The ARRL
Operating Manual

S0-AXS-131

Edited by
Robert J. Halprin, K1XA

The American Radio Relay League
Newington, CT USA 06111

Copyright © 1987 by

The American Radio Relay League, Inc.

Copyright secured under the
Pan-American Convention

International Copyright secured

This work is Publication No. 71 of the
Radio Amateur's Library, published by
the League. All rights reserved. No part
of this work may be reproduced in any
form except by written permission of the
publisher. All rights of translation are
reserved.

Printed in USA

Quedan reservados todos los derechos

ISBN: 0-87259-032-1

Third Edition

Contents

On the covers

Front (clockwise from top left): NØCP
with second op; N2FFY at the
mike; W6KQI and W6EKK at Field
Day; OH2MM with radio in tow;
WE6X tunes up aboard ship;

Back: KA5VLO garners Field Day
publicity; N5BGC hatches a packet
QSO; KC9KS/Field Day; KB6ISF at
the key

FOREWORD

Since 1966, the League has published manuals to satisfy the intense demand for specialized information on Amateur Radio operating procedures. This tradition has culminated in the third edition of the current *ARRL Operating Manual* series, the largest and most explicit volume devoted to on-the-air activity in the history of the Amateur Radio literature. To accomplish this remarkable objective, the multi-author approach—under one energetic editor—was utilized, as it had proved quite successful in previous editions. This was the most fitting way to set forth the great diversity of the Amateur Radio experience, and I hope you will agree with me that this goal has been accomplished with ingenuity and excellence.

As it happens, this new edition is very well-timed, now that the FCC has adopted the new Novice Enhancement rules. Amateur Radio is grooming itself for new life via Novice license-holders, presently 80,000 strong, the fundamentally most important segment of the Amateur Radio Service in terms of future vitality. You probably know people who have previously tried ham radio, only to drop out or become inactive because the Novice ticket simply didn't have enough to offer. You no doubt also know some who never got around to getting a ticket but *would have* if meaningful Novice operating privileges had been available. This book is for *everyone* interested in operating, but if you are a Novice or would-be Novice, this edition will be *particularly indispensable* to you in taking full advantage of those newly conferred privileges. With the contemporary operating know-how spelled out in these pages, you can't miss!

Please join us in the enormously entertaining world of Amateur Radio operating.

David Sumner, K1ZZ
Executive Vice President
March 1987

Introduction

LIGHTS, CAMERA, ACTION!

But wait—what's this Hollywood terminology doing in a book about Amateur Radio? Well, in serving as the Editor of this book, I began to visualize myself—within the framework of an admittedly unconventional imagination—as if I were directing a major motion picture, coming soon to a theater near you. That is, the megaphone-wielding principal in an epic production with a cast of thousands. Well, months and months of staring glassy eyed into a computer screen will do that to you!

A robust imagination (rather than delusions of grandeur) is healthy, and obviously *The ARRL Operating Manual* is something far short, say, of epics of the D. W. Griffith or Steven Spielberg eras. But I can dream, can't I? Moreover, in orchestrating a massive (at least by my modest standards) project like this, my control mechanism was a computer keyboard rather than a megaphone. And the refinements were made in computer files, rather than daily footage; the computer work-station functioned as the editing bench. Oh yes; there was a cast of thousands nevertheless—thousands of *bytes*, that is.

Countless hours were logged at the word processor, writing, rewriting and editing the text as necessary, and finally merging and moving the material (text, tables and graphics) into its correct place. This was followed by a generous amount of "industrial-strength" proofreading. Putting this book together was, at times during the process, an all-consuming passion, carefully nurtured in a very personal way. As such, I hope you won't find it too self-indulgent to express myself in this way. Perhaps my years of Amateur Radio message handling and contest training helped maintain the necessary stamina and persistence throughout!

Be that as it may, my purpose is not to engender sympathy cards; actually it was a project I gladly and eagerly undertook (as an adjunct to my ARRL HQ administrative responsibilities). More importantly, it was certainly a privilege to work with the many world-class radio amateurs, "stars" if you will, whose contributions enrich the pages of this book. Without their fine involvement, this edition of *The ARRL Operating Manual* could not have been assembled.

I could say that this, the third edition, *may* be the most complete book on Amateur Radio operating ever published, but alas that statement would not be accurate. No qualifiers are necessary! You, as the saying goes, are holding in your hands *the* most comprehensive volume of its kind ever published, further evidence of a revitalized ARRL publications program that responds to an increasingly diversified and dynamic Amateur Radio community. This is also the third edition for which I have officiated as Editor, but there is simply no comparison with the previous editions. In short, this sequel is better than the original! It was a wonderful experience in working toward this goal, and the best part is now I can share it with you.

Working on this book was irresistible because it is an extension of my profound interest in Amateur Radio operating. For some inexplicable reason perhaps, scanning the bands for new DXCC countries, being at the ground floor of a massive contest-style band opening into Europe or Japan, contacting a station 3000 miles away with 5 watts, or merely eavesdropping on a conversation on 75 phone or on an FM repeater, is just as intriguing and exciting now as it was 17 years ago when I first became licensed. If in the pages of this book, some of that intrigue is conveyed to you, this effort will be a success.

Consider this. In what other enterprise can you personally and self-sufficiently send and receive electronic impulses that traverse nations, that vault high mountain ranges and vast ocean seas, that leap into the outer perimeter of space? These signals are undaunted by political, territorial or generational boundaries. And to think, it's fun, rewarding and educational. It's the world's longest-running (electronic) happy hour, but with no downside, no morning after. And as the United States celebrates the bicentennial of the US Constitution, the remarkable document that has been the genesis of more individual freedoms ever known in human history, it is well to reflect, among its other provisions, on the Constitutional rights embodied in the "Commerce Clause" (Article 1, Section 8) and in the First Amendment. These liberties allow "We the People"—in this case, we Amateur Radio operators—to communicate with persons anywhere on the globe without governmental screening of that communication and without paying "tribute" to either a governmental or commercial entity. This characteristic of Amateur Radio is unique in the history of man.

Amateur Radio's major areas of interest are featured within these covers. The overall number of interest areas in the Amateur Radio hobby/service is staggering, and it is most gratifying that you are—with increasingly fewer government restraints—free to choose. Indeed, Amateur Radio may be in this respect a microcosm of our society; ours is the opportunity society and Amateur Radio is certainly the opportunity hobby. This book is designed to help you choose your own personal area of interest, and/or to assist in making your transition from one specialty to another as smooth as possible.

It would be extraordinary to find that one Renaissance man or woman with the sufficient breadth and depth of knowledge to address *all of* these important subjects. That's why I recruited an ensemble cast, if you will, of outstanding experts, to give you, the reader, the benefit of their experience. I've been through these materials about a dozen times during the editorial and production process, carefully evaluating and checking the content, and each time I learn something new that will enhance my own personal Amateur Radio operating. I hope you will derive enjoyment in equal measure, and perhaps you'll take a few minutes to fill out the feedback card in the back of this book to let me know the particulars of where the information hit (or even fell short) of the mark.

I read somewhere that because of the growth of specialization, knowing *more* about *less* will mean—played to its logical conclusion—knowing everything about nothing! Luckily, at least insofar as the Amateur Radio art is concerned, books like this (and others in the ARRL library) are available, providing under *one* cover a multidimensional presentation replete with practical, essential information. It will help you go from one Amateur Radio specialty to another with independence and adroitness. Or if you are satisfied where you are, it will perhaps cause you to approach your specialty with renewed determination.

In the course of the production work on this publication, the Federal Communications Commission was positioning itself to act on the League's "Novice Enhancement" proposal, a sweeping move to revitalize the entry-level amateur license. We waited with great anticipation, and perhaps a twinge of apprehension. Just as the finishing touches were being made on this book, the FCC did act, adopting new rules effective March 21, 1987 that closely mirror the League's original Novice Enhancement recommendations. This may be the most pivotal event in determining the future growth and viability of the Amateur Radio Service. It should prove to be a blockbuster.

The Novice license, Amateur Radio's "learners' permit," has traditionally been a code-only license (although there was a period running from the creation of the Novice ticket—circa 1951—and 1968, when Novices were permitted limited 2-meter phone operation along with the traditional HF CW privileges).

Up until now, the beginning Novice was limited, paradoxically, to the operating mode that required the *greatest* amount of operating skill.

To make the learner's permit more desirable (while simultaneously maintaining the present Novice HF code privileges), Novice Enhancement boldly departs from the past because it allows Novices (and Technicians) an ample segment of 10-meter phone (with all its exciting DX ramifications, once good propagation conditions reemerge), plus prime voice and data communication territory at 222 MHz and 1270 MHz.

The new privileges are as follows:

- 28.1-28.3 MHz: CW and RTTY, 200 watts PEP output max.
- 28.3-28.5 MHz: CW and SSB, 200 watts PEP output max.
- 222.10-223.91 MHz: all authorized modes, including operating through repeaters, 25 watts PEP output max.
- 1270-1295 MHz: all authorized modes, including operating through repeaters, 5 watts PEP output max.

By the time this book hits the streets, Novices will be tuning up their HF rigs (formerly in the CW-only configuration) for 10-meter SSB. As noted above, when good conditions return to 10 meters (and it shouldn't be long now until we enter into the next sunspot cycle), Novices will discover that the band is *the* most delightful amateur allocation for worldwide communication, bar none. The band is spacious, the antennas are unobtrusive and the intensity of the DX signals from around the world will be astonishing!

For you newcomers, you'll just have to trust me on this, but a couple of hundred watts and a modest antenna will be more than enough for page after page of DX contacts, many with beginners just like you. When 10 meters is open, you're just not going to want to be anywhere else. When I first obtained my Novice license, 10-meters was always dead; it was a curiosity, considered irrelevant by active hams. It wasn't until I found myself on a DXpedition to Senegal in 1977 that I experienced a bona fide 10-meter opening, one that lasted all day. It was incredible; wave after wave of strong signals from everywhere coming back to me. Despite the West African heat, I held forth at the operating position for hours, sustained by the potency of the wall-to-wall signals. Amateurs all over the world enjoyed these excellent 10-meter conditions through the early 1980s, and it appears a resurgence of the band—with the commencement of a new solar cycle—is almost at hand, just in time for Novices, too, to take full advantage, whether it be on SSB, or CW and RTTY, for that matter. Good times are straight ahead.

But this is in no way meant to minimize the importance of the new Novice VHF/UHF segments at 222 and 1270 MHz. On these bands, Novices can now hook up their computers to work RTTY and other digital modes. They can operate through existing FM repeaters or simplex to talk to other local hams. In short, no longer will the "socialization" of the new ham have to be conducted exclusively at five words-per-minute. And no longer will new and exciting aspects of Amateur Radio be off-limits to enthusiastic newcomers. Novices have joined the mainstream of Amateur Radio, as the first step towards greater personal achievement and license upgrading.

For Novices or would-be Novices preparing for an excursion into this new frontier, this manual—although substantially finalized prior to Novice Enhancement—should be a crucially important guidebook. To that end, you'll see specific references to Novice Enhancement sprinkled throughout this book. As a matter of fact, inside this book you'll find detailed descriptions of a wide range of operating activities that I hope *all* of you will be eager to try, regardless of where you are on the Novice to Extra Class continuum.

Keep in mind that the primary (if not the only) reason for attaining an Amateur Radio license is to operate the radio, to get on the air. This is the core of Amateur Radio.

This book attempts to suggest guidelines for desirable on-the-air conduct. Successful operating must be approached with enthusiasm, with a genuine desire to comport yourself on the air efficiently and intelligently. Amateur Radio operating should be enjoyed to its fullest, in a positive, future-oriented style. After all, having fun and becoming an able operator are *not* mutually exclusive! Your understanding and familiarity with established operating methods will make on-the-air operating more satisfying and entertaining for all concerned—especially yourself. And speaking of directing, ham radio makes you uniquely the program director of your own entertainment—no time constraints, no minimum box office receipts, no ratings, no previews of coming attractions and no commercials. *You* are in command. Within the substantial flexibility of the FCC rules, you are limited only by your own initiative, creativity and yes, your imagination.

Again, while not to sound like one of those (presumptively genuine) tearful, tuxedo-clad show-business "personalities" accepting a statuette from the "Academy," I am nonetheless grateful to the many persons who played significant roles in both the contours and the contents of this book. In addition to those distinguished men and women who are specifically recognized in the table of contents and elsewhere, I'd also like to particularly thank my good friend and colleague Mark Wilson, AA2Z, for his very helpful advice, and the ARRL HQ Production Department staff (notably Shelly Chrisjohn, WB1ENT, Debbie Strzeszkowski, Steffie Nelson, KA1IFB and Joel Kleinman, N1BKE) for a sincere effort and a job well done.

And now... *ROLL 'EM*!

Robert J. Halprin, K1XA
Editor
March 1987

Chapter 1

Shortwave Listening

*T*he year is 1969. This night is one of a series of deep summer Chicago nights throughout which the oppressive heat of day never finally dissipates. It is exactly this way in a second-floor bedroom where a boy sits alone at a radio.

He sits alone before an aging and modified National NC-100X receiver, its black wrinkled-enamel finish grayed with 33 years of dirt, the once-fine maroon trim about its bandswitch-indicator windows worn pale through decades of touch. There is no S-meter; the phosphoric green glow of the business end of a 6E5 "Magic Eye" indicator tube contracts and dilates in time with the rhythm imposed by propagation upon the signal to which the boy listens. Now and then a static crash pings the metal diaphragms of his Trimm headphones; the Magic Eye blinks shut until a timing capacitor discharges, the IF stages return to gain and the National's gaze is cast once again into the radio sky.

In the headset of the listener, the cicada-buzz and neighborhood sounds of day have been replaced with distant music. The frequency on which this music arrives is 5954 kHz; the station is TIQ, Radio Casino, out of Puerto Limon, Costa Rica. The boy at the radio likes this station; it is, for the moment, one of his favorites, for between the US pop tunes offered by Casino are the curiously accented English ramblings of the announcer, and between the music and the ramblings are the Sprite™ soft-drink ads—"tart and tingling"—and, after all, leaving a ring on the unsealed plywood of the radio bench is a heat-sweated can of Sprite itself. "This Radio Casino has something," he thinks. He begins to doze beneath the Trimm Featherweights. A static crash wakes him; there is another slow swallow of Sprite; the National is gently trimmed for drift. There is an additional pop tune, or three, another ramble, a further grumble of QRN. And then there is a vision of hands.

The first hand, or image of a hand—the boy alone at the radio does not have time to determine this, just as he is not certain if the hand is real or dreamed—is his own, busy in tuning the National to last week's logging of Ecos del Torbes, San Cristobal, Venezuela. But this hand is not alone. Contemporaneous with this he spies another, older hand as it rewires the radio's power supply. More are coming. Slowly, at first, and then with the dizzying rush of a rising exponential curve, there appear before his eyes every human hand in his radio's history—hands setting chassis punches, hands wiring, soldering, aligning each band's trimmer capacitors; hands bored, tired, checking in for work, hands calling their job complete. Three decades of hands simultaneously tune the NC-100X to every dial setting it has ever known; no hand's presence excludes another: They exist together in the same space, merging in, around and through the receiver. Twisting and fading in and out of view through the changing propagation of memory, the vision of hands spreads down the National's line cord to portend of the myriad lives implicit in the generation and transmission of energy....

QRN. It is 0605Z, forehead-sweat, and no TIQ. 'Spherics in the Featherweights remind the boy at the radio of a storm cell over Missouri and coming; the warm Sprite is flat, the ring in the benchtop swollen. He looks at his hands.[1]

The year is 1987. You've just read about a kid listening to a radio built in 1936 on a hot summer night in 1969. A bit overdramatic? Perhaps, but then, the hands tuning the dial that night were mine!

That old National *was* quite a receiver—or maybe "is," for it's still kicking as this is written in 1986, with a few modifications. It still "pulls them in"—"them" meaning mysterious radio signals from far away. Sure, a radio doesn't really pull signals in; that's wishful thinking—or poetic license—on the part of people who love radio.

When that particular NC-100X was built back in 1936 by the National Company of Malden, Massachusetts, radio was the thing. People gathered around the family radio and watched it—yes, they actually *watched* their radios—just as we watch television today. (Television had existed for some time in several experimental forms, but it didn't really create a stir until after World War II.) Common use of shortwave radio to bridge continents and oceans was less than 20 years old; radio was a big deal. It wasn't a fad. It provided people with something no other medium of communication had done before: Instantaneous contact with other places and people from the comfort of their own living rooms and kitchens.

Yet, with all the hoopla about radio back when the old National NC-100X was wired, you couldn't do as much with radio then as you can now. For instance, most of the signals you could hear in 1936 with an NC-100X—or with one of its competitors—would be of two types: Morse code (radiotelegraphy, or CW) or voice (radiotelephone). Even with civilization so excited over radio, that was about the extent of it, as far as most people were concerned. Efficient use of radio in aviation wouldn't take off until World War II. Television was an unfulfilled promise. What radar? What CB radio? What garage-door openers? What stereo music on FM? These didn't yet exist. But radio was king.

---

[1]The NC-100X incident originally appeared in the "It Came From a Radio" series, *Review of International Broadcasting*, June 1982. Used by permission.

Shortwave listening was a popular radio pastime then. Many family radios covered more than one band: You could receive "AM" on Band One, and if your set was classy enough, you might be able to tune in international broadcasts up to about 18 MHz—or, as it would have been put then, 18 Mc/s—using Bands Two and Three. Newspaper publishers worried that the ability of radio to report immediately on events would hurt their sales—and such worries were well-founded, as gripping radio coverage of World War II events proved.

Were you wondering what this has to do with Amateur Radio? You already know that this book is published by the American Radio Relay League, a nonprofit membership association dedicated to Amateur Radio. Here's what Amateur Radio has to do with the shortwave listening story.

Amateur Radio was going great guns back in the years when that National NC-100X was young, just as it is now. In the 1920s, hams had been first to prove that the shortwaves were good for something other than nothing by bridging the Atlantic Ocean, and this excitement hadn't diminished one iota right through the time the NC-100X rolled out the door in Malden in 1936. What was so amazing about such ham enthusiasm was that it had to be poured into relatively few radio options. You either sent code or you talked; those were your choices, pretty much. But radio in any "flavor" was a big deal, and those choices were enough to interest many people, especially if they'd heard ham radiotelephone on Band Two or Band Three of the family radio and itched to get in on the fun. Ham radio was a guaranteed blast: Talk to England from your bedroom? No problem. Toss together a bunch of tubes and parts on an orange-crate board and listen to buzzy whispers of Morse from Japan? "Why, just this morning..."

Let's get back to 1987. Radio is still as amazing as it ever was, and our options for enjoying it have exploded in number. When the old National was built, Morse code and voice were

The Super-Special Wide-Range Receiver is Here

"If the broadcast receiver were a very special one that could continue to tune higher in frequency (there are technical reasons why this is impractical without switching), you would find many different groups, or bands, of frequencies, used by many different services..." So began a discussion of the radio spectrum in the ARRL's *How to Become a Radio Amateur* of years gone by (since then replaced by *Tune in the World with Ham Radio*, also available from ARRL), the tuning dial of such a receiver being characterized in the drawing reproduced below. Times have changed. Even though, strictly speaking, some sort of bandswitching must be done as a receiver is tuned across wide stretches of the electromagnetic spectrum, many modern receivers and general-coverage Amateur Radio transceivers are set up so that you needn't manually operate a band switch unless you choose to: You can start tuning at the bottom of a receiver's tuning range (usually 100 kHz or so) and keep right on going—without taking your hand off the tuning knob—until you hit 30 MHz. The super-special wide-range receiver is here.

Although military radios had long afforded smooth general-coverage reception, and a few hard-to-get consumer receivers of the '60s made a stab at constant-tuning-rate general-coverage reception, it wasn't until the mid-'70s that frequency synthesis opened the door to the high-quality general-coverage reception available today in receivers from ICOM, Kenwood, Panasonic, Sony, Yaesu and other manufacturers. Within the span of 10 years, unbroken frequency coverage, constant tuning rate and optimum sensitivity throughout the spectrum below 30 MHz have become standard. Further, most new Amateur Radio transceivers incorporate no-compromise general-coverage receivers—and making the jump between amateur-bands-only and general-coverage reception is as easy as pushing a button or two.

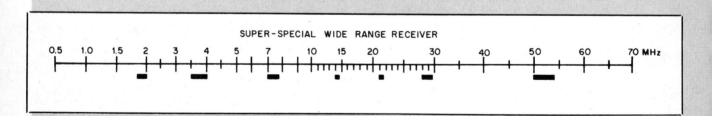

How They Do It Now

Here's a look at how five well-known Amateur Radio transceivers allow you to choose between ham-bands-only operation and smooth general-coverage reception. These examples only scratch the surface of what's available as this book is written, so check *QST* or see your dealer for news on the latest equipment.

Fig A—You can select which Megahertz slice of spectrum you want to tune with the Kenwood TS-430S by pressing 1 MHz STEP and using the UP and DOWN buttons under BAND.

Fig B—Band selection is simpler on the Kenwood TS-940S. Each ham band may be chosen by pressing its respective button. Entry into the general-coverage mode is automatic when you press the UP and DOWN buttons indicated by 1 MHz STEP. Of course, with the keypad frequency entry afforded by the 940S and its sibling TS-440S, you may key in a frequency digit by digit and get there directly. The Kenwood R-5000 shortwave listening receiver allows this, too.

Fig C—ICOM's IC-735 incorporates a quality general-coverage receiver. Use of its HAM button allows the operator to choose among ham bands only or step one Megahertz at a time through the spectrum below 30 MHz.

Fig D—The ICOM IC-751A shifts from ham-bands-only to general-coverage reception at a touch of its HAM/GENE button. On the receive-only front, ICOM's IC-R71A general-coverage shortwave listening receiver takes frequency selection a step further with the inclusion of direct frequency entry via keypad.

Fig E—Yaesu radios allow general-coverage reception, too. The FT-767GX allows UP and DOWN band selection as well as keypad frequency entry. The FT-757GX incorporates general-coverage reception without a keypad, much in the manner of the ICOM IC-735 (Yaesu's entry into the shortwave listening market is the FRG-8800 receiver).

the thing. By the time 1969 rolled around, humanity had so many uses for radio that they couldn't even be counted on the hands that appeared in a dream. The National was pulling in signals that couldn't have been understood without two or three refrigerator-sized racks of additional equipment. Radio amateurs had switched to single-sideband voice transmission decades before, and had long since taken the plunge into moonbounce, radioteletype, television and other exotic techniques—and the NC-100X was too unstable, unselective and limited in tuning range to do a very good job of receiving them.

Radio amateurs had done something else along the way, too. They'd gone over more and more to equipment covering only ham frequencies. This step was a natural one in the evolution of transmitting and receiving equipment into more and more efficient forms: If you're out to do some hammin', why waste dial space on much else of the radio realm? From the mid-1950s or so, up to the late 1970s, radio amateurs sacri-

ficed breadth of radio vision for superb equipment performance in the narrow chunks of the radio spectrum allocated to the Amateur Radio Service. Shortwave listeners tended to have a bigger picture; they continued to use general-coverage receivers, although the performance of many of these behemoths wasn't up to what amateurs were enjoying in the ham bands. The technology available allowed little else.

Well, it's 1987, and a revolution in shortwave listening is occurring: Almost every new ham transceiver incorporates a receiver covering the spectrum from 0 or 0.1 to 30 MHz *continuously*. And manufacturers of ham equipment are offering excellent general-coverage receivers covering the same frequencies for sale to shortwave listeners. You can buy a *portable* receiver covering more frequencies, with greater stability and sensitivity, than that NC-100X—for $150. Radioteletype and Morse code may be extracted from the air with the help of your home computer—the refrigerator-sized racks of equipment are gone.

(A)

(B)

(C)

(D)

Contemporary Amateur Radio transceivers (such as Kenwood's TS-430S and TS-940S, ICOM's IC-735 and IC-751A and Yaesu's FT-767GX) provide continuous frequency coverage at the touch of a button for ease of operation on the amateur bands as well as for general-coverage reception.

But what can you hear? That's the big question after the new receiver or transceiver is out of the carton and running, just as it's been since the early days of radio. (A car without a road map will get you nowhere fast!) After the relative order of life in the ham bands, radio amateurs may venture beyond Amateur Radio territory and wonder if all is chaos. Short-wave program listeners may figure nothing much happens outside of the international broadcasting bands and miss out on much of what's going on in their radio neighborhood. The hands on the tuning knobs need some idea of where to go to find what.

Over the next few pages, we'll discuss what kinds of radio signals you may hear as you tune, how to identify them, and why they operate where they do. We'll take a tour of the radio spectrum from 9 kHz to 30 MHz. And then we'll take a look at publications, listening clubs and other sources of information you'll need to get the most out of your radio listening. Enthusiasm for radio's excitement is easy to catch—and worth catching—whether you're on radio's receiving or transmitting end. That's why this chapter—and this book—is here.

ORDER IN THE RADIO SKY

The shortwave broadcast and amateur bands are pretty orderly places, for the most part. Similar stations are close together in clumps, and if you roam around the radio spectrum for a while, you notice right away that there are groups of similar stations at more than one spot on the dial. This isn't an accident: Radio stations, services and governments are following international rules and regulations to ensure that the radio spectrum is used in an orderly way. The body responsible for the promulgation of these rules is the International Telecommunication Union (ITU), an agency of the United Nations based in Geneva, Switzerland.

Once humankind had caught on to the utility of radio, a rush was on to grab frequencies. Beginning in 1903, periodic international radio conferences have been held to democratize the process. In January 1938 *QST*, in his introduction to a series of articles describing the business to be undertaken at a conference in Cairo later than year, A. L. Budlong, W1JFN (later W1BUD), put the reasons for international radio regs succinctly indeed: Why do we have to have international agreements on radio? Broadly speaking, there are three reasons:

1) Since stations of one nation are frequently in communication with stations of another nation, it is necessary to have agreements on such operating details as calling procedure, distress signals, call-letter assignments, methods of collecting tolls on radiograms, etc, unless utter confusion is to be encountered when any two stations try to do business over the air.

2) Because it is possible to operate radio stations throughout a wide range of frequencies, it is necessary to agree in advance where the various services will locate themselves in the spectrum, so that stations will know where to find each other.

3) Since radio signals are not confined to the borders of the country in which they originate, international agreements on allocations to services are also necessary in order to prevent chaotic conditions on the air and hopeless interference between services.[2]

That's it in a nutshell, and the reasons for the international regulation of radio haven't changed since then. Radio

[2]Budlong, A. L., "Cairo," *QST*, Jan 1938.

signals are classified by *service*—what they exist to do. Technical standards (frequency of operation, transmitter power, antenna radiation pattern and gain, and type, bandwidth, purity and stability of emissions) are established and applied to guarantee efficient use of the radio spectrum and minimize interference. Through their national telecommunication administrations, individual stations are assigned *call signs*—alphanumeric identifiers specific to each station's country of origin and service. You'll find a table of international call sign prefixes in Chapter 17 of this book. (For example, the call sign—in the Amateur Radio Service —AK7M; a listener with a good working knowledge of call sign structure knows immediately that AK7M belongs to an Amateur Radio Service station under the jurisdiction of the United States.) ITU allocates prefix blocks to nations; national telecommunication administrations (the Federal Communications Commission [FCC] in the US, and the Department of Communicatons [DOC] in Canada) assign call signs to stations.

Once you have the table of international call sign prefixes in hand, you still need to know what type of station— belonging to which service—you're likely to hear when you're listening at a certain spot in the spectrum. ITU allocates bands of frequencies to the different radio services; national administrations assign specific operating frequencies to stations under their jurisdiction.

The ITU has no enforcement powers in the same sense that the United Nations has security troops. But, like (at least theoretically) the nations in the UN, members of the ITU agree to be bound by its regulations, which are effective only if a consensus prevails. In practice, a number of ITU members have tended to "overlook" ITU regulations felt to be counterproductive to their interests. For instance, when you listen on the shortwave broadcasting bands, you'll hear jamming —growls, whirring, buzzes and whines transmitted deliberately by one nation to blot out programming from others. Such activity directly contravenes ITU regulations. Technology allows pinpointing the nations responsible for jamming, but politics keeps the transmitters grinding. ITU provisions for settling matters of harmful interference depend as much on the "honor" system for success as other ITU rules: Any station may operate on any frequency—even in radio territory improper for its service—as long as it does not cause harmful interference to stations operating in accordance with ITU frequency allocations. It is the responsibility of legitimate users of interfered-with frequencies to complain.

ITU has divided the world into three administrative regions to do its regulatory work (see Fig 1-1). Periodically, delegates from ITU's participant nations get together for a World Administrative Radio Conference (WARC), if sweeping changes need to be made to frequency allocations or technical standards worldwide. (The latest general WARC was held in Geneva in 1979.) Sometimes, such business might affect only one ITU region; then, a Regional Administrative Radio Conference is held. Conferences specific to service, frequency range or other areas of ITU concern may be called as needed.

The feeling that radio rules and regulations are the soul of dullness claims fewer victims among shortwave listeners with each WARC or RARC. What one year was just another radio acre of radioteletype or facsimile signals may swell with broadcast signals next year with a single stroke of a bureaucrat's pen. *The FCC Rule Book*, published by ARRL, goes into detail concerning how rulemaking proceeds at the FCC—and how you can get involved in the process. Radio amateurs, as licensees in a radio service, take special interest in their right

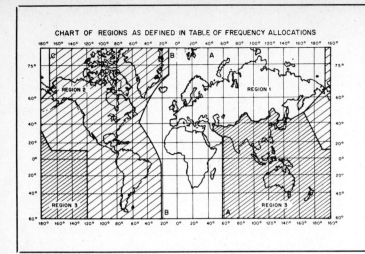

CHART OF REGIONS AS DEFINED IN TABLE OF FREQUENCY ALLOCATIONS

Note: Region 2 is defined as follows: On the east, a line (B) extending from the North Pole along meridian 10° of Greenwich to its intersection with parallel 72°; thence by Great Circle Arc to the intersection of meridian 50° W and parallel 40° N; thence by Great Circle Arc to the intersection of meridian 20° W and parallel 10° S; thence along meridian 20 degrees west to the South Pole. On the west, a line (C) extending from the North Pole by Great Circle Arc to the intersection of parallel 65°, 30′ N with the international boundary in Bering Strait; thence by Great Circle Arc to the intersection of meridian 165° East of Greenwich and parallel 50° N; thence by Great Circle Arc to the intersection of meridian 170° West and parallel 10° N; thence along parallel 10° N to its intersection with meridian 120° West thence along meridian 12° West to the South Pole.

Fig 1-1—ITU Region Map. Europe and Africa are Region 1, North and South America comprise Region 2 and the rest of the world makes up Region 3.

to petition their national telecommunication administrations for changes in the radio rules over and above just abiding by them—so you won't find the "regs" boring if you are a ham, talk to a ham or become one!

Secrecy and You

On October 21, 1986, President Reagan signed the Electronics Communications Privacy Act of 1986 (the ECPA), Public Law 99-508, which took effect January 19, 1987. The ECPA revises the Wiretap Act of 1968 to purportedly protect certain electronic communications by making it *illegal* to monitor communications defined as not "readily accessible" such as cellular telephones, electronic mail, various video/data communications and scrambled/encrypted transmissions. Perhaps as an unintended side effect, the ECPA creates a new set of rules governing the interception of radio communications additional to those already found in Section 705 of the Communications Act of 1934, as amended.

Secrecy of communications is the subject of 705. The message in this section is that it's not forbidden merely to *receive* a radio signal not intended for you; you also must not divulge or publish "the existence, contents, substance, purport, effect or meaning thereof." Additionally, you are not to use what you hear to your own benefit. Not to worry, though. The signals of several radio services transmitted for the use of the general public are *exempt* from Section 705 protection, and may be received and discussed freely: Broadcasting, Amateur Radio, CB or transmissions relating to ships, aircraft, vehicles or persons in distress. See *The FCC Rule Book,* published by ARRL, for more information.

As to the ECPA, it seems to parallel Section 705 to the extent that under its provisions, you can *legally* monitor among other things any amateur, CB or General Mobile Radio Service transmission, any marine or aeronautical mobile communication, any communication for the general use of the public or relating to ships, aircraft, vehicles or persons in distress, the radio portion of cordless telephone conversations, any public safety (including police/fire) radio communications system readily accessible to the general public, and satellite transmissions of network feeds (which are governed exclusively by Section 705), and more. For further details, see December 1986 *QST,* p 51.

THE SIGNALS YOU'LL HEAR

So bombarded are we with sound and vision broadcasting that we may never discover that there are other uses for radio. As you explore the shortwave spectrum, you'll hear some unusual noises. What are these signals, and what are they called?

You may already know a local broadcaster between 535 and 1605 kHz as an "AM" station. Fine; such stations mainly use AM, or amplitude modulation. But other forms of AM have different names. For instance (and assume that the AM station in question transmits monaural sound and not stereo), your local AMer transmits a carrier and two sidebands. This is the sort of emission many radio folks call "AM," and this term is used here to mean full-carrier double-sideband AM radiotelephone, since it's less of a mouthful. It was popular in the Amateur Radio bands until the 1950s, when a different species of AM became widespread: "Single sideband," or SSB. SSB emissions usually consist of a suppressed carrier ("suppressed" meaning greatly reduced in strength—preferably below easy audibility) and one sideband. Amateur Radio stations below 10 MHz usually use "lower sideband" (LSB), the sideband below the reference frequency afforded by the suppressed carrier; amateurs at 10 MHz and above use upper sideband, or USB. Most nonamateur users of SSB have settled on USB as a standard. A few stations, such as CHU at 3330 and 7335 kHz, and 14.670 MHz, transmit full-carrier SSB—one sideband (the upper) and a full-strength carrier. Shortwave broadcasters generally use AM, although they intend to switch to reduced-carrier SSB within the next 20 years or so. Reception of any amplitude-modulated signal with less than a full carrier will require that you use your receiver's beat-frequency oscillator (BFO) to supply the missing carrier at your receiver, so your radio's "mode" selector should be set to USB or LSB for such reception. Many shortwave-broadcast DXers use this technique even for weak *full*-carrier signals when reception conditions are poor. Shortwave broadcast "feeders" sometimes employ double-sideband AM with two independent (dissimilar) sidebands, each carrying unrelated programming; proper reception of these will require USB or LSB "mode" selection as well.

Another common form of amplitude modulation is CW (short for "continuous wave"), radiotelegraphy achieved by turning an otherwise-unmodulated carrier on and off. Your radio must be set to USB, LSB or CW to render CW intelligible. If the term "CW" sounds new to you, perhaps you've heard it referred to as "Morse" all along. A Morse variant is MCW (modulated CW), or Morse transmitted by keying a tone-modulated carrier. MCW may also be sent by keeping the carrier on and keying only the modulating tone. MCW is the emission of choice in the longwave beacon band.

Frequency modulation (FM) is used for voice communication below 30 MHz by some radio amateurs, and the few remaining 1600- to 1800-kHz cordless phones use FM as well. You'll have difficulty rendering it intelligible unless your receiver is set up for narrow-band FM (NBFM) detection, although switching to AM reception and tuning off to one side of a NBFM signal while trying various IF selectivities may give you enough audio to work with. The "AM-stereo" boom now sweeping the mediumwave broadcasting band means that "the AM band" is no longer just AM: The most popular AM stereo transmission system (Motorola) is transmitted using a combination of AM and PM (phase modulation, a relative of FM) in one signal.

Radioteletype (RTTY) below 30 MHz generally makes use of an FM emission in which printing commands are sent by switching rapidly between two frequencies ("mark" and "space"). For more information on RTTY, see Chapter 9. You'll find quite a bit of RTTY throughout the radio spectrum using your receiver's USB, LSB or CW "mode" selection, but if you want to demodulate it and read what's being transmitted, you'll need more than just your ears and a receiver. You'll *hear* RTTY as a steady musical "beedling" or whirring. Specialized multiplex RTTY emissions may be audible as tight clusters of RTTY signals within a kHz or two of spectrum; sometimes these sound like the drone of a cruising piston-engine aircraft if any of the multiplex channels are not carrying traffic. Specialized error-correcting forms of RTTY may be heard in the maritime mobile bands (TOR—Telex Over Radio and SITOR—Simplex TOR) and in bands allocated to the Amateur Radio Service (AMTOR). These sound like cricket chirps when listened to with your radio's BFO switch on.

Fascimile (FAX) is the transmission of fixed images over radio for storage in permanent form—the receiving party ends up with a *facsimile* of the original map or document. FAX receiving equipment is expensive, and not many shortwave listeners are equipped for FAX reception. But you'll certainly run into FAX transmissions between 9 kHz and 30 MHz; with your radio's BFO switched on, they sound like cyclic rapidly wobbling carriers punctuated with regular beeps (synchronization pulses). Radio amateurs may transmit FAX; they also transmit slow-scan television (SSTV) on the shortwaves, sending strings of still pictures at the rate of one every eight seconds. For further details on amateur image communication, see Chapter 16.

Progress in electronics has continued to give rise to new (and strange-sounding, to the listener equipped only for aural reception) methods of transmitting or gathering information by radio. Occasionally, you'll run into a signal sounding like repetitive flute music. This is a data emission called "Piccolo." Often, you'll hear a signal sounding like the tapping of a woodpecker, sometimes covering wide stretches of spectrum all at once. This is a form of over-the-horizon radar. Radio amateurs are using computers to communicate via packet radio ("packet" signifying that transmitted information is sent in precisely defined bundles); packet emissions below 30 MHz sound like short RTTY bursts without the repetitive "cricket" sound of AMTOR. See Chapter 10 for more information on packet.

You may run into data transmissions characterized by some listeners as sounding like metallic bonking mixed with the sound of escaping steam. Some secure data transmissions are disguised to appear as the background hiss you hear mixed with static throughout the shortwave spectrum. Even if you're unequipped to decipher their content—that's usually the idea behind the pains taken to hide and encrypt them—you'll find that they add mysterious spice to what erupts from your radio as you get to know the shortwave bands.

THE SERVICES YOU'LL HEAR

Now that we've covered the regulation of radio enough to know how services and stations receive their operating frequencies, and what kinds of signals you may expect to run across as you explore the bands, let's take a look at the different radio services populating the spectrum between 9 kHz and 30 MHz. Each service affords a particular type of listening to the DXer, and some hobbyists specialize in listening only to signals from one or two. They will be discussed alphabetically. Where possible, the structure of call signs for stations in the service under discussion will be given, but compliance with ITU rules concerning identification seems to vary with national administration.

Identification requirements for mobile radiotelephone stations and their associated land stations are generally eased as compared with those for nonvoice stations. These stations may, of course, use call signs, but they are also allowed the following substitutions: The official name of the ship or selective calling number (for ship stations); official registration mark or airline, followed by the flight identification number (for aircraft stations); the identity of the vehicle or any other appropriate indication (for land mobile stations); the geographical name of the place as it appears in the ITU's List of Coast Stations, followed preferably by the word "radio" or by any other appropriate indication (for coast stations); the geographical name of the place or airport name (for aeronautical stations); the geographical name of the place followed, if necessary, by any other appropriate indication (for base stations). Yes, means of identification vary!

Aeronautical Fixed Service stations operate between specific terrestrial points, carrying traffic intended to ensure safe air navigation and efficient air transport. Only two bands are allocated to this service in the 9 kHz to 30 MHz range; operation in these bands seems to concentrate on radioteletype and facsimile emissions. ITU rules do not specify particular call signs for aeronautical fixed operation, but do specify that the call signs of land and fixed stations shall consist of two characters and one letter, or two characters and one letter followed by not more than three digits; ITU also recommends that "as far as possible, the call signs of fixed stations shall consist of: two characters and one letter followed by two digits (other than the digits 0 and 1 in cases where they immediately follow a letter)."

Amateur and Amateur-Satellite Services. ITU's definition of Amateur Radio says that it exists "for the purpose of self-training, intercommunication and technical investigations carried out by amateurs, that is, by duly authorized persons interested in radio technique solely with a personal aim and without pecuniary interest." Implicitly, words like "radio fun, learning, friendship and emergency communication" are probably included in there somewhere. FCC terms it the Amateur Radio Service. The Amateur-Satellite Service uses space stations on earth satellites with the same aim as that of the Amateur Service. Amateur and experimental stations share the same call-sign guidelines: "One character and a single digit (other than 0 or 1), followed by a group of not more than three letters or two characters and a single digit (other than 0 or 1), followed by a group of not more than three letters."

Broadcasting Service. The transmissions of broadcasting stations are intended for direct reception by the general public. In the shortwave listening press, you'll likely come across references to "domestic" or "internal" broadcasting services—transmissions meant for reception within the country of origin—and "external," "international" or "overseas" services—those meant for reception outside a country's borders. Domestic broadcasters are more likely to use call signs than international stations. Whatever the intent, broadcasting is quite popular, but surprisingly, it wasn't the first use of radio on peoples' minds around the turn of the century. (To get an idea of one of radio's first jobs, see the paragraph on the maritime mobile service, below.) Today, we're saturated with television and radio programming around the clock. In the tour we're about to take through part of the

International Broadcasting: Another World

"Shortwave" is a general term usually applied to radio between 1600 kHz and 30 MHz. Amateur Radio and CB operators spend a lot of time on frequencies in this range; "Stepping Through the Megahertz," elsewhere in this chapter, hints that there's much more non-broadcast action on the shortwaves than that accounted for by Amateur Radio and CB. But judging from reportage in the mass news media, "shortwave" means international broadcasting to many people. Even when our domestic news networks report on comments made by Radio Moscow or Radio Tehran, however, the biggest news of all—that you can tune in such broadcasts on affordable and easy-to-operate radios—somehow never makes it to the headlines. Well, you can. Daily, strong shortwave stations broadcast many hours of enjoyable news and entertainment—in English—to listeners in North America.

International broadcasting is set apart from local AM, FM and TV by the highly directed nature of its transmissions. Most domestic stations in the US and Canada broadcast to a single target audience for all or most of each day. International shortwave broadcasters, on the other hand, target their transmissions to specific areas at scheduled times. Their broadcasts are usually between 15 minutes and several hours in duration, and they use directional antennas to concentrate transmitter power on their target areas. For instance, North Americans wishing to listen to English from Radio Netherlands have two chances each day—at 0230 and 0530 UTC—to hear RN's English transmissions for this area. At other times, RN is broadcasting to other parts of the world in English, Dutch, or one of seven additional languages. It's quite possible to listen in on transmissions not directed to your area, if you're willing to put up with poorer signal quality. Many shortwave listeners do this—it's called "DXing"!

Frequency agility also sets international broadcasters apart from locals. You expect to be able to find your favorite TV station on the same channel from week to week, and if you can't, they may expect to lose you as a viewer! International broadcasters are forced to shuffle frequencies with diurnal and seasonal changes in propagation—this in addition to using frequencies in several bands for each transmission ("parallel" frequencies, as shortwave listeners call them) to assure reception over a large target area during a given broadcast. Some international broadcasters, such as the Voice of America, BBC and Radio Moscow, transmit to some of their target areas for many hours at a stretch; at intervals throughout such broadcasts, you'll hear announcements made concerning the adding and dropping of frequencies in accordance with changing propagation. International broadcast enthusiasts are, by necessity, much more adept with the tuning dial than listeners to local stations!

Although quite a few radio *hobbyists* own receivers with accurate digital frequency displays, many people who depend on shortwave during travel or for reception of their countries' domestic services must make do with more economical, often dubiously calibrated, sets. It's not easy to return to the same dial setting time and again with a poorly calibrated radio—and even if you can, a frequency change by your favorite station may necessitate a search for its latest channel. To assist their listeners in finding a broadcast, many shortwave broadcasters transmit a tuning or *interval* signal ("IS" to long-time shortwave listeners) for several minutes prior to each broadcast. Listeners need only tune for a familiar repetitive musical phrase, chirping bird or barnyard animal cacophony at the appointed time to locate a broadcast. (Tuning signals are many: Radio Australia uses "Waltzing Matilda," Radio Canada plays the opening to "O, Canada," Radio Botswana transmits the sounds of cow bells and farm animals—complete with a crowing rooster—and Radio Polonia [Poland] plays an excerpt from Chopin's Revolutionary Etude, Op. 10, No. 12, in C Minor.) DXers wishing to identify a weak station broadcasting in an incomprehensible language need only identify the IS to be sure of what they're hearing.

What about shortwave programming? Sometimes, it is termed "propaganda," conjuring an image of little more than hard-line political harangues. While there are a few international broadcasters who still wear this decades-old face, blatant "propaganda" rarely applies to the program content of many modern stations—and some might suggest that propaganda heard is often softer sell than the advertisements we tolerate in our own domestic broadcasting! Shortwave is a good source of news and topical commentary from various points of view, it's true, but be advised that variations in reportage of a given event may vary from the comical to the alarming. Caveat emptor. Some people become "addicted" to news just as others wax fanatic about sports; international broadcasting is a natural haven for such news hounds.

Music, culture and language are what draw most confirmed shortwave listeners to their radios, however. From Radio Moscow's "Folk Box" to the Voice of America's "Jazz Hour"; from the BBC's "Anything Goes" to Radio Beijing's "Music From China"—not to mention bewitching indigenous music in the tropical broadcasting bands, while such broadcasting is, strictly speaking, not "international" even though the physics of radio makes it so—music lovers tend to gloat once they've discovered shortwave broadcasting. The folk and classical musics of many nations are regularly on aural display. Want to learn Chinese? Spanish? Dutch? German? Japanese? International broadcasting can help you with language lessions. Think next of what's "cultural" in your neighborhood—from dramatic performances to church services, from sporting events to discussions on history, medicine, the arts and science. You can hear such programming emanating from many points on the globe, all free for the tuning on the shortwaves.

Look for international broadcasting stations in bands beginning at 3900, 5950, 7100 and 9500 kHz, and at 11.650, 13.600, 15.100, 17.550, 21.450 and 25.670 MHz; see "Stepping Through the Megahertz" for more about them. Then, do a little browsing—and prepare not to be too surprised if your television set stays off a lot more than it used to once you've latched on to a few favorites in the world of international broadcasting.

radio spectrum, a distinction is made between broadcasting and *tropical* broadcasting. The tropical broadcasting bands are available for use only in tropical regions, where regional mediumwave coverage is restricted by thunderstorm static. Despite the regional nature of broadcasts in the tropical bands, great distances are often spanned by the transmitters there —much to the benefit of shortwave listeners, who may hear indigenous music and languages in the the tropical bands audible nowhere else on shortwave. Look elsewhere in this chapter for two sidebar features on broadcasting: "International Broadcasting: Another World" hints that you may be missing a lot if you think that international broadcasting is just like your local stations coming from farther away, and "DX Gold on 60 Meters" introduces you to tropical-band DXing.

The *CB Radio Service* resides on frequencies near 27 MHz, but you won't find mention of it in the ITU Radio Regulations, even as a footnote. Only voice emissions, SSB and AM, are allowed. The CB radio service offers short-range personal communication and remote control of models with minimal licensing requirements.

Fixed Service. Stations in this service communicate between points on the Earth's surface, and for this reason you'll sometimes find them referred to as point-to-point (PTP) stations. Since such point-to-point communication may be undertaken with any of a wide variety of aims in mind, the range of emissions you're apt to run into in fixed service bands is great. CW, SSB and RTTY emissions are there, as are facsimile and multiplex signals. Long-distance public telephone links are achieved by fixed stations. Fixed-service call signs are configured as they are for the aeronautical fixed service, above.

Industrial, Scientific and Medical (ISM) uses of radio don't really amount to a service. But note is made throughout the ITU Table of Allocations of small bands designated for ISM operation. A microwave oven is an ISM radio device, as is an industrial brazing or induction heating machine that makes use of radio—and so are RF-operated perimeter-protection systems and intrusion alarms. Diathermy—RF heating of body tissues for the purpose of physical therapy—was an early ISM use of radio.

Meteorological Aids Service. Only one band is allocated to this service below 30 MHz: 27.500-28.000 MHz. Any signals you may hear in this band will most likely be telemetry related to meteorology, including hydrological observations and exploration. This means possible reception of *radiosondes*—transmitters carried aloft by aircraft, balloons, kites or parachutes for the transmission of meteorological data.

Mobile Service. Communication in the mobile service may be between mobile stations—stations operating "while in motion or during halts at unspecified points," as the ITU Rules specify—or between mobile stations and land stations, a land station being "a station in the mobile service not intended to be used while in motion." (The idea behind this definition is that land stations not be confused with stations in the fixed service; land stations may not work each other except during emergencies, ensuring that routine point-to-point work remains the prerogative of fixed stations.) Call signs of land mobile stations may consist of two characters followed by four digits, or two characters and one or two letters followed by four digits; land stations may sport call signs consisting of two characters and one letter, or two characters and one letter followed by not more than three digits. At some points in the Table of Allocations, ITU specifies only "mobile" as a service allocation. More often, ITU instead names one of several mobile services of special importance, each with specific characteristics. Of particular interest to us below 30 MHz are the maritime mobile and aeronautical mobile services.

Communication in the maritime mobile service takes place between ship and coast stations, "coast station" being the name for a land station in this service. Maritime listening is popular with many shortwave listeners because of the allure and mystery of signals from faraway ships and ports. Adding to this attractiveness, some coast stations transmit periodic weather and news information using CW, USB, FAX and RTTY emissions, providing listeners with a window on weather and living conditions across the watery portions of the globe. Communication between ships and shore may be made using CW, USB, RTTY and TOR/SITOR emissions. Ship stations' call signs consist of two characters and two letters, or two characters, two letters and one digit; ship stations using only radiotelephony may use call signs consisting of two characters followed by four digits, or two characters and one letter followed by four digits. Call signs of coast stations are configured along the lines detailed for the aeronautical fixed service, above. Operation in the nature of the maritime mobile service was the first widespread practical use of radio early this century—it was some time before defined "services" existed as such.

The aeronautical mobile service comprises communication between aircraft, and between aircraft and aeronautical stations, "aeronautical station" being the term for a land station in this service. Survival craft stations and emergency position-indicating radiobeacon stations may also participate in the aeronautical mobile service on designated distress and emergency frequencies. Once upon a time, CW was heavily used in this service, but swifter voice communication (USB) has all but supplanted Morse for aeronautical use below 30 MHz. (Above 30 MHz, voice operation is still king, but AM is the emission of choice.) The aeronautical mobile bands are popular with shortwave listeners because of the excitement of catching transmissions of aircraft on the fly, as well as for the easily heard VOLMET (aviation weather) broadcasts by a number of aeronautical stations on shared frequencies. In ITU nomenclature, aeronautical mobile allocations bear one of two designators, "R" (denoting frequencies to be used by aircraft flying major world air routes, and their supporting ground stations) and "OR" (frequencies to be used by aircraft off the major routes). In practice, this means that "R" bands mostly carry traffic related to commercial aviation, while "OR" allocations carry communications related to military air transport. Aircraft stations may identify themselves with call signs consisting of two characters and three letters; aeronautical stations may use call signs along the lines of those detailed for the aeronautical fixed service, above.

Radio Astronomy Service. Radio astronomy is defined by ITU as "astronomy based on the reception of radio waves of cosmic origin." You won't hear radio astronomy stations on the air because radio astronomy involves only the use of receivers!

Radiodetermination, radionavigation and radiolocation. ITU defines radiodetermination as "the determination of the position, velocity and/or other characteristics of any object, or the obtaining of information relating to these parameters, by means of the propagation properties of radio waves."

Although you won't find "radiodetermination" listed as a service in the ITU Rules, it's an important word for us because we can't define radionavigation and radiolocation without it. Radionavigation, then, is "radiodetermination used for the purposes of navigation, including obstruction warning." You'll be meeting up with aeronautical and maritime mobile radionavigation services in our hike from 9 kHz to 30 MHz; these are made use of by aircraft and ships, respectively, for navigational radiodetermination. What about that last term above—"radiolocation"? The ITU's definition: "radiodetermination used for purposes other than radionavigation." Ground-based radar, for instance, is radiolocation—it makes use of radio to determine the height, distance, speed and identity of aircraft, or the characteristics of clouds or ocean currents. Longwave aeronautical and maritime beacons serve for radionavigation: Ships and aircraft use them for safe navigation—homing in on or avoiding the place associated with each beacon transmitter. Loran, in operation near 100 and 1900 kHz, stands for long range navigation. Some of the emissions used by radiodetermination stations are rather arcane—a number of stations use spread-spectrum techniques, for instance—but the MCW signals transmitted by aeronautical and maritime radionavigation beacons are quite popular with long- and mediumwave DXers.

Space Research Service. Space research receives a few narrow allocations below 30 MHz, each associated with standard frequency and time allocations (see below). ITU defines this service as "a radiocommunication service in which spacecraft or other objects in space are used for scientific or technological research purposes."

Standard Frequency and Time Signal Service. If you've ever tuned in WWV (Fort Collins, Colorado) or WWVH (Kekaha, Hawaii), you've already had contact with a service of direct practical value to all who may receive its signals. According to ITU, this service exists "for scientific, technical and other purposes, providing the transmission of specified frequencies, time signals, or both, of stated high precision, intended for general reception." Transmissions in this service, then, approach broadcasting more closely than any other non-broadcast service. A few stations transmit standard frequency and time signals on frequencies allocated to other services; CHU, Ottawa, Canada, is an example of this.

These are the services you'll run into in one form or another between 9 kHz and 30 MHz. And now you know not only what these services are called, but something of how to identify the stations operating in each. Next, more pieces of the puzzle: What are the various frequency allocations for each service operating in the spectrum below 30 MHz?

STEPPING THROUGH THE MEGAHERTZ

With some knowledge of the regulation of radio in place, let's start at 9 kHz and head up toward 30 MHz, noting frequencies and bands of interest as we go. There's one more regulatory tidbit worth mentioning, however. Some frequencies are allocated to more than one service. Which service takes precedence in such cases? Generally speaking, ITU takes care of this by categorizing services' allocations as primary, permitted and secondary. "Permitted" services haven't been singled out in the discussion to follow; such exceptions are few, and permitted services have been treated as primary. You will come across mention of primary, co-primary and secondary allocations, however. Co-primary services have equal rights in questions of frequency choice and harmful interference. Permitted services have the same

rights as primary services, except that primary services have prior choice of frequencies when frequency plans are prepared. Secondary services have a tougher row to hoe: They must not cause harmful interference to primary and permitted services, and can claim no protection from harmful interference on the part of primary and permitted services. Co-secondary services, where such exist, can claim protection from interference on the part of other secondary services.

The various stations mentioned here are just to whet your appetite; the idea is that you should do the exploring. In fact, it is recommended that you tune along with your radio as this discussion moves through the bands. Try to get a feel for the radio spectrum as a sort of landscape, with each station and band occupying the equivalent of a *place*. It's like learning your way around a neighborhood. Here, then, is a glimpse of the 9-kHz to 30-MHz radio landscape.

9 to 535 kHz

Even though there are radio uses for frequencies below 9 kHz, ITU "jurisdiction" over radio holds only for signals 9 kHz and up. The 9- to 14-kHz segment is allocated to radionavigation, and "radionavigation" means that the first denizen you meet as you tune up from 9 kHz is the Omega navigation system. The Omega network consists of eight transmitters worldwide, each transmitting on 10.2, 13.6 and 11.3 kHz in turn, for periods of 0.9, 1.0, 1.1 or 1.2 seconds in a sequence repeating every 10 seconds. During any given segment of the cycle, three stations are transmitting. With precise timekeeping at the receiving station, differences in arrival time and phase of signals in the Omega network may be calculated to provide the position of the receiving station to within 4 nautical miles.

The next segment, 14-190 kHz, is shared by a number of services. Depending on the ITU region and the specific frequency in question, the ITU allocation may be for the radionavigation, radiolocation, fixed, maritime mobile or standard-frequency-and-time-signal services. For instance, 19.95-20.05 kHz was once used by experimental station WWVL, Fort Collins, Colorado, to provide time signals allied with WWV and WWVH, but this service was discontinued. (WWVB still holds down the Fort with standard time and frequency signals at 60 kHz.) For many decades, station GBR in Rugby, England, has transmitted time signals at 16 kHz. And megawatt-powered stations NAA, Cutler, Maine, and NLK, Jim Creek (Oso), Washington, transmit encrypted radioteletype near 24 kHz in communications with submarines. Down here, there is no "skywave" propagation as such: Signals hug the Earth and even penetrate its surface.

Between 90 and 110 kHz you may run into the pulsed signals of the radiolocation signals of the Loran C system. From 110 to 160 kHz, several military CW coastal stations may be found: Try 124 kHz for CKN, Canadian Forces Radio, Vancouver, audible over most of North America.

The fixed service has "dibs" on the 160-190 kHz slot. Power companies use frequencies in this range for signaling and telemetry purposes, transmitting radio signals along the same long lines they use for 60-Hz ac energy. Certainly more mysterious and fun are the signals of the LOWfers, or low-frequency experimenters. FCC rules allow use of unlicensed transmitters between 160 and 190 kHz with transmitter power not to exceed 1 watt input, and antenna-plus-feedline length not to exceed 50 feet. LOWfer emissions will usually be CW, and will be weak if you're not nearby. But don't be fooled:

Distances of many hundreds of miles have been bridged by these radio experimenters (who often use their initials as call signs).

A relatively new denizen of the 160-190 kHz range is GWEN, the Ground Wave Emergency Network. These are military stations sending bursts of encoded data, and they're intended to provide coverage of the United States in the event that the ionosphere is disrupted by nuclear warfare. Another governmental station, on the air more sporadically, is WGU20, Gaithersburg, Maryland, on 179 kHz. When it's active, WGU20's voice time announcements may be heard over at least the eastern half of the US.

Things really start to perk up between 190 and 415 kHz. Right above 190, and in force above 200 kHz, appear aeronautical radionavigation beacons. These are very popular with DXers because they do exactly what you'd expect beacons to do: They send their identifiers ("idents" to the longwave DX fraternity) over and over in tone-modulated Morse, around the clock—so if you listen long enough, you should be able to piece together the ident. (Beacons don't identify with call signs; they send one- to three-character identifiers often reflective of their locations. For instance, the 362-kHz beacon at Boeing Field in Seattle sends ''BF,'' a 350-kHz beacon at Meigs Field in Chicago sends ''ME'' and a Nantucket beacon on 194 kHz sends ''TUK.'') Canadian longwave beacons usually send a long dash, or continuous tone, between idents, and usually run higher power than their US counterparts. US beacons may run up to a few hundred watts of output power. Aeronautical beacons at or near major air terminals may also send continuous weather voice broadcasts superimposed over their idents; just-mentioned beacons ME, BF and TUK are among them.

There are two additional residents of the longwave beacon spectrum: Maritime radionavigation beacons and longwave broadcasters. The maritime beacons usually appear in clusters on a given frequency in a region, transmitting their single-letter idents in turn in a carefully timed cycle. These often run very low power, so they're a treat to hear. The longwave

Mediumwave Broadcasting in the US and Canada

Once, if you said "broadcasting" it was understood that you had in mind the goings-on between 550 and 1600 kilocycles per second, commonly known as "the broadcast band" and perhaps a bit later, when shortwave broadcasting began to take off, "standard broadcasting." The very-high and ultra-high frequencies (VHF and UHF) had yet to be utilized for broadcasting purposes; as soon as they were (using frequency modulation, or FM), the broadcast band became the AM broadcast band. Once the terms longwave, mediumwave and shortwave came into use, the broadcast band also became known as the mediumwave band. Along the way, the band was expanded: In the US, the mediumwave band now spans 535 to 1605 kHz; in Canada, 525-1605 kHz.

Listening in the mediumwave band gives many people their first look into the mystery and excitement of radio from a distance. Even though US and Canadian mediumwave broadcasters assume high-quality reception only in local or regional target areas ("markets"), the nature of radio propagation at medium frequencies affords much longer-distance coverage if listeners are willing to put up with variable signal quality. The conversion of a listener from satisfaction with local programming to enthusiasm for distant radio is often triggered by the chance discovery that powerful mediumwave stations several states away be heard easily—at night—just by exploring with the tuning knob.

Technically, what's going on in the AM broadcast band? Well, AM—amplitude modulated—broadcasting, for one thing! But the "AM" label for mediumwave broadcasting is a bit misleading, for two reasons: (1) shortwave broadcasting makes use of amplitude modulation as well, so, precisely speaking, each shortwave broadcasting band is also an "AM" band—and television broadcasters use AM for the video portion of their TV signals; and (2) AM is no longer the only form of modulation used by broadcasters between 525 and 1605 kHz. The reason for this is the recent upswing in the popularity of *stereophonic* broadcasting on the mediumwaves, some of which systems make use of simultaneous AM and PM (phase modulation) of the broadcast signal.

In the US and Canada, mediumwave broadcasters operate on 107 carrier frequencies spaced every 10 kHz, beginning at 540 kHz and ending at 1600 kHz. The "extra" 5-kHz chunks below 540 and above 1605 allow for sideband energy, as does the 10-kHz channel spacing throughout the band. Typically—especially as stations improve their overall audio performance during the changeover from monaural to stereo operation—

mediumwave broadcasters may provide good sound reproduction out to 10 kHz or better.

During nighttime mediumwave bandscans, you'll discover that you can hear more than one station on most channels. How is this tolerated? Actually, the situation is much more organized than it seems. Mediumwave broadcasting stations may be classified with the application of three criteria: channel/service area, authorized operating time and antenna radiation pattern. The most heavily loaded frequencies, for instance, are local channels (1230, 1240, 1340, 1400, 1450 and 1490 kHz). Stations on these frequencies usually operate with non-directional antennas and no more than 1000 watts. (Mediumwave DXers call these the graveyard channels—take a listen to any of the local channels unoccupied by your local AM stations and find out why!) Regional channels hold much of the choicest mediumwave DX. Stations here intend regional coverage and use powers up to 5 kilowatts, often with directional antenna patterns, to accomplish it. Clear channels are occupied by relatively few high-powered stations (up to 50 kilowatts), many of which use nondirectional antennas. In Canada, the difficulty of achieving adequate coverage of large regions even by "clears" has given rise to a fourth class of stations: low-power relay transmitters, or LPRTs. These are unattended 20- or 40-watt transmitters operating with non-directional antennas and fed by telephone lines.

It's easy to be too simplistic in defining the characteristics of mediumwave broadcasters, however. For instance, under the category of authorized operating time, stations may be licensed to operate daytime only, nighttime only, limited time, specified hours, shared time (with another nearby station on the same frequency) or unlimited time. "Nighttime" means the hours from sunset to sunrise. If stations are to operate throughout the periods of vastly changing propagation afforded by sunrise and sunset, they must alter their antenna patterns to "protect" (minimize interference to) other broadcasters on the same (and sometimes even adjacent) channels. Limited-time stations are "limited" for this reason. Stations operating within specified hours have received expansions over the hours otherwise allowed limited-time stations; their particular hours of operation are specified on their station licenses. Protection afforded by changes in antenna radiation pattern and/or transmitter power is far more common. Quite a few unlimited-time stations need use a directional antenna only at night; some must use directional antennas all of the time, but with different patterns day and night. Relatively few unlimited-time stations are directional only at night. A singular solution to

broadcasting band stretches from 148.5 kHz to 283.5 kHz in ITU Region 1. Longwave broadcasters use transmitter powers in the megawatt range, but even so they're a DX delicacy in North America.

The longwave beacon band is divided into 1-kHz channels, and sometimes interference is fierce during extensive openings. Nighttime skywave propagation starts to take off here, so some pretty spectacular DX is possible if local electrical interference sources calm down and atmospheric noise is low. DXers on the coasts may log intercontinental DX with perseverance. Just about every recent general-coverage receiver, including Amateur Radio transceivers sporting general-coverage receive, will cover the band of the beacons. Give them a try.

Maritime coastal stations appear between 415 and 510 kHz, including the well-known international calling and distress channel at 500 kHz (600 meters). It's all CW down here, and you'll hear a lot of traffic between coastal stations and ships in the vicinity of 600 meters.

There are a few more aeronautical radionavigation beacon channels between 510 and 535 kHz, and these are particularly worth scouring for DX because skywave propagation really steps out here relative to that afforded by frequencies in the 190- to 405-kHz range. Give a listen for INE, Missoula, Montana, on 521 kHz; and SQM, Level Island (Sumner Strait), Alaska, 529 kHz. 530 kHz is the home of Travelers' Information Service (TIS) stations. Ever been driving near an airport or along a highway and seen a sign saying something like "Travelers' Information—Tune Your Radio to 530"? If you did as the sign suggested and heard a repeated tape-loop message, you heard a Travelers' Information Service station. This service is an FCC creation; it doesn't appear in the ITU Table of Allocations. It does appear again at 1610 kHz, as discussed below.

In the 525- to 535-kHz segment, you may also hear some Central and South American broadcasters—and one Canadian broadcaster: CJFT, Fort Erie, Ontario, on 530 kHz. The reason for this is that our next broadcast allocation—"AM"

the problem of protection is embodied by station CHYR, Leamington, Ontario, which transmits a 10-kilowatt directional signal on 710 kHz during the day—and moves to 730 kHz at night!

Another important term when authorized operating time is considered with respect to antenna pattern is *critical hours,* the two-hour periods after sunrise and before sunset. A daytime-only station, for instance, may have to use a special antenna pattern during critical hours even if it uses a non-directional pattern most of the day. Life may be most complicated for stations that must utilize *three* different directional patterns: One at night, another during the day, and a third during critical hours. Occasionally, operation during critical hours may require a special transmitter power as well.

Especially during the short days of the winter months, daytime-only stations may not be able to sign on at the usual beginning of their "broadcast day," 0600 local time. To ease this hardship, FCC may allow such a station to take to the air at 0600 with drastically reduced power, sunrise or not, under a Pre-Sunrise Authority (PSA). Since PSA operation entails use of a station's daytime antenna pattern, quite a few directional-at-night unlimited time stations make use of PSAs to cover the core of their markets better than they may with their normal nighttime patterns and powers. Sometimes, however, PSAs are granted but never used for reasons of economics: It may cost so much in hardware to step a one- or five-kilowatt transmitter down to PSA power—often on the order of only two to 10 watts—that some station managements decide against implementing their PSAs.

The mediumwave DXer and broadcaster operate at cross-purposes when it comes to the concept of medium-wave DX. Aside from the relatively few clear-channel stations that boast of their multistate coverage—possible only at night—many a station management would prefer putting an unvarying immense signal into only the receivers owned by its market audience. Time was when most broadcasters could be expected to close down sometime late on Sunday for equipment maintenance and return to the air early Monday morning. Mediumwave DXers learned that this was the best time for long-haul DX. But more and more stations began to stay on more of the time, perhaps leaving the air only one Monday morning a month for maintenance. Finally, as stereophonic FM broadcasting took more and more listeners away from the broadcast band, mediumwave broadcasters turned to a continuous operation—no time off for maintenance—as a means of keeping their market shares. Alternate main

transmitters keep each such station on the air; one transmits while the other is serviced. Mediumwave DXers call this NSP (No Silent Period) operation, and lament the continuous on-air presence of more and more NSP stations as frustrating their efforts to DX.

"AM stereo" is another effort by mediumwave broadcasters to recapture listeners lost to the crisp stereo sound and static-free reception afforded by VHF FM. Several such systems existed when FCC first gave the general go-ahead to AM stereo in the early 1980s —including a few that had existed, in experimental form, for decades—but at this writing, the contenders have been reduced to two, Motorola and Kahn. The Motorola system is a combination AM/PM system similar in concept to FM stereo, which uses modulation of a subcarrier to provide the necessary audio information for stereo. The Kahn system splits the transmitted signal into two independent sidebands corresponding to right and left stereo channels. Demodulation techniques for these competing systems have almost nothing in common, leaving receiver manufacturers to decide which systems to build into AM stereo/FM stereo receivers, if not both. Which system will "win"? No one can say at this time. Will the adoption of AM stereo make any significant dent in the FM market? Media watchers are doubtful, but stay tuned.

Whatever the fortunes of individual broadcasters on the mediumwave band, mediumwave DXers continue to enjoy the challenge of listening for distant stations between 525 and 1605 kHz. The problems inherent in mediumwave DXing are peculiar to the pursuit. Listening to a distant weak signal a few tens of kilohertz from a local 50-kilowatter puts the dynamic range of any receiver to a severe test. (You'll find that manufacturers intentionally reduce the sensitivity of general-coverage receivers and transceivers below 1600 kHz because of this problem.) Since fading and multipath effects occur much more slowly on the mediumwaves than on the shortwaves, the directional characteristics of multiturn directional loop antennas may routinely be used by mediumwave DXers in nulling unwanted stations and noise.

Your mediumwave DX efforts needn't be limited to intracontinental listening, by the way. With effort, many transpacific and transatlantic signals may be heard on "splits"—frequencies between the 10-kHz channels assigned to North American mediumwave broadcasters. Many stations in Europe, Africa and Asia operate with output powers of many hundreds of kilowatts, so if you should somehow tire of North American DXing (listening to the "domestics"), you'll have other worlds to conquer.

or "standard" broadcasting, also known as mediumwave—begins at 526.5 kHz in ITU Regions 1 and 3, and at 525 kHz in Region 2. It's lightly used by broadcasters in Region 2, however; ITU specifies that the output power of Region 2 broadcasters in this segment not exceed 1000 watts day and 250 watts night.

With this, so ends our brief jog through radio's "downstairs," below 535 kHz. Next is the medium-frequency (MF, 300-3000 kHz) range. The segment 525-1605 kHz—North American mediumwave broadcasting—deserves special treatment, and you can find detailed coverage of "the broadcast band" in the accompanying sidebar. After you've taken a look at the mediumwave band, proceed to the territory above 1605 kHz.

1605 to 2300 kHz

It's hard to know exactly where to chop the 1600-kHz to 30-MHz spectrum into inspectable bits, but stopping at 2300 kHz allows us to pull up short of bumping into our first "shortwave" broadcast band. This is still the medium frequencies; what most folks call shortwave doesn't really start until the high frequency (HF) range begins at 3000 kHz (3 MHz), but "shortwave" has always been a handily loose term. Skywave propagation is pretty good at night here, though, so worldwide DX is normal under the right conditions. This section concentrates on allocations in Region 2 unless additional allocations in other regions imply interesting DX.

At the 1979 WARC, the mediumwave broadcasting band was extended to include 1605-1705 kHz, contingent upon acceptance of this expansion by telecommunication administrations in Region 2. And that acceptance was quite a topic of debate as this chapter was written, for a number of reasons—one being that the Amateur Radio service might have to give up portions of its 160-meter (1800-2000 kHz) allocation to stations displaced from 1605-1705 kHz. In the US, FCC and broadcasters haven't been seeing eye to eye on who should be allowed access to this spectrum, should it become available. So, as of this writing, even though official ITU documents show broadcasting from 1605-1705 kHz, it isn't there yet. Let's take a look at what is.

Between 1605 and 1800 kHz, you may find signals from stations in the fixed, mobile, radiolocation and aeronautical radionavigation services. A few out-of-band broadcasters operate just above 1605 kHz: Try for the Caribbean Beacon, Anguilla, with 50 kilowatts on 1610 kHz, and Australian radio-for-the-print-handicapped stations on 1620, 1629 and 1692 kHz. More Travelers' Information Service stations may be found co-channel with the Caribbean beacon on 1610 kHz; since skywave propagation is better on 1610 than on the other TIS frequency (530 kHz), be prepared for some surprising TIS DX here when conditions allow. There are more denizens in the 1605-1800 kHz range, though: Unlicensed "pirate" broadcasters sometimes operate in this segment as well. Before the marketing of cordless telephones using 49 MHz for both base and handset transmitters, cordless phones used FM on frequencies between 1605 and 1800 kHz for their base-to-handset links. These phones still abound. Some hobbyists find such listening fun; if you do, remember that Section 705 of the Communications Act of 1934 forbids you to divulge what you hear.

Much of the radio action between the broadcast band and 1800 kHz, though—radiolocation—causes nothing more than odd sounds in our receivers. Two types of radiolocation

signals are commonly heard here: One sounds somewhat like crickets (you have to listen with you receiver's beat-frequency oscillator turned on) calling and answering on slightly different frequencies, and the other, a system marketed by Decca, results in signals sounding like repetitive "Js" in Morse code. Again, you need to have your radio set up for SSB or CW reception to hear this one. You may hear other whines and bleeps between 1605 and 1800 kHz; if so, chances are good that you're tuned into some sort of radiolocation emission.

If you recall the discussion of what's going on down at the longwave frequencies between 190 and 415 kHz, "aeronautical radio navigation" means "beacons" to you. That's what's going on when aeronautical radionavigation takes to the air between 1605 and 1800 kHz. Especially in Canada's far northern areas, and in Central and South America, quite a few tone-modulated CW beacons are still sending their idents in this medium-frequency range.

The range 1800-2000 kHz is "160 meters" to radio amateurs. Once upon a time, the amateur 160-meter band went all the way down to 1715 kHz, later shaved a bit to 1750 kHz. After WW II, though, pressures from other services for spectrum space meant shrinkage to the present band limits. Such inroads have continued right up to the present day, when amateurs must share 1900-2000 kHz with radiolocation. Still, 160 is a popular band, with intercontinental DX possible, especially right before dawn at the eastern area of a given propagation path. You'll know it's DX time on 160 if you begin to hear nonamateur signals from some of the services allocated these frequencies in Regions 1 and 3. A number of LOWfers (low-frequency experimenters; remember when we met them battling against adversity with 1 watt and 50-foot antennas between 160 and 190 kHz a while back?) use their amateur privileges and 160 meters for liaison if 1750-meter signals are down in the mud.

Frequencies between 2000 and 2300 kHz are allocated to the fixed, mobile or maritime mobile services. The segment 2173.5 to 2190.5 kHz is of particular interest because it includes the international distress and calling frequency of 2182 kHz. Set your radio for upper sideband reception and tune to 2182 kHz as darkness falls, especially if you live near a seacoast or inland waterway: 2182 should provide you some action before too long.

2300 to 2495 kHz

If this chapter concerned merely allocations in the US and Canada, discussion of 2300 to 2495 kHz would net little more than routine fixed and mobile allocations much like those between 2000 and 2300 kHz. But in equatorial and neo-equatorial areas of the world, 2300-2495 kHz is allocated to tropical broadcasting. Shortwave listeners call this the 120-meter band, and perseverance on this lowest-frequency "shortwave" band pays off in some real DX treats. Asia, Oceania, Africa and Central and South America are represented by low-powered stations intended for local coverage. Australia had just begun regional services on this band as this was being written. Indonesian stations are everywhere, and everywhere weak—challenge! High powered stations in the Democratic People's Republic of Korea [North Korea] (2300 kHz) and the People's Republic of China (2430 and 2490 kHz) may be quite audible around dawn, especially on the Pacific coast of North America. Weaker Chinese regional broadcasters may be found at 2340, 2415 and 2475 kHz, among others. 2380 kHz is used by the

Falkland Islands Broadcasting Station during winter in the Southern Hemisphere; this has been heard by quite a few patient DXers in North America. A number of Brazilian stations may be found at 120 meters, most operating with powers around 1 kilowatt. Plenty of DX down here if your local line noise cools down, you can put up with static (least prevalent in winter) and propagation goes your way.

2495 to 3500

Stations WWV, Fort Collins, Colorado, and WWVH, Kekaha, Hawaii, emanate standard-frequency-and-time signals centered on 2500 kHz; the allocation for this service spans 2495 to 2505 kHz. Other such stations use this frequency elsewhere in the world. Up to 2850 kHz, there's another nondescript fixed and mobile allocation. This isn't to say that there's no DX up here: Intercontinental propagation becomes more and more reliable as we head into the 3-MHz region, allowing us to hear voice, radioteletype and Morse transmissions from sites worldwide. 4XZ, Haifa, Israel, repeating its CW marker on 2800 kHz, is a good propagation indicator for listeners in eastern North America. Powerful out-of-band Asian broadcasters (China on 2600 and 2800 kHz, and drifty transmitters in North Korea near 2696, 2745, 2767, 2775, 2850 and 3015 kHz) make good propagation indicators for western listeners even if they never become strong enough to allow the recovery of audio.

Between 2850 and 3155 kHz, though, is our first aeronautical mobile allocation. This is used for "route" operation below 3025 kHz, and "off route" work above. If you set your receiver for upper sideband and tune through the "route" range during hours of darkness, you'll likely run into conversations between intercontinental flights and major air terminals such as New York, Oakland and Honolulu. "Off route" listening usually nets military flights. There are radioteletype and CW signals in there, too. You may hear 3023 kHz used by mobile and land stations participating in search-and-rescue operations. East Coast listeners, try for Italian CW station IDR8, Rome, sending its V marker on 3039 kHz.

Once again, you stand to miss out on something good if you look at the spectrum between 3155 and 3400 kHz only in terms of Region 2. In the US and Canada, this is another allocation to the fixed and mobile (excepting aeronautical "route") services. A number of Canadian radiotelephone circuits link remote northern camps with civilization in the 3200-2400 kHz range; Canada's standard-frequency-and-time station CHU transmits full carrier and upper sideband on 3330 kHz. A European treat worth tuning for are the wispy CW signals of time station OLB5, Prague, Czechoslovakia, on 3170 kHz. The best shortwave listening action in this radio neighborhood, though, is arguably the 90-meter tropical broadcasting band, 3200-3400 kHz. You wouldn't be argued with—much—if you thought that the broadcaster population was on the sparse side down at 120 meters. Such is not the case between 3200 and 3400 kHz! This is a premier DX band offering reception of many rare stations in Africa, Asia and Central and South America. Take a look at the *World Radio TV Handbook* for an idea of what's here—and don't be surprised if you're compelled to glance at your watch and fidget between now and the hours of darkness!

Aeronautical mobile has another "route" allocation at 3400-3500 kHz. You'll run into more air-to-ground conversations here. With perseverance, you may also intercept aviation weather ("VOLMET") transmissions in this allocation. Give 3485 kHz, upper sideband, a try; stations sharing this frequency (in North America) are located in Gander, Newfoundland, and New York City.

3500 to 4000 kHz

All 500 kHz of this band are allocated to the Amateur Radio Service in the US and Canada. This is the 80-meter band, and it makes for enjoyable listening whether or not you're a licensed radio amateur. US stations use CW and radioteletype emissions from 3500 to 3750 kHz, and you can find excellent code practice from two sources in this 250-kHz stretch: ARRL's code-practice and bulletin transmission station, W1AW, may be heard just above 3580 kHz, and the many ham CW conversations audible anytime are good practice, too. Beginning hams holding Novice and Technician class licenses may be heard using CW from 3700 to 3750 kHz, and their leisurely sending pace offers good practice for shortwave listeners getting into the business of copying code. Between 3750 and 4000 kHz, nearly all the amateur action is on lower-sideband radiotelephone. Hams using phone in the 3500-4000 kHz band call it "75 meters"; the CW and RTTY users call it "80."

There's more to this band than Amateur Radio, however. In ITU Regions 1 and 3, and in some parts of South America, stations in the fixed, mobile and aeronautical mobile services may be heard. Standard-time-and-frequency-station HD2IOA, Guayaquil, Ecuador, transmits on 3810 kHz. Of particular interest to the shortwave listener is the 3950-4000 kHz (75-meter) international broadcasting band available to stations in Regions 1 and 3. Especially in eastern North America, strong signals from well-known European broadcasters such as the BBC, Swiss Radio International and Deutsche Welle may be audible after dark. Choice domestic transmissions from Asian and African stations may be found between Region 2 Amateur Radio signals. Two rare shortwave broadcasting treats may be found on 75: Greenland on 3999 kHz and the Falkland Islands Broadcasting Service on 3958 kHz (Southern Hemisphere winters).

4000 to 5060 kHz

Just above 75 meters, you'll often run into stations that may remind you of Amateur Radio while using nonamateur call signs; you may hear such operation just outside of several HF amateur bands. These are stations in the Military Affiliate Radio System (MARS); often, the operators are amateurs doing their bit to help pass traffic for servicemen and women to their loved ones back home. Between the MARS stations and 4063 kHz, you may find a few upper-sideband signals from ships in the maritime mobile service. The real maritime mobile action, though, takes place between 4063 and 4438 kHz. This band is heavily used for contacts between ships and coastal stations on CW, voice, radioteletype and TOR/SITOR. (TOR and SITOR are error-correcting radioteletype systems coming more and more into use on the maritime bands. Their signals sound like short bursts of cricket chirps.) 4125 kHz is used as a distress and calling frequency supplemental to 2182 kHz in ITU Regions 1 and 2 south of latitude 15 degrees North, including Mexico, and in Region 3 south of latitude 25 degrees North. You can hear signals in this band around the clock, but darkness brings the best DX—and that DX may arrive from all over the globe if you monitor with patience.

As is the case throughout much of the shortwave spectrum allocated to nonbroadcast services, out-of-band broadcasters

DX Gold on 60 Meters

In the US and Canada, 60 meters (4750-5060 kHz, with 4995-5005 kHz set aside for standard frequency and time transmissions), the band containing what many listeners consider to be the most exotic shortwave broadcast DX of all isn't allocated to broadcasting—it's given over to fixed and mobile operation. During local daylight hours, you're likely to hear only a few radioteletype or voice transmissions. If you wait around for the onset of darkness, however, you'd better be prepared to make a grab for the logbook—things start to happen fast when the DX gold on 60 meters starts rolling in.

In eastern North America, 60-meter signals from across the Atlantic start to swell between 1900 and 2000 UTC. There's too much daylight yet in the Americas for us to hear the many Central and South American stations populating 60, but they'll have their chance after stations from Europe, the Middle East and Africa have come and gone. Although the ITU intends that broadcasting on 60 meters be confined to tropical regions, the USSR operates a number of transmitters here, and those in the Soviet west may be logged, if propagation allows, just before they sign off for the day, generally at 2000, 2100 or 2200 UTC. The Middle East is represented by Radio San'a of the Yemen Arab Republic, signing off at 2115 UTC on 4853 kHz.

The focus shifts to African stations. How many can you log before they sign off? With the earth's rotation into night being what it is, East African stations (Botswana [4820 kHz], Kenya [4915 and others], Swaziland [4980], Tanzania [4785], Uganda [varying near 5025]) will be afforded the shortest window by propagation before they leave the air at between 2000 and 2100 UTC (local midnight) or so (Mozambique [variable near 4735, among others] to 2200); Central Africans (Angola [several], Benin [4870], Central African Republic [varying near 5035], Equatorial Guinea [4925 or 5005], Ghana [4915], Lesotho [4800], Liberia [4760], Namibia [4935 or 4965], Niger [5020], Nigeria [4770 and 4990], Sao Tome & Principe [near 4805], Sudan [5039]) sign off at 2200 or 2300 (Chad [near 4904] at 2100); and West Africans sign off at around 2300 to 2400 (Burkina Faso [4815], Cameroon [4850 and variable near 4972], Gabon [4777 and 4810], Mauritania [4845], Senegal [4890], Togo [5047]). Use of 60 meters by the Republic of South Africa is variable, depending on season and target.

In the eastern half of the Americas, darkness is well on its way by 2300 UTC, and the closing African stations mingle with Spanish and Portuguese voices from Central and South America. Just as quite a few African states may be heard best (or only) on 60 meters, so it is with many American tropical broadcasters. You'll hear music varying from the lilting to the haunting, depending on whether it's rolling out of Venezuela or Colombia or coming from high in the Bolivian and Peruvian Andes. Brazilian stations are everywhere, sometimes three or four deep on a channel! Depending on station location and management, South American stations sign off for the night between about 0130 and 0500 UTC, clearing their channels for the reappearance of Africa...and for the appearance of Asia and Oceania.

Since 0600 is a usual sign-on time for local radio stations, the East Africans (ahead of UTC by three hours) you may have missed before 2100 UTC return to the air starting at 0300. During this sign-on cycle, your African loggings will be limited by the increasing illumination of the propagation path as sunrise overtakes them and approaches you. The Africans that faded in during your local afternoon will fade out with their brightening day. As

---

may also be found on 4-MHz maritime frequencies. Soviet and Mongolian outlets may be found between 4000 and about 4080 kHz, and a number of stations from the People's Republic of China and North Korea operate on frequencies throughout the 4-MHz maritime allocation.

Between 4438 and 5060 kHz, there are great differences in frequency allocations, depending on the area of the world in question. There is a strong Soviet broadcaster on 4485 kHz at Petropavlovsk-Kamchatskiy (on the Soviet Pacific coast) often offering fine music listening after dark in the North American West. And Australian standard-time-and-frequency station VNG, Lyndhurst, comes in nicely just before our North American dawn on 4500 kHz. Worldwide, 4650-4750 kHz is aeronautical mobile territory, with "route" operation below 4700 kHz and "off route" work above. If you confined your DX activity to gunning for stations operating under allocations made in Region 2, you might soon be yawning again: You'd gear up to hear just the good old fixed and mobile (but no aeronautical "route"!) services, kHz after kHz. There is a jewel of a DX preserve in this frequency range, however: The 60-meter tropical broadcasting band. Although its official limits are 4750-5060 kHz in Regions 1 and 3 (with a 10-kHz chunk centered at 5000 kHz set aside for standard-frequency-and-time stations like WWV and WWVH), your listening will reveal that band edges aren't that well defined in practice. Many broadcasters operate above and below the 60-meter band's limits. For more on what you may hear if you hit 60, see the accompanying sidebar, "DX Gold on 60 Meters." But, beware!—you may be in danger of becoming one of the shortwave broadcast DXers who consider

60 meters to be *the* premier shortwave broadcast DX band.

5060 to 5950 kHz

There're more fixed and mobile allocations between the top of the 60-meter band and 5450 kHz. Aeronautical mobile takes over from 5450 to 5730 kHz, with "off route" operation above 5680 kHz. Here you may find more air-to-ground communications and VOLMET aviation weather transmissions, using upper-sideband emission. As was the case with 3023 kHz, 5680 kHz may be used for search-and-rescue intercommunication between mobile stations, and between mobile and land stations. Several clandestine Central American broadcasters may also be intercepted in this range, as well as out-of-band Chinese broadcasting outlets.

Our friends "fixed" and "mobile" return again in the stretch 5730 to 5950 kHz, as well as more Chinese broadcasting stations. We're about to run into the 49-meter shortwave broadcast band at 5950 kHz. In practice, however, you'll find that some major shortwave broadcasters operate all the way down to 5900 kHz (the USSR, Czechoslovakia, Israel and Austria, among others), and a few even go lower. Then, tune up and across the 5950-kHz boundary and get ready with your receiver's input RF attenuator: You're in the 49-meter band.

5950 to 6200 kHz

Broadcasters had hoped to receive an expansion of the 5950-6200 kHz 49-meter band at the 1979 WARC, but they didn't. You'll find out for yourself how crowded this band is when you tune across it after dark: There are few, if any,

this occurs, you're left pretty much with a few 24-hour stations in the Americas for company—that is, until stations east of Brazil sign on again after 0800 or so.

With darkness still in place in the Americas, and heading west, Port Moresby, Papua New Guinea, soon pounds in on 4890 kHz—with only 10 kilowatts of transmitter power. Lively programming from the Solomon Islands may be heard on 5020. Soon, propagation steps out to Indonesia—and if you can't hear the 50-kilowatt station at Ujung Pandang on 4719 kHz, you probably won't hear any other Indonesians that day on 60!

Central American and the majority of South American stations begin to return to the air at 1000 and 1100 UTC, and this is a good chance to hear rare ones if their channels are normally blocked by Africans during your local evening—in addition to the likelihood that noise from local electronic interference sources, electric razors and televisions may be a bit easier on your ears as long as most of your neighbors are still asleep! Now's the time to hear the Far East, too—Chinese, Korean, Malaysian, Mongolian, Soviet and Vietnamese broadcasters may afford choice listening before your local sunrise washes them into oblivion.

The sunset-sunrise cycle just described pertains mainly to 60-meter reception in the eastern half of North America. Listeners on the West Coast will find African stations a much tougher challenge; East African stations may not even be audible in the period leading up to 2400 UTC because the western end of the path is still in daylight. On the other hand, during a few weeks in midwinter, West Coast listeners may be able to hear East African broadcasters with the help of *gray-line* propagation—propagation along earth's terminator between two points on the terminator—around 1500 UTC. Gray-line propagation works well in the 60-meter band, so you may be able to put it to good use if one of your 60-meter targets is on the air when you and your target are linked by the terminator at some time during the year.

Listeners in western North America may enjoy Asian reception that East Coast listeners only dream about—Asia is to West Coast 60-meter DXers as Africa is to East Coast folk. Any Asian audible on the East Coast is moderate to strong in the west. Many other stations, such as Burma on 4725 kHz, Nepal on 5005 kHz, and Indian and Chinese regional broadcasters, make regular showings in West Coast logs.

Receiver selectivity and interference rejection features are tested well by the conditions on 60 meters because so many of the stations there operate anywhere but smack on 5-kHz channel spacings, their closely-spaced carriers giving rise to strong heterodyne interference. Quite a few stations don't stay put, either, as some of the frequencies given above reflect (Mozambiquan and Vietnamese stations are among the most notorious).

QSLing most of the 60-meter broadcasters you hear will be a test of your patience at best; most are local stations operating in regions where catering to the whims of distant hobbyists is low to nonexistent on the priorities list. One 60-meter "beacon," Radio Botswana, easily identified as it transmits the sounds of cow bells and barnyard animals prior to its 0330 sign on, had a firm "no-QSL" policy for two reasons, one being tactless demands and falsified reports from DXers of years past. The other reason, related by its chief engineer, was that Radio Botswana is not interested in reports of reception outside of Botswana—that if anyone can hear the station outside of the country, it could be argued that he's not doing his job!

unoccupied usable frequencies. Several major European broadcasting organizations (BBC, Deutsche Welle, Radio France International and Radio Netherlands) operate relay stations in the Caribbean or South America, dumping enormous skywave signals into North America at night. (For the BBC, try 6175 and 5975 kHz; Deutsche Welle offers English on 6045 and other frequencies; Radio Netherlands is on 6165 kHz; and Radio France International uses 5950 kHz.) Enjoyable programming from the Voice of Free China can be heard via the strong 6065- and 5985-kHz transmitters made available by WYFR, Family Radio, Okeechobee, Florida. Radio Canada International has had a strong 49-meter presence for many years on 5960 kHz from its Sackville, New Brunswick, site.

But 49 meters isn't all "big boys." There are many choice DX morsels from Europe, Oceania, Asia, Africa and the Americas between the rock-crushing voices of the international broadcasters on the band—and a standard-frequency-and-time signal from YVTO, Observatorio Naval Cagigal, Caracas, Venezuela, on 6100 kHz! It'll take your skill and patience—along with a receiver capable of taking the strong signals some 49-meter users dish out—to find the DX.

6200 to 7000 kHz

The blend of signals available in this range is tantalizing. A few out-of-band broadcasters (some of the "legit" and some of them "pirate" or clandestine operators) hover just above the 6200-kHz upper limit of the 49-meter band, but the rightful users deserve your attention, too. The maritime mobile service has claim to 6200-6525 kHz, and here you'll find the mix of CW, RTTY, phone and TOR/SITOR signals you ran into in the 4-MHz maritime mobile band. Propagation brings stronger signals from distant places up here, though, as compared with 4 MHz, so try a little maritime mobile DXing here and now if you haven't done so already. 6 MHz is low enough so that you may still expect the best DX action during the hours of darkness. South of latitude 25 degrees N in ITU Region 3, 6215.5 kHz is used as a distress and calling frequency supplemental to 2182 kHz.

The 6525-to-6765-kHz segment "belongs" to the aeronautical mobile service. Again, air-to-ground communication is the thing. It's "route" operation below 6685 kHz, "off route" above. You may hear longer-distance VOLMET weather transmissions here than you may have at 3 or 5 MHz; Gander and New York Radios share 6604 kHz, upper sideband, while Honolulu, Hong Kong, Auckland and San Francisco share 6679 kHz. Most of these are audible all over North America, with patience.

Stations in the fixed service operate between 6765 and 7000 kHz, with a small band designated for industrial, scientific and medical (ISM) devices from 6765-6795 kHz. Many of the fixed stations you'll run into here are radioteletype and CW outlets, but you may occasionally run into shortwave broadcast "feeders"—signals carrying broadcast audio from one transmitter site to another, not intended for reception by listeners. Most feeders will be operating with a reduced carrier—"AM" or "diode" (rectification) detection of such signals will net you severely distorted audio, so you'll have to tune them in as single sideband. From time to time,

Listening to China

If the next 40 people you meet represent an accurate cross section of the citizenship of the human population of the world, 10 of them will hail from the People's Republic of China (mainland China). More people speak Standard Chinese (also known as Mandarin, or, on the Chinese mainland, putonghua ["common speech"]) than any other language in the world! But you can listen to broadcasts from China in English, Spanish, Portuguese, French, German—and in several Chinese dialects. Chinese domestic shortwave transmitters can be found on mysterious and easy-to-hear out-of-band frequencies. The sounds of Chinese flutes, cymbals, strings and vocalists arrive through a veil of polar propagational flutter.

China's radio broadcasting is handled by the Central People's Broadcasting Station, under the Central Broadcasting Administration. Your first brush with radio from China will likely be the English transmissions of China's overseas service, Radio Beijing (a name change from Radio Peking, "Beijing" more closely approaching the proper pronunication of the Chinese capital for English speakers). Radio Beijing broadcasts in over 40 languages, including transmissions in several Chinese dialects. Its English transmissions include programs on Chinese cooking and culture, such as "Listeners' Letterbox," "China Scrapbook," "Music Album" and "Music From China."

There's greater allure in China's use of shortwave frequencies for domestic use. The Central People's Broadcasting Station (CPBS) operates two domestic networks in Standard Chinese, using frequencies from 3220 kHz to 17.710 MHz to cover the whole of mainland China. Widely heard frequencies for the first network include 6665, 7504, 7516 and 9800 kHz, and 10.245, 11.330, 15.550, 15.590 and 17.605 MHz; the first network operates from 1958 to roughly 1735 UTC each day, with a few off periods during the week (apparently for equipment maintenance). Try for the second network on 6890 and 9020 kHz, and 10.260, 12.200, 15.030, 15.500 and 17.710 MHz; the second network operates from 2100-1600 UTC. The identification you'll hear is "Zhongyang renmin guangbo diantai" ("Central People's Broadcasting Station") in Standard Chinese, at the start or finish of programs on the hour, or at 15, 30 or 45 minutes past the hour.

There are three additional centralized shortwave domestic services in operation on the Chinese mainland. Taiwan is targeted by two special CPBS networks in the first of these services, often called "Taiwan 1" and "Taiwan 2" by shortwave listeners. Key Taiwan 1 outlets may be heard on 9380 and 9455 kHz, and 11.100 and 15.710 MHz; try for Taiwan 2 on 6790 and 9170 kHz, and 11.000 and 15.880 MHz. Another service for Taiwan, again mounted via two networks, is "The Voice of the Strait"

("Haixia zhi sheng guangbo diantai"). It is broadcast to Taiwan via powerful transmitters in China's Fujian province. Try for Haixia 1 on 2490, 3200, 3535, 4380, 5240, 5265, 6765 and 7850 kHz. Good bets for Haixia 2 are 3400, 4330, 5170, 7025 and 7280 kHz.

Finally, still under the auspices of the Central People's Broadcasting Station, are the Domestic Minority Language Services. Languages used include Korean, Tibetan, Kazakh, Uygur and Mongolian. Half-hour transmissions may be heard on a combination of these frequencies, depending on the season: 4190, 5145, 5420, 6430, 7035, 8566 and 9920 kHz, and 11.375, 11.675, 15.670 and 17.635 MHz.

Frequency usage in the services mentioned up to this point varies seasonally because of the relatively distant targets involved. Despite this, the Central services are relatively easy to hear, and serve well as propagation beacons because of their wide distribution throughout the shortwave spectrum (and, it's speculated, throughout China). China's Provincial People's Broadcasting Stations are far rarer DX targets even though they tend to stay put with regard to operating frequency. Not all of China's 21 provinces and five autonomous regions operate their own shortwave stations, but those that do may be found between 2000 kHz and 10 MHz, more often than not on in-band channels. There's mysterious Yunnan on 6937 kHz, and the Nei Menggu (Inner Mongolian) Autonomous Region on 6974; or try for Qinghai on 6260 and 9780 kHz. Sichuan may be heard on 7225 kHz when interference allows. Iffy propagation and varying interference may make necessary several years' effort in stalking an elusive Chinese regional.

Taiwan (the Republic of China) operates its own broadcasting administration, the Broadcasting Corporation of China (BCC). Aside from broadcasts intended for Taiwan itself, BCC is responsible for broadcasting to the mainland under the auspices of the Central Broadcasting System (Chungyang kuangpo tientai in Mandarin Chinese). You may be able to hear the three strongest BBC outlets on 7150 and 7250 kHz, and 11.905 MHz. Far more audible is Taiwan's international service, the Voice of Free China. While VOFC is audible direct from Taiwan, you'll have a hard time not hearing its transmissions as relayed via WYFR in Okeechobee, Florida, during North American evenings on 5985 kHz. Many SWLs consider VOFC's travel, folk tale and cooking programs listening treats.

The World Radio TV Handbook contains much more detail on the Chinese broadcasting scene. And remember that this sidebar only scratches the surface of what may be heard on radio from China—Amateur Radio is catching on there, and you may also listen to Chinese stations in the fixed, aeronautical mobile and maritime mobile bands.

---

you'll run into a feeder running *independent* sideband emission—with a different language on each sideband!

Long-time out-of-band broadcasters add a further slant to radio life between 6200 and 7000 kHz. As you might guess from our discussion of lower frequencies, the People's Republic of China and North Korea account for most such signals here. The latter may be found on 6576 and 6600 kHz, among other frequencies, and there are at least 20 Chinese outlets in the 6200-7000 kHz range. Of these, 6665 and 6790 are oft-heard domestic outlets, with Radio Beijing transmitters on 6933 and 6955 kHz sometimes putting in booming signals at the right time of year. Rare Laotian, Vietnamese and Chinese regionals may come through if propagation is good enough. This is not to mention flea-power South American out-of-banders, and several European "pirates," such as Radio Dublin, Ireland, on 6910 kHz. No, you're not likely to run into "radio boredom" between 6200 and 7000 kHz!

7000 to 7300 kHz

In Region 2, 7000-7300 is Amateur Radio's popular 40-meter band. You can find someone working DX, passing traffic or just plain conversing at any time of day on 40. Non-voice operation (CW, radioteletype and packet radio) holds sway from 7000 to 7150 kHz (except in Caribbean insular areas, where 7075-7100 kHz also supports voice work), including W1AW code practice and bulletin transmissions at 7080 kHz and the Novice/Technician segment from 7100 to 7150 kHz.

Most ham activity between 7150 and 7300 kHz is lower-sideband voice operation. 40-meter amateurs have something else to contend with, especially between 7100 and 7300 kHz: Strong shortwave broadcast signals in the 41-meter international broadcasting band.

In Regions 1 and 3, the amateur allocation ends at

7100 kHz, and shortwave broadcasters have exclusive use of 7100-7300 kHz. This makes for heavy interference to Region 2 amateur operations by broadcasters. According to the letter of the ITU regulations, 41-meter broadcasters are not to use the band for broadcasting to targets in Region 2. A quick spin of the dial any evening will show that this rule is not always obeyed. Worse, there are a number of broadcasters who routinely use frequencies in the exclusively amateur 7000-7100 kHz segment, such as Radio Beijing and Radio Tirana (Albania). Radio Tirana and Radio Moscow, among others, use 41 meters for broadcasting to North America.

7300 to 9500 kHz

Broadcasters also make heavy use of frequencies between 7300 and 7500 kHz—even our own Voice of America, WYFR-Okeechobee, and WRNO-New Orleans. The BBC has a long-standing channel on 7325 kHz; Radio Prague has been on 7345 kHz for many years. The USSR, Israel, China and Vietnam also operate in this segment. Perhaps the best-known resident here is CHU, Ottawa, Canada, transmitting a standard-time-and-frequency signal on 7335 kHz with full carrier and upper sideband. 7500 kHz is home to standard-frequency-and-time signals from VNG, Lyndhurst, Australia.

Actually, the 7300-8195 kHz stretch is allocated to the fixed service. Most of the CW and RTTY you hear in this segment are fixed transmitters. Still, on up toward 8 MHz, more out-of-band broadcasters may be found: A scattering of outlets from the People's Republic of China (7504, 7516, 7770, 7850 and 7935 kHz, among others), the Republic of Korea (7550 kHz), and Bulgaria (7670 kHz). You may also run into more broadcast feeders (touched upon in our discussion of 6200-7000 kHz). Standard-time-and-frequency signals from HD2IOA, Guayaquil, Ecuador, may be heard on 7600 kHz. Japanese standard-time-and-frequency station JJY, Tokyo, may be heard on 8000 kHz, in parallel with other JJY outlets on 2500 and 5000 kHz, and on 10 and 15 MHz.

There's much worth the listener's while between 8195 and 8815 kHz: It's the 8-MHz maritime mobile band, as heavily used by maritime interests as 40 meters is popular with amateurs. As is the case with lower-frequency maritime bands, CW, radiotelephone, radioteletype and TOR/SITOR are the emissions you'll surely hear here. Once you get used to the band, and if you've brushed up on your Morse code, you'll find maritime weather broadcasts to be of great value for code practice. 8364 kHz is designated for use by survival craft to communicate with aeronautical and maritime mobile stations engaged in rescue work. Several Asian out-of-band broadcast outlets, chiefly those of Radio Beijing and the Chinese domestic broadcast service, may be found squatting in the 8-MHz maritime mobile band.

At 8815 kHz, the focus of radio operations shifts from the sea to the air, for 8815 to 9040 kHz is another aeronautical service allocation. As usual, the higher portion of the band is reserved for "off route" operation, the dividing line being 8965 kHz. VOLMET aviation weather broadcasts are audible on this band, as they are at 3, 5 and 6 MHz. This segment harbors out-of-band broadcasters of its own: For decades, Israel has transmitted at 9009 kHz, Chinese domestic transmissions are heard on 9020 and 9030 kHz, and Radio Tehran is often strong on 9022 kHz.

The fixed service takes over again from 9040 to 9500 kHz. Another major broadcasting band resides at 9500-9900 kHz, so, as you might imagine, there is more use of this fixed band by broadcasting as we approach 9500 kHz. More Chinese

outlets may be found at 9064, 9080 and 9290 kHz; Radio Nacional de Espana booms in on 9360; there are a host of stations clustered from 9375 to 9390 kHz; and 9400-9500 kHz might as well be considered a broadcasting allocation! There's another long-time BBC outlet on 9410 kHz, Israel and Greece use frequencies in the 9420s and '30s, the Voice of America uses 9455, and Radio Cairo is on 9475 kHz. (If you're wondering why more out-of-band channels are listed than in-band ones, it's because out-of-band slots tend to remain unchanged over time, far more so than in-band channels. Stations mentioned in this chapter have been on out-of-band channels for many years, and will likely stay there. In band, where competition and interference are fierce, there's a lot of leapfrogging.)

9500 to 9900 kHz

This is the 31-meter shortwave broadcast band, and there is much congestion here. Stations tend to play "musical chairs" with their frequencies so much in such a major broadcasting band that it's tough to name stable residents. But there are some. Radio Japan has long used 9505 kHz. Radio Australia, with its near-ideal all-water transequatorial path to North America, is sometimes the only signal on the 31-meter band during geomagnetic storms; try for it on 9580 kHz around UTC noon. Radio Netherlands' Caribbean relay booms in on 9590 kHz. HCJB, the friendly Voice of the Andes, is one of the first stations heard by many short-wave listeners—try for it on 9745 kHz. Radio France International operates two 500-kilowatt powerhouses on 9790 and 9800 kHz. By the letter of the ITU regulations, the 9775-9900 kHz portion of 31 meters is not officially available to broadcasters, and wouldn't be until 1990 or so. But the fact is that these frequencies are heavily used by broadcasters. Yet, stations in the fixed service—the soon-to-be-erstwhile rightful citizens between 9775 and 9900 kHz—still ply their trade in Morse, radioteletype and voice.

9900 kHz to 10.100 MHz

The first 95 kHz of this segment is given over to the fixed service, but out-of-band broadcasters live here, too. A long-time BBC channel, 9915 kHz, is used for broadcasts to North America. India, Israel, Belgium, North Korea, Taiwan and the People's Republic of China are also represented.

Standard-frequency-and-time stations use 9995 kHz-10.005 MHz; you'll find WWV and WWVH here. "Route" aeronautical mobile takes over from 10.005 to 10.100 MHz. 10.051 MHz is another frequency shared by Gander and New York Radios for VOLMET activity. The Voice of Vietnam broadcasts out of band in this segment; its 10.010- and 10.040-MHz outlets are commonly audible in North America.

10.100 to 10.150 MHz

This is the 30-meter band—one of Amateur Radio's triumphs at the 1979 WARC in Geneva. US and Canadian amateurs are limited to nonvoice modes here, and since 30 is shared with the fixed service on a secondary basis, hams must avoid causing any interference to fixed stations. Since 30 meters is just about midway in frequency between the amateur 40- and 20-meter bands, it combines the characteristics of both, making it popular both for DX and intracontinental work.

10.150 to 11.650 MHz

The fixed service is allocated the 10.150-11.175 MHz

segment on a primary basis, with a secondary allocation for mobile. Out-of-band broadcasters are spread a bit thin in this range: There are several domestic outlets from the People's Republic of China (10.245, 10.260, 11.000, 11.040 and 11.100 MHz) and an All India Radio transmitter at 10.335 MHz. Radio Tirana transmits in Chinese on 10.510 MHz. Check this area for strong broadcast feeders as well. When propagation conditions allow, try for the singular DX target of station JJD, Tokyo, Japan, on 10.415 MHz (in parallel with 15.950 MHz). Every day at 0700 UTC, JJD transmits a GEOALERT message—a summary of solar and geomagnetic activity in Morse. The key to most of the codes and symbols used in GEOALERTs may be found on page 2.31 of the fourth edition of the Radio Society of Great Britain's *VHF/UHF Manual*, available from ARRL HQ. The RSGB book also lists frequencies for GEOALERTs from Meudon Observatory in France.

The aeronautical mobile service takes over again from 11.175-11.400 MHz. This time, "route" operations take the higher end of the band, the dividing line being 11.275 MHz. 11.290, 11.330 and 11.375 MHz are used for domestic broadcasting in the People's Republic of China. There is another small fixed service allocation from 11.400-11.650 MHz, but there is quite a bit of out-of-band broadcasting in this segment, especially between 11.500 and 11.650 MHz, including a few broadcast feeders.

11.650 to 12.050 MHz

Here is another major shortwave broadcasting band, also known by listeners as "25 meters." The segment 11.650-11.700 MHz might as well be considered as being part of the band, even though it hasn't officially been opened to broadcasting as a result of WARC-79. Broadcasters are using it, just the same! The same goes for the 11.975-12.050 MHz portion of the band.

12.050 to 12.230 MHz

Yes, indeed, broadcasters may be found above the 25-meter band: The USSR operates several transmitters here, and the BBC may be found with English on 12.095 MHz almost anytime propagation favors Britain. Israel and Syria operate here as well. Business is quite a bit slower if you're on the prowl for the Chinese domestic outlets so common on lower frequencies, although there is one at 12.200 MHz. Otherwise, this is more fixed service territory.

A DIGRESSION INTO THE "WAY" OF MULTIPLE FREQUENCY ALLOCATIONS

By now, you've caught on to a pattern in high-frequency spectrum usage. It's true that several services have only a few narrow allocations between 3 and 30 MHz: The aeronautical fixed service gets by with 21.870-21.924 and 23.200-23.350 MHz; radio astronomy has 13.360-13.410 MHz and 25.550-25.670 MHz; the meteorological aids service operates at 27.500-28.000 MHz. But most services are allocated frequencies from the bottom to the top of the HF spectrum because reliable communication may be assured only if stations may change frequency as propagation varies. If stations in a given service are to handle traffic over varying distances around the clock, they must have enough frequency agility to follow the maximum usable frequency (or optimum traffic frequency) as it varies over time. "Time" means not only time of day, but also time of year—and these two cycles

are superimposed upon the swing of the 11-year sunspot cycle. Very generally, with regard to communication via skywave, as the number of sunspots increases, so does the maximum usable frequency between two given points. Radio folk must be able to choose their operating frequencies with these factors in mind.

This is why shortwave broadcasters keep their listeners hopping with seasonal frequency changes. They'd love to be able to stay put in the manner of our local MF, VHF and UHF sound and television broadcasters; they know that a major component in shortwave's relative lack of popularity as compared with local broadcasting is that shortwave broadcasters can't be depended upon to park on a given channel for long. But they have no choice; one can't argue with propagation.

So that's the reason behind the multiple bands allocated to each service. Early on, Amateur Radio received more-or-less harmonically related bands at 1715, 3500 and 7000 kHz, and 14.000 and 28.000 MHz, for example (with continuation of the harmonic relationship on up into VHF); remnants of this approach may still be seen today, but gaps have been filled in with nonharmonic bands at 10, 18, 21 and 24 MHz. International broadcasting holds allocations from 3 through 26 MHz. Other services have chunks throughout the range as well. The aim is solid communication despite the vagaries of propagation; the charm of shortwave for many shortwave listeners and radio amateurs is that the actions of the ionosphere can't be predicted with absolute accuracy. In other words, chance + mystery = radio fun.

With the "why" of multiple bands for each service in mind, let's continue up toward 30 MHz in a speedier way. Below 7300 kHz, there were often variations in allocation from region to region; above 7300 kHz, the ITU Table of Frequency Allocations shows the same allocations for all three regions, with occasional variations taken care of in footnotes. We've covered enough ground to get a solid feel for what kind of signals may be encountered in spectrum allocated to a given service, and we've observed a reliable trend: Broadcasters tend to spill out of their allocations into neighboring bands. As to exactly which stations are where in each allocation, publications can help you with this—and you'll find information on them elsewhere in this chapter.

Preferably you've been tuning along on this allocations tour, as suggested earlier, and that you're getting to know who's operating where by treating different frequencies and bands as different places, each with singular residents, each with a specific propagational "feel." On we go.

12.230 MHz to 30.005 MHz

The maritime mobile service has the allocation from 12.230 to 13.200 MHz; this is a very busy band. Aeronautical mobile takes over from 13.200-13.360 MHz; "off route" operation is undertaken below 13.26 MHz. There's another fixed service band—shared with radio astronomy at 13.360-13.410 MHz, and with mobile services (except for the aeronautical mobile "route" service) in the 13.410-13.600-MHz stretch—from 13.360 to 13.600 MHz. Although it's not shown in the main ITU listings, a footnote tells us that 13.553-13.567 MHz is designated for use by ISM devices. The next segment, 13.600-13.800 MHz, is one of the prizes brought home from WARC-79 by the broadcasting service—22 meters, a never-before broadcasting band. As this is written, 22 meters still "belongs" to the fixed service, as was the case with WARC-79 extensions to the 25- and 31-meter bands. But powerful broadcasters are using 22, so have a listen; and, anyway, DX is DX, no matter what the service!

The fixed service has the allocation from 13.800-14.000 MHz; mobile services (again, excepting aeronautical mobile "route") may also use this band on a secondary basis. Then comes Amateur Radio's HF jewel: 14.000-14.350 MHz, or 20 meters—*the* DX band for radio amateurs. You'll usually find CW operation concentrated between 14.000 and 14.100 MHz, with digital modes most common in the higher half of this range. W1AW's 20-meter code practice frequency is 14.068 MHz. US hams use upper-sideband phone, slow-scan television and facsimile from 14.150 MHz up. The Amateur-Satellite service is allocated 14.000-14.250 MHz. In a few countries, the 14.250-14.350 MHz segment is allocated to the fixed service on a primary basis. Radio Tirana gets into the out-of-band broadcasting act with an unstable hummy transmitter on 14.320 MHz, used exclusively for Chinese-language transmissions. Above 20 meters, there's another fixed service enclave from 14.350 to 14.990 MHz; the mobile services (except for aeronautical "off route" stations) share this on a secondary basis. Try for Canada's standard-time-and-frequency CHU, Ottawa, in the clear with a full-carrier upper-sideband signal on 14.670 MHz.

Standard-frequency-and-time operations operate from 14.990 to 15.010 MHz, with a secondary space research chunk from 15.005-15.010 MHz. There's an "off route" aeronautical mobile allocation from 15.010 to 15.100 MHz; since this is just below the 19-meter shortwave broadcasting band, there is out-of-band broadcasting here, also. Try for these stations, among others: The Voice of Vietnam wanders a bit around 15.010 MHz; a Chinese home service outlet makes it into North America on 15.030 MHz with suitable propagation; Saudi Arabia makes heavy use of 15.060 MHz; the BBC has operated a World Service outlet for many years on 15.070 MHz; Radio Tehran is strong on 15.084 MHz; Israel and China use 15.095 MHz. The heavily congested 19-meter shortwave broadcast band lies between 15.100 and 15.600 MHz. The range 15.450-15.600 MHz was allocated to broadcasting at WARC-79; even though it has not been officially declared open for broadcasting, broadcasters are using it, mixing with fixed service denizens yet to be relocated.

As you might expect, some broadcasters, not to be outdone, go the extra mile and operate above 15.600 MHz anyway. Greece, Israel and Bangladesh operate between 15.600 and 15.635 MHz. Transmitters from the People's Republic of China may be found here on three frequencies: 15.670, 15.710 and 15.880 MHz. 15.670 MHz, shared with a BBC feeder, is activated during summer months for Beijing's Domestic Minorities Language Service—you won't hear Tibetan, Kazakh and Uygur on too many other shortwave stations! Otherwise, 15.600-16.360 MHz is used by stations in the fixed service. Another weird frequency for Radio Tirana Chinese broadcasts is 16.230 MHz.

Maritime mobile follows, with an allocation from 16.360-17.410 MHz. Another fixed service allocation stretches from 17.410-17.550 MHz. True to form, this fixed service block suffers incursions from the broadcasting band just above it: The 16-meter shortwave broadcast band covers 17.550-17.900 MHz. Since the 17.550-17.700 MHz portion is an addition resulting from WARC-79, you'll find a mix of fixed and broadasting stations there until the fixed service stations are relocated.

17.900-18.030 MHz is an aeronautical mobile allocation, with "route" operation below 17.970 MHz and "off route" above. 18.030-18.068 MHz is a fixed service allocation, with a small segment (18.052-18.068 MHz) shared on a secondary basis by space research.

Hams received three new HF bands at WARC-79, of which the 18.068-18.168 MHz band was one. It's allocated to the Amateur Radio and Amateur-Satellite Services. Hams are champing at the bit for access to this band, 17 meters, but until the present fixed-service users relocate, they must be content with listening here. For many years, the BBC has operated a 100-kilowatt transmitter on 18.08 MHz, usually carrying programs for Asia and Africa. Exactly what may be done with this transmitter when 17 meters is given over to Amateur Radio remains to be seen.

As this chapter is written, the fixed service has access to just about everything from 18.168-19.990 MHz. Two new maritime mobile allocations were dropped into this segment at WARC-79, however; this change has yet to be fully implemented. When the dust settles, the picture should look like this: 18.168-18.780 fixed, 18.780-18.900 maritime mobile, 18.900-19.680 fixed, 19.680-19.800 maritime mobile and 19.800-19.900 fixed.

A standard-frequency-and-time segment between 19.900 and 20.010 MHz straddles 20 MHz, with a chunk at 19.900-19.950 MHz allocated to space research on a secondary basis. Interestingly, 20 MHz holds a place in the history books where space research is concerned: Satellite Sputnik 1, launched by the Soviet Union in 1957, was easily received by shortwave listeners on 20.009 MHz (as was another transmitter at twice this frequency); so was China's first satellite, launched in 1970. But there was a difference: Sputnik merely beeped, while the Chinese vehicle broadcast music—"The East is Red"—and wailing telemetry!

The fixed service has another allocation at 20.010-21.000 MHz. Amateur Radio's 15-meter band, 21.000-21.450 MHz, is next. As was the case on lower bands, ARRL's W1AW transmits code practice here, on the frequency of 21.080 MHz. Novice and Technician class operators work quite a bit of DX in their 21.100-21.200 MHz subband. Amateur radiotelephone and slow-scan television operation—almost entirely suppressed-carrier upper sideband—fills the rest of the band, from 21.200 to 21.450 MHz. The entire 21.000-21.450 MHz stretch is also allocated to the Amateur-Satellite Service.

Broadcast has the next allocation: 21.450-21.850 MHz, the 13-meter international broadcasting band. This allocation is heavily used, but not so much by "small timers": 13 meters tends to be the haven for well-moneyed broadcasting organizations running high power. The fixed service follows with a tiny 21.850-21.870 MHz segment. The aeronautical fixed service has a little more space in its 21.870-21.924 MHz allocation. Above this is an aeronautical mobile "route" allocation from 21.924-22.000 MHz.

Next is a maritime mobile service allocation from 22.000 to 22.855 MHz. This is a good daytime DX band for shortwave listeners: During peak sunspot activity, Asian and African DX may parallel that being enjoyed by radio amateurs one MHz lower on 15 meters. 22 MHz is rounded out by the fixed service allocation from 22.855-23.200 MHz. 23.000-23.200 MHz is shared by mobile services (except for "route" aeronautical mobile) on a secondary basis. Aeronautical fixed and aeronautical mobile ("off route") share 23.200-23.350 MHz.

Stations in the fixed service operate between 23.350 and

24.890 MHz. Mobile services (except aeronautical mobile) share 23.350-24.000 MHz with fixed stations; land mobile stations share 24.000-24.890 MHz with the fixed service. Both of these sharing arrangements are on a co-primary basis.

Amateur Radio received a third new band at WARC-79 with the exclusive allocation to the Amateur Radio and Amateur-Satellite Services of 24.890-24.990 MHz—the 12-meter band. US amateurs were granted access to 12 meters in mid-1985. Look for amateur CW work from 24.890-24.920 MHz, CW and digital operation between 24.920 and 24.930 MHz, and upper sideband and slow-scan television between 24.930 and 24.990 MHz.

Twelve meters is the only HF amateur band sharing a border with a standard-time-and-frequency allocation: 24.990-25.010 MHz. Right now, though, there aren't any 25-MHz time stations to hear—WWV vacated this frequency several years back as a result of budget cuts and poor upper-HF propagation. Next is another fixed service allocation from 25.010-25.070 MHz, shared with co-primary mobile (except aeronautical mobile) services. Maritime mobile has the next allocation: 25.070-25.210 MHz. There's another co-primary fixed/mobile (except aeronautical mobile) allocation from 25.210-25.550 MHz.

Radio astronomy is rare among 3-30 MHz allocations; remember the 13.350-13.410 MHz allocation it shares with the fixed service? It has the band 25.210-25.670 MHz all to itself. Before WARC-79, the 25.600-25.670 MHz segment was in broadcasting hands; this addition to radio astronomy was one of only a few broadcasting losses at WARC-79.

The highest HF broadcasting band is the 11-meter band, 25.670-26.100 MHz. That this band is thinly populated as this is written is an understatement; even during years of high sunspot activity, there are a lot of wide open spaces on 11. Propagational unsuitability of the band is a major reason for this, but another is "hardware-related": Few broadcast-bands-only portable radios cover the 26-MHz range. This only makes the few stations using 11 meters all the rarer, however, and to the DXer, "rare" means "sought after." Engineers at HCJB, Quito, Ecuador, put a 100-watt 26.02-MHz signal on the air via a nondirectional antenna—until their modified Amateur Radio transmitter gave up the ghost, robbing propagation students of a valuable beacon. Speculation has been raised that 11 meters might serve well for broadcasting direct from satellites, but the consensus on this topic is that this will be a long way off, if at all. Reactivation of the HCJB experiment is more likely.

Broadcasting is not the only service making light use of upper-HF bands during years of poor propagation. The maritime mobile allocation at 26.100-26.175 MHz is another band where activity is limited. Nonetheless, DX is there for the digging, especially when propagation opens to Europe and Africa.

The fixed service operates between 26.174 and 28.000 MHz on a primary basis. This segment is shared on a co-primary basis with two other services. Mobile services, except for aeronautical mobile, may also use these frequencies up to 27.500 MHz; at that point, the "no aeronautical mobile" restriction falls away. 27.500-28.000 MHz is also shared by meteorological aids, the only such allocation in the 9-kHz-to-30-MHz range.

Mention of the 27-MHz range would not be complete without coverage of the CB Radio Service. You won't find CB radio mentioned in the ITU Rules and Regulations; its frequencies have been allocated on the initiative of national

entities. Legitimate CB stations operate between 26.960 and 27.410 MHz in 40 channels each 10 kHz wide, using full-carrier double-sideband AM phone and single sideband. The band 26.957 to 27.283 MHz is designated for ISM applications.

Our second-to-the-last stop is Amateur Radio's 10-meter band. Spanning 28.000-29.700 MHz, it's the largest amateur band below 30 MHz. Properly speaking, there are two amateur allocations covering the entirety of this range, one for the Amateur Radio Service and another for the Amateur-Satellite service. CW ham operation is the norm from 28.000-28.300 MHz, including Novice and Technician class operations from 28.100-28.200 MHz. With the FCC adopting new rules in response to the ARRL-initiated Novice Enhancement proposal, Novices and Technicians now have CW, SSB, and RTTY privileges at 28.300-28.500 MHz.

Ten-meter single-sideband and slow-scan television operation covers 29.300-29.700 MHz, with some important exceptions: satellites! The Soviet RS-series satellites transmit CW beacon signals between 29.300 and 29.500 MHz, and if you make a habit of listening for them from time to time, you'll have the thrill of hearing a satellite. Listen for RS-5 on 29.330, RS-6 on 29.410, RS-7 on 29.340 and RS-8 on 29.460 MHz.

Finally, the 29.700-30.005 segment is allocated to the fixed and mobile sources on a co-primary basis. And, with that, we've done it—we've made the hike from 9 kHz to 30 MHz. With this modest account of who's where on the long, medium and shortwaves, along with time spent tuning, reasonable propagation and a good radio, you're set for listening for a long time to come.

LISTENING REFERENCES AND RESOURCES

Earlier in this chapter, it was noted that the past few years have brought significant improvement in the quality and range of selection available in general-coverage receivers. These receivers wouldn't be of much value, however, if you couldn't back them up with reference material worthy of the quality of listening you may undertake with them. Shortwave fun quickly diminishes if you have no idea where (and when) to tune for broadcasts from a chosen country, or if you're hearing a coast station and wonder about its specific location. Luckily, the expansion of the shortwave receiver market has stimulated a boom in the publishing of frequency and programming guides for shortwave enthusiasts. The improvement of receivers and supporting documentation has had the further effect of broadening your options for participation in listening clubs throughout the world (or maybe the expansion of clubs has resulted in the expansion of the receiver/publication market!). Here's a summary of pertinent books, periodicals and listening clubs—at least enough so you're not left alone at the radio wondering where to turn (or tune) next.

Publications

In the field of publications supporting international broadcast listening, one name stands out above the rest: *The World Radio TV Handbook*, published yearly by (in the US) Billboard Publications, Inc, 1515 Broadway, New York NY 10036. If you've invested several hundred dollars or more in a shortwave receiver and intend to enjoy shortwave broadcasts on a regular basis—either as DX or for program content—you're not likely to get your money's worth out of the receiver without a copy of the *WRTH*. Shortwave—and

Don't Miss These Shortwave DX Programs

A quick way to get up to speed on what's happening in shortwave listening is to tune into the several excellent programs tailored particularly to the listening enthusiast. Here are the particulars of three such programs. Broadcast times are quite stable, but frequencies may vary seasonally; all times and dates are given in UTC (Coordinated Universal Time) unless otherwise stated. The latest information on broadcast times and frequencies may be obtained directly from the addresses given.

"Shortwave Listeners' Digest" is broadcast by Radio Canada International, PO Box 6000, Montreal PQ, HC3 3A8, Canada: Saturdays at 2130 on 15.150, 15.325 and 17.820 MHz; Sundays at 0100 (0000 when the US shifts to daylight time) on 5960 and 9755 kHz, and again on Sundays at 2300 on 9755 kHz and 11.710 MHz.

"Media Network" is broadcast by Radio Netherlands, PO Box 222, 1200 JG Hilversum, The Netherlands: Thursdays at 1050 on 6020 and 9650 kHz; Fridays at 0250 on 6165 and 9590 kHz, and again at 0550 on 6165 and 9715 kHz.

"World of Radio" is privately produced by Glenn Hauser, PO Box 490756, Ft Lauderdale FL 33349. It is broadcast by KCBI (Dallas), Fridays at 1930 and Sundays at 1730 on 11.735 MHz; and by WRNO Worldwide (New Orleans) on Fridays (EST/EDT) at 11 PM on 6185 kHz, Saturdays (EST/EDT) at 7:30 PM on 7355 or 9852.5 kHz, and Sundays (EST/EDT) at 9 AM on 9715 kHz.

---

major AM, FM and TV—stations for all of the countries of the world are listed, as well as addresses and personnel of stations and broadcasting organizations worldwide. What time is it in The Ivory Coast, or Lhasa? You can find out from the *WRTH*; each country listing shows local time relative to UTC. What kind of music or sounds serve as a station's interval signal? What does United Emirates Radio say when identifying in Arabic? You'd like to hear Mass from Vatican Radio in Standard Chinese—at what time and on which frequency should you listen? Want to subscribe to BBC's London Calling program guide, but don't know the address? The answers to all of these questions are in *The World Radio TV Handbook*. Each edition of the *WRTH* includes a master by-frequency listing for all shortwave stations of the world, as well as regional listings for all of the world's long- and mediumwave broadcasters. There's a listing of standard time and frequency stations. Names, addresses and publication data are listed for major listening clubs throughout the world. In addition, *WRTH* publishes shortwave equipment reviews and general-interest features on listening and broadcasting. *The World Radio TV Handbook's* 1986 edition (608 pages) was priced at $19.95; look for the *WRTH* at major bookstores, as well as dealers in Amateur Radio and shortwave listening equipment.

Radio Database International (edited by Lawrence Magne and Tony Jones), published by International Broadcast Services, Ltd, PO Box 300, Penn's Park PA 18943, is next. *RDI* covers shortwave broadcasters between 1600 kHz and 26.100 MHz. *RDI* details broadcast schedules frequency by frequency throughout the shortwave spectrum in bar-chart form: The X axis for each *RDI* page shows UTC hours 0 through 24; Y axis entries are broadcast frequencies in ascending order. Within each frequency block are horizontal entries for each broadcaster using that frequency, including transmitter site: At 7250 kHz, for instance, an entry for a 10-kilowatt All India Radio at Lucknow is shown as a bar

from 0700 to 0845 UTC. At a glance you learn also that Taiwan, the Federal Republic of Germany, Singapore, the USSR and the Vatican State also operate transmitters on 7250. Which of these are you hearing? That's where listening skill comes in—and *RDI* makes no claim to providing you with that particular commodity! The texture of the time bars indicates whether the language of a particular broadcast is English, German, French, Portuguese, Spanish, Arabic, Russian, Chinese, Japanese, multilingual or "other"—and jamming is also noted. *RDI* also includes coverage of unofficial broadcasters—"pirates" and clandestines few other mass-market publications have dared to cover. The 1987 edition (352 pages) of *Radio Database International* includes 24 pages of equipment reviews as well. *RDI* is available through ARRL ($13) as well as directly from the publisher.

RDI and the *WRTH* concentrate on broadcasting schedules and station information. What about coverage of what stations are saying—the contents of the broadcasts themselves? Two publications come to mind. *Review of International Broadcasting* (edited and published by Glenn Hauser, PO Box 490756, Ft Lauderdale, FL 33349; 40 pages per issue) includes program schedules, station information and "Listeners' Insights on Programming" (LIP)—a place for listeners to question and comment on the program content of the stations they're hearing. Since the cornerstone of *RIB* is participation by its readership—a worldwide readership —LIP is lively, indeed! Monthly BBC World Service program information and columns on US mediumwave programming, TV satellite reception ("Satellite Watch") and shortwave equipment ("Radio Equipment Forum") round out *Review of International Broadcasting*. A subscription for 10 issues of *RIB* costs $20 in North America; $2 for a sample issue. (*RIB's* companion publication, *DX Listening Digest*, concentrates less on programming and more on the nitty gritty of shortwave broadcast DXing and radio equipment; 10 40-page issues of *DXLD* cost $20 in North America, samples $2, from the same address as *RIB*, above).

The other is *Monitoring Times*, published by Grove Enterprises, PO Box 98, Brasstown NC 28902. In July 1986, *MT* merged with the shortwave listening magazine *International Radio*, going from 40 to 60 pages a month. Thus, *MT* includes *IR's* coverage of international broadcasting, as well as treatment of nonbroadcast and nonamateur services (including VHF/UHF scanning). A subscription in the US, Canada and Mexico costs $15 per year, with savings for multiple-year subscriptions.

The nonbroadcast coverage of *Monitoring Times* leads us to consideration of what many shortwave listeners term the "utilities." Earlier we ran into the fixed, aeronautical and maritime mobile services, among others—each dedicated to a use of radio for something other than broadcasting and Amateur Radio. These nonbroadcast, nonamateur services are the utilities. *Monitoring Times*, and to a lesser extent, *DX Listening Digest* (above), are nonclub periodicals covering the utilities. But what if you want to know which utility stations transmit where? Are there any frequency guides for the utilities?

Yes. Two excellent such books are available. *The Confidential Frequency List*, 6th Edition (Oliver P. Ferrell, editor; cover price $15.95) is published by Gilfer Associates, Inc, PO Box 239, Park Ridge NJ 07656. Most of its 304 pages are devoted to station listings, by frequency, from 4000 kHz to 27.998 MHz. For each station, frequency, emission, call sign, location, service and transmitter power are shown, with the addition of remarks as to agency of origin, time of scheduled transmissions, and so on. Special features on radioteletype, Piccolo (a flutelike data transmission mentioned earlier) and Cyrillic radioteletype are included.

The Guide to Utility Stations (Joerg Klingenfuss, editor) is published by Klingenfuss Publications, Panoramastrasse 81, Hagellock, D-7400 Tuebignen, Federal Republic of Germany. In the US, it's distributed by Universal Shortwave, 1280 Aida Dr, Reynoldsburg OH 43068, and Grove Enterprises (address for *Monitoring Times*, above), at a price depending on the distributor. *The Guide*'s 427 pages list utility stations, by frequency, from 1605 kHz to 29.400 MHz, showing call sign, location, emission and transmission times (if known) for each. Aside from much reprinted ITU information (such as the entire Table of Frequency Allocations from 9 kHz through 150.050 MHz!), *The Guide* offers its own listings of utility stations by call sign, utility stations identifying with other than official call signs, radioteletype news transmissions by country, and radioteletype news transmissions by time.

Longwave beacons are another story. The many beacons in the 200-415 kHz range operate on channels spaced every kilohertz, and official information on them requires sifting through the records of the FCC, the Federal Aviation Administration and other government agencies. And what about overseas beacons? The book for beacon DXers that takes care of all of this is the *Beacon Guide* by Ken Stryker. It lists over 6100 longwave beacons in the Americas, Asia and the Pacific by frequency order, cross-referenced by beacon ident. Associated information regarding beacon location and power is included. Longwave DXers who tend to slip below 200 kHz for a try at hearing LOWfer transmissions between 160 and 190 kHz may make use of the *Guide*'s separate listing of over 100 LOWfer stations. The *Beacon Guide* is published in three-ring loose-leaf format for easy updating, and is available from Ken Stryker, 6350 N Hoyne Ave, Chicago IL 60659. As of this writing, price was $10 postpaid in the US and $15 by airmail overseas.

Mediumwave broadcast DXers face a problem similar to that of beacon DXers, particularly with regard to DXing domestic broadcasting stations in the US and Canada. (*The World Radio TV Handbook* has space enough to list only the more powerful North American broadcasters.) What to do? Pick up a copy of the National Radio Club's *Domestic Log*. The *Domestic Log* lists, by frequency, all US and Canadian mediumwave stations, including locations, addresses, powers, antenna operations, network affiliations, schedules and formats. Like the *Beacon Guide*, the *NRC Domestic Log* is published in three-hole loose-leaf format for ease of updating. It's available for $10 postpaid in the US, Canada and Mexico to NRC nonmembers; NRC members get a price break. NRC publishes many more books and pamphlets covering all aspects of mediumwave DXing—be sure to ask about them if you order a copy of the *Domestic Log* from the NRC Publications Center, Box 164, Mannsville NY 13661.

This treatment of several key listening publications only scratches the surface of a widening field. Many of the best publications in shortwave listening are essentially homegrown newsletters; right now, radio monitoring supports only one mass-distribution magazine: *Popular Communications* (Tom Kneitel, K2AES, editor), 76 North Broadway, Hicksville NY 11801. A year of *PopComm* costs $16 in the US, and $20 in Canada. *PopComm* covers radio listening of all kinds—from longwave up through VHF and UHF scanner listening. The majority of listening publications, however, emanate from listening clubs—and the strength and value of clubs come from people.

Clubs

Aside from terrestrial static, and ionospheric and cosmic noise, most of the signals you'll hear lead you to other people—what they're saying, singing or doing, somewhere in the world. You don't have to wait for your radio to make the first move, however. All you need do is join one of the many radio listening clubs to tap in on the most important listening resource of all: people. For instance, if you're a long-wave enthusiast, the club for you is the Longwave Club of America. If shortwave program listening or DXing is your interest, the North American Short Wave Association (NASWA), the American Shortwave Listeners Club (ASWLC) and the Society for Preserving the Engrossing Enjoyment of DXing (SPEEDX)—to name a few—are what you're looking for. In the mediumwave department, the National Radio Club (NRC) and the International Radio Club of America (IRCA) are the places to go. Some clubs are national; some regional. The best way to find out about the many North American clubs is to get into contact with the Association of North American Radio Clubs (ANARC), PO Box 462, Northfield, MN 55057. An SASE and 25 cents will net you a copy of ANARC's Club List. A sample copy of ANARC's Newsletter may be had for an SASE and 60 cents. Clubs—it's the same way in shortwave listening as it is in Amateur Radio: They need you and you need them. More and more clubs are running computer bulletin boards for the dissemination of late-breaking tips even faster than the printed word allows. Between the yearly *RDI*, *WRTH* and utility guides releases, the clubs fill in with up-to-date information you won't find anywhere else.

"What Do I Do With Radio?"

There are as many approaches to the enjoyment of radio as there are people. Here are general trends of what people have done and are doing with the shortwave listening side of hobby radio.

Collectibles

Radio is a fine pursuit to be involved in if you like collecting things. If you hadn't done so beforehand, you began a "radio memories" collection as soon as you bought this book. Simple as this may seem, memories are the first collectibles you ever acquire. Many of the aural images presented by radio are vivid and singular. You may wish to use a tape recorder to help you capture them for later playback; shortwave is a natural hunting ground for armchair explorers eager to build collections of exotic music and speech.

Further, if you're not interested in owning "the latest" for reasons of taste or budgetary constraint, you may reap the hidden harvest of technological improvement in radio if you're willing to use "last year's" model. There's more to it than economics, however: For some radio enthusiasts, the older a set is, the more lovable it becomes. Often, their next step is the collection of radios themselves—another facet of radio fun that gains adherents as time goes by.

One of the oldest traditions in shortwave listening and Amateur Radio is that of the QSL card. "QSL" is one of the variants of the Q-Code used by radio operators world-wide (see Chapter 17) to facilitate brevity in transmission; it means "I acknowledge receipt." With the vagaries of language usage being what they are, it wasn't long before "QSL" signified the acknowledgement itself. Early on in the development of radio, then, acknowledgement of receipt of a message (or of an instance of reception or two-way contact) was done with a QSL card—a postcard-sized document, really, that verified the time, date, frequency, emission, location, signal strengths and call signs of the participants in an exchange of radio signals. This makes QSLs dandy collectibles, as you might imagine. Radio amateurs exchange QSLs regularly; a long-held maxim in amateur circles is that "the final courtesy of a QSO [contact] is a QSL." A careful approach may net you a QSL from a "utility" station. Many shortwave broadcasters acknowledge listeners' correct reception reports with QSL cards; some broadcasters go out of their way to produce beautiful and varied QSLs over periods of months with the idea of encouraging regular listening on the part of their audiences.

Especially in shortwave listening, however, in recent years the role of the QSL in broadcaster-audience relations has become less important. One reason for this is that some station managements became disgruntled at the discovery that many shortwave listeners were listening to broadcasts only long enough to guarantee receipt of that all-important QSL.

Another reason for the decline of shortwave broadcaster QSLing is the rising cost of staff time and postage. After all, stations exist primarily to transmit programming, not to mail out pieces of paper. Most importantly, however, the recent easing of popular access to shortwave reception, with so many quality portable and table model receivers becoming available for less money, has resulted in the rise of shortwave listening from arcane hobby to mainstream entertainment. People who routinely listen for news or entertainment generally are uninterested in QSLs.

Luckily, of course, whether you decide to quest for QSLs depends on your personal choice, and collecting them is still a popular choice among many shortwave listeners. A good return on the effort you put into QSLing requires skills over and above those used to gather data for reception reports: Enticing a QSL out of a rare Asian, African or South American regional may require years' patience and the skill of a diplomat. QSLs may lead to more than a collection of cards, however. They may also serve as the basis for competition.

Competition

Wherever skilled activities are undertaken by people, competition ensues. Shortwave listening is no different. Depending on the country list you use as a reference, there are between 150 and 300 possible countries to hear on the shortwave bands. Who has heard the most? Who has heard how many in how short a time? QSL cards serve as the evidence of such triumphs. If you're interested in gunning for shortwave listening awards, ask the Association of North American Radio Clubs (address above) for their Club List. Several clubs run awards programs that well test your listening skill. Some clubs run listening competitions at specific times of year as well; usually, the contents of your logbook serve in lieu of QSLs for these.

Shortwave listeners plying the ham bands become familiar with another sort of radio competition: Amateur Radio contesting. Hardly a weekend goes by during which an amateur contest isn't in progress somewhere between 1800 kHz and 29.7 MHz. It may be ARRL's International

QSL cards received by Amateur Radio station K1XA from shortwave listeners around the world.

DX Competition, wherein US and Canadian amateurs try to "work" [contact] as many foreign stations as possible; it may be one of the many state "QSO parties" in which radio amateurs try to work as many stations in a given state as possible. If listening to the amateur bands appeals to you, you'll hear more amateur stations in more places per minute during a contest than at any other time. Chapter 7 of this book covers amateur contesting in detail.

Imagination

Collectibles and competition are two means of enjoying radio that entail the use of radio as a means to an end. The end is a QSL card, the achievement of a certificate of accomplishment or your placement in a field of contest entrants. Many people in Amateur Radio and shortwave listening devote most of their time to collecting and competing, and they have a marvelous time of it.

On the other hand, shortwave listeners may, for example, at their option: Detail the activities of all stations between 11.500 and 11.700 MHz; concentrate on listening to broadcasters from South America; become experts on broadcasting in Cuba; characterize and catalogue the frequency variations of Vietnamese shortwave regionals; study propagational anomalies through variations of the received signal strength of coast stations ringing the Baltic Sea; listen to radioteletype press transmissions; enjoy every Sunday's "Happy Station" program from Radio Netherlands; become expert on Middle Eastern clandestine broadcasting; listen only to North American mediumwave stations; concentrate on longwave beacons; stick to nothing but listening to aeronautical mobile and ground stations; try to verify as many Chinese ship stations as possible; attempt to hear all states by listening to radio amateurs in the 30-meter band... These activities can pretty much be considered enjoying radio in and of itself; the list of things to do in this regard, of course, goes on. You need only invoke your imagination.

Alone at the Radio?

If it hasn't already been said to you somewhere before, let it be said here: "Welcome to shortwave listening!" There's a lot for you to do here. This chapter has aimed to provide you with basic knowledge concerning the "shortwaves"—that part of the radio landscape most accessible through the excellent reception afforded by recent general-coverage receivers and Amateur Radio transceivers.

The machinery we use to harness radio to our service is something of which we are, collectively, rightfully proud. Radio as a field of endeavor, radio as the integrator of complex and widespread telecommunication systems, radio as an example of our own ability to organize and regulate, and radio made real in the millions of transmitters and receivers in daily use throughout the world, is a fabulous subject for study and participation. But radio is not simply systems, history and hardware. The uniqueness of radio—at least, the magic of most of the radio you'll be listening to as a shortwave listener, and participating in as a radio amateur, should you decide to become one—is that it always leads you back to people. The dream-image of that old NC-100X on a sultry summer night in 1969 was one of the presence of people—past, present and future. The signals heard with that NC-100X—signals from Costa Rica, Venezuela, England, China—all portended of people.

Improvements in technology are hollow improvements unless they improve the lives of people. Radio continues to play a paramount role in the improvement of our lives because it, over all other technologies, stands best to cut down time and distance between people, and—it follows—to improve understanding and cooperation between us. You may participate *directly* in the radio that bridges the distance between people: You may listen to music, news and cultural programming from shortwave broadcast transmitters scattered across the globe. You may listen to ships at sea and planes in the air. Further, you may choose to get involved in Amateur Radio—radio where you get to use the shortwaves (and many more frequencies!) to respond with code, voice, keyboard and camera to the people whose signals you hear. These things are yours for the doing.

This idea of radio bringing people together isn't very new, if an article in September 1929 *QST* is any indication. The words are those of Eulalia M. Thomas, W8CNO, writing in "XYL":

> To me, one of the greatest things in amateur radio is the friendships one forms. Ofttimes there is naught but the thinnest strands of communication binding them. It is as the poet Dix has said, "My world is as wide as the realm of thought." But it is not only the "thoughts" of our game; it is the friendships that intrigue us. Only we, who have sat in a fast-chilling room, listening with queer, hypnotic fascination to a friend, a mile, a thousand—or even ten thousand miles away saying "73," can understand. We snap on our transmitter and out through the clear air go thin, etheric fingers; out through the infinite goes a handclasp of friendship. So we, sitting alone by our set, become suddenly warm and happy, for the Angel of Friendship has entered and we know that across the miles we have found a friend.[3]

This is the radio that this chapter, and the next 16, are about. Whether you listen or transmit, have fun—and remember that you're never really alone at the radio.

Beyond Voice and CW

More and more shortwave listeners are stepping beyond listening to "just" voice and radiotelegraphy transmissions into radioteletype and image communications. In-depth technical information on such modes appears in *The ARRL Handbook for the Radio Amateur.* Chapters elsewhere in this book cover the operational side of these techniques in detail: Antenna Orientation Chapter 4; RTTY Communications, Chapter 9; Packet Radio, Chapter 10; Satellites, Chapter 13; Emergency Communications, Chapter 14; and Image Communications, Chapter 16. You'll also probably be spending a lot of time with the reference material in Chapter 17—international call sign prefixes, sunrise/sunset information, Q-signals, and more. (And you won't really get the full benefit of this book unless you read it from cover to cover!)

---

[3]Thomas, E. M., "XYL," *QST*, Sept 1929.

Chapter 2

The Amateur Radio Spectrum

The electromagnetic spectrum is a limited resource. Every kilohertz of the radio spectrum represents precious turf that is blood sport to those who lay claim to it. Fortunately, the spectrum is a nondepletable resource—one that if misused can be restored to normal as soon as the misuse stops. Every day, we have a fresh chance to use the spectrum intelligently.

Simply because the radio spectrum has been used in a certain way, changes are not impossible. Needs of the various radio services evolve with technological innovation and growth. There can be changes both in the frequency-band allocations made to the Amateur Radio Service and how we use them. Also, to our benefit, signals transmitted from one spot on earth propagate only to certain areas, so specific frequencies may be reused numerous times throughout the globe. The Amateur Radio Service is richly endowed with a wide range of bands starting from 1.8 MHz and extending to any frequency above 300 GHz. Thus, we enjoy a veritable smorgasbord of bands with propagational "delicacies" of every type—direct wave, ground wave, sky wave, tropospheric scatter, meteor scatter—and can supplement this with moon reflections and active artificial satellites.

Why do you need to know any of this? Well, you don't. But you can be a better operator if you know the story of our frequency allocations, who the players are, and what influence you might have in the process.

INTERNATIONAL REGULATION OF THE SPECTRUM

Amateur Radio frequency band allocations don't just happen. Band allocation proposals must first crawl through a maze of national agencies and the International Telecommunication Union (ITU) with more adroitness than a computer-controlled mouse. Simultaneously, the proposals have to run the gauntlet of the competing interests of other spectrum users.

Treaties and Agreements

To bring some order to international relationships of all sorts, nations sign treaties and agreements. Otherwise (with respect to international communications) chaos, anarchy and bedlam would vie for supremacy over the radio spectrum. Pessimists think we already have some of that, but they haven't any idea of what we optimists know about how bad it could be *without* international treaties and agreements. The primary ones that affect the Amateur Service and the Amateur-Satellite Service to which the US is a party are:

• The International Telecommunication Convention, signed at Nairobi on November 6, 1982.

• The Radio Regulations annexed to the International Telecommunication Convention were signed at Geneva on December 6, 1979 and entered into force on January 1, 1982.

• Partial revisions of the Radio Regulations, relating to space and radio astronomy, were signed at Geneva on November 8, 1963 and July 17, 1971 and became effective on January 1, 1965 and January 1, 1973, respectively.

• The United States-Canada Agreement on Coordination and Use of Radio Frequencies above 30 MHz was agreed by an exchange of notes at Ottawa on October 24, 1962. A revision to the Technical Annex to the Agreement, made in October 1964 at Washington, was effected by an exchange of notes signed by the United States on June 16, 1965 and by Canada on June 24, 1965, and became effective on June 24, 1965. Another revision to this Agreement to add Arrangement E (between the Department of Communications of Canada and the National Telecommunications and Information Administration and the Federal Communications Commission of the United States on the use of the 406.1-430 MHz band in Canada-US Border Areas) was made by an exchange of notes signed by the United States on February 26, 1982 and Canada on April 7, 1982.

The International Telecommunication Union

The origins of the International Telecommunication Union (ITU) trace back to the invention of the telegraph in the 19th century. To establish an international telegraph network, it was necessary to reach agreement on uniform message handling and technical compatibility. Bordering European countries worked out some bilateral agreements. This eventually led to creation of the ITU at Paris in 1865 by the first International Telegraph Convention, which yielded agreement on basic telegraph regulations.

Starting with a conference in Berlin in 1906, attention focused on how to divide the radio spectrum by specific uses to minimize interference. In 1927, the ITU Washington Radio Conference of 27 maritime States was the beginning of international regulation of Amateur Radio. At that conference, the Table of Frequency Allocations was first devised, and six harmonically related bands from 1.715 and 60 MHz were allocated to amateurs. Some of the amateur bands were revised in this and subsequent conferences held in Cairo (1938), Atlantic City (1947) and Geneva (1959), in consideration of the needs of other radio services.

In 1932, a Telegraph and Telephone Conference and a Radiotelegraph Conference in Madrid merged into a single International Telecommunication Convention. The Radio Regulations resulting from this conference explicitly prohibited amateurs from handling international third-party

Glossary of Spectrum Management

Parenthetical abbreviations following definitions indicate the following sources:

(RR) ITU *Radio Regulations*

(NTIA) National Telecommunications and Information Administration *Manual of Regulations and Procedures for Federal Radio Frequency Management*

AIRS—ARRL Interference Reporting System—collection of information regarding out-of-band transmissions in the amateur bands (formerly called Intruder Watch).

Alligator—An unbalanced repeater with a "big mouth" (high-power transmitter) and "small ears" (insensitive receiver).

Adjacent Channel—The channel immediately above or below the reference channel.

Allocation (of a frequency band)—Entry in the Table of Frequency Allocations of a given frequency band for the purpose of its use by one or more (terrestrial or space) radiocommunication services or the radio astronomy service under specified conditions. This term shall also be applied to the frequency band concerned. (RR)

Note: Allocations are distributions of frequency bands to specific radio services. On the highest level, allocations are made by competent (legally qualified) World Administrative Radio Conferences. Certain specific uses for the spectrum are grouped together so that the probability of interference between major types of radio services is minimal. Domestic (US) allocations are made by the FCC and NTIA in coordination within the limits of the *Radio Regulations* and using broad principles... (Feller, *Planning an Electromagnetic Environment Model for Spectrum Management*, 1981, FCC)

Allotment (of a radio frequency or radio frequency channel)—Entry of a designated frequency channel in an agreed plan, adopted by a competent Conference, for use by one or more administrations for a (terrestrial or space) radiocommunication service in one or more identified countries or geographical areas and under specified conditions. (RR) Note: Allotments are distributions of frequencies to areas or to countries. They provide for the orderly use of certain frequency bands. (Feller, ibid)

Amateur-Satellite Earth Station—An earth station in the Amateur-Satellite Service.

Amateur-Satellite Service—A radiocommunication service using space stations on earth satellites for the same purposes as those of the Amateur Service. (RR)

Amateur-Satellite Space Station—A space station in the Amateur-Satellite service.

Amateur Service—A radiocommunication service for the purpose of self-training, intercommunication and technical investigation carried out by amateurs, that is, by duly authorized persons interested in radio technique solely with a personal aim and without pecuniary interest. (RR)

Amateur Station—A station in the Amateur Service. (RR)

AMSL—Height above mean sea level.

Amplitude Compandored Single Sideband (ACSSB)—A single-sideband transmission system which uses a pilot to tell the receiver how much expansion is required at each instant, in order to recover the dynamic range of the original signal.

Assigned Frequency—The center of the Assigned Frequency Band assigned to a station.

Assigned Frequency Band—The frequency band within which the emission of a station is authorized; the width of the band equals the necessary bandwidth plus twice the absolute value of the frequency tolerance. Where space stations are concerned, the assigned frequency band includes twice the maximum Doppler shift that may occur in relation to any point of the Earth's surface. (RR)

Assignment (of a radio frequency or radio frequency channel)—Authorization given by an administration for a radio station to use a radio frequency or radio frequency channel under specified conditions. (RR)

Note: Assignments are frequency distributions to specific stations. When making assignments, administrations authorize a station to use one or more specific frequencies under specified conditions. (Feller, ibid) With rare exceptions, the FCC does not assign specific frequencies within the Amateur Service but permits amateur stations to operate on any frequency within allocated frequency bands.

Attenuation Ratio—The magnitude of the path loss.

Authorized Bandwidth—Authorized bandwidth is... the necessary bandwidth (bandwidth required for transmission and reception of intelligence) and does not include allowance for transmitter drift or Doppler shift. (NTIA)

Automatic Transmitter Identification System (ATIS)—A device built into a transmitter that sends a unique identifier without action by the operator.

Availability—The characteristic denoting whether the resource is ready for immediate use, or the probability related to a period of time.

Balanced System—A system, usually a repeater, whose transmitting and receiving subsystems are designed for the same quality signal at the same range. (See *Alligator* and *Rabbit*.)

Band—(1) A range of frequencies. (2) One of the ranges of frequencies allocated to the Amateur Service.

Beacon, Engineering—A transmission or frequency used to convey technical information about the host system. In a satellite, it provides data on the health and welfare of the spacecraft.

Beacon (packet radio)—Periodic transmission of a frame giving station identification information—a early practice now deprecated as it adds to channel congestion.

Beacon, Propagation—A transmitting station that continuously or periodically emits a signal to permit listeners to determine whether a path is open and its quality.

Beam (antenna)—The major lobe of an antenna radiation pattern.

Beamwidth (antenna)—The width in degrees azimuth between the half-power points in the major lobe of the antenna radiation pattern.

Broadband—(See *Wideband*.)

Broadcast—(1) Dissemination of radio communications intended to be received by the public directly or by intermediary stations. (FCC Part 97); (2) To transmit a common message to all stations simultaneously rather than to make separate transmissions to each station—point-to-multipoint transmission.

Calling Frequency—A frequency used only for establishing contact, after which the stations change to another frequency to carry out their communication. Some band plans identify national and DX calling frequencies.

Carrier—A continuous radio frequency capable of being modulated by a baseband signal or subcarrier.

Carrier Power (of a radio transmitter)—The average power supplied to the antenna transmission line by a transmitter during one radio frequency cycle taken under the condition of no modulation. (RR)

Carrier-to-Noise Ratio—The ratio of the amplitude of the carrier vs the amplitude of noise, usually expressed in dB.

CCIR—International Radio Consultative Committee, an agency of the International Telecommunication Union.

CCITT—International Telegraph and Telephone Consultative Committee, an agency of the International Telecommunication Union.

Cell—A single service zone within a geographical grid. The network often has the structure of a honeycomb. This is the basis of the cellular telephone system which has base stations in the center of each cell to serve mobiles throughout the cell. Individual cells, if regular shapes, may be hexagonal, circular or elliptical. Cell shapes also may be modified according to terrain.

Cellular—A UHF radio system wherein groups of frequencies in a cell are reused in others in a controlled pattern and wherein a central computer is used to hand off control of a given mobile unit from cell to cell by means of landline trunks.

Centimetric Waves—Radio frequencies between 3 and 30 GHz. (SHF)

Channel—(1) A one- or two-way transmission path divided from any others in frequency or time. (2) A band of frequencies designated for a specific use.

Channel Access—Freedom to use, or protocol for using, a channel.

Channel Loading—The total occupancy of a channel by all users affected by each other's operation.

Channel Spacing—The frequency interval between successive (usually adjacent) channels, measured from center to center or between other characteristic frequencies.

Channel, Voice (Speech) Grade—A frequency channel that can faithfully pass frequencies in the 300-3000 Hz audio range.

Characteristic Frequency—A frequency which can be easily identified and measured in a given emission. A carrier frequency may, for example, be designated as the characteristic frequency. (RR) (See also *Reference Frequency*.)

CISPR—International Special Committee on Radio Interference, an organ of the International Electrotechnical Commission (IEC).

Class of Emission—The set of characteristics of an emission, designated by standard symbols, e.g., type of modulation, modulating signal, type of information to be transmitted, and also if appropriate, any additional signal characteristics. (RR)

Closed Repeater—A repeater where use by nonmembers is discouraged.

Cochannel Interference—Interference on the same channel.

Coded Squelch—A system wherein radio receivers are equipped with devices which allow audio signals to appear at the receiver output only when a carrier modulated with a specific signal is received. (NTIA)

Coordinated Station Operation—The repeater or auxiliary operation of an amateur station for which the transmitting and receiving frequencies have been implemented by the licensee in accordance with the recommendation of a frequency coordinator. (FCC Rules, section 97.3)

Coordination—The process of ascertaining from other users whether a proposed use of a radio frequency can occur without causing harmful interference.

Coordination Area—The area associated with an earth station outside of which a terrestrial station sharing the same frequency band neither causes nor is subject to interfering emissions greater than a permissible level. (RR)

Coordination Contour—The line enclosing the coordination area. (RR)

Coordination Distance—Distance on a given azimuth from an earth station beyond which a terrestrial station sharing the same frequency band neither causes nor is subject to interfering emissions greater than a permissible level. (RR)

Coordinated Station Operation—The repeater or auxiliary operation of an amateur station for which the transmitting and receiving frequencies have been implemented by the licensee in accordance with the recommendation of a frequency coordinator. (FCC Rules, section 97.3)

Critical Frequency—The frequency below which vertically directed radio waves are reflected back to earth.

Crossband—Two-way communication in which the frequencies lie in two distinctly separate frequency bands. (Compare In-band.)

Decametric—Radio frequencies between 3 and 30 MHz. (HF)

Decimetric Waves—Radio frequencies between 300 MHz and 3 GHz. (UHF)

Deviation (frequency)—The difference in frequency from the center of the channel to the upper and lower excursions of the carrier frequency.

Disaster Communication—Emergency communication relating to an unforeseen calamity.

Doppler shift—The change in apparent frequency resulting from relative motion between the source and observer.

Drift (frequency)—An undesired change in frequency, usually as a result of temperature change in frequency-determining components.

Duplex Operation—Operating method in which transmission is possible simultaneously in both directions of a telecommunication channel. (RR)

Duty Cycle—The ratio of the *on* time to the total time of an intermittent operation. (FM is said to have a 100% duty cycle, as its transmitter is at full power, when transmitting, regardless of the amount of modulation.)

Dynamic Range—The ratio to the maximum signal level (with a specified amount of distortion) to the noise level (or minimum signal level), usually expressed in dB.

Effective Radiated Power [ERP] (in a given direction)—The product of the power supplied to the antenna and its gain relative to half-wave dipole in a given direction. (RR)

Effective Isotropic Radiated Power (EIRP)—The product of the power supplied to the antenna and its gain relative to an isotropic antenna.

Electromagnetic Compatibility (EMC)—The ability of electronic equipment to function in its environment without introducing additional disturbances into that environment detrimental to other electronic equipment.

Emergency Communication—Any Amateur Radio communication directly relating to the immediate safety or life of individuals or the immediate protection of property. (FCC Rules, section 97.3)

Extremely High Frequencies (EHF)—Radio frequencies between 30 and 300 GHz. (Millimetric)

Facsimile—A form of telegraphy for the transmission of fixed images, with or without half-tones, with a view to their reproduction in a permanent form. In this definition the term telegraphy has the same general meaning as defined in the Convention. (RR)

Fading—Changes in signal strength generally caused by variations in the propagation medium.

Fading, Flat—Fading in which all frequencies within a channel fade together.

Fading, Selective—Fading in which frequencies within a channel fade somewhat independently.

Frequency Agility—The ability to shift the frequency of a transmitter and/or receiver rapidly to achieve a desired effect.

Frequency Coordinator—An individual or organization recognized in a local or regional area by amateur operators whose stations are eligible to engage in repeater or auxiliary operation which recommends frequencies and, where necessary, associated operating and technical parameters for amateur repeater and auxiliary operation in order to avoid or minimize potential interference. (FCC Rules, section 97.3)

(Glossary continued on next page)

Frequency Diversity—A form of diversity reception using simultaneous transmission on two or more frequencies to take advantage of the tendency that the different frequencies fade independently.

Frequency, Input—(repeater usage) The center frequency to which the repeater receiver is tuned.

Frequency Offset—The difference (in hertz) between two frequencies, such as: (a) between one used for transmitting and one used for receiving, (b) between a beat-frequency and the signal of interest in a receiver, or (c) between a specific transmission frequency and the nominal channel frequency.

Frequence Optimum de Travail (FOT)—A radio frequency 15% below the maximum usable frequency that will provide communications 90% of the days. (Also called *Optimum Working Frequency, OWF*).

Frequency, Output—(repeater usage) The center frequency to which the repeater transmitter is tuned.

Frequency Registration—The act of making a written record of frequency assignment or usage in a central registry or in a published form for distribution.

Frequency Reuse—The ability to use a frequency in a different geographical location without harmful interference.

Frequency Sharing—The common use of the same portion of the radio frequency spectrum by two or more users where a probability of interference exists. (NTIA)

Frequency-Shift Telegraphy—Telegraphy by frequency modulation in which the telegraph signal shifts the frequency of the carrier between predetermined values. (RR)

Frequency Tolerance—The maximum permissible departure by the center frequency of the frequency band occupied by an emission from the assigned frequency or, by the characteristic frequency of an emission from the reference frequency. The frequency tolerance is expressed in parts in 10^6 or in hertz. (RR)

Full Carrier Single-Sideband Emission—A single-sideband emission without suppression of the carrier. (RR)

Guard Band—A narrow band of frequencies surrounding a (usually weak-signal) channel (or band) that are not to be used in order to prevent interference.

Guard Channel (or Frequency)—A (calling) frequency designated to be monitored continuously or within scheduled times.

Geostationary Satellite—A geosynchronous satellite whose circular and direct orbit lies in the plane of the Earth's equator and which thus remains fixed relative to the Earth; by extension, a satellite which remains approximately fixed relative to the Earth. (RR)

Geosynchronous Satellite—An earth satellite whose period of revolution is equal to the period of rotation of the Earth about its axis. (RR)

Harmful Interference—Interference which endangers the functioning of a radionavigation service or of other safety services or seriously degrades, obstructs, or repeatedly interrupts a radiocommunication service operating in accordance with these Regulations. (RR) Interference which seriously degrades, obstructs or repeatedly interrupts the operation of a radiocommunication service. (FCC Rules, section 97.3)

Hectometric Waves—Radio frequencies between 300 kHz and 3 MHz. (MF)

High Band—VHF FM jargon for the 2-meter Amateur Radio band.

Highest Possible Frequency (HPF)—A radio frequency 15% above the maximum usable frequency, considered to be usable 10% of the days.

High Frequency (HF)—Radio frequencies between 3 and 30 MHz. (Decametric)

Image Frequency—(superheterodyne receiver) An undesired signal frequency that differs from the desired frequency by 2 times the intermediate frequency (IF).

In-band—Two or more frequencies within the same audio- or radio-frequency band.

Independent Sideband—Modulation of a carrier in which each sideband carries different intelligence.

Interference—Unwanted signals. (See *Harmful Interference*)

International Frequency Registration Board (IFRB)—An ITU agency that records frequency assignments made by different countries.

Inverse-Square Law—(radio) The power varies with distance inversely as the square of that distance, (a basic law of free-space propagation which says that as the distance is doubled, the signal will be one fourth or 6 dB weaker).

Jamming—Intentional radiation of a signal for the purpose of imparing the usefulness of radio reception.

Key Clicks—Undesired switching transients beyond the necessary bandwidth of a Morse code radio transmission caused by improperly shaped modulation envelopes.

Kilometric Waves—Radio frequencies between 30 and 300 kHz. (LF)

Left-Hand (or Anti-Clockwise) Polarized Wave—An elliptically or circularly-polarized wave, in which the electric field vector, observed in the fixed plane, normal to the direction of propagation, whilst looking in the direction of propagation, rotates with time in a left-hand or anti-clockwise direction. (RR)

Low Band—VHF FM jargon for the 6-meter Amateur Radio band. HF jargon for the 160, 80 or 40-meter bands.

Lowest Usable (High) Frequency (LUF)—The minimum radio frequency that will provide satisfactory ionospheric communication between two points 90% of the days.

Low Frequencies (LF)—Radio frequencies between 30 and 300 kHz. (Kilometric)

Machine—VHF FM jargon for a repeater.

Maximum Usable Frequency (MUF)—The highest radio frequency that will provide satisfactory ionospheric communication between two points at a given time 50% of the days.

Mean Power (of a radio transmitter)—The average power supplied to the antenna transmission line by a transmitter during an interval of time sufficiently long compared with the lowest frequency encountered in the modulation taken under normal operating conditions. (RR)

Media Access—Freedom to use, or protocol for using, media (such as a communication channel, writing to storage, or other resource).

Medium Frequencies (MF)—Radio frequencies between 300 kHz and 3 MHz. (Hectometric)

Meteor Burst Communications—Communications by the propagation of radio signals reflected by ionized meteor trails. (NTIA)

Metric Waves—Radio frequencies between 30 and 300 MHz. (VHF)

Microwaves—Radio frequencies above 1 GHz.

Millimetric Waves—Radio frequencies between 30 and 300 GHz. (EHF)

Modulation—Variation of one signal with another, usually to convey intelligence.

Multipath—Propagation of a radio signal via two or more distinct paths, such that signals (at certain frequencies and times) combine out of phase and produce a distorted output in the receiver.

Myriametric Waves—Radio frequencies between 3 and 30 kHz. (VLF)

Narrow-band—(1) A band of frequencies which are smaller than a reference bandwidth. (2) A band of frequencies which lie within the passband of a receiver or measuring device.

Necessary Bandwidth—For a given class of emission, the width of the frequency band which is just sufficient to ensure the transmission of information at the rate and with the quality required under specified conditions. (RR)

Net Control Station—A radio station responsible for real-time direction of a radio net, including maintenance of circuit discipline necessary to expeditious handling of traffic.

Net Manager—An individual responsible for the mission and resource management of a radio net.

Network Coordinating Agent (packet radio) —An entity to which a frequency coordinator delegates the responsibility for managing frequencies designated for packet-radio use.

NTIA—National Telecommunications & Information Administration of the US Department of Commerce.

Occupied Bandwidth—The width of a frequency band such that, below the lower and above the upper frequency limits, the mean powers emitted are each equal to a specified percentage B/2 of the total mean power of a given emission. Unless otherwise specified by the CCIR for the appropriate class of emission, the value of B/2 should be taken as 0.5%. (RR)

Open Repeater—A repeater where transient operators are welcome.

Optimum Working Frequency (OWF)—(Same as *Frequence Optimum de Travail, FOT*)

Out-of-band Emission—Emission on a frequency or frequencies immediately outside the necessary bandwidth which results from the modulation process, but excluding spurious emission. (RR)

Paging—One-way signaling to portable receivers for alerting or transmission of brief messages.

Pair (frequency)—Associated input and output frequencies.

Peak Envelope Power (of a radio transmitter)—The average power supplied to the antenna transmission line by a transmitter during one radio frequency cycle at the crest of the modulation envelope taken under normal operating conditions. (RR)

Period (of a satellite)—The time elapsing between two consecutive passages of a satellite through a characteristic point on its orbit. (RR)

Pilot—A signal to control the characteristics of a transmission. (For example, a 3.1-kHz pilot is used for automatic frequency control reference and to indicate the amount of expansion needed upon reception of an amplitude compandored single-sideband signal.)

Polarization—The alignment of the electric lines of force of a wave with respect to earth. If the lines are perpendicular to earth, the polarization is said to be vertical; if parallel to earth, horizontal. Polarization may also be elliptical or circular (right-hand or left-hand).

Polarization Diversity—A form of diversity reception using separate vertically and horizontally polarized antennas to take advantage of the tendency of vertical and horizontal components to rotate and thus fade independently.

Propagation—The motion of a radio wave.

Protection Ratio—The minimum value of the wanted-to-unwanted signal ratio, usually expressed in decibels, at the receiver input determined under specified conditions such that a specified reception quality of the wanted signal is achieved at the receiver output. (RR)

Quiet Zone—An area in which radio transmissions are restricted to reduce unwanted signals at sensitive receiving sites. The National Radio Quiet Zone is an area bounded by 39° 15′N on the north, 78° 30′W on the east, 37° 30′N on the south and 80° 30′W on the west. (FCC Rules, section 97.3)

Rabbit—An unbalanced repeater with ''big ears'' (sensitive receiver) and ''small mouth'' (low-power transmitter).

Radiation—The outward flow of energy from any source in the form of radio waves. (RR)

Radiocommunication—Telecommunication by means of radio waves. (RR)

Radiocommunication Service—A service . . . involving the transmission, emission and/or reception of radio waves for specific telecommunication purposes. In these regulations, unless otherwise stated, any radio-communication service relates to terrestrial radio-communication. (RR)

Raster—The pattern of spacing of channels in a specific group or band.

Reciprocity—(1) (radio propagation) The ability of radio signals to traverse the same path in both directions with the same results. (2) (mobile radio) Transmitting and receiving subsystems that have equal capability to work over the same path loss. (See *Balanced System*.)

Reduced Carrier Single-Sideband Emission—A single-sideband emission in which the degree of carrier suppression enables the carrier to be reconstituted and to be used for demodulation. (RR)

Reference Frequency—A frequency having a fixed and specific position with respect to the assigned frequency. The displacement of this frequency with respect to the assigned frequency has the same absolute value and sign that the displacement of the characteristic frequency has with respect to the center of the frequency band occupied by the emission. (RR) (See also *Characteristic Frequency*.)

Scanner—A radio receiver that can be programmed to sample different frequencies in a systematic sequence.

Selectivity—The ability of a radio receiver to discriminate against frequencies outside the desired passband.

Service Area—The area served by a radio station (typically a broadcast transmitter, base station or repeater).

Shift (frequency)—The difference between the frequency of the mark and space frequencies in frequency-shift-keyed modulation.

Sideband—The group of frequencies above or below the main carrier frequency which contain the product of modulation.

Silent Period—A designated time during which all stations must cease transmission and listen for distress traffic.

Simulcast—Simultaneous transmission on two or more frequencies or modulation modes. (Also known as quasi-synchronous transmission.)

Signal-to-Noise Ratio—The ratio of the amplitude of the signal vs noise, usually expressed in dB.

Single-Sideband Emission—An amplitude modulated emission with one sideband only. (RR)

Site Noise—Noise, external to the receiver, that limits reception at a location—a composite of co-channel interference, power-line noise, machine emissions, and spurious emissions from transmitters.

Spectrum Efficiency—The ratio of the information delivered to the spectrum resources (bandwidth, time and physical space) used.

Spectrum Engineering—The science of effective use of the radio spectrum to achieve the desired results economically.

Spectrum Management—Efficient and judicious use of the radio spectrum through application of techniques such as spectrum engineering, frequency coordination, monitoring and measurement, computer data base, and analysis.

Spectrum Metric—A unit of measure of spectrum-resource use.

Spectrum Occupancy—A frequency is said to be occupied when the signal strength exceeds a given threshold. Spectrum Occupancy is normally expressed in terms of probability or as a percent.

Splatter—A form of adjacent-channel interference caused by overmodulation on peaks.

Spread Spectrum—A signal structuring technique that employs direct sequence, frequency hopping or a hybrid

(Glossary continued on next page)

of these, which can be used for multiple access and/or multiple functions. This technique decreases the potential interference to other receivers while achieving privacy and increasing the immunity of spread spectrum receivers to noise and interference. Spread spectrum generally makes use of a sequential noise-like signal structure to spread the normally narrowband information signal over a relatively wide band of frequencies. The receiver correlates the signals to retrieve the original information signal. (NTIA)

Spurious Emission—Emission on a frequency or frequencies which are outside the necessary bandwidth and the level of which may be reduced without affecting the corresponding transmission of information. Spurious emissions include harmonic emissions, parasitic emissions, intermodulation products and frequency conversion products, but exclude out-of-band emissions. (RR)

Station—One or more transmitters or receivers or a combination of transmitters and receivers, including the accessory equipment, necessary at one location for carrying on a radiocommunication service, or the radio astronomy service. Each station shall be classified by the service in which it operates permanently or temporarily. (RR)

Subband—A part of a (frequency) band, usually having the same purpose or operator privileges.

Super High Frequencies (SHF)—Radio frequencies between 3 and 30 GHz. (Centimetric)

Suppressed Carrier Single-Sideband Emission—A single-sideband emission in which the carrier is virtually suppressed and not intended to be used for demodulation. (RR)

Telecommand—The use of telecommunication for the transmission of signals to initiate, modify or terminate functions of equipment at a distance. (RR)

Telecommunication—Any transmission, emission or reception of signs, signals, writing, images and sounds or intelligence of any nature by wire, radio, optical or other electromagnetic systems. (RR)

Telegraphy—A form of telecommunication which is concerned in any process providing transmission and reproduction at a distance of documentary matter, such as written or printed matter or fixed images, or the reproduction at a distance of any kind of information in such a form. For the purposes of the Radio Regulations, unless otherwise specified therein, telegraphy shall mean a form of telecommunication for the transmission of written matter by the use of a signal code. (RR)

Telemetry—The use of telecommunication for automatically indicating or recording measurements at a distance from the measuring instrument. (RR)

Telephony—A form of telecommunication set up for the transmission of speech or, in some cases, other sounds. (RR)

Television—A form of telecommunication for the transmission of transient images of fixed or moving objects. (RR)

Time-Bandwidth Product—The duration of a transmission multiplied by the bandwidth of the signal—a measure of how much bandwidth is used for how long.

Transmitter Power—The peak envelope power (output) present at the antenna terminals (where the antenna feedline, or if no feedline is used, the antenna, would be connected) of the transmitter. The term "transmitter" includes any external radio frequency power amplifier which may be used. Peak envelope power is defined as the average power during one radio frequency cycle at the crest of the modulation envelope, taken under normal operating conditions. (FCC Rules, section 97.3)

Trunking—A channel-utilization scheme whereby two stations automatically can select an unused channel out of a group.

Tuning, Tune-Up—Emission of a signal for the purpose of adjusting a radio transmitter.

Ultra High Frequencies (UHF)—Radio frequencies between 300 MHz and 3 GHz. (Decimetric)

Unwanted Emissions—Consist of spurious emissions and out-of-band emissions. (RR)

Very High Frequencies (VHF)—Radio frequencies between 30 and 300 MHz. (Metric)

Very Low Frequencies (VLF)—Radio frequencies between 3 and 30 kHz. (Myriametric)

Vestigial Sideband (VSB)—The sideband of an AM transmission that has been substantially suppressed and which rolls off gradually with distance from the carrier frequency.

WARC—World Administrative Radio Conference, held by the ITU.

Wideband—(1) A band of frequencies which are larger than a reference bandwidth. (2) A band of frequencies which exceed the passband of a receiver or measuring device.

Working Frequency—A frequency designated for exchange of communications after contact is made on a calling frequency or guard channel.

---

messages unless specifically permitted by the two countries involved. US amateurs may exchange third-party messages with the countries listed on page 5-8.

In Atlantic City in 1947, radical changes were made in the organization of the Union. The ITU became a specialized agency of the United Nations—the oldest such agency—and ITU headquarters was transferred from Berne to Geneva.

The internal structure of the ITU is shown in Fig 2-1.

Plenipotentiary Conferences

The ITU has Plenipotentiary Conferences every five or six years. "Plenipotentiary" is a two-dollar word that means the conference is fully empowered to do business. The conferences determine general policies, review the work of the Union, revise the Convention if necessary, elect the Members of the Union that are to serve on the Administrative Council, and elect the Secretary-General, the Deputy Secretary-General, Directors of the International Consultative Committees and the members of the International Frequency Registration Board. The last such conference was held in Nairobi in 1982.

International Consultative Committees

The International Radio Consultative Committee (CCIR) studies technical and operating questions, and issues recommendations relating to radio communication. The CCIR organization is shown in Fig 2-2. The Director, CCIR is Richard C. Kirby, WØLCT/HB9BOA. Amateurs have had a longstanding interest in the CCIR deliberative process. The League monitors, and makes input through, Study Group 8—Mobile, Radiodetermination and Amateur Services.

The International Telegraph and Telephone Consultative Committee (CCITT) studies technical, operating and tariff questions, and issues recommendations relating to telegraphy and telephony. The organization structure of the CCITT Study Groups is shown in Fig 2-3. Among other things, the international Morse code is specified in CCITT Recommendation F.1. Amateurs paid little attention to CCITT until the advent of packet radio, which sent people scurrying for international data-communications recommendations. The AX.25 packet-radio protocol, for example, is based on CCITT Recommendation X.25.

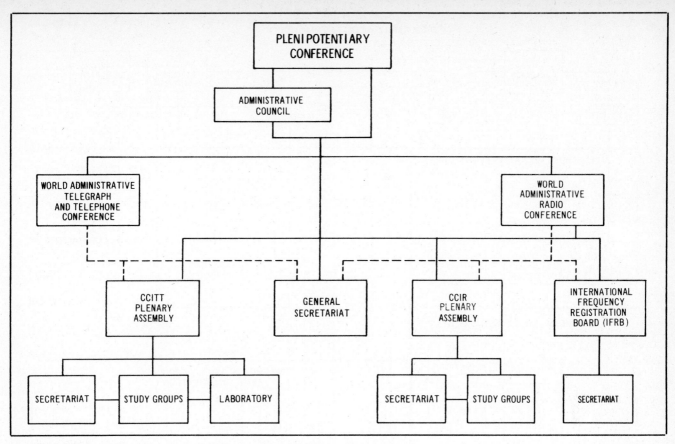

Fig 2-1—Organization chart of the International Telecommunication Union (ITU).

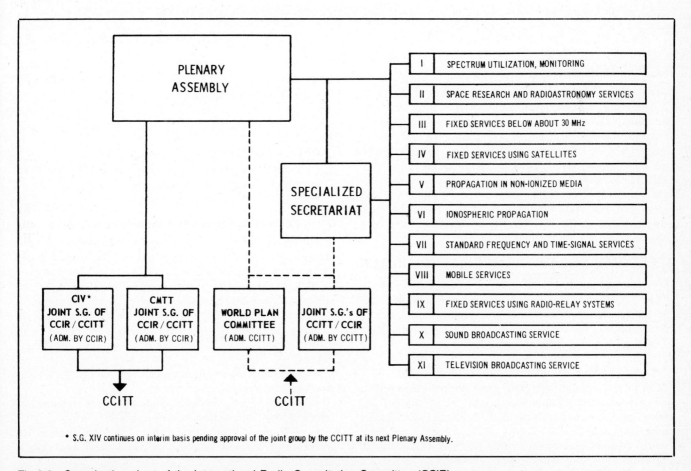

* S.G. XIV continues on interim basis pending approval of the joint group by the CCITT at its next Plenary Assembly.

Fig 2-2—Organization chart of the International Radio Consultative Committee (CCIR).

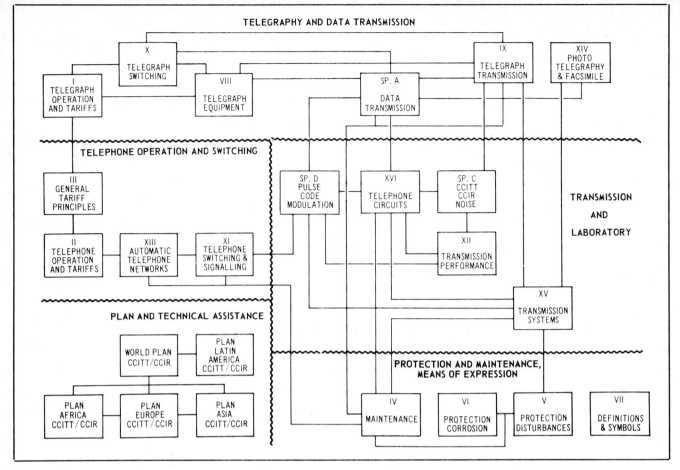

Fig 2-3—Organization chart of the International Telegraph and Telephone Consultative Committee (CCITT) Study Groups.

World Administrative Radio Conferences

The job of a World Administrative Radio Conference (WARC) is to revise the Radio Regulations. Conferences that had limited agendas considering space and radio astronomy matters and had some effect on the Amateur-Satellite Service were held in 1963, 1971 and 1977. The last conference to go over the complete set of ITU Radio Regulations was held in Geneva in 1979. WARC-79 will be covered in some detail later in this chapter. Specialized conferences are held about twice a year, but most have limited agendas and do not have the power to change the rules or the allocations of our two (Amateur and Amateur-Satellite) Services.

TELECOMMUNICATIONS REGULATION IN THE UNITED STATES

The Communications Act of 1934, as amended, provides for the regulation of interstate and foreign commerce in communication by wire or radio. This Act is printed in Title 47 of the US Code, beginning with Section 151.

Federal Communications Commission

The FCC is responsible, under the Communications Act, for regulating all telecommunications except that of the Federal government. This includes the Amateur Radio Service and the Amateur-Satellite Service, which are regulated under Part 97 of the Commission's rules. (Part 97 is available in *The FCC Rule Book,* published by the ARRL.)

The internal structure of the FCC is shown in Fig 2-4. The FCC components of most concern to Amateur Radio are:
- the five Commissioners
- the Private Radio Bureau (PRB)
- the PRB Special Services Division and its Personal Radio Branch
- the Licensing Division, Gettysburg, PA
- the Field Operations Bureau, its Regional Offices, Field Offices and Monitoring Stations, and
- the Office of Engineering & Technology

Federal Government Telecommunications

The functions relating to assignment of frequencies to radio stations belonging to, and operated by, the United States government were assigned to the Assistant Secretary of Commerce for Communications and Information (Administrator, National Telecommunications and Information Administration—NTIA) by Department of Commerce Organization Order 10-10 of May 9, 1978. See Figs 2-5 and 2-6 for NTIA organizational charts. Among other things, NTIA:
- coordinates telecommunications activities of the Executive Branch
- develops plans, policies and programs relating to international telecommunications issues
- coordinates preparations for US participation in international telecommunications conferences and negotiations
- develops, in cooperation with the FCC, a long-range plan

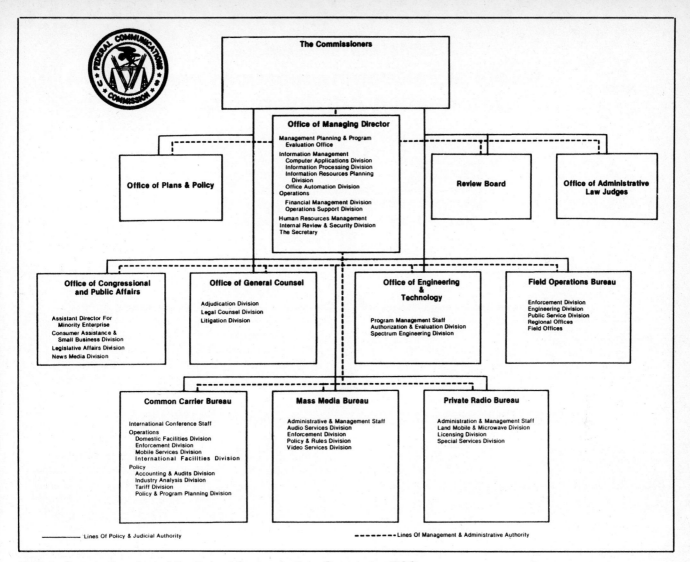

Fig 2-4—Organization chart of the Federal Communications Commission (FCC).

for improved management of all electromagnetic spectrum resources, including jointly determining the National Table of Frequency Allocations.

• conducts telecommunications research and development. The HF propagation predictions presented for your use in this manual (see Chapter 17) were prepared by NTIA's Institute for Telecommunication Sciences in Boulder, Colorado.

Interdepartment Radio Advisory Committee

The Interdepartment Radio Advisory Committee (IRAC) traces back to a mutual agreement of Government departments on June 1, 1922. Its status is as defined on December 10, 1964 and continued per Executive Order 12046 of March 27, 1978. The NTIA appoints the officers of the IRAC and chairmen of its subcommittees. It is composed of representatives appointed by the following member departments and agencies: Agriculture, Air Force, Army, Coast Guard, Commerce, Energy, Federal Aviation Administration, Federal Emergency Management Agency, General Services Administration, Health and Human Services, Interior, Justice, National Aeronautics and Space Administration, National Science Foundation, Navy, State, Treasury, US Information Agency, US Postal Service, and Veterans Administration.

The IRAC assists the Assistant Secretary of Commerce in assigning frequencies to US government radio stations and in developing and executing plans, procedures, and technical criteria. The FCC must coordinate with IRAC any actions that might affect the Federal government's use of the radio spectrum. The 220- to 225-MHz band has been a subject of joint FCC-IRAC study because it is shared by the Amateur Radio Service and the government. It is reasonable to expect that FCC dockets involving such sharing will take a bit longer to consider than those solely under the FCC's wing.

THE 16-YEARS WARC

Well, you've been a good class through the theory part. You'll want to stay awake for the history part, as it tells you how we really get and keep our frequency bands. The ITU-sponsored World Administrative Radio Conference of 1979 brought about numerous changes in Amateur Service and Amateur-Satellite band allocations. It is more than an intellectual exercise to retrace Amateur Radio WARC-79 preparations, the conference itself, and postconference actions.

As early as 1963, ARRL President Herbert Hoover, Jr., W6ZH, recognized that preparations had to begin for the next WARC. Efforts were already underway in Europe to strengthen Region 1 of the International Amateur Radio

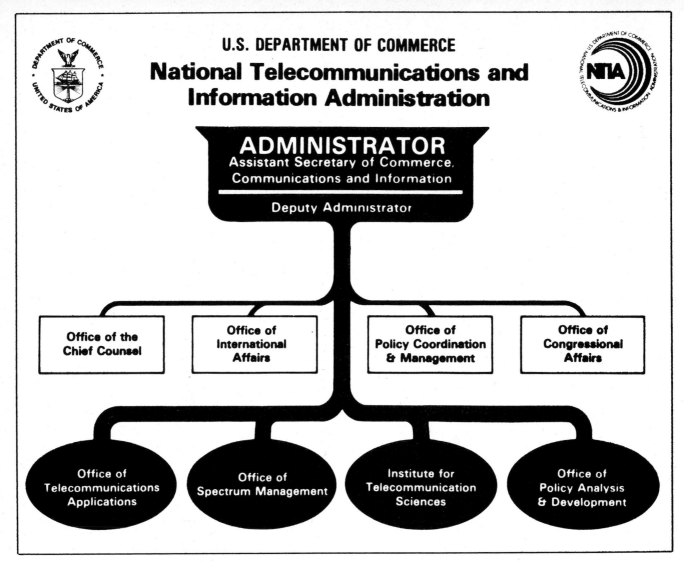

ADMINISTRATOR
Assistant Secretary of Commerce, Communications and Information

Deputy Administrator

Office of the Chief Counsel

Office of International Affairs

Office of Policy Coordination & Management

Office of Congressional Affairs

Office of Telecommunications Applications

Office of Spectrum Management

Institute for Telecommunication Sciences

Office of Policy Analysis & Development

Fig 2-5—Organization chart of the National Telecommunications and Information Administration (NTIA).

Union (IARU). Wearing his other hat as IARU president, Mr. Hoover launched a program to develop Amateur Radio in the new nations of Africa. These efforts continued under Mr. Hoover's successor, Robert W. Denniston, WØDX. The Intruder Watch was created, to document the interference from nonamateur stations suffered by amateurs in their exclusive bands. Bob also developed a personal rapport with a number of amateurs outside North America who were destined to play important roles in the IARU. When Noel B. Eaton, VE3CJ, became IARU president, the three regional organizations assumed a greater importance in the functioning of the Union; extensive travel was undertaken to improve the liaison between Headquarters and the member-societies and to bolster the societies' efforts to influence their governments. An International Working Group was established in 1976 to advise on WARC preparations. This group became the nucleus of the IARU Team which went to Geneva in 1979.

FCC preparations for WARC-79 were well underway in 1974. It was the Commission's responsibility to determine the long-term allocation requirements of non-Federal government users. This was done through a series of Notices of Inquiry (Docket 20271) soliciting comments, and through temporary Advisory Committees established for each service.

What is the IARU?

The International Amateur Radio Union is a federation of national Amateur Radio societies representing the interests of two-way Amateur Radio communications worldwide. Founded in Paris in 1925, the IARU is governed by a Constitution and Bylaws. It has three regional organizations: Region 1—Africa, Europe, USSR, Middle East (excluding Iran) and Mongolia; Region 2—North and South America, including Hawaii and the Johnston and Midway Islands; Region 3—the rest of Asia and Oceania. The International Secretariat is located in Newington, Connecticut. To date, the IARU has 125 member-societies.

Extensive Public Participation

The FCC made special efforts to encourage public participation in the formulation and discussion of WARC-related issues. No group was more active in this process than the nation's radio amateurs. Hundreds of individuals and clubs submitted comments at several stages of the proceeding, supplementing those of the League and the Commission's own Advisory Committee for Amateur Radio. The League's comments alone totaled nearly 200 pages.

The resultant 560-page Report and Order set forth the

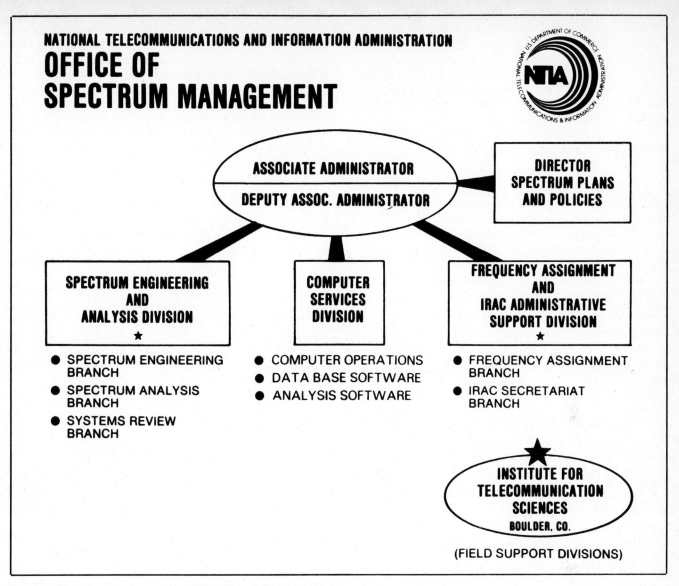

NATIONAL TELECOMMUNICATIONS AND INFORMATION ADMINISTRATION

OFFICE OF SPECTRUM MANAGEMENT

ASSOCIATE ADMINISTRATOR

DEPUTY ASSOC. ADMINISTRATOR

DIRECTOR SPECTRUM PLANS AND POLICIES

SPECTRUM ENGINEERING AND ANALYSIS DIVISION ★
- SPECTRUM ENGINEERING BRANCH
- SPECTRUM ANALYSIS BRANCH
- SYSTEMS REVIEW BRANCH

COMPUTER SERVICES DIVISION
- COMPUTER OPERATIONS
- DATA BASE SOFTWARE
- ANALYSIS SOFTWARE

FREQUENCY ASSIGNMENT AND IRAC ADMINISTRATIVE SUPPORT DIVISION ★
- FREQUENCY ASSIGNMENT BRANCH
- IRAC SECRETARIAT BRANCH

★ **INSTITUTE FOR TELECOMMUNICATION SCIENCES** BOULDER, CO.

(FIELD SUPPORT DIVISIONS)

Fig 2-6—Organization chart of the NTIA Office of Spectrum Management.

Commission's proposals. It was said to represent the combined thinking and agreement of the Commission and the various agencies of the Federal government. The document was forwarded to the Department of State with the recommendation that it be made the official US proposal for WARC-79.

IARU WARC-79 Objectives

Formal adoption of WARC objectives for Amateur Radio came at a series of meetings of the IARU Regions in 1975-76. IARU Region 3 led off in March 1975 by adopting objectives which essentially mirrored those of the four-man US study group. A month later Region 1 adopted very similar objectives, except that those below 10 MHz were more modest. Region 2 followed, in April 1976, by endorsing the objectives previously adopted by Region 3. WARC-79 was the first conference for which the world's amateurs adopted a common position in this way.

The IARU Geneva Team

In conjunction with the Region 2 Conference in April 1976, IARU President Eaton had convened a meeting with represen-

tatives of the three IARU regional organizations and several national amateur Societies. One idea to emerge from this meeting was need for a worldwide IARU working group to assist in guiding the preparatory work. This working group met several times over the following two years. One of their important products was a model position paper in English, French and Spanish which national Societies could use as a basis for their own formal submissions to their governments.

Even more important were the efforts by IARU Member-Societies to get Amateur Radio representatives on their national delegations. This was successful in 17 countries.

The basic approach, basic strategies and organization used in preparing Amateur Radio for WARC-79 proved to be sound. When the Final Acts of the Conference were signed in December 1979, Amateur Radio emerged not only with its present bands intact, but with important new allocations at 10, 18 and 24 MHz as well! Further, there were relaxation of restrictions on some of the existing amateur bands, and additional segments were earmarked for amateur satellite work. WARC-79 was a tremendous team effort on the part of many, many persons who had the faith and the vision to work successfully toward a common goal. The degree of success is attributable almost entirely to the intensive preparation conducted by the IARU. The IARU Geneva Team consisted of IARU President Noel Eaton, VE3CJ; Vice President Victor C. Clark, W4KFC; Secretary Richard L. Baldwin, W1RU; Thomas R. Clarkson, ZL2AZ; C. Eric Godsmark, G5CO; Bruce A. Johnson, WA6IDN; Shigetake Morimoto, JA1NET; Wojciech Nietyksza, SP5FM; David H. Rankin, 9V1RH/VK3QV; Pedro Seidemann, YV5BPG; Alberto Shaio, HK3DEU; Carl L. Smith, WØBWJ; R. F. Stevens, G2BVN; and David Sumner, K1ZZ.

Future WARC

The ARRL and IARU have begun preliminary discussions in preparation for a future WARC—whenever it may be held. Some preparatory work involves attending ITU meetings to get to know the issues and those people who will become decision-making delegates to a future WARC. WARC-preparatory travel to IARU societies has begun to reinforce the necessity of adequate liaison with government telecommunications people and to support the goals adopted by the IARU Administrative Council.

Bibliography

Palm, *The FCC Rule Book,* ARRL, Newington, CT.

FCC, *Rules and Regulations,* Parts 2 and 97.

Feller, *Planning an Electromagnetic Environment Model for Spectrum Management,* FCC, Washington, DC, Jun 26, 1981.

ITU, *Radio Regulations,* Geneva, 1982, revised in 1985.

NTIA, *Manual of Regulations and Procedures for Federal Radio Frequency Management,* revised to Jan 1985, US Department of Commerce.

Sumner, "FCC WARC Proposals Finalized," *QST,* Feb 79, p 55.

Sumner, "ITU Lays Technical Foundation for WARC-79," *QST,* Mar 79, p 56.

Sumner, "WARC Countdown—Mid-Conference Report from Geneva," *QST,* Dec 79, p 73.

Sumner, "WARC Countdown—Dateline Geneva: The Closing Days," *QST,* Jan 80, pp 62-63.

Baldwin and Sumner, "The Geneva Story," *QST,* Feb 80, pp 52-61.

AMATEUR FREQUENCY ALLOCATIONS AND BAND PLANS

There are bands of frequencies allocated to the Amateur Service extending from 1800 kHz to 250 GHz. Table 2-1 will give you an overview of the radio spectrum, some nomenclature and the Amateur Service bands allocated in the ITU *Radio Regulations.*

Table 2-2 is a more-detailed, band-by-band picture of the radio spectrum. The part labeled "Allocations" is an extract of the ITU *Radio Regulations* Table of Frequency Allocations. All footnotes that pertain to the Amateur Service and other services sharing the same frequency bands are included. The part labeled "Amateur Radio Band Plans" includes the band plans formally adopted by the IARU Regions and the ARRL. In most of the high-frequency bands, there are no ARRL band plans that were adopted by the ARRL Board of Directors.

There is a tendency for band plans to lag reality. This is due in part to the time needed to research, invite and digest comment from amateurs, arrive at a mix that will serve the diverse needs of the amateur community, and adopt a formal band plan. This is a process that can take a year or more on

the national level and a similar period in the IARU. Nevertheless, new communications modes or popularity of existing ones can make a year-or-two-old band plan look obsolete. Such revolutionary change has taken place recently with the popularity of packet radio, particularly in the 20- and 2-meter bands. As this is written, a similar change could be brewing for those bands in which Novices were given voice and digital privileges under FCC PR Docket 86-161 (Novice Enhancement) which became effective on March 21, 1987. Changes of this magnitude cause the new users to scramble for frequencies and some of the existing mode users to draw their wagons in a circle. The national Societies, their staffs and committees, and the IARU have the job of sorting out the contention for various frequencies and preparing new band plans. Fortunately, we have not exhausted all possible ways of improving our management of the spectrum so that all Amateur Radio interests can be accommodated.

The 160-Meter Band

Those who have a fascination for 160-meter DX should take a close look at page 5-22. It's a country-by-country listing of the frequency allocations in this band. The basic problem with allocations in this band has been competition with the Radiolocation Service. New pressures are possible as a result of planned expansion of the medium-frequency broadcast band in the 1605-1705 kHz range.

The 80-Meter Band

While US amateurs enjoy the use of 3500-4000 kHz, not all countries allocate such a wide range of frequencies to the 80-meter band. There are fixed, mobile and broadcast operations, particularly in the upper part of the band.

The 40-Meter Band

The 40-meter band has a big problem: International broadcasting occupies the 7100-7300 kHz (41-meter) band in many parts of the world. During the daytime when sunspots are high, broadcasting does not cause much interference to US amateurs. However, at night, especially when sunspots are low, the broadcast interference is heavy. Some countries allocate only the 7000-7100 kHz band to amateurs. Others, particularly in Region 2, allocate 7l00-7300 kHz as well, which at times is subject to interference. The result is that there is a great demand for frequencies in the 7000-7100 kHz segment. The effect is that there are two band plans overlaid on each other, ours that spreads out over 7000-7300 kHz and the foreign one that compresses everything into 7000-7100 kHz. An example of the problem is that US stations operate RTTY in the 7080-7100 kHz segment, and foreign stations typically use this segment for phone and the 7035-7045 kHz subband for RTTY. The two basic band plans coexist nicely until propagation permits stations in different regions to hear each other. The consolation is that 7000-7035 kHz is fine hunting grounds for domestic and DX CW contacts.

The 30-Meter Band

The relatively new 30-meter band is an excellent band for CW and digital modes. The only problem is that US amateurs must not cause harmful interference to the fixed operations outside the US. This restricts transmitter power output to 200 watts and is the reason for not having contests on this band.

The 20-Meter Band

The workhorse of DX is undoubtedly the 20-meter band. It offers excellent propagation to all parts of the world

Table 2-1

The Electomagnetic Spectrum with Amateur Service Frequency Bands by ITU Region

Wave-length	Frequency		Nomen-clature	Metric Band	Amateur Radio Bands by ITU Region		
					Region 1	Region 2	Region 3
1 mm	300 GHz						
			EHF	1 mm	241-250	241-250	241-250
			Milli-	2 mm	142-149	142-149	142-149
		M	metric	2.5 mm	119.98-120.02	119.98-120.02	119.98-120.02
		i		4 mm	75.5-81	75.5-81	75.5-81
		c		6 mm	47-47.2	47-47.2	47-47.2
1 cm	30 GHz	r					
		o	SHF	1.2 cm	24-24.25	24-24.25	24-24.25
		w	Centi-	3 cm	10-10.5	10-10.5	10-10.5
		a	metric	5 cm	5.65-5.85	5.65-5.925	5.65-5.85
		v		9 cm		3.3-3.5	3.3-3.5
10 cm	3 GHz	e					
		s	UHF	13 cm	2.3-2.45	2.3-2.45	2.3-2.45
			Deci-	23 cm	1240-1300	1240-1300	1240-1300
			metric	33 cm		902-928	
				70 cm	430-440	430-440	430-440
1	300 MHz						
			VHF	1.25 m		220-225	
			Metric	2 m	144-146	144-148	144-148
				6 m		50-54	50-54
10	30 MHz						
			HF	10 m	28-29.7	28-29.7	28-29.7
			Deca-	12 m	24.89-24.99	24.89-24.99	24.89-24.99
			metric	15 m	21-21.45	21-21.45	21-21.45
				17 m	18.068-18.168	18.068-18.168	18.068-18.168
				20 m	14-14.350	14-14.350	14-14.350
				30 m	10.1-10.150	10.1-10.150	10.1-10.150
				40 m	7-7.1	7-7.3	7-7.1
				80 m	3.5-3.8	3.5-4	3.5-3.9
100	3 MHz						
			MF	160 m	1.81-1.85	1.8-2	1.8-2
			Hecto-metric				
1000	300 kHz						
			LF Kilo-metric				
10,000	30 kHz						
			VLF Myria-metric				
100,000	3 kHz						

Note: This table should be used only for a general overview of where ITU Amateur Service and Amateur-Satellite Service frequencies fall within the radio spectrum. They do not necessarily agree with FCC allocations; for example, the 70-cm band is 420-450 MHz in the United States. Refer to subsequent tables for more specific information.

---

throughout the sunspot cycle and is virtually clean of interference from other services.

The 17-Meter Band

The 18068-18168 kHz band is not yet allocated to the Amateur Service in the US, pending reassignment of some fixed operations to other frequencies by July 1989. The ARRL has petitioned the FCC for early use of this band.

The 15- and 12-Meter Bands

The 21000-21450 and 24890-24990 kHz bands are excellent for DX when the sunspots are high. They also offer other openings throughout the sunspot cycle.

The 10-Meter Band

This is an exclusive amateur band worldwide. Its popularity is bound to rise sharply as a result of FCC PR Docket 86-161, better known as Novice Enhancement, which grants Novices and Technicians digital and voice privileges.

The VHF and Higher Bands

The 6-meter band is not universal, but the trend seems to be toward allocating it to amateurs as TV broadcasting vacates the 50-54 MHz band. Recent new 6-meter privileges for UK amateurs adds to worldwide interest in this band. It is also excellent for amateur exploitation of meteor-scatter communications using packet radio.

Two meters has been hot since the popularization of FM repeaters. It also has a chunk (145.8-146 MHz) devoted to amateur satellites. VHF packet radio also cut its teeth on the 2-meter band.

The 1.25-meter (220-225 MHz) band has had problems for

years. Because it was shared with other services in the US and its fate was unclear for some time, there has been some fear of losing it. This and the fact that the band is not worldwide combined to make manufacturers reluctant to offer much equipment for this band. It was a typical "chicken-and-egg" problem of how to get enough activity going with insufficient equipment in order to convince the equipment makers that they should build it. Novice Enhancement (PR Docket No. 86-161) is expected to bring a good number of Novices to the 222.1-223.91 MHz part of the band, probably much of it for FM voice operation. This is expected to prime the pump for heavy occupancy of the 1.25-meter band throughout the US and should create a viable market for new equipment. Unfortunately, within days after the Novice Enhancement announcement, the FCC issued a Notice of Proposed Rule Making (PR Docket 87-14) proposing to reallocate 220-222 MHz to land mobile services. The effect would be to compress 5 MHz of amateur usage into a 3-MHz band and perhaps relegate some operations to higher bands. The ARRL will vigorously oppose reallocation of the 220-222 MHz band.

The 70-cm band is prime UHF spectrum. The 430-440 MHz portion is virtually worldwide, but 420-430 and 440-450 MHz are not. The 435-438 MHz band is used for amateur satellites. Complexities of allocations make it difficult to come up with a completely satisfactory band plan.

The 33-cm band is new. While there is an ARRL Board-approved interim band plan, there is need for additional experience in that this band is shared with other services including ISM (industrial, scientific and medical). Also, there may be importation of radios built for the Japanese Personal Radio Service which is not channelized according to the US band plan.

Novice Enhancement brought along with it privileges in the 23-cm band (1270-1295 MHz), so some increased population is expected in this band. Also, manufacturers have shown interest in this band and now offer a variety of radios, which also should contribute to greater use.

The remaining microwave bands are the territory of amateur experimenters. It is important that amateurs use these bands and contribute to the state of the microwave art in order to retain them. Satellite and terrestrial relay services would dearly love to sink their transmitters into them.

FREQUENCY COORDINATION

From the earliest days of Amateur Radio, hams simply listened on a frequency to see if it was in use, and if not, went ahead and transmitted on that frequency. Amateurs operating on HF still use this simple listen-before-transmit procedure of randomly *assigning* themselves a frequency for use over a short period of time. This simple procedure has served us well over the years, but is no longer the only "rule of the road."

HF CW and phone nets historically have had problems finding frequencies on which to conduct their regular operations. Some have come up with strong arguments supporting the need for nets to have priority over a two-station contact on the same frequency. The rationale is that nets have many stations that would be difficult to move to another frequency and that the needs of the many translate to more public good than just a few rag chewers. The other side of the argument is that the two stations have a right to conduct their communication. The principle of priority becomes a bit clearer when one set of users has emergency traffic and the other does not. Nevertheless, no one (including the FCC and the ARRL) has expressed willingness to say who gets a specific HF channel and who does not. This is rooted in the knowledge that stations/needs come and go hour by hour, that ionospheric propagation changes hour by hour, and that most hams can bend a little in frequency and time to let both sets of users carry on.

Over time it became necessary to introduce a mild spectrum-management technique of frequency registration to let hams know where and when nets normally operate. This information is published annually in the ARRL *Net Directory*. This appearance of a listing in the *Net Directory* does not indicate an *assignment* by any authority except the entity responsible for the net. It is particularly useful when researching new frequencies for net or contest operations. It is of only limited use to individuals wishing to conduct random contacts, as the listen-before-transmit technique is more practical.

The Days Before Repeater Coordination

Two-meter FM repeaters became the rage in the 1960s. Just put your repeater on 34/94 (146.34 MHz input, 146.94 MHz output) and there was no problem...until someone else wanted to do the same thing nearby. "We got here first, so we are in the right; they are in the wrong. Why don't *they* find another frequency pair? What pair? That's *their* problem. Let's send two of our 250-pound guys over there and cut their coax!" No, that's not the way.

The early repeater contention problems were solved by having the latecomers find new frequencies. New transceivers typically were supplied with crystals for (146) 34, 52, 76 and 94. There were repeaters with inputs on 34 and outputs on 76 to avoid local simplex operation on 94. But, 76 became paired with a 16 input, and the 600-kHz offset was well on its way to becoming standard. Much of the early spadework in repeater band planning was done by the Texas VHF Society.

By early 1976, the repeater count in the US was more than 2000. The ARRL VHF Repeater Advisory Committee (VRAC) agreed on band plans which named repeater pairs in the 10-meter, 6-meter, 2-meter, 1.25-meter and 70-cm bands. The ARRL Board of Directors recommended that these band plans be published, which was done in October 1976 *QST*.

Many early tube-type amateur FM transceivers had seen service in the back of a police car or taxi. The land mobile services had been using 15-kHz deviation and had spaced their channels in a 30-kHz raster (30 kHz between channel center frequencies) but were in the process of converting to 5-kHz deviation and 15-kHz channel spacing. Amateurs followed suit by first setting up 30-kHz channel spacing with so-called *wideband* FM (15-kHz) then changing to 15-kHz channel spacing, which in that context was known as *narrow-band* FM. The new channels created in between the original 30-kHz spaced channels were called *splinter channels*. The change was accomplished in some cases by converting older rigs by backing off on the transmitter deviation and somehow tightening up the receiver bandwidth. The problem was solved by new solid-state rigs from Japan, first with a few crystals and room for more, later with synthesizers.

Synthesized transceivers were able to tune to any frequency

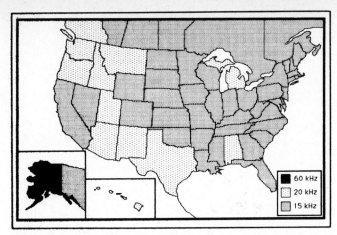

Fig 2-7—The state of the 146-148 MHz band plan in the US and Canada in the summer of 1986.

within the 2-meter band in 5-kHz steps. This opened up another possibility: It was now feasible to try some different channel spacings without causing people to buy several hundred dollars in new crystals. The technical imperative (It is possible, therefore do it!) was the mother of the 20-kHz channel-spacing plan and Clay Freinwald, K7CR, the father. The argument as to which is better (15 or 20) continues, with parts of the US on different spacing. A map showing 2-meter channel spacing throughout the US and Canada is shown in Fig 2-7.

This spreading out throughout the 2-meter and other bands to accommodate more repeaters had its limits, of course. It was clear that there would come a day when all the channels would be full, particularly in the metropolitan areas. What would happen then? Would latecomers just fire up new repeaters on top of established ones? Chaos! That was to be avoided at all costs, so amateurs actually began talking to each other. The prime movers were the repeater operators; they banned together and appointed *repeater frequency coordinators* to keep out the "upstarts." This was fine unless *you* were one of the upstarts.

Amateur Repeater Coordination Councils

When it looked like there would be contention for the finite number of repeater pairs, repeater operators established *repeater coordination councils*. Typically, the members of the council were the trustees of the area repeaters and perhaps others (if membership was open to them) who were planning repeaters or just concerned. The democratic processes began to work. The established repeater operators were able to keep their frequencies, and the newcomers somehow could be accommodated until all the pairs were full. Most areas had a *frequency coordinator,* who had the administrative job of keeping records on all existing repeaters. The frequency coordinator kept sufficient information, and theoretically had the skills to determine whether a proposed repeater would interfere with existing repeaters. If, after polling established repeater operators, there appeared to be no unacceptable interference, the frequency coordinator would *coordinate* the new repeater. This added two new terms to the Amateur Radio lexicon: *coordinated repeater* (as just explained) and *noncoordinated repeater* (ie, "sorry Jack"). The term "uncoordinated" wasn't used, as "Jack" would have taken it to mean that his repeater was messy in addition to being unwanted. "Aha, but the coordinator has a boss!" Jack can

become a member of the council, study up on *Roberts Rules of Order* (which the council founders blithely stipulated in their constitution and bylaws) and try to bring about democratically what he couldn't do administratively. Jack could even volunteer to be president of the council and influence the frequency coordinator into favorable action. Incidentally, you never fire a frequency coordinator who is doing the job; where would you get another competent person willing to work long hours for no pay?

But wait a minute? Is Repeater Coordination Council *X* legitimate? Don't they just represent the existing repeater owners, not prospective ones, not the repeater users, and not the other hams who might want to use the same frequencies for other purposes? Some have been challenged. Some have come and gone. A few frustrated individuals have started rival coordination entities. For a discussion of what happens when a frequency coordinator "doesn't," see the FM/RPT column in September 1986 *QST*.

Okay, "Let's ask the FCC," someone says. The FCC is not enthusiastic about this (and it is usually undesirable and not recommended from the amateur point of view). How would you like to be called in to settle a fight? After the FCC's in-boxes are empty and its phones stop ringing, you know what the bottom line will be: "Why don't you guys get together and work this thing out. *OUT*."

FCC PR Docket 85-22

Coordinating one's choice of frequencies for a repeater has never been a licensing requirement. But by early 1985, demand for frequencies had grown to the point where even an FCC predisposed to prefer "unregulation" decided something needed to be done. In February 1985, the FCC (in PR Docket 85-22) stated:

> Most of the reported cases of amateur repeater-to-repeater interference appear to involve one or more non-coordinated repeaters. In the past two years we have had to resolve amateur repeater-to-repeater interference disputes in which at least one non-coordinated repeater was involved and in which the parties to the dispute could reach no amicable solution. We attribute the growing number of instances of amateur repeater-to-repeater interference and the need for increased intervention in these matters to the mounting pressure which develops as to the desirable repeater frequencies become fully assigned.
>
> When we have intervened in such interference disputes, we have...favored the repeater which has been operating in accordance with the recommendation of a local frequency coordinator.
>
> Comments are also sought with regard to whether the FCC should recognize a single national frequency coordinator for the Amateur Radio Service. Such a coordinator could either be an individual national organization or an umbrella organization comprised of local coordination groups. We would expect this type of organization to promulgate coordination (and de-coordination) standards, to facilitate the use of advancing technology, to consider alternative frequency assignments, to consider frequency spacing and repeater separation distances and, in the case of an umbrella organization, to advise local coordinators.

In its comments to PR Docket 85-22, the League proposed these guidelines:

> Preferred status in instances of harmful repeater-to-repeater interference should be granted to amateur repeater operators who have implemented the recommendation of their local or regional volunteer frequency coordinator and are thereby coordinated.
>
> Frequency coordination should be strongly urged for all amateur stations in repeater, or auxiliary, operation in any geographical area which is served by a frequency coordinator.

The Commission should not consider alternatives to frequency coordination nor mandate methods of coordination.

The use of modern technological innovation...should be encouraged but not substituted for frequency coordination.

The Commission should not recognize a single entity, such as a national frequency coordinator for amateur repeater operation. Such coordination activities should be performed by local, or regional, volunteer frequency coordinators with appropriate support to these coordinators to be provided by the League.

In its Report and Order to PR Docket 85-22, the Commission stated:

We received sixty-four comments and reply comments to the *Notice*. Commenters included the...ARRL...and many leading amateur volunteer coordination entities.

Repeater operation in the Amateur service inherently requires operation on established frequencies. Amateur repeater operation is not frequency agile, as are other types of amateur station operation. As a result, most amateur operators have been willing to voluntarily cooperate to avoid interference to frequencies designed for repeater operation... in favor of the greater good, particularly since many amateur repeaters are open to all amateur operators who desire to use them. The cooperation has taken the form of adherence to the determinations of local frequency coordinators. While no amateur operator or amateur station "owns" a frequency, this type of coordination is the minimum joint effort by the amateur community needed to facilitate repeater operation...

Several coordinators urged us to establish some mechanism to officially recognize local or regional coordinators. Others were concerned about the potential for abuse of power at the local level. Another concern was the exclusive right to coordinate with a geographical area. It is essential that repeater coordinators respond to the broadest base of local amateurs, and consider the concerns not only of repeater owners but also of those users of spectrum affected by repeater operation. Their authority is derived from the voluntary participation of the entire amateur community; their recognition must be derived from the same source. We believe the new rules will assure that a coordinator is representative of all local amateur operators.

Several commenters urged us to abolish closed repeaters in the Amateur service, or, alternatively, to permit coordinators to relegate closed repeaters to secondary status or to give open repeaters preference when coordinating. We are not of the view, as were these commenters, that closed repeaters are any more or less desirable than open repeaters.

We proposed to make non-coordinated repeaters primarily responsible to resolve interference associated with coordinated repeaters. The ARRL commented that we should go even further and make non-coordinated repeaters solely responsible to resolve such interference, and require non-coordinated repeaters to cease operation if the interference is not resolved. Although the focus must be placed in the first instance upon the non-coordinated repeater to resolve such interference, we are adopting our proposed rules which continue to make the coordinated repeater secondarily responsible. This permits local coordinators and the FCC to consider technical alternatives, questions of equity, and spectrum efficiency in reaching the most reasonable solution.

There you have it—the rationale behind the current Part 97 rules that affect amateur repeater coordination.

The League's Role in Coordination

While the ARRL successfully argued before the FCC that there should not be a national frequency coordinator, there remains a significant role for the League in the frequency-coordination process. For one thing, the ARRL has for years published a *Repeater Directory*, which by 1987 had over 12,000 entries, primarily for North and South America. The data base used in preparation of the *Directory* will be accessible to frequency coordinators via telephone line.

The design of the National Data Base has been proposed by The Mid-Atlantic Repeater Council (TMARC) and is being circulated for comment as this is written. Here are the proposed fields:

Name	Bytes	Comments
ARRL File #	6	
List in *Repeater Directory*?	1	Yes/No
Frequency Band	4	29, 50, 144, etc
Frequency	7	in MHz
Call Sign	6	of repeater or link
Area/Region	16	
Location	24	city/town/mountain, etc
County	2	
State	2	
Latitude	7	degrees/minutes/seconds
Longitude	7	degrees/minutes/seconds
Antenna Height AGL	4	above ground level
Antenna HAAT	5	above average terrain
Antenna HAMSL	5	above mean sea level
Antenna Pattern	1	O = omni, B = beam, C = Cardioid
Antenna Bearing	3	degrees
Antenna Polarization	1	vertical/horizontal
Antenna Gain	2	in dB
Transmitter ERP	4	in watts
Transmitter Power Output	3	in watts
Use	2	AL = auxiliary link
		BD = base downlink
		BU = base uplink
		RI = repeater input
		RO = repeater output
		RX = replexer
		SX = simpatch
		TA = TV audio
		TV = TV video
Protected Coverage Area	3	radius in miles
Status	2	D = deleted
		NC = not coordinated
		OP = operational
		P = proposed
		T = under test
		U = under construction
		UK = unknown
Emission type	3	F3E/J3E/A3F/F2B etc
Access Mode	2	BT = burst tone
		CL = closed
		OP = open
		PL = subaudible tone
		TT = DTMF
Linked to	7/6/6	freq/call sign/ARRL file #
Notes	12	directory abbreviation codes
Sponsor	9	call sign/abbreviated name
Source	9	information originator
Coordinator	9	coordinator of record
Date Coordinated	4	year/month
Contact Person	6	call sign of contact
Additional Comments	35	

The League also undertook publishing of a *Repeater Coordinator's Newsletter* in June 1985 to enable coordinators to share their views and increase the dialog among them. This newsletter is available at no cost to individuals and groups recognized as coordinators in their area, and at nominal cost to others.

Bibliography

ARRL, *Repeater Coordinators' Newsletter*.

Horzepa, S., "FM/RPT: When the Frequency Coordinator Doesn't...," *QST*, Sep 86.

McCoy, L., "ARRL Repeater Band Plans," *QST*, Oct 76.

PROPAGATION BEACONS

NCDXF Beacons

The following borrows liberally from June 1983 *QST* and from the Irish Radio Transmitters Society *Amateur Radio Yearbook, 1986-87.*

The worldwide beacon network on 14.100 MHz is probably the most useful group of beacons on any HF band. There are nine beacons which transmit in sequence every 10 minutes. Each beacon transmits for 1 minute, and the last minute in each 10-minute sequence is silent. The complete sequence is shown in Table 2-3.

The unique, and very interesting feature, of these beacons is that the transmitter power is varied from 100 watts to 1/10 of a watt in four steps during every one-minute transmission period. The transmission starts with QST DE 4U1UN/B (in New York), then four continuous tones are sent, for about nine seconds each, at 100, 10, 1 and 1/10 of a watt. Then the beacon signs off at full power (100 watts). If you listen for a while, you'll be pleasantly surprised at how effective 1/10 of a watt into a vertical can be.

Besides monitoring, the beacons also can be used for several do-it-yourself practical observations or projects. The beacons:

• provide an "in-band" time/frequency standard.

• are a means of comparing antennas or receivers by switching back and forth during the same nine-second dash. (Check DX versus short-range characteristics of different antennas. When a beacon signal is steady, rotate the beam antenna [from center to side] to check beamwidth.)

• enhance propagation investigation. (Record the lowest signal level copied from each beacon [including "no copy"] hour-to-hour and day-to-day.) How do your observations and interpretations of the beacon's reception compare with published propagation forecasts, or correlate with various A and K indices, WWV broadcasts, sunspot activity or just generally observed band activity? Is there any correlation at all? Can beacon signals be used to improve forecasts of conditions?

The beacon net was organized and financed by the Northern California DX Foundation (NCDXF). The overall beacon transmitter concept and RF power-level switching was conceived by Mike Villard, W6QYT, and designed by Dave Leeson, W6QHS. Jack Curtis, K6KU, designed the clock, the microprocessor and the programming components. The engineering, production and packaging was done by Cam Pierce, K6RU. NCDXF asks that all listeners making regular obser-

Table 2-3
NCDXF Beacons and 10-minute Sequence
[Entire sequence repeats every 10 minutes]

Minute	Call	Location
00	4U1UN/B	United Nations, New York City
01	W6WX/B	Stanford University, CA
02	KH6O/B	Honolulu Community College, HI
03	JA2IGY	JARL, Mt Asama, Japan
04	4X6TU/B	Tel Aviv University, Israel
05	OH2B	Helsinki Technical University, Finland
06	CT3B	ARRM (Madeira Radio Society), Madeira Island
07	ZS6DN/B	Transvaal, South Africa
08	LU4AA/B	Radio Club Argentino, Buenos Aires
09	Silent	

[Note: Minutes 09-14 tentatively planned for a 15-minute sequence.]

vations send them to Al Lotze, W6RQ, who analyzes such reports and correlates them with solar and geomagnetic data. Send reports to W6RQ at 46 Cragmont Ave, San Francisco, CA 94116 USA.

28-MHz and Higher Beacons

The IARU is in the process of improving the International Beacon Project (IBP). The aim is to enhance the power and quality of those stations selected to receive IARU sponsorship and, with the cooperation of NCDXF, to extend the power-stepping time-shared networks to 15 and 10 meters. There will also be regional networks and continuous-duty stations on 10 meters to investigate the special properties of that band. Details of new stations and frequencies will be announced from time-to-time in *QST*. The old 10-meter beacon subband (28200-28300) is being phased out by 1990 in favor of the new 28190-28225 kHz subband. See Table 2-4 for an all-source listing of amateur propagation beacons. Note: You are urged to avoid transmitting on beacon frequencies; interference reduces the value of the beacon service to other amateurs.

Bibliography

IRTS, *Amateur Radio Yearbook, 1986-87*, Irish Radio Transmitters Society, PO Box 462, Dublin 9, Ireland.

Troster and Pierce, "Worldwide Beacon Net: The Possibilities Abound," *QST*, June 1983, p 27.

Tynan, "The World Above 50 MHz—Beacons: Where Do We Go from Here?," *QST*, July 1976, p 66.

Table 2-4
Consolidated Beacon List

[This list is correct and complete to the best of our knowledge. Please send any updates to the attention of the Editor, *ARRL Operating Manual*.]

Freq	Call Sign	Location	Grid Square	Mode	Ant	Pol	ERP (W)	Remarks
14.100	4U1UN/B	UN, New York NY USA	FN30	A1A	Omni		100	T + 0 min
14.100	W6WX/B	Stanford Univ CA USA		A1A	Omni		100	T + 1 min
14.100	KH6O/B	Honolulu HI USA		A1A	Omni		100	T + 2 min
14.100	JA2IGY	Mt Asama Japan		A1A	Omni		100	T + 3 min
14.100	4X6TU/B	Tel Aviv Israel		A1A	Omni		100	T + 4 min
14.100	OH2B	Helsinki Finland		A1A	Omni		100	T + 5 min
14.100	CT3B	Funchal Madeira Island		A1A	Omni		100	T + 6 min

Table 2-4 continued on next page

Freq	Call Sign	Location	Grid Square	Mode	Ant	Pol	ERP (W)	Remarks
14.100	ZS6DN/B	Transvaal RSA		A1A	Omni		100	T + 7 min
14.100	LU4AA/B	Buenos Aires Argentina		A1A	Omni		100	T + 8 min
21.150	IBP							
28.050	PY2GOB	Sao Paulo Brazil			Vert		15	
28.175	VE3TEN	Ottawa ON Canada						Continuous
28.195	IY4M	Bologna Italy	JN54qk	A1A	Omni		20	Robot
28.200	IBP							
28.200	GB3SX	Crowboro UK	JO01bb	F1A	N/S		8	H + 25&55
28.200	KF4MS	St Petersburg FL USA			Omni		75	Continuous
28.2025	ZS5VHF	Natal RSA			Omni		5	
28.205	DLØIGI	Mt Predigtstuhl FRG		F1A	Omni		100	Continuous
28.2075	W8FKL	Venice FL USA			Omni		10	Continuous
28.208	WA1IOB	Marlboro MA USA			Omni		75	Continuous
28.210	3B8MS	Mauritius			Omni			Continuous
28.210	K4KMZ	Elizabethtown KY USA			Omni		20	Intermittent
28.212	ZD9GI	Gough Island		F1A	Omni			Continuous
28.215	GB3RAL	Slough UK	IO91rl	F1A	Omni		20	Continuous
28.2175	WB9YMY	Oklahoma City OK USA			Omni		4	Continuous
28.220	5B4CY	Zyyi Cyprus		F1A	Omni		26	Continuous
28.222	W9UXO	Chicago IL USA			Omni		10	Continuous
28.2225	HG2BHA	Tapolca Hungary		F1A	Omni		10	Continuous
28.2275	EA6AU	Mallorca		A1A	Omni		10	Continuous
28.230	ZL2MHF	Mt Climie New Zealand		F1A	Omni		50	Continuous
28.232	W7JPI	Sonoita, AZ USA			Yagi-NE		5	Continuous
28.2325	KD4EC	Jupiter, FL USA			Omni		7	Continuous
28.235	VP9BA	Southampton Bermuda		F1A	Omni		5	Continuous
28.2375	LA5TEN	Oslo Norway		A1A	Omni		10	Continuous
28.240	OA4CK	Lima Peru	FH17mw	A1A			10	Continuous
28.2405	5Z4ERR	Kenya						
28.2425	ZS1CTB	Capetown RSA		F1A	Omni		20	
28.245	A92C	Bahrain		F1A	NE/SW			
28.2475	EA2HB	San Sebastian Spain			Omni		6	Intermittent
28.248	K1BZ	Belfast ME USA			Omni		5	Continuous
28.250	Z21ANB	Bulawayo Zimbabwe		F1A	Omni		15	Continuous
28.253	WB4JHS	Durham, NC USA			Omni		7	Intermittent
28.255	LU1UG	G'ral Pico Argentina	FF84dh		Omni		5	
28.2575	DKØTE	Konstanz FRG		F1A	Omni		40	Continuous
28.260	VK5WI	Adelaide Australia		A1A	Omni		10	Continuous
28.262	VK2RSY	Dural Australia		A1A	Omni	V	25	Continuous
28.264	VK6RWA	Perth Australia		A1A				Continuous
28.266	VK6RTW	Albany Australia						Continuous
28.268	W9KFO	Eaton IN USA			Omni		0.75	Intermittent
28.270	ZS6PW	Pretoria RSA			Yagi to "G"		10	Continuous
28.270	VK4RTL	Townsville Australia						Continuous
28.2725	9L1FTN	Freetown Sierra Leone			Omni		10	Intermittent
28.275	AL7GQ	Jackson MS USA			Loop		1	Continuous
28.2775	DFØAAB	Kiel FRG	JO54ni	A1A	Omni		10	Continuous
28.280	YV5AYV	Caracas Venezuela	FK60ni	F1A	Sweep		10	
28.280	LU8EB	Argentina					5	
28.285	VP8ADE	Adelaide Is UK					8	Continuous
28.286	KA1YE	Henrietta NY USA		A1A	Omni		2	
28.287	H44SI	Solomon Is			Omni		15	Continuous
28.287	W8OMV	Ashville NC USA			Omni		5	
28.288	W2NZH	Moorestown NJ USA		A1A	Omni		3	Intermittent
28.290	VS6TEN	Mt Matilda Hong Kong		A1A	Omni		10	Continuous
28.2925	LU2FFV	San Jorge Argentina			Omni		5	
28.299	PY2AMI	Sao Paulo Brazil	GG67ig	A1A	Omni		10	Continuous
28.300	ZS1LA	Still Bay RSA		F1A	NW		20	Continuous
28.315	ZS6DN	Irene RSA			Omni		100	Continuous
28.888	W6IRT	N Hollywood CA USA		A1A	Omni		5	Intermittent
28.890	WD9GOE	Freeburg IL USA						
28.992	DLØNF	FRG		A1A	E/W		1	
50.003	PY1RO	Rio de Janeiro Brazil		A1A				
50.005	H44HIR	Honiara Solomon Is		A1A				
50.005	PY1AA	Brazil		A1A				
50.005	ZS2SIX	Cape Province RSA		A1A	Omni		10	
50.006	GB3RMK	Inverness UK		F1A	N/S		30	
50.010	JA2IGY	Mie Japan		A1A			10	
50.010	ZS1STB	Still Bay RSA		F1A	N		50	
50.010	ZS6STB	Vereeniging RSA						
50.013	P29BPI	Papua New Guinea		A1A			30	
50.015	SZ2DH	Athens Greece						>1300Z
50.020	GB3SIX	Anglesey Wales UK	IO73tj	F1A	W		100	
50.025	ZS6SIX	Kempton Park RSA						

Freq	Call Sign	Location	Grid Square	Mode	Ant	Pol	ERP (W)	Remarks
50.025	5Z4YV	Kenya						
50.025	6Y5RC	Jamaica		F1A	NW		40	
50.030	XE3VV	Yucatan Mexico	EL50ex	A1A	Omni	V	5	
50.030	ZS6PW	Pretoria RSA		A1A	N/NNW			1000-2000Z
50.033	LU8YYO	Argentina						
50.035	EL2CA	Monrovia Liberia						
50.035	ZB2VHF	Gibraltar	IM76he	A1A	WNW		100	
50.038	FY7THF	French Guiana		F1A	Omni	V	100	
50.041	WA8KGG	OH USA						
50.045	OX3VHF	Julianehaab Greenland	GP60qq	A1A	Omni	V	20	
50.048	VE6ARC	Alberta Canada		A1A	Omni	V	50	
50.050	GB3NHQ	London UK	IO91vq	F1A	Omni	H	15	
50.050	LU2DH	Argentina						
50.050	ZS6LN	Petersburg RSA						
50.055		Oslo Norway						Proposed
50.555	WA9FEF	Chicago IL USA		A1A				
50.060	K1NFE	Burlington CT USA	FN31		Omni	V	25	
50.060	GB3RMK	Rosemarkie Scotland UK	IO77ud	F1A	N/S		20	
50.060	PY2AA	Sao Paulo Brazil			Omni	V	25	
50.060	WA8DNQ	Cincinnati OH USA		A1A	Omni	H	1	
50.060	ZS6DN	Pretoria RSA			N		100	
50.062	W3VD	Laurel MD USA	FM19	A1A	Omni	V	0.1	
50.064	N4PZ	Sarasota FL USA		A1A	Omni	V	0.5	
50.064	WA5UUD	New Orleans LA USA						
50.064	N7DB	Boring OR USA	CN85				30	
50.065	W5VAS	Metairie LA USA		A1A	Omni	H	1	
50.065	WB5ZRL	New Orleans LA USA		A1A	Omni	H	2	
50.065	W0JR/KA0CDN	Denver CO USA	DM79	A1A	Omni	H	20	
50.067	WA6IJZ	Oxnard CA USA	DM04		Omni	V	7-70	
50.069	W0BJ	N Platte NE USA		A1A	Omni	H	6	
50.070	W2CAP/1	Cape Cod MA USA	FN41	A1A	Omni	V	10	
50.070	KS2T	Toms River NJ USA	FM29vx	A1A	Omni	H	10	
50.070	WA2YTM	Rochester NY USA						
50.070	KA4VEY	Harvest AL USA			Omni	V	10	
50.070	WB0CGH/5	Lewisville TX USA	EM13	A1A	Omni	H	0.5	
50.070	WA7ECY	Troutdale OR USA			Omni	V	10	
50.070	K0HTF	Des Moines IA USA	EN3ldx	AIA	Omni		2	
50.070	4U1ITU	Geneva Switzerland						Proposed
50.071	WA2YTM	Victor NY USA	FN12	A1A	Omni	H	15	
50.072	W9KFO	Eaton IN USA						
50.075	VS6SIX	Hong Kong		A1A	Omni	V	10	
50.075	N5JM	New Orleans LA USA	EL49	A1A	Omni	V	2	
50.077	N0LL	Smith Center KS USA	EM09	A1A	Omni	H	25	
50.080	TI2NA	San Jose Costa Rica		A1A				
50.080	W1AW	Newington CT USA	FN31	A1A	W	H	50	
50.080	ZS5TR	Durban RSA						
50.080	ZS5VHF	RSA		A1A	Omni	H	10	
50.085	VE2STL	Val Belair PQ Canada	FN46				3	
50.088	VE1SIX	NB Canada						
50.096	HD1QRC	Quito Ecuador						
50.100	HC2FG	Guayaquil Ecuador		A1A				
50.100	PY5YD	Brazil						
50.109	JD1YAA	Minami Tori-shima						
50.110	KG6DX	Guam						
50.110	ZS6LN	RSA		A1A			100	Intermittent
50.125	ZS3AK	South West Africa						
50.246	LU8MBL	Argentina						
50.440	K1NFE	Hartford CT USA						
50.500	5B4CY	Cyprus	KM64pr	F1A	Omni	V	15	
50.945	ZS1SIX	Cape Province RSA		F2A	Omni		8	
51.020	ZL1UHF	Nihotupu New Zealand		F1A			25	
51.030	ZL2MHB	Fernhill New Zealand		F2A			15	
51.225	ZL2VHT	Inglewood New Zealand		F1A			15	
52.033	P29BPL	Papua New Guinea						
52.100	ZK2SIX	Niue Is						
52.150	VK0CK	Macquarie Is						
52.200	VK8VF	Darwin NA Australia			Omni	V	15	
52.250	ZL2VHM	Pahiatua Track New Zealand		F1A			8	
52.300	VK6RPH	Perth WA Australia						
52.300	VK6RTV	Perth WA Australia						
52.310	ZL3MHF	Christchurch New Zealand		F2A			20	
52.320	VK6RTT	Carnarvon WA Australia						
52.325	VK2RHV	Newcastle NSW Australia						

Table 2-4 continued on next page

Freq	Call Sign	Location	Grid Square	Mode	Ant	Pol	ERP (W)	Remarks
52.330	VK3RGG	Geelong Victoria Australia		F1A			4	On trial
52.350	VK6RTU	Kalgoorlie WA Australia						
52.370	VK7RST	Hobart Tasmania Australia						
52.420	VK2RSY	Sydney NSW Australia				H	25	
52.425	VK2RGB	Gunnedah NSW Australia						
52.435	VK3RMV	Hamilton Victoria Australia						
52.440	VK4RTL	Townsville QLD Australia						
52.450	VK5VF	Mt Lofty SA Australia						
52.460	VK6RPH	Perth WA Australia						
52.465	VK6RTW	Albany WA Australia						
52.470	VK7RNT	Lanceston Tasmania Australia						
52.490	ZL2SIX	Blenheim New Zealand		F2A			10	
52.500	JA2IGY	Japan						
52.500	ZL2VHM	Palmerston North New Zealand						
52.510	ZL2MHF	Mt Climie New Zealand		F1A			2	
70.030	GB3CTC	UK	IO71gy					
70.050	GB3BUX	UK	IO93be					
70.060	GB3ANG	UK	IO86mn					
70.112	5B4CY	Cyprus	KM64pr					
70.120	ZB2VHF	Gibraltar	IM76he					
70.130	EI4RF	Ireland	IO63sn					
144.019	VK6RBS	Busselton Australia						
144.139	5B4CY	Cyprus	KM64ht					
144.145	ZB2VHF	Gibraltar	IM76he					
144.157	EA3URE	Spain	JN00et					
144.175	KP4EOR	Puerto Rico						
144.400	VK4RTT	Mt Mowbullan Australia						
144.410	VK1RCC	Canberra Australia						
144.420	VK2RSY	Sydney NSW Australia				H	25	
144.465	VK6RTW	Albany Australia						
144.480	VK8VH	Darwin Australia						
144.550	VK5RSE	Mt Gambier Australia						
144.565	VK6RPB	Port Hedland Australia						
144.800	VK5VF	Mt Lofty Australia						
144.800	OH8VHF	Finland	KP25tc					
144.810	IS0A	Sardinia	JN40sx					
144.825	I0A	Italy	JN61es					
144.840	IT9G		JM68					
144.850	DL0UB	FRG	JO62ql					
144.855	LA5VHF	Norway	JP99kp					
144.860	LA1VHF	Norway	JO94gm					
144.865	HB9HB	Switzerland	JN37md					
144.867	EA1VHF	Spain	IN53ug					
144.870	LA2VHF	Norway	JP53eg					
144.875	SK2VHF	Sweden	JP94tf					
144.885	OY6VHF	Faroe Is	IP62na					
144.890	LA4VHF	Norway	JP20oq					
144.895	FX0THF		JN08ml					
144.900	OH6VHF	Finland	KP02tg					
144.910	DL0PR	FRG	JO44jh					
144.915	GB3CTC	UK	IO70oj					
144.920	SK7VHF	Sweden	JO65sn					
144.925	GB3VHF	UK	JO01dh					
144.930	OZ7IGY	Denmark	JO55vo					
144.945	SP4VHG	Poland	JO71sw					
144.950	SK1VHF	Sweden	JO97cg					
144.950	VK2RCW	Sydney Australia					25	
144.960	SK4MPI	Sweden	JP70nj					
144.965	GB3LER	UK	IP90jd					
144.975	GB3ANG	UK	IO86mn					
144.975	DL0SG	FRG	JN68eq					
144.980	SP3VHC	Poland	JO94gm					
144.984	ON4VHF	Belgium	JO20fp					
144.985	Y41B	GDR	JO53rp					
145.000	VK6RPH	Perth Australia						
145.810	OSCAR 10	Mode B General Beacon						
145.987	OSCAR 10	Mode B Eng Beacon						
220.051	K9XI	Mokena IL USA	EN61					

Freq	Call Sign	Location	Grid Square	Mode	Ant	Pol	ERP (W)	Remarks
220.051	WB2IEY	Canandaigua NY USA	FN12					
220.052	KA0ADV/2	Rochester NY USA	FN12nh				15	
220.055	WB4FQR	Woodbridge VA USA	FM18				13	
220.055	WB9VMY/5	Calumet OK USA	EM15				20	
220.058	WB2ELB	Buffalo NY USA	FN03					
220.085	K5BYS	Lewisville TX USA	EM12		Omni	H	0.2	
432.057	VK6RBS	Busselton Australia						
432.0715	N3CX	Pennsburg PA USA	FN20		Omni	H	1	
432.0715	K4MSK	Lookout Mt GA USA	EM85md	A1A	Omni	H	1	
432.160	VK6RPR	Nedlands Australia						
432.410	VK1RBC	Canberra Australia						
432.420	VK2RSY	Sydney NSW Australia				H	15	
432.425	VK3RMB	Ballarat Australia						
432.440	VK4RBB	Brisbane Australia						
432.855	LA5UHF	Norway	JP99kp					
432.855	SK3UHF	Sweden	JP92gw					
432.860	LA1UHF	Norway	JO59jw					
432.865	OZ2UHF	Denmark	JO46jd					
432.870	FX4UHF		IN93eh					
432.870	LA2UHF	Norway	JP53eg					
432.880	LA3UHF	Norway	JO48bc					
432.885	OY6UHF	Faroe Is	IP62na					
432.890	LA4UHF	Norway	JO29pj					
432.895	OZ4UHF	Denmark	JO75je					
432.900	OH3UHF	Finland	KP11un					
432.910	GB3MLY	UK	IO93eo					
432.925	SK6UHF	Sweden	JO67bf					
432.930	OZ7IGY	Denmark	JO55vo					
432.935	OK0EA	Czechoslovakia	JO70ss					
432.960	SK4UHF	Sweden	JO79kh					
432.970	GB3CTC	UK	IO70oj					
432.975	SK5UHF	Sweden	JP80sa					
432.980	GB3ANG	UK	IO86mn					
432.983	OZ2ALS	Denmark	JO45vb					
432.984	HB9F	Switzerland	JN36xn					
436.020	OSCAR 10	Mode L Eng Beacon						
436.040	OSCAR 10	Mode L General Beacon						
903.076	N3CX	Pennsburg PA USA	FN20		Omni	H	3	
1296.171	VK6RBS	Busselton Australia						
1296.2145	N3CX	Pennsburg PA USA	FN20		Omni	H	2	
1296.2145	K4MSK	Lookout Mt GA USA	EM85md	A1A	Omni	H	2	
1296.410	VK1RBC	Canberra Australia						
1296.420	VK2RSY	Sydney NSW Australia			Omni	H	5	
1296.480	VK6RPR	Nedlands Australia						
1296.810	GB3NWK	UK	JO01bi					
1296.830	GB3BPO	UK	JO02pb					
1296.854	DB0JO	FRG	JN36xn					
1296.870	GB3AND	UK	IO91gf					
1296.885	ON5UHF	Belgium	JO10un					
1296.890	GB3DUN	UK	IO91sv					
1296.910	GB3CLE	UK	IO82rl					
1296.920	PA0QHN	Netherlands	JO22fh					
1296.920	DB0VC	FRG	JO54bh					
1296.925	SK6UHG	Sweden	JO57tq					
1296.930	GB3MLE	UK	IO91eo					
1296.930	OZ7IGY	Denmark	JO55vo					
1296.975	DB0JU	FRG	JO31bu					
1296.990	GB3EDN	UK	IO85jw					
2304.020	N3CX	Pennsburg PA USA	FN20		SW	H	0.1	
2304.3575	K4MSK	Lookout Mt GA USA	EM85md	A1A	SW	H	0.1	
3456.572	K4MSK	Lookout Mt GA USA	EM85md	A1A	SW	H	0.001	
5760.9295	K4MSK	Lookout Mt GA USA	EM85md	A1A	SW	H	8	
10000.265	WB6DNX	Santiago Peak CA USA	DM13fr					
10369.716	K4MSK	Lookout Mt GA USA	EM85md	A1A				
10300	VK6RVF	Roleystone Australia						

Sources: IARU, IRTS, RSGB, G3DME, K2OLG, VK5LP, W3XO, W6ISQ

Table 2-2

Amateur Frequency Allocations and Band Plans

The "Allocations" part of this table is an extract of the ITU *Radio Regulations* Table of Frequency Allocations. Services listed in capital letters denote Primary allocations and have priority over secondary services listed in normal characters. Stations of a secondary service may not cause harmful interference to Primary service stations, nor may they claim protection from harmful interference from Primary stations. All footnotes pertaining to the Amateur Radio Service are included. Most of the US is in ITU Region 2 (some of the US possessions in the Pacific are in ITU Region 3).

The "FCC Part 97 Privileges" portion of this table indicates frequency privileges for US amateurs under Federal Communications Commission regulations. The part labeled "Amateur Radio Band Plans" includes voluntary band plans formally adopted by the IARU and/or ARRL.

Allocations							
INTERNATIONAL			UNITED STATES				
Region 1 kHz	Region 2 kHz	Region 3 kHz	Band kHz	National Provisions	Government Allocation	Non-Government Allocation	Remarks
1800-1810 RADIOLOCATION 485 486	1800-1850 AMATEUR	1800-2000 AMATEUR FIXED MOBILE except aeronautical mobile RADIONAVIGATION Radiolocation	1800-1900			AMATEUR	
1810-1850 AMATEUR 490 491 492 493	489						
1850-2000 FIXED MOBILE except aeronautical mobile 484 488 495	1850-2000 AMATEUR FIXED MOBILE except aeronautical mobile RADIOLOCATION RADIONAVIGATION 489 494	489	1900-2000	US290	RADIOLOCATION	RADIOLOCATION AMATEUR	

484—Some countries of Region 1 use radiodetermination systems in the bands 1606.5-1625 kHz, 1635-1800 kHz, 1850-2160 kHz, 2194-2300 kHz, 2502-2850 kHz and 3500-3800 kHz. The establishment and operation of such systems are subject to agreement obtained under the procedures set forth in Article 14. The radiated mean power of these stations shall not exceed 50 W.

485—Additional allocation: in Angola, Bulgaria, Hungary, Mongolia, Nigeria, Poland, the German Democratic Republic, Chad, Czechoslovakia and the U.S.S.R., the bands 1625-1635 kHz, 1800-1810 kHz and 2160-2170 kHz are also allocated to the fixed and land mobile services on a primary basis subject to agreement obtained under the procedure set forth in Article 14.

486—In Region 1, in the bands 1625-1635 kHz, 1800-1810 kHz and 2160-2170 kHz (except in the countries listed in No. 485 and those listed in No. 499 for the band 2160-2170 kHz), existing stations in the fixed and mobile, except aeronautical mobile, services (and stations of the aeronautical mobile (OR) service in the band 2160-2170 kHz) may continue to operate on a primary basis until satisfactory replacement assignments have been found and implemented in accordance with Resolution 38.

487—In Region 1, the establishment and operation of stations of the radiolocation service in the bands 1625-1635 kHz, 1800-1810 kHz and 2160-2170 kHz shall be subject to agreement obtained under the procedure set forth in Article 14 (see also No. 486). The radiated mean power of radiolocation stations shall not exceed 50 W. Pulse systems are prohibited.

488—In the Federal Republic of Germany, Denmark, Finland, Hungary, Ireland, Israel, Jordan, Malta, Norway, Poland, the German Democratic Republic, the United Kingdom, Sweden, Czechoslovakia and the U.S.S.R., administrations may allocate up to 200 kHz to their amateur service in the bands 1715-1800 kHz and 1850-2000 kHz. However, when allocating the bands within this range to their amateur service, administrations shall, after prior consulations with administrations of neighboring countries, take such steps as may be necessary to prevent harmful interference from their amateur service to the fixed and mobile services of other countries. The mean power of any amateur station shall not exceed 10 W.

489—In Region 2, Loran stations operating in the band 1800-2000 kHz shall cease operation by December 31, 1982. In Region 3, the Loran system operates either on 1850 kHz or 1950 kHz, the bands occupied being 1825-1875 kHz and 1925- 1975 kHz respectively. Other services to which the band 1800-2000 kHz is allocated may use any frequency therein on condition that no harmful interference is caused to the Loran system operating on 1850 kHz or 1950 kHz.

490—Alternative allocation: in the Federal Republic of Germany, Angola, Austria, Belgium, Bulgaria, Cameroon, the Congo, Denmark, Egypt, Spain, Ethiopia, France, Greece, Italy, Lebanon, Luxembourg, Malawi, the Netherlands, Portugal, Syria, the German Democratic Republic, Somalia, Tanzania, Tunisia, Turkey and the U.S.S.R., the band 1810-1830 kHz is allocated to the fixed and mobile, except aeronautical mobile, services on a primary basis.

491—Additional allocation: in Saudi Arabia, Iraq, Israel, Libya, Poland, Romania, Chad, Czechoslovakia, Togo and Yugoslavia, the band 1810-1830 kHz is also allocated to the fixed and mobile, except aeronautical mobile, services on a primary basis.

492—In Region 1, the use of the band 1810-1850 kHz by the amateur service is subject to the condition that satisfactory replacement assignments have been found and implemented in accordance with Resolution 38, for frequencies to all existing stations of the fixed and mobile, except aeronautical mobile, services operating in this band (except for the stations of the countries listed in Nos. 490, 491 and 493). On completion of satisfactory transfer, the authorization to use the band 1810-1830 kHz by the amateur service in countries situated totally or partially north of 40°N shall be given only after consultation with the countries mentioned in Nos. 490 and 491 to define the necessary steps to be taken to prevent harmful interference between amateur stations and stations of other services operating in accordance with Nos. 490 and 491.

493—Alternative allocation: in Burundi and Lesotho, the band 1810-1850 kHz is allocated to the fixed and mobile, except aeronautical mobile, services on a primary basis.

494—Alternative allocation: in Argentina, Bolivia, Chile, Mexico, Paraguay, Peru, Uruguay and Venezuela, the band 1850-2000 kHz is allocated to the fixed, mobile, except aeronautical mobile, radiolocation and radionavigation services on a primary basis.

US290—In the band 1900-2000 kHz, amateur stations may continue to operate on a secondary basis to the Radiolocation Service, pending a decision as to their disposition through a future rule making proceeding in conjunction with implementation of the Standard Broadcasting Service in the 1625-1705 kHz band.

FCC Part 97 Privileges

| License Class | Terrestrial location of the amateur radio station | | | |
	Region 1 kHz	Region 2 kHz	Region 3 kHz	Limitations
General Advanced Extra	1810-1850	1800-2000	1800-2000	3,5,21

Amateur Radio Band Plans

Region 1 kHz	Region 2 kHz	Region 3 kHz	ARRL* kHz
1800-1838 CW	1800-1830 CW/RTTY		1800-1840 CW/RTTY/NB
1838-1840 CW/RTTY	1830-1840 CW/RTTY DX CW Window		1830-1850 Intercontinental QSOs only
1840-1842 RTTY/Phone/CW	1840-1850 Phone/CW DX Phone Window		1840-2000 Phone SSTV WB CW
1842-2000 Phone/CW	1850-2000 Phone		

*Adopted by the ARRL Board of Directors January 1986.

§97.7(g)
Limitations . . .

(3) Where, in adjacent regions or subregions, a band of frequencies is allocated to different services of the same category, the basic principle is the equality of right to operate. Accordingly, the stations of each service in one region or subregion must operate so as not to cause harmful interference to services in the other regions or subregions. (See International Telecommunication Union Radio Regulations, RR 346 (Geneva, 1979).)

(5) Amateur stations in the 1900-2000 kHz, 220-225 MHz, 420-450 MHz, 902-928 MHz, 1240-1300 MHz, 2300-2310 MHz, 2390-2450 MHz, 3.3-3.5 GHz, 5.650-5.925 GHz, 10.0-10.5 GHz, 24.05-24.25 GHz, 76-81 GHz, 144-149 GHz and 241-248 GHz bands must not cause harmful interference to stations in the Government radiolocation service and are not protected from interference due to the operation of stations in the Government radiolocation service.

(21) Amateur stations in the 1900-2000 kHz, 10.45-10.50 GHz, 76-81 GHz, 144-149 GHz and 241-248 GHz bands must not cause harmful interference to stations in the non-government radiolocation service and are not protected from interference due to the operation of stations in the non-government radiolocation service.

INTERNATIONAL			UNITED STATES				
Region 1 kHz	Region 2 kHz	Region 3 kHz	Band kHz	National Provisions	Government Allocation	Non-Government Allocation	Remarks
3500-3800 AMATEUR 510 FIXED 　MOBILE except 　　aeronautical 　　mobile 484	3500-3750 AMATEUR 510 509 511	3500-3900 AMATEUR 510 FIXED MOBILE	3500-4000	510		AMATEUR	
3800-3900 FIXED AERONAUTICAL 　MOBILE (OR) LAND MOBILE	3750-4000 AMATEUR 510 FIXED MOBILE except 　aeronautical 　mobile (R)						
3900-3950 AERONAUTICAL 　MOBILE (OR) 513		3900-3950 AERONAUTICAL 　MOBILE BROADCASTING					
3950-4000 FIXED BROADCASTING	511 512 514 515	3950-4000 FIXED BROADCASTING 516					

484—Some countries of Region 1 use radiodetermination systems in the bands 1606.5-1625 kHz, 1635-1800 kHz, 1850-2160 kHz, 2194-2300 kHz, 2502-2850 kHz and 3500-3800 kHz. The establishment and operation of such systems are subject to agreement obtained under the procedures set forth in Article 14. The radiated mean power of these stations shall not exceed 50 W.

509—Additional allocation: in Honduras, Mexico, Peru and Venezuela, the band 3500-3750 kHz is also allocated to the fixed and mobile services on a primary basis.

510—For the use of the bands allocated to the amateur service at 3.5 MHz, 7.0 MHz, 10.1 MHz, 14.0 MHz, 18.068 MHz, 21.0 MHz, 24.89 MHz and 144 MHz in the event of natural disasters, see Resolution 640.

511—Additional allocation: in Brazil, the band 3700-4000 kHz is also allocated to the radiolocation service on a primary basis.

512—Alternative allocation: in Argentina, Bolivia, Chile, Ecuador, Paraguay, Peru and Uruguay, the band 3750-4000 kHz is allocated to the fixed and mobile, except aeronautical mobile, services on a primary basis.

513—Alternative allocation: in Botswana, Lesotho, Malawi, Mozambique, Namibia, South Africa, Swaziland, Zambia and Zimbabwe, the band 3900-3950 kHz is allocated to the broadcasting service on a primary basis. The use of this band by the broadcasting service is subject to agreement obtained under the procedure set forth in Article 14 with neighboring countries having services operating in accordance with the Table.

514—Additional allocation: in Canada, the band 3950-4000 kHz is also allocated to the broadcasting service on a primary basis. The power of broadcasting stations operating in this band shall not exceed that necessary for a national service within the frontier of this country and shall not cause harmful interference to other services operating in accordance with the Table.

515—Additional allocation: in Greenland, the band 3950-4000 kHz is also allocated to the broadcasting service on a primary basis. The power of the broadcasting stations operating in this band shall not exceed that necessary for a national service and shall in no case exceed 5 kW.

516—In Region 3, the stations of those services to which the band 3995-4005 kHz is allocated may transmit standard frequency and time signals.

FCC Part 97 Privileges

License Class	Terrestrial location of the amateur radio station			
	Region 1 kHz	Region 2 kHz	Region 3 kHz	Limitations
Novice Technician	3700-3750	3700-3750	3700-3750	1,3,32
General	3525-3750 —	3525-3750 3850-4000	3525-3750 3850-3900	3,32
Advanced	3525-3750 3775-3800	3525-3750 3775-4000	3525-3750 3775-3900	3,32
Extra	3500-3800	3500-4000	3500-3900	3,32

§97.7(g)
Limitations. . .
    (1) Novice and Technician class radio operators are limited to the use of international Morse code when communicating in this band.
    (3) Where, in adjacent ragions or subregions, a band of frequencies is allocated to different services of the same category, the basic principle is the equality of right to operate. Accordingly, the stations of each service in one region or subregion must operate so as not to cause harmful interference to services in the other regions or subregions. (See International Telecommunication Union Radio Regulations, RR 346 (Geneva, 1979).)
    (32) Amateur stations in these bands may be used for communications related to relief operations in connection with natural disasters. See Appendix 6 to this Part.

Amateur Radio Band Plans

Region 1 kHz	Region 2 kHz	Region 3 kHz	ARRL kHz
3500-3510 DX CW	3500-3510 DX CW		CW
3510-3560 CW Contest preferred segment	3510-3525 CW		
	3525-3605 CW (Phone permitted)		
3560-3580 CW			
3580-3600 CW/RTTY			3590 DX RTTY
3600-3620 RTTY/Phone/CW Contest preferred segment	3605-3645 RTTY/CW (Phone permitted)		3605-3645 RTTY
3620-3650 Phone/CW Contest preferred segment			
3650-3700 Phone/CW	3645-3750 CW (Phone permitted)		3645-3750 CW
3700-3730 Phone/CW Contest preferred segment			
3730-3740 SSTV/Phone/CW Contest preferred segment			
3740-3775 Phone/CW Contest preferred segment	3750-3775 Phone (CW permitted)		3750-3790 Phone
3775-3800 DX Phone	3775-3800 DX Phone (CW permitted)		
			3790-3800 Phone DX Window
	3800-3840 Phone (CW permitted)		3800-4000 Phone
	3840-3850 SSTV/Phone (CW permitted)		3845 SSTV
	3850-4000 Phone (CW permitted)		

INTERNATIONAL			UNITED STATES				
Region 1 *kHz*	Region 2 *kHz*	Region 3 *kHz*	Band *kHz*	National Provisions	Government Allocation	Non-Government Allocation	Remarks
7000-7100	AMATEUR 510 AMATEUR-SATELLITE 526 527		7000-7100	510		AMATEUR AMATEUR-SATELLITE	
7100-7300 BROADCASTING	7100-7300 AMATEUR 510 528	7100-7300 BROADCASTING	7100-7300	510 528		AMATEUR	

510—For the use of the bands allocated to the amateur service at 3.5 MHz, 7.0 MHz, 10.1 MHz, 14.0 MHz, 18.068 MHz, 21.0 MHz, 24.89 MHz and 144 MHz in the event of natural disasters, see Resolution 640.

526—Additional allocation: in Angola, Iraq, Kenya, Rwanda, Somalia and Togo, the band 7000-7050 kHz is also allocated to the fixed service on a primary basis.

527—Alternative allocation: in Egypt, Ethiopia, Guinea, Libya, Madagascar, Malawi and Tanzania, the band 7000-7050 kHz is allocated to the fixed service on a primary basis.

528—The use of the band 7100-7300 kHz in Ragion 2 by the amateur service shall not impose constraints on the broadcasting service intended for use within Region 1 and Region 3.

FCC Part 97 Privileges

License Class	Terrestrial location of the amateur radio station			
	Region 1 kHz	Region 2 kHz	Region 3 kHz	Limitations
Novice Technician	7050-7075	7100-7150	7050-7075	1,3,32
General	7025-7100	7025-7150 7225-7300	7025-7100	3,32
Advanced	7025-7100	7025-7300	7025-7100	3,32
Extra	7000-7100	7000-7300	7000-7100	3,32

§97.7(g)

Limitations. . .

(1) Novice and Technician class radio operators are limited to the use of international Morse code when communicating in this band.

(3) Where, in adjacent regions or subregions, a band of frequencies is allocated to different services of the same category, the basic principle is the equality of right to operate. Accordingly, the stations of each service in one region or subregion must operate so as not to cause harmful interference to services in the other regions or subregions. (See International Telecommunication Union Radio Regulations, RR 346 (Geneva, 1979).)

(32) Amateur stations in these bands may be used for communications related to relief operations in connection with natural disasters. See Appendix 6 to this Part.

Amateur Radio Band Plans

Region 1 kHz	Region 2 kHz	Region 3 kHz	ARRL kHz
7000-7035 CW	7000-7035 CW	7000-7025 CW	7000-7080 CW
		7025-7030 NB/CW	
		7030-7035 NB/CW/Phone	
7035-7045 RTTY/SSTV/CW	7035-7040 DX RTTY/CW	7035-7100 Phone/CW	
	7040-7080 Phone (CW permitted)		7040 DX RTTY
7045-7100 Phone/CW	7080-7100 DX Phone (CW permitted)		7080-7100 RTTY
	7100-7166 Phone (CW permitted)	7100-7300 Phone/CW (Secondary operation)	7100-7150 CW
			7150-7300 Phone
	7166-7176 SSTV/Phone (CW permitted)		7171 SSTV
	7176-7300 Phone (CW permitted)		

INTERNATIONAL			UNITED STATES				
Region 1 kHz	Region 2 kHz	Region 3 kHz	Band kHz	National Provisions	Government Allocation	Non-Government Allocation	Remarks
10100-10150	FIXED Amateur 510		10100-10150	US247 510		AMATEUR	

510—For the use of the bands allocated to the amateur service at 3.5 MHz, 7.0 MHz, 10.1 MHz, 14.0 MHz, 18.068 MHz, 21.0 MHz, 24.89 MHz and 144 MHz in the event of natural disasters, see Resolution 640.

US247—The band 10100-10150 kHz is allocated to the fixed service on a primary basis outside the United States and Possessions. Transmissions of stations in the amateur service shall not cause harmful interference to this fixed service use and stations in the amateur service shall make all necessary adjustments (including termination of transmission) if harmful interference is caused.

FCC Part 97 Privileges

License Class	Terrestrial location of the amateur radio station			
	Region 1 kHz	Region 2 kHz	Region 3 kHz	Limitations
General Advanced Extra	10100-10150	10100-10150	10100-10150	28,32

Amateur Radio Band Plans

Region 1 kHz	Region 2 kHz	Region 3 kHz	ARRL kHz
10100-10140 CW	10100-10140 CW	10100-10150 CW/NB (Secondary operation)	10100-10140 CW
10140-10150 RTTY/CW	10140-10150 RTTY/CW		10140-10150 RTTY

§97.7(g)
Limitations. . .

(28) Amateur stations in the 10100-10150 kHz band must not cause harmful interference to stations authorized by other nations in the fixed service. Amateur stations shall make all necessary adjustments (including termination of transmission) if harmful interference is caused.

(32) Amateur stations in these bands may be used for communications related to relief operations in connection with natural disasters. See Appendix 6 to this Part.

INTERNATIONAL			UNITED STATES				
Region 1 kHz	Region 2 kHz	Region 3 kHz	Band kHz	National Provisions	Government Allocation	Non-Government Allocation	Remarks
14000-14250	AMATEUR 510 AMATEUR-SATELLITE		14000-14250	510		AMATEUR AMATEUR-SATELLITE	
14250-14350	AMATEUR 510 535		14250-14350	510		AMATEUR	

510—For the use of the bands allocated to the amateur service at 3.5 MHz, 7.0 MHz, 10.1 MHz, 14.0 MHz, 18.068 MHz, 21.0 MHz, 24.89 MHz and 144 MHz in the event of natural disasters, see Resolution 640.

535—Additional allocation: in Afghanistan, China, the Ivory Coast, Iran and the U.S.S.R., the band 14250-14350 kHz is also allocated to the fixed service on a primary basis. Stations of the fixed service shall not use a radiated power exceeding 24 dBW.

FCC Part 97 Privileges

License Class	Terrestrial location of the amateur radio station			
	Region 1 kHz	Region 2 kHz	Region 3 kHz	Limitations
General	14025-14150 14225-14350	14025-14150 14225-14350	14025-14150 14225-14350	32
Advanced	14025-14150 14175-14350	14025-14150 14175-14350	14025-14150 14175-14350	32
Extra	14000-14350	14000-14350	14000-14350	32

§97.7(g)
Limitations. . .
　　(32) Amateur stations in these bands may be used for communications related to relief operations in connection with natural disasters. See Appendix 6 to this Part.

Amateur Radio Band Plans

Region 1 kHz	Region 2 kHz	Region 3 kHz	ARRL kHz
14000-14060 CW Contest preferred segment	14000-14070 CW	14000-14070 CW	14000-14070 CW
14060-14075 CW	14070-14099.5 RTTY/CW	14070-14099.5 NB/CW	14070-14099.5 RTTY
14075-14099 RTTY/CW			
14099-14101 Beacons	14099.5-14100.5 Beacons	14099.5-14100.5 Beacons	14099.5-14100.5 Beacons
14101-14125 Phone/CW	14100.5-14150 Phone/CW permitted International traffic	14100.4-14225 Phone/CW	14100.5-14150 CW
14125-14225 Phone/CW Contest preferred segment	14150-14225 Phone/CW permitted		14150-14350 Phone
14225-14235 SSTV/CW Contest preferred segment	14225-14235 SSTV/Phone/ CW permitted	14225-14235 SSTV/CW	14230 SSTV
14235-14300 Phone/CW Contest preferred segment	14235-14250 Phone/CW permitted	14235-14350 Phone/CW	
	14250-14340 Phone/CW permitted International traffic		
14300-14350 Phone/CW			14313 Maritime Mobile
	14340-14350 Phone Emergency (CW permitted) International traffic		

INTERNATIONAL			UNITED STATES				
Region 1 kHz	Region 2 kHz	Region 3 kHz	Band kHz	National Provisions	Government Allocation	Non-Government Allocation	Remarks
18068-18168	AMATEUR 510 AMATEUR-SATELLITE 537 538		18068-18168	US248 510		AMATEUR AMATEUR-SATELLITE	

510—For the use of the bands allocated to the amateur service at 3.5 MHz, 7.0 MHz, 10.1 MHz, 14.0 MHz, 18.068 MHz, 21.0 MHz, 24.89 MHz and 144 MHz in the event of natural disasters, see Resolution 640.

537—The band 18068-18168 kHz is allocated to the fixed service on a primary basis subject to the procedure described in Resolution 8. The use of this band by the amateur and amateur-satellite services shall be subject to the completion of satisfactory transfer of all assignments to stations in the fixed service operating in this band and recorded in the Master Register, in accordance with the procedure described in Resolution 8.

538—Additional allocation: in the U.S.S.R., the band 18068-18168 kHz is also allocated to the fixed service on a primary basis for use within the boundary of the U.S.S.R., with a peak envelope power not exceeding 1 kW.

US248—Until reaccommodation actions of the international Telecommunication Union are completed, the band 18068-18168 kHz and 24890-24990 kHz are allocated as an alternative allocation to the fixed service. In the interim, assignments to stations in the fixed service shall be made in accordance with the policy set forth in 8.2.13 of the NTIA Manual of Regulations and Procedures and Part 2, Section 2.102 of the FCC Rules and Regulations. However, assignments to the fixed service in these bands shall be terminated no later than 1 July 1989.

FCC Part 97 Privileges

License Class	Terrestrial location of the amateur radio station			
	Region 1 kHz	*Region 2 kHz*	*Region 3 kHz*	*Limitations*
(none)	18068-18168	18068-18168	18068-18168	Allocation pending reassignment of fixed stations

Amateur Radio Band Plans

Region 1 kHz	*Region 2 kHz*	*Region 3 kHz*	*ARRL kHz*
18068-18100 CW	18068-18100 CW	18068-18100 CW	18068-18100 CW
18100-18110 RTTY/CW	18100-18110 RTTY/CW	18100-18110 NB	18100-18110 RTTY
18110-18168 Phone/CW	18110-18168 Phone/CW	18110-18168 Phone/CW	18110-18168 Phone/CW

INTERNATIONAL			UNITED STATES				
Region 1 kHz	Region 2 kHz	Region 3 kHz	Band kHz	National Provisions	Government Allocation	Non-Government Allocation	Remarks
21000-21450	AMATEUR 510 AMATEUR-SATELLITE		21000-21450	510		AMATEUR AMATEUR-SATELLITE	

510—For the use of the bands allocated to the amateur service at 3.5 MHz, 7.0 MHz, 10.1 MHz, 14.0 MHz, 18.068 MHz, 21.0 MHz, 24.89 MHz and 144 MHz in the event of natural disasters, see Resolution 640.

License Class	Terrestrial location of the amateur radio station			
	Region 1 kHz	Region 2 kHz	Region 3 kHz	Limitations
Novice Technician	21100-21200	21100-21200	21100-21200	1,32
General	21025-21200 21300-21450	21025-21200 21300-21450	21025-21200 21300-21450	32
Advanced	21025-21200 21225-21450	21025-21200 21225-21450	21025-21200 21225-21450	32
Extra	21000-21450	21000-21450	21000-21450	32

§97.7(g)
Limitations. . .
(1) Novice and Technician class radio operators are limited to the use of international Morse code when communicating in this band.
(32) Amateur stations in these bands may be used for communications related to relief operations in connection with natural disasters. See Appendix 6 to this Part.

Amateur Radio Band Plans

Region 1 kHz	Region 2 kHz	Region 3 kHz	ARRL kHz
21000-21080 CW	21000-21070 CW		
21080-21120 RTTY/CW	21070-21100 RTTY/CW	21070-21125 NB/CW	21070-21100 RTTY
21120-21149 CW	21100-21149.5 CW		21100-21149.5 CW
		21125-21149.5 CW	
21149-21151 Beacons	21149.5-21150.5 Beacons	21149.5-21150.5 Beacons	21149.5-21150.5 Beacons
21151-21335 Phone/CW	21150.5-21200 Phone/CW permitted International traffic	21150.5-21335 Phone/CW	21150.5-21200 CW
	21200-21300 Phone/CW permitted		21200-21450 Phone
	21300-21335 Phone/CW permitted International Traffic		
21335-21345 SSTV/CW	21335-21345 SSTV/Phone (CW permitted)	21335-21345 SSTV/CW	21340 SSTV
21345-21450 Phone/CW	21345-21440 Phone (CW permitted)	21345-21450 Phone/CW	
	21440-21450 Phone Emergency (CW permitted)		

INTERNATIONAL			UNITED STATES				
Region 1 kHz	Region 2 kHz	Region 3 kHz	Band kHz	National Provisions	Government Allocation	Non-Government Allocation	Remarks
24890-24990	AMATEUR 510 AMATEUR-SATELLITE 542 543		24890-24990	US248 510		AMATEUR AMATEUR-SATELLITE	

510—For the use of the bands allocated to the amateur service at 3.5 MHz, 7.0 MHz, 10.1 MHz, 14.0 MHz, 18.068 MHz, 21.0 MHz, 24.89 MHz and 144 MHz in the event of natural disasters, see Resolution 640.

542—Additional allocation: in Kenya, the band 23600-24900 kHz is also allocated to the meteorological aids service (radiosondes) on a primary basis.

543—The band 24890-24990 kHz is allocated to the fixed and land mobile services on a primary basis subject to the procedure described in Resolution 8. The use of this band by the amateur and amateur-satellite services shall be subject to the completion of the satisfactory transfer of all assignments to fixed and land mobile stations operating in this band and recorded in the Master Register, in accordance with the procedure described in Resolution 8.

US248—Until reaccommodation actions of the International Telecommunication Union are completed, the band 18068-18168 kHz and 24890-24990 kHz are allocated as an alternative allocation to the fixed service. In the interim, assignments to stations in the fixed service shall be made in accordance with the policy set forth in 8.2.13 of the NTIA Manual of Regulations and Procedures and Part 2, Section 2.102 of the FCC Rules and Regulations. However, assignments to the fixed service in these bands shall be terminated no later than 1 July 1989.

FCC Part 97 Privileges

License Class	Terrestrial location of the amateur radio station			
	Region 1 kHz	Region 2 kHz	Region 3 kHz	Limitations
General Advanced Extra	24890-24990	24890-24990	24890-24990	29,32

Amateur Radio Band Plans

Region 1 kHz	Region 2 kHz	Region 3 kHz	ARRL* kHz
24890-24920 CW	24890-24920 CW	24890-24920 CW	24890-24920 CW
24920-24930 RTTY/CW	24920-24930 RTTY/CW	24920-24930 NB/CW	24920-24930 CW/Digital
24930-24990 Phone/CW	24930-24990 Phone/CW	24930-24990 Phone/CW	24930-24990 Phone/SSTV/CW

*Adopted by the ARRL Board of Directors July 1985.

§97.7(g)
Limitations. . .

(29) Until July 1, 1989, amateur stations in this band must not cause harmful interference to stations authorized by other nations in the fixed and mobile services. Amateur stations must make all necessary adjustments (including termination of transmission) if harmful interference is caused.

(32) Amateur stations in these bands may be used for communications related to relief operations in connection with natural disasters. See Appendix 6 to this Part.

Allocations							
INTERNATIONAL			UNITED STATES				
Region 1 MHz	Region 2 MHz	Region 3 MHz	Band MHz	National Provisions	Government Allocation	Non-Government Allocation	Remarks
28-29.7	AMATEUR AMATEUR-SATELLITE		28.00-29.70			AMATEUR AMATEUR-SATELLITE	

FCC Part 97 Privileges

License Class	Region 1 kHz	Region 2 kHz	Region 3 kHz	Limitations
Novice Technician	28100-28500	28100-28500	28100-28500	1
General Advanced Extra	28000-29700	28000-29700	28000-29700	

§97.7(g)
Limitations. . .
(1) Novice and Technician class radio operators are limited to the use of international Morse code when communicating in this band.

Amateur Radio Band Plans

Region 1 kHz	Region 2 kHz	Region 3 kHz	ARRL kHz
28000-28050 CW	28000-28070 CW	28000-28050 CW	28000-28070 CW
28050-18150 RTTY/CW	28070-28150 RTTY	28050-28150 NB/CW	28070-28150 RTTY
28150-28200 CW	28150-28190 CW	28150-28190 CW	28150-28190 CW
	28190-28200 CW New beacon band	28190-28200 CW New beacon band	28100-28200 CW New beacon band
28200-28300	28200-28300 CW Old beacon band until 1990	28200-28300 CW Old beacon band until 1990	28200-28300 CW Old beacon band until 1990
28300-28675 Phone/CW	28300-28675 Phone CW permitted	28300-28675 Phone/CW	28300-29300 Phone
	28670-28690		
28675-28685 SSTV/CW	SSTV/Phone CW permitted	28675-28685 SSTV/CW	
28685-29300 Phone/CW	28690-29300 Phone CW permitted	28685-29300 Phone/CW	
29300-29550 Satellites	29300-29510 Satellites	29300-29510 Satellites	29300-29510 Satellites
	29510-29700	29510-29700	29510-29590
29550-29700 Phone/CW	FM Phone and repeaters	WB (6 kHz)/CW	Repeater inputs
			29600 FM simplex calling frequency
			29610-29700 Repeater outputs

INTERNATIONAL			UNITED STATES				
Region 1 MHz	Region 2 MHz	Region 3 MHz	Band MHz	National Provisions	Government Allocation	Non-Government Allocation	Remarks
47-68 BROADCASTING							
	50-54 AMATEUR		50-54			AMATEUR	
553 554 555 559 561	556 557 558 560						

553—Additional allocation: in Hungary, Kenya, Mongolia, Czechoslovakia and the U.S.S.R., the bands 47-48.5 MHz and 56.5-58 MHz are also allocated to the fixed and land mobile services on a secondary basis.

554—Additional allocation: in Albania, the Federal Republic of Germany, Austria, Belgium, Bulgaria, Denmark, Finland, France, Gabon, Greece, Israel, Italy, Lebanon, Liechtenstein, Luxembourg, Mali, Malta, Morocco, Nigeria, Norway, the Netherlands, Poland, the German Democratic Republic, the United Kingdom, Senegal, Sweden, Switzerland, Tunisia, Turkey and Yugoslavia, the band 47-68 MHz, and in Romania, the band 47-58 MHz, are also allocated to the land mobile service on a permitted basis. However, stations of the land mobile service in the countries mentioned in connection with each band referred to in this footnote shall not cause harmful interference to, or claim protection from, existing or planned broadcasting stations of countries other than those mentioned in connection with the band.

555—Additional allocation: in Angola, Cameroon, the Congo, Madagascar, Mozambique, Somalia, Sudan, Tanzania, Chad and Yemen (P.D.R. of), the band 47-68 MHz is also allocated to the fixed and mobile, except aeronautical mobile, services on a permitted basis.

556—Alternative allocation: in New Zealand, the band 50-51 MHz is allocated to the fixed, mobile and broadcasting services on a primary basis, the band 53-54 MHz is allocated to the fixed and mobile services on a primary basis.

557—Alternative allocation: in Afghanistan, Bangladesh, Brunei, India, Indonesia, Iran, Malaysia, Pakistan, Singapore and Thailand, the band 50-54 MHz is allocated to the fixed, mobile and broadcasting services on a primary basis.

558—Additional allocation: in Australia, China and the Democratic People's Republic of Korea, the band 50-54 MHz is also allocated to the broadcasting service on a primary basis.

559—Alternative allocation: in Botswana, Burundi, Lesotho, Malawi, Namibia, Rwanda, South Africa, Swaziland, Zaire, Zambia and Zimbabwe, the band 50-54 MHz is allocated to the amateur service on a primary basis.

560—Additional allocation: in New Zealand, the band 51-53 MHz is also allocated to the fixed and mobile services on a primary basis.

561—Additional allocation: in Botswana, Burundi, Lesotho, Malawi, Mali, Namibia, Rwanda, South Africa, Swaziland, Zaire, Zambia and Zimbabwe, the band 54-68 MHz is also allocated to the fixed and mobile, except aeronautical mobile, services on a primary basis.

FCC Part 97 Privileges

| License Class | Terrestrial location of the amateur radio station | | | Limitations |
	Region 1 MHz	Region 2 MHz	Region 3 MHz	
Technician General Advanced Extra	—	50-54	50-54	3

§97.7(g)
Limitations. . .
(3) Where, in adjacent regions or subregions, a band of frequencies is allocated to different services of the same category, the basic principle is the equality of right to operate. Accordingly, the stations of each service in one region or subregion must operate so as not to cause harmful interference to services in the other regions or subregions. (See International Telecommunication Union Radio Regulations, RR 346 (Geneva, 1979).)

Amateur Radio Band Plans

Region 1 MHz	Region 2 MHz	Region 3 MHz	ARRL* MHz
		50.0-50.1 CW/Beacons	50.0-50.1 CW/Beacons
			50.06-50.08 Automatically controlled beacons
		50.1-54 CW/Phone/WB	50.1-50.6 SSB/AM phone
			50.11 SSB DX calling
			50.2 SSB national calling (QSY up for local, down for long-distance QSOs)
			50.4 AM calling
			50.6-51.0 Experimental and special modes
			50.80-50.98 Radio Control Ten channels on 20-kHz raster
			51.0-51.1 Pacific DX Window
			51.1-52.0 FM Simplex
			52.0-54.0
			52.00-52.05 Pacific DX Window
			FM repeater** and simplex
			52.490 Simplex 52.510 Simplex 52.525 National Simplex Frequency

*Adopted by the ARRL Board of Directors April 1983.

**ARRL repeater frequency pairs (input/output)

52.010/53.010	52.330/53.330	52.710/53.710
52.030/53.030	52.350/53.350	52.730/53.730
52.050/53.050	52.370/53.370	52.750/53.750
52.070/53.070	52.390/53.390	52.770/53.770
52.090/53.090	52.410/53.410	52.790/53.790
52.110/53.110	52.430/53.430	52.810/53.810
52.130/53.130	52.450/53.450	52.830/53.830
52.150/53.150	52.470/53.470	52.850/53.850
52.170/53.170	52.550/53.550	52.870/53.870
52.190/53.190	52.570/53.570	52.890/53.890
52.210/53.210	52.590/53.590	52.910/53.910
52.230/53.230	52.610/53.610	52.930/53.930
52.250/53.250	52.630/53.630	52.950/53.950
52.270/53.270	52.650/53.650	52.970/53.970
52.290/53.290	52.670/53.670	52.990/53.990
52.310/53.310	52.690/53.690	

Simplex frequencies: 52.490 52.510 52.525†

†National Simplex Frequency

INTERNATIONAL			UNITED STATES				
Region 1 MHz	Region 2 MHz	Region 3 MHz	Band MHz	National Provisions	Government Allocation	Non-Government Allocation	Remarks
144-146	AMATEUR 510 AMATEUR-SATELLITE 605 606		144-146	510		AMATEUR AMATEUR-SATELLITE	
146-149.9 FIXED MOBILE except aeronautical mobile (R)	146-148 AMATEUR 607	146-148 AMATEUR FIXED MOBILE 607	146-148			AMATEUR	

510—For the use of the bands allocated to the amateur service at 3.5 MHz, 7.0 MHz, 10.1 MHz, 14.0 MHz, 18.068 MHz, 21.0 MHz, 24.89 MHz and 144 MHz in the event of natural disasters, see Resolution 640.

605—Additional allocation: in Singapore, the band 144-145 MHz is also allocated to the fixed and mobile services on a primary basis. Such use is limited to systems in operation on or before 1 January 1980, which in any case shall cease by 31 December 1995.

606—Additional allocation: in China, the band 144-146 MHz is also allocated to the aeronautical mobile (OR) service on a secondary basis.

607—Alternative allocation: in Afghanistan, Bangladesh, Cuba, Guyana and India, the band 146-148 MHz is allocated to the fixed and mobile services on a primary basis.

FCC Part 97 Privileges				
	Terrestrial location of the amateur radio station			
License Class	Region 1 MHz	Region 2 MHz	Region 3 MHz	Limitations
Technician General Advanced Extra	144-146	144-148	144-148	3,32

§97.7(g)
Limitations. . .

(3) Where, in adjacent regions or subregions, a band of frequencies is allocated to different services of the same category, the basic principle is the equality of right to operate. Accordingly, the stations of each service in one region or subregion must operate so as not to cause harmful interference to services in the other regions or subregions. (See International Telecommunication Union Radio Regulations, RR 346 (Geneva, 1979).)

(32) Amateur stations in these bands may be used for communications related to relief operations in connection with natural disasters. See Appendix 6 to this Part.

Amateur Radio Band Plans

Region 1 MHz	Region 2 MHz	Region 3 MHz	ARRL* MHz
144.0-144.15 CW exclusive		144.0-144.025 EME	144.00-144.05 EME (CW)
144.00-144.01 EME			
144.05 CW calling			144.05-144.06 Propagation Beacons
144.10 CW random meteor scatter			144.06-144.10 General CW and weak signals
			144.10-144.20 EME and weak-signal SSB
144.20 SSB random meteor scatter			144.20 National calling
			144.20-144.30 General SSB operation
144.30 SSB calling			144.30-144.50 New OSCAR subband
144.40 SSB meteor scatter reference frequency			
144.50 SSTV calling			144.50-144.60 Linear translation input
144.60 RTTY calling 144.675 Data calling 144.70 FAX calling			144.60-144.90** FM repeater inputs
144.845-144.99 Regional Beacons			144.90-145.10 Weak signal and FM simplex****
145.00-145.175 Repeater inputs R0 through R7			145.10-145.20 Linear translator outputs
145.20-145.575 Channelized FM simplex			145.20-145.50** FM repeater outputs
145.5 FM calling			145.50-145.80 Miscellaneous and experimental modes
145.60-145.775 Repeater outputs R0 through R9		145.80-146.00 Satellite	145.80-146.00 OSCAR subband
145.80-146.00 Space communications			146.01-146.37** Repeater inputs
			146.40-146.58 Simplex***
			146.61-147.39** Repeater outputs
			147.42-147.57 Simplex***
			147.60-147.99** Repeater inputs

Region 1:

Repeater (input/output):				Simplex channels:	
145.00	R0	145.60	R0	145.25	S10
145.025	R1	145.625	R1	145.275	S11
145.05	R2	145.65	R2	145.30	S12
145.075	R3	145.675	R3	145.325	S13
145.10	R4	145.70	R4	145.35	S14
145.125	R5	145.725	R5	145.375	S15
145.15	R6	145.75	R6	145.40	S16
145.175	R7	145.775	R7	145.425	S17
				145.45	S18
				145.475	S19
				145.50	S20

145.50 S20 — Most common simplex calling freq; once contact established move off frequency

145.525 S21
145.55 S22
145.575 S23

U.K. & European repeaters require a 1750 Hz (±25 Hz) tone burst of about 200 ms to open the repeater and in practice it is best to transmit this tone at the start of each transmission, especially in the U.K.

ARRL:

*This band plan has been proposed by the ARRL VHF-UHF Advisory Committee.

**Repeater frequency pairs (input/output):

144.51/145.11	144.89/145.49	147.75/147.15
144.53/145.13	146.01/146.61	147.78/147.18
144.55/145.15	146.04/146.64	147.81/147.21
144.57/145.17	146.07/146.67	147.84/147.24
144.59/145.19	146.10/146.70	147.87/147.27
144.61/145.21	146.13/146.73	147.90/147.30
144.63/145.23	146.16/146.76	147.93/147.33
144.65/145.25	146.19/146.79	147.96/147.36
144.67/145.27	146.22/146.82	147.99/147.39
144.69/145.29	146.25/146.85	
144.71/145.31	146.28/146.88	
144.73/145.33	146.31/146.91	
144.75/145.35	146.34/146.94	
144.77/145.37	146.37/146.97	
144.79/145.39	146.40 or 147.60/147.00†	
144.81/145.41	146.43 or 147.63/147.03†	
144.83/145.43	146.46 or 147.66/147.06†	
144.85/145.45	147.69/147.09	
144.87/145.47	147.72/147.12	†local option

Some areas use 146.40-146.60 and 147.40-147.60 MHz for either simplex or repeater inputs and outputs. Frequency pairs in those areas are:

147.415/146.415	147.46/146.46	147.505/146.505
147.43/146.43	147.475/146.475	147.595/146.595
147.445/146.445	147.49/146.49	

For repeaters located east of the Continental Divide, the ARRL suggests standardizing on upright 15-kHz splinter channels: that is, all repeater frequency pairs are low in/high out between 146 and 147 MHz, and high in/low out between 147 and 148 MHz. Recognizing the existence of a significant minority of inverted split repeaters east of the Divide, and wishing to minimize expense and inconvenience of these repeater groups, the ARRL recommends that these changes be completed by April 1989.

146.025/146.625	146.295/146.895	147.735/147.135
146.055/146.655	146.325/146.925	147.765/147.165
146.085/146.685	146.355/146.955	147.825/147.225
146.115/146.715	146.385/146.985	147.855/147.255
146.145/146.745	147.615/147.015	147.885/147.285
146.175/146.775	147.645/147.045	147.915/147.315
146.205/146.805	147.675/147.075	147.945/147.345
146.235/146.835	147.705/147.105	147.975/147.375
146.265/146.865		

***Simplex frequencies

‡146.40	‡146.505	147.42	147.525
‡146.415	#146.52	147.435	147.54
‡146.43	146.535	147.45	147.555
‡146.445	146.55	147.465	147.57
‡146.46	146.565	147.48	147.585
‡146.475	146.58	147.495	
‡146.49	146.595	147.51	

#National Simplex Frequency
‡May also be a repeater (input or output). See repeater pairs listing.

Several states have chosen to re-align the 146-148 MHz band, using 20 kHz spacing between channels. This choice was made to gain additional repeater pairs.

The transition from 30 to 20 kHz spacing is taking place on a case by case basis as the need for additional pairs occurs. Typically the repeater on an odd numbered pair will shift to 10 kHz, up or down, creating a new set on an even numbered channel. For example, the pair of 146.13/.73 would change to 146.12/.72 or 146.14/.74, while the pairs of 146.10/.70 and 146.16/.76 would be left status quo.

****The channels 145.01, .03, .05, .07 and .90 MHz are widely used for packet radio, and have been established by a number of frequency coordinators. Some have recognized Network Coordinating Agents to coordinate packet-radio use of these frequencies.

INTERNATIONAL			UNITED STATES				
Region 1 MHz	Region 2 MHz	Region 3 MHz	Band MHz	National Provisions	Government Allocation	Non-Government Allocation	Remarks
223-230 BROADCASTING Fixed Mobile 622 628 629 631 632 633 634 635	220-225 AMATEUR FIXED MOBILE Radiolocation 627 225-235 FIXED MOBILE	223-230 FIXED MOBILE BROADCASTING AERONAUTICAL RADIONAVIGATION Radiolocation 636 637	220-225	US243 627	Radiolocation FIXED MOBILE G2	AMATEUR FIXED MOBILE	This allocation is subject to further discussion between the FCC and NTIA as part of their long range planning activities

622—Different category of service: in the Federal Republic of Germany, Austria, Belgium, Denmark, Spain, Finland, France, Israel, Italy, Liechtenstein, Luxembourg, Monaco, Norway, the Netherlands, Portugal, the United Kingdom, Sweden, Switzerland, and Yemen (P.D.R. of), the band 223-230 MHz is allocated to the land mobile service on a permitted basis (see No. 425). However, the stations of the land mobile service shall not cause harmful interference to, nor claim protection from, broadcasting stations, existing or planned, in countries other than those listed in this footnote.

623—Additional allocation: in the Congo, Ethiopia, Gambia, Guinea, Kenya, Libya, Malawi, Mali, Uganda, Senegal, Sierra Leone, Somalia, Tanzania and Zimbabwe, the band 174-223 MHz is also allocated to the fixed and mobile services on a secondary basis.

625—Additional allocation: in Australia and Papua New Guinea, the bands 204-208 MHz and 222-223 MHz are also allocated to the aeronautical radionavigation service on a primary basis.

626—Additional allocation: in China, India and Thailand, the band 216-223 MHz is also allocated to the aeronautical radionavigation service on a primary basis and to the radiolocation service on a secondary basis.

627—In Region 2, the band 216-225 MHz is allocated to the radiolocation service on a primary basis until 1 January 1990. On and after 1 January 1990, no new stations in that service may be authorized. Stations authorized prior to 1 January 1990 may continue to operate on a secondary basis.

628—Additional allocation: in Somalia, the band 216-225 MHz is also allocated to the aeronautical radionavigation service on a primary basis, subject to not causing harmful interference to existing or planned broadcasting services in other countries.

629—Additional allocation: in Oman, the United Kingdom and Turkey, the band 216-235 MHz is also allocated to the radiolocation service on a secondary basis.

630—Additional allocation: in Japan, the band 222-223 MHz is also allocated to the aeronautical radionavigation service on a primary basis and to the radiolocation service on a secondary basis.

631—Different category of service: in Spain and Portugal, the band 223-230 MHz is allocated to the fixed service on a permitted basis (see No. 425). Stations of this service shall not cause harmful interference to, or claim protection from, broadcasting stations of other countries, whether existing or planned, that operate in accordance with the Table.

632—Additional allocation: in Saudi Arabia, Bahrain, the United Arab Emirates, Israel, Jordan, Oman, Qatar and Syria, the band 223-235 MHz is also allocated to the aeronautical radionavigation service on a permitted basis.

633—Additional allocation: in Spain and Portugal, the band 223-235 MHz is also allocated to the aeronautical radionavigation service on a permitted basis until 1 January 1990, subject to not causing harmful interference to existing or planned broadcasting stations in other countries.

634—Additional allocation: in Sweden, the band 223-235 MHz is also allocated to the aeronautical radionavigation service on a permitted basis until 1 January 1990, subject to agreement obtained under the procedure set forth in Article 14, and on condition that no harmful interference is caused to existing and planned broadcasting stations in other countries.

635—Alternative allocation: in Botswana, Lesotho, Namibia, South Africa, Swaziland and Zambia, the bands 223-238 MHz and 246-254 MHz are allocated to the broadcasting service on a primary basis subject to agreement obtained under the procedure set forth in Article 14.

636—Alternative allocation: in New Zealand, Western Samoa and the Niue and Cook Islands, the band 225-230 MHz is allocated to the fixed, mobile and aeronautical radionavigation services on a primary basis.

637—Additional allocation: in China, the band 225-235 MHz is also allocated to the radio astronomy service on a secondary basis.

US243—In the band 220-225 MHz, stations in the radiolocation service have priority until 1 January 1990.

G2—In the bands 216-225, 420-450 (except as provided by US217), 890-902, 928-942, 1300-1400, 2300-2450, 2700-2900, 5650-5925, and 9000-9200 MHz, the Government radiolocation is limited to the military services.

	FCC Part 97 Privileges			
	Terrestrial location of the amateur radio station			
License Class	Region 1 MHz	Region 2 MHz	Region 3 MHz	Limitations
Novice	—	222.1-223.91	—	5
Technician General Advanced Extra	—	220-225*	—	3,4,5

*At presstime, the FCC issued a Notice of Proposed Rule Making proposing to reallocate the bottom 2 megahertz (220-222 MHz) exclusively to the Land Mobile Service. Watch QST for further developments.

§97.7(g)
Limitations. . .
(3) Where, in adjacent regions or subregions, a band of frequencies is allocated to different services of the same category, the basic principle is equality of right to operate. Accordingly, the stations of each service in one region or subregion must operate so as not to cause harmful interference to services in the other regions or subregions. (See International Telecommunication Union Radio Regulations, RR 346 (Geneva, 1979).)

(4) This band is allocated to the amateur, fixed and mobile services in the United States on a co-primary basis. The basic principle which applies is the equality of right to operate. Amateur, fixed and mobile stations must operate so as not to cause harmful interference to each other.

(5) Amateur stations in the 1900-2000 kHz, 220-225 MHz, 420-450 MHz, 902-928 MHz, 1240-1300 MHz, 2300-2310 MHz, 2390-2450 MHz, 3.3-3.5 GHz, 5.650-5.925 GHz, 10.0-10.5 GHz, 24.05-24.25 GHz, 76-81 GHz, 144-149 GHz and 241-248 GHz bands must not cause harmful interference to stations in the Government radiolocation service and are not protected from interference due to the operation of stations in the Government radiolocation service.

Amateur Radio Band Plans

Region 1 MHz	Region 2 MHz	Region 3 MHz	ARRL* MHz
			220.00-220.05 EME
			220.5-220.06 Propagation beacons
			220.06-220.10 Weak-signal CW
			220.10 Calling frequency
			221.10-220.50 General weak signal, rag chewing, and experimental communications
			220.50-221.90 Experimental and control links
			221.90-222.00 Weak signal guard band
			220.00-222.05 EME
			222.05-222.06 Propagation beacons
			222.06-222.10 Weak-signal CW
			222.10 Calling frequency
			222.10-222.30 General operation, CW or SSB, etc
			222.34-223.38 Repeater inputs**
			223.34-223.90 Simplex*** and repeater outputs** (local option)
			223.40 suggested Packet frequency
			223.94-224.98 Repeater outputs**

*This band plan is under review by the ARRL VHF-UHF Advisory Committee.

**Repeater frequency pairs (input/output):

222.32/223.92	222.54/224.14	222.76/224.36	222.98/224.58	223.20/224.80
222.34/223.94	222.56/224.16	222.78/224.38	223.00/224.60	223.22/224.82
222.36/223.96	222.58/224.18	222.80/224.40	223.02/224.62	223.24/224.84
222.38/223.98	222.60/224.20	222.82/224.42	223.04/224.64	223.26/224.86
222.40/224.00	222.62/224.22	222.84/224.44	223.06/224.66	223.28/224.88
222.42/224.02	222.64/224.24	222.86/222.46	223.08/224.68	223.30/224.90
222.44/224.04	222.66/224.26	222.88/224.48	223.10/224.70	223.32/224.92
222.46/224.06	222.68/224.28	222.90/224.50	223.12/224.72	223.34/224.94
222.48/224.08	222.70/224.30	222.92/224.52	223.14/224.74	223.36/224.96
222.50/224.10	222.72/224.32	222.94/224.54	223.16/224.76	223.38/224.98
222.52/224.12	222.74/224.34	222.96/224.56	223.18/224.78	

***Simplex frequencies:

223.42	223.52	223.62	223.72	223.82
223.44	223.54	223.64	223.74	223.84
223.46	223.56	223.66	223.76	223.86
223.48	223.58	223.68	223.78	223.88
223.50‡	223.60	223.70	223.80	223.90

‡National simplex frequency

INTERNATIONAL			UNITED STATES				
Region 1 MHz	Region 2 MHz	Region 3 MHz	Band MHz	National Provisions	Government Allocation	Non-Government Allocation	Remarks
420-430	FIXED MOBILE except aeronautical mobile Radiolocation 651 652 653		420-450	US7 US87 US217 US228 664 668	RADIOLOCATION	Amateur	
430-440 AMATEUR RADIOLOCATION 653 654 655 656 657 658 659 661 662 663 664 665	430-440 RADIOLOCATION Amateur 653 658 659 660 663 664				G2 G8 G105	NG135	
440-450	FIXED MOBILE except aeronautical mobile Radiolocation 651 652 653 666 667 668						

651—Different category of service: in Australia, the United States, India, Japan and the United Kingdom, the allocation of the bands 420-430 MHz and 440-450 MHz to the radiolocation service is on a primary basis (see No. 425).

652—Additional allocation: in Australia, the United States, Jamaica and the Philippines, the bands 420-430 MHz and 440-450 MHz are also allocated to the amateur service on a secondary basis.

653—Additional allocation: in China, India, the German Democratic Republic, the United Kingdom and the U.S.S.R., the band 420-460 MHz is also allocated to the aeronautical radionavigation service (radio altimeters) on a secondary basis.

654—Different category of service: in France, the allocation of the band 430-434 MHz to the amateur service is on a secondary basis (see No. 424).

655—Different category of service: in Denmark, Libya, Norway and Sweden, the allocation of the bands 430-432 MHz and 438-440 MHz to the radiolocation service is on a secondary basis (see No. 424).

656—Alternative allocation: in Denmark, Norway and Sweden, the bands 430-432 MHz and 438-440 MHz are allocated to the fixed and mobile, except aeronautical mobile, services on a primary basis.

657—Additional allocation: in Finland, Libya and Yugoslavia, the bands 430-432 MHz and 438-440 MHz are also allocated to the fixed and mobile, except aeronautical mobile, services on a primary basis.

658—Additional allocation: in Afghanistan, Algeria, Saudi Arabia, Bahrain, Bangladesh, Brunei, Burundi, Egypt, the United Arab Emirates, Ecuador, Ethiopia, Greece, Guinea, India, Indonesia, Iran, Iraq, Israel, Italy, Jordan, Kenya, Kuwait, Lebanon, Libya, Liechtenstein, Malaysia, Malta, Nigeria, Oman, Pakistan, the Philippines, Qatar, Syria, Singapore, Somalia, Switzerland, Tanzania, the band 430-440 MHz is also allocated to the fixed service on a primary basis and the bands 430-435 MHz and 438-440 MHz are also allocated to the mobile, except aeronautical mobile, service on a primary basis.

659—Additional allocation: in Angola, Bulgaria, Cameroon, the Congo, Gabon, Hungary, Mali, Mongolia, Niger, Poland, the German Democratic Republic, Romania, Rwanda, Chad, Czechoslovakia and the U.S.S.R., the band 430-440 MHz is also allocated to the fixed service on a primary basis.

660—Different category of service: in Argentina, Colombia, Costa Rica, Cuba, Guyana, Honduras, Panama and Venezuela, the allocation of the band 430-440 MHz to the amateur service is on a primary basis (see No. 425).

661—In Region 1, except in the countries mentioned in No. 662, the band 433.05-434.79 MHz (centre frequency 433.92 MHz) is designated for industrial, scientific and medical (ISM) applications. The use of this frequency band for ISM applications shall be subject to special authorization by the administration concerned, in agreement with other administrations whose radiocommunication services might be affected. In applying this provision, administrations shall have due regard to the latest relevant CCIR Recommendations.

662—In the Federal Republic of Germany, Austria, Liechtenstein, Portugal, Switzerland and Yugoslavia, the band 433.05-434.79 MHz (centre frequency 433.92 MHz) is designated for industrial, scientific and medical (ISM) applications. Radiocommunication services of these countries operating within this band must accept harmful interference which may be caused by these applications. ISM equipment operating in this band is subject to the provisions of No. 1815.

663—Additional allocation: in Brazil, France and the French Overseas Departments in Region 2, and India, the band 433.75-434.25 MHz is also allocated to the space operation service (Earth-to-space) on a primary basis until 1 January 1990, subject to agreement obtained under the procedure set forth in Article 14. After 1 January 1990, the band 433.75-434.25 MHz will be allocated in the same countries to the same service on a secondary basis.

664—In the bands 435-438 MHz, 1260-1270 MHz, 2400-2450 MHz, 3400-3410 MHz (in Regions 2 and 3 only) and 5650-5670 MHz, the amateur-satellite service may operate subject to not causing harmful interference to other services operating in accordance with the Table (see No. 435). Administrations authorizing such use shall ensure that any harmful interference caused by emissions from a station in the amateur-satellite service is immediately eliminated in accordance with the provisions of No. 2741. The use of the bands 1260-1270 MHz and 5650-5670 MHz by the amateur-satellite service is limited to the Earth-to-space direction.

665—Additional allocation: in Austria, the band 438-440 MHz is also allocated to the fixed and mobile, except aeronautical mobile, services on a secondary basis.

666—Additional allocation: in Canada, New Zealand and Papua New Guinea, the band 440-450 MHz is also allocated to the amateur service on a secondary basis.

667—Different category of service: in Canada, the allocation of the band 440-450 MHz to the radiolocation service is on a primary basis (see No. 425).

668—Subject to agreement obtained under the procedure set forth in Article 14, the band 449.75-450.25 MHz may be used for the space operation service (Earth-to-space) and the space research service (Earth-to-space).

US7—In the band 420-450 MHz and within the following areas, the peak envelope power output of a transmitter employed in the amateur service shall not exceed 50 watts, unless expressly authorized by the Commission after mutual agreement, on a case-by-case basis, between the Federal Communications Commission Engineer in Charge at the applicable District Office and the Military Area Frequency Coordinator at the applicable military base:

(a) Those portions of Texas and New Mexico bounded on the south by latitude 31°45′ North, on the east by 104°00′ West, on the north by latitude 34°30′ North, and on the west by longitude 107°30′ West;

(b) The entire State of Florida including the Key West area and the areas enclosed within a 200-mile radius of Patrick Air Force Base, Florida (latitude 28°21′ North, longitude 80°43′ West), and within a 200-mile radius of Eglin Air Force Base, Florida (latitude 30°30′ North, longitude 86°30′ West);

(c) The entire State of Arizona;

(d) Those portions of California and Nevada south of Latitude 37°10′ North, and the areas enclosed within a 200-mile radius of the Pacific Missile Test Center, Point Mugu, California (latitude 34°09′ North, longitude 119°11′ West).

(e) In the State of Massachusetts within a 160-kilometer (100 mile) radius around locations at Otis Air Force Base, Massachusetts (latitude 41°45′ North, longitude 70°32′ West).

(f) In the State of California within a 240-kilometer (150 mile) radius around locations at Beale Air Force Base, California (latitude 39°08′ North, longitude 121°26′ West).

(g) In the State of Alaska within a 160-kilometer (100 mile) radius of Clear, Alaska (latitude 64°17′ North, longitude 149°10′ West). (The Military Area Frequency Coordinator for this area is located at Elmendorf Air Force Base, Alaska.)

(h) In the State of North Dakota within a 160-kilometer (100 mile) radius of Concrete, North Dakota (latitude 48°43′ North, longitude 97°54′ West). (The Military Area Frequency Coordinator for this area can be contacted at: HQ SAC/SXOE, Offutt Air Force Base, Nebraska 68113.)

US87—The frequency 450 MHz, with maximum emission bandwidth of 500 kHz, may be used by Government and non-Government stations for space telecommand at specific locations, subject to such conditions as may be applied on a case-by-case basis.

US217—Pulse-ranging radiolocation systems may be authorized for Government and non-Government use in the 420-450 MHz band along the shorelines of Alaska and the contiguous 48 States. Spread spectrum radiolocation systems may be authorized in the 420-435 MHz portion of the band for operation within the contiguous 48 States and Alaska. Authorizations will be granted on a case-by-case basis; however, operations proposed to be located within the zones set forth in US228 should not expect to be accommodated. All stations operating in accordance with this provision will be secondary to stations operating in accordance with the Table of Frequency Allocations.

US228—Applicants of operation in the band 420 to 450 MHz under the provisions of US217 should not expect to be accommodated if their area of service is within the following geographic areas:

(a) Those portions of Texas and New Mexico bounded on the south by latitude 31°45′ North, on the east by 104°00′ West, on the north by latitude 34°30′ North, and on the west by longitude 107°30′ West;

(b) The entire State of Florida including the Key West area and the areas enclosed within a 200-mile radius of Patrick Air Force Base, Florida (latitude 28°21′ North, longitude 80°43′ West), and within a 200-mile radius of Eglin Air Force Base, Florida (latitude 30°30′ North, longitude 86°30′ West);

(c) The entire State of Arizona;

(d) Those portions of California and Nevada south of Latitude 37°10′ North, and the areas enclosed within a 200-mile radius of the Pacific Missile Test Center, Point Mugu, California (latitude 34°09′ North, longitude 119°11′ West).

(e) In the State of Massachusetts within a 160-kilometer (100 mile) radius around locations at Otis Air Force Base, Massachusetts (latitude 41°45' North, longitude 70°32' West).

(f) In the State of California within a 240-kilometer (150 mile) radius around locations at Beale Air Force Base, California (latitude 39°08' North, longitude 121°26' West).

(g) In the State of Alaska within a 160-kilometer (100 mile) radius of Clear, Alaska (latitude 64°17' North, longitude 149°10' West). (The Military Area Frequency Coordinator for this area is located at Elmendorf Air Force Base, Alaska.)

(h) In the State of North Dakota within a 160-kilometer (100 mile) radius of Concrete, North Dakota (latitude 48°43' North, longitude 97°54' West). (The Military Area Frequency Coordinator for this area can be contacted at: HQ SAC/SXOE, Offutt Air Force Base, Nebraska 68113.)

(i) In the States of Alabama, Florida, Georgia and South Carolina within a 200-kilometer (124 mile) radius of Warner Robins Air Force Base, Georgia (latitude 32°38' North, longitude 83°35' West).

(j) In the State of Texas within a 200-kilometer (124 mile) radius of Goodfellow Air Force Base, Texas (latitude 31°25' North, longitude 100°24' West).

G2—In the bands 216-225, 420-450 (except as provided by US217), 890-902, 928-942, 1300-1400, 2300-2450, 2700-2900, 5650-5925, and 9000-9200 MHz, the Government radiolocation is limited to the military services.

G8—Low power Government radio control operations are permitted in the band 420-450 MHz.

G105—In the band 420-460 MHz, Radio Altimeter operations are limited to the military services and to existing equipments which may continue to operate until January 1, 1985 on the condition that harmful interference is not caused to stations of services operating in accordance with the Table of Frequency Allocations.

NG135—In the 420-430 MHz band the Amateur service is not allocated north of line A. (def. §2.1). All amateur radio stations shall operate north of line A in accordance with the Agreement between the United States and Canada.

FCC Part 97 Privileges

License Class	Terrestrial location of the amateur radio station			
	Region 1 MHz	Region 2 MHz	Region 3 MHz	Limitations
Technician General Advanced Extra	430-440	420-450	420-450	3,5,6,7,10,30

§97.7(g)

Limitations. . .

(3) Where, in adjacent regions or subregions, a band of frequencies is allocated to different services of the same category, the basic principle is the equality of right to operate. Accordingly, the stations of each service in one region or subregion must operate so as not to cause harmful interference to services in the other regions or subregions. (See International Telecommunication Union Radio Regulations, RR 346 (Geneva, 1979).)

(5) Amateur stations in the 1900-2000 kHz, 220-225 MHz, 420-450 MHz, 902-928 MHz, 1240-1300 MHz, 2300-2310 MHz, 2390-2450 MHz, 3.3-3.5 GHz, 5.650-5.925 GHz, 10.0-10.5 GHz, 24.05-24.25 GHz, 76-81 GHz, 144-149 GHz and 241-248 GHz bands must not cause harmful interference to stations in the Government radiolocation service and are not protected from interference due to the operation of stations in the Government radiolocation service.

(6) No amateur station shall operate north of Line A (see §97.3(i)) in the 420-430 MHz band.

(7) The 420-430 MHz band is allocated to the Amateur service in the United States on a secondary basis, but is allocated to the fixed and mobile (except aeronautical mobile) services in the International Table of Allocations on a primary basis. Therefore, amateur stations by other nations in the fixed and mobile (except aeronautical mobile) services and are not protected from interference due to the operation of stations authorized by other nations in the fixed and mobile (except aeronautical mobile) services.

(10) The 430-440 MHz band is allocated to the Amateur service on a secondary basis in ITU Regions 2 and 3. Amateur stations in this band in ITU Regions 2 and 3 must not cause harmful interference to stations authorized by other nations in the radiolocation service and are not protected from interference due to the operation of stations authorized by other nations in the radiolocation service. In ITU Region 1 the 430-440 MHz band is allocated to the Amateur service on a co-primary basis with the radiolocation service. As between these two services in this band in Region 1 the basic principle which applies is the equality of right to operate. Amateur stations authorized by the United States and radiolocation stations authorized by other nations in Region 1 must operate so as not to cause harmful interference to each other.

(30) Amateur stations in the 449.5-450 MHz band must not cause interference to and are not protected from interference due to the operation of stations in the space operation service, the space research service, or for space telecommand.

Amateur Radio Band Plans

Region 1 MHz	Region 2 MHz	Region 3 MHz	ARRL* MHz
			420.00-426.00 ATV repeater or simplex with 421.25-MHz video carrier, control links and experimental
		430.0-431.9	426.00-432.00 ATV simplex with 427.25-MHz video carrier
		431.9-432.24 EME	432.00-432.07 EME
			432.07-432.08 Propagation beacons
			432.08-432.10 Weak-signal CW
			432.10 Calling frequency
			432.10-432.125 Mixed mode and weak signal
			432.125-432.175 OSCAR inputs
		432.24-435.00	432.175-433.00 Mixed mode and weak signals
			433.00-435.00 Auxiliary/repeater links
		435.00-438.00 Satellite	435.00-438.00 Satellite only (internationally)
		438.00-450.00	438.00-444.00 ATV repeater input with 439.25-MHz video carrier and repeater links
			442.00-445 Repeater inputs and outputs (local option)
			445.00-447.00 Shared by auxiliary and control links, repeaters and simplex (local option)
			446.0 National simplex frequency
			447.00-450.00 Repeater inputs and outputs (local option)

*This band plan is under review by the ARRL VHF-UHF Advisory Committee.

The Amateur Radio Spectrum   2-47

INTERNATIONAL			UNITED STATES				
Region 1 MHz	Region 2 MHz	Region 3 MHz	Band MHz	National Provisions	Government Allocation	Non-Government Allocation	Remarks
890-942 FIXED MOBILE except aeronautical mobile BROADCASTING 703 Radiolocation		890-942 FIXED MOBILE BROADCASTING Radiolocation	902-928	US215 US218 US267 US275 707	RADIOLOCATION	AMATEUR	(ISM 915 ± 13 MHz)
	902-928 FIXED Amateur Mobile except aeronautical mobile						
704	Radiolocation 705 707	706			G11 G59		

704—Additional allocation: in Bulgaria, Hungary, Mongolia, Poland, the German Democratic Republic, Romania, Czechoslovakia and the U.S.S.R., the band 862-960 MHz is also allocated to the aeronautical radionavigation service on a permitted basis until 1 January 1998. Up to this date, the aeronautical radionavigation service may use the band, subject to agreement obtained under the procedure set forth in Article 14. After this date, the aeronautical radionavigation service may continue to operate on a secondary basis.

705—Different category of service: in the United States, the allocation of the band 890-942 MHz to the radiolocation service is on a primary basis (see No. 425) and subject to agreement obtained under the procedure set forth in Article 14.

706—Different category of service: in Australia, the allocation of the band 890-942 MHz to the radiolocation service is on a primary basis (see No. 425).

707—In Region 2, the band 902-928 MHz (center frequency 915 MHz) is designated for industrial, scientific and medical (ISM) applications. Radiocommunication services operating within this band must accept harmful interference which may be caused by these applications. ISM equipment operating in this band is subject to the provisions of No. 1815.

US215—Emissions from microwave ovens manufactured on and after January 1, 1980, for operation on the frequency 915 MHz must be confined within the band 902-928 MHz. Emissions from microwave ovens manufactured prior to January 1, 1980, for operation on the frequency 915 MHz must be confined within the band 902-940 MHz. Radiocommunications services operating within the band 928-940 MHz must accept any harmful interference that may be experienced from the operation of microwave ovens manufactured before January 1, 1980.

US218—The band segments 902-912 MHz and 918-928 MHz are available for Automatic Vehicle Monitoring (AVM) Systems subject to not causing harmful interference to the operation of Government stations authorized in these bands. These systems must tolerate any interference from the operation of industrial, scientific, and medical (ISM) devices and the operation of Government stations authorized in these bands.

US267—In the band 902-928 MHz, amateur radio stations shall not operate within the States of Colorado and Wyoming, bounded by the area of: latitude 39° N to 42° N and longitude 103° W to 108° W.

US275—The band 902-928 MHz is allocated on a secondary basis to the amateur service subject to not causing harmful interference to the operations of Government stations authorized in this band or to the Automatic Vehicle Monitoring (AVM) systems. Stations in the amateur service must tolerate any interference from the operations of industrial, scientific and medical (ISM) devices, AVM systems, and the operations of Government stations authorized in this band.

G11—Government fixed and mobile radio services, including low power radio control operations, are permitted in the band 902-928 MHz on a secondary basis.

G59—In the bands 902-928 MHz, 3100-3300 MHz, 3500-3700 MHz, 5250-5350 MHz, 8500-9000 MHz, 9200-9300 MHz, 13.4-14.0 GHz, 15.7-17.7 GHz and 24.05-24.25 GHz, all Government non-military radiolocation shall be secondary to military radiolocation, except in the sub-band 15.7-16.2 GHz, airport surface detection equipment (ASDE) is permitted on a co-equal basis subject to coordination with the military departments.

FCC Part 97 Privileges

License Class	Terrestrial location of the amateur radio station			
	Region 1 MHz	Region 2 MHz	Region 3 MHz	Limitations
Technician General Advanced Extra	—	902-928	—	3,5,8,9

§97.7(g)
Limitations. . .

(3) Where, in adjacent regions or subregions, a band of frequencies is allocated to different services of the same category, the basic principle is the equality of right to operate. Accordingly, the stations of each service in one region or subregion must operate so as not to cause harmful interference to services in the other regions or subregions. (See International Telecommunication Union Radio Regulations, RR 346 (Geneva, 1979).)

(5) Amateur stations in the 1900-2000 kHz, 220-225 MHz, 420-450 MHz, 902-928 MHz, 1240-1300 MHz, 2300-2310 MHz, 2390-2450 MHz, 3.3-3.5 GHz, 5.650-5.925 GHz, 10.0-10.5 GHz, 24.05-24.25 GHz, 76-81 GHz, 144-149 GHz and 241-248 GHz bands must not cause harmful interference to stations in the Government radiolocation service and are not protected from interference due to the operation of stations in the Government radiolocation service.

(8) In the 902-928 MHz band, amateur stations shall not operate within the States of Colorado and Wyoming, bounded by the area of: latitude 39° N to 42° N, and longitude 103° W to 108° W. This band is allocated on a secondary basis to the Amateur service subject to not causing harmful interference to the operations of Government stations authorized in this band or to the Automatic Vehicle Monitoring (AVM) systems. Stations in the Amateur service are not protected from any interference due to the operation of industrial, scientific and medical (ISM) devices, AVM systems or Government stations authorized in this band.

(9) In the 902-928 MHz band, amateur stations shall not operate in those portions of the States of Texas and New Mexico bounded on the south by latitude 31°41′N, on the east by longitude 104°11′W, on the north by latitude 34°30′N, and on the west by longitude 107°30′W.

Amateur Radio Band Plans

Region 1 MHz	Region 2 MHz	Region 3 MHz	ARRL* MHz
			902-904 NB, weak-signal
			902.0-902.8 SSTV, FAX, experimental
			902.8-903.0 reserved for EME, CW expansion
			903.0-903.05 EME exclusive
			903.7-903.8 CW beacons
			903.1 CW/SSB calling
			903.4-903.6 Linear translater inputs
			903.6-903.8 Linear translater outputs
			903.8-904 Experimental beacons exclusive
			904-906 (2,8) Digital communications
			906-907 (6) NB FM simplex services, 25-kHz channels
			906.5 National simplex
			907-910 (2,4) FM repeater inputs paired with 919-922 MHz, 119 pairs, every 25 kHz, eg, 907.025, 050, 075, etc.
			908/920 Uncoordinated Pair
			910-916 (2,3) ATV
			916-918 (2,8) Digital communications
			918-919 NB FM control links and remote bases
			919-922 (2,4) FM repeater outputs paired with 907-910 MHz
			922-928 (5,7) WB experimental, simplex ATV, spread spectrum

*Adopted by the ARRL Board of Directors October 1984.

Canadian amateurs note:
The Amateur Service will continue to have secondary status in the band 902-928 MHz throughout Canada, using any of the following emissions: NØN, A1A, A2A, A3C, A3E, C3F, F1A, F2A, F2B, F1C, F3E.
Before operation in this band, Canadian amateur licensees are required to consult with their DOC District Office to ensure interference will not be caused to other services operating in the area as per Section 45 of the General Radio Regulations Part II, given in TRC 25.
Government of Canada shipborne radiolocation service is permitted within 150 km of the East and West Coasts, Arctic Ocean, Hudson Bay, James Bay and up the St. Lawrence River as far as Rimourski on pre-coordinated channels in the 902-928 MHz band.

Footnotes:
1) Deleted
2) Coordinated frequency assignments required.
3) ATV assignments should be made according to modulation type, e.g. VSB-ATV, SSB-ATV or combinations. Coordination of multiple uses of a single channel in a local area can be achieved through isolation by means of cross-polarization and directional antennas.
4) Coordinated assignments at 100 kHz until allocations are filled, then assign 50 kHz until allocations are filled, before assigning 25-kHz channels.
5) Simplex services only, permanent users shall not be coordinated in this segment. High altitude repeaters for other unattended fixed operations are not permitted.
6) Voice and non-voice operation.
7) Spread-spectrum requires FCC authorization.
8) Consult FCC R&R §97.69 for allowable data rates and bandwidths.

INTERNATIONAL			UNITED STATES				
Region 1 MHz	Region 2 MHz	Region 3 MHz	Band MHz	National Provisions	Government Allocation	Non-Government Allocation	Remarks
1240-1260	RADIOLOCATION RADIONAVIGATION-SATELLITE(Space-to-Earth) 710 Amateur 711 712 713 714		1240-1300	664 713 714	RADIOLOCATION	Amateur	
1260-1300	RADIOLOCATION Amateur 664 711 712 713 714				G56		

664—In the bands 435-438 MHz, 1260-1270 MHz, 2400-2450 MHz, 3400-3410 MHz (in Regions 2 and 3 only) and 5650-5670 MHz, the amateur-satellite service may operate subject to not causing harmful interference to other services operating in accordance with the Table (see No. 435). Administrations authorizing such use shall ensure that any harmful interference caused by emissions from a station in the amateur-satellite service is immediately eliminated in accordance with the provisions of No. 2741. The use of the bands 1260-1270 MHz and 5650-5670 MHz by the amateur-satellite service is limited to the Earth-to-space direction.

710—Use of the radionavigation-satellite service in the band 1215-1260 MHz shall be subject to the condition that no harmful interference is caused to the radionavigation service authorized under No. 712.

711—Additional allocation: in Afghanistan, Angola, Saudi Arabia, Bahrain, Bangladesh, Cameroon, China, the United Arab Emirates, Ethiopia, Guinea, Guyana, India, Indonesia, Iran, Iraq, Israel, Japan, Jordan, Kuwait, Lebanon, Libya, Malawi, Morocco, Mozambique, Nepal, Nigeria, Oman, Pakistan, the Philippines, Qatar, Syria, Somalia, Sudan, Sri Lanka, Chad, Thailand, Togo and Yemen (P.D.R. of), the band 1215-1300 MHz is also allocated to the fixed and mobile services on a primary basis.

712—Additional allocation: in Algeria, the Federal Republic of Germany, Austria, Bahrain, Belgium, Benin, Burundi, Cameroon, China, Denmark, the United Arab Emirates, France, Greece, India, Iran, Iraq, Kenya, Liechtenstein, Luxembourg, Mali, Mauritania, Norway, Oman, Pakistan, the Netherlands, Portugal, Qatar, Senegal, Somalia, Sudan, Sri Lanka, Sweden, Switzerland, Tanzania, Turkey and Yugoslavia, the band 1215-1300 MHz is also allocated to the radionavigation service on a primary basis.

713—In the bands 1215-1300 MHz, 3100-3300 MHz, 5250-5350 MHz, 8550-8650 MHz, 9500-9800 MHz and 13.4-14.0 GHz, radiolocation stations installed on spacecraft may also be employed for the earth exploration-satellite and space research services on a secondary basis.

714—Additional allocation: in Canada and the United States, the bands 1240-1300 MHz and 1350-1370 MHz are also allocated to the aeronautical radionavigation service on a primary basis.

G56—Government radiolocation in the bands 1215-1300, 2900-3100, 5350-5650 and 9300-9500 MHz is primarily for the military services; however, limited secondary use is permitted by other Government agencies in support of experimentation and research programs. In addition, limited secondary use is permitted for survey operations in the band 2900-3100 MHz.

FCC Part 97 Privileges				
	Terrestrial location of the amateur radio station			
License Class	Region 1 MHz	Region 2 MHz	Region 3 MHz	Limitations
Novice	1270-1295	1270-1295	1270-1295	5,22
Technician General Advanced Extra	1240-1300	1240-1300	1240-1300	5,11,22

§97.7(g)
Limitations. . .

(5) Amateur stations in the 1900-2000 kHz, 220-225 MHz, 420-450 MHz, 902-928 MHz, 1240-1300 MHz, 2300-2310 MHz, 2390-2450 MHz, 3.3-3.5 GHz, 5.650-5.925 GHz, 10.0-10.5 GHz, 24.05-24.25 GHz, 76-81 GHz, 144-149 GHz and 241-248 GHz bands must not cause harmful interference to stations in the Government radiolocation service and are not protected from interference due to the operation of stations in the Government radiolocation service.

(11) In the 1240-1260 MHz band amateur stations must not cause harmful interference to stations authorized by other nations in the radionavigation-satellite service and are not protected from interference due to the operation of stations authorized by other nations in the radionavigation-satellite service.

(22) Amateur stations in the 1240-1300 MHz, 10.0-10.5 GHz, 24.05-24.25 GHz, 76-81 GHz, 144-149 GHz and 241-248 GHz bands must not cause harmful interference to stations authorized by other nations in the radiolocation service and are not protected from interference due to the operation of stations authorized by other nations in the radiolocation service.

Region 1 MHz	Region 2 MHz	Region 3 MHz	ARRL* MHz
		1240-1260	1240-1246 (2,3) ATV #1
			1246-1248 (2) NB FM point-to-point links and digital duplex with 1258-1260
			1248-1252 (2,8,9) Digital communications
			1252-1258 (2,3) ATV #2
			1258-1260 (2) NB FM point-to-point links and digital, duplexed with 1246-1252
		1260-1270 Satellite	1260-1270 (5,6) Satellite uplinks ref WARC 1979 WB experimental, simplex ATV
		1270-1296	1270-1276 (2,4,9,10) Repeater inputs, FM and linear, paired with 1282-1288, 239 pairs, every 25 kHz, eg, 1270.025, 050, 075, etc.
			1271/1283 uncoordinated test pair
			1276-1282 (2,3) ATV #3
			1282-1288 (2,4,9,10) Repeater outputs paired with 1270-1276
			1288-1294 (6) WB experimental, simplex ATV
			1294-1295 (7) NB FM simplex, 25-kHz channels
			1294.5 National FM simplex calling
			1295-1297 NB weak-signal (no FM)
			1295.0-1295.8 SSTV, FAX, ACSSB experimental
			1295.8-1296.0 Reserved for EME, CW expansion
		1296-1297 EME	1296.0-1296.05 EME exclusive
			1296.07-1296.08 CW beacons
			1296.1 CW/SSB calling
			1296.4-1296.6 Crossband linear translator input
			1296.6-1296.8 Crossband linear translator output
			1296.8-1297.0 Experimental beacons (exclusive)
		1297-1300	1297-1300 Digital communications

*Adopted by the ARRL Board of Directors January 1985.

Footnotes:
1) Deleted
2) Coordinated assignments required.
3) ATV assignments should be made according to modulation type (for example), VSB-ATV, SSB-ATV, or combination. Coordination of multiple users of a single channel in a local area can be achieved through isolation by means of cross polarization and directional antennas. DSB ATV may be used, but only when local and regional activity levels permit. The excess bandwidths from such users are secondary to the assigned services.
4) Coordinate assignments with 100-kHz channels, beginning at the lower end of the segment until allocations are filled, then assign 50-kHz channels until allocations are filled before assigning 25-kHz channels.
5) Wide bandwidth experimental users are secondary to the satellite service and may be displaced upon the installation of any new satellites. Users are EIRP-limited to the noise floor of the satellites in service and may suffer interference from satellite uplinks.
6) Simplex services only; permanent users shall not be coordinated in this segment. High-altitude repeaters or other unattended fixed operations are not permitted.
7) Voice and non-voice operations.
8) Consult 47 C.F.R. 97.69 (FCC regulations) for allowable data rates and bandwidths.
9) Provide guard bands at the higher frequency end of segments, as required, to avoid interference to ATV.
10) 1274.0-1274.2 and 1286.0-1286.2 MHz are optionally reserved for contiguous Linear Translators supporting multiple narrow bandwidth users. These may also be duplexed with other non-coordinated band segments.

INTERNATIONAL			UNITED STATES				
Region 1 MHz	Region 2 MHz	Region 3 MHz	Band MHz	National Provisions	Government Allocation	Non-Government Allocation	Remarks
2300-2450 FIXED Amateur Mobile Radiolocation	2300-2450 FIXED MOBILE RADIOLOCATION Amateur		2300-2310	US253	RADIOLOCATION Fixed Mobile G2	Amateur	
			2310-2390		RADIOLOCATION MOBILE Fixed G2	MOBILE	
664 752	664 751 752		2390-2450	664 752	RADIOLOCATION G2	Amateur	(ISM 2450 ± 50 MHz)

664—In the bands 435-438 MHz, 1260-1270 MHz, 2400-2450 MHz, 3400-3410 MHz (in Regions 2 and 3 only) and 5650-5670 MHz, the amateur-satellite service may operate subject to not causing harmful interference to other services operating in accordance with the Table (see No. 435). Administrations authorizing such use shall ensure that any harmful interference caused by emissions from a station in the amateur-satellite service is immediately eliminated in accordance with the provisions of No. 2741. The use of the bands 1260-1270 MHz and 5650-5670 MHz by the amateur-satellite service is limited to the Earth-to-space direction.

751—In Australia, the United States and Papua New Guinea, the use of the band 2310-2390 MHz by the aeronautical mobile service for telemetry has priority over other uses by the mobile services.

752—The band 2400-2500 MHz (center frequency 2450 MHz) is designated for industrial, scientific and medical (ISM) applications. Radio services operating within this band must accept harmful interference which may be caused by these applications. ISM equipment operating in this band is subject to the provisions of No. 1815.

G2—In the bands 216-225, 420-450 (except as provided by US217), 890-902, 928-942, 1300-1400, 2300-2450, 2700-2900, 5650-5925, and 9000-9200 MHz, the Government radiolocation is limited to the military services.

| License Class | Terrestrial location of the amateur radio station | | | |
	Region 1 MHz	Region 2 MHz	Region 3 MHz	Limitations
Technician General Advanced Extra	2300-2310 2390-2450	2300-2310 2390-2450	2300-2310 2390-2450	3,5,12,13 3,5,13,14

Region 1 MHz	Region 2 MHz	Region 3 MHz	ARRL MHz
			2300-2310
			2304.1 Calling frequency
			2310-2390 (Not authorized)
			2390-2450

§97.7(g)
Limitations. . .

(3) Where, in adjacent regions or subregions, a band of frequencies is allocated to different services of the same category, the basic principle is the equality of right to operate. Accordingly, the stations of each service in one region or subregion must operate so as not to cause harmful interference to services in the other regions or subregions. (See International Telecommunication Union Radio Regulations, RR 346 (Geneva, 1979).)

(5) Amateur stations in the 1900-2000 kHz, 220-225 MHz, 420-250 MHz, 902-928 MHz, 1240-1300 MHz, 2300-2310 MHz, 2390-2450 MHz, 3.3-3.5 GHz, 5.650-5.925 GHz, 10.0-10.5 GHz, 24.05-24.25 GHz, 76-81 GHz, 144-149 GHz and 241-248 GHz bands must not cause harmful interference to stations in the Government radiolocation service and are not protected from interference due to the operation of stations in the Government radiolocation service.

(12) In the United States, the 2300-2310 MHz band is allocated to the Amateur service on a co-secondary basis with the Government fixed and mobile services. In this band, the fixed and mobile services must not cause harmful interference to the Amateur service.

(13) In the 2300-2310 MHz and 2390-2450 MHz bands, the Amateur service is allocated on a secondary basis in all ITU Regions. In ITU Region 1, stations in the Amateur service must not cause harmful interference to stations authorized by other nations in the fixed service, and are not protected from interference due to the operation of stations authorized by other nations in the fixed service. In ITU Regions 2 and 3, stations in the Amateur service must not cause harmful interference to stations authorized by other nations in the fixed, mobile and radiolocation services, and are not protected from interference due to the operation of stations authorized by other nations in the fixed, mobile and radiolocation services.

(14) Amateur stations in the 2400-2450 MHz band are not protected from interference due to the operation of industrial, scientific and medical devices on 2450 MHz.

INTERNATIONAL			UNITED STATES				
Region 1 MHz	Region 2 MHz	Region 3 MHz	Band MHz	National Provisions	Government Allocation	Non-Government Allocation	Remarks
3300-3400 RADIOLOCATION 778 779 780	330-3400 RADIOLOCATION Amateur Fixed Mobile 778 780	3300-3400 RADIOLOCATION Amateur 778 779	3300-3500	US108 664 778	RADIOLOCATION	Amateur Radiolocation	
3400-3600 FIXED FIXED-SATELLITE (Space-to-Earth) Mobile Radiolocation 781 782 785	3400-3500 FIXED FIXED-SATELLITE (Space-to-Earth) Amateur Mobile Radiolocation 784 664 783				 G31		

664—In the bands 435-438 MHz, 1260-1270 MHz, 2400-2450 MHz, 3400-3410 MHz (in Regions 2 and 3 only) and 5650-5670 MHz, the amateur-satellite service may operate subject to not causing harmful interference to other services operating in accordance with the Table (see No. 435). Administrations authorizing such use shall ensure that any harmful interference caused by emissions from a station in the amateur-satellite service is immediately eliminated in accordance with the provisions of No. 2741. The use of the bands 1260-1270 MHz and 5650-5670 MHz by the amateur-satellite service is limited to the Earth-to-space direction.

778—In making assignments to stations of other services, administrations are urged to take all practicable steps to protect the spectral line observations of the radio astronomy service from harmful interference in the bands 3260-3267 MHz, 3332-3339 MHz, 3345.8-3352.5 MHz and 4825-4835 MHz. Emissions from space or airborne stations can be particularly serious sources of interference to the radio astronomy service (see Nos. 343 and 344 and Article 36).

779—Additional allocation: in Afghanistan, Saudi Arabia, Bahrain, Bangladesh, China, the Congo, the United Arab Emirates, India, Indonesia, Iran, Iraq, Israel, Japan, Kuwait, Lebanon, Libya, Malaysia, Oman, Pakistan, Qatar, Syria, Singapore, Sri Lanka and Thailand, the band 3300-3400 MHz is also allocated to the fixed and mobile services on a primary basis. The countries bordering the Mediterranean shall not claim protection for their fixed and mobile services from the radiolocation service.

780—Additional allocation: in Bulgaria, Cuba, Hungary, Mongolia, Poland, the German Democratic Republic, Romania, Czechoslovakia and the U.S.S.R., the band 3300-3400 MHz is also allocated to the radionavigation service on a primary basis.

781—Additional allocation: in the Federal Republic of Germany, Israel, Nigeria and the United Kingdom, the band 3400-3475 MHz is also allocated to the amateur service on a secondary basis.

782—Different category of service: in Austria, the allocation of the band 3400-3500 MHz to the radiolocation service is on a primary basis (see No. 425), subject to the agreement of the Administrations of the following countries: Hungary, Italy, the German Democratic Republic, Czechoslovakia and Yugoslavia. Such use is limited to ground-based stations. However, this Administration is urged to cease operations by 1985. After this date, this Administration shall take all practicable steps to protect the fixed-satellite service and coordination requirements shall not be imposed on the fixed-satellite service.

783—Different category of service: in Indonesia, Japan, Pakistan and Thailand, the allocation of the band 3400-3500 MHz to the mobile, except aeronautical mobile, service is on a primary basis (see No. 425).

784—In Regions 2 and 3, in the band 3400-3600 MHz, the radiolocation service is allocated on a primary basis. However, all administrations operating radiolocation systems in this band are urged to cease operations by 1985. Thereafter, administrations shall take all practicable steps to protect the fixed-satellite service and coordination requirements shall not be imposed on the fixed-satellite service.

785—In Denmark, Norway and the United Kingdom, the fixed, radiolocation and fixed-satellite services operate on a basis of equality of rights in the band 3400-3600 MHz. However, these Administrations operating radiolocation systems in this band are urged to cease operations by 1985. After this date, these Administrations shall take all practicable steps to protect the fixed-satellite service and coordination requirements shall not be imposed on the fixed-satellite service.

US108—Within the bands 3300-3500 MHz and 10000-10500 MHz, survey operations, using transmitters with a peak power not to exceed five watts into the antenna, may be authorized for Government and non-Government use on a secondary basis to other Government radiolocations operations.

G31—In the bands 3300-3500 MHz, the Government radiolocation is limited to the military services, except as provided by footnote US108.

| License Class | Terrestrial location of the amateur radio station | | | |
	Region 1 MHz	Region 2 MHz	Region 3 MHz	Limitations
Technician General Advanced Extra	—	3300-3500	3300-3500	3,5,15,16,17

§97.7(g)
Limitations. . .

(3) Where, in adjacent regions or subregions, a band of frequencies is allocated to different services of the same category, the basic principle is the equality of right to operate. Accordingly, the stations of each service in one region or subregion must operate so as not to cause harmful interference to services in the other regions or subregions. (See International Telecommunication Union Radio Regulations, RR 346 (Geneva, 1979).)

(5) Amateur stations in the 1900-2000 kHz, 220-225 MHz, 420-250 MHz, 902-928 MHz, 1240-1300 MHz, 2300-2310 MHz, 2390-2450 MHz, 3.3-3.5 GHz, 5.650-5.925 GHz, 10.0-10.5 GHz, 24.05-24.25 GHz, 76-81 GHz, 144-149 GHz and 241-248 GHz bands must not cause harmful interference to stations in the Government radiolocation service and are not protected from interference due to the operation of stations in the Government radiolocation service.

(15) Amateur stations in the 3.332-3.339 GHz, 3.3458-3.3525 GHz, 119.98-120.02 GHz, 144.68-144.98 GHz, 145.45-145.75 GHz, 146.82-147.12 GHz and 343-348 GHz bands must not cause harmful interference to stations in the radio astronomy service. Amateur stations in the 300-302 GHz, 324-326 GHz, 345-347 GHz, 363-365 GHz and 379-381 GHz bands must not cause harmful interference to stations in the space research service (passive) or Earth exploration-satellite service (passive).

(16) In both ITU Regions 2 and 3 the 3.3-3.5 GHz band is allocated to the Amateur service on a secondary basis. In the 3.3-3.4 GHz band amateur stations must not cause harmful interference to stations authorized by other nations in the radiolocation service, and are not protected from interference due to the operation of stations authorized by other nations in the radiolocation service. In the 3.4-3.5 GHz band amateur stations must not cause harmful interference to stations authorized by other nations in the fixed and fixed-satellite services, and are not protected from interference due to the operation of stations authorized by other nations in the fixed and fixed-satellite services.

(17) In the United States the 3.3-3.5 GHz band is allocated to the amateur service on a co-secondary basis with the non-government radiolocation service.

INTERNATIONAL			UNITED STATES				
Region 1 MHz	Region 2 MHz	Region 3 MHz	Band MHz	National Provisions	Government Allocation	Non-Government Allocation	Remarks
5650-5725	RADIOLOCATION Amateur Space Research (Deep space) 664 801 803 804 805		5650-5850	664 806 808	RADIOLOCATION	Amateur	(ISM 5800 ± 75 MHz)
5725-5850 FIXED-SATELLITE (Earth-to-space) RADIOLOCATION Amateur 801 803 805 806 807 808	5725-5850 RADIOLOCATION Amateur 803 805 806 808				G2		
5850-5925 FIXED FIXED-SATELLITE (Earth-to-space) MOBILE 806	5850-5925 FIXED FIXED-SATELLITE (Earth-to-space) MOBILE Amateur Radiolocation 806	5850-5925 FIXED FIXED-SATELLITE (Earth-to-space) MOBILE Radiolocation 806	5850-5925	US245 806	RADIOLOCATION G2	Amateur FIXED-SATELLITE (Earth-to-space)	

664—In the bands 435-438 MHz, 1260-1270 MHz, 2400-2450 MHz, 3400-3410 MHz (in Regions 2 and 3 only) and 5650-5670 MHz, the amateur-satellite service may operate subject to not causing harmful interference to other services operating in accordance with the Table (see No. 435). Administrations authorizing such use shall ensure that any harmful interference caused by emissions from a station in the amateur-satellite service is immediately eliminated in accordance with the provisions of No. 2741. The use of the bands 1260-1270 MHz and 5650-5670 MHz by the amateur-satellite service is limited to the Earth-to-space direction.

801—Additional allocation: in the United Kingdom, the band 5470-5850 MHz is also allocated to the land mobile service on a secondary basis. The power limits specified in Nos. 2502, 2505, 2506 and 2507 shall apply in the band 5725-5850 MHz.

802—Between 5600 MHz and 5650 MHz, ground-based radars used for meteorological purposes are authorized to operate on a basis of equality with stations of the maritime radionavigation service.

803—Additional allocation: in Afghanistan, Saudi Arabia, Bahrain, Bangladesh, Cameroon, the Central African Republic, China, the Congo, the Republic of Korea, Egypt, the United Arab Emirates, Gabon, Guinea, India, Indonesia, Iran, Iraq, Israel, Japan, Jordan, Kuwait, Lebanon, Libya, Madagascar, Malaysia, Malawi, Malta, Niger, Nigeria, Pakistan, the Philippines, Qatar, Syria, Singapore, Sri Lanka, Tanzania, Chad, Thailand and Yemen (P.D.R. of), the band 5650-5850 MHz is also allocated to the fixed and mobile services on a primary basis.

804—Different category of service: in Bulgaria, Cuba, Hungary, Mongolia, Poland, the German Democratic Republic, Czechoslovakia and the U.S.S.R., the allocation of the band 5670-5725 MHz to the space research service is on a primary basis (see No. 425).

805—Additional allocation: in Bulgaria, Cuba, Hungary, Mongolia, Poland, the German Democratic Republic, Czechoslovakia and the U.S.S.R., the band 5670-5850 MHz is also allocated to the fixed service on a primary basis.

806—The band 5725-5875 MHz (centre frequency 5800 MHz) is designated for industrial, scientific and medical (ISM) applications. Radiocommunication services operating within this band must accept harmful interference which may be caused by these applications. ISM equipment operating in this band is subject to the provisions of No. 1815.

807—Additional allocation: in the Federal Republic of Germany and in Cameroon, the band 5755-5850 MHz is also allocated to the fixed service on a primary basis.

808—The band 5830-5850 MHz is also allocated in the amateur-satellite service (space-to-Earth) on a secondary basis.

US245—The Fixed-Satellite Service is limited to International inter-Continental systems and subject to case-by-case electromagnetic compatibility analysis.

G2—In the bands 216-225, 420-450 (except as provided by US217), 890-902, 928-942, 1300-1400, 2300-2450, 2700-2900, 5650-5925, and 9000-9200 MHz, the Government radiolocation is limited to the military services.

FCC Part 97 Privileges

License Class	Terrestrial location of the amateur radio station			
	Region 1 MHz	Region 2 MHz	Region 3 MHz	Limitations
Technician General Advanced Extra	5650-5850	5650-5850	5650-5850	3,5,18,19,20

§97.7(g)
Limitations. . .

(3) Where, in adjacent regions or subregions, a band of frequencies is allocated to different services of the same category, the basic principle is the equality of right to operate. Accordingly, the stations of each service in one region or subregion must operate so as not to cause harmful interference to services in the other regions or subregions. (See International Telecommunication Union Radio Regulations, RR 346 (Geneva, 1979).)

(5) Amateur stations in the 1900-2000 kHz, 220-225 MHz, 420-450 MHz, 902-928 MHz, 1240-1300 MHz, 2300-2310 MHz, 2390-2450 MHz, 3.3-3.5 GHz, 5.650-5.925 GHz, 10.0-10.5 GHz, 24.05-24.25 GHz, 76-81 GHz, 144-149 GHz and 241-248 GHz bands must not cause harmful interference to stations in the Government radiolocation service and are not protected from interference due to the operation of stations in the Government radiolocation service.

(18) In the 5.650-5.725 GHz band, the Amateur service is allocated in all ITU regions on a co-secondary basis with the space research (deep space) service. In the 5.725-5.850 GHz band the Amateur service is allocated in all ITU regions on a secondary basis. In the 5.650-5.850 GHz band amateur stations must not cause harmful interference to stations authorized by other nations in the radiolocation service, and are not protected from interference due to the operation of stations authorized by other nations in the radiolocation service. In the 5.725-5.850 GHz band amateur stations must not cause harmful interference to stations authorized by other nations in the fixed-satellite service in ITU Region 1, and are not protected from interference due to the operation of stations authorized by other nations in the fixed-satellite service in ITU Region 1. In the 5.850-5.925 GHz band the Amateur service is allocated in ITU Region 2 on a co-secondary basis with the radiolocation service. In the 5.850-5.925 GHz band amateur stations must not cause harmful interference to stations authorized by other nations in the fixed, fixed-satellite and mobile services, and are not protected from interference due to the operation of stations authorized by other nations in the fixed, fixed-satellite and mobile services.

(19) In the United States, the 5.850-5.925 GHz band is allocated to the Amateur service on a secondary basis to the non-government fixed-satellite service. In the 5.850-5.925 GHz band amateur stations must not cause harmful interference to stations in the non-government fixed-satellite service and are not protected from interference due to the operation of stations in the non-government fixed-satellite service.

(20) Amateur stations in the 5.725-5.875 GHz band are not protected from interference due to the operation of industrial, scientific and medical devices on 5.8 GHz.

INTERNATIONAL			UNITED STATES				
Region 1 *kHz*	Region 2 *kHz*	Region 3 *kHz*	Band *kHz*	National Provisions	Government Allocation	Non-Government Allocation	Remarks
10-10.45 FIXED MOBILE RADIOLOCATION Amateur	10-10.45 RADIOLOCATION Amateur	10-10.45 FIXED MOBILE RADIOLOCATION Amateur	10-10.45	US58 US108 828	RADIOLOCATION	Amateur Radiolocation	
828	828 829	828			G32	NG42	
10.45-10.5	RADIOLOCATION Amateur Amateur-Satellite		10.45-10.5	US58 US108	RADIOLOCATION	RADIOLOCATION Amateur Amateur-Satellite	
	830				G32	NG42 NG134	

828—The band 9975-10025 MHz is also allocated to the meteorological-satellite service on a secondary basis for use by weather radars.

829—Additional allocation: in Costa Rica, Ecuador, Guatemala and Honduras, the band 10-10.45 GHz is also allocated to the fixed and mobile services on a primary basis.

830—Additional allocation: in the Federal Republic of Germany, Angola, China, Ecuador, Spain, Japan, Kenya, Morocco, Nigeria, Sweden, Tanzania and Thailand, the band 10.45-10.5 GHz is also allocated to the fixed and mobile services on a primary basis.

US58—In the band 10000-10500 MHz, pulsed emissions are prohibited, except for weather radars on board meteorological satellites in the band 10000-10025 MHz. The amateur service and the non-Government radiolocation service, which shall not cause harmful interference to the Government radiolocation service, are the only non-Government services permitted in this band. The non-Government radiolocation service is limited to survey operations as specified in footnote US108.

US108—Within the bands 3300-3500 MHz and 10000-10500 MHz, survey operations, using transmitters with a peak power not to exceed five watts into the antenna, may be authorized for Government and non-Government use on a secondary basis to other Government radiolocations operations.

G32—Except for weather radars on meteorological satellites in the band 9975-10025 MHz and for Government survey operations (see footnote US108), Government radiolocation in the band 10000-10500 MHz is limited to the military services.

NG42—Non-Government stations in the radiolocation service shall not cause harmful interference to the amateur service.

NG134—In the band 10.45 GHz non-Government stations in the radiolocation service shall not cause harmful interference to the amateur and amateur-satellite services.

| License Class | Terrestrial location of the amateur radio station | | | |
	Region 1 GHz	Region 2 GHz	Region 3 GHz	Limitations
Technician General Advanced Extra	10.0-10.5	10.0-10.5	10.0-10.5	5,21,22,23,31

Region 1 GHz	Region 2 GHz	Region 3 GHz	ARRL GHz
			10.00-10.5
			10.364* Calling frequency

*Adopted by ARRL Board of Directors January 1987.

§97.7(g)
Limitations. . .

(5) Amateur stations in the 1900-2000 kHz, 220-225 MHz, 420-450 MHz, 902-928 MHz, 1240-1300 MHz, 2300-2310 MHz, 2390-2450 MHz, 3.3-3.5 GHz, 5.650-5.925 GHz, 10.0-10.5 GHz, 24.05-24.25 GHz, 76-81 GHz, 144-149 GHz and 241-248 GHz bands must not cause harmful interference to stations in the Government radiolocation service and are not protected from interference due to the operation of stations in the Government radiolocation service.

(21) Amateur stations in the 1900-2000 kHz, 10.45-10.50 GHz, 76-81 GHz, 144-149 GHz and 241-248 GHz bands must not cause harmful interference to stations in the non-government radiolocation service and are not protected from interference due to the operation of stations in the non-government radiolocation service.

(22) Amateur stations in the 1240-1300 MHz, 10.0-10.5 GHz, 24.05-24.25 GHz, 76-81 GHz, 144-149 GHz and 241-248 GHz bands must not cause harmful interference to stations authorized by other nations in the radiolocation service and are not protected from interference due to the operation of stations authorized by other nations in the radiolocation service.

(23) In the 10.00-10.45 GHz band in ITU Regions 1 and 3 amateur stations must not cause harmful interference to stations authorized by other nations in the fixed and mobile services, and are not protected from interference due to the operation of stations authorized by other nations in the fixed and mobile services.

(31) In the United States, the 10.0-10.5 GHz band is allocated to the Amateur service on a co-secondary basis with the non-government radiolocation service.

INTERNATIONAL			UNITED STATES				
Region 1 GHz	Region 2 GHz	Region 3 GHz	Band GHz	National Provisions	Government Allocation	Non-Government Allocation	Remarks
24-24.05	AMATEUR AMATEUR-SATELLITE 881		24-24.05	US211 881		AMATEUR AMATEUR-SATELLITE	
24.05-24.25	RADIOLOCATION Amateur Earth Exploration-Satellite (Active) 881		24.05-24.25	US110 881	RADIOLOCATION Earth Exploration Satellite (Active) G59	Amateur Radiolocation Earth Exploration Satellite (Active)	(ISM 24.125 ± 125 MHz)

881—The band 24-24.25 GHz (centre frequency 24.125 GHz) is designated for industrial, scientific and medical (ISM) applications. Radiocommunication services operating within this band must accept harmful interference which may be caused by these applications. ISM equipment operating in this band is subject to the provisions of No. 1815.

US211—In the bands 1670-1690, 5000-5250 MHz, and 10.7-11.7, 15.1365-15.35, 15.4-15.7, 22.5-22.55, 24-24.05, 31.0-31.3, 31.8-32, 40.5-42.5, 84-86, 102-105, 116-126, 151-164, 176.5-182, 185-190, 231-235, 252-265 GHz, applicants for airborne or space station assignments are urged to take all practicable steps to protect radio astronomy observations in the adjacent bands from harmful interference; however, US74 applies.

G59—In the bands 902-928 MHz, 3100-3300 MHz, 3500-3700 MHz, 5250-5350 MHz, 8500-9000 MHz, 9200-9300 MHz, 13.4-14.0 GHz, 15.7-17.7 GHz and 24.05-24.25 GHz, all Government non-military radiolocation shall be secondary to military radiolocation, except in the sub-band 15.7-16.2 GHz airport surface detection equipment (ASDE) is permitted on a co-equal basis subject to coordination with the military departments.

FCC Part 97 Privileges

License Class	Terrestrial location of the amateur radio station			
	Region 1 GHz	Region 2 GHz	Region 3 GHz	Limitations
Technician General Advanced Extra	24.00-24.25	24.00-24.25	24.00-24.25	3,5,22,24,26

§97.7(g)
Limitations. . .

(3) Where, in adjacent regions or subregions, a band of frequencies is allocated to different services of the same category, the basic principle is the equality of right to operate. Accordingly, the stations of each service in one region or subregion must operate so as not to cause harmful interference to services in the other regions or subregions. (See International Telecommunication Union Radio Regulations, RR 346 (Geneva, 1979).)

(5) Amateur stations in the 1900-2000 kHz, 220-225 MHz, 420-450 MHz, 902-928 MHz, 1240-1300 MHz, 2300-2310 MHz, 2390-2450 MHz, 3.3-3.5 GHz, 5.650-5.925 GHz, 10.0-10.5 GHz, 24.05-24.25 GHz, 76-81 GHz, 144-149 GHz and 241-248 GHz bands must not cause harmful interference to stations in the Government radiolocation service and are not protected from interference due to the operation of stations in the Government radiolocation service.

(22) Amateur stations in the 1240-1300 MHz, 10.0-10.5 GHz, 24.05-24.25 GHz, 76-81 GHz, 144-149 GHz and 241-248 GHz bands must not cause harmful interference to stations authorized by other nations in the radiolocation service and are not protected from interference due to the operation of stations authorized by other nations in the radiolocation service.

(24) In the United States, the 24.05-24.25 GHz band is allocated to the Amateur service on a co-secondary basis with the non-government radiolocation and Government and non-government Earth exploration-satellite (active) services.

(26) Amateur stations in the 24.00-24.25 GHz band are not protected from interference due to the operation of industrial, scientific and medical devices on 24.125 GHz.

INTERNATIONAL			UNITED STATES				
Region 1 GHz	Region 2 GHz	Region 3 GHz	Band GHz	National Provisions	Government Allocation	Non-Government Allocation	Remarks
47-47.2	AMATEUR AMATEUR-SATELLITE		47-47.2			AMATEUR AMATEUR-SATELLITE	

FCC Part 97 Privileges				
	Terrestrial location of the amateur radio station			
License Class	Region 1 GHz	Region 2 GHz	Region 3 GHz	Limitations
Technician General Advanced Extra	47.0-47.2	47.0-47.2	47.0-47.2	

INTERNATIONAL			UNITED STATES				
Region 1 GHz	Region 2 GHz	Region 3 GHz	Band GHz	National Provisions	Government Allocation	Non-Government Allocation	Remarks
75.5-76	AMATEUR AMATEUR-SATELLITE		75.5-76			AMATEUR AMATEUR-SATELLITE	
76-81	RADIOLOCATION Amateur Amateur-Satellite 912		76-81	912	RADIOLOCATION	RADIOLOCATION Amateur Amateur-Satellite	

912—In the band 78-79 GHz radars located on space stations may be operated on a primary basis in the Earth exploration-satellite service and in the space research service.

FCC Part 97 Privileges

| License Class | Terrestrial location of the amateur radio station | | | |
	Region 1 GHz	Region 2 GHz	Region 3 GHz	Limitations
Technician General Advanced Extra	75.5-81	75.5-81	75.5-81	5,21,22

§97.7(g)
Limitations. . .

(5) Amateur stations in the 1900-2000 kHz, 220-225 MHz, 420-450 MHz, 902-928 MHz, 1240-1300 MHz, 2300-2310 MHz, 2390-2450 MHz, 3.3-3.5 GHz, 5.650-5.925 GHz, 10.0-10.5 GHz, 24.05-24.25 GHz, 76-81 GHz, 144-149 GHz and 241-248 GHz bands must not cause harmful interference to stations in the Government radiolocation service and are not protected from interference due to the operation of stations in the Government radiolocation service.

(21) Amateur stations in the 1900-2000 kHz, 10.45-10.50 GHz, 76-81 GHz, 144-149 GHz and 241-248 GHz bands must not cause harmful interference to stations in the non-government radiolocation service and are not protected from interference due to the operation of stations in the non-government radiolocation service.

(22) Amateur stations in the 1240-1300 MHz, 10.0-10.5 GHz, 24.05-24.25 GHz, 76-81 GHz, 144-149 GHz and 241-248 GHz bands must not cause harmful interference to stations authorized by other nations in the radiolocation service and are not protected from interference due to the operation of stations authorized by other nations in the radiolocation service.

INTERNATIONAL			UNITED STATES				
Region 1 GHz	Region 2 GHz	Region 3 GHz	Band GHz	National Provisions	Government Allocation	Non-Government Allocation	Remarks
116-126	EARTH EXPLORATION-SATELLITE (Passive) FIXED INTER-SATELLITE MOBILE 909 SPACE RESEARCH (Passive) 722 915 916		116-126	US211 US263 722 909 915 916	FIXED INTER-SATELLITE MOBILE EARTH EXPLORATION SATELLITE (Passive) SPACE REA SPACE RESEARCH (Passive)	FIXED MOBILE INTER-SATELLITE EARTH EXPLORATION-SATELLITE (Passive) SPACE RESEARCH (Passive)	(ISM 122.5 GHz ± 500 MHz)

722—In the bands 1400-1727 MHz, 101-120 GHz and 197-220 GHz, passive research is being conducted by some countries in a program for the search for intentional emissions of extra-terrestrial origin.

909—In the bands 54.25-58.2 GHz, 59-64 GHz, 116-134 GHz, 170-182 GHz and 185-190 GHz, stations in the aeronautical mobile service may be operated subject to not causing harmful interference to the inter-satellite service (see No. 435).

915—The band 119.98-120.02 GHz is also allocated to the amateur service on a secondary basis.

916—The band 122-123 GHz (centre frequency 122.5 GHz) is designated for industrial, scientific and medical (ISM) applications. The use of this frequency band for ISM applications shall be subject to special authorization by the administration concerned in agreement with other administrations whose radiocommunication services might be affected. In applying this provision administrations shall have due regard to the latest relevant CCIR Recommendations.

US211—In the bands 1670-1690, 5000-5250 MHz, and 10.7-11.7, 15.1365-15.35, 15.4-15.7, 22.5-22.55, 24-24.05, 31.0-31.3, 31.8-32, 40.5-42.5, 84-86, 102-105, 116-126, 151-164, 176.5-182, 185-190, 231-235, 252-265 GHz, applicants for airborn or space station assignments are urged to take all practicable steps to protect radio astronomy observations in the adjacent bands from harmful interference; however, US74 applies.

US263—In the frequency bands 21.2-21.4, 22.21-22.5, 36-37, 50.2-50.4, 54.25-58.2, 116-126, 150-151, 174.5-176.5, 200-202 and 235-238 GHz, the Space Research and the Earth Exploration-Satellite Services shall not receive protection from the Fixed and Mobile Services operating in accordance with the Table of Frequency Allocations.

FCC Part 97 Privileges				
	Terrestrial location of the amateur radio station			
License Class	*Region 1 GHz*	*Region 2 GHz*	*Region 3 GHz*	*Limitations*
Technician General Advanced Extra	119.98-120.02	119.98-120.02	119.98-120.02	15,25

§97.7(g)
Limitations. . .

(15) Amateur stations in the 3.332-3.339 GHz, 3.3458-3.3525 GHz, 119.98-120.02 GHz, 144.68-144.98 GHz, 145.45-145.75 GHz, 146.82-147.12 GHz and 343-348 GHz bands must not cause harmful interference to stations in the radio astronomy service. Amateur stations in the 300-302 GHz, 324-326 GHz, 345-347 GHz, 363-365 GHz and 379-381 GHz bands must not cause harmful interference to stations in the space research service (passive) or Earth exploration-satellite service (passive).

(25) The 119.98-120.02 GHz band is allocated to the Amateur service on a secondary basis. Amateur stations in this band must not cause harmful interference to stations operating in the fixed, inter-satellite and mobile services, and are not protected from interference caused by the operation of stations in the fixed, inter-satellite and mobile services.

INTERNATIONAL			UNITED STATES				
Region 1 GHz	Region 2 GHz	Region 3 GHz	Band GHz	National Provisions	Government Allocation	Non-Government Allocation	Remarks
142-144	AMATEUR AMATEUR-SATELLITE		142-144			AMATEUR AMATEUR-SATELLITE	
144-149	RADIOLOCATION Amateur Amateur-Satellite 918		144-149	918	RADIOLOCATION	RADIOLOCATION Amateur Amateur-Satellite	

918—The bands 140.69-140.98 GHz, 144.68-144.98 GHz, 145.45-145.75 GHz and 146.82-147.12 GHz are also allocated to the radio astronomy service on a primary basis for spectral line observations. In making assignments to stations of other services to which the bands are allocated, administrations are urged to take all practicable steps to protect the radio astronomy service from harmful interference. Emissions from space or airborne stations can be particularly serious sources of interference to the radio astronomy service (see Nos. 343 and 344 and Article 36).

FCC Part 97 Privileges

| License Class | Terrestrial location of the amateur radio station | | | |
	Region 1 GHz	Region 2 GHz	Region 3 GHz	Limitations
Technician General Advanced Extra	142-149	142-149	142-149	5,15,21,22

§97.7(g)
Limitations. . .

(5) Amateur stations in the 1900-2000 kHz, 220-225 MHz, 420-450 MHz, 902-928 MHz, 1240-1300 MHz, 2300-2310 MHz, 2390-2450 MHz, 3.3-3.5 GHz, 5.650-5.925 GHz, 10.0-10.5 GHz, 24.05-24.25 GHz, 76-81 GHz, 144-149 GHz and 241-248 GHz bands must not cause harmful interference to stations in the Government radiolocation service and are not protected from interference due to the operation of stations in the Government radiolocation service.

(15) Amateur stations in the 3.332-3.339 GHz, 3.3458-3.3525 GHz, 119.98-120.02 GHz, 144.68-144.98 GHz, 145.45-145.75 GHz, 146.82-147.12 GHz and 343-348 GHz bands must not cause harmful interference to stations in the radio astronomy service. Amateur stations in the 300-302 GHz, 324-326 GHz, 345-347 GHz, 363-365 GHz and 379-381 GHz bands must not cause harmful interference to stations in the space research service (passive) or Earth exploration-satellite service (passive).

(21) Amateur stations in the 1900-2000 kHz, 10.45-10.50 GHz, 76-81 GHz, 144-149 GHz and 241-248 GHz bands must not cause harmful interference to stations in the non-government radiolocation service and are not protected from interference due to the operation of stations in the non-government radiolocation service.

(22) Amateur stations in the 1240-1300 MHz, 10.0-10.5 GHz, 24.05-24.25 GHz, 76-81 GHz, 144-149 GHz and 241-248 GHz bands must not cause harmful interference to stations authorized by other nations in the radiolocation service and are not protected from interference due to the operation of stations authorized by other nations in the radiolocation service.

INTERNATIONAL			UNITED STATES				
Region 1 GHz	Region 2 GHz	Region 3 GHz	Band GHz	National Provisions	Government Allocation	Non-Government Allocation	Remarks
241-248	RADIOLOCATION Amateur Amateur-Satellite 922		241-248	922	RADIOLOCATION	RADIOLOCATION Amateur Amateur- Satellite	(ISM 245 GHz ± 1 GHz)
248-250	AMATEUR AMATEUR-SATELLITE		248-250			AMATEUR AMATEUR-SATELLITE	

922—The band 244-246 GHz (centre frequency 245 GHz) is designated for industrial, scientific and medical (ISM) applications. The use of this frequency band for ISM applications shall be subject to special authorization by the administration concerned in agreement with other administrations whose radiocommunication services might be affected. In applying this provision administrations shall have due regard to the latest relevant CCIR Recommendations.

License Class	Terrestrial location of the amateur radio station			
	Region 1 GHz	Region 2 GHz	Region 3 GHz	Limitations
Technician General Advanced Extra	241-250	241-250	241-250	5,21,22,27

§97.7(g)
Limitations. . .

(5) Amateur stations in the 1900-2000 kHz, 220-225 MHz, 420-450 MHz, 902-928 MHz, 1240-1300 MHz, 2300-2310 MHz, 2390-2450 MHz, 3.3-3.5 GHz, 5.650-5.925 GHz, 10.0-10.5 GHz, 24.05-24.25 GHz, 76-81 GHz, 144-149 GHz and 241-248 GHz bands must not cause harmful interference to stations in the Government radiolocation service and are not protected from interference due to the operation of stations in the Government radiolocation service.

(21) Amateur stations in the 1900-2000 kHz, 10.45-10.50 GHz, 76-81 GHz, 144-149 GHz and 241-248 GHz bands must not cause harmful interference to stations in the non-government radiolocation service and are not protected from interference due to the operation of stations in the non-government radiolocation service.

(22) Amateur stations in the 1240-1300 MHz, 10.0-10.5 GHz, 24.05-24.25 GHz, 76-81 GHz, 144-149 GHz and 241-248 GHz bands must not cause harmful interference to stations authorized by other nations in the radiolocation service and are not protected from interference due to the operation of stations authorized by other nations in the radiolocation service.

(27) Amateur stations in the 244-246 GHz band are not protected from interference due to the operation of industrial, scientific and medical devices on 245 GHz.

Where, in adjacent Regions or sub-Regions, a band of frequencies is allocated to different services of the same category... the basic principle is the equality of right to operate. Accordingly, the stations of each service in one Region or sub-Region must operate so as not to cause harmful interference to services in the other Regions or sub-Regions.

ITU Resolution No. 640

Relating to the International Use of Radiocommunications, in the Event of Natural Disasters, in Frequency Bands Allocated to the Amateur Service

The World Administrative Radio Conference, Geneva, 1979.

considering

a) that in the event of natural disaster normal communication systems are frequently overloaded, damaged, or completely disrupted;

b) that rapid establishment of communication is essential to facilitate worldwide relief actions;

c) that the amateur bands are not bound by international plans or notification procedures, and are therefore well adapted for short-term use in emergency cases;

d) that international disaster communications would be facilitated by temporary use of certain frequency bands allocated to the amateur service;

e) that under those circumstances the stations of the amateur service, because of their widespread distribution and their demonstrated capacity in such cases, can assist in meeting essential communication needs;

f) the existence of national and regional amateur emergency networks using frequencies throughout the bands allocated to the amateur service;

g) that, in the event of a natural disaster, direct communication between amateur stations and other stations might enable vital communications to be carried out until normal communications are restored;

recognizing

that the rights and responsibilities for communications in the event of a natural disaster rest with the administration involved;

resolves

1. that the bands allocated to the amateur service which are specified in No. 510 may be used by administrations to meet the needs of international disaster communications;

2. that such use of these bands shall be only for communications in relation to relief operations in connection with natural disasters;

3. that the use of specified bands allocated to the amateur service by non-amateur stations for disaster communications shall be limited to the duration of the emergency and to the specific geographical areas as defined by the responsible authority of the affected country;

4. the disaster communications shall take place within the disaster area and between the disaster area and the permanent headquarters of the organization providing relief;

5. that such communications shall be carried out only with the consent of the administration of the country in which the disaster has occurred;

6. that relief communications provided from outside the country in which the disaster has occurred shall not replace existing national or international amateur emergency networks;

7. that close cooperation is desirable between amateur stations and the stations of other radio services which may find it necessary to use amateur frequencies in disaster communications;

8. that such international relief communications shall avoid, as far as practicable, interference to the amateur service networks;

invites administrations

1. to provide for the needs of international disaster communications;

2. to provide for the needs of emergency communications within their national regulations.

Chapter 3

Basic Operating

Congratulations on obtaining your first license or on getting back into the exciting world of ham radio. A wide vista of opportunities awaits you. The only limitations are your imagination and your interest in different operating modes and techniques. Of course, ham radio does have a few basic rules and regulations that must be followed. But within these parameters, you will be able to enjoy one of the most fascinating hobbies in the world.

Think of your participation in Amateur Radio as a neverending journey—always something new to explore, always something new to do. This chapter will help you get started on your journey.

LET'S GET STARTED

You have your license; now it's time to use it. Getting on the air can seem to be an overwhelming task, and your head is probably spinning with all kinds of questions ranging from "What kind of antenna should I use?" to "How do I make my first contact?"

Every newcomer needs help. Even experienced "Old Timers" once had those seemingly embarrassing questions when they began. After all, everyone has to start somewhere. In this chapter, we will try to give you some building blocks to use as a foundation for success in ham radio.

Who Are Hams?

Because of the nature of the diversity of activities found within ham radio, almost everyone can find something to pique his/her interests. The 14-year-old high-school student can keep in contact with friends across town. A 70-year-old retired engineer can chase those last elusive countries that will put him on the DXCC (DX Century Club) Honor Roll. A Midwestern homemaker can show off her skills at contesting by racking up consistent QSO rates of 60 contacts per hour in the ARRL November Sweepstakes. You just never know who you might run into.

Amateur Radio is a Service!

The element that binds all these people together is ham radio, and the underlying basis for its existence is *service*. A combination of operator skills and a willingness by those operators to use those skills and whatever talents or equipment that may be necessary to serve the public in times of need is what this service is all about.

If you have read the Public Service column in *QST*, you no doubt have read of the exploits of hams from all walks of life who have selflessly donated their time and given of themselves to aid their neighbors with emergency communications when all conventional means of communications have broken down. You have probably seen a piece on the 6 o'clock

news about a local radio amateur who provided the only lines of communication to a remote area of the country when extraordinary circumstances (tornado, flood, hurricane, major fire) eliminated the usual commercial means of communication. Many, many more hams provide this service than are given recognition by the media, but that's all part of being a ham—the intrinsic reward is the satisfaction of doing a job well.

You may or may not be called upon to provide a service to your community. But by being prepared, honing your on-the-air operating skills to their sharpest, maximizing your station/equipment to obtain the best from it, and being prepared to pitch in should the need arise, you will be ready. And in so doing, you will derive untold hours of satisfaction from the exciting hobby we know as Amateur Radio.

Experience? Get On the Air

Actual on-the-air operating experience is the best teacher. In this section, we will try to give you enough of the basics to be able to make it through your first QSO. If nothing else, read the following material for additional reference notations. Learn by doing and don't be afraid to ask questions of someone who might be able to help. After all, we're in this hobby together, and assistance is only as far away as the closest ham.

How do you develop good operating habits? This *ARRL Operating Manual* is an excellent place to start. An entire chapter is devoted to each of the major Amateur Radio

Melissa, KA8WSQ, is on the air from Williamston, Michigan.

I Made Shep Laugh

Let me tell you about the night I spoke to Shep—humorist Jean Shepherd, K2ORS, that is. I remember clearly that warm, starry night, walking along Riverside Drive in Coral Springs, Florida. The weather was perfect for a relaxing walk with my 2-meter hand-held for company. The sky was clear, humidity low, while the temperature was around 70 degrees as the moon hovered over the horizon.

I called some friends on the 61 repeater, then clicked the digits over to the 82 machine in Boca Raton. Just then a police cruiser pulled over; the officer had a few questions about my hand-held rig. After a short demonstration of the repeater system, the officer was very impressed, thought the radio was kind of cute, that I was too...so I made a date with her!

The evening would get more interesting. I remember as I turned onto Cardinal Road and approached the bike path I heard "K2ORS/mobile 4 on frequency" crisp and clear. For a moment I dreamed about 2-land and about New Jersey where I once lived...about the Route 3 Drive-in, the Flagship on Route 22. Then I felt a chill. That was Shep calling—Jean Shepherd! The radio and TV personality, the screenwriter and the author of *In God We Trust, Wanda Hickey's Night of Golden Memories* and many others. Shep, it's you, it's you!

I called him. "K2ORS, WB4RXB, over." And he answered me! He told me his name was Jean, but I knew him more as Shep. But as I leaned on the bike-path sign, excited, I quickly realized that if I rambled on about how much I enjoyed his creative stories he would clear off the air fast. So I acted like a "dildock" (one of those guys Shep says is on the slow team), and after the usual exchange of information, I casually asked him what he did for a living.

Shep casually sidestepped my question (undoubtedly out of modesty) and asked if I was walking alone. I said yes and then got the nutty idea that I might try to make him laugh. So I carefully pressed the transmit bar on my radio and said, "You know, Jean, here's a comment on our society. I'm leaning on a bike-path sign that's been partially run over by a car!" I almost couldn't get the words out I was so nervous. But as I let go of the transmit button, there was Shep, laughing. I made Shep laugh! Ah. . . I felt great.

What an extraordinary hobby Amateur Radio is. Here I was ambulating down a quiet, surburban street, speaking with an award-winning contemporary humorist, with our conversation traveling at the speed of light. I also remember that he would pause every few transmissions to let anyone jump in to join the conversation or use the frequency. Amazingly, there was none. His courtesy was the mark of a good operator.

He was kind enough to listen to a humorous story about my first code contact. I explained that about 10 years ago I was a night-school instructor teaching basic electronics. One student, who was also a chemistry teacher in a nearby town, offered to teach me organic chemistry if I would help him work toward a Novice license. A friendship began, and soon Stan received his Novice ticket.

Stan decided to build his Novice station from a kit, primarily because he enjoyed the challenge of low power and Morse code. Personally, I didn't share his affection for CW; after I passed my General, I sort of placed the code on a mental shelf somewhere between Algebra and my

---

activities, everything from working DX to space communications—each chapter written by a ham with considerable experience in that area of hamdom. Experience, even through the words of others, is a pretty powerful teacher. Notice the table of contents; you'll be amazed at the diversity and amount of good solid reference material that you have at your fingertips.

Amateur Radio Publications

The American Radio Relay League (ARRL), publisher of this book, and the only nonprofit organization to serve the over 400,000 Amateur Radio operators (hams) in the United States, produces an entire line of reference manuals and specialty publications, as well as the monthly journal, *QST*. In fact just about anything you may need to know—whether it be from the technical or operating sides of the hobby—can be found in an ARRL publication or obtained directly from the HQ staff. The League's basic beginner's publication is *Tune in the World with Ham Radio,* which contains everything you need to master—including two Morse code instructional cassette tapes—to get geared up for your Novice amateur license exam. You only need write to ARRL HQ, 225 Main St, Newington, CT 06111, for complete information on all of the League's services.

QST Magazine

The ARRL journal *QST*, published monthly since 1914, is an excellent source of technical, operating, regulatory and general Amateur Radio information. Since it is published each month, *QST* is a very timely source of up-to-date information, and perhaps the most significant benefit of League membership. Write to the ARRL at the address given above for current membership rates.

On-the-Air Bulletins

Up-to-the-minute information on everything of immediate interest to amateurs—from Federal Communications Commission policy-making decisions to propagation predictions, to news of DXpeditions—is transmitted on a timely basis in W1AW bulletins. W1AW is the Amateur Radio station maintained at ARRL HQ in Newington, Connecticut. W1AW bulletins are transmitted at regularly scheduled intervals on CW (Morse code), phone (voice) and RTTY (teletype) on various frequencies, in addition to a compre-

Meet KA7WRG, Gary Anderson, of Kent, Washington.

first visit to the dentist.

Well, we carefully assembled the kit over a cool, summer weekend. We then set up a dipole antenna between the maple trees in the backyard and hooked the rig up to a homebrew 12-volt power supply. By Sunday afternoon, the station was operational.

Unbeknownst to Stan, who thought I was a certified electronics genius, I was about to make my first CW contact. As I demonstrated to him how to tune and load the rig, he surely must have been congratulating himself on getting an expert like me to show him the ropes.

I cranked the output power all the way to 2 watts and sent CW. I couldn't believe my ears, but as if by magic, my call was returned. I was rusty on my code, and Stan assisted me with the translation. The other station must have been going at least 15 WPM. QRS, I sent. Then Stan said, "He wants to know what class of license you have." I proudly sent GENERAL.

Then eagerly, we both deciphered the next (and apparently final) message—YOU LID, YOU'RE ON THE WRONG FREQUENCY. "That, Jean, was my first CW contact."

Shep laughed again. "Good story, Mitch. I've had similar experiences and know how you must have felt. You know (he said with a chuckle), some guys make it and then there's guys like us!"

This reminded me that Shep would occasionally include a humorous story about ham radio on his nightly radio program on WOR-AM, 710 kHz, in New York City. Between 10:15 and 11 PM, he would share with his listeners the rich memories of an Indiana youth seen through adult eyes. The show's format included contemporary topics also, and Shep's story-telling skill was just as impressive as the material itself. With his story-telling gift, Shep, a radio amateur since his teens, could capture the interest of even an unsuspecting listener tuning across the radio dial. Because of all the years of listening enjoyment that he provided, I was glad I could reciprocate in a small way by sharing one of my stories with him.

I asked Shep if he had time to listen to a poem. Not one I had written, but one written by Don Marquis, a newspaperman during the '20s. See, Don Marquis is one of Shep's favorite authors. He would read some of the newsman's creations on the air during his radio program. This had piqued my curiousity, and as a result, I had hunted down a collection of Marquis' work in book form years ago and still treasure it.

Shep said yes, and told me he would try to identify the poem I planned to recite. I pressed the transmit switch:

Life is not all jazz and joy,
Smiles and sunny weather!
Every gold has its alloy
To hold the stuff together!
If luck is good, why man alive,
Welcome it and cheer it!
But if the drink's two-seven-five,
Try to grin and bear it!

After he waited for the courtesy beep on the repeater, Shep answered. "Mitch, that sounds like Don Marquis."

I mentioned to him that someone I knew years ago interested me in Don Marquis. I did my best to hold back a chuckle. I'm sure he realized then that *he* was that someone.

I was sad to hear that Shep needed gasoline and had to clear. You know something . . . he thanked me for the contact as he said goodbye. That really made me feel good. I returned the compliment with gusto, and almost added "Excelsior!"—*Mitch Cohen, WB4RXB, Margate, Florida*

---

hensive Morse-code practice agenda. A detailed schedule of W1AW transmissions appears regularly in *QST* and may also be obtained on request from ARRL HQ (please include a self-addressed, stamped envelope).

Local Amateurs/Clubs

Don't overlook the greatest source of information of them all—the experience of your fellow hams. There is nothing a ham likes better to do than share the "vast" wealth of experience he/she has in the hobby. Being human, most hams like to brag a little about their on-the-air exploits. A general question on a particular area of operation to a ham who has experience in that area is likely to bring you all kinds of data. A dedicated DXer will talk for hours on techniques for working that rare DX QSO (ham terminology for long-distance contact). Similarly, a traffic handler will be only too happy to give you hints on efficient net procedures.

For the new ham (or potential ham), the problem may not be in asking the proper questions, but in finding another amateur of whom to ask the questions. That's where Amateur Radio clubs play an important role.

As a potential ham or a new ham, you might not know of an amateur in your immediate area. A good choice would be to try to find the time and meeting place for a nearby Amateur Radio club. What better source of information for the not-yet-licensed and/or the newly licensed than a whole club full of experienced hams? If you have no local ham contacts (in person) yet, don't despair, it's the ARRL to the rescue again. Simply write to the Club Services Department at ARRL HQ for a referral to an ARRL-affiliated Amateur Radio club (ARC) near you. The League is more than happy to supply this information to you. If you are not yet licensed, your local

Amateur Radio club is the place where Morse code and electronics theory courses will be given to prepare you for the

---

How to Apply for an Amateur Radio License

If you are applying for a Novice class license, you must first find two volunteer examiners who are over 18 years of age, hold a US General class or higher amateur license and are not related to you. You must also obtain FCC Form 610, which is available from any FCC district office or from ARRL HQ (SASE, please). If you have difficulty locating volunteer examiners, drop a note to the Club Services Department at ARRL HQ for assistance.

The volunteer examiners will first give you a five-word-per-minute code test (both sending and receiving). If you pass the code test, they will then give you the written theory-and-regulations exam, based on an FCC-approved public list of questions (available from the Volunteer Examiner Department at ARRL HQ). If you are successful, they will fill in the section on your Form 610 certifying that you have successfully completed the code and written requirements. After you fill out the remaining pertinent sections of the form, the volunteer examiners will send the form to the FCC in Gettysburg, Pennsylvania, for issuance of your license.

If you are applying for a Technician, General, Advanced or Extra Class license, you must appear before a team of three accredited Volunteer Examiners. If you want the latest list of test sessions near you, send an SASE to the Volunteer Examiner Department at ARRL HQ. The tests are typically given by local Amateur Radio Clubs and are assembled from a standard list of questions issued by the FCC. These questions and answers appear in the ARRL *License Manuals* series. If you are physically unable to travel to an examination point, you may make special arrangements for taking your exam.

Complete details on the Volunteer Examining program are available from ARRL HQ.

examination for the entry-level Amateur Radio license, the Novice class.

Listen, Listen, Listen...!

"Experience is the best teacher." The most efficient way to learn to do something is to actually pitch right in and give it a try. Since there are rules and an established order of protocol in the Amateur Radio Service, it would be time well spent in just plain, old-fashioned listening. Use a receiver (yours or one borrowed from a fellow amateur) and your ears. Tune the amateur frequency bands, and just listen, listen, listen. Copy as many QSOs as you can. Learn how the operators there conduct themselves, see what works for other operators and what doesn't, and incorporate those good points in your own operating habits. In other words, "copy the best and forget the rest."

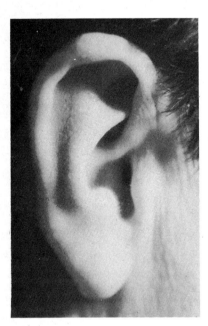

This is about the most important tool the Amateur Radio operator has at his disposal. Of course the other half of this matched set and the "gray matter" that lives between them are pretty important also.

Especially when going onto a new band or trying a new mode, take the time to listen to how those already established operators conduct themselves. Analyze the different types of operators you hear. Try to recognize good operating habits and incorporate them into your own style. You'll be surprised how much operator savvy you can pick up just by listening. A little common sense goes a long way in helping decide what is and what isn't good operating practice.

On-the-Air Experience for Newcomers

When it is time for you to take the plunge, fire up the rig and make your first QSO, go for it! Nothing can compare with actual on-the-air experience. Remember, everyone on the air today had to make that first QSO at some time or other. They had the same tentativeness as you might have now, but they made it through fine. So will you.

It's the nature of the ham to be friendly, especially towards other members of this wonderful fraternity that we all joined when we passed that examination. So trust the operator at the other end of your first QSO to understand your feelings and help all that she/he can. After all, that operator was in your shoes once, too!

But, it's not quite time yet to fire up the rig and give it a go. A few more basics are in order. We'll quickly cover rules and regulations, move on to a few tips on setting up your first station, give you a few of the basics in operating procedures, let you in on some on-the-air activities, then let you go to make that first QSO.

RULES AND REGULATIONS

The government of the United States grants us, as radio amateurs, certain privileges. In order to retain these privileges, we must conform to the rules and regulations in Part 97 of the Federal Communications Commission (FCC) rules.

This chapter will summarize only the most obvious rules as they pertain to day-to-day operation. This is by no means a complete discussion of all the FCC rules and regulations. For more complete information on the FCC rules as they pertain to Amateur Radio, along with the complete text of Part 97, see the *The FCC Rule Book,* published by ARRL. (Contact ARRL HQ for availability and price.) See also Spectrum Management, Chapter 2 of this book.

Table 3-1
Amateur Operator Licenses*

Class	Code Test	Written Examination	Privileges
Novice	5 WPM (Element 1A)	Elementary theory and regulations. (Element 2)	CW: 3700-3750, 7100-7150, 21,100-21,200 kHz. CW and RTTY: 28,100-28300 kHz. CW and SSB: 28,300-28,500 kHz. } 200-watts PEP output maximum. All authorized modes: 222.10-223.91 MHz (25-watts PEP output maximum). All authorized modes: 1270-1295 MHz (5-watts PEP output maximum).
Technician	5 WPM (Element 1A)	Elementary theory and regulations, general theory and regulations. (Elements 2 and 3A)	All amateur privileges above 50.0 MHz plus Novice privileges.
General	13 WPM (Element 1B)	Elementary theory and regulations, general theory and regulations (Elements 2, 3A and 3B)	All amateur privileges except those reserved for Advanced and Amateur Extra Class. 1500-watts PEP output maximum.
Advanced	13 WPM (Element 1B)	General theory and regulations, plus intermediate theory (Elements 2, 3A, 3B and 4A)	All amateur privileges except those reserved for Amateur Extra Class. 1500-watts PEP output maximum.
Amateur Extra	20 WPM (Element 1C)	General theory and regulations, intermediate theory, plus advanced techniques (Elements 2, 3A, 3B, 4A and 4B)	All amateur privileges. 1500-watts PEP output maximum.

*A licensed radio amateur will be required to pass only those elements that are not included in the examination for the amateur license currently held.

Your License

You must have a license issued by the FCC to operate an Amateur Radio station in the United States, in any of its territories or possessions, or from any vessel or aircraft registered in the United States. The FCC sets no minimum or maximum age requirement for obtaining a license, nor does it require that an applicant be a United States citizen.

An Amateur Radio license incorporates two kinds of authorization. For the operator, the license grants operator privileges, issued after passing a code test and a written examination that assesses the applicant's knowledge of Amateur Radio regulations, radio theory and the proper operation of transmitting equipment. The license also authorizes an Amateur Radio station—operation of transmitting equipment at a specified location. It also authorizes portable and mobile operation as permitted under the regulations.

Alien Amateur Radio operators, if licensed by a country with which the United States has signed a reciprocal-operating agreement, may apply to the FCC for permission to operate without having to pass the FCC amateur examinations.

The United States has five classes of operator licenses: Novice, Technician, General, Advanced, and Extra. See Table 3-1. The Novice license is the entry-level license requiring a minimum of technical knowledge and of Morse-code proficiency. The Novice license authorizes operation on six of the amateur frequency bands, with a variety of modes. Each successive class of license requires more technical knowledge and/or Morse-code proficiency. But each license rewards the licensee with more privileges, including expanded frequency allocations and more modes of operation. See Fig 3-1.

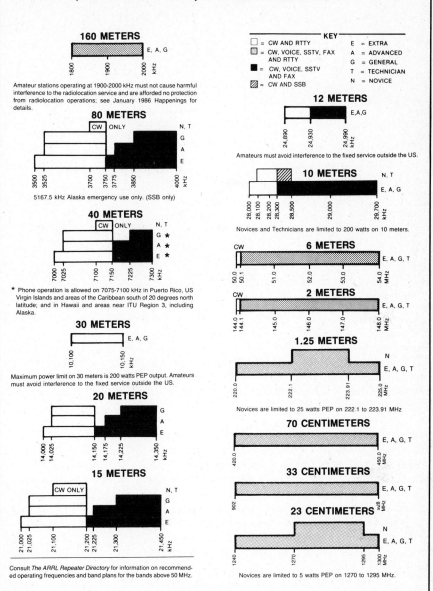

US Amateur Subband Allocations, 1.8 to 1300 MHz

Power Limits: All US amateur are limited to 200 watts PEP output in the Novice segments below 28,100 kHz and in the 30-meter band. On all other segments, 1500 watts PEP output is permitted. In addition, there are ERP limitations for stations in repeater operation. (See 97.67, FCC rules.) At all times the power level should be kept down to that necessary to maintain communications.

Fig 3-1—The popular HF and VHF amateur bands at a glance.

License Renewal/Modification

Once you have a license, you can keep it for your lifetime, but it must be renewed every 10 years. To renew an Amateur Radio license, simply:

1) Fill out FCC Form 610 (available from any FCC office or from ARRL HQ), and attach a photocopy, or the original, of your license.

2) Mail the completed application to the FCC, PO Box 1020, Gettysburg, PA 17325.

3) Keep copies of everything, if possible, as proof of filing before expiration. If you file your renewal application before the license-expiration date, you may continue to operate beyond the expiration date until the new license arrives. After expiration, there is a five-year grace period under which you may still renew without retesting. However, after two years

of the grace period has elapsed, you will lose your call sign, and will be assigned a new one. After this five-year grace period is over, you must be reexamined for a new license.

4) Note that 10-year-term licenses, which have been issued to all amateur licensees renewing since January 1984, have only a two-year grace period before *both* the license and call sign expire.

5) You may apply to have your license renewed at any time during the term of the license. FCC suggests that the application be made at least 90 days before expiration.

6) If you are simply modifying your license (eg, changing your address), you must still fill out Form 610. Incidentally, your license will also automatically be renewed for 10 years at this time.

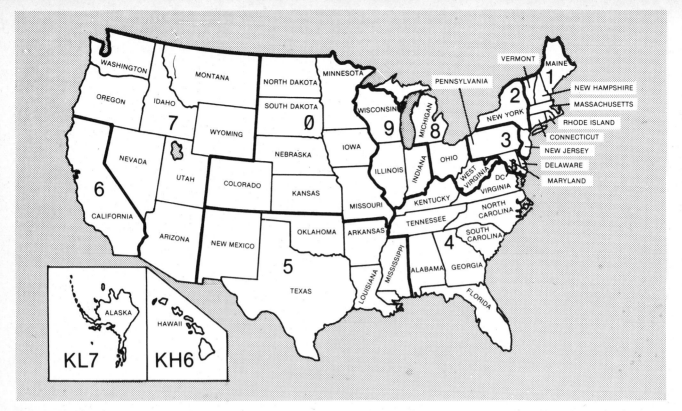

Fig 3-2—The 10 US call districts. See Table 3-2 for a list of FCC-allocated call-sign prefixes for US territory outside the 48 contiguous states.

If your license is lost, mutilated or destroyed, you must apply to the FCC for a duplicate. Send a letter to the FCC, Gettysburg, PA 17325, giving the circumstances and details under which the license was lost, mutilated or destroyed. After receiving your letter, the FCC will issue you a duplicate license bearing the same expiration date as the original license. If, after applying, you find the original license, either the original or the duplicate must be returned to the FCC.

In the unlikely event that FCC personnel should arrive while you were operating an Amateur Radio station, would you be able to produce the original or a photcopy of your license? The amateur rules require that you have a copy of your license or the original license in your personal possession or posted in a conspicuous place near your operating position. See Section 97.82 of the rules for further details.

If you are a guest operating at a station, you must also have the original or a photocopy of the station license whose call letters you are using. This station's license should be posted in a conspicuous place at the operating position, or it could be in your personal possession. If you give another ham permission to use your station while you are away, remember to leave behind a photocopy of your station license.

Even if operations are being conducted from a club station, the operator must be able to produce the club station license or a photocopy of that license in addition to his or her own operator license. The next time your club plans to operate from a remote location (Field Day, for example), be sure everybody brings along their operator licenses and that the club has the club station license.

US AMATEUR CALL SIGNS

The International Telecommunication Union (ITU) radio regulations outline the basic principles used in forming amateur call signs. According to these regulations, an amateur call sign must consist of one or two letters (sometimes the first or second may be a number) as a prefix, followed by a number and then a suffix of not more than three letters. Refer to Chapter 17 for a complete listing of international call-sign prefixes.

Every US Amateur Radio station call sign is a combination of a 1- or 2-letter prefix, a number and a 1, 2 or 3-letter suffix. The first letter of every US Amateur Radio call sign is always an A, K, N or W. For example, in the call sign W1AW, the W is the prefix, 1 is the number, and AW is the suffix.

For Amateur Radio stations located within the continental United States, the number in the call sign designates the geographic area in which the station was *originally* licensed. The call-sign districts for the continental US are illustrated in Fig 3-2. For Amateur Radio stations under FCC jurisdiction, but not located within the continental US, a prefix/number combination designates the geographic area in which the station was *originally* licensed. The prefix/numeral designators for US Amateur Radio districts outside the continental US are shown in Table 3-2.

The FCC used to require the number (or in the case of US stations located outside the continental US, the prefix/number) to correspond to the call-sign district in which the station was located. In 1978, however, the FCC relaxed the rules and now allows hams to retain their present call sign even if the station location is moved permanently to another call-sign district. It's routine to hear station call signs on the air that *do not* correspond to the call-sign district in which the station is located.

Some amateurs make their locations clear by including a second designator at the end of their call signs. For example, the voice identification, "This is W9KDR, permanent W1" or "fixed W1" means that the station, though originally licensed in the ninth call district, is now licensed to an address

Table 3-2
FCC-Allocated Prefixes for Areas Outside the Continental US

Prefix	Location
AH1, KH1, NH1, WH1	Baker, Howland Is
AH2, KH2, NH2, WH2	Guam
AH3, KH3, NH3, WH3	Johnston I
AH4, KH4, NH4, WH4	Midway I
AH5K, KH5K, NH5K, WH5K	Kingman Reef
AH5, KH5, NH5, WH5	(except K suffix) Palmyra, Jarvis Is
AH6, KH6, NH6, WH6	Hawaii
AH7, KH7, NH7, WH7	Kure I
AH8, KH8, NH8, WH8	American Samoa
AH9, KH9, NH9, WH9	Wake I
AH0, KH0, NH0, WH0	Northern Mariana Is
AL7, KL7, NL7, WL7	Alaska
KP1, NP1, WP1	Navassa I
KP2, NP2, WP2	Virgin Is
KP4, NP4, WP4	Puerto Rico
KP5	Desecheo

in the first call district. When identifying on CW, W9KDR/1 could be used to clarify the station's location.

It is important to make a distinction between clarifying one's *permanent* location and one's *portable* location, especially on voice. Using CW, the slant bar / (DAH-DIDIDAHDIT) followed by a number, could designate either portable or permanent operation. To clarify a location status during voice identification, the operator could say "portable" when she means portable and "permanent" when she means permanent.

On-The-Air Identification Requirements

The amateur rules specifically prohibit unidentified transmissions, but what is required for proper station identification? An amateur station must be identified at the end of a transmission or series of transmissions, and at intervals not to exceed 10 minutes during a single transmission or a series of transmissions of more than 10 minutes' duration. The identification must be given using the International Morse code or, when identifying on phone, in the English language. Note: At the end of an exchange of third-party communications (that is, on behalf of anyone other than the control operator) with a station located in a foreign country, you must also give the call sign of the station with which third-party traffic was exchanged.

Temporary Designator

Temporary designators must be used after a call sign under some circumstances. These designators are used for "instant upgrading," an FCC rule that permits holders of an amateur license who successfully pass an examination for a higher class of license to use the new privileges immediately after leaving the test session. The accredited Volunteer Examiner issues the amateur a certificate of successful completion, which authorizes the new privileges for one year or until the issuance of the new, upgraded, permanent license (whichever comes first). The temporary amateur permit also sets forth a two-letter designator for the upgraded class of license. For example, "temporary AA" means the operator has upgraded to Advanced.

When operating under the authority of a temporary amateur permit with privileges authorized by the permit, but which exceed the privileges of the licensee's permanent station license, the station must be identified in the following manner:

1) On phone, by the transmission of the station call sign, followed by the word "temporary" followed by the special

identifier shown on the certificate, appropriate to the newly earned class of license.

2) On CW, by the transmission of the station call sign, followed by the slant bar, followed by the special identifier shown on the temporary amateur permit, eg, W5NOW/AA.

Additional Identification Requirements

To meet the identification requirements of the amateur rules, the call sign must be transmitted on each frequency being used. If identification is made by an automatic device used only for identification by CW, the code speed must not exceed 20 words per minute. While the FCC does not require use of a specific list of words to aid in sending one's call sign on phone, it does encourage the use of a nationally or internationally recognized standard phonetic alphabet as an aid for correct phone identification. The ITU phonetic alphabet appears in Chapter 17.

Linda, KA4AAT, and Ken, WA4SFQ, shared operating and logging during Field Day.

Permitted Communications

The US amateur rules clearly set forth the kinds of stations with which amaters may communicate. These stations are:

1) other amateur stations;

2) stations in other services licensed by the FCC and with US Government stations for civil defense purposes, in accordance with the Radio Amateur Civil Emergency Service (RACES), in emergencies and, on a temporary basis, for test purposes; and

3) any station that is authorized by the Commission to communicate with amateur stations.

Additionally, amateur stations may be used for transmitting signals, communications or energy to receiving equipment for the measurement of emissions, temporary observations of transmission phenomena, radio control of remote objects and similar experiments, and for the objectives set forth in the basis and purpose of the Amateur Radio Service.

Every amateur should operate with one eye on the basis and purpose of the Amateur Radio Service, as set forth in Section 97.1 of the amateur rules. Adhere to these fundamental principles and you can't go wrong!

§97.1 Basis and Purpose

The rules and regulations in this part are designed to provide an amateur radio service having a fundamental purpose as expressed in the following principles:

(a) Recognition and enhancement of the value of the amateur service to the public as a voluntary noncommercial communication service, particularly with respect to providing emergency communications.

(b) Continuation and extension of the amateur's proven ability to contribute to the advancement of the radio art.

(c) Encouragement and improvement of the amateur radio service through rules which provide for advancing skills in both the communication and technical phases of the art.

(d) Expansion of the existing reservoir within the amateur radio service of trained operators, technicians, and electronics experts.

(e) Continuation and extension of the amateur's unique ability to enhance international goodwill.

Third-Party Communications

Third-party traffic is defined in the US amateur rules as "Amateur radio communication by or under the supervision of the control operator at an amateur radio station to another amateur radio station on behalf of anyone other than the control operator" (97.3[v]); in other words, sending or receiving any communication on behalf of anyone other than yourself or the control operator of the other station. If you allow anyone to say "hello" into the microphone at your station, that is third-party traffic. Sending or receiving formal radiogram messages is third-party traffic (see Chapter 15). Autopatching and phone patching (interconnecting your radio to the telephone system) are also third-party traffic. You, the control operator of the station, are the first party; the control operator at the other station is the second party; and anyone else participating in the two-way communication from either station is a third party.

Certain kinds of third-party traffic are strictly prohibited. All amateurs should know that they must not handle the following kinds of third-party traffic:

1) international third-party traffic, except with countries that allow it;

2) third-party traffic involving material compensation, either tangible or intangible, direct or indirect, or to a third party, station licensee, a control operator, or any other person;

3) except for an emergency communication, third-party traffic consisting of business communications on behalf of any party.

Business communications of any type are specifically prohibited by 97.110, except for emergency communications. 97.3(bb) defines business communications as "any transmission or communication the purpose of which is to facilitate the regular business or commercial affairs of any party." Emergency communication is defined in 97.3(w) as "Any amateur communication directly relating to the immediate safety of human life or the immediate protection of property."

Third-party traffic may be handled by US amateur stations with US possessions and territories where amateur stations are also licensed by the FCC. This message handling is subject to the same criteria as domestic traffic in the states. There must be no pecuniary (ie, financial) interest, direct or indirect, to any party. Note: Only amateur stations identified by authorized call signs having a one- or two-letter prefix beginning with AA-AL, K, W or N are licensed by the US Government. Third-party traffic may be handled freely with these stations.

In general, third-party traffic is prohibited with amateur stations in the rest of the world (see Chapter 5 for a list of countries that do permit third-party traffic handling). This is not through reluctance on the part of the US Government, but because of prohibitions on the part of the other governments. The amateur at the other end is commonly forbidden to handle any third-party traffic. US amateurs must abide by this restriction and must have no participation in the handling of third-party traffic in such cases.

Active participation by a third party: One of the best ways to get a nonham interested in Amateur Radio is to allow him or her to speak into a microphone and experience first-hand the excitement of amateur communication. While the US amateur rules allow active participation by third parties, the control operator must make certain that the communications abide by all the restrictions pertaining to third-party traffic. For example, the third party may not talk about business and may not communicate with any country that has not signed a third-party agreement with the US. Also, the control operator may not leave the controls of the station while the third party is participating in amateur communications. The amateur rules allow third-party participation only if "a control operator is present and continuously monitors and supervises the radio communication to ensure compliance with the rules."

Third-party participation is not the same as designating another licensed amateur to be control operator of a station. A control operator, designated by a station licensee, may operate the station to the extent authorized by the control operator's class of license.

FCC recently (late 1985) modified 97.114 to close a loophole that previously allowed amateurs who have had their licenses suspended or revoked to participate in Amateur Radio communications as third parties. Section 97.114 now reads as follows:

(a) Subject to the limitations specified in paragraphs (b) and (c) of this section, an amateur radio station may transmit third-party traffic.

Guest Operating

An FCC rules interpretation allows the person *in physical control* of an Amateur Radio station to use his or her own call sign when guest operating at another station. Of course, the guest operator is bound by the frequency privileges of his or her own operator's license, no matter what class of license the station licensee may hold. For example, Brandy, KA1KPI, a Novice, is visiting Mac, KZ1A, an Extra Class licensee. Brandy may use her own call sign at KZ1A's station, but she must stay within the Novice subbands.

When a ham is operating from a club station, the club call is usually used. Again, the operator may never exceed the privileges of his or her own operator's license. The club station trustee and/or the club members may decide to allow individual amateurs to use their own call signs at the club station, but it is optional. In cases where it is desirable to retain the identity of the club station (W1AW at ARRL HQ, for example), the club may require amateurs to use the club call sign at all times.

In the rare instance where the guest operator at a club station holds a higher class of license than the club station trustee, the guest operator must use the club call sign and his/her own call sign. For example, if Billy, KR1R, who holds an Amateur Extra Class license, visits the Norfolk Novice Radio Club station, KA4CVX, and wants to operate the club station outside the Novice subbands, Billy would sign KA4CVX/KR1R on CW and "KA4CVX, KR1R controlling" on phone. (Of course, this situation would prevail *only* if the club requires that the club call sign be used at all times.)

(b) The transmission or delivery of the following third-party traffic is prohibited:

(1) International third-party traffic except with countries which have assented thereto;

(2) Third-party traffic involving material compensation, either tangible or intangible, direct or indirect, to a third party, station licensee, a control operator or any other person;

(3) Except for emergency communications as defined in this part, third-party traffic consisting of business communications on behalf of any party.

(c) The licensee of an amateur radio station may not permit any person to participate in traffic from that station as a third party if:

(1) The control operator is not present at the control point and is not continuously monitoring and supervising the third-party participation to ensure compliance with the rules.

(2) The third party is a prior amateur radio licensee whose license was revoked; suspended for less than the balance of the license term and the suspension is still in effect; suspended for the balance of the license term and relicensing has not taken place; surrendered for cancellation following notice of revocation, suspension or monetary forfeiture proceedings; or who is the subject of a cease and desist order which relates to amateur operation and which is still in effect.

Logging Eliminated

The FCC has eliminated the requirement that amateurs keep logs of routine operating activities. Also, repeater-station trustees no longer need to tape record autopatch and other third-party messages. Amateurs may continue to keep voluntary logs, and FCC Engineers-In-Charge may still mandate log-keeping in specific cases. Details of nonroutine operations, such as repeater data when certain power limits are exceeded, auxiliary and remote control operations, and names and call signs of guest control operators are still required to be kept as part of the official station records.

Authorized Power

Generally, an Amateur Radio transmitter may be operated with a PEP output of up to 1500 watts. However, there are circumstances under which less power must be used. These are:

1) Under all circumstances, an amateur is required to use the minimum amount of transmitter power necessary to carry out the desired communications.

2) When operating in the 80, 40 and 15-meter Novice subbands, *all amateurs* may use no more than 200-watts PEP output. The Novice subbands are as follows: *80 meters*—3.700-3.750 MHz; *40 meters*—7.100-7.150 MHz (7.050-7.075 MHz when the station is not within ITU Region 2—see Fig 3-3); *15 meters*—21.100-21.200 MHz; *10 meters*—28.100-28.500 MHz; *1.25 meters*—222.10-223.91 MHz; and *.23 meters*—1270-1295 MHz. When operating in the 10-meter Novice subband, Novice and Technician control operators *only* may use no more than 200-watts PEP output; in the 1.25-meter Novice subband, Novices *only* may use no more than 25-watts PEP output; and in the .23-meter Novice subband, Novices *only* may use no more than 5-watts PEP output.

3) For stations in repeater operation, the authorized effective radiated power (ERP) shall not exceed certain levels, in accordance with Section 97.67 of the FCC rules.

Note: Operation on the band 10.1 to 10.150 MHz is, at this writing, limited to 200 watts PEP output using CW or RTTY only.

Amateur operation in a foreign country: Operation may not be conducted from within the jurisdiction of a foreign government unless that foreign government has granted permission for such operation. Some countries have allowed US amateurs to operate while visiting; however, not all countries permit foreigners this privilege. Every country has the right to determine who may and who may not operate an Amateur Radio station within its jurisdiction. There are severe penalties in some countries for operating transmitting equipment without the proper authority. See Chapter 6 for details on overseas operating.

Canada-US reciprocal operating: The most liberal agreement between two countries allowing reciprocal operating for their radio amateurs is between Canada and the US. No formal application or permit is required for US amateurs to oper-

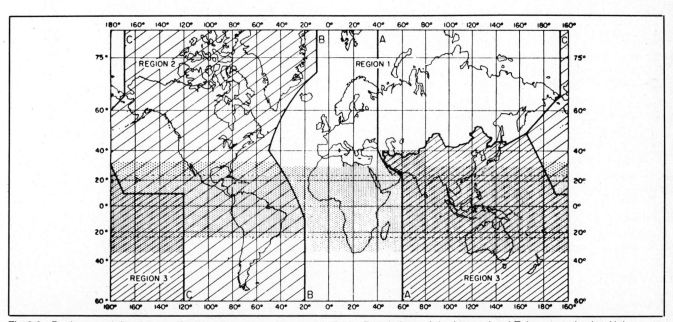

Fig 3-3—Regions as defined in the Table of Frequency Allocations, Radio Regulations of the International Telecommunication Union (ITU). The shaded part represents the Tropical Zone.

ate amateur equipment in Canada, nor is any formal paperwork required for Canadian amateurs to operate in the United States. It's automatic. US radio amateurs may operate in Canada with the mode and frequency privileges authorized them in the United States, and Canadian radio amateurs may operate in the US with the mode and frequency privileges authorized to them in Canada. In all cases, however, visitors must stay within the band and mode restrictions of the host country.

Other reciprocal operating agreements: Except for Canada, US amateurs must obtain a written permit to be allowed to operate in any foreign country. The administrations of some countries have signed reciprocal operating agreements with the US. Such agreements facilitate the application procedures for getting permission to operate. These agreements, because they are reciprocal, also allow foreign amateurs the opportunity to receive permission to operate while in the US. If you want to operate a station in a foreign country, ARRL HQ may be able to help you obtain forms and information. Write to ARRL HQ's Information Services for details. Many countries require a US licensee to hold a General or higher-class license to qualify for a permit to operate. See Chapter 6 for a list of the countries that have signed reciprocal operating agreements with the US.

Radio Frequency Interference

The FCC authorizes radio amateurs to use up to 1500 watts PEP output and many different bands of the radio spectrum. However, these privileges do not come without some effort on the part of the license applicant. A prerequisite for getting on the air is passing an examination that tests the applicant's technical knowledge. One area of special importance is knowing about radio frequency interference (RFI), its causes and its cures. The FCC expects radio amateurs to identify and solve most RFI problems, but RFI is not always a simple matter.

Much has been written about RFI. For example, the material in the ARRL *Radio Frequency Interference* book examines this subject from both a legal and technical standpoint. Also, *The ARRL Handbook for the Radio Amateur* devotes an entire chapter to solving radio frequency interference problems. If you are the recipient of interference complaints, these two books will be of interest to you.

A growing number of amateurs are unjustly accused of causing RFI. This occurs when an amateur transmitter is operating properly, and the responsibility for the interference falls with the electronic device experiencing the interference. For example, some television sets receive signals they are not supposed to because the manufacturer decided that reducing the per-unit cost was more important than incorporating adequate shielding in the circuit design. In these cases, the transmitter is operating with a "clean" signal; it is the television set that must be brought up to today's engineering standards. This usually means that the only effective cure must be performed by adding a filter at the television set. Trying to make your neighbor understand that it is his equipment that is at fault is usually no easy task, however.

Public Law 97-259 now allows the FCC to require standards for unwanted-signal-rejection filtering in electronic home-entertainment devices. Also, some manufacturers have made some progress in dealing with this problem by providing free assistance to owners of their products who experience RFI. However, the onus for causing the interference continues to fall on amateurs.

Citizens Band operators are also increasingly involved in RFI complaints. FCC figures show that only five percent of RFI complaints can be attributed to CB operation. The CB situation is exacerbated because the typical CB operator is usually less prepared to solve an RFI problem. The FCC figures also show that over 80% of RFI complaints concerning home-entertainment devices would never have occurred if the manufacturers had included proper shielding or filtering.

If you are accused of causing RFI:

1) Check your log. Were you operating at that time? (A complete log, although no longer required by the FCC, is very useful in interference situations.)

2) Check with your own nonamateur equipment. If you are not interfering with your own TV set, chances are the problem lies with your neighbor's receiver and not with your transmitter.

3) Solicit the cooperation of your neighbor in testing to determine the exact cause of the interference.

4) Check with your local radio club for a TVI committee or other assistance.

5) Request RFI assistance from the manufacturer of the home-entertainment device.

6) Read the ARRL *Radio Frequency Interference* book and

Note to Canadian Readers

For purely practical reasons, this chapter has been written to reflect the regulations established by the Federal Communications Commission for US amateurs. Canadian amateurs are regulated by the Department of Communications (DOC). Though the privileges and responsibilities of amateurs in Canada and the US are generally similar, there are differences. Several sources of Canadian information are listed below.

There are three classes of operator certificates in Canada: the Amateur Radio Operator's Certificate, the Amateur Radio Operator's Advanced Certificate and the Amateur Digital Radio Operator's Certificate. Examinations include written tests on radio theory and DOC regulations for all three classes of Certificates, and Morse code receiving tests for Amateur (10 words per minute) and Advanced (15 words per minute) Certificates. There is no code requirement for the Digital Certificate, but Advanced theory, Advanced regulations, plus a digital theory exam must be written. There is an annual fee of $20 for the station license (the initial fee is $26), which is renewable before April 1 each year. There is no minimum age requirement.

A Telecommunications Regulation Circular, TRC-24, is available without charge from local offices of DOC. Entitled "Information for the Guidance of Candidates Preparing to Attend Examination for an Amateur Radio Operator's Certificate, or an Amateur Radio Operator's Advanced Certificate, or an Amateur Digital Radio Operator's Certificate," it describes the method of periodic examinations and the study curriculum to qualify for examination for all three classes of certification. It is available in English and French. Anyone planning to take examinations is advised to obtain a copy.

The *Canadian Amateur Radio Licencing Manual,* by Ralph Zbarsky, VE7BTG, published by the Canadian Radio Relay League (CRRL), is written expressly for those wishing to become licensed in Canada. It is organized into a looseleaf, three-ring binder format and is easily identified by its yellow cover with the CRRL-ARRL diamonds on the front cover. Inquiries should be directed to the Canadian Radio Relay League, Box 7009, Station E, London, ON N5Y 4J9, or to Canadian Amateur Radio Licensing Manual, 3275 West 22nd Ave, Vancouver, BC V6L 1N1.

The CRRL also publishes the *CRRL Questions and Answers Book,* with 12 chapters corresponding to the 12 subject areas in DOC's TRC-24. Contact CRRL for more information and ordering details.

the appropriate chapter in *The ARRL Handbook for the Radio Amateur.*

7) Be prompt, courteous and helpful; Amateur Radio's reputation is at stake, as well as your own.

8) For further assistance, write ARRL HQ. Include as much detail as you can about the symptoms and circumstances, but avoid emotional commentary.

On the Horizon

A basic appreciation of the rules will provide amateurs with increased levels of enjoyment. The rules have evolved over a long time and will continue to evolve. In the last few years, deregulation of the Amateur Radio Service has accelerated, and for the most part, this deregulation has had a desirable impact. The FCC proposes additional rules changes periodically; check *QST* for news of late-breaking events.

The rules are dynamic because Amateur Radio is dynamic—constantly changing to meet and create new communications technologies to better serve the public and society.

A New Era for Amateur Radio

September 14, 1982, was a red-letter day for Amateur Radio, as President Reagan signed into law Public Law 97-259 that has opened the door to sweeping changes in the Amateur Radio Service. The law amends the Communications Act of 1934 in several critical areas:

1) FCC now has the authority to regulate the susceptibility of electronic equipment to radio-frequency interference, authority that is needed to stem the flow of electronic devices that cannot function normally in the presence of RF energy. This provision is a significant victory for amateurs weary of their long-standing battle with such devices in the households of uninformed neighbors having TVI troubles.

2) The Amateur Service is exempted from the "secrecy of communications" provisions of Section 605 [now Section 705] of the Communications Act, thus clearing the way for a more active role on the part of amateurs in policing their own bands.

3) FCC is now authorized to use volunteers in monitoring for rules violations. Amateur volunteers are able to issue advisory notices to apparent violators and to convey information to FCC personnel (but are not authorized to take enforcement actions themselves). ARRL has expanded its Official Observer program to take a leadership role in volunteer monitoring through the Amateur Auxiliary to the FCC's Field Operations Bureau, administered by the League's Section Managers.

4) FCC is authorized to use volunteers in preparing and administering amateur exams, a necessary response to the effects of budget cuts on the FCC's ability to prepare and supervise exams. Amateur Radio license exams are now administered by accredited radio amateurs acting as Volunteer Examiners. The work of these volunteers is coordinated by a Volunteer Examiner Coordinator (VEC). The ARRL is a VEC, serving each of the FCC call districts, plus any other location worldwide. Write to the Volunteer Examiner Department at ARRL HQ for a list of test sessions near you.

5) The term of amateur station licenses is made 10 years to reduce the administrative burden on the FCC and its licensees.

The law is a milestone in Amateur Radio. It has been a catalyst for a new, dynamic service, with amateurs playing a more integral role in processes that affect them. It is an enabling law in every sense of the word. As ARRL Executive Vice President David Sumner noted in a *QST* editorial, "Where protecting the future of Amateur Radio is concerned, it boils down to this: If we, the amateur community, won't do it, no one will."

Rules of the Game

Basically, that's a bare-bones description of the rules that govern how the game of ham radio is played, only a short summation of the FCC rules in Part 97. A copy of the entire amateur rules can be found in *The FCC Rule Book*, published by ARRL.

Now that we know the basics of the rules of the game, it's time to get down to the real fun—"How to play the ham radio game!"

YOUR STATION

If you haven't already done it, think about exactly what kind of station you would like to have.

There are so many different possible station configurations it seems an unlikely task to be able to choose the "best" station/antenna combination for you. Many hams face constraints such as having a limited budget to spend for ham gear or having restrictions on the size and location of the antenna system.

The best approach to selecting the "best" station arrangement for you is to do some exhaustive (but oh so much fun) research. Utilize all the sources of information on ham gear at your disposal, some of which are:

1) *Radio club members.* Ask members of the local radio club what their personal preferences in gear and antennas are, but be prepared for a great volume of input. Every amateur has an opinion on which equipment/antenna is the "best" of them all. Years of experimentation usually go into finding just the right station equipment to meet a particular amateur's needs. Listen and take note of each ham's choices and reasons for selecting a particular kind of gear; there's a lot of experience, time, effort and money behind each of those choices.

2) *Hands-on experience.* Try to use as many different pieces of gear as possible before you decide. Ask a nearby ham friend or one or more of your fellow radio club members if you can use their station (remember to abide by the FCC rules for guest operators as described earlier). Also, use the club station. Note what you like and what features you don't care for in each of the stations that you use.

3) *Advertisements.* QST is chock-full of ads for all the newest up-to-date equipment as well as some premium used gear. Read the ads, and don't be afraid to contact the manufacturers of the gear for further information. Compare specifications and prices to get the best deal.

Buying Your Station Equipment

After you've made your final choice (it's not really final, as most hams will trade station equipment several times during their ham careers), consider the sources where you might get the best deal whether it be new or used, transceiver or separates. Several possible sources are:

1) *Local amateurs—private sale.* Many hams will have spare used gear that they may be willing to part with at a reasonable price. Be sure you know what a particular rig is going for on the open market before settling on a final price.

2) *Hamfests/flea markets.* Many radio clubs run conventions called hamventions, hamfests or flea markets. Usually one of the big attractions of these events are the flea markets or equipment sales. Much used gear, usually in passable shape, can be found at reasonable rates. Local distributors and manufacturers of new ham gear sometimes show up at these events to sell equipment at fair prices.

3) *Local electronics dealers.* If you are lucky enough to live

Your First Station

One of the biggest nail biters faced by the new amateur is deciding which and how much equipment is necessary to get on the air without unnecessary expense. The following overview may help you decide. (A detailed treatment of used equipment and getting on the air quickly and effectively may be found in ARRL's *Tune In the World with Ham Radio.*)

1) *Used separate transmitter/receiver combination.* Perhaps the least expensive route is second-hand transmitters and receivers. Most of these rigs sell used for under $100 each (see *QST* Ham Ads). Some manufacturers such as Heath, Kenwood, Drake and Collins offered "twins"—matching transmitter/receiver combinations that are almost as easy to use as transceivers. A disadvantage is that more controls need to be adjusted when changing frequency. Also, older equipment is somewhat large and heavy compared with present-day, solid-state gear.

Separate transmitter. For CW operation, consider a second-hand Johnson Viking I or II transmitter, or a Viking Ranger I or II with built-in VFO. Another popular VFO-controlled transmitter is the Hallicrafters HT-32 and HT-37 series. Other available used transmitters include the Heath SB-400, SB-401 and DX-60; Collins 32V and 32S series; Drake 2NT and T4X series; and Kenwood T599.

Separate receiver. Some pretty good receivers can be purchased for around $100, and they will do the job nicely. The Hallicrafters SX-71, SX-99, SX-100, SX-101; National HRO-50T1 and HRO-60; Collins 75A and 75S series; Hammarlund HQ-170 and HQ-180; Heath SB-300, SB-301 and SB-303; Drake 2B, 2C, R4 series and R7 series; and Kenwood R599 are examples.

2) *Transceivers.* The advantage of a transceiver is that one knob adjusts transmitter and receiver main tuning simultaneously. These rigs will cost more than older separates, partly because both the transmitter and the receiver are contained in a single box, which may also contain a power supply. Worthy of consideration are the Drake TR3 and TR4; Swan 350, 400 and 500C; Heath HW-100, HW-101, SB-100, SB-101 and SB-102; Kenwood TS-511S, TS-520 or TS-820; Yeasu FT-101 series and FT-102. Such transceivers can be purchased in the $300 range (see *QST* Ham Ads again). The older Kenwood TS 520 and Yaesu FT-101 are also good choices.

3) *New equipment.* Most of the new gear is compact, has solid-state circuitry, and is capable of better overall performance than older, used gear. It would be inappropriate to recommend any single item, but be prepared to pay from $600 to $1500 for a brand-new transceiver.

4) *Homemade equipment.* Beginners usually lack the skill and experience to design or build their first station, but if you have a background in electronics, don't pass up the chance to build ("home brew") your own. Many good circuits appear in *The ARRL Handbook for the Radio Amateur* and *Solid-State Design for the Radio Amateur.* Keep in mind, though, that you might want to enjoy the operating aspects of Amateur Radio for a while before tackling a major construction project.

Now let's address other accessory items that might be needed in your first station:

SWR indicator. This device is handy for adjusting an antenna for the lowest obtainable SWR when the antenna is first erected. Thereafter, it serves only as a monitor to keep tabs on the condition of the antenna system. If a Transmatch is used with a multiband wire antenna, an SWR indicator is essential for insuring that the Transmatch is adjusted for an SWR if 1:1. If a multiband trap or single-band 50-ohm antenna is used, an SWR meter may not be necessary.

Transmatch (antenna tuner/antenna coupler). Many of these devices exist on the market today, and a newcomer can easily get a false impression that it is impossible to operate without a Transmatch. If your station antennas present an SWR of less than, say, 2:1 in a 50-ohm system, you can forget about Transmatches. But if you elect to use one antenna for several bands (assuming that antenna traps aren't used), a Transmatch becomes a necessary tool. Under such conditions, the Transmatch improves the impedance match between the antenna or its feed line and the transmitter. Some Transmatches (depending upon the circuit used) will attenuate harmonic emissions from the transmitter, thereby aiding in the reduction of TV and other forms of interference. Transmatches can also aid reception by rejecting out-of-band signals. A manufactured Transmatch will sell for $75 to $1500, depending on its power-handling capability and operating features. A homemade unit is easy to build and shouldn't cost more than $50 if surplus parts are used. (*The ARRL Handbook* gives directions for building Transmatches.) Check the radio flea markets for inexpensive components.

Keys, keyers and paddles. There is a certain mystique connected with using a hand or straight key. If we have reasonable rhythm in our souls, we can send good CW, up to 20 words per minute, with a straight key. A new key can be purchased for less than $6, and a surplus one (military J-38 replica) costs even less. A paddle (for use with an electronic keyer) can, on the other hand, cost as much as $60. Keyers are prevalent on the amateur bands, and are great for sending good, smooth CW. They are, however, only as good as the commands sent by the operator. Some of the most dreadful CW heard on the amateur bands today is being sent by means of poorly operated electronic keyers. Keyers vary in price from a few dollars to a few hundred dollars. One economical approach is to start out with a straight key, and then switch to an electronic keyer as soon as you become reasonably adept with the code. Some hams even use home computers such as the Commodore 64™ to send CW. Software for this application has been published in various amateur journals.

Station Setup

Now that you have an idea of what equipment to have, think about how to arrange it. Don't stack your gear. Allow room for passage of air around and through the equipment. Your CW key should be far enough away from the table edge to permit your sending arm to lie flat from elbow to fingertips. To reduce arm tension and ensure good CW sending, locate the key at an angle of 20-30 degrees from a line that is perpendicular to the front table edge.

Be certain to have an effective earth ground routed to your station. Use as many grounds as you can locate, bonding them together electrically and bringing the connection plate to the operating position.

Connections from the various ground points to the shack should be as short as possible. Since RF currents flow mainly on the surfaces of conductors, ground conductors such as shield braid from RG-8 coaxial cable or wide strips of flashing copper are best. (Large, short conductors help reduce unwanted lead inductance and make for better RF grounding.) The leads between the common ground point in the shack and the various pieces of station gear should be similarly short, and large in surface area.

Amateur Radio is fun. Get started on the right track, and your enjoyment of this wonderful hobby can keep growing.

J. D. Cale, KA8WTZ, earned his Novice license at age 11. *(N8FEB photo)*

Larry, KA8WSP, upgraded to General shortly after this picture was taken.

near an electronics distributor who handles a line or two of ham gear, so much the better. The dealer can usually answer any questions that you may have, will usually have a demonstration unit and will be only too happy to assist you in purchasing your new gear.

4) *Mail order.* There are many mail order ham equipment distributors to choose from. Some deal in new equipment, some deal in used equipment and some deal in both. Check the ads in the ham publications, such as those found in *QST*, for the equipment that you want. The ads are also an excellent place to find out the going price of a particular piece of gear.

The Choice is Yours

Take your time in deciding which gear to get, and even consider homebrewing your gear. There's a lot of satisfaction to be had in telling the operator on the other end of your QSO (contact) that the "rig here is homebrew." *The ARRL Handbook for the Radio Amateur* has extensive coverage of construction and basic-to-advanced electronics theory information.

Be sure that you will be satisfied (at least for a time) with the gear that you have selected.

THE ANTENNA

Now that you've got the rig capable of producing watts, it's time to think about installing an antenna to put those watts where they will do the most good for you—into the ether (on the airwaves).

Again, knowledge is your greatest ally. Gather all the information you can on the different types of antennas. Get down to basics, learn antenna theory from the ground up (see *The ARRL Handbook* and *The ARRL Antenna Book*). Once you know how a particular antenna works, and what pattern of radiation you can expect through its use, you can intelligently assess its value to you.

A good rule of thumb to follow is "always erect as much antenna as possible." The better your antenna array, the better will be your radiated signal. A good antenna system will make up for inadequacies or shortcomings in station equipment; a less-sensitive receiver "hears better" with a good antenna system and a "bigger" antenna system will make a QRP (low power) station sound a lot louder at the receiving station's end.

You may even wish to experiment with new antenna designs of your own. Why not? Erect several antennas for the same band, experiment and see which works best for you.

Simple Novice Antenna

The antenna is a very important part of any amateur's station; indeed, the antenna can make or break your Amateur Radio activities. Many newcomers and veterans alike choose the dipole antenna for getting on the air easily and efficiently. This elementary antenna consists of a half wavelength of antenna wire usually extending in a straight line with a feed line and an insulator at its center. See Fig 3-4.

For Novice activities, a single- or multiband dipole system is a good selection. Each has its advantages. The single-band dipole is very easy to erect and has a low *SWR* (standing wave ratio—a measure of how well an antenna is tuned to the transmitted frequency) for its desired cut-to-length operating range. The single-band dipole allows you, however, to operate only on one band. The multiband system is more expensive, but it gives you the flexibility of operating on more than one band for both local and DX communications. Keep in mind, though, that this multiband system also passes *harmonics* (unwanted multiples of the transmitted frequency) more easily, something all amateurs try to avoid. Precise transmitter tuneup procedures will help prevent this.

Based on these considerations, choose the type of dipole you wish to erect.

Where?

Think about your antenna location next. Remember that your antenna should be as high and as far away from surrounding trees and structures as possible. Never put an antenna around or even near power lines! The dipole will require one support at each of its ends (perhaps trees, poles or even house or garage eaves), so survey your potential antenna site with that in mind. If you find that space is so limited that you can't put up a straight-line dipole, don't give up. You can bend or coil-up the ends of the dipole and still make plenty of contacts. You can put up an antenna under almost any circumstances, but you may need to use your imagination and other reference sources to find an antenna suitable to your particular needs. Check *The ARRL Antenna Book, Tune in the World with Ham Radio* and *QST* for many workable antenna ideas.

Antenna Parts List

1) *Antenna wire.* No. 12 or No. 14 hard-drawn Copperweld® (copper-clad steel) preferably, so that the antenna won't stretch, and it will be strong enough to support itself as well as the weight of the feed line connected at its center. If you plan to erect a single-band dipole, buy just enough wire for a single band, plus some extra. If you will put up a multiband dipole, buy enough wire for the total lengths of all bands you wish to use, plus some extra for each antenna.

2) *Insulators.* One center and two end antenna insulators for the single-band dipole; only one center, then two spreader

Build Your Own Antenna

To connect the coax to the center insulator, cut a few inches of the outer covering off the coax. Next, separate the copper braid from the inner conductor and insulation. After you've done that, twist the strands of copper together to form a single wire. Remove about half the insulation covering the inner conductor and bend it away from the twisted strands of copper braid. Loop the cable over the insulator as shown and solder the braid to one half the antenna and the inner conductor to the other. Be sure to tape all connections securely for waterproofing; the braid can soak up water like a sponge or wick, making the coax useless after a while. B and C show how the wire is connected to various types of insulators at the ends, and D shows the connection of the feed line at the center.

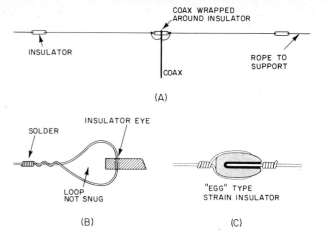

(A)

(B)   (C)

Examples of simple but effective wire antennas. A horizontal dipole is shown at A. The legs can be drooped to form an "inverted V" as shown at B. A sloping dipole (sloper) is illustrated at C. The feed line should come away from the sloper at a 90° angle for best results. If the supporting mast is metal, the antenna will have some directivity in the direction of the slope.

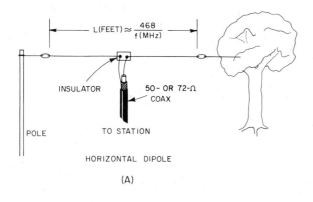

$$L(\text{FEET}) \approx \frac{468}{f(\text{MHz})}$$

INSULATOR

50- OR 72-Ω COAX

POLE

TO STATION

HORIZONTAL DIPOLE

(A)

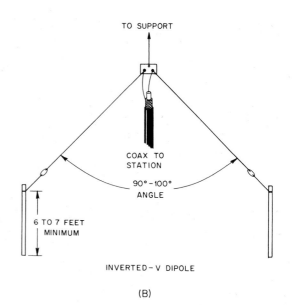

TO SUPPORT

COAX TO STATION

90° - 100° ANGLE

6 TO 7 FEET MINIMUM

INVERTED – V DIPOLE

(B)

Using a PL-259 Connector

If you are using RG-58 or RG-59 with a PL-259 connector, you will need to use an adapter, as shown here. This material courtesy of Amphenol Electronic Components, RF Division, Bunker Ramo Corp.

2) Fan braid slightly and fold back over cable.

1) Cut end of cable even. Remove vinyl jacket ¾"—don't nick braid. Slide coupling ring and adapter on cable.

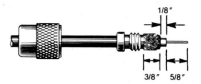

3) Position adapter to dimension shown. Press braid down over body of adapter and trim to 3/8". Bare 5/8" of center conductor. Tin exposed center conductor.

Fig 3-4—Building your own antenna.

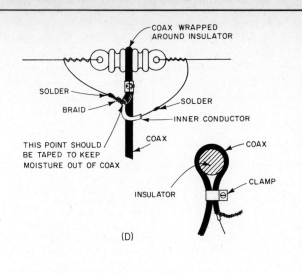

(D)

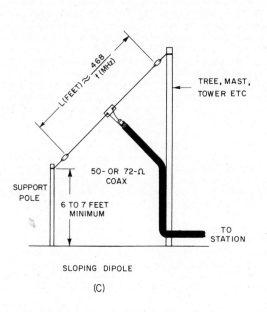

SLOPING DIPOLE

(C)

4) Screw the plug assembly on adapter. Solder braid to shell through solder holes. Solder conductor to contact sleeve.

5) Screw coupling ring on plug assembly.

Fig 3-4, continued

insulators for each individual antenna of your multiband system.

3) *Clamp.* Large enough to fit over two widths of your coaxial cable.

4) *Coax.* Feed line that consists of a center conductor surrounded by an insulating dielectric, which in turn is surrounded by a braid called the shield. You need RG-58/U (or /AU or /BU or polyfoam) or RG-8U. Look for coax with heavy braid shield such as Belden, Saxton or Times Wire and Cable. Stay away from cheap cable from unknown manufacturers and from deals on "surplus cable."

5) *Connector.* Connects the feed line (coax) to your rig. This will probably be a PL-259, standard on most rigs. If your radio needs another kind, check your radio's instruction manual for installation information. Connectors may also be needed for connections between your SWR-meter cable and your rig.

6) *Electrical tape and coax sealant.* This is needed to cover and tape antenna ends and joints to make them waterproof. Otherwise water can get into coax and "short" it, ruining your antenna system.

7) *Rope.* You need enough to tie the ends of the antenna to a supporting structure. If you make a multiband system, you may wish to use the same rope to attach ends of shorter antennas to longer antennas above.

8) *SWR meter.* This is essential for your station and especially for antenna adjustment. SWR meters are readily available and inexpensive, making them easier to buy than build.

9) *Notebook.* This is handy for noting antenna lengths, SWR readings, antenna adjustments and any problems and/or changes in your antenna system.

Gather all the parts you'll need for your chosen antenna. Most everything is available from your local electronics store or from suppliers advertising in *QST*. When this is done, the fun—actually putting together your antenna—can begin.

Assembly is quite simple. Your dipole consists of two lengths of wire, each approximately ¼ wavelength of your chosen operating frequency. These two wires are connected in the center, at an insulator, to the feed line. In our antenna, the feed line is coaxial cable, and it brings the signals to and from your radio. If you are making a multi-band dipole, think of it as several dipoles simply connected to a common center insulator and having a common feed line. The total length of the half-wave dipole is calculated by using this formula: antenna length in feet = 468/freq in MHz. (The Novice-band information below has approximate lengths already calculated.) Now measure the antenna wire, keeping it as straight as possible. Leave some extra, about 8 inches, for length adjustment and connections on each wire, and then cut your antenna.

Antenna Lengths in Feet
(cut to center of Novice band)

	½ wavelength	¼ wavelength
80 m	125'7.7"	62'9.84"
40 m	65'8.2"	32'19.1"
15 m	22'1.5"	11'0.72"
10 m	16'7.5"	8'3.8"

Remember to add about 8 in. to each end of the dipole for tuning adjustment.

Putting It Together

Carefully assemble your antenna according to the corresponding single- or multiband dipole diagrams, paying special attention to waterproofing the coax. Don't solder the antenna ends until later, when retuning is completed. Just tight-wrap them for now. Route the coax to your station, remembering to keep it as inobtrusive as possible. Cut the coax to a length that will leave some excess for strain relief so your rig won't be pulled around during strong winds! Install the connector(s) according to the diagrams. Now connect your SWR bridge between the feed line (coax) and the transmitter.

Your antenna must be "tuned" to your desired operating frequency to save wear and tear on your rig's final-amplifier stage. To do this, use your SWR meter, following the directions that came with it. If it shows an SWR of 2:1 or less at

---

Less is More . . . Or Enjoying QRP

Are you a newly licensed radio amateur trying to get on the air with a minimum dollar investment? Are you a certified Old-Timer with 5BWAS, 5BWAC and 5BDXCC under your belt and have found that hamming has lost some of its excitement? Are you somewhere in between Novice and veteran, and want to introduce your ham palate to a new spice? If you fit into any of these categories, read on.

Operating QRP (low power, defined by ARRL as 10 watts input or 5 watts output) is a popular modus operandi of thousands of our ham radio brethren. The thrill of communicating at low power levels is perhaps best described as being similar to the excitement experienced during your first QSO. Interestingly enough, the level of enjoyment proportionally goes up as your power goes down!

Many amateurs believe that QRP is a relatively new avenue of hamming that started in the '70s with the introduction of the Ten-Tec PM series and the Heathkit HW-7 transceivers. In reality, QRP dates back at least 50 years prior to the birth of these rigs. In the early days of Amateur Radio, particularly during the Depression years in the '30s, amateurs didn't have a lot of money to invest in their hobby. "Make do with what you have" and professional scrounging were the order of the day to acquire the necessary parts to build up a station. This, more often than not, limited the station transmitter to one of the QRP variety.

A look at page 45 of May 1933 *QST* reveals an article entitled "Low Power Records," listing success stories hams of the day were enjoying using QRP. Actual QSOs, not just hello-goodbye contacts, at 30 milliwatts input were being enjoyed by these pioneer QRPers. Low-power operating is, therefore, nothing new and might well be one of the few remaining common bonds with a simpler time.

The first thing anyone contemplating a jump into the QRP sport should do is cast off the notion that you must run high power (QRO). A more enlightened attitude is needed. Consider your QRP operating as an adventure, a challenge, a unique and very personal voyage on the airwaves—riding a leaf instead of a supersonic jet. It's a gentle form of communication; think "heart and soul," not "blood and guts." Shoot down your DX prey with a pea-shooter rather than a double-barrel shotgun. A positive frame of mind will set the stage for an enjoyable time with QRP. And here are some important ground rules:

1) Listen, listen, listen.
2) Call other stations; don't call CQ.
3) Expect less-than-optimum signal reports.
4) Be persistent and patient.
5) Know when to quit.

Listen to the bands and try to figure out what the prevailing propagation is. Is the skip short or long? Who's working who? Is there much interference, static or fading (QRM, QRN or QSB) present? A quick analysis of the band conditions should be the first thing you do when sitting down for a session of QRP operating. Listening will help you decide what band to operate on. A thorough study of the radio-wave propagation information in this and other ARRL publications will help you understand what you hear. Always listen, listen, listen!

Calling CQ should be done as a last resort. Most hams prefer to answer a strong signal, which you probably will not have. You will be much better off answering a CQ. Try answering someone like this: AK7M DE KA1CV/QRP or AK7M DE KB1MW/2W K. This tells AK7M why you aren't doing a meltdown on his headphones!

Don't be discouraged if you receive signal reports like RST 249 or RS 33. With less than 5 watts output, you can't expect to be overloading the receiver front ends out there in DX-land. Have faith, for you will get more than your fair share of very respectable reports. The ultimate ego gratification of a 599 or 59 will be yours if you keep at it!

If at first you don't succeed, then try again. And again. This QRP stuff is a game of persistence, so don't give up if you don't get an answer to your call on the first try. Don't be discouraged. Make up your mind that instant success just isn't part of the plan—that's what makes it so much fun.

If band conditions are inadequate, if your dipole has become the prey of a falling tree limb, if your neighbor has just started up a QRN-generating appliance, if the dog is barking, the baby is crying and your spouse is out shopping, give it up! The last thing a QRPer needs is such distractions. Go tend the baby and see what the hound is upset about. Know when to quit. Sometimes you will have days like this, so know when to take a break from hamming. Try again tomorrow (or at least wait until your spouse returns from the store).

There is a small selection to choose from if you plan to purchase a QRP-only transceiver. The Heathkit QRP radio, the HW-9, is CW-only, covering the lower portions of the popular HF bands, with WARC-band options. The HW-7 and HW-8 are earlier versions and can be found at reasonable prices in the *QST* Ham Ads or at flea markets.

The Ten-Tec Argonaut is considered the rig of rigs for QRP, featuring full CW and SSB coverage on 80-10 meters. The old Ten-Tec PM series transceivers are more basic, barebones units and should be good to get you started. Since the Argonaut and PMs are no longer manufactured, the used market is the only place to obtain them.

Some of the Japanese multimode HF transceivers can be purchased with a low-power amplifier (Japan has a 10-watt amateur license class). These are full-feature transceivers, with all the advantages of modern technology, *sans* the full-power amplifier.

If you already have the typical amateur transceiver running 100-200 watts, never fear! All that has to be done is reduce the output power by cranking down the "drive" or "RF output" control, adjusting the power output using a wattmeter.

Many amateurs have as much fun building QRP transmitters as they do operating them! Let's face it—nobody has the resources to build a deluxe, professional-quality transceiver. A two- or three-stage CW transmitter, on the other hand, is certainly within the technical ability of most of us. See the ARRL publication *Solid-State Design for the Radio Amateur,* if you have the urge to roll your own!

A very useful tool for the QRP station is the wattmeter. A commercially built unit can be had for a reasonable price. A basic single meter unit with switchable forward and reverse power is a good way to start. In time you may want to add another meter to eliminate the need to switch back and forth between forward and reverse power. Save that switch, though, and use it to change the power range of the meter. This way you can have one range for a 5-watt full scale and the other a 1-watt scale. You can calibrate this wattmeter with a VTVM (vacuum-tube voltmeter), a simple homebrew dummy load and an RF

---

your desired operating frequency, your antenna system is "tuned," and you can go ahead and operate. If your SWR is greater than 2:1, you must "retune" your antenna to obtain a lower SWR. Disconnect the transmitter, and try shortening the antenna a few inches on each end. Keep notes! Reconnect the transmitter, and check the SWR. If the SWR

gets lower, continue shortening the ends until you get the lowest possible SWR reading. Be careful not to shorten it too much! If the SWR instead gets higher, attach an extra length of wire to each end with an alligator clip. Then shorten the extra length a little at a time until you find the correct length for the antenna. For a multiband dipole system, this process

probe. Not only will you be saving some hard-earned bucks, but you will be gaining experience in designing, building, modifying and calibrating test equipment. Quite a bit to get out of such a seemingly simple project.

With QRP, your antenna is going to be much more important and instrumental in your success than if you run QRO. Running 100 watts into a random wire will net you plenty of solid contacts. But when you reduce power into that same hunk of wire, your signal effectiveness will decrease, too. As a result, the old axiom of putting up the biggest antenna you can muster, as high and as in the clear as possible, means more to the QRPer than someone running 100 or 1000 watts. The important thing is to optimize your antenna to your own personal circumstances. Many operators have reported amazing results using less-than-optimum antenna systems, but this is not to say that you should be lax in your antenna installation. By running QRP, you are already reducing your effective radiated power (ERP); no need for a further (unintentional) reduction by cutting corners on your antenna system.

For 160-30 meters, dipoles generally will work well. Height is always nice, so do the best you can get away with. Loops, end-fed wires and verticals can also be used on these bands. The popular HF DX bands, 20-10 meters, deserve some serious thought as to rotatable gain antennas—Yagis or quads. Although this train of thought usually leads to a considerable outlay of cash, you will benefit in several ways. A 1-watt signal to a 10-dB-gain Yagi will give you an effective radiated output of about 10 watts! That's like having an amplifier that needs no power to run. A directional antenna is a reciprocal device as well. That is, it complements received signals and the transmitted signal. Listening to Europeans is so much more fun when you don't have to hear them along with signals from other unwanted directions.

The VHF and UHF bands are ripe for QRP operating and some serious antenna-design work. Amateurs with even very limited space can put up a high-gain Yagi for 6 or 2 meters or above. The "headaches" of the HF bands (QRM, QRN, QSB and so on) are virtually nonexistent on VHF/UHF. The creative QRPer should be able to have a lot of success on these bands.

If you can't swing a big Yagi, either physically or financially, just get a decent wire antenna in the air. They all work, just some better than others. Again, the important thing is to put up the best system you possibly can.

Now that you're hooked, what to do about it? General operating and ragchewing is always fun, but don't limit yourself to just one kind of operating. Study the rest of this *Operating Manual*, and you should have enough info to attack several fronts with your mighty QRP signal! Here are a few suggestions.

Chase DX. Don't plan on busting a pileup that is 250 stations deep, but very respectable DX can be worked with an extremely modest station and antenna. If you are cagey enough and have a few tricks up your sleeve (see Chapter 5 on DX operating techniques), then DX will indeed be yours.

Get in a contest. During the course of a contest weekend, activity on the bands is very high. What better time to try out a new transmitter or antenna? Go ahead and jump in; the water's fine. But how to compete against the super contest stations? You don't. You can compete with other QRPers or, better yet, compete against yourself. Set a personal goal—100 contacts, 30 states or countries

or whatever—and keep a record of your results. Try to beat your previous record during the next contest. Once you get involved with QRP contesting, though, watch out—it's addicting!

Chase VHF/UHF grid squares. First, pack up your equipment and drive up to a mountain or hilltop location. On VHF, if the band is open, your 1 watt will sound like 100 watts! With a couple of watts output to a gain antenna at a high elevation, you should be able to bend a few S-meter needles.

Go portable. Put the rig in your knapsack, take a hike up a hill or mountain and set up your rig. Throw a wire over a tree limb and load 'er up. Working QRP from a scenic spot is hard to beat. Battery powered and away from all man-made electrical noise, you'll be able to hear what your receiver really can do. You can operate Field Day—not once a year, but every time you go for a hike!

Try RTTY, SSB or SSTV. Low-power operating does not automatically mean you have to operate CW exclusively. If you like RTTY, then you'll love it using QRP. The same goes for SSB, SSTV or other modes. Further, operating SSB is an excellent way to spread the QRP word. More people seem to be "reading the mail" in the voice portions of the bands, so what better place to be a salesman for the QRP movement?

You might be wondering how low can you go. Conduct minimum-power-level experiments and log your results. Get on the air with a fellow QRPer and, while monitoring your output power, see just how low in power you can go while still maintaining communications. You will be amazed at the possibilities. If the band conditions are right, there will be no difference between 1 watt and 5 watts; you may even be able to communicate at 10 or 20 milliwatts. That's less power than some flashlights run. This type of threshold experimentation can be one of the most ego-gratifying experiences in your Amateur Radio career.

On the bands, CW QRP activity centers around 1810, 3560, 7040 (7030 for QRP DX), 10,106, 14,060, 21,060 and 28,060 kHz. Phone operation is around 3985, 7285, 14,285, 21,385 and 28,885 kHz. Novices should check 3710, 7110, 21,110 and 28,110 kHz.

QRP operating can result in earning some awards with a special QRP endorsement. There are many hams who have WAS and WAC under their belt using no more than 2 watts. DXCC has even been achieved by a few hardy QRP souls. And rumor has it that one amateur is attempting to work all US counties with 250 mW. Now that's a challenge!

There are several clubs that cater to the QRPer. The QRP Amateur Radio Club International and the G-QRP group are two of the more well-known organizations, the latter being based in England. Each club publishes a quarterly newsletter for members with operating news, contest results and construction articles, as well as sponsoring QRP awards, QRP contests and regular QRP nets. This is a great way to "tune in" to the QRP world.

Two fine books now on the market are devoted to the topic of QRP. The *QRP Notebook* by Doug DeMaw, W1FB, is available from your local radio bookstore or ARRL HQ. *The Joy of QRP* by Adrian Weiss, WØRSP, is published by Milliwatt Press. These books can be quite helpful for any amateur interested in QRP operating.

So what's holding you back? Join in on the fun and thrill of communicating with low power. Amaze your fellow amateurs and outdo the "big gun" across town with your operating accomplishments. Be a rebel; *Quit Running Power!—Jeff Bauer, WA1MBK*

must be done for each antenna. Generally, if your SWR reading gets better at a higher frequency, then lengthen the antenna ends. If the SWR gets better at a lower frequency, shorten the antenna ends.

Problems and Cures

An SWR reading of greater than 4:1 probably means that something is wrong beyond merely length misadjustments. Check to see if your coax is "open" (circuit no longer continuous so current won't flow) or "shorted" (a conductor has touched another conductor so current will flow through the shorter path instead of the desired one). Check to see that your antenna isn't touching anything and that all your connections are sound.

When all systems are "go," get on the air and operate, remembering to use good on-the-air procedures. As you settle into that first QSO with your new antenna, enjoy those feelings of pride, accomplishment and fun that will naturally follow. After all, that's what Amateur Radio is all about.

Almost Time

Good show! You've chosen and acquired your rig, assembled your station and erected that antenna. The next logical step is to fire that baby up and put some RF out—but wait. Bear with us for just a little longer, while we explore just a sampling of operating procedures and on-the-air operating

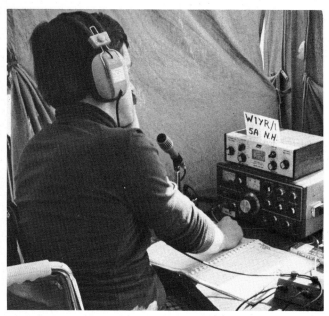

Vic, K1JUL, operated Field Day with the Nashua (New Hampshire) Area Radio Club, W1YR.

activities. Stop looking at that shiny new digital frequency readout beckoning you so seductively and read on.

BASIC OPERATING—AT LAST

We're finally getting down to the real nitty-gritty. You have learned a little of the rules used to play the game of ham radio. You should also have a solid foundation in how to go about selecting a rig and station equipment and erecting a decent antenna system.

Now it's time to learn a little about how to go about playing the ham radio game. Amateurs strive for uniformity. For everyone to be able to communicate with everyone else effectively, operating procedures should be as uniform as

Ed, KB6JOH, is a Novice class operator from Castro Valley, California. (K6SIK photo)

possible. Makes sense, doesn't it? If we all use nearly the same procedures, our communications will be accomplished more quickly to the benefit of all.

A brief explanation follows, describing the major modes of ham radio communication and a few of the procedures and conventions in use on the air today. This will give you a good foundation upon which to build your system of good operating habits, while adhering to the rules set forth for Amateur Radio. Operating procedures and conventions are generally a matter of common sense, courtesy and efficiency.

Experience is the best teacher. Spend as much time as possible listening to other operators on the bands. Find out what works and what doesn't. Copy the good habits of the good communicators and incorporate those habits into your own operating style. Building a foundation of good operating habits this early in your Amateur Radio career will stand you in good stead as an effective communicator for as long as you pound that CW key or speak into that microphone.

QRP vs. QRO

In the ham radio community, there are both the QRO (use power output up to the full legal limit) advocates and the QRP (10 watts input or 5 watts output) enthusiasts. Simply a matter of taste? Probably, but the FCC rules state that an operator "will use no more than the minimum" amount of power to maintain solid communications. Most hams run around 200 watts PEP, basically what the typical transceiver puts out. And there are times when it is necessary to run the full legal power limit to establish and maintain solid communications.

QRP operation does produce some significant benefits. For example, low-powered QRP transmitters are generally simpler in design and easier to homebrew (make yourself). QRP operation forces the operator to develop operating skills to a higher degree. The QRPer will quickly learn how to handle calling in large crowds of high-power stations to produce the maximum results. QRP operation forces the operator to maximize station design/layout to reap maximum results from the hardware (rig, antenna, etc) at hand. The QRP operator will generally learn that an improved antenna system will result in a more effective radiated signal.

If nothing else, QRP operation is becoming more popular because of the challenge that it presents to one's operating skills. QRP operators also have little or no trouble with interference complaints from neighbors! For further details on QRP, see accompanying sidebar. Note: The International Amateur Radio Union (IARU) has designated June 17 as World QRP Day

CW Operating

CW (continuous wave, Morse code) is the universal language, the common bond of hamdom. Every licensed ham has to take and pass a CW receiving test to get her/his license. A ham should take pride in mastering the code, and that is easy enough to do. All it takes is practice to master the art of sending good code on a hand key, bug (semi-automatic key) or electronic keyer. You should practice to get that smooth rhythm, practice to get that smooth spacing between words and characters, and practice to learn the sound of whole words and phrases, rather than just individual letters.

CW is an expedient mode of communications. CW transmitters are simpler devices than their phone counterparts, and a CW signal can usually get through very heavy QRM (interference on the band) much more effectively than a phone signal given equal output power, antenna arrays and propagation conditions. Therefore, it is appropriate that there be shortcuts and abbreviations used during a CW QSO. Most of the abbreviations hams use have developed within the ham fraternity; some of them are borrowed from the old-time telegraph operators. Q signals are among the most useful of these abbreviations. See Chapter 17 for a list of Q signals and their meanings.

Learn the Q signals and use any other standard abbreviations in common use on the ham bands (remember how important listening to other operators can be?). With practice, you'll find that as your CW proficiency rises, you will be able to communicate almost as quickly on CW as you can on the voice modes. For more information on CW operating, see *Morse Code: The Essential Language*, published by ARRL.

It may not seem that way to you now, but your CW sending and receiving speed will rise very quickly with on-the-air practice. For your first few QSOs, carefully choose to answer the calls from stations that are sending at a speed you can copy (perhaps another first timer on the band?). Courtesy on the ham bands dictates that an operator will slow his/her code speed to accommodate another operator. Don't be afraid to call someone who is sending just a bit faster than you can copy comfortably. That operator will generally slow down to meet your CW speed. Helping each other is the name of the game in ham radio.

To increase your speed, you may continue to copy the code practice sessions from W1AW, the ARRL HQ station. W1AW transmits flawless computer-generated CW, a pleasure to copy and an accurate indication of your copying ability. It might be a good idea to spend some time sending in step (on a code-practice oscillator—not on the air, of course) with W1AW transmissions; it may help develop your sending ability.

Phone Operating Procedures

All license classes above the Novice license already had some voice operating privileges, and now that Novice Enhancement is a reality, Novices are authorized phone on 28.3-28.5, 222.10-223.91 and 1270-1295 MHz. So now, or after you upgrade, you might want to give phone operation a try.

Learning phone procedures is exactly like learning CW operating procedure. Listen to what others are doing, and incorporate their good habits into your own operating style.

Use common sense in your day-to-day phone QSOs, too:
1) Listen before transmitting. Ask if the frequency is in

Correct CW Procedures

The best way to establish a contact, especially at first, is to listen until you hear someone calling CQ. CQ means "I wish to contact any amateur station." Avoid the common operating pitfall of calling CQ endlessly; it clutters up the air and drives off potential new friends. The typical CQ would go like this: CQ CQ CQ DE K5RC K5RC K5RC K. The letter K is an invitation for any station to go ahead.

If you hear a CQ, wait until the ham indicates he is listening (by ending with the letter K), then call him, thus: K5RC K5RC DE K3YL K3YL $\overline{AR}$ ($\overline{AR}$ is equivalent to "over").

In answer to your call, the called station will begin the reply by sending K3YL DE K5RC R....

That R ("roger") means that he has received your call correctly. That's all it means—received. It does not mean (a) correct, (b) I agree, (c) I will comply, etc. It is not sent unless everything from the previous transmission was received correctly. Perhaps K5RC heard someone calling him but didn't quite catch the call because of interference (QRM) or static (QRN). In this case, he might come back with QRZ? DE K5RC $\overline{AR}$ ("Who is calling me?").

The QSO—During the contact, it is necessary to identify your station only once every 10 minutes. Keep the contact on a friendly and cordial level, remembering that the conversation is not private and many others, including nonamateurs, may be listening. It may be helpful at the beginning to have a fully-written-out script typical of the first couple of exchanges in front of you at the operating position. A typical first-transmission might sound like this: K3YL DE K5RC R TNX CALL. RST 599 599 QTH HOUSTON TX NAME TOM. HW? DE K5RC $\overline{KN}$. This is the basic exchange that begins most QSOs. Once these basics are exchanged, the conversation via the airwaves can turn in any direction. Many people talk about their jobs, other hobbies, families, travel experiences, and so on.

Both on CW and phone, it is possible to be informal, friendly and conversational; this is what makes the Amateur Radio QSO enjoyable. During the contact, when you want the other station to take a turn, the recommended signal is $\overline{KN}$, meaning that you want only the contacted station to come back to you. If you don't mind someone else signing in, just K ("go") is sufficient.

Ending the QSO—When you decide to end the contact or the other ham expresses the desire to end it, don't keep talking. Briefly express your pleasure at having worked the other operator and then sign out: $\overline{SK}$ DE K5KG. If you are leaving the air, add CL to the end, right after your call sign.

These ending signals, which indicate to the casual listener the "status of the contact," establish Amateur Radio as a cordial and fraternal hobby. At the same time, they foster orderliness and denote organization. These signals have no legal standing; FCC regulations say little about our internal procedures. The procedures we ourselves adopt are even more important than that, because they indicate that we are not just a bunch of hobbyists playing around in random fashion, but that we are an established communications service (that we take pride in) with distinctive procedures tailored to our special needs.

Voice Equivalents to Code Procedure

Voice	Code	Meaning
over	$\overline{AR}$	after call to specific station
end of message	$\overline{AR}$	(self-explanatory)
wait, stand by	$\overline{AS}$	(self-explanatory)
roger	R	all received correctly
go	K	any station transmit
go only	$\overline{KN}$	addressed station only
clear	$\overline{SK}$	end of contact
closing station	CL	going off the air

Phone Operation for the Novice

Great news for Novices (and Technicians)! The FCC has approved the League's Novice Enhancement proposal, PR Docket 86-161, giving the green light to phone operation on 10 meters (specifically 28.3-28-5 MHz) as well as all authorized modes (including voice) on 222.10-223.91 MHz and 1270-1295 MHz. (Novices also gained an expanded CW/RTTY segment on 10 meters, 28.1 to 28.3 MHz.) See page 3-4 for a detailed listing of Novice privileges. For Novices, the 10-meter power limit is 200-watts output, with 25-watts output on 222 MHz and 5-watts output on 1270 MHz.

All this is *in addition to* the present Novice privileges (CW on 80, 40, 15 and 10 meters) which remain in full force. The FCC adopted the Novice Enhancement rules changes on January 28, 1987, released the Report and Order on February 10, 1987, with an effective date for the new privileges of March 21, 1987, at 0001 UTC.

This without a doubt is the most stirring moment in the history of the Novice license. Aside from the basic operating information presented here, the Novice is particularly directed to several other chapters of this book for guidance in using these exciting new privileges: DXing (Chapter 5), radioteletype (Chapter 9), packet radio (Chapter 10), repeaters (Chapter 11) and VHF (Chapter 12). While these chapters provide detailed information, below are some general principles of phone (voice) operation that the Novice should take into consideration as he or she prepares to enjoy this enhanced authorization.

Operating techniques and procedures vary from band to band; for example, Novices accustomed to 80-meter CW may understandably be a bit uncertain when approaching 10-meter SSB operating or when using a VHF repeater for the first time. As outlined in this book, each set of techniques is appropriate for the particular circumstances, and a good way to become familiar with them is to spend time listening to the good operators, noting how and why their methods succeed. Be discriminating; don't simply mimic whatever you have heard previously when going on the air for the first time. Follow the lead of the good operators *only*.

Whatever band, mode or type of operating you are undertaking, there are three fundamental things to remember. The first is that courtesy costs very little and is very often amply rewarded by bringing out the best in others. The second is that the aim of each radio contact should be 100% effective communication. The good operator is never satisfied with anything less. The third is that the "private" conversation with another station is actually in public. Keep in mind that many amateurs are uncomfortable discussing so-called controversial subjects over the air. Also, never give any information on the air that might be of assistance to the criminally inclined, for example, when you are going to be out of town and so forth.

Correct phone operation is more challenging than it first may appear, even though it does not require the use of code or special abbreviations and prosigns. This may be because operators may have acquired some imperfect habits in their pronunciation, intonation and phraseology even before entering Amateur Radio! To these might be added a whole new set of clichés and mannerisms derived from listening to below-par operators.

The use of proper procedure to get best results is very important. Voice operators say what they want to have understood, while CW operators have to spell it out or abbreviate. Since the speed of transmission on phone is generally between 150 and 200 words per minute, the matter of readability and understandability is critical to good communications. The good voice operator uses operating habits that are beyond reproach.

It is important to speak clearly and not too quickly, *not just* when talking to, for example a DX station who does not fully understand our language, but at all times. This is an excellent practice.

Avoid using CW abbreviations (including "HI") and the Q-code on phone, although QRZ (for "who is calling?") has become an accepted practice on HF phone. [HF, ie high frequency, refers to the 160-10 meter amateur bands.] Otherwise, plain language should be used; keep jargon to

Joyce Miller, N1CXV, of Manchester, Connecticut, is a Technician class licensee who takes full advantage of the Novice privileges that run concurrently with the Tech license. *(K1XA photo)*

a minimum. In particular, avoid the use of "we" when "I" is meant and "handle" when "name" is meant. Also, don't say "that's a roger" when you mean "that's correct." No doubt you will hear many more. Taken individually, any of these saying are almost harmless, but when combined together give a false-sounding "radioese" which is actually less effective than plain language in most cases and otherwise reflects poorly on the Amateur Service.

Procedure

Essentially there are two ways to initiate a voice contact: *call CQ* (a general call to any station) or *answer a CQ*. If activity on a band seems low and you have a reasonable signal, a CQ call may be worthwhile.

Before calling CQ, it is important to find a frequency that appears unoccupied by any other station. This may not be easy, particularly in crowded band conditions. Listen carefully—perhaps a weak DX station is on frequency. If you're using a beam antenna, rotate it to make sure the frequency is clear.

Always listen before transmitting. Make sure the frequency isn't being used *before* you come barging in. If, after a reasonable time, the frequency seems clear, ask if the frequency is in use, followed by your call. "Is the frequency in use? N2EEC." If, as far as you can determine, no one responds, you are ready to make your call.

As in CW operation, CQ calls should be kept short. Long calls are considered undesirable operating technique. For one thing, interference may be unwittingly caused to stations that were already on the frequency but whom you didn't hear in the initial check. Also, stations intending to reply to the call may become impatient and move to another frequency. If no one comes back, try again. If two or three calls produce no answer, it may be that interference is present on frequency and a new frequency should be sought, or that the particular band is not open or that your station capabilities are more suited to *calling other stations* rather than expecting to be called.

An example of a CQ call would be:

"CQ CQ Calling CQ. This is N2EEC, November-Two-Echo-Echo-Charlie, November-Two-Echo-Echo-Charlie, calling CQ and standing by."

There is no need to say what band is being used, and certainly no need to add "tuning for any possible calls, dah-di-dah!" or "K someone please!" and the like.

When replying to a CQ, both call signs should be given clearly, and the calling station may want to sign his call using the ITU phonetic alphabet (see Chapter 17) if warranted by the working conditions—that is, if your call is hard to understand or if conditions are poor. For example, phonetics are necessary when calling into a DX pileup (or initially in most HF contacts) but not necessary when calling into an FM repeater.

When calling a specific station, it is good practice to keep calls short and to say the call sign of the station called once or twice only, followed by your call repeated several times, pronounced carefully and clearly. Phonetics are appropriate when signal strengths are not good or in the presence of interference. This calling procedure should be repeated as required until a reply is obtained or the station has come back to someone else. VOX (voice operated switch) operation is helpful because, if properly adjusted, it enables you to listen between words, so that you know what is happening on the frequency. Speak in distinct, measured tones:

"N2EEC N2EEC, this is W2GD, Whiskey-Two-Golf-Delta, Over."

Once contact has been established, it is no longer necessary to use the phonetic alphabet or sign the other station's call. By FCC regulations, you need only sign your call every ten minutes or at the conclusion of the contact. (The exception is handling international third-party traffic; you must sign both calls in this instance.) A normal two-way conversation can thus be enjoyed, without the need for continual identification. Use of the word "Over" may be added at the end of a transmission to indicate a reply from the other station is expected. When FM is used, it is self-evident when you or the other station has stopped transmitting because the carrier drops.

Signal reports on phone are two-digit numbers using the RS portion of the RST system (no tone report is required). The maximum signal report would be "59," that is readability 5, strength 9. On FM repeaters, RS reports are not appropriate—when a signal has fully captured the repeater, this is termed "full quieting."

Conducting the Contact

Aside from signal strength, it is customary (as in a typical CW QSO described elsewhere in this chapter) to exchange name, location and information on equipment being used, the latter often consisting only of the model number of the transceiver and antenna. Once these routine details are out of the way, you can discuss virtually *any* appropriate topic, particularly when you are conversing with another stateside amateur.

As good conditions return to 10 meters, worldwide communication on a daily basis will be commonplace. Ten meters has proved to be an outstanding DX band when conditions are right. A particular advantage of 10 for DX work is that effective beam-type antennas tend to be small and light, making for relatively easy installation.

Keep in mind that while many overseas amateurs have an exceptional command of English (which is especially remarkable since very few US amateurs have any foreign-language capability whatsoever), they may not be intimately familiar with many of our colloquialisms and so on. Because of the language differences, some DX stations are more comfortable with the "bare-bones" type contact, and you should be sensitive to their preferences. A further point is that in unsettled conditions, it may be necessary to keep the whole contact short in case fading or interference occur. The good operator takes these factors into account when expanding on a basic contact.

When the time comes to end the contact, end it. Thank the other operator (once) for the pleasure of the contact and say goodbye: "This is N2EEC, clear." This is all that is required. Unless the other amateur is a good friend, there is no need to start sending best wishes to everyone in the household including the family dog! Nor is this the time to start digging up extra comments on the contact which will require a "final final" from the other station (there may be other stations waiting to call in). Please understand that during a band opening on 10 meters or on VHF, it is crucial to keep contacts brief so as many stations as possible can work the DX coming through.

Additional Recommendations

Listen with care. It is very natural to answer the loudest station that calls, but with a little digging, if need be, answer the best signal. Not all amateurs can run a kilowatt, but there is no reason every amateur cannot have a signal of the highest quality. Do not reward with a contact the operator who has cranked-up his transmitter gain to the point of splattering if another station is calling.

Use VOX or push-to-talk (PTT). If you use VOX, don't defeat its purpose by saying "aah" to keep the relay closed. If you use PTT, let go of the mic button every so often to make sure you are not "doubling" with the other station. Don't filibuster.

Keep microphone gain constant. Don't "ride" the mic gain. Try to speak in an even amplitude the same distance from the microphone, keeping the gain down to eliminate room noise. Follow the manufacturer's instructions for use of the microphone; some require close talking, while some need to be turned at an angle to the speaker's mouth. Speech processing (sometimes built in to contemporary transceivers or available as an outboard accessory) is often a mixed blessing. It can help you cut through the interference and static, but if too much is used, the audio quality suffers greatly. Tests should be made to determine the maximum level that can be used effectively, and this should be noted or marked on the control. Be ready to turn it down or off if it is not really required during a contact.

The speed of voice transmission (with perfect accuracy) depends almost entirely on the skill of the two operators concerned. Use a rate of speech that allows perfect understanding as well as permitting the receiving operator to record the information. Further voice-operating hints appear elsewhere in this chapter.

V/UHF

In general, "weak-signal" VHF/UHF operating practices are not very different that those common on HF. Pileups can occur in good conditions, and because these openings can be short-lived, it is good procedure to limit the contact to exchange of signal reports or grid squares and location details. Other details can be left to the QSL card. As mentioned above, in this way more stations get a chance to work the DX station.

When conditions are good, *call CQ sparingly,* if at all, unless you have an extremely potent signal. The old adage that "if you can't hear 'em, you can't work 'em" is just as true at VHF as HF. All too often, stations can be heard calling CQ DX on the same frequency as distant stations that they cannot hear.

One feature of VHF operating which is rarely found on the HF bands is the use of *calling frequencies.* The idea is that each frequency provides a "meeting place" for operators using the same mode. Once contact has been set up, a change to another frequency (the working frequency) is arranged so that others can use the calling frequency. See Chapter 12 for details.

Repeaters

A repeater is a device that receives a signal on one frequency and simultaneously transmits (repeats) it on another frequency. In a sense, it is a robot transmitter. Often located atop a tall building or high mountain, V/UHF repeaters greatly extend the operating coverage of amateurs using mobile and hand-held transceivers.

To use a repeater, you must have a transceiver with the capability of transmitting on the receiver's *input frequency* (the frequency that the repeater listens on) and receiving on the repeater's *output frequency* (the frequency the repeater transmits on). This capability can be acquired by dialing the correct frequency and selecting the proper *offset* (frequency difference between input and output).

When you have the frequency capability (and most rigs today are fully synthesized), all you need do is key the microphone button and you turn on (access) the repeater. Some repeaters have limited access, requiring the transmission of a subaudible tone, a series of tones or bursts to gain access. However, the vast majority of repeaters are open. Most repeaters briefly transmit a carrier after a user has stopped transmitting to inform the user that he is actually accessing a repeater.

After acquiring the ability to access a repeater, you should become acquainted with the operating practices that are inherent to this unique mode of Amateur Radio. See Chapter 11 for specifics on operator practices.

Further notes on a variety of other operating procedures for specialized modes/activities can be found in the relevant chapters noted above. *(Tnx RSGB Operating Manual)*

use before making a call on any particular frequency.

2) Give your call sign as needed, using the approved ITU (International Telecommunication Union) Phonetics. Those phonetics are given in Chapter 17.

3) Make sure your signal is "clean." Do not keep your microphone gain turned up too high.

4) Keep your transmissions as short as possible to give as many operators as possible a chance to use the frequency spectrum.

An efficient phone operator can pass a lot of information in a very short period of time. Strive to become an efficient phone operator. See accompanying sidebar on "Phone Operation for the Novice" for more details

FM and Repeaters

Generally, FM (frequency modulation) is a voice mode of operation, available to Technician class operators and higher (although Novices are now able to operate FM on 222.10-223.91 MHz and 1270-1295 MHz, under Novice Enhancement). This manual has an excellent chapter on FM and repeater operation; see Chapter 11.

Specialized Communications

Specialized communications, including satellite communications, the digital modes of communications, including radioteletype, AMTOR and packet radio, all have procedures distinct to each particular mode. This manual thoroughly covers these modes in other chapters, while additional information can be found in *The ARRL Handbook*.

Your First QSO

That's it in a nutshell. You now have more than enough information to get started on making your first QSO. There's no doubt that you can do it and do it well. See accompanying sidebar.

Here are a few more hints, tips, suggestions and a little helpful advice. You might want to hang on for a few minutes and read a few more pages before rushing to fire up your rig.

Your First Contact

Are your palms sweaty, knees weak, hands shaking and is there a queasy herd of butterflies (wearing spikes, perhaps) performing maneuvers in your stomach? Chances are you're facing your first Amateur Radio QSO. If so, take heart! Although few may admit it, the vast majority of hams felt the same way before firing up the rig for that first contact.

Although nervousness is natural, there are some preparations that can make things go smoothly so that a few of those butterflies are released. Practicing QSOs with a code-practice oscillator is a good way to become familiar with the typical QSO format. You can practice solo, with a friend, or perhaps with several members of the local Amateur Radio Club.

Another thing that some people find helpful is to write down information that you will use during the QSO in advance. While spelling in code may pose no problem while you are calm, nerves make it tough for some to remember where they were in a word or, for that matter, how to spell it.

One nervous Novice made notes of her own QSO data on index cards. One card contained her QTH, another her name (no kidding!); but after a half dozen contacts, she gradually forgot to use the cards and could spell her address in perfect code—asleep.

So will you.

If you are concerned that you'll have trouble copying the code, a good security blanket to have is an experienced operator in your radio room shack during your QSOs. You'll find that after your initial nervousness wears off that you are able to do just fine by yourself, but you will have honored a ham friend (perhaps your teacher or your "Elmer") by allowing him the privilege of sharing your initial QSO. Indeed you may also want your Elmer to help you write up an equipment tune-up checklist so that you can put a properly adjusted signal on the air every time, even when Elmer isn't around.

Establishing Contact

You may choose to answer someone who is soliciting a QSO by calling CQ, the abbreviation used by a station willing to

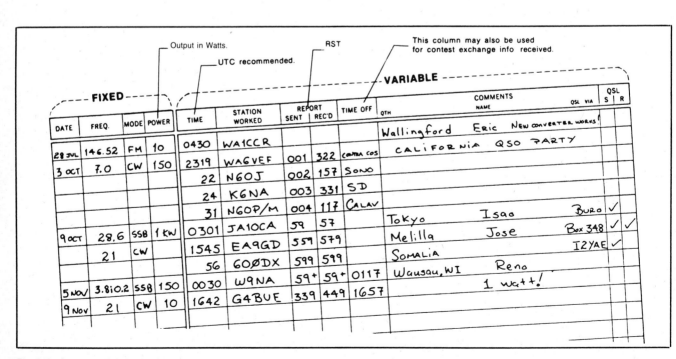

Fig 3-5—An example of a well-kept log.

talk to anyone who answers, or you may choose to call CQ yourself. If you answer someone else's CQ, try to find someone who is sending CW at a rate that you can copy. If you choose to send a CQ yourself, be careful to send only as fast as you are comfortably able to copy. A long CQ is unnecessary; try a 3 × 3 (CQ CQ CQ DE AB1U AB1U AB1U K) [substitute your own call for AB1U]. This means "Calling anyone interested in establishing contact with me DE (from) AB1U (your call sign) K (go ahead, any station transmit)."

On voice, a simple "CQ CQ, this is AB1U, Alfa Bravo One Uniform, over," will suffice.

QSO Format

About the only things common to all QSOs are:

1) An exchange of personal data. Names, QTHs (locations), signal reports and QSL-card information (if appropriate).

2) An honest signal report. The signal report allows the operators to assess how well they are hearing each other and adjust their operating procedures as necessary. See Chapter 17 for details on the RST signal-reporting system.

After an exchange of personal data and a signal report, the scope or the direction that the QSO takes is limited only by the imagination and willingness of the participants to carry it anywhere they choose. This leads to many fascinating QSOs; you never know what direction a QSO will take.

LOGGING AND QSLING

Although the FCC does not require logging except for certain specialized occurrences (see *FCC Rule Book*), you can record the history and a lot of pleasant memories of your on-the-air progress if you choose to keep an accurate log such as the one shown in Fig 3-5. *The ARRL Logbook*, on sale at your local radio bookstore or directly from ARRL HQ, provides a good method for maintaining your log data at your fingertips.

The log entry should include:

1) The call sign of the station worked.

2) The date and time of the QSO. Always use UTC (aka GMT or Zulu time) when entering the date and time. In fact, use UTC whenever you need a time or date in your hamming activities. The use of UTC helps all hams avoid confusion

---

UTC Explained

Ever hear of Greenwich Mean Time? How about Coordinated Universal Time? Do you know if it is light or dark at 0400 hours? If you answered no to any of these questions, you'd better read on!

Keeping track of time can be pretty confusing when you are talking to other hams around the world. Europe, for example, is anywhere from 4 to 11 hours ahead of us here in North America. Over the years, the time at Greenwich, England, has been universally recognized as *the* standard time in all international affairs, including ham radio. (We measure longitude on the surface of the earth in degrees east or west of the Prime Meridian, which runs approximately through Greenwich, and which is halfway around the world from the International Date Line.) This means that wherever you are, you and the station you contact will be able to reference a common date and time easily. Mass confusion would occur if everyone used their own local time.

Coordinated Universal Time (abbreviated UTC from the French *Universelle Tempes Coordinaté*) is the name for what used to be called Greenwich Mean Time.

Twenty-four-hour time lets you avoid the equally confusing question about AM and PM. If you hear someone say he made a contact at 0400 hours UTC, you will know immediately that this was 4 hours past midnight, UTC, since the new day always starts just after midnight. Likewise, a contact made at 1500 hours UTC was 15 hours past midnight, or 3 PM (15 − 12 = 3).

Maybe you have begun to figure it out: Each day starts at midnight, 0000 hours. Noon is 1200 hours, and the afternoon hours merely go on from there. You can think of it as adding 12 hours to the normal PM time— 3 PM is 1500 hours, 9:30 PM is 2130 hours, and so on. However you learn it, be sure to use the time everyone else does—UTC. See chart below.

The photo shows a specially made clock, with an hour hand that goes around only once every day, instead of twice a day like a normal clock. Clocks with a digital readout that show time in a 24-hour format are quite popular as a station accessory.

UTC	EDT/AST	CDT/EST	MDT/CST	PDT/MST	PST
0000*	2000	1900	1800	1700	1600
0100	2100	2000	1900	1800	1700
0200	2200	2100	2000	1900	1800
0300	2300	2200	2100	2000	1900
0400	0000 *	2300	2200	2100	2000
0500	0100	0000 *	2300	2200	2100
0600	0200	0100	0000 *	2300	2200
0700	0300	0200	0100	0000 *	2300
0800	0400	0300	0200	0100	0000 *
0900	0500	0400	0300	0200	0100
1000	0600	0500	0400	0300	0200
1100	0700	0600	0500	0400	0300
1200	0800	0700	0600	0500	0400
1300	0900	0800	0700	0600	0500
1400	1000	0900	0800	0700	0600
1500	1100	1000	0900	0800	0700
1600	1200	1100	1000	0900	0800
1700	1300	1200	1100	1000	0900
1800	1400	1300	1200	1100	1000
1900	1500	1400	1300	1200	1100
2000	1600	1500	1400	1300	1200
2100	1700	1600	1500	1400	1300
2200	1800	1700	1600	1500	1400
2300	1900	1800	1700	1600	1500
2400	2000	1900	1800	1700	1600

Time changes one hour with each change of 15° in longitude. The five time zones in the U.S. proper and Canada roughly follow these lines.

*0000 and 2400 are interchangeable. 2400 is associated with the date of the day ending, 0000 with the day just starting.

Fig 3-6—UTC Explained.

The N7BH World Time Finder

I eliminated the chance for error in local-to-UTC time and date conversions by keeping a two-function clock set to UTC. This does not help to determine the date and time in other time zones around the world, however. To solve this problem. I devised a circular slide rule that will provide the conversion when set to local time. Notice that for every 15° of longitude around the earth, the time changes one hour. Thus there are 24 Time Zones, each 1 hour different from the next, around the world.

This World Time Finder converts time and date between all time zones. The outer scale is in hours using the 24-hour format. At midnight, the date becomes either "tomorrow" or "yesterday" according to the direction of time conversion indicated by the plus-minus arrows.

The moveable second scale indicates the zone identification letter. This letter is used to identify the location of time being used. For example, 2100T would indicate 8 PM Mountain Standard Time or Pacific Daylight Time. Note that adapting daylight time has the same effect as shifting one time zone to the east, and thus one hour closer to UTC. If you pass the International Date Line in doing a time conversion, you again gain or lose a day, depending on which direction you are going.

You can make your own World Time Finder by cutting the two patterns from Chapter 17 of this book. Put each scale between two pieces of clear plastic, or laminate them between sheets of clear CONTACT® paper (a sticky-backed plastic material sold in many department stores). Punch a hole in the center of each scale, and use a 1/8-inch rivet with washers for the center pin.

The World Time Finder is operated by aligning the time zone letter for your area with the current hour. In the example shown, at zone U (PST) it is 2000 hours (8 PM). Zone U is 120° from the Prime Meridian, or 8 hours earlier than UTC. If the zone U date is July 1, then the time and date at UTC is 0400 (4 AM) on July 2. In Japan (zone 1), which is 135° from the Prime Meridian and 9 hours later than UTC, it would be 1300 (1 PM) on July 2.

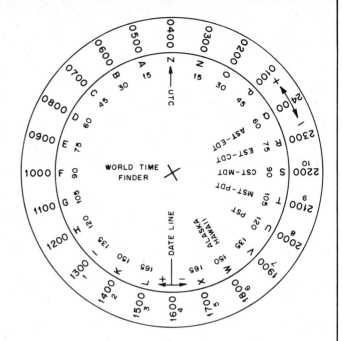

Note that this time conversion can be done by going in either a clockwise or counterclockwise direction around the chart. Converting time by going in a counterclockwise direction from zone U to UTC, you must pass midnight on the outer scale, thus gaining a day. If you perform the conversion by going in a clockwise direction around the scale, you must pass the International Date Line on the center scale, again gaining a day. The gain or loss of date is established easily by the + or − sign next to the arrows.

Fig 3-7—The N7BH World Time Finder.

through conversion to local time. Figs 3-6 and 3-7 explain the UTC system. Please study this material carefully.

3) The frequency or frequency band on which the QSO took place.

4) The emission mode of communication.

5) Signal reports sent and received.

6) Any miscellaneous data, such as the other operator's name or QTH, that you care to record.

The FCC has promulgated a rather elaborate system of emission designators (see Chapter 2 on spectrum management). For logging simplicity, the following common abbreviations are suggested instead:

Abbreviation	Explanation
CXR	steady, unmodulated pure carrier
CW	telegraphy on pure continuous wave
MCW	tone-modulated telegraphy
SSB	single-sideband suppressed carrier
AM, DSB	double-sideband with full, reduced or suppressed carrier
FAX	facsimile
FM	frequency- or phase-modulated telephony
RTTY	radioteletype
AMTOR	time diversity radioteletype
P	pulse
TV or SSTV	television or slow-scan television

Computer Logging

If you are into personal computing, and have your own machine or access to a personal computer, there are many log-keeping programs that will take your input, format your log file and even give you a printed output. The On Line column in *QST* lists many electronic logging programs and information on how to obtain them. Numerous computer programs for contest applications appear in Chapter 7.

A well-kept log will help you preserve your fondest ham radio memories for years, while serving as an accurate bookkeeping system should you embark upon a quest for ham radio awards or a complete collection of QSL cards.

QSLing

The QSL card is the final courtesy of a QSO. It confirms specific details about your two-way contact with another ham. Whether you want the other station's QSL as a memento of an enjoyable QSO or for your collection of the cards necessary to apply for some ham radio operating award, it is wise to have your own QSL cards and know how to fill them out. That way, when you send your card to the other station, it will result in the desired outcome (his card sent to you).

Your QSL

Your QSL card makes a statement about you. It may also hang in ham radio shacks all over the world. So you will want to choose carefully the style of QSL that you have printed.

There are many QSL vendors listed in the Ham Ad section of *QST* each month. A nominal fee will bring you many

The Final Courtesy

QSL cards are a tradition in ham radio. Exchanging QSLs is fun, and they can serve as needed confirmations for many operating awards. You'll probably want your own QSL cards, so look in *QST* for companies that sell them, or you may even want to make your own. Your QSL should be attractive, yet straightforward. All necessary QSO information should be on the card; in fact, putting all the information on one side of the card will make answering your QSL a relatively simple matter.

A good QSL card should be the standard 3.5- × 5.5-in. size and should contain the following information:

1) Your call. If you were portable or mobile during the contact, this should be indicated on the card.

2) The geographical location of your station. Again, portables/mobiles should indicate where they were during the contact.

3) The call of the station you worked. This isn't as simple as it sounds. Errors are very common here.

4) Date and time of the contact. Use UTC for both and be sure you convert the time properly. It is best to write out the date *in words* to avoid ambiguity, ie, May 10, 1987 or 10 May 1987, rather than 5/10/87.

5) Frequency. The band in wavelength (meters) or approximate frequency in kHz or MHz is required.

6) Mode of operation. Use accepted abbreviations, but be specific.

7) Signal report.

8) Leave no doubt that the QSL is confirming a two-way contact by using language such as "confirming two-way QSO with," or "2 X" or "2-Way" before the other station's call. Other items, such as your rig, antenna and so on, are optional.

9) If you make any errors filling out the QSL, destroy the card and start over. Do not make corrections or mark-overs on the card, as such cards are not acceptable for awards purposes (see "PSE QSL . . . Correctly").

Now comes the problem of how to get your QSLs to the DX station. Sending them directly can be expensive, so many amateurs use the ARRL's Overseas QSL Service. This is an outgoing service that allows ARRL members to send DX QSL cards to foreign countries at a minimum of cost and effort. An unlimited number of your cards may be sent for distribution 12 times a year, and the fee is just $1 per pound or portion thereof (155 QSLs average a pound) for ARRL members. This amounts to quite a bargain!

To receive QSL cards from DX (overseas) stations, the ARRL sponsors incoming QSL Bureaus, provided free for all amateurs throughout the United States and Canada. Each call area has its own bureau staffed totally by volunteers. To expedite the handling of your QSL cards (both incoming and outgoing), be sure to follow the bureau's requirements at all times. See Chapter 5 for further details on the ARRL QSL Bureau System.

These typical QSL cards will guarantee you many happy returns.

---

samples from which to choose or you may design and/or print your own style. The choice is up to you. See the accompanying sidebar for more information on QSLs.

That just about finishes our ever so slight foray into operating procedures. Your knowledge on these subjects and other topics in operating will grow quickly with your on-the-air experience. Enjoy the learning process. Ham radio is such a diverse activity that the learning process never stops—there is always something new to learn. Here's a quick tour of some of the many on-the-air activities that you may enjoy.

ON-THE-AIR ACTIVITIES

There are so many activities open to us as radio amateurs that it's hard to decide which to tell you about first. Among the most popular are ragchewing, traffic handling, DXing, contesting and awards hunting.

"Chewing the rag" refers to getting on the air and spending minutes (or hours!) in interesting conversation on virtually any and every topic imaginable. Without a doubt, the most popular operating activity is ragchewing.

Traffic handling is part of that public-service responsibility of the Amateur Radio Service that we spoke of at the beginning of this chapter. Hams handle third-party traffic (messages for nonhams) in both routine situations and in times of disaster. Public-service communications make Amateur Radio the valuable public resource that it is. Traffic handling is covered in great detail in Chapter 15.

Nets are regular gatherings of hams who share a mutual interest and use the net (short for "network") to further that interest. Nets most commonly meet to pass traffic or participate in one of the many other ham activities from awards chasing and DXing to just plain old ragchewing.

There are several nets dedicated to Novice or slow speed CW operation that can help a newcomer sharpen operating skills. The *ARRL Net Directory,* a compilation of public-service nets, is available from ARRL HQ for $1.

DX (ham shorthand for long distance) is one of those terms in ham radio that holds a special fascination for many hams. DX is one of those ambiguous concepts that is correctly de-

Fourteen-year-old Brian Clark, KA8WAG, operated his first contest—the 1986 Novice Roundup.

PSE QSL . . . Correctly

Thousands of QSLs come through the ARRL DXCC and Awards Desks each day. Most pass inspection and are used to qualify for an award, but some don't make it.

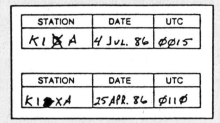

Fig 3-8—Two examples of call-sign alterations. *Who* changed the call isn't important. Neither card is acceptable for awards credit.

The most common cause for rejection is the altered card, the QSL that has letters in the call sign either crossed out or written over (see Fig 3-8). Even if the alteration is made with the best of intentions, the card won't make the grade. Imagine if this were a card for a WAS or WAC application. Worse, what if it's the card that could knock your DXCC count to 99. If your call appears correctly in another place (say in the address area), the card is still unacceptable. Your call (unaltered) must appear accurately in the QSO report. A good rule of thumb is, when you accidentally make a writing error while filling out a QSL, destroy the card and start over.

Believe it or not, the second most common mistake occurs when the QSLer puts his or her call in place of yours in the QSO report. This confirms a self QSO! In this case, if your call appears correctly somewhere else on the card, it's okay.

Sometimes, the QSLer may write your call sign wrong, even if he's worked you under the right call. This is common with 2 × 1 calls. For example, suppose while on a DXpedition to Peter I island, you work KR1R, but you mistakenly make out the card to K1RR. KR1R now has a card made out to K1RR. Even if you send the card to KR1R's address, it still doesn't count for KR1R. We don't even want to think what would happen if K1RR should get hold of the card!

Once in a great while, the QSLer forgets to put your call in the QSO report. In this case, if your call is somewhere else on the card, the card is good. If not, no credit.

Occasionally, a QSL will come through without the band or mode indicated. Is this the end of the world? Only if that card is part of a Five-Band DXCC, Five-Band WAC or Five-Band WAS, or a specialty-mode WAC, WAS or DXCC that requires that information. For example, all cards submitted for a CW endorsement must confirm a code contact. If the card lacks any mode indication but an RST report is provided, this information is acceptable to show that it was a CW contact, if all else fails.

Sometimes, the date is important. For 5BWAS, cards must be dated on or after January 1, 1970. For CW DXCC, cards must be dated on or after January 1, 1975. So before submitting for these special awards, check the date!

Okay, so you've checked over your cards and found some with one or two of these errors. What now? You've got two choices: You can send the card(s) back to the station(s) worked and ask for a replacement, or you can get on the air and work that country or state again (of course, getting a replacement card for, say, FO0XX or 3C0A is much easier than waiting to work the next DXpedition!). If you decide to obtain a replacement card, do so as quickly as possible. Sometimes logs get tossed out, and the QSO information is lost forever. Obtaining a valid replacement card first and having your application pass with flying colors is much better than having your application hang in limbo because one or more of your cards was obviously not acceptable. Finally, if someone asks you for a replacement card, return the new card promptly.

QSLing is the final courtesy of the QSO and, no doubt, for most of us it's fun to exchange cards. If we all take the time to check over our cards before we send them out, we can avoid the hassles mentioned above. This makes QSLing and award collecting that much more enjoyable. —*Frank Vesci, NG1J, ARRL Awards Manager (originally published in Sep 1986 QST)*

In Pursuit of . . . DX

"I'll never forget the thrill of my first DX contact. It was on CW in the Novice portion of 15 meters. I heard a G3 (a station from England) calling CQ, and no one answered right away so I decided to give it a try. Success! With a lump in my throat and sweat on my palms, I managed to complete my first DX QSO and become hopelessly hooked. DX is great!"

DX is Amateur Radio shorthand for long distance; furthermore, DX is universally understood by hams to be a station in a foreign country. Chasing DX has steadily become one of the most popular activities in our hobby, and you—the newly licensed amateur—can get in on the fun!

Within your Novice frequency allocations, you can work DX just as easily as the Extra Class ham can. You don't need a super kilowatt station and a huge antenna installation. The beauty of DXing is that your operating skills can overcome deficiencies in your station equipment. Good operating habits and techniques are every bit as important as an elaborate station set-up.

Jumping into the world of DX can be tantalizing but challenging. New operating techniques, new jargon, new ways of thinking about radio—all of these must be mastered by the successful DXer. Don't worry; anyone willing to invest the time and effort can learn the skills necessary to become a top-notch DXer.

A large part of DXing is listening. Listen to how others work DX and copy their good habits. Learn all you can from the established DXers; they are a gold mine of information. Corner a DXer at your next local radio club meeting and get all the tips you can. Don't be afraid to ask for help in person or on the air. DXers will usually go out of their way to help someone who is genuinely interested. This book, plus several others published by ARRL, is chock-full of DX operating tips.

When you feel confident enough to try working DX, find out where to locate the DX stations. Listen for them, and get to know where they congregate. Study the propagation characteristics of the bands. Your Novice studies taught you that 10 meters is a good DX band when it is open, that 15 meters is probably the most reliable Novice DX band, and that time of day also plays a part in considering when and where to work DX. Armed with this information, the DXer need only be in the right place at the right time.

A Hobby with No Barriers

One of the outstanding characteristics of hamming is that anyone can enjoy it. But Amateur Radio has a special appeal to persons with visual or physical disabilities: it provides a means of people-to-people contact on a basis of absolute equality. Ham radio has always been a natural way of making new friends, renewing old acquaintances and visiting—via radio—people and places all over the world. And it is perfect for those with limited mobility. There are hundreds of individuals with disabilities active in Amateur Radio. They have shown that with the proper training, patience and determination, almost any physical disability can be overcome.

The Handi-Hams

A nationwide group specializes in helping persons with disabilities become a part of the hobby—the Handi-Hams. Bruce Humphrys, KØHR, the former director of the Handi-Hams, described the program in January 1980 *QST:* "Courage Center is a comprehensive rehabilitation facility for persons with physical, speech, hearing and visual handicaps, based in Golden Valley, Minnesota. Aside from the Handi-Ham System, Courage Center offers a tremendously wide array of services for handicapped people . . . the Handi-Ham System is one of the Center's services—serving more than 1000 people all over the world. The System provides three direct (and many indirect) services: (1) educational material, fraternity and close personal attention; (2) Amateur Radio equipment on loan; and (3) specially designed devices for ease of station control. The System relies on a trained cadre of radio amateurs to help provide these services. We not only help new students to get their first license, but also help handicapped licensed hams to upgrade."

Every year, the Handi-Hams hold radio camps for students with disabilities who are studying for new or upgraded ham tickets. FCC exams are conducted after each session.

Further information can be obtained by contacting the Courage Handi-Hams, 3915 Golden Valley Rd, Golden Valley, MN 55422, tel 612-588-0811.

Getting the Ticket

All candidates for an amateur license must prove their ability to send and receive the International Morse code,

and must pass a test on radio regulations and theory. Blind applicants may use a typewriter or a Braille-writer for the code test, or call out the words to be recorded by someone else. Similarly, answers to the written exam may be dictated to a sighted person, under the provisions of Section 97.26(g) of the US amateur regulations. In addition, the FCC does not deny anyone the right to take the examination because of a disability that prevents travel to an examination point. In such cases, the examination may be conducted (by accredited Volunteer Examiners) at a location convenient to the applicant. All that's required is that the application be accompanied by a physician's written statement certifying the nature of the applicant's disability. The applicant must also include a statement describing any special procedures required in the administration of the exam, and/or the names of persons assisting the applicant in the dictation and transcribing of test questions and answers.

ARRL Involvement

The ARRL Program for the Disabled provides sources of information that enable persons with disabilities to prepare for FCC amateur exams. ARRL HQ serves as an information center and clearinghouse for this purpose, and maintains a current reference file containing lists of operating aids and clubs, and other pertinent information. In addition, *QST* publishes technical and general-interest articles of special interest to persons with disabilities as well as a comprehensive booklet on the Disabled Program that includes reprints of informative *QST* articles (contact ARRL HQ for information on obtaining a copy).

Membership in the League is available at a reduced rate without *QST* to persons who are blind. This gives already licensed sightless members an opportunity to take an active role in ARRL activities, vote in ARRL Director and SM elections, run for League offices and otherwise share in the privileges of full membership. Unlicensed blind persons, too, may show their support for the work of the ARRL through an associate membership. For more information, write the Program for the Disabled Coordinator at ARRL HQ.

QST is available on a flexible disk, directly from the Library of Congress, Division for the Blind and Handicapped, 1291 Taylor St, NW, Washington, DC 20542.

Keep these tips in mind. If you hear a lone DX station calling CQ, work it. If there is a pileup (a lot of stations trying to work the same station), don't jump right in. Listen for five or six contacts so that you can figure out how the DX station is handling the pileup. Remember to keep your calls short, and don't call the DX station after he has gone back to someone else. Try not to send the DX station's call more than once, if at all. It's better to sign just your call. If the DX station is working "tailenders" (stations who call immediately at the end of the DX station's last QSO), try it. If the station is not working tailenders, don't disrupt the proceedings by trying it. If the DX station is working only stations in a call area other than your own, perhaps conditions are not yet quite right for you. Listen, and follow the DX station's instructions to the letter. Be courteous and fair to all the stations on frequency; using good operating procedures is every bit as important as logging that DX.

Check the How's DX column in *QST*. It contains tips on propagation and news of interest to DXers. And don't forget to listen to the weekly W1AW DX bulletin for the latest in DX operations. Good DX!

fined in different ways by different people. In the beginning, to most amateurs DX is the lure of seeing how far away you can establish a QSO—the greater the distance, the better. DX can be a personal achievement, bettering some previous "best distance worked," involving a set of self-imposed parameters. DX can also be competing on a larger scale, trying to break the pileups on a rare DX (foreign) station and aiming for one of the DX-oriented awards. DXing can be a full-time goal for some hams and a just-for-fun challenge for others. No matter if you turn into a serious DXer or just have a little fun, DXing is one of the more fascinating aspects of Amateur Radio. See Chapter 5 for more particulars on DXing.

Contesting

Contesting is to Amateur Radio what the Olympic Games are to worldwide amateur athletic competition: a showcase to display talent and learned skills, as well as a stimulus for further achievement through competition. Increased operating skills and greater efficiency are the predominant end results of Amateur Radio contesting, whether the operator is a serious competitor or a casual participant.

Don't believe it? Tune across the band, any band, and listen for the most efficient operators. Chances are better than even that they are avid contesters or at least enjoy casual participation is some contests.

The contest operator is also likely to have one of the better signals on the band—not necessarily the most elaborate station equipment, but a signal enhanced by the most efficient use of station components available. Contest operation encourages optimization of station and operator efficiency.

Nearly every contest has competitors vying to see which one can work the most stations (depending on the rules of the particular contest) in a given timeframe. In some contests, the top-scoring stations have consistently worked 100 or more stations per hour for the entire 48-hour contest period.

The ARRL contest program is so diverse that it holds some appeal for almost every operator—the beginning contester and the old hand, the newest Novice and oldest Extra-Classer, "Top Band [160 meters] Buff" and microwave enthusiast. Complete contest entry rules and results appear in *QST*.

Two of the ARRL-sponsored contests hold particular appeal for the newly licensed ham: Novice Roundup and Field Day.

The Novice Roundup is competition geared for the beginning (Novice and Technician class) amateur. Usually held in late January through early February, the Novice Roundup (NR) is just what the doctor ordered for the bashful Novice waiting to make her first QSO. Since the format of each QSO is exactly the same, you need only write down your QSO data, transmit it to the station you contact, and receive similar data, in the same order, in return. In the Novice Roundup, you increase code speed through operation, and work stations needed for the WAS (Worked All States) Award and other achievements. There are certificates for ARRL section winners, and a good time is had by all.

Always held on the fourth full weekend of June, Field Day (FD) sees upwards of 25,000 hams taking their stations to the fields and operating under simulated emergency conditions. There is plenty of opportunity for a newly licensed ham to gain some supervised operating time at a club-sponsored Field Day site. Don't miss this contest.

See Chapter 7 for further details on the exciting world of contesting.

A lot of fun can be had in the pursuit of one or more of the many awards offered for various operating achievements. Generally, an award is based on some particular operating achievement such as working and confirming (getting QSL cards from) all 50 United States on various bands and various modes.

Amateur Radio Societies from around the world offer different awards (see Chapter 8) and even local radio clubs offer special awards for those who pursue such things. Many of the awards are very handsome certificates or intricately designed plaques very much in demand by awards chasers.

In addition to the international awards programs described in Chapter 8, ARRL administers an extensive awards regimen, including the premier operating award in Amateur Radio—DXCC. See Chapter 8 for details.

Well, this is the end of the chapter, but just the beginning of your Amateur Radio career. Ham radio is a diverse, exciting and dynamic hobby. There's a lot of fun to be had, and we're glad we could be here at the beginning to share some with you.

It's time for you to go and make that first QSO now, because that's what ham radio is all about. Have fun, and we hope to work you on the air soon.

Chapter 4

Antenna Orientation

Anyone laying out a fixed directive array does so in order to put his signal into certain parts of the world; in such cases, it is essential to be able to determine the bearings of the desired points. Too, amateurs with a rotatable directive array like to know where to aim if they are trying to pick up certain countries. And even amateurs with the single wire are interested in the directive pattern of the lobes when the wire is operated harmonically at the higher frequencies, and often are able to vary the direction of the wire to take advantage of the lobe pattern.

Which Direction?

It is probably no news to most people nowadays that true direction from one place to another is not what it appears to be on the old Mercator school map. On such a map, if one starts "west" from Pratt in central Kansas, he winds up in the neighborhood of Beijing, People's Republic of China. Actually, as a minute's experiment with a strip of paper on a small globe will show, a signal starting due west from Pratt never hits China at all but rather passes over Perth, in Western Australia.

"The shortest distance between two points is a straight line" is true only on a flat surface. The determination of the shortest path between two points on the surface of a sphere is a bit more complicated. Imagine a plane that intersects two points on the surface and the center of the sphere. The intersection of the plane and the sphere describes a circle on the surface of the sphere that is defined as a great circle. The shortest distance between the points follows the path of the great circle. The direction or bearing from your location to another point on the earth is the direction of a great circle as it passes through your location on its way to the other point.

If, therefore, we want to determine the direction of some distant point from our own location, the ordinary Mercator projection alone is utterly useless. True bearing, however, may be found in several ways: by using a special type of world map that does show true direction from a specific location to other parts of the world, by working directly from a globe or by using mathematics.

DETERMINING TRUE NORTH

Determining the direction of distant points is of little use to amateurs erecting a directive array unless they can put up the array itself in the desired direction. This, in turn, demands a knowledge of the direction of *true* north (as against magnetic north), since all our directions from a globe or map are worked in terms of true north.

A number of ways may be available to amateurs for determining true north from their location. Frequently, the streets of a city or town are laid out, quite accurately, in north-south and east-west directions. A visit to the office of your city or county engineer will enable you to determine whether or not this is the case for the street in front of or parallel to your own lot. Or from such a visit it is often possible to locate some landmark, such as a factory chimney or church spire, that lies true north with respect to your house. If you cannot get true north by such means, three other methods are available: compass, pole star and sun.

By Compass

Get as large a compass as you can; it is difficult, though not impossible, to get satisfactory results with the "pocket" type. In any event, the compass *must* have not more than 2 degrees per division.

It must be remembered that the compass points to *magnetic* north, not true north. The amount by which magnetic north differs from true north in a particular location is known as *variation*. Your city engineer's office or the flight office at a nearby airport can tell you the magnetic variation for your locality. The information is also available from US Geological Survey topographic maps for your locality; these may be on file in your local library. When correcting your "compass north," do so *opposite* to the direction of the variation. For instance, if the variation for your locality is 12 degrees west (meaning that the compass points 12 degrees west of north), then true north is found by counting 12 degrees *east* of north as shown on the compass.

When taking the bearing, make sure that the compass is located well away from ironwork, fencing, pipes, etc. Place the instrument on a wooden tripod or support of some sort, at a convenient height as near eye level as possible. Make yourself a sighting stick from a flat stick about 2 feet long with a nail driven upright in each end (for use as "sights") and then, after the needle of the compass has settled down, carefully lay this stick across the face of the compass—with the necessary allowance for variation— to line it up on true north. *Be sure you apply the variation correctly*.

This same sighting-stick and compass rig can also be used in laying out directions for supporting poles for antennas in other directions—provided, of course, that the compass dial is graduated in degrees.

By the Pole Star

Many amateurs in the Northern Hemisphere use the pole star, Polaris, in determining the direction of true north. An advantage is that the pole star is never more than $0.8°$ from true north, so that in practice no corrections are necessary. Disadvantages are that some people have difficulty identifying the pole star, and that because of its comparatively high angle above the horizon at high northerly latitudes, it is not always easy to "sight" on it accurately. Polaris is not visible in the Southern Hemisphere. In any event, if visible, it is a handy check on the direction secured by other means.

Table 4-1
Time Correction for Various Dates of the Year
To get time of true noon, apply to clock time as indicated by the sign.

Date		Min	Date		Min
Jan	1	+ 4	Jul	10	+ 5
	10	+ 8		20	+ 6
	20	+11		30	+ 6
	30	+13			
Feb	10	+14	Aug	10	+ 5
	20	+14		20	+ 3
	28	+13		30	+ 1
Mar	10	+10	Sep	10	− 3
	20	+ 8		20	− 7
	30	+ 4		30	− 10
Apr	10	+ 1	Oct	10	− 13
	20	− 1		20	− 15
	30	− 3		30	− 16
May	10	− 4	Nov	10	− 16
	20	− 4		20	− 14
	30	− 3		30	− 11
Jun	10	− 1	Dec	10	− 7
	20	+ 1		20	− 2
	30	+ 4		30	+ 3

By the Sun

With some slight preparation, the sun can be used easily for determination of true north. One of the most satisfactory methods is described below. The method is based on the fact that exactly at noon, local time, the sun bears due south, so at that time the shadow of a vertical stick or rod will bear north. The resulting shadow direction, incidentally, is *true* north.

Two corrections to your Standard Time must be made to determine the exact moment of local true noon.

The first is a longitude correction. Standard Time is time at some particular meridian of longitude: EST is based on the 75th meridian, CST on the 90th meridian, MST on the 105th meridian and PST on the 120th meridian. From an atlas or perhaps Table 4-3, determine the difference between your longitude and the longitude of your time meridian. Getting this to the nearest ¼ degree of longitude is close enough. Example: Newington, Connecticut, which runs on 75th meridian time (EST) is at 72.75° longitude, or a difference of 2.25°. Now for each ¼° of longitude, figure one minute of time; thus 2.25° is equivalent to nine minutes of time (there are 60 "angle" minutes to a degree, so that each degree of longitude equals four minutes of time). *Subtract* this correction from noon if you are *east* of your time meridian; *add* if you are *west*.

To the resulting time, apply a further correction for the date from Table 4-1. The resulting time is the time, by Standard Time, when it will be true noon at your location. Put up your vertical stick (use a plumb bob to make sure it is actually vertical), check your watch with Standard Time, and, at the time indicated from your calculations, mark the position of the shadow. That is true north.

In the case of Newington, if we wanted correct time for true noon on October 20: First, subtracting the longitude correction—because we are east of the time meridian—we get 11:51 AM; then, applying the further correction of − 15

minutes, we get 11:36 AM EST (12:36 PM EDST) as the time of true noon at Newington on October 20.

AZIMUTHAL MAPS

While the Mercator projection does not show true directions, it is possible to make up a map that will show true bearings for all parts of the world from any single point. Three such maps are reproduced in this chapter. Fig 4-1 shows directions from Washington, DC, Fig 4-2 gives directions from San Francisco and Fig 4-3 (a simplified version of the ARRL Amateur Radio map of the world) gives directions from the approximate center of the United States—Wichita, Kansas.

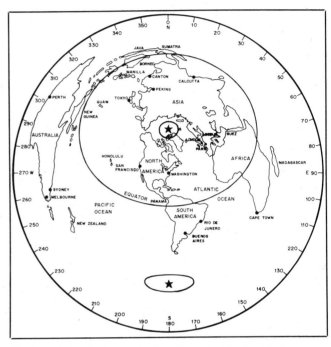

Fig 4-1—Azimuthal map centered on Washington, DC.

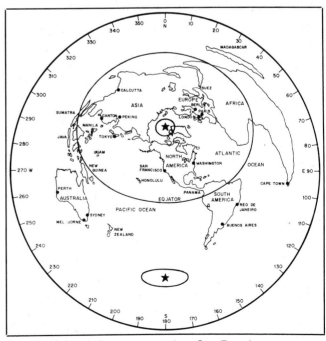

Fig 4-2—Azimuthal map centered on San Francisco, California.

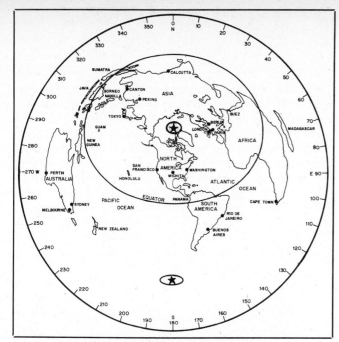

Fig 4-3—Azimuthal map centered on Wichita, Kansas. Copyright by Rand McNally & Co, Chicago. Reproduction License No. 394.

For anyone living in the immediate vicinity (within 150 miles) of any of these three reference points, the directions as taken from maps will have a high degree of accuracy. However, one or the other of the three maps will suffice for any location in the United States for all except the most accurate work; simply choose the map whose reference point is nearest you. Greatest errors will arise when your location is to one side or the other of a line between the reference point and the destination point; if your location is near or on the resulting line, there will be little or no error.

By tracing the directional pattern of the antenna system on a sheet of tissue paper, then placing the paper over the azimuthal map with the origin of the pattern at one's location, the "coverage" of the antenna will be readily evident. This is a particularly useful stunt when a multi-lobed antenna, such as any of the long single-wire systems, is to be laid out so that the main lobes cover as many desirable directions as possible. Often a set of such patterns will be of considerable assistance in determining what length antenna to put up, as well as the direction in which it should run.

The current edition of the ARRL Amateur Radio Map of the World, entirely different in concept and design from any other radio amateurs' map, contains a wealth of information especially useful to amateurs. A special type of azimuthal projection made by Rand-McNally to ARRL specifications, it gives great-circle bearings from the geographical center of the United States, as well as the great-circle distance measurement in miles and kilometers, within an accuracy of two percent. The map shows principal cities of the world, local time zones, WAC divisions, index of DXCC countries and amateur prefixes throughout the world. The map is large enough to be easily readable from the operating position, 31 × 41 inches, and is printed in six colors on heavy paper. The map is available from ARRL HQ; write for details.

The *Radio Amateur's Callbook* includes great-circle maps and tables, and another Callbook publication, *The Radio Amateur's World Atlas*, features a polar-projection world map, maps of the continents and world amateur prefixes. The maps are in color.

Bill Johnston, N5KR, offers computer-calulated and -drawn great-circle maps; an extensive selection of these fine maps for various areas of the world appears in Chapter 17. An 11- × 14-inch map can be custom-made for your location. Write to 1808 Pomona Dr, Las Cruces, NM 88001.

WORKING FROM A GLOBE

Bearings for beam-heading purposes may be determined easily from an ordinary globe with nothing more complicated than a small school protractor of the type available in any school-supply or stationery store. For best results, however, the globe should be at least 8 inches in diameter. A thin strip of paper may be used for a straightedge to determine the great-circle path between your location and any other location on the earth's surface. The bearing from your location may be determined with the aid of a protractor. For convenience, a paper-scale circle calibrated in degrees of bearing may be made and affixed over the point representing your location on the globe. The 0° mark of this scale should point toward the North Pole.

A SIMPLIFIED DIRECTION FINDER

A simplified direction finder may be made by removing a globe from its brazen meridian (semi-circular support) and remounting it in the manner shown in Fig 4-4. Drill a hole that will accept the support at your location on the globe, and another hole directly opposite the first. This second hole will have the same latitude as yours but will be on the other side of the equator (north latitude vs. south latitude). Its longitude will be opposite in direction from yours from the Greenwich or 0° meridian, east vs. west, and will be equal to 180° minus your longitude. For example, if your location is 42° N lat, 72° W long, the point opposite yours on the globe is 42° S lat, 180 − 72 or 108° E long.

Fig 4-4—A simple direction finder made by modifying a globe. Bearing and distance to other locations from yours may be determined quickly after modification, no calculations being required.

Once the holes are drilled, remount the globe with your location in the position formerly occupied by the north pole. By rotating the globe until the distant point of interest lies beneath the brazen meridian, this support may be used to indicate the great-circle path. A new "equator" calibrated in a manner to indicate the bearing may be added with India ink, as shown in Fig 4-4, or a small protractor-like scale may be added at the top of the globe over your location. A distance scale can be affixed to the brazen meridian so that both the bearing and distance to other locations may be readily determined (12,500 miles or 40,000 km to the semicircle).

DIRECTION AND DISTANCE BY TRIGONOMETRY

The methods to be described will give the bearing and distance as accurately as one cares to compute them. All that is required is a table of latitude and longitude information, such as is found in the table at the end of this chapter, and a calculator or computer with trigonometric functions. The latitude and longitude for any other location can be taken from a map of the area in question.

Fig 4-5 will help you to visualize the nature of the situation. That sketch represents the path between points situated relatively such as Pratt, Kansas, USA (at point A), and Perth, Western Australia (at point B). In using these equations, northerly latitudes are taken as positive, and southerly latitudes are taken as negative. Also, westerly longitudes are taken as positive, and easterly longitudes are taken as negative. *In all calculations, the appropriate signs are to be retained. All additions and subtractions throughout the procedure are to be made algebraically*. Thus, if a negative-value number is subtracted from a positive-value number, the resultant will be positive, and it will be the *sum* of the two absolute values, and so on.

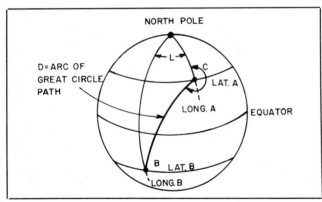

Fig 4-5—The various terms used in the equations for determining bearing and distance. North latitudes and west longitudes are taken as positive, while south latitudes and east longitudes are taken as negative.

The Calculations

The two equations we'll be using for these calculations are:

$$\cos D = \sin A \sin B + \cos A \cos B \cos L \qquad \text{(Eq 1)}$$

$$\cos C = \frac{\sin B - \sin A \cos D}{\cos A \sin D} \qquad \text{(Eq 2)}$$

where

A = *your* latitude in degrees
B = latitude of the other location in degrees

L = *your* longitude minus that of the other location (algebraic difference)
D = distance along path in degrees of arc
C = true bearing from north if the value for sin L is positive. If sin L is negative, true bearing is 360 − C.

The term *cos* is an abbreviation for cosine, and the term *sin* is an abbreviation for sine. A knowledge of the meanings of these terms isn't necessary for their use here.

The actual calculating procedure uses, first, Eq 1 to determine the angular value for D, in degrees. From this value the path-length distance may be determined in miles or kilometers. Next, Eq 2 is used to determine the bearing angle.

Using the Pratt-to-Perth example mentioned earlier, refer to Fig 4-5 to see how the equations are used. From the table at the end of the chapter, it can be seen that the location of Pratt is 37.7° N lat, 98.7° W long. Similarly, Perth is located at 32° S lat, 115.9° E long. Your location is in Pratt. Values for use in the equations are as follows:

A = lat A = +37.7°
B = lat B = −32°
L = long A − long B
  = 98.7° − (− 115.9°) = 214.6°

Solving Eq 1, cos D = sin 37.7° sin (− 32°) + cos 37.7° cos (− 32°) cos 214.6°. D = 151.21°. Each degree along the path equals 60 nautical miles. Therefore, 151.2° of arc is equivalent to 60 × 151.2 = 9072 nautical miles. To convert to statute miles, multiply degrees by 69.041. If the distance is desired in kilometers, multiply degrees by 111.111. This means that the Pratt-to-Perth distance is 10,440 miles or 16,801 kilometers.

Solving Eq 2

$$\cos C = \frac{\sin (− 32°) − \sin 37.7° \cos 151.2°}{\cos 37.7° \sin 151.2°}$$

C = 89.1°. Because the sin of L (214.6°) is negative, however, the correct value for C is 360° − 89.1° = 270.9°. Thus, the true bearing from Pratt to Perth is 270.9° and the distance is 10,440 statute miles. If the bearing from Perth were desired, it would be necessary only to work through Eq 2, interchanging latitude values for A and B. Because of the way L is defined, sin L will be positive in this case, and it will not be necessary to subtract from 360° to get the true bearing at Perth, which is 68.9°.

These equations give information for the great-circle bearing and distance for the shortest path. For long-path work, the bearing will be 180° away from the answers obtained.

The equations described above may be used for any two points on the earth's surface—both locations in the northern hemisphere, both locations in the southern hemisphere, either or both on the equator, and so on. The equations themselves are exact, not being based on any approximations. However, there are some cases where practical limitations exist in the accuracy of the results obtained from Eq 2, in relation to the number of significant figures used during calculations. (Round-off errors in calculators and computers during computations will effectively reduce the number of significant figures in the resulting answers.) These cases are where both locations are near or at exact opposite points on the earth (antipodes), where the locations are close together, or where *your* location is at or near one of the poles. (At the poles, all directions are either south or north, anyway.) More specifically, these situations exist when lat A is near ±90°, or where D is near 0° or 180°.

Table 4-2 is a BASIC language program for calculating

Table 4-2

BASIC Language Program for Determining Bearings and Distances

```
10 REM * * * BEARINGS/BAS * * *
15 REM A=YOUR LAT.
20 REM B=OTHER STATION LAT.
25 REM C=BEARING ANGLE
30 REM D=DEGREES OF ARC
35 REM E=INTERMEDIATE VALUE
40 REM K=CONVERSION CONST., ARC TO KILOMETERS
45 REM L=DIFF  IN LONGITUDES
50 REM L1=YOUR LONG.
55 REM L2=OTHER STATION LONG.
60 REM M=CONVERSION CONSTANT, DEGREES TO RADIANS
65 REM N=CONVERSION CONST., ARC TO NAUT. MI.
70 REM S=CONVERSION CONST., ARC TO STATUTE MI.
75 REM V=STRING VARIABLE
100 CLS
110 PRINT"PROGRAM TO CALCULATE GREAT CIRCLE DISTANCES AND BEARINGS"
120 PRINT
130 PRINT"BY J. HALL AND C. HUTCHINSON, ARRL HQ., JULY 1981"
140 PRINT"THIS PROGRAM IS NOT COPYRIGHTED AND MAY BE REPRODUCED FREELY"
150 DEFDBLA,B,C,D,E,L,M:DEFSTRV
160 D=1:K=111.11:M=57.29577951308238:N=60:S=69.041:V="######.#"
170 PRINT
180 PRINT"ENTER NEGATIVE VALUES FOR SOUTHERLY LATITUDES"
190 PRINT"ENTER NEGATIVE VALUES FOR EASTERLY LONGITUDES"
200 PRINT:IFD<>1THEN230
210 INPUT"YOUR LATITUDE (DEGREES AND DECIMAL)";A:A=A/M
220 INPUT"YOUR LONGITUDE (DEGREES AND DECIMAL)";L1
230 INPUT"OTHER LATITUDE (DEGREES AND DECIMAL)";B:B=B/M
240 INPUT"OTHER LONGITUDE (DEGREES AND DECIMAL)";L2
250 L=(L1-L2)/M
260 E=SIN(A)*SIN(B)+COS(A)*COS(B)*COS(L)
270 D=-ATN(E/SQR(1-E*E))+1.57079
280 C=(SIN(B)-SIN(A)*E)/(COS(A)*SIN(D))
290 IFC>=1THENC=0:GOTO310ELSEIFC<=-1THENC=180/M:GOTO310
300 C=-ATN(C/SQR(1-C*C))+1.57079
310 C=C*M
320 IFSIN(L)<0THENC=360-C
330 PRINT"THE BEARING IS ";:PRINTTAB(29)USINGV;C;:PRINT" DEGREES"
340 PRINT"THE GREAT CIRCLE DISTANCE IS ";
350 PRINTUSINGV;K*D*M;:PRINT" KILOMETERS"
360 PRINTTAB(29)USINGV;N*D*M;:PRINT" NAUTICAL MILES"
370 PRINTTAB(29)USINGV;S*D*M;:PRINT" STATUTE MILES"
380 PRINT
390 D=0:PRINT"TO CONTINUE PRESS ENTER"
400 INPUT"TO CALCULATE BEARING FROM A DIFFERENT LOCATION ENTER 1";D
410 GOTO170
```

bearings and distances by computer. The program is written for a Radio Shack TRS-80® Level II computer. There are numerous versions of BASIC, and it may be necessary for you to modify portions of the program for use with your computer system. Statements 15 through 75 indicate by way of remarks what variables are used in the program and the information they represent.

Table 4-3 shows latitude and longitude for various US and Canadian cities and other areas of the world. The data are arranged alphanumerically by call prefixes; DX information generally follows the ARRL DXCC List.

Acknowledgments

Data for Table 3 was compiled from various sources, including *The World Almanac and Book of Facts 1981*, Newspaper Enterprise Associates, Inc, New York. Special thanks for latitude and longitude data for Table 4-3 go to Jer-ry Cross, N4NO, Chuck Hutchinson, K8CH, Daryl Kiebler, WB8EUN, and Michael Kaczynski, W1OD.

BIBLIOGRAPHY

Source material and more extended discussions of topics covered in this chapter can be found in the references given below.

Davis, "A Simplified Direction Finder," Hints and Kinks, May 1972 *QST*.

Hall, "Bearing and Distance Calculations by Sleight of Hand," August 1973 *QST*.

Klopf, "A Bearing and Distance Calculator," March 1971 *QST*.

Norton, *Norton's Star Atlas and Reference Handbook*, Gail and Inglis, London, England; also published in the US by Sky Publishing Corp, Cambridge, Massachusetts.

Table 4-3

Latitude and Longitude of Various US/Canadian Cities and DX Locations

Pref	State/Province/Country/City	Lat	Long
VE1	New Brunswick, Saint John	45.3 N	66.1 W
	Nova Scotia, Halifax	44.6 N	63.6 W
	Prince Edward Island, Charlottetown	46.2 N	63.1 W
VE2	Quebec, Montreal	45.5 N	73.6 W
	Quebec City	46.8 N	71.2 W
VE3	Ontario, London	43.0 N	81.3 W
	Ottawa	45.4 N	75.7 W
	Sudbury	46.5 N	81.0 W
	Toronto	43.7 N	79.4 W
VE4	Manitoba, Winnipeg	49.9 N	97.1 W
VE5	Saskatchewan, Regina	50.5 N	104.6 W
	Saskatoon	52.1 N	106.7 W
VE6	Alberta, Calgary	51.0 N	114.1 W
	Edmonton	53.5 N	113.5 W
VE7	British Columbia, Prince George	53.9 N	122.8 W
	Prince Rupert	54.3 N	130.3 W
	Vancouver	49.3 N	123.1 W
VE8	Northwest Territories, Yellowknife	62.5 N	114.4 W
	Resolute	74.7 N	95.0 W
VY1	Yukon, Whitehorse	60.7 N	135.1 W
VO1	Newfoundland, St John's	47.6 N	52.7 W
VO2	Labrador, Goose Bay	53.3 N	60.4 W
W1	Connecticut, Hartford	41.8 N	72.7 W
	Maine, Bangor	44.8 N	68.8 W
	Portland	43.7 N	70.3 W
	Massachusetts, Boston	42.4 N	71.1 W
	New Hampshire, Concord	43.2 N	71.5 W
	Rhode Island, Providence	41.8 N	71.4 W
	Vermont, Montpelier	44.3 N	72.6 W
W2	New Jersey, Atlantic City	39.4 N	74.4 W
	New York, Albany	42.7 N	73.8 W
	Buffalo	42.9 N	78.9 W
	New York City	40.8 N	74.0 W
	Syracuse	43.1 N	76.2 W
W3	Delaware, Wilmington	39.7 N	75.5 W
	District of Columbia, Washington	38.9 N	77.0 W
	Maryland, Baltimore	39.3 N	76.6 W
	Pennsylvania, Harrisburg	40.3 N	76.9 W
	Philadelphia	39.9 N	75.2 W
	Pittsburgh	40.4 N	80.0 W
	Scranton	41.4 N	75.7 W
W4	Alabama, Montgomery	32.4 N	86.3 W
	Florida, Jacksonville	30.3 N	81.7 W
	Miami	25.8 N	80.2 W
	Pensacola	30.4 N	87.2 W
	Georgia, Atlanta	33.8 N	84.4 W
	Savannah	32.1 N	81.1 W
	Kentucky, Lexington	38.0 N	84.5 W
	Louisville	38.2 N	85.8 W
	North Carolina, Charlotte	35.2 N	80.8 W
	Raleigh	35.8 N	78.6 W
	Wilmington	34.2 N	77.9 W
	South Carolina, Columbia	34.0 N	81.0 W
	Tennessee, Knoxville	36.0 N	83.9 W
	Memphis	35.1 N	90.1 W
	Nashville	36.2 N	86.8 W
	Virginia, Norfolk	36.9 N	76.3 W
	Richmond	37.5 N	77.4 W
W5	Arkansas, Little Rock	34.7 N	92.3 W
	Louisiana, New Orleans	29.9 N	90.1 W
	Shreveport	32.5 N	93.7 W
	Mississippi, Jackson	32.3 N	90.2 W
	New Mexico, Albuquerque	35.1 N	106.7 W
	Oklahoma, Oklahoma City	35.5 N	97.5 W
	Texas, Abilene	32.5 N	99.7 W
	Amarillo	35.2 N	101.8 W
	Dallas	32.8 N	96.8 W
	El Paso	31.8 N	106.5 W
	San Antonio	29.4 N	98.5 W
W6	California, Los Angeles	34.1 N	118.2 W
	San Francisco	37.8 N	122.4 W
W7	Arizona, Flagstaff	35.2 N	111.7 W
	Phoenix	33.5 N	112.1 W
	Idaho, Boise	43.6 N	116.2 W
	Pocatello	42.9 N	112.5 W
	Montana, Billings	45.8 N	108.5 W
	Butte	46.0 N	112.5 W
	Great Falls	47.5 N	111.3 W
	Nevada, Las Vegas	36.2 N	115.1 W
	Reno	39.5 N	119.8 W
	Oregon, Portland	45.5 N	122.7 W
	Utah, Salt Lake City	40.8 N	111.9 W
	Washington, Seattle	47.6 N	122.3 W
	Spokane	47.7 N	117.4 W
	Wyoming, Cheyenne	41.1 N	104.8 W
	Sheridan	44.8 N	107.0 W
W8	Michigan, Detroit	42.3 N	83.0 W
	Grand Rapids	43.0 N	85.7 W
	Sault Ste Marie	46.5 N	84.4 W
	Traverse City	44.8 N	85.6 W
	Ohio, Cincinnati	39.1 N	84.5 W
	Cleveland	41.5 N	81.7 W
	Columbus	40.0 N	83.0 W
	West Virginia, Charleston	38.4 N	81.6 W
W9	Illinois, Chicago	41.9 N	87.6 W
	Indiana, Indianapolis	39.8 N	86.2 W
	Wisconsin, Green Bay	44.5 N	88.0 W
	Milwaukee	43.0 N	87.9 W
WØ	Colorado, Denver	39.7 N	105.0 W
	Grand Junction	39.1 N	108.6 W
	Iowa, Des Moines	41.6 N	93.6 W
	Kansas, Pratt	37.7 N	98.7 W
	Wichita	37.7 N	97.3 W
	Minnesota, Duluth	46.8 N	92.1 W
	Minneapolis	45.0 N	93.3 W
	Missouri, Columbia	39.0 N	92.3 W
	Kansas City	39.1 N	94.6 W
	St Louis	38.6 N	90.2 W
	Nebraska, North Platte	41.1 N	100.8 W
	Omaha	41.3 N	95.9 W
	North Dakota, Fargo	46.9 N	96.8 W
	South Dakota, Rapid City	44.1 N	103.2 W
A2	Botswana, Gaborone	24.8 S	25.9 E
A3	Tonga, Nukualofa	21.1 S	175.2 W
A4	Oman, Muscat	23.6 N	58.6 E
A5	Bhutan, Thimbu	27.3 N	89.4 E
A6	United Arab Emirates, Abu Zaby	24.5 N	54.2 E
A7	Qatar, Ad-Dawhah	25.3 N	51.5 E
A9	Bahrein, Al-Manamah	26.2 N	50.6 E
AP	Pakistan, Karachi	24.9 N	67.1 E
AP	Islamabad	33.7 N	73.2 E
BV	Taiwan, Taipei	25.1 N	121.5 E
BY	People's Rep of China, Beijing	40.0 N	116.4 E
	Harbin	45.8 N	126.7 E
	Shanghai	31.2 N	121.5 E
	Fuzhou	26.1 N	119.3 E
	Xian	34.3 N	108.9 E
	Chongqing	29.8 N	106.5 E
	Chengdu	30.7 N	104.1 E
	Lhasa	29.7 N	91.2 E
	Urumqi	43.8 N	87.6 E
	Kashi	39.5 N	76.0 E
C2	Nauru	0.5 S	166.9 E
C3	Andorra	42.5 N	1.5 E
C5	The Gambia, Banjul	13.5 N	16.7 W
C6	Bahamas, Nassau	25.1 N	77.4 W
C9	Mozambique, Maputo	26.0 S	32.6 E
	Mozambique	15.1 S	40.7 E
CE	Chile, Santiago	33.5 S	70.8 W
CEØ	Easter Island	27.1 S	109.4 W
CEØ	San Felix	26.3 S	80.1 W
CEØ	Juan Fernandez	33.6 S	78.8 W
CM,CO	Cuba, Havana	23.1 N	82.4 W

Pref	State/Province/Country/City	Lat	Long
CN	Morocco, Casablanca	33.6 N	7.5 W
CP	Bolivia, La Paz	16.5 S	68.4 W
CT	Portugal, Lisbon	38.7 N	9.2 W
CT2, CU	Azores, Ponta Delgada	37.7 N	25.7 W
CT3	Madeira Islands, Funchal	32.6 N	16.9 W
CX	Uruguay, Montevideo	34.9 S	56.2 W
D2	Angola, Luanda	8.8 S	13.2 E
D4	Cape Verde, Praia	14.9 N	23.5 W
D6	Comoros, Moroni	11.8 S	43.7 E
DA-DL	Germany, Fed Rep of (W), Bonn	50.7 N	7.0 E
DU	Philippines, Manila	14.6 N	121.0 E
EA	Spain, Madrid	40.4 N	3.7 W
EA6	Balearic Is, Palma	39.5 N	2.6 E
EA8	Canary Is, Las Palmas	28.4 N	14.3 W
EA9	Ceuta & Melilla, Ceuta	35.9 N	5.3 W
	Melilla	35.3 N	3.0 W
EI	Ireland, Dublin	53.3 N	6.3 W
EL	Liberia, Monrovia	6.3 N	10.8 W
EP	Iran, Tehran	35.8 N	51.8 E
ET	Ethiopia, Addis Ababa	9.0 N	38.7 E
	Asmera	15.3 N	38.9 E
F	France, Paris	48.8 N	2.3 E
FG	Guadeloupe	16.0 N	61.7 W
FG, FS	St Martin	18.1 N	63.1 W
FH	Mayotte	13.0 S	45.3 E
FK	New Caledonia, Noumia	22.3 S	166.5 E
FM	Martinique	14.6 N	61.0 W
FO	Clipperton	10.3 N	109.2 W
FO	Fr Polynesia, Tahiti	17.6 S	159.5 W
FP	St Pierre & Miquelon, St Pierre	46.7 N	56.0 W
FR	Glorioso	11.5 S	47.3 E
FR	Juan de Nova	17.0 S	42.8 E
	Europa	22.3 S	40.4 E
FR	Reunion	21.1 S	55.6 E
FR	Tromelin	15.9 S	54.4 E
FT8W	Crozet	46.0 S	52.0 E
FT8X	Kerguelen	49.3 S	69.2 E
FT8Y	Antarctica, Dumont D'Urville	66.6 S	140.0 E
FT8Z	Amsterdam & St Paul Is, Amsterdam	37.7 S	77.6 E
FW	Wallis & Futuna Is, Wallis	13.3 S	176.3 W
FY	Fr Guiana, Cayenne	4.9 N	52.3 W
G	England, London	51.5 N	0.1 W
GD	Isle of Man	54.3 N	4.5 W
GI	Northern Ireland, Belfast	54.6 N	5.9 W
GJ	Jersey	49.3 N	2.2 W
GM	Scotland, Glasgow	55.8 N	4.3 W
	Aberdeen	57.2 N	2.1 W
GU	Guernsey	49.5 N	2.7 W
GW	Wales, Cardiff	51.5 N	3.2 W
H4	Solomon Islands, Honiara	9.4 S	160.0 E
HA	Hungary, Budapest	47.5 N	19.1 E
HB	Switzerland, Berne	47.0 N	7.5 E
HB0	Liechtenstein	47.2 N	9.6 E
HC	Ecuador, Quito	0.2 S	78.0 W
HC8	Galapagos Is	0.5 S	90.5 W
HH	Haiti, Port-Au-Prince	18.5 N	72.3 W
HI	Dominican Republic, Santo Domingo	18.5 N	70.0 W
HK	Colombia, Bogota	4.6 N	74.1 W
HK0	Malpelo Is	4.0 N	81.1 W
HK0	San Adreas	12.5 N	81.7 W
HL, HM	Rep of Korea, Seoul	37.5 N	127.0 E
HP	Panama, Panama	9.0 N	79.5 W
HR	Honduras, Tegucigalpa	14.1 N	87.2 W
HS	Thailand, Bangkok	13.8 N	100.5 E
HV	Vatican City	41.9 N	12.5 E
HZ, 7Z	Saudi Arabia, Dharan	26.3 N	50.0 E
	Mecca	21.5 N	39.8 E
	Riyadh	24.6 N	46.7 E
I	Italy, Rome	41.9 N	12.5 E
	Trieste	45.7 N	13.8 E
	Sicily	37.5 N	14.0 E
IS	Sardinia, Cagliari	39.2 N	9.1 E
J2	Djibouti, Djibouti	11.6 N	43.2 E
J3	Grenada	12.0 N	61.8 W
J5	Guinea-Bissau, Bissau	11.9 N	15.6 W
J6	St Lucia	13.9 N	61.0 W
J7	Dominica	15.4 N	61.3 W
J8	St Vincent	13.3 N	61.3 W
JA-JS	Japan, Tokyo	35.7 N	139.8 E
	Nagasaki	32.8 N	129.9 E
	Sapporo	43.1 N	141.4 E
JD	Minami Torishima	24.3 N	154.0 E
JD	Ogasawara, Kazan Is	27.5 N	141.0 E
JT	Mongolia, Ulan Bator	47.9 N	106.9 E
JW	Svalbard, Spitsbergen	78.8 N	16.0 E
JX	Jan Mayen	71.0 N	8.3 W
JY	Jordan, Amman	32.0 N	35.9 E
KC4	Antarctica, Byrd Station	80.0 S	120.0 W
	McMurdo Sound	77.7 S	166.7 E
	Palmer Station	64.8 S	64.0 W
KC6	Micronesia, Ponape	6.9 N	158.3 E
KC6	Belau, Yap	9.5 N	138.2 E
	Koror	7.3 N	134.5 E
KG4	Guantanamo Bay	19.9 N	75.2 W
KH1	Baker, Howland Is	0.5 N	176.0 W
KH2	Guam, Agana	13.5 N	144.8 E
KH3	Johnston Is	17.0 N	168.5 W
KH4	Midway Is	28.2 N	177.4 W
KH5	Palmyra Is	5.9 N	162.1 W
KH5K	Kingman Reef	7.5 N	162.8 W
KH6	Hawaii, Hilo	19.7 N	155.1 W
	Honolulu	21.3 N	157.9 W
KH7	Kure Is	28.4 N	178.4 W
KH8	American Samoa, Pago Pago	14.3 S	170.8 W
KH9	Wake Is	19.3 N	166.6 E
KH0	Mariana Is, Saipan	15.2 N	145.8 E
KL7	Alaska, Adak	51.8 N	176.6 W
	Anchorage	61.2 N	150.0 W
	Fairbanks	64.8 N	147.9 W
	Juneau	58.3 N	134.4 W
	Nome	64.5 N	165.4 W
KP1	Navassa Is	18.4 N	75.0 W
KP2	Virgin Islands, Charlotte Amalie	18.3 N	64.9 W
KP4	Puerto Rico, San Juan	18.5 N	66.2 W
KP5	Desecheo Is	18.3 N	67.5 W
KX6	Marshall Islands, Kwajalein	9.1 N	167.3 E
LA-LJ	Norway, Oslo	60.0 N	10.7 E
LU	Argentina, Buenos Aires	34.6 S	58.4 W
LX	Luxembourg	49.6 N	6.2 E
LZ	Bulgaria, Sofia	42.7 N	23.3 E
OA	Peru, Lima	12.1 S	77.1 W
OD	Lebanon, Beirut	33.9 N	35.5 E
OE	Austria, Vienna	48.2 N	16.3 E
OH	Finland, Helsinki	60.2 N	25.0 E
OH0	Aland Is	60.2 N	20.0 E
OJ0	Market Reef	60.3 N	19.0 E
OK	Czechoslovakia, Prague	50.1 N	14.4 E
ON	Belgium, Brussels	50.9 N	4.4 E
OX, XP	Greenland, Godthaab	64.2 N	51.7 W
	Thule	76.6 N	68.8 W
OY	Faroe Islands, Torshavn	62.0 N	6.8 W
OZ	Denmark, Copenhagen	55.7 N	12.6 E
P2	Papua New Guinea, Madang	5.2 S	145.6 E
	Port Moresby	9.4 S	147.1 E
PA-PI	Netherlands, Amsterdam	52.4 N	4.9 E
PJ, P4	Netherlands Antilles, Willemstad	12.1 N	69.0 W
PJ5-8	St Maarten and Saba, St Maarten	17.7 N	63.2 W
PY	Brazil, Brasilia	15.8 S	47.9 W
	Rio De Janeiro	23.0 S	43.2 W
	Natal	6.0 S	35.2 W
	Manaus	3.1 S	60.2 W
	Porto Alegre	30.1 S	51.2 W
PY0	Fernando De Noronha	3.9 S	32.4 W
PY0	St Peter & St Paul Rocks	1.0 N	29.4 W
PY0	Trindade & Martin Vaz Is, Trindade	20.5 S	29.3 W
PZ	Suriname, Paramaribo	5.8 N	55.2 W
S2	Bangladesh, Dacca	23.7 N	90.4 E
S7	Seychelles, Victoria	4.6 S	55.5 E
S9	Sao Tome	0.3 N	6.7 E
SJ-SM	Sweden, Stockholm	59.3 N	18.1 E
SP	Poland, Cracow	50.0 N	20.0 E
	Warsaw	52.2 N	21.0 E
ST	Sudan, Khartoum	15.6 N	32.5 E
ST0	Southern Sudan, Juba	5.0 N	31.6 E

Table 4-3 (continued)

Pref	State/Province/Country/City	Lat	Long		Pref	State/Province/Country/City	Lat	Long
SU	Egypt, Cairo	30.0 N	31.4 E		UL	Kazakh, Alma-ata	43.3 N	76.9 E
SV	Greece, Athens	38.0 N	23.7 E		UM	Kirghiz, Frunze	42.9 N	74.6 E
SV5	Dodecanese, Rhodes	36.4 N	28.2 E		UO	Moldavia, Kishinev	47.0 N	28.8 E
SV9	Crete	35.4 N	25.2 E		UP	Lithuania, Vilna	54.5 N	25.5 E
SV/A	Mount Athos	40.2 N	24.3 E		UQ	Latvia, Riga	57.0 N	24.1 E
T2	Tuvalu, Funafuti	8.7 S	178.6 E		UR	Estonia, Tallinn	59.4 N	24.8 E
T3Ø	West Kiribati, Bonriki	1.4 N	173.2 E		V2	Antigua & Barbuda, St Johns	17.1 N	61.8 W
T31	Central Kiribati, Kanton	2.8 S	171.7 W		V3	Belize, Belmopan	17.3 N	88.8 W
T32	East Kiribati, Christmas Is	1.9 N	157.4 W		V4	St Christopher & Nevis	17.3 N	62.6 W
T5	Somalia, Mogadishu	2.1 N	45.4 E		V8	Brunei, Bandar Seri Begawan	4.9 N	114.9 E
T7	San Marino	43.9 N	12.3 E		VE1, CY	Sable Is	43.8 N	60.0 W
TA	Turkey, Ankara	39.9 N	32.9 E		VE1, CY	St Paul Is	47.2 N	60.1 W
	Istanbul	41.2 N	29.0 E		VK	Australia, Canberra (VK1)	35.3 S	149.1 E
TF	Iceland, Reykjavik	64.1 N	22.0 W			Sydney (VK2)	33.9 S	151.2 E
TG	Guatemala, Guatemala City	14.6 N	90.5 W			Melbourne (VK3)	37.8 S	145.0 E
TI	Costa Rica, San Jose	9.9 N	84.0 W			Brisbane (VK4)	27.5 S	153.0 E
TI9	Cocos Is	5.6 N	87.0 W			Adelaide (VK5)	34.9 S	138.6 E
TJ	Cameroon, Yaounde	3.9 N	11.5 E			Perth (VK6)	31.9 S	115.8 E
TK	Corsica	42.0 N	9.0 E			Hobart, Tasmania (VK7)	42.9 S	147.3 E
TL	Central African Republic, Bangui	4.4 N	18.6 E			Darwin (VK8)	12.5 S	130.9 E
TN	Congo, Brazzaville	4.3 S	15.3 E		VK	Lord Howe Is	31.6 S	159.1 E
TR	Gabon, Libreville	0.4 N	9.5 E		VK9	Christmas Is	10.5 S	105.7 E
TT	Chad, N'Djamena	12.1 N	15.0 E		VK9	Cocos-Keeling Is	12.2 S	96.8 E
TU	Ivory Coast, Abidjan	5.3 N	4.0 W		VK9	Mellish Reef	17.6 S	155.8 E
TY	Benin, Porto Novo	6.5 N	2.6 E		VK9	Norfolk Is	29.0 S	168.0 E
TZ	Mali, Bamako	12.7 N	8.0 W		VK9	Willis Is	16.3 S	149.5 E
UA	Russia, European, Leningrad (UA1)	59.9 N	30.3 E		VKØ	Heard Is	53.0 S	73.4 E
	Archangel (UA1)	64.6 N	40.5 E		VKØ	Macquarie Is	54.7 S	158.8 E
	Murmansk (UA1)	69.0 N	33.1 E		VP2E	Anguilla	18.3 N	63.0 W
	Moscow (UA3)	55.8 N	37.6 E		VP2M	Montserrat	16.8 N	62.2 W
	Kuibyshev (UA4)	53.2 N	50.1 E		VP2V	British Virgin Is, Tortola	18.4 N	64.6 W
	Rostov (UA6)	47.5 N	39.5 E		VP5	Turks & Caicos Islands, Grand Turk	21.4 N	71.2 W
UA1P	Franz Josef Land	80.0 N	53.0 E		VP8	Falkland Islands, Stanley	51.7 S	57.9 W
UA2	Kaliningrad	55.0 N	20.5 E		VP8	So Georgia Is	54.3 S	36.8 W
UA9,Ø	Russia, Asiatic, Novosibirsk (UA9)	55.0 N	82.9 E		VP8	So Orkney Is	60.6 S	45.5 W
	Perm (UA9)	58.0 N	56.3 E		VP8	So Sandwich Is, Saunders Is	57.8 S	26.7 W
	Omsk (UA9)	55.0 N	73.4 E		VP8	So Shetland Is, King George Is	62.0 S	58.3 W
	Norilsk (UAØ)	69.3 N	88.1 E		VP9	Bermuda	32.3 N	64.7 W
	Irkutsk (UAØ)	52.3 N	104.3 E		VQ9	Chagos, Diego Garcia	7.3 S	72.4 E
	Vladivostok (UAØ)	43.2 N	131.9 E		VR6	Pitcairn Is	25.1 S	130.1 W
	Petropavlovsk (UAØ)	53.0 N	158.7 E		VS6	Hong Kong	22.3 N	114.3 E
	Khabarovsk (UAØ)	48.5 N	135.1 E		VU	India, Bombay	19.0 N	72.8 E
	Krasnoyarsk (UAØ)	56.0 N	92.8 E			Calcutta	22.6 N	88.4 E
	Yakutsk (UAØ)	62.0 N	129.7 E			New Delhi	28.6 N	77.2 E
	Wrangel Island (UAØ)	71.0 N	179.5 W			Bangalore	13.0 N	77.6 E
	Kyzyl (UAØY)	51.7 N	94.5 E		VU7	Andaman Islands, Port Blair	11.7 N	92.8 E
UB	Ukraine, Kiev	50.4 N	30.5 E		VU7	Laccadive Is	10.0 N	73.0 E
UC	Byelorussia, Minsk	53.9 N	27.6 E		XE	Mexico, Mexico City (XE1)	19.4 N	99.1 W
UD	Azerbaijan, Baku	40.4 N	49.9 E			Chihuahua (XE2)	28.7 N	106.0 W
UF	Georgia, Tbilisi	41.7 N	44.8 E			Merida (XE3)	21.0 N	89.7 W
UG	Armenia, Ierevan	40.3 N	44.5 E		XF4	Revilla Gigedo	19.0 N	111.5 W
UH	Turkoman, Ashkhabad	38.0 N	58.4 E		XT	Burkina Faso, Onagadougou	12.4 N	1.6 W
UI	Uzbek, Bukhara	39.8 N	64.4 E		XU	Kampuchea, Phnom Penh	11.7 N	104.8 E
	Tashkent	41.2 N	69.3 E		XW	Laos, Viangchan	18.0 N	102.6 E
UJ	Tadzhik, Samarkand	39.7 N	66.8 E					
	Dushanbe	39.1 N	68.8 E					

Pref	State/Province/Country/City	Lat	Long	Pref	State/Province/Country/City	Lat	Long
XX9	Macao	22.2 N	113.6 E	3CØ	Pagalu Is	1.5 S	5.6 E
XZ	Burma, Rangoon	16.8 N	96.0 E	3D2	Fiji Is, Suva	18.1 S	178.4 E
Y2-9	German Dem Rep (E), Berlin	52.5 N	13.4 E	3D6	Swaziland, Mbabane	26.3 S	31.1 E
YA	Afghanistan, Kandahar	31.0 N	65.8 E	3V	Tunisia, Tunis	36.8 N	10.2 E
	Kabul	34.4 N	69.2 E	3W	Vietnam, Ho Chi Minh City (Saigon)	10.8 N	106.7 E
YB-YD	Indonesia, Jakarta	6.2 S	106.8 E		Hanoi	21.0 N	105.8 E
	Medan	3.6 N	98.7 E	3X	Republic of Guinea, Conakry	9.5 N	13.7 W
	Pontianak	0.0	109.3 E	3Y	Bouvet	54.5 S	3.4 E
	Jayapura	2.6 S	140.7 E	3Y	Peter I Is	68.8 S	90.6 W
YI	Iraq, Baghdad	33.0 N	44.5 E	4S	Sri Lanka, Colombo	7.0 N	79.9 E
YJ	Vanuatu, Port Vila	17.7 S	168.3 E	4U	ITU Geneva	46.2 N	6.2 E
YK	Syria, Damascus	33.5 N	36.3 E	4U	United Nations HQ	40.8 N	74.0 W
YN, HT	Nicaragua, Managua	12.0 N	86.0 W	4W	Yemen, Sanaa	15.4 N	44.2 E
YO	Romania, Bucharest	44.4 N	26.1 E	4X, 4Z	Israel, Jerusalem	31.8 N	35.2 E
YS	El Salvador, San Salvador	13.7 N	89.2 W	5A	Libya, Tripoli	32.5 N	12.5 E
YU	Yugoslavia, Belgrade	44.9 N	20.5 E		Benghazi	32.1 N	20.0 E
YV	Venezuela, Caracas	10.5 N	67.0 W	5B	Cyprus, Nicosia	35.2 N	33.4 E
YVØ	Aves Is	15.7 N	63.7 W	5H	Tanzania, Dar es Salaam	7.0 S	39.5 E
Z2,ZE	Zimbabwe, Harare	17.8 S	31.0 E	5N	Nigeria, Lagos	6.5 N	3.4 E
ZA	Albania, Tirane	41.3 N	19.8 E	5R	Madagascar, Antananarivo	18.9 S	47.5 E
ZB	Gibraltar	36.1 N	5.4 W	5T	Mauritania, Nouakchott	18.1 N	16.0 W
ZC4	British Cyprus	34.6 N	33.0 E	5U	Niger, Niamey	13.5 N	2.0 E
ZD7	St Helena	16.0 S	5.9 W	5V	Togo, Lome	5.8 N	1.2 E
ZD8	Ascension Is	8.0 S	14.4 W	5W	Western Samoa, Apia	13.5 S	171.8 W
ZD9	Tristan da Cunha	37.1 S	12.3 W	5X	Uganda, Kampala	0.3 N	32.5 E
ZF	Cayman Is	19.5 N	81.2 W	5Z	Kenya, Nairobi	1.3 S	376.8 E
ZK1	No Cook Is, Manihiki	10.4 S	161.0 W	6W	Senegal, Dakar	14.7 N	17.5 W
ZK1	So Cook Is, Rarotonga	21.2 S	159.8 W	6Y	Jamaica, Kingston	18.0 N	76.8 W
ZK2	Niue	19.0 S	168.9 W	7O	Yemen, People's Dem Rep, Aden	12.8 N	45.0 E
ZK3	Tokelaus, Atafu	8.4 S	172.7 W	7P	Lesotho, Maseru	29.3 S	27.5 E
ZL	New Zealand, Auckland (ZL1)	36.9 S	174.8 E	7Q	Malawi, Lilongwe	14.0 S	33.8 E
	Wellington (ZL2)	41.3 S	174.8 E		Blantyre	15.8 S	35.0 E
	Christchurch (ZL3)	43.5 S	172.6 E	7X	Algeria, Algiers	36.7 N	3.0 E
	Dunedin (ZL4)	45.9 S	170.5 E	8P	Barbados, Bridgetown	13.1 N	59.6 W
ZL5	Antarctica, Scott Base	77.9 S	166.4 E	8Q	Maldive Is	4.4 N	73.4 E
ZL7	Chatham Is	44.0 S	176.5 W	8R	Guyana, Georgetown	6.8 N	58.2 W
ZL8	Kermadec Is	29.3 S	177.9 W	9G	Ghana, Accra	5.5 N	0.2 W
ZL9	Auckland & Campbell Is, Auckland	50.7 S	166.5 E	9H	Malta	36.0 N	14.4 E
	Campbell Is	52.5 S	169.1 E	9J	Zambia, Lusaka	15.4 S	28.3 E
ZP	Paraguay, Asuncion	25.3 S	57.7 W	9K	Kuwait	29.5 N	47.8 E
ZS	South Africa, Cape Town (ZS1)	33.9 S	18.4 E	9L	Sierra Leone, Freetown	8.5 N	13.2 W
	Port Elizabeth (ZS2)	34.0 S	25.7 E	9M2	West Malaysia, Kuala Lumpur	3.2 N	101.6 E
	Bloemfontein (ZS4)	29.2 S	26.1 E	9M6,8	East Malaysia, Sabah, Sandakan	5.8 N	118.1 E
	Durban (ZS5)	29.9 S	30.9 E		(9M6)		
	Johannesburg (ZS6)	26.2 S	28.1 E		Sarawak, Kuching (9M8)	1.6 N	110.3 E
ZS2	Prince Edward & Marion Is, Marion	46.8 S	37.8 E	9N	Nepal, Katmandu	27.7 N	85.3 E
ZS3	Namibia, Windhoek	22.6 S	17.1 E	9Q	Zaire, Kinshasa	4.3 S	15.3 E
1AØ	SMOM	41.9 N	12.4 E		Kisangani	0.5 N	25.2 E
1S	Spratly Is	8.8 N	111.9 E		Lubumbashi	11.7 S	27.5 E
3A	Monaco	43.7 N	7.4 E	9U	Burundi, Bujumbura	3.3 S	29.3 E
3B6	Agalega	10.4 S	56.6 E	9V	Singapore	1.3 N	103.8 E
3B7	St Brandon	16.3 S	59.8 E	9X	Rwanda, Kigali	2.0 S	30.1 E
3B8	Mauritius	20.3 S	57.5 E	9Y	Trinidad & Tobago, Port of Spain	10.5 S	61.3 W
3B9	Rodriguez Is	19.7 S	63.4 E	J2/A	Abu Ail	14.1 N	42.8 E
3C	Equatorial Guinea, Bata	1.8 N	9.8 E				
	Malabo	3.8 N	8.8 E				

Chapter 5

DXing

Almost every radio amateur at one time or another becomes interested in DX, surely the oldest activity of Amateur Radio. After all, Guglielmo Marconi, the first amateur, went on the first DXpedition—to W1 land! And there hardly breathes an operator who is not interested in working a station farther away than his normal range, even if only through the local repeater.

But for many amateurs, pushing a high-frequency (HF) signal as far as it will go, and working for contacts with stations in countries never before logged, is a primary or *the* primary interest in Amateur Radio. DXing is one of the most exciting and satisfying aspects of Amateur Radio, and one that virtually any amateur with a working HF station can pursue.

Is your station capable of working DX? If your station can work both coasts on a consistent basis on 40, 20 or 15 meters, you are capable of working DX, especially on CW. You do not have to own a *B-I-G* linear, nor do you need a Yagi on a tower. Of course, such things help, and can at times help a lot, but one of the really nice things about DXing is that it is based on the art of the *possible*.

To be more specific, if you are running one of the typical 150-watt transceivers, and a decent dipole, inverted V, vertical or better on the bands from 40-10 meters, you can work DX. And, with such equipment, you can be assured that you will be able to work 95% of the stations that you can hear. Much DX has been worked with similar antennas and 5 watts or less. Of course, if you hear a really rare one buried in a howling pileup, your chances of snagging a QSO are slim, but if you trip across the same station a day later calling CQ in solitary splendor, your chances of nailing him are very good.

Before going further, let's agree on a definition of what DX really is, at least for purposes of this chapter. Although the classic definition of DX is "distance," today it generally means amateur stations in foreign countries. Typically, this suggests countries thousands of miles away, but not always! For example, for nearly half the amateurs in the United States, the French-owned St Pierre and Miquelon Islands off the coast of Newfoundland are closer than Seattle, yet these islands, using the prefix FP, are considered DX by most amateurs, while the W7 in Seattle is not. And again, Clipperton Island, with the prefix FO, lying off the coast of Mexico, is considered to be a very rare DX country, even though it is closer to San Francisco than New York City is. So, for most HF work, any station in a foreign country is considered to be DX.

Certainly, a JT1 in Ulan Bator, Mongolia, is going to raise a good deal more excitement on the bands than an HI8 in the Dominican Republic, but both are considered DX. Again, working a rare country excites some amateurs, while other amateurs are equally pleased with a friendly chat with a G3 in London. But, it has been noticed more than once that the amateur claiming to welcome equally a contact with England or with Mongolia already has a card from both England and Mongolia hanging on the wall!

Which raises the question of why people try to work DX. For some, it is a simple matter of satisfaction. They find a ragchew with a station in another land interesting and rewarding in its own right, and have no interest beyond that. For others, a contact with a country never before worked provides a thrill unmatched by any other QSO. Likely, such an amateur will avidly pursue the QSL card as a trophy for the wall and for the premier award of DXing, the ARRL DX Century Club (DXCC).

And for you? If you don't know the answer, be glad. If the thought of a contact with a foreign land excites your fancy, you are in for a lot of fun seeking the answer.

TUNING THE BANDS

Let's slip the headphones on and have a look round the bands, and see if a worthwhile bit of DX to snag can be located. It's a late Friday afternoon in April. With the implicit blessing of my boss, I head home early. I grab a cup of coffee, settle the headphones down over my ears, and turn the gear on. The antenna rotator shows that the tribander is pointed at 45 degrees, the nominal short path bearing for Europe from the Midwest. I move the transceiver to 20 meters, switch to

John, OE3ZOC (NK4N), is joined by a friend who likes the sound of CW.

the CW filter, set the AGC (automatic gain control) to "fast," the RF gain on full, move the dial to the bottom of the CW band, and start tuning slowly up the band.

The first few kilohertz are fairly empty of activity, but as I move a bit higher up the band, the activity seems to build. There are several loud stations coming in, and a brief listen to each confirms my expectation that they are Americans and Canadians. One is calling CQ, another is working the West Coast. Nothing interesting there. I keep tuning higher.

There—there's a station calling CQ. It's an OH6 from Finland. Hmmm. I listen carefully to his signal. He is S8, and nice and steady with a clean note.

[Whenever you are checking out a band, especially 20 meters, the quality of signal from a northern European, such as an SM, OH, LA, or a UA1, will tell you a lot about the band in a hurry. If the signal is a bit fluttery, with a trace of auroral buzz to it, there are at least some signs of disturbance. It is therefore unlikely that there will be a transpolar opening, depending on how strong the auroral effect is on the north European signal. On the other hand, if the signals from the high latitude Europeans are strong and clean, with any luck there will be an excellent transpolar opening into such areas as Soviet central Asia, India, Pakistan, Sri Lanka and so on, at the right time for propagation to those areas. From late spring through early fall, this will be an hour or so after sunrise and an hour or so after sunset.]

I keep moving. There's a weak signal signing with a W4. Okay, it's a 4X4, Israel. Another one I don't need. There's another CQ. Right, it's an EI3, Ireland. I don't need that either.

I come to a W2 calling CQ DX, and listen to see what he raises by way of answer. His signal is strong and has a definite bit of backscatter trace to it, a raspy, weaker overlay, spread a few hertz above and below his signal. The upper layers of the ionosphere reflect a portion of his signal back toward me. The uneven reflections yield a Doppler shift "hash" mixed with cleaner bounces of his signal, causing the rasp. [The effect is more generally heard on the signals of stations using Yagis or quads pointed to the same area that the receiving station antenna is aimed at. And, generally, the more the backscatter from nearby stations (for which, read "North American"), the better conditions are likely to be.]

There, the W2 signs. I listen. Yes, there are several stations

Angela, HC1HW, and Alfred, HC1HC, are an active husband-and-wife team from Quito, Ecuador.

calling him. I listen to them. I hear a DK3, West Germany. And there—there's an ON8 in Belgium. Ah, the W2 starts coming back to the DK3. Not surprising; the DK3 is dead on the W2's frequency, and started his call almost immediately after the W2 signed. And his speed is virtually identical to the W2's. On the other hand, the ON8 started his call several seconds later, is two or three hundred hertz above the W2's frequency, and is calling rather slower than the W2 sent his CQ.

I smile to myself; the act of analyzing why the DK3 got the QSO instead of the ON8 is a good way to learn how to be effective at calling other stations, DX or otherwise. But the lessons learned are no good unless I am smart enough to practice the lessons for myself. But I don't need a QSO with a DK3, and today I am after more exotic game. I tune on.

Hmmm. I come across six or eight stations calling someone. I listen; this might be interesting. Ah, there—someone just below the the callers' frequency comes back to a WB8, someone with a good, clean signal, with no backscatter or flutter. He just sends the WB8's call, and 5NN K, the 5NN being CW shorthand for 599. I hear the W8 coming back to him, via backscatter, R TU 5NN 5NN 73 BK. The other station responds, R TU QRZ? and that's all he sends. All the other stations start calling again.

I continue listening carefully and with interest, but I make no effort to call. I want to see who the DX station is before I commit myself to the mini-pileup. The DX station puts away a W3, then a VE3. Then, as he signs with the VE3, he sends, QSL VIA I2CRG I2CRG NW QRZ UP 1 UP1 DE 5L2EQ K.

A 5L2? What in the heck is that? I glance quickly at my DXCC Countries List [which appears in Chapter 17] and don't find it; 5L is not a standard amateur call sign, apparently. I check the back of the log book for the ITU call sign allocations table [which also appears in Chapter 17], and look up 5L there. Yes, there it is; 5LA-5MZ is assigned to Liberia, in West Africa. [5L also is listed in the prefix cross-reference section of the Countries List.] Darn. I need a lot of African countries, especially the "Terrible Ts," but Liberia I don't need. I have an EL2 in the log and on the wall. Well, ELs aren't particularly rare, but the unusual prefix does help explain the mini-pileup. No doubt some of the stations calling want the special prefix for one of the prefix hunter's awards, while others calling just like the idea of a fast QSO with an out-of-the-ordinary station. But I'm after bigger game. I tune on.

I come across UA2FAM from Soviet Kaliningrad signing clear with another W9. I check my list; I have worked several UA2s but have no QSL as yet to show for it. I check my QSL list; UA2FAM is not one of the stations I have worked, so I better give him a call. The W9 is finishing up the QSO. He is not very strong, coming in on weak groundwave, but I hear him well enough. I set my keyer speed to match his. Next I throw the VOX switch to OFF on my transceiver, check to insure that my RIT (receiver incremental tuning) control is in the OFF position, as it should always be when tuning, and touch my paddle to send a string of dits that are heard only in my monitor and don't go out over the air.

I move the transceiver dial until the W9's signal is identical in tone to my monitor, then flip on the RIT, and retune the signal by adjusting the RIT control until he is back where he was when I first heard him. My CW signal is now zero beat with the W9, so I know I will transmit on the same frequency, where, after all, I know the UA2 is listening. I turn the VOX back on.

The W9 finishes, and the UA2 returns with the traditional "dit dit." I wait a brief moment to see that the UA2 does not send a CQ or a QRZ. I hear nothing, so I give him a call, UA2FAM DE W9KNI W9KNI AR. I listen; hah, I got him! I log the details of the QSO; the UA2, like most Soviet amateurs, keeps the contact short, and after the promise of an exchange of QSLs, I am soon tuning the band again. Though the UA2 wasn't a new country for me, if he is the first UA2 that I receive a QSL from, it will be as good as the first QSO with a new one.

I look out the window; darkness is gathering. It will be dinnertime shortly, but the band seems very good, and I don't want to miss the opportunity it offers. I arrange with the family for a quick dinner a bit later, and return to the rig with a fresh cup of coffee.

I glance at my frequency, rather high in the CW portion of the band. I quickly tune up higher a few more kilohertz then, having found nothing of interest, I return to the bottom edge again, and start tuning higher.

The very bottom of the band is as quiet as before, but about three kilohertz in from the edge, things start to pick up. There—a UL7 calling CQ. UL7s are relatively common on 20 meters, but they are a harbinger of rarer DX, coming through as they do on the same path as the rarer UHs, UJs and UMs, as well as a path close to the path for India and the islands of the Indian Ocean.

There's a CQ—not real strong, but enough for copy. I dig for the call, but have trouble getting it at first. I adjust the frequency a bit, and he gets clearer. Yes, it's UM8FZ. And that would be a new one for me! Quickly, I start lining him up. Oops; my RIT is still on from the last QSO. But it must be close to the transmit frequency; I trimmed only a couple hundred hertz for that UA2. (While looking for DX, never tune a transceiver with the RIT on unless you have it set at a specific tone for a specific reason.)

I turn off the RIT; the tone of the UM8 shifts, but he's still inside my crystal filter's passband, so I "get away with" my mistake. I switch off the VOX and hit the paddle, and adjust the frequency of the transceiver until the UM8 is at the same tone as the monitor. There, that's set. Good thing he's making that CQ a long call. I switch the VOX and the RIT back on, but before I retune the RIT to make his tone more suitable to my ear, he finishes his call. No matter, I'm ready for him. I call, UM8FZ DE W9KNI W9KNI W9KNI AR, adjusting my keyer speed to match the speed of the UM8 as I start my transmission. I finish my call, and listen. Darn! He's already coming back to someone else—a WA5.

I copy all the data that he is sending the WA5. That way, if I have trouble with QRM when I get him, if I get lucky, then I'd still have his name and QTH. Yes, he is copying the WA5 RST 579, his QTH is Frunze, and his name is Victor. He turns it back to the WA5. I wince, as a wave of RF sends my S-meter slamming into the peg. Guess we must have short skip. And if the WA5 is anywhere near as strong for the UM8 as he is in my receiver, it's no wonder he beat me out. I temporarily switch my RIT out again, and find the WA5 within a hundred hertz of my monitor note; that means I was close to being on the same frequency, anyhow. The WA5 is giving the UM8 a 579 report. I can only say if he's hearing the UM8 a real 579, he's doing a lot better than I am. The UM8 is really 559 at best here.

There, the WA5 is describing his station. Yes, a kW and a five-element Yagi at 30 meters. Okay, that's 100 feet, give or take, against my little three-element tribander at 45, and

his kW against my 100 watts. No wonder I got beat out. I decide not to move my transmit frequency; I was close enough to the WA5's frequency that there's no reason to move. But there, they are signing now; maybe it's my turn this time. I call the UM8 again, then listen.

As the transceiver goes back to receive, I hear another station for a moment, a KA4. Guess he must have also been calling the UM8. There...there's the UM8 coming back.

Phooey. Beaten again. This time the UM8 is coming back to a K8. Should I wait around again, or should I go tuning again, looking for another new one? Maybe I would find another UM8, as far as that goes. But if I do, I'll probably lose any chance of snagging Victor. I decide to ride it out a bit longer. Besides, I think his signal is starting to come up a bit; the path is getting better. I glance at my antenna rotator. Darn, it's still at 45 degrees. I check out the beam heading on my *Second Op*. Okay, I should be at 13 degrees. I turn the antenna more north, and the UM8 seems to pick up another half S-unit, bringing him to a solid 569 now. I decide to stay after him.

The UM8 signs clear with the K8, and I wait for the UM8's "dit dit." But, two stations start calling the UM8 while I wait, and they are both on almost exactly the same frequency as the K8 was—and where my transmit frequency is. If I call there, all I'm going to do is get QRMed by those other stations. Quickly, blindly, I move my frequency up about 300 Hz. I know that I am moving away from where the UM8 was listening, but with my not-overly-strong signal, there's no way I'm likely to blow those other callers away. So a fast move seems likely to be my best bet. I make my call a little longer, as well, hoping to give the UM8 more time to tune off those other stations and find me.

I listen again, moving the RIT back down to the UM8's frequency. Yes, there are still two stations calling him, right on his frequency. I can make out that the UM8 is in there. He must have come back to someone, but there's no way I'm going to copy him with those fellows still calling him dead on his frequency. But there, one of them stops transmitting, and then almost immediately so does the other station...569 BT QTH HR FRUNZE FRUNZE ES NAME IS VICTOR VICTOR QSL SURE VIA BURO HW COPY W9KNI DE UM8FZ KN. All right—I got him!

R UM8FZ DE W9KNI R FB VICTOR ES TNX QSO ES FIRST UM8 BT UR RST 569 569 HR QTH NR CHICAGO NR CHICAGO ES NAME IS BOB BOB QSL SURE VIA BURO ES TNX AGN NEW COUNTRY DR VICTOR HW CPY UM8FZ DE W9KNI KN.

R W9KNI DE UM8FZ R R FB BOB ES TNX REPORT FM NR CHICAGO ES GLAD BE UR FIRST UM8 BT QSL SURE QRU NW 73 ES GOOD DX DSW SK W9KNI DE UM8FZ TU.

I start breathing again; a real "goody" in the log. I look over at the UTC (Coordinated Universal Time) clock, and get him noted in. Let's see here, the clock shows 0039 Zulu (ie, UTC). I must have started my QSO with Victor at about 0034, and I enter that time into the log. But let's see, 0034 is after midnight Zulu time; that means I have to put *tomorrow's date* in the log for the date of the QSO.

[A fellow who does a lot of QSL managing for rare DX stations told me once that more stations are not found in a DX log because of either using local time instead of UTC or because they got the date wrong than for any other reason. Although he checks for wrong dates and thus salvages a lot of stations their QSLs, he thinks a lot of managers don't bother or just don't have the time to check. Ever since he told me that, I've been careful to see that the date is rendered correctly in my log book, being sure that I don't accidentally enter the *local date* when it's already *tomorrow* in UTC.]

Well, that's a good one. I listen on my frequency. There are four or five stations calling the UM8 now. There, he comes back to the KA4 that was calling him on the same frequency as whoever the other station was. The UM8 gives him a 589. That's two S-units better than my report was! But I got the UM8 first, because I moved away the QRM. Hmm. Maybe I learned something here. I start tuning up the band again. Conditions seem pretty good tonight; maybe I'll get lucky again.

The Azores is well represented by Manuel, CT2CB.

20 METERS—THE QUEEN OF THE DX BANDS

There is a lot of DX to be worked on *any* of the HF bands; 20-meter capability is not necessary to achieve a high country total, but it surely helps! No matter what point in the sunspot cycle, there is DX to be worked on 20 at some hour every day, with only an occasional lapse of a day or so courtesy of massive solar storms. Twenty meters is the DX crossroads of the Amateur Radio world. At the high point of the solar cycle, 20 is often open around the world around the clock —the DXer's ultimate dream. And, at the bottom of the cycle, especially in the summer months, 20 is often the *only* band where everyone has a real chance of working DX. If there is going to be a path open to a remote DX spot that is to be workable during difficult conditions, 20 meters will almost always be the band that opens the path.

All of this, of course, means there are a lot of serious DXers on 20, many with very large antenna arrays and using maximum power levels. But even the station with modest capabilities can work a great deal of DX on 20 meters. There assuredly are people on the DXCC Honor Roll who got there using 150 watts or less and simple antennas, such as dipoles or verticals, with 20 meters their primary medium. You can, too, especially on CW.

The best times of the year on 20 are the spring and fall, during which time the sunrise and sunset times are the best for long-haul DX, but with openings to all parts of the world at various times. In summer, 20 is poor in the four to six hours around local noon time, and often open the rest of the time. In winter, 20 is excellent from sunrise to sunset, and often some hours into the evening. And, always watch for long-

path openings around sunrise and sunset! [See the propagation charts and sunrise-sunset tables which appear in Chapter 17.]

Amateurs interested in learning how to become effective DXers and who hold the appropriate class of license should spend a major part of their operating time on 20 meters. Although all of the HF bands, including 160 meters, offer at least some DX opportunities, the reliability of 20 meters and the quality of the competition make it the ideal DX learning ground.

Back to the Bands

Well, that UM8 was a good one! The thrill of a new country always makes me feel good. I run upstairs from my basement shack to tell the family of my triumph, grab another cup of coffee and head back to the shack. I'm going hunting again! We have a good band opening, and I'm going to keep after it.

I slowly wind the transceiver knob higher up the band, looking for the weak, watery signals that spell DX. There's one, not really so weak, giving out a report. I listen; when he says QTH ALMA ATA, I tune on. He's a UL7, and one that I have hanging on the wall. The funny thing is that Alma Ata is all of 10 miles from the border of UM8, but it counts as a UL7. I keep moving.

I come across a loud signal, FB KHANRI QSL SURE VIA UR MGR ES TNX NEW COUNTRY 73. Needless to say, thanking the other station for the new country makes me all ears! A new country for whoever the loud station is might make it a new country for me, too. He continues, 73 ES GOOD DX S̄K̄ S21AC DE WB5FZH. Wow! S2, Bangladesh, a very rare country indeed. I listen carefully, as I look over my gear. Yes, the RIT is off, the keyer is at the right speed, and so on. And there the S21 is, coming back to the WB5, FB BILL TNX QSO ES 73 NW QRZ? DE S21AC ĀR̄. By the time he finishes sending QRZ, I'm all ready for him. Sure hope no one else has noticed him.

I pause for a brief moment after he signs—and my mouth drops open. The frequency suddenly sounds as if it were the focal point of a volcano. My poor S-meter is buried against the pin. There must be hundreds of stations calling the S21 all at once, all obviously desperate for the QSO. I shake my head sadly, knowing that there's no hope for me in this one. My barefoot transceiver and low little tribander just aren't about to beat out that mob. I don't even hang around to see who snags the S21, knowing that there's no way it is going to be me. After all, DXing is supposed to be the art of the possible. A happier thought occurs to me. With all of the interest in the S21, maybe there's some other good bit of DX lurking around somewhere—without a pileup.

The strength of the pileup is so strong that I have to move up better than 5 kHz for my receiver to be clear of the pounding QRM. I start tuning again, listening carefully to each station. It seems that I was right about the S21 pileup attracting all the attention on the band; there are very few W stations around. Now, if I could only find a good one to catch without the competition! There's a weak CQ, one with a bit of warble to it. I dig him out of the mud, but it's only a UA9. Of course, "only" a UA9 means I have UA9 confirmed. I keep on looking.

The sound of RTTY tells me that I have tuned up past 14,070 kHz and into the digital area. I have a very quick look on up to 14,100 to see that there are no stray CW signals that should be inspected, then move the rig back to the bottom of the band. Almost immediately, I come across a fairly strong signal, but it's one clearly of DX origin, as indicated by its watery sound (marking a station on the other side of the North Pole). OK DR SHANTI QSL SURE VIA BURO 73 ES DSW SK 4S7WP DE UKØQAC K. Pay dirt! The UKØ is working a 4S7, Sri Lanka. But can I hear the 4S7?

I listen intently. Yes, there he is—and good copy, too, a clean 559 signal standing out against a relatively quiet band. As he signs clear with the Siberian, I quickly line up on him. I swing the antenna a touch to dead north, make sure my RIT is off, switch the VOX off, tune the receiver while keying the monitor so the monitor note is the same as the 4S7's (which is the same as the UKØ's), switch the RIT and VOX back on and wait. Yes, there's the "di dit" from the UKØ. I listen on my frequency for a moment; no loud signals calling, but another watery signal starts calling the 4S7, probably another Siberian.

I start my call. 4S7WP 4S7WP DE W9KNI W9KNI W9KNI AR. I make my call longer than normal, figuring that the other station calling is probably louder in Sri Lanka than I am. If all goes well, the curiosity of the 4S7 will make him want to see who the weaker station calling is. I listen. There is a pause, then I hear that most glorious of musical symphonies—my call coming back from a new country! My long call must have worked. W9KNI DE 4S7WP GM OM ES TNX QSO BT UR RST...

And so it goes. This DXing style is how the amateur with less than an outstanding signal gets a respectable DXCC country-total: by *tuning, listening carefully* to every station and being *ready to pounce* quickly whenever a new one is found. And, by recognizing a lost cause when he sees one—knowing when the game is over, and that it's time to move on to the next signal.

There is one other important lesson in tuning. The DXer quickly learns to tell if a signal is from a DX station or from a local. But, if the station *heard* is not DX, the DXer must always be aware that the station heard might be *working* rare DX. On a good DX band, most QSOs are between two stations that are DX to each other. So, if you hear a VE or W in QSO on 20 meters if the band is open for DX, chances are good that that station is working DX, maybe even rare DX.

There are some telltale signs to watch for when listening to QSOs. When you find a loud station promising to QSL direct or through a manager, or asking for QSL instructions, be alert! If you hear a loud signal with a fist or voice that sounds nervous or excited, take note. There is a reason, and you want to find out what it is. Chances are good that someone is in QSO with a new one, and possibly one that would be new for you.

TUNING TECHNIQUES

You will have noticed references to turning off the RIT and VOX as part of the routine of zeroing the other station's frequency. For the CW operator and would-be DXer, there is no skill more important to learn than how to put your transmitter or transceiver exactly on another station's frequency. The frustration and downfall of many a potential DXer comes from a failure to understand the proper technique. DXers using split transmitter-and-receiver combinations can ignore this section, as can those with rigs having specific CW zero-beating switches. If you are fortunate enough to own such a rig, be sure to study the manual and affirm that you truly understand how to use it.

If you own one of the common American or Japanese transceivers that lack a specific zero-beating function, the following techniques will enable you to match, or zero beat, the frequency of a station. First, see that the RIT, and if you have it, the XIT (transmitter incremental tuning) is turned off. Remember, you should never engage in general tuning with the RIT on when you may need to zero beat a station, such as during a DX-chasing session. The RIT is to be used *after* you are zeroed, not before!

If your transceiver has passband tuning, be sure the passband tuning is at the center of its range or not far off, so you can be positive you are not listening to the wrong side of the BFO (beat frequency oscillator) from where the transmitter offset is placed. Never tune with the passband on the wrong side, unless you fancy a career as an SWL. But be warned. DXCC is not offered to SWLs.

Next, turn off the VOX. On most rigs, you can now use your keyer to send a string of dits that will be heard in the CW monitor only, while the receiver continues to operate. The audio frequency (AF) of the monitor matches the offset of the CW note from your transceiver's transmitter in almost any transceiver built today. Adjust the transceiver frequency until the audio tone of the station you wish to zero beat is identical in tone to the transceiver's CW monitor note. You are now at zero beat with the other station, and when you transmit, you are exactly on that station's frequency.

For your own good, please be sure you understand this material. For working anyone on CW, DX or otherwise, the technique is of vital importance. And if you have been finding that stations that should, by all rights, be coming back to you are not, review your zero-beating technique. Especially if you are listening on the wrong side of the BFO, or using the RIT and failing to turn it off, you could well be calling DX as much as 2 kilohertz off, when the DX station is listening with a 500-hertz filter. [Additional information on zero-beating techniques can be found in *The Complete DX'er,* referenced at the end of this chapter.]

The Morning Shift

I brush the sleep out of my eyes as I settle into the operating chair, shivering slightly in my heavy robe. The glare of the rising sun reflects off the snow and into my basement window. I turn the antenna to point into the southwest, set the transceiver at the bottom edge of 20-meter SSB and begin tuning up the band. It is almost immediately obvious that the band is in fairly decent shape; the low background rumble in my receiver indicates a high activity level on the band. Immediately, I come across a station speaking with a definite accent, almost surely Australian. Right, he signs over to a WB3. Yes, it's a VK6 in Perth, Western Australia. He's up late. I keep tuning.

There's a strong signal, but with a pure Midwest American accent. He's talking about the weather in a very calm and unexcited voice. I go on. I trip over a small pile-up of stations. Must be some DX there; yes, I hear a station coming back. Wonder where he is? Oh, right, he's a P29 in Papua New Guinea. Not very strong, either. But then I remember that my antenna is pointed southwest for the long path into the Middle East, rather than a bit north of west as would be correct for a P29. Since I already have a P29, I keep on moving.

There's a station passing out a report, and the operator has a different accent. There, he signs. Ah, it's a 3B8 on Mauritius in the Indian Ocean. I have one confirmed from a QSO on 40-meter CW (one of my prouder moments on that band), but I do not have it on SSB. I check the bearing on my

DX Operating Procedures and Guidelines

For many years DX operating procedures have consisted of split-frequency operation and transceive-type operation with good operator control. Recently, several other methods have emerged which, if not used properly, cause frustration, unhappiness and undue interference. These operations are known as lists, DX nets and roulette.

Below are the different DX operating procedures and how and when to use them. Following these recommendations should increase everyone's pleasure in working DX, and in being DX operators, resulting in smooth, efficient operation which will allow as many amateurs as possible to make good DX contacts.

Split-Frequency Operation—Best Procedure of All

Split-frequency operation is, by far, the best method to use, and results in less confusion, interference and frustration than other procedures. In split-frequency operation, the DX station transmits on one frequency and receives on another. This method give the highest QSO rate, allowing more amateurs to contact the DX station. Anyone considering going on a DXpedition should plan to take proper equipment along so that split-frequency operation can be used. Any DX operator planning to purchase a transmitter/transceiver should purchase one that has split-frequency capability or has the provisions for use of an external VFO.

It is recommended that a DX station operating phone transmit outside the US phone subbands, when possible, and "listen up" (listen higher in frequency) for US stations and "listen down" (listen lower in frequency) for other non-US stations. Examples: Phone—DX station transmits on 14.190 MHz, listens for US stations from 14.200 to 14.230 MHz and listens for others from 14.180 to 14.160 MHz. CW—DX station transmits on 14.020 MHz and listens up from 14.025 to 14.035 MHz.

No more bandwidth should be specified than the absolute minimum that permits rapid separation of the calling stations. Not everyone on the band is interested in calling the DX station, and the interests of others should be respected.

Transceive Operation—Requires Experience and Control

In this method of DX operation, the DX station transmits and receives on the same frequency. This takes less bandwidth than split-frequency operation, but more so than operating split, its success depends on operator control and experience. There are three important points that lead to a successful transceive operation: (1) Experience and maintaining proper control; (2) A good to excellent signal; and (3) Knowing when and how to divide a pileup.

Handling a pileup is an art, and one must maintain firm control of the situation and know what to do to regain control should it be lost. There are some DX operators who can handle a pileup no matter how big it gets. But, there are times when it is advantageous to divide the pileup in order to reduce it to manageable limits. There are many good methods now in use to divide a pileup. The most common method is by call districts.

Example: The DX station requests calls from the first US call district. After working a specific number (usually three to 10), the DX station requests calls from the second US call district, moving on to the other areas until he or she is finished with W0s. Then he requests non-US stations, working a specific number of these.

Whatever sequence is used, it can be repeated as long as the DX station so desires. However, once a sequence is started, it should be continued until completed. This gives everyone a fair chance to have an opportunity to contact the DX station.

If the DX station's time is limited, the number of QSOs per call area can be reduced to, for example, three QSOs per district instead of 10. Whichever way the DX station decides to divide a pileup, he must be very firm and not

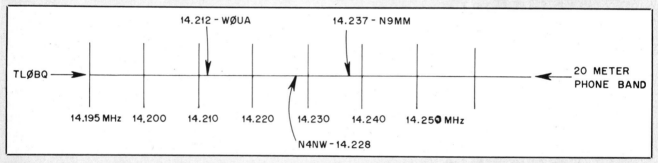

Fig 5-1—An example of split-frequency operation. We are trying to work TL0BQ, Central African Republic, who is transmitting on 14.195 MHz. TL0BQ says he is listening for stations to call him from 14.210 to 14.240 MHz. We listen a bit and find: W0UA calling at 14.212, N4NW calling at 14.228 and N9MM calling at 14.237. We write these calls and frequencies down. Then, when TL0BQ works one of the stations that we are using for a benchmark, say N9MM at 14.237, we can tune our transmitter quickly to this frequency and have a good shot at working TL0BQ.

great-circle map and glance at the antenna rotator. I move the antenna a hair to the north, so that my bearing is a little south of west, giving me the correct bearing for a long-path shot at 3B8.

He signs with his last QSO, and I call, "W9KNI, W9KNI, OVER." I listen. There are several other stations calling him, but I hear him through them, although I cannot yet copy him because of the QRM (interference). There, the frequency clears. "5 and 8, 5 and 8, KB9IY go ahead, please." I listen; yes, there's KB9IY. He lives not too far from me, and is very loud. He runs a linear, and I'm not disappointed at getting beaten out. He is dead on the frequency of the 3B8, so I don't need to move. He gives the 3B8 a report, and signs clear all in one transmission. That's good; apparently the 3B8 is

interested in working a lot of stations, so I have a better chance of getting him. He clears with KB9IY, and calls QRZ.

I call him. "Whiskey Nine Kilowatt, Norway, India, W9KNI, Over," and listen again. He's already back to someone. Yes, another W8. He turns it over to the W8, who is weak, and coming through on backscatter. I am able to copy him well enough to realize that he is nearly half a kilohertz lower than the 3B8's frequency, and I confirm it by tuning him in with the RIT control temporarily engaged. Quickly, I note the frequency displayed when I am on his frequency, kill the RIT and move my dial so that my displayed frequency is now the same frequency as the W8's. Just as I do, he turns it back to the 3B8. I move the RIT back to copy the 3B8 properly. He signs clear with the W8, and I call, now

QSO stations out of turn, even though some will call out of turn. To do so will encourage others to follow, leading to on-the-air bickering, and, in short order, chaos on the frequency. To prevent stations from calling out of turn, the DX station should announce often that he or she is dividing the pileup and the method being used to do so.

Another important factor in a successful transceive operation is that the DX station should have a good to excellent signal, so that calling stations will know who the DX station has acknowledged. Weak DX stations tend to be overcome by their own pileups and no one calling can be sure who has been acknowledged, if anyone. When this happens, QSO rates go down and confusion takes over. Then, the best approach for the DX station is silence until the pileup reduces itself to a low level or goes away. The DX operator regains control by announcing that he or she is going to divide the pileup and the method that he is going to use. Again, whichever method the DX operator uses, he must stick to it and give everyone a fair chance to contact him.

DX Networks—More Orderly, But Slower

DX nets have been set up for many purposes. Some are for certain areas, club members, and so on. Some are for working DX only. They range from fully "open" nets for all to those controlled as to who can check in and when. For the most part, the DX operators who participate in these DX nets are permanent residents. They check in regularly and, with a reasonable effort, almost anyone can obtain a contact with them. If the DX station desires to participate in these DX nets, he or she has the right to do so.

Any net operated for the sole purpose of working DX should be operated in such a manner that all non-DX check-ins are given a fair chance to work the DX. Many of the guidelines used in transceive operation can be applied to DX net operations. Rare or semi-rare DX stations with limited time should refrain from using DX nets because of the resulting low QSO rate. DX nets are helpful to those DX operators who do not have the proper equipment to operate split frequency and/or the experience to operate transceive only.

Lists—Limited Use Only

List-style DX operations are, in some cases, the only way to work a DX station. However, it should be understood that lists should be used only when other DX operating procedures are not possible. Here are some guidelines that can make list operations run smoothly, be fair to almost everyone, and reduce interference to all.

Conditions under which lists might be used: (1) DX operator cannot speak English or understand it well or at all; (2) DX operator is inexperienced; or (3) DX station has a very weak signal because of poor antenna or location, low power or poor propagation.

List operation guidelines: Take a list for a DX station only if he asks or desires to do so. Do not pressure him. If it is determined that he is inexperienced in DX operating procedures, one could suggest a list operation, but leave the decision up to the DX station.

Take lists only for real-time operation. Old lists taken on a previous day or on different bands should not be used. Many on these lists do not show up, leading to frustration to those on frequency. It wastes time.

Prearranged lists should not be used. Lists that are not accessible to all who can hear the DX station should be avoided.

Only stations designated by the DX station to take lists should do so. When taking a list for a DX station, propagation differences should be taken into account. If necessary, the MC (master of ceremonies or list-taker) should direct one or two alternates in different areas to pick up the stations whom the MC cannot copy because of propagation. This gives all a fair chance to get on the list.

The MC should have good to excellent copy on the DX station, so that control can be maintained at all times. Lists should conform to the DX station's desires as to content, number per call area, length and so on. It is much better to take a short list, about 20 stations or so, run them, then come back and pick up a second group of 20, repeating this procedure for as long as the DX station is willing to do so, than to take a list of 300 stations in one swoop.

Never relay signal reports and call-sign corrections. If it is really a QSO, the DX station should be able to copy the call and report of the station he is working without the help of the MC.

The MC should be clear and concise in giving instructions as to what he and/or the DX station desires. The instructions should be repeated often to reduce confusion and to inform those who have just arrived on frequency what is going on.

When working a DX station in a list operation, conform to the DX station's desires as to length of QSO. If the DX station is giving only signal reports, do likewise and do not give your name, QTH or a weather report. If the DX station gives reports and his name, be courteous and give him the report and your name; but in any case, conduct the QSO clearly and quickly, so that others can also have a chance to work him. Think of others.

Remember, list operations should be used only when required, not as a replacement for other unaided DX operating procedures.

Roulette—Should Not Be Used

At times, a DX station may appear on a predetermined frequency and announce that he will listen for callers within a certain band segment and will answer callers on their own frequency. This shotgun technique is referred to as roulette. This method is not at all recommended for DX operating and should be highly discouraged. It causes confusion, encourages bootleggers, has a very poor QSO rate, results in questionable QSOs and produces a lot of unnecessary QRM.

These suggestions are aimed at increasing your DX operating pleasure.—*John Kanode, N4MM (originally published in Sep 1979 QST)*

---

transmitting on the W8's frequency.

I dump my call in, "Whiskey Nine Kilowatt, Norway, India, W9KNI, Over." The smoke clears as he is giving a report—and my chest swells with pride when he turns it over to me. A new one on phone! I'm surely glad I went to the trouble to dig that W8 out of the mud enough to know he was off the 3B8's frequency. He was probably off because he left his RIT on by accident, and then got lucky when the 3B8 tuned off enough to find him. But when I found him off frequency, my chances of nailing the 3B8 increased a great deal; I transmitted on the frequency I knew that the 3B8 was listening on, rather than that of the stations who stayed dead on the 3B8's frequency. I log in the details and run up the stairs to grab another cup of

coffee. And quickly, I am tuning the band again.

As I move up the band, I hear a good deal of DX, but nothing really juicy. I was rather hoping to nail one of the Middle East stations that I need, such as a 9K2 or an HZ1, but all I hear are 4X4s, a loud UA6 and several other East Europeans. Obviously, the band is open to the Middle East, but there are no players for me to chase. I finish tuning up past 14,300 and move back down to the bottom of the phone portion of the band.

There I find a French station calling a W6, apparently on schedule, and about where I found the VK6 before. Just above him, I run into a pileup, with all stations on a single frequency. The QRM clears rapidly, and I hear the DX station come back to a W4 with a report. The DX station has a big signal,

International Third-Party Traffic—Proceed with Caution

Occasionally, DX stations may ask you to handle a third-party message to a friend or relative in the States. This is all right as long as the US has an official Third-Party Traffic Agreement with that particular country, and the traffic is noncommercial and of a personal, unimportant nature. During an emergency, our State Department will often work out a special temporary agreement with the country involved, but in normal times, never handle traffic without first making sure it is legally permitted.

US Amateurs May Handle Third-Party Traffic With:

C5	The Gambia	V2	Antigua and
CE	Chile		Barbuda
CO	Cuba	V3	Belize
CP	Bolivia	V4	St Christopher
CX	Uruguay		and Nevis
EL	Liberia	VE	Canada
GB*	Great Britain	VK	Australia
HC	Ecuador	VR6**	Pitcairn Island
HH	Haiti	XE	Mexico
HI	Dominican Republic	YN	Nicaragua
HK	Colombia	YS	El Salvador
HP	Panama	YV	Venezuela
HR	Honduras	ZP	Paraguay
J3	Grenada	3D6	Swaziland
J6	St Lucia	4U1ITU	ITU, Geneva
J7	Dominica	4U1VIC	VIC, Vienna
J8	St Vincent	4X	Israel
JY	Jordan	6Y	Jamaica
LU	Argentina	8R	Guyana
OA	Peru	9G	Ghana
PY	Brazil	9L	Sierra Leone
TG	Guatemala	9Y	Trinidad and
TI	Costa Rica		Tobago

*Third-party traffic permitted between US amateurs and special-events stations in the United Kingdom having the prefix GB only, with the exception that GB3 stations are not included in this agreement.
**Since 1970, there has been an informal agreement between the United Kingdom and the US, permitting VR6TC and US amateurs to exchange messages concerning medical emergencies, urgent need for equipment or supplies, and private or personal matters of island residents.

Please note that the Region 2 Division of the International Amateur Radio Union (IARU) has recommended that international traffic on the 20- and 15-meter bands be conducted on the following frequencies:

14.100-14.150 MHz
14.250-14.350 MHz
21.150-21.200 MHz
21.300-21.450 MHz

The IARU is the alliance of Amateur Radio Societies from around the world; Region 2 comprises Member-Societies in North, South and Central America, and the Caribbean.

Note: At the end of an exchange of third-party traffic with a station located in a foreign country, an FCC-licensed amateur must also transmit the call sign of the foreign station as well as his own call.

running well over S9, which would help explain the size of the pileup. The W4 responds with thanks, but without signing the DX station's call. But then the DX station comes back, saying "Roger, 73. QRZ 4s from HZ1AB."

The frequency explodes with stations calling the HZ1. This is one that I need. But if he is going by call areas, he's sure to be on the 9th call area sooner or later. For a new country like that, I'm going to sit and wait, and see if I can learn anything about this fellow's operating habits in the meantime. And as he keeps working 4s, I begin to realize that he is the loud station with the American Midwest accent that I heard earlier when I first came onto the band. I realize that I blew it, since the pileup on him then was probably far smaller, a time when I would have had a much better chance of getting him into the log.

Well, you live and learn. At least I hope I do. I should have never passed over that station without seeing who and where he was. As I listen, he describes his station; stacked rhombics at 150 feet and a full-bore linear. No wonder he's so loud. Finally, he goes on to the W5s, and a new pileup jumps into being on the frequency. I listen as he takes six W5s, and moves on to the 6s. I cringe mentally when he calls for the W6s, expecting the frequency to be obliterated. But, strangely, there are no W6s calling. The band must not be open out there yet. Finally, he works one lone W6 operating portable 4, and then goes on to the W7s.

After only a few W7s, he moves on to the W8s, and the situation returns to normal—seemingly hundreds of eager DXers, all tying to score. I begin to get nervous about my chances. With the fellow working people as long as I have been listening to him, and with the size of the pileup calling him, by now there's sure to be a load of W9s in wait. And I don't have the signal to beat out hordes of fellow W9s.

After six W8s, our turn comes; my fears prove to be well-founded. I pause for a moment before calling, and see my S-meter move over to the pin. I call frantically, shouting into the microphone like everyone else, and, like for most of the W9s calling, to no avail. After six QSOs, he's on to the suffering W0s. I can't complain; I had my chance. But, I mentally kick myself. I know that I had a much better chance earlier, and I failed to take advantage of it.

PROPAGATION-DETERMINING TOOLS

Great-circle bearings and long-path propagation are two of the more interesting technical elements of DXing. Most amateurs probably were taught about great-circle routes in school. Strings were pulled across globes to demonstrate that the shortest route between two points on earth was rarely a single compass bearing, but rather a constantly changing bearing going across a globe. Remember? Our signals travel through the ionosphere in the same manner, seeking the shortest route, which is a great-circle bearing.

Special maps utilizing an azimuthal equidistant projection possess the unique property of showing great-circle bearings and paths as a straight line, which is of course the path our signals take in normal circumstances. Also, they show distances accurately from the center of the map to all points, something that more normal projections are not readily capable of. [A set of fine azimuthal maps produced by Bill Johnston, N5KR, appears in Chapter 17. N5KR can supply a similar map centered on your QTH as well as beam-heading charts. Contact him at 1808 Pomona Dr, Las Cruces, NM 88001, for details.] Several maps also of special interest to DXers are offered by the League and by the Radio Amateur Callbook Company. These maps are constructed around the geographical center of the United States, which offers accuracies within 2% for stations anywhere in the contiguous 48 states or the southern parts of Canada. To use these maps for determining the short-path bearings to any point on the globe, simply draw a real or imaginary line from the center of the map to the point where you wish to work. Extend the line on to the edge of the map, and read the bearing off from that point. That is the correct direction to point your antenna.

The antenna set up at Switzerland's HB9CXZ.

An alternative approach is to make or purchase a list of computerized beam headings from your location. Computer programs for this are in several ARRL publications, or customized lists are available from several suppliers advertising in *QST*. The Radio Amateur Callbook includes lists of bearings for a number of major American cities. Also, the N6RJ *Second Op*, available at many dealers, offers a convenient and quick access tool for determining bearings for your antenna, along with a host of other useful pieces of information for the DXer.

Long-path propagation, mentioned above, is one of the more fascinating aspects of DXing. Of course, when we transmit, our signals travel in a straight-line great-circle path. What happens when our signals travel to the point on the opposite side of the globe, 12,000 + miles away? Do they stop dead, or fall off the edge? Of course not; they keep on going in the same direction exactly as they were, sometimes a great deal farther still, strong enough for communication. And, often, when they travel in such a fashion, the propagation to that farther point via short-path propagation is nil. Indeed, thanks to seasonal and diurnal variations, the long-path propagation to some points is often more reliable at certain seasons than short-path propagation!

For example, from the American Midwest in the winter months, propagation on 20 meters to the Middle East and southern Asia is best in the hours after sunrise, with the station antenna pointed southwest, on the reverse bearing for the Middle East, and south for southern Asia. In summer months, southern Africa is often best worked in the morning hours with the antenna aimed straight west. Although long-path propagation is most consistently usable on 20 meters, especially for stations with relatively modest antennas, the phenomenon can be used to advantage on all bands from 160 to 6 meters. The bearing for long-path propagation is 180 degrees opposite that of the short-path bearing.

Sometimes, both the long and the short paths are open, especially in the spring and fall seasons. If you are listening to a DX station on the short path and notice an echo on his signal, try turning the antenna to the long path. If long-path propagation is superior, signals can show dramatic improvement. Needless to say, your opportunity to work him is equally enhanced!

There are two other propagation anomalies that occur reliably enough such that DXers should be aware and take advantage of them. They are "grayline" propagation enhancement and "crooked-path" propagation. Grayline enhancement takes advantage of the fact that propagation conditions are best in twilight conditions, thanks to solar absorption dropping greatly in twilight while upper-atmosphere ionization has not yet weakened enough to cause a reduction in the MUF (maximum useable frequency).

At any given moment, there is a circle or band of twilight girdling the globe somewhere. All stations in this twilight band are capable of communicating more easily with other stations in the twilight band—especially if they can point a directional gain antenna down the band. Stations in the twilight band will often find communication possible over paths as long as 8000-10,000 miles on three or even four bands at the same time. [A comprehensive table of sunrise-sunset times for each DXCC country appears in Chapter 17.] Very useful tools demonstrating grayline times anywhere on the globe are also available from Xantec Inc, PO Box 834, Madison Square Station, New York, NY 10159.

Crooked-path propagation is a unique phenomenon that also can be useful to DXers, especially those with directive antennas. Sometimes, signals will be observed coming in from directions that are unrelated to short-path or long-path bearings. Almost always, such signals are weak and watery, but they are workable. Often, crooked-path propagation is combined with grayline enhancement. For example, in December and January, at sunrise in the American Midwest, it is sunset in India and Sri Lanka. Aiming a 40- or 20-meter Yagi southwest, pointing down the grayline belt, will often turn up workable signals from that part of the world, even though the great-circle bearings are far removed from the actual antenna bearing. Indeed, the southwest crooked-path opening on 40 meters in winter can be one of the most exciting DX activities of a season.

Again, crooked-path openings can be found on other occasions. Another example can occur during the late fall and early spring contest seasons. Stations in North America, anxious to add multipliers to their score, will look for Europeans on 10 meters when the only band opening is to southern Africa. By pointing directional antennas at the bearing to southern Africa, the DXer will sometimes hear faint Europeans who can be worked to advantage, while turning the antenna to Europe results in the loss of reception. Note that this opening is usually very weak, and one that requires better-than-normal stations at both ends. Crooked-path openings are generally more useful on the lower frequencies.

Checking 15 Meters

It's 10 AM on a fine, rainy April morning. It's fine because I can sit in the shack most of the day chasing DX instead of feeling guilty about undone yard work. I have the antenna

Sixth Annual DXCC Countries-Needed Survey Results (information courtesy of The DX Bulletin)

World	N America: West	N America: Midwest	N America: East
ZA	ZA	ZA	ZA
7O	7O	XZ	7O
3Y	3Y	3Y	VU7/A
VU7/A	VU7/A	7O	XV
XV	A6	VU7/A	3Y
YA	XZ	XV	YA
4W	YA	4W	4W
XZ	5A	YA	XU
5A	4W	XW	XW
S2	S9	XU	S2
XW	SV/A	S2	1S
A6	XV	5A	5A
XU	FR/G	A6	A5
S9	3CØ	S9	XZ
A5	S2	1S	S9
SV/A	A5	VP8/S.GA	A6
1S	5U	SV/A	VP8/S.GA
VP8/S.GA.	TT	A5	VP8/S.SA
ZS2M	VP8/S.SA	FR/G	ZS2M
VP8/S.SA	ZS2M	VU7/L	BY
FR/G	FR/T	C9	C9
C9	1AØ	ZS2M	SV/A
3CØ	J2/A	VP8/S.SA	FR/G
5U	VP8/S.GA.	3CØ	FR/J
FR/J	FR/J	5U	3CØ
FR/T	XW	BY	VU7/L
VU7/L	3B6	FR/T	ET
ET	HKØ/M	FR/J	5U
BY	VKØ/H	YI	FR/T
3B6	5X	ET	3B6
TN	C9	UG	9M6/8
T5	ET	TN	KH5K
J2/A	T5	T5	FT8Z
TT	XU	J2/A	KH1
5X	TN	TT	VK9/M
YI	5R	3B6	TN
1AØ	PYØ/T	FT8X	ZL9
VK/H	TA	VK9/C	T5
KH5K	VU7/L	1AØ	VKØ/M
FT8X	PYØ/P&P	5X	JD/M
5R	D6	ZL9	YI
ZL9	FH	VKØ/H	VK9/C
PYØ/P&P	1S	STØ	BV
STØ	STØ	PYØ/P&P	VK9/W
KH1	TZ	TA	FT8Z
VK9/C	9U	CEØX	VKØ/H
HKØ/M	3V	VKØ/M	KC6/W
VKØ/M	TY	FH	5X
FT8Z	SU	5R	D2
D2	TJ	VK9/X	J2/A
VK9/M	5V	TY	XX9
FH	FT8Z	EP	STØ
9M6/8	3C	KH5K	5R
VK9/W	J5	KH1	JD/O
PYØ/T	YI	JD/M	T31
TY	9N	FT8Z	PYØ/P&P
D6	UG	D2	VK9/X
UG	CEØX	JD/O	TT
JD/M	EP	PYØ/T	D6
TZ	D2	TZ	FH

need. I start working my way up from the bottom of the CW band.

One of the nice things about 15 and 10 meters is that there usually is not much short skip, so that the great majority of stations heard, especially the stronger ones, are DX rather than locals. Of course, that makes things tougher on the "tune for the pileup" crowd, but for DXers willing to listen carefully to the weak signals, it can be a real advantage. As I move up the band, I unearth loads of DX signals, mostly Europeans, as on SSB, but again, nothing that I need. But at 21,055 my luck changes. I hear a station, about 569, running antenna tests with another station. The station doing the tests carries the funny little noises propagation can add to a signal that has come a long way; subtle little shifts of frequency caused by slight Doppler effects from slightly different paths and reflection areas propagating the signal. On top of that, he has a distinctive bit of chirp on the first letter of each word; probably a power-supply regulation problem. Then, after several repetitions of NW ANT 1 ANT 1 NW ANT 2 ANT 2, he signs, WA4RNE DE 5R8AP KN. Ohhh boy, a 5R8 on Madagascar, a real good one! An all-time new one!

I listen carefully for the WA4. Yes, there he is, about 559, almost dead zero with the 5R8. The WA4 is telling the 5R8 that the two antennas appear about the same, which was the same report that I would have sent, given the chance. It's a chance that I hope to get. I check my frequency out, getting zero beat with the WA4, and can only sit and wait for the end of the QSO. The 5R8 and the WA4 seem to be friends, chatting back and forth easily, and I begin to suspect that they have a regular schedule. I make notes on my little blackboard of the time, frequency and date; if they are meeting a schedule, next week I'll be meeting it as well, unless I get lucky first.

After what seems like an "interminable" wait (actually only about 10 minutes), they sign clear. I listen for a moment before I start my call to see if I have much competition on the frequency. I discover there can't be more than 15 or 20 stations calling that I can hear, with doubtless more that I can't. Not knowing what else to do, I call, but I have a nasty feeling about how effective my call is going to be. I listen; there, yes, he's coming back to someone, yes, an F3 in France. Well, that's not really surprising; I expect a 5R8 is pretty good DX almost anywhere. He is sending to the F3 in French. He now turns it back to the French station. I look, and almost immediately find the F3, about 500 Hz above the 5R8's frequency.

That's interesting; apparently the 5R8 is willing to tune off his frequency. He listened up a half kHz. I can hear the F3 pretty well; he's about the same strength as the 5R8. That means that likely everyone in the States at least can hear the F3 as well as I do, and I have no doubt that the majority of the pileup will be zero-beating the F3 station's frequency. And I'm virtually positive that at least several of the stations calling will be significantly louder in Madagascar than I am. In short, for me to call on the F station's frequency is probably a waste of time.

Hmmm. The 5R8 tuned higher to avoid the bulk of the pileup. He has several choices, assuming that he intends to stick on the frequency and deal with the pileup. He can tune higher yet to work the next station, he can try to fish someone out of the pile on the French station's frequency or he can tune lower. If all goes well, he will avoid working any more stations on his frequency, or the QRM there will become unbearable. I decide that my best bet is to call him higher, perhaps another 500 Hz above the French station's frequency.

on Europe, and I'm looking across 15 meters, hoping for a pleasant surprise—one of the elusive Middle East stations that I need. I have already checked out the phone band, and now I'm down at the other end, having a look on CW.

The band is good; there were plenty of Europeans, with northern Africans mixed in, on SSB, but nothing that I need. Ten meters is essentially dead, with only a few Central and South Americans coming through, and again nothing that I

I move my transmit frequency, and wait for the next call.

The 5R8 signs with the F3, and at least 40 of us call. I give him a 1 × 3 (5R8AL DE W9KNI W9KNI W9KNI $\overline{AR}$) and listen. Yes, he's coming back to someone, but the call is covered up by those still calling him on his frequency. He gives a report, his name and QSL manager, and turns it over to a KA1. I let my breath out, as probably do the 40 other operators in unison. I look for the KA1. Yes, there he is. My transmit frequency is almost exactly the same as that of the successful KA1. I was on the right track. I guessed right, but so did the KA1. When it came down to a horse race, I got beat.

What to do now, for my next call? That one is easy; chances are the 5R8 will again tune a bit higher. But now that pattern is more obvious to a lot of the operators calling him in the pileup, and I have no doubt that a number of them will be calling the 5R8 another half kHz above the KA1. So if I want to have my signal stand out in solitary splendor, I need to try something a bit different.

I could call 1 kHz above the KA1, figuring that to move me above the new pileup group. Or, I could go back down to where the F station made the previous contact, figuring the 5R8 to tune down from the pileup. This is more likely if he doesn't have an external VFO and has to rely on his RIT to tune through the pileup. But I obviously don't know what he's using for equipment. I decide to try calling 1 kHz higher.

He finishes with the KA1. I pause a moment and listen on my transmit frequency before I start my call. It appears to be quiet. Maybe this time it's my turn. I give him a 1 × 3 call again, and listen. And again, he's back to someone, and again buried by stations still blindly calling him on his own frequency. That distinctive chirp on the opening character of a word makes his signal stand out from the rest, and that helps find him. The smoke clears, and the whole lot of us wait to see who made it through. Darn, it's a WB2.

I look for the WB2's frequency. There he is, down where the F8 station worked him two QSOs ago. That suggests that he is probably going to tune a fairly narrow range above his own frequency, probably using RIT, unless he also tunes below his own frequency. If I'm right, and he is using an RIT control, that's his only other option. But he hasn't yet worked anyone below himself. I decide to try calling him above his frequency again, about a half kHz above the WB2, and I move my transmit frequency. There, he's clearing with the WB2; he's stepping up the tempo a bit.

I give him another 1 × 3 call; obviously he knows his own call, and there is no doubt about who the pileup on and above his frequency is for. I listen. Once again he comes back, but to who? We all wait.

And, I wait in vain. A K3 puts him away that time. I find the K3's frequency. Darn, almost dead zero with me. I guessed right again, but couldn't pull the trigger. Let's set up a half kHz above the K3; that's about the frequency where he worked the KA1, and the highest frequency that he worked anyone on. I begin to be distinctly unhappy that all he has worked so far has been East Coast stations; that does not bode well. It suggests that the East Coast has a definite propagation edge on the Midwest. But there, good, he's going faster; he's clearing with the K3. I call again.

Okay, he's back. The pileup sweats out the wait to see who got him; hello, it's a WB0. Where is he? I look all around the area above the 5R8, but can't find the WB0. Darn! I look below the 5R8, and there, nearly 1 kHz below the 5R8's frequency is the WB0, "fat, dumb and happy" with a new one. I grunt as I move my transmitter a half kHz below the

PT7BZ is a very active Brazilian amateur.

WB0. Sure hope that I'm the only operator to find the WB0. If I am, my chances ought to be pretty good now. I give a 1 × 3 call.

There, he's back. The gang on his frequency have finally learned enough to try something different; I can hear him spelling out the W5's call. Phooey. Okay, where's the W5? I look below again; he and I were real close in frequency. I move down 1 kHz. Wait a minute. What's the 5R8 saying now? Darn! He's going QRT. Shucks. Yup, there he goes with a CL.

I give him one last long call, hoping against hope that he will decide that he needs a W9 in the log worse than he needs to go to sleep! But even as I call, I know it's futile. And I'm right. I sigh, and go over to the little blackboard to add some notes on the 5R8. That was the first time that I have ever heard him. Although I didn't get him, I learned a lot about him, what his signal sounds like, and what his pileup operating habits are. Next time I'll be ready. It looks like he may well have a sked with that WA4 next Sunday, and I'll be there, ready and waiting. And my experience of how he deals with a pileup will give me a big edge over people who haven't a clue. There's hope!

[Except in years near the bottom of the sunspot cycle, 15 meters is (after 20 meters) the best amateur DX band presently available. Every band has its own "flavor," and in some respects, 15 is somewhat bland. It lacks the reliability of 20 meters and the seductive appeal of 10 meters. But, it is a good, solid performer, with more room and fewer of the commercial intruders that plague 20 meters at times. Antennas can be smaller and lower and still be effective, and somehow the competition never seems quite so rough. And, short skip is rarer, indicating that most strong stations on the band are DX. Fifteen rarely opens before sunrise in the best of times, and is usually disappointing in mid-summer, except in the early evening hours. But in the better parts of the sunspot cycle, it is a terrific DX band. And, it is a great band for Novice licensees to get their first taste of CW DX. Some Novices have even attained DXCC on 15 meters, and no doubt more will.]

Working a New One

I'm just getting ready for bed on a mid-August evening when the phone rings. My heart skips a beat. It's too late for a casual call, so either there is a family emergency or there's a new one on!

DX, Pileups and Common Sense

A couple of recent DXpeditions have generated more than the usual amount of mail. The common thread in the letters is that we have a real problem, and something needs to be done about it. The problem, as defined by most of the correspondents, is that individual DXpeditions are taking up altogether too much of the bands as they spread out calling stations so as to make it easier to pick out call signs from the pile. The problem usually is associated with split-frequency SSB operation, particularly on 20 meters, but the CW band has seen its abuses as well.

The reason this is a problem, of course, is that not everyone on the band shares the single-minded goal of contacting that particular DX station at that particular moment. Nets, ragchews and a host of other worthwhile activities are going on at the same time, and these operators are understandably miffed when they are set upon by a horde of people shouting their call signs ad infinitum, sometimes apparently without listening on their transmitting frequency. It's not deliberate or malicious interference, but it's certainly thoughtless, and much of it is unnecessary. From there tempers flare, there's retaliation, things escalate—and we have another incident that wreaks havoc on Amateur Radio's image and does nothing good for our collective blood pressure.

The one individual in the best position to do something about the problem is, of course, the DXpedition operator. He or she specifies the range of listening frequencies and the stations (divided by call sign numerically, alphabetically or geographically) whose calls will be answered, and otherwise manages the pileup. A talented operator can keep hundreds, even thousands, of stations in line while making contacts at the rate of four or more per minute, without taking up more than a few kilohertz of the band. It takes a rare combination of good nature, fairness and operating skill, but it can be done—and the result is a well-ordered pileup that is a joy to behold, fun to be in and doesn't get in anyone else's way.

Is it too much to hope for that operators on rare DX-peditions will all measure up to that ideal? Probably it is, since the people with the time and money for such ventures aren't automatically the best operators (though there are a number who do mighty well). For the rest of us who may find ourselves on the other end of a massive pileup, though, a little common sense will go a long way. Nearly everyone in a pileup will follow instructions and not get out of line if things are being handled fairly and if contacts are being made at a reasonable rate. Nearly everyone not in a pileup will take pains to avoid causing interference as long as it's not "polluting" too much of the band.

How much is too much? Reasonable people will differ, but 10 kHz strikes us as being sufficient for nearly any situation, and over the years we've been on both ends a few times. Clearly, there's no justification for "listening 200 to 350," sending CW stations into RTTY territory, or being so nonspecific with instructions that the whole band gets clobbered. An expedition to a rare spot is a significant event in Amateur Radio, but everything else that goes on in our bands can't be expected to stop for it.

Those of us who stay at home aren't completely helpless, either. Expeditions to rare spots generally require financial backing, and the talent and temperament of operators being sent certainly deserves more than a passing thought when sponsorship is considered. DX organizations ought to be especially sensitive to this.

Will things get better? We certainly hope so. If not, the non-DX part of the amateur community is going to press the League, if not the FCC, for measures that will protect its interests. A little self-restraint by DXers will go a long way toward avoiding restraints designed by others.—*David Sumner, K1ZZ (originally published in April 1983 QST)*

"Bob, it's me. Chuck. There's a station portable C9 on 14,195, listening above 200, with an OH for a QSL manager. Okay?"

"Right. Fine, Chuck. Did you get him yet?"

"Yes, I caught him calling CQ, I was lucky. I got to call Pete and Dick and the rest of the club."

"Right. Thanks for the call. 'Bye."

With that, I'm racing down the stairs to the shack like a madman. A C9! Mozambique. That would be an all-time new one. Sure hope he's for real. There have been a few stations claiming to be active from there the last few years, but none of them ever amounted to anything. Okay, that should be a good path this time of year.

As the gear and the linear warm, I swing the antenna straight east for the C9. If that's not the exact bearing, it certainly is close enough. I move the transceiver down to 195; yes, there he is, nice and loud, a good S8.

"Roger, QSL via OH3QT. QRZ? Listening 14210 to 225, from OH3QT, portable C-Ninety-Seven."

Okay, he wants to do it that way. I kick the external VFO in, and move it up to 14,210 to have a look.

Wow! I guess that I am hardly the only operator to get a phone call out of the night. I run smack into a wall of QRM. Most of it is not too loud; long skip must be in. But I can't really copy any of the stations calling, and there sure seems to be a lot of them. I glance over at the linear. Yes, the ready light is on, and it is tuned for 20-meter SSB. I'm ready. I move the VFO experimentally to 14,220. I flip the VFO control switch back to the transceiver VFO, and listen back on the C9's frequency. He's calling QRZ ("Who's calling me") again.

I simply sign my call, "Whiskey Nine Kilowatt, Norway, Italy, W9KNI, over." There, he's back to another W9. Yes, it's W9DWQ, a member of my DX club. When I hear the OH3 say the W9-part of the call sign, I'm thrilled for a tiny moment, thinking that I had gotten him. But it was W9DWQ. Anyhow, 'DWQ will be strong enough that I can easily find his transmit frequency.

I flip the switch to the external VFO, and look for 'DWQ. Yes, there he is on 221. I leave my VFO on him, and switch back to the OH3/C97. He clears with 'DWQ, and I call again. There, he's back, to a WA5. He turns it over to the WA5; I flip to the external VFO, starting on my own frequency. Nope, not on my frequency, anyhow. I tune a bit higher up, and find the WA5 about 2 kHz above my own frequency. Okay, the OH3 moved up 2 after he worked W9DWQ, so I need to move 2 up from the WA5's frequency. Hmm, that puts me right on 225, the OH3's upper calling limit.

The OH3 signs with the WA5, and I call. Nope, a KB8 got him. Hope I can find the KB8 on backscatter. I flip back to the external VFO. There he is, right on my frequency. So I guessed right, but got beaten. Now what? If the OH follows his own instructions, he'll work another station on the same frequency, which so far he hasn't done since I have been listening to him. Thus far he's tuned back down a couple kHz after each QSO, or he's just tuned back to 210, and then started moving up. The old shell game, eh? Right. Well, I'll make my call on 210, and see what happens.

I call, then listen. He's back to another W9. It's WD9AHJ; he's in my club, too, and lives nearby. Yes, I find him at 223. Okay, I guessed wrong, but Ron didn't. Good for him. I'll try 221.

I set up on 221 and make my call. Phooey. He's back to a WD4. He turns it over to the WD4. Well, I was within a few hundred hertz of him, so I was about right. I'll go to 219. I leave the external VFO in for a moment and can tell when the OH3 goes QRZ. The roar in my headphones is as perfectly orchestrated as if Georg Solti was directing the Chicago Symphony on frequency. I listen for an instant and hear several stations right on my frequency. I slide down half a kHz; it sounds clearer, and I call.

Bingo! All right! "W9KNI, you're 5-by-9. W9KNI go ahead."

"Roger, from W9KNI. You're 5-and-9, thanks for the new one, over."

"W9KNI, Roger, QSL. QRZ from Ocean-Henry-Three-Quebec- Tango portable-Charlie-Ninety-Seven, listening 210 to 225."

I sit in my chair, stupidly grinning from ear to ear. I turn on the 2-meter rig and go to the DX repeater frequency, just in time to hear W9DWQ congratulating me.

"Roger. Thanks, Ed. Excuse me, I better get this into the log book before I forget what time it is. I'll be back in a minute and try to help with the spotting."

I neglected to do that once on an all-time new one. The next day, I had to ask around about what time the other fellows in the local "wolf pack" had worked the station, so that I could guesstimate for my log book. I was lucky; the guess was close enough that the DX station found me in his log. The QSO time on the QSL I got back was eight minutes off from the time in my log, and under the circumstances, I wasn't complaining at all! I was lucky as it was. But now I make sure that I always write the QSO time/date down *immediately*.

I get the QSO into the log book along with the QSL information, make sure that I have the UTC date entered correctly, and start trying to spot the frequencies of the stations working the OH3/C9 to help others in our DX club get through.

QSLING

As the old saying goes, "The job ain't done 'till the paperwork is finished." For the competitive DXer, the paperwork consists of getting the QSL in hand from the rare DX station. And, as any experienced DXer will tell you, sometimes getting the QSL is as difficult or more so than getting the QSO was in the first place. But, it is a part of the game, one that the successful DXer must be good at.

Of course, getting confirmations from three quarters or more of the new ones that you work is easy. The DX station has a stateside QSL manager or handles his own cards in a prompt and orderly fashion. And the DXer who cares enough to go after the card in a proper manner will soon be enjoying the thrill of finding a new one in his mailbox.

With the exception of Soviet stations, there are almost always at least two routes that can be used in pursuit of the QSL. The most inexpensive is, of course, the ARRL Outgoing Bureau, which can be surprisingly effective (see accompanying sidebar). Of course the return trip does take a while, with total round-trip times commonly one or two years. It is the route that many DXers use for the bulk of their cards, except for the all-time new ones or DX stations with US QSL

QSLs from some recent successful DXpeditions. *(Karl Townsend photo)*

managers. (For Soviet stations, all cards must go through the Soviet Bureau, the famous Box 88, Moscow, whether sent via the ARRL Outgoing Bureau or mailed direct.)

Almost every DXer is in the DXCC chase, but many DXers also pursue other DX awards which require possession of valid QSL cards as proof of contact. (For details on DXCC and many other operating awards, see Chapter 8.) A DXer interested in even a small number of different operating awards will often find the expense of a direct airmail QSL with provision for return air postage to be more than he or she is comfortably prepared to pay. In such situations, QSLing via the League's Outgoing Bureau is the best and most cost-effective way to go.

But, for the DXer intent on gaining the DXCC Honor Roll as fast as humanly possible, the direct airmail QSL with return envelope and provision for return postage is considered appropriate (unless the DX station has a QSL manager). For this, the DXer will need a set of current *Callbooks*. The *Callbooks* include all the reference information that the DXer needs for QSLing, as well as a good deal of useful reference information. If the DX station asks for QSL direct or via CLBK (*Callbook* address), the address information is almost always available in the *Callbook*.

To provide for return postage from a foreign land, preferably via airmail, the DXer has three choices:

He can affix unused postage stamps from the foreign country on his return envelope in an amount appropriate for an airmail reply. Such stamps can often be purchased from a stamp dealer.

The second choice is to enclose International Reply Coupons (IRCs), which can be purchased at any post office. In almost any country of the world, an IRC can be exchanged at the local post office for enough local stamps to pay for postage for the return of a QSL card via surface mail. Several IRCs can be exchanged for enough postage for an airmail return of the QSL, but the number of IRCs needed varies from country to country. A table published in the *Callbook* gives the information on how many IRCs are needed in most countries for an airmail reply. The present price of an IRC is 80 cents US.

The third choice is the famous "green stamp," a euphemism for a dollar bill. Almost everywhere, a dollar exchanged into the local currency is more than enough to pay for an airmail response and certainly is cheaper than the two

The ARRL Outgoing QSL Service

One of the greatest bargains of League membership is being able to use the ARRL Outgoing QSL Bureau to conveniently send your DX QSL cards overseas to foreign QSL bureaus. Your ticket for using this service is your *QST* address label and just $1 per 100 cards. You can't even get a deal like that at your local warehouse supermarket! And the potential savings over the substantial cost of individual QSLing is equal to many times the price of your annual dues. Your cards are sorted promptly by the Outgoing Bureau staff, and cards are on their way overseas usually within a week of arrival at ARRL HQ. More than one million cards are handled by the Bureau each year!

QSL cards are shipped to QSL bureaus throughout the world, which are typically maintained by the national Amateur Radio Society of each county. While no cards are sent to individuals or individual QSL managers, keep in mind that what you might lose in speed is more than made up in the convenience and savings of not having to address and mail QSL cards separately. (In the case of DXpeditions and/or active DX stations that use US QSL managers, a better approach is to QSL directly to the QSL manager. The various DX newsletters, the W6GO *QSL Manager Directory* and other publications, are good sources of up-to-date QSL manager information.)

As postage costs become increasingly prohibitive, don't go broke before you're even halfway toward making DXCC. There's a better and cheaper way—"QSL VIA BURO" through the ARRL Outgoing QSL Bureau!

How To Use the ARRL Outgoing QSL Service

1) Presort your DX QSLs alphabetically by call-sign prefix (AP, C6, CE, DL, F, G, JA, LU, PY, 5N, 9Y and so on).

2) Enclose the address label from your current copy of *QST*. The label shows that you are a current ARRL member.

3) Enclose payment of $1 per pound (or less) of cards —approximately 155 cards weigh 1 pound. In other words, $1 is the minimum charge whether you send 1 card or 150 cards. Please pay by check (or money order) and write your call sign on the check. Do not send cash.

4) Include only the cards, address label and check in the package. Wrap the package securely and address it to

At typical day's work at the ARRL Overseas QSL Bureau at HQ!

the ARRL Outgoing QSL Service, 225 Main St, Newington CT 06111.

5) Family members may also use the service by enclosing their QSLs with those of the primary member. Include the appropriate fee with each individual's cards and indicate "family membership."

6) Blind members who do not receive *QST* need only include the appropriate fee along with a note indicating the cards are from a blind member.

7) ARRL affiliated-club stations may use the service when submitting club QSLs by indicating the club name. Club secretaries should check affiliation papers to ensure that affiliation is current. In addition to club station QSLs through this service, affiliated clubs may also "pool" their members' individual QSL cards to save even more. Each club member using this service must also be a League member. Cards should be sorted "en masse" by prefix, and a *QST* label enclosed for each ARRL member sending cards.

Recommended QSL-Card Dimensions

The efficient operation of the worldwide system of QSL bureau requires that cards be easy to handle and sort.

---

or more IRCs required for an airmail return. But there are two problems with the use of the dollar bill. The first one is that in some countries, possession of US currency is illegal, largely because of currency restrictions, but illegal nonetheless. Sending a dollar to DX stations in such countries can reportedly get the DX operator in serious trouble. Never send a "green stamp" unless you are absolutely sure that it will not cause problems for the DX station. If you are not sure, either ask or don't send the "stamp."

Secondly, theft from the mails is a major problem in some countries. And once a postal employee with larceny in his heart discovers that envelopes addressed to "Amateur Radio DX1DX" often have money in them, you can imagine how many of the envelopes with QSLs actually get through. Indeed, in some countries the chance of getting a QSL through to a station in any way, except as a post card or via registered mail at $6-plus per toss, can become remote. The DXer should feel pleased, however, that he is helping to stimulate the local economy! Although IRCs can also disappear, they are generally more successful for your cause, since they are harder to convert into the local currency and are easier to conceal in an envelope. Be sure that your envelope going overseas with either IRCs or "green stamps" is opaque in bright sunshine.

The easiest and usually most reliable way to get a QSL is from a DX station's manager. Not every DX station with a manager announces that fact on the air, so the DXer out to run up a score will find the *QSL Manager's Directory* a blessing. The *Directory*, complete with update service, is available from W6GO, Box 700, Rio Linda, CA 95673. Write for details.

When sending a QSL to a DX station's QSL manager, there are a few tips that will make that volunteer's job easier. And, since he or she is a volunteer, whatever you can do to make the job easier is the least that you can do. First, don't even bother to send a QSL (to a QSL manager or any DX station) unless you show the QSO time and date in UTC. Be sure that the self-addressed stamped envelope (SASE) that you supply the manager is large enough to accommodate a normal-size QSL—unless you like folded QSLs. And be sure that the return envelope has adequate postage.

Another technique that will help the manager is to write on the rear of the return envelope the call of the station that the QSL is for, as well as the date, time and band of the QSO. That way, the manager can arrange the cards for easiest processing while he waits for the logs to come in from the DX station.

Cards of unusual dimensions, either much larger or much smaller than normal, slow the work of the bureaus, most of which is done by unpaid volunteers. A review of the cards received by the ARRL Bureau indicates that most fall in the following range: height = 2¾ to 4¼ in. (70 to 110 mm), width = 4¾ to 6¼ in. (120 to 160 mm). Cards in this range can be easily sorted, stacked and packaged. Cards outside this range create problems; in particular, the larger cards often cannot be handled without folding or otherwise damaging them. In the interest of efficient operation of the worldwide QSL bureau system, it is recommended that cards entering the system be limited to the range of dimensions given. Note: IARU Region 2 has suggested the following dimensions as optimum—height = 3½ in. (90 mm), width = 5½ in. (140 mm).

Countries Not Served by the Outgoing QSL Bureau

Approximately 270 DXCC countries are served by the ARRL Outgoing QSL Bureau, as detailed in the ARRL DXCC Countries List. This includes nearly every active country. As noted previously, cards are forwarded from the ARRL Outgoing Bureau to a counterpart bureau in each of these countries. In some cases, there is no bureau in a particular country and cards, therefore, cannot be forwarded. For this reason, the ARRL Outgoing Bureau cannot forward cards to the following countries:

A5	Bhutan
A6	United Arab Emirates
A7	Qatar
BV	Taiwan
C9	Mozambique
D6	Comoros
ET	Ethiopia
HZ	Saudi Arabia
J5	Guinea-Bissau
KC4	US bases in Antarctica
KC6	Belau
KC6	Micronesia
KH1	Baker and Howland Is
KH3	Johnston I
KH5	Palmyra and Jarvis Is
KH7	Kure I
KH9	Wake I
KP1	Navassa I
KP5	Desecheo I.
P5	North Korea
S9	Sao Tome and Principe
SU	Egypt
T2	Tuvalu
T3	Kiribati
T5	Somalia
TJ	Cameroon
TL	Central African Rep
TN	Congo
TT	Chad
TY	Benin
TZ	Mali
V4	St Christopher and Nevis
VP2E	Anguilla
VR6	Pitcairn Island
XT	Burkina Faso
XU	Kampuchea
XW	Laos
XX9	Macao
XZ	Burma
YA	Afghanistan
ZA	Albania
ZD7	St Helena
ZD9	Tristan de Cunha
ZK3	Tokelau
3C	Equatorial Guinea
3V	Tunisia
3W	Vietnam
3X	Guinea
4W	North Yemen
5A	Libya
5H	Tanzania
5R	Madagascar
5U	Niger
5X	Uganda
7O	South Yemen
7Q	Malawi
8Q	Maldives
9G	Ghana
9N	Nepal
9U	Burundi

Note: The ARRL QSL Service should not be used to exchange QSL cards within the 48 contiguous states.

Which brings up another point. If you are slow in getting your card from the manager, don't harass him for the card. The QSL manager is at the mercy of the mails, of the DX station and a number of other factors over which he has no control. He is a volunteer, doing a job that is a lot of work. Treat him kindly. After a wait of some months, or if you have not gotten the card while everyone else in town has been showing off theirs for weeks, mail a polite note to the manager explaining the circumstances, and enclose an SASE for his reply. And don't forget the QSO information; maybe what was lost was your card to the manager. For more on QSL managers, see the accompanying sidebar.

Eavesdropping on 40

I settle into the old captain's chair in front of the rig, settling the headphones into place, and move the antenna selector switch to my new southwest 40-meter sloper. It sure would be nice to have one of those "shorty 40" Yagis, but the conventional wisdom of the local DX gang seems to be that a set of good slopers is as good a simple wire DX antenna as you can get. With a bit of help from my son, we were able to get them up yesterday, before the winter storms.

I glance at my clock. Okay, its just after 6 AM local, 1200 UTC. I warm one hand on my coffee cup while the other hand starts tuning the receiver. I'm out to see if my new southwest sloper is any good for the grayline DX I've been hearing about on 40. I was on 40 last night, but results from testing the north sloper were inconclusive. But then I am not so sure that the band was very good either. And, to get the feed line for one of the slopers, I had to take the old inverted V down, so I have no A/B test that I can run; only a set of three slopers.

I set the receiver down at 7000 and begin tuning slowly up the band. I'm sure glad that I got my Extra Class license last fall; it has opened up a whole new world of DX opportunities for me, especially on the CW bands. It has done more for my DXCC country total than buying a linear, and as much for my DXing as putting up a tribander.

There's a K1 calling someone. He's about an hour further into daylight than I am, so I'll get some idea of how good a band opening is by hearing what he's working, figuring that my chance comes a bit later. I listen; yes, there's someone coming back. It's a JA7, weak and watery sounding. I tune on.

I hear a weak CQ, just barely above the noise. I listen carefully. Okay, it's a UA9. I switch the north sloper in for a moment to see if he gets any stronger, but he almost

The ARRL Incoming QSL Bureau System
Purpose

Within the US and Canada, the ARRL DX QSL Bureau System is made up of numerous call area bureaus that act as central clearing houses for QSLs arriving from foreign countries. These "incoming" bureaus are staffed by volunteers. The service is free and ARRL membership is not required.

How It Works

Most countries have "outgoing" QSL bureaus that operate in much the same manner as the ARRL Outgoing QSL Service. The member sends his cards to his outgoing bureau where they are packaged and shipped to the appropriate countries.

A majority of the DX QSLs are shipped directly to the individual incoming bureaus where volunteers sort the incoming QSLs by the first letter of the call-sign suffix. One individual may be assigned the responsibility of handling one or more letters of the alphabet.

Claiming Your QSLs

Send a 5- × 7½-in. self-addressed, stamped envelope (SASE) to the bureau serving your call-sign district. Print your call sign neatly in the upper-left corner of the envelope. A preferred way to send envelopes is to affix a first-class stamp and clip extra postage to the envelope. Then, if you receive more than 1 oz of cards, they can be sent in the single package.

Some incoming bureaus sell envelopes or postage credits in addition to the normal SASE handling. They provide the proper envelope and postage upon the prepayment of a certain fee. The exact arrangements can be obtained by sending your inquiry with an SASE to your area bureau. A list of bureaus appears below.

Helpful Hints

Good cooperation between the DXer and the bureau is important to ensure a smooth flow of cards. Remember that the people who work in the area bureaus are volunteers. They are providing you with a valuable service. With that thought in mind, please pay close attention ot the following DOs and DON'Ts.

DOs

Do keep self-addressed 5- × 7½-in. envelopes on file at your bureau, with your call in the upper left corner, and affix at least one unit of first-class postage.

Do send the bureau enough postage to cover envelopes on file and enough to take care of possible postage-rate increases.

Do respond quickly to any bureau request for envelopes,

Remember to keep a supply of envelopes on file with your incoming QSL bureau (see text for address). Affix one unit of first-class postage to your self-addressed, stamped envelope, put your call sign in the upper-left corner of the envelope and clip additional postage to the envelope if you are expecting many cards.

stamps or money. Unclaimed card backlogs are the bureau's biggest problem

Do notify the bureau of your new call as you upgrade. Send envelopes with new call in addition to envelopes with old call.

Do include an SASE with any information request to the bureau.

Do notify the bureau in writing if you don't want your cards.

DON'Ts

Don't expect DX cards to arrive for several months after the QSO. Overseas delivery is very slow. Many cards coming from overseas bureaus are over a year old.

Don't send your outgoing DX cards to your call-area bureau.

Don't send envelopes to your "portable" bureau. For example, AA2Z/1 sends envelopes to the W2 bureau, not the W1 bureau.

ARRL Incoming DX QSL Bureau Addresses

First Call Area: All calls
W1QSL Bureau
Mt Tom Amateur Repeater Assn
Box 216, Forest Park Station
Springfield, MA 01108
Second Call Area: All calls*
ARRL 2nd District QSL Bureau
NJDXA, PO Box 599
Morris Plains, NJ 07950

---

disappears. That's interesting. I thought that the slopers were supposed to have a cardioid pattern, which would certainly include a UA9 on a north-facing sloper, unless he isn't coming through on the short path. Yes, I guess he could be coming through on a crooked path, since it's sunset somewhere in Siberia. Interesting.

I tune on; I don't need a UA9, but the experience of the one that I just heard intrigues me. There are three or four stations calling someone. Ah, there he is coming back. Good grief! It's VS6DO! I've heard him on 40 before, but never better than Q2 to Q3, and there he is an honest 559. Maybe there is something to these slopers! I decide to try to put him in the log. I have a QSL from VS6 on 20 meters, but nailing one on 40 would be real satisfaction for me. I find the station that he's working, a W4, and set my transmit frequency just above the W4. The VS6 is working people quickly, passing out reports and QSL information, a K7 manager. If I can nail him, it means an easy QSL to boot!

The W4 signs, the VS6 goes QRZ, and I give a 1 × 2 call. I listen. He's back to a WA8. He turns it to the WA8. I listen; yes, there he is, down about a half kHz. I move my transmit frequency down a bit, closer to the WA8 but not on top of him. The VS6 clears with him and calls QRZ again. I call him, giving another 1 × 2, and wait to see what's happening. Hah! I got him! I note the details in the log, grinning from ear to ear. Guess that new sloper is okay!

I start moving up the band again. I come across a VK with good signal strength clearing with a WB5. I find a mini pileup on an FK8, but I already have one on 40. Then I strike paydirt—a weak signal calling CQ. I listen carefully, and dig out his call sign. It's VU2TEC! I've never before heard a VU2 on 40 with my old inverted V, and here's one calling CQ, RST 449. Guess that new sloper is doing the job. I set up on his frequency, and when he ends his call, I'm ready. I pause a moment, and, hearing nothing, call him with a 1 × 3. I listen, anxiously. There, he's coming back, but to a UKØ. Darn.

Third Call Area: All calls
  C-CARS, PO Box 448
  New Kingstown, PA 17072-0448
Fourth Call Area: All single-letter prefixes
  Mecklenburg Amateur Radio Club
  PO Box DX
  Charlotte, NC 28220
Fourth Call Area: All two-letter prefixes (AA4, KB4, NC4,
  WD4, etc)
  Sterling Park Amateur Radio Club
  Call Box 599
  Sterling Park, VA 22170
Fifth Call Area: All calls*
  ARRL W5 QSL Bureau
  PO Box 44246
  Oklahoma City, OK 73144
Sixth Call Area: All calls*
  ARRL Sixth (6th) District DX QSL Bureau
  PO Box 1460
  Sun Valley, CA 91352
Seventh Call Area: All calls
  Willamette Valley DX Club, Inc
  PO Box 555
  Portland, OR 97207
Eighth Call Area: All calls
  Columbus Amateur Radio Assn
  Radio Room, 280 E Broad St
  Columbus, OH 43215
Ninth Call Area: All calls*
  Northern Illinois DX Assn
  Box 519
  Elmhurst, IL 60126
Zero Call Area: All calls*
  WØQSL Bureau
  Ak-Sar-Ben Radio Club
  PO Box 291
  Omaha, NE 68101
Puerto Rico: All calls*
  Radio Club de Puerto Rico
  PO Box 1061
  San Juan, PR 00902
US Virgin Islands: All calls
  Virgin Islands ARC
  Ron Hall, KP2N
  PO Box 3987
  St Thomas
  Virgin Islands 00801
Hawaiian Islands: All calls*
  John H. Oka, KH6DQ
  PO Box 101
  Aiea, Oahu, HI 96701

Alaska: All calls*
  Alaska QSL Bureau
  4304 Garfield St
  Anchorage, AK 99503
Guam: MARC
  Box 445
  Agana, GU 96910
SWL: Mike Witkowski
  4206 Nebel St
  Stevens Point, WI 54481
QSL Cards for Canada may be sent to:
  CRRL DX QSL Bureau System
  Kennebecasis Valley Amateur Radio Club
  Box 51
  St John, NB E2L 3X1

QSL cards may also be sent to the individual bureaus:

VE1*	L. J. Fader, VE1FQ	
	PO Box 663	
	Halifax, NS B3J 2T3	
VE2	A. G. Daemen, VE2IJ	
	2960 Douglas Ave	
	Montreal, PQ H3R 2E3	
VE3	The Ontario Trilliums	
	PO Box 157	
	Downsview, ON M3M 3A3	
VE4	Larry Lazar, VE4SL	
	30 Bathgate Bay	
	Winnipeg, MB R3T 0L2	
VE5	B. J. Madsen, VE5FX	
	739 Washington Dr	
	Weyburn, SK S4H 3C7	
VE6*	Norm Waltho, VE6VW	
	General Delivery	
	9714 - 94th St	
	Morinville, AB T0G 1P0	
VE7*	Alex Ivisc, VE7CNE	
	1107-7434 Kingsway	
	Burnaby, BC V3N 3B7	
VE8*	Rolf Ziemann, VE8RZ	
	2888 Lanky Ct	
	Yellowknife, NWT X1A 2G4	
VY1	W. L. Champagne, VY1AU	
	PO Box 4597	
	Whitehorse, YU Y1A 2R8	
VO1, VO2	Roland Peddle, VO1BD	
	PO Box 6	
	St John's, NF A1C 5H5	

*These bureaus sell envelopes or postage credits. Send
an SASE to the bureau for further information.

Well, not really a surprise. A UKØ is certainly much closer than a W9, and much stronger.

I wait out the QSO with the UKØ. When the VU2 turns it over, I try to hear the UKØ so that I can be sure that I transmit on the UKØ's frequency after he has finished the QSO. After all, that is where the VU2 certainly is listening. But, I can find no trace of the UKØ. I keep careful watch. There, the VU2 is back. Ah, good, he's promising to QSL and is starting to sign clear. I'm nervous; I've worked at least a dozen VUs on 20, but I've never heard one on 40, and for me this is real DX.

The VU signs clear with the UKØ, and I don't know what to do. Is the VU copying the closing transmission of the UKØ? If so, will he give a simple "di dit" at the end of the QSO, or will he call QRZ, or will he call CQ again? I wait, unsure, listening to try to hear his "di dit." Ah, there he is. . .R 73 DR IVAN S̄K̄ NW QRZ W9? DE VU2TEC K̄N̄. All right! I call him nervously, giving a 1 × 3, and turning it back to him. There

it is, beautiful music, R W9KNI DE VU2TEC...

[Forty Meters is one of the best DX bands available, an important complement to 20 meters, offering consistent DX openings especially when 20 meters is dead. Although the primary DX activity on 40 is on CW, there is a fair amount of DX worked on SSB as well. But, because the bulk of the serious DX activity takes place in the first 25 kHz of 40 meters, an Extra Class license is an essential for the serious DXer. Not so long ago, any form of Yagi on 40 meters was rare, but now many DXers have two-element Yagis, making the pileups more competitive. Those with such antennas are especially enthusiastic about the band, but there are still a lot of DXers using various forms of wire antennas on 40 to good advantage, including stations who have achieved the Worked All Zones award on 40 monoband. Often, DX signals are weak on 40, and at other times display good signal strength, but against a background of high noise levels. And, many of the DXers on 40 are very proficient at the game. But these

So You Want To Be a QSL Manager!!

There have been many fine articles directed at new and old-time DXers on the proper ways to use the services of QSL managers—and service is exactly what a QSL manager provides.

Let's determine which DX station will be interested in your efforts. First, there are the well-equipped stations who have the time to work quite a few stateside and other DX stations, but simply do not have the time to spend filling out QSLs. Second, there are fine DX operators who make many QSLs, but simply have no desire to send or receive any, even though they have the time and financial means to do so. Third, many DX operators find that QSLing is a financial burden. We often do not realize the financial sacrifice so many of our DX friends have had to make just to get a transceiver and a good antenna.

What personal qualifications are necessary to make a good QSL manager? A good manager will be a DXer. This will give him or her the insight into, and understanding of, the feelings of those he'll be trying to help. A DXer will also know what is happening on the bands and will probably subscribe to one or more of the DX newsletters. Time is also very important. If you are pressed for time by your occupation or family, it is best that you do not consider this added demand on your time. And, remember, time spent on QSL manager duties takes that much time away from your own operating activities.

Being a QSL manager is not a job that you start and then abruptly stop. It takes months and years to become well-known and to have your call published in the several QSL manager directories. Don't start unless you are prepared to stay with it for a long time.

About the expenses involved: Being a QSL manager will cost you money if you intend to do a good job. Regardless of the SASEs received and an occasional dollar bill from some thoughtful DXer, you will find that you will have to lay out some money. You will always have the cost of mailing cards you have received to the DX station. There is also the cost of printing the QSLs for the DX station who just can't afford the expense. In some cases, you will have to send postage money to the DX station so he can send you the logs. Expenses are probably the main reason why many QSL managers ''throw in the towel'' after just a few months or a year on the job. This can leave hundreds of DXers waiting for QSL cards that they will never receive, and can also give managers in general a bad name. If spending a few extra dollars will be a financial burden to you, don't consider becoming a QSL manager.

A good manager must be able to deal with detailed paperwork and be very organized. Now, if you are still interested in becoming a QSL manager, how do you go about finding a DX station who is interested in using your services? You must ''sell yourself'' and your abilities to the DX station. After all, you will be his representative to the DX community, and he has every right to be able to feel

The left photo shows what was sent to the QSL manager of 3Y1VC (Bouvet I). Note the self-addressed envelope and the IRCs for return postage. On the right: 3Y1VC's QSL was received promptly.

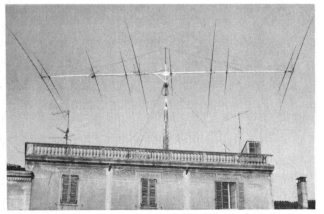

This impressive antenna installation belongs to I2VRN: a four-element Yagi for 40 meters and a six-element Yagi for 20 meters, both full size.

factors combine to make 40 a superb DX band, yet one where the working of a new one is always a special thrill.]

CONTEST!

As I settle into the operating chair, I'm a touch anxious; it's 2350 Zulu, 10 minutes until the start of the CW weekend of the ARRL DX Contest. I have decided to start out on 20 meters. A listen across 20 confirms that the band is open. I glance at the antenna rotator; yes, I'm pointed into the northwest, the great-circle direct path for Southeast Asia. My strategy is clear and uncluttered; I am only interested in new countries. I look at the clock; Okay, 2354. Time to run for a fresh cup of coffee.

It's just turned 2359. I listen across the a few kilohertz near the bottom of the band; everywhere I hear carriers, as stations tune their linears for the last watt of power. Then, with about 30 seconds left before the start, there suddenly seem to be

that he can trust you and depend on you.

Type up a description of yourself with a friendly offer to be a QSL manager, stating why you wish to be a manager and why you would make a good job of the required duties for as long as required. When you QSL a DX station who does not have a manager, and for whom you'd like to assume those duties, mail your typed description along with a picture of yourself and your QSL to him. If you have made a good presentation of yourself, in time you will have made some good contacts as well as having made some good DX friends.

Say you got the manager's job. Now you are ready to start. Requests are coming in for QSLs for "your" DX station. The DX station is telling his contacts to QSL to your call and, finally, the log sheets of QSLs that he has sent to you arrive. Get into it!

Probably, no two managers organize their work the same way, but the first thing they all have to do is open the mail. See who the card is for, note the date and call on the back of the envelope and file all requests by station (if you manage for more than one station) and date.

You will need some supplies. Stock up on envelopes and postage stamps. Remember that all envelopes must meet the postal code for size. From the local stationery store, get a glue stick, adhesive tape, a roll of 3-inch gummed paper tape, scissors, a moistener to wet the glue on the envelopes, and a rubber stamp with your name, call and address.

Most DXers will make it easy for you to send them their QSLs. They will have their call sign and all QSO information printed on the same side of their QSL. They will also have the correct date and time in UTC. And most importantly, they will enclose an SASE with their QSL request. The DXers will also keep in mind that it may take several months for you to get the logs from the DX station for verification.

Overseas stations should send you a self-addressed envelope with two IRCs for airmail return or one IRC for seamail return. Some will only send you an IRC or two without the self-addressed envelope. For these, take your scissors and cut out the return address on their envelope and, using your glue stick, attach that name and return address to one of your stock of envelopes. Never rewrite the names and addresses unless you absolutely have to. About 25 percent of the overseas stations will QSL you via the ARRL QSL Bureau for your call area. Be sure to inform your district QSL Bureau of the calls of the stations you manage. Keep plenty of proper-size envelopes and postage at your bureau. Return overseas cards received via the bureau through the ARRL outgoing QSL Bureau. This will save you the expense of sending direct to the overseas bureaus. Be sure to answer all the SWL (shortwave listener) QSLs received. These SWL cards mean a lot to prospective amateurs in many countries.

About 10 percent of the stateside stations will fail to send you an SASE or even a postage stamp. For these, you must pay the postage yourself or else return the DX QSL via the bureau and hope the requesting station has an envelope at his or her district QSL bureau. Otherwise, he'll never get that QSL.

Carefully check the requests for cards with your DX station's logs. Very often the time and or date on the QSL is off considerably, so make every effort to find the QSO in the logs. If you can't find the QSO, you can do either of two things: send the card back to the station making the request with the notation "not in log, please check time and date," or you can send the card direct to the DX station for him to decide whether to QSL. The best thing to do is to return the card to the station making the request. This is the most disappointing thing that you, as manager, will have to do. Operators requesting QSLs from your DX station really believe they have made the contact, but in the QRM of the pileup, it is easy to be mistaken. This happens to all DXers at some time or another. Don't just disregard the request; let the DXer know that he is not in the log.

Carefully fill out all QSLs. Be sure that the date, frequency and mode are correct. Use a ballpoint or felt-tipped pen. Sign your name or initials and your call sign, then stamp the back of the QSL with your rubber stamp to add authenticity.

Some DX stations that you manage for will tell you they do not want any cards received for them. It's best to keep these cards, so store them away in a dry place in case the DX station changes his mind at some future time. For those DX stations who want their incoming cards sent to them, wrap the cards in heavy wrapping paper (grocery bags work fine) seal securely with gummed wrapping tape in packets that weigh less than two pounds (approximately 310 QSLs). Send them via seamail with the required customs declaration stating that the contents are "used radio postcards" with no monetary value. This is the most economical way to send the cards to your DX station. Airmail costs over twice as much.

Keep up with your QSLing. Send out the QSLs every day if you have the logs for confirmation. If you get far behind, the job will seem more like drudgery than a pleasant chore. If necessary, enlist the aid of your XYL/OM or an amateur friend or two to help out. Remember, some anxious DXer is waiting for the mail every day.

In spite of all the work, it will never cease to be rewarding when you tune in on your DX friends and hear them say to QSL via your call sign, or most of all, when you open an envelope and find a note from an appreciative DXer who simply says "thanks, OM."

Now, get with it, do a good job, and good luck!—*Don Brickey, W7OK*

---

dozens of stations calling CQ TEST. Fascinated, I turn the selectivity to the 8 kHz position, and hear about six stations calling CQ TEST all at once. As I listen, I watch the clock. Magically, as the clock goes to 0000, the CQ calls are ended, and there is a brief moment of silence. Then, stations lucky enough to have had an answer to their CQs start acknowledging with their contest exchanges, while others start calling CQ again.

A few minutes at the bottom of the band shows me that the big gun contest stations are all intent on staking out their turf in the bottom 15 or 20 kHz, but as I tune higher, things start to loosen up. The QRM seems deafening at first, but I find that I can copy plenty of weaker stations in between the big guns. I slowly work up the band, examining every signal, trying to determine if it comes from a country that I need.

I hear JAs by the score, Siberians by the bunch. I trip across

an HL4 (Korea) passing out reports to an eager following, and a little higher a DU (Philippines) station is proving to be most popular. But I don't need any of them. Then, I come across a tremendous pileup. I listen carefully, wondering what manner of DX could attract such a throng. The pile seems to be several kilohertz across, and I cannot find anyone inside it that would be the target. I tune back down a moment, trying to find the lower side of the pileup, and in doing so find AH2G (Guam) passing out a report. His signal is terrific, well over S9. Could he be the target? I open up the selectivity as he signs with a station. Wow! As one, the whole pileup calls him.

I am mesmerized for a few minutes. I listen to a couple rounds, then start calling him myself. After about 10 minutes with no joy, I come back to my senses, and laugh. After all, I don't need it. I look at the AH2 card on the wall for reassurance. Just because he's loud, and apparently a superb

160 Meters (1.8-2.0 MHz)

Devotees of DXing on 160 meters, aka Top Band, tend to be viewed by others as gluttons for punishment. Why else would anyone stay up to all hours of the night (and early morning) just to get an earful of static? Those who see it that way should probably move back to the safer regions of 20 meters. But for those who really are up to the challenge, 160 meters is irresistible.

Top Band was at one time an exclusive club. It became known as the "gentleman's band" because some of the shenanigans often heard in 20-meter pileups just did not happen on 1.8 MHz. That "gentleman's band" concept continues today, but the level of competition has increased manifold with the proliferation of transceivers with 160-meter capability. Participants are also discovering that acres of real estate aren't a necessity to compete. The two most common antennas found a high inverted V, for those with towers of 80 feet and higher, and the inverted L, the favorite of the "tribander at 50 feet" crowd, who can locate the antenna next to the tower and extend the horizontal section to a support such as a tree.

DXing operating habits on 160 have also undergone some dramatic change with the burgeoning population and worldwide frequency allocations in a state of flux. To keep pace, the ARRL Board of Directors adopted a modified band plan in 1986. This plan recommends that CW, RTTY and other narrow-band modes be used from 1800 to 1840 kHz, while CW, SSB, SSTV and other wideband modes should be used from 1840 to 2000 kHz. Because the 1830-1850 segment is the most common international allocation in the band, amateurs are encouraged to use it only for intercontinental QSOs.

Transceive operation, both CW and phone, is commonplace between 1830 and 1850 kHz. The segment 1850-1855 kHz, is often used by Eastern European countries that are restricted to that portion of the band. General phone operation is by convention used above 1855.

The other band segment of significance is 1907.5 to 1912.5 kHz, the window for Japanese stations. JA stations will announce where they are listening, usually on the low end. VK and ZL stations will be found on CW at our sunrise, near the bottom of the band, just up from 1800 kHz, best during the northern hemisphere's winter season. As of January 1, 1985, Soviet 160-meter allocations are 1830 to 1930 kHz on a secondary basis. 1830 to 1860 is CW only, 1860 to 1900 is CW and LSB, and 1900 to 1930 CW, LSB and AM.

160 meters is strictly a nighttime DX band. Knowledge and use of grayline sunset and sunrise information is probably of more use on this band than any other. The equinoxes can be especially productive. DXing on Top Band offers a greater challenge than any other HF band. There is no better reward than to have another sleepless night produce a "new one." Where it once took years to accumulate 100 countries for Top Band DXCC, DXers are now doing it in a single operating season—meaning fall through spring.

80/75 Meters (3.5-4.0 MHz)

It's called 80 meters in the CW portion of the band and 75 meters in the phone part of the band. We'll refer to it simply as "80."

80 (like 160) is a nighttime band with best DX times being an hour before sunset and an hour after sunrise. Those hours between sunset and sunrise are not a vast wasteland, either. More-than-adequate possibilities to work some medium- and long-haul DX exist during the hours of darkness. Again as on 160, long-haul DX paths between two given areas are enhanced when sunset at one end of the DX path corresponds with sunrise in the area at the other end. This is called shadow edge or grayline DXing. See the sunrise/sunset tables in Chapter 17. It's hard to imagine that the daytime short-hop ragchewing and net operation that seems to be the mainstay of 80 meters gives way to a world of some "heavy-duty" DXing by night. Several 80-meter DXers have worked from 200 to well over 300 countries. If they can do it, so can you, with a little time and effort. On 80, an Advanced or Extra Class license can be the difference between success and failure in your DXing efforts, although a respectable country total can be amassed with one of the other license classes. [To gauge propagation conditions to Eastern Europe, check the Russian beacon continuously sending the letter U in Morse code on 3636.3 kHz.]

Eighty-meter WAC (Worked All Continents) and racking up 100 countries for your Five-Band DXCC are well within

operator, is not enough reason for me to depart from my game plan. I start tuning again, up the band.

After passing two more JAs calling relatively weak CQs, I hit big, in the form of V85AK, S5 and calling CQ. Brunei, a new one, if I can nail him. I set up as he calls his CQ, hoping that no one else will find him. He signs, and I drop my call in, a ×2 shot. I listen; darn, he's back to a W5. The 5 is just about on the same frequency as I am, so I hold my ground and do not move. There, he finishes, and the V85 calls a QRZ. I wince, as I anticipate a pileup forming that will dwarf that of the AH2. I pause for a moment before I begin my call, waiting to hear just how loud the pileup is.

There is so much silence that I almost blow my chance. I call, and listen incredulously as he comes back, W9KNI R TU 579150 579150 W9KNI K̄N.

I barely manage to respond, R V85AK DE W9KNI R TU NEW COUNTRY 559ILL 559ILL QSL 73 DE W9KNI K̄N.

He comes right back, R W9KNI FB QSL OK BOX 73 BANDAR S̄K QRZ DE V85AK K.

Wow, just like that! I log him in, and listen. There is no one calling him at all now. There he goes again, calling CQ. Okay, now he has a taker, a WA6. Hmm. I glance back down the band for a moment. Yes, that AH2 has a bigger pileup than ever now, while the relatively rare V85 goes un-challenged, begging for QSOs. I guess it's the AH2's big signal. But you'll get no complaint from me!

Contests for the DXer

Contests offer the DXer an unparalleled opportunity to snag new countries and to practice pileup techniques. (But remember—if you are "practicing" in a pileup for a station whose country you *don't* need, you just may be missing one a few kilohertz away that you *do* need.) There are two particularly important contests for the DXer. Rules for all major contests are published in the Contest Corral column of *QST*.

The best contest for country hunting is the CQ World Wide Contest, with the phone section held the fourth weekend of October, and CW the fourth weekend of November, which is usually, but not always, the weekend after the US Thanksgiving holiday.

The ARRL DX Contest is a close second in importance, with the CW weekend the third full weekend of February, and the phone weekend the first full weekend of March. This contest is of particular interest to American and Canadian DXers, because DX stations can work only Ws and VEs. Unlike the CQ contest, DX stations are not allowed to work each other for points.

the grasp of any operator with even a modest station/antenna setup and a little bit of operating savvy. Antennas for 80 can range from simple dipoles and trapped verticals to elaborate delta loops and quad configurations. The rig should have split-frequency capabilities if you are interested in working 75-meter phone DX. To work some of the countries that can't transmit in the US phone band, an external receiver or external VFO would definitely work to your advantage.

The low end of the US phone band (3.750-3.800 MHz) is a hotbed of DX activity. Extras can operate from 3.750 and up, while the Advanced portion starts at 3.775 MHz. The segment 3.790-3.800 MHz is considered a "DX Window." Note that many DX stations will operate split frequency, transmitting below 3.750 and listening above 3.800 MHz for callers. So the knowledgable 80-meter DXer should listen well below 3.800 for a DX station who may wish to work the US using split-frequency operation. (Check also for SSB DX calling CQ around 3.640, listening in the American band.) On CW, the majority of DX stations work in the 3.500- to 3.525-MHz range. An accompanying table shows the 80-meter phone frequency allocations from around the world.

For those interested in serious DXing on 80 meters, there is a definitive work on the subject: *Low-Band DXing* by John Devoldere, ON4UN, published by ARRL. The book is probably the most complete work on the topic, written by an operator with 300 countries worked on 80!

40 Meters (7.0-7.3 MHz)

Forty, like the other low bands, 160 and 80 meters, is basically a nighttime and early-morning DX band. Again, grayline-DXing methods prove most fruitful.

On 40, though, the DXer faces a new enemy. Propagation and band characteristics are not the only obstacles to overcome. Forty meters is (like 160 and 80) a shared band; shortwave broadcast stations have assigned frequencies on 40 also. Thus, as the band conditions begin to change and become more favorable for working DX, it seems as if every broadcast station in existence is trying to claim a piece of 40 meters as its own. The bc stations use several kilohertz of bandwidth and legally run tens and hundreds of kilowatts of power. Strangely enough, it's possible to tell when 40 is open to certain areas by the strength of a received bc signal.

At 7 MHz, rotatable/steerable antenna arrays start to become more manageable in size, and some of the simpler wire and vertical antennas that work well on 80 meters do a creditable job on 40. By running near or at full legal transmitter power to an antenna array, one can put out quite a respectable signal on 40 meters. The beam antennas make it possible to explore different signal paths utilized on the higher bands, such as the long-path signal. Long path is a phenomenon in which a signal will peak if you steer your antenna so that it is facing approximately 180° away from the azimuthal heading (beam heading, or short-path beam heading) as determined from a great-circle projection centered on your QTH. For example, if you usually point your antenna at 120° to work a station in Africa, a long-path opening would cause the same African station's signal to peak up when you rotated your antenna to a bearing of about 300°. It's worth checking for long-path signals, especially when signals seem to be down early in the morning or early evening.

As they do on most of the other bands, the CW DXers like to stay in the lower 25 kHz of the 40-meter band. But since there is such strong competition for space on 7 MHz, DX stations can regularly be found on CW much higher in the band—as high as 7.050 MHz or so.

Working DX on 40 phone will quickly bring home the need to have the ability to work split frequency. The US phone band extends from 7.150-7.300 MHz, and South and Central Americans as well as Africans can be found and worked here transceive (on their own frequency, not split). European and some Asian stations *must* be worked split-frequency on 40. The Europeans can be found transmitting betwee 7.040 and 7.100 MHz, listening for US amateurs in the US phone segment.

When the band is quiet and not crowded, it is not necessary to have a "killer" signal to work DX easily. Many QRPers (operators who run a maximum of 10 watts input or 5 watts output) are quite successful in achieving respectable country totals on this band. QRPers on 40, as on most of the other bands, tend to be CW operators and gather at or near 7.040 MHz. Forty meters, for all its quirks, is quite an interesting band and has a lot to offer for a serious or casual DXer.

Another popular event on the contest calendar is the All Asia Contest, with phone and CW dates announced in the major magazines. This contest is for Asian stations "against the world," meaning that non-Asians can work only Asian stations, and Asians can work only non-Asian stations. It is a fine contest—unless you already have all the Asian countries worked and confirmed. Of course, there are a number of other contests that will be of great interest, such as the CQ WPX Contest and the IARU HF World Championship, especially to the DXer with the smaller country total.

Countries with enough amateurs to organize and promote a DX contest are usually countries that are not considered rare. Still, many contests, such as the WAE (Worked All Europe), offer an occasional pleasant surprise (many amateurs still need ZA) and, in any case, offer good operating practice and the chance of new band-countries. Details and dates on such contests are published regularly in *QST*.

Contests, including the minor ones, are especially fine opportunities for the QRP operator to make a mark. An alert and aggressive operator in a contest will regularly find even rare stations calling CQ, with the chance of a one-on-one shot at a QSO. If you can't get a QSO under such circumstances, something is usually wrong.

Persons seriously interested in contests from the view of

JA1ASO's all-band, all-mode station in Tokyo.

competing for a score or just simply in learning more about contests should refer to Chapter 7 for further information. Remember, even if you're not particularly interested in

Worldwide Top-Band Frequency Allocations

This list shows international frequency allocations on 160 meters. Information was garnered from International Amateur Radio Union (IARU) annual reports submitted by 91 of the 124 IARU Member-Societies. Some other countries, with or without IARU representation, may give their amateurs access to 160 meters. Since this information reflects administering government regulations, these frequencies may apply to several DXCC countries within that jurisdiction. For example, the listing for G (United Kingdom) applies to G, GD, GI, GJ, GM, GU and GW.

Country	Privileges (kHz)	Notes	Country	Privileges (kHz)	Notes
A2	1820-2000	32 W PEP Max	SP	1810-1930	10 W input max (except 1830-1850)
A4	1800-2000	10 W max for CW	T7	1810-1850	
A9	1810-1850		TA	1810-1850	30 W max ERP
AP	1800-2000		TG	1800-2000	
BY	1800-2000		TI	1800-1850	
C2	1800-2000		TR	1810-1850	
C3	1810-1875	Ph only: 1825-1875	U	1830-1930	Low power ph only: 1860-1930
C6	1800-2000		V2	1800-2000	
CP	1800-1850		V3	1800-2000	
CT	1830-1850	CW & RTTY 60 W Max	V8	1800-2000	
DL	1815-1835	SSB only: 1832-1835	VE	1800-2000	
	1850-1890		VK	1800-1866	Ph: 1825-1866
DU	1800-2000	SSB only: 1900-2000		1874-1875	
EA	1830-1850		VP2M	1800-2000	CW & SSB only
EI	1800-2000		VP9	1800-1825	CW & ph
F	1830-1850			1875-1900	CW & ph
FR/R	1810-1850		VS6	1800-2000	CW only
FG,FR,FS,FW,FY	1800-2000		VU	1820-1860	
FO	1800-2000		W,K,N,AA-AK	1800-2000	See ARRL recommended band plan
G	1810-2000	10 W max	XE	1800-2000	
H4	1800-2000		Y2-9	1810-1950	10 W max ph only: 1900-1950
HA	1830-1850		YB	1800-1875	SSB only: 1835-1875
HB	1810-1850		YK	1830-1850	
HK	1800-2000		YJ	1800-2000	SSB:1850-2000
HL	1800-1825		YN	1800-2000	
HP	1800-2000		YS	1800-2000	
HR	1800-2000		YU	1810-1850	SSB: 1830-1850
I	1830-1850		ZB2	1800-2000	
J2	1810-1850		ZL	1803-1857	
J3	1800-2000			1863-1950	
JA	1907.5-1912.5	CW only	ZP	1800-2000	
LA	1802-1850	CW only, 15 W max	ZS	1810-1850	
LU	1800-1810	CW only	3A	1830-1850	
	1810-1850	CW & Ph	3B8	1810-2000	10 W max
LX	1810-1850		3D2	1800-1850	
OA	1800-2000		4S	1800-2000	
OE	1810-1950	100 W max SSB: 1840-1850	4X	1800-2000	
OH	1820-1845	10 W max	5B	1800-2000	Ph only: 1900-2000
	1915-1955	10 W max	5N	1800-2000	10 W max
OK	1750-1950	10 W output max ph only: 1820-1950	5W	1800-2000	Ph only: 1850-2000
ON	1830-1850	CW, 10 W max	5Z	1830-1850	
OY	1830-1850	CW, 10 W max	6W	1810-1850	
OZ	1830-1850	CW, 10 W max	7P	1800-2000	10 W max
P2	1800-1866		9H	1810-2000	10 W max
	1874-2000		9J	1810-1850	10 W dc input
PA	1825-1835	CW & SSB 10 W output max	9K	1810-1850	
	1835-1850	CW only	9L	1800-2000	10 W max
PJ2-4,P4	1800-2000		9M	1800-2000	
PP-PY	1800-1850		9V	1800-2000	10 W input max
PZ	1800-2000		9Y	1800-2000	
SM	1830-1845	CW: 10 W input max			

competing in the contest from a scoring standpoint, you should nevertheless be aware of the operating opportunity to snag new countries and get experience under fire.

CONCLUDING NOTES

The scope of this book obviously does not permit an exhaustive study of DXing. It is designed more as a basic introduction that points you in the proper direction. On that basis, here—briefly—are a few important subjects.

First, the reader will have noted that only the bands 40, 20, and 15 were especially covered. These major bands are useful DX bands for new DXers almost anywhere in North America, a situation perhaps not as true of 80 or 160 meters. Which is not to say that DX cannot be worked on those bands, but for the newcomer operating with 150 watts from the American Midwest, for example, it is a more challenging proposition. Chasing DX on 160 meters especially requires that good operators follow established practices that are unique to the band. See accompanying sidebar on these "low" bands.

Ten meters is a wonderful and very popular band when the sunspot cycle favors it, and a band where even low-powered

80-Meter Phone Allocations Throughout the World

Area	Frequency (MHz)
Africa	3.600-3.800
Asia (most)	3.600-3.900
Australia	3.535-3.700, 3.794-3.800
Canada	3.725-4.000
Europe (except USSR)	3.600-3.800
Japan	3.525-3.575, 3.791-3.805
Pacific	3.600-3.900
South America	3.500-3.800
USA and possessions	3.750-4.000
USSR	3.600-3.650

IARU Region 1 has officially recommended that 3.790-3.800 MHz be used only for intercontinental contacts.

Note: There are several exceptions, including license-class restrictions, to this table.

stations with simple antennas can work the world. [Novices will certainly find this out via their new 10-meter SSB privileges—28.3 to 28.5 MHz—courtesy of Novice Enhancement.] But DXers who concentrate on 10 meters are people who will have to find another hobby every few years, and most nights. The band is a very seductive one when it is wide open—and very dead when closed. Any serious DXer must be operational on the band; but, the DXer intent on running up a country total will be better off concentrating on the other bands, at least until the upswing of the next sunspot cycle.

Thanks to the ARRL and other IARU societies, the last WARC conference brought us some exciting new bands to play with at 10, 18 and 24 MHz. These bands are just becoming significant to the DXer, offering outstanding additional opportunities for the future. [Contacts on 24 MHz *only* are valid for DXCC, but remember that contacts made on any of the three WARC bands are *not* valid for 5BDXCC.] Now that 10 and 24 MHz are available to us, they offer operators who wish it a quiet sanctuary from weekend contests.

Each of the new bands is located—in terms of frequency—approximately midway between two of the existing amateur bands, and, as might be expected, their propagation characteristics are intermediate between those of their neighbors. During the first year, 10 MHz was not usually "MUF-limited" except for a short period around dawn in midwinter. As solar activity declines, there will be an increasing tendency for the MUF on more northerly paths to fall below 10 MHz during darkness. Apart from this, absorption will be the main factor

affecting DXing, which will generally be best when much of the path is in darkness or, if not, is reasonably close to twilight. DX openings will therefore follow the same pattern as those on 40 meters, but because absorption is lower, they will start earlier and end later than on 40 and peak signals will be stronger. As to 18 (not yet available to US amateurs) and 24 MHz, while there will be detailed differences with different levels of solar and geomagnetic activity, the propagation characteristics of these two bands will show marked similarities to those of their immediate neighbors, sometimes behaving like the next lower band and sometimes resembling the next higher. Under present conditions, 24 MHz will remain open for DX on days when 28 MHz is not useful, while the 18-MHz band should prove good for daytime DX even during the solar minimum.

Another aspect of DXing not referenced above are the DX newsletters. Four are published in the US at present, each with its own viewpoint and following. They are:

The DX Bulletin
816 Fourth St, Suite 1001
Santa Rosa, CA 95404

The junior op at GW4BLE calling into his very first DX pileup.

The International Amateur Radio Union

The International Amateur Radio Union (IARU), founded in 1925, is the federation of national Amateur Radio Societies around the world. The IARU consists of Member-Societies (currently 124), an Administrative Council, three Regional Organizations (one in each ITU Region) and an International Secretariat (ARRL has been serving this function since the formation of the IARU).

The major objectives of the IARU are: representation of the interests of Amateur Radio at and between conferences and meetings of international telecommunications organizations; encouragement of agreements between national Amateur Radio societies on matters of common interest; enhancement and promotion of Amateur Radio; and encouragement of international goodwill and friendship. Shown here is Lou, PAØLOU, Chairman of IARU Region 1. Lou exemplifies the spirit of dedication of these volunteer IARU officials.

WARC-Bands Frequency Allocations

The following information on international frequency allocations for 10, 18 and 24 MHz was supplied by International Amateur Radio Union (IARU) Member-Societies. Other countries, with or without IARU representation, may give their amateurs access to one or more of these bands.

Country	10 MHz	18 MHz	24 MHz
A2	10.100-10.150	18.068-18.168	24.890-24.990
A3	10.100-10.150		24.890-24.990
A4		18.068-18.168	24.890-24.990
A9		18.068-18.168	24.890-24.990
BY	10.100-10.150	18.068-18.168	24.890-24.990
C3	10.100-10.150	18.068-18.168	24.890-24.990
C6	10.100-10.150	18.680-18.168	24.890-24.990
CT	10.100-10.150	18.068-18.168	24.890-24.990
DA-DL	10.100-10.150	18.068-18.168	24.890-24.990
DU	10.100-10.150		
EA	10.100-10.150		
EI	10.100-10.150	18.068-18.168	24.890-24.990
F	10.100-10.150	18.068-18.168	24.890-24.990
G	10.100-10.150	18.068-18.168	24.890-24.990
H4	10.100-10.150		
HB	10.100-10.150	18.068-18.168	24.890-24.990
HK	10.100-10.150	18.068-18.168	24.890-24.990
HL	10.100-10.150		
HP	10.100-10.150	18.068-18.168	24.890-24.990
HR	10.100-10.150	18.068-18.168	24.890-24.990
I	10.100-10.150	18.068-18.168	24.890-24.990
J2	10.100-10.150	18.068-18.168	24.890-24.990
J3	10.100-10.150	18.068-18.168	24.890-24.990
J7	10.100-10.150		
JA-JS	10.100-10.150		
K,W,N, AA-AK	10.100-10.150		24.890-24.990
LA	10.100-10.150	18.068-18.168	24.890-24.990
LU	10.1005-10.1030	18.0730-18.0765	24.890-24.990
	10.1190-10.1215	18.0835-18.0895	
	10.1435-10.1465	18.0965-18.1085	
		18.1215-18.149	
		18.1515-18.1675	
LX	10.100-10.150	18.068-18.168	24.890-24.990
OA	10.100-10.150	18.068-18.168	24.890-24.990
OE	10.100-10.150	18.068-18.168	24.890-24.990
OK	10.100-10.150	18.068-18.168	24.890-24.990
ON	10.100-10.150	18.068-18.168	24.890-24.990
OY	10.100-10.150	18.068-18.168	
OZ	10.100-10.150	18.068-18.168	24.890-24.990
P2	10.100-10.150		24.890-24.990
PA	10.100-10.150	18.068-18.168	24.890-24.990
PJ2-4, P4	10.100-10.150	18.068-18.168	24.890-24.990
SM	10.100-10.150	18.068-18.168	24.890-24.990
SV	10.100-10.150		
T7	10.100-10.150	18.068-18.168	24.890-24.990
TA	10.100-10.150	18.068-18.168	24.890-24.990
TI	10.100-10.150	18.068-18.168	24.890-24.990

Country	10 MHz	18 MHz	24 MHz
TR	10.100-10.150	18.068-18.168	24.890-24.990
U	10.100-10.150		
V2	10.100-10.150	18.068-18.168	24.890-24.990
V3	10.100-10.150		
V8	10.100-10.150	18.068-18.168	24.890-24.990
VE	10.100-10.150		
VK	10.100-10.126	18.068-18.071	24.890-24.896
	10.1350-10.1375	18.080-18.101	24.904-24.990
	10.1456-10.150	18.110-18.121	
		18.135-18.141	
		18.150-18.156	
		18.165-18.168	
VP2M	10.100-10.150		
VP9	10.100-10.150		
VS6	10.100-10.150		
VU		18.068-18.169	24.890-24.990
Y2-9	10.000-10.150	18.068-18.168	24.890-24.990
YB	10.100-10.150		24.890-24.990
YJ	10.100-10.150	18.068-18.168	24.890-24.990
YK	10.100-10.150	18.068-18.168	24.890-24.990
YN	10.100-10.150		
YS	10.100-10.150	18.068-18.168	24.890-24.990
YU	10.100-10.150	18.068-18.168	24.890-24.990
ZB2	10.100-10.150		
ZF	10.100-10.150	18.068-18.168	24.890-24.990
ZL	10.100-10.127	18.068-18.168	
	10.133-10.150		
ZS	10.100-10.150	18.068-18.168	24.890-24.990
3A	10.100-10.150	18.068-18.103	24.890-24.990
		18.117-18.129	
		18.130-18.135	
		18.136-18.165	
		18.166-18.168	
3B8	10.100-10.150	18.068-18.168	24.890-24.990
3D2	10.100-10.150		
4S	10.100-10.150	18.068-18.168	24.890-24.990
4X	10.100-10.150	18.068-18.168	24.890-24.990
5B	10.100-10.150	18.068-18.168	24.890-24.990
5N	10.100-10.150	18.068-18.168	24.890-24.990
5W	10.100-10.150		
6W	10.100-10.150	18.068-18.168	24.890-24.990
7X	10.100-10.150	18.068-18.168	24.890-24.990
9H	10.100-10.150		
9J	10.100-10.150	18.068-18.168	24.890-24.990
9K	10.100-10.150	18.068-18.168	24.890-24.990
9L	10.100-10.150	18.068-18.168	24.890-24.990
9M	10.100-10.150	18.068-18.168	24.890-24.990
9V	10.100-10.150	18.068-18.168	24.890-24.990
9Y	10.100-10.150	18.068-18.168	24.890-24.990

QRZ DX
PO Box 834072
Richardson, TX 75083

The Long Island DX Bulletin
Box 173
Huntington, NY 17743

Inside DX
436 N Geneva St
Ithaca, NY 14850

All of these bulletins will be happy to quote you subscription rates and send you sample copies upon receipt of your SASE with postage for three ounces. A subscription to a DX newsletter is an excellent investment for the operator seriously interested in working new countries, as it gives information on DXCC, DXpeditions, news of stations from rare spots and a wealth of intelligence on the happenings of the bands, all timely enough to be helpful.

Another subject covered generally has been propagation. The committed DXer will learn through experience the propagation patterns of his favorite bands, but a study of the physical phenomena that make DX propagation possible will enhance both the DXer's interest and knowledge of when to be laying for the A5 or another needed country. The chapter on antenna orientation in this book provides a good introduction. See *The ARRL Handbook for the Radio Amateur* for a more detailed treatment. The definitive book on the subject for the DXer is *The Shortwave Propagation Handbook,* second edition, by George Jacobs, W3ASK, and Theodore Cohen, N4XX, available from *CQ* magazine. See also the propagation charts and sunrise-sunset tables in Chapter 17.

Of course, a truly comprehensive discussion of such a complex matter as DXing is impossible in the space allocated to one chapter, and there are significant subjects that have not been covered at all, such as buying gear for maximum DX benefit, antennas, site selection and CQ calls. [See Chapter 8 for a complete survey of international awards.] Those amateurs interested in a more intensive book-length treatise on DXing should obtain *The Complete DX'er* by Bob Locher, W9KNI, available from most amateur dealers, the ARRL or from the publisher, Idiom Press, Box 583, Deerfield, IL 60015. Another good reference is *DX POWER: Effective Techniques for Radio Amateurs,* by Eugene B. Tilton, K5RSG, copublished by ARRL and Tab Books.

Good hunting!

Chapter 6

Overseas DXing/DXpeditions

"**Y**ou want to go on a what?" asks your spouse. "D-X-pedition," you enunciate very clearly. "What in the world is that?" comes the not unexpected response.

"Well, honey, you know all those DX stations all over the world that I've been chasing the last couple of years...."

"Uh huh."

"Well, I would like to try being chased myself for a change."

She grins and replies, "Come now, you're too old for that!"

"Seriously, this year instead of taking the kids to Disney World, I want to go on a DXpedition!"

Now she knows you are serious. The grin is gone and in its place is a hurt abandoned look. "You want to leave me and the children while you go off and get hurt playing Indiana Jones?"

My goodness, you think, this woman can sure make you feel guilty. "No, darlin'," you soothe, "I want to leave the kids at your mother's and I want us to spend a week on some beautiful tropical island."

"But what about your DXpedition?"

"That will be my DXpedition!"

"It doesn't sound so bad; you'll just be talking on your radio at night like you do now?"

"Well," not wanting to limit your activity, "maybe a *little* more than that; after all, I will be on vacation!"

Now that the stage has been set, let's plan a DXpedition: where to go, when to go, what to take, what to expect, and some obligations. If everything else goes as it should, you will then be free to enjoy being the much sought-after DX. While these ideas are written primarily for amateurs in the US, they could apply anywhere with little difficulty. This chapter will attempt to follow the same chronological flow as would be done in actually planning and executing a DXpedition.

There are five general categories: (1) planning and preparation, (2) travel and arrival, (3) documentation and assembly of the station, (4) DX operating, and (5) post-DXpedition requirements.

There are some things about operating which deserve a little thought before going further. At a certain level of competency, all ham radio operators know how to operate, and all DXers know how to DX. There is no denying this. One purpose of this manual, particularly this chapter and the previous chapter on domestic DXing, is to help you become a better operator. Whether you actually go on a DXpedition or not, the philosophies expressed here are aimed at improving the quality of operating.

At this point, it is appropriate to define some terms. For the purpose of this chapter, a DX station, shortened to "the DX" or just "DX," is an Amateur Radio station, on land, in an ARRL DXCC country. DX can be both a noun and a verb. For example: DXers DX when they pursue the elusive DX. DXing is the activity of trying to contact DX stations and it is pursued by DXers. "To DX" is to engage in DXing. You should realize that the DX (the hunted) can also be and often is a DXer (the hunter). Certainly there are quite proper domestic uses of the term "DX" which won't be addressed in this chapter.

Let us begin with the most logical point, choosing your destination.

PLANNING AND PREPARATION

Where to go

Most DXers know which countries are the rarest and consequently the ones from which an Amateur Radio operation would generate a really large pileup. Many DXers never consider going on a DXpedition because they can't be the one to put on an operation from one of those very rare DXCC countries. This is a sad mistake! While it is certainly normal to pursue lofty goals, you should note that, with only minor exceptions, there are few differences in pileups other than their size. Because of the many facets of DXing, it is quite possible to conduct a truly exciting DXpedition from a country that already has an active ham population! This is because many resident hams prefer rag chewing to pileups, and will welcome your efforts to take the pressure off of them!

Many relatively-easy-to-contact DXCC countries are quite rare on certain modes (eg, CW or RTTY) or bands, such as 160, 80 or 40 meters. Likewise, by choosing operating times and frequencies carefully, some nice operating can be had from what might be considered common countries which are actually rare in other geographical areas. Some examples are the Caribbean area (rare to Japanese amateurs), Europe (difficult to contact from the West Coast of the US) and the Pacific Islands (which offer some challenge to be worked from the East Coast). From time to time, the DX newsletters will poll their subscribers and list, in descending order, the most needed or the rarest DXCC countries (see the 1985 *DX Bulletin* survey in Chapter 5). Numerous DX clubs maintain similar lists of the countries needed by their members, usually in conjunction with a telephone or 2-meter FM DX alert system. Both of these can be good sources of information to help you decide which location to choose for your DXpedition. Just remember that if all of your information comes from one geographical location, then the results will be distorted, favoring that area.

Another person to help you choose a DX location is your travel agent. In the present competitive status of airline fares, a wide variety of distant locations are within reach of some rather modest budgets. The travel agent charges you nothing, but receives a commission on the sale of your ticket. This is an excellent source of information and savings which you should not ignore. If your trip is in conjunction with business, a little research is still in order. You might be able to change from the most direct routing and include a stopover in a semi-rare DXCC country and not increase the cost of your air travel at all. In addition, your business destination might be located only a short distance from a relatively good DX operating location. The business traveller/DXer arriving in Hong Kong will find some DXing fun there, but might experience a "major league" DXpedition in nearby Macao (XX9), just a short and inexpensive trip by hydrofoil, jetfoil or ferry.

When to Go

If your sojourn is employment related, then you might not have a choice of times. But if you do have a choice, then you will probably want to choose a time during the Northern Hemisphere's winter. With the Earth closer to the sun during the winter months, the F-layer of the ionosphere (the one more likely to bring DX) will be more favorably charged. In the summer, those long days allow the D layer (commonly blamed for absorbing signals) to build up. The third, and possibly the most significant, reason for planning a DXpedition for the wintertime is to get away from snow and cold weather!

Another appropriate time for a DXpedition is during a major contest. You don't have to be a serious competitor to enjoy yourself in a contest. But even so, an examination of the published results will make it clear that it is quite easy to enter a category that will ensure the DXpedition a nice operating award. The results of a previous contest can also be used to help select the destination, but they can also be very misleading since there is no certainty that the same countries will be activated from year to year (check the DX newsletters to find out who is going where).

Local weather conditions at the chosen DX location can either be devastating or richly rewarding to the operation. Some geographical regions have no winter or summer; instead they have only a wet season and a dry season. The rainy season can ruin QSO totals on lower frequency bands such as 160 and 80 meters, not to mention multiplying the risk of contracting malaria in susceptible areas many times over!

If this is to be a specialty-band-related trip, it is important to remember that the lower the frequency, the more critical the selection of the time for the operation. It would be prudent to examine such DX aids as the sunrise/sunset tables and propagation tables that appear in this book, along with *The DX Edge*, SM6BGG's package of maps and tables, or any similar product. This would help to ensure that the goals for the operation are realistic, and that every consideration has been taken to optimize the results.

Learning What to Expect

Assuming that a tentative choice of one (or more) of the DXCC countries has been made as the destination, it is time now to research and investigate so as to not receive any surprises which might ruin your adventure. If the choice is a place where tourism blossoms, there will be relatively few pitfalls, and these should be very minor. If you receive a tourist visa from the embassy of the destination country late in the year, and it is numbered in single, double or triple digits, then it is safe to assume that tourism there is rare (usually for a good reason) and that some difficulty will be encountered.

Licensing is essential, of course, but it might occur at the beginning of the planning stage or after arrival at the DX location. It is easy to see the possibilities of failure if licensing is attempted only after arrival.

Some countries can be activated only in this most precarious way, because it is simply impossible with certain countries to reach the correct person via telephone or to receive an answer to a letter. In nations such as these, it is very important to reach the person who actually has the authority to grant the permission. A lesser official cannot say "yes," but if he says "no," then it will be very difficult to get a reversal of that decision.

ARRL HQ maintains a great deal of information regarding licensing in other countries. This should adequately cover the countries where most tourists go and, surprisingly, it can be of significant benefit in the more exotic places. For example, the difference in receiving permission to operate from a former European colony might very well hinge on having a license from the former colonial power. On the other hand, depending upon the relationship between the two countries, that same license might be a ticket to sure failure. A little history study at your local library might make a significant difference. It will also increase your enjoyment of the trip!

Begin your research anywhere convenient, but do not accept either a glowing or a gloomy report at face value without evaluating the source and possible motives for the report. The US State Department will give a conservative outlook, since some of their duties encompass trying to rescue US citizens in distress. Representatives of the host nation will never volunteer anything which might be construed as derogatory about their own nation. Your evaluation should also weigh quite heavily on your own international experience (or lack thereof). Do not choose a so-called developing country as a destination unless you are a seasoned traveler. (A possible exception to this would be when a very dependable, healthy and resourceful person is going to meet your flight and assist you in-country.)

Other amateurs who have operated from the particular country are an important source of information. They understand what you want to do (it's unlikely that anyone else will understand or care). If the ill-fated Spratly Islands DXpedition had contacted the previous successful expedition (1S1DX) before them, 1S might be much lower on the list of DXCC countries needed and, far more important, that awful tragedy might never have occurred. (When the 1S1DX group approached one of the few islands of the Spratly Group large enough to land on, they discovered that it was occupied. Not only was it occupied, but the inhabitants were not at all sociable. Gunfire has never been a sign of open friendship! Luckily, no one was hit, and several days later the DXpedition located an uninhabited tiny patch of sand and coral large enough to install some antennas and still have room from which to operate. Several years later another group went back to the same original island, apparently without benefit of the knowledge gained by the 1S1DX crew. They were greeted by more accurate gunfire. The boat was hit, and the gasoline tanks exploded. Three persons were killed, and the survivors drifted for days in the South China Sea before being rescued by a passing merchant vessel.) Some good sources for this information would be any QSL route listing, including *QST*'s How's DX column, the *Callbook*, The W6GO/K6HHD QSL

Operating Overseas—A Brief Tutorial

Do you need a permit to operate in a country with which the United States holds a reciprocal-operating agreement?

Yes, you do, except in the case of Canada, where your FCC license is automatically valid. Generally, the existence of such an agreement relieves you only from taking an examination for a license and minimizes the paperwork necessary for a permit. On the other hand, the absence of such an agreement does not necessarily mean you cannot be licensed.

Where do you apply for a permit, and how?

Send a business-size SASE to ARRL Information Services and indicate which country you're planning to visit. You'll be sent instructions on how to apply and (if available) an application form. If sufficient up-to-date information isn't on hand, you will at least receive the address of the licensing administration or the national Amateur Radio Society (usually our International Amateur Radio Union—IARU—Sister Society) of the country from which information can be obtained.

If you do not have a specific application form, write a letter requesting a permit. This should include the purpose of your trip, dates and places of your stay, passport information and the equipment you intend to use. You should also attach a copy of your FCC license and, in some cases, a letter attesting to your character signed by your local chief of police (or equivalent).

If a certified copy of your license is required, a notary public should be able to provide you with one if he can compare a photocopy with your original (and is given the phone number of the FCC Consumer Assistance Line, 717-337-1212, to obtain confirmation of the validity of the license). If you're a League member and the foreign administration concerned honors this, you may request ARRL HQ to write the appropriate administration attesting to your membership and license. Be sure to provide HQ with a photocopy of your license and an SASE.

How is an application fee paid for?

Some administrations charge for permits. It is best to send a bank draft drawn on a bank that does business in the country concerned. Or draw an international postal money order and send a copy of the receipt, unless otherwise instructed. Personal checks are rarely accepted in foreign countries.

Is it possible to apply on short notice, less than one month in advance of my projected trip?

You may be taking a chance. But if you do, and if you enclose a fee (or if you're sure there is no charge), you should furnish your address in the host country where you would like your permit, if granted, to be mailed. If you're not sure whether a fee is charged, you should promise to pay any charge when you pick up the permit upon arrival in that country.

I've applied well in advance, but haven't received a permit. What to do?

Write or call the administration. If you prefer calling but do not have the telephone number, call their embassy or consulate in the United States and you will be furnished the number.

I don't have a permit, but want to take a transceiver into a foreign country.

For most countries, you cannot take in such equipment without a license or permit. If you've already applied for the permit and expect to pick it up on arrival in the country, show the customs officer a copy of your appli-cation. If you intend to apply for a permit later or are just carrying the equipment before or after operating in some other country, show your FCC license (and permit of the other country). This will help you verify that you are carrying communication equipment lawfully. It is quite likely, however, that your gear will be held at customs until you obtain a permit or are leaving the country. Do not try to take in any equipment secretly.

I'd like to operate "on the way" (ship or plane). How can I do this?

ARRL HQ can furnish you information on how to conduct international maritime-mobile operation. If your ship isn't of US registry, let ARRL know the nationality. You'll need a reciprocal-operating permit before you seek the captain's permission. Amateur operation aboard a commercial passenger aircraft isn't allowed, since it might interfere with the navigation control system.

Is ARRL's information accurate and up-to-date?

In the case of countries that normally license foreign amateurs, ARRL contacts the administrations (or IARU Sister Societies) at least every other year to confirm the information on file. Various DX news sources are followed closely to ascertain the current licensing situation. It isn't possible to be 100% up-to-date, but HQ does try.

How do I find answers to other questions?

ARRL's Information Services keeps track of any changes regarding reciprocal-operating and/or third-party-traffic agreements. Additionally, you can obtain a list of amateur frequencies and subbands outside of the United States. For any information request, please enclose an SASE and ask either specific questions or for a copy of the general information package.

There are cautions to keep in mind, of course. Inquiries about regulatory matters (privileges, etc) should be addressed to the administration concerned. Queries about purely amateur matters (repeaters, local clubs, etc) should go to the national Amateur Radio Society. Subjects of general matters (importing a general-coverage receiver, the limit of territorial waters, etc), should go to the embassy or consulate here in the United States.

While signing an FCC call /MM, /AM or /VE (Canada), you must observe the FCC-authorized portions that are part of the amateur frequencies of the Region concerned. You must never use the 50-MHz band while cruising the Mediterranean, operate on 18 MHz before FCC releases the band to US licensees, work SSB on 14.110 MHz, etc. But, this does not apply when you are authorized by a foreign administration (except Canada) to sign a portable FCC call, such as W1AW/D2A or G4/W1AW. You are then bound only by the rules and regulations of the host country.

Further specific information regarding travel, etc, within certain countries (particularly information generally not available from a travel agency or from the diplomatic mission of the country concerned) may be obtained from the US State Department (telephone 202-655-4000). Travel advisories, with the primary purpose of alerting the public to adverse conditions in specific countries, can be obtained from the Citizens Emergency Center, Room 4811 NS, Department of State, Washington, DC 20520 (telephone 202-647-5225).

Have a good trip!

Note to Canadian readers: For information on how to obtain an amateur license or operating permit in most foreign countries, contact CRRL, Box 7009, Station E, London, ON N5Y 4J9 (tel 519-225-2188).

Managers List and several other Amateur Radio magazines and newsletters (see Chapter 5 for a discussion of DX publications).

Here's to Your Health

Quite possibly the most important research to be undertaken for a DXpedition is medical. This advice is not intended to frighten you, but to impress upon you the need to be careful. If you are cautious, there is little chance of harm. However, disregard prudent health precautions and you are assured of discomfort and possibly far worse. There are two major sources of information: Your family doctor and the Centers for Disease Control (CDC) in Atlanta, Georgia. The CDC is part of the Federal government and offers much good information about conditions in other parts of the world. Call the CDC for information about your destination, even if thousands of tourists go there annually. Phone calls are inexpensive, and your health is everything! Some suggestions of subjects to inquire about include Cholera, yellow fever, malaria and chloroquin-resistant malaria. Your local health department can help out with some vaccinations but the really serious ones will have to be given by your doctor. Don't put this off until the last minute. Cholera and yellow fever shots are quite unpleasant, and traveling with a sore arm or hip can do little to improve your spirits. Furthermore, cholera is a two-shot series that requires one to four weeks between vaccinations.

Here are some good guidelines that, while they might not apply everywhere, won't hurt you anywhere.

1) Do not swim, bathe or walk barefooted in or near fresh water. Parasites can cause debilitating disease.

2) Do not drink water that has not been brought to a rolling boil for at least 30 minutes or treated with water-purification tablets. Remember: Ice cubes are water.

3) Do not eat food that may have been in contact with untreated water or handled by anyone without good personal hygiene habits.

4) Do not get too romantic. Romance may be good for the soul, but exotic QTHs have some very exotic levels of disease!

5) Do not stand under mango trees. Parasitic larvae can drop onto your hair and scalp.

6) Do take some form of anti-malaria medicine, and do use insect repellant, especially if in an area where malaria and falaria (river blindness) might be prevalent.

7) Do take a first-aid kit. Talk to your family doctor and the Centers for Disease Control about its contents. A common side effect of the primary anti-malarial used is diarrhea. Water purification tablets can kill the good little creatures inside of you, resulting in diarrhea. Deaths from cholera result from dehydration caused by vomiting and diarrhea. (The most common travelers' ailment is diarrhea. Your doctor can prescribe an effective medicine to combat it.) Drink lots of (safe) water as well as control diarrhea to prevent dehydration. Sunscreen is another valuable asset to include in the first-aid kit; sunburn is an insidious and serious problem.

Some additional suggestions for the first-aid kit include: Cough drops (especially for phone operators), lip balm, aspirin, an ear syringe, adhesive bandage material, antibiotic ointment and disinfectant.

Your Equipment

The selection and preparation of equipment can make the difference between success and failure, at least as far as your own evaluation of dollars spent as compared to QSOs. Almost all DXpeditions go through some stage of air travel. The maximum allowable checked luggage for flights originating from the US is three bags: Two checked and one carry-on. The total dimensions (length + width + height) of the checked luggage cannot exceed 109 inches. One single bag cannot exceed 62 inches in total dimensions. The maximum allowable weight is 70 pounds per bag. Your carry-on baggage cannot exceed 45 pounds, and it must be of a size to fit

underneath the airplane seat. Working within these parameters, make the most intelligent choice of equipment for your DXpedition possible. (Note: The non-US baggage weight limitation is 44 lb, so if your flight originates outside of the US and has a destination outside of the US, it will be necessary to figure within these more restrictive confines.)

A DXpeditioner will have many equipment requirements for any operation. Think of a DXpedition as going out on Field Day, with the single exception that after you leave home it will not be possible to acquire anything else—either from home or from a store. Of course, radio, antennas, documents and personal effects are very broad and general classifications, but everything needed could be lumped together under one of these. Let us take them one at a time.

The Radio

Selection of the rig to be used is quite important, especially if there is any doubt about the stability of the current source to be used. The power supply should have a tapped transformer in the not-unlikely event that the voltage source will be somewhat different than the standard choices of 120/240 volts offered by some manufacturers. (From personal experience, using a Yaesu FT-101B in The Gambia in 1978: The power source was an old, old 208-volt generator, and there was a significant voltage drop along the line. My VOM read 190 volts! The Yaesu's power output on 160 meters, which was marginal to begin with, was about 20 watts. By careful selection of the voltage taps, however, normal operation was restored.)

No one in their right mind would trust an unprotected radio to the cargo hold of even the finest, most conscientious air carrier. The radio then automatically becomes carry-on luggage. Very seldom does carry-on luggage get weighed, so it is more of a weight saving to carry the heavier items and pack the lighter ones. Indeed, it is more prudent to carry a piece of equipment which has a built-in power supply than to carry the radio and pack the supply and VFO. Many of the Amateur Radio manufacturers are now building solid-state equipment which is compact enough to carry both radio and power supply in a large briefcase. A built-in second VFO or frequency memory circuit goes a long way toward conserving both space and weight. Vince Thompson, K5VT, and Erik Sjolund, SMØAGD, both fine DXpedition operators, have chosen this solid-state type of equipment (others prefer the older hybrid transceivers). Vince made his choice because he prefers to keep his radio out of sight so as to not invite questions from customs officials, while Erik likes the memory features. Erik discovered several years ago that when operating split frequency, it was easier to work the pileup when everyone was on the same frequency or spread out on several specific frequencies.

If the DXpedition will use more than one radio, these radios should be checked, while operating in close proximity to each other, for interference problems. ARRL Field Day is an excellent testing ground for this aspect of equipment performance. When this problem exists, it still might be possible to work around it by using the radio which suffers the most interference on the lower frequency or sometimes on a band which has no harmonic relationship to the band where the interfering radio is operating. Since the intermediate frequency (IF) mixer might be involved, the solution might be as simple as finding the right band combinations by trial and error.

''Bells and whistles'' might be nice to have on your home

Modification of the Hy-Gain TH3 Jr

To modify a TH3 Jr to fit into a 38-inch cardboard box, it is necessary to make a total of 10 cuts and splices. There are six 6-foot main element halves, two 6-foot boom halves and two trap assemblies, which are longer than 3 feet. A drill, an inexpensive tubing cutter, a hacksaw, a tape measure, an old pocket knife and a smooth file are all the tools required. Several feet of different size tubing is needed for making the splices.

Tubing and pipe are different. Tubing is measured on the outside diameter (OD) and pipe on the inside diameter (ID). Tubing dimensions are expressed in nominal wall thickness in thousandths of an inch, and pipe dimensions by its "schedule." The result is that tubing ID is determined by tubing OD minus its wall thickness times 2. Commercial antennas are constructed of tubing, so to modify one to fit the dimension of your box, you will need aluminum tubing only. You will cut any piece which is too long (greater than 36 inches) into a desired length. Then during assembly at your destination, splice the two pieces together with a short aluminum sleeve and clamp it, returning that antenna part to its original dimension.

A good choice of tubing is 6061T6. The "6061" refers to the alloy and the "T6" refers to the tensile strength. This alloy is very common and inexpensive, and T6 is an excellent strength for antennas. The wall thickness for inner splices is not critical, but it should not be really thin. If some outside splices are used (the TH3 Jr uses both inside and outside splices with the same size tubing), the wall thickness can be critical. Never use 0.058-in. or larger for outside splices. The tubing increases in size steps of 1/16 of an inch, or 0.0625-in., so if you want to splice a ¾-inch tube on the inside, select a piece of 5/8-inch tubing with about 0.035-in. wall.

It will be necessary to sand down the splice (a long and difficult task) or have it "turned down" in a machine shop when splicing thick-walled tubing. This is because there is a "wall" on each side of a diameter, so the ID of ¾-inch (0.058-in.) wall is calculated by 0.750-in. (¾-inch) minus 0.116-in. (2 × 0.058-in). You might note that 1/8 of an inch is 0.125-in., so it appears that 0.058-in. wall thickness will give a side clearance of 0.009-in. Not really; it gives 0.0045-in. on each side, and, since wall thicknesses are described as nominal, it might be less. Aluminum is soft, and this whole operation is not very precise anyway. The result of using 0.058-in. wall thickness is a probability that two pieces of aluminum will get stuck together. Once this happens, the chances of getting them apart are somewhere between slim and none. Most commercial antennas are made from 0.035-in. wall and should be no trouble to modify. The exception to this is the boom of the Hy-Gain TH3 Jr. It is thick-walled; and the two splice pieces must be reduced in OD.

Once the antenna and splicing sleeves have been cut, each piece must be "deburred." The small blade of an old pocket knife will do fine for the ID, but a smooth file will be required for the OD of the cut. Rough edges should be completely eliminated since they can cut you or cause the same sticking together of antenna parts. Slots will be made in the ends of antenna material (except the boom) to be spliced if it is an inside splice or in the splice material if it

is an outside splice. This will enable you to clamp the slotted aluminum tightly onto the other piece. Cut twice with a hacksaw very close to each other (1/16 inch) and pull the little piece between the cuts outward (not inward), creating a small slot. Pulling the piece outward will make it easier to deburr. Deburring the slots is more difficult, but do it thoroughly. (Alternately, it is just as good to drill screw holes through the antenna part and the splice, and attach them with screws; but then each splice end becomes "matched," and it is necessary to do something to identify each antenna piece and its matching splice. Drilled splices can need deburring more so than slotted splices, if the drill size was too small for the screw size you use for the job.)

It will be preferable to splice the boom sections using screws, since the thickness of the material makes it difficult to compress. Simply insert the splice half way into one of the boom pieces to be joined, and drill a hole through both pieces. Install a sheetmetal screw in the hole. Next, slide the other piece of boom over the splice material, and drill a hole through that section of boom and the splice. Lock these together with another sheetmetal screw.

Aluminum tubing can be purchased several ways. Occasionally, ads in the ham magazines will feature a company selling antenna aluminum to the public. Otherwise, look in the Yellow Pages of the telephone book under "Metals." If you choose slotted splices, the local auto parts or hardware store will be able to supply you with hose clamps. If you choose a Hy-Gain antenna, replace the supplied clamps with hose clamps.

Using the same techniques, a short mast can easily be constructed. Stores that sell tubing usually have a minimum, so there can be an added incentive to building yourself a 12-foot or so mast. This gives you something to attach to another structure and costs little in space and weight.

It is advisable to splice mast sections by screws. This will insure that the mast sections don't slide together under compression, causing a binding problem on disassembly and future use. Some of those spare clamps can readily attach your mast to a railing or flag pole too large for your antenna boom-to-mast clamp.

station radio, but an excess of internal options can complicate your life at a DX location which has a poor earth (or ground) connection for the electrical power source. Most of the rest of the world uses 220 volts ac which can transfer power without the third wire as is required in the US. It is common for technicians in developing nations to quit wiring as soon as power is established and completely ignore that very important third wire.

The Antennas

While it is not the purpose of this chapter to endorse any particular product, some (either by accident or design) are more ideally suited for DXpeditions. Therefore, the first antennas discussed will be triband beams. Two different models were tried simply because of their small size and their modest acquisition cost at a local hamfest. These antennas are the Mosley TA-33 Jr and the Hy-Gain TH3 Jr. No electrical comparison was made, and the assumption is that they are equivalent antennas. The way both of them come from the factory is not very acceptable for a DXpedition, especially one on a budget, though.

The longest pieces of both antennas are 6 feet or 72 inches.

The container for this antenna should stay under a total dimension of 62 inches to avoid airline excess baggage charges. Already the length is 10 inches too long, without consideration for width and height. A box 38 inches long, 12 inches wide and 12 inches high would meet this 62-inch dimension requirement, and this would require only one splice in each section of antenna longer than 3 feet or 36 inches. This allows for 1-inch-thick ends of the box. (A box this size is large enough to hold other luggage or equipment as well as the antenna.) The Mosley antenna requires several more splices than the Hy-Gain antenna, which means more work at the operating site; it also needs more different sizes of splicing material. The detailed instructions included in this chapter (see accompanying sidebar) are for modifying the Hy-Gain TH3 Jr. The two Mosley TA-33 Jr antennas modified in this manner have found their way to the Caribbean Sea and Indian Ocean areas and are serving Amateur Radio well there.

You can see in the picture that Hy-Gain used a total of eight 6-foot-long pieces of aluminum in the construction of their antenna. This is probably to take advantage of the fact that aluminum tubing is sold in 12-foot lengths. To modify any antenna with splices requires a few simple hand tools and only a moderate amount of knowledge of tubing. For a box in which to put that antenna, you might try a commercial box-manufacturing firm. Look in the Yellow Pages of the phone book under "Boxes."

Any antenna constructed of aluminum tubing can be modified easily to fit standard luggage dimensions. The most common variety, aside from Yagis, is the vertical. Probably the easiest to adapt and smallest would be the trap vertical. These antennas certainly have their place, but it is probably not on a DXpedition. Other types of verticals might perform well with a suitable ground plane, but unless installing dipoles and a beam is not possible, a trap vertical is not recommended.

Simplest and easiest to build, hardest to see and a moderately good performer is the trusty dipole. Refer to *The ARRL Antenna Book* for a wide variety of inexpensive wire antennas. A ZL-Special would make an excellent rotatable antenna for a DXpedition, especially if bamboo flourishes at your destination. An exceptionally good antenna is a three-element beam constructed of wire; though probably not rotatable, it will perform almost as well as one made of aluminum tubing. An inverted L antenna should work nicely on lower frequency bands, but necessitates your taking along an impedance-matching device. There are several small antenna tuners on the market, but this represents still another package to take along.

Any wire antennas taken on a DXpedition should be made of insulated wire. Insulation will not adversely affect their performance, but you might get your antenna much closer to high-voltage electrical wires than you would under saner circumstances. Additionally, uninsulated antenna wire could hardly double for anything else except guy wire, where insulated wire could have several alternate uses. Plan to take some extra wire also; you might need to construct another antenna or add another guy wire.

Miscellaneous Supplies

Plenty of coaxial cable (figure on about 100 feet per antenna) is recommended for several reasons. It might be necessary to locate the antennas quite some distance from the actual station. If more than one radio is operating at the same time and there is an interference problem, it might help to roll an RF choke in the coax at several places. This is accomplished by simply rolling a coil of six to eight turns of coax about a foot in diameter. (Antenna currents along the outside of the feed line can cause an interference problem, if only by having the effect of bringing the RF field closer to the other transceiver.) The last and possibly the most important reason is the possibility that you might meet a local amateur who has great need for such a simple commodity!

Miscellaneous supplies and equipment should include tools to assemble everything you are taking with you, connectors for joining two coaxial fittings, extension cord(s), a multiple electrical outlet, a foreign voltage converter (with adapter plugs), wire cutters, a pocket knife, tools for opening up your radio, your radio's owner's manual, a 220-volt soldering iron (120-volt will do if you absolutely can't find a 220-volt iron and are very careful to unplug it just as soon as it gets hot enough to melt solder), solder, nylon cord (to support wire antennas and to turn your triband beam), strapping tape, several hose clamps (a variety of sizes, along with the strapping tape can be used to secure antennas, lights and numerous other things), a Boy Scout compass and a map (a great circle centered on your destination might be hard to find so a Polar projection will do nicely. To find approximate directions, just orient it with your location directly underneath the North Pole).

Documents

The necessary and nice-to-have papers are relatively few, but as usual there are some simple items which might make things go much smoother for you.

Your passports are especially important. Yes, *passports*—if your journey includes travel into countries which are at odds with each other. The US State Department will issue more than one passport if you need it. Reserve your short-term passport for entry into countries such as the Republic of South Africa (RSA). (Never present a passport with an RSA stamp or visa at almost any other African nation's border or embassy.)

Try to have all of the visas for all of the countries you might visit before you go. There are numerous visa services in both Washington, DC, and New York City, which can assist you and expedite the handling of your papers. These companies can "walk through" a passport application in days instead of the weeks it normally takes. They do charge a fee for each service they offer. Visa service companies normally have travel agents as their clients, but they will take individuals, if you are willing to pay up front.

Extra passport pictures can make the difference if you wish to alter your travel plans in mid-stream or if it was not possible to get all of the necessary visas before beginning your travel. For example, suppose you planned a trip to Nigeria (5N) and you entered the country, but then left Nigeria for another country. If you wanted to return to Nigeria, you would need an additional Nigerian visa since they don't (at this writing) give multiple-entry visas. Also, never pack your passport in your suitcase or leave it in your hotel room.

Your amateur licenses and several photocopies of them should be with your important papers. If all goes well, you will have obtained operating permission before beginning a DXpedition. Take several photocopies of whatever you have. When customs inspectors see your equipment, they will also want to see some papers. They will go away with your papers, and since you just got there, you might not be able to find or identify the person to whom you gave them. You won't worry as much if you have more copies. (Photocopy machines are a rarity in parts of the world generally associated with DXpeditions.)

The Magic Adapter Cord

Several years ago, I acquired a new electronic keyer. It is small and rugged, making it ideal for DXpeditions. A telephone call to the manufacturer led to the discovery that they did not have a 220-volt power supply available for it.

In my search for information about the country of Chad, N4HX (previously N4HX/TT8) advised that a DXpedition there might have to be totally self-sufficient. It was decided to purchase a generator in Nigeria; therefore, its output would be 220 volts. Lamps might be needed, but the only inexpensive, lightweight cord available in the US are of the 120-volt variety.

My solution to this quandary is admittedly cheap and dirty, but it works quite well. I took two extension cords, cut off the plugs, and then wired them back together in a "Frankenstein" manner. One plug would supply 220 volts in series to both 120-volt socket heads. One wire from each side of both socket heads is connected together where the plug had been attached. Each remaining wire of the two socket heads is connected to a side of the plug, which then plugs into the 220-volt power source; the result being that neither socket head will work unless something is plugged into the other socket head. The only restriction here is that whatever is plugged into one socket head must be equal or nearly equal to one-half the total load. In this manner, two 120-volt items will share the 220 volts with no ill effects.

Using Ohm's law, it's evident that a 120-volt, 60-watt light bulb will draw 0.5 ampere. Two bulbs in series still draw 0.5 ampere, reducing the power consumed in each by one-half. Therefore, if operated from 220 volts, they will use nearly the same amount of power as if they were in parallel on 120 volts. A nice safety feature is that when one unit is unplugged, there is no danger, since doing so removes the voltage from the other unit.

Ohm's law can also be used to understand why the keyer still works. It is placed in parallel with one of the series-connected 60-watt bulbs, since double-socket extension cords are used. The keyer is described by the manufacturer as requiring anything between 9 and

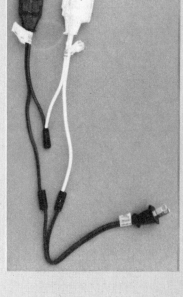

15 volts. This tolerance is easily met in the described fashion. Beware: If you use multiple-socket extension cords, the fail-safe feature is defeated when a third item is added. Now if the wrong item is unplugged first (the bulb which is in parallel with the keyer), a possible destructive voltage is applied to the most expensive item in the chain—the keyer. In all probability, the other bulb would burn out before the transformer since it glows so-o-o bright!

The resistance of a 120-volt, 60-watt light bulb is 240 ohms, while the resistance of a 120-volt, 8-watt keyer power supply is a whopping 1791 ohms. The series resistance of the parallel circuit is approximately 212 ohms. The single series bulb drops about 117 volts and uses 57 watts. The remaining bulb and keyer have to share only 103 volts and consume about 44 watts and 6 watts, respectively. The keyer is operating from just a little over 10 volts (assuming the provided transformer has a ratio of 10 to 1). A more even balance could be had by using an 80-watt bulb in series with the combination of the 60-watt bulb and 8-watt keyer, but then this additional factor would require more care and more spares.

Your International Certificate of Vaccination (which you should obtain from your local health department), if lost, might mean that you would have to be reinoculated. It is theorized that dirty medical needles in Africa are responsible for the spread of disease in that continent! You do not want to have to be reinoculated!

Personal Effects

Your personal effects include all of your clothing, which should be appropriate to the climate where you are heading. The desert requires long sleeves, not short, to protect your arms from sunburn in the daytime and keep you warm at night (deserts get very cold). The only price you pay is a little sweat. Cotton is preferable to polyester for coolness and comfort. Revealing or skimpy clothing for women is not recommended for travel in Islamic countries (a woman in a conservative one-piece bathing suit is considered to be nude). Always take very comfortable and broken-in walking shoes.

Immersion water heaters can be of great service to you. Don't expect to boil water for 30 minutes with them, but it is quite nice to make instant coffee or instant soup which you should also take along. If no 220-volt models are to be found, it is possible to put two 120-volt heaters in series with no ill effects. (This is a simple trick using the adapter cord described in the accompanying sidebar.)

In addition to instant drinks and soups, a small amount of snacks such as granola bars might be useful in case

mealtime coincides with a band opening that is too good to miss. Food might just not be affordable because of artificially controlled exchange rates, or, as in some countries, there may simply not be enough food. If a person can go without any food for a couple of months and survive, then you should have no major difficulty living on instant soup, powdered milk and granola bars for a couple of weeks.

A small battery-operated or key-wound alarm clock will help you keep your log accurately, help you meet schedules and/or avoid sleeping through band openings, and keep you from missing your plane when it comes time to go home! If your clock operates off of ac voltage, make sure that it has its own crystal time base. (Clocks which use the line voltage's 60-cycle frequency are slightly better than useless if the power source is an overloaded 50-cycle generator.) A wristwatch alarm might not be loud enough to wake a tired and weary DXer.

Here is a list of other items which ought to be considered for taking along as well: toilet tissue, laundry powder, a universal bathtub stopper (flat round rubber), eating utensils, candles, matches (or disposable lighter), flashlight and spare batteries (or disposable flashlights), water-purification tablets and a couple of collapsible water containers. Many of the things you might need will be found where camping supplies are sold in your hometown. Of course, the rarer your destination, the more important the list becomes.

Plenty of logbooks or logsheets are a must! You might want

Countries with which the US Shares Reciprocal Operating Agreements

A2	Botswana	HI	Dominican	OZ	Denmark	YB	Indonesia
C6	Bahamas		Republic	PA	Netherlands	YN	Nicaragua
CE	Chile	HK	Colombia	PJ	Netherlands	YU	Yugoslavia
CP	Bolivia	HP	Panama		Antilles	YV	Venezuela
CT	Portugal	HR	Honduras	PY	Brazil	YS	El Salvador
CX	Uruguay	I	Italy	PZ	Suriname	ZL	New Zealand
DL	Fed Rep of	J3	Grenada	SM	Sweden	ZP	Paraguay
	Germany	J6	St Lucia	SV	Greece	ZS	South Africa
DU	Philippines	JA	Japan	T2	Tuvalu	3A	Monaco
EA	Spain	JY	Jordan	T3	Kiribati	3D	Fiji
EI	Ireland	LA	Norway	TF	Iceland	4X	Israel
EL	Liberia	LU	Argentina	TG	Guatemala	5B	Cyprus
F	France*	LX	Luxembourg	TI	Costa Rica	6Y	Jamaica
G	United Kingdom**	OA	Peru	V3	Belize	8P	Barbados
H4	Solomon Islands	OE	Austria	VE	Canada	8R	Guyana
HB	Switzerland	OH	Finland	VK	Australia	9K	Kuwait
HC	Ecuador	ON	Belgium	VU	India	9L	Sierra Leone
HH	Haiti					9Y	Trinidad & Tobago

*includes all French overseas departments/territories
**includes the following: J7, J8, V2, VP2M, VP2V, VP5, VP8, VP9, VS6, YJ, ZB2, ZD7, ZF.

to design your own logs, similar to a contest log, containing the minimum amount of information for the maximum number of contacts. Several cheap black (not blue) ink ballpoint pens will ensure that your logs will photocopy accurately, but using sharpened pencils is often preferred since this permits erasures and corrections while you're operating (make sure you obtain an inexpensive, battery-powered pencil sharpener to go with the pencils). And by photocopying your logs when you return, you and your QSL manager can each have a set.

TRAVEL AND ARRIVAL

As previously mentioned, a travel agent will help you immensely and will not charge for the services. The travel agent will be paid by the airlines and other service companies that receive your business. Some knowledge is recommended on your part to ensure that the agent provides the service you desire.

As discussed earlier, the 70 pounds maximum per bag is for flights to and from the US. This higher limitation is required by the US Government, while, outside the US, the limitation is set by the International Airline Transport Authority (IATA). Membership by airlines in IATA is voluntary, and it should be realized that excess-baggage charges are a significant portion of international airline revenues.

For these reasons, it might be impossible to avoid excess-baggage charges on your return trip from the DXpedition. Even checking the luggage straight through to the US might not be enough, unless only one carrier is used. Sometimes that doesn't work either. Plan on taking enough money with you to cover the possibility of excess-baggage charges, just in case.

A woman's purse would never be considered as "carry-on" or "hand" baggage. A giant purse will hold many breakable things such as a VSWR bridge, VOM, spare tubes, microphone—and her normal purse. A man's toiletry kit would be in the same category. What a perfect place to carry all of those heavy tools that would otherwise skyrocket those excess-baggage charges! It may be cumbersome and uncomfortable, but it will keep more cash in your pocket!

The fewer airlines and fewer connections involved in any trip, the less likely are travel-connected problems. It is also a good idea to pack in a balanced way, so that, if only one

piece of luggage arrives at your destination, you will have at least your minimum personal requirements and enough equipment to put one station on the air.

As risky as it might seem, it might be advisable to take at least part of your money in cash. There are places in the world where traveler's checks and credit cards are not honored. If you take a significant amount of cash, purchase a money belt or similar pickpocket-proof way of carrying the bulk of your money. Of course, keep out enough for taxis, tips and other incidental expenses. Several international airports in the Western world have bad reputations for thieves, and some third-world airports can be downright dangerous for the unwary.

Strapping tape wrapped tightly around your suitcase might keep thieves from looking inside. Protecting your person might require a little more imagination. One large infamous airport has been the scene of such a regularly occurring ruse that it is worthwhile to mention here. You are standing in the terminal and an official-looking person walks up and demands to see some identification. Dutifully you show your passport and other papers. The person examines them carefully, returns them and leaves. (Or the person reads your name from a luggage tag.) A few minutes later another person arrives and asks if you are Mr, Mrs or Ms (insert your name) and, when given an affirmative answer, says that he has been sent to pick you up. If this person is legitimate, then he will have something from the party who was supposed to meet you. If he can't produce something to satisfy you, then don't go with him; he is probably a thief, who obtained your name from the "official."

Don't misunderstand: International travel is not dangerous in and of itself. There are some hot spots around the world to be avoided, and many places where close attention should be paid to activities around you. It should also be noted that it is highly illegal to take photographs at the airports of nearly all of the so-called developing nations. Many of these countries are overly sensitive about their security and about possible spies. Carrying a radio makes you doubly suspicious.

Upon arrival at your destination, you and your equipment will have to go through customs and immigration. If operating permission has not already been secured, be prepared to leave your radio equipment "in bond" at the airport or some

(text continues on page 6-13)

Electric Current Abroad

[Information courtesy of the US Department of Commerce.]

The characteristics of electric current—type (alternating or direct), number of phases, frequency (hertz) and voltage—found in major foreign cities are listed in this report. In addition, the stability of the frequency and the number of wires to a commercial or residential installation are given where available.

The current characteristics and other data furnished relate to domestic and commercial service only. It does not include special commercial installations involving relatively high voltage requirements, nor does it refer to any industrial installations.

Persons who are planning to use or ship appliances abroad should acquaint themselves with the characteristics of the electric supply available in the area in which the appliance is to be used. It may be less expensive to buy a new appliance than to purchase the auxiliary equipment needed to make presently owned appliances usable. A transformer may be obtained locally to correct the voltage if required; however, if the operation of an appliance will require exact timing or speed and if the frequency of the foreign electricity supply is other than that for which a United States appliance has been designed, it is advisable that an appliance designed especially for the foreign frequency to be used. Auxiliary equipment to change frequency is too bulky and expensive for the average residential or commercial installation.

Some foreign hotels have a special circuit, providing approximately 120 volts, for the convenience of guests using electric shavers and other low-wattage appliances made in the US. Such circuits have markers at the convenience outlets indicating the voltage.

Among the nominal voltages indicated in this report, the lower voltages shown are used primarily for lighting and smaller appliances, while the higher voltages are used primarily for air conditioners, heating and other large appliances.

Readers are reminded that the list of characteristics presented here were compiled over a period of months from a large number of sources. There is, consequently, some possibility of errors and omissions for which the Department of Commerce cannot assume responsibility.

Readers are further reminded that the information presented here should not be taken as final in the case of industrial or highly specialized commercial installations. The Department of Commerce regrets that it is impossible for it to maintain complete data on every foreign industrial installation. It is recommended that for special equipment required for commercial use or any heavy equipment required for industrial use, the current characteristics for the area of installation be obtained from the end user.

[Editor's Note: There are often different *sizes* in use within these plug types, and the traveler might even find a mixture of these types being used in developing countries. Former French colonies tend to use French-style plugs. Voltage adapters that cover almost anything are usually on sale at major European airports.]

Key to Terms Used

Type of Current—ac indicates alternating current and dc, direct current.

Frequency—Shown in number of hertz (cycles per second). Even if voltages are similar, a 60-hertz US clock or tape recorder will not function properly on 50-hertz current.

Number of phases—1 and 3 are the conventional phases which may be available.

Nominal voltages—Direct current nominal voltages, for example, are: 110/220 and 120/240. The lower voltage is always one-half of the higher voltage. On a direct current installation, the lower voltage requires 2 wires while the higher voltage requires 3 wires.

Alternating current nominal voltage: Alternating current is normally distributed either through 3-phase wye ("star") or delta ("triangle"), 4-wire secondary distribution systems. In the wye or star distribution system the nominal voltage examples are: 120/208, 127/220, 220/380 and 230/400. The

TYPE A
FLAT BLADE ATTACHMENT PLUG

TYPE B
FLAT BLADES WITH ROUND GROUNDING PIN

TYPE C
ROUND PIN ATTACHMENT PLUG

TYPE D
ROUND PINS WITH GROUND

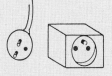

TYPE E
ROUND PIN PLUG AND RECEPTACLE WITH MALE GROUNDING PIN

TYPE F
"SCHUKO" PLUG AND RECEPTACLE WITH SIDE GROUNDING CONTACTS

TYPE G
RECTANGULAR BLADE PLUG

TYPE H
OBLIQUE FLAT BLADES

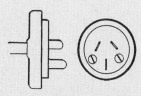

TYPE I
OBLIQUE FLAT BLADES WITH GROUND

TYPE J
OBLIQUE FLAT BLADES WITH GROUND

higher voltage is 1.732 (the square root of 3) times the lower voltage. In a delta or triangle system, 110/220 and 230/460 are examples of nominal voltages. The higher voltage is always double the lower voltage.

The higher voltage is obtained by using 2- or 3-phase wires and the neutral wire while the lower voltage is the voltage between the neutral wire and one phase wire. The higher voltage may be single or 3 phase while the lower voltage is always single phase and used primarily for lighting and for small appliances.

In this report, the term nominal voltage is used to denote the reported voltage in use in the majority of residential and commercial establishments in the city or area named.

Type of attachment plug in use—Attachment plugs used throughout the world come in various forms, dimensions and configurations too numerous to describe in this report. This report does, however, attempt to point out the basic and most commonly used types of plugs by country. Adapters may be purchased to change from the American type to other types. The availability of these adapters is shown for each country immediately after the key to the type of plug used.

Number of wires to the consumer—The number of wires which may be used by the consumer is shown. Normally, a single phase, 220/380 volt system or 127/220 system will have two wires if only the lower voltage is available (one phase wire and the neutral). It will have three wires if both the higher and lower voltages are available (two phase wires and the neutral) and where three phase motors will be used, four wires will be available for the higher voltage (the three phases wires and the neutral wire).

Frequency stability—A "Yes" in the column indicates that the frequency stability is stable and that service interruptions are rare.

Types of Electric Plugs by Country

Country	Plug type shown	Adapters readily available
Afghanistan	D	Yes
Algeria	C & D	No
American Samoa	A,B,F & I	Yes
Angola	C	—
Antigua	A,B & G	Yes
Argentina	C & G	—
Australia	I	No
Austria	F	No
Azores	C & D	Yes
Bahamas	A & B	Yes
Bahrain	D & G	Yes
Bangladesh	C & D	No
Barbados	A & B	—
Belgium	C & E	No
Belize	A & B	—
Benin	D	—
Bermuda	A,B,G & I	No
Bolivia	A,B,C & F	—
Botswana	D & G	Yes
Brazil	A,B,C,D,E,F, G,H,I & J	Yes
Bulgaria	C & F	No
Burma	D & G	—
Burundi	C,E & F	No
Cambodia [Kampuchea]	No information	
Cameroon	C & E	—
Canada	A & B	—
Canary Islands	C & E	Yes
Cape Verde, Rep of	C & F	No
Cayman Islands	A & B	—
Central African Rep	C	—
Chad	D,E, & F	No
Channel Islands	C & G	Yes
Chile	C	Yes
China, People's Rep of	C & J	No
Colombia	A & C	Yes
Congo	C	—
Costa Rica	A,D,I & J	Yes
Cyprus	G	Yes
Czechoslovakia	E	No
Denmark	C	No
Djibouti	C & E	No
Dominica	G	—
Dominican Rep	A & J	No
Ecuador	A & C	—
Egypt	C	No
El Salvador	A,B,G,I & J	No
England	C & G	Yes
Equatorial Guinea	C	No
Ethiopia	C,D & F	—
Faeroe Islands	C	No
Fiji	I	Yes
Finland	C & F	No
France	C,E,F, & G	Yes
French Guiana	C,D & E	No
Gabon	C & E	—
Gambia, The	G	Yes
German Dem Rep	C & F	Yes
Germany, Fed Rep of	F	No
Ghana	C,D & G	No
Gibraltar	C & G	Yes
Greece	C,D & F	No
Greenland	C	No
Grenada	C,D & G	Yes
Guadeloupe	C,D & E	Yes
Guam	A & B	—
Guatemala	A & I	—
Guinea	C & E	No
Guinea-Bissau	C,D & F	No
Guyana	A,B,C,D & G	Yes
Haiti	A,B,I & J	—
Honduras	A & B	—
Hong Kong	D	No
Hungary	F	—
Iceland	C & F	Yes
India	C & D	—
Indonesia	C,E, & F	Yes
Iran	No Information	—
Iraq	C,D & G	No
Ireland	F & G	—
Isle of Man	C & G	Yes
Israel	C,D & J	Yes

Electric Current Abroad—Characteristics

This is a simplified version (with footnotes omitted) of the detailed listing appearing in *Electric Current Abroad,* published by the US Department of Commerce. For further particulars (all such information is subject to change at any moment), obtain the latest edition of this booklet from the Commerce Department.

[1]	[2]	[3]	[4]	[5]	[6]
Afghanistan					
Kabul	ac 50	1,3	220/380	2,4	Yes
Algeria					
Algiers	ac 50	1,3	127/220 220/380 120/240	2,4	Yes
American Samoa	ac 60	1,3	240/480	2,3,4	Yes
Angola	ac 50	1,3	220/380	2,4	Yes
Antigua	ac 60	1,3	230/400	2,3,4	Yes
Argentina					
Buenos Aires	ac 50	1,3	220/380	2,4	Yes
	dc		220/440	2,3	Yes
Australia					
Canberra	ac 50	1,3	240/415	2,3,4	Yes

[1]	[2]	[3]	[4]	[5]	[6]
Austria					
Vienna	ac 50	1,3	220/380	2,4	Yes
Azores	ac 50	1,3	220/380	2,3,4	Yes
Bahamas					
Nassau	ac 60	1,3	120/240 120/208	2,3,4	Yes
Bahrain	ac 50	1,3	230/400	2,3,4	Yes
Bangladesh					
Dacca	ac 50	1,3	220/380	3,4	No
Barbados					
Bridgetown	ac 50	1,3	115/230 115/200	3,4	Yes
Belgium					
Brussels	ac 50	1,3	127/220 220/380	2,3,4	Yes
Belize	ac 60	1,3	110/220	2,3,4	Yes
Benin	ac 50	1,3	220/380	2,4	Yes
Bermuda	ac 60	1,3	120/240 120/208	2,3,4	Yes

Country	Plug type shown	Adapters readily available	Country	Plug type shown	Adapters readily available
Italy	C & F	Yes	Portugal	C & D	—
Ivory Coast	C	—	Puerto Rico	A & B	—
Jamaica	A & B	—	Qatar	D & G	Yes
Japan	A	—	Romania	C & F	—
Jerusalem	C,D & J	Yes	Rwanda	C & J	No
Jordan	C,F & G	Yes	St Kitts-Nevis	D & G	Yes
Kenya	D & G	—	St Lucia	G	—
Korea	A & F	Yes	St Vincent	G	—
Kuwait	C,D & G	Yes	Saudi Arabia	A,B,C,D,E,F,G, H,I & J	No
Laos	A & C	Yes			
Lebanon	C	—	Scotland	C & G	Yes
Lesotho	C	No	Senegal	C & F	No
Liberia	A & G	No	Seychelles	D & G	No
Libya	D	—	Sierra Leone	D & G	—
Luxembourg	C & F	Yes	Singapore	C & G	Yes
Macao	C & D	—	Somalia	C & D	Yes
Madagascar	C & E	No	South Africa, and		
Madeira	C & D	Yes	Southwest Africa/Namibia	D	—
Majorca Island	C & E	Yes	Spain	C & E	Yes
Malawi	G	—	Sri Lanka	D	No
Malaysia	G	Yes	Sudan	C,F & G	No
Maldives	D	—	Suriname	C & F	Yes
Mali	C & E	—	Swaziland	D	—
Malta	G	No	Sweden	C & F	No
Martinique	C,D & E	Yes	Switzerland	C & F	No
Mauritania	C	—	Syria	C	Yes
Mauritius	C,D & G	—	Tahiti	A	—
Mexico	A	—	Taiwan	A,B,H,I & J	Yes
Monaco	C,D,E & F	Yes	Tanzania	D & G	—
Montserrat	A,B & G	Yes	Thailand	A & C	No
Morocco	C,D,E & F	—	Togo	C & E	—
Mozambique	C & F	—	Tonga	D & I	Yes
Nepal	D	Yes	Trinidad & Tobago	A,B,D,G & I	Yes
Netherlands	C & F	No	Tunisia	C	—
Netherlands Antilles	A,B,C,D & F	Yes	Turkey	C,E & F	—
New Caledonia	No Information		Uganda	G	—
New Zealand	I	Yes	USSR	C & I	No
Nicaragua	A	—	United Arab Emirates	D & G	Yes
Niger	C	No	Upper Volta		
Nigeria	D & G	—	[Burkina Faso]	C	—
Northern Ireland	C & G	Yes	Uruguay	C & I	Yes
Norway	C & F	No	Venezuela	I & J	—
Okinawa	A	—	Virgin Islands (American)	A & B	—
Oman	D & G	Yes	Wales C & G		Yes
Pakistan	C & D	No	Western Samoa	No Information	
Panama	A & B	—	Yemen (Aden)	A & D	—
Papua New Guinea	H & I	No	Yemen Arab Rep	C & D	—
Paraguay	C	No	Yugoslavia	C & F	—
Peru	A	—	Zaire	E	Yes
Philippines	A,B,C,D & J	Yes	Zambia	G	—
Poland	C,E & F	No	Zimbabwe	D & G	No

[1]	[2]	[3]	[4]	[5]	[6]	[1]	[2]	[3]	[4]	[5]	[6]
Bolivia						Canary Islands	ac 50	1,3	127/220 220/380	2,3,4	Yes
La Paz	ac 50	1,3	115/230 220/380	2,3	Yes	Cape Verde	ac 50	1,3	220/380	2,3,4	No
Botswana						Cayman Islands	ac 60	1,3	120/240	2,3	Yes
Gaborone	ac 50	1,3	220/380	2,4	Yes	Central African Republic					
Brazil						Bangui	ac 50	1,3	220/380	2,4	Yes
Brazilia	ac 60	1,3	220/380	2,3,4	Yes	Chad	ac 50	1,3	220/380	2,4	No
Bulgaria						Channel Islands	ac 50	1,3	240/415	2,4	Yes
Sofia	ac 50	1,3	220/380	2,4	No	Chile					
Burkina Faso						Santiago	ac 50	1,3	220/380	2,3,4	Yes
Ouagadougou	ac 50	1,3	220/380	2,4	No	China, People's Rep of					
Burma						Beijing	ac 50	1,3	220/380	2,4	No
Rangoon	ac 50	1,3	230/400	2,4	No	Colombia					
Burundi	ac 50	1,3	220/380	2,4	No	Bogota	ac 60	1,3	150/240	2,3,4	Yes
Cambodia [Kampuchea]							ac 60	1,3	110/220	2,3,4	Yes
Phnom-Penh	ac 50	1,3	220/380	2,3,4	No	Congo, Rep of					
Cameroon						Brazzaville	ac 50	1,3	220/380	2,4	Yes
Yaounde	ac 50	1,3	127/220 220/380	2,4	Yes	Costa Rica					
						San Jose	ac 60	1,3	120/240	2,3,4	Yes

[1]	[2]	[3]	[4]	[5]	[6]
Cyprus	ac 50	1,3	240/415	2,4	Yes
Czechoslovakia					
Prague	ac 50	1,3	220/380	2,3,4	Yes
Denmark					
Copenhagen	ac 50	1,3	220/380	2,3,4	Yes
Djibouti	ac 50	1,3	220/380	2,4	Yes
Dominica	ac 50	1,3	230/400	2,4	N.A.
Dominican Republic					
Santo Domingo	ac 60	1,3	110/220	2,3	Yes
Ecuador					
Quito	ac 60	1,3	120/208 127/220	2,3,4	Yes
Egypt					
Cairo	ac 50	1,3	220/380	2,3,4	No
El Salvador					
San Salvador	ac 60	1,3	115/230	2,3	Yes
Equatorial Guinea	ac 50	1	220	2	No
Ethiopia					
Addis Ababa	ac 50	1,3	220/380	2,4	Yes
Faeroe Islands	ac 50	1,3	220/380	2,3,4	Yes
Fiji	ac 50	1,3	240/415	2,3,4	Yes
Finland					
Helsinki	ac 50	1,3	220/380	2,4	Yes
France					
Paris	ac 50	1,3	115/230	2,4	Yes
French Guiana					
Cayenne	ac 50	1,3	220/380	2,3,4	No
Gabon					
Libreville	ac 50	1,3	220/380	2,4	Yes
Gambia, The					
Banjul	ac 50	1,3	220/380	2,4	No
German Democratic Rep					
Berlin	ac 50	1,3	220/380	2,4	No
Germany, Federal Rep of					
Bonn	ac 50	1,3	220/380	2,4	Yes
Ghana					
Accra	ac 50	1,3	220/440	2,3,4	No
Gibraltar	ac 50	1,3	240/415	2,4	Yes
Greece					
Athens	ac 50	1,3	220/380	2,4	Yes
Greenland					
Godthaab	ac 50	1,3	220/380	2,3,4	Yes
Grenada	ac 50	1,3	230/400	2,4	N.A.
Guadeloupe	ac 50	1,3	220/380	2,3,4	Yes
Guam	ac 60	1,3	110/220 120/208	3,4	Yes
Guatemala					
Guatemala City	ac 60	1,3	120/240	2,3,4	Yes
Guinea					
Conakry	ac 50	1,3	220/380	2,3,4	No
Guinea-Bissau	ac 50	1,3	220/380	2,3,4	No
Guyana					
Georgetown	ac 50	1,3	110/220	2,3,4	Yes
Haiti					
Port-au-Prince	ac 60	1,3	110/220	2,3,4	No
Honduras					
Tegucigalpa	ac 60	1,3	110/220	2,3	No
Hong Kong	ac 50	1,3	200/346	2,3,4	Yes
Hungary					
Budapest	ac 50	1,3	220/380	2,4	Yes
Iceland					
Reykjavik	ac 50	1,3	220/380	2,3,4	Yes
India					
New Delhi	ac 50	1,3	230/400 230/415	2,4	Yes
Indonesia					
Jakarta	ac 50	1,3	220/380	2,4	Yes
Iran					
Tehran	ac 50	1,3	220/380	2,3,4	Yes
Iraq					
Baghdad	ac 50	1,3	220/380	2,4	Yes
Ireland (Eire)					
Dublin	ac 50	1,3	220/380	2,4	Yes
Isle of Man					
Douglas	ac 50	1,3	240/415	2,4	Yes
Israel					
Tel Aviv	ac 50	1,3	230/400	2,4	Yes
Italy					
Rome	ac 50	1,3	127/220 220/380	2,4	Yes
Ivory Coast					
Abidjan	ac 50	1,3	220/380	2,4	Yes
Jamaica					
Kingston	ac 50	1,3	110/220	2,3,4	Yes
Japan					
Tokyo	ac 50	1,3	100/200	2,3	Yes
Jordan					
Amman	ac 50	1,3	220/380	2,3,4	Yes
Kenya					
Nairobi	ac 50	1,3	240/415	2,4	Yes
Korea	ac 60	1,2	220/380	2,3,4	Yes
Kuwait	ac 50	1,3	240/415	2,4	Yes
Laos					
Vientiane	ac 50	1,3	220/380	2,4	Yes
Lebanon					
Beirut	ac 50	1,3	110/190	2,4	No
Lesotho					
Maseru	ac 50	1,3	220/380	2,4	Yes
Liberia					
Monrovia	ac 60	1,3	120/240 120/208	2,3,4	No
Libya					
Tripoli	ac 50	1,3	127/220	2,4	Yes
Luxembourg	ac 50	1,3	120/208 220/380	3,4,5	Yes
Macao	ac 50	1,3	220/380	2,3	Yes
Madagascar					
Tananarive	ac 50	1,3	127/220 220/380	2,3,4	Yes
Madeira					
Funchal	ac 50	1,3	220/380	2,3,4	Yes
	dc		220/440	2,3	Yes
Majorca Island					
Palma de Majorca	ac 50	1,3	127/220 220/380	2,3,4	Yes
Malawi					
Blantyre	ac 50	1,3	230/400	3,4	Yes
Malaysia					
Kuala Lumpur	ac 50	1,3	240/415	2,4	Yes
Maldives	ac 50	1,3	230/400	2,4	Yes
Mali, Rep of					
Bamako	ac 50	1,3	220/380	2,4	No
Malta					
Valletta	ac 50	1,3	240/415	2,4	Yes
Mexico					
Mexico, DF	ac 60	1,3	127/220	2,3,4	No
Monaco					
Monte-Carlo	ac 50	1,3	127/220 220/380	2,4	Yes
Montserrat	ac 60	1,3	230/400	2,4	NA
Morocco					
Rabat	ac 50	1,3	127/220	2,4	Yes
Mozambique	ac 50	1,3	220/380	2,3,4	Yes
Nepal					
Kathmandu	ac 50	1,3	220/440	2,4	No
Netherlands					
The Hague	ac 50	1,3	220/380	2,4	Yes
Netherlands Antilles					
Curacao	ac 50	1,3	223/380	2,3,4	Yes
New Caledonia					
Noumea	ac 50	1,3	220/380	2,3,4	Yes
New Zealand					
Auckland	ac 50	1,3	230/400	2,3,4	Yes
Nicaragua					
Managua	ac 60	1,3	120/240	2,3,4	Yes
Niger					
Niamey	ac 50	1,3	220/380	2,3,4	No
Nigeria					
Lagos	ac 50	1,3	230/415	2,4	Yes
Norway					
Oslo	ac 50	1,3	230	2,3	Yes
Okinawa	ac 60	1	120/240 100/200	2,3	Yes
Oman					
Muscat	ac 50	1,3	240/415	2,4	Yes
Pakistan					
Rawalpindi	ac 50	1,3	230/400	2,3,4	No
Panama					
Panama City	ac 60	1,3	115/230 126/208	2,3,4	Yes
Papua New Guinea	ac 50	1,3	240/415	2,4	Yes
Paraguay					
Asuncion	ac 50	1,3	220/380	2,4	Yes

[1] country/city	[2]	[3]	[4]	[5]	[6]
Peru					
Lima	ac 60	1,3	220	2,3	Yes
Philippines					
Manila	ac 60	1,3	115/230 110/220	2,3,4	Yes
Poland					
Warsaw	ac 50	1,3	220/380	2,4	No
Portugal					
Lisbon	ac 50	1,3	220/380	2,3,4	Yes
Puerto Rico					
San Juan	ac 60	1,3	120/240	2,3,4	Yes
Qatar					
Doha	ac 50	1,3	240/415	2,4	No
Romania					
Bucharest	ac 50	1,3	220/380	2,4	Yes
Rwanda	ac 50	1,3	220/380	2,4	Yes
St Kitts and Nevis	ac 60	1,3	230/400	2,4	Yes
St Lucia	ac 50	1,3	240/416	2,4	NA
St Vincent	ac 50	1,3	230/400	2,4	NA
Saudi Arabia					
Jidda	ac 60	1,3	127/220	2,4	No
Senegal					
Dakar	ac 50	1,3	127/220	2,3,4	No
Seychelles	ac 50	1,3	240	2,3	Yes
Sierra Leone					
Freetown	ac 50	1,3	230/400	2,4	No
Singapore	ac 50	1,3	230/400	2,4	Yes
Somalia					
Mogadishu	ac 50	1,3	220/380	2,4	No
South Africa and Southwest Africa/Namibia					
Cape Town	ac 50	1,3	220/380	2,4	Yes
Spain					
Madrid	ac 50	1,3	127/220 220/380	2,3,4	Yes
Sri Lanka					
Colombo	ac 50	1,3	230/400	2,4	Yes
Sudan					
Khartoum	ac 50	1,3	240/415	2,4	Yes
Suriname	ac 60	1,3	115/230	2,3,4	Yes
Swaziland	ac 50	1,3	230/400	2,4	Yes
Sweden					
Stockholm	ac 50	1,3	220/380	2,3,4	Yes
Switzerland					
Bern	ac 50	1,3	220/380	2,3,4	Yes
Syria					
Damascus	ac 50	1,3	220/380	2,3	No
Tahiti					
Papeete	ac 60	1,3	127/220	2,3,4	No
Taiwan					
Taipei	ac 60	1,3	110/220	2,3,4	Yes
Tanzania					
Dar es Salaam	ac 50	1,3	230/400	2,3,4	Yes
Thailand					
Bangkok	ac 50	1,3	220/380	2,3,4	Yes
Togo					
Lome	ac 50	1,3	127/220 220/380	2,4	Yes
Tonga	ac 50	1,3	240/415	2,3,4	NA
Trinidad and Tobago	ac 60	1,3	115/230 230/400	2,3,4	Yes
Tunisia					
Tunis	ac 50	1,3	127/220 220/380	2,4	Yes
Turkey					
Ankara	ac 50	1,3	220/380	2,3,4	Yes
Uganda					
Kampala	ac 50	1,3	240/415	3,4	Yes
USSR					
Moscow	ac 50	1,3	220/380	NA	No
United Arab Emirates					
Abu Dhabi	ac 50	1,3	240/415	2,3,4	No
United Kingdom					
London	ac 50	1	240/480	2,3	Yes
		3	240/415	4	Yes
Wales	ac 50	1,3	240/415	2,4	Yes
Northern Ireland	ac 50	1,3	220/380 230/400	2,4	Yes
Uruguay					
Montevideo	ac 50	1,3	220	2,3	Yes
Venezuela					
Caracas	ac 60	1,3	120/240	2,3,4	Yes
Viet Nam					
Saigon-Cholon	ac 50	1,3	120/208 220/380	2,4	No
Virgin Islands (US)	ac 60	1,3	120/240	2,3,4	Yes
Western Samoa	ac 50	1,3	230/400	2,3,4	Yes
Yemen (Aden)	ac 50	1,3	230/400	2,4	Yes
Yemen (Arab Rep)	ac 50	1,3	220/380	2,4	No
Yugoslavia					
Belgrade	ac 50	1,3	220/380	2,4	Yes
Zaire, Rep of					
Kinshasa	ac 50	1,3	220/380	2,3,4	No
Zambia	ac 50	1,3	220/380	2,4	Yes
Zimbabwe					
Harare	ac 50	1,3	220/380	2,3,4	Yes

KEY
[1] country/city
[2] type and frequency of current
[3] number of phases
[4] nominal voltage
[5] number of wires
[6] frequency stability—stable enough for electric clocks

(continued from page 6-8)

customs warehouse. This is usually quite safe, as long as you get a receipt for whatever is left there. Some countries, such as the Bahama Islands, may require you to post a sizeable cash bond even though you have operating permission. This is to ensure that you are not importing merchandise to sell and are just trying to avoid import taxation. You won't be helping any local amateurs at all if you get caught illegally importing Amateur Radio equipment into their country!

Some countries, especially the poorer developing nations, have a problem with the authorities soliciting bribes. The residents of these countries should be entitled to good government, whether or not they get it. Don't contribute to the problem by paying bribes!

Many hotels and private residences in the Caribbean actively solicit hams to stay at their facilities for DXpedition vacations. One hotel in Freeport, Bahamas, has a ham radio suite on the top floor with specially drilled holes to accommodate coax from the rooftop. Some hotels will also gladly provide postcards for QSL cards. Almost all hotels will try their best to help you, as long as other guests are not disturbed and your antennas don't present a safety hazard.

Documentation and Station Assembly

If you received operating permission before your arrival, you should already have several extra copies. Many hotels will be far less reluctant to allow an Amateur Radio antenna installation on their roof if they are given a copy of the permission which they can present to any inquiring government security personnel. If the permission is to be granted after arrival, explain to the hotel what your plans are, emphasizing that nothing will be put up before permission is in hand. Most places find this quite acceptable. After receiving the permission, try to find a way to duplicate it. If it is not possible to photocopy the permission, then never let it leave your sight. Do not leave it in the hotel room. Chances are that it would be quite safe but, if not, then you might never get it replaced. Hotels, banks and your embassy are the three most likely places to find a photocopy machine. The US Embassy is generally not in the business of photocopying for tourists. However, if there is no other machine around, then the Embassy probably would not mind making a copy or two for you, since they would be interested in being apprised of your whereabouts.

Documentation for the ARRL DXCC Desk is required only if the DXpedition is to a country or place where there has previously been permission denied or which has been inactive for so long as to indicate that Amateur Radio is not permitted there. Some places are rare because of their inaccessibility rather than the difficulty in obtaining permission to operate. Acceptable documentation in these instances might include an original photograph showing members of the DXpedition at some clearly identifiable landmark that proves the DXpedition was really there. Receipts for passage or passport stamps are also good "documentation" for this purpose. See Chapter 8 for details on DXCC accreditation.

If operating permission is granted at the location and photocopies are obtained, it might not be a bad idea to just mail a copy right then to the DXCC Desk at ARRL HQ in Newington. That way the ARRL has a copy of the permission and a canceled stamp showing the date and place from which it was mailed. The HQ staff isn't concerned about documentation for operations from countries that normally permit Amateur Radio operations. You can have a great time operating from Jamaica, for example, and not have to worry about providing documentation to the ARRL DXCC Desk.

Another piece of documentation which could be of significance is a currency exchange transaction record if it is required in the particular country. Some nations protect their currency from the free trade market with a fixed rate of exchange. The purpose of this is to control their economy by preventing too much of their nation's wealth from leaving.

Because of this control, there will be people who have more of the local currency than they need and who want to exchange it for a currency which can be used in other countries. They will be willing to exchange it at a more favorable rate than the bank. Therefore, the country has rules requiring you to keep your copy of the transaction record until you leave the country. You can be made to account for all the money you brought into the country. You are supposed to have all the foreign exchange currency (dollars, pounds, yen, etc) which you declared upon entry minus that which shows up on the transaction record as having been exchanged. Some countries have very severe penalties for even first offenders. You certainly won't do yourself any good, or ham radio any good for that matter, if you get caught dealing in the black market merely to save a few dollars!

With permission in hand and the nod from the hotel, you must now assemble the station and get it on the air. Assuming there is no one else on the DXpedition or that you know locally to assist, it could be quite difficult to put up the antennas. One thing about developing countries is that human labor is usually plentiful and inexpensive. The hotel will probably be happy to provide you with young men for climbing trees and otherwise installing antennas. For this service, you will rightfully be expected to pay. Try to find out from the hotel manager or some other person familiar with the local economy an appropriate amount to pay. It is important that you pay a fair, even generous amount, but it could be even more important that you not create jealousies by paying too much.

OPERATING TECHNIQUES

Like it or not, the actions of the sought-after DX stations (DXpeditioners and residents alike) have a great and profound influence on the operating techniques of DXers all over the world. If you go on a DXpedition, you have a responsibility to make every effort to be a positive influence on the sport of DXing.

DXers will generally do whatever is necessary to work DX. A "good" DXer will draw the line at illegal or unethical operating practices. Some poor operating techniques get used so much that a gray area exists in the minds of many DXers as to what is or isn't ethical (good sportsmanship and fair play). It is easy to understand how this came about. Isn't it true that the object of any competition is to come away with the prize? DX operators should never "reward" poor operating since it reinforces undesirable behavior.

If a station became active from the mythical DXCC country of Outer Baldonia, and the operator of that station could be active only for one hour a day, there would be frenzy upon the bands. It would be impossible for that operator to work all the DXers who want to contact that country. How then will the competition be conducted? Within the minor limitations of the DXCC rules, the decision will be made by the DX station. Whatever DXers hear other people doing that works is what they will do. Therefore, when you come on the air from whatever DXCC country you choose for your DXpedition, the responsibility of good sportsmanship and fair play rests squarely in your hands. Possibly more important than the operating during that one operation, people will learn (or reinforce earlier learning) about how to work other DX stations. If all this chapter is able to do is start DX operators thinking about how their operating affects DXing, then it will be worthwhile.

The next few paragraphs are some thoughts and reasoning about preferences in operating practices. There are several methods of conducting a DX operation which might seem to be inconsistent with other ways. In fact, the different skill levels of the individual operators (or other circumstances) make what would be totally inappropriate for one ideal for the other. Because of this, there can be few hard and fast rules.

Finally, now after all of the research, licensing, planning, preparation, traveling and hassles, you have the first antenna in the air and it is time for the fun to begin! You rub your hands together with anticipation as you hurry back to your hotel room.

You flip on the switch, and your receiver leaps to life. In seconds, you are ready to operate. After making the initial contacts, a number of interested DXers have found you. Soon there are quite a few people calling you. You have a pileup. Now a decision must be made as to how to handle it.

Assuming that you have two VFOs or a frequency memory circuit that can function as a VFO, you simply say, "Okay everyone, I will transmit here and listen up 5 to 10 kHz." (On CW, you might send QSX UP 5.) Almost all DXers would prefer to listen on one frequency to the station they are working while actually transmitting on another. This makes it easier to hear the station being worked. Here are some suggestions regarding split-frequency operation:

1) Always try to operate within the band limitations of the DXers you are working. This way, if the DXer makes a mistake and calls on your frequency, he or she will not have transmitted "out of band." This automatically eliminates the interference from the "helpful" operator who feels a need to warn the out-of-band offender. It might also prevent that

DXer from getting into trouble with his or her own licensing authority.

2) Try to keep your transmit frequency as close as possible to the frequency or frequencies to which you are listening. This will make it easier for people who are spinning the dial between frequencies trying to work you without a separate VFO. Of course, stay far enough away so that your pileup doesn't interfere with your own transmissions.

3) It may even be a good idea to listen on only one frequency. Erik Sjolund, SMØAGD, discovered that on SSB the overlapping interference caused by stations above and below the station he was trying to hear was much worse than having all of the stations on one frequency or several separated specific frequencies. By having all the stations piled up in one spot, it is relatively easy to take the loudest stations first. The only difficulty results when there are several stations of equal or nearly equal signal strength.

4) The amount of total spectrum used should be reasonable. Never use more than you actually need. It would be hard to imagine a situation where more than 20 kHz would be needed. As an alternative to spreading the pileup out more, it might be worthwhile to reduce your own output power. This might reduce the number of people hearing your signal and thus reduce the size of the pileup. This certainly ought to be tried if the radio has only one VFO. Remember that non-DXers have a right to use the band as well. They probably won't realize what you did for them, but they will surely realize it if you don't do it for them!

5) It might become necessary to bring the pileup back to your frequency for a while in the event that adjacent activity starts to move in, keeping the pileup from hearing you.

6) The pileup can also be divided several ways. This is difficult or perhaps impossible to do and maintain good sportsmanship and fair play. The most common method is to go by call districts when working the US. But no two call districts have the same number of amateurs. If the same number of contacts are made from each district, then the more populous 4th and 6th districts are "cheated," and it is easy to see how some people in these areas, or other areas for that matter, could rationalize poor operating habits to compensate for being "cheated."

Since the FCC chose to have us keep our call signs when we move into another district, a station in this situation gets two chances to call: Once when the district is called that matches the number in the call sign, and once again when the district is called where that stateside DXer actually lives. In the latter instance, he or she usually identifies as "portable." If the DXer is ethical and only calls with the district where he or she lives, then there is a good possibility, paradoxically enough, that he will be ignored for calling out of turn! A poor reward for exhibiting good sportsmanship and fair play!

Another way to divide the pileup is by the last letter of the call sign. But is it proper to ask a station to identify by transmitting only part of its identification? Probably not. It may also be technically illegal, depending upon the identification rules of the country (or countries) concerned.

7) Try to keep your instructions to the pileup down to a minimum. The more the DX station talks, the fewer stations he or she is working.

8) Try to end each QSO the same way. Some operators prefer TU or simply K on CW. On phone, the call sign and QRZ? are commonly used. The important part is making it clear when it is time to call.

9) Remember that many, many more people will hear you than will hear the station you are working. If a DXer operates poorly or unethically and it makes you angry, try not to chastise that person on the air. For one thing, it may have been an honest misunderstanding; for another, you will seem like the "heavy" to many people listening to the one-sided conversation. It usually won't help control the pileup, wastes your time and robs you of enjoyment.

10) People will ask you to do things for them. They will ask to do things for you. Some will be genuinely helpful, others will have less commendable motives. You will be asked to listen on other bands, make schedules for friends and friends of friends, work different modes, check into a particular net or operate from a list. Do those things which coincide with your goals for your DXpedition. Do not do those which are at odds with your goals or which you feel do not promote good sportsmanship and fair play. Always be polite but firm. Be assured, it will be difficult!

11) Make thoughtful decisions about the ground rules you lay down for the pileup, then modify them only if absolutely necessary. If the DX station calls for 7s and then answers a W2, the other operators in the pileup have no reasonable expectation that stations will be answered in proper turn in the future. In that case, they feel that they must now call no matter which number is in their call sign. Suddenly, a no-holds-barred situation is in effect.

These suggestions apply to transceive operating as well. There are some additional complications with the single-VFO situation which are accentuated by not being able use separate transmit and receive frequencies.

Typical Operating Situations

The Tail-ender

If done properly, this technique (on the part of the DXers in the pileup) is extremely helpful to the DX station. Some years ago, proper "tail-ending" was the rule and not the exception. Unfortunately, in recent years improper tail-ending has worked so well for some people that now many employ this very poor operating technique. The result is that it disrupts and slows the entire operation.

Let us examine a QSO in order to see the effect. The DX station is Igor, L3ØA, from the Republic of Outer Baldonia. The pileup calls and Igor answers, "K1XA, 5 by 9!"

"QSL, Igor! You are also 5 by 9. L3ØA, this is K1XA."

"WB4ZNH."

"WB4ZNH, 5 by 9!" Igor was able to identify the station and give a signal report and get one back without interruption. He was also saved the time and trouble of asking QRZ.

Now let's tail-end wrong. The pile-up calls and Igor answers, "K1XA, 5 by 9!"

"WB4ZNH!"

"...Igor! You are..."

"WB4 Zulu November Hotel!" "...XA."

"Please WB4ZNH, stand by! K1XA, again my report please." Igor was able to identify the station and give a signal report, but WB4ZNH called and prevented Igor from receiving his report from K1XA. Then Igor made a severe tactical error. He identified the offending station, which means that WB4ZNH has "broken" the pileup and will most likely be the next station worked. But what is the cost? The extra time spent in this exercise reduced the number of total DXers worked by 33%! Where Igor contacted two, he could have worked three. Worse yet, all of the amateurs in that pileup now realize that if they call while another station is talking, they stand a better chance of getting through and getting that precious QSL card from Outer Baldonia. They will also attempt to use this disruptive tactic in the future in other DX pileups.

If Igor had a separate VFO, he could merely turn the dial and start over, but even then some people will find the station he is working before the QSO is completed.

Igor had several alternatives. Possibly the best approach would have been to ignore WB4ZNH and have just asked K1XA to repeat his signal report. In any case, WB4ZNH should not be rewarded with a QSO until his operating practices conform to the standard of good sportsmanship and fair play. That is easy to say but difficult to do without the aid of the second VFO. Some DX stations dealing with this problem attempt to punish the offender by keeping a "blacklist" of stations which will not receive a QSL card, ever. This is excessive punishment that will more likely damage the reputation of the DX station more than that of the original offender. Other DX stations simply tell the offender to come back tomorrow for they will not receive a QSO that day. This is a lot more reasonable than the blacklist, but uses up a lot of time.

Inadequate Operators

Fortunately when you consider our numbers, there really aren't many; it just seems like there are, sometimes. No matter what your personal opinion of a particular amateur might be, never fail to work any DXer if that person is operating correctly. It can be difficult to bring yourself to have a QSO with someone for whom you have little regard, but you will feel better in the long run. You will be setting the example, so make it a good one.

Intentional Interference

It is embarrassing to admit that this negative aspect of DXing exists, but there are ways to minimize its effects. For your own sense of respect for the hobby and for DXing, it is very important to remember for every amateur calling when they should be listening and every amateur actively intentionally causing trouble, there are thousands (yes, thousands) of good operators quietly enduring the same agony as you.

The best way to deal with intentional interference is don't deal with it. Ignore it. Above all, don't quit operating. That might be the goal of the person interfering, and he must not be allowed to control any aspect of Amateur Radio. Whatever the ratio of troublemakers to good people, there seems to be an equal spread throughout society.

This is another argument for not splitting the pileup: If the call-sign number is the determining factor of the pileup division, then 9/10ths of the troublemakers (if they are also DXers) are free to engage in mischief. If they are calling in the pileup, then chances are they are not causing trouble.

[It is interesting to note that, on numerous occasions when DXers stopped answering my calls, someone in the pileup would say "Interference on your frequency," and I would be totally unable to hear the signal causing the problem. It would seem to substantiate the previous conclusion.]

Talking to Your Friends

It will be necessary or desirable sometimes to have a QSO much longer than the "QSL QRZ?" type. DX operators are hams, too. When it comes time for the schedule with your QSL manager back home, or whatever the reason, and you are taking a break from DXing, every attempt should be made to inform the DXers there on frequency as to what is going to transpire. Remember, some of these folks have possibly been waiting hours and even days to talk with you. When

you're at the DX end, you are most certainly entitled to conduct the operation as you please, but that does not relieve the obligation to be polite and thoughtful.

It is always important to give respect if one is to expect to receive respect. Tell the pileup that you are leaving only when it is time to go. Any earlier than that creates a certain frantic frenzy which deteriorates operating efficiency. Thoughtfulness dictates that they be told when and where to expect your return to the game. If you see that it won't be possible to begin operating then, at least try to go there at the prescribed time and announce a new time. If the non-DXing QSO occurs on the pileup frequency, then be careful to acknowledge only those who are a party to that conversation. Otherwise, try to avoid lengthy discussions on the pileup frequency. It is much better to have a prearranged frequency for this purpose.

Multiple Operators and Duplicate Contacts

If your goals include contacting the maximum number of DXers possible, then some attention must be paid to "insurance" contacts. To avoid getting their QSL card back marked "Sri, not in log," most DXers will try to make more than one contact; if any mistake occurred on one QSO, they will have the "insurance" of another contact. Most of these DXers will be considerate of others and make the additional contact only on another mode or band. Depending on the situation, this seems to be quite ethical operating. These same operators will have no compunction whatsoever about working another operator of the same DXpedition if another call sign is involved. To passively combat this wasteful duplication of contacts, try to work out an operating schedule where no two operators work the same bands or band/modes. In almost all cases, it is not practical to keep track of duplicate contacts during a DXpedition. The only opportunity to influence future operating is during the QSLing stage of the operation, and you should make your intentions clear.

POST-DXPEDITION CHORES

Finally, as always, all good things must come to an end. It is important that, when you leave this DX location, the government or licensing authority will have no reason at all to be reluctant about permitting another such operation. It is always good to express your gratitude. While you're there, the rules laid down by the licensing authority should have been adhered to without exception.

If the DXpedition operators are good amateurs, then there will be some new friends left behind. This is always kind of sad; but remember, this might be the spark to create another ham radio operator, or DXer, if the new friend is already an amateur!

If there are local amateurs, be sensitive to their needs, and, if possible, leave behind for their use any equipment or hardware which might be difficult for them to obtain.

A QSL manager is a good idea for all but the very smallest of DXpeditions. Select a QSL manager who is capable of handling the expected influx of QSL cards. It is important that the QSL manager have a positive outlook with similar ideals as your own. The manager should be as accessible to you as possible.

While the logs and your memory are still fresh, go over them and make any notes that will aid the QSL manager. In the heat of the pileup, your penmanship can deteriorate considerably. This is a good time to add clarity to the logs.

If all goes well, plenty of photographs will have been taken. Start to work on a slide program about your DXpedition. There are many DXers who will never have the opportunity to do what you have done, and the closest they will ever come to it is through your pictures. (A cassette narration is also nice for those beyond your traveling range. A CW sidetone makes an excellent signal to advance to the next slide.)

QSL cards should be printed as soon as possible. Most people don't stop to think that, in many cases, the DXpedition won't even know the call sign to be used until a day or two before the actual operating begins. Also, it is not until after the fact, when your QSO totals are in, that a guess of the quantity of QSL cards to be ordered can be made. Surprisingly enough, unless the DXpedition is to a fairly rare spot, QSL requests will usually only equal about 50% of the QSO total.

Not only is it poor form to request money on the air, but the DXCC rules require ethical behavior when it comes to QSLing procedures. Donations should never be a condition for getting a QSL. DXpeditions can be very expensive, so they should be undertaken only with the expectation that the operators themselves will bear all of the expense. Some DXers are under the mistaken impression that donations sent with the QSL cards are a nice source of income. It is okay if they think that, but if you are planning a DXpedition, it is important to realize that if you are lucky, "green stamps" might cover the cost of QSL cards and postage, but nothing more.

The cards should be sorted as they arrive (another good reason to have a manager, since the operators will still be on DXpedition when the first cards arrive). In the first category should be placed the cards which include return postage. The second contains the cards without postage and cards from the QSL bureau. The third and final category is the duplicate-contact people.

Some sort of "first-in, first-out" schedule should be established. The QSL cards for the operators with excessive duplicate contacts should go out last, and there should be included a brief and matter-of-fact note explaining the delay. While in the DXpedition country, you can buy some low-denomination stamps as souvenirs to be included with the cards for those people who were especially helpful during the operation or assisted in some other way.

Some operations choose to have their QSL manager in a country with a lower population, since stateside or European amateurs are more likely to send a donation where international postage rates apply rather than domestic ones. Typically, these operations then send the cards to some domestic point for distribution. While this is not really unethical, it is considered by many to be poor form.

Conclusion

DX operating and DXpeditions can be very satisfying and rewarding activities. These should not be taken lightly, since many people will look at you in the spotlight and form very distinct opinions about Amateur Radio, DXing and you personally. With just a little effort, though, all three can benefit.

Chapter 7

Contests

The word "contest" brings to mind a number of images. Extravaganzas sponsored by various magazine publishers, where "you may have already won $100,000" may be most familiar. If you've ever had any dealings with the legal system or the legal profession, the term "contest your claim" may hold special significance (particularly depending on the outcome of that contest). Many sporting events are zealously described as "titanic contest(s) between close rivals."

To many radio amateurs, this same word describes an event encompassing none of these images, but at the same time, all of them. An Amateur Radio contest is an operating event, held over a predefined time period where the goal is... to enjoy yourself. Of course everyone has their own definition of enjoyment; that diversity of definition only adds to the appeal.

K1XA is one of the more active CW contesters in the Hartford, Connecticut, area. (*NA1F photo*)

The first Amateur Radio contests were operating periods set aside specifically to attempt trans-oceanic communication between North America and Europe, just after World War I. Through this coordination effort, many of the participants successfully made some of the very first intercontinental contacts. Ultimately, this activity evolved into the ARRL International DX Contest. Although the technology and results have changed "somewhat," the object is still the same: Contact as many stations in as many areas of the globe as possible during the contest period.

Two other early operating activities were conceived as intense exercises to train radio amateurs in traffic handling and emergency communications. These are now known as the ARRL Sweepstakes and Field Day. The only real difference between the two is that Field Day emphasizes portable operation as might be required at a disaster site.

These examples demonstrate the two primary motivations for establishing radio contests: advancing the state of the art in radio amateur communications and advancing the expertise level of the Amateur Radio operator. Today, these same two goals provide the inspiration for sponsoring organizations to continue support of Amateur Radio contests. Many of the hams who take part in contests use them as opportunities to work a few new states or collect a few new countries. Others test the effectiveness of a new antenna or other piece of gear. Some work at increasing code speed. There are even some who take all of this very seriously and are out to *win*. In the final analysis, each participant gains expertise, learns a little more about propagation, and becomes a better operator, more equipped to perform effectively in the case of an emergency or to explore some new area of the Amateur Radio art, all while having a good time.

PHILOSOPHY OF EFFECTIVE CONTESTING

For many hams, participation in contests is very casual. If they happen to be on the air and hear a flurry of activity that can only be the result of a contest or big DXpedition, they'll figure out what's going on and pass out some contacts to the hams who are seriously going at it. Indeed, if it were not for these casual operators who are willing to get on the air and work a few contesters, most contests would be extremely slow and boring after the first few hours (see accompanying sidebar "Contesting for Non-Contesters"). For casual contesters, who may not have the time or inclination to compete (at least on that particular weekend), little forethought or planning is required. Getting on the air and knowing the contest exchange (or at least be willing to have someone tell you the exchange) is adequate preparation.

By contrast, serious contesting, at any level, requires some advance preparation. Preparation for a contest entails a number of considerations, and truly is an interdisciplinary subject. Unlike many forms of competition, the success of a radio amateur contest effort is a synergistic product of the station, the operator and a dose of luck as well. An inefficient operator at the controls of a giant station at a great location may not fare as well as a superior operator at a mediocre station. An experienced, competent operator at an outstanding station might completely dominate his or her class, or may have an electrical outage three hours into the contest and have to give up. One thing is certain, though; an inexperienced operator at a poorly prepared station not only should expect dismal results, but shouldn't expect to have fun, either.

Setting Goals

The very first step when entering competition is to set a goal for yourself. As mentioned earlier, a goal might be to improve your code speed by 10 words per minute or to work the last two states you need for the Worked All States award. These would certainly be very worthwhile achievements, and a contest is an excellent arena in which to attain them.

If you're more oriented toward the competitive aspects of contests, then you should establish a competitive goal for yourself. The goal should be reasonably attainable, or you'll just get discouraged when you don't achieve it. Don't set the goal too low or you won't progress, or, more importantly, draw any satisfaction out of achieving it.

The best types of goals are relative goals; that is, how well you can do compared to a particular station in your area who might have similar capabilities to your own. Absolute goals, such as making 250 contacts on 28 MHz during a DX contest, may prove to be totally unreasonable; the band might not even open during periods of low sunspot activity or might be so good that stations using a converted CB set and a mobile whip antenna are capable of making 500 contacts. By setting your sights on an absolute goal, you probably are preventing yourself from doing the best that you can.

Another reason for setting a relative goal as opposed to an absolute goal is the wide disparity between different geographical areas. For example, during a DX contest, because of the vast differences in propagation to population centers around the world, the only real relationship between a station competing from the East Coast of the United States and one from the West Coast during a DX contest is the coincidence that both stations happen to be active during the same time period.

The immense Amateur Radio population in Japan is readily accessible to the average western United States station, so typically the West Coast results show a higher number of contacts than do East Coast results. The East Coast stations, who may only have a short period of useful propagation to Japan, usually have good propagation to Europe. While the amateur population in Europe is less than that of Japan, the population is distributed among a large number of countries, each counting as separate multipliers.

Since the final score is the product of the number of contacts and the number of multipliers, the bottom-line score (with a lot of multipliers and a fair number of contacts) of an East Coast station may be similar to the bottom-line score of a West Coast station (with a lot of contacts, but relatively few multipliers). However, the scores are not comparable, since the conditions are so different. Are football championships comparable to baseball championships? The obvious answer is no; each is an achievement unto itself. By setting your contest goals relative to a nearby station, you eliminate the difficulty of competing with something you have no control over—propagation.

The ideal goal is to compete with a station similar in capability to yours, operated by someone who has similar experience and talent to your own. The absolute optimum goal would be to compete with someone you know personally; after the contest is over you can then compare notes, learn from each other, and spur each other on to new levels of achievement in the next contest.

Compete in a class that is commensurate with your own

abilities as well as your station's. For example, if you live on a suburban lot and at present only have a three-element 15-meter monoband beam antenna, it is pointless to enter the all-band category in a DX contest. Similarly, if you don't own (or wish to own) a high-power linear amplifier for the HF bands, compete in a low-power category.

The contest sponsors make it easy for you to be judged only against your "peers." Not only are there multiple classes (such as monoband 15 meters, or transmitter power below 150 watts) that you can enter, but results are geographically processed. You only compete against other stations in your own geographical area (ARRL section, state, or call area, depending on the contest). The contest sponsors give awards based on these classes and geographical distributions.

Finally, be realistic. Don't try to set the contest world afire in your first attempt. Most of the top competitors not only share your drive and desire, but also have vast experience coupled with that desire. After all, there is a reason they got to the top. Don't forget, as you yourself advance, that the competition isn't standing still either. In the event you don't find yourself challenged enough, set higher goals or choose a new station to compete with.

If you find that you consistently win all the contests in every category you enter by a wide margin, please contact the editor of this book for possible employment in writing the contest chapter of the next edition!

JAØJHA operates both phone and CW contests; he particularly enjoys 20 meters.

The Fundamentals

Once you have set your goal, the next step is to decide exactly what tools you need to do to attain it. First, the fundamentals.

What kind of contest are you going to enter? Most contests fall into four somewhat overlapping categories:

Domestic—contact as many stations as possible during the contest period in the United States and Canada. The ARRL Sweepstakes (SS) is an example of a domestic contest.

DX—contact as many stations during the contest period outside your own country. The ARRL International DX Contest and the *CQ* Magazine World Wide DX Contest are examples of DX contests.

VHF—contact as many stations as possible on the amateur bands above 50 MHz during the contest period. The ARRL VHF QSO Parties and the *CQ* Magazine VHF WPX Contest are examples of VHF contests.

Specialty—contact as many stations as possible during the contest period. Usually all contacts must be made with oper-

Program 1

1A0,1A0,1A0,EU	6C,YK,YK,AS	9B,EP,EP,AS	AP,AP,AP,AS
1S,1S,1S,AS	6D,XE,XE,NA	9C,EP,EP,AS	AS,AP,AP,AS
3A,3A,3A,EU	6E,XE,XE,NA	9D,EP,EP,AS	AT,*VU.VU/AN.VU/L.,*VU.VU/AN.VU/L.,AS
3B6,3B6,3B6,AF	6F,XE,XE,NA	9E,ET,ET,AF	AU,*VU.VU/AN.VU/L.,*VU.VU/AN.VU/L.,AS
3B7,3B7,3B7,AF	6G,XE,XE,NA	9F,ET,ET,AF	AV,*VU.VU/AN.VU/L.,*VU.VU/AN.VU/L.,AS
3B8,3B8,3B8,AF	6H,XE,XE,NA	9G,9G,9G,AF	AW,*VU.VU/AN.VU/L.,*VU.VU/AN.VU/L.,AS
3B9,3B9,3B9,AF	6I,XE,XE,NA	9H,9H,9H,EU	AX,VK,VK,OC
3C,3C,3C,AF	6J,XE,XE,NA	9I,9J,9J,AF	AY,LU,LU,SA
3C0,3C0,3C0,AF	6O,T5,T5,AF	9J,9J,9J,AF	AZ,LU,LU,SA
3D2,3D2,3D2,OC	6T,ST,ST,AF	9K,9K,9K,AS	B,BY,BY,AS
3D6,3D6,3D6,AF	6U,ST,ST,AF	9L,9L,9L,AF	BV,BV,BV,AS
3E,HP,HP,NA	6V,6W,6W,AF	9M2,9M2,9M2,AS	C2,C2,C2,OC
3G,CE,CE,SA	6W,6W,6W,AF	9M4,9M2,9M2,AS	C3,C3,C3,EU
3V,3V,3V,AF	6Y,6Y,6Y,NA	9M6,9M8,9M8,OC	C4,5B,5B,AS
3Y,*3Y/B.3Y/P.,*3Y/B.3Y/P.,AF	6Z,EL,EL,AF	9M8,9M8,9M8,OC	C5,C5,C5,AF
3Z,SP,SP,EU	7A,YB,YB,OC	9N,9N,9N,AS	C6,C6,C6,NA
4A,XE,XE,NA	7B,YB,YB,OC	9O,9Q,9Q,AF	C8,C9,C9,AF
4B,XE,XE,NA	7C,YB,YB,OC	9P,9Q,9Q,AF	C9,C9,C9,AF
4C,XE,XE,NA	7D,YB,YB,OC	9Q,9Q,9Q,AF	CA,CE,CE,SA
4D,DU,DU,OC	7E,YB,YB,OC	9R,9Q,9Q,AF	CA0,*CE0/EI.CE0/SF.CE0/JF.,*CE0/EI.CE0/SF.CE0/JF.,SA
4E,DU,DU,OC	7F,YB,YB,OC	9S,9Q,9Q,AF	CB,CE,CE,SA
4F,DU,DU,OC	7G,YB,YB,OC	9T,9Q,9Q,AF	CB0,*CE0/EI.CE0/SF.CE0/JF.,*CE0/EI.CE0/SF.CE0/JF.,SA
4G,DU,DU,OC	7H,YB,YB,OC	9U,9U,9U,AF	CC,CE,CE,SA
4H,DU,DU,OC	7I,YB,YB,OC	9V,9V,9V,AS	CC0,*CE0/EI.CE0/SF.CE0/JF.,*CE0/EI.CE0/SF.CE0/JF.,SA
4I,DU,DU,OC	7J,JA,JA,AS	9X,9X,9X,AF	CD,CE,CE,SA
4K,*KC4.VP8/Sh.,*KC4.VP8/Sh.,SA	7K,JA,JA,AS	9Y,9Y,9Y,SA	CD0,*CE0/EI.CE0/SF.CE0/JF.,*CE0/EI.CE0/SF.CE0/JF.,SA
4M,YV,YV,SA	7L,JA,JA,AS	9Z,9Y,9Y,SA	CE,CE,CE,SA
4N,YU,YU,EU	7M,JA,JA,AS	A,W,W,NA	CE0,*CE0/EI.CE0/SF.CE0/JF.,*CE0/EI.CE0/SF.CE0/JF.,SA
4O,YU,YU,EU	7N,JA,JA,AS	A2,A2,A2,AF	CE9,*VP8/Sh.KC4.,*VP8/Sh.KC4.,SA
4P,4S,4S,AS	7O,7O,7O,AS	A3,A3,A3,OC	CL,CO,CO,NA
4Q,4S,4S,AS	7P,7P,7P,AF	A4,A4,A4,AS	CM,CO,CO,NA
4S,4S,4S,AS	7Q,7Q,7Q,AF	A5,A5,A5,AS	CN,CN,CN,AF
4T,OA,OA,SA	7R,7X,7X,AF	A6,A6,A6,AS	CO,CO,CO,NA
4U1I,4U/I,4U/I,EU	7S,SM,SM,EU	A7,A7,A7,AS	CP,CP,CP,SA
4U1U,4U/U,4U/U,NA	7T,7X,7X,AF	A8,EL,EL,AF	CQ,CT,CT,EU
4U1V,OE,4U/V,EU	7U,7X,7X,AF	A9,A9,A9,AS	CQ2,CT2,CT2,EU
4V,HH,HH,NA	7V,7X,7X,AF	AH1,KH1,KH1,OC	CQ3,CT3,CT3,AF
4W,4W,4W,AS	7W,7X,7X,AF	AH2,KH2,KH2,OC	CR,CT,CT,EU
4X,4X,4X,AS	7X,7X,7X,AF	AH3,KH3,KH3,OC	CR2,CT2,CT2,EU
4Z,4X,4X,AS	7Y,7X,7X,AF	AH4,KH4,KH4,OC	CR3,CT3,CT3,AF
5A,5A,5A,AF	7Z,HZ,HZ,AS	AH5,KH5,KH5,OC	CS,CT,CT,EU
5B,5B,5B,AS	8A,YB,YB,OC	AH5K,KH5K,KH5K,OC	CS2,CT2,CT2,EU
5H,5H,5H,AF	8B,YB,YB,OC	AH6,KH6,KH6,OC	CS3,CT3,CT3,AF
5I,5H,5H,AF	8C,YB,YB,OC	AH7,KH7,KH7,OC	CT,CT,CT,EU
5J,HK,HK,SA	8D,YB,YB,OC	AH8,KH8,KH8,OC	CT2,CT2,CT2,EU
5K,HK,HK,SA	8E,YB,YB,OC	AH9,KH9,KH9,OC	CT3,CT3,CT3,AF
5L,EL,EL,AF	8F,YB,YB,OC	AH0,KH0,KH0,OC	CT9,CT3,CT3,AF
5M,EL,EL,AF	8G,YB,YB,OC	AL7,KL7,KL7,NA	CU,CT2,CT2,EU
5N,5N,5N,AF	8H,YB,YB,OC	AM,EA,EA,EU	CV,CX,CX,SA
5O,5N,5N,AF	8I,YB,YB,OC	AM6,EA6,EA6,EU	CW,CX,CX,SA
5R,5R,5R,AF	8J,JA,JA,AS	AM8,EA8,EA8,AF	CX,CX,CX,SA
5T,5T,5T,AF	8K,JA,JA,AS	AM9,EA9,EA9,AF	CY,VE,VE,NA
5U,5U,5U,AF	8L,JA,JA,AS	AN,EA,EA,EU	CY0,*CY0/S.CY0/SP.,*CY0/S.CY0/SP.,NA
5V,5V,5V,AF	8M,JA,JA,AS	AN6,EA6,EA6,EU	CZ,VE,VE,NA
5W,5W,5W,OC	8N,JA,JA,AS	AN8,EA8,EA8,AF	D2,D2,D2,AF
5X,5X,5X,AF	8O,A2,A2,AF	AN9,EA9,EA9,AF	D3,D2,D2,AF
5Y,5Z,5Z,AF	8P,8P,8P,NA	AO,EA,EA,EU	D4,D4,D4,AF
5Z,5Z,5Z,AF	8Q,8Q,8Q,AS	AO6,EA6,EA6,EU	D5,EL,EL,AF
6A,SU,SU,AF	8R,8R,8R,SA	AO8,EA8,EA8,AF	D6,D6,D6,AF
6B,SU,SU,AF	8S,SM,SM,EU	AO9,EA9,EA9,AF	D7,HL,HL,AS

D8,HL,HL,AS	FG,*FG.FG/FS.,*FG.FG/FS.,NA	J3,J3,J3,NA
D9,HL,HL,AS	FH,FH,FH,AF	J4,SV,SV,EU
DA,DL,DL,EU	FK,FK,FK,OC	J5,J5,J5,AF
DB,DL,DL,EU	FM,FM,FM,NA	J6,J6,J6,NA
DC,DL,DL,EU	FO,*FO/CI.FO/FPo.,*FO/CI.FO/FPo.,OC	J7,J7,J7,NA
DD,DL,DL,EU	FP,FP,FP,NA	J8,J8,J8,NA
DF,DL,DL,EU	FR,*FR/G.FR/JN.FR/R.FR/T.,*FR/G.FR/JN.FR/R.FR/T.,AF	JA,JA,JA,AS
DG,DL,DL,EU	FS,FS,FS,NA	JB,JA,JA,AS
DH,DL,DL,EU	FT8W,FT8W,FT8W,AF	JC,JA,JA,AS
DI,DL,DL,EU	FT8X,FT8X,FT8X,AF	JD1,*JD/MT.JD/O.,*JD/MT.JD/O.,OC
DJ,DL,DL,EU	FT8Y,VP8/Sh,VP8/Sh,SA	JE,JA,JA,AS
DK,DL,DL,EU	FT8Z,FT8Z,FT8Z,AF	JF,JA,JA,AS
DL,DL,DL,EU	FW,FW,FW,OC	JG,JA,JA,AS
DS,HL,HL,AS	FY,FY,FY,SA	JH,JA,JA,AS
DT,HL,HL,AS	G,G,G,EU	JI,JA,JA,AS
DU,DU,DU,OC	GB,G,G,EU	JJ,JA,JA,AS
DV,DU,DU,OC	GD,GD,GD,EU	JK,JA,JA,AS
DW,DU,DU,OC	GI,GI,GI,EU	JL,JA,JA,AS
DX,DU,DU,OC	GJ,GJ,GJ,EU	JM,JA,JA,AS
DY,DU,DU,OC	GM,GM,*GM.GM/Sh.,EU	JN,JA,JA,AS
DZ,DU,DU,OC	GU,GU,GU,EU	JO,JA,JA,AS
EA,EA,EA,EU	GW,GW,GW,EU	JP,JA,JA,AS
EA6,EA6,EA6,EU	H2,5B,5B,AS	JQ,JA,JA,AS
EA8,EA8,EA8,AF	H3,HP,HP,NA	JR,JA,JA,AS
EA9,EA9,EA9,AF	H4,H4,H4,OC	JS,JA,JA,AS
EB,EA,EA,EU	H5,ZS,ZS,AF	JT,JT,JT,AS
EB6,EA6,EA6,EU	H6,YN,YN,NA	JU,JT,JT,AS
EB8,EA8,EA8,AF	H7,YN,YN,NA	JV,JT,JT,AS
EB9,EA9,EA9,AF	H8,HP,HP,NA	JW,JW,*JW.JW/B.,EU
EC,EA,EA,EU	H9,HP,HP,NA	JX,JX,JX,EU
EC6,EA6,EA6,EU	HA,HA,HA,EU	JY,JY,JY,AS
EC8,EA8,EA8,AF	HB,HB,HB,EU	JZ,YB,YB,OC
EC9,EA9,EA9,AF	HB0,HB0,HB0,EU	K,W,W,NA
ED,EA,EA,EU	HC,HC,HC,SA	KA1,*JD/MT.JD/O.,*JD/MT.JD/O.,OC
ED6,EA6,EA6,EU	HC8,HC8,HC8,SA	KB6,KH1,KH1,OC
ED8,EA8,EA8,AF	HD,HC,HC,SA	KC4,KC4,KC4,SA
ED9,EA9,EA9,AF	HD8,HC8,HC8,SA	KC6,KC6,KC6,OC
EE,EA,EA,EU	HE,HB,HB,EU	KG4,KG4,KG4,NA
EE6,EA6,EA6,EU	HE0,HB0,HB0,EU	KG6,KH2,KH2,OC
EE8,EA8,EA8,AF	HF,SP,SP,EU	KG6R,KH0,KH0,OC
EE9,EA9,EA9,AF	HF0,*KC4.VP8/Sh.,*KC4.VP8/Sh.,SA	KG6S,KH0,KH0,OC
EF,EA,EA,EU	HG,HA,HA,EU	KG6T,KH0,KH0,OC
EF6,EA6,EA6,EU	HH,HH,HH,NA	KH1,KH1,KH1,OC
EF8,EA8,EA8,AF	HI,HI,HI,NA	KH2,KH2,KH2,OC
EF9,EA9,EA9,AF	HJ,HK,HK,SA	KH3,KH3,KH3,OC
EG,EA,EA,EU	HK,HK,HK,SA	KH4,KH4,KH4,OC
EG6,EA6,EA6,EU	HK0,*HK0/M.HK0/SA.,*HK0/M.HK0/SA.,NA	KH5,KH5,KH5,OC
EG8,EA8,EA8,AF	HL,HL,HL,AS	KH5K,KH5K,KH5K,OC
EG9,EA9,EA9,AF	HN,YI,YI,AS	KH6,KH6,KH6,OC
EH,EA,EA,EU	HO,HP,HP,NA	KH7,KH7,KH7,OC
EH6,EA6,EA6,EU	HP,HP,HP,NA	KH8,KH8,KH8,OC
EH8,EA8,EA8,AF	HQ,HR,HR,NA	KH9,KH9,KH9,OC
EH9,EA9,EA9,AF	HR,HR,HR,NA	KH0,KH0,KH0,OC
EI,EI,EI,EU	HS,HS,HS,AS	KJ6,KH3,KH3,OC
EJ,EI,EI,EU	HT,YN,YN,NA	KL7,KL7,KL7,NA
EL,EL,EL,AF	HU,YS,YN,NA	KM6,KH4,KH4,OC
EP,EP,EP,AS	HV,HV,HV,EU	KP1,KP1,KP1,NA
EQ,EP,EQ,AS	HW,F,F,EU	KP2,KP2,KP2,NA
ET,ET,ET,AF	HY,F,F,EU	KP4,KP4,KP4,NA
F,F,F,EU	HZ,HZ,HZ,AS	KP5,KP5,KP5,NA
FB8W,FT8W,FT8W,AF	I,I,I,EU	KP6,KH5,KH5,OC
FB8X,FT8X,FT8X,AF	IS,IS,IS,EU	KS6,KH8,KH8,OC
FB8Y,VP8/Sh,VP8/Sh,SA	IT,I,IT,EU	KV4,KP2,KP2,NA
FB8Z,FT8Z,FT8Z,AF	J2,J2,J2,AF	KW6,KH9,KH9,OC

KX6,KX6,KX6,OC	OQ,ON,ON,EU	RA1P,UA/FJL,UA/FJL,EU	S9,S9,S9,AF
L,LU,LU,SA	OR,ON,ON,EU	RA2,UA2,UA2,EU	SA,SM,SM,EU
LA,LA,LA,EU	OS,ON,ON,EU	RA3,UA/Eu,UA/Eu,EU	SB,SM,SM,EU
LB,LA,LA,EU	OT,ON,ON,EU	RA4,UA/Eu,UA/Eu,EU	SC,SM,SM,EU
LC,LA,LA,EU	OX,OX,OX,NA	RA6,UA/Eu,UA/Eu,EU	SD,SM,SM,EU
LD,LA,LA,EU	OY,OY,OY,EU	RA9,UA/As,UA/As,AS	SE,SM,SM,EU
LE,LA,LA,EU	OZ,OZ,OZ,EU	RB,UB,UB,EU	SF,SM,SM,EU
LF,LA,LA,EU	P2,P2,P2,OC	RC,UC,UC,EU	SG,SM,SM,EU
LG,LA,LA,EU	P3,5B,5B,AS	RD,UD,UD,AS	SH,SM,SM,EU
LH,LA,LA,EU	P41,PJ/SA,PJ/SA,SA	RF,UF,UF,AS	SI,SM,SM,EU
LI,LA,LA,EU	P42,PJ/SA,PJ/SA,SA	RG,UG,UG,AS	SJ,SM,SM,EU
LJ,LA,LA,EU	P43,PJ/SA,PJ/SA,SA	RH,UH,UH,AS	SK,SM,SM,EU
LK,LA,LA,EU	P44,PJ/SA,PJ/SA,SA	RI,UI,UI,AS	SL,SM,SM,EU
LL,LA,LA,EU	P46,PJ/NA,PJ/NA,NA	RJ,UJ,UJ,AS	SM,SM,SM,EU
LM,LA,LA,EU	P47,PJ/NA,PJ/NA,NA	RL,UL,UL,AS	SN,SP,SP,EU
LN,LA,LA,EU	P48,PJ/NA,PJ/NA,NA	RM,UM,UM,AS	SO,SP,SP,EU
LO,LU,LU,SA	P49,PJ/SA,PJ/SA,SA	RN0,UA/As,UA/As,AS	SP,SP,SP,EU
LP,LU,LU,SA	P5,HL,HL,AS	RN1,UA/Eu,UA/Eu,EU	SQ,SP,SP,EU
LQ,LU,LU,SA	P6,HL,HL,AS	RN1N,UA/Eu,UN,EU	SR,SP,SP,EU
LR,LU,LU,SA	P7,HL,HL,AS	RN1P,UA/FJL,UA/FJL,EU	SS,SU,SU,AF
LS,LU,LU,SA	P8,HL,HL,AS	RN2,UA2,UA2,EU	ST,ST,ST,AF
LT,LU,LU,SA	P9,HL,HL,AS	RN3,UA/Eu,UA/Eu,EU	ST0,ST0,ST0,AF
LU,LU,LU,SA	PA,PA,PA,EU	RN4,UA/Eu,UA/Eu,EU	SU,SU,SU,AF
LV,LU,LU,SA	PB,PA,PA,EU	RN5,UA/Eu,UA/Eu,EU	SV,SV,SV,EU
LW,LU,LU,SA	PC,PA,PA,EU	RN6,UA/Eu,UA/Eu,EU	SV5,SV5,SV5,EU
LX,LX,LX,EU	PD,PA,PA,EU	RN9,UA/As,UA/As,AS	SV9,SV9,SV9,EU
LY,UP2,UP2,EU	PE,PA,PA,EU	RO,UO,UO,EU	SW,SV,SV,EU
LZ,LZ,LZ,EU	PF,PA,PA,EU	RP,UP,UP,EU	SW5,SV5,SV5,EU
M1,T7,T7,EU	PG,PA,PA,EU	RQ,UQ,UQ,EU	SW9,SV9,SV9,EU
N,W,W,NA	PH,PA,PA,EU	RR,UR,UR,EU	SX,SV,SV,EU
NH1,KH1,KH1,OC	PI,PA,PA,EU	RT,UB,UB,EU	SX5,SV5,SV5,EU
NH2,KH2,KH2,OC	PJ0,PJ/SA,PJ/SA,SA	RV0,UA/As,UA/As,AS	SX9,SV9,SV9,EU
NH3,KH3,KH3,OC	PJ1,PJ/SA,PJ/SA,SA	RV1,UA/Eu,UA/Eu,EU	SY,SV,SV,EU
NH4,KH4,KH4,OC	PJ2,PJ/SA,PJ/SA,SA	RV1N,UA/Eu,UN,EU	SY5,SV5,SV5,EU
NH5,KH5,KH5,OC	PJ3,PJ/SA,PJ/SA,SA	RV1P,UA/FJL,UA/FJL,EU	SY9,SV9,SV9,EU
NH5K,KH5K,KH5K,OC	PJ4,PJ/SA,PJ/SA,SA	RV2,UA2,UA2,EU	SZ,SV,SV,EU
NH6,KH6,KH6,OC	PJ6,PJ/NA,PJ/NA,NA	RV3,UA/Eu,UA/Eu,EU	SZ5,SV5,SV5,EU
NH7,KH7,KH7,OC	PJ7,PJ/NA,PJ/NA,NA	RV4,UA/Eu,UA/Eu,EU	SZ9,SV9,SV9,EU
NH8,KH8,KH8,OC	PJ8,PJ/NA,PJ/NA,NA	RV6,UA/Eu,UA/Eu,EU	T2,T2,T2,OC
NH9,KH9,KH9,OC	PJ9,PJ/SA,PJ/SA,SA	RV9,UA/As,UA/As,AS	T30,T30,T30,OC
NH0,KH0,KH0,OC	PP,PY,PY,SA	RW0,UA/As,UA/As,AS	T31,T31,T31,OC
NL7,KL7,KL7,NA	PP0,*PY0/FN.PY0/SP.PY0/T.,*PY0/FN.PY0/SP.PY0/T.,SA	RW1,UA/Eu,UA/Eu,EU	T32,T32,T32,OC
NP1,KP1,KP1,NA	PQ,PY,PY,SA	RW1N,UA/Eu,UN,EU	T4,CO,CO,NA
NP2,KP2,KP2,NA	PQ0,*PY0/FN.PY0/SP.PY0/T.,*PY0/FN.PY0/SP.PY0/T.,SA	RW1P,UA/FJL,UA/FJL,EU	T5,T5,T5,AF
NP4,KP4,KP4,NA	PR,PY,PY,SA	RW2,UA2,UA2,EU	T6,YA,YA,AS
NP5,KP5,KP5,NA	PR0,*PY0/FN.PY0/SP.PY0/T.,*PY0/FN.PY0/SP.PY0/T.,SA	RW3,UA/Eu,UA/Eu,EU	T7,T7,T7,EU
NP6,KH5,KH5,OC	PS,PY,PY,SA	RW4,UA/Eu,UA/Eu,EU	TA,TA,TA,AS
OA,OA,OA,SA	PS0,*PY0/FN.PY0/SP.PY0/T.,*PY0/FN.PY0/SP.PY0/T.,SA	RW6,UA/Eu,UA/Eu,EU	TB,TA,TA,AS
OB,OA,OA,SA	PT,PY,PY,SA	RW9,UA/As,UA/As,AS	TC,TA,TA,AS
OC,OA,OA,SA	PT0,*PY0/FN.PY0/SP.PY0/T.,*PY0/FN.PY0/SP.PY0/T.,SA	RY,UB,UB,EU	TD,TG,TG,NA
OD,OD,OD,AS	PU,PY,PY,SA	RZ0,UA/As,UA/As,AS	TE,TI,TI,NA
OE,OE,OE,EU	PU0,*PY0/FN.PY0/SP.PY0/T.,*PY0/FN.PY0/SP.PY0/T.,SA	RZ1,UA/Eu,UA/Eu,EU	TF,TF,TF,EU
OF,OH,OH,EU	PV,PY,PY,SA	RZ1N,UA/Eu,UN,EU	TG,TG,TG,NA
OG,OH,OH,EU	PV0,*PY0/FN.PY0/SP.PY0/T.,*PY0/FN.PY0/SP.PY0/T.,SA	RZ1P,UA/FJL,UA/FJL,EU	TH,F,F,EU
OH,OH,OH,EU	PW,PY,PY,SA	RZ2,UA2,UA2,EU	TI,TI,TI,NA
OH0,OH0,OH0,EU	PW0,*PY0/FN.PY0/SP.PY0/T.,*PY0/FN.PY0/SP.PY0/T.,SA	RZ3,UA/Eu,UA/Eu,EU	TI9,TI9,TI9,NA
OI,OH,OH,EU	PX,PY,PY,SA	RZ4,UA/Eu,UA/Eu,EU	TJ,TJ,TJ,AF
OJ,OJ,OJ,EU	PX0,*PY0/FN.PY0/SP.PY0/T.,*PY0/FN.PY0/SP.PY0/T.,SA	RZ6,UA/Eu,UA/Eu,EU	TK,TK,TK,EU
OK,OK,OK,EU	PY,PY,PY,SA	RZ9,UA/As,UA/As,AS	TL,TL,TL,AF
OL,OK,OK,EU	PY0,*PY0/FN.PY0/SP.PY0/T.,*PY0/FN.PY0/SP.PY0/T.,SA	S2,S2,S2,AS	TN,TN,TN,AF
OM,OK,OK,EU	PZ,PZ,PZ,SA	S4,ZS,ZS,AF	TO,F,F,EU
ON,ON,ON,EU	RA0,UA/As,UA/As,AS	S6,9V,9V,AS	TR,TR,TR,AF
OO,ON,ON,EU	RA1,UA/Eu,UA/Eu,EU	S7,S7,S7,AF	TS,3V,3V,AF
OP,ON,ON,EU	RA1N,UA/FJL,UN,EU	S8,ZS,ZS,AF	TT,TT,TT,AF

```
TU,TU,TU,AF              V3,V3,V3,NA                        XF,XE,XE,NA      ZK3,ZK3,ZK3,OC
TY,TY,TY,AF              V4,V4,V4,NA                        XF4,XF4,XF4,NA   ZL,ZL,ZL,OC
UA0,UA/As,UA/As,AS       V8,V8,V8,OC                        XG,XE,XE,NA      ZL5,VP8/Sh,VP8/Sh,SA
UA1,UA/Eu,UA/Eu,EU       VA,VE,VE,NA                        XG4,XF4,XF4,NA   ZL7,ZL7,ZL7,OC
UA1N,UA/Eu,UN,EU         VB,VE,VE,NA                        XH,XE,XE,NA      ZL8,ZL8,ZL8,OC
UA1P,UA/FJL,UA/FJL,EU    VC,VE,VE,NA                        XH4,XF4,XF4,NA   ZL9,ZL9,ZL9,OC
UA2,UA2,UA2,EU           VD,VE,VE,NA                        XI,XE,XE,NA      ZM,ZL,ZL,OC
UA3,UA/Eu,UA/Eu,EU       VE,VE,VE,NA                        XI4,XF4,XF4,NA   ZP,ZP,ZP,SA
UA4,UA/Eu,UA/Eu,EU       VF,VE,VE,NA                        XQ,CE,CE,SA      ZR,ZS,ZS,AF
UA6,UA/Eu,UA/Eu,EU       VG,VE,VE,NA                        XS,BY,BY,AS      ZR3,ZS3,ZS3,AF
UA9,UA/As,UA/As,AS       VH,VK,VK,OC                        XT,XT,XT,AF      ZS,ZS,ZS,AF
UB,UB,UB,EU              VI,VK,VK,OC                        XU,XU,XU,AS      ZS3,ZS3,ZS3,AF
UC,UC,UC,EU              VJ,VK,VK,OC                        XV,3W,3W,AS      ZT,ZS,ZS,AF
UD,UD,UD,AS              VK,VK,VK,OC                        XW,XW,XW,AS      ZT3,ZS3,ZS3,AF
UF,UF,UF,AS              VK0,*VK0/H.VK0/M.KC4.,*VK0/H.VK0/M.KC4.,SA   XX9,XX9,XX9,AS   ZU,ZS,ZS,AF
UG,UG,UG,AS              VK9,*VK9/W.VK9/C.VK9/CK.VK9/MR.VK9/N.,       XY,XZ,XZ,AS      ZU3,ZS3,ZS3,AF
UH,UH,UH,AS                *VK9/W.VK9/C.VK9/CK.VK9/MR.VK9/N.,OC       XZ,XZ,XZ,AS      ZV,PY,PY,SA
UI,UI,UI,AS              VL,VK,VK,OC                        Y2,Y2,Y2,EU      ZY,PY,PY,SA
UJ,UJ,UJ,AS              VM,VK,VK,OC                        Y3,Y2,Y2,EU      ZZ,PY,PY,SA
UL,UL,UL,AS              VN,VK,VK,OC                        Y4,Y2,Y2,EU
UM,UM,UM,AS              VO,VE,VE,NA                        Y5,Y2,Y2,EU
UN0,UA/As,UA/As,AS       VP2E,VP2E,VP2E,NA                  Y6,Y2,Y2,EU
UN1,UA/Eu,UA/Eu,EU       VP2M,VP2M,VP2M,NA                  Y7,Y2,Y2,EU
UN1N,UA/Eu,UN,EU         VP2V,VP2V,VP2V,NA                  Y8,Y2,Y2,EU
UN1P,UA/FJL,UA/FJL,EU    VP8,*VP8/F.VP8/SG.VP8/SO.VP8/SS.VP8/Sh.,     Y9,Y2,Y2,EU
UN2,UA2,UA2,EU             *VP8/F.VP8/SG.VP8/SO.VP8/SS.VP8/Sh.,SA     YA,YA,YA,AS
UN3,UA/Eu,UA/Eu,EU       VP9,VP9,VP9,NA                     YB,YB,YB,OC
UN4,UA/Eu,UA/Eu,EU       VQ9,VQ9,VQ9,AF                     YC,YB,YB,OC
UN6,UA/Eu,UA/Eu,EU       VR6,VR6,VR6,OC                     YD,YB,YB,OC
UN9,UA/As,UA/As,AS       VS6,VS6,VS6,AS                     YE,YB,YB,OC
UO,UO,UO,EU              VT,*VU.VU/AN.VU/L.,*VU.VU/AN.VU/L.,AS        YF,YB,YB,OC
UP,UP,UP,EU              VU,*VU.VU/AN.VU/L.,*VU.VU/AN.VU/L.,AS        YG,YB,YB,OC
UQ,UQ,UQ,EU              VV,*VU.VU/AN.VU/L.,*VU.VU/AN.VU/L.,AS        YH,YB,YB,OC
UR,UR,UR,EU              VW,*VU.VU/AN.VU/L.,*VU.VU/AN.VU/L.,AS        YI,YI,YI,AS
UT,UB,UB,EU              VY,VE,VE,NA                        YJ,YJ,YJ,OC
UV0,UA/As,UA/As,AS       VZ,VK,VK,OC                        YK,YK,YK,AS
UV1,UA/Eu,UA/Eu,EU       W,W,W,NA                           YM,TA,TA,AS
UV1N,UA/Eu,UN,EU         WH1,KH1,KH1,OC                     YN,YN,YN,NA
UV1P,UA/FJL,UA/FJL,EU    WH2,KH2,KH2,OC                     YO,YO,YO,EU
UV2,UA2,UA2,EU           WH3,KH3,KH3,OC                     YP,YO,YO,EU
UV3,UA/Eu,UA/Eu,EU       WH4,KH4,KH4,OC                     YQ,YO,YO,EU
UV4,UA/Eu,UA/Eu,EU       WH5,KH5,KH5,OC                     YR,YO,YO,EU
UV6,UA/Eu,UA/Eu,EU       WH5K,KH5K,KH5K,OC                  YS,YS,YS,NA
UV9,UA/As,UA/As,AS       WH6,KH6,KH6,OC                     YT,YU,YU,EU
UW0,UA/As,UA/As,AS       WH7,KH7,KH7,OC                     YU,YU,YU,EU
UW1,UA/Eu,UA/Eu,EU       WH8,KH8,KH8,OC                     YV,YV,YV,SA
UW1N,UA/Eu,UN,EU         WH9,KH9,KH9,OC                     YV0,YV0,YV0,SA
UW1P,UA/FJL,UA/FJL,EU    WH0,KH0,KH0,OC                     YW,YV,YV,SA
UW2,UA2,UA2,EU           WL7,KL7,KL7,NA                     YW0,YV0,YV0,SA
UW3,UA/Eu,UA/Eu,EU       WP1,KP1,KP1,NA                     YX,YV,YV,SA
UW4,UA/Eu,UA/Eu,EU       WP2,KP2,KP2,NA                     YX0,YV0,YV0,SA
UW6,UA/Eu,UA/Eu,EU       WP4,KP4,KP4,NA                     YY,YV,YV,SA
UW9,UA/As,UA/As,AS       WP5,KP5,KP5,NA                     YY0,YV0,YV0,SA
UY,UB,UB,EU              WP6,KH5,KH5,OC                     YZ,YU,YU,EU
UZ0,UA/As,UA/As,AS       XA,XE,XE,NA                        Z2,Z2,Z2,AF
UZ1,UA/Eu,UA/Eu,EU       XA4,XF4,XF4,NA                     ZA,ZA,ZA,EU
UZ1N,UA/Eu,UN,EU         XB,XE,XE,NA                        ZB2,ZB2,ZB2,EU
UZ1P,UA/FJL,UA/FJL,EU    XB4,XF4,XF4,NA                     ZC4,ZC4,ZC4,AS
UZ2,UA2,UA2,EU           XC,XE,XE,NA                        ZD7,ZD7,ZD7,AF
UZ3,UA/Eu,UA/Eu,EU       XC4,XF4,XF4,NA                     ZD8,ZD8,ZD8,AF
UZ4,UA/Eu,UA/Eu,EU       XD,XE,XE,NA                        ZD9,ZD9,ZD9,AF
UZ6,UA/Eu,UA/Eu,EU       XD4,XF4,XF4,NA                     ZF,ZF,ZF,NA
UZ9,UA/As,UA/As,AS       XE,XE,XE,NA                        ZK1,*ZK1/NC.ZK1/SC.,*ZK1/NC.ZK1/SC.,OC
V2,V2,V2,NA              XE4,XF4,XF4,NA                     ZK2,ZK2,ZK2,OC
```

Program 2

```
10 '    DUPER.BAS version 1.3  -  19 August 1986  by Clarke Greene K1JX
20 '
30 '    This Microsoft (tm) BASIC program will build a sorted "Dupe Sheet"
40 '
50 '    This file will be produced:
60 '
70 '          <filename> - this is a sorted duplicate listing ready for printing
80 '
90 '
100 '    Depending on the version of BASIC for your particular machine, the CLS (Clear Screen) command must
110 '    be changed.  Consult your own computer's BASIC documentation for more information.
120 '
130 '
140 '    If compiling (a VERY good idea for several orders of magnitude improvement in speed), use O and E switches
150 '
160 '
170 '  Define arrays and variables
180 DEFINT A-Z : OPTION BASE 1
190 DIM ENTRY$(2000)
200 BLANK$=" " : BL$="" : BS$=CHR$(8) : CTRLE$=CHR$(5) : CR$=CHR$(13) : DEL$=BS$+CHR$(32) : ESC$=CHR$(27)
210 TRUE= -1
220 DUPE$=CHR$(7)+"    ** Duplicate QSO **"
230 DUPFORM$="        \          \ \          \ \          \ \          \ \          \"
240 '
250 '  Print message to user
260 CLS
270 PRINT TAB(26) "Interactive Contest Log Duper"
280 PRINT : PRINT
290 PRINT TAB(5) "What is your station's callsign?  ";
300  INPUT; "", MYCALL$ : IF LEN(MYCALL$)=0 THEN 300 ELSE PRINT : PRINT
310 '
320 '  Clear array
330 FOR I=1 TO 2000
340  ENTRY$(I)=BL$
350  NEXT I
360 '
370 '  Initialize variables
380 QSOS=0 : DUPES=0
390 CLS
400 '
410 '  Main user entry loop
420 PRINT : PRINT
430 PRINT TAB(5) "Enter callsign [Esc to end] >  ";
440 THISENTRY$=BL$ : CHAR$=BL$
450 WHILE CHAR$<>CR$ AND CHAR$<>ESC$
460  CHAR$=INKEY$ : IF LEN(CHAR$)=0 GOTO 530
470  IF CHAR$=CR$ OR CHAR$=ESC$ GOTO 530                    ' if the character is an <ESC> or <CR>, jump to exit loop
480  IF CHAR$=BS$ AND LEN(THISENTRY$)>0 THEN THISENTRY$=LEFT$(THISENTRY$,LEN(THISENTRY$)-1) : PRINT DEL$;: GOTO 520
490  IF ASC(CHAR$)<47 GOTO 530                              ' ignore invalid characters
500  IF ASC(CHAR$)>96 THEN GOSUB 1340                       ' capitalize character if necessary
510  THISENTRY$=THISENTRY$+CHAR$                            ' add character to string
520  PRINT CHAR$;                                           ' echo character to screen
530  WEND
540 IF CHAR$=ESC$ GOTO 740                                 ' if the user wants to quit, jump to close log
550 IF LEN(THISENTRY$)=0 GOTO 440
560 '
570 '  Check for dupes
580 DUPE.QSO=NOT TRUE : NOTE$=BL$                           ' clear note field
590 FOR I=1 TO QSOS
600  IF THISENTRY$=ENTRY$(I) THEN NOTE$=DUPE$ : DUPE.QSO=TRUE : I=QSOS
610  NEXT I
620 '
```

```
630 '   Print result of dupe search to screen
640  PRINT NOTES$ : PRINT
650  PRINT TAB(9) "Type Ctrl-E to change the last entry,"
660  PRINT TAB(9) "or any other key to continue.  ";
670  ANS$=INPUT$(1)
680   IF ANS$=CTRLE$ GOTO 410                                           ' if ^E was input, go back and edit entry
690 '
700 '   Adjust variables and loop
710  IF DUPE.QSO THEN DUPES=DUPES+1 : GOTO 410
720  QSOS=QSOS+1 : ENTRY$(QSOS)=THISENTRY$ : GOTO 410
730 '
740 '   Get filename from user
750 CLS
760  PRINT
770 PRINT TAB(5) "What is the name of the file you want to save the dupe sheet in?"
780  PRINT : PRINT TAB(8) "> ";
790  INPUT "", OUTFILE$ : IF LEN(OUTFILE$)=0 THEN 790 ELSE PRINT
800 '
810 '   Routine to prevent overwriting existing file
820 ON ERROR GOTO 1310
830 OPEN OUTFILE$ FOR INPUT AS #1                                        ' try opening file
840 CLOSE
850 PRINT CHR$(7) : PRINT TAB(5) "That file already exists - do you want to overwrite it? <Y/N>  ";
860  ANS$=INPUT$(1) : PRINT
870  IF ANS$<>"Y" AND ANS$<>"y" THEN 740 ELSE PRINT
880 ON ERROR GOTO 0
890 PRINT : PRINT TAB(5) "What frequency band is this dupe sheet for?  ";
900  INPUT; "", BAND$ : IF LEN(BAND$)=0 THEN 900 ELSE PRINT
910 '
920 '   Build dupe sheet
930 PRINT : PRINT TAB(5) "Preparing dupe sheet...  ";
940 '
950 '   Sort callsigns for dupe sheet
960 M=QSOS\2
970 WHILE M>0
980  FOR I=M+1 TO QSOS
990    J=I-M
1000   WHILE J>0
1010    IF ENTRY$(J)>ENTRY$(J+M) THEN SWAP ENTRY$(J),ENTRY$(J+M) : J=J-M ELSE J=0
1020   WEND
1030   NEXT I
1040  M=M\2
1050  WEND
1060 '
1070 '   Enter dupe sheet into file
1080 OPEN OUTFILE$ FOR OUTPUT AS #1
1090  IF QSOS MOD 250=0 THEN LASTPAGE=QSOS\250 ELSE LASTPAGE=QSOS\250+1
1100 FOR PAGE=1 TO LASTPAGE
1110  PRINT #1, SPC(20-(LEN(MYCALL$)+LEN(BAND$))/2); MYCALL$; " -- Dupe Sheet for "; BAND$; " MHz Band -- Page"; STR$(PAGE)
1120  PRINT #1, BL$ : PRINT #1, BL$
1130  FOR ROW=1 TO 50
1140   E=(PAGE-1)*250+ROW
1150    PRINT #1, USING DUPFORM$; ENTRY$(E); ENTRY$(E+50); ENTRY$(E+100); ENTRY$(E+150); ENTRY$(E+200)
1160   NEXT ROW
1170  PRINT #1, CHR$(12)                                                 ' go to next page
1180  NEXT PAGE
1190 CLOSE
1200 PRINT "done"
1210 '
1220 '   Print results
1230 PRINT : PRINT TAB(8) "Valid QSOs: "; QSOS
1240 PRINT : PRINT TAB(8) "Duplicate QSOs: "; DUPES
1250 PRINT : PRINT : PRINT
1260 PRINT TAB(5) "Type C to continue with another band for this contest,"
```

```
1270 PRINT : PRINT TAB(5) "or any other key to Exit  ";
1280 ANS$=INPUT$(1)
1290  IF ANS$="C" OR ANS$="c" THEN 320 ELSE CLS : SYSTEM
1300 '
1310 '  Error trap for existing file
1320 RESUME 880
1330 '
1340 '  Subroutine to capitalize character
1350 ALPHA=ASC(CHAR$)
1360 WHILE ALPHA >96
1370   ALPHA=ALPHA-32
1380   WEND
1390 CHAR$=CHR$(ALPHA)
1400 RETURN
```

Program 3

```
10 '   DXTEST.BAS version 1.2 -  13 August 1986  by Clarke Greene K1JX
20 '
30 '   This Microsoft (tm) BASIC program will build a complete log package for the ARRL International DX Test.
40 '
50 '   The file containing the log entries must be an ASCII file in the following format:
60 '           (each band requires a separate log entry file)
70 '
80 '        TIME        CALLSIGN        RCV'D REPORT  (each log entry must be followed by a carriage return)
90 '
100 '   At least one space must be between each field of each log entry. Only a changed digit in the time field must
110 '   be present; for example, if the contest begins at 1800Z and the first contact is made at 1802Z and the second
120 '   contact is made at 1805Z, then only 5 need be entered in the time field. If the third contact is made at
130 '   1812Z, then 12 should be entered in the time field. If the next contact is made at 1812Z, then no number need be
140 '   entered in the time field (however, be sure to enter a space to indicate separation between fields).
150 '
160 '   These files will be produced:
170 '           <filename>.LOG - this is a complete log ready for printing
180 '           <filename>.DUP - this is a sorted duplicate listing ready for printing
190 '           <filename>.SUM - this is a summary sheet ready for printing
200 '
210 '
220 '   Depending on the version of BASIC for your particular machine, the CLS (Clear Screen) command must
230 '   be changed.  Consult your own computer's BASIC documentation for more information.
240 '
250 '
260 '   If compiling (a VERY good idea for several orders of magnitude improvement in speed), use O and E switches
270 '
280 '   This program also uses a prefix library file (DXPREFIX.LIB), which MUST be on the same disc (and in the
290 '   same subdirectory) as this program.
300 '
310 DEFINT A-Z : OPTION BASE 1
320 DIM ENTRY$(1500), MULT$(175), PFX$(900), CTRY$(900), WIERDPFX$(50), WIERDCTRY$(50), AMBCTRY$(10), Q(175)
330 BLANK$=" " : BL$="" : SLANT$="/" : TRUE=-1
340 DUPES$="- Duplicate QSO -" : NEWCTRY$=" - Mult. #" : INVALID$="- Invalid QSO -"
350 ' Define format strings for printouts
360 LOGFORM$="  \       \ \ \     \        \ \   \   \   \   \                      \"
370 DUPFORM$="    \         \ \         \ \         \ \         \ \         \"
380 SUMFORM$="    \         \ \         \ \         \ \         \ \         \"
390 '
400 CLS
410 PRINT TAB(27) "ARRL DX Test Log Processor" : PRINT : PRINT
420 '
430 ' Read Prefix table file
440 PRINT TAB(5)  "Reading prefix library...  ";
450 I=0                                                ' initialize array subscript
```

```
460 OPEN "DXPREFIX.LIB" FOR INPUT AS #1
470   WHILE NOT EOF(1)
480     I=I+1
490     INPUT #1, PFX$(I), CTRY$(I), DUMMY$, DUMMY$                    ' DUMMY$ is a dummy variable for unused data
500   WEND
510 CLOSE
520 TABLESIZE=I                                                        ' prefix table length
530 PRINT "done"
540 '
550 ' Get user input
560 PRINT : PRINT TAB(5) "What is the station callsign?   ";
570   INPUT "", MYCALL$
580 PRINT : PRINT TAB(5) "What is the two letter abbreviation for the station's state?   ";
590   INPUT "", MYSTATE$
600   IF LEN(MYSTATE$)<>2 THEN PRINT CHR$(7);: GOTO 580
610 PRINT : PRINT TAB(5) "What is the beginning date of the contest? <dd/mm/yr> ";
620   INPUT "", STARTDATE$
630   MARK=INSTR(STARTDATE$,"/") : IF MARK=0 THEN MARK=INSTR(STARTDATE$,"-")
640   STARTDAY=VAL(LEFT$(STARTDATE$,MARK-1))
650   STARTDATE$=RIGHT$(STARTDATE$,LEN(STARTDATE$)-MARK)
660   MARK=INSTR(STARTDATE$,"/") : IF MARK=0 THEN MARK=INSTR(STARTDATE$,"-")
670   MON=VAL(LEFT$(STARTDATE$,MARK-1))
680     IF MON=2 THEN MON$=" Feb.  " : RST$="599" ELSE MON$=" Mar.  " : RST$="59"
690     SENT$=RST$+MYSTATE$
700     YR$=RIGHT$(STARTDATE$,LEN(STARTDATE$)-MARK)
710 PRINT : PRINT TAB(5) "What is the GMT starting time for the contest?  ";
720   INPUT "", STARTGMT$
730 PRINT : PRINT TAB(5) "What file is the log extract located in?  ";
740   INPUT "", INFILE$ : GOSUB 2420                                   ' check to see if file is valid
750   IF INSTR(INFILE$,".")<>0 THEN OUTFILE$=LEFT$(INFILE$,INSTR(INFILE$,".")-1) ELSE OUTFILE$=INFILE$
760 PRINT : PRINT TAB(5) "What frequency band is the log extract for?  ";
770   INPUT "", BAND$
780 '
790 CLS
800 PRINT : PRINT TAB(5) "Duping and counting...  ";
810 '
820 ' Clear arrays
830 FOR I=1 TO 1500
840   ENTRY$(I)=BL$
850   NEXT I
860 FOR I=1 TO 175
870   MULT$(I)=BL$
880   Q(I)=1
890   NEXT I
900 '
910 ' Initialize variables
920 RAWTOTAL=0 : QSOS=0 : DUPES=0 : INVALIDS=0 : MULTNR=0
930 DAY=STARTDAY : PREVIOUSGMT$=STARTGMT$
940 '
950 ' Open files for data input and .LOG output
960 OPEN INFILE$ FOR INPUT AS #1
970 OPEN OUTFILE$+".LOG" FOR OUTPUT AS #2
980 '
990 ' Collect log data, process, and enter into output file
1000  WHILE NOT EOF(1)
1010    LINE INPUT #1, THISENTRY$                                      ' read entire line from disc file
1020    IF THISENTRY$=BL$ GOTO 1590                                    ' jump ahead if line is blank
1030    WHILE ASC(RIGHT$(THISENTRY$,1))<48
1040     THISENTRY$=LEFT$(THISENTRY$,LEN(THISENTRY$)-1)                ' strip off trailing spaces,etc
1050    WEND
1060    IF LEN(THISENTRY$)>0 THEN RAWTOTAL=RAWTOTAL+1 ELSE GOTO 1590
1070 '
1080 ' Separate received report from THISENTRY$
1090    RCVD$=BL$                                                      ' initialize RCVD$ to be null string
```

```
1100    WHILE ASC(RIGHT$(THISENTRY$,1))>=48
1110    RCVD$=RIGHT$(THISENTRY$,1)+RCVD$
1120    THISENTRY$=LEFT$(THISENTRY$,LEN(THISENTRY$)-1)              ' parse last character of string
1130    WEND
1140    IF LEN(RCVD$)<(LEN(RST$)+3) THEN RCVD$=RST$+RCVD$          ' if signal report is left out, append std. report
1150    WHILE ASC(RIGHT$(THISENTRY$,1))<48
1160    THISENTRY$=LEFT$(THISENTRY$,LEN(THISENTRY$)-1)             ' strip off trailing spaces,etc
1170    WEND
1180    '
1190    '    Separate GMT from THISENTRY$
1200    WHILE ASC(LEFT$(THISENTRY$,1))<48
1210    THISENTRY$=RIGHT$(THISENTRY$,LEN(THISENTRY$)-1)            ' strip off leading spaces
1220    WEND
1230    IF INSTR(THISENTRY$,BLANK$)<>0 THEN GMT$=LEFT$(THISENTRY$,INSTR(THISENTRY$,BLANK$)-1) ELSE GMT$=BL$
1240    THISENTRY$=RIGHT$(THISENTRY$,(LEN(THISENTRY$)-LEN(GMT$)))
1250    WHILE LEFT$(THISENTRY$,1)=BLANK$
1260    THISENTRY$=RIGHT$(THISENTRY$,LEN(THISENTRY$)-1)            ' strip off leading spaces
1270    WEND
1280    '   Fill in missing time data
1290    GMT$=LEFT$(PREVIOUSGMT$,(4-LEN(GMT$)))+GMT$
1300    THEDATE$=BL$ : IF GMT$<PREVIOUSGMT$ THEN DAY=DAY+1 : THEDATE$=STR$(DAY)+MON$
1310    '
1320    '   Check for dupes
1330    DUPE.QSO=NOT TRUE : NOTE$=BL$                              ' blank note
1340    FOR I=1 TO QSOS
1350    IF THISENTRY$=ENTRY$(I) THEN NOTE$=DUPE$ : DUPES=DUPES+1 : DUPE.QSO=TRUE : I=QSOS
1360    NEXT I
1370    IF DUPE.QSO GOTO 1530                                      ' skip over prefix search if this entry is a dupe
1380    QSOS=QSOS+1 : ENTRY$(QSOS)=THISENTRY$
1390    '
1400    '   Determine prefix and search prefix library for contact country
1410    IF INSTR(THISENTRY$,SLANT$)>0 THEN GOSUB 2520 ELSE THISPFX$=LEFT$(THISENTRY$,4)
1420    GOSUB 2650 : IF NOT INLIST THEN GOSUB 2790
1430    IF THISCTRY$="W" OR THISCTRY$="VE" THEN NOTE$=INVALID$ : INVALIDS=INVALIDS+1 : GOTO 1530
1440    IF ASC(THISCTRY$)<48 THEN GOSUB 3030                       ' resolve ambiguous prefix
1450    '
1460    '   Search multiplier table for new multiplier
1470    NEWMULT=TRUE                                               ' initially call contact a new multiplier
1480    FOR I=1 TO MULTNR
1490    IF MULT$(I)=THISCTRY$ THEN Q(I)=Q(I)+1 : NEWMULT=NOT TRUE : J=MULTNR
1500    NEXT I
1510    IF NEWMULT THEN MULTNR=MULTNR+1 : MULT$(MULTNR)=THISCTRY$ : NOTE$=THISCTRY$+NEWCTRY$+STR$(MULTNR)
1520    '
1530    '   Write entry to file
1540    IF (RAWTOTAL-1) MOD 50=0 THEN GOSUB 3250                   ' print header if this is the beginning of a page
1550    PRINT #2, USING LOGFORM$; THEDATE$; GMT$; THISENTRY$; SENT$; RCVD$; NOTE$
1560    IF RAWTOTAL MOD 50=0 THEN PRINT #2, CHR$(12)               ' print a form feed if this is the end of a page
1570    '   Reset variables for next entry
1580    PREVIOUSGMT$=GMT$ : GMT$=BL$
1590    WEND
1600    IF RAWTOTAL MOD 50<>0 THEN PRINT #2, CHR$(12)              ' if a form feed hasn't been printed, print one
1610    CLOSE
1620    PRINT "done"
1630    '
1640    '   Build dupe sheet
1650    PRINT : PRINT TAB(5) "Preparing dupe sheet...   ";
1660    '   Sort callsigns for dupe sheet
1670    M=QSOS\2
1680    WHILE M>0
1690    FOR I=M+1 TO QSOS
1700    J=I-M
1710    WHILE J>0
1720    IF ENTRY$(J)>ENTRY$(J+M) THEN SWAP ENTRY$(J),ENTRY$(J+M) : J=J-M ELSE J=0
1730    WEND
```

```
1740    NEXT I
1750    M=M\2
1760   WEND
1770   '
1780   '  Enter dupe sheet into file
1790   OPEN OUTFILE$+".DUP" FOR OUTPUT AS #1
1800    IF QSOS MOD 250=0 THEN LASTPAGE=QSOS\250 ELSE LASTPAGE=QSOS\250+1
1810   FOR PAGE=1 TO LASTPAGE
1820    PRINT #1, SPC(20-(LEN(MYCALL$)+LEN(BAND$))/2); MYCALL$; " -- Dupe Sheet for ";
1830     PRINT #1, BAND$; " MHz Band -- Page"; STR$(PAGE)
1840    PRINT #1, BL$ : PRINT #1, BL$
1850    FOR ROW=1 TO 50
1860    E=(PAGE-1)*250+ROW
1870    PRINT #1, USING DUPFORM$; ENTRY$(E); ENTRY$(E+50); ENTRY$(E+100); ENTRY$(E+150); ENTRY$(E+200)
1880    NEXT ROW
1890    PRINT #1, CHR$(12)                                    ' go to next page
1900    NEXT PAGE
1910   CLOSE
1920   PRINT "done"
1930   '
1940   '  Build summary listing
1950   PRINT : PRINT TAB(5) "Preparing summary sheet...  ";
1960   '  Sort multipliers for summary sheet
1970   M=MULTNR\2
1980   WHILE M>0
1990    FOR I=M+1 TO MULTNR
2000    J=I-M
2010    WHILE J>0
2020     IF MULT$(J)>MULT$(J+M) THEN SWAP MULT$(J),MULT$(J+M) : SWAP Q(J),Q(J+M) : J=J-M ELSE J=0
2030    WEND
2040    NEXT I
2050    M=M\2
2060   WEND
2070   '
2080   '  Append number of qsos per country onto country prefixes
2090   FOR I=1 TO MULTNR
2100    MULT$(I)=MULT$(I)+SPACE$(6-LEN(MULT$(I)))+" -"+STR$(Q(I))
2110    NEXT I
2120   '
2130   '  Enter summary sheet into file
2140   OPEN OUTFILE$+".SUM" FOR OUTPUT AS #1
2150    PRINT #1, SPC(14-(LEN(MYCALL$)+LEN(BAND$))/2); MYCALL$; " -- Summary Sheet for "; BAND$;
2160     PRINT #1, " MHz Band - "; YR$; " ARRL DX Test"
2170    PRINT #1, BL$
2180    PRINT #1, TAB(15); "Country Listing and number of contacts per Country"
2190    PRINT #1, BL$ : PRINT #1, BL$
2200    IF MULTNR MOD 5=0 THEN LASTROW=MULTNR\5 ELSE LASTROW=MULTNR\5+1
2210    FOR ROW=1 TO LASTROW
2220     PRINT #1, USING SUMFORM$; MULT$(ROW); MULT$(ROW+LASTROW); MULT$(ROW+LASTROW*2); MULT$(ROW+LASTROW*3); MULT$(ROW+LASTROW*4)
2230     NEXT ROW
2240    PRINT #1, BL$ : PRINT #1, BL$ : PRINT #1, BL$
2250    PRINT #1, TAB(18) "Total Valid QSOs - "; STR$(QSOS-INVALIDS); TAB(45)"Dupes - "; STR$(DUPES)
2260    PRINT #1, TAB(18) "Countries - "; STR$(MULTNR)
2270   CLOSE
2280   PRINT "done"
2290   '
2300   '  Print results
2310   CLS : PRINT CHR$(7)
2320   PRINT : PRINT TAB(5) "Results for the "; BAND$; " MHz band"
2330   PRINT : PRINT TAB(8) "Valid QSOs: "; QSOS-INVALIDS
2340   PRINT : PRINT TAB(8) "Duplicate QSOs: "; DUPES
2350   PRINT : PRINT TAB(8) "Countries: "; MULTNR
2360   PRINT : PRINT : PRINT
2370   PRINT TAB(5) "Type C to continue with another band for this contest,"
```

```
2380 PRINT : PRINT TAB(5) "or any other key to Exit   ";
2390 ANS$=INPUT$(1)
2400  IF ANS$="C" OR ANS$="c" THEN CLS : GOTO 730 ELSE CLS : SYSTEM
2410 '
2420 '  Subroutine to trap missing file
2430 ON ERROR GOTO 2480
2440 OPEN INFILE$ FOR INPUT AS #1                              ' try opening file
2450 ON ERROR GOTO 0
2460 CLOSE
2470 RETURN
2480 PRINT CHR$(7) : PRINT TAB(4) "That file does not exist - type X to Exit or any other key to continue ";
2490 ANS$=INPUT$(1) : IF ANS$="X" OR ANS$="x" THEN CLS : SYSTEM
2500 PRINT
2510 RESUME 730
2520 '
2530 '  Subroutine to determine prefix from portable designator
2540  MARK=INSTR(THISENTRY$,SLANT$)
2550  IF MARK>3 THEN THISPFX$=RIGHT$(THISENTRY$,LEN(THISENTRY$)-MARK) ELSE THISPFX$=LEFT$(THISENTRY$,MARK-1)
2560  IF LEN(THISPFX$)>1 GOTO 2630                             ' have prefix - return
2570  IF ASC(THISPFX$)>58 OR ASC(THISPFX$)<47 THEN THISPFX$=LEFT$(THISENTRY$,4) : GOTO 2630 ' (local portable designator)
2580  K=2                                                      ' find position of first numeral in call
2590   WHILE (ASC(MID$(THISENTRY$,K,1))>57 OR ASC(MID$(THISENTRY$,K,1))<48) AND K<LEN(THISENTRY$)
2600    K=K+1
2610   WEND
2620   THISPFX$=LEFT$(THISENTRY$,K-1)+THISPFX$                 ' new prefix = portable number in old prefix
2630  RETURN
2640 '
2650 '  Subroutine to search prefix library for standard country prefix
2660   K=4 : INLIST=NOT TRUE : SAVEDPFX$=THISPFX$
2670   WHILE K>0 AND INLIST=NOT TRUE
2680    THISPFX$=LEFT$(THISPFX$,K)
2690    LOW=1 : HIGH=TABLESIZE : INLIST=NOT TRUE              ' initial values for binary sort
2700    WHILE LOW<=HIGH AND INLIST=NOT TRUE
2710     L=(LOW+HIGH)\2
2720     IF THISPFX$=PFX$(L) THEN INLIST=TRUE : THISCTRY$=CTRY$(L)
2730     IF THISPFX$<PFX$(L) THEN HIGH=L-1 ELSE LOW=LOW+1
2740    WEND
2750    K=K-1
2760   WEND
2770  RETURN
2780 '
2790 '  Subroutine to search unusual prefix list
2800  IF NRWIERDPFX=0 GOTO 2910                                ' if the supplementary prefix list is empty, skip ahead
2810   K=4
2820   WHILE K>0
2830    SAVEDPFX$=LEFT$(SAVEDPFX$,K)
2840    FOR J=1 TO NRWIERDPFX
2850     IF SAVEDPFX$=WIERDPFX$(J) THEN INLIST=TRUE : THISCTRY$=WIERDCTRY$(J) : J=NRWIERDPFX : K=1
2860    NEXT J
2870    K=K-1
2880   WEND
2890   IF INLIST THEN RETURN                                   ' if the prefix was found, return
2900 '
2910 '  Routine to get prefix definition and continent from user for prefix not found in library
2920  PRINT CHR$(7) : PRINT
2930  PRINT TAB(5) "The prefix for "; THISENTRY$; " can't be found in the prefix library."
2940  PRINT : PRINT TAB(8) "What is the callsign prefix?  ";
2950   INPUT "", HELDPFX$
2960  PRINT : PRINT TAB(8) "What standard prefix is that equivalent to?  ";
2970   INPUT "", THISPFX$
2980   GOSUB 2650 : IF NOT INLIST GOTO 2920
2990   NRWIERDPFX=NRWIERDPFX+1 : WIERDPFX$(NRWIERDPFX)=HELDPFX$ : WIERDCTRY$(NRWIERDPFX)=THISCTRY$
3000  PRINT : PRINT TAB(5) "Back to duping and counting...  ";
3010  RETURN
```

```
3020 '
3030 '  Subroutine to resolve ambiguous prefix with user interaction
3040  THISCTRY$=RIGHT$(THISCTRY$,LEN(THISCTRY$)-1)                    ' strip initial delimiter
3050    J=0
3060    WHILE LEN(THISCTRY$)>0
3070      J=J+1
3080      MARK=INSTR(THISCTRY$,".")
3090      AMBCTRY$(J)=LEFT$(THISCTRY$,MARK-1)                         ' put multipiler name into array
3100      THISCTRY$=RIGHT$(THISCTRY$,LEN(THISCTRY$)-MARK)
3110    WEND
3120  PRINT CHR$(7) : PRINT
3130  PRINT TAB(5) "The prefix for "; THISENTRY$; " could indicate several different countries."
3140  PRINT : PRINT TAB(8) "The possiblities are:" : PRINT
3150  FOR K=1 TO J
3160   PRINT TAB(11) STR$(K); ". "; AMBCTRY$(K)                      ' print choices to screen
3170   NEXT K
3180  PRINT : PRINT TAB(8) "Type the number of the correct country. > ";
3190   INPUT "", CHOICE$
3200   K=VAL(CHOICE$) : IF K > J OR K < 1 THEN PRINT CHR$(7); : GOTO 3180
3210  THISCTRY$=AMBCTRY$(K)
3220  PRINT : PRINT TAB(5) "Back to duping and counting...  ";
3230  RETURN
3240  '
3250  '  Subroutine to print log sheet header
3260  PRINT #2, BL$
3270  PRINT #2, TAB(5); MYCALL$; "  "; BAND$; " MHz Log"; TAB(70); "Page"; STR$(RAWTOTAL\50+1)
3280  PRINT #2, BL$
3290  PRINT #2, "   Date    Time    Callsign      Sent    Rcv'd         Notes"
3300  PRINT #2, "  "; STRING$(74,61)
3310  THEDATE$=STR$(DAY)+MON$
3320  RETURN
```

Program 4

```
10 '   CQWWLOG.BAS version 1.2 -  13 July 1986  by Clarke Greene K1JX
20 '
30 '   This Microsoft (tm) BASIC program will build a complete log package for the CQ Worldwide DX Contest.
40 '
50 '   The file containing the log entries must be an ASCII file in the following format:
60 '           (each band requires a separate log entry file)
70 '
80 '        TIME          CALLSIGN          RCV'D REPORT  (each log entry must be followed by a carriage return)
90 '
100 '   At least one space must be between each field of each log entry. Only a changed digit in the time field must
110 '   be present; for example, if the contest begins at 1800Z and the first contact is made at 1802Z and the second
120 '   contact is made at 1805Z, then only 5 need be entered in the time field. If the third contact is made at
130 '   1812Z, then 12 should be entered in the time field. If the next contact is made at 1812Z, then no number need be
140 '   entered in the time field (however, be sure to enter a space to indicate separation between fields).
150 '
160 '   These files will be produced:
170 '           <filename>.LOG - this is a complete log ready for printing
180 '           <filename>.DUP - this is a sorted duplicate listing ready for printing
190 '           <filename>.SUM - this is a summary sheet ready for printing
200 '
210 '
220 '   Depending on the version of BASIC for your particular machine, the CLS (Clear Screen) command must
230 '   be changed.  Consult your own computer's BASIC documentation for more information.
240 '
250 '
260 '   If compiling (a VERY good idea for several orders of magnitude improvement in speed), use O and E switches
270 '
280 '   This program also uses a prefix library file (DXPREFIX.LIB), which MUST be on the same disc (and in the
```

```
290 '   same subdirectory) as this program.
300 '
310 '  Define arrays and variables
320 DEFINT A-Z : OPTION BASE 1
330 DIM ENTRY$(1500), MULT$(175), PFX$(900), CTRY$(900), CNT$(900), WIERDPFX$(50), WIERDCTRY$(50), WIERDCNT$(50), AMBCTRY$(10)
340 DIM Q(175), ZONE(40)
350 BLANK$=" " : BL$="" : SLANT$="/" : TRUE=-1
360 DUPE1$="  - Duplica" : DUPE2$="te QSO -"
370 '  Define format strings for printouts
380 LOGFORM$=" \      \ \ \    \          \ \ \  \ \ \  \        \\          \ #"
390 DUPFORM$="    \          \ \        \ \        \ \        \ \          \"
400 SUMFORM$="    \          \ \        \ \        \ \        \ \          \"
410 FOOTFORM$=" ##          ##         ##        ###"
420 '
430 CLS
440 PRINT TAB(26) "CQWW DX Contest Log Processor" : PRINT : PRINT
450 '
460 '  Read Prefix table file
470 PRINT TAB(5)  "Reading prefix library...   ";
480 I=0                                                      ' initialize array subscript
490 OPEN "DXPREFIX.LIB" FOR INPUT AS #1
500   WHILE NOT EOF(1)
510   I=I+1
520   INPUT #1, PFX$(I), DUMMY$, CTRY$(I), CNT$(I)           ' DUMMY$ is a dummy variable for data not used
530   WEND
540 CLOSE
550 TABLESIZE=I                                              ' prefix table length
560 PRINT "done"
570 '
580 '  Get user input
590 PRINT : PRINT TAB(5) "What is the station callsign?   ";
600   INPUT "", MYCALL$
610   THISENTRY$=MYCALL$ : IF INSTR(THISENTRY$,SLANT$)>0 THEN GOSUB 3100 ELSE THISPFX$=LEFT$(THISENTRY$,4)
620   GOSUB 3230 : IF NOT INLIST THEN GOSUB 3370
630   MYCTRY$=THISCTRY$ : MYCNT$=THISCNT$ : IF MYCNT$="NA" THEN MYCNTPTS=2 ELSE MYCNTPTS=1
640 PRINT : PRINT TAB(5) "What is the station's WAZ zone?   ";
650   INPUT "", MYZONE$
660   IF VAL(MYZONE$)<1 OR VAL(MYZONE$)>40 THEN PRINT CHR$(7);: GOTO 640
670   IF VAL(MYZONE$)<10 AND LEN(MYZONE$)=1 THEN MYZONE$="0"+MYZONE$
680 PRINT : PRINT TAB(5) "What is the beginning date of the contest? <dd/mm/yr> ";
690   INPUT "", STARTDATE$
700   MARK=INSTR(STARTDATE$,"/") : IF MARK=0 THEN MARK=INSTR(STARTDATE$,"-")
710   STARTDAY=VAL(LEFT$(STARTDATE$,MARK-1))
720   STARTDATE$=RIGHT$(STARTDATE$,LEN(STARTDATE$)-MARK)
730   MARK=INSTR(STARTDATE$,"/") : IF MARK=0 THEN MARK=INSTR(STARTDATE$,"-")
740   MON=VAL(LEFT$(STARTDATE$,MARK-1))
750    IF MON=10 THEN MON$=" Oct.  " : RST$="59" ELSE MON$=" Nov.  " : RST$="599"
760    SENT$=RST$+MYZONE$
770   YR$=RIGHT$(STARTDATE$,LEN(STARTDATE$)-MARK)
780 PRINT : PRINT TAB(5) "What is the GMT starting time for the contest?  ";
790   INPUT "", STARTGMT$
800 PRINT : PRINT TAB(5) "What file is the log extract located in?  ";
810   INPUT "", INFILE$ : GOSUB 2910                          ' check to see if file is valid
820   IF INSTR(INFILE$,".")<>0 THEN OUTFILE$=LEFT$(INFILE$,INSTR(INFILE$,".")-1) ELSE OUTFILE$=INFILE$
830 PRINT : PRINT TAB(5) "What frequency band is the log extract for?  ";
840   INPUT "", BAND$
850 '
860 '  Build log file
870 CLS
880 PRINT : PRINT TAB(5) "Duping and counting...   ";
890 '
900 '  Clear arrays
910 FOR I=1 TO 1500
920  ENTRY$(I)=BL$
```

```
930  NEXT I
940  FOR I=1 TO 175
950   MULT$(I)=BL$
960   Q(I)=1
970  NEXT I
980  FOR I=1 TO 40
990   ZONE(I)=0
1000  NEXT I
1010 '
1020 '  Initialize variables
1030 RAWTOTAL=0 : QSOS=0 : DUPES=0 : CTRYNR=0 : ZONENR=0 : TOTPOINTS=0
1040 PGQSOS=0 : PGZONES=0 : PGCTRY=0 : PGPTS=0
1050 DAY=STARTDAY : PREVIOUSGMT$=STARTGMT$
1060 '
1070 '  Open input file and output .LOG file
1080 OPEN INFILE$ FOR INPUT AS #1
1090 OPEN OUTFILE$+".LOG" FOR OUTPUT AS #2
1100 '
1110 '  Input data, process, and enter into output file
1120  WHILE NOT EOF(1)
1130   LINE INPUT #1, THISENTRY$                              ' read entire line from disc file
1140    IF THISENTRY$=BL$ GOTO 1840                           ' jump ahead if line is blank
1150    WHILE ASC(RIGHT$(THISENTRY$,1))<48
1160     THISENTRY$=LEFT$(THISENTRY$,LEN(THISENTRY$)-1)       ' strip off trailing spaces,etc
1170    WEND
1180    IF LEN(THISENTRY$)>0 THEN RAWTOTAL=RAWTOTAL+1 ELSE GOTO 1840
1190 '
1200 '  Separate received report from THISENTRY$
1210   RCVD$=BL$                                              ' initialize RCVD$ to be null string
1220   WHILE ASC(RIGHT$(THISENTRY$,1))>=48
1230    RCVD$=RIGHT$(THISENTRY$,1)+RCVD$
1240    THISENTRY$=LEFT$(THISENTRY$,LEN(THISENTRY$)-1)        ' parse last character of string
1250   WEND
1260   IF LEN(RCVD$)<(LEN(RST$)+2) THEN RCVD$=RST$+RCVD$      ' if no RST was typed, append std report
1270   WHILE ASC(RIGHT$(THISENTRY$,1))<48
1280    THISENTRY$=LEFT$(THISENTRY$,LEN(THISENTRY$)-1)        ' strip off trailing spaces,etc
1290   WEND
1300 '
1310 '  Separate GMT from THISENTRY$
1320   WHILE ASC(LEFT$(THISENTRY$,1))<48
1330    THISENTRY$=RIGHT$(THISENTRY$,LEN(THISENTRY$)-1)       ' strip off leading spaces
1340   WEND
1350   IF INSTR(THISENTRY$,BLANK$)<>0 THEN GMT$=LEFT$(THISENTRY$,INSTR(THISENTRY$,BLANK$)-1) ELSE GMT$=BL$
1360   THISENTRY$=RIGHT$(THISENTRY$,(LEN(THISENTRY$)-LEN(GMT$)))
1370   WHILE LEFT$(THISENTRY$,1)=BLANK$
1380    THISENTRY$=RIGHT$(THISENTRY$,LEN(THISENTRY$)-1)       ' strip off leading spaces
1390   WEND
1400 '  Fill in missing time data
1410   GMT$=LEFT$(PREVIOUSGMT$,(4-LEN(GMT$)))+GMT$
1420   THEDATE$=BL$ : IF GMT$<PREVIOUSGMT$ THEN DAY=DAY+1 : THEDATE$=STR$(DAY)+MON$
1430 '
1440 '  Check for dupes
1450   DUPE.QSO=NOT TRUE : POINTS=3
1460   FOR J=1 TO QSOS
1470    IF ENTRY$(J)=THISENTRY$ THEN NEWZONE$=DUPE1$ : NEWCTRY$=DUPE2$ : DUPES=DUPES+1 : POINTS=0 : DUPE.QSO=TRUE : J=QSOS
1480   NEXT J
1490    IF DUPE.QSO GOTO 1790                                 ' skip over prefix search if this entry is a dupe
1500   QSOS=QSOS+1 : ENTRY$(QSOS)=THISENTRY$
1510 '
1520 '  Determine zone and search zone table for new multiplier
1530   NEWZONE$=BL$
1540   THISZONE$=RIGHT$(RCVD$,LEN(THISZONE$)-LEN(RST$))       ' allow for 1 digit zone entry
1550   J=VAL(THISZONE$) : IF J<1 OR J>40 THEN GOSUB 3020
1560    IF ZONE(J)=0 THEN ZONENR=ZONENR+1 : NEWZONE$="Zone #"+STR$(ZONENR) : PGZONES=PGZONES+1
```

```
1570    ZONE(J)=ZONE(J)+1
1580 '
1590 '   Determine prefix and search prefix library for contact country and continent
1600    IF INSTR(THISENTRY$,SLANT$)>0 THEN GOSUB 3100 ELSE THISPFX$=LEFT$(THISENTRY$,4)
1610    GOSUB 3230 : IF NOT INLIST THEN GOSUB 3370
1620    IF ASC(THISCTRY$)<48 THEN GOSUB 3610                        ' resolve ambiguous prefix
1630 '
1640 '   Search multiplier table for new country
1650    NEWMULT=TRUE : NEWCTRY$=BL$
1660    FOR J=1 TO CTRYNR
1670     IF MULT$(J)=THISCTRY$ THEN Q(J)=Q(J)+1 : NEWMULT=NOT TRUE : J=CTRYNR
1680     NEXT J
1690     IF NEWMULT THEN CTRYNR=CTRYNR+1 : MULT$(CTRYNR)=THISCTRY$ : NEWCTRY$=THISCTRY$+" #"+STR$(CTRYNR) : PGCTRY=PGCTRY+1
1700 '
1710 '   Determine point value for QSO
1720    IF THISCTRY$=MYCTRY$ THEN POINTS=0 : GOTO 1750             ' contacts in your own country are worth 0 points
1730    IF THISCNT$=MYCNT$ THEN POINTS=MYCNTPTS
1740 '
1750 '   Update page totals
1760    PGQSOS=PGQSOS+1 : PGPTS=PGPTS+POINTS
1770    TOTPOINTS=TOTPOINTS+POINTS
1780 '
1790 '   Write entry to file
1800    IF (RAWTOTAL-1) MOD 50=0 THEN GOSUB 3830                  ' print header if this is the beginning of a page
1810    PRINT #2, USING LOGFORM$; THEDATE$; GMT$; THISENTRY$; SENT$; RCVD$; NEWZONE$; NEWCTRY$; POINTS
1820    IF RAWTOTAL MOD 50=0 THEN GOSUB 3900                      ' print footer if this is the end of a page
1830    PREVIOUSGMT$=GMT$ : GMT$=BL$
1840   WEND
1850 IF RAWTOTAL MOD 50<>0 THEN PRINT#2, CHR$(12)                 ' if a form feed hasn't been printed, print one now
1860 CLOSE
1870 PRINT "done"
1880 '
1890 '   Build dupe sheet
1900  PRINT : PRINT TAB(5) "Preparing dupe sheet...   ";
1910 '   Sort callsigns for dupe sheet
1920 M=QSOS\2
1930 WHILE M>0
1940  FOR I=M+1 TO QSOS
1950   J=I-M
1960    WHILE J>0
1970     IF ENTRY$(J)>ENTRY$(J+M) THEN SWAP ENTRY$(J),ENTRY$(J+M) : J=J-M ELSE J=0
1980     WEND
1990    NEXT I
2000  M=M\2
2010 WEND
2020 '
2030 '   Enter dupe sheet into file
2040 OPEN OUTFILE$+".DUP" FOR OUTPUT AS #1
2050   IF QSOS MOD 250=0 THEN LASTPAGE=QSOS\250 ELSE LASTPAGE=QSOS\250+1
2060 FOR PAGE=1 TO LASTPAGE
2070  PRINT #1, SPC(20-(LEN(MYCALL$)+LEN(BAND$))/2); MYCALL$; " -- Dupe Sheet for ";
2080   PRINT #1, BAND$; " MHz Band -- Page"; STR$(PAGE)
2090  PRINT #1, BL$ : PRINT #1, BL$
2100  FOR ROW=1 TO 50
2110   E=(PAGE-1)*250+ROW
2120    PRINT #1, USING DUPFORM$; ENTRY$(E); ENTRY$(E+50); ENTRY$(E+100); ENTRY$(E+150); ENTRY$(E+200)
2130   NEXT ROW
2140  PRINT #1, CHR$(12)                                          ' go to next page
2150  NEXT PAGE
2160 CLOSE
2170 PRINT "done"
2180 '
2190 '   Build summary listing
2200 PRINT : PRINT TAB(5) "Preparing summary sheet...   ";
```

```
2210 '  Sort countries for summary sheet
2220 M=CTRYNR\2
2230 WHILE M>0
2240 FOR I=M+1 TO CTRYNR
2250   J=I-M
2260   WHILE J>0
2270    IF MULT$(J)>MULT$(J+M) THEN SWAP MULT$(J),MULT$(J+M) : SWAP Q(J),Q(J+M) : J=J-M ELSE J=0
2280    WEND
2290   NEXT I
2300   M=M\2
2310  WEND
2320 '
2330 '  Append number of qsos per country onto country prefixes
2340 FOR I=1 TO CTRYNR
2350 MULT$(I)=MULT$(I)+SPACE$(6-LEN(MULT$(I)))+" -"+STR$(Q(I))
2360  NEXT I
2370 '
2380 '  Enter country listing into file
2390 OPEN OUTFILE$+".SUM" FOR OUTPUT AS #1
2400 PRINT #1, SPC(12-(LEN(MYCALL$)+LEN(BAND$))/2); MYCALL$; " -- Summary Sheet for "; BAND$;
2410   PRINT #1, " MHz Band - "; YR$; " CQWW DX Contest"
2420 PRINT #1, BL$
2430 PRINT #1, TAB(15); "Country Listing and number of contacts per Country"
2440 PRINT #1, BL$ : PRINT #1, BL$
2450 IF CTRYNR MOD 5=0 THEN LASTROW=CTRYNR\5 ELSE LASTROW=CTRYNR\5+1
2460 FOR ROW=1 TO LASTROW
2470   PRINT #1, USING SUMFORM$; MULT$(ROW); MULT$(ROW+LASTROW); MULT$(ROW+LASTROW*2); MULT$(ROW+LASTROW*3); MULT$(ROW+LASTROW*4)
2480  NEXT ROW
2490 '
2500 '  Build listing of zones worked and contacts per zone
2510  J=0
2520 FOR I=1 TO 40
2530   IF ZONE(I)>0 THEN J=J+1 : MULT$(J)="Zone"+STR$(I)+" -"+STR$(ZONE(I))
2540  NEXT I                                    ' put zone count into array
2550 FOR I=J TO 40
2560   MULT$(I)=BL$
2570  NEXT I                                    ' blank out remainder of array
2580 '
2590 '  Enter zone listing
2600 PRINT #1, BL$
2610 PRINT #1, TAB(18); "Zone Listing and number of contacts per Zone"
2620 PRINT #1, BL$
2630 IF ZONENR MOD 5=0 THEN LASTROW=ZONENR\5 ELSE LASTROW=ZONENR\5+1
2640 FOR ROW=1 TO LASTROW
2650   PRINT #1, USING SUMFORM$; MULT$(ROW); MULT$(ROW+LASTROW); MULT$(ROW+LASTROW*2); MULT$(ROW+LASTROW*3); MULT$(ROW+LASTROW*4)
2660  NEXT ROW
2670 '
2680 '  Enter summary into file
2690 PRINT #1, BL$ : PRINT #1, BL$
2700 PRINT #1, "     Total Valid QSOs - "; STR$(QSOS); "       Dupes - "; STR$(DUPES)
2710 PRINT #1, "     QSO points - "; STR$(TOTPOINTS)
2720 PRINT #1, "     Zones - "; STR$(ZONENR)
2730 PRINT #1, "     Countries - "; STR$(CTRYNR)
2740 CLOSE
2750 PRINT "done"
2760 '
2770 '  Print results
2780 CLS : PRINT CHR$(7)
2790 PRINT : PRINT TAB(5) "Results for the "; BAND$; " MHz band"
2800 PRINT : PRINT TAB(8) "Valid QSOs: "; QSOS
2810 PRINT : PRINT TAB(8) "Duplicate QSOs: "; DUPES
2820 PRINT : PRINT TAB(8) "QSO points: "; TOTPOINTS
2830 PRINT : PRINT TAB(8) "Zones: "; ZONENR
2840 PRINT : PRINT TAB(8) "Countries: "; CTRYNR
```

```
2850 PRINT : PRINT : PRINT
2860 PRINT TAB(5) "Type C to continue with another band for this contest,"
2870 PRINT : PRINT TAB(5) "or any other key to Exit   ";
2880 ANS$=INPUT$(1)
2890  IF ANS$="C" OR ANS$="c" THEN CLS : GOTO 800 ELSE CLS : SYSTEM
2900 '
2910 '  Subroutine to trap missing file
2920 ON ERROR GOTO 2970
2930 OPEN INFILE$ FOR INPUT AS #1                                ' try opening file
2940 ON ERROR GOTO 0
2950 CLOSE
2960 RETURN
2970 PRINT CHR$(7) : PRINT TAB(4) "That file does not exist - type X to Exit or any other key to continue ";
2980 ANS$=INPUT$(1) : IF ANS$="X" OR ANS$="x" THEN CLS : SYSTEM
2990 PRINT
3000 RESUME 800
3010 '
3020 '  Subroutine to clear up impossible zone number
3030  PRINT CHR$(7) : PRINT
3040  PRINT TAB(5) "The zone for "; THISENTRY$; " ["; THISZONE$; "] must be incorrect."
3050  PRINT : PRINT TAB(8) "What is the correct zone number?  ";
3060   INPUT "", THISZONE$ : J=VAL(THISZONE$)
3070   IF J<1 OR J>40 GOTO 3030
3080  PRINT : PRINT TAB(5) "Back to duping and counting...  ";
3090  RETURN
3100 '
3110 '  Subroutine to determine prefix from portable designator
3120  MARK=INSTR(THISENTRY$,SLANT$)
3130  IF MARK>3 THEN THISPFX$=RIGHT$(THISENTRY$,LEN(THISENTRY$)-MARK) ELSE THISPFX$=LEFT$(THISENTRY$,MARK-1)
3140  IF LEN(THISPFX$)>1 GOTO 3210                               ' have prefix - return
3150  IF ASC(THISPFX$)>58 OR ASC(THISPFX$)<47 THEN THISPFX$=LEFT$(THISENTRY$,4) : GOTO 3210 ' (local portable designator)
3160  K=2                                                        ' find position of first numeral in call
3170  WHILE (ASC(MID$(THISENTRY$,K,1))>57 OR ASC(MID$(THISENTRY$,K,1))<48) AND K<LEN(THISENTRY$)
3180   K=K+1
3190   WEND
3200  THISPFX$=LEFT$(THISENTRY$,K-1)+THISPFX$                    ' new prefix = portable number in old prefix
3210  RETURN
3220 '
3230 '  Subroutine to search prefix library for standard country prefix and continent
3240   K=4 : INLIST=NOT TRUE : SAVEDPFX$=THISPFX$
3250   WHILE K>0 AND INLIST=NOT TRUE
3260    THISPFX$=LEFT$(THISPFX$,K)
3270    LOW=1 : HIGH=TABLESIZE : INLIST=NOT TRUE                 ' initial values for binary sort
3280    WHILE LOW<=HIGH AND INLIST=NOT TRUE
3290     L=(LOW+HIGH)\2
3300     IF THISPFX$=PFX$(L) THEN INLIST=TRUE : THISCTRY$=CTRY$(L) : THISCNT$=CNT$(L)
3310     IF THISPFX$<PFX$(L) THEN HIGH=L-1 ELSE LOW=LOW+1
3320     WEND
3330    K=K-1
3340   WEND
3350   RETURN
3360 '
3370 '  Subroutine to search unusual prefix list
3380  IF NRWIERDPFX=0 GOTO 3480                                  ' if the supplementary prefix list is empty, skip ahead
3390   K=4
3400   WHILE K>0
3410    SAVEDPFX$=LEFT$(SAVEDPFX$,K)
3420    FOR J=1 TO NRWIERDPFX
3430     IF SAVEDPFX$=WIERDPFX$(J) THEN INLIST=TRUE : THISCTRY$=WIERDCTRY$(J) : THISCNT$=WIERDCNT$(J) : J=NRWIERDPFX : K=1
3440     NEXT J
3450    K=K-1
3460    WEND
3470   IF INLIST THEN RETURN                                     ' if the prefix was found, return
3480 '  Routine to get prefix definition and continent from user for prefix not found in library
```

```
3490   PRINT CHR$(7) : PRINT
3500   PRINT TAB(5) "The prefix for "; THISENTRY$; " can't be found in the prefix library."
3510   PRINT : PRINT TAB(8) "What is the callsign prefix?  ";
3520     INPUT "", HELDPFX$
3530   PRINT : PRINT TAB(8) "What standard prefix is that equivalent to?  ";
3540     INPUT "", THISPFX$
3550     GOSUB 3230 : IF NOT INLIST GOTO 3490
3560   NRWIERDPFX=NRWIERDPFX+1 : WIERDPFX$(NRWIERDPFX)=HELDPFX$
3570   WIERDCTRY$(NRWIERDPFX)=THISCTRY$ : WIERDCNT$(NRWIERDPFX)=THISCNT$
3580   PRINT : PRINT TAB(5) "Back to duping and counting...  ";
3590   RETURN
3600   '
3610   ' Subroutine to resolve ambiguous prefix with user interaction
3620   THISCTRY$=RIGHT$(THISCTRY$,LEN(THISCTRY$)-1)                ' strip initial delimiter
3630     J=0
3640     WHILE LEN(THISCTRY$)>0
3650     J=J+1
3660     MARK=INSTR(THISCTRY$,".")
3670     AMBCTRY$(J)=LEFT$(THISCTRY$,MARK-1)                ' put multipiler name into array
3680     THISCTRY$=RIGHT$(THISCTRY$,LEN(THISCTRY$)-MARK)
3690     WEND
3700   PRINT CHR$(7) : PRINT
3710   PRINT TAB(5) "The prefix for "; THISENTRY$; " could indicate several different countries."
3720   PRINT : PRINT TAB(8) "The possiblities are:" : PRINT
3730   FOR K=1 TO J
3740     PRINT TAB(11) STR$(K); ". "; AMBCTRY$(K)                ' print choices to screen
3750     NEXT K
3760   PRINT : PRINT TAB(8) "Type the number of the correct country. > ";
3770     INPUT "", CHOICE$
3780     K=VAL(CHOICE$) : IF K > J OR K < 1 THEN PRINT CHR$(7); : GOTO 3760
3790   THISCTRY$=AMBCTRY$(K)
3800   PRINT : PRINT TAB(5) "Back to duping and counting...  ";
3810   RETURN
3820   '
3830   ' Subroutine to print log sheet header
3840   PRINT #2, "   "; MYCALL$; " "; BAND$; " MHz Log"; TAB(72); "Page"; STR$(RAWTOTAL\50+1)
3850   PRINT #2, " Date    Time   Callsign        Sent    Rcvd    New Zone   New Country   Pt."
3860   PRINT #2, " "; STRING$(78,45)
3870   THEDATE$=STR$(DAY)+MON$
3880   RETURN
3890   '
3900   ' Subroutine to print log sheet footer
3910   IF RAWTOTAL MOD 50=0 GOTO 3950                          ' if at the end of a page, jump ahead
3920   FOR J=1 TO 50-(RAWTOTAL MOD 50)
3930     PRINT #2, BL$
3940     NEXT J                                               ' fill last page with blank lines
3950   PRINT #2, " "; STRING$(78,45)
3960   PRINT #2, "    Totals for this page:  Valid QSOs - ";
3970     PRINT #2, USING FOOTFORM$; PGQSOS; PGZONES; PGCTRY; PGPTS
3980   PRINT #2, CHR$(12)
3990   PGQSOS=0 : PGZONES=0 : PGCTRY=0 : PGPTS=0                ' reset page counts
4000   RETURN
```

Program 5

```
10 '   IARULOG.BAS version 1.2 -  13 August 1986  by Clarke Greene K1JX
20 '
30 '   This Microsoft (tm) BASIC program will build a complete log package for the IARU HF Championship.
40 '
50 '   The file containing the log entries must be an ASCII file in the following format:
60 '           (each band requires a separate log entry file)
70 '
80 '           TIME          CALLSIGN          RCV'D REPORT  (each log entry must be followed by a carriage return)
90 '
100 '  At least one space must be between each field of each log entry. Only a changed digit in the time field must
110 '  be present; for example, if the contest begins at 1800Z and the first contact is made at 1802Z and the second
120 '  contact is made at 1805Z, then only 5 need be entered in the time field. If the third contact is made at
130 '  1812Z, then 12 should be entered in the time field. If the next contact is made at 1812Z, then no number need be
140 '  entered in the time field (however, be sure to enter a space to indicate separation between fields).
150 '
160 '  These files will be produced:
170 '           <filename>.LOG - this is a complete log ready for printing
180 '           <filename>.DUP - this is a sorted duplicate listing ready for printing
190 '           <filename>.SUM - this is a summary sheet ready for printing
200 '
210 '
220 '  Depending on the version of BASIC for your particular machine, the CLS (Clear Screen) command must
230 '  be changed.  Consult your own computer's BASIC documentation for more information.
240 '
250 '
260 '  If compiling (a VERY good idea for several orders of magnitude improvement in speed), use O and E switches
270 '
280 '  This program also uses a prefix library file (DXPREFIX.LIB), which MUST be on the same disc (and in the
290 '  same subdirectory) as this program.
300 '
310 ' Define arrays and variables
320 DEFINT A-Z : OPTION BASE 1
330 DIM ENTRY$(1500), MULT$(100), PFX$(900), CNT$(900), WIERDPFX$(50), WIERDCNT$(50), Q(100)
340 BLANK$=" " : BL$="" : SLANT$="/" : TRUE=-1
350 DUPE$="- Duplicate QSO -" : NEWZONE$="     Mult. #"
360 CONTINENTS$="AFASEUNAOCSA"
370 ' Define format strings for printouts
380 LOGFORM$="  \    \ \ \  \          \  \  \  \       \ #  \             \"
390 DUPFORM$="     \       \  \       \  \      \  \      \  \           \"
400 SUMFORM$="     \       \  \       \  \      \  \      \  \        \"
410 '
420 CLS
430 PRINT TAB(24) "IARU HF Competition Log Processor" : PRINT : PRINT
440 '
450 ' Read Prefix table file
460 PRINT TAB(5)  "Reading prefix library...  ";
470 I=0                                              ' initialize array subscript
480 OPEN "DXPREFIX.LIB" FOR INPUT AS #1
490   WHILE NOT EOF(1)
500   I=I+1
510   INPUT #1, PFX$(I), DUMMY$, DUMMY$, CNT$(I)    ' DUMMY$ is a dummy variable for unused data
520   WEND
530 CLOSE
540 TABLESIZE=I                                      ' prefix table length
550 PRINT "done"
560 '
570 ' Get user input
580 PRINT : PRINT TAB(5) "What is the station callsign?  ";
590  INPUT "", MYCALL$
600   THISENTRY$=MYCALL$ : IF INSTR(THISENTRY$,SLANT$)>0 THEN GOSUB 2690 ELSE THISPFX$=LEFT$(THISENTRY$,4)
610   GOSUB 2820 : IF NOT INLIST THEN GOSUB 2960              ' if the prefix can't be found in table, look elsewhere
```

```
620    MYCNT$=THISCNT$
630 PRINT : PRINT TAB(5) "What is the station's zone?  ";
640  INPUT "", MYZONE$
650   IF VAL(MYZONE$)>=1 AND VAL(MYZONE$)<=75 GOTO 680
660    PRINT CHR$(7) : PRINT TAB(8) "Is '"; MYZONE$; "' correct? <Y/N>  ";
670    INPUT "", ANS$ : IF ANS$<>"Y" AND ANS$<>"y" GOTO 630
680   IF VAL(MYZONE$)<10 AND LEN(MYZONE$)=1 THEN MYZONE$="0"+MYZONE$
690 PRINT : PRINT TAB(5) "What is the beginning date of the contest? <dd/mm/yr> ";
700  INPUT "", STARTDATE$
710   MARK=INSTR(STARTDATE$,"/") : IF MARK=0 THEN MARK=INSTR(STARTDATE$,"-")
720   STARTDAY=VAL(LEFT$(STARTDATE$,MARK-1))
730   STARTDATE$=RIGHT$(STARTDATE$,LEN(STARTDATE$)-MARK)
740   MARK=INSTR(STARTDATE$,"/") : IF MARK=0 THEN MARK=INSTR(STARTDATE$,"-")
750   MON$=" July  "
760   YR$=RIGHT$(STARTDATE$,LEN(STARTDATE$)-MARK)
770 PRINT : PRINT TAB(5) "What is the GMT starting time for the contest?  ";
780  INPUT "", STARTGMT$
790 PRINT : PRINT TAB(5) "What file is the log extract located in?  ";
800  INPUT "", INFILE$ : GOSUB 2590                          ' check to see if file is valid
810  IF INSTR(INFILE$,".")<>0 THEN OUTFILE$=LEFT$(INFILE$,INSTR(INFILE$,".")-1) ELSE OUTFILE$=INFILE$
820 PRINT : PRINT TAB(5) "What frequency band is the log extract for?  ";
830  INPUT "", BAND$
840 PRINT : PRINT TAB(5) "Which mode is the log extract for?"
850  PRINT TAB(5) "  [Type 1 for CW or 2 for Phone]  ";
860  INPUT "", MODE
870  IF MODE=1 THEN RST$="599" ELSE RST$="59"
880  SENT$=RST$+MYZONE$
890 '
900 '  Build log file
910 CLS
920 PRINT : PRINT TAB(5) "Duping and counting...  ";
930 '
940 '  Clear arrays
950 FOR I=1 TO 1500
960  ENTRY$(I)=BL$
970  NEXT I
980 FOR I=1 TO 100
990  MULT$(I)=BL$
1000  Q(I)=1
1010  NEXT I
1020 '
1030 '  Initialize variables
1040 RAWTOTAL=0 : QSOS=0 : DUPES=0 : MULTNR=0 : TOTPOINTS=0
1050 DAY=STARTDAY : PREVIOUSGMT$=STARTGMT$
1060 '
1070 '  Open input file and ouput .LOG file
1080 OPEN INFILE$ FOR INPUT AS #1
1090 OPEN OUTFILE$+".LOG" FOR OUTPUT AS #2
1100 '
1110 '  Input data, process, and enter into output file
1120  WHILE NOT EOF(1)
1130   LINE INPUT #1, THISENTRY$                          ' read entire line from disc file
1140    IF THISENTRY$=BL$ GOTO 1740
1150    WHILE ASC(RIGHT$(THISENTRY$,1))<48
1160     THISENTRY$=LEFT$(THISENTRY$,LEN(THISENTRY$)-1)    ' strip off trailing spaces,etc
1170    WEND
1180   IF LEN(THISENTRY$)>0 THEN RAWTOTAL=RAWTOTAL+1 ELSE GOTO 1740
1190 '
1200 '  Separate received report from THISENTRY$
1210   RCVD$=BL$                                           ' initialize rcvd field to blank
1220   WHILE ASC(RIGHT$(THISENTRY$,1))>=48
1230    RCVD$=RIGHT$(THISENTRY$,1)+RCVD$
1240    THISENTRY$=LEFT$(THISENTRY$,LEN(THISENTRY$)-1)
1250   WEND
```

```
1260    IF LEN(RCVD$)<(LEN(RST$)+2) OR ASC(RCVD$)>=65 THEN RCVD$=RST$+RCVD$
1270    WHILE ASC(RIGHT$(THISENTRY$,1))<48
1280     THISENTRY$=LEFT$(THISENTRY$,LEN(THISENTRY$)-1)                ' strip off trailing spaces,etc
1290    WEND
1300  '
1310  '  Separate GMT from THISENTRY$
1320    WHILE ASC(LEFT$(THISENTRY$,1))<48
1330     THISENTRY$=RIGHT$(THISENTRY$,LEN(THISENTRY$)-1)               ' strip off leading spaces
1340    WEND
1350    IF INSTR(THISENTRY$,BLANK$)<>0 THEN GMT$=LEFT$(THISENTRY$,INSTR(THISENTRY$,BLANK$)-1) ELSE GMT$=BL$
1360    THISENTRY$=RIGHT$(THISENTRY$,(LEN(THISENTRY$)-LEN(GMT$)))
1370    WHILE LEFT$(THISENTRY$,1)=BLANK$
1380     THISENTRY$=RIGHT$(THISENTRY$,LEN(THISENTRY$)-1)               ' strip off leading spaces
1390    WEND
1400  '  Fill in missing time data
1410    GMT$=LEFT$(PREVIOUSGMT$,(4-LEN(GMT$)))+GMT$
1420    THEDATE$=BL$ : IF GMT$<PREVIOUSGMT$ THEN DAY=DAY+1 : THEDATE$=STR$(DAY)+MON$
1430  '
1440  '  Check for dupes
1450    DUPE.QSO=NOT TRUE : NOTE$=BL$                                  ' blank note
1460    FOR I=1 TO QSOS
1470     IF ENTRY$(I)=THISENTRY$ THEN NOTE$=DUPE$ : DUPES=DUPES+1 : POINTS=0 : DUPE.QSO=TRUE : I=QSOS
1480    NEXT I
1490    IF DUPE.QSO GOTO 1690                                         ' skip over prefix search if this entry is a dupe
1500    QSOS=QSOS+1 : ENTRY$(QSOS)=THISENTRY$
1510  '
1520  '  Determine "zone" and search through multiplier table
1530    THISZONE$=RIGHT$(RCVD$,LEN(RCVD$)-LEN(RST$))
1540    NEWMULT=TRUE                                                  ' initially call contact a new multiplier
1550    FOR I=1 TO MULTNR
1560     IF MULTS$(I)=THISZONE$ THEN Q(I)=Q(I)+1 : NEWMULT=NOT TRUE : J=MULTNR
1570    NEXT I
1580    IF NEWMULT THEN MULTNR=MULTNR+1 : MULTS$(MULTNR)=THISZONE$ : NOTE$=NEWZONE$+STR$(MULTNR)
1590    IF ASC(THISZONE$)>=65 OR THISZONE$=MYZONE$ THEN POINTS=1 : GOTO 1660
1600  '
1610  '  Determine prefix and search prefix library for contact continent
1620    IF INSTR(THISENTRY$,SLANT$)>0 THEN GOSUB 2690 ELSE THISPFX$=LEFT$(THISENTRY$,4)
1630    GOSUB 2820 : IF NOT INLIST THEN GOSUB 2960                    ' if the prefix can't be found in table, look elsewhere
1640    IF THISCNT$=MYCNT$ THEN POINTS=3 ELSE POINTS=5
1650  '
1660  '  Total QSO points
1670    TOTPOINTS=TOTPOINTS+POINTS
1680  '
1690  '  Write entry to file
1700    IF (RAWTOTAL-1) MOD 50=0 THEN GOSUB 3240                      ' print header if this is the beginning of a page
1710    PRINT #2, USING LOGFORM$; THEDATE$; GMT$; THISENTRY$; SENTS$; RCVD$; POINTS; NOTE$
1720    IF RAWTOTAL MOD 50=0 THEN PRINT #2, CHR$(12)                  ' print a form feed if this is the end of a page
1730    PREVIOUSGMT$=GMT$ : GMT$=BL$
1740   WEND
1750  IF RAWTOTAL MOD 50<>0 THEN PRINT #2, CHR$(12)                   ' if a form feed hasn't been printed, print one now
1760  CLOSE
1770  PRINT "done"
1780  '
1790  '  Build dupe sheet
1800   PRINT : PRINT TAB(5) "Preparing dupe sheet...  ";
1810  '  Sort callsigns for dupe sheet
1820  M=QSOS\2
1830  WHILE M>0
1840   FOR I=M+1 TO QSOS
1850    J=I-M
1860    WHILE J>0
1870     IF ENTRY$(J)>ENTRY$(J+M) THEN SWAP ENTRY$(J),ENTRY$(J+M) : J=J-M ELSE J=0
1880    WEND
1890   NEXT I
```

```
1900  M=M\2
1910  WEND
1920  '
1930  '  Enter dupe sheet into file
1940 OPEN OUTFILE$+".DUP" FOR OUTPUT AS #1
1950  IF QSOS MOD 250=0 THEN LASTPAGE=QSOS\250 ELSE LASTPAGE=QSOS\250+1
1960 FOR PAGE=1 TO LASTPAGE
1970  PRINT #1, SPC(20-(LEN(MYCALL$)+LEN(BAND$))/2); MYCALL$; " -- Dupe Sheet for ";
1980   PRINT #1, BAND$; " MHz Band -- Page"; STR$(PAGE)
1990  PRINT #1, BL$ : PRINT #1, BL$
2000  FOR ROW=1 TO 50
2010   E=(PAGE-1)*250+ROW
2020    PRINT #1, USING DUPFORM$; ENTRY$(E); ENTRY$(E+50); ENTRY$(E+100); ENTRY$(E+150); ENTRY$(E+200)
2030   NEXT ROW
2040  PRINT #1, CHR$(12)                                      ' go to next page
2050  NEXT PAGE
2060 CLOSE
2070 PRINT "done"
2080  '
2090  '  Build summary listing
2100 PRINT : PRINT TAB(5) "Preparing summary sheet...  ";
2110  '  Sort multipliers for summary sheet
2120 M=MULTNR\2
2130 WHILE M>0
2140  FOR I=M+1 TO MULTNR
2150   J=I-M
2160    WHILE J>0
2170     IF MULT$(J)>MULT$(J+M) THEN SWAP MULT$(J),MULT$(J+M) : SWAP Q(J),Q(J+M) : J=J-M ELSE J=0
2180    WEND
2190   NEXT I
2200  M=M\2
2210  WEND
2220  '
2230  '  Append number of qsos per zone onto zone numbers
2240 FOR I=1 TO MULTNR
2250  MULT$(I)=MULT$(I)+SPACE$(6-LEN(MULT$(I)))+" -"+STR$(Q(I))
2260  NEXT I
2270  '
2280  '  Enter summary sheet into file
2290 OPEN OUTFILE$+".SUM" FOR OUTPUT AS #1
2300  PRINT #1, SPC(13-(LEN(MYCALL$)+LEN(BAND$))/2); MYCALL$; " -- Summary Sheet for "; BAND$;
2310   PRINT #1, " MHz Band - "; YR$; " IARU HF Championship"
2320  PRINT #1, BL$
2330  PRINT #1, TAB(11); "Multiplier Listing and number of contacts per multiplier"
2340  PRINT #1, BL$ : PRINT #1, BL$
2350  IF MULTNR MOD 5=0 THEN LASTROW=MULTNR\5 ELSE LASTROW=MULTNR\5+1
2360  FOR ROW=1 TO LASTROW
2370   PRINT #1, USING SUMFORM$; MULT$(ROW); MULT$(ROW+LASTROW); MULT$(ROW+LASTROW*2); MULT$(ROW+LASTROW*3); MULT$(ROW+LASTROW*4)
2380   NEXT ROW
2390  PRINT #1, BL$ : PRINT #1, BL$ : PRINT #1, BL$
2400  PRINT #1, TAB(17); "Total Valid QSOs - "; STR$(QSOS); TAB(45); "Dupes - "; STR$(DUPES)
2410  PRINT #1, TAB(17); "Total QSO points - "; STR$(TOTPOINTS)
2420  PRINT #1, TAB(17); "Multipliers - "; STR$(MULTNR)
2430 CLOSE
2440 PRINT "done"
2450  '
2460  '  Print results
2470 CLS : PRINT CHR$(7)
2480 PRINT : PRINT TAB(5) "Results for the "; BAND$; " MHz band"
2490 PRINT : PRINT TAB(8) "Valid QSOs: "; QSOS
2500 PRINT : PRINT TAB(8) "Duplicate QSOs: "; DUPES
2510 PRINT : PRINT TAB(8) "Total QSO points: "; TOTPOINTS
2520 PRINT : PRINT TAB(8) "Multipliers: "; MULTNR
2530 PRINT : PRINT : PRINT
```

```
2540 PRINT TAB(5) "Type C to continue with another band for this contest,"
2550 PRINT : PRINT TAB(5) "or any other key to Exit    ";
2560 ANS$=INPUT$(1)
2570  IF ANS$="C" OR ANS$="c" THEN CLS : GOTO 790 ELSE CLS : SYSTEM
2580 '
2590 '  Subroutine to trap missing file
2600 ON ERROR GOTO 2650
2610 OPEN INFILE$ FOR INPUT AS #1                              ' try opening file
2620 ON ERROR GOTO 0
2630 CLOSE
2640 RETURN
2650 PRINT CHR$(7) : PRINT TAB(4) "That file does not exist - type X to Exit or any other key to continue ";
2660 ANS$=INPUT$(1) : IF ANS$="X" OR ANS$="x" THEN CLS : SYSTEM
2670 PRINT
2680 RESUME 790
2690 '
2700 '  Subroutine to determine prefix from portable designator
2710  MARK=INSTR(THISENTRY$,SLANT$)
2720  IF MARK>3 THEN THISPFX$=RIGHT$(THISENTRY$,LEN(THISENTRY$)-MARK) ELSE THISPFX$=LEFT$(THISENTRY$,MARK-1)
2730  IF LEN(THISPFX$)>1 GOTO 2800                             ' have prefix - return
2740  IF ASC(THISPFX$)>58 OR ASC(THISPFX$)<47 THEN THISPFX$=LEFT$(THISENTRY$,4) : GOTO 2800 ' (local portable designator)
2750  K=2                                                      ' find position of first numeral in call
2760  WHILE (ASC(MID$(THISENTRY$,K,1))>57 OR ASC(MID$(THISENTRY$,K,1))<48) AND K<LEN(THISENTRY$)
2770   K=K+1
2780  WEND
2790  THISPFX$=LEFT$(THISENTRY$,K-1)+THISPFX$                  ' new prefix = portable number in old prefix
2800  RETURN
2810 '
2820 '  Subroutine to determine station's continent from prefix
2830  K=4 : INLIST=NOT TRUE : SAVEDPFX$=THISPFX$
2840  WHILE K>0 AND INLIST=NOT TRUE
2850   THISPFX$=LEFT$(THISPFX$,K)
2860   LOW=1 : HIGH=TABLESIZE : INLIST=NOT TRUE               ' initial values for binary sort
2870   WHILE LOW<=HIGH AND INLIST=NOT TRUE
2880    L=(LOW+HIGH)\2
2890    IF THISPFX$=PFX$(L) THEN INLIST=TRUE : THISCNT$=CNT$(L)
2900    IF THISPFX$<PFX$(L) THEN HIGH=L-1 ELSE LOW=LOW+1
2910   WEND
2920   K=K-1
2930  WEND
2940  RETURN
2950 '
2960 '  Subroutine to search unusual prefix list
2970  IF NRWIERDPFX=0 GOTO 3080                                ' if the supplementary prefix list is empty, skip ahead
2980   K=4
2990   WHILE K>0
3000   SAVEDPFX$=LEFT$(SAVEDPFX$,K)
3010   FOR J=1 TO NRWIERDPFX
3020    IF SAVEDPFX$=WIERDPFX$(J) THEN INLIST=TRUE : THISCNT$=WIERDCNT$(J) : J=NRWIERDPFX : K=1
3030   NEXT J
3040   K=K-1
3050   WEND
3060   IF INLIST THEN RETURN                                   ' if the prefix was found, return
3070 '
3080 '  Routine to get prefix definition and continent from user for prefix not found in library
3090  PRINT CHR$(7) : PRINT
3100  PRINT TAB(5) "The prefix for "; THISENTRY$; " can't be found in the prefix library."
3110  PRINT : PRINT TAB(8) "What is the callsign prefix?  ";
3120   INPUT "", THISPFX$
3130   NRWIERDPFX=NRWIERDPFX+1 : WIERDPFX$(NRWIERDPFX)=THISPFX$
3140  PRINT : PRINT TAB(8) "What is the continent? [AF, AS, EU, NA, OC, SA]  ";
3150   INPUT "", THISCNT$
3160   FOR J=1 TO 11 STEP 2
3170    IF THISCNT$=MID$(CONTINENTS$,J,2) THEN INLIST=TRUE : J=11
```

```
3180      NEXT J                                           ' check for valid continent name
3190    IF NOT INLIST THEN PRINT CHR$(7);: GOTO 3140
3200   WIERDCNT$(NRWIERDPFX)=THISCNT$
3210 PRINT : PRINT TAB(5) "Back to duping and counting...  ";
3220 RETURN
3230 '
3240 ' Subroutine to print log sheet header
3250 PRINT #2, BL$
3260 PRINT #2, TAB(5); MYCALL$; "  "; BAND$; " MHz Log"; TAB(70); "Page"; STR$(RAWTOTAL\50+1)
3270 PRINT #2, BL$
3280 PRINT #2, "   Date    Time  Callsign         Sent   Rcvd    Pt.        Notes"
3290 PRINT #2, "  "; STRING$(74,61)
3300 THEDATE$=STR$(DAY)+MON$
3310 RETURN
```

Program 6

```
10 '   SSLOG.BAS version 1.3 -  13 August 1986  by Clarke Greene K1JX
20 '
30 '   This Microsoft (tm) BASIC program will build a complete log package for the ARRL Sweepstakes Contest.
40 '
50 '   The file containing the log entries must be an ASCII file in the following format:
60 '           (each band requires a separate log entry file)
70 '
80 '   <*BAND>  <+TIME ON -TIME OFF>   <TIME>   RCV'D NR   RCV'D PREC   RCV'D CALLSIGN   RCV'D CHECK   RCV'D SEC.
90 '           (each log entry must be followed by a carriage return)
100 '
110 '   At least one space must be between each field of each log entry.
120 '
130 '   Any time a band change is made, the first field in the log entry should be an asterisk, followed by the band.
140 '
150 '           for example:   *14  would indicate a band change to 14 MHz.
160 '
170 '   To indicate on and off times, enter the time of going on or off, preceded immediately by a plus <+> to
180 '   indicate ON or a minus <-> to indicate OFF. (MUST be 4 digits)
190 '
200 '           for example:   +0134  would indicate an on time of 0134
210 '
220 '
230 '   Only a changed digit in the time field must be present; for example, if the contest begins at 1800Z and the
240 '   first contact is made at 1802Z and the second contact is made at 1805Z, then only 5 need be entered in the
250 '   time field. If the third contact is made at 1812Z, then 12 should be entered in the time field. If the next
260 '   contact is made at 1812Z, then no number need be entered in the time field (however, be sure to enter a space
270 '   to indicate separation between fields).
280 '
290 '
300 '           A short set of log entries might be:
310 '
320 '       *14 +0100  0104    434 B WB1AVA 78 CT
330 '                      5    2  A K1JX   68 CT
340 '       *21            9   99  A WA1NEV 73 CT
350 '                         765 B W3ZZ   52 MDC
360 '           -0123    14  948 B W1VD    70 CT
370 '
380 '   This log extract shows an operating period beginning at 0100 on the 14 MHz band.
390 '   A band change to 21 MHz was made at 0109, and the period ended at 0123.
400 '
410 '   These files will be produced:
420 '           <filename>.LOG - this is a complete log ready for printing
430 '           <filename>.DUP - this is a sorted duplicate listing ready for printing
440 '           <filename>.SUM - this is a summary sheet ready for printing
450 '
```

```
460 '
470 '    Depending on the version of BASIC for your particular machine, the CLS (Clear Screen) command must
480 '    be changed.  Consult your own computer's BASIC documentation for more information.
490 '
500 '
510 '    If compiling (a VERY good idea for several orders of magnitude improvement in speed), use O and E switches
520 '
530 '
540 '  Define arrays and variables
550 DEFINT A-Z : OPTION BASE 1
560 DIM ENTRY$(2000), MULT$(75), Q(75)
570 MULT$(1)="CT" : MULT$(2)="EMA" : MULT$(3)="ME" : MULT$(4)="NH" : MULT$(5)="RI" : MULT$(6)="VT" : MULT$(7)="WMA"
580 MULT$(8)="ENY" : MULT$(9)="NLI" : MULT$(10)="NNJ" : MULT$(11)="SNJ" : MULT$(12)="WNY"
590 MULT$(13)="DE" : MULT$(14)="EPA" : MULT$(15)="MDC" : MULT$(16)="WPA"
600 MULT$(17)="AL" : MULT$(18)="GA" : MULT$(19)="KY" : MULT$(20)="NC" : MULT$(21)="NFL" : MULT$(22)="SC" : MULT$(23)="SFL"
610  MULT$(24)="TN" : MULT$(25)="VA" : MULT$(26)="WI"
620 MULT$(27)="AR" : MULT$(28)="LA" : MULT$(29)="MS" : MULT$(30)="NM" : MULT$(31)="NTX" : MULT$(32)="OK" : MULT$(33)="STX"
630  MULT$(34)="WTX"
640 MULT$(35)="EB" : MULT$(36)="LAX" : MULT$(37)="ORG" : MULT$(38)="SB" : MULT$(39)="SCV" : MULT$(40)="SDG"
650  MULT$(41)="SF" : MULT$(42)="SJV" : MULT$(43)="SV" : MULT$(44)="PAC"
660 MULT$(45)="AZ" : MULT$(46)="ID" : MULT$(47)="MT" : MULT$(48)="NV" : MULT$(49)="OR" : MULT$(50)="UT" : MULT$(51)="WA"
670  MULT$(52)="WY" : MULT$(53)="AK"
680 MULT$(54)="MI" : MULT$(55)="OH" : MULT$(56)="WV"
690 MULT$(57)="IL" : MULT$(58)="IN" : MULT$(59)="WI"
700 MULT$(60)="CO" : MULT$(61)="IA" : MULT$(62)="KS" : MULT$(63)="MN" : MULT$(64)="MO" : MULT$(65)="NE"
710  MULT$(66)="ND" : MULT$(67)="SD"
720 MULT$(68)="MAR" : MULT$(69)="PQ" : MULT$(70)="ON" : MULT$(71)="MB" : MULT$(72)="SK" : MULT$(73)="AB"
730  MULT$(74)="BC" : MULT$(75)="NWT"
740 BANDCHG$= "*" : BLANK$=" " : BL$="" : CLOCKOFF$="-" : CLOCKON$="+" : TRUE=-1
750 DUPES$="- Duplicate -" : NEWSECTION$="  Mult. #"
760 '  Define format strings for printouts
770 LOGFORM$="\ \  \    \ \      \ \ \ ####  \   \  \        \ \\ \ \ \          \"
780 DUPFORM$="      \        \  \        \  \      \  \        \  \      \"
790 SUMFORM$="  \      \  \      \  \      \  \      \  \      \  \      \"
800 '
810 '  Get information from user
820 CLS
830 PRINT TAB(23) " Sweepstakes Contest Log Processor " : PRINT : PRINT
840 PRINT : PRINT TAB(5) "What is the station callsign?  ";
850  INPUT "", MYCALL$
860 PRINT : PRINT TAB(5) "What is your precedence (A or B)?  ";
870  INPUT "", MYPREC$
880   IF MYPREC$<>"A" AND MYPREC$<>"B" THEN PRINT CHR$(7);: GOTO 860
890 PRINT : PRINT TAB(5) "What is your check?  ";
900  INPUT "", MYCHECK$
910   IF LEN(MYCHECK$)<>2 THEN PRINT CHR$(7);: GOTO 890
920 PRINT : PRINT TAB(5) "What ARRL section is the station in?  ";
930  INPUT "", MYSECTION$
940   SECTION$=MYSECTION$ : GOSUB 3730 : IF NOT INLIST THEN PRINT CHR$(7);: GOTO 920
950 PRINT : PRINT TAB(5) "What is the beginning date of the contest? <dd/mm/yr> ";
960  INPUT "", STARTDATE$
970   MARK=INSTR(STARTDATE$,"/") : IF MARK=0 THEN MARK=INSTR(STARTDATE$,"-")
980    STARTDAY=VAL(LEFT$(STARTDATE$,MARK-1))
990    STARTDATE$=RIGHT$(STARTDATE$,LEN(STARTDATE$)-MARK)
1000   MARK=INSTR(STARTDATE$,"/") : IF MARK=0 THEN MARK=INSTR(STARTDATE$,"-")
1010    MON$=" Nov"
1020    YR$=RIGHT$(STARTDATE$,LEN(STARTDATE$)-MARK)
1030 PRINT : PRINT TAB(5) "What is the GMT starting time for the contest?  ";
1040  INPUT "", STARTGMT$
1050   IF LEN(STARTGMT$)<>4 THEN PRINT CHR$(7);: GOTO 1030
1060 PRINT : PRINT TAB(5) "What file is the log extract located in?  ";
1070  INPUT "", INFILE$
1080   GOSUB 3600                                        ' check to see if file exists
1090 IF INSTR(INFILE$,".")<>0 THEN OUTFILE$=LEFT$(INFILE$,INSTR(INFILE$,".")-1) ELSE OUTFILE$=INFILE$
```

```
1100 '
1110 '  Collect log extract from input file and assign data to arrays
1120 CLS
1130 PRINT : PRINT TAB(5) "Duping and counting...  ";
1140 '
1150 '  Initialize variables
1160 RAWTOTAL=0 : QSOS=0 : DUPES=0 : MULTNR=0 : TOTAL.TIME=0
1170 DAY=STARTDAY : PREVIOUSGMT$=STARTGMT$                          ' (re)set starting time and day
1180 ONTIME$=STARTGMT$
1190 ON.TIME=VAL(LEFT$(ONTIME$,2))*60+VAL(RIGHT$(ONTIME$,2))        ' set on-time to default value of contest beginning
1200 '
1210 '  Open input data file and output .LOG file
1220 OPEN INFILE$ FOR INPUT AS #1
1230 OPEN OUTFILE$+".LOG" FOR OUTPUT AS #2
1240 '
1250 '  Input data, process, and enter into output file
1260  WHILE NOT EOF(1)
1270   LINE INPUT #1, THISENTRY$                                    ' read entire line from disc file
1280    IF THISENTRY$=BL$ GOTO 2520                                 ' jump ahead if this is a blank line
1290    WHILE ASC(RIGHT$(THISENTRY$,1))<48
1300     THISENTRY$=LEFT$(THISENTRY$,LEN(THISENTRY$)-1)             ' strip off trailing spaces,etc
1310    WEND
1320    IF LEN(THISENTRY$)>0 THEN RAWTOTAL=RAWTOTAL+1 ELSE GOTO 2520
1330 '
1340 '  Look for band change
1350    BAND$=BL$
1360    MARK=INSTR(THISENTRY$,BANDCHG$) : IF MARK=0 GOTO 1450       ' if there isn't a bandchange, jump ahead
1370    MID$(THISENTRY$,MARK,1)=BLANK$                              ' blank out *
1380    MARK=MARK+1                                                 ' point to next character
1390    WHILE ASC(MID$(THISENTRY$,MARK,1))>=46 AND ASC(MID$(THISENTRY$,MARK,1))<=57
1400     BAND$=BAND$+MID$(THISENTRY$,MARK,1)                        ' parse bandchange string
1410     MID$(THISENTRY$,MARK,1)=BLANK$
1420     MARK=MARK+1
1430    WEND
1440 '
1450 '  Look for on-time indication
1460    ONTIME$=BL$ : TIME.NOTE$=BL$
1470    MARK=INSTR(THISENTRY$,CLOCKON$) : IF MARK=0 GOTO 1590
1480    MID$(THISENTRY$,MARK,1)=BLANK$
1490    MARK=MARK+1
1500    WHILE ASC(MID$(THISENTRY$,MARK,1))>=46 AND ASC(MID$(THISENTRY$,MARK,1))<=57
1510     ONTIME$=ONTIME$+MID$(THISENTRY$,MARK,1)                    ' parse on-time string
1520     MID$(THISENTRY$,MARK,1)=BLANK$
1530     MARK=MARK+1
1540    WEND
1550    ONTIME$=LEFT$(PREVIOUSGMT$,(4-LEN(ONTIME$)))+ONTIME$        ' complete missing time string
1560    ON.TIME=VAL(LEFT$(ONTIME$,2))*60+VAL(RIGHT$(ONTIME$,2))
1570    TIME.NOTE$="on  "+ONTIME$
1580 '
1590 '  Look for off-time indication
1600    OFFTIME$=BL$
1610    MARK=INSTR(THISENTRY$,CLOCKOFF$) : IF MARK=0 GOTO 1750
1620    MID$(THISENTRY$,MARK,1)=BLANK$
1630    MARK=MARK+1
1640    WHILE ASC(MID$(THISENTRY$,MARK,1))>=46 AND ASC(MID$(THISENTRY$,MARK,1))<=57
1650     OFFTIME$=OFFTIME$+MID$(THISENTRY$,MARK,1)                  ' parse off-time string
1660     MID$(THISENTRY$,MARK,1)=BLANK$
1670     MARK=MARK+1
1680    WEND
1690    OFFTIME$=LEFT$(PREVIOUSGMT$,(4-LEN(OFFTIME$)))+OFFTIME$     ' complete missing time string
1700    TIME.NOTE$="off "+OFFTIME$
1710    OFF.TIME=VAL(LEFT$(OFFTIME$,2))*60+VAL(RIGHT$(OFFTIME$,2))
1720    IF OFF.TIME<ON.TIME THEN OFF.TIME=OFF.TIME+1440
1730    TOTAL.TIME=TOTAL.TIME+OFF.TIME-ON.TIME                      ' calculate cumulative operating time
```

```
1740 '
1750 ' Separate received section from THISENTRY$
1760   SECTION$=BL$                                             ' initialize section field to blank
1770   WHILE ASC(RIGHT$(THISENTRY$,1))>=48
1780    SECTION$=RIGHT$(THISENTRY$,1)+SECTION$
1790    THISENTRY$=LEFT$(THISENTRY$,LEN(THISENTRY$)-1)
1800   WEND
1810   WHILE ASC(RIGHT$(THISENTRY$,1))<48
1820    THISENTRY$=LEFT$(THISENTRY$,LEN(THISENTRY$)-1)          ' strip off trailing spaces,etc
1830   WEND
1840 '
1850 ' Separate received check from THISENTRY$
1860   CHECK$=BL$                                               ' initialize check field to blank
1870   WHILE ASC(RIGHT$(THISENTRY$,1))>=48
1880    CHECK$=RIGHT$(THISENTRY$,1)+CHECK$
1890    THISENTRY$=LEFT$(THISENTRY$,LEN(THISENTRY$)-1)
1900   WEND
1910   WHILE ASC(RIGHT$(THISENTRY$,1))<48
1920    THISENTRY$=LEFT$(THISENTRY$,LEN(THISENTRY$)-1)          ' strip off trailing spaces,etc
1930   WEND
1940 '
1950 ' Separate callsign from THISENTRY$
1960   CALLSIGN$=BL$                                            ' initialize callsign field to blank
1970   WHILE ASC(RIGHT$(THISENTRY$,1))>=47
1980    CALLSIGN$=RIGHT$(THISENTRY$,1)+CALLSIGN$
1990    THISENTRY$=LEFT$(THISENTRY$,LEN(THISENTRY$)-1)
2000   WEND
2010   WHILE ASC(RIGHT$(THISENTRY$,1))<48
2020    THISENTRY$=LEFT$(THISENTRY$,LEN(THISENTRY$)-1)          ' strip off trailing spaces,etc
2030   WEND
2040 '
2050 ' Separate received prec from THISENTRY$
2060   PREC$=RIGHT$(THISENTRY$,1)                               ' 1 character prec
2070    THISENTRY$=LEFT$(THISENTRY$,LEN(THISENTRY$)-1)
2080   WHILE ASC(RIGHT$(THISENTRY$,1))<48
2090    THISENTRY$=LEFT$(THISENTRY$,LEN(THISENTRY$)-1)          ' strip off trailing spaces,etc
2100   WEND
2110 '
2120 ' Separate received number from THISENTRY$
2130   NR$=BL$                                                  ' initialize number field to blank
2140   WHILE ASC(RIGHT$(THISENTRY$,1))>=48
2150    NR$=RIGHT$(THISENTRY$,1)+NR$
2160    THISENTRY$=LEFT$(THISENTRY$,LEN(THISENTRY$)-1)
2170   WEND
2180   WHILE ASC(RIGHT$(THISENTRY$,1))<48
2190    THISENTRY$=LEFT$(THISENTRY$,LEN(THISENTRY$)-1)          ' strip off trailing spaces,etc
2200   WEND
2210   NR$=SPACE$(4-LEN(NR$))+NR$+"  "+PREC$                    ' concatenate number and prec.
2220 '
2230 ' Separate GMT from THISENTRY$
2240   IF LEN(THISENTRY$)=0 GOTO 2290
2250   WHILE ASC(LEFT$(THISENTRY$,1))<48
2260    THISENTRY$=RIGHT$(THISENTRY$,LEN(THISENTRY$)-1)         ' strip off leading spaces
2270   WEND
2280 ' Fill in missing time data
2290   GMT$=LEFT$(PREVIOUSGMT$,(4-LEN(THISENTRY$)))+THISENTRY$
2300   THEDATE$=BL$ : IF GMT$<PREVIOUSGMT$ THEN DAY=DAY+1 : THEDATE$=STR$(DAY)+MON$ : THEDATE$=RIGHT$(THEDATE$,LEN(THEDATE$)-1)
2310 '
2320 ' Check for dupes
2330   DUPE.QSO=NOT TRUE : NOTE$=BL$                            ' clear note field
2340   FOR I=1 TO QSOS
2350    IF CALLSIGN$=ENTRY$(I) THEN NOTE$=DUPE$ : DUPES=DUPES+1 : DUPE.QSO=TRUE : I=QSOS
2360   NEXT I
2370   IF DUPE.QSO GOTO 2450                                    ' if contact is a dupe, skip over multiplier search
```

```
2380    QSOS=QSOS+1 : ENTRY$(QSOS)=CALLSIGN$
2390 '
2400 '   Check for multipliers
2410    GOSUB 3730 : IF NOT INLIST THEN GOSUB 3800
2420    Q(INDEX)=Q(INDEX)+1
2430    IF Q(INDEX)=1 THEN MULTNR=MULTNR+1 : NOTE$=NEWSECTION$+STR$(MULTNR)
2440 '
2450 '   Write entry to file
2460    IF (RAWTOTAL-1) MOD 50=0 THEN GOSUB 3890              ' print the header if this is the beginning of a page
2470    PRINT #2, USING LOGFORM$; BAND$; THEDATE$; TIME.NOTE$; GMT$; RAWTOTAL; NR$; CALLSIGN$; CHECK$; SECTION$; NOTE$
2480    IF RAWTOTAL MOD 50=0 THEN PRINT #2, CHR$(12)          ' print a form feed if this is the end of the page
2490 '
2500 '   Update variables for next entry
2510    PREVIOUSGMT$=GMT$ : GMT$=BL$
2520 WEND
2530 IF RAWTOTAL MOD 50 <>0 THEN PRINT #2, CHR$(12)           ' if a form feed wasn't printed, print one now
2540 CLOSE
2550 '
2560 '   Calculate total operating time
2570 IF OFFTIME$<>BL$ GOTO 2610                               ' if an off-time was specified, jump ahead
2580    OFF.TIME=VAL(LEFT$(PREVIOUSGMT$,2))*60+VAL(RIGHT$(PREVIOUSGMT$,2))
2590    IF OFF.TIME<ON.TIME THEN OFF.TIME=OFF.TIME+1440
2600    TOTAL.TIME=TOTAL.TIME+OFF.TIME-ON.TIME                ' calculate cumulative operating time
2610 TOTAL.HRS$=STR$(TOTAL.TIME\60)
2620 TOTAL.MIN$=STR$(TOTAL.TIME MOD 60)
2630    TOTAL.MIN$=RIGHT$(TOTAL.MIN$,LEN(TOTAL.MIN$)-1)       ' delete leading space
2640    WHILE LEN(TOTAL.MIN$)<2
2650     TOTAL.MIN$="0"+TOTAL.MIN$                            ' append leading zero
2660     WEND
2670    TOTAL.MIN$=":"+TOTAL.MIN$
2680 PRINT "done"
2690 '
2700 '   Build dupe sheet
2710 PRINT : PRINT TAB(5) "Preparing dupe sheet...  ";
2720 '
2730 '   Sort callsigns for dupe sheet
2740 M=QSOS\2
2750 WHILE M>0
2760  FOR I=M+1 TO QSOS
2770   J=I-M
2780    WHILE J>0
2790     IF ENTRY$(J)>ENTRY$(J+M) THEN SWAP ENTRY$(J),ENTRY$(J+M) : J=J-M ELSE J=0
2800     WEND
2810   NEXT I
2820  M=M\2
2830  WEND
2840 '
2850 '   Enter dupe sheet into file
2860 OPEN OUTFILE$+".DUP" FOR OUTPUT AS #1
2870  IF QSOS MOD 250=0 THEN LASTPAGE=QSOS\250 ELSE LASTPAGE=QSOS\250+1
2880 FOR PAGE=1 TO LASTPAGE
2890  PRINT #1, SPC(22-(LEN(MYCALL$))/2); MYCALL$; " -- Sweepstakes Dupe Sheet -- Page"; STR$(PAGE)
2900  PRINT #1, BL$ : PRINT #1, BL$
2910  FOR ROW=1 TO 50
2920   E=(PAGE-1)*250+ROW
2930    PRINT #1, USING DUPFORM$; ENTRY$(E); ENTRY$(E+50); ENTRY$(E+100); ENTRY$(E+150); ENTRY$(E+200)
2940   NEXT ROW
2950  PRINT #1, CHR$(12)                                      ' go to next page
2960  NEXT PAGE
2970 CLOSE
2980 PRINT "done"
2990 '
3000 '   Build summary listing
3010 PRINT : PRINT TAB(5) "Preparing summary sheet...  ";
```

```
3020 '
3030 '  Append QSO count to section name
3040  FOR I=1 TO 75
3050   IF Q(I)=0 THEN MULT$(I)=MULT$(I)+SPACE$(4-LEN(MULT$(I)))+"- None" : GOTO 3070
3060   MULT$(I)=MULT$(I)+SPACE$(4-LEN(MULT$(I)))+"-"+STR$(Q(I))
3070    NEXT I
3080 '
3090 '  Enter summary sheet into file
3100 OPEN OUTFILE$+".SUM" FOR OUTPUT AS #1
3110  PRINT #1, SPC(16-(LEN(MYCALL$))/2); MYCALL$; " -- Summary Sheet for "; YR$; " Sweepstakes Contest"
3120  PRINT #1, BL$ : PRINT #1, BL$
3130  PRINT #1, TAB(15) "Section Listing and number of contacts per Section"
3140  PRINT #1, BL$ : PRINT #1, BL$
3150  PRINT #1, "       1           2           3           4           5           6"
3160  PRINT #1, BL$
3170  PRINT #1, USING SUMFORM$; MULT$(1); MULT$(8); MULT$(13); MULT$(17); MULT$(27); MULT$(35)
3180  PRINT #1, USING SUMFORM$; MULT$(2); MULT$(9); MULT$(14); MULT$(18); MULT$(28); MULT$(36)
3190  PRINT #1, USING SUMFORM$; MULT$(3); MULT$(10); MULT$(15); MULT$(19); MULT$(29); MULT$(37)
3200  PRINT #1, USING SUMFORM$; MULT$(4); MULT$(11); MULT$(16); MULT$(20); MULT$(30); MULT$(38)
3210  PRINT #1, USING SUMFORM$; MULT$(5); MULT$(12); BL$; MULT$(21); MULT$(31); MULT$(39)
3220  PRINT #1, USING SUMFORM$; MULT$(6); BL$; BL$; MULT$(22); MULT$(32); MULT$(40)
3230  PRINT #1, USING SUMFORM$; MULT$(7); BL$; BL$; MULT$(23); MULT$(33); MULT$(41)
3240  PRINT #1, USING SUMFORM$; BL$; BL$; BL$; MULT$(24); MULT$(34); MULT$(42)
3250  PRINT #1, USING SUMFORM$; BL$; BL$; BL$; MULT$(25); BL$; MULT$(43)
3260  PRINT #1, USING SUMFORM$; BL$; BL$; BL$; MULT$(26); BL$; MULT$(44)
3270  PRINT #1, BL$ : PRINT #1, BL$ : PRINT #1, BL$
3280  PRINT #1, "       7           8           9           0           VE"
3290  PRINT #1, BL$
3300  PRINT #1, USING SUMFORM$; MULT$(44); MULT$(54); MULT$(57); MULT$(60); MULT$(68); BL$
3310  PRINT #1, USING SUMFORM$; MULT$(45); MULT$(55); MULT$(58); MULT$(61); MULT$(69); BL$
3320  PRINT #1, USING SUMFORM$; MULT$(46); MULT$(56); MULT$(59); MULT$(62); MULT$(70); BL$
3330  PRINT #1, USING SUMFORM$; MULT$(47); BL$; BL$; MULT$(63); MULT$(71); BL$
3340  PRINT #1, USING SUMFORM$; MULT$(48); BL$; BL$; MULT$(64); MULT$(72); BL$
3350  PRINT #1, USING SUMFORM$; MULT$(49); BL$; BL$; MULT$(65); MULT$(73); BL$
3360  PRINT #1, USING SUMFORM$; MULT$(50); BL$; BL$; MULT$(66); MULT$(74); BL$
3370  PRINT #1, USING SUMFORM$; MULT$(51); BL$; BL$; MULT$(67); MULT$(75); BL$
3380  PRINT #1, USING SUMFORM$; MULT$(52); BL$; BL$; BL$; BL$; BL$
3390  PRINT #1, USING SUMFORM$; MULT$(53); BL$; BL$; BL$; BL$; BL$
3400  PRINT #1, BL$ : PRINT #1, BL$ : PRINT #1, BL$
3410  PRINT #1, TAB(18); "Total Valid QSOs - "; STR$(QSOS); TAB(45); "Dupes - "; STR$(DUPES)
3420  PRINT #1, TAB(18); "Sections - "; STR$(MULTNR)
3430  PRINT #1, TAB(18); "Total operating time - "; TOTAL.HRS$; TOTAL.MIN$
3440  PRINT #1, CHR$(12)
3450 CLOSE
3460 PRINT "done"
3470 '
3480 '  Print results
3490 CLS : PRINT CHR$(7)
3500 PRINT : PRINT TAB(8) "Results" : PRINT TAB(8) "-------"
3510 PRINT : PRINT TAB(8) "Valid QSOs: "; QSOS
3520 PRINT : PRINT TAB(8) "Duplicate QSOs: "; DUPES
3530 PRINT : PRINT TAB(8) "Sections: "; MULTNR
3540 PRINT : PRINT TAB(8) "Total operating time: "; TOTAL.HRS$; TOTAL.MIN$
3550 PRINT : PRINT : PRINT
3560 PRINT TAB(29) "Press any key to Exit  ";
3570  ANS$=INPUT$(1)
3580 CLS : SYSTEM
3590 '
3600 '  Subroutine to trap missing file
3610 ON ERROR GOTO 3670
3620 OPEN INFILE$ FOR INPUT AS #1                    ' try opening file
3630 ON ERROR GOTO 0
3640 CLOSE
3650 RETURN
```

```
3660 '
3670 '  Error trap for missing file
3680 PRINT CHR$(7) : PRINT TAB(4) "That file does not exist - type X to Exit or any other key to continue ";
3690 ANS$=INPUT$(1) : IF ANS$="X" OR ANS$="x" THEN CLS : SYSTEM
3700 PRINT
3710 RESUME 1060
3720 '
3730 '  Subroutine to search multiplier list
3740 INLIST=NOT TRUE
3750 FOR I=1 TO 75
3760  IF SECTION$=MULT$(I) THEN INDEX=I : INLIST=TRUE : I=75
3770  NEXT I
3780 RETURN
3790 '
3800 '  Subroutine to clear up impossible section
3810  PRINT CHR$(7) : PRINT
3820  PRINT TAB(5) "The section for "; CALLSIGN$; " ["; SECTION$; "] must be incorrect."
3830  PRINT : PRINT TAB(8) "What is the correct section?  ";
3840   INPUT "", SECTION$
3850   GOSUB 3730 : IF NOT INLIST GOTO 3810
3860   PRINT : PRINT TAB(5) "Back to duping and counting...  ";
3870   RETURN
3880 '
3890 '  Subroutine to print log sheet header
3900 PRINT #2, MYCALL$; " Sweepstakes Log"; TAB(73); "Page"; STR$(RAWTOTAL\50+1)
3910 PRINT #2, TAB(27) "Sent:  :--- "; MYPREC$; " "; MYCALL$; " "; MYCHECK$; " "; MYSECTION$
3920 PRINT #2, TAB(34) "V"
3930 PRINT #2, "Band Date   On/Off   Time  Nr   Nr Pr Call        Ck  Sec      Notes"
3940 PRINT #2, STRING$(80,61)
3950 THEDATE$=STR$(DAY)+MON$ : THEDATE$=RIGHT$(THEDATE$,LEN(THEDATE$)-1)
3960 RETURN
```

Program 7

```
10 '   VHFLOG.BAS version 1.3 -  13 August 1986  by Clarke Greene K1JX
20 '
30 '   This Microsoft (tm) BASIC program will build a complete log package for any of the ARRL VHF Contests.
40 '
50 '   The file containing the log entries must be an ASCII file in the following format:
60 '           (each band requires a separate log entry file)
70 '
80 '       TIME          CALLSIGN          RCV'D GRID SQUARE   (each log entry must be followed by a carriage return)
90 '
100 '   At least one space must be between each field of each log entry. Only a changed digit in the time field must
110 '   be present; for example, if the contest begins at 1800Z and the first contact is made at 1802Z and the second
120 '   contact is made at 1805Z, then only 5 need be entered in the time field. If the third contact is made at
130 '   1812Z, then 12 should be entered in the time field. If the next contact is made at 1812Z, then no number need be
140 '   entered in the time field (however, be sure to enter a space to indicate separation between fields).
150 '
160 '   These files will be produced:
170 '           <filename>.LOG - this is a complete log ready for printing
180 '           <filename>.DUP - this is a sorted duplicate listing ready for printing
190 '           <filename>.SUM - this is a summary sheet ready for printing
200 '
210 '
220 '   Depending on the version of BASIC for your particular machine, the CLS (Clear Screen) command must
230 '   be changed.  Consult your own computer's BASIC documentation for more information.
240 '
250 '
260 '   If compiling (a VERY good idea for several orders of magnitude improvement in speed), use O and E switches
```

```
270 '
280 '
290 '  Define arrays and variables
300 DEFINT A-Z : OPTION BASE 1
310 DIM ENTRY$(1500), GRID$(1500), MULT$(250), Q(250), MONTH$(12)
320 BLANK$=" " : BL$="" : TRUE=-1
330 MONTH$(1)="Jan." : MONTH$(2)="Feb." : MONTH$(3)="Mar." : MONTH$(4)="Apr."
340 MONTH$(5)="May" : MONTH$(6)="June" : MONTH$(7)="July" : MONTH$(8)="Aug."
350 MONTH$(9)="Sept." : MONTH$(10)="Oct." : MONTH$(11)="Nov." : MONTH$(12)="Dec."
360 DUPES$="- Duplicate QSO -" : DIFGRID$="Station in new Grid" : NEWGRID$=" - Mult. #"
370 '  Define format strings for printouts
380 LOGFORM$="  \      \ \ \      \        \      \ \    \ \    \          \"
390 DUPFORM$="   \        \ \      \ \        \ \      \ \        \ \        \"
400 SUMFORM$="   \        \ \        \ \        \ \        \ \        \"
410 '
420 '  Get information from user
430 CLS
440 PRINT TAB(27) " VHF Contest Log Processor " : PRINT : PRINT
450 PRINT : PRINT TAB(5) "What is the station callsign?  ";
460  INPUT "", MYCALL$
470 PRINT : PRINT TAB(5) "What Grid Square is the station in?  ";
480  INPUT "", MYGRID$
490   IF LEN(MYGRID$)<>4 THEN PRINT CHR$(7);: GOTO 470
500 PRINT : PRINT TAB(5) "What is the beginning date of the contest? <dd/mm/yr> ";
510  INPUT "", STARTDATE$
520   MARK=INSTR(STARTDATE$,"/") : IF MARK=0 THEN MARK=INSTR(STARTDATE$,"-")
530    STARTDAY=VAL(LEFT$(STARTDATE$,MARK-1))
540    STARTDATE$=RIGHT$(STARTDATE$,LEN(STARTDATE$)-MARK)
550   MARK=INSTR(STARTDATE$,"/") : IF MARK=0 THEN MARK=INSTR(STARTDATE$,"-")
560    MON$=MONTH$(VAL(LEFT$(STARTDATE$,MARK-1)))
570    YR$=RIGHT$(STARTDATE$,LEN(STARTDATE$)-MARK)
580 PRINT : PRINT TAB(5) "What is the GMT starting time for the contest?  ";
590  INPUT "", STARTGMT$
600 PRINT : PRINT TAB(5) "What file is the log extract located in?  ";
610  INPUT "", INFILE$
620  GOSUB 2250                                          ' check to see if file exists
630  IF INSTR(INFILE$,".")<>0 THEN OUTFILE$=LEFT$(INFILE$,INSTR(INFILE$,".")-1) ELSE OUTFILE$=INFILE$
640 PRINT : PRINT TAB(5) "What frequency band is the log extract for?  ";
650  INPUT "", BAND$
660 '
670 '  Collect log extract from input file and assign data to arrays
680 CLS
690 PRINT : PRINT TAB(5) "Duping and counting...  ";
700 '
710 '  Clear arrays
720 FOR I=1 TO 1500
730  ENTRY$(I)=BL$
740  GRID$(I)=BL$
750  NEXT I
760 FOR I=1 TO 250
770  MULT$(I)=BL$
780  Q(I)=1
790  NEXT I
800 '
810 '  Initialize variables
820 RAWTOTAL=0 : QSOS=0 : DUPES=0 : MULTNR=0
830 DAY=STARTDAY : PREVIOUSGMT$=STARTGMT$                        ' (re)set starting time and day
840 '
850 '  Open input data file and output .LOG file
860 OPEN INFILE$ FOR INPUT AS #1
870 OPEN OUTFILE$+".LOG" FOR OUTPUT AS #2
880 '
890 '  Input data, process, and enter into output file
900  WHILE NOT EOF(1)
```

```
910    LINE INPUT #1, THISENTRY$                              ' read entire line from disc file
920     IF THISENTRY$=BL$ GOTO 1400                           ' jump ahead if this is a blank line
930     WHILE ASC(RIGHT$(THISENTRY$,1))<48
940      THISENTRY$=LEFT$(THISENTRY$,LEN(THISENTRY$)-1)       ' strip off trailing spaces,etc
950      WEND
960     IF LEN(THISENTRY$)>0 THEN RAWTOTAL=RAWTOTAL+1 ELSE GOTO 1400
970  '
980  '   Separate grid from THISENTRY$
990     THISGRID$=RIGHT$(THISENTRY$,4)
1000    THISENTRY$=LEFT$(THISENTRY$,LEN(THISENTRY$)-4)
1010    WHILE ASC(RIGHT$(THISENTRY$,1))<48
1020     THISENTRY$=LEFT$(THISENTRY$,LEN(THISENTRY$)-1)       ' strip off trailing spaces
1030     WEND
1040  '
1050  '   Separate GMT from THISENTRY$
1060    WHILE ASC(LEFT$(THISENTRY$,1))<48
1070     THISENTRY$=RIGHT$(THISENTRY$,LEN(THISENTRY$)-1)      ' strip off leading spaces
1080     WEND
1090    IF INSTR(THISENTRY$,BLANK$)<>0 THEN GMT$=LEFT$(THISENTRY$,INSTR(THISENTRY$,BLANK$)-1) ELSE GMT$=BL$
1100    THISENTRY$=RIGHT$(THISENTRY$,(LEN(THISENTRY$)-LEN(GMT$)))
1110    WHILE LEFT$(THISENTRY$,1)=BLANK$
1120     THISENTRY$=RIGHT$(THISENTRY$,LEN(THISENTRY$)-1)      ' strip off leading spaces
1130     WEND
1140  '  Fill in missing time data
1150    GMT$=LEFT$(PREVIOUSGMT$,(4-LEN(GMT$)))+GMT$
1160    THEDATE$=BL$ : IF GMT$<PREVIOUSGMT$ THEN DAY=DAY+1 : THEDATE$=STR$(DAY)+BLANK$+MON$
1170  '
1180  '   Check for dupes
1190    DUPE.QSO=NOT TRUE : NOTES$=BL$                        ' clear note field
1200    FOR I=1 TO QSOS
1210     IF THISENTRY$=ENTRY$(I) AND THISGRID$<>GRID$(I) THEN NOTES$=DIFGRID$
1220     IF THISENTRY$=ENTRY$(I) AND THISGRID$=GRID$(I) THEN NOTES$=DUPES$ : DUPES=DUPES+1 : DUPE.QSO=TRUE : I=QSOS
1230     NEXT I
1240    IF DUPE.QSO GOTO 1340                                 ' if contact is a dupe, skip over multiplier search
1250    QSOS=QSOS+1 : ENTRY$(QSOS)=THISENTRY$ : GRID$(QSOS)=THISGRID$
1260  '
1270  '   Check for multipliers
1280    NEWMULT=TRUE                                          ' initially set multiplier flag true
1290    FOR I=1 TO MULTNR
1300     IF MULT$(I)=THISGRID$ THEN Q(I)=Q(I)+1 : NEWMULT=NOT TRUE : J=MULTNR
1310     NEXT I
1320    IF NEWMULT THEN MULTNR=MULTNR+1 : MULT$(MULTNR)=THISGRID$ : NOTES$=THISGRID$+NEWGRID$+STR$(MULTNR)
1330  '
1340  '   Write entry to file
1350     IF (RAWTOTAL-1) MOD 50=0 THEN GOSUB 2380             ' print the header if this is the beginning of a page
1360     PRINT #2, USING LOGFORM$; THEDATE$; GMT$; THISENTRY$; MYGRID$; THISGRID$; NOTES$
1370     IF RAWTOTAL MOD 50=0 THEN PRINT #2, CHR$(12)         ' print a form feed if this is the end of the page
1380  '  Update variables for next entry
1390     PREVIOUSGMT$=GMT$ : GMT$=BL$
1400    WEND
1410  IF RAWTOTAL MOD 50 <>0 THEN PRINT #2, CHR$(12)          ' if a form feed wasn't printed, print one now
1420  CLOSE
1430  PRINT "done"
1440  '
1450  '   Build dupe sheet
1460  PRINT : PRINT TAB(5) "Preparing dupe sheet...  ";
1470  '
1480  '   Sort callsigns for dupe sheet
1490  M=QSOS\2
1500  WHILE M>0
1510   FOR I=M+1 TO QSOS
1520    J=I-M
1530    WHILE J>0
1540     IF ENTRY$(J)>ENTRY$(J+M) THEN SWAP ENTRY$(J),ENTRY$(J+M) : J=J-M ELSE J=0
```

```
1550    WEND
1560    NEXT I
1570    M=M\2
1580   WEND
1590   '
1600   ' Enter dupe sheet into file
1610 OPEN OUTFILE$+".DUP" FOR OUTPUT AS #1
1620  IF QSOS MOD 250=0 THEN LASTPAGE=QSOS\250 ELSE LASTPAGE=QSOS\250+1
1630 FOR PAGE=1 TO LASTPAGE
1640  PRINT #1, SPC(20-(LEN(MYCALL$)+LEN(BAND$))/2); MYCALL$; " -- Dupe Sheet for "; BAND$; " MHz Band -- Page"; STR$(PAGE)
1650  PRINT #1, BL$ : PRINT #1, BL$
1660  FOR ROW=1 TO 50
1670   E=(PAGE-1)*250+ROW
1680    PRINT #1, USING DUPFORM$; ENTRY$(E); ENTRY$(E+50); ENTRY$(E+100); ENTRY$(E+150); ENTRY$(E+200)
1690   NEXT ROW
1700  PRINT #1, CHR$(12)                                      ' go to next page
1710  NEXT PAGE
1720 CLOSE
1730 PRINT "done"
1740 '
1750 ' Build summary listing
1760 PRINT : PRINT TAB(5) "Preparing summary sheet...  ";
1770 '
1780 ' Sort multipliers for summary sheet
1790 M=MULTNR\2
1800 WHILE M>0
1810  FOR I=M+1 TO MULTNR
1820   J=I-M
1830    WHILE J>0
1840     IF MULT$(J)>MULT$(J+M) THEN SWAP MULT$(J),MULT$(J+M) : SWAP Q(J),Q(J+M) : J=J-M ELSE J=0
1850    WEND
1860   NEXT I
1870  M=M\2
1880 WEND
1890 '
1900 ' Append QSO count to Grid Square
1910  FOR I=1 TO MULTNR
1920   MULT$(I)=MULT$(I)+" - "+STR$(Q(I))
1930   NEXT I
1940 '
1950 ' Enter summary sheet into file
1960 OPEN OUTFILE$+".SUM" FOR OUTPUT AS #1
1970  PRINT #1, SPC(12-(LEN(MYCALL$)+LEN(BAND$)+LEN(MON$))/2); MYCALL$; " -- Summary Sheet for "; BAND$;
1980    PRINT #1, " MHz Band - "; MON$; " "; YR$; " VHF Contest"
1990  PRINT #1, BL$
2000  PRINT #1, TAB(13); "Grid Square Listing and number of contacts per Grid"
2010  PRINT #1, BL$ : PRINT #1, BL$
2020  IF MULTNR MOD 5=0 THEN LASTROW=MULTNR\5 ELSE LASTROW=MULTNR\5+1
2030  FOR ROW=1 TO LASTROW
2040    PRINT #1, USING SUMFORM$; MULT$(ROW); MULT$(ROW+LASTROW); MULT$(ROW+LASTROW*2); MULT$(ROW+LASTROW*3); MULT$(ROW+LASTROW*4)
2050   NEXT ROW
2060  PRINT #1, BL$ : PRINT #1, BL$ : PRINT #1, BL$
2070  PRINT #1, TAB(18); "Total Valid QSOs - "; STR$(QSOS); TAB(45); "Dupes - "; STR$(DUPES)
2080  PRINT #1, TAB(18); "Grid Squares - "; STR$(MULTNR)
2090  PRINT #1, CHR$(12)
2100 CLOSE
2110 PRINT "done"
2120 '
2130 ' Print results
2140 CLS : PRINT CHR$(7)
2150 PRINT : PRINT TAB(5) "Results for the "; BAND$; " MHz band"
2160 PRINT : PRINT TAB(8) "Valid QSOs: "; QSOS
2170 PRINT : PRINT TAB(8) "Duplicate QSOs: "; DUPES
2180 PRINT : PRINT TAB(8) "Grid Squares: "; MULTNR
```

```
2190 PRINT : PRINT : PRINT
2200 PRINT TAB(5) "Type C to continue with another band for this contest,"
2210 PRINT : PRINT TAB(5) "or any other key to Exit   ";
2220 ANS$=INPUT$(1)
2230  IF ANS$="C" OR ANS$="c" THEN CLS : GOTO 600 ELSE CLS : SYSTEM
2240 '
2250 '  Subroutine to trap missing file
2260 ON ERROR GOTO 2320
2270 OPEN INFILE$ FOR INPUT AS #1                           ' try opening file
2280 ON ERROR GOTO 0
2290 CLOSE
2300 RETURN
2310 '
2320 '  Error trap for missing file
2330 PRINT CHR$(7) : PRINT TAB(4) "That file does not exist - type X to Exit or any other key to continue ";
2340 ANS$=INPUT$(1) : IF ANS$="X" OR ANS$="x" THEN CLS : SYSTEM
2350 PRINT
2360 RESUME 600
2370 '
2380 '  Subroutine to print log sheet header
2390 PRINT #2, BL$
2400 PRINT #2, TAB(5); MYCALL$; " "; BAND$; " MHz Log"; TAB(70); "Page"; STR$(RAWTOTAL\50+1)
2410 PRINT #2, BL$
2420 PRINT #2, "    Date    Time     Callsign        Sent      Rcvd         Notes"
2430 PRINT #2, "   "; STRING$(74,61)
2440 THEDATE$=STR$(DAY)+BLANK$+MON$
2450 RETURN
```

Program 8

```
10 '   VHFWPX.BAS version 1.3 -  13 August 1986  by Clarke Greene K1JX
20 '
30 '   This Microsoft (tm) BASIC program will build a complete log package for the CQ WPX VHF Contest.
40 '
50 '   The file containing the log entries must be an ASCII file in the following format:
60 '           (each band requires a separate log entry file)
70 '
80 '       TIME        CALLSIGN          RCV'D GRID SQUARE   (each log entry must be followed by a carriage return)
90 '
100 '   At least one space must be between each field of each log entry. Only a changed digit in the time field must
110 '   be present; for example, if the contest begins at 1800Z and the first contact is made at 1802Z and the second
120 '   contact is made at 1805Z, then only 5 need be entered in the time field. If the third contact is made at
130 '   1812Z, then 12 should be entered in the time field. If the next contact is made at 1812Z, then no number need be
140 '   entered in the time field (however, be sure to enter a space to indicate separation between fields).
150 '
160 '   These files will be produced:
170 '           <filename>.LOG - this is a complete log ready for printing
180 '           <filename>.DUP - this is a sorted duplicate listing ready for printing
190 '           <filename>.SUM - this is a summary sheet ready for printing
200 '
210 '
220 '   Depending on the version of BASIC for your particular machine, the CLS (Clear Screen) command must
230 '   be changed.  Consult your own computer's BASIC documentation for more information.
240 '
250 '
260 '   If compiling (a VERY good idea for several orders of magnitude improvement in speed), use O and E switches
270 '
280 '
290 '   Define arrays and variables
300 DEFINT A-Z : OPTION BASE 1
310 DIM ENTRY$(1500), PFX$(250)
```

```
320 BLANK$=" " : BL$="" : SLANT$="/" : TRUE=-1
330 DUPE$="- Duplicate QSO -" : NEWPFX$="- Mult. #"
340 ' Define format strings for printouts
350 LOGFORM$="   \        \ \ \     \         \      \ \   \ \      \                \"
360 DUPFORM$="      \       \     \       \    \ \    \   \        \ \            \"
370 SUMFORM$="         \ \        \ \         \ \         \ \        \ \"
380 '
390 ' Get information from user
400 CLS
410 PRINT TAB(23) " VHF WPX Contest Log Processor " : PRINT : PRINT
420 PRINT : PRINT TAB(5) "What is the station callsign?  ";
430  INPUT "", MYCALL$
440 PRINT : PRINT TAB(5) "What Grid Square is the station in?  ";
450  INPUT "", MYGRID$
460  IF LEN(MYGRID$)<>4 THEN PRINT CHR$(7);: GOTO 440
470 PRINT : PRINT TAB(5) "What is the beginning date of the contest? <dd/mm/yr> ";
480  INPUT "", STARTDATE$
490   MARK=INSTR(STARTDATE$,"/") : IF MARK=0 THEN MARK=INSTR(STARTDATE$,"-")
500    STARTDAY=VAL(LEFT$(STARTDATE$,MARK-1))
510    STARTDATE$=RIGHT$(STARTDATE$,LEN(STARTDATE$)-MARK)
520   MARK=INSTR(STARTDATE$,"/") : IF MARK=0 THEN MARK=INSTR(STARTDATE$,"-")
530    MON$=" July  "
540    YR$=RIGHT$(STARTDATE$,LEN(STARTDATE$)-MARK)
550 PRINT : PRINT TAB(5) "What is the GMT starting time for the contest?  ";
560  INPUT "", STARTGMT$
570 PRINT : PRINT TAB(5) "What file is the log extract located in?  ";
580  INPUT "", INFILE$
590  GOSUB 2210                                            ' check to see if file exists
600  IF INSTR(INFILE$,".")<>0 THEN OUTFILE$=LEFT$(INFILE$,INSTR(INFILE$,".")-1) ELSE OUTFILE$=INFILE$
610 PRINT : PRINT TAB(5) "What frequency band is the log extract for?  ";
620  INPUT "", BAND$
630 '
640 ' Build log listing
650 CLS
660 PRINT : PRINT TAB(5) "Duping and counting...  ";
670 '
680 ' Clear arrays
690 FOR I=1 TO 1500
700  ENTRY$(I)=BL$
710  NEXT I
720 FOR I=1 TO 250
730  PFX$(I)=BL$
740  NEXT I
750 '
760 ' Initialize variables
770 RAWTOTAL=0 : QSOS=0 : DUPES=0 : MULT=0
780 DAY=STARTDAY : PREVIOUSGMT$=STARTGMT$                  ' reset starting time and day
790 '
800 ' Open input file and output .LOG file
810 OPEN INFILE$ FOR INPUT AS #1
820 OPEN OUTFILE$+".LOG" FOR OUTPUT AS #2
830 '
840 ' Input data, process, and enter into output file
850  WHILE NOT EOF(1)
860   LINE INPUT #1, THISENTRY$                            ' read entire line from disc file
870    IF THISENTRY$=BL$ GOTO 1440                         ' jump ahead if the line is blank
880    WHILE ASC(RIGHT$(THISENTRY$,1))<48
890     THISENTRY$=LEFT$(THISENTRY$,LEN(THISENTRY$)-1)     ' strip off trailing spaces,etc
900     WEND
910    IF LEN(THISENTRY$)>0 THEN RAWTOTAL=RAWTOTAL+1 ELSE GOTO 1440
920 '
930 ' Separate grid from ENTRY$
940   GRID$=RIGHT$(THISENTRY$,4)
950    THISENTRY$=LEFT$(THISENTRY$,LEN(THISENTRY$)-4)
```

```
960    WHILE ASC(RIGHT$(THISENTRY$,1))<48
970      THISENTRY$=LEFT$(THISENTRY$,LEN(THISENTRY$)-1)                ' strip off trailing spaces
980    WEND
990  '
1000 '   Separate GMT from ENTRY$
1010   WHILE ASC(LEFT$(THISENTRY$,1))<48
1020     THISENTRY$=RIGHT$(THISENTRY$,LEN(THISENTRY$)-1)               ' strip off leading spaces
1030   WEND
1040   IF INSTR(THISENTRY$,BLANK$)<>0 THEN GMT$=LEFT$(THISENTRY$,INSTR(THISENTRY$,BLANK$)-1) ELSE GMT$=BL$
1050   THISENTRY$=RIGHT$(THISENTRY$,(LEN(THISENTRY$)-LEN(GMT$)))
1060   WHILE LEFT$(THISENTRY$,1)=BLANK$
1070     THISENTRY$=RIGHT$(THISENTRY$,LEN(THISENTRY$)-1)               ' strip off leading spaces
1080   WEND
1090 ' Fill in missing time data
1100   GMT$=LEFT$(PREVIOUSGMT$,(4-LEN(GMT$)))+GMT$
1110   THEDATE$=BL$ : IF GMT$<PREVIOUSGMT$ THEN DAY=DAY+1 : THEDATE$=STR$(DAY)+MON$
1120 '
1130 '   Check for dupes
1140   DUPE.QSO=NOT TRUE : NOTE$=BL$                                   ' clear note field
1150   FOR I=1 TO QSOS
1160     IF THISENTRY$=ENTRY$(I) THEN NOTE$=DUPE$ : DUPES=DUPES+1 : DUPE.QSO=TRUE : I=QSOS
1170   NEXT I
1180   IF DUPE.QSO GOTO 1370                                          ' if the QSO was dupe, don't bother looking at prefix
1190   QSOS=QSOS+1 : ENTRY$(QSOS)=THISENTRY$
1200 '
1210 '   Determine station prefix
1220   THISPFX$=LEFT$(THISENTRY$,2)                                   ' the prefix is always at least two characters long
1230   IF ASC(MID$(THISENTRY$,3))>=48 AND ASC(MID$(THISENTRY$,3))<=57 THEN THISPFX$=LEFT$(THISENTRY$,3)
1240   MARK=INSTR(THISENTRY$,SLANT$)                                  ' look for portable designator
1250   IF MARK=0 GOTO 1300                                           ' if not portable, jump ahead
1260   IF MARK>3 THEN PORTPFX$=RIGHT$(THISENTRY$,LEN(THISENTRY$)-MARK) ELSE PORTPFX$=LEFT$(THISENTRY$,MARK-1)
1270     IF LEN(PORTPFX$)=1 AND PORTPFX$>="A" GOTO 1300
1280     IF LEN(PORTPFX$)=1 THEN THISPFX$=LEFT$(THISPFX$,LEN(THISPFX$)-1)+PORTPFX$ : GOTO 1300
1290     IF LEN(PORTPFX$)>1 AND RIGHT$(PORTPFX$,1)>="A" THEN THISPFX$=PORTPFX$+"0" ELSE THISPFX$=PORTPFX$
1300 ' Search multiplier table for prefix
1310   NEWPFX=TRUE
1320   FOR J=1 TO MULT
1330     IF THISPFX$=PFX$(J) THEN NEWPFX=NOT TRUE : J=MULT
1340   NEXT J
1350   IF NEWPFX THEN MULT=MULT+1 : PFX$(MULT)=THISPFX$ : NOTE$=THISPFX$+SPACE$(4-LEN(THISPFX$))+NEWPFX$+STR$(MULT)
1360 '
1370 '   Print log entry to file
1380   IF (RAWTOTAL-1) MOD 50=0 THEN GOSUB 2340                       ' print the header if this is the beginning of a page
1390   PRINT #2, USING LOGFORM$; THEDATE$; GMT$; THISENTRY$; MYGRID$; GRID$; NOTE$
1400   IF RAWTOTAL MOD 50=0 THEN PRINT #2, CHR$(12)                   ' print a form feed if this is the end of the page
1410 '
1420 '   Set variables for next pass
1430   PREVIOUSGMT$=GMT$ : GMT$=BL$
1440 WEND
1450 IF RAWTOTAL MOD 50 <>0 THEN PRINT #2, CHR$(12)                   ' if a form feed wasn't printed, print one now
1460 CLOSE
1470 PRINT "done"
1480 '
1490 ' Sort callsigns for dupe sheet
1500 PRINT : PRINT TAB(5) "Preparing dupe sheet...  ";
1510 M=QSOS\2
1520 WHILE M>0
1530   FOR I=M+1 TO QSOS
1540     J=I-M
1550     WHILE J>0
1560       IF ENTRY$(J)>ENTRY$(J+M) THEN SWAP ENTRY$(J),ENTRY$(J+M) : J=J-M ELSE J=0
1570     WEND
1580   NEXT I
1590   M=M\2
```

```
1600  WEND
1610  '
1620  '  Enter dupe sheet into file
1630  OPEN OUTFILE$+".DUP" FOR OUTPUT AS #1
1640  IF QSOS MOD 250=0 THEN LASTPAGE=QSOS\250 ELSE LASTPAGE=QSOS\250+1
1650  FOR PAGE=1 TO LASTPAGE
1660    PRINT #1, SPC(20-(LEN(MYCALL$)+LEN(BAND$))/2); MYCALL$; " -- Dupe Sheet for "; BAND$; " MHz Band -- Page"; STR$(PAGE)
1670    PRINT #1, BL$ : PRINT #1, BL$
1680    FOR ROW=1 TO 50
1690      E=(PAGE-1)*250+ROW
1700      PRINT #1, USING DUPFORM$; ENTRY$(E); ENTRY$(E+50); ENTRY$(E+100); ENTRY$(E+150); ENTRY$(E+200)
1710    NEXT ROW
1720    PRINT #1, CHR$(12)                                   ' go to next page
1730  NEXT PAGE
1740  CLOSE
1750  PRINT "done"
1760  '
1770  '  Enter summary sheet into file
1780  PRINT : PRINT TAB(5) "Preparing summary sheet...  ";
1790  '  Sort multipliers for summary sheet
1800  M=MULT\2
1810  WHILE M>0
1820    FOR I=M+1 TO MULT
1830      J=I-M
1840      WHILE J>0
1850        IF PFX$(J)>PFX$(J+M) THEN SWAP PFX$(J),PFX$(J+M) : J=J-M ELSE J=0
1860      WEND
1870    NEXT I
1880    M=M\2
1890  WEND
1900  '
1910  '  Enter summary sheet into file
1920  OPEN OUTFILE$+".SUM" FOR OUTPUT AS #1
1930    PRINT #1, SPC(12-(LEN(MYCALL$)+LEN(BAND$))/2); MYCALL$; " -- Summary Sheet for "; BAND$;
1940    PRINT #1, " MHz Band -  "; YR$; " VHF WPX Contest"
1950    PRINT #1, BL$
1960    PRINT #1, TAB(33); "Prefix Listing"
1970    PRINT #1, BL$ : PRINT #1, BL$
1980    IF MULT MOD 5=0 THEN LASTROW=MULT\5 ELSE LASTROW=MULT\5+1
1990    FOR ROW=1 TO LASTROW
2000      PRINT #1, USING SUMFORM$; PFX$(ROW); PFX$(ROW+LASTROW); PFX$(ROW+LASTROW*2); PFX$(ROW+LASTROW*3); PFX$(ROW+LASTROW*4)
2010    NEXT ROW
2020    PRINT #1, BL$ : PRINT #1, BL$ : PRINT #1, BL$
2030    PRINT #1, TAB(18); "Total Valid QSOs - "; STR$(QSOS); TAB(45); "Dupes - "; STR$(DUPES)
2040    PRINT #1, TAB(18); "Total Prefixes - "; STR$(MULT)
2050    PRINT #1, CHR$(12)
2060  CLOSE
2070  PRINT "done"
2080  '
2090  '  Print results to screen
2100  CLS : PRINT CHR$(7)
2110  PRINT : PRINT TAB(5) "Results for the "; BAND$; " MHz band"
2120  PRINT : PRINT TAB(8) "Valid QSOs: "; QSOS
2130  PRINT : PRINT TAB(8) "Duplicate QSOs: "; DUPES
2140  PRINT : PRINT TAB(8) "Total Prefixes: "; MULT
2150  PRINT : PRINT : PRINT
2160  PRINT TAB(5) "Type C to continue with another band for this contest,"
2170  PRINT : PRINT TAB(5) "or any other key to Exit  ";
2180  ANS$=INPUT$(1)
2190  IF ANS$="C" OR ANS$="c" THEN CLS : GOTO 570 ELSE CLS : SYSTEM
2200  '
2210  '  Subroutine to trap missing file
2220  ON ERROR GOTO 2280
2230  OPEN INFILE$ FOR INPUT AS #1                           ' try opening file
```

```
2240 ON ERROR GOTO 0
2250 CLOSE
2260 RETURN
2270 '
2280 ' Error trap for missing file
2290 PRINT CHR$(7) : PRINT TAB(4) "That file does not exist - type X to Exit or any other key to continue ";
2300 ANS$=INPUT$(1) : IF ANS$="X" OR ANS$="x" THEN CLS : SYSTEM
2310 PRINT
2320 RESUME 570
2330 '
2340 ' Subroutine to print log sheet header
2350 PRINT #2, BL$
2360 PRINT #2, TAB(5); MYCALL$; " "; BAND$; " MHz Log"; TAB(70); "Page"; STR$(RAWTOTAL\50+1)
2370 PRINT #2, BL$
2380 PRINT #2, "      Date     Time     Callsign        Sent    Rcvd        Notes"
2390 PRINT #2, "   "; STRING$(74,61)
2400 THEDATE$=STR$(DAY)+MON$
2410 RETURN
```

(continued from page 7-9)

you don't need as narrow an electronic filter as previously. If you can attain this level of proficiency, you gain an added advantage over most of the competition. Since your receive filter is wider, "chasing" contacts is much easier because you can effectively tune the band faster than someone with a narrow filter.

Strategy plays an important role in successful contest operating. The decision to be on a particular band at a particular time can be crucial to the outcome of a competition, particularly when you make the right move and the competition doesn't. Your strategic decisions should be made on the basis of hard facts and educated choices (in reality, a lot of luck will be involved, too). To have a full complement of facts available during the contest, you have to do yet more research before the contest. There are two areas where you can gather information prior to the contest that will be of value during the contest. The first is propagation.

During the weeks prior to the contest, study the bands. At what times do they open to where? From what direction? For how long? Keep records of your observations. The most significant times to observe are the few days just prior to the contest, and 28 days before the contest period (the sun rotates once every 28 days; propagation trends tend to follow the same pattern). Once you have completed your study, plot your results in graphical form. Use the resulting presentation in preparing a tentative operating plan for the contest. Of course, like the weather, you truly can't predict band openings with a high degree of reliability, so be prepared to modify your strategy according to the propagation you actually encounter. Be flexible.

Speaking of the weather, you should also follow the weather forecasts for the days immediately preceding the contest. Most enhanced VHF propagation conditions are the result of slowly moving weather systems; weather-map study can help you predict potential favorable directions in a VHF contest. HF contest enthusiasts should also follow the weather; passing storm fronts cause QRN (static), particularly on the lower HF bands. If you expect a front to pass through your area during the contest, plan to emphasize the higher frequency bands during that period to avoid the QRN. Plan your lower frequency operation around the QRN. Some television stations now show foreign weather conditions as well as local forecasts; for DX contest operating, use the foreign weather information to plan your lower frequency operation around QRN at the DX end, so the DX stations can hear you.

The second area where you can gather useful contest information before the contest is in the study of activity patterns. Having a knowledge of certain stations' operating habits can truly be a blessing with regard to finding rare contest multipliers. Is there a particular gathering frequency for groups of stations in a particular geographical area (such as for severe weather reporting in tropical areas)? Does a station located in a rare multiplier habitually get on the air just before his or her bedtime every evening, perhaps way up in frequency on the band? Is a particular station willing to change bands with you for a new multiplier? Is someone going to a rare area specifically for the contest? All this information can be put to good use if you document it carefully and put the data in a form that you can understand well into the contest.

Sometime before the contest, prepare your logging materials for contest use. Obviously you can't fill in contact data in advance, but you can put in some constant information like your own call sign, the contest starting date, a consecutive number for each log page, the dupe sheet frequency band information and the like. Not only will this step save time during the contest, it will also lead to less confusion during the contest, allowing you to concentrate on making contacts.

It's also a good idea before the contest to have your station all set up for the event. Adjust the VOX controls on the transmitter, set your clock, calibrate your receiver, replace the burned-out light bulb over the operating position, program the memory keyer, note the optimum band markings for quick tuneup on each band, and any other maintenance tasks that might need doing. You certainly don't want to perform maintenance during the contest; you have enough to do just in operating and fixing the inevitable visits from Murphy.

Be sure to read and thoroughly understand the latest contest rules for the contest in question. Perhaps the rules have changed since last year or your recollection of the existing rules is rather fuzzy. In either case, being clear on the rules might prevent you from making a fatal mistake that might make your hard work during the contest weekend for naught. Finally, prepare yourself mentally for the contest. As most sports psychologists will tell you, your mental attitude often is the determining factor in how well you perform in a competitive activity. Convince yourself that you can do well by putting in the effort. Picture in your mind what you're going to do in each situation you might encounter during the contest.

Run through it over and over until it becomes second nature. You have to determine for yourself how to get motivated, since everyone is different. One thing is for sure, though; if you aren't *prepared* to do your best, *you won't*.

OPERATING THE CONTEST

The big moment finally arrives. Now what? Regardless of whether you are operating a VHF contest, an HF DX contest, the Novice Roundup, or whatever, there is one fundamental guideline that you should follow in a contest: Do what gains you the most points. "Another obvious suggestion," you're probably thinking. It may be obvious, but if you truly follow it, you'll end up with the highest score possible for your station on that weekend.

For example, early in the contest you come across a rare multiplier. He is just casually working a few stations at a rate of about 12 an hour. He is on an "oddball" frequency, away from most of the contest activity. What should you do? If you decide that the only way you'll work the multiplier during the whole weekend is to stick it out and spend the time, you'll be missing the highest activity point in the contest. The time might be well spent. Of course, you could also spend 45 minutes and never work him; think of the contacts you lost and will never recover.

Decisions like this need to be made almost constantly during a contest. Your decision should be based on considering that simple fundamental. If you feel that propagation might be favorable right then, that your station is loud enough, and that the addition of another multiplier is more beneficial to the final score than running up your contact total at that point,

---

Two's Not Company; It's a Multiop

In almost every major contest there are two broad entry categories: single-operator, where all operating, logging and other station functions are performed by a single person during the contest; and multioperator, where the operating functions are divided among a group of people.

Most amateurs, particularly newcomers to contests, think of multiop participation as a sort of serial relief activity; one ham operates the station for a while until he or she is tired or otherwise wants a break. Then, another operator takes over. Nobody gets too tired, hungry or otherwise uncomfortable. This definition of multiop contest operating has its obvious advantages, particularly at a family or club station where several hams must share the same equipment. This approach might be likened to a relay race where several participants pool their efforts to race a longer distance faster than they could as individuals.

Another more ambitious approach to multiopping is a parallel team concept. Just as in any other team activity, each participant plays a specific role in the overall activity and several participate at once. In the multioperator, multi-transmitter (ie, unlimited number of transmitters) category, this suggests several operating positions operating on several bands simultaneously so as to cover all the active bands. In the multioperator, single transmitter category, only one transmitter can be active at one time; however, there is no restriction on the number of listening positions used or the number of people used for logging functions. For example, one operator might actually be making contacts on 20 meters, while a second "spots" on 20 meters with a second receiver, scanning the band for stations to work. When the spotter finds one, he somehow informs the "main" operator, who moves to work the spotted station (this is especially effective for searching for new multipliers). A third operator might be scanning 15 meters and making a "map" of the active stations there (a map is a listing of call signs of stations heard along with the frequencies they are transmitting on). When the main station changes bands to 15 meters, a fast start can be made by contacting the stations on the map. The operator scanning 15 meters also provides valuable information in making the decision to change bands as well. Yet another operator might be keeping the duplicate listing for the main operator; the "duper" tells the main operator when a station has already been worked and to go on to the next one. That's four operators, all working at once, with only one active station! Their combined effort allows more productive use of the contest period and yields a higher score. These same ideas can be even further expanded until all the bands are covered simultaneously.

All of these functions can really increase the overall score, but they don't relieve the work level; indeed, they *add* to it. Nobody said this would be easy. The logical question might be: Aside from producing a higher score, in a category where the scores are higher to begin with, why go to all the extra effort for multiopping? The answer to that question probably has as many answers as there are

contest participants. Multiops provide an opportunity for hams with a marginal station or no station at all to participate in a contest. Multiops also provide a good opportunity for beginners to get some hands-on experience while learning from established participants. For some (college students, for example), a multiop at the club station may be the best way to parcel out contest operating time fairly.

For many, participating in a multiop adds another dimension to contesting. Even the most ardent single-op has to admit that sitting alone in the station for up to 48 hours, while listening through headphones and speaking through a microphone or key to persons unseen, is a solitary and perhaps lonely activity. The thrill of having a rare station respond to your CQ, or of breaking through a moderate pileup (the one that only has 80% of the active stations in the world calling . . . usually out of turn) tends to get lost when you're by yourself.

Having some other witnesses to your good fortune (or, more precisely, astonishing feat), particularly those who will not only marvel at but appreciate what has transpired, enhances the experience immensely for *everyone* involved. Somehow, describing the feeling you got when the Antarctica station came back to you on 160 meters, or when you worked your 75th section, is lost on your car-pool partners Monday morning. Sharing the experience live, up-close and in-person can make all the headaches, sore throats, dizziness and sleeplessness secondary to those few magical moments, the ones remembered long after the contest results are published and forgotten.

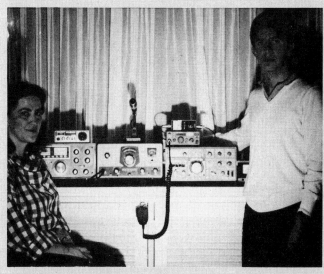

HK3IJM and HK3BPB (Bogota, Colombia) were a multi-single team in a recent ARRL DX Contest.

then by all means try to work the multiplier. Your gamble may or may not work. You might decide that you can more easily work a number of stations in the same time that it might take to work the multiplier, more than enough contacts to compensate for the "lost" multiplier. Your best choice then probably is to forget about the multiplier; he might be easier to work later. Decisions like these are the very essence of contesting, and only you can decide based on your knowledge of propagation, your station's capabilities, contest strategy and your own capabilities. Indeed, preparation, experience and raw talent determine your ability to make the right decision when needed.

There are some operating hints that might make your strategic decisions easier to make. First, learn the relative point value of a new multiplier compared to a new QSO. Simply dividing the number of QSOs made at that point in the contest by the number of multipliers worked at that point in the contest gives a ratio that represents the number of contacts a multiplier is worth at that point. You can do this during the contest with a calculator or slide rule, but arithmetic exercises when you're already fatigued likely will be very time consuming, and may not yield accurate results. Instead, a prepared chart, showing the number of QSOs along one axis, the number of multipliers along the other axis and the QSO/multiplier ratio at the intersections, is much easier and more reliable to use during a contest (in fact, add that to your list of items to prepare before the contest).

At each point in a contest, the QSO/multiplier ratio changes depending on the contest format and your own performance. For example, during a DX contest, where there is a relatively high number of multipliers available, the QSO/multiplier ratio might be around three. That means each new multiplier worked at that point is equivalent to making three (non-multiplier) QSOs, as far as your point total is concerned. In the November Sweepstakes, in which there is a finite number of multipliers available (75), the QSO/multiplier ratio might be over 10. In that circumstance, it would require making 10 QSOs to equal the point value of one new multiplier.

By applying your knowledge of propagation, your intimate familiarity with your station's capabilities, and your best estimate of how the propagation and activity is evolving in the contest, you can use the QSO/multiplier ratio information to decide whether to look for contacts or to try jumping into a pileup for a new multiplier. During high activity and good propagation periods, making contacts is probably a better tack, since you might never be able to recover the band-opening time later on. Spending 10 minutes in a pileup when the band is hot and you have the potential to make a lot of contacts is a poor choice compared to spending 10 minutes in a pileup when the activity or propagation has slackened, and you only might be able to make two contacts during that same 10 minutes.

The key principle is to utilize your band-opening time wisely, as you have little or no control over propagation and have to work what you can when you can. This strongly implies that any unusually good propagation should be taken advantage of immediately, since it may never return. At VHF especially, where band openings tend to be short-lived (anywhere from a second to a couple of hours), enhanced propagation is a sure sign to stop whatever you're doing and work what you can during the opening. You may not get another chance. The value of band-opening time is inversely proportional to its duration.

As an example, consider 6-meter scatter propagation during a VHF contest morning. A meteor burst might make a 1000-mile-distant station audible (and potentially workable) for 20 seconds. Unless you make the contact during that time period, you're out of luck. You should call quickly and succinctly and exchange contest reports the same way. If you make yourself understood the first time, there is no need for repeats that you might not have propagation for. There is no time to give the other station's call (which he or she presumably already knows) five times, your call eight times and the instruction to go ahead six times. By the third "go ahead," the burst is over, the contact is lost, and the operator at the other end is frustrated over the failure to complete the contact successfully. The next time there is a meteor burst, that other operator might tell you to "go away," and then try to work someone quicker on the draw than you were.

HF contesters are just as guilty of bad (or no) timing as VHF contesters. How many times have you heard the din of a pileup go on for two or three minutes after the station being called has already come back to someone? Who gained there? The moral: Don't repeat unless asked to.

There are two general approaches to making contacts—calling CQ (and "running" people/getting responses) and hunting. Each has its own advantages and disadvantages, as you might expect. Calling CQ effectively requires getting the attention of other stations and motivating them to call you. Getting attention at the other end is most easily accomplished by wielding a big signal. Big signals always seem to get answered (as do signals emanating from rare multipliers). Of course, you may not be fortunate enough to posssess a big signal, so calling CQ may not be as productive for you. Here again, knowledge of your station's capabilities is required to make that decision.

GW4BLE's nutritionally unconventional contest food: orange soda, chocolate bars and tomato ketchup!

Motivating respondents to call is usually easy early in the contest while you still haven't drained the pool of active contesters. The contesters are already motivated to call; they want the QSO. The strain begins *after* you have worked the serious competitors, and the casual operator is all that remains. Here you have to be prepared to answer questions about the contest, and perhaps tell the operator at the other end what

Contesting for Noncontesters

It's the cool of a picture-perfect summer evening. You're sitting on the porch, enjoying the ambience of your serene, residential neighborhood, and everything is grand and glorious. Then it happens—the roar of a dozen throbbing Harley-Davidsons invading your quiet environment. Trouble in paradise.

Isn't this analogous to what most hams think about the onslaught of contesters during an otherwise carefree ragchew weekend on the amateur bands? And why not? After all, ragchewing, the purest and mellowest form of communication, is the bread-and-butter of Amateur Radio operating, while contesting is life in the fast lane. Contesting is basically seeing how many stations you can work in a limited amount of time, as specified in the particular contest rules. Usually, a premium is placed on working as many different geographical areas (which could be states, ARRL sections, grid squares, DXCC countries and so forth) as possible. These geographical areas are called "multipliers." The final contest score is generally the number of QSOs (or QSO points) times the number of multipliers. Contesting is like putting your normal hamming into fast forward!

Sure, it's all been said before. That is, there is always some substantial portion of the low-band frequency spectrum free and clear for noncontest activities. But how about a different approach? Instead of heading for the hills, why not participate, even if only briefly, in contests? In other words, if you can't lick 'em, join 'em. Contests are fun, and there's always room for the noncontester to jump in for a few contacts.

These competitions come in all shapes and sizes, and you don't have to spend all weekend neglecting your family or other responsibilities to join in. Relatively few contesters even come close to operating the maximum allowable number of hours in the contest period, so there is no fundamental difference between a contester who puts in an hour or two and a noncontester who does the same. Most amateurs don't play contests to win. Only a small group of enthusiasts (see The Contester's Persona, below) go all out, so virtually everyone else taking part is in the same boat. Indeed, it's the multitude of casual participants that keep the contest program alive and well.

The Contester's Persona

To understand contesting, it may be useful to get "into the head" of the serious contester, the kind of amateur sufficiently warped to devote a good deal of the weekend to the event. Strange as it may seem, spending the weekend exchanging signal reports and other so-called information machine-gun style is considered fun, even though no real communication—in the conversational sense—occurs. Some suggest contests are even dehumanizing, and they may be right, particularly if you've seen a contester slumped over his radio, haggard and needing sleep (if not a shower) during one of the major contests—a certified zombie.

For some reason, weird or otherwise, contesters get their kicks from pushing themselves and their gear to the limit, sharpening operating skills and station efficiency. It's the same kind of personal satisfaction people derive when they compete in marathons, work out with weights, play tennis or racquetball, and so on. Contesting makes a person feel fulfilled and stimulated, even though exhaustion overpowers everything else, at least in the short term. It doesn't seem that sitting in a chair in front of a radio would be that strenuous, but try doing it for 24 hours or so. In fact, it only takes an hour before the hallucinations set in!

For most contesters, life is perfect when conditions are good, when there are plenty of contacts and multipliers to be had, and particularly when a contester is able to stake out one prime frequency, and "CQ-in" one contact after another (ie, "running"). At this point, the adrenalin is really flowing—and whatever delirium that may have set in as a result of disruption of the normal sleep pattern and the relentless pounding of signals inside headphones is no no longer a factor.

But contesting—like life—is not perfect. Depending on the contest, there are slow hours, and the "rate" (number of contacts made per hour) plummets. Eyelids become as heavy as anvils.

Contesters, in their self-indulgent way, want each contest to be like an all-you-can-eat shrimp-and-salad bar. They want to keep going back for more, and their "reservoir" can never be filled. So why not help them out? Keep those contesters busy, awake and satiated.

The True Story

Obviously, the principal question is what can contests and contesters do for you, the noncontester. Initially, let's lay to rest the myth about being "in the contest." You can spontaneously work as many, or as few, stations as you want (as previously indicated, only a small percentage of the competitors play to win). And you go at your own pace. You're free to choose; working stations in the contest is simply one of many fun things to do while you're just tuning around the bands.

You do not need to register in advance, fill out forms or get an entry number and official jersey. Nor do you need "official logs." You can operate as much as you want, and you are under no obligation whatsoever to send in a log to the organization sponsoring the contest.

Indeed, no one is "in the contest" until he or she submits an official log. Keep in mind, however, that should you decide to submit a log, the use of the official log and summary forms (typically available for an SASE) is tremendously appreciated by the log checkers. Further, you can submit your log for noncompetitive purposes (called a "check log"), to aid the contest checkers in cross-checking other logs for accuracy. Either way, don't hesitate to submit even a modest score; but you are absolutely under no obligation to do so—no salesman will call! Also, you need not be a League member, a subscriber to CQ Magazine, or the lead singer of Kool and the Gang to participate. But if you do send in an entry, be sure to retain a complete photocopy of all submitted materials for your records and for QSLing.

So much for myths. Here are some real advantages to operating contests: It gives you something more purposeful to do when you're otherwise scanning the bands (often looking for something to do). It helps you continually gauge the effectiveness of your antennas and your operating prowess, and the combination of the two—in other words, what you can do with what you've got!

If you're interested in obtaining the numerous awards that are out there in radioland, you might remember that making the requisite contacts for these awards under routine operating conditions is often labor intensive. In the fast-paced environment of a major contest, it's incredible how quickly you can rack up those states, DXCC countries, prefixes, grid squares or whatever your heart desires (in the framework of the particular contest, of course) to qualify for those neat certificates.

Table 7-1 summarizes the basics of the major contests. A "major" contest is one that sports a high activity level, presumably a more desirable training exercise than one with fewer participants. But you might find that such ample activity levels are too much of a good thing, so lower-profile contests may be more to your liking, particularly when you're new to this form of operating. These run the gamut from state QSO parties and other special-interest contests to European, South American and other international encounters. Each contest, regardless of size, yields some valuable experience and enjoyment toward becoming contest-ready.

If you happen to live in a "rare" state, ARRL section (see Chapter 17), grid square or even a rare country, you can have a lot of fun and at the same time make many contesters happy by giving them that avidly sought after multiplier. The amateur world will beat a path to your

Table 7-1

Major Contests

Month	Contest	Scope	Exchange	For more Information
Jan	ARRL VHF Sweepstakes	Primarily W/VE	Grid-square locator	Dec QST
Jan	CQ Worldwide 160-Meter Contest (CW)	International	W/VE: signal report and state/province; DX: signal report and country	Dec CQ; contest Corral, Jan QST
Jan	73's 160-Meter World SSB Championship	International	W/VE: signal report and state/province; DX: signal report and country	Dec 73; Contest Corral, Jan QST
Jan	73's 75-Meter World SSB Championship	International	See above	See above
Jan	73's 40-Meter World SSB Championship	International	See above	See above
Jan	73's 20-Meter World SSB Championship	International	See above	See above
Jan	73's 15-Meter World SSB Championship	International	See above	See above
Feb	ARRL Novice Roundup	Novices/Techs work others	Signal report and ARRL section	Jan QST
Feb	ARRL DX Contest (CW)	W/VE stns work DX stns	W/VE: signal report and state/province; DX: signal report and power	Dec QST
Feb	CQ Worldwide 160-Meter Contest (phone)	International	See above	See above
Mar	ARRL DX Contest (phone)	International	See above	Dec QST
Mar	Spring RTTY Contest	International	UTC, signal report and consecutive QSO serial number	Contest Corral Mar QST
Mar	CQ WPX Contest (phone)	International	Signal report and consecutive QSO serial number	Jan CQ; Contest Corral, Feb QST
May	USSR CQ-M Contest (phone and CW)	International	Signal report and consecutive QSO serial number	Contest Corral, Apr QST
May	CQ WPX Contest (CW)	International	See above	See above
Jun	ARRL VHF QSO Party	International	See above	See above
Jun	All Asian DX Contest (phone)	Asian stns work others	Signal report and age	Contest Corral Jun QST
Jun	ARRL Field Day	Primarily W/VE	Transmitter "class" and ARRL Section	May QST
Jul	IARU HF World Championship (phone and CW)	International	Signal report and ITU zone	May QST
Jul	CQ VHF WPX Contest	International	Consecutive QSO serial number and call sign	Feb CQ; Contest Corral, Jul QST
Aug	Worked All Europe (CW)	EU stns work others	Signal report and consecutive QSO serial number	Contest Corral, May QST
Sep	Worked All Europe (phone)	See above	See above	See above
Sep	ARRL VHF QSO Party	Primarily W/VE	Grid-square locator	Aug QST
Sep	CRRL Can-Am Contest	W/VE	Signal report, consecutive QSO serial number and "multiplier-area" abbreviation	Aug QST
Oct	CQ Worldwide DX Contest (phone)	International	Signal report and CQ zone	Sep CQ; Contest Corral, Oct QST
Nov	ARRL Sweepstakes (CW)	W/VE	See Oct QST	
Nov	ARRL Sweepstakes (phone)	W/VE	See Oct QST	
Nov	CQ Worldwide DX Contest (CW)	International	See above	See above
Dec	ARRL 160-Meter Contest (CW)	International	W/VE: signal report and ARRL Section; DX: signal report	Nov QST
Dec	ARRL 10-Meter Contest (phone and CW)	International	W/VE: signal report and state/province; DX: signal report and consecutive QSO serial number	Nov QST

door—you'll have instant celebrity status! Find a good frequency (logically on the band that is most open at the time), call CQ and let the good times roll. You'll be surprised how fast you can fill up those log pages, and you'll understand quickly why contesters derive so much pleasure from running contacts. (Note that, as discussed below, it's more desirable to be hunting-and-pouncing rather than CQing at the outset of your contesting endeavors.)

On some bands, contests are about the only times you can expect to work anybody. This is particularly true with 10 meters under the present sunspot conditions, as well as 15 meters to a certain extent, and the VHF (non-FM) and some UHF bands. So operating during contests can be an efficient, if not the only, way to gain some significant experience on these frequencies.

The predictability of contest operation is another advantage, primarily on code. Since you can pretty well anticipate the component parts of the exchange, you'll probably feel more relaxed in CW contesting, even if the code is not necessarily your mode of preference. And predictability notwithstanding, if increased code speed is what you're after, consider that one local Novice nearly doubled his code speed over the course of one contest weekend!

The bottom line is that contests—even during less-desirable band conditions—set the stage for all sorts of people and places to show up, and they all can be worked with reasonable effort. If you put yourself on a band under contest-activity periods, good things are bound to happen.

Hunting-and-Pouncing

When you first start out, it is preferable to concentrate on hunting-and-pouncing—tuning for stations calling CQ CONTEST and responding to them. This is a great way to acclimate yourself to the contest environment; you probably shouldn't try to run stations until you have sufficient contest experience under your belt. Tune around and listen until you feel comfortable with the goings on, then pick out a contester who sounds like he knows what he's doing, listen to the give-and-take and, when you're ready, go for it!

Indeed, you may find that hunting-and-pouncing—the thrill of the hunt—is so attractive that it's all you'll ever want to do in contests. It is certainly understandable that you may not see the point of running strings of "stereotyped" QSOs when you can be selective as to what you work. Further, you may prefer to collect multipliers in a given contest on a one-of-a-kind basis, such as making one contact in each ARRL section for the "clean sweep" in the ARRL Sweepstakes, or aiming to make one contact per DXCC country heard in the *CQ* Worldwide or ARRL DX Contest. So you may find quality rather than quantity more suited to your personal operating tastes, and that's fine.

Contest Protocol

Are you convinced there's something for you in contesting? If contesting is now in your future, it's important to keep in mind some matters of protocol.

When you are working your way up or down the band hunting-and-pouncing, please don't stop a contester's 100-contact hour to ask what the contest is. Listen around first for a few minutes to get the particulars on your own, including the required contest exchange, or check *QST*'s Contest Corral or *CQ*'s Contest Calendar (see Table 7-1). Don't upset a contester's rhythm, particularly when he's running strings of stations during a peak hour, with Dragnet-style questions. Think of it this way: Would you run out into the middle of a marathon and clothesline one of the runners to ask him who is sponsoring the marathon, how many miles it is, or where he bought his groovy running shoes? Of course not. A contester is running a radio marathon, so similar courtesy should prevail.

Once you've determined the appropriate exchange (each contest requires different information to be transmitted), send the information once and only once—that is, after you've signed your call only once when responding to a

CQ 'Test. Remember: FCC rules don't mandate signing the other station's call. However, if you're the one running stations, you'll obviously have to sign the other station's call so people know who you've come back to! But when you start out, you'll be tuning and replying to CQs; in this instance, signing your own call once is sufficient.

Unnecessary "fills" (repeats) needlessly slow everyone down. Generally, the aim is to maximize the number of contacts made during each operator's own personally limited operating time. So repeat only when asked.

On the other hand, a valid contest exchange requires you to copy accurately the information sent to you by the other station. So if you need a fill, ask for it! Every attempt should be made to get each bit of information correctly; the objective is a valid two-way QSO between the parties. A shortcut you can use to avoid asking the station for fills (which is often considered annoying and intrusive) is to listen to subsequent QSOs the station is making on that frequency. In this way, you can listen to him send his call and his exchange several times to make sure your log entry is accurate. This obviously pertains to stations who are calling CQ and who will be "holding court" there for a while. In short, if you miss part of the exchange at first, or even the other station's call, no harm done. Just listen to subsequent QSOs until you get it.

If you've replied to a contester who's having great success running stations on a particular frequency, your excessive repeats may threaten him with the loss of that productive frequency. A contester jealously tries to maintain his CQ frequency like a family heirloom. It's a very delicate balance, working stations efficiently while keeping the frequency clear on receive. If you painstakingly repeat your information (and it is painful to the contester who's sitting there helplessly waiting for you to finish), some aggressive individual—who may not hear you all that well—may assume that this choice frequency is unoccupied. He then will swoop in and start banging out CQ 'TEST at 40 dB over 9. This causes extreme heartburn for the innocent contester who has been on that frequency for a substantial amount of time working stations one after another.

Be prepared, also, for the brevity of the contester's transmission to you. Don't be put off by a contester's seemingly impersonal nature. In the heat of battle, the most you'll get in addition to the exchange is an R (roger) TU (thank you), or CFM (confirmed). That's about it, the bare essentials. Often (on CW), the entire "script" is programmed into a memory keyer, with no wasted words. On phone, a tape loop is often used to preserve those vocal cords.

At all times, the avid contester has his eye on the clock, trying to maintain the highest rate possible. So don't be offended if he doesn't stop to chat; your contact is, nevertheless, greatly appreciated, because they all add up. And there's plenty of time to engage in dialogue after the fray.

Most contest rules allow for working a station once per band (typically, the major contests have separate and distinct weekends for phone and CW, which are considered separate contests). The ARRL Sweepstakes is an exception—you can work a station only once, regardless of band. The point here is that once you've had your shot, don't keep calling a station you've already logged. Working a station on a band on which you've already had a complete two-way QSO is called a "dupe," which counts 0 points. A good method to avoid this is to use a "dupe sheet," such as ARRL's CD-77, which is a convenient way to write in the calls by prefix. At a glance, you can tell whom you've already contacted.

Another thing to consider is that participation in contests does not require a "big station." You can join in the fun with any kind of gear, including low power. And it should be mentioned that many of the remarkable scores listed in various contest write-ups are made with standard tribanders and routine wire antennas. Indeed, your present station may be better equipped than those in use by contest regulars.

Making Contact

Suppose one Saturday you're casually tuning across the low end of 20 meters and you hear Mike, KH6ND, calling CQ 'TEST. Having read this material (and committed its contents to memory!), you decide that it would be real slick to work Hawaii to wrap up your Worked All States award. So you listen around the band for a minute or two, and rapidly determine that this particular contest merely requires the exchange of signal report and state. And you're off:

And the contact, and your WAS, is history. Hawaii is now in your logbook, and you don't need to be Laurence Olivier to remember all your lines in this little drama. It couldn't be easier; KH6ND also gets another contact for his contest effort. So everyone wins.

Here's how it would go on phone:

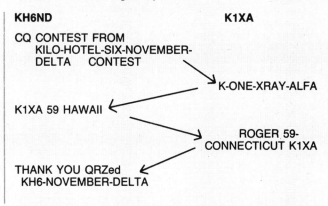

KH6ND   **K1XA**

CQ TEST TEST
    DE KH6ND KH6ND K
                    →K1XA

K1XA 599 HI K ←

                 → R 599 CT K1XA K

R   TU KH6ND QRZ ←

KH6ND   **K1XA**

CQ CONTEST FROM
    KILO-HOTEL-SIX-NOVEMBER-
    DELTA   CONTEST
                 → K-ONE-XRAY-ALFA

K1XA 59 HAWAII ←

                 → ROGER 59-
          CONNECTICUT K1XA

THANK YOU QRZed ←
    KH6-NOVEMBER-DELTA

Other contests require a somewhat more elaborate exchange. Regardless, you have an open invitation to jump right in, whether it's a domestic or DX contest, phone or CW, single band or all band, RTTY or VHF, and so on. The most popular contests—where you might encounter the most activity—are listed in Table 7-1.

The rules for ARRL contests are given feature treatment in *QST*, as are the results, while the rules for numerous other important contests are summarized in *QST*'s monthly Contest Corral column and in other Amateur Radio periodicals. Also, January *QST* provides a calendar of the major ARRL contests for the upcoming year.

So, next time the commotion on your favorite frequency

seems as tumultuous as a rock video, remind yourself that contesting is part of what makes our Amateur Radio hobby rich and unique. But don't stop there, because there is no real substitute for actually sitting down in front of a radio and making real contacts during an actual contest. Why not broaden your Amateur Radio experience, and have some fun, by personally savoring the friendly competition of contesting. And when you do, going at it at your own pace and within your available spare time, don't be surprised when you get so charged up and feeling good that you'll get carried away by all the action!

—*Robert Halprin, K1XA* [originally published
in Sep 1985 *QST*]

---

information you need for a valid contest QSO.

There is one operating strategy to keep in mind here: Don't scare anyone off. Try to be as polite as you can, and give the information at a reasonable pace. Sending an exchange at 50 words per minute or speaking as fast as you can may be fine when you are in contact with another contest devotee, but is totally lost on most everyone else. Not only will they not be able to acknowledge your transmitted information, they (as well as anyone else listening) might become discouraged or even incensed, disappearing without giving you your QSO.

There seemingly are as many hunting techniques as there are contest operators. While everyone has a favorite approach, there is one generalization about the successful versions; they are all organized. The operator starts at one end of the band and methodically scans to the other end, while trying to contact all the stations not already worked in between. Maintenance of an updated duplicate check sheet (dupe sheet) is a necessity for hunting. It is not only embarrassing but a waste of time to call someone already worked. Some operators maintain a "map" of the band that lists the call signs of the stations holding forth on the various frequencies. A map such as this allows the hunting operator to skip over frequencies that he or she knows are occupied by someone already in the log. This can save considerable time, especially on bands that are clogged with big pileups; spending 10 minutes to work a station that turns out to be a duplicate

contact doesn't add much excitement to the contest.

Successful contest operating can be summarized very easily: Always spend your time where you can produce the most points and treat that time as if it is precious—because it is! Always strive to make your log entries an accurate and complete representation of what was sent to you. Accurate communication is the very basis for the contest; inaccurate communication can be the basis for disqualification.

AFTER THE CONTEST

Afterwards, every participant faces the same task, regardless of his or her performance in the contest—preparing the log for submission. The proficient and ambitious contester adds a second task—analyzing the results. The first task shouldn't be taken lightly. Many contests have been won or lost as a result of accurate or inaccurate log checking before submission. Log preparation is a lot like preparing income-tax returns. The accuracy of your submission has a lot to do with the outcome. While you may be able to get away with "creative logging" for a year (or contest) or two, eventually you will be revealed. From that point on, your return (or log) will be carefully scrutinized on a routine basis. What is probably worse about contest logs than income-tax returns is that once you are disqualified, you will never have any credibility with your fellow contesters. That in itself might be a more severe punishment than a few measly years in prison for

Manually Logging/Duping Your Log During the Contest

Operating a contest includes effective and efficient logging and duping procedures while the contest is in full swing. Each contester has his own view on what method works best. Several computer-logging programs are provided elsewhere in this chapter, but many, if not most, contesters opt for the traditional paper-and-pencil approach during the contest period itself, since real-time computer contest logging is a bit awkward (computers are great for preparing an elegant entry after the contest is over, however). The following information provides the generally accepted approach for handwritten logging.

Use of a pencil with a good eraser is practically mandatory for clean logs and dupe sheets. Remember that the contest log checker must look over your log when it is submitted, and a clean log is more desirable than a sloppy one. You will never initially be able to copy every piece of information perfectly, so erasing is fairly frequent. And to keep a sharp point on your pencils, a small battery-operated pencil sharpener is also very helpful.

In contests where more than a few contacts are to be made, keeping a crosscheck (dupe) sheet of each station as worked is the most convenient way of avoiding duplicate contacts (which don't count). For most contests, this means keeping a separate dupe sheet for each band; for others, only one sheet is needed. Many contest sponsors provide dupe sheets for the asking (ARRL's form CD-77 and CD-175, for example), although you may use a different format if that system works best for you. For domestic US contests, the usual procedure is to group each call area together and further break down the call area by suffixes. Whether you choose to enter the entire call sign (advisable for at least the one- or two-letter suffixed calls, since each suffix might have a number of different possible prefixes), or the suffix only, is a matter of preference.

For DX contests, the duping procedure is to break the world down by prefixes and enter the suffix only. Since the prefix is already listed, there is usually no need to enter the full call sign.

But the call sign must first be entered into the log—in other words, you have to work 'em before you can dupe 'em! Note, though, that unlike the typical station log, in which detailed QSO information is often entered, a number of shortcuts may be employed while you are logging during the contest.

Since the date will change only once or twice during an entire contest, there is no need to enter it for each QSO. For entering the time of a QSO when many contacts are made in an hour, the general approach is to enter only the minute of each QSO, leaving off the hour except for the first QSO in each hour. When running strings of stations in an hour or series of hours, some contesters log the time only every five minutes.

When the required exchange includes a QSO number, these should be preentered on your log sheets before the contest begins. The other parts of the transmitted exchange usually remain the same and need only be entered at the top of each page (or just the first page), with arrows or the equivalent pointing downward to show that they remained the same for each QSO. The most important parts of each QSO are the call sign of the station worked (shortcuts here are not acceptable; the complete call must be logged) and the exchange received.

The exchange received usually should be logged completely. One exception is in the case of repetitive signal reports. In most contests, the trend is that most operators give out 599 or 59, the maximum signal strength, for each and every contact. If, for example, everyone gives you a 599 report, the same procedure as for the transmitted exchange applies, and you have to log only the variations. It is also desirable to indicate on either your summary sheet or at the top of your first log sheet "all reports 599 [59] unless otherwise noted." As far as keeping track of new multipliers on the log sheet itself, don't bother! Leave it until rest periods are taken or after the contest is over. The same applies to the number of QSO points earned for each QSO.

Every effort should be made to get each bit of information correct. If it means asking for the call sign a number of times to be sure, do it! Any miscopied information may result in the removal of credit for that contact from your score by the contest log checkers. Too many errors may cause your log to be disqualified.

Examples of efficient logging and duping procedures using standard forms are illustrated in Figs 7-1 through 7-4.

Logging on only one side of each sheet permits you to scan previous pages more easily when necessary, and gives you a good psychological lift when you see all those log pages you are rapidly filling up. Using one side of your dupe sheet allows the entire sheet to be visible at once and scanned quicker as well. Some contesters use two dupe sheets for each band, pasting each sheet to the side of a manilla folder to keep things organized.

As mentioned above, instead of keeping track of multipliers on the log sheets, most contesters use some other pieces of paper. For example, during DX contests, a scan of your dupe sheet (assuming you have kept an accurate sheet) shows quite accurately which country multipliers have been worked. It is helpful in this regard to circle the suffix of the first contact in each country worked (see Fig 7-2) so you can see how you're doing at a glance.

In a contest such as the *CQ* Worldwide, where the score is also based on working zone multipliers, some extra record-keeping is necessary for those who really want to stay organized. One simple method is to use a *CQ* WW log sheet, and number each line to represent the 40 *CQ* zones. Then let each column represent a different band, and simply check the appropriate "box" when a new band-zone is worked. Another way is to take a piece of paper

BAND MHZ	TIME UTC 16 FEB 85	STATION	COMPLETE EXCHANGE SENT	COMPLETE EXCHANGE RCVD	1.8	3.5	7	14	21	28	POINTS	
14	1219	SM0TW	599 CT	599 500								
	20	YU4EZC		100								
		DL4IW		100								
	21	PA0CF		100								
	22	FE6HME		100								
	23	U39AW2		200								
		YU2EU		100								
	24	UT4UH		100								
		DJ0YI		KW								
	25	DL9HC		500								
	27	EV4AP		200								
		RB5GW		200								
	28	G6ZY/EA6		150				EA6				
	29	G3RHI		150								
21	32	U2LLW2		200					UA			
		HA5KDQ		KW					HA			
		HA5JI		200								
	33	OK1EV		300					OK			
	34	Y42YG		100					Y			
		HA7KLG		200								
		4N4TN		20					YU			
	35	YU2TO		300								
	36	I1BAYA		100					I			
	37	OK3FON		200								
		OK3MM		KW								
		OK1MHI		150								
	38	YU2LLL		100								
	39	Y22UE		100								
		DL23AW		100					DL			
	40	HG5A		KW								
		HA6KNX		100								
		HA1WD		200								
	41	DF2US		100								
		YU4AAI		100								
	42	RW3AY		200								
		OK1ALU		500								
		Y33XB		500								
		OK2RU		150								
	43	OK2XA		100								
		UA6ED		200								
		Y36SG		50								
	44	Y25SE		100								
		DK2MG		200								
	45	L22TU		50					L2			
		SP2JKC		200					SP			
		UB5ICK		200					UB			
	46	DK0FN		500								
		OE1TKW		100					OE			
	47	HB9ALO		100					HB			
		DL8FBD		200								
				TOTALS					1	12		

PAGE: 9
CALLSIGN: K1TO
☒ CW
☐ PHONE

ARRL INTERNATIONAL DX CONTEST

1. Separate summary sheets and logs required for each mode.
2. Check entry for duplicates on each band.
3. Dupe/check sheets must be included with every entry of 500 or more QSOs.

W/VE: These points per QSO.
DX: These points per W/VE QSO.

CD-55(R1179)
Printed in U.S.A.

Fig 7-1—A page out of the log of K1TO's CW DX Contest effort.

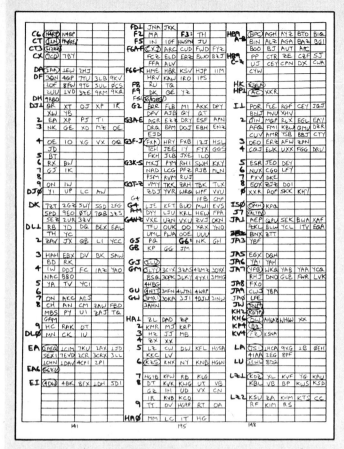

Fig 7-2—Keeping an accurate dupe sheet is the best way to avoid duplicate QSOs.

of paper and write the bands horizontally across the top, and number 1-40 below each band heading, crossing each zone off as worked. Whatever method of keeping track of your contacts and multipliers you are comfortable with—particularly in the heat of battle—is the *best* one. Also, many of the contest clubs to which many amateurs belong also provide very workable dupe and check sheets to their members.

Keeping track of which multipliers have been worked and which have not is essential so you can reach such decisions as to whether to stay in a pileup to work a new one, or turn the beam toward the Pacific in each of KH6, VK, ZL and other rarer stations, or go to 80 meters looking for a Vermont or Wyoming station.

When the contest is over, it is highly recommended that your log be reduped from scratch (and this is where a computer can provide great assistance), and otherwise checked for accuracy, each new multiplier numbered, operating time calculated, sloppy log entries cleaned up and so on. There is no need to *rewrite* your log; it doesn't have to be an artistic success, as long as it is accurate and your summary sheet and dupe sheets are complete, readable and generally in good shape.

Since the deadline for submitting logs is usually many weeks after the contest, there is absolutely no excuse for turning in a log with duplicate contacts unremoved. If it means double- and triple-checking an exceptionally large log, do it! When removing duplicates from your log, just draw a light line through the contact information and put a 0 in the points column. Never erase duplicates.

Reduping your log may, on the positive side, reveal that a contact originally thought to be a duplicate is actually a valid contact after all. Also, you may find new multipliers that you missed during the contest. This kind of checking can reveal scoring-boosts that might have otherwise gone unnoticed. Log checkers often find the same thing and probably do as much *increasing* of scores as decreasing.

Remember, when you put a reasonable effort into the

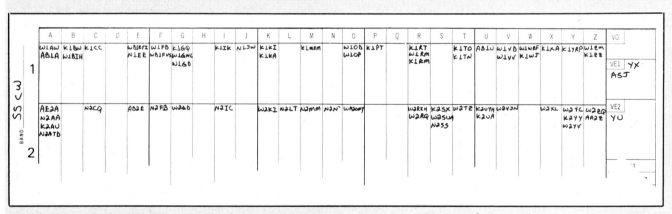

Fig 7-3—Dupe sheets, such as ARRL's CD-77, are mandatory in many stateside contests.

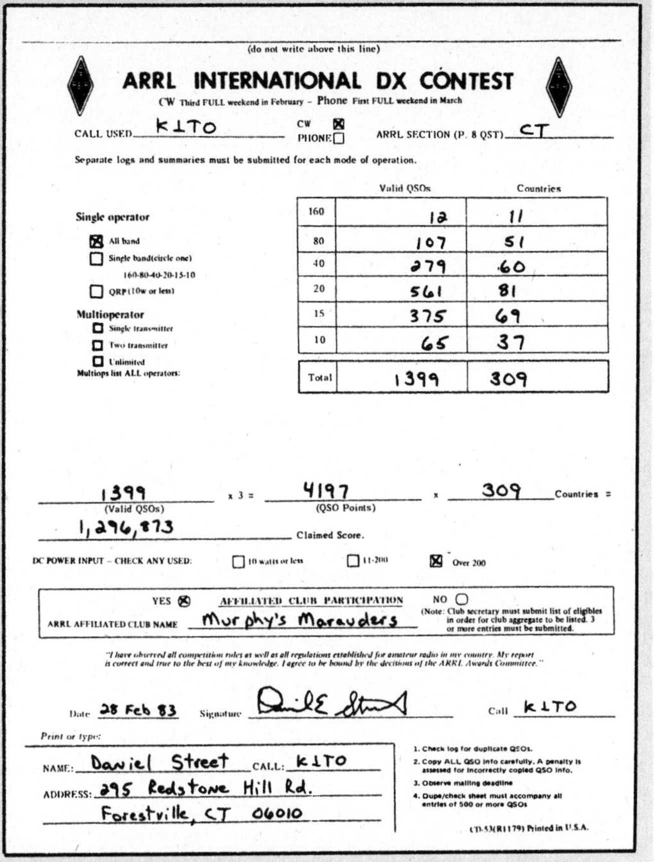
(do not write above this line)

ARRL INTERNATIONAL DX CONTEST

CW Third FULL weekend in February – Phone First FULL weekend in March

CALL USED K1TO   CW [X] PHONE [ ]   ARRL SECTION (P. 8 QST) CT

Separate logs and summaries must be submitted for each mode of operation.

Single operator
[X] All band
[ ] Single band (circle one) 160-80-40-20-15-10
[ ] QRP (10w or less)

Multioperator
[ ] Single transmitter
[ ] Two transmitter
[ ] Unlimited
Multiops list ALL operators:

	Valid QSOs	Countries
160	12	11
80	107	51
40	279	60
20	561	81
15	375	69
10	65	37
Total	1399	309

1399 (Valid QSOs) x 3 = 4197 (QSO Points) x 309 Countries =
1,296,873 Claimed Score.

DC POWER INPUT – CHECK ANY USED: [ ] 10 watts or less [ ] 11-200 [X] Over 200

YES (X) AFFILIATED CLUB PARTICIPATION NO ( )
ARRL AFFILIATED CLUB NAME Murphy's Marauders
(Note: Club secretary must submit list of eligibles in order for club aggregate to be listed. 3 or more entries must be submitted.

"I have observed all competition rules as well as all regulations established for amateur radio in my country. My report is correct and true to the best of my knowledge. I agree to be bound by the decisions of the ARRL Awards Committee."

Date 28 Feb 83   Signature Daniel E Street   Call K1TO

Print or type:
NAME: Daniel Street   CALL: K1TO
ADDRESS: 295 Redstone Hill Rd.
Forestville, CT 06010

1. Check log for duplicate QSOs.
2. Copy ALL QSO info carefully. A penalty is assessed for incorrectly copied QSO info.
3. Observe mailing deadline
4. Dupe/check sheet must accompany all entries of 500 or more QSOs

CD-53(R1179) Printed in U.S.A.

Fig 7-4—Your contest activity culminates in filling out the summary sheet, which at a glance chronicles your achievement.

contest, you should also put in a reasonable effort *after* the contest. Once the contest has ended, by no means is the mission completed. A number of contests have been won and lost long after the equipment has been shut off—in the paperwork. Your goal should be to have a clean and accurate contest submission. And it is a good idea to make a complete photocopy of your contest entry for your personal files before you mail it away to the sponsoring organization.

Standard Contest Guidelines

1) Make sure your log details the date, time, band, call sign and complete exchange sent and received for each QSO claimed for contest credit.
2) Your summary sheet should indicate your score, including how you figured it, and a declaration that you followed FCC regulations (or the regulations of the licensing authority in your country) and the contest rules. Your name, call sign and complete address should be typed or printed in block letters.
3) Crossband, crossmode and repeater contacts are usually not permitted. Contacts with the same station on different bands are usually permitted.
4) Your log should be checked carefully for duplicate QSOs and, if more than 200 contacts are made, dupe sheets should be included with your entry.
5) Your log may be considered a checklog or be disqualified if it is incomplete or if too many errors are detected by the contest committee.
6) Avoid standard net frequencies.
7) International contests generally offer awards to top scorers from each US call area and each country; state QSO parties to each state/province.
8) Your summary sheet should include the following statement: "I have observed all competition rules as well as regulations established for Amateur Radio in my country." The declaration should be signed and dated.

defrauding the government!

A proper log submission consists of a photocopy of the original log sheets, with invalid and duplicate contacts so marked. Don't erase these contacts, just mark them so that the log checkers can identify and verify them by crosschecking with other logs. It is virtually imperative to "re-dupe" a log before submission. Any duplicate check-sheet made during the contest is bound to be inaccurate and should only be considered an operating aid. In these modern times, computers are increasingly being used to eliminate duplicate contacts from contest logs prior to submission (be sure to double check for keyboarding errors, however).

A contest log might be considered the history of a "lost" weekend. While there is no explicit description whatsoever of the events of the weekend, a clear record is available for the experienced to interpret. Your best source of information to use while preparing for the next contest is your record of the last contest. What mistakes did you make? What bands were the most or least productive and why? What equipment worked well? What equipment failed? Responding to these questions should keep you busy until the next contest. And you can be sure that the next contest will be more productive and satisfying than the last.

Chapter 8

Operating Awards

Awards hunting is a significant part of the life-support system of Amateur Radio operating. It's a major motivating force of so many of the contacts that occur on the bands day after the day. It takes skillful operating to qualify, and the reward of having a beautiful certificate or plaque on your ham-shack wall commemorating your achievement is very gratifying. (On the other hand, you don't necessarily have to seek them actively; just pull out your shoebox of QSLs on a cold, winter afternoon, and see what gems you already have on hand.) Aside from expanding your Amateur Radio-related knowledge, it's also a fascinating way to learn about the geography, history or political structure of another country, or perhaps even your own.

This chapter provides detailed information on awards sponsored by ARRL and other major US organizations. Following that is perhaps the most comprehensive collection of international awards information ever gathered under one cover, obtained by the Editor through the good offices of International Amateur Radio Union (IARU) Societies, the sister associations of the League. In so doing, the awards data base has been expanded like never before.

There are some basic considerations to keep in mind when applying for awards. Always carefully read the rules, so that your application complies fully. Use the standard award application if possible, and make sure your application is neat and legible, and indicates clearly what you are applying for. Official rules and application materials are available directly from the organization sponsoring the particular award; always include an SASE (self-addressed, stamped envelope) or, in the case of international awards, a self-addressed envelope with IRCs (International Reply Coupons, available from your local Post Office) when making such requests. Sufficient return postage should also be included when directing awards-related correspondence to Awards Managers, many (if not most) of whom are volunteers. So above all, be patient!

As to QSL cards, if they are required to be included with your application, send them the safest possible way and always include sufficient return postage for their return the same way. (A return-postage chart for ARRL awards appears in Chapter 17 of this book). It is vital that you check your cards carefully before mailing them; make sure each card contains your call and other substantiating information (band, mode, and so on). Above all, don't send cards that are altered or have mark-overs, even if such modifications are made by the amateur filling out the card. Altered cards, even if such alterations are made in "good faith," are not acceptable on this no-fault basis. If you are unsure about a particular card, don't submit it. Secure a replacement.

None of the above is meant to diminish your enthusiasm for awards hunting. Just the opposite, since this chapter has been painstakingly put together to make awards hunting even more enjoyable. These are just helpful hints to make things even more fun for all concerned. Chasing awards is a robust facet of hamming that makes each and every QSO a key element in your present or future Amateur Radio success.

ARRL AWARDS

To make Amateur Radio QSOing more enjoyable and to add challenge, the League sponsors awards for operating achievement, some of which are among the most popular awards in ham radio. Please note that with the exception of RCC and Code Proficiency, applicants in the US and possessions, Canada and Puerto Rico must be League members to participate.

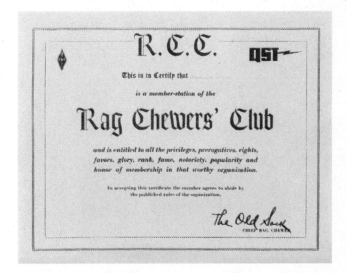

Rag Chewers Club (RCC)

RCC is often the first operating award a new ham aims for. This is designed to encourage friendly contacts of more substance than the hello-goodbye type QSO. RCC has just one requirement: "Chew the rag" over the air for at least one solid half hour. If you want to obtain the RCC certificate, report the QSO to ARRL HQ, with a business-size SASE, and you'll soon be issued the distinctive blue award. If you want to nominate someone else for membership, send the nomination to him or her, not to ARRL HQ. If this person wants the award, he or she can send the nomination to HQ. This way, no one gets an unwanted certificate, and confirmed ragchewers can still nominate those they think are qualified. RCC is a free ARRL service to all amateur licensees.

Code Proficiency (CP)

You don't have to be a ham to earn this one. But you do have to copy one of the W1AW qualifying runs. (A current W1AW operating schedule is available from ARRL HQ for an SASE.) Twice a month, five minutes worth of text is transmitted at the following speeds: 10-15-20-25-30-35 WPM. Twice yearly, W1AW transmits 40-WPM runs. To qualify, just copy one minute solid. Your copy can be written, printed or typed. Underline the minute you believe you copied perfectly and send this text to ARRL HQ along with your name, call (if licensed) and complete mailing address. Your copy is checked directly against the official W1AW transmission copy, and you'll be advised promptly if you've passed or failed. If the news is good, you'll soon receive either your initial certificate or an appropriate endorsement sticker. Please include an SASE with your submission.

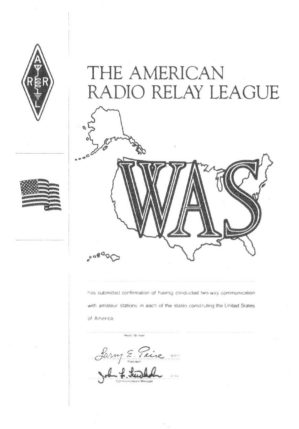

THE AMERICAN RADIO RELAY LEAGUE

WAS

has submitted confirmation of having conducted two-way communication with amateur stations in each of the states constituting the United States of America.

Worked All States (WAS)

The WAS (Worked All States) award is available to all amateurs worldwide who submit proof with written confirmation of having contacted each of the 50 states of the United States of America. The WAS awards program includes 10 different and separately numbered awards as listed below. In addition, endorsement stickers are available as listed below.

Two-way communications must be established on amateur bands with each state. Specialty awards and endorsements must be two-way ($2\times$) on that band and/or mode. There is no minimum signal report required. Any or all amateur bands, with the exception of 10 MHz (contacts on 10 MHz are not valid for WAS) may be used for general WAS. The District of Columbia may be counted for Maryland.

Contacts must all be made from the same location, or from locations no two of which are more than 50 miles apart which is affirmed by signature of the applicant on the application. Club-station applicants, please include clearly the club name and call sign of the club station (or trustee).

Contacts may be made over any period of years. Contacts must be confirmed in writing, preferably in the form of QSL cards. Written confirmations must be submitted (no photocopies). Confirmations must show your call and indicate that two-way communication was established. Applicants for specialty awards or endorsements must submit confirmations that clearly confirm two-way contact on the specialty mode/band. Contacts made with Alaska must be dated January 3, 1959 or later, and with Hawaii dated August 21, 1959 or after.

Specialty awards (numbered separately) are available for OSCAR satellites, SSTV, RTTY, 432 MHz, 220 MHz, 144 MHz, 50 MHz, and 160 meters. In addition, the "75-Meter 2-letter Extra Class" award is available to any Extra Class amateur who has worked all states consisting of 1×2, 2×1 or 2×2 call signs on the Extra Class portion of 75 meters. Endorsement stickers for the basic mixed mode/band award and any of the specialty awards are available for SSB, CW, EME, Novice, QRP, packet and any single band. The Novice endorsement is available for the applicant who has worked all states as a Novice licensee. QRP is defined as 10-watts input (or 5-watts output) as used by the applicant only and is affirmed by signature of the applicant on the application.

Contacts made through "repeater" devices or any other power relay method cannot be used for WAS confirmation. A separate WAS is available for OSCAR contacts. All stations contacted must be "land stations." Contact with ships (anchored or otherwise) and aircraft cannot be counted.

A W/VE applicant must be an ARRL member to participate in the WAS program. DX stations are exempt from this requirement.

HQ reserves the right to "spot call" for inspection of cards (at ARRL expense) of applications verified by an HF Awards Manager. The purpose of this is not to question the integrity of any individual, but rather to ensure the overall integrity of the program. More difficult-to-be-attained specialty awards (220-MHz WAS, for example) are more likely to be so called. Failure of the applicant to respond to such a spot check will result in nonissuance of the WAS certificate.

Disqualification: False statements on the WAS application or submission of forged or altered cards may result in disqualification. ARRL does not attempt to determine who has altered a submitted card; therefore do not submit any marked-over cards. The decision of the ARRL Awards Committee in such cases is final.

Application Procedure (please follow carefully): Confirmations (QSLs) and application form (MCS-217) may be submitted to an approved ARRL Special Service Club HF Awards Manager. ARRL Special Service Clubs appoint HF Awards Managers whose names/addresses are on file at HQ. If you do not know of an HF Awards Manager in your local area, call a club officer to see if one has been appointed or contact HQ. If you can have your application so verified locally, you need not submit your cards to HQ. Otherwise, send your application and cards to HQ, as indicated on the application form (reproduced in Chapter 17).

Be sure that when cards are presented for verification (either locally or to HQ) they are sorted alphabetically by state, as listed on the back of application form MCS-217.

All QSL cards sent to HQ must be accompanied by sufficient postage for their safe return (postage guidelines appear in Chapter 17).

Five-Band WAS (5BWAS)

This award, for working all states on five amateur bands, is designed to foster more uniform activity throughout the bands, encourage the development of better antennas and generally offer a challenge to both newcomers and veterans. The basic WAS rules apply, including cards being checked in the field by Awards Managers; in addition, 5BWAS carries a start date of January 1, 1970. Unlike WAS, 5BWAS is a one-time-only award; no band/mode endorsements are available. Contacts made on 10/18/24 MHz are not valid for 5BWAS. In addition to the 5BWAS certificate, a 5BWAS plaque is available at an additional charge.

VHF/UHF Century Club (VUCC)

The VHF/UHF Century Club Award is awarded for contact with a minimum number of "Maidenhead" 2° × 1° grid-square locators per band as indicated below. Grid squares are designated by a combination of two letters and two numbers. More information on grid squares can be found in January 1983 *QST*, pp 49-51 (reprint available from HQ on request), *The ARRL World Grid Locator Atlas* and the ARRL *Grid Locator for North America* (see latest *QST* for prices and ordering information). The VUCC certificate and endorsements are available to League members in Canada, the US and possessions, and Puerto Rico, and to other amateurs worldwide. Only those contacts dated January 1, 1983 and later are creditable for VUCC purposes. (A VUCC lapel pin is available for a nominal charge of $2 US.)

The minimum number of squares needed to qualify initially for each individual band award is as follows: 50 MHz—100; 144 MHz—100; 220 MHz—50; 432 MHz—50; 902 MHz—25; 1296 MHz—25; 2.3 GHz—10; 3.4 GHz—5; 5.7 GHz—5; 10 GHz—5; 24 GHz—5; and 47 GHz—5. Certificates for 220 and 432 MHz are designated as Half Century, 902 and 1296 MHz as Quarter Century, and those above as SHF Awards. Individual band awards are endorsable in the following increments: 50 and 144 MHz—25 credits; 220 and 432 MHz—10 credits; 902 MHz and above—5 credits. There are no specialty endorsements such as "CW only," etc.

Separate bands are considered as separate awards. No cross-band contacts permitted. No contacts through active "repeater" or satellite devices, or any other power relay methods, are permitted. Contacts with aeronautical mobiles (in the air) *do not* count. Contacts with stations shipboard (located on a body of water) *do* count.

Stations who claim to operate from more than one grid square simultaneously (such as from the intersection of four grid squares) must be physically present in more than one square to give multiple square credit with a single contact. This requires the operator to know precisely where the intersection lines are located and placing the station exactly on the boundary to meet this test. To achieve this precision work requires either current markers permanently in place or the precision work of a professional surveyor. Operators of such stations should be prepared to provide some evidence of meeting this test if called upon to do so. Multiple QSL cards are not required.

For VUCC awards on 50 through 1296 MHz, all contacts must be made from a location or locations within the same grid square or locations in different grid squares no more than 50 miles apart. For SHF awards, contacts must be made from a single location, defined as within a 300-meter-diameter circle.

Application Procedure (please follow carefully): Confirmations (QSLs) and application forms (MCS-259 and MCS-260) must be submitted to an approved VHF Awards Manager for certification. ARRL Special Service Clubs appoint VHF Managers whose names are on file at HQ. If you do not know of an Awards Manager in your area, HQ will give you the name of the closest Manager. Do not send cards to HQ.

For the convenience of the Awards Manager in checking cards, applicants may indicate in pencil (pencil only) the grid-square locator on the address side of cards that do not clearly indicate the grid locator. The applicant affirms that he/she has accurately determined the proper locator from the address information given on the card by signing the affirmation statement on the application.

Cards must be sorted alphabetically by field and numerically from 00 to 99 within that field.

Where it is necessary to mail cards for certification, postage equal to the amount needed to send them must be included for return of cards along with a separate self-addressed mailing label. In addition, US and Canadian applicants will enclose $1 US for any initial application to cover postage and packaging of the certificate. For endorsements, enclose only an SASE with two units of First-Class postage. When mailing cards, registered or certified mail is recommended.

Enclosed with the initial VUCC certificate from HQ will be a photocopy of the original list of grid squares for which the applicant has received credit (MCS-259). When applying for endorsement, the applicant will indicate in red on that photocopy those new grid squares for which credit is sought, and submit cards for certification to an Awards Manager. A new, updated photocopy listing will be returned with the endorsement sticker. Thus, a current list of grid squares worked is always in the hands of the VUCC award holder, available to the VHF Manager during certification, and a permanent historical record always maintained at HQ. Reminder: For initial application, enclose $1 and for endorsements, enclose an SASE (with two units of postage). For endorsement applications, it is necessary to submit only those MCS-259s that indicate new grid squares worked since the previous submission (indicated in red).

Disqualification: Altered/forged confirmations or fraudulent applications submitted may result in disqualification of the applicant from VUCC participation by action of the ARRL Awards Committee. The applicant affirms he/she has abided by all the rules of membership in the VUCC and agrees to be bound by the decisions of the ARRL Awards Committee.

Decisions of the ARRL Awards Committee regarding interpretation of the rules here printed or later amended shall be final.

Operating Ethics: Fair play and good sportsmanship in operating are required of all VUCC members.

Worked All Continents (WAC)

In recognition of international two-way Amateur Radio communication, the International Amateur Radio Union (IARU) issues Worked All Continents certificates to Amateur Radio stations of the world. Qualificiation for the WAC award is based on examination by the International Secretariat, or a Member-Society, of the IARU, that the applicant has received QSL cards from other amateur stations in each of the six continental areas of the world (see the ARRL DXCC Countries List in Chapter 17 for a complete listing of continents). All contacts must be made from the same country or separate territory within the same continental area of the world. All QSL cards (no photocopies) must show the mode and/or band for any endorsement applied for.

WAC certificates are issued as follows:

- Basic Certificate (mixed mode)
- CW Certificate
- Phone Certificate
- SSTV Certificate
- RTTY Certificate
- FAX Certificate
- Satellite Certificate
- 5-Band Certificate

WAC endorsement stickers are issued as follows:

- 6-Band sticker
- QRP Sticker
- 1.8-MHz Sticker
- 3.5-MHz Sticker
- 50-MHz Sticker
- 144-MHz Sticker
- 430-MHz Sticker
- Any higher band

Contacts made on 10/18/24 MHz or via satellites are void for the 5-band certificate and 6-band sticker. All contacts for the QRP sticker must be made on or after January 1, 1985, while running a maximum power of 5-watts output or 10-watts input.

For amateurs in the United States or countries without IARU representation, applications should be sent to the IARU International Secretariat, PO Box AAA, Newington CT 06111, USA. After verification, the cards will be returned, and the award sent soon afterward. There is no application fee; however, sufficient return postage for the cards, in the form of a self-addressed, stamped envelope or IRCs, is required. US amateurs must have current ARRL membership. [A sample application form and a return-postage chart appear in Chapter 17.] All other applicants must be members of their national Amateur Radio Society affiliated with IARU and must apply through the Society only.

DX Century Club (DXCC)

DXCC is the premier operating award in Amateur Radio. The initial DXCC certificate (a nominal fee of $2 US is charged for the DXCC lapel pin) is available to League members in Canada, the US and possessions, and Puerto Rico, and all amateurs in the rest of the world. There are six separate DXCC awards available:

Mixed (general type)—contacts must be made using any mode since November 14, 1945.

Phone—contacts must be made using radiotelephone since November 14, 1945. Confirmations for cross-mode contacts for this award must be dated before October 1, 1981. Confirmations need not indicate two way (2×) to be credited.

CW—contacts must be made using CW since January 1, 1975. Confirmations for cross-mode contacts for this award must be dated before October 1, 1981. Confirmations need not indicate two-way (2×) to be credited.

RTTY—contacts must be made using radioteletype since November 15, 1945. Confirmations for cross-mode contacts for this award must be dated before October 1, 1981. Confirmations need not indicate two-way (2X) to be credited.

160 meter—contacts must be made on 160 meters since November 15, 1945.

Satellite—contacts must be made using satellites since March 1, 1965 (non-endorsable).

Confirmations (QSL cards) must be submitted directly to ARRL HQ for all countries claimed. Confirmations for a total of 100 or more countries must be included with first application. Contacts made on all amateur bands, with the exception of 10 MHz, are valid for DXCC. **[Contacts made on 10/18/24 MHz are not valid for 5-Band DXCC.]**

The ARRL DXCC Countries List criteria (see below) will be used in determining what constitutes a "country."

Confirmations must be accompanied by a list of claimed DXCC countries and stations to aid in checking and for future reference (The required DXCC application materials are available from ARRL HQ for an SASE).

Endorsement stickers for affixing to certificates will be awarded as additional credits are granted. These stickers are in exact multiples of 25 (ie, 125, 150, 175) between the 100 and 250 levels, in multiples of 10 between 250 and 300, and in multiples of 5 above the 300 level (Note: Satellite DXCC is non-endorsable). Confirmations for additional countries may only be submitted for further credits in increments that will at least bring the new total up to the next endorsement level. Exception: Once per year, any DXCC participant having an accredited DXCC total of 250 or more may make a submission without regard to the number of cards submitted.

All contacts must be made with amateur stations working in the authorized amateur bands or with other stations licensed to work amateurs.

In cases of countries where amateurs are licensed in the normal manner, credit may be claimed only for stations using regular government-assigned call letters. No credit may be claimed for contacts with stations in any countries in which amateurs have been temporarily closed down by special government edict where amateur licenses were formerly issued in the normal manner.

All stations contacted must be "land stations." Contacts with ships (anchored or otherwise) and aircraft cannot be counted.

All stations must be contacted from the same DXCC "country."

Contacts may be made over any period of years since November 15, 1945 for the mixed, phone, 160-meter and RTTY DXCCs; January 1, 1975 for the CW DXCC, and from March 1, 1965 for the Satellite DXCC, provided only that all contacts were made from the same DXCC country and by the same station licensee; contacts may have been under different call letters or in the same area (or country), if the licensee for all was the same. (You may feed one DXCC from several calls held simultaneously.)

Any altered or forged confirmations submitted for DXCC credit may result in disqualification of the applicant. The eligibility of any DXCC applicant who was ever barred from DXCC to reapply, and the conditions for such application, shall be determined by the ARRL Awards Committee. Any holder of the DXCC Award submitting forged or altered confirmations must forfeit his right to continued DXCC participation.

a) Fair play and good sportsmanship in operating are required of all DXCC members. In the event of specific objections relative to poor operating ethics, an individual may be disqualified from DXCC by action of the ARRL Awards Committee.

b) Credit for contacts with individuals who have displayed continued poor operating ethics may be disallowed by action of the ARRL Awards Committee.

c) For (a) and (b) above, operating includes confirmation procedures.

League membership is required of DXCC applicants in the US and possessions, Canada and Puerto Rico. All new DXCC applications must contain sufficient postage in the form of US currency, check or money order. For DXCC endorsements, sufficient funds for return postage are also required. A chart showing suggested sums for return postage is available on request from ARRL HQ (and is reproduced in Chapter 17).

ARRL membership is not required of foreign applicants. Each new DXCC application and endorsement application must contain suffient funds for return postage. A chart showing suggested sums for return postage is available on request from ARRL HQ (and is reproduced elsewhere in this book).

Decisions of the ARRL Awards Committee regarding DXCC rules interpretations (either currently in effect or later amended) shall be final.

Official DXCC application forms are required. These may be obtained from the DXCC Desk at ARRL HQ. Please include a business-size SASE.

5BDXCC

A Five-Band DXCC Award has been established to encourage more uniform DX activity throughout the amateur bands, encourage the development of more versatile antenna systems and equipment, provide a challenge for DXers, and enhance amateur-band occupancy. The basic DXCC rules apply, although the starting date for valid QSOs is January 1, 1969.

The 5BDXCC certificate is issued after the applicant submits a minimum of 500 QSLs representing two-way contact with 100 different DXCC countries on each of five Amateur Radio bands. Phone and CW segments of the band do not count as separate bands for this award. Confirmations made on any legal mode are acceptable, but no cross-mode or cross-band contacts are acceptable. Contacts made on 10/18/24 MHz are not valid for 5BDXCC. All QSLs must be checked by the ARRL HQ DXCC Desk. 5BDXCC is non-endorsable. In addition to the 5BDXCC certificate, a 5BDXCC plaque is available at an extra charge.

Countries List Criteria

The ARRL DXCC Countries List (reproduced in Chapter

17) is the result of approximately 40 years of progressive changes in DXing. The full list will not necessarily conform completely with these criteria since some of the listings were set up and recognized from pre-WW II. While the general policy has remained the same, specific mileages and additional points have, over the past 25 years, been added to the criteria. These specific mileages in Point 2(a) and Point 3 (see below) have been used in considerations made April 1960 and after. The specific mileage in Point 2(b) has been used in considerations made April 1963 and after.

Any land area in the world, with the exception of such land that would come under Points 4 and 5, can be placed in one or more of the following categories. When the area in question meets at least one of the points in the criteria, it may be considered eligible as a separate entity, ie, a country, for the ARRL DXCC Countries List.

1) *Government*. An area by reason of Government constitutes a separate entity.

2) *Separation by Water*. An island or a group of islands, not having its own government, is considered as a separate entity under the following conditions:

    a) Islands situated offshore from their governing area must be geographically separated by a minimum of 225 statute miles of open water. This point is concerned with islands offshore from the mainland *only*. This point is *not* concerned with islands which are part of an island group or are geographically located adjacent to an island group.

    b) Islands forming part of an island group or which are geographically located adjacent to an island, or island group, which have a common government, will be considered as separate entities provided there is at least 500 statute miles of open water separation between the two areas in question.

3) *Separation by foreign land*. In the case of a country, such as that covered by Point 1, which has a common government but which is geographically separated by land which is foreign to that country, if there is a complete separation of the country in question by a minimum of 75 miles of foreign land, the country is considered as two separate entities. This 75 statute miles of land is a requirement which is applicable to land areas *only*. In cases of areas made up of a chain of islands, there are no minimum requirements concerned with the separation by foreign land.

4) *Unadministered Area*. Any area which is unadministered will not be eligible for consideration as a separate entity.

5) (a) Any area which is classified as a Demilitarized Zone, Neutral Zone or Buffer Zone will not be eligible for consideration as a separate entity.

    (b) The following will not be eligible for consideration as a separate entity from the host country: Embassies, consulates and extraterritorial legal entities of all nature, including, but not limited to, monuments, offices of the United Nations agencies or related organizations, other intergovernmental organizations or diplomatic missions.

DXCC Accreditation Criteria

During the course of more than 40 years of DXCC administration, standards have evolved in the acceptance of confirmations for DXCC credit. These criteria codify long-standing practice. The intent is to assure that DXCC credit is given only for contacts with operations that are conducted appropriately in two respects: (1) properly licensed, and (2) physically present in the country to be credited. The following points should be of particular interest to DXpeditions:

1) The vast majority of operations are accredited routinely without any requirement for submission of documents.

2) In some instances, especially DXpeditions and in countries that previously have evidenced some reluctance to license amateur stations or allow access, authenticating documents may be requested for review prior to accreditation. Such supporting documents could included the following: (a) Photocopy of license or operating authorization. (b) For amateurs foreign to the country, photocopy of passport entry and exit stamps. (c) For offshore islands, a landing permit and/or signed statement of the transporting ship's captain showing all pertinent data (dates, etc). (d) For some locations where special permission is known to be required to gain access legally, evidence of this permission having been given may be required.

The purpose of these accreditation requirements is to: (1) preserve the program's continued integrity; and (2) ensure that the DXCC program does not encourage amateurs "to bend the rules" in their enthusiasm, thus jeopardizing the future development of Amateur Radio. Every effort will be made to apply these criteria in a uniform manner in conformity with these objectives.

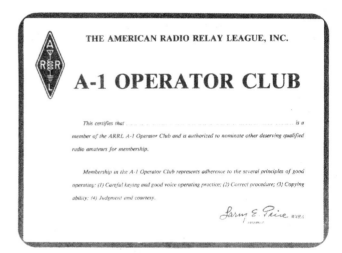

A-1 Operator Club (A-1 OP)

Membership in this elite group attests to superior competence and performance in the many facets of Amateur Radio operation: CW, phone, procedures, copying ability, judgment and courtesy. You must be recommended for the certification independently by two amateurs who already are A-1 Ops. This honor is unsolicited; it is earned through the continuous observance of the very highest operating standards.

Old-Timers' Club (OTC)

In recognition of amateurs who have held an amateur license 20-or-more years (lapses permitted), a suitable award is available—OTC. If you qualify as an "old-timer," you'll find the necessary paperwork pretty easy. Drop a note to HQ (with an SASE) with the date of your first amateur license and your call then and now. HQ will verify the information, and if you're eligible, you'll soon receive your OTC certificate by return mail.

Contests

The League sponsors exciting on-the-air contests (see Chapter 7) for developing operating skills from 160 meters to 10 GHz, and participation certificates are available for all levels of activity, from casual participation to "all-out war"!

Contests are also the "fast-track" way to make those needed QSOs for ARRL awards.

Plaques

•Those who qualify for either 5BWAS and/or 5BDXCC are eligible for a handsome individually engraved 9- × 12-inch walnut plaque. Further information, including required fee, is included in the 5BWAS or 5BDXCC application materials.

•ARRL International DX Contest Awards Program. Beautiful plaques can be won for specific achievements in the ARRL DX Contest. Details appear in *QST*.

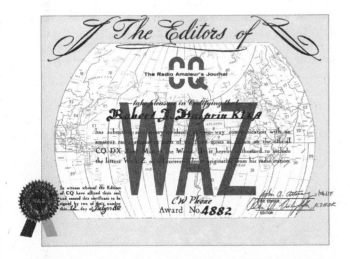

RULES FOR CQ MAGAZINE'S DX AWARDS

Worked all Zones (WAZ)

The CQ WAZ Award will be issued to any licensed amateur station presenting proof of contact with the 40 zones of the world. This proof shall consist of proper QSL cards, which may be checked by any of the authorized *CQ* checkpoints or

sent directly to the WAZ Award Manager, Leo Haijsman, W4KA, 1044 Southeast 43rd St, Cape Coral, FL 33904. Many of the major DX clubs in the United States and Canada and most national Amateur Radio societies abroad are authorized *CQ* check points. If in doubt, consult the WAZ Award Manager or the *CQ* Magazine DX Editor. Any legal type of emission may be used, providing communication was established after November 15, 1945.

The official *CQ* WAZ Zone Map and the printed zone list will be used to determine the zone in which a station is located. [A DXCC Countries List, which includes *CQ* Zones, appears in Chapter 17 of this book.] Confirmation must be accompanied by a list of claimed zones, using *CQ* form 1479, showing the call letters of the station contacted within each zone. The list should also clearly show the applicant's name, call letters and complete mailing address. The applicant should indicate the type of award for which he or she is applying, such as All SSB, All CW, or Mixed. In remote locations and in foreign countries, a handwritten list may be submitted and will be accepted for processing, provided the above information is shown.

All contacts must be made with licensed, land-based, amateur stations operating in authorized amateur bands, 160-10 meters. All contacts submitted by the applicant must be made from within the same country. It is recommended that each QSL clearly show the station's zone number. When the applicant submits cards for multiple call signs, evidence should be provided to show that he or she also held those call letters. Any altered or forged confirmations will result in permanent disqualification of the applicant.

A processing fee ($4 for subscribers—a recent *CQ* mailing label must be included; $10 for nonsubscribers) and a self-addressed envelope with (sufficient postage or IRCs to return the QSL cards by the class of mail desired and indicated) must accompany each application. IRCs equal in redemption value to the processing fee are acceptable. Checks can be made out to Leo Haijsman, WAZ Award Manager.

In addition to the conventional certificate for which any and all bands and modes may be used, specially endorsed and numbered certificates are available for phone and single-sideband operation. The phone certificate requires that all contacts be two-way phone, and the SSB certificate requires that all contacts be two-way SSB.

If, at the time of the original application, a note is made pertaining to the possibility of a subsequent application for an endorsement or special certificate, only the missing confirmations required for that endorsement need be submitted with the later application, providing a copy of the original authorization signed by the WAZ Manager is enclosed.

Decisions of the *CQ* DX Awards Advisory Committee on any matter pertaining to the administration of this award will be final.

All applications should be sent to the WAZ Award Manager, W4KA, after the QSL cards have been checked by an authorized *CQ* checkpoint. [Contact W4KA for a list of checkpoints in your area.] Zone maps, printed rules and application forms are available from the WAZ Award Manager. Send a business-size (4- × 9-inch), self-addressed envelope with two units of First-Class postage, or a self-addressed envelope and 3 IRCs.

Single Band WAZ

Effective January 1, 1973, special WAZ Awards will be issued to licensed amateur stations presenting proof of contact with the 40 zones of the world on one of the six high-frequency bands, 160-10 meters. Contacts for a Single Band WAZ award must have been made after 0000 hours UTC January 1, 1973. For 160-meter WAZ, contacts are valid from January 1, 1975. Proof of contact shall consist of proper QSL cards checked by the *CQ* DX Editor, the WAZ Manager or an authorized *CQ* checkpoint. Single-band certificates will be awarded for both two-way phone, including SSB and two-way CW (for 160-meter WAZ, separate awards are *not* offered for the different modes). The Single Band WAZ program is governed by the same rules and uses the same zone boundaries.

5 Band WAZ

Effective January 1, 1979, the *CQ* DX Department, in cooperation with the *CQ* DX Awards Advisory Committee, announced a most challenging DX award—5 Band WAZ. Applicants who succeed in presenting proof of contact with the 40 zones of the world on the five HF bands—80, 40, 20, 15 and 10 meters (for a total of 200)—will receive a special certificate in recognition of this achievement.

These rules are in effect as of July 1, 1979, and supersede all other rules. Five Band WAZ will be offered for any combination of CW, SSB, phone or RTTY contacts, mixed mode only. Separate awards will not be offered for the different modes. Contacts must have been made after 0000 UTC January 1, 1979. Proof of contact shall consist of proper QSL cards checked only by WAZ Award Manager W4KA. The first plateau will be a total of 150 zones on a combination of the five bands. Applicants should use a separate sheet for each frequency band, using *CQ* form 1479.

A regular WAZ or Single Band WAZ is a prerequisite for a 5 Band WAZ certificate. All applications should show the applicant's WAZ number. After the 150 zone certificate is earned, the final objective is 200 zones for a complete 5 Band WAZ. The applicant has a choice of paying a fee for a plaque and/or applying for an endorsement sticker commemorating this achievement.

All applications should be sent to the WAZ Award Manager, W4KA. The 5 Band Award is governed by the same rules as the regular WAZ Award and uses the same zone boundaries.

The Prefix Award Program

WPX

The *CQ* WPX Award recognizes the accomplishments of confirmed QSOs with the many prefixes used by amateurs throughout the world. Separate distinctively marked certificates are available for two-way SSB, CW and mixed modes, as well as the VPX Award for shortwave listeners and the WPNX Award for Novices.

All applications for WPX certificates (and endorsements) must be submitted on the official application form, *CQ* 1051A. This form can be obtained by sending a self-addressed stamped, business-size (4- × 9-inch) envelope to the WPX Award Manager, Norm Koch, K6ZDL, PO Box 1351, Torrance, CA 90505. All QSOs must be made from the same country. All call letters must be in strict alphabetical order and the entire call letters must be shown. All entries must be clearly legible. Certificates are issued for the following modes and number of prefixes. Cross-mode QSOs are not valid for the CW or 2× SSB certificates. Mixed (any mode): 400 prefixes confirmed. CW: 300 prefixes confirmed. Separate applications are required for each mode. Cards need not be sent, but they must be in the possession of the applicant. Any

and all cards may be requested by the WPX Award Manager or by the *CQ* DX Committee. The application fee for each certificate is $4 for subscribers (subscribers must include a recent *CQ* mailing label) and $10 for nonsubscribers, or the equivalent in IRCs. All applications and endorsements should be sent to the WPX Award Manager.

Prefix endorsements are issued for each 50 additional prefixes submitted. Band endorsements are available for working the following numbers of prefixes on the various bands: 1.8 MHz—50, 3.5 MHz—175, 7 MHz—250, 14 MHz—300, 21 MHz—300, 28 MHz—300. Continental endorsements are given for working the following numbers of prefixes in the respective continents: North America—160, South America—95, Europe—160, Africa—90, Asia—75, Oceania—60. Endorsement applications must be submitted on *CQ* form 1051A. Use separate applications for each mode and be sure to specify the mode of your endorsement application. For prefix endorsements, list only additional call letters confirmed since the last endorsement application. A self-addressed, stamped envelope or proper IRCs for surface or airmail return is required, and $1 or 5 IRCs for each endorsement sticker.

Prefixes: The two or three letter/numeral combinations which form the first part of any amateur call will be considered the prefix. Any difference in the numbering, lettering, or order of same shall constitute a separate prefix. The following would be considered different: W2, WA2, WB2, WN2, WV2, K2 and KN2. Any prefix will be considered legitimate if its use was licensed or permitted by the governing authority in that country after November 15, 1945. A suffix would designate portable operation in another country or call area and would count only if it is the normal prefix used in that area. For example, K4IIF/KP4 would count as KP4. However, KP4XX/7 would not count as KP7, since this is nor a normal prefix. Suffixes such as /M, /MM, /AM, /A and /P are not counted as prefixes. An exception to this rule is granted for portable operation within the issued call area. Thus, contacts with a special prefix such as WS2JRA/2 counts for WS2; however, WS2JRA/3 would count for W3. All calls without numbers will be assigned an arbitrary 0 plus the first two letters to constitute a prefix. For example, RAEM counts as RA0, AIR as AI0, UPOL as UP0. All portable suffixes that contain no numerals will be assigned an arbitrary 0. For example, W4BPD/LX counts as LX0, and WA6QGW/PX counts as PX0.

WPNX

The WPNX Award can be earned by US Novices who work 100 different prefixes prior to upgrading. The application may be submitted after receiving the higher license, providing the actual contacts were made as a Novice. Prefixes worked for the WPNX Award may be used later for credit toward the WPX Award. The rules for the WPNX Award are the same as for the WPX Award except that only 100 prefixes must be confirmed. Applications are sent to the WPX Award Manager.

WPX Honor Roll

The WPX Honor Roll recognizes those operators and stations that maintain a high standing in confirmed, current prefixes. The rules therefore, reflect the belief that Honor Roll membership should be accessible to all active radio amateurs and not to be unduly advantageous to the "old-timers." With the exceptions listed below, all general rules for WPX apply toward Honor Roll credit. A minimum of 600 prefixes is required to be eligible for the Honor Roll.

Only current prefixes may be counted toward WPX Honor Roll standings, those prefixes to be listed and updated annually in *CQ* or to be available from the WPX Award Manager. Special-issue prefixes (ie, OF, OS, 4A, etc) will be considered current for as long as they are assigned to a particular country and deducted as credit for Honor Roll standings one year after cessation of their use or assignment. Honor Roll applicants must submit their list of current prefixes (entire call required) separate from their regular WPX applications. Use regular form 1051 and indicate "Honor Roll" at the top of the form. Forms may be obtained by sending a business-size SASE or one IRC (foreign stations send extra postage or IRC if airmail desired) to the WPX Award Manager. A separate application must be made for each mode. Endorsements for the Honor Roll may be made for 10 or more prefixes. An SASE or IRC should be included. For prefixes by countries, see Chapter 17.

WPX Award of Excellence

This is the ultimate award for the prefix DXer. The requirements are 1000 prefixes mixed mode, 600 prefixes SSB, 600 prefixes CW, all six continental endorsements, and the 5 band endorsements 80-10 meters. A special 160-meter endorsement bar is also available. The WPX Plaque fee is $50, and the 160-meter bar is $4.50.

The CQ DX Awards Program

The *CQ* CW DX Award and *CQ* SSB DX Award are issued to any amateur station submitting proof of contact with 100 or more countries on CW or SSB. Applications should be submitted on the official *CQ* DX Award application (form 1067B). All QSOs must be 2 × SSB or 2 × CW. Cross-mode or one-way QSOs are not valid for the *CQ* DX Awards. QSLs must be listed in alphabetical order by prefix, and all QSOs must be dated after November 15, 1945. QSL cards must be verified by one of the the authorized checkpoints for the *CQ* DX Awards or must be included with the application. If the cards are sent directly to the *CQ* DX Awards Manager, Billy Williams, N4UF, PO Box 9673, Jacksonville, FL 32208, postage for return by First-Class mail must be included. If certified or registered mail return is desired, sufficient postage should be included. Country endorsements for 150, 200, 250, 275, 300, 310 and 320 countries will be issued.

Any altered or forged confirmations will result in permanent disqualification of the applicant. Fair play and good sportsmanship in operating are required for all amateurs working toward *CQ* DX Awards. Continued use of poor ethics will result in disqualification of the applicant. A fee of $4 for subscribers (subscribers must include a recent *CQ* mailing label with their application) or $10 for nonsubscribers, or the equivalent in IRCs, is required for each award to defray the cost of the certificate and handling. An SASE or 1 IRC is required for each endorsement.

The ARRL DXCC Countries List (see Chapter 17) constitutes the basis for the *CQ* DX Award country status. Deleted countries will not be valid for the *CQ* DX Awards. Once a country has lost its status as a current country, it will automatically be deleted from our records. All contacts must be with licensed land-based amateur stations working in authorized amateur bands. Contacts with ships and aircraft cannot be counted. Decisions of the *CQ* DX Advisory Committee on any matter pertaining to the administration of these awards shall be final.

To promote multiband usage and special operating skills,

special endorsements are available:

- A 28-MHz endorsement for 100 or more countries confirmed on the 28-MHz band.
- A 3.5/7-MHz endorsement for 100 or more countries confirmed using any combination of the 3.5- and 7-MHz bands.
- A 1.8-MHz endorsement for 50 or more countries confirmed on the 1.8-MHz band.
- A QRPp endorsement for 50 or more countries confirmed using 5-watts input or less.
- A Mobile endorsement for 50 or more countries confirmed while operating mobile. The call-area requirement is waived for this endorsement.
- An SSTV endorsement (*CQ* SSB DX Award only) for 50 or more countries confirmed using two-way slow scan TV.
- An OSCAR endorsement for 50 countries confirmed via amateur satellite.

(After the basic award is issued, only a listing of confirmed QSOs is required for these seven special endorsements. However, specific QSLs may be requested by Award Manager N4UF.)

The *CQ* DX Honor Roll will list all stations with a total of 275 countries or more. Separate Honor Rolls will be maintained for SSB and CW. To remain on the Honor Roll, a station's country total must be updated annually.

USA-CA Rules and Program

The United States of America Counties Award, also sponsored by *CQ*, is issued for confirmed two-way radio contacts with specified numbers of US counties under rules and con-

ditions below. [Note: A complete list of US counties appears in Chapter 17.]

The USA-CA is issued in seven (7) different classes, each a separate achievement as endorsed on the basic certificate by use of special seals for higher class. Also, special endorsements will be made for all one band or mode operations subject to the rules.

Class	Countries Required	States Required
USA-500	500	Any
USA-1000	1000	25
USA-1500	1500	45
USA-2000	2000	50
USA-2500	2500	50
USA-3000	3000	50

USA 3076-CA for ALL counties and
Special Honors Plaque [$40]

USA-CA is available to all licensed amateurs everywhere in the world and is issued to them as individuals for all county contacts made, regardless of calls held, operating QTHs or dates. All contacts must be confirmed by QSL, and such QSLs must be in one's possession for identification by certification officials. Any QSL card found to be altered in any way disqualifies the applicant. QSOs via repeaters, satellites, moonbounce and phone patches are not valid for USA-CA. So-called "team" contacts, wherein one person acknowledges a signal report and another returns a signal report, while both amateur call signs are logged, are not valid for USA-CA. Acceptable contact can be made with only one station at a time.

The National Zip Code & Directory of Post Offices will be the official guide in determining identity of counties of contacts as ascertained by name of nearest municipality. Publication No. 65, Stock No. 039-000-00264-7, is available at your local Post Office or from the Superintendent of Documents, US Government Printing Office, Washington, DC 20402, for $9, but will be shipped only to US or Canada.

Unless otherwise indicated on QSL cards, the QTH printed on cards will determine county identity. For mobile and portable operations, the postmark shall identify the county unless information stated on QSL cards makes other positive identity. In the case of Cities, Parks or Reservations not within counties proper, applicants may claim any one of adjoining counties for credit (once).

The USA-CA program will be administered by a *CQ* staff member acting as USA-CA Custodian, and all applications and related correspondence should be sent directly to the custodian at his or her QTH. Decisions of the Custodian in administering these rules and their interpretation, including future amendments, are final.

The scope of USA-CA makes it mandatory that special Record Books be used for application. For this purpose, *CQ* has provided a 64-page 4¼ - × 11-inch Record Book which contains application and certification forms and which provides record-log space meeting the conditions of any Class award and/or endorsement requested. A completed USA-CA Record Book constitutes medium of basic application and becomes the property of *CQ* for record purposes. On subsequent applications for either higher classes or for special endorsements, the applicant may use additional Record Books to list required data or may make up own alphabetical list conforming to requirements. Record Books are to be obtained directly from *CQ*, 76 North Broadway, Hicksville, NY 11801,

for $1.25 each. It is recommended that two be obtained, one for application use and one for personal file copy.

Make Record Book entries necessary for county identity and enter other log data necessary to satisfy any special endorsements (band-mode) requested. Have the certification form provided signed by two licensed amateurs (General class or higher) or an official of a national-level radio organization or affiliated club verifying the QSL cards for all contacts as listed have been seen. The USA-CA Custodian reserves the right to request any specific cards to satisfy any doubt whatever. In such cases, the applicant should send sufficient postage for return of cards by registered mail. Send the original completed Record Book (not a copy) and certification forms and handling fee. Fee for nonsubscribers to *CQ* is $10 US or 40 IRCs; for subscribers, the fee is $4 or 12 IRCs. (Subscribers, please include recent *CQ* mailing label.) Send applications to USA-CA Custodian, Dorothy Johnson, WB9RCY, 333 South Lincoln Ave, Mundelein, IL 60060.

For later applications for higher-class seals, send Record Book or self-prepared list per rules and $1.25 or 6 IRCs handling charge. For application for later special endorsements (band/mode) where certificates must be returned for endorsement, send certificate and $1.25 or 8 IRCs for handling charges. Note: At the time any USA-CA award certificate is being processed, there are no charges other than the basic fee, regardless of number of endorsements or seals; likewise, one may skip lower classes of USA-CA and get higher classes without losing any lower awards credits or paying any fee for them. Also note: IRCs are not accepted from US stations.

[The Mobile Emergency and County Hunters Net meets on 14,336 kHz SSB every day and on 3866 kHz evenings during the winter. The CW County Hunters Net meets on 14,066.5 MHz daily.]

GERATOL NET

The ARRL offers many endorsements to the popular Worked All States (WAS) award. One of them is the "75-Meter 2-letter Extra Class" Award, available to any Extra Class amateur who has worked all states consisting of 1 × 2, 2 × 1 or 2 × 2 call signs on the Extra Class portion of 75 meters.

During the winter of 1977, a separate award—the "Unbelievable Operating Achievement Award" (otherwise known as the GERATOL Net Award)—was established by N4BA, K5BG, W7RQ and WØGX. On March 10, 1978, an informal net went into operation to provide a meeting place and to assist others with the completion of both the ARRL WAS and the new award developed by N4BA and company.

Around this time, someone suggested that since most of those working toward the unique 2-letter WAS award were senior citizens, the name "Geritol Net" might be appropriate! That name stuck; the spelling has been changed to the acronym "GERATOL" (Greetings Extra Radio Amateurs—Tired of Operating Lately?). In many ways, this describes the essence of the GERATOL Net. More than one "GERATOLLER" will tell you that the fellowship and challenge has revitalized their amateur operations and allowed a welcome escape from the mundane "Roger 59" DX world.

The GERATOL Net is quite active during the good 75-meter propagation months, from October through April. As many as 48 states have checked into the net during the course of an evening, including Alaska and Hawaii. Using three net-control shifts, it convenes around 0100 UTC Friday

and Saturday evenings, and routinely runs until after 0700 (informal nets occur throughout the year on most any night). Meeting around 3767 kHz, the net still pursues the original purpose developed by its founders. No dues or membership fees are required. DX stations are welcome to check in and work any W or VE, but W/VEs are not allowed to call DX stations (for obvious reasons).

To qualify for the GERATOL Award, you first must qualify for the ARRL 75-Meter 2-letter WAS Award. Once you have secured the ARRL WAS Award, you are eligible to apply for your GERATOL Award. Include your ARRL WAS certificate serial number and date, your name as you want it to appear on the GERATOL certificate, and $2 to defray printing and mailing costs, and send to Jim Hardee, KT5F, 12731 South 23rd East Ave, Broken Arrow, OK 74012.

Once you have secured the basic GERATOL certificate, 12 interesting and challenging endorsements are available to add to it. The honor system is used, so no QSL cards are required. A brief description of each follows:

1) WAS with 1 × 2 call signs, K prefixes only.
2) 48 states with 1 × 2 call signs, N prefixes only.
3) 48 states with 1 × 2 call signs, W prefixes only.
4) WAS with 1 × 2 calls signs, with the suffix of each containing each operator's initials.
5) WAS with 1 × 2 call signs, with the suffixes containing the postal abbreviations of the states.
6) WAS with 1 × 2 call signs, with the suffixes containing double letters twice thorough the alphabet.
7) WAS with 2 × 1 call signs.
8) WAS with other stations that have already qualified for the ARRL 75-Meter 2-letter WAS award.
9) WAS with 2 × 1 call signs, with the suffixes containing each letter twice through the alphabet.
10) WAS with Extra Class 1 × 3 call signs.
11) WAS with 2 × 1 call signs, A prefixes.
12) WAS with 2 × 1 call signs, K prefixes.

For further information on endorsements, and on any GERATOL Net details, send your request with a business-size SASE to William J. Stopka, W9IH, 5016 N Natchez Ave, Chicago, IL 60656.

The GERATOL Net also is involved in a parallel operation, the Canadian 2/80 Award. Requirements for the award consist of working the following Canadian provinces and territories:

VO1 New Foundland	VE3 Ontario
VO2 Labrador	VE4 Manitoba
VE1 New Brunswick	VE5 Saskatchewan
VE1 Nova Scotia	VE6 Alberta
VE1 Prince Edward Island	VE7 British Columbia
VE2 Quebec	VE8 Northwest Territories

Contacts may be made anywhere on 75/80 meters on any mode. All QSL cards must show a contact date of 0000 UTC November 1, 1972 or later. Stations worked must have 2-letter call-sign suffixes. One you have all 12 QSL cards, submit them to Norm Smith, VE3EJQ, 2/80 Contest Coordinator, 265 Homewood Ave, Willowdale, ON M2R 2N4, Canada.

So take a shot of GERATOL and join the group on 75 meters to experience a new operating challenge.—*John S. Wilcox, KS4B, PO Box 475, Leonardtown, Maryland 20650*

3905 CENTURY CLUB

Shortly after getting my General ticket, I was tuning through the bands one night and stumbled upon a net. They seemed like a friendly group, so I checked in to see what it was all about. I was immediately deluged with calls from other net check-ins who wanted nothing more than a signal report and a subsequent QSL card. Strange group, I thought, but they all seemed to be having loads of fun, so I stuck with it. Net control that night was a sultry-voiced YL named "Lisa" from Arkansas. By the time the net ended, I had worked an additional 10 states toward my WAS and felt that I was more than welcome to come back again. So began my involvement with the 3905 Century Club.

The 3905 Century Club is basically a Worked All States net, but it is much, much more. It all began back in 1976, when the ARRL celebrated the US Bicentennial by establishing a special Bicentennial WAS. A group of hams decided to help one another qualify for the award by forming a WAS net on 3.905 MHz. It was so successful and so many friendships were made that the net has continued and flourished ever since.

Over the years, the net has expanded to four separate nets:

•3.903 MHz—the original net has shifted 2 kHz down, as a compromise to another group at 3.907 MHz. It meets 365 days a year at 0100Z, with a late net following at 0500Z.

•7.233 MHz—the 40-meter version, at 0100Z during Daylight Savings Time (one hour earlier during Standard Time).

•1.865 MHz—operates Friday and Saturday evenings from autumn through spring.

•7.082 MHz—the RTTY counterpart, this net meets every Sunday afternoon at 1900Z.

As was mentioned above, the 3905 Century Club is more than just a WAS net; it's a social experience. Hams who join nets seem to be a gregarious bunch. I have made many everlasting friendships with members of the net. And of vital interest, the QSL return rate by 3905 Century Club members is excellent. During 1985, for example, 92% of my contacts resulted in QSL cards.

Much of this high return rate has to do with the fact that the net operates a QSL bureau for each net. KE5XU runs the 40-meter bureau, and WA9CNV runs the other three. The turnaround time is exceptional, too. I have at times received cards within five days of the QSO.

The Club also has its own awards program. There are some 35 separate awards and endorsements, including Worked All State Capitals, the Night Owl Award and a DX Award (we do get check-ins from many parts of the world).

Every summer, the club has an "eyeball" at some attractive stateside location. I have found that an eyeball with a ham friend who I've never met before is one of the most enjoyable experiences of my life.

So, for those of you who have not had the opportunity to encounter the 3905 Century Club, join us. If you wish to find out more about the Club, simply send an SASE to the Club Information Officer, Mike Sheehan, KB1GN, 27 Border Rd, Holbrook, MA 02343, who will be happy to send you a complete information package.

Hope to see you on the net.—*John O'Keefe, N3DPF, Brackney, Pennsylvania*

QRP AMATEUR RADIO CLUB INTERNATIONAL (ARCI)

The objective of the QRP ARCI, Inc awards program is to demonstrate that power is no substitute for skill, encouraging full enjoyment of Amateur Radio while running the minimum power necessary to complete a QSO and reduce QRM on crowded amateur bands. Requirements for the following awards are set forth below:

QRP-25. Issued to any amateur for working 25 members of the QRP ARCI while those members were running a power output of 5 watts or less. Endorsements are offered for 50, 100 and every 100 thereafter. To apply, send full log info with the list of the member-number in order of the member-number.

WAC-QRP. Issued to any amateur for confirmed contacts with stations in all six continents. A power output of 5 watts or less must have been used by the applicant for all QSOs. To apply, send full log info with photocopies of QSLs or a confirmed list.

WAS-QRP. Issued to any amateur for confirmed QSOs with each of the 50 states, while running a power output of 5 watts or less. Basic award issued for 20 states, endorsement seals for 30, 40 and 50. To apply, send full log data with photocopies of QSLs or a confirmed list.

1000-Mile-Per Watt (km/W). Issued to any amateur transmitting from or receiving the transmissions of a low-power station such that the Great Circle bearing distance between the two, divided by the power output of the low-power station, equals or exceeds 1000-miles-per-watt. Additional certificates can be earned with different modes/bands. To apply, send full log info and photocopies of QSLs or certified list including RST and power used on both sides, band, mode and exact (as nearly as possible) QTH on both sides.

QRP-Net (QNI-25). Issued once to those members completing 25 check-ins into any individual QRP net. Subsequent 25 QNIs in another net will earn an endorsement seal. Net Managers send a list of those qualifying for the QNI certificate to the QRP ARCI Awards Manager at the end of each month. Apply for a certificate or endorsement after the end of the month in which you qualified. Applicants do not need to send in a copy of their log to verify their claim to 25 QNIs.

Notes:

1) A two-way QRP seal will be issued for all the above awards, except QRP-25, if log data indicates the power output on both sides of all QSOs was 5 watts or less.

2) Other endorsement seals available—one band, one mode, natural power, Novice. Please specify endorsements desired.

3) Certified list (GCR). QRP ARCI will accept as satisfactory proof of confirmed QSOs and that the QSLs are on hand as claimed by the applicant if the list is signed by (a) a radio club official, or (b) two Amateur Radio operators of General class or higher, or (c) a notary public, or (d) a CPA (Certified Public Accountant). If you must send QSLs, be sure to include postage for their return.

4) QRP ARCI member-numbers are not published. The Awards Manager will accept as satisfactory proof for any of the club awards a QSO with a club member giving his membership number and power output in the log data. If the QRP number and power level are not given, a QSL is required for confirmation.

5) Fee structure

    (a) original certificate and seal—$2 or 10 IRCs.

    (b) subsequent endorsement seals—$1 or five IRCs.

    (c) return postage for QSLs or other application material to be returned.

6) Send applications and the required fee to QRP ARCI Awards Manager, Leo Delaney, KC5EV, 2106A Courtney, Austin, TX 78745.

TEN-TEN

Amateurs operating on 10 meters are often bewildered by requests for "10-10 numbers." 10-10 numbers are assigned by the 10-10 International Net, Inc. A number is available to any amateur who works ten 10-10 members and submits the log data to the appropriate 10-10 Call Area Manager. Once the log data has been received and approved, the applicant will be issued his own 10-10 number, which can be used to exchange with others on 10 meters to obtain a "Bar" for each 100 numbers collected. 10-10 numbers are issued for life, and well over 40,000 10-10 numbers have been issued internationally.

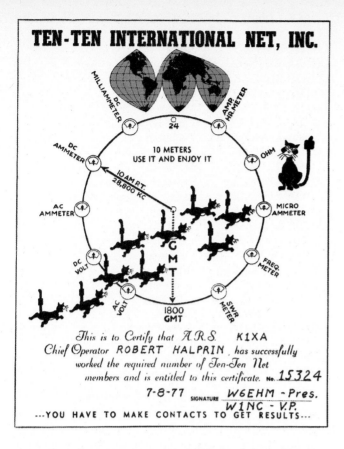

The purpose of 10-10 is to promote interest and activity on the 10-meter band. 10-10 holds a net meeting each day, except Sunday, on 28.800 MHz at 1800Z. All amateurs, 10-10 members or not, are invited to check in on the daily net. In addition to holding both CW and SSB contests twice each year, 10-10 has more than 220 Chapters which hold net meetings at least once each week. Most 10-10 Chapters issued certificates for working members of that Chapter. If chasing certificates and awards is your "thing," obtaining a 10-10 number is highly recommended. For information on how to join 10-10, including an application form and the latest list of 10-10 Chapters, meeting times and frequencies, send a business-size SASE to Chuck Imsande, W6YLJ, 18130 Bromley St, Tarzana, CA 91356.

Beside the collection of "Bars" for each 100 10-10 members worked, 10-10 offers some additional awards for contacts on 10 meters. 10-10 Worked All States (10-10 WAS) is awarded in recognition of confirmed contacts with members in all 50 states on 10 meters. QSL card confirmations must indicate the 10-10 numbers of stations worked. For information and application, send an SASE to Susan Brackeen, KA1CAD, PO Box 7081, Johnston, RI 02919. The 10-10 Countries Award is issued for confirmed contacts with 10-10 members in 25 different countries on the 10-meter band. Endorsement may also be earned for additional countries worked. For information on this award and an application form, send an SASE to Alan Sherman, K1AS, RR 4, Box 422, Danielson, CT 06239.

WORLD ITU ZONE AWARD

This award will be issued to amateurs who work the prescribed number of ITU (International Telecommunication Union) zones of the world. There are 90 ITU zones. QSL cards are suggested, but not required. If QSLs are not submitted, a certified list of contacts (verified by a local club official or two licensed amateurs) shall be submitted.

The ITU Award is issued in the following classes:

Class D—40 zones
Class C—50 zones
Class B—65 zones
Class A—75 zones
Class AA—90 zones

The official ITU zone list [see Chapter 17 for an ITU zone list] and map will be the criteria for determining zones.

Fees are $2 US postpaid for the basic award, including endorsements requested at the same time, and 50 cents US postpaid for each future endorsement.

Applications should be sent to John Lee, K6YK, 3654 Three Oaks Rd, Stockton, CA 95205.

INTERNATIONAL AWARDS

The following awards information was obtained from International Amateur Radio Union (IARU) Societies around the world, who kindly provided the rules of the Amateur Radio awards they sponsor. [Not all of the IARU Societies are, of course, fortunate enough to have the resources necessary for awards sponsorship.] In addition to the HF awards described below, many IARU Societies also sponsor localized and/or VHF-only awards, or VHF-endorsed versions of their HF awards. Further, shortwave listeners (SWLs) are eligible for many, if not most, of these awards as well. Since awards rules have been necessarily summarized for publication, further details can be secured by writing directly to the Awards Manager(s) of the particular Society or Societies concerned. When addressing correspondence to an Awards Manager, it is a good idea to enclose a self-addressed envelope and at least 2 IRCs (International Reply Coupons).

As a general principle, QSLs are not required to be included with the application. Ordinarily, a list of the appropriate QSLs with a signed certification by an official of a national-level Amateur Radio Society or two officers of a local Amateur Radio Club stating that the valid QSLs are in possession of the applicant will be honored by the IARU Society sponsoring the award. This list should show the call sign, date, time, frequency, mode and signal report for each contact. This is often referred to as a "GCR" (General Certification Rules) list. Awards Managers may reserve the right to inspect QSL cards on rare occasions, and any decisions made by the Awards Managers as to the validity of applications are final.

When applying for a particular award, type or neatly print your name, call sign and mailing address. State precisely what award and/or endorsement you are applying for, and enclose the appropriate application forms and the required fee (usually in IRCs) to defray postage and stationery costs. You should also state that you have complied with the rules of the particular award as well as the regulations established for Amateur Radio in your country.

[Note: At presstime, the selling price of IRCs is 80 cents US.]

ARGENTINA (LU)

Radio Club Argentino (RCA) awards will be issued to any recognized radio amateur holding an official license. Each request must be accompanied by a log-extract or list of the appropriate contacts for the award. The list must be accompanied by the corresponding QSL cards or by a statement from an officer of the national Amateur Radio Society stating that the necessary cards have been checked. All QSOs must be made November 20, 1945 or later, except where otherwise noted. Contacts with mobile stations will not be accepted (except for the CEMA and CEMARA awards). The fee for each certificate is 10 IRCs (or $5 US). The fee for endorsements is 4 IRCs ($2 US). Awards applications should be sent to: Radio Club Argentino, Awards Manager Sr. Ricardo Schroder, LU8AEJ, Casilla de Correo 97, (1000) Buenos Aires, Argentina.

101 (101 Paises)

Awarded to all amateurs for QSOs with 101 different DXCC countries. Endorsements for 121, 141, 161, etc. Different certificates issued for CW, Phone and Mixed modes—one band or mixed bands. Five bands—"laureated." One LU contact required.

TPA (Todos Paises America)

Awarded to amateurs for QSOs made on any band or mode with 21 American Republics and Canada (22 contacts): CE, CO, CP, CX, HC, HH, HI, HK, HP, HR, LU, OA, PY, TG, TI, VE, W, XE, YN, YS, YV, ZP.

CCC (Cinco Continentes Comunicados)

Awarded to amateurs for QSOs with America, Africa, Asia, Australia and Europe (North American stations must contact South American stations for America; South Americans must contact North Americans). All contacts must be made on two bands, once on each band. "Laureated"—making the required contacts on three or more bands. Different certificates are available for CW and phone.

CAA (Certificado Antartico Argentino)

Awarded to amateurs for working at least one LU Antarctic base (LU-Z). LU stations must work three bases. Endorsement stickers are issued for working additional bases, and separate certificates are issued for CW and phone. All bands.

CEMA (Certificado Moviles Argentinas)

Awarded to amateurs for working 25 LU mobile stations, any band or mode.

CEMARA (Certificado Moviles Armada Argentina)

The Servicio de Radioaficionados de la Armada (SARA) and the Radio Club Argentino jointly issue this award to amateurs who work 25 Maritime Mobile stations, 10 of which must be LUs (including five from the Argentine Navy). The 15 other Maritime Mobiles stations can be from any country. Starting date: January 1, 1960. All bands. Separate awards issued for CW, AM or SSB. No mixed modes.

CA (Certificado Argentino)

Awarded to non-LU amateurs who work 100 LU stations. Any mode or band.

RA (Republica Argentina)

Awarded to all amateurs working 18 LU stations to form the words Republica Argentina with the first letter of each call-sign suffix. Example: LU1RA, LU6EAM, LU2PB, LU5US, LU9BBA, LU3LA, LU2IO, LU9CAM, LU4AA, and so on. All modes/bands. Starting date: January 1, 1965.

TRA (Toda Republica Argentina)

Awarded to amateurs working all 25 Argentine districts:

LU1AA-LU9CZZ	Federal Capital
LU1DA-LU9EZZ	Buenos Aires
LU1FA-LU9FZZ	Santa Fe
LU1GA-LU9GOZ	Chaco
LU1GP-LU9GZZ	Formosa
LU1HA-LU9HZZ	Cordoba
LU1IA-LU9IZZ	Misiones
LU1JA-LU9JZZ	Entre Hios
LU1KA-LU9KZZ	Tucuman
LU1LA-LU9LZZ	Corrientes
LU1MA-LU9MZZ	Mendoza
LU1NA-LU9NZZ	Santiago del Estero
LU1OA-LU9OZZ	Salta
LU1PA-LU9PZZ	San Juan
LU1QA-LU9QZZ	San Luis
LU1RA-LU9RZZ	Catamarca
LU1SA-LU9SZZ	La Rioja
LU1TA-LU9TZZ	Jujuy
LU1UA-LU9UZZ	La Pampa
LU1VA-LU9VZZ	Rio Negro
LU1WA-LU9WZZ	Chubut
LU1XA-LU9XOZ	Santa Cruz
LU1XP-LU9XAA	Tiera del Fuego
LU1YA-LU9YZZ	Neuquen
LU1ZA-LU9ZZZ	Antarctic Bases

LU 10 DL (LU 10 doble letras)

Awarded to amateurs working 10 stations (at least one must be an LU) with double letter prefixes from 1 to 0 without repeating the pair of letters (eg, PY1AA, CP2FF, LU5BB, etc). Any band/mode. Starting date: January 1, 1965.

RCA (Radio Club Argentino)

Awarded to amateurs working stations (at least one must be an LU) to form the words Radio Club Argentino using only the first two letters after the call-sign suffix. Examples: CT1RA, OK3DI, PY2OC, W8LU, LU3BAC. Any band/mode. Starting date: January 1, 1965.

AUSTRALIA (VK)

The *Worked All VK Call Areas (WAVKCA) Award* is offered by the Wireless Institute of Australia (WIA) for working and confirming contacts with the various call areas of the Commonwealth of Australia. All VK amateurs and all amateurs who are members of their IARU-affiliated National Society are eligible for the award. Twenty-two (22) QSOs are required (77 for VK amateurs) as follows:

Territory	Call Area	*Required* *QSOs/QSLs*
Australian Antarctica Heard Island Macquarie Island	VK0	1
Australian Capital Territory	VK1	1
State of New South Wales	VK2	3
State of Victoria	VK3	3
State of Queensland Thursday Island	VK4	3
State of South Australia	VK5	3
State of Western Australia	VK6	3
State of Tasmania King Island Flinders Island	VK7	3
Northern Territory	VK8	1
[All VK9 Territories]	VK9	1

Contacts must have been made from January 1, 1946 on. Any band/mode, but crossband contacts not allowed. Maritime and aeronautical mobile contacts do not count, but land-mobile and -portable contacts are valid provided the QSL card shows the station location at time of contact. Send a certified log extract (in lieu of the actual QSLs), verified by an official of the applicant's IARU Society or two licensed amateurs, application and 5 IRCs to: The Federal Awards Manager, Wireless Institute of Australia, Mr. K.D. Hall, VK5AKH, St Georges Square Rectory, South Australia 5014, Australia.

In addition, approximately 60 awards are issued by local clubs in Australia. Contact the Federal Awards Manager for further information.

AUSTRIA (OE)

Except where otherwise noted, applications for Austrian awards should be sent to the Austrian IARU Society (OVSV) as follows: OVSV HQ, International Awards Manager, Theresiengasse 11, A-1180 Vienna, Austria.

WAOE (Worked All Austria)

North American amateurs must work eight of the nine OE call areas on any band (different requirements apply to European applicants). Awards are issued as mixed and SSB. A separate 160-meter award, WAOE-160, is also available, for working four OE call areas on 160. Send verified list of QSLs and 10 IRCs.

OE 100

This certificate is issued for working at least 100 different OE stations. Contacts must be made on any mode and band that OE stations can legally operate. All contacts must be made from the same country and must be made after April 1, 1954. Endorsement stickers are available for 200, 300 . . . different OE contacts. Send certified list and 10 IRCs (2 IRCs for each endorsement).

WLOE (Worked District Locators in Austria)

Each OE call district has its own three-digit "District Locator" (ADL). For the basic award, 30 different ADLs must be worked in at least six different OE call districts (OE1-OE9). ADLs in the block 001-099 count as a separate call district. Contacts from January 1, 1968 are valid.

The WLOE award is available for mixed HF and CW HF. Endorsement stickers are issued for each additional 10 ADLs and a special 160-meter sticker for working 10 ADLs on 160. QSLs are not required for contacts made during the WAOE 160-Meter International Contest.

To apply, send a certified list of QSLs and 10 IRCs (4 IRCs for each endorsement sticker). Note: Only OVSV members have an ADL number and only those stations count for the award.

WPXZ 15 (Worked Prefix Zone 15 Awards)

This award is offered for working stations in five classes:

Class 1—15 countries and 50 prefixes
Class 2—12 countries and 40 prefixes
Class 3— 8 countries and 30 prefixes

Contacts must be made with the following countries (by prefix): HA, HV, I, IS, IT, OH, OK, T7, TK, UA2, UR, UP, UQ, YU, ZA, 1AØ, 4U1VIC, 9H. Starting date: January 1, 1958. Send verified list of QSLs and 10 IRCs to OVSV—LV OE1, Eisvogelgasse 4, A-1060 Vienna, Austria. Endorsement stickers—2 IRCs.

WIEN Diplom (Vienna Award)

This award requires contacting Vienna districts as follows: Class 1—all 23 districts, Class 2—15 districts. Vienna stations (OE1) should be asked to clearly mark their district on their QSL card. All bands/modes. Send certified list and 10 IRCs to OVSV—LV OE1, Eisvogelgasse 4, A-1060 Vienna, Austria.

WDRA (Worked Danube River)

This award is issued for working countries adjoining the Danube River. To qualify, work 15 DL stations, seven OE stations (one of which must be an OE1), five HA stations, three YO stations, three YU stations, two OK stations, and one each from LZ and UO5. Send certified list and 10 IRCs to OVSV—LV OE1, Eisvogelgasse 4, A-1060 Vienna, Austria.

ACA (Austrian Large Cities Award)

For this award, you qualify for one point for each Austrian large city worked. The award is issued in three classes:

Class 1—100 points
Class 2—70 points
Class 3—40 points

Send certified list and 10 IRCs to OVSV—LV OE1, Eisvogelgasse 4, A-1060 Vienna, Austria.

AMRS (Austrian Military Radio Section)

This award is issued for working members of the Austrian Military Radio Section, a subdivision of OVSV, for a total of 20 points. Each AMRS club station counts as one point, AMRS club stations count two points, and stations in United Nations duty (presently OE————/YK and some 5B stations from the United Nations OE contingent in Cyprus) count three points. To apply, send certified list to Werner Hafner, OE8HFL, Hausergasse 30/1/7/16, A-9500 Villach, Austria.

A variety of localized awards are also available. Contact OVSV HQ for details.

BELGIUM (ON)

The Union Belge des Amateurs-Emetteurs (UBA) sponsors the *Worked All Belgian Provinces (WABP)* award for amateurs who have worked and confirmed contacts with each of the nine ON provinces on two bands. To apply, send a certified list of QSLs and 10 IRCs (or $3 US) to: ON5KL, UBA-HF Awards Manager, Van Campenhout Mat, Hospicestraat 175, B-9080 Moerbeke-Waas, Belgium.

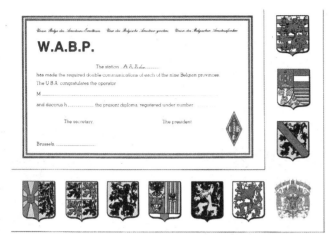

ON Provinces:

AN—Antwerp	LG—Liege	OV—East Flanders
BT—Brabant	LM—Linbourg	NR—Nanur
HT—Hainaut	LU—Luxembourg	WV—West Flanders

BERMUDA (VP9)

Applications, lists of contacts, QSLs and other required materials for Bermuda awards should be sent to: The Awards Manager, Radio Society of Bermuda, Box 275, Hamilton 5, Bermuda.

Bermuda 100 Club

This award is a mahogany plaque with the Bermuda Coat of Arms embossed thereon in full color. An engraved plate for affixing to the plaque is provided for each additional 25 stations worked. It is suitably imprinted with the name and call sign of the recipient. The rules are as follows: 100 different VP9 call signs must be worked on any band or mode from the same QTH (note: A reissued VP9 call sign held by two different operators counts as two call signs if both operators have been worked. However, only one contact credit is allowed for working the same station mobile, portable, etc); there is no time limit, and the award is free. A letter of application, together with a list of contacts and the relevant QSL cards, is required.

Worked All Bermuda Award

This award is a print of an antique map of Bermuda, signed by the Governor. The award is issued to licensed amateurs who submit proof (valid QSL cards) of contacts with all nine Bermuda parishes: Hamilton, Devonshire, Paget, Pembroke, Sandys, Smith's, Southampton, St George's and Warwick. The following rules apply: Applications must be accompanied by valid QSL cards from each parish, only one of which can be from a mobile or other non-fixed station; there is no time limit for the award and it is free; VP9 stations are not eligible; all contacts must be made from the same QTH or within a 25-mile radius thereof. A letter of application listing stations contacted and all nine QSL cards required.

BRAZIL (PY)

To apply for these awards sponsored by the Liga de Amadores Brasileiros de Radio Emissao (LABRE), the relevant log information must be checked by the Awards Manager of the applicant's National Society or by two licensed amateurs. All contacts must be with land stations. Send the application and 10 IRCs for each award to: Awards Manager, LABRE HQ, PO Box 07-0004, 7000-Brasilia, DF, Brazil.

WAB (Worked All Brazil)

This award is available for working and confirming contacts with Brazilian stations in all 23 Brazilian states and the capital city, Brasilia. A special ribbon (to attach to the certificate) is available for working and confirming the two Brazilian Federal Territories, Amapa and Roraima.

WAO (Worked All Oceans)

This award is available for working and confirming contacts with nine Brazilian Geographic Regions, as follows:

1st Region—PY1-PP1
2nd Region—PY2-PP2-PT2
3rd Region—PY3
4th Region—PY4
5th Region—PY5-PP5
6th Region—PY6-PP6
7th Region—PY7,PP7,PR7,PS7,PT7
8th Region—PY8,PP8,PR8,PS8,PT8,PU8,PV8,PW8
9th Region—PY9,PT9,PY0 and 21 countries of the Atlantic Ocean

WAA (Worked All America)

This award is available for working and confirming contacts with 45 countries in the American Geographic Area; one contact must be with Brazil.

DBDX (Brazilian DX Award)

This award is available for working and confirming contacts with a minimum of 20 different DXCC countries. One contact must be Brazil. Endorsement stickers are available for additional countries in groups of 10. All contacts must be made on 160, 80 or 40 meters. Four kinds of certificates are issued—phone/CW, phone only, CW only, and RTTY.

BULGARIA (LZ)

The Bulgarian Federation of Radio Amateurs (BFRA) offers six certificates for two-way contacts on CW, SSB/AM or mixed modes. Applications should include a list of claimed QSOs verified by two licensed radio amateurs or local Amateur Radio Club authorities indicating stations worked, date, time, band and mode, together with 10 IRCs. Send all applications to: Bulgarian Federation of Radio Amateurs, PO Box 830, Sofia 1000, Bulgaria.

People's Republic of Bulgaria Award

To qualify, make 20 QSOs with different LZ stations, 10 with LZ1 and 10 with LZ2 irrespective of band (Europeans must work 5 QSOs each with LZ1 and LZ2 on 80 and 40 meters or 20 QSOs with different LZ stations). Contacts must be made after January 1, 1965.

5-Band LZ Award

To qualify, work 10 LZ1 stations and one LZ2 station on all bands, 80-10 meters. Contacts must be made after January 1, 1979.

W 100 LZ Award

To qualify, work 100 QSOs with different LZ stations during one calendar year. Contacts must be made after January 1, 1979.

W 28 Z ITU Award

To qualify, work these countries in ITU Zone 28: DL, DL7/West Berlin, HA/HG, HB9, HBØ, HV, I, IS, LZ, OE, OK, SP, SV, SV5, SV9, SV/A, T7, TK, Y2, YO, YU, ZA, 1AØ, 9H, 4U1ITU.

This award is issued in three classes:
Class 1—28 QSOs with different stations in 20 countries.
Class 2—28 QSOs with different stations in 16 countries.
Class 3—28 QSOs with different stations in 10 countries.

All classes require mandatory QSOs with five LZ stations. All contacts must be made after January 1, 1979.

Black Sea Award

To qualify, work 60 different amateur stations in countries bordering the Black Sea. QSOs with LZ, TA, YO, UA6 and UB are mandatory. All contacts must be made after January 1, 1979.

SOFIA Award

To qualify, 100 total points must be earned by working

amateur stations in Sofia, the capital of Bulgaria. Points are allocated as follows: 80 meters—15 points, 40 meters—five points, 20 meters—one point, 15 meters—two points, 10 meters—three points (for Europeans, two points are earned for working Sofia stations on 80, 40, 15 and 10 meters, and one point for 20-meter contacts). All contacts must be made after January 1, 1979. Stations may be worked only once per band, regardless of mode. (The more active stations in Sofia are LZ1KAA, KAB, KDP, KPG, KSA, KSF, KVV, KWF, AB, AD, AM, AP, AQ, AU, BC, FF, FN, IA, JW, KX, LB, MS, NP, QG, QI, QP, SS, UA, UO, WD, WJ, WV, XL, XX, ZQ.)

CHILE (CE)

WACE (Worked All Chile)

This award is issued by the Radio Club de Chile to all amateurs who can submit satisfactory evidence of having established two-way contact with CE stations in each of the 10 zones in which the Chilean territory is divided. The application must contain a list of the stations worked, including date, signal report and mode. QSL cards are not necessary if the list is verified and signed by a recognized radio club or national Amateur Radio organization. All contacts must be made from the same "country." Eight IRCs required. Send the application to: Awards Manager, Radio Club de Chile, Casilla 13630, Santiago, Chile.

Diploma Republica de Chile (Republic of Chile Award)

To qualify, applicants must make contact with 16 different CE stations from any Chilean zone so as to form the phrase "Republica de Chile" with the last letter of the suffix of their call signs. All contacts are valid starting January 1, 1986 on any band and mode. Applicants may send QSL cards or preferably a list of QSOs duly certified by an IARU Member-Society. QSLs or QSO list must be arranged in order to spell Republica de Chile. Fee: 8 IRCs. Send applications to the Radio Club de Chile Awards Manager at the address above.

COLOMBIA (HK)

Unless otherwise noted, please send all applications to: Liga

Colombiana de Radioaficionados (LCRA), PO Box 584, Bogota, Colombia.

CHK Award

Contact and confirm different HK stations as follows: North and South American stations must contact 50 HK stations, elsewhere 25 different HK stations. Send log data and 8 IRCs.

ZHK Award

Contact and confirm the 10 Colombian call areas as follows: North and South American stations must work 9 call areas, elsewhere 8 call areas. Log data required.

HK5 Diploma

American stations must contact 12 HK5 stations, others must contact 8 HK5s. Contacts are valid after January 1, 1957.

City of Cali Foundation Diploma

This diploma is awarded for contacts made on July 25 (Sectional Five Anniversary) with HK5VD, 5J5VD, 5K5VD and five other HK5 stations. Applications to LCRA, Sectional Cali, Apartado 6149, Cali, Colombia.

100 HK3 Diploma

One hundred contacts with HK3 stations are required.

COSTA RICA (TI)

All applications should be sent to: Awards Manager, Radio Club de Costa Rica (RCCR), PO Box 999, Heredia, Costa Rica.

Diploma FRACAP

The principal reasons for the creation of this diploma are to increase the international awareness of FRACAP (the Federation of Radio Amateurs of Central America and Panama) and to raise the interest and knowledge of the geographical location of its member-countries—Costa Rica, El Salvador, Guatemala, Honduras, Nicaragua and Panama. The requirements are two-way confirmed contacts (any band/mode) with each of the six FRACAP countries. Each contact must be made with an amateur who is a member of a FRACAP-affiliated radio club. Contacts are valid after August 16, 1985. The application must be accompanied by either the QSL cards or a list verified by the radio club of which the applicant is a member. Ten IRCs (or $5 US) required.

TTI Diploma

To qualify, two-way QSOs (any mode) must be made with seven of the eight Costa Rican call areas: TI2—San Jose, TI3—Cartago, TI4—Heredia, TI5—Alajuela, TI6—Limon, TI7—Guanacaste, TI8—Puntarenas, TI9—Isla del Coco. A QSO with the official RCCR club station, TI0RC, can be used to replace one call area. Verified list of QSLs and 10 IRCs (or $4 US) required.

CUBA (CO)

To apply for awards sponsored by the Federacion de Radioaficionados de Cuba (FRC), send a certified list of QSLs and 10 IRCs (or $2 US) for each award. All bands/modes are acceptable. Address applications to: FRC, Awards Department, PO Box 1, Habana 1, Cuba.

America Award

To qualify, the applicant must work 45 countries and islands in America for Class III; 50 countries for Class II; and 51+ for Class I. A contact with Cuba is mandatory; KG4 contacts are not acceptable.

Cuba Award

To qualify, the appplicant must work all eight Cuban call

sign districts, CM1/CO1-CM8/CO8. Cuban Amateur Radio Club stations (three-letter suffixes) can be substituted for up to three missing districts. Contacts are valid after January 1, 1959.

Caribbean Award

To qualify, the applicant must work 20 or more of the 32 countries in the Caribbean, including CO, HK, HP, HR, TI, TG, V3, XE, YN and YV. CO contacts are mandatory; KG4 contacts are not acceptable. Contacts are valid after January 1, 1959.

Cuba DX Group

To qualify, the applicant must work four official members of the Cuba DX Group for a total of four points (one point per member). Aspiring members of the DX Group count a ½ point. Contacts are valid after September 1, 1980.

CYPRUS (ZC4)

The *Cyprus Award* Certificate, sponsored by the Cyprus Amateur Radio Society, is awarded to any licensed amateur outside Cyprus who makes a specified number of two-way

contacts with licensed amateurs on the island of Cyprus. To reduce as far as possible any advantage accruing to stations by reason of their geographical location and to encourage activity on the less-frequently used bands, the certificate will be awarded on a points basis determined by *CQ* Zone location and bands used as shown below:

Zone	1.8	3.5	7	14	21	28
	Points Scored Per Contact					
20	8	2	1	1	2	4
1, 2, 3, 6, 7, 10, 12, 19, 24, 25, 26, 27, 28, 29, 30, 31 and 32	16	8	4	2	4	8
All Other Zones	8	4	2	1	2	4

The total number of points required to win the Award depends on the bands used. If all contacts are made on only one band, 32 points are required. If the contacts are made on any two bands, 24 points are required. If the contacts are made on any three bands, 16 points are required, and any four bands, 12 points are required. Any mode may be used. Contacts must be made after April 1, 1973. Contacts with any one Cyprus station can only count once per band.

To claim the award, copies of log entries should be submitted under the following heading: DATE/TIME UTC, STATION WORKED, FREQUENCY BAND, SIGNAL REPORTS IN AND OUT. Each log sheet should be headed with the applicant's call sign, zone number and complete mailing address, preferably typed or printed in block capitals. Either the appropriate QSLs or a statement from the applicant's National Society certifying that the QSL cards have been produced to them is required. In countries without a National Society, a similar statement signed by two other amateurs will suffice. Log sheets and 10 IRCs should be sent to: Awards Manager, Cyprus Amateur Radio Society, PO Box 1267, Limassol, Cyprus.

CZECHOSLOVAKIA (OK)

The following awards are issued by the Central Radio Club of Czechoslovakia (CRCC). All certificates are issued free of charge for members of clubs and associations accepting this rule reciprocally. The fee for all others is 5 IRCs (10 IRCs for the P75P). For endorsement stickers, send a confirmed list and 1 IRC, and indicate the number and issue date of the

basic award. QSL cards may be sent with the application; cards need not be sent when the national-level Amateur Radio Society or club has confirmed possession of the listed QSLs by the applicant. [Cards for the P75P must contain locations (QTH) of listed stations.] Applications should be sent to: Central Radio Club, Awards Manager, PO Box 69, 113 27 Praha 1, Czechoslovakia.

S6S (Six Continents)

To qualify, work and confirm contacts with at least one amateur station located in each of the six continents (as defined by IARU) since January 1, 1950; all CW, all phone or all RTTY. Endorsement stickers for the basic certificate are available for single bands (80, 40, 20, 15, 10).

P75P (Worked 75 Zones)

To qualify, work and confirm contacts with at least one amateur fixed station located in each of the ITU zones since January 1, 1960, in three classes: 1st Class—70 zones, 2nd Class—60 zones, 3rd class—50 zones.

ZMT

To qualify, work and confirm contacts with at least one amateur station since April 26, 1949, in each of the following

39 areas: OK1, OK2, OK3, HA, LZ, UA1, UA2, UA3, UA4, UA6, UA9, UA0, UB, UC, UD, UF, UG, UH, UI, UJ, UL, UM, UN, UO, UP, UQ, UR, SP (three different districts), Y2 (three different regions, as determined by the last letter of the call sign), YO (three different districts), YU (three different districts). A special certificate will be issued for ZMT contacts made during a 24-hour period.

100 OK

To qualify, work and confirm contacts with at least 100 different OK stations since January 1, 1954. Endorsement stickers are available for every additional 100 stations confirmed up to 500.

OK SSB Award

To qualify, work and confirm SSB QSOs with different OK stations for a total of 25 points. Each QSO on 10, 15 and 20 meters counts one point; two points for each QSO made on 80 or 40 meters. No date limitation.

[Note: An opportunity for qualifying for the S6S, ZMT and 100 OK certificates is the annual OK DX Contest. When all the necessary contacts were made during the contest, the contest log can replace QSL cards. The OK DX Contest log must be included with the application.]

DENMARK (OZ)

The following awards are sponsored by Eksperimenterende Danske Radioamatorer (EDR):

Cross Country Award

This award is issued for all CW or all phone. European amateurs must make 50 points, all others 40 points. The call sign is used as the basis for the award. Each call prefix, OZ1 to OZ9, and OX3, must be contacted. Three contacts with each call prefix on permitted on each band (two contacts permitted for European stations), with the exception of OX3 (nine contacts permitted). Each contact counts one point, with the exception of 70 cm, where each contact counts two points. Starting date: April 1, 1970. Send a list certified by the Awards Manager of the National Society and 5 IRCs (or $2 US) to: EDR Awards Manager, Tage Eilman, OZ1WL, PO Box 213, 5100 Odense C, Denmark.

Greenland Award

This award is issued in three categories—CW, phone or mixed, in three classes:

Class 1—five different locations and 15 different stations.
Class 2—four different locations and 10 different stations.
Class 3—three different locations and five different stations.

Only two-way CW or phone QSOs with OX3 stations made after January 1, 1978 count. One contact per band with the same station permitted. All amateur bands (3.5 to 1296 MHz) are valid. Crossband QSOs do not count. Minimum report 33 (phone), 338 (CW). Contacts with portable or mobile stations are not valid. A detailed list of contacts (certified by the Awards Manager of a National Society) and 5 IRCs (or $2 US) should be sent to OZ1WL (see address above).

Fairytale Award

Radio amateurs in Odense, the native town of Hans Christian Andersen, the fairytale writer, issue the Fairytale Award. Amateurs outside of Scandinavia are eligible for the award after making nine two-way CW QSOs with nine different OZ stations; each OZ call area (1-9) must be represented, a minimum of three must be from Odense (separate rules apply to those in LA/OZ/SM). A contact with OZ3FYN, the EDR Odense Division club station, may be used to replace one call area. Only two-way CW QSOs since December 6, 1967, the centenary of Hans Christian Andersen's being appointed an honorary citizen of Odense, count for the award. All amateur bands (3.5 to 1296 MHz) are valid; cross-band QSOs do not count, nor do mobile/portable contacts. Minimum signal report is 338. To apply, send a certified list of QSOs and 6 IRCs to: E. Hansen, OZ7XG, 14 Sophus Bauditz Vej, DK 5210, Odense, Denmark.

OZ Prefix Award

The Copenhagen Division of EDR, in commemoration of the 50th anniversary of its foundation, issues the OZ Prefix Award, available to licensed amateurs and SWLs under the following rules. Amateurs outside of Denmark and Europe must work one station in each OZ call area (1-9) for a total of nine QSOs. [Europeans must work two stations in each, OZ stations three stations in each.] A contact with club station OZ5EDR can be used as a "joker" (i.e. "wild card") to replace another contact. All bands/modes; special endorsements for CW, SSB, RTTY, single band, and so on. Send a certified list of QSOs and 10 IRCs to: Allis Andersen, OZ1ACB, Kagsaavej 34, DK-2730 Herlev, Denmark.

DJIBOUTI (J2)

The Association des Radio Amateurs Djibouti (ARAD) issues the *Award J2 Djibouti* in two classes:
First Class—Eight QSOs with J2 stations on two or more bands, any mode.
Second Class—15 QSOs, with five on CW.
The same station can be worked on more than one band.
Send a list of certified QSOs, a copy of the log or copies of the QSLs, and 12 IRCs ($5 US) to: ARAD, BP 1076, Djibouti.

DOMINICAN REPUBLIC (HI)

The Radio Club Dominicano (RCD) offers two awards. The *HI Award* is granted to those amateurs who confirm 25, 50 or 100 HI QSOs; for CW operators, 10 QSOs. Asian amateurs get a five-QSO bonus towards the HI Award. The *9-Circuits Award* is granted to those amateurs who confirm all nine HI circuits (call areas HI1-HI9). Applicants must submit QSL cards together with a list of QSOs and 12 IRCs (or $5 US) to help cover costs. The application should be sent to: Radio Club Dominicano, PO Box 1157, Santo Domingo, Dominican Republic.

ECUADOR (HC)

The Guyaquil Radio Club (GRC) sponsors the *WHC Certificate*, a diploma issued to any amateur who works at least five of the eight HC zones and provides proof through photocopies of the appropriate QSL cards. Applications can be sent to: Guyaquil Radio Club, PO Box 5757, Guyaquil, Ecuador.

FEDERAL REPUBLIC OF GERMANY (WEST GERMANY—DL)

The Deutscher Amateur Radio Club (DARC) sponsors an extensive array of awards, as detailed below. In addition to the individual Awards Managers, general correspondence about the DARC and DIG awards programs can be directed to: Eberhard Warnecke, DJ8OT, PO Box 101244, D-5620 Velbert 1, Fed Rep of Germany.

DLD Award

The DLD awards program provides recognition for working West German DOKs (Districts/Ortsverbands-Kenner) [Districts Area-Indicator]. The DOK number must be shown in print on the QSL (DARC stamps are permissible). The applicant must submit QSLs for each DLD Award, and at least 100 different DOKs must be worked on one band. A special DLD application is available on request.

Send completed application, QSL cards and appropriate IRCs to: DLD Awards Manager, Hans Peter Guenther, DL9XW, Am Strampel 22, D-4460 Nordhorn, Fed Rep of Germany.

1) The basic DLD is issued in two classes, DLD 100 and DLD 200. Radio amateurs in non-European countries get the DLD Award for QSOs on 80, 20, 15 and 10 meters, and

80-meter QSOs count for two DOKs. German and European amateurs get the DLD 100 and DLD 200 for 80-meter QSOs only. All QSOs must be made since January 1, 1956. Fifteen IRCs required.

2) The DLD 40-Meter Award is awarded in two classes, DLD 100/40 Meter and DLD 200/40 Meter. QSOs must be made on 40 meters only since May 7, 1959. Fifteen IRCs required.

3) DLD Proficiency Badges are awarded in three classes:
• bronze—DLD 300 (300 confirmed DOKs) [15 IRCs]
• silver—DLD 400 (400 confirmed DOKs) [20 IRCs]
• gold—DLD 500 (500 confirmed DOKs) [25 IRCs]

4) Holders of DLD 500 may apply for stickers which are issued in four classes:
• green—600 confirmed DOKs
• red—700 confirmed DOKs
• silver—800 confirmed DOKs
• gold—900 confirmed DOKs
[For each sticker, 10 IRCs are required.]

5) Holders of DLD 200, DLD 200/40 Meters, DLD 500 and the 600-900 stickers may apply for the highest level, DLD 1000, if they work and confirm the remaining DOKs on 80/20/15/10 meters (non-Europeans) or 80 and 40 meters (Europeans). The DLD 1000 requires 15 IRCs.

6) DLD 10 Meter is awarded for confirming two-way QSOs with 50 different DOKs on 10-meters since January 1, 1976. Stickers are awarded for an additional 25 DOKs. For the basic DLD 10 Award, 15 IRCs are required; for the endorsement sticker, 10 IRCs are required.

For further details and an official application, send a large SASE and 1 IRC to: DARC DLD-H Awards Manager, Herman Gerls, DL6ME, Reinhaeuser Landstr 103, D-3400 Goettingen, Fed Rep of Germany.

EU-FAX-D (Europa Faksimile Diplom)

This EU-FAX-D, to promote FAX communication, is issued in three classes:

a) EU-FAX-D 3—QSLs from at least five different countries (regardless of band used) and a minimum of 10 prefix-points are required.

b) EU-FAX-D 2—QSLs from 10 countries and 20 prefix points.

c) EU-FAX-D 1—QSLs from 20 countries and 40 prefix points.

European countries are determined by the Worked-All-Europe list [see below] and each official European prefix counts as one prefix-point per band. All amateur bands where FAX is permitted may be used. All QSLs must confirm two-way FAX contacts and must be dated January 1, 1980 or later. For each certificate, 15 IRCs are required.

Send a QSL list confirmed by your official radio club and the necessary IRCs to: DARC FAX Manager, Hans-Juergen Schalk, DJ8BT, Hammarskjoeldring 174, D-6000 Frankfurt 50, Fed Rep of Germany.

EURD (Europaeisches-RTTY-Diplom)

The EURD, to promote RTTY activity, is issued in four classes:

a) EURD III—QSLs from at least 20 different countries (regardless of band used) and a minimum of 100 prefix-points are required.

b) EURD II—QSLs from at least 30 countries and 150 prefix-points.

c) EURD I—QSLs from at least 40 countries and 200 prefix-points.

d) EURD Trophy—QSLs from at least 50 countries and 250 prefix-points.

European countries are determined by the Worked-All-Europe list [see below], and each official European country counts for one prefix-point per band. All amateur bands may be used, and all QSLs must confirm RTTY (which comprises all digital modes, such as ASCII and AMTOR). QSLs must be dated January 1, 1965, or later. For each certificate, 15 IRCs are required.

Send a QSL list confirmed by the local radio club and the necessary IRCs to: DARC-RTTY Manager, Heinz Moestl, DE8BUS, PO Box 1123, D-6473 Gedern 1, Fed Rep of Germany.

[Worked-All-Europe Countries List: CT1, CT2, C3, DL, EA, EA6, EI, F, G, GD, GI, GJ, GM, GU, GW, HA, HB, HBØ, HV, I, IS, IT, JW, JX, LA, LX, LZ, OE, OH, OHØ, OJØ, OK, ON, OY, OZ, PA, SM, SP, SV, SV/A, SV5, SV9, TA1, TF, TK, T7, UA1/3/4/6, UA1/FJL, UN/UK1N, UA2, UB, UC, UO, UP, UQ, UR, YO, YU, Y2, ZA, ZB2, 1AØ, 3A, 4U1ITU, 4U1VIC, 9H]

SSTV-AD (Slow Scan Television Activity Diploma)

The SSTV-AD is issued to enhance the development of SSTV on the amateur bands. It is issued at four levels:

Basic Award—25 points   [20 IRCs required]
Sticker 50—50 points   [10 IRCs required]
Sticker 75—75 points   [10 IRCs required]
SSTV Trophy—100 points   [free]

Each QSO counts one point; contacts with club stations DFØBUS or DLØATV count 10 points. Stations can be worked only once. QSLs must show two-way SSTV and be dated January 1, 1984 or later. To qualify for the SSTV Trophy, the Basic Award and the 50 and 75 stickers must be earned first.

Send a confirmed list of QSLs and the necessary IRCs to DE8BUS (see address above).

DLYL

Stations outside of Europe work 10 West German licensed YLs (Europeans must work 25). (A QSL from a YL operating from a club station counts extra if this QSL and the QSL from the YL's personal station show different dates.) All bands and modes are allowed, no date limit. Send QSLs or photocopies of QSLs showing all necessary dates, along with a list of calls, first names and date, and $3 US (or the equivalent) to: DLYL Custodian, Ursula Buerger, DL3LS, 12 Fuerbergerstrasse, D-5630 Remscheid, Fed Rep of Germany. US applicants may apply to the DLYL "subcustodian" Karla A. Holmes, WA1UVJ, 2 Belfast St., Nashua, NH 03063.

AFZ Diplom

To qualify for this diploma, the applicant must submit a log confirmed by two licensed amateurs that documents that the applicant has worked one of the DARC special stations (DFØAFZ or DBØAFZ) and 20 different West German prefixes. [Note: The special station must be worked first.] The DOK numbers of each station must be logged, and contacts must be made on two HF bands. Send the QSO list, the QSLs and an application, along with 24 IRCs ($4 US) to DL9XW (see address above).

DARC DX AWARDS

All contacts for DARC DX Awards must be made from the same country. DARC DX Awards are based on the Worked-All-Europe Countries List (see above) and the ARRL DXCC List (see Chapter 17). All amateur bands for which the applicant holds a valid license may be used. Application forms (mandatory) are available from the address below for 3 IRCs. QSL cards must also be submitted with the application. Applications should be sent to: DARC DX Awards, Walter Geyrhalter, DL3RK, Box 1328, D-8950 Kaufbeuren, Fed Rep of Germany.

Special Note: US and Canadian amateurs should submit applications for DARC DX Awards to Ralph Hirsch, K1RH, 172 Newton Rd, Woodbridge CT 06525. Official application forms (which are mandatory) are available from K1RH; please include $1 US (or 3 IRCs) and a business-size SASE with each request. QSL cards must be submitted with the application. The service charge is $5 US (or 15 IRCs) per award. Endorsements are $3 US (or 10 IRCs).

WAE Award (Worked All Europe)

This certificate is awarded for contacts with European countries on different bands. The WAE is issued in two divisions: two-way telegraphy (CW) and two-way telephony (SSB/AM/FM).

Each European country counts one point on each band.

For stations outside Europe, contacts on 80 meters and 160 meters count two points. A country can not be worked on more than five bands.

Classes: WAE III—at least 40 countries and 100 points
           WAE II—at least 50 countries and 150 points
           WAE I—at least 55 countries and 175 points

Holders of WAE I get a special WAE badge.

EU-DX-D (Europa DX Diplom)

The EU-DX-D is an award that may be claimed annually and is issued in the following classes: Telegraphy, 2 × SSB and mixed modes. For the mixed award, at least 30% of the contacts must be made in a second mode. A minimum of 50 points is required for the EU-DX-D per year; 20 points must be obtained by contacts with European countries and 30 points by contacts with countries outside Europe. All bands can be used. Each country counts one point (two points on 160 and 80 meters). Stickers are available for each additional block of four European and six non-European points within the same calendar year. Each year's score may be added to obtain the EU-DX-D 500 badge and the EU-DX-D 1000 trophy. There is no limit to the number of years.

Europa Diplom

This certificate is awarded for working amateurs in European countries for a total score of at least 100 points. Annual score: Each confirmed European country counts one point per year on each amateur band. Total score: sum of the annual score for the year of application and the five preceding years. Each certificate holder with an actual score of at least 300 points becomes eligible for the Europa Diplom Honor Roll. For US/Canadian amateurs, QSL cards can be submitted to K1RH twice a year; K1RH must receive the cards before June 15 or December 15 of each year to be considered for listing in the Europa Diplom Honor Roll published twice yearly in *CQ-DL*, the official journal of DARC.

Europa 300 Trophy

Holders of the Europa Diplom may obtain the Europa 300 Trophy. Applicants must achieve 300 country points counting each country on each band only once in all the years. There is a charge of $10 US (or 30 IRCs) for the trophy when requested simultaneously with the Europa Diplom.

DIG AWARDS PROGRAM

Each of the DIG (Diploma Interests Group German) awards has a different Awards Manager as noted below.

WGLC (Worked German Large Cities)

The WGLC recognizes contacts with amateurs in German "large cities" (population 100,000 or larger). It is available in three classes:
Class 3—DX stations must work 10 cities
    (EU stations 20 cities)
Class 2—DX stations must work 20 cities
    (EU stations 40 cities)
Class 1—DX stations must work 30 cities
    (EU stations 60 cities)
In the application, each city can be listed only once. The award is available for contacts on all HF bands and all modes, or for CW only. Starting date: January 1, 1962. Send a certified list of QSLs and 10 IRCs to H. W. Schutte, DB3OR, PO Box 810660, D-3000 Hannover 81, Fed Rep of Germany.

The following cities count for WGLC: Aachen, Augsburg, Bergisch-Gladbach, Berlin, Bielefeld, Bochum, Bonn, Bottrop, Braunschweig, Bremen, Bremerhaven, Darmstadt, Dortmund, Dusseldorf, Duisburg, Erlangen, Essen, Frankfurt/Main, Freiburg, Furth, Geisenkirchen, Gottingen, Hagen, Hamburg, Hamm, Hannover, Heidelberg, Heilbronn, Herne, Hildesheim, Kaiserlautern, Karlsruhe, Kassel, Kiel, Koblenz, Koln, Krefeld, Leverkusen, Ludwigshafen, Lubeck, Mainz, Mannheim, Monchen-Gladbach, Mulheim/Ruhr, Munchen, Munster, Neuss, Nurnberg, Oberhausen, Offenbach, Oldenburg, Osnabruck, Paderborn, Pforzheim, Recklinghausen, Regensburg, Remscheid, Rheydt, Saarbrucken, Salzgitter, Siegen, Solingen, Stuttgart, Trier, Ulm, Wanne-Eickel, Wiesbaden, Wilhelmshaven, Witten, Wolfsburg, Wurzburg, Wuppertal.

DIG Diplom 77

To qualify, it is necessary to work 77 DIG members in at least seven different countries, but only 49 DIG members in the applicant's own country. The diploma is available for CW only and mixed modes/bands. Send certified list of QSLs and 10 IRCs to Heinz Louis, DK4KW, Oberforstbacher 419, D-5100 Aachen, Fed Rep of Germany.

WDIGM (Worked DIG Members)

The WDIGM is issued in three classes as follows:
Class 3—DX stations work 15 DIG members
    EU stations work 50 DIG members
Class 2—DX stations work 30 DIG members
    EU stations work 75 DIG members
Class 1—DX stations work 50 DIG members
    EU stations work 100 DIG members
The diploma is available for contacts on all bands and modes and CW only. Endorsement stickers are available for each 100 additional DIG members up to 2000. Send a certified list of QSLs and 10 IRCs (one IRCs for endorsements) to Werner Theis, DH1PAL, Tilsiter 16, D-5350 Euskirchen, Fed Rep of Germany.

IAPA (International Airport Award)

This award is in recognition of working amateurs in 50 different cities with an international airport. Only one contact from the applicant's home country counts. All six contacts have to be worked, and only contacts with fixed stations on or after January 1, 1973, are valid. All bands/modes or CW only. Send a certified list of QSLs and 10 IRCs to Walter Hymmen, DL8JS, PO Box 1925, D-4980 Bunde, Fed Rep of Germany.

TMA (Two-Mode Award)

Work 50 different DXCC countries, including Germany, and all six continents on CW, and work the same 50 countries on phone. Send certified list of QSLs and 10 IRCs to DK4KW (see address above).

EU-PX-A (European Prefixes Award)

Work 100 European stations with different prefixes (eg, DL1, DL2, etc). The diploma is available for mixed bands/modes or CW only. Endorsement stickers are available for additional 50 prefixes. Send certified list of QSLs and 10 IRCs (SASE for stickers) to Alfons Niehoff, DJ8VC, Ernst-Hase-Weg 6, D-4407 Emsdetten, Fed Rep of Germany.

One Million Award

The 1,000,000 Award can be earned for working amateurs

in towns with different postal code numbers (ZIP codes) in West Germany. The total addition of the different postal code numbers must be at least 1,000,000. Each postal code number may be used only once. Postal code numbers with less than four digits are to be completed with zeros (eg, 41 becomes 4100). List postal codes in numerical order in the application. The diploma is available for mixed bands/modes or CW only. Send certified list of QSLs and 10 IRCs to Dieter Petring, DL1YCA, Bruderstabe 52, D-4972 Lohne 2, Fed Rep of Germany.

Familia Award

Contacts with at least two licensed amateurs in the same family count one point for each contact. The applicant must make 100 points. Starting date: January 1, 1980. The diploma is available for mixed bands/modes or CW only. Send a certified list of QSLs and 10 IRCs (or $3 US) to DK4KW (see address above).

WDXS (Worked DX Stations)

The WDXS diploma is available in four classes:
Class 4—DX stations work 200 EU stations
      (at least 10 contacts on 80/40 meters)
    EU stations work 200 DX stations
      (at least 20 contacts on 80/40 meters)
Class 3—DX stations work 500 EU stations
      (at least 25 contacts on 80/40 meters)
    EU stations work 500 DX stations
      (at least 50 contacts on 80/40 meters)
Class 2—DX stations work 1000 EU stations
      (at least 50 contacts on 80/40 meters)
    EU stations work 1000 DX stations
      (at least 50 contacts on 80/40 meters)
Class 1—DX stations work 2000 EU stations
      (at least 100 contacts on 40 meters and 20 contacts on 80 meters)
    EU stations work 2000 DX stations
      (at least 100 contacts on 40 meters and 20 contacts on 80 meters)

All contacts must be made January 1, 1964 or after. The diploma is available for mixed bands/modes or CW only. Send a certified list of QSLs and 10 IRCs to DL8JS (see address above).

DIG Trophy

This one-time award can be earned by those amateurs who have previously qualified for at least four different DIG diplomas and 500 points from QSOs/QSLs with DIG members. Applications are available from DL9XW (see address above).

DIG CW Plaque

This one-time award can be earned by those amateurs who have previously qualified for at least three different DIG-CW diplomas and 25 points from two-way CW QSOs/QSLs with DIG members. Applications are available from DL9XW (see address above).

DIG Trophy 1000

This trophy is the premier award issued by DIG for those amateurs who have qualified for all the DIG diplomas at the highest class and the DIG trophy and both DIG plaques (CW and VHF). In addition, the applicant must be in possession of QSLs from at least 1000 DIG members. Applications are available from DL9XW (see address above).

General Rules for DIG trophies and plaques

All QSLs from DIG members count one point, holders of DIG trophies and DIG plaques (as indicated in the DIG membership list) count two points, club stations DL0DIG, DF0DIG, DK0DIG, OE1XDC and PI4DIG count three points. DIG members must have their DIG number on their QSLs (or mention their DIG number on the air). The fee for each of these trophies/plaques (which includes engraving) is 40 IRCs. There is no fee for the DIG Trophy 1000.

The DIG membership list is published in the first quarter of each year and can be obtained from DIG Secretary Eberhard Warnecke, DJ8OT, PO Box 101244, D-5620 Velbert 1, Fed Rep of Germany. Please include 4 IRCs.

FINLAND (OH)

General rules for OH awards, sponsored by the Finnish IARU Society, Suomen Radioamatooriliitto (SRAL), are as follows:

All contacts must be made with Finnish fixed stations after June 6, 1947 (after February 1, 1967 for OHA 500). Do not send QSLs; send a list of QSLs checked by two amateurs or the Awards Manager of the national Amateur Radio Society. This list must include the call sign, date, time, band, mode and report for each QSO/QSL. Eight IRCs (or $2 US) are required for each award. For further information and application forms, contact Jukka Kovanen, OH3GZ, SRAL Awards Manager, Varuskunta 47 as 11, SF-11310, Riihimaki 31, Finland.

Finnmaid—Work three OH YLs.

OHA—Work 15 OH stations in at least five call areas.

OHA 100—Work 100 OH stations in all 10 call areas using at least two bands.

OHA 300—Work at least 300 OH stations.

OHA 500—Work at least 500 OH stations.

OHA 600—Work at least 600 OH stations.

FRANCE (F)

The following material is translated from *Les Diplomes Francais* by F6AXP, published by Reseau des Emetteurs Francais (REF).

General application requirements:

Each application must be dated and contain a signed statement that all awards rules have been complied with. The application must be addressed to the appropriate Awards Manager.

The application must include a detailed list of QSOs, indicating call sign, date, frequency and mode for each QSO.

The applicant must provide evidence in the form of QSLs or a certified list of QSOs/QSLs verified by an official of the applicant's National Society. A self-addressed envelope must be provided when sending QSLs. When applying for an endorsement, the number of the basic certificate and mode must be supplied.

DDFM (French Metropolitan Departments Diploma)

This award is issued in these classes:

- DDFM phone
- DDFM CW
- DDFM mobile or special mention

For applicants outside of France, it is necessary to work and confirm 40 different Departments on the same band. Endorsement stickers are available in multiples of ten; an "excellence" endorsement sticker is issued for working all 96 departments on one band. QSLs are required with all applications and must clearly indicate the Department of the station contacted. Contacts are valid from June 30, 1957 (January 1, 1982, for the new WARC bands). For the basic award, 10 IRCs are required; five IRCs are required for each endorsement. For further details, see REF form C3-01-7 (2/82). Awards Manager: Max Anouzet, F6FWH, 8, allee du Parc, 63110 Beaumont, France. A 5-band award *(5BDDFM)* is also available for working 300 Departments using a combination of five bands (at least 10 contacts must be made per band). A personalized and numbered plaque is issued to each 5BDDFM recipient; 65 IRCs are required for both the return of the cards and the plaque. Contact F6FWH for further details.

French Departments	Province
01 Ain	Rhone-Alpes
02 Aisne	Picardie
03 Allier	Auvergne
04 Alpes Haute Provence	Provence-Cote d'Azur
05 Alpes (Hautes)	Provence-Cote d'Azur
06 Alpes-Maritimes	Provence-Cote d'Azur
07 Ardeche	Rhone-Alpes
08 Ardennes	Champagne
09 Ariege	Midi-Pyrenees
10 Aube	Champagne
11 Aude	Languedoc-Roussillon
12 Aveyron	Midi-Pyrenees
13 Bouches-du-Rhone	Provence-Cote d'Azur
14 Calvados	Basse Normandie
15 Cantal	Auvergne
16 Charente	Poitou-Charente
17 Charente-Maritime	Poitou-Charente
18 Cher	Centre
19 Correze	Limousin
2A Corse Sud	Corse
2B Corse (Haute)	Corse
21 Cote d'Or	Bourgogne
22 Cotes du Nord	Bretagne
23 Creuse	Limousin
24 Dordogne	Aquitaine
25 Doubs	Franche-Comte
26 Drome	Rhone-Alpes
27 Eure	Haute Normandie
26 Eure-et-Loir	Centre
29 Finistere	Bretagne
30 Gard	Languedoc-Roussillon
31 Haute-Garonne	Midi-Pyrenees
32 Gers	Midi-Pyrenees
33 Gironde	Aquitaine
34 Herault	Languedoc-Roussillon
35 Ille-et-Vilaine	Bretagne
36 Indre	Centre
37 Indre-et-Loire	Centre
38 Isere	Rhone-Alpes
39 Jura	Franche-Comte
40 Landes	Aquitaine
41 Loir-et-Cher	Centre
42 Loire	Rhone-Alpes
43 Haute-Loire	Auvergne
44 Loire-Atlantique	Pays de Loire
45 Loiret	Centre
46 Lot	Midi-Pyrenees
57 Lot-et-Garonne	Aquitaine
48 Lozere	Languedoc-Roussillon
49 Maine-et-Loire	Pays de la Loire
50 Manche	Basse-Normandie
51 Marne	Champagne
52 Marne (Haute)	Champagne
53 Mayenne	Pays de Loire
54 Meurthe-et-Moselle	Lorraine
55 Meuse	Lorraine
56 Morbihan	Bretagne
57 Moselle	Lorraine
58 Nievre	Bourgogne
59 Nord	Nord
60 Oise	Picardie
61 Orne	Basse-Normandie
62 Pas-de-Calais	Nord
63 Puy-de-Dome	Auvergne
64 Pyrenees-Atlantique	Aquitaine
65 Pyrenees (Hautes)	Midi-Pyrenees
66 Pyrenees-Orientales	Languedoc-Roussillon
67 Bas-Rhin	Alsace
68 Haut-Rhin	Alsace
69 Rhone	Rhone-Alpes
70 Saone (Haute)	Franche-Comte
71 Saone-et-Loire	Bourgogne
72 Sarthe	Pays de Loire
73 Savoie	Rhone-Alpes
74 Haute-Savoie	Rhone-Alpes
75 Ville de Paris	Ile-de-France
76 Seine-Maritime	Haute-Normandie
77 Seine-et-Marne	Ile-de-France
78 Yvelines	Ile-de-France
79 Deux-Sevres	Poitou-Charentes
80 Somme	Picardie
81 Tarn	Midi-Pyrenees
82 Tarn-et-Garonne	Midi-Pyrenees
83 Var	Provence-Cote d'Azur
84 Vaucluse	Provence-Cote d'Azur
85 Vendee	Pays de Loire
86 Vienne	Poitou-Charente
87 Haute-Vienne	Limousin
88 Vosges	Lorraine
89 Yonne	Bourgogne
90 Territoire de Belfort	Franche-Comte
91 Essone	Ile-de-France
92 Hauts-de-Seine	Ile-de-France
93 Seine-Saint-Denis	Ile-de-France
94 Val-de-Marie	Ile-de-France
95 Val-d'Oise	Ile-de-France

[Note: Many attractive certificates are also available for contacts with individual Departments. For particulars, see Chapter II of *Les Diplomes Francais*; for information on how to obtain this booklet, contact REF, 2 Square Trudaine, 75009 Paris, France.]

DPF (French Provinces Diploma)

This award, for working a required number of French provinces, is issued in these classes:

- DPF CW
- DPF phone

- DPF "Special Mention" for any single band or mode.

For the DPF HF, the applicant must work and confirm 25 provinces. Contacts are valid from January 1, 1951 (WARC-band contacts valid from January 1, 1982). QSLs must be submitted with the application, and the cards must clearly indicate the provinces of the stations contacted. For further details, see REF form C3-01-5 (2/82). To cover the postage for returning the QSLs and the certificate, 10 IRCs are required. Awards Manager: F6FWH (see address above).

5BDPF (5-Band Province Diploma)

To qualify, work and confirm contacts with stations in 22 different provinces on five HF bands. For further details, see REF form C3-01-8 (2/82). A personalized and numbered plaque is sent to each successful applicant. For return of QSLs and shipping for the plaque, 65 IRCs are required. Awards Manager: F6FWH (see address above).

French Provinces	Includes These Departments (By Number)
1—Alsace	67, 68
2—Aquitaine	24, 33, 40, 47, 54
3—Auvergne	3, 15, 43, 63
4—Basse-Normandie	14, 50, 61
5—Bourgogne	21, 58, 71, 89
6—Bretagne	22, 29, 35, 56
7—Centre	18, 28, 36, 37, 41, 45
8—Champagne	8, 10, 51, 52
9—Corse	2A, 2B
10—Franche-Comte	25, 39, 70, 90
11—Haute-Normandie	27, 76
12—Languedoc-Roussillon	11, 30, 34, 48, 66
13—Limousin	19, 23, 87
14—Lorraine	54, 55, 57, 88
15—Midi-Pyrenees	9, 12, 31, 32, 46, 65, 81, 82
16—Nord	59, 62
17—Pays-de-Loire	44, 49, 53, 72, 85
18—Picardie	2, 60, 80
19—Poitou-Charentes	16, 17, 79, 86
20—Provence Cote-d'Azur	4, 5, 6, 13, 83, 84
21—Ile de France	75, 77, 78, 91, 92, 93, 94, 95
22—Rhone-Alpes	1, 7, 26, 38, 42, 69, 73, 74

DUF (French-speaking "Universe" Diploma)

This award, for working stations in the French-speaking world, is available in four classes:

DUF1—for confirmed contacts with five countries in three continents.

DUF2—for confirmed contacts with eight countries in four continents.

DUF3—for confirmed contacts with 10 countries in five continents.

DUF4—for confirmed contacts with 20 countries in six continents.

QSL cards must accompany the application and must clearly indicate the country and continent. Contacts are valid from April 1, 1946, unless otherwise indicated. For more details, see REF form C3-01-9 (2/82). Awards Manager: Edmond Dubois, F9IL, Aubencheul-au-Bac, 59265, Aubigny-au-Bac, France.

5BDUF

A personalized and numbered plaque is awarded to those qualifying for 5BDUF, which requires confirmed contacts with 30 countries in six continents on bands higher than 7 MHz and confirmed contacts with 15 countries in at least five continents on 7 MHz and each band below. Contacts are valid from January 1, 1981, and QSLs that clearly indicate

the country and continent must accompany the application; 65 IRCs, to cover cost of returning the QSLs and sending the plaque, must also be included. For further details, see REF form C3-01-10 (2/82). Awards Manager: F9IL (see address above).

DUF Countries [simplified list]:

Europe—C3, DA, F, TK, 3A.

Africa—CN, D6 (after 5 July 1975), FH (after 5 July 1975), FR/R, FR/E (after 1 August 1968), FR/G (after 25 June 1960), FR/J (after 25 June 1960), FR/T (after 25 June 1960), J2, TJ (after 1 January 1960), TL (after 13 August 1960), TN (after 15 June 1960), TR (after 17 August 1960), TT (after 11 August 1960), TU (after 7 August 1960), TY (after 1 August 1960), TZ (after 20 June 1960), XT (after 5 August 1960), 3V, 3X (after 1 October 1958), 5R (after 14 October 1958), 5T (after 20 June 1960), 5U (after 3 August 1960), 5V (after 27 April 1960), 6W (after 20 June 1960), 7X (after 1 July 1962).

Asia—XU, XV, XW.

North America—FG, FM, FS, FO, FP.

South America—FY.

Oceania—FK, FO, FW, YJ.

Austral Continent—FT8W, FT8X, FT8Y, FT8Z.

DDTOM (French Overseas Departments and Territories Diploma)

This award is issued for working the following overseas departments (FG, FS, FM, FP, FR) and the following overseas territories (FH, FK, FO, FW, 3Y). All contacts are valid from January 1, 1982, and QSLs and 10 IRCs must accompany the application. For further details, see REF form C3-01-13 (2/82). Awards Manager: Max Pomel, F6AXP, PO Box 73, 63370 Lempdes, France.

DTA (Southern French Lands Diploma)

To qualify for this award, three confirmed contacts are required with FT8W, FTX, FT8Y and FT8Z. The *DTA Excellence* diploma is issued for four confirmed contacts with each country. QSLs and 10 IRCs must accompany the application. For further details, see REF form C3-01-14 (2/82). Awards Manager: F6AXP (see address above).

DTC (CW Class C Diploma)

This award is issued for making 1000 CW QSOs. All bands and all countries are valid. The following endorsements are available:

DTC 1—2000 additional CW QSOs.

DTC 2—4000 additional CW QSOs (for a total of 5000).

DTC 3—9000 additional CW QSOs (for a total of 10,000).

DTC 4—14,000 additional CW QSOs (for a total of 15,000).

DTC REF HONOR—20,000 CW QSOs.

To apply, send certified list of QSOs (no QSLs) and 10 IRCs (30 IRCs for DTC HONOR). Awards Manager: Patrick Beunier, F6HWH, La Tuilerie F 150 Rue Dauphine, 03150 Varennes Sur Allier, France.

F-CW-500

This award is issued for working 500 different French metropolitan stations on CW. Endorsements are available in 100-QSO increments. Applications must include a certified list of QSOs in alphanumerical order by call (with complete

QSO data) and 10 IRCs. Awards Manager: F6HWH (see address above).

YL De France Diploma

This award is issued in three classes:

Class 1—work five French YLs, plus one YL contact on three different continents.

Class 2—work 100 YLs, including five French YLs, on three different continents.

Class 3—work 500 YLs, including five French YLs, on six continents.

QSL cards and 20 IRCs (to defray costs of returning the QSLs and sending the certificate) must be included in each application. Awards Manager: Gilda Le Gall, F6FMO, Ecole Publique, 56490 Guilliers, France.

GABON (TR)

Association Gabonaise des Radio-Amateurs (AGRA) issues three different certificates as follows:

DDG1—for confirmed contacts with eight different TR stations; any mode, any band.

DDG2—for confirmed contacts with 12 different TR stations on at least three different HF bands, any mode.

DDG Special—for one confirmed TR contact on each of five different HF bands; two of the five bands must be 160, 80 and/or 40 meters.

Minimum signal report: 33(9).

To apply, send a certified list of QSLs signed by two amateurs or an official of the applicant's National Society and 10 IRCs to: AGRA Diplome Manager, PO Box 1826, Libreville, Gabon.

GERMAN DEMOCRATIC REPUBLIC (EAST GERMANY—Y2)

There are generally no mode restrictions for GDR awards. When applying, the applicant must include his call sign, name and address, award and/or endorsement desired, and a certified list of stations worked as well as other QSO data that may be required for the specific award applied for. QSOs are valid after January 1, 1980.

Fees: Basic Class—10 IRCs (Air Mail 12 IRCs);
each sticker—2 IRCs (Air Mail 3 IRCs),
except for the SOP pennant and the
Y2-CA-Trophy which are 20 IRCs
(Air Mail 25 IRCs).

Send applications to: The Radioclub of the GDR, Awards Bureau, PO Box 30, Berlin, 1055 GDR, German Democratic Republic.

WA-Y2 (Worked All Y2)

This award is issued in four classes for confirmed contacts with different GDR counties indicated by the last letter of the call sign suffix:

A,U	Rostock	H,V	Halle
B	Schwerin	I,Q	Erfurt
C	Neubrandenberg	J,Y	Gera
D,P	Potsdam	K	Suhl
E	Frankfurt/Oder	L,R	Dresden
F,X	Cottbus	M,S	Leipzig
G,W	Magdeburg	N,T	Karl Marx Stadt
		O	Berlin

The four classes are as follows:

I—Basic	= 20 points,	10 countries
II—1st sticker	= 40 points,	13 countries
III—2nd sticker	= 75 points,	15 countries
IV—3rd sticker	= 120 points,	15 countries

Each county counts one point per band (a maximum of five bands may be used). Four bonus points can be earned for working the same Y2 station on four bands, five points for five bands, but only once per county and only once for Classes II, III and IV. For contacts with 12 Y2 counties on 160 meters, 30 meters, 17 meters and 12 meters, separate band-stickers are issued. All call signs Y2-Y9 are valid, and call signs (including special call signs) ending in Z, /p, /m, and /a can be counted for the county in which the stations are located.

Y2-KK (Y2-Kreiskennerdiplom)

This award is issued in four classes for confirmed contacts with different GDR districts. The GDR is subdivided into 229 districts; these districts are indicated by the "KK" (Kreiskenner district locator) on QSL cards. A KK consists of a letter (A-O, indicating the county) and a two-digit number.

The four classes are as follows:

I—Basic	= 100 points
II—1st sticker	= 150 points
III—2nd sticker	= 200 points
IV—3rd sticker	= 225 points

For applicants outside of Europe, each KK counts four points (two points for European applicants), irrespective of the band or mode used.

SOP (Sea of Peace)

The Sea of Peace pennant is issued for confirmed contacts with the following different countries/areas bordering the Baltic Sea as well as Norway and Iceland during the period July 1-July 31: Y2____A (Rostock County), DL, LA, OH1/2/5/6/8, OHØ, OJØ, SM1/2/3/5/6/7/Ø, SP1/2, TF, UA1, UA2, UP, UQ, UR. Applicants outside Europe must work 10 of these areas (European applicants must work 15 areas), and a contact with Rostock County is mandatory. The SOP is a one-time award; there are no endorsements.

Y2-CA (Y2 Certificate Hunters Award)

This award is issued in four classes for confirmed contacts with Y2 members of the Certificate Group (CG). For applicants outside of Europe, each CG member counts two points (one point for European applicants). Send an SAE and one IRC for Y2-CG membership list.

The four classes are as follows:

I—Basic	= 50 points,	8 counties
II—1st sticker	= 100 points,	10 counties
III—2nd sticker	= 150 points,	12 counties
IV—3rd sticker	= 200 points,	15 counties
Trophy	= 250 points,	15 counties

Y2-DX-A (Y2 DXer Award)

This award is issued in four classes for confirmed contacts with Y2 amateurs who are in the "Y2 Honor Roll." Each contact counts one point.

The four classes are as follows:

I—Basic	= 50 points
II—1st sticker	= 75 points
III—2nd sticker	= 100 points
IV—3rd sticker	= 125 points

GIBRALTAR (ZB2)

The *ZB2 Award* is issued to amateurs who work at least five different ZB2 stations on any band at any time. QSLs are not necessary; certified log entries are acceptable. Send log and $3 US to defray costs. The *ZB2BU Award* is issued to amateurs who work Gibraltar Amateur Radio Society club station ZB2BU on three different bands (one contact per band is sufficient). Otherwise, same rules as apply to the ZB2 Award. Send applications to: Awards Manager, Gibraltar Amateur Radio Society, PO Box 292, Hargraves Parade, Gibraltar.

GREECE (SV)

The *RAAG Award* is offered by the Radio Amateur Association of Greece for contacts made after January 1, 1975. Applicants must submit certified log extracts (no QSLs) of contacts with at least seven Greek stations from the nine SV call areas (SV1-SV9). There are no band or mode limitations. Send application and 8 IRCs (or $2 US) to: RAAG Award Manager, PO Box 3564, 10210 GR Athens, Greece.

GUATEMALA (TG)

The Club de Radioaficionados de Guatemala sponsors the *WATG Award* for working all TG zones (consisting of TG4-TG0). Send either the QSL cards or photocopies to the Awards Manager, along with 5 IRCs (or $2 US) to Awards Manager, Club de Radioaficionados de Guatemala (CRAG), PO Box 115, Guatemala City, Guatemala.

HONG KONG (VS6)

To apply for awards sponsored by the Hong Kong Amateur Radio Transmitting Society (HARTS), send application and certified log extracts (no QSL cards required) to: Awards Manager, HARTS, GPO Box 541, Hong Kong.

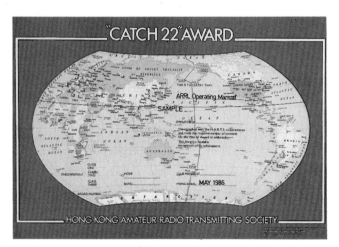

Catch 22 Award

To qualify, the applicant must submit verified evidence of two-way contacts with amateur stations located in the 22nd Parallel of Latitude North. The award is issued in three classes:

Class 3—contacts with at least 15 countries.
Class 2—contacts with at least 20 countries.
Class 1—contacts with at least 25 countries.

A VS6 contact is mandatory. Contacts must be made after January 1, 1980. Band/mode endorsement stickers are available. Fee: $7 US; endorsements $1 US.

22nd Parallel countries: A4, A6, BV, BY, C6, CN, CO, HZ, KH6, S2, ST, SU, TT, TZ, VS6, VU, XE, XW, XX9, XZ, 3W, 5A, 5T, 5U, 7X.

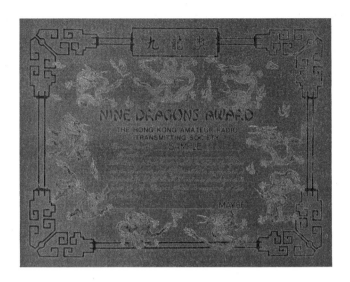

Nine Dragons Award

To qualify, work one country in zone 18, 19 and 24-30, for a total of nine zones. The zone 24 contact must be with a VS6. Contacts are valid after January 1, 1979. Fee: $3 US or 25 IRCs.

Firecracker Award

To qualify, work six different VS6 stations. Contacts are valid after January 1, 1964. Fee: $2 US or 10 IRCs.

HUNGARY (HA)

For awards sponsored by the Hungarian Radioamateur Society (MRASZ), all amateur bands and modes may be used, except repeaters. Each station may be contacted only once on any band and any mode. Contacts may be made from any location within the same DXCC country. Applications must include a log showing the name, call sign and QTH of applicant, and the following information: station worked, UTC, band, mode, report. With the exception of Szeged Festival, Dunaferr and HCS awards, each application must be accompanied by a statement from the applicant's National Society, or from any two amateurs, that the valid QSL cards listed are in the possession of the applicant. Applications must be sent to the individual Awards Managers as indicated below. General correspondence may be sent to: Hungarian Radioamateur Society, Awards Committee, PO Box 22, Tiszakecske H-6061, Hungary.

BD (Balaton Diploma)

This award is issued by the Radioclub Siofok for QSOs made on or after January 1, 1967. The applicant must make 15 points, including at least one contact with a Siofok club member (Europeans must make 30 points, including at least two club members). Each club member counts five points. Fee: 10 IRCs.

Soifok club members (counting five points) are as follows: HA/HG 3s KGJ KHL GI GJ GQ HE HL HQ HZ IG IK IQ IS NG; HA4XW; HA6NP; HA8UA.

The following stations, residing in the Lake Balaton area, count three points: HA/HG 1s KXX XA XX ZY; 2s KRQ RC RQ SH Y YRC, 3s KHB KHO GG GO HK HO HU.

The following stations, in Zala, Veszprem and Somogy county, count one point: HA/HG 1s KRA-KRZ, KXA-KXZ, KZA-KZZ, RA-RZ, XA-XZ, ZA-ZZ, DRA-DZZ; 2s KPA-KTZ, PA-TZ, ENA-EZZ; 3s KGA-KIZ, GA-IZ, FLA-FSZ.

Awards Manager: Jozsef Turjanyi, HA3GJ, PO Box 78, Siofok H-8601, Hungary.

BPA (Budapest Award)

This award is issued by the Radioamateur Society of Budapest for QSOs with HA5/HG5 stations made on or after January 1, 1959. Requirements:

- European applicants must work 75 different HA5/HG5 stations
- Applicants outside Europe must work 25 different HA5/HG5 stations.

There is no restriction as to the use of repeaters, satellites or moon reflections over 500 km. Fee: 10 IRCs.

Awards Manager: Verebes Janosne, HA5YR, PO Box 64, Budapest H-1475, Hungary.

DD (Dunakanyar Diploma)

This award is issued by the Radioamateur Society of Pest County for QSOs with HA7/HG7 stations made on or after January 1, 1970. The applicant must confirm five HA7/HG7 contacts (Europeans must confirm 20). Fee: 6 IRCs.

Awards Manager: HA7PL, PRASZ Awards Manager, PO Box 36, Budapest H-1387, Hungary.

HCS (Hungarian Castle Series)

This award, granted for QSOs made on or after January 1, 1968, is issued by directly by the MRASZ in three classes:
Class I, Gold—for all QSLs, numbers 1-36.
Class II, Silver—for QSLs numbers 1-24 or 12-36.
Class III, Bronze—for QSLs numbers 1-12 or 13-24 or 25-36.

QSL numbers by district:
HA1/HG1   7-22-25-31.
HA2/HG2   6-8-12-15-21-23-30-32-35.
HA3/HG3   3-14-23-30-32-33-35.
HA4/HG4   17-23-30-32-35.
HA5/HG5   1-13-36.
HA6/HG6   4-10-11-34.
HA7/HG7   2-5-19.
HA8/HG8   16-20-24.
HA9/HG9   18-27-28-29.
HA0/HG0   9-26-29.

Applicants must include the QSL number "slice," cut from each QSL card, along with the application and the appropriate number of IRCs (5 for bronze, 8 for silver, 10 for gold).

Awards Manager: Janos Retkes, HA8UB, PO Box 22, Tiszakecske H-6061, Hungary.

Pannonia Award

This award is issued by the Radioamateur Society of Gyor-Sopron County for contacts made on or after January 1, 1966. The applicant must submit proof of two QSOs each with HA/HG1, HA/HG2, HA/HG3, and HA/HG4, for a total of eight contacts, on more than one band. Fee: 10 IRCs.

Awards Manager: Radioclub HA1KSA, PO Box 79, Gyor H-9001, Hungary.

Savaria Award

This award is issued by the Radioamateur Society of VAS County for contacts with HA1/HG1 stations made on or after January 1, 1976. The applicant must confirm 10 different HA1/HG1 QSOs (Europeans must confirm 20). Fee: 10 IRCs. Awards Manager: Savaria Radioclub, Puskas T.u.7, Szombathely H-9700, Hungary.

Szeged Festival Award

This award is issued yearly by the Radioamateur Society of Csongrad County for QSOs made between July 1 and August 31. The applicant must make five points (Europeans, 10 points). Fee: 5 IRCS.

Stations in Szeged count two points: HA/HG8 CA CB CD CH CP CT CV CX CZ DC DE DF DP DQ DR DT DZ EK EL KCC KCK KDA.

Stations in Csongrad County (HA/HG8 CA-FZ, KCA-KFZ, LSA-LZZ) count one point.

Awards Manager: Imre Kelemen, HA8CH, PO Box 673, Szeged H-6701, Hungary.

VTA (Videoton Award)

This award is issued by the Videoton Radioclub for QSOs with HA4/HG4 stations made on or after January 1, 1969. Contacts must be confirmed by special QSL cards representing broadcast (BC) and TV receivers and computers respectively (there are three groups of special cards, 3-4-3).

Classes:

Gold—all ten cards [fee: 5 IRCs]
Silver—complete set of any two groups.
     [fee: 3 IRCs]
Bronze—one complete set of any group
     [fee: 2 IRCs]

Awards Manager: Halmi Belane, HA4XP, Berkes F.ltp.40, Szekesfehervar H-8000, Hungary.

HRD (Hungarian Rummy Diploma)

The Rummy awards are issued by the Radio Amateur

Society of Somogy County. Contacts are valid on or after September 1, 1972 [for HRD-108 and HCD, contacts are valid on or after April 4, 1980]. Fee: 5 IRCs.

The *basic HRD award* is issued in three degrees:

Bronze degree——"hand rummy," collecting 14 cards in accordance with the rules of the game.
Silver degree——full collection of one of the four series, plus one joker of the same color.
Gold degree——full pack, 54 cards.

The *HRD-108* award requires two packs of QSL cards from 108 different stations. For the *Hungarian Canasta Diploma (HCD),* three canastas, 21 cards, have to be confirmed from different call signs in accordance with the rules of the game. The canasta contains seven cards of the same figure; two of them can be equivalent (eg, seven cards of figure 5, seven cards of figure 8 and seven cards of kings). Not more than three cards can be substituted by the four jokers and the eight "little jokers" in one canasta.

Allocation of the HFD cards:

HA/HG Call Areas	Spades/Hearts R/a	Diamonds/Clubs R/b
1	A	J
2	2	J
3	3	J
4	4	J
5	5	Q
6	6	Q
7	7	K
8	8	K
9	9	K
0	10	K
???	red and black Joker = Y =	

Awards Manager: Dr Janos Mihalyfy, HA3GA, PO Box 173, Kaposvar H-7401, Hungary.

WHD (Worked Hungarian Districts)

This award is issued directly by MRASZ for QSOs made with HA/HG stations on or after January 1, 1958. The applicant must confirm 2 QSOs with at least five different HA call districts on two bands, for a total of 10 contacts (Europeans must confirm 2 QSOs with at least eight different districts). Fee: 5 IRCs.

Awards Manager: HA8UB (see address above).

HADXC (HUNGARIAN DX CHAPTER) AWARDS

The general rules for HADXC awards are essentially the same as for MRASZ awards. The fee for all HADXC awards is 12 IRCs. Send all correspondence to: HADXC Awards Manager, HA8UB (see address above).

WHADXCA (Worked Hungarian DX Chapter Award)

Applicants must work HADXC members on or after January 1, 1986, for a total of 15 points (each club member counts one point, and club station HG3DXC counts two points). Europeans must make 20 points.

WAHA-HF (Worked All Hungarian Award, HF)

This award, for contacts made on or after January 1, 1980, is issued in three classes:

Class III—250 points required
Class II—500 points required
Class I—1000 points required

Contacts with club station HG3DXC counts 10 points,

HADXC members count five points, and all other Hungarian stations count one point.

HADXC Members: HG1S, HA1TJ, HG1W, HA1YA, HG1Z, HA2RD, HG3DXC, HA3GJ, HA3NU, HA4XH, HA4XT, HA4XX, HA4YD, HG5A, HA5AM, HA5DW, HA5FM, HA5HO, HA5MY, HG6N, HA6NF, HG6V, HG7B, HA8ET, HG8U, HA8UB, HG9R, HA9RU, HA0LZ, HA0MM.

ICELAND (TF)

The *Icelandic Radio Amateurs Award (IRAA)* is available to amateurs outside of Iceland. Only contacts with amateurs who are Icelandic citizens operating from Icelandic territory are valid. There is no date limit for QSOs. QSL cards or certified photocopies of the QSL cards must be submitted together with a complete list of the log entries for the contacts. Mixed-mode and crossband contacts are not valid, except for amateur satellites. Each TF station may be contacted only once per band per mode. [TF Novices run 5 watts input and are permitted to operate on three bands, 3500-366 kHz, 7000-7040 kHz and 21100-21150 kHz, and are identified by a three-letter suffix ending in N.] Total points needed to qualify for the award depend upon the applicant's ITU zone:

I. ITU Zones 5, 9, 18-20, 27-29	=	98 points
II. ITU Zones 1-4, 6-8, 21-26, 30, 31, 36, 37	=	48 points
III. ITU Zones 10-13, 32-35, 38-40, 46-48	=	28 points
IV. ITU Zones 14-16, 41-45, 49-75	=	18 points

Points by band and mode are calculated according to the following chart:

Band	Novice	CW	RTTY	SSTV	SSB	Via Satellite
1.8		10	8	8	6	
3.5	32	8	6	6	4	
7	24	6	5	5	3	
10		5	4	0	0	
14		3	2	2	1	8
18		4	3	3	2	
21	16	5	4	4	3	
24		6	5	5	4	
28		7	6	6	5	

Send application materials and 14 IRCs to: Islenzkir Radioamatorar (IRA), Awards Manager, PO Box 1058, 101 Reykjavik, Iceland.

INDIA (VU)

The Amateur Radio Society of India issues the *WRI Award* (Worked Republic of India). Amateurs must work and produce QSLs as evidence of making at least 50 contacts with VU stations since January 26, 1950, all bands/modes. In lieu of sending QSLs, an executive of an IARU Member-Society may certify in writing that the cards are valid. Sufficient postage must be enclosed for return mail. Applications should be sent to: The Amateur Radio Society of India (ARSI), Foreign Mails and QSL Service, PO Box 3005, New Delhi 3, India.

INDONESIA (YB)

The general rules for awards sponsored by ORARI (Organisasi Amatir Radio Indonesia) are as follows: Awards are issued for two-way SSB, two-way CW, mixed mode or single mode, mixed or single band. To be valid, contacts must be made on or after July 9, 1968. Only contacts with land stations are valid. Applications must be accompanied by a list of QSL cards in alphabetical order indicating station worked, dates, bands and modes, and a statement from the applicant's National Society, club station or from any two amateurs, stipulating that the correct QSL cards of the contact listed are in the possession of the applicant. Include $8 (US) or 16 IRCs with each appplication. Send applications to: ORARI National QSL & Awards Bureau, PO Box 96, Jakarta 10002, Indonesia.

JA (Jakarta Award)

To qualify, work a total of 20 stations, including at least one club station in Jakarta (Ø call area), the capital of Indonesia. [YBØ club stations: YBØZAA, YBØZAB, YBØZAD, YBØZAE, YBØZAF, YBØZBA, YBØZBB, YBØZCA, YBØZCB, YBØZCD, YBØZCE, YBØZDB, YBØZDC, YBØZDD, YBØZDE, YBØZDG, YBØZEA, YBØZEE, YBØZZ.]

WAIA (Worked All Indonesia Award)

To qualify, work two stations in each of the Indonesian call areas (1-Ø), a total of 20 contacts.

WTEA (Worked the Equator Award)

To qualify, confirmed contacts are required with the DXCC countries along the equator: C2, HC, HC8, HK, KH1/KB6, PP-PY, PYØ (St Peter & Paul), S9, T3Ø, T31, T32, T5, TN, TR, YB5, YB7, YB8, 5X, 5Z, 8Q, 9Q [YB5, YB7, YB8 contacts are mandatory]. The WTEA is issued in three classes:

Class I—for confirmed contacts with 15 countries
Class II—for confirmed contacts with 12 countries
Class III—for confirmed contacts with 8 countries

IRELAND (EI)

The Irish Radio Transmitters Society (IRTS) has just introduced its first award, *WEIC*. The WEIC Award is available to licensed amateurs who have worked stations located in at least 20 of the 26 different counties of Ireland (EI/EJ). Contacts from January 1, 1982 are valid; no band or mode endorsements. Applications for the WEIC Award must be accompanied by 10 IRCs, a QSO list and a statement from the applicant's National Society Awards Manager that correctly filled-in QSL cards are in the possession of the applicant. If this is not possible, the applicant must submit all QSLs concerned.

Counties of Ireland: Carlow, Cavan, Clare, Cork, Donegal, Dublin, Galway, Kerry, Kildare, Kilkenny, Laois, Leitrim, Limerick, Longford, Louth, Mayo, Meath, Monaghan, Offaly, Roscommon, Sligo, Tipperary, Waterford, Westmeath, Wexford, Wicklow.

Send applications to: Irish Radio Transmitters Society, PO Box 462, Dublin 9, Ireland.

ISRAEL (4X)

The following awards are sponsored the Israel Amateur Radio Club (IARC). Applications must be accompanied by a certified list of QSLs (checked and signed by two amateurs) together with 7 IRCs and should be sent to: Mark Stern, 4Z4KX, Israel Awards Manager, Israel Amateur Radio Club, PO Box 4099, 61040 Tel-Aviv, Israel.

The 4 × 4 = 16 Award

This award can be obtained by sending a certified list of at least 16 QSLs that confirm contacts with 16 different Israeli stations. No restrictions as to modes or bands, but contacts must be made on at least four different bands.

The Israel Award

This award can be obtained by making 25 points, as follows: Each QSO with an Israeli station above 10 MHz counts one point, and each contact below 10 MHz counts two points. Each station may be worked only once per band. No restrictions as to modes or bands, but contacts must be made after January 1, 1983.

The Jerusalem Award

Send a certified list of four contacts with stations located in Jerusalem. Each certificate will be signed by Teddy Kollek, the mayor of Jerusalem. Applications and seven IRCs should be sent to Dr Milton Gordon, 4X6AA, Box 4079, 91040 Jerusalem, Israel.

The Tel-Aviv Award

This award is issued in commemoration of the 75th anniversary of the city of Tel-Aviv. To obtain this award, 15 points are required, as follows: Stations in Tel-Aviv count one point, stations located in Jaffo count two points, special-events station 4X75TA counts five points. No restrictions as to bands or modes, but each station may be worked only once on each band, and all contacts must be made after January 1, 1984. Each certificate will be signed by Shlomo Lahat, the mayor of Tel-Aviv. Applications with certified log extract and $3 US should be sent to Shlomo Mussali, 4X6LM, PO Box 8225, 61081 Tel Aviv-Jaffo, Israel.

ITALY (I)

To apply for the following awards sponsored by Associazione Radioamatori Italiani (ARI), send all applications to: ARI Awards Manager, Via Scarlatti 31, 20124 Milano, Italy.

WAIP (Worked All Italian Provinces)

The WAIP is awarded for confirmed contacts with fixed stations in at least 60 Italian provinces (75 provinces for Italian amateurs). Contacts must be made since January 1, 1949, and the same station may be worked more than once if located in different provinces. Endorsements are available for CW, phone, mixed and RTTY. Minimum signal report 33(8). A log extract and 10 IRCs (or $3 US) must be included with the application; QSLs are not necessary if the log extract is certified by the Awards Manager of the applicant's National Society or two amateurs.

Italian provinces are as follows:

Agrigento, Alessandria, Ancona, Aosta, Arezzo, Ascoli P, Asti, Avellino, Bari, Belluno, Benevento, Bergamo, Bologna, Bolzano, Brescia, Brindisi, Cagliari, Caltanissetta, Campobasso, Caserta, Catania, Catanzaro, Chieti, Como, Cosenza, Cremona, Cuneo, Enna, Ferrara, Firenze, Foggia, Forli, Frosinone, Genova, Gorizia, Grosseto, Imperia, Isernia, L'Aquila, La Spezia, Latina, Lecce, Livorno, Lucca, Macerata, Mantova, Massa, Matera, Messina, Milano, Modena, Napoli, Novara, Nuoro, Padova, Palermo, Parma, Pavia, Perugia, Pesaro, Pescara, Piacenza, Pisa, Pistoia, Pordenone, Potenza, Ragusa, Ravenna, Reggio C, Reggio E, Rieti, Roma, Rovigo, Salerno, Sassari, Savona, Siena, Siracusa, Sondrio, Taranto, Teramo, Terni, Torino, Trapani, Trento, Treviso, Trieste, Udine, Varese, Venezia, Vercelli, Verona, Vicenza, Viterbo.

CDM (Certificato del Mediterraneo)

The CDM is awarded for confirmed contacts with fixed stations in at least 22 Mediterranean countries excluding Italy, plus at least 30 amateur stations located on the Italian mainland. Contacts must be made since June 1, 1952, and stations may be worked only once. Endorsements are available for phone, CW, mixed and RTTY. Minimum signal report—33(8). A log extract and 10 IRCs (or $3 US) must be included with application; QSLs are not necessary if the log extract is certified by either the Awards Manager of the applicant's National Society or two amateurs.

CDM countries: EA, EA6, EA9, CN, F, ISØ, IT, I/Trieste (before December 31, 1957), OD5, SU, SV, SV5, SV9, TA, TK, YK, YU, ZA, ZB, ZC4/5B, 3A, 3V, 4X, 5A, 7X, 9H

IIA (Italian Islands Award)

Applicants must work and confirm contacts with Italian islands in the following archipelagos: IA1—Liguri Group, IA5—Tuscan Arch, IBØ—Ponziano Arch, IC8—Napoli Arch, ID9—Eolie Arch, IE9—Ustica, IF9—Egadi Arch, IG9—Pelagie Arch, IH9—Pantelleria, IJ7—Cheradi Arch, IL7—Tremiti Arch, IMØ—Maddalena Arch, ISØ—Sardinia, IT9—Sicily. Contacts must be made starting January 1, 1970, on SSB, CW, RTTY or mixed modes.

To qualify, applicants must make 10 points (Italian applicants must make 40 points, all other Europeans 20 points). In general, each island counts one point, but some count two-or-more points. Further details can be obtained from the ARI Awards Manager for a self-addressed envelope and one IRC. Point values usually appear on the QSL cards. Note: Each island counts one-or-more points even if in the same

archipelago, and islands can be worked more than once if a different mode is used. *IIA Honor Roll* requires 60 points. *Five-Band IIA* requires 10 contacts with 10 different Italian islands on each of five HF bands. QSLs are normally not required, but must be available on request. Application and log extract must be accompanied by 30 IRCs (or $8 US).

JAPAN (JA)

General information for awards issued by the Japan Amateur Radio League (JARL): Each application must be accompanied by a list of QSL cards, stating the call signs of each station worked, date, band and mode. Each list must be accompanied by a statement from the applicant's National Society or from any two amateurs that the valid QSL cards are in the possession of the applicant. If such a statement is not available, the applicant must submit the actual QSL cards. Eight IRCs are required for each award (2 additional IRCs for airmail delivery). If QSL cards are submitted, sufficient funds for their return must also be included. A list of cities and guns, along with the Asian Countries list, is available for 3 IRCs. Only contacts with land stations (including mobile stations on a river or lake) are valid. Contacts with KA stations are not acceptable. For all JARL awards, contacts are acceptable if made July 29, 1952, or later, with the following exceptions:

- Satellite endorsement—December 15, 1972
- SSTV endorsement—April 10, 1973
- RTTY endorsement—August 8, 1968
- Okinawa Prefecture contacts—May 15, 1972

Send all applications to: Japan Amateur Radio League, Awards Section, 1-14-1 Sugamo, Toshima, Tokyo 170, Japan.

AJD (All Japan Districts)

Work and confirm amateur stations in each of the Japanese call areas (1-Ø). Endorsements available: 160 meters, 80/75 meters, 40 meters, 30 meters, 20 meters.

WAJA (Worked All Japanese Prefectures Award)

Work and confirm amateur stations in each of the 47 prefectures (administrative districts) of Japan listed immediately below. The list of QSLs should be arranged in order of WAJA reference number in lieu of prefecture name. Endorsements: 15 meters, 10 meters.

Prefectures:

Call Area 1—Tokyo, Kanagawa, Chiba, Saitama, Ibaraki, Tochibi, Gunma, Yamanashi

Call Area 2—Shizuoka, Gifu, Aichi, Mie

Call Area 3—Kyoto, Shiga, Nara, Osaka, Wakayama, Hyogo

Call Area 4—Okayama, Shimane, Yamaguchi, Tottori, Hiroshima

Call Area 5—Kagawa, Tokushima, Ehime, Kochi

Call Area 6—Fukuoka, Saga, Nagasaki, Kumamoto, Oita, Miyazaki, Kogoshima, Okinawa

Call Area 7—Aomori, Iwate, Akita, Yamagata, Miyagi, Fukushima

Call Area 8—Hokkaido

Call Area 9—Toyama, Fukui, Ishikawa

Call Area Ø—Niigata, Nagano

JCC (Japan Century Cities)

Work and confirm amateur stations in each of at least 100 different Japanese cities. JCC-200, 300, 400, 500 and 600 will be issued as separate awards. The list of QSLs should be arranged in order of JCC reference number in lieu of city name. Endorsements: CW, AM, SSB, FM, SSTV, RTTY, ATV, FAX.

JCG (Japan Century Guns)

Same rules, including endorsements, as JCC (above), with guns replacing cities. [Note: Japan has, as administrative districts, 47 prefectures, which are divided into cities, towns and villages. The gun, not being an administrative district, is a regional congregation of towns and villages.]

ADXA (Asian DX Award)

Work and confirm amateur stations in at least 30 Asian countries, including Japan, in accordance with the ARRL DXCC List and the Asian Countries List. The QSL list should be arranged in order by country per the Asian Countries List. Endorsements: 160 meters, 80 meters, SSTV, RTTY, FAX.

WACA (Worked All Cities Award)

Work and confirm amateur stations in each of all the cities of Japan that are in existence on the day when the final contact claimed for the award is made. The QSL list should be arranged in order of JCC reference number. A satellite endorsement is available.

WAGA (Worked All Guns Award)

Same rules as WACA (above), with guns replacing cities.

JORDAN (JY)

To apply for these Royal Jordanian Radio Amateur Society (RJRAS) awards, applications must be accompanied by photocopies of the QSL cards (or certification) and 10 IRCs. Send applications to: Awards Manager, PO Box 1055, Amman, Jordan.

The Silver Award

This award, issued by King Hussein, JY1, requires proof of working six different JY prefixes. Photocopies of QSL cards (or certification) and 10 IRCs required.

The Coral Award

This award, issued by King Hussein, JY1, is for amateurs who visit Jordan and make a QSO from Aqaba.

The Arabian Knights Award

This award, issued by the Arab Radio Amateur League (ARL) and presented by JY1, is for working 10 Arab countries, one of which must be either JY1 or JY2, since January 1, 1971. [The Arab countries are A4, A6, A7, A9, CN, HZ, J2, JY, OD, ST, SU, T5, YI, YK, 3V, 4W, 5A, 5T, 7O, 7X, 9K.]

AKA (All Korea Award)

The AKA Award is issued for confirmed contacts with HL stations in each of the seven Korean call areas (1,2,3,4,5,8,0).

KDN (Korean District Number Award)

The KDN Award is issued for confirmed contacts with HL stations located in each of the 50 different cities, guns or gus in Korea. The award will be issued in multiples of 50 (KDN 50, 100, 150, etc) upon submission of the QSL cards accompanied by a list prepared in order of KDN reference numbers.

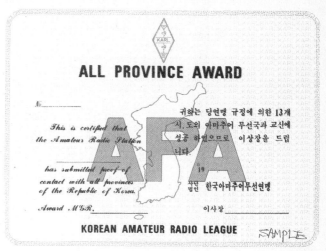

KENYA (5Z)

The Radio Society of Kenya (RSK) issues the *Kenyan Award* for contacts made after December 31, 1977. Ten points are necessary to qualify, as follows: Each contact with a 5Z station who must be a member of RSK = two points, and a contact with club station 5Z4RS = five points. All bands/modes. To apply, send a log photocopy, witnessed and signed by a responsible official of the local radio society/club or the licensing authority, and 10 IRCs (or $5 US). A self-addressed, adhesive label must be enclosed with the letter of application, which should be addressed to RSK and marked "Kenyan Award" at the top left-hand corner of the envelope. Send application to: The Radio Society of Kenya, PO Box 45681, Nairobi, Kenya.

KOREA (HL)

General rules for Korean Amateur Radio League (KARL) Awards:

Eight IRCS are required for each award, 4 IRCs for each endorsement. If QSL cards are submitted, IRCs for return postage must be provided. Separate endorsements for specific bands/modes, and QRP, are available. Contacts with HL stations (except HL9) are acceptable on or after February 3, 1959. Contacts with HL9s are not acceptable. All contacts must be made from the same call area. Mail application to: Korean Amateur Radio League, CPO Box 162, Seoul 100, Korea.

HLA (HL Award)

The HLA is issued for confirmed contacts with HL stations (except HL9s) in the following classes.

Class K—5 QSLs required
Class O—10 QSLs required
Class R—20 QSLs required
Class E—30 QSLs required
Class A—50 QSLs required

Endorsement stickers are available in multiples of 50 upon submission of the QSL cards.

APA (All Province Award)

The APA is issued for confirmed contacts with HL stations located in each of the different special cities and provinces in Korea. The area codes for each city and/or province are as follows:

Area Code	Province and/or City
1	City of Seoul
2	Inchon City, Kyonggi-do, Kangwon-do
3	Chungchongnam-do, Chungchongbuk-do
4	Chollanam-do, Chollabuk-do, Cheju-do
5	Pusan City, Taegu City, Kyongsangnam-do, Kyongsangbuk-do

KUWAIT (9K)

The *Kuwait Award* is issued by the Kuwait Amateur Radio Society (KARS) for working 10 different 9K stations. There is no date limitation, any band/mode may be used, and all contacts must be made from the same location or within a radius of 100 km of the original location. Send 5 IRCs and certified list of contacts to: Awards Manager, Kuwait Amateur Radio Society, PO Box 5240, Safat, Kuwait, State of Kuwait.

LIBERIA (EL)

For awards sponsored by the Liberia Radio Amateur Association (LRAA), 20 IRCs (or $5 US) is required. QSL cards need not be submitted; instead, satisfactory evidence is a certified list of claimed contacts, signed by an officer of the applicant's National Society, or two amateurs of at least General class. This list must include each station worked, date, time, band, mode and signal report. Send all applications to: Awards Manager, Liberia Radio Amateur Association, PO Box 987, Monrovia, Liberia.

Worked All Liberia Award

Work and confirm two-way contacts with at least one amateur station in each of Liberia's nine counties on at least three amateur bands since April 1, 1964.

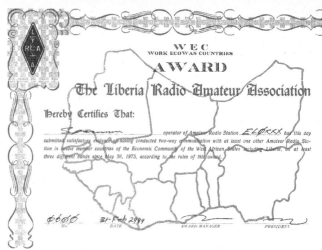

Six Counties Award

Work and confirm two-way contacts with at least one amateur station in six counties in Liberia on at least two different bands since April 1, 1964.

West African Countries Award

Work and confirm two-way contacts with at least one amateur station in eight West African countries and at least five amateur stations in Liberia on at least two different bands since January 1, 1962.

Work ECOWAS Countries Award

Work and confirm two-way contacts with at least one amateur station in each of the 16 member-countries of the Economic Order of West African States (ECOWAS), including Liberia, on at least three different bands since May 28, 1975. [ECOWAS countries: C5, EL, D4, J5, TU, TY, TZ, XT, 3C, 5N, 5T, 5U, 5V, 6W, 9G, 9L.]

LUXEMBOURG (LX)

Applications for awards issued by the Reseau Luxembourgeois des Amateurs d'Ondes Courtes (RL) should be sent to: Awards Manager, A. Engelhardt, LX2EA, RL, PO Box 1352, L-1013 Luxembourg.

LX Award

To commemorate 50 years of Amateur Radio activity in the Grand Duchy of Luxembourg, the LX Award is issued for contacts with LX amateurs since January 1, 1951. Applicants outside of Europe must make 20 points (contacts with LX stations on 160/80/40/30 meters count two points, and one point on 20/17/15/12/10 meters). Europeans must make 30 points (at least 20% on 160/80/40 meters).

An LX station can be worked once on each band in different modes. Fifteen points (10 for Europeans) can be counted for working the same LX station on five bands. No mode restrictions. No QSLs required; send a detailed list showing the calls of the LX stations worked, and date, time, band and mode, countersigned by two licensed amateurs or the Awards Manager of the National Society. Ten IRCs (or $4 US) is required for each award.

European Community Award

This is the official RL diploma to commemorate the 25th anniversary of the European Community. Each contact with a station with one of the European Community member-countries, made on or after the date of the country's entry into the European community (see list below), counts one point. Non-European applicants must make 50 points; each member-country must be worked at least once, and at least three LX stations must be worked (Europeans need 100 points, must work each member-country at least once, 5 LX stations, and no more than 10 contacts are allowed with stations in the applicant's own country). No band/mode restrictions.

Each station may be counted only once. No more than 20% of the QSO points may be obtained by contacts with the same member-country. A contact with special station LX0RL may replace a missing contact with any of the member-countries. Contacts made via active earthbound reflectors or repeaters are not valid. To apply, send a list of QSLs (confirmed by two licensed amateurs, by one club official or by a notary) and 10 IRCs (or $4 US).

Date of entry into the European Community by prefix:

March 25, 1957—DL, F, I, LX, ON, PA
January 1, 1973—EI, G, OZ
January 1, 1981—SV
January 1, 1986—CT, EA

MALAYSIA (9M)

The Malaysian Amateur Radio Transmitters' Society (MARTS) issues the *Worked All Malaysia Award* to amateurs who work and confirm 10 9M2 contacts, and one each with 9M6 and 9M8. The award and endorsements are issued for two-way SSB, two-way RTTY, two-way CW, mixed or single band, mixed or single mode. All contacts must be made starting August 31, 1957. Send a certified list of QSLs and 10 IRCs ($5 US) to: MARTS Awards Manager, Eshee Razak, 9M2FK, PO Box 13, Penang, Malaysia.

MALTA (9H)

To apply for diplomas issued by the Malta Amateur Radio League (MARL), send a certified list of QSL cards and 15 IRCs (or $3 US) to: Malta Amateur Radio League (MARL), PO Box 575, Valletta, Malta.
[Note: For European applicants, send 12 IRCs or $2 US]

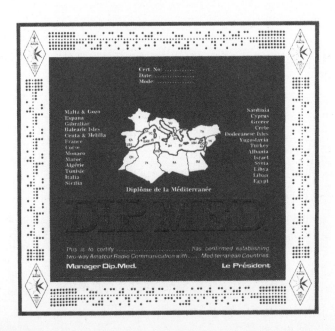

Dip-MED Award

Work and confirm a minimum of 15 of the 26 Mediterranean countries (see list below). A 9H QSO is mandatory.
[Mediterranean countries, by prefix: CN, EA, EA6, EA9, F, I, IS, IT, OD, SU, SV, SV5, SV9, TA, TK, TZ, YK, YU, ZA, ZB, 3A, 3V, 4X, 5A, 5B, 7X.]

9H Diploma

To qualify for the 9H Diploma, applicants outside Europe need 5 points (Europeans need 10 points). Each QSO with a 9H1 station counts one point, each QSO with a 9H4 counts two points. A QSO with club station 9H1MRL counts two points.

Golden Jubilee Award

To qualify, it is necessary to work four Maltese stations.

MONACO (3A)

The Association des Radio-Amateurs de Monaco (ARM) sponsors the *Principality of Monaco Award* for working and confirming contacts with three amateur stations in Monaco. Only contacts with permanent stations in Monaco are valid, and contacts must be made after January 1, 1980. All bands/modes, except repeater contacts. Send photocopy of log or the appropriate QSL cards, along with 10 IRCs (or $6

US) to: Awards Manager, ARM, PO Box 2, MC 98001, Monaco.

NETHERLANDS (PA)

Applications for awards issued by VERON (Vereniging voor Experimenteel Radio Onderzoek in Nederland) should be sent to: Traffic Bureau, VERON, A. Sanderse, PAØMOD, Obdammerdijk 2, 1713 RA Obdam, The Netherlands.

PACC

Work and confirm contacts with 100 different PA/PI stations. Endorsement stickers are available for 200-900 contacts. Contacts must be made June 1, 1945 and after. Certified list of QSLs and 8 IRCs required.

PAMC

Same rules as above, except 1000 different contacts required.

QSL Region Certificate

Work and confirm each of the 50 Dutch QSL Regions on the HF bands. Contacts September 1, 1980 and after are valid, and the QSL Region has to be mentioned on the QSL card. Certified list and 8 IRCs required.

NETHERLANDS ANTILLES (PJ)

Vereniging voor Experimenteel Radio Onderzoek in Nederlandse Antillen (VERONA) sponsors the *Curacao Certificate*, issued to any amateur who works and confirms contacts with at least three licensed amateurs on the island

of Curacao. All contacts after January 1, 1967 are valid. All bands/modes, signal reports must show at least readibility 3. The applicant can make the necessary contacts from any location in the same call area. Send certified list of QSOs (certified by two officers of a local Amateur Radio club or by three licensed amateurs in the absence of a club) and $3 US to: VERONA, PO Box 383, Curacao, Netherlands Antilles.

NEW ZEALAND (ZL)

To apply for these many fine awards sponsored by the New Zealand Association of Radio Transmitters (NZART), QSL cards are not required as NZART stresses the Honor System. It is sufficient merely to certify that the QSL was legitimately made. $1 US is required for each award (except WAP and 5X5 which are $2); 1 IRC and a self-addressed envelope required for each separate endorsement. Appropriate application forms may be obtained from the Awards Manager for 1 IRC. Please address all applications and correspondence to: NZART Awards Manager, Jock White, ZL2GX, 152 Lytton Rd, Gisborne, New Zealand.

ENZART

This prestige award, to honor NZART, is issued in three classes:

Basic Award—200 points (endorsements available for 250, 300, and 350 points).

Honour Award—400 points (with endorsements for each additional 50 points).

Honour Plaque—500 points.

It is essential to clearly determine that the operator contacted is a financial member of NZART and on the voting roll of the branch concerned. Points are allocated as follows: Phone QSO, 1 point; CW QSO, 2 points; club station, 2 points; YL operator, 1 point; mobile QSO, 1 point; AMSAT QSO, 1 point; Branch President, 1 point. All contacts must be made November 1, 1976 or later. Contacts claimed with a Branch Station must be from a club-room or other permanent headquarters, or alternatively when the station is being operated in a contest, from an exhibition, or other activity of a similar kind. (Contacts as a result of call-sign "swapping" at a member's QTH are not eligible.) Contacts with mobiles are valid only when the mobile station is within 30 miles of his Branch's club-rooms or his own QTH. The same station may be claimed in several point categories, but each must be for a separate contact, and the contacts must be not less than 24 hours apart. Contacts must be "phone to phone" or "CW to CW," and contacts made on awards nets are not valid.

ZL applicants must be members of NZART, and all applicants must certify that each contact made fulfills award requirements. No QSL cards need be submitted; applications must be made on the special NZART Check-list (available for 10 cents and a large SASE). Each certificate requires $1 US; endorsements are free, but an SASE is required.

WAZL (Worked All New Zealand)

To qualify for the WAZL award, work 35 branches of NZART (ZLs must work 45). Special endorsements are available if WAZL is completed within a 12-month period. Note: Mobiles operating out of their own Branch area must use the Branch number of the area in which they are operating.

NZART Branches

01 Ashburton
02 Auckland
03 Western Suburbs
04 Cambridge
05 Christchurch
06 Dannevirke
07
08 East Southland
09 Egmont
10 Franklin
11 Gisborne
12 Hamilton
13 Hastings-Havelock Nth
14 Hawera
15 Hawke's Bay Central
16 Horowhenua
17 Huntly
18 Hutt Valley
19 Inglewood
20 Manawatu
21 Manakau
22 Marlborough
23 Marton
24 Motueka
25 Napier
26 Nelson
27 New Plymouth
28 Northland
29 North Shore
30 Otago
31 Pahiatua
32 Rahotu Coastal
33 Rotorua
34 South Cantebury
35 South Otago
36 South Westland
37 Southland
38 Taumarunui
39 Tauranga
40 Te Awamuta
41 Thames Valley
42 Titahi Bay
43 Waihi
44 Matamata Radio Club
45 Waimarino
46 Wairarapa
47 Waitara
48 Wanganui
49 Westland
50 Wellington
51 Eastern Bay of Plenty
52 Northern Hawkes Bay
53 Te Puke
54 Patea
55 Waitomo
56 Christchurch West
57 Tokoroa
58 Helensville
59 Mangakino
60 Taupo
61 Central Otago
62 Reefton-Buller
63 Upper Hutt
64 North Otago
65 Papakura
66 Auckland VHF
67 Kawerau
68 North Canterbury
69 Kapiti
70 Fielding
71 Rodney
72 Opotiki
73 North Wairoa
74 Wellington VHF
75 Queenstown
76 Kaikoura
77 Te Aroha & Districts
78 Far North
79 Howick & Districts
80 Hibiscus Coast
81 Waikato VHF Group
82 Southern Wairarapa

NZA (New Zealand Award)

The NZA is available to all radio amateurs other than ZLs. Required contacts: 35 contacts with ZL1, plus 35 contacts with ZL2, plus 20 contacts with ZL3, plus 10 contacts with ZL4,

plus one contact with a ZL "territory" (either NZ Antarctica, Chatham, Kermadec or Campbell). This makes a minimum of 101 required contacts. Twenty ordinary ZL contacts can be substituted for the territory contact, if desired. All contacts must be dated from December 8, 1945.

NZC (New Zealand Counties)

The basic NZC award requires contacts with 20 different New Zealand counties. Endorsements for 40, 60, 80 and 100 counties are available, with a special certificate for all 112 counties. A counties map and checking sheet are available from NZART. Initial certificate with any or all endorsements to 100 costs $1 US. Separate endorsements require 1 IRC plus self-addressed envelope. The special checking sheet must be used and are available from NZART for 1 IRC. The "NZC 224," to recognize the outstanding achievement of "double 112" (different stations contacted in the relevant counties for each NZC 112), is available to those who already hold two NZC 112 awards. Cost of the plaque is $12.

5X5 (Five by Five)

This premier award has been instituted to recognize the increasing interest in five-band DX operation. The initial award requires that the same station be contacted on five bands; first endorsement after another five has been contacted (a total of 10), with the 20 endorsement requiring another 10, and so on to 100. Certified list with full QSO data and $2

(for initial application and for each endorsement) required. Contacts must date from 1945.

NZLA (NZ Lakeside Award)

To qualify for the basic award, contacts must be made with stations operating on the shores of 10 freshwater lakes in New Zealand. Endorsements for each additional lakes up to 40; Honour Award for 50 lakes, with further endorsements for each additional 10. Stations worked must be located within one kilometer of the water or in a town on the lake shore. Complete rules are with the check sheets (available from ZL2GX for an IRC and a self-addressed envelope) which must be used when applying. Contacts must be dated May 1, 1976 or later.

WAP (Worked All Pacific)

Contacts required with 30 different Oceania countries (as indicated on the ARRL DXCC List). Fee $1.50 US.

Five Band WAP Plaque

To qualify, contact any 30 Oceania countries, each on five different bands, for a total of 150 contacts. $8 US required.

ZLA Award

To qualify, contacts are required with Auckland City/ZL1, Wellington City/ZL2, Christchurch/ZL3, Dunedin City/ZL4, Antarctica/ZL5, Chatham Island/ZL7, Kermadec Island/ZL8 and Campbell Island/ZL9.

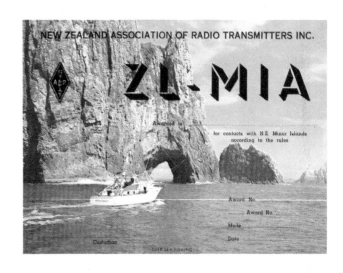

MIA (Minor Islands Award)

To qualify, five contacts with minor islands off the New Zealand coast are required (eg, Kapiti Island, Portland Island, Great Barrier Island, Little Barrier Island, Waiheke Island, D'Urville Island, Rangitoto Island, Kawau Island, Motutapu Island). The so-called islands in the Auckland area which are connected to the mainland by a causeway are not eligible.

ZL Awards

ZL1—the basic award requires contacts with 125 ZL1 stations in the post-war era. Endorsements are available for 175 and 250.

ZL2—the basic award requires contacts with 100 different ZL2 stations in the post-war era. Endorsements available for 150 and 200.

ZL3—the basic award requires contacts with 50 different ZL3 stations in the post-war era. Endorsements available for 75 and 100.

ZL4—the basic award requires contacts with 25 different ZL4 stations in the post-war ear. Endorsements available for 35 and 50.

[Certificates can be endorsed for any band/mode.]

Tiki Award

To qualify, work five different ZLs, each on five different bands, in the post-war era.

CJC (Captain James Cook Award)

To perpetuate the memory of this world-famous navigator and seaman, the CJC award is issued in three classes:

•Basic "Sailor Class"—requires contacts with G in Yorkshire, FO in Fr Polynesia, ZL2, VK2, KH6.

•"Officer Class"—requires all "Sailor Class" contacts plus ZL1, ZL3, ZL4, VK3, VK4, P2 and any Antarctic station.

•"Command Class"—requires all "Officer Class" contacts plus any five of the following, VE2, VO, YJ, FK, CE8, KL7, A3.

Antipodes Award

This award recognizes all-round achievement as follows:
1) *Technical*

(a) The applicant's station (receiver, transmitter, antenna) must be home built, completely of amateur design and construction, which the applicant has personally developed and constructed.

(b) Commercially constructed equipment or kits from any source are not valid.

2) *Operating*

The applicant must make three QSOs with stations in the general "Antipodes continental area" (the opposite side of the Earth) from the applicant.

[Note: The following awards are not officially sponsored by NZART but are supported by NZART branches.]

IARU Region 3 Operating Award

Contacts made after April 5, 1982, are valid; a certified list in lieu of QSL cards required. The basic award requires seven countries, the Silver Star endorsement requires 15 countries, and the Gold Star endorsement requires 20 countries. Any mode/band endorsement is available. Send $1 US ($2 airmail) and application to ZL2GX. Eligible countries: A3, DU, H4, HL, HS, JA, KH1/2/8/9/Ø, P2, VK, VS6, VU, YB, ZL, 3D2, 4S, 5W, 9M, 9V. Applications to NZART Awards Manager ZL2GX (see address above).

Regional Awards

Auckland Branch Certificate

For overseas stations only. To qualify, work eight members of the Auckland Branch (02) since January 1, 1957. Branch station ZL1AA counts as two members. No QSLs required. (Note: not all Auckland stations are members of the Auckland Branch.) Send certified list and $1 US to Awards Manager, Auckland Branch, Box 18-003, Glen Innes, Auckland, New Zealand.

Worked New Zealand Novices

For the Basic Award, QSO 10 ZL Novices, including one each from ZL1, ZL2, ZL3, ZL4. Endorsements for 25 (two QSOs from each call area), 50 (five QSOs from each call area), 100 (ten QSOs from each call area). All contacts must be made on or after October 1, 1979. Certificates may be endorsed all CW or phone. No QSLs required; send certified list with relevant QSO data. [Note: The ZL Novice band is 3525-3575 kHz.] Send applications to Awards Manager, Auckland Branch (see address above). One dollar for each certificate required, no charge for endorsements.

Christchurch Award

Work five stations in Christchurch. Award fee: $1 US. Applications to Awards Manager, Box 1733, Christchurch, New Zealand.

Bay of Plenty Award

Work five stations in the Bay of Plenty area (Opotiki, Rotorua, Tauranga, Whakatane). Send certified list of QSOs and 8 IRCs to ZL1BJP, 36 Western Rd, Otumoetai, Tauranga, New Zealand.

Horowhenua Award

Work stations in the Branch 15 area (Levin, Otaki, Waikanae, Shannon) with bonus points in certain categories. Involved rules; contact E. Tombs, ZL2IG, Ihakara, RD 1, Levin, New Zealand for details. Please include self-addressed envelope and 1 IRC.

Kapiti Award

Work stations in the Kapiti Branch (69) as follows: ITU Region 1 applicants need five contacts, Region 2 applicants seven contacts, and Region 3 applicants nine contacts [except for VK/ZL, which requires 15 contacts on 160 through 10 meters]. Contacts with club station ZL2KB count double (two contacts). Net and repeater contacts invalid. One contact with any one club member permitted. Special endorsements for single band, all phone, all CW. Contacts must be made January 1, 1976 or later. Certified list and 3 IRCs (or 50 cents US) to Box 1608, Paraparaumu Beach, New Zealand.

Manawatu Award

To qualify, work five stations who are members of the Manawatu Branch of NZART. Certified list and 5 IRCs to ZL2AFT, 431 Albert St, Palmerston, New Zealand.

Papakura Radio Club "21" Award

To qualify, work 20 members of the Papakura Radio Club, plus ZL1VK. Contacts must date from January 1, 1977. Band/mode endorsements. Apply to Ms C. Johnson, ZL1AJL, 63 Red Hill, Papakura, New Zealand.

Gisborne Award

To qualify, work two stations in Gisborne (Branch 11). Apply to ZL2GX (see address above).

NIGERIA (5N)

Unless otherwise noted, contacts for these awards sponsored by the Nigerian Amateur Radio Society (NARS) must be made on or after January 1, 1980. To apply, send a certified log (that must be verified by the applicant's National

Society) and 10 IRCs (or $4 US) to: Awards Manager, NARS, PO Box 2873, Lagos, Nigeria.

WNS (Worked All Nigeria States Award)

To qualify, the applicant must work one amateur station from each of the 19 states in the Federal Republic of Nigeria:

5N0—Lagos State and Federal Territory Abuja
5N1—Ogun State and Oyo State
5N2—Kwara State and Niger State
5N3—Ondo State and Bendel State
5N4—Anambra State and Rivers State
5N5—Imo State and Cross River State
5N6—Benue State and Plateau State
5N7—Bauchi State and Gongola State
5N8—Bornu State and Kano State
5N9—Kaduna State and Sokoto State

5NDX (Outstanding Nigeria DX Award)

This award is issued in three classes:

First Class—work 100 5N stations.
Second Class—work 50 5N stations.
Third Class—work 20 5N stations.

NARS President's station (5N0AAJ) counts as five stations, XYL stations count as two stations and club stations count as two stations.

NOA (Nigerian October Award)

The month of October has special significance in Nigerian history, and in commemoration, 5N amateurs are authorized to use a special prefix every year in the month of October (eg, 5N20 for 1980, 5N21 for 1981, 5N27 for 1987). To qualify for the award, work at least five different 5N stations using the special prefix in the month of October of that particular year. This award is effective October 1, 1980.

WNZ (Worked All Nigeria Zones Award)

To qualify, work one station in each of the 10 5N call areas. Usual rules apply.

NORWAY (LA)

The following awards are sponsored by the Norsk Radio Relae Liga [Norwegian Radio Relay League], NRRL:

WALA (Worked All Norway)

Contacts with LA and LB stations made after January 1, 1950, are valid for this award. Applicants must work and confirm contacts with each of the 19 Norwegian counties ("flykers") on any band (Scandinavian applicants must make two contacts, on separate bands, with each county). All contacts must be made from the same QTH, within a radius of 100 km. Contacts via repeaters or satellites are not valid. Contacts may be made on all legal modes and may be endorsed as appropriate. Cross-band contacts are not valid. Contacts with Arctic stations (JW or JX) count and may be substituted for W, X and/or Y counties. Applications should include a list of stations worked and QSL cards or a certified list of QSLs (verified by the applicant's National Society) that includes date and time in UTC, call sign, signal report and QTH of station worked, along with other relevant information in the case of endorsements. Send application and 10 IRCs (or $2 US) to either the Norwegian Radio Relay League, PO Box 21 Refstad, Oslo 5, Norway, or NRRL Awards Manager

Erik Jahnsen, LA7AJ, Kaupangruta 21, N-3250 Larvik, Norway.

Norwegian Counties:

A	Oslo	S	Sogn ug Fjordane
B	Ostfold	T	More og Romsdal
C	Akershus	U	Sor-Trondelag
D	Hedmark	V	Nord-Trondelag
E	Oppland	W	Nordland
F	Buskerud	X	Troms
Z	Vestfold	Y	Finnmark
H	Telemark	JW	Svalbard/Bear Island
I	Aust-Agder	JX	Jan Mayen
K	Vest-Agder		
L	Rogaland		

WNC (Worked Norwegian Cities)

The WNC certificate is issued in three classes:

Class 3—work 5 cities (Europeans, 10 cities)
Class 2—work 10 cities (Europeans, 20 cities)
Class 1—work 15 cities (Europeans, 30 cities)

No date, band or mode limitations, but QSOs with LF, LH and LJ stations are not valid. Send a certified list verified by two amateurs and 10 IRCs (or $2 US) to Larvik Society of NRRL, Awards Manager, PO Box 59, N-3251 Larvik, Norway.

Valid Norwegian Cities:

Arendal	Kristiansand S
Bergen	Kristiansand N
Bodo	Kragero
Drammen	Larvik
Egersund	Lillehammer
Fredrikstad	Mandal
Gjovik	Molde
Grimstad	Mosjoen
Hammerfest	Moss
Halden	Mo i Rana
Hamar	Namsos
Harstad	Narvik
Haugesund	Notodden
Holmestrand	Oslo
Horten	Porsgrunn
Kongsberg	Sarpsborg
Kongsvinger	Sandnes

Sandefjord	Tonsberg
Stavanger	Tromso
Skien	Vardo
Steinkjer	Alesund
Trondheim	

W 100 LA (Worked One Hundred LA Amateurs)

To qualify, applicants must work 100 different LA/LB stations. Contacts made after January 1, 1984 are valid. The awards is issued for CW, phone or mixed modes. All valid amateur bands may be used (10, 18 and 24 MHz will not be available before January 1, 1989). LF, LH and LJ stations do not count. A list showing full details of the contacts should be certified by the Awards Manager of the applicant's National Society. The application must contain call sign, date, band, RST and modes, and must be accompanied by 10 IRCs (or $2 US). Send applications to Awards Manager, Stavangergruppen av NRRL, PO Box 354, 4001 Stavanger, Norway.

PAPUA NEW GUINEA (P2)

The Papua New Guinea Amateur Radio Society (PNGARS) sponsors *the Bird of Paradise Award* to encourage contacts with amateurs in Papua New Guinea, to advance the art of radio communication, and to help foster international friendship and goodwill. Applicants outside of Oceania must contact at least five amateurs in P2, with at least three of them located in different provinces. Amateurs in Oceania (ITU zones 27-32) must contact seven stations in P2, at least five located in different provinces. The National Capital District (Port Moresby) is a separate province. The PNGARS official station, P29PNG, which is only activated on special occasions, may be substituted for any province. Contacts must be made after September 16, 1975 (the date of independence). Any band/mode; special endorsements for specific bands and modes are available if specifically requested. QSL cards are not required; send certified long entries and $3 US to offset printing and postage costs.

The PNG provinces are as follows:

National Capital District	Chimbu
East New Britain	Gulf
West New Britain	Manus
Southern Highlands	Madang
Eastern Highlands	Northern

Western Highlands	Western
New Ireland	East Sepik
Moroba	Milne Bay
Central	North Solomons
Enge	West Sepik

Send all applications to: The Awards Committee, PNGARS, PO Box 204, Port Moresby, Papua New Guinea.

PARAGUAY (ZP)

To qualify for awards sponsored by the Radio Club Paraguayo, all contacts must be made after May 15, 1952. Send a certified list of QSLs and 5 IRCs to: Radio Club Paraguayo, Awards Manager, PO Box 512, Asuncion, Paraguay.

All Mediterranean Counties Awards

This award is issued for confirmed contacts with Mediterranean countries as follows:

Class A—41 countries
Class B—30 countries
Class C—20 countries

A ZP contact is mandatory for each class.
Countries: A2, A5, C3, CP, HA, HB, HBØ, HV, JT, LX, OE, OK, T7, TL, TT, TZ, UC, UD, UG, UH, UI, UJ, UL, UM, UO, XT, XW, YA, Z2, ZP, 3D6, 4U1ITU, 5U, 5X, 7P, 8Q, 9J, 9N, 9U, 9X.

Tropics of Cancer and Capricorn Award

This award is issued for confirmed contacts with countries touched by the Tropic of Cancer and Capricorn as follows:

Class A—28 countries
Class B—20 countries
Class C—12 countries

A ZP contact is mandatory for each class.
Countries: Tropic of Cancer—A4, A6, BV, BY, C6, KH6, S2, SU, TZ, VU, XE, XZ, 5A, 5T, 5U, 7X, 7Z. Tropic of Capricorn—A2, CE, C9, LU, PY, VK, ZP, ZS, ZS3, 5R.

All Zone 11 Prefixes Award

This award is issued for confirmed contacts with stations in CQ Zone 11 as follows:

Class A—30 different prefixes
Class B—19 different prefixes
Class C—12 different prefixes

[Prefixes consist of ZP1-ZP9, PY1-PYØ, and the special prefixes issued for the *CQ* WPX Contest.]

Diploma Sud-America

This award is issued for confirmed contacts with countries located in ITU Zones 12, 13, 14, 15, 16 and 73 as follows:

Class A—33 countries and six ITU zones
Class B—25 countries and six ITU zones
Class C—18 countries and five ITU zones

A ZP contact is mandatory for each class.
Countries:
Zone 12—CP1, CP8, CP9, FY, HC, HC8, HK, HKØ/Malpelo, OA, PZ, YV, 8R
Zone 13—PY6, PY7, PY8, PYØ/Fernando, PYØ/St Peter & Paul

Zone 14—CE1, CE2, CE3, CE4, CE5, CEØ/San Felix, CEØ/Juan Fernandez, CP2, CP3, CP4, CP5, CP6, CP7, CX, LU-A/U/Y, ZP
Zone 15—PY1, PY2, PY3, PY4, PY5, PY9, PYØ/Trindade
Zone 16—CE6, CE7, CE8, VP8/Falklands, LU-V-W/X
Zone 73—KC4USP/Palmer, LU-Z, CE9AA/AM, VP8/S Georgia, VP8/S Orkney, VP8/S Sandwich, VP8/S Shetland

DP (Diploma Paraguay)

This award is issued for confirmed contacts with five different ZP stations (South Americans must contact 15 ZP stations).

WAZP (Worked All ZP)

This award is issued for confirmed contacts with one station in each of the nine ZP call areas.

CRCP (Certificado Radio Club Paraguayo)

This award is issued for confirmed contacts with 15 different ZP stations (South Americans must contact 50 different stations).

ZP 100, ZP 150, and ZP 200 Awards

These awards are issued for confirmed contacts with 100, 150 and 200 different ZP stations.

ZP3 Award

This award is issued for confirmed contacts with two different ZP3s (CE, CX, LU and PY must work five different ZP3s).

DDP (Diploma Departamentos del Paraguay)

This award is issued for confirmed contacts with the capital of Paraguay and the different departments in which Paraguay is divided, as follows:

Class A—20 contacts
Class B—16 contacts
Class C—12 contacts

Paraguay Departments, by prefix:

ZP1—Boqueron, Chaco, Nueva Asuncion
ZP2—Alto Paraguay, Pte Hayes
ZP3—Amambay, Concepcion
ZP4—Canendiyu, San Pedro
ZP5—Asuncion (nation's capital)
ZP6—Central, Cordilleras, Paraguari
ZP7—Caaguazu, Caazapa, Guaira
ZP8—Misiones, Neembucu
ZP9—Alto Parana, Itapua

PHILIPPINES (DU)

The Philippine Amateur Radio Association (PARA) sponsors the UN-DU Award in commemoration of the formation of the United Nations in 1945, of which the Philippines is a charter member.

For US applicants, the procedure is as follows:

The applicant must work and confirm 100 or more QSOs with amateurs in member-nations of the UN, including the applicant's own country. All QSOs must be dated on or after the date of the particular country's admission to the UN.

Send K6EDV the original QSL cards, photocopies of the QSL cards and an alphabetical list to be certified by K6EDV,

Pete Peterson, 845 Ramona Dr, Santa Rosa CA 95404. [Application forms, including the UN admission date for each country, are available from K6EDV for an SASE.] All material sent to K6EDV must include sufficient return postage; all submitted QSL cards shall be returned by K6EDV after they have been checked and the application has been certified. When returned, the applicant then sends the alphabetical list and the photocopies of the QSL cards to the Philippine Amateur Radio Association, PO Box 4083, Manila, Philippines. American amateurs can send the original QSLs to PARA if desired; all other amateurs must send original QSLs directly to PARA.

The fee for the UN-DU Award is $16 US, which also covers return air mail postage. This amount must be by certified check or postal money order, made out ot PARA, and sent directly to PARA with the application.

The initial application is for 100 countries; endorsements can be submitted to K6EDV in groups of five. The original QSLs and photocopies are required when applying for endorsements.

The UN-DU Award is granted for mixed modes and for single modes (SSB, CW, RTTY, SSTV, satellite).

POLAND (SP)

To apply for awards issued by the Polski Zwiazek Krotkofalowcow, PZK, please include 10 IRCs for each award to cover the cost of each certificate and postage. All contacts, mixed or single band, mixed or single mode, are valid (exception—satellite and/or repeater contacts are invalid). To apply, send a certified list of QSLs (in lieu of QSL cards) and the IRCs to: PZK Awards Manager, PO Box 320, PL 00-950 Warszawa 1, Poland.

The POLSKA Award

This award is available in three classes:

Class 1—for confirmed contacts with all provinces (49) in Poland.
Class 2—for confirmed contacts with 35 provinces.
Class 3—for confirmed contacts with 20 provinces.

All contacts are valid from June 1, 1975, and must be confirmed by QSL cards. The applicant's QSL list must be in alphabetical order by province abbreviation (see below). When applying for a higher class, please supply the number of your

diploma of the next lower class and a list of your additional contacts.

Abbreviations, and prefixes, of the 49 Polish provinces:

BB/SP9	KI/SP7	OL/SP4	SK/SP7	ZG/SP3
BK/SP4	KL/SP3	OP/SP6	SL/SP1	
BP/SP8	KN/SP3	OS/SP5	SU/SP4	
BY/SP2	KO/SP1	PI/SP3	SZ/SP1	
CH/SP8	KR/SP9	PL/SP5	TA/SP9	
CI/SP5	KS/SP8	PO/SP3	TG/SP7	
CZ/SP9	LD/SP7	PR/SP8	TO/SP2	
EL/SP2	LE/SP7	PT/SP7	WA/SP5	
GD/SP2	LG/SP6	RA/SP7	WB/SP6	
GO/SP3	LO/SP4	RZ/SP8	WL/SP2	
JG/SP6	LU/SP8	SE/SP5	WR/SP6	
KA/SP9	NS/SP9	SI/SP7	ZA/SP8	

The AC 15 Z Award

This award is issued for working and confirming contacts with at least 23 countries/call areas located in *CQ* Zone 15: HA, HV, I, IS, IT, OE (two call areas), OH (three call areas), OHØ, OJØ, OK, SP (four call areas), T7, TK, UA2, UP, UQ, UR, YU (three call areas), ZA, 9H.

Contacts with four call areas in Poland are mandatory. All contacts are valid from January 1, 1985. The list of QSLs must be in alphabetical order by country.

The W 21 M Award

This award is issued for working and confirming contacts with at least 16 countries situated on the meridian 21 east: A2, D2, JW, LA, OH, OHØ, OK, SP, SV, TL, TT, UA2, UP, UQ, UR, YO, YU, ZS, ZS3, 5A, 9Q. Contacts with Poland are mandatory. All contacts are valid from January 1, 1985. The list of QSLs must be in alphabetical order by country.

SP DX Award

The SP DX Club issues a certificate attesting to honorary SP DX Club membership. To qualify, work 10 (15 for European applicants) regular SPDX members after October 1, 1959. Send certified list and 10 IRCs to: SP DX Club of PZK, Awards Manager SP9PT, PO Box 131, 44-201 Rybnik, Poland.

ROMANIA (YO)

Awards issued by the Federatia Romana de Radioamatorism (Romanian Radioamateur Federation), FRR, are issued for CW, AM, SSB, RTTY or mixed modes; awards are also issued for single bands (80-10 meters) and for any combination of bands. Contacts are valid after August 23, 1949. [Note: Special requirements apply to European applicants; contact FRR for details.] Applications must include a certified list of QSLs (no cards) and 7 IRCs sent to: Romanian Radioamateur Federation, PO Box 22-50, R-71100 Bucharest, Romania.

Romania Award

This award is issued for working at least 30 different YO stations, each in a different YO county, Bucharest (capital city) included. All eight YO districts (YO2-YO9) must also be represented.

County		District	County		District
AB	Alba	YO5	HR	Harghita	YO6
AR	Arad	YO2	IL	Ialomita	YO9
AG	Arges	YO7	IS	Iasi	YO8

County		District	County		District
BC	Bacau	YO8	MM	Maramures	YO5
BH	Bihor	YO5	MH	Mehedinti	YO5
BN	Bistrita-Nasaud	YO5	MS	Mures	YO6
BT	Botosani	YO8	NT	Neamt	YO8
BR	Braila	YO4	OT	Olt	YO7
BV	Brasov	YO6	PH	Prahova	YO9
BZ	Buzau	YO9	SJ	Salaj	YO5
CL	Calarasi	YO9	SM	Satu Mare	YO5
CS	Caras-Severin	YO2	SB	Sibiu	YO6
CJ	Cluj	YO5	SV	Suceava	YO8
CT	Constanta	YO4	TR	Teleorman	YO9
CV	Covasna	YO6	TM	Timis	YO2
DB	Dimbovita	YO9	TL	Tulcea	YO4
DJ	Dolj	YO7	VS	Vaslui	YO8
GL	Galati	YO4	VL	Vilcea	YO7
GJ	Gorj	YO7	VR	Vrancea	YO4
GR	Giurgiu	YO9			
HD	Hunedoara	YO2			

Bucuresti Award

This award is issued for working 10 different YO3 stations. The Jubiliar Bucuresti award is issued for working five different YO3 stations annually, between August 20 and October 30. During this time there are various activities celebrating the national day of Romania. All YO stations located in the city of Bucharest use the YO3 prefix. Application must be mailed by December 31 of the same year.

YO-AD (YO—All Districts)

This award is issued in three classes for working all YO districts:

Class I—work all eight YO districts, three QSOs/district
Class II—work six YO districts, two QSOs/district
Class III—work three YO districts, one QSO/district

YO-AM (YO—Alma Mater)

This award is issued for working YO stations located in the university centers of Romania. The following cities are university centers: Bucharest, Cluj, Timisoara, Brasov, Lasi, Constanta, Oradea, Baia Mare, Ploiesti, Sibiu, Bacau, Pitesti, Galati, Suceava, Petrosani, Hunedoara, Tirgu Mures, Craiova, Resita. North American applicants must make 10 points; contacts on 80 meters = 8 points, 40 meters = 3 points, 20 meters = 1 point, 15 meters = 2 points, 10 meters = 4 points. A station can be worked on more than one band.

YO-BZ (YO—Balkans Zone of Peace)

This award is issued for working stations in LZ, SV, TA, YO, YU and ZA in three classes:

Class I—work at least four countries, including 10 call districts and three YO call districts.
Class II—work at least three countries, including eight call districts and three YO call districts.
Class III—work at least two countries, including six call districts and three YO call districts.

YO-DC (YO—Double Call)

This award is issued for working 26 "double-letter" call sign suffixes using all letters from AA to ZZ respectively. This includes three-letter suffixes, such as AAT, BBZ, BCC. Applicants outside of Europe must work at least two YO stations with double calls.

YO-CM (YO—Chess Mate)

This award is issued for working all 64 stations as follows: On the first horizontal row of a chessboard, fill in the call signs of eight different YO stations you have worked. On each corresponding vertical row, fill in the call signs of seven other DX stations (YO okay) whose suffixes will include at least one similar letter, and always the same, from the suffix of the first listed YO call sign. For example, the first list YO station is YO3JP; on the following seven vertical squares, you might list YO3AAJ, JA1BJA, DL1CJ, YO3JW, I1SJ, SP5JJ, W8ACJ (the letter J being the "key" of this series). If the first YO call signs listed on the first horizontal row belong to YO DX Club members, you will be eligible for the special *YO—CM Master award*. In this case, you may use the same YO DX Club member call sign several times if you have worked it on several bands.

YO-DR (YO—Danube River)

This award is issued for working different stations located in countries along the River Danube (DL, HA, LZ, OE, OK, YO, YU, UA) on two bands. Two QSOs are required with each country except YO (three QSOs required). At least three QSOs must be with stations located in cities on the Danube River.

YO-DX-C (YO—DX Club Members)

This award is issued for working two YO DX Club members or honorary members. A membership list is available on request for 1 IRC. Working YODXC members during the annual YO DX Contest on the first weekend in August can qualify you as an honorary member.

YO-LC (YO—Large Cities)

This award is issued for working YO stations located in the large cities in three classes:

Class I—work 20 cities
Class II—work 10 cities
Class III—work 5 cities

YO large cities:
YO2—Arad, Deva, Hunedoara, Lugoj, Petrosani, Resita, Timisoara
YO3—Bucharest
YO4—Braila, Constanta, Focsani, Galati, Tecuci, Tulcea
YO5—Alba Iulia, Baia Mare, Bistrita, Cluj, Dej, Oradea, Satu Mare, Sighetul Marmatiei, Turda, Zalau
YO6—Brasaov, Medias, Miercurea Ciuc, Odorheiul Secuiesc, Sfintu Gheorghe, Sibu, Sighsoara, Targu Mures
YO7—Craiova, Pitesti, Rimnicu Vilcea, Slatina, Tirgu Jiu, Drobeta-Turnu Severin
YO8—Bacau, Birlad, Botosani, Lasi, Oras, Gheorghe Gheorghiu-Dej, Piatra Neamt, Roman, Suceava
YO9—Alexandria, Buzau, Calarasi, Giurgiu, Ploiesti, Slobozia, Tirgoviste, Turnu Magurele

YO-NC (YO—Namesake Calls)

This award is issued for working five different stations having the same one- or two-letter suffix as the applicant. For instance, KN5A may apply for the award if he worked LA1A, BV2A, HP1A, OZ3A, EL2A; N1JW may apply if he worked ON4JW, UA0JW, YO3YW, DL6JW, VK9JW. Stations with three-letter suffixes need to work only three different stations with the same suffix.

YO-5 On 5 (YO—5 Continents on 5 Bands)

This award is issued for working one station in each continent (other than the continent of the applicant) on five bands. One continent may be replaced by a YO station.

Single-Band Awards:
YO-10 × 10—work 10 different YO stations on 10 meters
YO-15 × 15—work 15 different YO stations on 15 meters
YO-20 × 20—work 20 different YO stations on 20 meters
YO-40 × 40—work 40 different YO stations on 40 meters
YO-80 × 80—work 80 different YO stations on 80 meters

YO-100

Work 100 different YO stations, regardless of band. This award is available in further multiples of 100 (200, 300...).

YO-20 Z (YO—Zone 20)

This award is issued in three classes for working Zone 20 countries: LZ, JY, OD, SV, TA, YK, YO, 4X, 5B/ZC.

Class I—work six countries
Class II—work four countries
Class III—work two countries

For all classes, a YO contact is mandatory.

YO-25 M (YO—25 Meridian)

This award is issued in three classes for working the following countries situated on 25 east meridian: A2, LA, LZ, OH, ST, SU, SV, TL, UA, YO, Z2, ZS, 5A, 9J, 9Q, 9U, 9X.

Class I—work 12 countries
Class II—work eight countries
Class III—work five countries

For all classes, a YO contact is mandatory.

YO-45 P (YO—45 Parallel)

This award is issued in three classes for working the following countries situated on 45 north parallel: BY, F, FP, I, JA, JT, UA, VE, W, YO, YU.

Class I—work eight countries
Class II—work six countries
Class III—work five countries
For all classes, a YO contact is mandatory

SENEGAL (6W)

The *Senegalese Award* is issued by the Association des Radio-Amateurs du Senegal (ARAS) to amateurs who contact five different 6W stations. The request for the award must be forwarded with either a certified list of QSL cards or photocopies of the QSL cards and 10 IRCs to: Awards Manager, 6W1JN, ARAS, PO Box 971, Dakar, Senegal.

SOUTH AFRICA (ZS)

To qualify for awards sponsored by the South African Radio League (SARL), the applicant must submit a list of QSLs endorsed by the Awards Manager of his National Society and 10 IRCs. The minimum CW report is 338, or R3 S3 on phone. Send applications to: Awards Manager, South African Radio League, PO Box 3911, 8000 Cape Town, South Africa.

AAA (All Africa Award)

This award is issued to amateurs working and confirming two-way contacts with each of the six ZS call areas (ZS1-6), as well as one contact with A2, 7P, 3D6, plus one contact each from 25 different areas of the remaining groups of country prefixes (this list is available from the SARL Awards Manager for a self-addressed envelope and 1 IRC). All

contacts must be made after November 1945, and must be land stations only (islands do not count).

WAZS (Worked All ZS Call Areas)

This award is issued to amateurs working and confirming two-way contacts with 100 South African amateur stations as follows: ZS1—16 contacts, ZS2—10 contacts, ZS3—three contacts, ZS4—10 contacts, ZS5—16 contacts, ZS6—45 contacts. Contacts may be made on CW, phone or mixed modes; endorsements are available for all CW or all phone. All contacts must be made since January 1, 1958.

SPAIN (EA)

To apply for awards sponsored by the Union de Radioaficionados Espanoles (URE), applications must be accompanied by a log sheet certified by the Awards Manager of an IARU Society and 5 IRCs unless otherwise noted. Send all applications to: URE, PO Box 220, 28080 Madrid, Spain.

Espana Award

To qualify, the applicant must meet the following requirements:

Phone—work 200 different EA stations, including nine call districts and in at least 25 provinces.

CW—work 100 different EA stations, including eight call districts and 20 provinces.

Mixed—work 125 different EA stations, including eight call districts and 20 provinces. At least 50 CW contacts required.

The same station can be counted only once. Repeater and

satellite contacts are not valid, nor are contacts with or from mobiles. Contacts must be made after January 1, 1952. Application must be accompanied by log sheet containing QSO details arranged in call district order and alphabetical order.

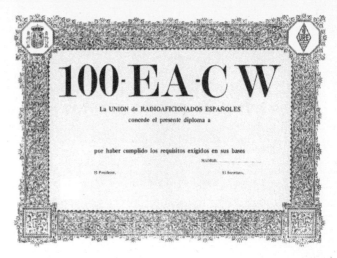

TPEA Award

This award is issued for working and confirming all 52 Spanish provinces. All HF bands; no mobile contacts. Repeater and satellite contacts are not acceptable. Contacts must be made after March 1, 1979. The special *5-Band TPEA Plaque* may be obtained by those amateurs contacting all 52 Spanish provinces on five bands (80-10 meters). Applications must be accompanied by the QSL cards and 25 IRCs in addition to the log sheet.

Provinces of Spain:
   EA1—Asturias (O), Avila (AV), Burgos (BU), Cantabria (S), La Coruna (C), Leon (LE), Lugo (LU), Orense (OR), Palencia (P), Pontevedra (PO), La Rioja (LO), Salamanca (SA), Segovia (SG), Soria (SO), Valladolid (VA), Zamora (ZA)
   EA2—Alava (VI), Guipuzcoa (SS), Huesca (HU), Navarra (NA), Teruel (TE), Vizcaya (BI), Zaragoza (Z)
   EA3—Barcelona (B), Gerona (GE), Lerida (L), Tarragona (T)
   EA4—Bardajoz (BA), Caceres (CC), Ciudad Real (CR), Cuenca (CU), Guadalajara (GU), Madrid (M), Toledo (TO)
   EA5—Albacete (AB), Alicante (A), Castellon (CS), Murcia (MU), Valencia (V)
   EA6—Baleares (PM)
   EA7—Almeria (AL), Cadiz (CA), Cordoba (CO), Granada (GR), Huelva (H), Jaen (J), Malaga (MA), Sevilla (SE)
   EA8—Las Palmas (GC), Tenerife (TF)
   EA9—Ceuta (CE), Melilla (ML)

CIA Award

This award is issued in two categories:
Gold—work 20 Iberoamerican countries, plus Portugal and Spain, for a total of 22 required QSOs.
Silver—work 15 Iberoamerican countries, plus Portugal and Spain, for a total of 17 required QSOs.
[Iberoamerican countries: CE, CO, CP, CX, HC, HI, HK, HP, HR, HT, KP4, LU, OA, PY, TG, TI, YS, YV, XE.]

100-EA-CW Award

This award is issued for making 100 points, as follows:
• Applicants in CQ Zones 14, 15, 16, 30 and 33 must make 100 CW contacts with EA stations (each QSO = one point).
• Applicants in CQ Zones 4, 5, 8, 9, 11, 17, 18, 21, 22, 34, 35, 36, and 37 must make 50 CW contacts with EA stations (each QSO = two points).
• Applicants in all other CQ Zones must make 25 CW QSOs (each QSO = four points).

Each applicant must make contacts in at least seven of the EA call districts, and the contacts must be made on at least three bands. The same station may be worked on different bands if there are three days between contacts (except during the EA DX CW Contest). Contacts are valid from January 1, 1966. A silver medal is awarded for making 500 points, and a gold medal is awarded for making 1000 points.

EA DX100 Award

To qualify for this award, the applicant must work at least 100 countries; contacts with EA, EA6, EA8 and EA9 are mandatory. Separate awards are issued for phone and CW; all HF bands can be used. Contacts are valid from April 1, 1949. All DXCC and WAE (see below) countries are valid; deleted countries are not valid. Endorsements are issued in 25 or 50 country increments; endorsements are issued in single increments once 300 accredited countries have been attained. QSL cards must accompany the log sheet (which must be filled out in alphabetical order).

TD-EA-CW Award

The Villarreal section of URE sponsors this award, which is available for working and confirming the nine EA call districts on CW (9 contacts). Contacts are valid from January 1, 1976, and can be made on any authorized amateur HF band. Contacts made through repeaters, satellites and the like are not valid; contacts made from or with mobiles are not valid. Contacts must be made from the same DXCC country with the same call sign. Endorsements: *5B-TD-EA-CW* (work all nine call districts on five bands, 45 contacts); *160-TD-EA-CW* (work all nine call districts on 160 meters). Send log and $3 (or the equivalent in IRCs) to Delegacion Local de URE, La Mura 67, Villarreal (Castellon), Spain.

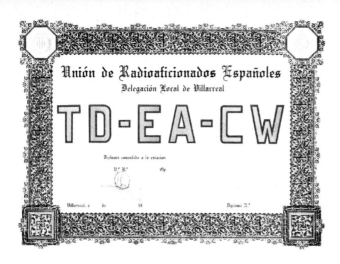

SRI LANKA (4S)

The Radio Society of Sri Lanka (RSSL) issues the *Worked 4S7 Award* for working 10 different 4S7 stations. Further details can be obtained from RSSL, PO Box 907, Colombo, Sri Lanka.

SURINAME (PZ)

The Vereniging van Radio Amateurs in Suriname, VRAS, issues one operating award for making two-way contacts with three PZ stations. Certified log data and 10 IRCs (or $5 US) for defrayment of postage and printing of the certificate should be sent to VRAS, PO Box 1153, Paramaribo, Suriname.

SWEDEN (SM)

The Foreningen Sveriges Sandareamatorer (SSA) sponsors the following operating awards:

WASM (Worked All SM)

Applicants must work one station in each of the eight Swedish call areas (European applicants must work two stations in each). The Swedish call areas are SM1 to SM7, inclusive, and SMØ. The contacts may be made with SM, SK and SL stations, and all stations have to be land-based. All contacts must be made after World War II. In lieu of QSL cards, a certified list of QSLs (checked by the Awards Manager of the applicant's IARU Society) may be submitted with the application. To cover expenses, please include 11 IRCs. Send applications to: SSA Diploma Manager, Ostmarksgatan 43, S-123 42 Farsta, Sweden.

WASM II (Worked All SM Laens)

This award, for working Swedish laens (counties), is issued in two classes:

Class A (LA, OH, OZ, SM amateurs only)—two-way contacts with amateurs in the 25 laens on two different bands (a total of 50 contacts) required.

Class B (rest of the world)—two-way contacts with amateurs in the 25 laens required (25 contacts). All contacts must have been made since January 1, 1953. Contacts can be made on any amateur band, and on CW or phone or any combination of both. Swedish laens are designated as follows:

Swedish Laens:

A—City of Stockholm (SM5 and SMØ)
B—Stockholm (SM5 and SMØ)
C—Uppsala (SM5)
D—Sodermanlands (SM5)
E—Ostergotlands (SM5)
F—Jonkopings (SM7)
G—Kronobergs (SM7)
H—Kalmar (SM7)
I—Gotlands (SM1)
K—Blekinge (SM7)
L—Kristianstads (SM7)
M—Malmohus (SM7)
N—Hallands (SM6)
O—Goteborgs och Bohus (SM6)
P—Alvsborgs (SM6)
R—Skaraborgs (SM6)
S—Varmlands (SM4)
T—Orebo (SM4)
U—Vastmanlands (SM5)
W—Kopparbergs (SM4)
X—Gavleborgs (SM3)
Y—Vasternorrlands (SM3)
Z—Jamtlands (SM3)
AC—Vasterbottens (SM2)
BD—Norrbottens (SM2)

Applications for WASM II must be accompanied by a verified list of claimed stations and 7 IRCs. Send applications to: WASM II Manager, SM6ID, Karl O. Friden, Morup 1084, 311 03 Langas, Sweden.

SCA (Swedish Communes Award)

In a parlimentary resolution in 1962, the Swedish government divided the country into 279 communes. The SCA is issued in six different classes to encourage two-way contacts with as many of the 279 communes as possible as follows:

- SCA-50 for 50 confirmed communes
- SCA-100 for 100 confirmed communes
- SCA-150 for 150 confirmed communes
- SCA-200 for 200 confirmed communes
- SCA-250 for 250 confirmed communes
- SCA-ALL for 279 confirmed communes

All contacts must be confirmed with QSL cards, and contacts with fixed, portable or mobile stations from January 1, 1971 (unless otherwise stated) will be valid.

Club SK5AJ has complied a commune Record book (which is the basis of the application for the award) containing complete rules, a systematically ordered commune list with space for complete QSO data, and application forms (including a space for certification from two licensed amateurs that they have inspected the QSLs). The Record Book is available for 5 IRCs from: Club SK5AJ, Box 46, S-591 01 Motala, Sweden.

When applying for the SCA, send the completed record book and 12 IRCs to SK5AJ at the address above.

Field Award

The Field Diploma is issued for verified contacts with amateurs in fields [grid locators] in Sweden, as defined by the Maidenhead locator system adopted as of January 1, 1985. Contacts on or after that date are valid for the diploma. The Field Award is issued in four classes:

Platinum—all 324
Gold—300 fields verified

Silver—200 fields verified
Bronze—100 fields verified

All bands/modes, no endorsements. QSLs should clearly state the grid locator or longitude/latitude. Send a certified list (verified by the applicant's National Society) and 10 IRCs (or $4 US) to: Field Award Manager, SSA, Ostmarksgatan 43, S-123 42 Farsta, Sweden.

[Note: *The ARRL Amateur Radio Grid Locator Atlas* is available from League HQ or your local radio bookshop for $4].

SWITZERLAND (HB)

The *Helvetia Award* is issued by the USKA (Union of Swiss Short Wave Amateurs) for working and confirming contacts with all 26 Swiss cantons since January 1, 1979. The award is issued in four categories: mixed phone and CW, CW only, RTTY, and SSTV. Cross-mode and cross-band contacts are not valid.

The cantons, preceded by their abbreviation, are as follows:

AG—Aargau	NW—Nidwalden
AI—Appenzell Inner Rhoden	OW—Obwalden
AR—Appenzell Outer Rhoden	SG—St Gall
BE—Berne	SH—Schaffhausen
BL—Basle-Country	SO—Solothurn
BS—Basle-City	SZ—Schwyz
FR—Fribourg	TG—Thurgau
GE—Geneva	TI—Ticino

GL—Glaris	UR—Uri
GR—Grisons	VD—Vaud
JU—Jura	VS—Valais
LU—Lucerne	ZG—Zug
NE—Neuchatel	ZH—Zurich

QSLs should indicate the canton and they should be sent, along with a list of the QSLs (with complete QSO data), to: Kurt Bindschedler, HB9MX, Strahleggweg 28, 8400 Winterthur, Switzerland.

The award is free, but sufficient IRCs should be included to cover return postage for QSLs.

THAILAND (HS)

The Radio Amateur Society of Thailand (RAST) sponsors the *Siam Award*. To qualify, contact 10 different HS stations, any mode. To apply, send certified list and $5 US (for postage) to: Awards Manager, Radio Amateur Society of Thailand, GPO Box 2008, Bangkok 10501, Thailand.

UNITED KINGDOM (G)

The Radio Society of Great Britain (RSGB) sponsors awards in the HF and VHF categories, administered by separate Awards Managers. The following general rules apply to HF awards:

Applicants from the UK, Channel Islands and Isle of Man must be RSGB members and, as proof of membership, should provide a recent address label from the RSGB journal, *Radio Communication*. All other applicants need not be members of RSGB, but should include a label if they are members. Each application from within the UK must be accompanied by QSL cards. All others must include QSLs in the case of a special plaque or cup; otherwise, a statement from the applicant's National Society, or a statement by two officers of a local society affiliated with the National Society, that the necessary cards have been checked will be accepted. Each application from a nonmember of RSGB must include 12 IRCs or $4 US per certificate. (At press time, RSGB was in the process of revising this fee schedule; write to the RSGB HF Awards Manager for details.) Applicants submitting QSL cards must also include sufficient payment to cover their return.

For RSGB HF Awards, all contacts must be made by the holder of the call sign on bands below 30 MHz. Contacts may be made from any location in the same geographical call area or, if no such area exists, then from the same country. Except where otherwise indicated, credit will be given for confirmed contacts made on or after November 15, 1945. Contacts with land-mobile stations will be accepted provided that the exact location of each station at the time of contact is clearly stated on the evidence submitted. By decision of the RSGB's HF Committee, credit will not be given for contacts made on the 10, 18 and 24-MHz bands. (This decision will be reviewed when the bands become freely available to amateurs worldwide and restricted power limits are removed.) Credit will be given for contacts made entirely on one mode or a combination of modes. Certificate endorsements for single modes and/or single bands are available. No credit for cross-band or cross-mode contacts.

Send the appropriate fee, QSL cards (if necessary) and application to: RSGB HF Awards Manager, Peter Miles, G3KDB, PO Box 73, Lichfield, Staffs WS13 6UJ, England.

Additional information on HF awards as well as a 5-Band Commonwealth checklist can be obtained by sending a self-addressed envelope to: Radio Society of Great Britain (RSGB), Lambda House, Cranborne Road, Potters Bar, Herts, EN6 3JW, England.

CCC (Commonwealth Century Club)

To apply, work and confirm contacts with amateurs in at least 100 British Commonwealth call areas on the current list [see below] since January 1, 1984. An engraved plaque is available to all recipients upon payment of a contributory charge. Applicants working all of the Commonwealth call areas on the current list will be eligible for a suitably engraved cup (charge to be determined).

5BCC (5-Band Commonwealth Century Club)

This award, available in five classes, is available to any amateur who has worked the requisite number of stations in the call areas listed since November 15, 1945 using all five bands (80-10 meters). Each station should be located in a different call area per band. The five classes are as follows:

5BCCC Supreme—500 stations
Class 1—450 stations
Class 2—450 stations, minimum 50 per band
Class 3—300 stations, minimum 40 per band
Class 4—200 stations, minimum 30 per band

Certificates are issued for each class. In addition, those qualifying for Class 1 are eligible for an engraved plaque, while those qualifying for 5BCC receive an engraved cup (additional fees apply). Applicants are required to pay the standard fees for any lower classes of certificates not already held.

Commonwealth Call Areas:

A2, A3, C2, C5, C6, G, GD, GI, GJ, GM, GU, GW, H4, J3, J6, J7, J8, P2, S2, S7, T2, T30, T31, T32, V2, V3, V4, V8, VE1/Mar, VE1/Sable, VE1/St Paul, VE2, VE3, VE4, VE5, VE6, VE7, VE8, VK1, VK2, VK3, VK4, VK5, VK6, VK7, VK8, VK9L, VK9N, VK9X, VK9Y, VK9, VK9Z, VK0/H, VK0/M, VK0/A, VO1, VO2, VP2E, VP2M, VP2V, VP5, VP8/Falklands, VP8/S Georgia, VP8/S Orkney, VP8/S Sandwich, VP8/S Shetland, VP9, VQ9, VR6, VS6, VY1, VU, VU7/A, VU7/L, YJ, Z2, ZB, ZC4, ZD7, ZD8, ZD9, ZF, ZK1/N, ZK1/S, ZK2, ZK3, ZL1, ZL2, ZL3, ZL4, ZL7, ZL8, ZL9, 3B6/3B7, 3B8, 3B9, 3D2, 3D6, 4S, 5B, 5H, 5N, 5W, 5X, 5Z, 6Y, 7P, 7Q, 8P, 8Q, 8R, 9G, 9H, 9J, 9L, 9M2, 9M6/9M8, 9V, 9Y. [Contacts with 5B valid only after March 12, 1961; contacts with 8Q valid only after July 8, 1982.]

WITUZ (Worked ITU Zones)

To qualify, work and confirm contacts with land-based amateur stations located in at least 70 of the 75 ITU zones. QSOs must be made January 1, 1984 or later to be valid. Those qualifying for WITUZ are eligible for an engraved plaque. For those working all 75 ITU zones, a suitably engraved cup is awarded. Additional fees apply for both the plaque and cup.

5BWITUZ (5 Band Worked ITU Zones)

This award, available in five classes, is available to any amateur who has confirmed contacts since November 15, 1945 with the requisite number of land-based amateur stations located in the 75 ITU zones, using all bands (80-10 meters).

Each station should be located in a different ITU zone per band. The five classes are as follows:

Supreme—350 stations
Class 1—325 stations
Class 2—300 stations, minimum 50 per band
Class 3—250 stations, minimum 40 per band
Class 4—200 stations, minimum 30 per band

Certificates are issued for each class. Class 1 winners are eligible for an engraved plaque, and Supreme winners are eligible for an engraved cup. Additional fees apply. Applicants are required to pay the standard fees for any lower classes of certificates not already held. [Note: Minami Torishima (JD) lies outside the 75 ITU zones; a confirmed contact will be accepted as credit for one missing zone for WITUZ and as one missing zone per band for 5BWITUZ.]

Islands on the Air (IOTA)

The IOTA award program was created by Geoff Watts, a leading British SWL, in the mid '60s. In March 1985, it was taken over by RSGB at his request. In all, the IOTA program consists of five separate awards. Any amateur may qualify by working and confirming since December 1, 1964 the requisite number of amateur stations located on islands worldwide and regional. Many of the islands are DXCC countries in their own right; others are not, but by meeting particular eligibility criteria, they also count for credit. IOTA is an evolving program, with new islands being added to the list when they are activated for the first time. The following awards are available:

- IOTA Africa (AF)
- IOTA Antarctica (AN)
- IOTA Asia (AS)
- IOTA Europe (EU)
- IOTA North America (NA)
- IOTA Oceania (OC)
- IOTA South America (SA)
- IOTA Arctic Islands (AI)
- IOTA British Isles (BI)
- IOTA West Indies (WI)
- IOTA Century Club Award 100 (CC100)
- IOTA Century Club Award 200 (CC200)
- IOTA Century Club Award 300 (CC300)
- IOTA Century Club Award 400 (CC400)
- IOTA World Diploma (WW)

The 14-page *Directory of Islands,* available for 8 IRCs or $3 US, lists all islands that count for IOTA and gives full information on the above awards. Requests for *Directories* and IOTA awards applications, which in all cases must be accompanied by QSL cards, should be addressed to the IOTA Awards Manager, Roger Balister, G3KMA, La Quinta, Mimbridge, Chobham, Woking, Surrey, GU24 8AR, England.

IARU Region 1 Award

This award is available in three classes for confirmed contacts with stations located in the requisite number of countries whose National Societies are members of IARU Region 1. The three classes for contacts are as follows:

Class 1—all IARU member-countries on list
Class 2—45 member-countries
Class 3—30 member-countries

IARU Region 1 member countries: A2, A4, A9, C3, C5, CN, CT, DL, EA, EI, EL, F, G, HA, HB, I, J2, JY, LA, LX, LZ, OD, OE, OH, OK, ON, OY, OZ, PA, SM, SP, SV, T7, TF, TR, TU, UA, Y2, YO, YU, Z2, ZB, ZS, 3A, 3B8, 4X, 5B, 5N, TZ, 6W, 7P, 7X, 9G, 9H, 9J, 9K, 9L.

A special version of this award is available in the same three classes for confirmed contacts on 10 meters since July 1, 1983.

28-MHZ Counties Award

This award is available for confirmed 28-MHz contacts since April 1, 1983 with amateur stations located in 40 counties/regions of the UK, Channel Islands and the Isle of Man. Stickers are available for confirming 60 and all 177 counties/regions.

[Information on the RSGB's extensive VHF awards program can be obtained from VHF Awards Manager Jack Hum, G5UM; please include 1 IRC and a self-addressed envelope. Address all correspondence to: RSGB VHF Awards Manager, 27 Ingarsby Lane, Houghton-on-the-Hill, Leicester LE7 9JJ, England.]

USSR (U)

The Radio Sport Federation of the USSR (RSF) and the E. T. Krenkel Central Radio Club of the USSR jointly sponsor the awards listed below. To apply, enclose a certified list of confirmed QSOs (endorsed by an official of your National Amateur Radio Society or club official) and 14 IRCs for each diploma. Send applications to: Radio Sport Federation of the USSR (RSF), PO Box 88, Moscow, USSR.

Diploma R-100-O

This award is issued for confirmed contacts with 100 amateur stations operating in 100 different oblasts (regions) of the USSR [a list of Russian Oblasts appears in Chapter 17]. The diploma is issued in three grades: the 1st-grade diploma is awarded for making QSOs on 160 and 80 meters; the 2nd-grade diploma for QSOs on 40 meters; and the 3rd-grade diploma for QSOs on any band. QSOs are valid from January 1, 1957 and can be made on any band/mode. Endorsement stickers are available for working 150 regions and all regions.

Diploma R-150-C

This award is issued for confirmed contacts with 150 different amateur stations and territories of the world, including 15 Union Republics of the USSR. QSOs are valid from June 1, 1956, and may be made on any HF band and on any mode. Endorsement stickers are granted for QSOs with 200, 250, 300 and 325 different countries of the world.

Diploma R-15-R

This award is issued for confirmed contacts with 15 different Union Republics of the USSR. QSOs are valid from July 1, 1958, and can be made by CW or phone on any HF band.

Diploma W-100-U

This award is issued for confirmed contacts with 100 different amateur stations in the USSR, including five radio stations in the Sverdlovsk region, the birthplace of Russian scientist/inventor A. S. Popov. QSOs are valid from January 1, 1959, and can be made on CW and phone. Endorsement stickers are available for working 300, 500 and 1000 different amateur stations.

Diploma Cosmos-RS

This award, in commemoration of Astronaut Yuri Gagarin's space flight, is issued for confirmed contacts via satellites: 1st-grade diploma for 100 contacts, 2nd-grade diploma for 200 contacts, and 3rd-grade diploma for 300 contacts. QSOs made on any mode from December 17, 1981, are valid.

The Diploma RAEM

This award, in honor of E. T. Krenkel, polar explorer and first president of the RSF, is issued for confirmed contacts with amateur stations operating beyond the North and South polar circles. To obtain the diploma, 68 points are required. Each QSO with RAEM counts 15 points; each QSO with stations in the Arctic and Antarctic counts 10 points; each QSO with stations in the Arctic Islands counts five points; and each QSO with stations beyond the North polar circle counts two points. Valid QSOs must be made since December 24, 1972 on any HF band, CW only.

Diploma R-6-K

This award is issued for confirmed contacts with each of the six continents, plus three QSOs with European Russia and three with Asiatic Russia, for a total of 12 QSOs. The 1st-grade diploma is issued for QSOs made on 160 and 80 meters; 2nd grade for 40 meters; and 3rd grade for any other band(s). Valid QSOs must be made since May 7, 1962 on CW, SSB, and phone on any HF band.

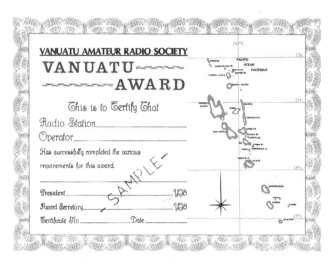

VANUATU (YJ)

The Vanuatu Amateur Radio Society (VARS) issues the *Vanuatu Award* for working at least six YJ8 stations who are members of VARS. Contacts made from July 30, 1980, Vanuatu Independence Day, are acceptable. Contacts may be made by any mode. Endorsements for single band, single mode, or additional stations worked are available. Two contacts with any one YJ8 station will be accepted provided that the contacts are made on different days, bands or modes. Send a certified log extract and 10 IRCs (or $2 US) to: Awards Manager, VARS, PO Box 665, Port Vila, Vanuatu.

Chapter 9

RTTY Communications

Radioteletype (RTTY) or radioteleprinter communications is probably the most-used form of *digital communications*. Thanks to the computer age, this mode has experienced a great resurgence of popularity in recent years. There have been many changes in the last few years, and as a result, there are several different forms of RTTY. From the late 1940s, when radioteletype was first developed, until 1980, RTTY was synonymous with communications using the International Telegraph Alphabet Number 2 (ITA2), also known as the Baudot or Murray code. Amateurs today also use the American National Standard Code for Information Interchange (ASCII), the AMTOR code and some codes for exchanging picture information. (Packet Radio, another popular form of digital communications, is discussed in Chapter 10.)

In the early years of radioteletype, amateurs used surplus teletypewriter machines. These machines were slow, noisy and often dirty, with excess oil from the mechanism dripping out onto the floor. In 1975, affordable microcomputers were introduced, and many radio amateurs became involved with these new microcomputer systems. It became possible to use a video-display terminal (VDT) for keyboarding and displaying RTTY communications.

Since computers use a code called the American National Standard Code for Information Interchange, or ASCII for short, to represent the characters typed at the keyboard, it was necessary to convert between ASCII and the Baudot code used for radioteletype. So, many RTTY enthusiasts got busy and wrote code-conversion programs for their computers. Some programmers couldn't stop with writing a program to convert from ASCII to ITA2 codes, and they also wrote programs to convert the ASCII codes to Morse code. Since then, amateurs have made increasing use of microcomputers for RTTY and CW operation. Today, many amateur RTTY operators have microprocessor-based equipment. The noisy teletypewriters have taken a back seat. A high-speed computer printer can be used when hard copy (a paper print-out) is desired, although much operation simply involves reading the information on the video screen. Today, the microcomputer is largely responsible for an increased interest in radioteletype, and for the heavy population of amateurs on the RTTY frequencies.

In early 1980, the FCC legalized amateur use of ASCII, along with packet-switching techniques (such as are used in packet radio systems) for radioteletype operation. In 1982, acting on a petition from the League, the FCC permitted amateurs to use AMTOR (Amateur Teleprinting Over Radio), an error-free RTTY system designed for the Maritime Mobile Service and introduced to Amateur Radio by J. P. Martinez,

G3PLX. (Operating notes by Martinez and Paul Newland, AD7I, were helpful in preparing the AMTOR operating section of this chapter. Also, some of the Baudot RTTY operating information was taken from a May 1985 *QST* article, "A RTTY Operator's Guide," by Bill Snyder, WØLHS.)

By 1984, AMTOR had gained considerable popularity with amateurs, and a number of manufacturers were producing AMTOR systems. AMTOR is a system that is governed by International Radio Consultative Committee (CCIR) Recommendation 476-2 or 476-3. See Fig 9-1 for the W1AW AMTOR installation.

Fig 9-1—The W1AW AMTOR operating position.

Packet-radio techniques are being pioneered by Amateur Radio experimenters. By early 1984, packet-radio hardware and software were available to amateurs from three radio clubs and three manufacturers. For more information on packet radio, see Chapter 10.

RTTY enthusiasts have long been fascinated by automatic operation. Many had their receivers tuned to a certain frequency and set their equipment to copy any activity on that channel. An auto-start circuit was used to turn on the printer motor when a signal was received and, more importantly, turn it off when there was no activity. A few used a stunt-box feature of some teleprinters to turn on the printer only when a specific character sequence (for example a call sign) was received. A more recent example of automation is the computer-based message system (CBMS) that answers callers and repeats messages stored in its memory. Under present FCC rules, such stations may operate above 50 MHz under repeater rules, but in the HF bands must be under operator control to ensure against transmitter malfunction and retransmission of impermissible traffic.

Some amateurs have suggested that RTTY is the best mode for traffic handling; that's what commercial and governmental communications used for years. In those services, tape relay was the norm until the early 1960s when it was replaced by computer-switched networks. With few exceptions, neither tape-relay nor computer-switched networking has been used for Amateur Radio message communications. For one thing, not many operators have had the necessary equipment to receive and automatically retransmit message traffic. Packet radio, however, is developing toward a computer-switched network that will cover North America and a number of overseas countries by the end of this decade. This shows great promise for automatically relaying messages, and may prove quite popular with traffic handlers.

In this chapter, the operating techniques used on the various RTTY modes will be covered, along with how the pieces of equipment in a typical RTTY station connect. This chapter will give you all the information you need to sound like a "pro," even on your first contact. The technical aspects of digital communications are covered in the *ARRL Handbook for the Radio Amateur* and the RSGB *Teleprinter Handbook*, both available from ARRL HQ or your local radio bookstore.

Operating Frequencies

On the HF bands, RTTY frequencies are usually found at the high end of the CW portion of each band. Traditionally, the 20-meter RTTY subband has received the heaviest use during daylight hours and 80 meters gets most of the activity at night. The increased popularity of RTTY has also brought an increased use of the 40-meter RTTY subband. The HF frequencies on which you will most likely find RTTY operation are listed in Table 9-1.

There is some digital communications activity on the VHF bands, particularly on 2 meters. The 146.10/146.70 MHz repeater pair is one used for RTTY operation in many areas. Most simplex packet-radio repeater operation is on 2 meters, with some localities using 220 MHz. Check the current *ARRL Repeater Directory* for digital repeaters in your area.

Table 9-1

Common RTTY Frequencies

US RTTY Operation (kHz)	IARU Region 1 (Europe/Africa) Band Plan (kHz)
1800-1840	
(1830-1840 for intercontinental contacts)	
3590 RTTY DX 3605-3645	3580-3620
7040 RTTY DX 7080-7100	7035-7045
10,140-10,150	10,140-10,150
14,070-14,099.5	14,080-14,100
18,100-18,110*	18,100-18,110
21,070-21,100	21,080-21,120
24,920-24,930	24,920-24,930
28,070-28,150	28,050-28,150

145.880 MHz (OSCAR RTTY)
*Pending FCC approval for US amateur use.

BAUDOT RADIOTELETYPE

Radioteletype communication using ITA2 is widely used on the Amateur Radio HF bands in most areas of the world. The technical details of this code are beyond the scope of this book, but if you want to know more about the actual code and how the characters are formed, you can find all of the details in Chapter 19 of *The ARRL Handbook*. Check the portions of the bands listed in Table 9-1 every weekend and most evenings for active RTTY operators. If you are not equipped to copy RTTY, you can still get some idea of what is going on by listening. You may hear an RTTY station transmitting short bursts from a keyboard or a steady stream of characters coming from a digital memory. Some operators use teletypewriter machines, but many have gone to computer-based systems. It is an enjoyable mode that is good for ragchewing and handling traffic.

Setting Up

To get on RTTY using ITA2, the first thing you will need is a teleprinter or a computer-based terminal. Teletype® Model 15 or 19 teletypewriters may still be found for $50 or just for hauling them away. Model 28 or Model 32 machines may be available for $100 or so, depending on their condition. If you obtain one of these (or European) machines, be prepared to do most of your own maintenance. So, be sure to get a good set of manuals.

Most radioteletype operators are now using computer-based terminals. One type of terminal is designed for amateur RTTY service, such as those manufactured by Hal Communications and Microlog. Another approach is to build your own or purchase a program and electrical interface for a home computer. Check the *QST* Product Review column, and articles and ads for information about these products.

You will also need a modem, a contraction of modulator-demodulator (also called "terminal unit" or "TU"). The current standard frequency shift for HF RTTY operation is 170 Hz. Most US amateur RTTY modems demodulate audio tones of 2125 Hz for mark and 2295 Hz for space (sometimes called "high tones"). The IARU Region 1 standard is 1275 Hz mark and 2125 Hz space, also known as the "low tones." Both high and low tones can be used interchangeably on the HF bands because only the frequency shift is important.

If you want to tune to an exact RTTY frequency, remember that the SSB radio will display the frequency of its (suppressed) carrier, not the frequency of the mark signal. For example, if you want to operate on 14,083 kHz and you are using a 2125-Hz AFSK mark frequency, your SSB radio (suppressed-carrier) frequency should be 14,083 kHz + 2.125 kHz = 14,085.125 kHz. (On lower sideband, the audio frequency subtracts from the carrier frequency to give the actual frequency of the transmitted signal.) Many modern transceivers include an FSK mode. In this case, the relationship of the rig's frequency readout to the transmitted mark frequency varies from one model to the next. Consult the operator's manual of your radio for details.

It is normal to use the lower sideband mode for RTTY on HF SSB radio equipment. This makes the MARK tone come out higher in frequency than the SPACE tone, even though the audio tones used to produce the signal had a higher frequency SPACE tone. If you hear someone say that the lower sideband is inverted, this is what they mean.

Some modern transceivers have two positions on the sideband-selection knob, labeled NORMAL and REVERSE. This labeling may cause some uncertainty when you use the rig for RTTY. On 80 and 40 meters, normal SSB operation uses the lower sideband, so setting the selection switch to NORMAL will put you on the lower sideband for RTTY operation. Upper sideband is used on the other HF phone bands, however, so you will have to set the switch to the REVERSE position in order to select the lower sideband for RTTY operation on those bands.

If you are not using the correct sideband, then your signal will be "upside down" (the SPACE tone will be the higher-frequency tone) and other operators will have to change their normal operating setup in order to copy your signals. If other operators switch their modems to work with inverted tones, it will probably affect their transmitter MARK/SPACE tone relationships, and then you may have to switch your system to copy them! So be sure you always select the lower sideband and transmit signals that are "right-side up."

On VHF, the most common practice is to use AFSK. This makes high- and low-tone modems unable to talk to each other as they can on HF. In most areas, 170-Hz shift is used for VHF RTTY, but some hams still use the older standard of 850 Hz (2125-Hz mark and 2975-Hz space).

In choosing a modem, or a complete RTTY system with a built-in modem, look for versatility. Currently it would be good to have a modem with shift capabilities of 170, 425 and 850 Hz (both high and low tones). The 425-Hz shift is useful in copying commercial stations.

Some modems have an autostart feature. Mechanical teletypewriters cannot print until their motors are running. Rather than letting the motor run continuously, some RTTY stations use an autostart circuit that senses the presence of a mark signal and turns on the motor. In this way, the equipment can print a record of any activity on the frequency even when the operator is not in the station.

What about a rig? Almost any multimode HF transceiver can be used for RTTY. The main thing is stability. Frequency error, for whatever cause, should be minimized—ideally on the order of no more than ± 10 Hz to make use of minimum receiver bandwidth. Many solid-state transceivers will do this after the temperature has stabilized. A transceiver with a frequency display of 10-Hz resolution is also helpful, although that is not a necessity. Older tube-type radios can be stabilized by operating the filaments of the frequency-determining stages (eg, oscillators) from a separate power supply that is left on at all times.

Ideally, the receiver used for 170-Hz-shift RTTY should have the minimum practical bandwidth, preferably between 270 and 340 Hz. Many CW filters have bandwidths around 500 Hz and would be good for 170-Hz RTTY reception, but many receivers (in transceivers) can only place the SSB filter in the circuit in the SSB or RTTY modes. Some RTTY operators have modified their receivers to use the CW filter when the RTTY mode is selected. A switch can be added so it is possible to select the CW filter with the mode switch set for SSB operation. It is desirable to still have the option of using the standard SSB IF filter for modems designed for higher signaling speeds of up to 1200 bauds. (US amateurs can only use speeds of 300 bauds or less below 28 MHz.) In determining the amount of bandwidth you need, add in the amount of frequency drift expected. If you are going to be able to manually fine tune the receiver as needed, however, you won't have to worry about the extra bandwidth needed because of any drift. The bandwidth of the receiver and demodulator must include the anticipated frequency drift for unattended operation.

You can have some fun with RTTY using a transmitter with 100 watts output. But for reliable RTTY communication, it is desirable to have a higher-power amplifier available when needed. Bear in mind that your transmitter must be able to withstand 100% duty cycle when transmitting conventional Baudot and ASCII radioteletype, as well as AMTOR Mode B. AMTOR Mode A and packet radio have shorter transmission blocks. They are on 100% of the time when transmitting but are in receive condition more than half the time when averaged over more than a few seconds. (One hundred watts output is usually sufficient for AMTOR operation; more power is desirable for Baudot and ASCII codes, which don't include the error-correction feature.)

Receiver-tuning accuracy is important. Thus, a tuning aid can be a great help in proper receiver adjustment. Most modems have some type of tuning indicator, possibly just flashing LEDs. Some have oscilloscopes that produce patterns such as those shown in Fig 9-2. The mark signal is displayed

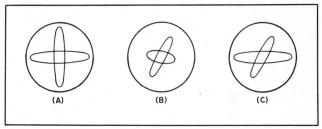

Fig 9-2—Tuning indicator oscilloscope patterns. Part A shows a properly tuned RTTY signal. At B the receiver is slightly off frequency. The received signal displayed at C has a different shift than what the demodulator is set for. The mark (horizontal) line is normal, but the space line is tilted and smaller in amplitude, indicating that it is coming in at the wrong frequency.

as a horizontal line and the space signal as a vertical line. Theoretically this should appear as a "+" sign on the 'scope screen, but in practice it may look more like a pair of crossed bananas! To avoid the high cost of an oscilloscope, some manufacturers are offering LED displays that imitate the 'scope display.

If your modem has separate MARK- and SPACE-tone outputs, you can connect a standard oscilloscope for use as a tuning indicator. Old oscilloscopes found at flea markets usually can do the job without modification. (That is if you happen to find one in working order!) One tone goes to the vertical scope plate, while the other goes to the horizontal plate. When you have tuned in an RTTY signal properly, you will see the two ellipses on the screen.

Mastering The Green Keys

No one can disagree that touch typing (typing without looking at the keys) is the best way to go. Touch typing is mainly a matter of finger placement and practice. Fig 9-3 shows the keyboard of a teletype machine. The "home" position for the fingers is with the left forefinger on the F key, and the other fingers of the left hand on D, S and A. The four right fingers are on J, K, L and CAR RET. The thumbs are used for the space bar only. You hit Q, A, Z and FIGS with the little finger of the left hand. The left forefinger's territory covers R, T, F, G, V and B. Similarly the right forefinger is used for Y, U, H, J, N and M and the right little finger for P, CAR RET and blanks. The main part

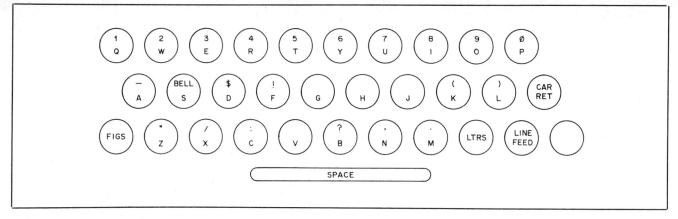

Fig 9-3—ITA2 keyboard. The arrangement resembles that of a regular typewriter, but it has only three rows of keys instead of four. All letters are capitalized; there are no lower case letters. The characters appearing above the alphabet letters are produced when preceded by the FIGS key, shifting the carriage.

of the keyboard is just like a typewriter or computer-keyboard layout. Instead of two SHIFT keys, however, the teletype keyboard has one key labeled FIGS and one labelled LTRS. The FIGS key is used to type the numbers and other symbols shown near the top of the keys. Notice that not all keys have a FIGS character. To return to the letters, you must type a LTRS character. The letters are printed only as upper-case letters. There are no lower-case characters on the teletype printer.

One fact you'll soon discover is that it's impossible to type faster than 60 words per minute on a 60-WPM (45.45-baud) teleprinter. Other machines operate at speeds of 67, 75 or 100 WPM, corresponding to 50, 56 and 75 bauds, respectively. (The baud rate is just another common way to refer to the speed of transmitting the teletype characters.) If you try to type two letters too quickly, the second key will lock up until the first letter is finished. To send on this type of keyboard, you have to develop a steady rhythmic typing technique. This does not apply to most paper-tape perforators and to computer-based keyboards.

On a mechanical teleprinter, you have to hit two keys, Carriage Return (CR) and Line Feed (LF), to start a new line. It's a good idea to get into the habit of typing the end-of-line sequence of 2CR LF LTRS at the end of each line (2CR LF FIGS if the next line begins with a figures-case character). Two carriage returns are needed by some older printers because their heavy carriages take some time to return to the first printing position. It's also necessary to send the end-of-line sequence at the end of each transmission to ensure that the other station's printer or screen is not mid line and then prints "garbage" over the last line if it thinks that it received a CR but no LF. Most computerized RTTY setups require only that a CR be entered to send the entire end-of-line sequence. If you are using a computerized system, be sure the program is set up to generate the end-of-line sequence automatically.

Many RTTY operators are reluctant to use the figures-case characters and only use the letters case. This practice has two roots. One is that HF RTTY signals are subject to garbling, particularly because of fading signals or the use of inferior demodulators. The other is that some of the surplus teletypewriters used by RTTY operators had weather symbols in figures case. Most of these machines are out of service, and it is fairly safe to assume that active RTTY operators are using either ITA2 or a US variation of it. The FIGS case for

the letters D, F, G, H, J, S, V and Z may produce different results on different machines. For that reason, it is prudent to avoid these combinations unless you are sure what the other station is using for equipment. On the other hand, on a good circuit there is no good reason why the other standard figures-case characters shouldn't be used.

The standard punctuation symbols are "-" (FIGS A), "?" (FIGS B), ":" (FIGS C), "(" (FIGS K), ")" (FIGS L), "." (FIGS M), "," (FIGS N), and "/" (FIGS X). Also, there is a trick that will allow you to use the BELL function, regardless of what terminal is at the distant station. If you want the distant station to hear 5 BELLS, send FIGS JJJJJSSSSS LTRS. That will not only sound the audible signal but show 5 apostrophes on the screen or printer.

When using figures case, you should be aware that some printers have a feature called unshift on space. If you send a group of numbers and a space, the machine will revert to letters-case characters after the space if it has the unshift-on-space feature. To avoid this problem, it is good practice to always send a FIGS just before each group of figures that are preceded by a space.

Some teleprinters have a "nonoverline" feature. When a teleprinter or terminal with this feature receives a CR, it acts as though it received both a CR and LF. Others have a "wrap-around" feature that automatically moves to the first printing position of the new line when the old line is full. If you have these features, don't assume that everyone else has them. Send the 2CRs and LF anyway so any type of printer or screen can display what you send.

When transmitting RTTY with the ITA2 code, you are usually in control of the receiving station's copy format. Because of the variety of mechanical and electronic devices in use, your message may be displayed on a video-display terminal (VDT) or a printer.

It is a good practice to keep the number of characters per line to a maximum of 69, the CCITT international telex standard. That line length can be handled by virtually any teleprinter designed for the ITA2. Many US teleprinters will print 72 or 74 characters, but some foreign printers take only 69. Computer screens and printers vary widely and may have 22 (Commodore VIC 20), 32 (Sinclair ZX81), 40 (Apple II and Commodore 64), 64 (some video boards), 80 (common for word processing) or 132 (some more-expensive displays and printers) characters per line.

After a mistake, the "oops" signal is XXXXX (some operators may use EEEEE). There is no need to overdo this, though, because generally the other operator will figure out what you meant to say anyway, and a page filled with XXXXX can be confusing.

Station Identification

When you are transmitting RTTY, simply send your call sign at not more than 10-minute intervals, just as you would on any other mode. You can send a CW ID if you wish—it does give stations not equipped to receive RTTY a means of finding out your call sign. If you send a CW ID, you will drive some RTTY operators crazy if you send "CW ID TO FOLLOW" at the end of a transmission. Most operators know that if you're going to send a CW ID, you'll do it at the end of your RTTY transmission.

FCC rules permit a station engaged in digital communications using ITA2, AMTOR or ASCII, to identify in the digital code used for the communication. Identification may also be given in Morse code. Above 50 MHz, identification may be given using one of the methods just mentioned or by voice (where radiotelephone operation is permitted on the same frequency being used for the RTTY communication). Identifying the station you are communicating with, in addition to your own station, is not legally required, but it will help those monitoring your QSO to determine band conditions.

Special Character Sequences

ITA2 has a limited character set that provides only capital letters, numerals and limited punctuation. Unlike ASCII, which has a variety of control characters, ITA2 has only a few machine functions: space, carriage return, line feed, figures, letters and blank. Because of the lack of control characters, commercial and government services have evolved a number of character sequences to be used for special purposes. Two that you may have seen are ZCZC and NNNN. Both of these sequences are from CCITT Recommendation S.4, which specifies the special signals given in Table 9-2.

Computer software for receiving RTTY should recognize any of the combinations shown in Table 9-2, even when they are received in the wrong case. Those writing software should carefully research the CCITT recommendations before using these character sequences. Where relays may be energized or machines (such as reperforators) need time to come up to full speed, it is prudent to include a 500 ms delay after the signal to allow the mechanical action to occur. To avoid misinterpretations and possible conflicts in the use of these signals,

please write to: Chairman, ARRL Digital Communications Committee c/o ARRL HQ.

Stored-Message Library

Although many RTTY operators now use computers, some use paper tape to store a repertoire of text or graphics. See Fig 9-4. Even if you are working someone using a computer, don't be surprised to hear older tape-related terms such as "brag tape" in common usage.

When typing to tape or computer memory, there is no need to use XXXXX to correct errors. If you make a mistake while punching tape, backspace the perforator over the error and type LTRS. Since all five holes are punched out for this character, it effectively erases the error. Watch out, though; if you were in FIGS case, you are now in LTRS case. On computer keyboards, you will normally have a delete or backspace key to correct mistakes.

A stored message consisting of repetitive RYs, as shown below, is useful for local and on-the-air testing of ITA2 equipment.

RYRYRYRYRYRYRYRYRYRYRYRYRYRYRY
RYRYRYRYRYRYRYRYRYRYRYRYRY DE W1AW

The use of the RY sequence goes back to the early Teletype-machine days. To set the mechanical machines, it was necessary to send a string of Rs and Ys while the receiving operator adjusted a knob called the "rangefinder." The

Table 9-2
Special Control Sequences Used With ITA2

ZCZC	start-of-message signal
+ + + +	end-of-input signal (+ is FIGS Z in ITA2 keyboards)
NNNN	end-of-message signal (also for restoring the waiting signal per CCIR Rec. U.22)
HHHH	to prevent transmission of LTRS delay signals (per CCIR Rec. U.22)
CCCC	for remotely switching on a reperforator or equivalent device
SSSS	for remotely switching on a terminal
FFFF	for remotely turning off a reperforator
KKKK	ready-for-test signal (per CCIR Recs. R.75 or R.79 bis)
KLKL	for remotely switching on a reader
XXXXX	error signal
LF 4CR	operator recall on RTTY telex circuit

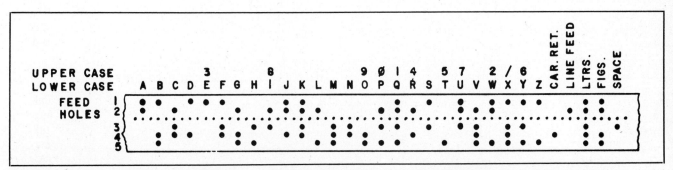

Fig 9-4—Teleprinter letter code as it appears on perforated tape. The start and stop pulses are not on the tape, but are automatically generated in the machine. The dots (perforations) indicate mark pulses. Numerals, punctuation marks and other arbitrary symbols are secured by carriage shift (FIGs key). Where undesignated positions appear in the "upper case" line of the above chart, characters may differ on various machines.

military and commercial services also used RYRYRYRY as a channel-holding signal during idle periods. They even typed the line over and over without sending a line feed, to save paper. The letters R and Y both alternate between MARK and SPACE tones in the Baudot code, so it was natural to use these letters for the range-setting operation. (If we represent the SPACE tone with a 0 and the MARK tone with a 1, then R is 01010 and Y is 10101. So you can see that a string of RY characters sends alternating SPACE and MARK tones.)

Use the RY sequence sparingly. RYs have a musical sound that is easily recognizable, and is useful for tuning a signal in properly. This makes it good for rousing attention when you are sending a CQ call, or when calling another station on schedule, but it is unnecessary during routine contacts. Don't use RY when you are working DX.

The RY sequence is good for exercising mechanical teleprinters, but it does not test to see whether every different character can be printed. The mechanical printers need a test signal that can check the operation of the code bars and vanes that decode the electronic signal pulses to print each character. To do this, a test message is needed. The Quick Brown Fox sentence meets this need:

THE QUICK BROWN FOX JUMPS OVER THE LAZY DOG 01234567879

There are other variations of this sentence, and there are other sentences that have been made up to include all of the letters of the alphabet, but this is probably the most common one. Such a message can be stored on paper tape or on computer memory for use at any time that you are asked to send a test message for someone who is testing his or her equipment.

Making Contact

There are two ways of establishing contact with other amateur stations using radioteletype. You can either answer someone else's call at the end of a contact or when he or she calls CQ, or you can try calling CQ yourself. When you call CQ, you depend on someone else to tune your signal properly, and if you are a newcomer to the mode, that may be an advantage. By listening for other calls, you have a good chance of finding stations that you might not contact otherwise. You will want to use both methods at various times, but whichever one you choose, learn the proper procedure and follow it.

Calling CQ

Since it is a little easier to explain, we will start by describing the procedure for calling CQ on RTTY. Of course the first thing you must do is locate a clear frequency on which to make your call. Courtesy should be the rule. Ask if the frequency is busy by sending "QRL," or by asking in clear text. Sometimes it is possible that there is a QSO on frequency, but you may only be able to hear one side of it. So the frequency may sound clear to you while the other station is transmitting. Once you have established that you have a clear frequency, it is time to make the CQ call.

The general CQ call, like RY, has a recognizable, musical sound. Take advantage of this fact, and send CQ in a pattern something like this:

CQ CQ CQ CQ CQ CQ CQ CQ DE WØLHS CQ CQ CQ CQ CQ CQ DE WØLHS

There are enough CQs in a row to create the musical pattern, and the call sign sent once will not disturb the recognizable sound of the transmission. If your computer system has a "CQ key" with a pattern similar to this, use it, but if not, store a CQ message in one of the message memories.

Many operators add their name and QTH to the last line of the CQ call. This is a good idea, but keep it simple. If you live in North Dakota, for example, you could add your QTH, because there are lots of hams who need that state for WAS! That will increase the likelihood of receiving an answer to your call.

You should not corrupt a CQ call by getting tricky with added punctuation and/or fancy screen displays. This would destroy the musical pattern, and you lose the advantage of having a casual listener being able to recognize the general call and tune you in.

There are all kinds of "arty" CQ calls around. Some are very clever. But if you want to attract attention, you can't beat the musical notes of a repeated CQ. Leave all the CQCQCQCQCQCQCQCQ; SEEKYOU, SEEKYOU; C-Q-C-Q-C-Q; and other fancy stuff in the dummy load. Don't clutter the airwaves with it.

Calling Another Station

You have just heard 9K2KA in Kuwait calling CQ. He is booming in with an S9 signal, and this would be a new country for you. So you jump in and excitedly run a full string of RYs to get his attention, then you send his call ten times to make sure he knows you are calling him, and then sign your call 15 times just to be sure he gets it right! But, alas, when you stand by after doing all that, you find Adnan working another station. And chances are, you will have a bunch of RTTY DX operators mad at you for being so long-winded.

Keep your calls short. Forget the RYs; send the called station's call sign only once, and your call no more than five times. Then stand by and listen. If he doesn't answer, try another 1 × 5 call. The DX station has a better chance of picking out someone to answer if he has a pile of short calls rather than long-winded strings of RYs and call signs. Some rare DX stations have had to go QRT because of such bedlam getting on their nerves. Signing your call as little as three times has also proven successful. The point is that RTTY QSOs can be pleasant and abundant, but only if everyone keeps their calls short.

Line Feeds and Other Things

When another station answers your call, gives you a report and then turns it back to you, what should you do? Well, always send a line feed (L/F) and a carriage return (C/R) first. This will put his computer and or printer into the "letters" mode, and eliminate the possibility of part of your message being garbled by numbers and/or punctuation marks.

Next, send his call and your call once each. On your first exchange, send your name and QTH message from the computer memory, but on subsequent transmissions do not send this line: it is unnecessary. If you have a type-ahead buffer on your system, then you can compose some of your reply before it is actually transmitted. This saves time, especially if you are not a fast typist.

If you give another station a signal report of RST 599 or RST 519, it means you are copying him solid. If he gives you a like report, then it is unnecessary to repeat everything twice. Give honest reports—if the copy is solid, then the readability report is five. Adapt your operating technique to the reported conditions. If copy is marginal, repeats or extra spaces are in order. But with solid copy, you can zip right along. It's that simple.

In the days when almost everyone used surplus Teletype machines, it was customary to send the "LTRS" character as a "diddle" during pauses between typing, to make certain that the receiving station was not typing a string of jibberish instead of the desired text. The letters key is a nonprinting character that does not advance the carriage, and so the print head stands still and does not waste paper during these idle moments. The letters key clears the possibility of text being garbled by numerals and punctuation. There are still a few of these machines being used, so adapt your computer operating to those amateurs using the machines. They will appreciate it.

Another caution is to not send a string of carriage returns to clear the screen. The other operator may be copying on a printer, and it will waste a lot of his paper. The last two characters of each transmission should be carriage returns, however. The purpose of this is to move the other station's printer (or screen cursor) to the left side of the page, and thus avoid leaving the print head or cursor stranded in the middle of a line of type.

Date and Time Generators

One of the little goodies of a computerized RTTY station is the date and time generator. Unless you are bragging to the world that you have one of these gadgets in your tool kit, don't use it! When computers first started to show up on the ham bands, it was considered quite classy to put the date and time on every transmission, but that is old hat now, and mostly a waste of time. There are legitimate uses for including a date and time in your transmission, such as for message-storage operations on a "mailbox" system, and for auto-start timing, so let that be your guide.

Signing Off

You have now completed your QSO with 9K2KA and are about to sign off. What prosign do you use? Well, the standard prosigns mean the same thing on RTTY as they do on CW, and should be adequate. Table 9-3 summarizes the meaning of the most commonly used prosigns.

Contests

RTTY contests are fun. The pace is much slower than the frantic operations on CW and SSB, but they are very enjoyable. If you are not entering the contest because you feel that your typing is not up to speed, perish the thought. Jump right in and join the fun!

Read the rules carefully and do as they say. You might have a potentially winning score tossed out of the competition because you did not bother to log the time that the other stations send you in the exchange if the rules call for both times sent and received to be logged, for example.

RTTY contests are run on a gentlemanly basis, and renewing friendships is part of the excitement. So there is nothing wrong with pausing for a chat after the contest exchange has been made. There are enough 'tests during the year to make each one a special event for all entrants.

DX Operations

Chasing DX is just as much fun on RTTY as on any other mode. The leading DXers in this mode have a country count of about 240 at the present time, so there is still room at the top! Less than 100 DXCC certificates with a RTTY endorsement have been issued so far by the ARRL, but the pace is picking up. Many countries have only one active RTTY station, so the chase is quite spirited. When a new country appears on the screens, the word spreads like wildfire. Avid DXers begin alerting each other with the news. This great spirit of cooperation is one of the really rewarding facets of RTTY DX operating.

Radiograms by Baudot RTTY

If you have a carload of messages to send to another station, RTTY is a lot faster than CW or radiotelephone. Net operation is a bit more cumbersome on RTTY than on CW or SSB unless all stations have fast break-in and are exactly on the same frequency. Information about RTTY traffic nets can be found in the *ARRL Net Directory*. When stations have only one or two messages each, the traffic is normally handled on the net frequency.

RTTY message format is very similar to the CW traffic-handling procedure covered earlier in this book. A typical message with all ITA2 machine functions is shown in Fig 9-5.

The 10LTRS at the beginning and the 12LTRS at the end of a message are needed in the event that paper tape is used anywhere in the network, for origination, relay or receiving. The 5SP 2CR LF LTRS functions before the start of the heading are to synchronize mechanical printers and get them to the first printing position of the next line.

The start-of-heading (SOH) sequence is ZCZC, one of the CCITT Recommendation S.4 signals listed in Table 9-2. This is now almost universally used on networks using the ITA2 code to signal computer switches and terminal receiving equipment that there is a message coming.

The preamble and address are handled the same as on CW, with machine functions added as needed. Note the use of BT before and after the text in this example. It is added as a step toward compatibility with computer message processing.

The text is typed no more than 69 characters per line. The punctuation marks - ? : ( ) . , and / may be used in the text. An apostrophe can be used by sending both FIGS J and FIGS S—one or the other will print the apostrophe ('), and the other will ring the bell, regardless of the type of terminal. Spell out QUOTE or UNQUOTE (for ''), SEMICOLON (for ;), AND (for &), POUND (for # or £), NUMBER (for #), EQUALS or DOUBLE HYPHEN (for =), and PLUS or CROSS (for +).

After the signature line, a confirmation line should be used

Table 9-3

Common Prosigns Used On RTTY

K—Invitation to transmit.
KN—Invitation to the addressed station only to transmit.
SK—Signing off. End of contact.
CL or CLEAR—I am shutting my station down.
SK QRZ—Signing off and listening on this frequency for any other calls. The idea is to indicate which station is remaining on the frequency.

Some operators use various other combinations of prosigns:

SK KN—Signing off, but listening for one last transmission from the other station.
SK SZ—Signing off and listening on this frequency for any other calls.

```
↓↓↓↓↓↓↓↓↓↓↓→→→→→<<≡↓                                                      Leader
ZCZC→NR→↑345↓→R→WA↑0↓XYZ→↑15↓→SIOUX→FALLS→SD→↑1030↓Z→JAN→↑3<<≡↓          SOH and Preamble
MRS→SAM→SWARZ<<≡↑                                                         Address
1234↓→MAIN→ST<<≡↓
ANYTOWN→IL→↑64578<<≡↑
617-555-9683<<≡↓                                                         Telephone Number
BT<<≡↓                                                                   Start of Text
HAVING→A→WUNNERFUL→TIME↑.↓→→WISH→YOU→WERE→HERE↑.↓→→SEE→YOU→AND→THE→KIDS<<≡↓  Text
SOON↑.↓→→LOVE<<≡↓
BT<<≡↓                                                                   End of Text
MARY<<≡↓                                                                 Signature
CFM→SWARZ→↑1234→↑64578→↑617-555-9683↓→WUNNERFUL<<≡≡≡≡≡≡≡≡               Confirmation




NNNN↓↓↓↓↓↓↓↓↓↓↓↓                                                         End of Message

                    Legend:   ↓LTRS    ↑FIGS    <CR    ≡LF    →SP
```

Fig 9-5—A sample message format using the ITA2 code.

if the message contains any figures, proper names or unusual words. At the end of the confirmation line, type 2CR 8LF LTRS NNNN 12LTRS to provide a separation between messages.

Long messages are normally frowned upon for handling by CW or phone. Longer messages can be sent, however, if RTTY will be used over the entire route to the message destination. Type no more than 54 lines per page, including heading and end of message. Using the example shown in Figure 9-5, at the end of the first page type 2CR 4LF LTRS PAGE 2 WA0XYZ NR 345 2CR 4LF LTRS. Follow the same procedure for any more pages except for the last one. The last page will end with the confirmation line, 2CR 8LF LTRS NNNN 12LTRS.

Remember that you must identify your station every 10 minutes. If that must happen in the middle of a long message, you can use the following procedure to allow tape reperforators or other storage devices to turn off before the ID is sent, and turn on immediately after the ID to resume copying:

After an end of line sequence (2CR LF LTRS) type FFFF DE (your call sign) CCCC 2CR LF LTRS and resume sending your message. The CCIR Recommendation S.4 signal FFFF turns a reperforator or other storage device off, and CCCC turns it back on. Stations not equipped to recognize these character sequences will simply print them as part of the message. They can be edited out after receipt.

AMTOR OPERATION

Amateur Teleprinting Over Radio (AMTOR) is a more robust transmission system than ITA2 because of its error-detection and -correction properties. In many respects, AMTOR operation, including sending radiograms, is the same as for ITA2. AMTOR uses the same character set as ITA2, but the coding is different. Instead of transmitting 5 bits of information, AMTOR sends 7. If we use B to represent the higher radio frequency, and Y to represent a frequency 170 Hz below it, each AMTOR character has a constant ratio of 4 Bs to 3 Ys. For a technical overview of AMTOR, see the current issue of *The ARRL Handbook*. AMTOR technical standards are found in Section 97.69 of the FCC rules and CCIR Recommendations 476-2 and 476-3.

At the moment, most AMTOR activity is around 14,075 and 3637.5 kHz, which are used as calling frequencies. Operators usually move off the calling frequency to free it for others. However, there are some automatically controlled stations that just stay on these frequencies. At this writing, automatic control of an AMTOR station in the US requires an FCC Special Temporary Authorization.

AMTOR is permitted in the US and many other countries, but check the Amateur Radio rules for your country and license class before obtaining an AMTOR system. Some countries may not allow AMTOR operation without a special permit.

AMTOR Modes

CCIR Recommendations 476-2 and 476-3 specify two modes of operation: Mode A (ARQ—Automatic Repeat Request) and Mode B (FEC—Forward Error Correction). Mode B is further subdivided into Collective B-Mode and Selective B-Mode. Amateurs have added Mode L for monitoring AMTOR transmissions.

Stations in contact using AMTOR are classified either as Information Sending Stations (ISS) or Information Receiving Stations (IRS). The stations change identities as information is exchanged during a QSO. The station that originates the communication is also called the master station and the other station is called the slave station. The master/slave relationship does not change during the contact.

Mode A

Mode A is a synchronous system in which the ISS transmits blocks of three characters to the IRS. The ISS sends the block in 210 ms, listens for replies from the IRS for 240 ms, then sends three more characters. The ISS keeps the three characters in memory until it receives a receipt from the IRS. When the IRS acknowledges, the ISS moves on to the next three characters. If the IRS does not acknowledge, the ISS keeps sending the three characters until they are acknowledged. AMTOR keeps trying until it gets it right!

Mode A is used only for contacts between two stations. It should not be used for calling CQ, in nets or roundtables. This is the mode that has the characteristic "chirp" on the air. It has a 47% duty cycle, allowing transmitters to operate at full CW power output. Mode L is used by stations to monitor, or listen to, Mode A contacts.

Collective B-Mode

Collective B-Mode is basically a "broadcast" mode, although most hams have an aversion to that word because FCC rules say that we are not allowed to "broadcast" in the entertainment sense of the word. In this mode, there is only one sending station, called the Collective B Sending Station (CBSS) and any number of Collective B Receiving Stations (CBRSs). The CBSS sends each character twice: the first transmission of a character, called DX, is followed by four other characters, and the repetition, called RX, is sent. This allows for time-diversity reception with a spacing of 280 ms. The CBRSs examine DX and RX and print whichever one has the correct 4B/3Y ratio. A space or error symbol is printed whenever both DX and RX are mutilated.

This is the mode that W1AW uses for bulletins and the one that should be used for calling CQ, for nets and for making multi-way contacts. It has a 100% duty cycle, so you will have to reduce your transmitter power output to a level that will withstand this duty cycle (25% of peak power on many SSB transmitters).

Selective B-Mode

Selective B-Mode is intended for transmissions to a single station or group of stations. It is similar to Collective B-Mode except that the signals have a 3B/4Y ratio (inverted with respect to Mode A and Collective B-Mode). Only stations set up to accept this mode and recognize the specific selective-calling signal are intended to receive Selective B-Mode transmissions. The Selective B-Mode Sending Station is called SBSS, and the Receiving Station is called SBRS. As with Collective B-Mode, a space or error signal will be printed if both DX and RX are incorrectly received.

AMTOR Selective-Calling Identities

CCIR Recommendations 476-2 and 476-3 provide for four-letter identities to be used in the "call" signal, which consists of two transmission blocks. The first block of three characters has the "signal repetition" (RQ) character in the second position and the first two letters of the station identity in positions 1 and 3. The second block has the last two letters of the station identity in positions 1 and 2 and the RQ character in position 3.

Amateurs can't simply use their call signs as their selective calling identities because amateur call signs consist of letters and numerals. Some calls have up to six characters. Most AMTOR stations use the first letter and the last three letters of their call sign for the selective-calling identities. Using this algorithm, the calling identity for W1AW would be WWAW. This scheme doesn't provide unique identities for every call sign (for example, WWAW would also be used for W2AW, W2WAW and others), so be aware that someone may ask you to choose another identity if an early bird (chirp) got yours.

Making Contact on AMTOR

If you try to make your first contact using Mode A (ARQ) and there is a problem somewhere in the system, you will not make contact. It takes a two-way contact to make Mode A work at all.

Try your first contact using Collective B-Mode. Use this mode for calling CQ, and include your selective-calling identity. Use of Mode B should ensure that you can get the circuit working one way. In Mode B, you can check to confirm that you are using the correct frequency-shift polarity for both receiving and transmitting.

After making contact using Mode B, ask the other station to go into the monitor (listen) mode and try to receive your Mode A transmission. If there is a problem with Mode A, the most likely cause is that your transceiver is taking too long to change between receive and transmit. If that's the case, check with other AMTOR operators having the same equipment set-up. If you are using commercial AMTOR gear, the manufacturer will probably have the information on which transceivers will switch fast enough and which require specific modification. Another source of modification information is a users' group for your brand of transceiver.

If the other station (in monitor mode) is copying your Mode A transmission, ask the station to make a Mode A call to your selective-calling identity. The other station is acting as the master and your station the slave in this case. Finally, reverse the roles and try being the master station.

Because of the timing cycle specified in CCIR Recommendations 476-2 and 476-3, it is not practical to contact AMTOR stations on the opposite side of the earth using Mode A. Mode B can be used for communication with stations too distant for Mode A contacts.

If you need to change frequency during a contact, either to move off a calling frequency or to sidestep QRM, it's best to conclude contact on the original frequency and start over on the new one. AMTOR operators call this a "cold switch," or "cold QSY." Changing frequency without discontinuing contact, called a "hot switch" or "hot QSY," may disrupt other QSOs while you are changing frequency.

ASCII OPERATION

ASCII operation differs from ITA2 operation in that it has a larger character set. Besides a number of control characters, ASCII also provides both upper- and lower-case letters and a more complete set of punctuation marks.

ASCII does away with letters and figures cases, which sometimes cause ITA2 characters to print in the wrong case. ASCII is a character set which was designed for both computer and data communications uses.

Because of printers with faster carriage-return action and elimination of the LTRS/FIGS problem, only CR LF need be sent as an end-of-line sequence. There are other ASCII format effectors (control signals that affect where print occurs on the page) if you want to make use of its entire capability. For more information about the complete ASCII coded character set, and for the technical details about methods and transmission rates used to transmit the ASCII code, see the latest *ARRL Handbook*.

When using ASCII to communicate with overseas stations, avoid the use of so-called "national" characters because they are different in other countries. In fact, ASCII is only the US version of the CCITT International Alphabet Number 5 (IA5), also called ISO-646-006. (The ASCII characters to be avoided are: # $ [ ] △ { | } and ~).

To take advantage of the expanded character set, an ASCII test message could be as follows:

ThE QuicK BrowN FoX JumpS OveR ThE LazY DoG
1234567890

By using some upper-case letters, the message shows that the upper and lower-case letters are printing properly. Other

```
↓↓↓↓↓↓↓↓↓→→→→→<≡                                              Leader
ZCZC→NR→345→R→WA0XYZ→15→SIOUX→FALLS→SD→1030Z→JAN→3<≡           SOH and Preamble
Mrs→Sam→Swarz<≡                                               Address
1234→Main→St<≡
Anytown→IL→64578<≡
617-555-9683<≡                                                Telephone Number
BT<≡                                                          Start of Text
Having→a→wunnerful→time.→→Wish→you→were→here.→→See→you→and→the→kids<≡   Text
soon.→→Love<≡
BT<≡                                                          End of Text
Mary<≡                                                        Signature
CFM→1234→64578→617-555-9683→wunnerful<≡≡≡≡≡≡≡≡                 Confirmation


NNNN↓↓↓↓↓↓↓↓↓↓↓                                               End of Message

                   Legend:    ↓DEL    <CR    ≡LF    →SP
```

Fig 9-6—A sample message format using ASCII.

variations of this message are, of course, possible. By sending it once with all capital letters and once with all lower-case letters, both complete sets can be verified.

When ASCII is used, it is possible that the way the transmitting station formats a message will not be the way that the message is presented to the receiving operator. Computer-based systems may be used to provide split-screen displays, simply file messages away for later use, or process incoming messages according to an agreed protocol.

Radiograms by ASCII

Essentially the same message format is used for radiograms sent by ASCII as is used with ITA2. This is desirable for several reasons, not the least of which is that the same message may be transmitted by one code, then the other, on the way to its destination, and if the same format is used, it can be done without reformatting the computer file. An ASCII example of the same message as used in Fig 9-5 is shown in Fig 9-6.

COMPUTER-BASED MESSAGE SYSTEMS

No discussion of modern radioteletype operation could be complete without some information about auto-start and/or mailbox operation. Since the mid 1970s, a number of Computer-Based Message Systems (CBMSs) have appeared on the ham bands. Possibly you have heard them called one of these terms: MSO (Message Storage Operation), bulletin board or mailbox.

All of these systems have some features in common. They will automatically respond to calls on their operating frequency if the calling station uses the correct character sequence. A remote station can write messages to be stored in the system. A station may request a listing of the messages on file. Stations can subsequently read remotely entered messages plus any bulletin messages. Some systems may provide for messages to have password protection to restrict access to the information.

Message handling in this manner is third-party traffic, and the system operator (SYSOP) is required to observe appropriate rules concerning message content. Also the SYSOP is responsible for maintaining control of the transmitter and removing it from the air in the event of malfunction.

While there are many advantages to CBMS operation, there are some drawbacks, as well. Imagine yourself operating in a worldwide RTTY contest. After you find a clear frequency, you ask twice to see if it is busy, then call CQ CONTEST. Immediately you receive an answer from a European station, and contest messages are exchanged. Then, just as you are about to sign off, a US station comes up on the frequency and says, "QSY, QSY, YOU ARE RIGHT ON TOP OF W7***'S MAILBOX."

We operate on a first-come, first-served basis on the ham bands. We must all realize, however, that there are skip-distance problems and other occurrences that will inadvertently cause QRM to other stations, but we learned to live with this years ago. Because auto-start and mailbox operations must necessarily operate on fixed frequencies, perhaps subbands can be designated for these operations. Then the general conversational RTTY/AMTOR contacts will not bother the automated operations, and vice versa. If you know that a mailbox system operates on a certain frequency, try to avoid that frequency unless you are checking in to the mailbox. As was stated earlier, always be courteous.

RTTY GRAPHICS

If you aren't afraid of getting hooked on something, you might try RTTY graphics. Currently there are three ways of transmitting graphics via RTTY. The oldest is forming pictures by using the different printing characters on a teleprinter. Another is to use the graphics capability that comes with many home computers. The third is to use an emerging computer-graphics system with the generic name of videotex.

Teleprinter Art

Actually, it takes more persistence than raw artistic talent to produce teleprinter art. You can start with a picture or sketch, then make a photocopy to slip in your printer. If it is not the right size, you can get it reduced or enlarged in well-equipped photocopy shops.

Now, turn on the reperforator or open a computer file, and type over the picture with the printer. For dark areas, use dense letters such as W or M, and use H, X, I and punctuation marks for lighter areas.

Patience is required to do the necessary corrections. When you are done with the first draft, you will probably have to go back and do some touching up. When you get the final version done, be sure to sign it with your call sign. As a courtesy to the originator, when you relay a picture be sure not to omit the author's call. An example of what can be done by talented keyboard artists appears in Fig 9-7.

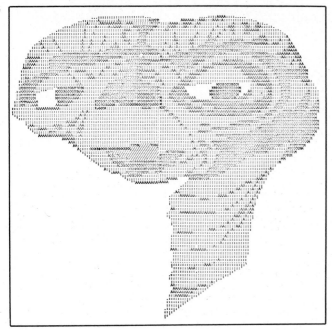

Fig 9-7—"The Extra Terrestrial," by WA3CHN, as received by AG9H. Courtesy of Argonne Amateur Radio Club and *RTTY Journal*.

Keeping RTTY art on paper tape can create a storage problem if you are a compulsive collector. Older tapes tend to dry out and become easily worn. Many RTTY art collectors have adopted the Phillips cassette. However, to get the audio tape to play at full RTTY transmission speed, you will have had to use a tape reader or a computer file to make the original recording. If you have a computer, you're better off using digital storage such as magnetic disk or digital tape.

Printer Art from a TV Camera

Some computer hobbyists have developed ways of connecting their fast-scan video cameras to their computers to produce printer art. Virtually all have been done in ASCII, but there is no reason why an ITA2 version can't be implemented. You may have seen these types of computer-generated photographs at shops in tourist areas.

Personal-Computer Graphics

Many personal computers have a graphics capability. Unfortunately, you can't display someone else's graphics unless you have a compatible computer. Obviously, similarly equipped computers of the same model number with the same graphics software can communicate with each other. It doesn't necessarily follow that so-called "look alikes" have graphics that are compatible with those of the original computer.

Nevertheless, amateur RTTY operators using the same computers and software can get together and have some fun exchanging graphics. It's worthwhile to keep in mind that computer graphics are not necessarily legal on all bands. There has been a long-established practice of sending teleprinter art via Amateur Radio HF bands using the ITA2 character set to print the pictures. There is no problem with this, and there isn't anything in the FCC rules to prevent you from doing the same

thing with AMTOR or the ASCII character set. But once you jump out of these character sets, you are using another code. These other codes would be permissible above 50 MHz under existing rules but not below 50 MHz unless you obtain a Special Temporary Authorization (STA) from the FCC.

Videotex

There is some Amateur Radio interest in transmitting graphics using videotex standards. There are two major videotex standards:
• The North American Presentation-Level Protocol Syntax (NAPLPS)
• International Telegraph and Telephone Consultative Committee (CCITT) Recommendation S.100, "International Information Exchange for Interactive Videotex."

Both of these videotex standards are essentially extensions of the ASCII coded character set. The 7-bit ASCII code allows up to 128 combinations, which are used for letters, numbers, punctuation and control signals. It would be possible to double that number of combinations by adding another bit, making it an 8-bit coding scheme. The other way of getting more combinations is used by videotex—that is using the ASCII SO (shift out) character to jump out of the ASCII character set into one or more different character (or graphic) sets. When you leave ASCII, you have to identify what graphic set you're going into by specific ESCape sequences. You get back from these graphic sets by invoking the ASCII SI (shift in). CCITT Recommendation S.100 has several graphic operations. The options are called *alphamosaic, alphageometric,* and *alpha-dynamically redefinable character sets (DRCS).* CCITT Recommendation S.100 provides a rich repertoire of graphic options for future Amateur Radio applications.

NAPLPS, on the other hand, has had some Amateur Radio experimentation. VE3FTT and VE3GQW gave an account of their NAPLPS on-the-air experiments using the Telidon system in September 1983 *QST.* A black-and-white rendering of the graphics capability of this color-transmission system is shown in Fig 9-8. Using the Telidon system, the operator can select from seven geometric shapes: dot, line, rectangle, arc, circle, polygon and text. Then the operator chooses the fill patterns, colors and pixel densities. For some additional ideas on videotex, see *Byte,* February 1983, pp 203-255 and July 1983, pp 40-129.

Fig 9-8—A black-and-white photo of a color graphic image sent via videotex. Thanks VE3FTT.

Chapter 10

Packet Radio

Packet radio is hot!
A conservative estimate puts 25,000 packet radio terminal node controllers (TNCs) in the hands of Amateur Radio operators today. Meanwhile, Amateur Radio equipment manufacturers work night and day to keep up with the demand for new TNCs as 145.01 MHz becomes one of the most active amateur frequencies from one end of the radio spectrum to the other.

Why has this relatively new amateur mode of digital communications caught on so fast?

It's child's play! Learn a few commands and you too can experience the new adventure in Amateur Radio called packet radio. *(WA1LOU photo)*

Packet radio is communications for the computer age. In the 1980s, a computer in a ham shack is as common as a 2-meter hand-held was in the 1970s. As computers appeared in more and more shacks, computer programs were written to perform amateur modes of communications (primarily, CW and RTTY). The problem is that these modes of communications do not begin to use the true power of the computer.

Writing computer programs to perform CW and RTTY today is like building a mechanical horse for transportation at the turn of the century. It could be done, but what a waste of technology! Rather than building a mechanical horse, some inventors used the technology of the day to build the horseless buggy. And instead of writing computer programs to emulate keyers and teleprinters, some farsighted hams developed a new amateur mode of communication that unleashes the power of the computer. That mode is packet radio.

Being a child of the computer age, packet radio has computer age features that you would expect:

• It is data communications; high speed and error-free packet-radio communications lends itself to the transfer of large amounts of data.

• It is fast, much faster than the highest speed CW or RTTY.

• It is error-free; no "hits" or "misses" caused by propagation variances or electrical interference.

• It is spectrum efficient; multiple communications may be conducted by multiple stations on the same frequency at the same time.

• It is networking; any packet-radio station can command other packet stations to create a network for packet radio communications.

• It is message storage; packet-radio bulletin boards (PBBS) provide storage of messages for later retrieval.

So, the many hams with computers in their shacks are naturally attracted to packet radio. Now, they are able to unleash the power of their computers, too.

THE FUNCTION OF A TNC

A packet-radio TNC could be considered a very intelligent modem. Whereas a telephone modem permits a computer to communicate by means of the telephone network, a packet radio "modem" permits a computer to communicate by means of the packet-radio network.

Similarly, as an "intelligent" telephone modem augments its simple modem functions by including a wide range of built-in commands to facilitate computer telephone communications, the built-in intelligence of a packet-radio modem facilitates packet-radio communications. The advantage of this built-in intelligence is that it allows the computer to use communication programs of the "plain vanilla" variety. The software does not have to provide the intelligence because it is already in the modem, thus freeing the computer to perform more important tasks.

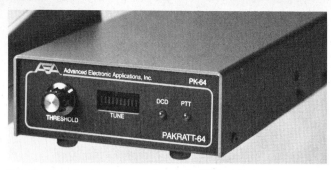

Besides functioning as a packet-radio TNC, AEA's PK-64 transmits and receives Morse, Baudot, ASCII and AMTOR. The "PAKRATT"® is specifically designed for interfacing with the Commodore 64®, SX-64 and C-128 computers.

To initiate computer telephone communications, you command an intelligent modem to address (dial the telephone number of) the other computer. Similarly, to initiate computer packet-radio communications, you command a TNC to address (make a connection with the call sign of) the other Amateur Radio station.

Both a telephone modem and a packet-radio modem are connected to a communications medium: the telephone modem to the telephone line and the packet-radio modem to the radio. And just as a telephone modem is normally connected to the RS-232-C interface of a computer, a packet-radio modem is also connected to that same computer interface.

One of the first commercial TNCs, the GLB PK-1 has a unique command set and is very popular in dedicated digipeater applications.

This chapter describes what to do after you connect a TNC to your radio and computer.

A Few Words about Protocols, Commands, and this Chapter

The manner in which packet-radio communications is conducted is called a "protocol." The protocol consists of a standard set of rules and procedures that are programmed into each TNC so all TNCs will communicate in a compatible manner. Although there have been several protocols used from time to time, the standard in wide use today is called the AX.25 protocol.

After the Canadian Department of Communications authorized amateur packet radio in 1978, the VADCG protocol and a companion TNC were developed by the Vancouver Amateur Digital Communications Group

(VADCG). Two years later, the FCC authorized amateur packet radio in the US and after extensive experimentation, American packet-radio developers agreed on a protocol: the AX.25 protocol.

The original TNC that was designed and kitted by VADCG only supported VADCG protocol (the only protocol that existed at the time). These TNCs are commonly known as "VADCG" TNCs.

When AX.25 was implemented for the first time [by Tucson Amateur Packet Radio Corporation (TAPR) in their "TNC 1"], the VADCG protocol was also supported to be compatible with the VADCG TNCs that, at that time, out-numbered the new TAPR TNCs. The TNC 1 also had a new design and command set to support AX.25. The same design and commands were later used in "TNC 1 clones," most notably AEA's PKT-1 and Heathkit's HD-4040.

The Heathkit® HD-4040, like its forefather, the TAPR TNC 1, is a kit that can be assembled in a few hours.

In 1985, TAPR introduced an AX.25-only TNC, the "TNC 2," which used an expanded TNC 1 command set. The TNC 2 also had its clones, most notably AEA's PK-80, GLB's TNC2A, MFJ's MFJ-1270, and Pac-Comm's TNC200.

The TNC 2 command set is the most popular in packet radio today. Therefore, this chapter uses TNC 2 commands throughout its description of packet-radio operation. If your TNC is not a TNC 2 or an exact TNC 2 clone, equivalent commands may be found in Table 10-1.

The TNC 2 (and TNC 1) is configured and controlled by entering commands into the keyboard of your computer or terminal. Commands may only be entered when the TNC's command prompt (cmd:) is displayed by your computer or terminal. When this prompt is displayed, type the desired command and follow it with a carriage return. For example, to command the TNC to disconnect, you type "DISCONNE" and a carriage return at the command prompt. In this chapter, this operation will be represented as:

cmd: DISCONNE <CR>

(A)

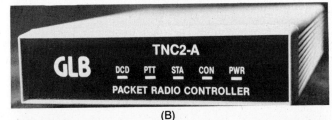

(B)

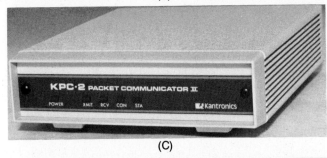

(C)

(D)

TNC 2 clones, the AEA PK-80, the GLB TNC2-A, the Kantronics Packet Communicator II, and the MFJ Model MFJ-1270, are all preassembled and ready to packet. Another TNC 2 clone, the Pac-Comm TNC-200, is available as a kit, as well as preassembled.

To save keystrokes, many TNC 1 and 2 commands may be used in their shorthand version. For example, instead of typing DISCONNE for the disconnect command, you can simply type the letter D. In this chapter, each command will be printed partially in uppercase characters and partially in lowercase characters. The uppercase and lowercase characters together represent the longhand version of the command, whereas the uppercase characters alone represent the shorthand version of the command. For example, the disconnect command will be represented as:

Disconne

where Disconne is the longhand version of the command and D is the shorthand version.

Tailoring The TNC To You

Before you transmit your first packet, you must configure the TNC to your needs. These include the requirements of your station equipment, specifically your computer or terminal and your radio, and the requirements of the radio spectrum you will use, either HF or VHF/UHF.

Terminal Parameters

Whether you are using a terminal or a computer running communications software that emulates a terminal (hereinafter also referred to as a "terminal"), you should configure the TNC to be compatible with the terminal's operating parameters.

Baud Rate—Set the baud rate of your TNC's serial port to the baud rate of your terminal (your terminal is connected to the TNC's serial port). On the TNC 2, the baud rate (300, 1200, 2400, 4800 or 9600 baud) is set by means of a rear panel DIP switch (positions 1 through 5) while the TNC is powered off. (All other parameters, except radio baud rate, are set while the TNC is powered on.)

Echo—The echo function causes the TNC to print on your terminal's display each character that you type into your terminal's keyboard. This TNC function should be turned on or off depending on whether your terminal does or does not provide this function. (Echo on is the TNC default.)

If two characters are displayed for each one that you type into the keyboard (you type HELLO and HHEELLLLOO is displayed), turn off the echo function by typing:

cmd: Echo OFF <CR>

If nothing is displayed when you type into the keyboard, turn on the echo function by typing:

cmd: Echo ON <CR>

Character Length—Set the TNC to the character length used by your terminal. (7 bits per character is the default value.)

To select 7 bits per character, type:

cmd: AWlen 7 <CR>

To select 8 bits per character, type:

cmd: AWlen 8 <CR>

Parity—Set the TNC to the parity used by your terminal. (Even parity is the TNC default).

To select even parity, type:

cmd: PARity 3 <CR>

To select odd parity, type:

cmd: PARity 1 <CR>

To select no parity, type:

cmd: PARity 0 <CR>

or

cmd: PARity 2 <CR>

Screen Length—Set the TNC to the maximum number of characters that is displayed on each line of your terminal (80 characters per line is the TNC default), by typing:

cmd: Screenln n <CR>

where n is number 0 to 255, representing the number of characters per line displayed by your terminal. This will cause the TNC to insert a <CR> automatically if you receive a line longer than your terminal screen's line length. If your terminal provides the feature automatically, or if you simply wish to disable this feature, set Screenln to 0.

Line Feeds—The automatic line feed function causes the TNC to insert a line feed after each carriage return that is received in incoming packets. This TNC function should be

Table 10-1

Command Compatibility

The following table lists the user commands of the various TNCs that are in use today. Each line lists equivalent or nearly equivalent commands. Note that in some instances, a command in one TNC is equivalent to multiple commands in another TNC (for example, the GLB PK1 AC, SD, and SV commands are equivalent to the TAPR TNC 2 CONNECT command).

TAPR TNC 1/2 (1)	AEA PK-64 (4)	GLB PK1 & PK1L (5)	KANTRONICS PC II
8BITCONV	---	---	---
ABAUD (2)	---	---	ABAUD
ABIT (2)	---	---	---
---	AUTOCR-R	---	---
---	AUTOCR-T	---	---
AUTOLF	AUTOLF-R	OL	AUTOLF
AWLEN	---	---	---
AX25 (2)	---	SX	---
AX25L2V2 (3)	AX25L2V2	---	AX25L2V2
AXDELAY	AXDELAY	---	AXDELAY
AXHANG	AXHANG	---	AXHANG
---	BAND HF/VHF	---	CCITT
BEACON	BEACON	BT	BEACON
---	---	---	---
BKONDEL	---	---	BKONDEL
BTEXT	BTEXT	BK BR BS	BTEXT
BUDLIST (3)	MONREJ	---	BUDLIST SUPLIST
CALIBRA	---	DH DL DO	CALIBRATE
CALSET	---	---	---
CANLINE	---	SY	CANLINE
CANPAC	---	---	CANPAC
CHECK (3)	CHECK	SQ	CHECK
CLKADJ (3)	---	---	DAYTWEAK
CMDTIME	---	SK	CMDTIME
CMSG (3)	CONMAG	---	CMSG
COMMAND	---	ME SY	COMMAND
CONMODE	CONMODE	ML	CONMODE
CONNECT	CONNECT	AC SD SV	CONNECT
CONOK	---	---	CONOK
CONPERM	---	---	---
CONSTAMP (3)	CONSTAMP	---	CSTAMP
CONVERS	CONVERS	MS	CONVERS
CPACTIME	CPACTIME	---	CPACTIME
CR	---	---	CR
CSTATUS (3)	---	---	STATUS
CTEXT (3)	CTEXT	---	CTEXT
CWID (2)	---	---	---
---	DAYSTAMP	---	---
DAYTIME (3)	---	---	DAYTIME
DAYUSA (3)	---	---	DAYUSA
DEBUG (2)	---	---	---
DELETE	---	SY	DELETE
DIGIPEAT	DIGIPEAT	MR	DIGIPEAT
DISCONNE	DISCONNECT	AD	DISCONNECT
DISPLAY	---	S	DISPLAY
DWAIT	DWAIT	---	DWAIT
ECHO	ECHO	SE	ECHO
---	---	---	EQUALIZE
ESCAPE	---	SY	ESCAPE
---	---	---	EXCARDET
FLOW	---	---	FLOW
FRACK	FRACK	ST	FRACK
FULLDUP	FULLDUP	---	FULLDUP
---	HARDTIME	---	---
HBAUD (2)	---	SJ	HBAUD
HEADERLN (3)	HEADERLN	---	HEADERLN
---	---	---	HELP
---	BAND HF/VHF	---	HF
---	BAND HF/VHF	---	HFTONES
HID (3)	---	---	HID
ID (3)	---	---	ID
ID (2)	---	MI	---
IDTEXT (2)	---	---	---
LCALLS (3)			
MTO MFROM (2)	BUDCALLS SUPCALLS	MTO MFROM	OA
LCOK	LCOK	SU	LCOK
LFADD	AUTOLF-T	MF	LFADD
MALL	MALL	---	MV
MAXFRAME	MAXFRAME	SH	MAXFRAME
MCOM (3)	MONITOR	OD	MCOM
MCON	MONCON	OC	MCON
---	MONDIGI	---	---
MFILTER (3)	MFILT	---	FILTER
MHCLEAR (3)	---	---	MHCLEAR
MHEARD (3)	---	---	MHEARD
---	---	---	MODEMENA
MONITOR	MONITOR	OO	MONITOR
MRPT (3)	MONRPT	---	MRPT
MSTAMP (3)	MONSTAMP	---	MSTAMP
MYALIAS (3)	---	---	MYALIAS
MYCALL (3)	MYCALL	SC	MYCALL
MYVADR (2)	---	SA	---
NEWMODE (3)	---	---	NEWMODE
NUCR	---	---	NUCR
NULF	---	---	NULF
NULLS	---	ON	NULLS
PACLEN	PACLEN	SL	PACLEN
PACTIME	PACTIM	SK	PACTIME
PARITY	---	---	PARITY
PASS	---	---	PASS
PASSALL (3)	PASSALL	SG	---
PERM (2)	---	---	PERM
---	PRINTER	---	---
---	PRNCMD	---	---
PROGRAM (2)	---	---	---
RECONNECT (3)	---	---	RECONNECT
REDISPLA	---	---	REDISPLA
---	---	---	REPEATER
RESET	---	SI	RESET
RESPTIME	RESPTIME	---	RESPTIME
RESTART	---	---	---
RETRY	RETRY	SN	RETRY
---	---	---	RING
SCREENLN	---	OW	SCREENLN
SENDPAC	---	SY	SENDPAC
---	SQUELCH	SB	---
START	---	---	START
---	---	---	STATSHRT
STOP	---	---	STOP
STREAMCA (3)	---	---	STREAMCA
STREAMDBL (3)	---	---	STREAMDB
---	---	---	STREAMEV
STREAMSW (3)	---	---	STREAMSW
---	---	---	SWDETENA
TRACE	---	---	TRACE
TRANS	TRANSPARENT	MS MX	TRANS
TRFLOW (3)	---	---	TRFLOW
TRIES (3)	---	---	---
TXDELAY	TXDELAY	SF SP	TXDELAY
TXFLOW	---	---	TXFLOW
UNPROTO	UNPROTO	BD BV MU	UNPROTO
USERS (3)	CONMAX	---	MAXUSERS
VDIGIPEA (2)	---	---	---
VRPT (2)	---	---	---
---	WRAP	---	---
XFLOW	---	---	XFLOW
XMITOK	XMITOK	AS	XMITOK
XOFF	XOFF	---	XOFF
XON	XON	---	XON

(1) TNC 1 compatibles: AEA PKT-1, Heathkit HD-4040. TNC 2 compatibles: AEA PK-80, GLB TNC2A, MFJ MFJ-1270, Pac-Comm TNC-200.

(2) TNC 1-only commands (not TNC 2)

(3) TNC 2-only commands (not TNC 1)

(4) Only usable with Commodore C-64 and C-128 computers

(5) Unique commands: AA AH AI AR AT AW AX BC I K MA MC MD MK MM MT OB OF OI OM OP OQ OU SR SS SZ

Note: TNC manufacturers' addresses:

AEA, P.O. Box C-2160, Lynnwood, WA 98036
GLB, 151 Commerce Pkwy., Buffalo, NY 14224
Heathkit, Benton Harbor, MI 49022
Kantronics, 1202 E. 23rd St., Lawrence, KS 66046
MFJ, P.O. Box 494, Mississippi State, MS 39762
Pac-Comm, 4040 W. Kennedy Blvd., Suite 620, Tampa, FL 33609
TAPR, P.O. Box 22888, Tucson, AZ 85734-2888

Packet-Radio Software

Today, nearly all computers in use in the ham shack may be programmed to function as terminals. The software that does this transformation may turn a computer into a bare-bones, dumb terminal or a very sophisticated data communications device. The drawback with such software is that it is primarily designed to operate with telephone modems, not radio modems (TNCs). As a result, these programs include features that are applicable only to telephone data communications, features that are useless to the packet-radio user. At the same time, these programs lack features that the packet-radio user would find useful.

Do not despair! Now, there is data-communications software available that is optimized for packet-radio operation. This software was written with packet radio in mind, so the user does not have to be compromised by telephone modem software any longer. A description of some of the packet-radio software that is now available follows.

MacPacket—this program for the Apple® Macintosh computer creates an amiable interface for the TAPR and Kantronics TNCs. As you would expect from a Macintosh program, MacPacket has windows (for transmitted and received data) and pull-down menus (to access all of the TNC commands). One of the nicest features of the program is its ability to store the routes of the stations you contact frequently. All you have to do is select the connect command from a pull-down menu and type the call sign of the station you wish to contact. The program looks up the route for that station and a connection may be initiated.

MacPacket is available in two flavors, MacPacket/TAPRterm for the TAPR TNCs (and clones) and MacPacket/KANterm for the Kantronics TNCs. Each is available from Brincomm Technology, 3155 Resin St, Marietta, GA 30066.

TNC64—This program is the result of a project conducted by the Texas Packet Radio Society to provide packet-radio terminal programs for popular computers. The Society chose the most popular computer in ham radio, the Commodore 64™, as the object of its initial effort, and the result is TNC64. The program supports all TAPR TNC 1 and TNC 2 (and clone) terminal control functions which have been made selectable from plain-language menus. Easy-to-understand screen-prompts guide the user through the many features of the program.

The program's 50,000-character capture buffer is its best feature. All data that pass between the TNC and the computer are captured in the buffer, which can be viewed on a monitor, copied to disk, printed or erased. When the buffer is full, its contents are automatically saved to disk. Each disk can save the contents of three full buffers, a total of 150,000 characters.

TNC64 is available from the Texas Packet Radio Society, PO Box 831566, Richardson, TX 75083.

Model 100 Mailbox—Radio Shack's TRS-80® Model 100 lap-top computer has been programmed to function as an automatic-answering mailbox for packet-radio applications by Dick Roux, N1AED. The computer responds to connects with a stored message that you create and then proceeds to store received messages in a file with the time and date of the connection. The file may be read or printed at any time, it may be cleared from memory, or more messages may be appended to it.

The program also has the ability to handshake with the W0RLI message forwarding system to allow you to have your mail forwarded directly to your Model 100. Another feature of the program is a call sign look-up table, which you may create to add the personal touch. As a result, when K7UGA connects with you, your mailbox's greeting is prefaced with ''Hello Barry...''

N1AED has also written a companion program that sets a Kantronics TNC to the parameters required by the mailbox and then automatically loads the mailbox program. Both programs may be obtained from N1AED (25 Greenfield Dr, Merrimack, NH 03054) by sending him an SASE for printed listings of the programs or a blank cassette tape and a post-paid cassette mailer for recorded copies of the programs. The programs may also be downloaded from the CP/M files sections of many PBBS.

The Software Approach—Bob Richardson, W4UCH, has written Z80® assembly-language programs for the Radio Shack TRS-80® Model I, III and 4 computers that emulate the functions of a TNC. All you need is an external modem. The programs are menu driven and provide access to all the functions required for packet-radio data communications.

Bob has written programs for both the VADCG and AX.25 protocols. They are available from Richcraft Engineering, Ltd, Drawer 1065, 1 Wahmeda Industrial Park, Chautauqua, NY 14722.

```
 File  Edit  Service  Options  Mode

 00:11:58      [✉][^s][💾]   9600-N-8-1-FULL    [O]^s [O]^q [O]^c

*** DISCONNECTED

cmd:mon on
was  OFF
cmd:c wallou v kalion
*** CONNECTED to WA2FTC-1
Can't CONNECT, Link state is: CONNECTED to WA2FTC-1 VIA W1AW-5
HI guy.>
hi guy...finally got the tnc1 hooked up to the 2 meter box>
ok, was just abt to ask you which TNC u are using....
did you ever put a msg on  cis re: the MFJ & mac?>
NO. HAVE NOT CHECKED IN TO CIS SINCE THE PROBLEM OCCURRED....
TRIED TO LOG ON AT 2400 BAUD EARLIER, BUT WHEN THAT FAILED, I
HUNG UP AND WILL LOG ON LATER (OR TOMORROW) AT 1200 BAUD AND
POSE THE PROBLEM (UNLESS I FIND THERE NO LONGER IS A PROBLEM)!!!>>>
well I have a cure for your 2400 bit/s problem...
delete the three letters on the front of the modem...
and it should work fine!!>
NICE GUY! BY THE WAY, DONT SELL ALL YOUR STOCK YET...
THEY ARE SHIPPING MORE STUFF THIS QUARTER THAN THEY'VE SHIPPED IN ANY
QUARTER DURING THE LAST TWO YEARS!!!!>>>
see what getting rid of me did for the co.>
YES. IT TOOK A FEW MONTHS TO HAVE ANY EFFECT, BUT IT WAS SURELY WORTH IT!!!>>>
█
```

Typically, the packet-radio operator uses communications software designed for telephone data communications. Here, an Apple® MacIntosh running Red Ryder® 7.0, displays a window interspersed with received data and transmitted data. The software includes unnecessary telephone-specific commands like dial and redial.

```
 File  Edit  Command  Parameters  Routing

                      MacPacket

yes I saw...
did you come and conquer also?
what u running now?>
MacPacket
what???>
oh>
are you xeroxing now?
no....
think pwr sply bad...
have ibm pc and wa7mbl code for tnc>
nice
i mean bbs>
i mean nice
what's nice.?>
a city in france
ha ha>
picture this.........
no...you mean hi hi...

════════════════════ Send ════════════════════
are you xeroxing now?
nice
i mean nice
a city in france
no...you mean hi hi...
```

The interface between a packet-radio operator and a TNC is friendlier when communications software designed for packet radio is used. Here, an Apple® Macintosh running MacPacket, displays an upper window containing received data and a lower window containing transmitted data. The pull-down menus across the top of the screen access TAPR TNC commands.

Packet Radio Quick Start

The following steps will get you on the air as quickly as possible:

Installation . . .

1) Connect a standard RS-232-C modem cable between the serial ports of your TNC and computer or terminal.

2) Connect the radio port of your TNC to your transceiver's audio output (SPKR), audio input (MIC), press-to-talk (PTT) and ground connections. The TNC audio-output level should be adjusted for proper transmitter deviation.

3) Connect the TNC's power cable between the power connector of your TNC and a suitable power source.

4) If your TNC is a TAPR TNC 2 or TNC 2 clone, set the rear panel switch of your TNC to the baud rate to be used by your radio and computer or terminal with the TNC powered off.

Operation . . .

5) Power up your TNC.

6) Type a carriage return <CR> to display the command prompt (cmd:).

7) If your TNC is a TAPR TNC 1 or TNC 1 clone, set your TNC to the baud rate to be used with your radio by typing:

cmd: HBAUD n <CR>

where n is the radio baud rate. Also, set your TNC to the baud rate of your computer or terminal by typing:

cmd: ABAUD n <CR>

where n is the computer or terminal baud rate.

8) Install your call sign in your TNC by typing:

cmd: MYCALL WA1LOU <CR>

where WA1LOU is your call sign.

9) Initiate a contact by typing:

cmd: CONNECT K1WJ <CR>

where K1WJ is the call sign of the station to contact. To use a digital repeater to initiate a contact, type:

cmd: CONNECT K1WJ VIA W1AW-5 <CR>

where K1WJ is the call sign of the station to contact and W1AW-5 is the call sign of the digipeater. When the message:

*** CONNECTED to K1WJ

is displayed, a contact is established. Anything that you type into your computer or terminal is sent to the other station after typing <CR>.

10) To end the contact, enter command mode (<CTRL-C> on most TNCs), then type:

cmd: DISCONNE <CR>

When the message:

*** DISCONNECTED

is displayed, your contact is ended.

Note: While this method will get you on the air, be sure to read your TNC's manual.

---

turned on or off depending on whether your terminal does or does not require this function. (Automatic line feed on is the TNC default.)

If a blank line follows each line of received text, turn off the automatic line feed function by typing:

cmd: AUtolf OFF <CR>

If each line of received text is printed over the previously received line of text, turn on the automatic line feed function by typing:

cmd: AUtolf ON <CR>

Lowercase and Uppercase Characters—This TNC function

normally sends both lowercase and uppercase characters received in incoming packets to your terminal; however, if your terminal does not accept lowercase characters, this TNC function may be set to automatically translate all received lowercase characters to uppercase characters.

If your terminal accepts both lowercase and uppercase characters, type:

cmd: LCok ON <CR>

If your terminal accepts only uppercase characters, type:

cmd: LCok OFF <CR>

Those are the most important parameters that you need to set to configure your TNC for compatibility with your terminal. By setting these parameters correctly, your terminal should now display text from your TNC in a clear, legible manner. If not, you should recheck these parameters and also check the physical connection between your TNC and terminal.

(In addition to the previously described terminal parameter commands, there are other terminal-related commands that allow you to select ASCII characters that perform various functions: input editing, flow control, etc. Refer to the sidebar entitled "Control Characters" for more information.)

Radio Parameters

The TNC also includes commands that customizes it to the needs of your radio equipment and the radio spectrum that you are using.

Baud Rate—Set the radio baud rate of the TNC to the data rate you will use (300 bauds for HF, 1200 or 9600 bauds for

---

Control Characters

Control characters, that is, characters that are typed while simultaneously pressing your terminal's control <CTRL> key and one other key, perform a number of useful functions. Some of the TNC default selections for these characters are the same as those commonly used by other devices. If your terminal is unable to generate a particular character, or if you simply wish to select a different character to perform the same function, the following commands allow you to change the default character to another character of your choosing. (Note that the following list also includes control characters <CR> and <DELETE> which are *not* used in conjunction with the <CTRL> key.)

Default	Character	Command Character Function
<CR>	SEndpac	Send previous keyboard input as a packet
<CTRL-C>	COMmand	Enter command mode (from converse mode)
<CTRL-Q>	STArt	Restart sending to terminal (from TNC)
<CTRL-Q>	XON	Restart sending to TNC (from terminal)
<CTRL-R>	REDisplay	Redisplay typed line before <CR>
<CTRL-S>	STOp	Stop sending to terminal (from TNC)
<CTRL-S>	XOff	Stop sending to TNC (from terminal)
<CTRL-V>	PASs	Insert special character
<CTRL-X>	CANline	Cancel typed line before <CR>
<CTRL-Y>	CANPac	Cancel typed packet before <CR>
<DELETE>	DELete	Delete typed character
<|>	STReamsw	Change streams

VHF/UHF). On the TNC 2, the radio baud rate is set by means of a rear panel DIP switch (positions 6 through 8) while the TNC is powered off. (All other parameters, except terminal baud rate, are set while the TNC is powered on.)

Receive-to-Transmit Turnaround—When a transceiver is switched from the receive mode to the transmit mode, there is a short delay after the switching begins and before intelligence can be sent over the air. Although this delay is only measured in milliseconds (ms), it is critical in packet radio because a TNC is capable of enabling a transmitter and sending intelligence instantaneously. Therefore, some delay must be programmed into the TNC to be compatible with the receive-to-transmit turnaround time of your transceiver (the receiving station requires a few milliseconds to synchronize the transmitted signal as well). This delay may be varied from 0 to 1200 ms by typing:

cmd: TXdelay n <CR>

where n is number 0 to 120 representing a delay in 10 ms increments [30 (300 ms) is the TNC default]. [For example, to set the delay to 500 ms, set n to 50 (500 ms ÷ 10 = 50).]

Your transceiver's receive-to-transmit turnaround time may be specified in your transceiver's manual. If not, you will have to experiment to find a delay that is suitable. In general, if your transceiver uses solid-state switching, you may set the delay relatively low. If your transceiver uses a mechanical relay for switching, you should set the delay higher. Also, synthesized transceivers may require a greater delay than crystal-controlled transceivers. If you are using an external amplifier, you must also take its receive-to-transmit switching delay into account.

Digital Repeater Delays—If you contemplate packet-radio operation through a digital repeater, you must counteract the increased delays that occur when a digital repeater packet collides with another packet. When such a collision occurs, the station originating the packet, not the digital repeater, must initiate resending the packet, and this process consumes a lot of time.

To avoid collisions with digital repeater packets, you may use the Dwait command to insert a delay that causes your TNC to wait the selected Dwait time before it begins sending a packet. (A digital repeater does not have this delay, so it always transmits first, thus reducing collisions with non-repeater stations that do have this delay.) This delay is selected by typing:

cmd: DWait n <CR>

where n is the number 0 to 250 representing a delay in 10 ms increments [16 (160 ms) is the TNC default]. [For example, to set the delay to 1200 ms, set n to 120 (1200 ms ÷ 10 = 120).]

Voice Repeater Delays—If you plan on operating packet radio through a standard VHF/UHF voice repeater, you should add additional delay to the TNC's receive-to-transmit turnaround time to compensate for the relatively long receive-to-transmit delay that is usually experienced when keying up a voice repeater. This additional delay is selected by typing:

cmd: AXDelay n <CR>

where n is number 0 to 180 representing a delay in 10 ms increments (0 is the TNC default). [For example, to set the delay to 1000 ms, set n to 100 (1000 ms ÷ 10 = 100).]

The Axdelay time is not needed by a second station when the repeater continues transmitting after a station stops transmitting through the repeater. This repeater feature is commonly referred to as a "squelch tail" or hang time, and packet stations can take advantage of this feature by configuring their TNCs to start transmitting before the end of the squelch tail. The length of the squelch tail is programmed into the TNC by typing:

AXHang n <CR>

where n is number 0 to 20 representing the length of the squelch tail in 100 ms increments (0 is the TNC default). [For example, to set the squelch tail length to 500 milliseconds, set n to 5 (500 ms ÷ 100 = 5).]

Usually, a repeater's key-up delay and squelch tail length are not readily obtainable information, so you may have to experiment with the Axdelay and Axhang commands until you find the appropriate timing for each parameter. Of course, if other stations have been sending packets through the same repeater, you can ask them for advice!

Station Identification—Perhaps one of the most important TNC commands is the one that inserts your call sign in the TNC. To program your TNC with your call sign, type:

cmd: MYcall WA1LOU <CR>

where WA1LOU is the call sign of your station.

An optionally specified secondary station identification (SSID), number 1 through 15, may be appended to your call sign by typing a hyphen and the desired number immediately after the call sign in the Mycall command (for example, WA1LOU-4). SSIDs are normally used to differentiate digital repeaters, packet bulletin-board systems (PBBSs), and other secondary packet-radio station operations from individual packet-radio stations that use the same call sign. [For example, my individual packet-radio station is identified as WA1LOU (or WA1LOU-0 because 0 is assumed to be the SSID unless specified otherwise), whereas my PBBS is identified as WA1LOU-4. On the TNC 1, the PERM command should be used to make this setting, as well as the others mentioned above, permanent.]

Making a Connection

Now that your TNC is tailored to your station, you can send your first packet.

To contact another station, type:

cmd: Connect K3OX <CR>

where K3OX is the call sign of the station you wish to contact.

If K3OX's packet-radio station is on the air and receives your connect request, your station and his will exchange packets to set up a connection between the two stations. When the connection is completed, your terminal will display:

*** CONNECTED to K3OX

and your TNC automatically switches to the Converse Mode.

Now, everything you type into the terminal keyboard is packetized and sent to the other station. A packet is sent whenever you enter a carriage return <CR> (which is the TNC default for the Sendpac character) or whenever the byte length (character count) of your keyboard input equals the number of bytes selected with the Paclen command. (The TNC default is 128 bytes.)

The byte length of a packet may be changed in the Command Mode by typing:

cmd: Paclen n <CR>

where n is number 0 to 255 (0 selects 256 bytes and 1 through 255 selects 1 through 255 bytes, respectively).

When you are finished conversing with the other station, return to the Command Mode by typing <CTRL-C> (which is the TNC default for the Command character). When the command prompt (cmd:) is displayed, type:

cmd: Disconne <CR>

and your station will exchange packets with the other station to break the connection between stations. When the connection is broken, your terminal will display:

*** DISCONNECTED

If, in mid-contact, you wish to enter a command and then continue with the contact, enter the Command Mode by typing <CTRL-C>. When the command prompt (cmd:) is displayed, enter the desired command. When you are ready to exit the Command Mode and return to your contact, at the command prompt, type:

cmd: CONVers <CR>

and you are back in the Converse Mode. Anything you type now is packeted and sent to the other station. Note that CONVers may also be abbreviated with the single letter K.

If, for some reason, the other station does not respond to your initial connect request, your TNC will resend the request until the number of attempts equals the number selected with the Retry command plus one. (The TNC default is 10 attempts.) This may be changed by typing:

cmd: REtry n <CR>

where n is number 0 to 15, representing the maximum number of retries.

When the number of attempts exceeds the Retry command selection, your TNC ceases sending connect requests and your terminal will display:

*** retry count exceeded
*** DISCONNECTED

(Retry not only sets the maximum number of times that a connect request is sent before the TNC gives up, but also sets the number of times any packet is sent without receiving an acknowledgement before the TNC quits.) A TNC can reject a connect request if it is busy or the operator has set CONOK off. If this happens when you try to connect, your TNC will display:

*** K3OX busy
*** DISCONNECTED

PACKET RADIO REPEATING

When you are unable to make a connection with another station, the problem may be caused by terrain or propagation preventing your signal from being received by the other station. Packet radio has the ability to circumvent this problem by calling on other packet-radio stations to relay your signal to the intended station. All that is required is knowledge of which on-the-air packet-radio stations can send and receive signals between your station and the station you are intending to contact. Once the existence of a suitable intermediary station is known, type:

cmd: Connect K3OX Via W1AW-5

where K3OX is the call sign of the station to connect to and W1AW-5 is the call sign of the station that will act as the intermediary.

When W1AW-5 receives your connect request, it stores your

request in memory until the frequency is quiet and then it retransmits your request to K3OX on the same frequency. This action is called "digipeating," a contraction of "digital repeating." If K3OX's packet-radio station is on the air and receives the relayed connect request, your station and his will exchange packets through W1AW-5 to set up a connection. Once the connection is established, your terminal will display:

*** CONNECTED to K3OX VIA W1AW-5

Your station then continues to use the facilities of the digital repeater until the connection is broken.

Digital and voice repeaters both repeat, but the similarity ends there. Notice that digital repeaters differ from typical voice repeaters in a number of ways. A digital repeater ("digipeater") usually receives and transmits on the same frequency (whereas a voice repeater receives and transmits on different frequencies). As a result, a digipeater operates simplex; that is, it does not receive and transmit at the same time (as compared to a voice repeater, which simultaneously transmits whatever it receives). When a digipeater receives a packet, it stores the packet in memory (in the TNC's random-access memory, RAM) temporarily until the frequency is clear. It then retransmits the packet. Also, a digipeater repeats packets *only* when it is asked to repeat packets (versus a voice repeater that repeats everything that it receives on its input frequency).

If one digipeater is insufficient to establish a connection, as many as eight stations can be called upon for digipeater operation in your connect request! Additional digipeaters are added to the connect command separated by commas. For example, by typing "Connect K3OX Via W1AW-5,

Go to the Source

Active packet-radio amateurs must keep tabs on the ever-changing world of packet radio, and they have a number of good sources of information.

ARRL HQ (225 Main St, Newington, CT 06111) provides packet-radio coverage in *QST;* the magazine's computer column, On Line, is a regular source of packet-radio information. ARRL also publishes *Gateway,* the ARRL packet-radio newsletter, 25 times a year, and also annually publishes the papers presented at the ARRL Amateur Radio Computer Networking Conference.

Both the US and Canadian "national" packet-radio organizations publish newsletters that cover each organization's activities. TAPR (PO Box 22888, Tucson, AZ 85734-2888) publishes monthly as part of *Packet Radio Magazine,* and VADCG (9531 Odlin Rd, Richmond, BC V6X 1E1, Canada) publishes *The Packet.*

Florida Amateur Digital Communications Association (FADCA, 812 Childers Loop, Brandon, FL 33511) publishes *Packet Radio Magazine* each month. The magazine includes newsletters from other amateur packet-radio organizations, as well as a lot of good original material.

Amateur Radio Research and Development Corporation (AMRAD, PO Drawer 6148, McLean, VA 22106-6148) publishes the monthly *AMRAD Newsletter,* which devotes a lot of space to packet radio.

Get ***CONNECTED to Packet Radio by Jim Grubbs, K9EI (available from ARRL or QSKY Publishing, PO Box 3042, Springfield, IL 62708) is a recently published book that provides a good introduction for packet-radio newcomers.

HamNet is a ham radio group on *CompuServe* [5000 Arlington Centre Blvd, Columbus, OH 43220, telephone 800-848-8199 (in Ohio, 614-457-0802)] that is accessible, with a modem, over the telephone. The group has a section devoted to packet radio with a roster of members that is a who's who of the packet-radio community.

WA2FTC-1'' after the command prompt (cmd:), your TNC will send the K3OX connect request to W1AW-5 which relays it to WA2FTC-1. Then, WA2FTC-1 relays it to K3OX.

It is recommended that you not use more than one or two digipeaters at any one time, especially during the prime-time operating hours (evenings and weekends) because multiple digipeater throughput decreases dramatically as activity increases. Each time you use a digipeater, you are competing with other stations attempting to use the same digipeater. Each station that you compete with has the potential of generating a packet that may collide with your packet (which causes your TNC to resend the packet). The more digipeaters you use, the more stations you compete with, greatly increasing the chance of a packet collision. As a result, it may be difficult to get one packet through multiple digipeaters and it is often impossible to carry on a meaningful conversation in this way.

Any packet-radio station can act as a digipeater. This occurs automatically without any intervention by the operator of the station being used as a digipeater. You do not need his permission, only his cooperation, because he can disable his station's digipeater function by means of the Digipeat command. (In the spirit of Amateur Radio, most packet-radio operators leave the digipeater function enabled, disabling it only under special circumstances.)

Similar to VHF/UHF voice repeaters, some stations are set up as dedicated digipeaters. They are usually set up in good radio locations by packet-radio clubs. Besides location, the other advantage of a dedicated digipeater is that it is always there (barring a calamity). Stations do not have to depend on the whims of other packet-radio stations, which may or may not be on the air when their digipeater functions are most needed.

If the path between stations is marginal (that is, there are a lot of retransmissions of packets and possibly disconnection after the retry counter runs out), you may switch to another path without disconnecting. For example, after WA1LOU makes a direct connection (a connection without using a digital repeater) with K8KA, the high number of retries makes it obvious that a more reliable path is needed. To change paths in the middle of a contact without disconnecting, WA1LOU transfers to the Command Mode by typing <CTRL-C>. When the command prompt is displayed, he types:

cmd: RECOnnect K8KA Via W1AW-5 <CR>

His TNC switches the connection to the new path (via digital repeater W1AW-5) without breaking the previously established connection and the contact can continue without missing a beat and with a lot fewer retries.

Digital Repeater, Where Art Thou?

How do you know which stations are within communications range so that they may be used as digital repeaters?

Some packet-radio operators have compiled maps that graphically represent the packet-radio links in various regions of the country. These maps are usually posted on PBBS and are very useful in determining the best paths between stations. Usually, the maps include only dedicated digipeaters, PBBS and mailbox stations.

Besides maps, monitoring the frequency should give you an idea of which active stations are within radio range. The TNC has a number of commands to facilitate monitoring.

The Monitor command causes the TNC to display on your terminal the contents of each packet received by your station

while in the Command Mode, as well as the call sign of the station originating each packet and the call sign of the station that is the intended recipient of that packet. A monitored packet will be displayed as:

WA1LOU > K3OX: DATA

where WA1LOU is the call sign of the station originating the packet, K3OX is the call sign of intended destination of the packet, and DATA is the contents of the packet. (Monitor ON is the TNC default.) To enable (or disable) this function, type:

cmd: Monitor ON (or OFF) <CR>

The Mrpt command makes the monitoring function more revealing by displaying the call sign(s) of the station(s) that repeat a monitored packet, in addition to the call signs of the originating and destination stations. With Mrpt on, a monitored packet is displayed as:

WA1LOU > K3OX,KG1O-9*,W1AW-5:DATA

where WA1LOU is the station originating the packet, K3OX is the intended destination of the packet, KG1O-9 and W1AW-5 are the stations repeating the packet, and DATA is the packet's contents. The asterisk (*) indicates that KG1O-9 transmitted the packet you are receiving. (Mrpt ON is the TNC default.) To enable (or disable) this function, type:

cmd: MRpt ON (or OFF) <CR>

Other monitoring commands that may be useful are:

MAll (default: on)—monitor both connected and unconnected packets;

MCOM (default: off)—monitor connect requests, disconnect requests, connect/disconnect acknowledgments, and nonconnect/disconnect acknowledgments, as well as packets containing data;

MCon (default: on)—display packets received from other stations on frequency, while connected to another station;

MHeard—list the last 18 stations received by typing:

cmd: MHeard <CR>

MHClear—clear the received stations that are logged for recall by the Mheard command by typing:

cmd: MHClear <CR>

MStamp (default: off)—stamp the date and time of each monitored packet. To use this function, the date and time must be entered into the TNC using the DAytime command, by typing:

cmd: DAytime yymmddhhmm <CR>

where yy is the last two digits of the year (00-99), the first mm is two digits representing the month (00-12), dd is the day of the month (00-31), hh is the hour (00-23) and second mm is the minute of the hour (00-59). Note that the TNC has a 24-hour, rather than a 12-hour, clock. For example, to enter March 8, 1991, 1:30 PM, type:

cmd: DAytime 9103081330 <CR>

where 91 is the last two digits of 1991, 03 represents the third month (March), 08 represents the eighth day of the month, 13 represents the 13th hour of the day, and 30 represents the 30th minute of the hour.

The Budlist and Lcalls commands may be used together to limit your monitoring. Use the Lcalls command to list a maximum of eight stations that you do or do not wish to monitor by typing:

cmd: LCAlls aaaaaa,bbbbbb...,hhhhhh
                        \<CR\>

where aaaaaa, bbbbbb..., hhhhhh are the call signs of stations you do or do not wish to monitor.

Next, use the Budlist command to limit your monitoring according to the Lcalls list of stations. Your terminal will display only packets originating from stations in the Lcalls list by typing:

cmd: BUDlist ON \<CR\>

Your terminal will display packets of all stations except the stations in the Lcalls list (the TNC default) by typing:

cmd: BUDlist OFF \<CR\>

Surely your monitoring needs will be met with such a wide range of monitoring commands at your disposal.

VHF/UHF Operating

Today, most amateur packet-radio activity occurs at VHF, on 2 meters. 1200 baud is the most commonly used data rate with frequency modulated AFSK tones of 1200 and 2200 Hz. This is referred to as the "Bell 202" telephone modem standard.

The majority of TNCs are optimized for VHF/UHF FM operation, so getting one on the air is a simple matter of turning on your radio and tuning to your favorite packet radio frequency.

145.010 MHz is "the" packet-radio frequency. Most activity hovers on or near "01" including digital repeater, PBBS and direct-connect contacts (refer to the sidebar entitled "The Frequencies of Packet Radio" for information concerning other packet-radio hotspots). In some areas, an effort has been undertaken to establish a band plan for packet radio. Generally, these plans partition different types of packet-radio operations to different frequencies with digipeater, PBBS and direct-connect activity typically being the object of this division. If there is a packet-radio band plan in effect in your area, it is recommended that you comply with it. If there is no established band plan in your area, the following rule of thumb should be followed:

If you are conducting a direct connect (*sans* digital repeater) contact, move your contact to an unused frequency. It is very inefficient to try to exchange packets on a frequency where other stations, especially digipeater stations, are also exchanging packets. The competition slows down your station's ability to

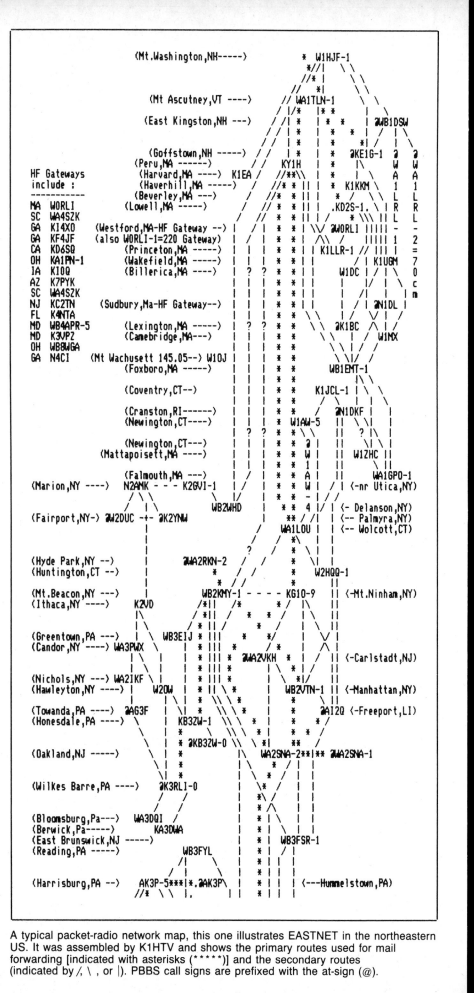

A typical packet-radio network map, this one illustrates EASTNET in the northeastern US. It was assembled by K1HTV and shows the primary routes used for mail forwarding [indicated with asterisks (*****)] and the secondary routes (indicated by /, \ , or |). PBBS call signs are prefixed with the at-sign (@).

The Frequencies of Packet Radio

Today, most packet-radio activity is clustered around a handful of HF and VHF frequencies. A summary of the locale of the packet-radio action follows:

80 meters—3.607 and 3.630 MHz (LSB)

40 meters—7.087 to 7.097 MHz (LSB), plus or minus interference from international shortwave broadcasters. 7.093 and 7.097 MHz are popular PBBS/Gateway frequencies.

30 meters—10.147 MHz (LSB)

20 meters—14.103 MHz (LSB) was the old packet-radio watering hole on 20. Today, more activity occurs on 14.105 and 14.107 MHz to avoid interfering with the DX beacons on 14.100 MHz.

2 meters—145.010 MHz (FM) is "the" packet-radio frequency. Because of the 145.010 logjam, 145.030, 145.050, 145.070, and 145.090 MHz contain more and more activity. In certain areas of the country, you may find packet-radio activity on unused simplex voice channels and on voice and RTTY repeaters.

exchange packets and, in return, you are also slowing down all of the other stations. You should use a frequency occupied by a digipeater *only* if you are using that digipeater.

VHF/UHF packet-radio operation is similar to VHF/UHF FM voice operation. Under normal propagation conditions, you may communicate only with stations that are in line of sight of your station. By using a repeater (voice or digital), this limitation is reduced. If a repeater is within range of your station, you may communicate with other stations that are also within range of the same repeater, even though those stations may be beyond your own line of sight. The AX.25 protocol expands on this by permitting the simultaneous use of eight digipeaters to relay a packet from one station that is line of sight of the first digipeater in the link to a station within radio coverage of the last digipeater in the chain.

Besides digipeaters, some packet-radio operations are conducted using voice repeaters. Such operations are usually conducted in the absence of voice activity and with the blessing of the sponsor of the voice repeater. It is imperative that you monitor a voice repeater before initiating packet-radio operation. One way to wear out your welcome fast is to send a few packets in the middle of a voice conversation! To take full advantage of the voice repeater, you should set your TNC's Axdelay and Axhang parameters as described earlier in this chapter. Finally, when entering the Connect command, do not include Via and the call sign of the voice repeater.

HF Operating Procedures

HF packet radio is very different from VHF/UHF packet radio. 300 baud is used rather than 1200 baud, and the AFSK tones are applied to an SSB (as opposed to FM) transmitter, producing a pseudo-FSK signal. 200-Hz shift is used, with 1600 and 1800 Hz the most common tones.

The majority of TNCs are optimized for VHF/UHF FM operation, so getting on HF requires more effort than getting on V/UHF. For starters, the modems in some TNCs must be modified and recalibrated for HF operation or, TNC permitting, an external modem may be used that is compatible with HF packet-radio operation. (A few TNC modems have the capability of switching between V/UHF and HF modem requirements.)

Modem modification in most TNCs is simple. For example, the TNC 1 requires changing the values of a resistor and capacitor, followed by the recalibration of the modem.

Adding an external HF modem (such as the AEA PM-1) is a simple matter with some TNCs and a major project for others. It is best to consult the documentation that accompanied your TNC to decide how to go about getting your TNC on HF.

A complete data-communications station; the TAPR TNC 2 is used on 2 meters, the unenclosed TAPR TNC 1 is used on HF, and the 1200-baud intelligent modem is used on the telephone lines. The three data-communication devices are connected/disconnected to/from the Apple® Macintosh computer's serial port by means of a four-position switch. (WA1LOU photo)

While some TNC parameters may be set to whatever suits you for VHF/UHF operation, the setting of these same parameters may be much more critical for HF operation. The reason is that HF operation is affected by interference, both atmospheric and man-made, and by the fickleness of propagation, to a much greater extent than V/UHF FM operation. Low signal-to-noise ratio also makes HF packet more difficult. Fortunately, certain TNC parameters may be set to compensate for these conditions. Some of these critical parameters were discussed earlier in this chapter. They are Axdelay, Axhang, Dwait, Paclen and Txdelay. One other, Maxframe, will be discussed here for the first time.

To increase throughput and decrease congestion on frequency, your packets should be short, and no more than one unacknowledged packet should be outstanding. The maximum number of outstanding unacknowledged packets is set with the Maxframe command (the TNC default is 4), so set Maxframe to 1 by typing:

cmd: MAXframe 1 < CR >

Shorten the length of your packets by using the Paclen command, as described earlier in this chapter. A value of 40 is a good place to start.

HF packet is adversely affected by spurious noise, which may be considered by the TNC as a valid signal. The TNC timing parameters are reset each time such a "signal" is detected, thus causing your TNC to delay transmitting a packet for no legitimate reason. To avoid this, simply disable all of the timing parameters by setting Axhang, Dwait and Txdelay to zero, as described earlier in this chapter. Axdelay should be set to the required receiver-transmitter turnaround time.

Whereas tuning a VHF/UHF receiver is simply a matter of setting the dial to the desired frequency, HF packet-radio tuning is much more critical. It is recommended that you tune your receiver very slowly, in as small an increment as possible (10-Hz increments are desirable) until your terminal begins displaying packets. When tuning, do not change frequency until a whole packet is received. If you shift frequency in midpacket, that packet will not be received properly and will not be displayed on your terminal even if you were on the correct frequency before or after the frequency shift.

Some TNCs and external modems have tuning indicators that make tuning a lot easier. Kits are also available to allow you to add a tuning indicator to a TNC without one.

The same rule of thumb that applies to V/UHF packet-radio operation also applies to HF operation. Move to an unused frequency if there is other packet-radio activity on the frequency you are presently using. Often, one frequency is used as a calling frequency where stations transmit packets to attract the attention of other stations who may wish to contact them. Once a contact/connection is established, the stations should move to another frequency where there is less activity, thus clearing the calling frequency and increasing their own stations' throughput.

The Dreaded Beacon

All TNCs have a beacon function. This function allows a station to send an unconnected packet at regular intervals. These unconnected packets usually contain a message to the effect that the station originating the beacon is on the air and is ready, willing and able to carry on a packet-radio contact.

The purpose of the beacon function is to generate activity when there is none. This purpose was legitimate when there was little packet-radio activity. Back in the early 1980s, when a new packet-radio station appeared on the air, it was a rare occurrence. Without beacons, that new station might believe that he was the only packet-radio station that was active in the area. Similarly, packet-radio stations already on the air would not be aware of the new station's existence. It would be very discouraging to build a TNC (they were all kits in the early days), get on the air and find no one to contact. The beacon function was the solution to the problem; it made stations aware of each other's existence.

Today, beacons are usually unnecessary. There is absolutely no need to resort to beaconing in order to make your existence known. On HF and 2 meters, there is plenty of activity in most areas. If you are getting on the air for the first time, monitor 14.103 or 145.010 MHz for a few minutes and you will quickly have a list of other stations that are on the air. When one of these stations disconnects, send a connect request to that station. After a few connections, your existence on the air will be known.

Instead of beaconing, leave a message announcing your existence on your local PBBS. This is more effective than beaconing because your message will be read even when your station is off the air (you can't beacon while you are off the air).

Beacons only add congestion to already crowded packet-radio channels, so do the packet-radio community a favor and disable your TNC's beacon function by typing:

cmd: Beacon Every 0 < CR >

PACKET RADIO BBS OPERATING PROCEDURES

Circa 1982, there was a little warehouse in Texas full of surplus Xerox® 820-1 computers. The "820" is a single-board computer using the Zilog Z80® central processing unit (CPU). It runs the CP/M® operating system (version 2.2) and features 64 kbytes of RAM, one parallel port, two serial ports, a disk controller and an 80 × 24 video display generator. As surplus, the computer could be purchased for as little as $50!

A typical WØRLI PBBS station, WA2FTC-1; its hub is a Xerox® 820-I computer board (mounted in the IBM® PC clone cabinet). This PBBS includes a 2-meter-to-HF Gateway. (WA2FTC photo)

Hank Oredson, WØRLI, put this bargain computer to good use by writing software that permitted it to function as a BBS—not a telephone BBS, but a packet-radio BBS. Shortly, WØRLI PBBSs began appearing throughout packet radioland, not solely because the 820 was inexpensive, but also because Hank's software was easy to install. The major features in later versions, and the reason WØRLI systems have become popular, were the two-TNC gateway and the automatic forwarding of messages. (There are other BBS systems in use in the US today.) A detailed discussion of the WØRLI BBS follows.

The mailbox function of the PBBS allows you to post mail on the system for later retrieval by the addressee. You can also retrieve mail from the system that is addressed to you or anyone else.

The mailbox does not limit your message posting to stations that frequent the local PBBS. If you know of a station that checks into a distant PBBS, you can post a message to that station on the local PBBS and it will be routed automatically through the packet-radio network to that distant PBBS for retrieval by that station.

Besides storing individual messages, large files of interest to the general packet-radio population may be stored as text files. Here you find packet-radio network maps, programs, W1AW bulletins, newsletters and so on.

The gateway function allows you to make connections on other frequencies. For example, from 2 meters you can send packets on 20 meters.

To use a WØRLI PBBS, you must find one. On HF, a number of PBBSs are active in the packet-radio subbands at various times of the day (refer to the sidebar entitled "The Frequencies of Packet Radio"). On VHF/UHF, there is liable to be at least one active PBBS station in an area where there is any amount of packet-radio activity.

Logging on—Once you locate a PBBS, you must log onto it in order to use it. Logging onto a PBBS is as simple as initiating a contact with any other packet-radio station; you simply connect. For example:

cmd: Connect W1AW-4 < CR >

where W1AW-4 is the call sign of the PBBS. After you are connected to the PBBS, the PBBS preamble is displayed on your terminal followed by a request for commands from you. (The sidebar entitled "WØRLI PBBS Command Set" lists the commands associated with WØRLI's software.)

Logging off—Whenever you are finished using a PBBS, you log off the system by using the B (for "bye") command. When the PBBS receives the B command, it logs you off the system and disconnects.

Reading the Mail—To read the messages that are posted on the PBBS, you must know what messages are on the

WØRLI PBBS Command Set

The commands associated with the WØRLI packet radio bulletin-board system are simple, yet powerful. The following commands are available with version 11.2 of the WØRLI software.

B	Bye—log off PBBS
D x	Download CP/M file named x
G	Access Gateway function and use the Gateway commands:

	B	Bye—log off PBBS
	C x	Send connect request to station x on Gateway
	J	List last 10 stations heard by or connected to PBBS
	M	Monitor Gateway frequency
	R	Return to PBBS command mode
	U	Send unconnected messages (CQs) on Gateway
	X	Toggle between short and extended command menu
	<CTRL-W>	Abort Gateway commands C or U

H	Help—obtain command descriptions
I	Obtain information about PBBS
J	List last 10 stations heard by or connected to PBBS
K n	Delete message numbered n
KT n	Delete message numbered n and send acknowledgment to sender that the message was retrieved
L	List all messages stored on PBBS since your last log-on
L n	List messages starting with message numbered n to most recently stored message
LA	List all ARRL bulletin messages
LB	List all bulletin messages
LL n	List last n number of messages stored on PBBS
LM	List all messages addressed to you
LT	List all NTS messages
L@ x	List all messages to be forwarded to PBBS x
L> x	List all messages addressed to x
L< x	List all messages stored by x
N	Enter your name in PBBS log
R n	Read message numbered n
RM	Read all messages addressed to you
S	Store a message
SB	Store a bulletin message
S x	Store a message addressed to x
S x @ y	Store a message addressed to x to be forwarded to PBBS y
SB	Store a bulletin message
SP	Store a private message
T	Talk to PBBS sysop
U x	Upload CP/M file named x
W	List all CP/M files
WA:	List all CP/M files on PBBS disk drive A
WB:	List all CP/M files on PBBS disk drive B
W*.xyz	List all CP/M files named with suffix xyz
X	Toggle between short and extended command menu

system. If there is a message stored on the PBBS that is specifically addressed to you, you are informed of that fact when you log onto the system and you can read it by using the RM (for "read mine") command. When the PBBS receives the RM command, it retrieves each message that is addressed to you and displays it on your terminal. After you have read a message addressed to you, use the K (for "kill") command to delete the message from the system (and conserve the PBBS storage facilities) by typing:

$$K \ n \ <CR>$$

where n is the number of the message to be deleted.

To read other messages that are on the PBBS, use the L (for "list") command to obtain a list of all of the messages that have been stored on the PBBS since the last time you logged onto the system. The PBBS lists each message by its message number and includes an indication (Y or N for yes or no) as to whether or not the addressee has read the message, the size of the message (in bytes), the addressee, the addressor, the call sign of the destination BBS, if any, the date that the message was posted, and the title of the message. If you wish to read a particular message, use the R (for "read") command by typing:

$$R \ n \ <CR>$$

where n is the number of the message to be read.

Sending Mail—To send a message by means of a PBBS, use the S (for "send") command. When the PBBS receives the S command, it asks you for the call sign of the addressee. To post the message on the local PBBS, simply type the call sign of the station to receive the message followed by <CR>. If you want to have your message forwarded to a station at another PBBS, type the call sign of the station to receive the message, followed by a space, the at-sign (@), another space, the call sign of the destination PBBS, and <CR> (for example, WB9FLW @ K4NTA <CR> where WB9FLW is the call sign of the station to receive the message and K4NTA is the call sign of the destination PBBS for mail forwarding).

Next, the PBBS asks you for the title of the message. Type a short (fewer than 38 characters) title that represents the contents of the message followed by <CR>. Now, you type the actual contents of the message. After you have typed the last line of the message, type a <CRTL-Z> and <CR> to indicate the end of the message. When the PBBS receives the <CTRL-Z>, it stores the message for later retrieval. It is good operating practice to use the P (private) qualifier when storing a message that is not of general interest. For example:

$$SP \ WA1LOU \ <CR>$$

The P qualifier will prevent the display of the message during the L command except when WA1LOU is using the system.

File Retrieval—Retrieving files that are stored under CP/M® requires knowledge of what files exist. Use the W (for "what's there?") command to obtain a list of the files.

Once you have chosen a file to retrieve, use the D (for "download") command to command the PBBS to send the file to you. (If your terminal is actually a computer running communications software, it may have the capability to save files to tape or disk for later reading, printing or uploading. You may consider using the software's downloading capability at this time.) To download the file, type:

$$D \ x \ <CR>$$

where x is the name of the file to be downloaded.

File Storage—If you wish to upload a file to the PBBS, use the U (for "upload") command, by typing:

$$U \ xxxxxxxx.xxx \ <CR>$$

where xxxxxxxx.xxx is the name of the file to be uploaded. The name may contain 1 to 8 characters before the period and 0 to 3 characters after the period. No spaces should be embedded in the name. An example of legal filenames are FILENAME.123, FILENAME, FILE.1 and FILE.

The only efficient way to upload a file is to use a computer running communications software with the capability of

Packet-Radio Talk

Definitions of the words used in the amateur packet-radio world follow.

AFSK—abbreviation for audio-frequency-shift keying.

ASCII—abbreviation for American National Standard Code for Information Interchange, the standard used to encode alphanumeric and control characters for data communications; it consists of seven information bits.

audio-frequency-shift keying—a type of modulation in which a tone is changed from one audio frequency to a second audio frequency to represent the on and off states of the digital information that is being modulated.

AX.25—an amateur packet-radio link-layer protocol.

baud—a unit of signaling speed equal to one pulse (event or symbol) per second.

Bell 103—a 300-baud modem standard using 200-Hz frequency-shifted tones centered at 1170 and 2125 Hz.

Bell 202—a 1200-baud modem standard using the tones 1200 Hz and 2200 Hz, used for VHF packet radio.

BBS—abbreviation for bulletin-board system.

bit—binary digit, a signal that is either on/one or off/zero; bits are combined to represent alphanumeric and control characters for data communications.

buffer—a portion of computer memory that is set aside to temporarily store data that is being uploaded or downloaded.

bulletin-board system—a computer system where messages and files can be stored for other users.

byte—a group of bits, usually eight in number.

carriage return—a key on a terminal keyboard that is used to indicate the end of a line of typed information and commands a terminal display to begin printing at the left-hand margin of the display.

character bits—the bits that represent an alphanumeric or control character.

<CR>—abbreviation for carriage return.

collision—a condition that occurs when two or more transmissions occur at the same time and causes interference to the intended receivers.

connect—to establish a communications link between two packet-radio stations.

data rate—the speed at which information is transferred, usually expressed in bauds or bits per second.

default—the state of a TNC parameter at the time of initially powering up the TNC or after resetting it.

digipeater, digital repeater—a packet-radio station that repeats the transmissions of other packet-radio stations.

download—to receive files from a PBBS or another packet-radio station.

duplex—a mode of communications in which you transmit on one frequency and receive on another frequency.

echo—a terminal and/or TNC function that prints on the terminal's display each character typed into the terminal keyboard.

FO-12—A JAMSAT/JARL amateur satellite with packet-radio capabilities.

frequency-shift keying—a type of modulation in which a carrier is changed from one frequency to a second frequency to represent the on and off states of the digital information that is being modulated.

FSK—abbreviation for frequency-shift keying

full duplex—a mode of communications in which you transmit and receive simultaneously.

Gateway—a PBBS function that allows packet-radio stations to make transmissions and connections on another frequency.

---

uploading files from disk, tape or memory. Manually uploading a long file by typing it at your terminal is very inefficient and ties up the PBBS for a long time. After you have finished uploading the file, type <CRTL-Z> and <CR> to indicate the end of the file.

Before uploading a long file, it is recommended that you check with the PBBS system operator (sysop) to make sure that there is enough room on disk to contain your file. If space is lacking, you could crash the system by attempting to upload a file that exceeds the available space.

To ask the sysop about the available file space or to ask him about anything else, you can leave him a message on the PBBS. For an immediate response, you can page him by using the T (for "talk to the sysop") command. When the PBBS receives the T command, it summons the sysop by ringing the bell of the sysop's terminal. If the sysop hears the bell, he can place the PBBS in the Converse Mode and answer your questions.

Gateway Operation—The PBBS gateway allows you to make a packet-radio connection from one frequency to another. This is accomplished by means of a separate TNC connected to each serial port of the Xerox® 820 computer with one TNC connected to a transceiver on one frequency and the other TNC connected to a transceiver on another frequency. WØRLI's software routes packets between the two ports in the gateway mode. The majority of gateways currently in operation are set up to interconnect 2 meters and an HF band (usually 7, 10 or 14 MHz).

Before using a gateway, it is advisable to check the level of activity on the gateway frequency. This can be accomplished by using the J (for "just heard" or "just con-

nected") command. When the PBBS receives the J command, it sends a list of the stations that were most recently heard by the PBBS or connected to the PBBS on the frequency you are using and the gateway frequency. The times of each hearing/connection are also listed to give you an indication of which stations are on the gateway frequency and how recently they were there.

Once you have determined which stations might be workable, use the G (for "gateway") command to access the gateway function. When the PBBS receives the G command, it requests additional commands from you.

The R (for "return to PBBS") command disables the gateway function and returns you to the PBBS command mode.

The C (for "connect") command followed by a station call sign sends a connect request to the specified station on the gateway frequency. <CTRL-W> aborts the connect request and returns you to the gateway command menu.

The U (for "unprotocol") command links you with the TNC on the gateway frequency and allows you to send unconnected, unprotocol packets (CQs) on the gateway frequency. <CTRL-W> returns you to the gateway command menu.

The M (for "monitor") command allows you to monitor activity on the gateway frequency. To exit the monitor mode, type any character and <CR>.

Using the gateway function, it is possible to make a connection with another PBBS that is operating on the gateway frequency and use its facilities including its gateway. You can also use VHF or UHF stations on the gateway frequency as digipeaters (HF digipeater operation is illegal under

half duplex—a mode of communications in which you transmit and receive alternately.

line feed—a control signal that commands a terminal display to begin printing on the next line.

log off—to inform a PBBS that you are finished using the system.

log on—to inform a PBBS that you wish to begin using the system.

modem—modulator-demodulator, an electronic device that permits computers and terminals to use analog communication media for data communications.

PACSAT—a proposed AMSAT packet-radio satellite.

packet-radio bulletin-board system—a BBS that is accessed via packet radio.

parity—a method of checking the accuracy of a received character by adding an extra bit to the character; the number of the character-bits set is even or odd depending on the type of parity in use (even or odd parity).

PBBS—abbreviation for packet-radio bulletin-board system.

protocol—a recognized set of procedures.

RAM—abbreviation for random-access memory.

random-access memory—a data storage device that can be written to and read from.

read-only memory—a data storage device that can be read from only.

ROM—abbreviation for read-only memory.

RS-232-C—an Electronic Industries Association (EIA) standard for data terminal equipment to data communication equipment interfacing that specifies the interface signals and their electrical characteristics.

RTTY—abbreviation for radioteletype.

secondary station identifier—a number that follows a packet-radio station's call sign to differentiate between two or more packet-radio stations operating under the same call sign.

simplex—a mode of communications in which you transmit and receive on the same frequency.

SSID—abbreviation for secondary station identifier.

stop bit—one or two extra bits that follow a character to indicate its end.

TAPR—abbreviation for Tucson Amateur Packet Radio Corporation.

terminal node controller—a device that assembles and disassembles packets.

TNC—abbreviation for terminal node controller.

Tucson Amateur Packet Radio Corporation [TAPR]—an Amateur Radio organization that was instrumental in the development of packet radio in the US (TAPR's address is PO Box 22888, Tucson, AZ 85734-2888).

turnaround time—the time required to switch between the receive and transmit modes of operation of a half-duplex circuit.

upload—to send files to a PBBS or other packet-radio station.

VADCG—addreviation for Vancouver Amateur Digital Communications Group.

Vancouver Amateur Digital Communications Group—an Amateur Radio organization that was instrumental in the development of packet radio in Canada (VADCG's address is 9531 Odlin Rd, Richmond, BC V6X 1E1, Canada).

present FCC rules). Either way, once a connection is established, the gateway is invisible to you; it's as if you were directly connected to the station on the gateway frequency.

PBBS Procedures—Those are the basic commands for using WØRLI's PBBS. A few basic operating rules need to be mentioned also.

If the PBBS does not respond to a command immediately, be patient and do not resend the command. One of the features of packet radio is that whatever you send is received perfectly at the other end. To achieve this result sometimes takes a number of attempts, especially if there is a lot of activity on frequency. If you do send a command twice, the PBBS will eventually receive the command twice, and will respond to the command twice! If the response is long (downloading a file, for example), your repeated command will waste a lot of PBBS time.

Do not perform operations that consume a lot of time during the prime-time operating hours. The demand for time on a PBBS is very high during prime time; therefore, do not use the gateway or the file transfer functions during these hours. Downloading or uploading a file takes time under the best conditions; during prime time, it may take large multiples of that time to perform the same file transfer because of the volume of other activity on frequency that is contending with your file transfer. Use good judgment before deciding to perform a time-consuming task.

To save time, you can send more than one command at one time by preceding the <CR> that follows each command with the pass character. <CTRL-V> is the TNC default for the pass character which, in this case, prevents each <CR> from causing the commands to be sent individually. For

example, to send the RM, L and W commands at one time, you would type:

RM <CTRL-V> <CR> L <CTRL-V> <CR> W <CR>

The last <CR> is not preceded by a <CTRL-V> because you want the last <CR> to force the transmission of this complete packet. The <CTRL-V>s preceding the other <CR>s prevent those <CR>s from forcing transmission of each command individually. Without the <CTRL-V>s, RM would be sent in one packet, L would be sent as a second packet and W would be sent as a third packet.

Public Service Communications

Emergency Communications—Although packet radio is relatively new, it has already served the public by providing communications during emergencies that have occurred during the past few years. With portable computers and dc-powered TNCs and radio equipment, packet-radio stations have been quickly set up in the middle of disasters of various proportions. These stations have been able to get all-important messages on the air and to their destinations quickly, and just as important, accurately. And accuracy is very important in emergency communications. When a packet-radio station sends NEED A SURGEON, you can be sure that the station receiving the message will not have a fish delivered to the originating station. Since each packet-radio station is a self-contained digipeater, packet-radio networks can be set up just as quickly in times of emergencies.

Those who have already gained experience in packet-radio emergency communications have indicated that the ideal emergency station setup includes two computers. One is used

A portable packet-radio station consisting of a TAPR TNC 2, a Radio Shack TRS-80® Model 100 laptop computer, and a Tempo S1 2-meter hand-held transceiver. Ac or dc can power this station. *(WA1LOU photo)*

to compose the actual messages and the other is used (with a TNC and radio) to get those messages on the air. Using the same computer for both composition and transmission of messages is too inefficient in times of emergencies.

The potential of packet radio for emergency communications is just beginning to be explored now. However, its value is already so impressive that various governmental agencies involved with emergencies of different degrees have allocated funds to purchase packet-radio equipment to be available for hams in times of emergencies.

Traffic Handling—Making packet-radio communications most effective in emergency communications takes a lot of work before an actual emergency occurs, and some hams are working very hard now on establishing a means of handling traffic efficiently through the packet-radio network. As more and more packet-radio stations appear on the air each day, the packet-radio network grows. Each new station that appears on the air has the potential of being an outlet for emergency traffic. The message forwarding system of WØRLI PBBS software is a key part of the packet-radio traffic system. That software has the ability (the KT, kill traffic command) to send an acknowledgment to the originating station when packet traffic is retrieved by a station at the destination PBBS for local delivery. This system is now being fine tuned and will soon be a running at full speed, ready to handle volumes of emergency traffic when the need arises.

Space Age Communications

Amateur Satellites—Packet-radio experiments in space via man-made satellites are laying the groundwork for a future packet-radio network that will rely on digital packet transponders in orbit around the Earth.

Experiments conducted in 1984 demonstrated the practicality of space packet-radio communications when connections across the North American continent were successfully established through the OSCAR 10 satellite. As much as 50 kbytes of data from W3IWI's BBS were successfully transferred via the satellite.

In 1985, UoSAT-OSCAR 11 was the medium for messages transferred between packet-radio stations in Honolulu, Los Angeles and Surrey, England. This experiment proved that worldwide store-and-forward packet messages via low earth orbiting satellites was a viable concept. PBBSs in space are accessible on VHF and UHF frequencies through Fuji-OSCAR 12 (FO-12) and in the future via PACSAT with 1 to 4 megabytes of on-board memory allocated to the store-and-forward function. Also, the Rudak transponder on the Phase IIIC spacecraft will supply very long-range digipeater service.

Meteor Scatter—An elite group of VHF and UHF enthusiasts make successful contacts by bouncing their signals off the ionized trails of meteors as they enter the Earth's atmosphere. The amount of voice or CW information that can be transferred via one meteor trail is small and, as a result, a number of meteors are necessary before two stations can successfully exchange their call signs and signal reports.

While one meteor trail has a limited potential for transferring voice or CW information, that same trail can transfer more information if it is sent by means of packet radio. And some hams have already begun experimenting with this mode of VHF communications. According to the 1987 edition of *The ARRL Handbook,* well before the Perseids meteor showers in early August 1984, Ralph Wallio, WØRPK, in central Iowa held 6-meter packet-radio schedules with Bob Carpenter, W3OTC, in Maryland. W3OTC received about 2% of the packets sent by WØRPK. Four nights later, these two stations had what is believed to be the first two-way amateur packet-radio meteor-scatter contact.

Since meteors are constantly entering the Earth's atmosphere, meteor-scatter packet-radio can be a reliable form of communications once enough stations begin to take advantage of this mode and once 9600-baud modems become readily available for packet-radio communications. Then, the amount of data that can be transferred via one meteor trail will be astounding!

The Future

Amateur packet radio is in its infancy. Most of its life is in the future, and the amateurs involved in packet radio today will determine that future. They are the pioneers who will be responsible for laying the groundwork and making decisions for an amateur packet-radio network that will encompass the Earth. You too can become a pioneer; add a TNC to your station now and join the packet-radio adventure.

Chapter 11

FM and Repeaters

Beer and pizza, Abbott and Costello, FM and repeaters: there are some combinations that make it and some that do not. In Amateur Radio, the combination of frequency modulation and repeaters has made it in a big way. The FM and repeater modes of communication have been the most popular modes in ham radio since the early 1970s, and they are still in the forefront of the hobby today.

The reason for the popularity of the modes is their versatility. FM and repeaters offer something for everyone. Today, virtually all active amateurs have at least one transceiver in their possession that is capable of FM and repeater operation. Otherwise, they are missing something (often, a lot). The HF DXer can swap DX information with his fellow DXers on the local DX club repeater. The traffic handler can pass traffic between the local and section levels by means of the local repeater outlets. The packet radio and radioteletype enthusiasts can communicate digitally by means of repeaters dedicated to their specific modes. Even if you want to communicate with a shuttle astronaut traveling through space, 2-meter FM is the place to do it. And if you just like to talk (and who in our hobby does not like to talk?),

then strike up a conversation on any repeater. Adding to the versatility of FM and repeater operation is its ability to make it easy to communicate while walking or driving, or from the comfort of your home.

Licensing

With the exception of the new Novice VHF/UHF sub-bands, you need at least a Technician class license (or higher) to operate in the FM and repeater modes in the VHF and UHF

Base, Mobile or Portable

Transceivers capable of FM and repeater operation are available in three basic varieties: base, mobile and portable. The primary differences between the three are their size and how they are powered. The base transceiver is the largest of the three and, because it is intended for operation from the ham shack, it is usually ac-powered. The mobile transceiver is intended for operation from within a vehicle and is therefore more compact than a base transceiver and is dc-powered (from the vehicle's battery). The portable transceiver is intended to be operated while in the palm of your hand, so it is designed to be so compact that it can be hand held. It is also dc-powered, but from a self-contained, rechargeable battery rather than an external source of power.

Although there is a great deal of difference in the sizes of base, mobile and portable transceivers, size is not a factor as far as the capabilities of the different transceivers are concerned. Today's FM and repeater transceivers are full of LSIs and microprocessors that are capable of filling a hand-held transceiver with as many different bells and whistles as are contained in a base transceiver. Memory, tone encoding, programmable scanning and LCD frequency display are features that can be found in portable, mobile and base transceivers alike. To squeeze all of these capabilities into these smaller transceivers, microswitches are used on portable transceivers, and controls are often mounted on the microphones of mobile transceivers.

The only thing that is forsaken in the smaller-sized transceiver is the power output of its transmitter. Whereas the typical portable transceiver is capable of transmitting with an output of 0.5 to 5 watts and the typical mobile transceiver 10 to 30 watts, some base transceivers are capable of transmitting with an output of 100 watts!

The type of antenna you use is dependent on the installation of your transceiver. Obviously, a mobile installation will require an antenna that is able to withstand being attached to vehicles traveling at speeds up to 55 miles per hour. A portable installation should have an antenna that is not too obtrusive, while a base installation has a little more leeway. In base installations, the higher and bigger, the better; you are limited only by what your spouse and/or neighbor will tolerate. No matter what kind of installation you have, all antennas for the FM repeater modes have one thing in common: vertical polarization.

Whether you're out walking or mobiling, ham radio can be operated everywhere, through VHF repeaters. All you need is a hand-held transceiver, such as the one used here by Joyce, N1CXV. *(K1XA photo)*

Amateur Radio spectrum, ie, above 50 MHz. A General class license (or higher) is required to operate 10-meter FM and repeaters. Note that Technician class hams may also use VHF/UHF repeaters that transmit on frequencies not normally permitted for use by the Technician class (eg, a 450-MHz repeater that also has an output on 20 meters or an orbiting amateur satellite that repeats 2-meter transmissions on 10 meters). Under the exciting new Novice Enhancement rules changes put into effect by the FCC at ARRL urging, Novices now have considerable VHF/UHF privileges as follows: 222.10-223.91 MHz (25 watts maximum, all authorized modes) and 1270-1295 MHz (5 watts maximum, all authorized modes). Novice Enhancement, therefore, means that Novices can operate through all existing repeaters in those two subbands! For further details, see Chapter 3.

OPERATING

FM repeaters provide the means to communicate efficiently. However, if the communicators (Amateur Radio operators) communicate inefficiently, the function of the repeater is wasted. Therefore, prior to making your first transmission in the FM and repeater world of communications, you should be aware of some of the fundamental operating techniques that are unique to it.

The Great Frequency Hunt—Users

If you are planning to use FM repeaters for amateur communications, you will need to know what frequencies are being used by the repeaters in your area. The best source of this information is the ARRL *Repeater Directory,* which lists the frequencies (along with other informative data) of over 10,000 repeaters in the US and Canada. You may also wish to contact your area frequency coordinator (listed in the *Directory*), who can tell you the frequencies of the *active* repeaters in your area.

Following is a list of the 10 most popular 2-meter repeater frequency pairs in the US and Canada. With these channels programmed in your transceiver, the chances are very good that you will be able to find an accessible repeater wherever your travels take you.

The top-10 frequency pairs (listen on the second one) are:

146.34/146.94	146.22/146.82
146.16/146.76	146.07/146.67
146.25/146.85	146.37/146.97
146.13/146.73	146.31/146.91
146.28/146.88	146.40/147.00

Finding a Repeater

To use a repeater, you must know that one exists. There are various ways to determine the existence of a repeater. The local hams should be able to provide you with information about local repeater activity or you can consult a repeater listing. Various clubs, often those associated with the local frequency coordinator, publish statewide and regional directories of repeaters. Each spring, the ARRL publishes the *Repeater Directory,* a comprehensive listing of repeaters throughout the US, Canada and other parts of the repeater world. Besides identifying local repeater activity, the *Directory* is handy for finding repeaters during vacations and business trips. Once you find a repeater to use, listen and familiarize yourself with its operating procedures before making your first transmission.

Your First Transmission

Making your first transmission on a repeater is as simple as your call sign. If the repeater is quiet, pick up your

---

The Great Frequency Hunt—Builders

Amateurs who are contemplating putting a repeater on the air may wonder how to obtain the frequency pairs for their repeaters. Frequency coordination is the recommended method. Across the US and Canada, volunteer individuals and groups have taken on the task of recommending frequencies for proposed repeaters within their "jurisdiction." These frequency coordinators keep extensive records on repeater operation in their area, as well as adjacent areas. With this information, they are usually able to recommend the best pair of frequencies for the proposed repeater. Please note that under §97.85 of the FCC's rules, in cases of repeater-to-repeater interference, the two repeater stations are equally and fully responsible for resolving the interference *unless* one repeater is coordinated and the other is not. In that case, the uncoordinated repeater has *primary responsibility* for resolving the interference.

Because of the amount of work involved in finding the best frequency pair, and the fact that this work is done on a volunteer basis, there is a time lag between the request for a frequency pair and receipt of the coordinated channel. So if you plan to request a pair of frequencies, do it early. Consult the latest edition of the ARRL *Repeater Directory* for the listing of frequency coordinators.

---

microphone, press its switch, and transmit your call sign— "K1ZZ" or "K1ZZ listening"—to attract someone's attention. After you stop transmitting, you will usually hear a short, unmodulated carrier transmitted by the repeater to let you know that the repeater is working. If someone is listening and is interested in talking to you, he will call you after your initial transmission. Some repeaters have specific rules for making yourself heard, but, in general, your call sign is all you need to do the trick.

Don't call CQ to initiate a conversation. It takes a lot longer to complete a long CQ than to simply transmit your call sign. (In some areas, a solitary "CQ" is permissible.) Efficient communications is our goal. You are not on HF, trying to attract the attention of someone who is casually tuning his receiver across the band. In the FM and repeater mode, stations are either monitoring their favorite frequency or they are not. Except for scanner operation, there is not much tuning across the repeater bands.

If you want to join a conversation that is already in progress, transmit your call sign during the break that occurs between transmissions. The station that transmits after you drop in your call sign will acknowledge you. Don't use the word "break" to join a conversation (unless it is the operating practice in your area). In some areas, "break" indicates that there is an emergency and that all stations should stand by for the station with emergency traffic.

If you want to call another station and the repeater is inactive, simply call the other station, eg, "K1XA, this is K1WJ." If the repeater is active, but the conversation that is in progress sounds like it is about to end, wait until it is over before calling another station. However, if the conversation sounds like it is going to continue for a while, transmit your call sign between their transmissions. After you are acknowledged, ask to make a quick call. Usually, the other stations will acquiesce. Make your call short. If your friend responds to your call, ask him to stand by on frequency until the present conversation is over or, if possible, ask him to meet you on another repeater or a simplex frequency.

Acknowledging Stations

If you are in the midst of a conversation and a station

The Ins and Outs of Repeater Operating

Repeater users should know how repeaters work. Users who are not familiar with the workings of a repeater may unintentionally misuse the equipment and interfere with other users.

A repeater is like any other Amateur Radio station. It has a receiver, a transmitter and an antenna system. The difference is that a repeater's receiver and transmitter are tuned to different frequencies and the output of the receiver is connected (through a "carrier operated relay" or COR) to the input of the transmitter; thus, anything the receiver hears is repeated by the transmitter.

Your transceiver's transmitter and receiver are also tuned to different frequencies. Your transmitter is tuned to the repeater's receive—or "input"—frequency, so whatever you transmit is received by the repeater. Meanwhile, your receiver is tuned to the repeater's transmit—or "output"—frequency, so that whatever the repeater transmits (repeats) is received by your transceiver.

To tune the majority of FM transceivers, you turn its frequency control (actually the receiver frequency control) to the transmit/output frequency of the desired repeater. Your transmitter's frequency automatically follows the tuning of your receiver and is adjusted to the repeater's receive/input frequency. Usually, you must also select a switch that chooses between the "simplex" and "duplex" mode. In the simplex mode, your transmitter and receiver are tuned to the same frequency, while in the duplex mode, your transmitter and receiver are tuned to different frequencies, as required for repeater operation.

The frequency separation between the transmitter and receiver differs with each repeater band. On 2 meters, the standard separation between the transmitter and receiver frequencies is 600 kHz; on 220 MHz, the separation is 1.6 MHz; and on 450 MHz, the separation is 5 MHz. A 2-meter repeater with an input frequency of 146.34 MHz has an output of 146.94 MHz (146.34 MHz + 600 kHz = 146.94 MHz [146.34 + 0.6 = 146.94]).

Transmitter-receiver frequency separation is usually built into your transceiver, so you do not have to be concerned with it when tuning. However, you do have to be concerned with whether the output frequency is higher or lower than the input frequency. In some segments of the 2-meter band, for example, the input frequency is 600 kHz *lower* than the output frequency (as in our 146.34/146.94 repeater example above), while in other segments of the band, the input frequency is 600 kHz *higher* than the output frequency. For example, a repeater with an output frequency of 147.18 MHz has an input frequency of 147.78 MHz.

Block diagrams of a typical FM repeater and FM transceivers illustrate the relationship of the freq X, the repeater's "input frequency," and freq Y, the repeater's "output frequency." Transceiver A transmits on freq X and is received by the repeater's receiver, which is also tuned to freq X. The repeater's transmitter transmits (repeats) the received signal on freq Y and is received by transceiver B's receiver, which is also tuned to freq Y.

Usually, when you select the duplex mode on your transceiver, you also have to select whether your transmitter frequency will be higher or lower than your receiver frequency. You would set your transmitter frequency lower than your receiver frequency to use the 146.34/146.94 repeater. However, to use the 147.78/147.18 repeater, you would set your transmitter frequency higher than your receiver frequency. Often, the simplex/duplex switch—designated "+/−"—performs this function; the "+" position sets the transmitter frequency higher than the receiver frequency, while the "−" position sets the transmitter lower than the receiver frequency. The position between the "+" and "−" often selects the simplex mode.

transmits his call sign between transmissions, the next station in queue to transmit should acknowledge that station and permit him to make a call or join the conversation. It is impolite not to acknowledge him and, furthermore, it is impolite to acknowledge him but not let him speak. You never know; the calling station may need to use the repeater immediately. He may have an emergency on his hands, so let him make a transmission promptly.

The Pause That Refreshes

A brief pause before you begin each transmission allows other stations to participate in the conversation, so don't key your microphone as soon as someone else releases his. If your exchanges are too quick, you can prevent other stations from getting in. Again, there may be an emergency, so leave that pause.

The "courtesy beepers" found on some repeaters force users to leave spaces between transmissions. The beeper sounds a second or two after each transmission to permit new stations to transmit their call signs in the intervening time

period. The conversation may continue only after the beeper sounds. If a station is too quick and begins transmitting before the beeper sounds, the repeater will indicate the violation, sometimes by shutting itself down.

Brevity

Keep each transmission as short as possible. Short transmissions permit more people to use the repeater. All repeaters promote this by having timers that "time-out," ie, shutting down the repeater whenever someone transmits too long. With this in the back of their minds, most users keep their transmissions brief.

Learn the length of the repeater's timer and stay well within its limits. The length may vary with each repeater; some are as short as 15 seconds and some are as long as three minutes. Others automatically vary their length depending on the amount of traffic on frequency; the more traffic, the shorter the timer. The other purpose of a repeater timer is to prevent extraneous signals from keeping the repeater on continuously, which could damage the repeater's transmitter.

Identification

You must transmit your call sign at the end of each transmission or series of transmissions and at least every 10 minutes during the course of any communication. You do not have to transmit the call sign of the station to whom you are transmitting.

Never transmit without identification. For example, keying your microphone to turn on the repeater without identifying is illegal. If you do not want to engage in conversation, but simply want to check if you are able to access a particular repeater, simply say "N1ED testing." Thus, you have accomplished what you wanted to do, legally.

Go Simplex

After you have made a contact on a repeater, move the conversation to a simplex frequency, if possible. The repeater is not a soapbox; you may like to listen to yourself, but others, who may need to use the repeater, will not be as appreciative.

The function of a repeater is to provide communications between stations that would not normally be able to communicate because of terrain and/or equipment limitations; see Fig 11-1. It logically follows that if stations are able to communicate without the need of a repeater, they should not use a repeater. In other words, when it is possible to use a simplex frequency for communications, use simplex, so the

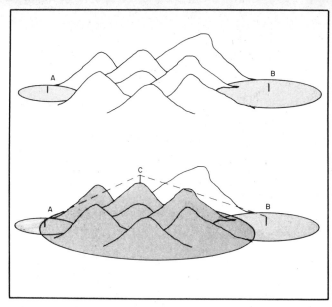

Fig 11-1—In the upper diagram, stations A and B cannot communicate because their coverage is limited by the mountains between them. In the lower diagram, stations A and B can communicate because the coverage of each station falls within the coverage of repeater C, which is located on one of the mountains.

Band Plans

"Band plan" is a term you may encounter during your sojourns in the world of repeaters. It refers to an agreement among concerned VHF and UHF operators as to how each VHF and UHF Amateur Radio band should be utilized. The goal of a band plan is to systematically minimize interference between the various modes sharing each band. Besides FM repeater and simplex activity, you can find CW, SSB, AM, OSCAR and R/C (radio control) operations, among others, on these bands.

Most of you who use FM or plan on using FM will probably be using 2 meters, so a description of the 2-meter band plan will be useful. The band is divided for various modes according to the following proposal by the ARRL VHF-UHF Advisory Committee:

Frequency (MHz)	Operation
144.00-144.05	EME (CW)
144.05-144.06	Propagation beacons
144.06-144.10	General CW and weak signals
144.10-144.20	EME and weak signal SSB
144.20	National calling frequency
144.20-144.30	General SSB operation
144.30-144.50	New OSCAR subband
144.50-144.60	Linear translator inputs
144.50-144.90	FM repeater inputs (local option)
144.90-145.10	Weak signal and FM simplex*
145.10-145.20	Linear translator outputs
145.10-145.50	FM repeater outputs (local option)
145.50-145.80	Miscellaneous and experimental modes
145.80-146.00	OSCAR subband
146.01-146.37	Repeater inputs (local option)
146.40-146.58	Simplex
146.61-147.39	Repeater outputs (local option)
147.42-147.57	Simplex
147.60-147.99	Repeater inputs (local option)

*The channels 145.01, .03, .05, .07 and .09 MHz are widely used for packet radio, and have been established by a number of frequency coordinators. In some cases, local conditions may dictate variations from the national band plan. In these cases,

the written determination of the regional frequency coordinating body shall prevail and be considered good amateur operating practice in that region.

In the 144.51-145.49 MHz range, the spacing between the repeater channels is 20 kHz. The first input frequency is 144.51 MHz and the next is 144.53 MHz (144.51 MHz + 20 kHz = 144.53 MHz).

In the 146.01-147.99 MHz range, channel spacing is either 20 or 30 kHz, depending on local policies.

Also, in those areas that utilize the 30-kHz spacing plan, some repeaters will be set up on "splinter" (15-kHz spacing) frequencies, or between two repeater channels. For example, spacing would normally place repeater inputs on 146.01 and 146.04 MHz, but an additional repeater input can be established on 146.025 MHz.

According to the above band plan, the spacing between each repeater input and output is 600 kHz. For example, the input frequency of 146.34 MHz has an output frequency of 146.94 MHz (146.34 MHz + 600 kHz = 146.94 MHz).

The simplex frequencies between 144.90 and 145.10 are currently being used for packet operation, while those between 146.40 and 146.60, and between 147.40 and 147.60 MHz, are channelized for voice-simplex operation. (Note: Select frequencies between 146.40 to 146.60 and 147.40 to 147.60 may also be a repeater input or output.)

A common variation to the band plan is the reversal of the input and output frequencies of those repeaters operating on splinter frequencies, such as 146.625 and 146.315 (not 146.85 and 147.15) MHz. The purpose of this reversal is to reduce cochannel interference.

Under this band plan, repeater pairs on 146 MHz (ie, 146.01/61) will normally have a low input/high output split; that is, the lower frequency (146.01) would be the input and the higher frequency (146.61) would be the output. Conversely, repeater pairs on 147 MHz (ie, 147.81/21) will normally have a high input/low output.

For more information on band plans and repeater operating in your local area, contact your frequency coordinator, listed in the front of the ARRL *Repeater Directory*.

repeater can be made available for stations that need its facilities. (Besides, communications on a simplex frequency offer a degree of privacy that is impossible to achieve on a repeater, and you can usually carry on an extensive conversation on a simplex channel without interruption.)

You would be surprised how well one can communicate with a modest equipment installation on a simplex frequency. There is almost never any need for fixed stations in the same community to use a repeater to communicate with each other. Similarly, portable operators walking around the same ham radio flea market or convention floor should be able to stay in touch on a simplex frequency.

When selecting a simplex frequency, make sure that it is a simplex frequency designated for FM simplex operation (check the band plans listed in the ARRL *Repeater Directory*). Each band is subdivided for specific modes of operation, such as satellite communications and weak-signal CW and SSB. If you select a simplex frequency indiscriminately, you may interfere with stations operating in other modes (and you may not be aware of it).

Fixed Stations and Prime Time

Fixed stations should stay off the repeaters during rush hours. At these times, mobile stations have preference. Originally, repeaters were intended to enhance mobile communications. During mobile operating prime time, fixed stations should generally yield to mobile stations. (Some repeaters have a specific policy to this effect.) But when you're operating as a fixed station, don't abandon the repeater completely; monitor mobile activity. Your assistance may be needed in an emergency.

Support

Support your local repeater. Many repeaters are run by clubs whose members pool their resources to operate the repeater. Some members provide the actual equipment that makes up the repeater, other members provide funds to pay the electric, phone, maintenance and site-rental bills, and others provide manpower at the repeater site whenever their assistance is needed. If you are a regular user of a particular repeater, you should help to support the effort to keep the repeater on the air. Without support, the repeater may not be there when you need it.

Limited Access

Most of the repeaters that you are likely to encounter are open to all users. There are no restrictions to use of the repeater's facilities. However, limited-access repeaters do exist. Their use is restricted to exclusive groups. Although such operations go against the grain of the spirit of our hobby, "closed" repeaters are legal according to section 97.85(a) of the FCC regulations, which states that "Provisions to limit automatically the access of a station in repeater operation may be incorporated but are not mandatory."

There are also some repeaters that are of the "open" variety, but require the use of special codes or subaudible tones to gain access. The reason for limiting the access to some "open" repeaters is to prevent interference. In cases where extraneous transmissions often cause the repeater to be turned on, limited access is the only means to counteract the interference.

Minimum Power

Use only the minimum amount of power necessary to maintain communications through the repeater to avoid the

Donna, N2FFY, particularly enjoys 2-meter FM during "drive time," the daily commute to and from the office. *(WA4PFN photo)*

possibility of accessing distant repeaters on the same frequency. The use of minimum power is not only a courtesy to the distant repeaters, but is an FCC requirement.

Autopatch

An autopatch is a device that allows repeater users to make telephone calls through the repeater. In most repeater autopatch systems, the user simply has to generate the standard telephone company tones to access and dial through the system. Usually, this is accomplished by interfacing a telephone tone pad with the user's transceiver. Tone pads are often mounted on the front of a portable transceiver or on the front of a fixed or mobile transceiver's microphone. Whatever is used, the same autopatch operating procedures apply.

Is the call necessary? If it is an emergency situation, there is no problem. Any other reason falls into a gray area. As a result, autopatch use, except for emergency situations, is expressly forbidden by some repeater groups.

Never use an autopatch where regular telephone service is available. How many times have you heard the OM use the autopatch to call the XYL to let her know that he has just left the office, or vice versa? Why didn't he make that call from the office before he left?

Never use the autopatch for anything that could be construed as business communications. The FCC strictly forbids any business communications in Amateur Radio. Don't use the autopatch to call the local radio store to find out if they have a modified trimmer capacitor in stock, and don't call the local pizza palace to order a large mozzarella. However, you may call a business if the call is related to an emergency situation, such as calling an ambulance service for the removal of injured persons from an accident or a towing service for the removal of a disabled vehicle from a hazardous location.

ARRL Phone Patch and Autopatch Guidelines

Radio amateurs in the United States enjoy a great privilege: The ability to interconnect their individual stations and repeaters with the public telephone system. The wisdom of the federal government in permitting, and even defending, this freedom has been demonstrated time and again. There is no way to calculate the value of the lives and property that have been saved by the intelligent use of phone patch and autopatch facilities in emergency situations. The public interest has been well served by amateurs with interconnect capabilities.

As with any privilege, this one can be abused, and the penalty for abuse could be the loss of the privilege for all amateurs. What constitutes abuse of phone patch and autopatch privileges? In the absence of specific regulations governing their use, the answer to this question depends on one's perspective. Consider these facts: To other amateurs, phone patching activities that result in unnecessary frequency congestion or which appear as a commercialization of Amateur Radio operation are an abuse of their privilege to engage in other forms of amateur activities.

To the telephone company, which needs to protect its massive investment in capital equipment, anything that endangers its equipment, its personnel or its revenues is an abuse.

To the Federal Communications Commission, which is responsible for the efficient use of the radio spectrum by the services it regulates, any radiocommunication that could be handled more appropriately by wire is an unnecessary use of a valuable resource.

To the commercial suppliers of radiocommunication for business purposes (Radio Common Carriers), competition from a noncommercial service constitutes a possible threat to their livelihood.

At one time or another, threats to radio amateurs' interconnect privileges have come from each of these sources. And threats may come from another quarter: The governments of certain nations that prohibit amateurs from handling third-party messages internationally in competition with government-owned telecommunications services. If illegal phone patching to and from their countries cannot be controlled, they reason, the solution may be to ban *all* international third-party traffic by amateurs and to permit no such special arrangements.

The question facing amateurs is this: Should phone patches and autopatches be subject to reasonable voluntary restraints, thereby preserving most of our traditional flexibility, or should we risk forcing our government to define for us specifically what we can and cannot do? Experience has shown very clearly that when specific regulations are established, both innovation and flexibility are likely to suffer.

The Amateur Radio Service is not a common carrier, and its primary purpose is not handling of routine messages on behalf of nonamateurs. However, third-party communications as an incidental part of Amateur Radio adds an important dimension to our public service capability.

It is the policy of the American Radio Relay League to safeguard the prerogative of amateurs to interconnect their stations, including repeaters, to the public telephone system. An important element of this defense is encouraging amateurs to maintain a high standard of legal and ethical conduct in their patching activities. It is to this end that these guidelines are addressed. They are based on several sets of standards that have been in use for several years on a local or regional basis throughout the country. The ideas they represent have widespread

---

Never use an autopatch to avoid a toll call. Autopatch operation is a privilege granted by the FCC; abuses of autopatch privileges may lead to their loss. (The League's phone patch and autopatch guidelines appear in the accompanying sidebar.)

Now that you have a legitimate reason to use the autopatch, how do you use it? First, you must *access* (turn on) the autopatch, usually by pressing a designated key or combination of keys on the telephone tone pad. The # and * signs are common access keys. When you hear a dial tone, you know that you have successfully accessed the autopatch. Now, simply punch in the telephone number you wish to call.

Once a call is established, do not speak at the same time as the party at the other end of the line (you also may have to tell that party not to speak at the same time as you speak) because an autopatch only provides communications in one direction at any particular time. (This is "half duplex": while you are transmitting, you obviously cannot hear the other end of the conversation, whereas on a telephone you can talk and listen at the same time. Thus, a telephone is "full duplex.") To inform the party at the other end of the line that you are finished talking and that party can speak, use the word "over." Many autopatches have timers that terminate the connection after a certain period of time, so keep your telephone conversation short and end the autopatch as quickly as possible.

Turning off the autopatch is similar to accessing it. A key or combination of keys must be punched to return the repeater to normal operation. Autopatch users should consult the repeater group sponsoring the autopatch for specific information about access and deaccess codes, as well as timer specifics.

Traffic

Traffic handling is a natural for repeater operation. Where else can you find as many local outlets for the traffic coming down from the section-level nets. Everyone who has a transceiver capable of repeater operation is a potential traffic handler. As a result, local traffic nets have flourished on repeaters.

The procedures for handling traffic by means of a repeater are basically the same as handling traffic anywhere else (consult Chapter 15 for full details). However, there are two things that are unique to repeater traffic handling. Repeaters have timers that shut down the repeater if a transmission is too long. Therefore, when you relay a message by means of a repeater, remember to release your PTT switch during natural breaks in the message to reset the timer. If you read the message in one long breath without resetting the timer, the repeater will likely shut down in the middle of your message. As a result, you will have to repeat the message and hold up the net in the process.

The audio quality of the FM mode of communications is excellent. It lacks the noise and interference common to other modes. Therefore, the use of phonetics and repetition is not necessary when relaying a message by means of a repeater. The only time that phonetics are necessary is to spell out an unfamiliar word or words with similar sounding letters. On the rare occasion that a receiving station missed something, he can ask you to fill in the missing information. Then, and only then, do you have to repeat something.

The efficiency provided by repeaters has made repeater traffic nets very popular. If there are a number of repeaters in your area, there is likely to be one or more active repeater traffic nets (consult the latest edition of the ARRL *Net*

support within the amateur community. All amateurs are urged to observe these standards carefully so our traditional freedom from government regulation may be preserved as much as possible.

1) International phone patches must be conducted only when there is a special third-party agreement between the countries concerned. The only exception is when the *immediate safety of life or property* is endangered.

2) Phone patches or autopatches involving the business affairs of any party must not be conducted *at any time.* The content of any patch should be such that it is clear to any listener that business communications is *not* involved. Particular caution must be observed in calling any business telephone. Calls to place an order for a commercial product must not be made, nor may calls be made to one's office to receive or to leave business messages. However, calls made in the interests of highway safety, such as for the removal of injured persons from the scene of an accident or for the removal of a disabled vehicle from a hazardous location, are permitted.

3) All interconnections must be made in accordance with telephone company tariffs. If you have trouble obtaining information about them from your telephone company representatives, the tariffs are available for public inspection at your telephone company office.

4) Phone patches and autopatches should *never* be made solely to avoid telephone toll charges. Phone patches and autopatches should *never* be made when normal telephone service could just as easily be used.

5) Third parties should not be retransmitted until the responsible control operator has explained to them the nature of Amateur Radio. Control of the station must *never* be relinquished to an unlicensed person. Permitting a person you don't know very well to conduct a patch in a language you don't understand amounts to relinquishing

control.

6) Phone patches and autopatches must be terminated *immediately* in the event of any illegality or impropriety.

7) Autopatch facilities must not be used for the broadcasting of information of interest to the general public. If a repeater has the capability of transmitting information, such as weather reports, which is of interest to the general public, such transmissions must occur only when requested by a licensed amateur and must not conform to a specific time schedule. The retransmission of radio signals from other services is not permitted in the Amateur Radio Service. However, the retransmission of tape material from other sources is permitted.

8) Station identification must be strictly observed. In particular, US stations conducting international phone patches must identify in English at least once very 10 minutes, and must also give their call sign and the other station's call sign at the end of the exchange of transmissions.

9) In selecting frequencies for phone patch work, the rights of other amateurs must be considered. In particular, patching on 20 meters should be confined to the upper portion of the 14,200-14,350 kHz segment in accordance with the IARU Region 2 recommendation, Miami, April 1976.

10) Phone patches and autopatches should be kept as *brief as possible,* as a courtesy to other amateurs; the amateur bands are intended to be used primarily for communication among radio amateurs.

11) If you have any doubt as to the legality or advisability of a patch, *don't make it.*

Compliance with these guidelines will help ensure that our interconnection privilege will continue to be available in the future, which will in turn help us contribute to the public interest.

Directory to find a local repeater traffic net). If there is no repeater traffic net in your area, fill the need by starting one yourself. Contact your ARRL Section Manager or Section Traffic Manager for details.

RTTY

Radioteletype (RTTY) operation is popular in the FM repeater world because V/UHF repeater operation offers certain advantages over HF. On FM, there is minimal noise and interference, therefore making it possible to send and receive error-free RTTY messages. Also, above 50 MHz, any digital code is permissible, whereas HF RTTY operators are limited to three digital codes: Baudot/Murray, ASCII and AMTOR. [Note: Although any digital code is permissible above 50 MHz, station identification—at the end of each communication and every 10 minutes during any communication—must be made using voice (where permitted), CW, Baudot/Murray, ASCII or AMTOR.]

The operating procedures for RTTY repeater operation are similar to RTTY procedures on HF (consult Chapter 9 for full details). However, one difference is the length of transmissions. RTTY operators must keep their transmissions short enough so that they do not time-out the repeater. On many RTTY repeaters, this situation is alleviated by circuitry that disables the timer when an RTTY signal is detected.

Also, some repeaters that are used for RTTY may also be used for voice communications. Under these circumstances, RTTY operators must keep an ear open to be sure that they do not disrupt voice communications by indiscriminately transmitting an RTTY message.

The operating procedures for RTTY repeater operation are

simple, so the transition between HF and VHF/UHF green keys should be easy. That's why so many RTTY operators have made the move up and above 50 MHz.

Packet Radio

FM and repeater packet-radio operation is a digital mode of communications that has grown tremendously in popularity during the past few years. Its operating procedures differ from the other digital modes because any packet-radio station may be called upon to function as a repeater when necessary. For example, station A wishes to communicate with station B, but because of the distance between the two stations and the lack of an intermediary repeater, communication is impossible. However, if stations A and B were using packet radio and were aware that another packet-radio station, station C, was approximately half the distance between them, they could call on station C, assuming that his station was on the air, to act as a repeater between A and B. And if station C was not adequate, station D could also be used as a repeater to get from A to B via C. Yes, more than one repeater can be used in packet radio. Isn't that convenient? (For a full explanation of packet radio operation, consult Chapter 10.)

Just as there are dedicated voice repeaters, there are dedicated packet repeaters that perform the same function as their voice counterparts. That is, they act as intermediaries between stations that otherwise could not intercommunicate. What is interesting is that packet-radio repeaters (also known as digital repeaters or *digipeaters*) use the same frequency for transmission and reception. Furthermore, more than one, and sometimes several, conversations can be conducted on the same frequency (by means of the same repeater). How's that

FM and Repeat/Speak

Definition of the words used everyday in the FM and repeater world of Amateur Radio.

access code—one or more numbers and/or symbols that are keyed into the repeater by means of a telephone tone pad to activate a repeater function, such as an autopatch.

autopatch—a device that interfaces a repeater to the telephone system to permit repeater users to make telephone calls.

break—the word used to interrupt a conversation on a repeater *only* to indicate that there is an emergency situation.

carrier operated relay (COR)—a device that causes the repeater to transmit in response to a received signal.

channel—the pair of frequencies (input and output) used by a repeater.

closed repeater—a repeater whose access is limited to a select group.

control operator—the Amateur Radio operator who is designated to "control" the operation of the repeater, as required by FCC regulations.

courtesy beeper—an audible indication that a repeater user may go ahead and transmit.

coverage—the geographic area within which the repeater provides communications.

digipeater, digital repeater—a packet-radio repeater.

duplex—a mode of communication in which you transmit on one frequency and receive on another frequency.

frequency coordinator—an individual or group that is responsible for assigning channels to new repeaters without interference to existing repeaters.

full duplex—a mode of communication in which you transmit and receive simultaneously.

full quieting—a received signal that contains absolutely no noise.

half duplex—a mode of communication in which you transmit at one time and receive at another time.

hand-held—a portable transceiver that is small enough to fit in the palm of your hand.

input frequency—the frequency of the repeater's receiver (and your transceiver's transmitter).

key-up—to turn on a repeater by transmitting on its input frequency.

machine—the complete repeater system (slang).

magnetic mount, magmount—a mobile antenna with a magnetic base that permits quick installation and removal from a motor vehicle.

NiCd—a nickel-cadmium battery that may be recharged over and over, often used to power portable transceivers.

open repeater—a repeater whose access is not limited to any select group.

output frequency—the frequency of the repeater's transmitter (and your transceiver's receiver).

over—the word used to indicate the end of a voice transmission.

radio direction finding (RDF)—the art and science of locating a hidden transmitter.

Repeater Directory—an annual ARRL publication that lists repeaters in the US, Canada and other areas.

separation—the difference (in kHz) between a repeater's transmitter and receiver frequencies.

simplex—a mode of communication in which you transmit and receive on the same frequency.

time-out—to cause the repeater or a repeater function to turn off because you have transmitted for too long.

timer—a device that measures the length of each transmission and causes the repeater or a repeater function to turn off after a transmission has exceeded a certain length.

tone pad—a device that is used to generate the standard telephone dialing signals.

vertical polarization—a plane that is 90 degrees relative to the horizon of the earth that is used for RF emissions in the V/UHF FM repeater mode.

---

for efficient repeater communications!

The operating procedures for packet-radio communications are different than any other form of Amateur Radio communications because you command your packet-radio equipment, a terminal node controller (TNC), to perform various tasks for you. The commands differ slightly depending on which type of TNC you use (and there are a lot of commands available to do a lot of things besides communications). However, they basically work the same way.

To call another packet-radio station, you simply enter a one- or two-letter command followed by the call sign of the station you wish to contact; for example, C WA2FTC (using the Tucson Packet Amateur Radio [TAPR] command set). Once you have entered the command and call sign, your TNC will attempt to contact the other station automatically.

To use a digipeater to contact another station, you use another command with the repeater's call sign, for example, C WA2FTC V W1AW-5 (using the TAPR command set again in this example, contact with station WA2FTC is being attempted via digipeater W1AW-5). Again, once you have entered a command, your TNC will do your bidding. (As with voice FM and repeater operation, it is advisable in packet-radio operation to use a digipeater only if its services are required. If you can communicate directly, move to a frequency where there are no digipeaters so that the digipeater is clear for those who need it.)

You can call CQ on packet radio by entering a CQ message in your TNC and commanding the TNC to transmit the message over the air. It is highly recommended that you do not send a CQ message through several digipeaters because it is difficult to maintain a conversation via several digipeater hops while those same digipeaters are being used for single-hop communications by other stations.

If your TNC establishes contact with another station (or if someone else's TNC establishes contact with your TNC), your TNC will display a message to that effect (*** CONNECTED TO WA2FTC in TAPR parlance). Once contact is established, anything you type at your keyboard will be transmitted to the other station after entering a control character or carriage return (depending on your TNC). During the prime-time operating hours, it is good procedure to keep your messages short because you are sharing the channel with a lot of other stations. Short transmissions are more likely to make it through to the other station than long transmissions. (Short transmissions are less likely to collide with the transmissions emanating from another conversation on the same frequency; a collision causes your TNC to repeat the transmission.)

To turn the train of thought of a conversation over to the other station, end your last transmission with K, BK, >, or anything else that lets the other station know that you are done (things are very loose on packet radio). To end a conversation, you again command your TNC to do the dirty work (the simple command D, for disconnect, does the job in TAPR TNCs). When that task is completed, the TNC will again indicate it (***** DISCONNECTED is TAPR's indication).

Packet radio is unique and so are its operating procedures. However, the end justifies the unique means many times over.

Satellite Gateway

A satellite gateway is a specially equipped station that interfaces your local FM repeater to an Amateur Radio satellite. This makes it much easier to try out a ham satellite without having to plunk down cash for some new equipment. It's literally your "gateway" to OSCAR operation by allowing you to use the simplest ham radio station, a hand-held transceiver, to work stations around the world.

Operating through a gateway is fairly easy. The Operations Controller (OC) is in charge of what goes into the repeater and, in general, keeps things in order. A "list" operation works well to ensure minimum confusion by those trying to use the gateway by means of the repeater. The OC should ensure that he has instructed each potential communicator to keep each contact within a maximum length, to speak clearly, and preferably, to say "over" to avoid doubling. Good discipline in gateway operation is essential. The signals from the gateway will be heard over a large portion of the earth, so efforts to minimize confusion are highly desirable.

The OC can update the list every half hour or so to permit new communicators to join in. It is also helpful for the OC to suggest to individuals using the repeater while it is in the gateway mode that they may find listening easier by using a set of headphones instead of their transceiver's speaker.

It is permissible to call CQ through a repeater to work someone by means of a satellite. More stations will respond to "CQ from WA1LOU via gateway WB2NOM" than "QRZ OSCAR"! Beyond these considerations, operating through a satellite gateway is straightforward. Common sense dictates the rest.

FM, the Fun Mode

These are some of the basic operating procedures for FM and repeater operation. Monitor your local repeater and listen to how these procedures (and variations thereof) work in actual practice; you may discover that the local repeater has its own special operating procedures. A little reading (of this chapter) and a little listening (to the local repeater) will prepare you to communicate efficiently and have a lot of fun in the FM and repeater world.

Direction Finding

Radio direction finding (RDF), also known as fox hunting, bunny hunting and hidden transmitter hunting, has become a popular radiosport in the V/UHF FM mode. Someone hides a transmitter in the woods, and the troops, with direction-finding equipment in hand, go out and try to be the first to find the hidden transmitter. Many hamfests regularly include a hidden transmitter hunt as part of their agenda, and the competition is vigorous.

RDF also has a more serious side. It has been used successfully by repeater operators to track down repeater jammers. It has also been used to track down stolen transceivers that suddenly pop up on the air operated by individuals who appear very unfamiliar with Amateur Radio operating procedures.

Locating a hidden transmitter is an art and, like any other art, practice makes perfect. (*The ARRL Handbook for the Radio Amateur* devotes an entire chapter to RDF.) So, if you have the opportunity to participate in a fox hunt, take advantage of it to hone your direction-finding skills for the real thing.

Chapter 12

VHF/UHF Operating

The radio spectrum between 30 and 3,000 MHz is one of the greatest resources available to the radio amateur. The VHF and UHF amateur bands are a haven for ragchewers and experimenters alike; new modes of emission, new antennas and state-of-the-art equipment are all developed in this territory. Plenty of commercial equipment is available for the more popular bands, and "rolling your own" is very popular as well. Propagation conditions may change rapidly and seemingly unpredictably, but the keen observer can take advantage of subtle clues to make the most of the bands.

Most North American hams are already well acquainted with 2-meter or 450-MHz FM. For many, channelized repeater operation is their first exposure to VHF or UHF. However, FM is only part of the story! This chapter will discuss the "other end" of the VHF/UHF bands, the low end, where almost all weak-signal SSB and CW activity takes place. It will discuss the bands from 50 MHz up through microwaves, but will emphasize the lower VHF frequency bands because the vast majority of amateur activity occurs there.

HOW ARE THE BANDS ORGANIZED?

One of the keys to using this immense resource properly is knowing how the bands are organized. For example, if you are unaware of the calling frequencies on most of the VHF bands, you might spend weeks tuning around before you find someone to talk to! But, knowing the best frequencies and times, you will have little trouble making plenty of contacts, working DX, and enjoying the "world above 50 MHz." See Fig 12-1 for suggested band use in the VHF/UHF range.

Currently, activity on VHF/UHF is concentrated in the two lower VHF bands, 6 and 2 meters (50 and 144 MHz). The number of active stations on these bands is about equal. Above the 2-meter band, there are considerably more active stations on 432 MHz than any other. Because it is not available worldwide, 220 MHz is not inhabited as much as it could be. The 902-928 MHz band has recently become available to US amateurs and its use is just beginning; 1296 MHz has quite a few occupants near major cities. For US amateurs, the 2300-MHz band is split into two parts (2300-2310 and 2390-2450 MHz) and many European amateurs have access only to the frequencies above 2320 MHz. Higher frequencies, above 3300 MHz, are the territory of the true experimenters because little commercial equipment is available for those bands. See accompanying sidebar on the microwave bands.

The VHF/UHF bands are roomy. That is, there is lots of frequency spectrum available for ragchewing, experimenting or working DX. Although some of the more popular frequencies occasionally get very crowded, most VHF/UHF allocations are available all the time. This means that whatever you like to do—chat with your friends across town, test amateur television or bounce signals off the moon—there is usually plenty of spectrum available. VHF/UHF is a great resource!

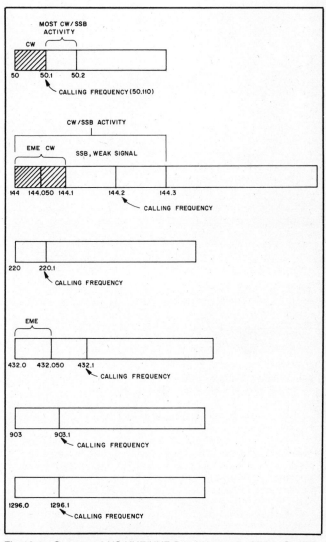

Fig 12-1—Suggested US VHF/UHF Band Usage (in MHz). On 220 MHz and above, activity is usually centered around the calling frequency. [Note: At presstime, the FCC had released a Notice of Proposed Rule Making that proposed to reallocate 220-222 MHz exclusively to the Land Mobile Service and 222-225 exclusively to the Amateur Radio Service. Watch QST for further developments.]

Microwaves

The region above 3000 MHz is known as the microwave spectrum. Although European amateurs have made tremendous use of this rich resource, few Americans have done so. Little commercial equipment is available and entirely different construction techniques are required if amateurs are to build their own equipment. Nevertheless, the microwave bands are receiving a lot of attention from the experimentally minded among us. We can expect a considerable amount of progress as more amateurs try out the microwave bands.

10 GHz is the most popular microwave band, and VHF enthusiasts—such as Cheryl, KA1IXI—find 10-GHz mountaintopping particularly effective. *(NA1L photo)*

The most popular microwave band in the USA is the 10-GHz (3 cm) band. Integrated QRP transmitter/receiver/horn antenna assemblies using Gunn diodes are available commercially. They provide a reliable range of several kilometers using wideband FM. Using narrowband techniques (SSB/CW) and high-gain dish antennas, hams have made QSOs of over 1000 miles. These are particularly remarkable since the power levels involved were less than 10 watts. Other microwave bands are less popular and have little activity at the present time. However, as designs for SSB/CW gear become more widely available, activity is certain to spread.

Most microwave activity is by prearranged schedules (at least in the US). This is partly because activity is so sparse and partly because antenna beamwidths are so narrow. With a 5° beamwidth, you would have to call CQ 72 times to cover a complete circle!

The key to enjoyable use of this resource is to know how everyone else is using it, and to follow their lead. Basically, this means to listen first. Pay attention to the segments of the band already in use, and follow the operating practices that experienced operators are using. This way, you won't

interfere with ongoing use of the band by others, and you'll "fit in" right away.

All of the bands between 50 and 1296 MHz have widely accepted calling frequencies (see Table 12-1). These are the frequencies most operators monitor most of the time, where you just park the receiver when you're doing something else around the shack. If someone wants to call you, or someone

Table 12-1
North American Calling Frequencies

Band (MHz)	Calling Frequency
50	50.110 DX calling
	50.200 US, local
144	144.200 SSB calling
	144.100, 144.110 CW
	144.010 EME calling
200	220.100 CW/SSB
432	432.100 CW/SSB
	432.010 EME calling
902	903.1
1296	1296.100 CW/SSB
2304	2304.1
10,000	10,368

calls CQ, he or she will use the calling frequency and you will already be "on frequency" to hear them call.

The most important thing to remember about the calling frequencies is that they are not for ragchewing. After all, if a dozen other stations want to have a place to monitor for calls, it's really impolite to carry on a long-winded conversation on that frequency, isn't it?

In most areas of the country, everyone uses the calling frequency to establish a contact, and then the two stations move up or down a few tens of kHz to chat. This way, everyone can share the calling frequency without having to listen to each other's QSOs. You can easily tell if the band is open by monitoring the call signs of the stations making contact on the calling frequency—you sure don't need to hear their whole QSO!

ACTIVITY NIGHTS

Although you can scare up a QSO on 50 or 144 MHz almost any evening (especially during the summer), in some areas of the country there is not always enough activity to make it easy to find someone to talk with. Therefore, informal "activity nights" have been established in many parts of the country so you will know when to expect some activity. Each band has its own night.

There is a lot of variation in activity nights from place to place, so don't trust the following list to be entirely accurate. Check with someone in your area to find out about local activity nights.

Common Activity Nights

Band (MHz)	Day	Local Time
50	Sunday	6:00 PM
144	Monday	7:00 PM
220	Tuesday	8:00 PM
432	Wednesday	9:00 PM
902	Friday	9:00 PM
1296	Thursday	10:00 PM

Activity nights are particularly important for 220 MHz and above. On these bands, there is often little activity during most "normal" propagation nights. If you have just finished a new transverter or antenna for one of these bands, you will have a much better chance to try them out during the band's weekly activity night. That doesn't mean there is no activity on other nights, especially if the band is open. It may just take longer to get someone's attention during other times.

Local VHF/UHF nets often meet during activity nights. Two national organizations, SMIRK (Six Meter International Radio Klub) and SWOT (Sidewinders on Two), run nets in many parts of the country. These nets provide a meeting place for active users of the 50- or 144-MHz bands. In addition, the Southeastern VHF Society runs a net on 432 MHz. For those whose location is far away from the net control's location, the nets may provide a means of telling if your station is operating up to snuff, or if propagation is enhanced. Furthermore, you can sometimes catch a rare state or grid square checking into the net. For information on the meeting times and frequencies of the nets which are run by SMIRK and SWOT, ask other local occupants of the bands in your area or write to these organizations directly: SMIRK c/o Ray Clark, K5ZMS, 7158 Stone Fence Dr, San Antonio, TX 78227; SWOT c/o Harry Arsenault, K1PLR, 704 Curtiss Dr, Garner, NC 27529.

Donna, N2FFY, of Randolph, Massachusetts (outside of Boston) shares this formidable VHF setup with Dennis, WA4PFN. (WA4PFN photo)

WHERE AM I?

One of the first things you will notice when you tune the low end of any VHF band is that most QSOs include an exchange of "grid squares." What are grid squares? Well, first of all, grid squares aren't squares. They're more like rectangles, and they're just a way of dividing up the surface of the Earth. Grid squares are a shorthand means of describing your general location anywhere on the Earth. (For example, instead of trying to tell a distant station that I'm in Canton, New York, I tell them I'm in "grid square FN24." It sounds strange, but FN24 is a lot easier to locate on a map than my small town!)

Grid squares were developed by an international group at a conference in Maidenhead, England, so they are sometimes called "Maidenhead" grid squares.

Grid squares are based on latitude and longitude. The world is first divided into 324 very large areas called "fields," which cover 10° of latitude by 20° of longitude. Each field is divided into 100 "squares" from which grid squares get their name. Grid squares are 1° × 2° in size. When the full grid square designator is used, even smaller "sub-square" divisions are made, down to areas only a few kilometers wide. For most of us, knowing about "squares" is sufficient.

Grid squares are coded with a 2-letter/2-number/2-letter code (such as FN24KP). Most people just use the first four characters (such as FN24). This handy designator uniquely identifies the grid square; no two have the same identifying code. This is the biggest advantage of the Maidenhead system over the older QRA-locator system used for decades in Europe. The European system is not applicable worldwide because the same QRA designator might be used for more than one place.

There are several ways to find out your own grid square identifier. The best place to start is with the QST article "VHF/UHF Century Club Awards," published in January 1983 QST, pp 49-51 (reprints available from ARRL HQ for an SASE). This article explains grid squares very simply. The hardest part is finding your QTH on a good map that has latitude and longitude on it; the rest is easy. Most high-quality road maps have this on the margins. Or, you could go a step further and purchase the topographic map of your immediate

area from the US Geological Survey. For maps, write to US Geological Survey, Map Distribution, Federal Center, Building 41, Denver, CO 80225. Once you have your latitude and longitude, the rest is a snap!

The ARRL publishes a map (see Fig 12-2) of the continental United States and most populated areas of Canada, on which grid squares are marked. This map is available from ARRL for $1. It really helps to keep a copy of the map handy to help locate stations as you work them. If you are keeping track of grid squares for the purpose of VUCC (the VHF/UHF Century Club award—see below), you can color in each square as you work it. Some operators use a light color when they work the grid for the first time, then color it more darkly when they receive a QSL. Others use color or pattern schemes to indicate different propagation modes. [Note: The ARRL also publishes a World Grid Locator Atlas, available for $4.]

If you have a computer, you might try running this BASIC program, GRIDLOC. It will determine anybody's grid square if you know their latitude and longitude. Or, it will go the other way, and tell latitude and longitude if you know the grid square. In addition, it calculates the true (great circle) and reverse bearing and distance between you and some other location. It is based on programs written by Ron Dunbar, WØPN, with lots of modifications. GRIDLOC runs as-is on any computer which runs PC-DOS or MS-DOS (such as the IBM® PC, Zenith Z-150, Compaq®, and so on). It will probably run on most other computers with minor changes in punctuation or syntax because it is a very "basic" BASIC program. To adapt it to most other computers, you will probably want to change or delete the statements that affect the screen, such as COLOR and LOCATE. This program is reproduced in this chapter and is also available on a 5¼-in. disk (standard IBM® PC format) from W9IP. Also included are programs for real-time tracking of the moon and sun, as well as predicting meteor-shower peaks. For more information, send an SASE to M. R. Owen, W9IP/2, 21 Maple St, Canton, NY 13617.

PROPAGATION

Normal Conditions

If you are new to the world above 50 MHz, you might wonder what sort of range is considered "normal." To a large

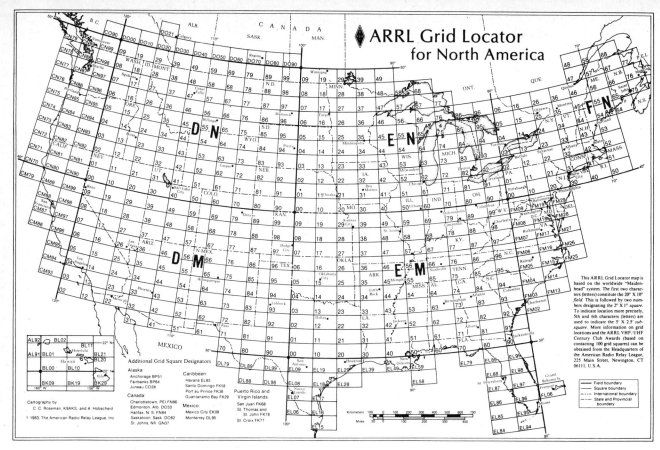

Fig 12-2—ARRL Grid Locator Map

```
1Ø REM *** GRIDLOC: A PROGRAM TO RELATE LATITUDE/LONGITUDE TO
2Ø REM *** MAIDENHEAD GRID SQUARES.  BASED ON PROGRAMS BY WA5IED AND
3Ø REM *** WØPN.  THIS VERSION BY M.R. OWEN, W9IP/2.  WRITTEN FOR THE
4Ø REM *** IBM-PC AND COMPATIBLE COMPUTERS (BASICA OR GW-BASIC).
5Ø REM ***              Last update 1-3Ø-87
6Ø REM *** The following two lines must be changed to indicate your
7Ø REM *** latitude and longitude (in decimal form).  NS$ must be either
8Ø REM *** "N" or "S" and EW$ must be either "E" or "W".
9Ø MYLATD=44.56: NS$="N"
1ØØ MYLOND=75.17: EW$="W"
11Ø IF NS$="S" THEN MYLATD=-MYLATD
12Ø IF EW$="W" THEN MYLOND=-MYLOND
13Ø PI=3.14159
14Ø DEG=PI/18Ø!: DEG$=CHR$(248)
15Ø REM PC BASIC DOESN'T HAVE ARCCOS, SO WE DEFINE IT HERE
16Ø DEF FNACOS(X)=1.57Ø796-ATN(X/SQR(1.ØØØØØ1-X*X))
17Ø CLS: KEY OFF
18Ø LOCATE 5,22: PRINT " ****** GRIDLOC MAIN MENU ******"
19Ø LOCATE 9,33
2ØØ PRINT "OPTIONS:"
21Ø LOCATE 11,2Ø
22Ø PRINT "<A>  CONVERT LAT/LON TO GRID SQUARE"
23Ø LOCATE 13,2Ø
24Ø PRINT "<B>  CONVERT GRID SQUARE TO LAT/LON"
25Ø LOCATE 18,25
26Ø PRINT "WHICH OPTION (A OR B)?"
27Ø WHICH$=INKEY$: IF WHICH$="" THEN 27Ø
28Ø GOSUB 238Ø
29Ø IF WHICH$<>"A" AND WHICH$<>"B" THEN BEEP: GOTO 26Ø
3ØØ IF WHICH$="A" THEN GOSUB 43Ø ELSE GOSUB 1Ø8Ø
```

```
31Ø GOSUB 154Ø
32Ø LOCATE 24,5
33Ø PRINT "Press A to convert lat/lon to grid square,"
34Ø PRINT "        B to convert grid square to lat/lon,"
35Ø PRINT "        Q to quit,
36Ø PRINT "        Any other key to return to GRIDLOC menu"
37Ø Z$=INKEY$: IF Z$="" THEN 37Ø
38Ø WHICH$=Z$
39Ø GOSUB 238Ø
4ØØ IF WHICH$="Q" OR WHICH$="q" THEN END
41Ø IF WHICH$="A" OR WHICH$="B" THEN 3ØØ
42Ø ABC=Ø: GOTO 17Ø
43Ø REM ROUTINE TO CALCULATE "MAIDENHEAD" GRID LOCATORS
44Ø CLS
45Ø IF ABC=1 THEN 52Ø
46Ø LOCATE 6,2Ø: PRINT "FORMAT FOR LATITUDE AND LONGITUDE"
47Ø LOCATE 11,7: PRINT "Choose one:  <A> Degrees, Minutes, Seconds
48Ø LOCATE 13,21: PRINT "<B> Decimal degrees (ie: 34.6)
49Ø LOCATE 15,21: PRINT "<C> Degrees, decimal minutes (ie: 36, 14.2)
5ØØ DMS$=INKEY$: IF DMS$="" THEN 5ØØ
51Ø ABC=1
52Ø CLS: LOCATE 5,15: PRINT "*** LAT/LON TO GRID SQUARE CONVERSION ***"
53Ø IF DMS$="B" OR DMS$="b" THEN GOSUB 197Ø
54Ø IF DMS$="C" OR DMS$="c" THEN GOSUB 216Ø
55Ø ON ERROR GOTO 246Ø
56Ø LOCATE 12,2Ø: PRINT "Example: 34,33,12,N"
57Ø LOCATE 1Ø,1Ø
58Ø PRINT "Enter latitude in degrees, minutes, seconds + N or S"
59Ø LOCATE 12,5Ø: INPUT"", L1,L2,L3,SN$
6ØØ IF L1>9Ø OR L2>6Ø OR L3>6Ø THEN PRINT "INPUT ERROR: TRY AGAIN": GOTO 58Ø
61Ø GOSUB 25ØØ  'CHECK FOR N OR S
62Ø IF SN$="S" OR SN$="s" THEN L1=-L1: L2=-L2: L3=-L3
63Ø LOCATE 16,2Ø: PRINT "Example 125,14,36,E"
64Ø LOCATE 14,1Ø
65Ø PRINT "Enter longitude in degrees, minutes, seconds + E or W"
66Ø LOCATE 16,5Ø: INPUT"", G1,G2,G3,WE$
67Ø IF G1>18Ø OR G2>6Ø OR G3>6Ø THEN PRINT "INPUT ERROR: TRY AGAIN": GOTO 65Ø
68Ø GOSUB 253Ø  'CHECK FOR E OR W
69Ø IF WE$="E" OR WE$="e" THEN G1=-G1: G2=-G2: G3=-G3
7ØØ GOSUB 194Ø
71Ø L9=L1: G9=G1: REM THIS IS TO LINK THIS ROUTINE WITH DISTANCE CALC.
72Ø G4= (G1*-1)+18Ø
73Ø C=INT (G4/2Ø)
74Ø M1$=CHR$(C+65)
75Ø R=ABS(G1/2Ø)
76Ø R=INT(((R-INT(R))*2Ø+.ØØ1))
77Ø C=INT(R/2)
78Ø IF G1>=Ø THEN C=ABS(C-9)
79Ø M3$=CHR$(C+48)
8ØØ M=ABS(G1*6Ø)
81Ø M=((M/12Ø)-INT(M/12Ø))*12Ø
82Ø M=INT(M+.ØØ1)
83Ø C=INT(M/5)
84Ø IF G1>=Ø THEN C=ABS(C-23)
85Ø M5$=CHR$(C+65)
86Ø L4=L1+9Ø
87Ø C=INT(L4/1Ø)
88Ø M2$=CHR$(C+65)
89Ø R=ABS(L1/1Ø)
9ØØ R=INT(((R-INT(R))*1Ø+.ØØ1))
91Ø C=INT(R)
92Ø IF L1<Ø THEN C=ABS(C-9)
93Ø M4$=CHR$(C+48)
94Ø M=ABS(L1*6Ø)
95Ø M=((M/6Ø)-INT(M/6Ø))*6Ø
```

```
960 C=INT(M/2.5)
970 IF L1<0 THEN C=ABS(C-23)
980 M6$=CHR$(C+65)
990 M$=M1$ + M2$ + M3$ + M4$ + M5$ +M6$
1000 CLS: LOCATE 8,15
1010 PRINT "-------------------------------------"
1020 LOCATE 14,15: PRINT "-------------------------------------"
1030 LOCATE 11,20
1040 PRINT "GRID LOCATOR :"
1050 LOCATE 11,37: COLOR 15
1060 PRINT M1$+M2$;" "; M3$+M4$;" "; M5$+M6$: COLOR 7
1070 RETURN
1080 REM ROUTINE TO CALCULATE LATITUDE AND LONGITUDE FROM
1090 REM MAIDENHEAD GRID SQUARE LOCATORS.
1100 CLS
1110 LOCATE 5,15: PRINT "*** GRID SQUARE TO LAT/LON CONVERSION ***"
1120 LOCATE 12,20: PRINT "Examples: FN24  or  FN24KP"
1130 LOCATE 10,18
1140 INPUT " GRID SQUARE (AAnnAA) : ", A$
1150 L1=LEN(A$)
1160 FOR I=1 TO L1
1170 IF ASC(MID$(A$,I,1))>90 THEN MID$(A$,I,1)=CHR$(ASC(MID$(A$,I,1))-32)
1180 NEXT I
1190 IF L1<4 THEN A$=A$+"44"
1200 IF L1<6 THEN A$=A$+"LL"
1210 FOR K=1 TO 6
1220 A(K)=ASC(MID$(A$,K,1))
1230 NEXT K
1240 L9=-90+(A(2)-65)*10+A(4)-48+(A(6)-64.5)/24
1250 L9= L9+.0005
1260 IF L1<4 THEN L9=L9+.5
1270 IF L1<6 THEN L9=L9+.02
1280 G9=-180+(A(1)-65)*20+(A(3)-48)*2+(A(5)-64.5)/12
1290 G9=(G9+.0005)*-1
1300 IF L1<4 THEN G9=G9-1!
1310 IF L1<6 THEN G9=G9-.041
1320 LATDEG=INT(L9)
1330 LATDECIMAL=(CINT(100*(L9-LATDEG)))/100
1340 LAT=LATDEG+LATDECIMAL
1350 LONDEG=INT(G9)
1360 LONDECIMAL=(CINT(100*(G9-LONDEG)))/100
1370 LON=LONDEG+LONDECIMAL
1380 DEG$=CHR$(248)
1390 IF LAT>0 THEN D$="N" ELSE D$="S"
1400 IF LON>0 THEN L$="W" ELSE L$="E"
1410 CLS
1420 LOCATE 11,15: PRINT "-------------------------------------------"
1430 LOCATE 12,17
1440 PRINT "GRID SQUARE:":COLOR 15: LOCATE 12,32: PRINT A$: COLOR 7
1450 LOCATE 12,40: IF L1<6 THEN PRINT "(MIDDLE OF GRID)"
1460 LOCATE 14,20:PRINT "LATITUDE: "
1470 LOCATE 16,19:PRINT "LONGITUDE:"
1480 COLOR 15
1490 LOCATE 14,30: PRINT USING "###.##& &";ABS(LAT);DEG$;D$
1500 LOCATE 16,30: PRINT USING "###.##& &";ABS(LON);DEG$;L$
1510 COLOR 7
1520 LOCATE 17,15: PRINT "-------------------------------------------"
1530 RETURN
1540 REM ROUTINE TO DETERMINE THE BEARING AND DISTANCE
1550 REM BETWEEN ANY TWO POINTS ON THE EARTH.
1560 HISLATD=L9
1570 HISLOND=G9
1580 HISLOND=HISLOND*-1
1590 DIFLOND=MYLOND-HISLOND
```

```
1600 REM *** DIFFERENCE IN LONGITUDES MUST FALL BETWEEN -18Ø AND +18Ø
1610 IF DIFLOND<-18Ø THEN DIFLOND=DIFLOND+36Ø
1620 IF DIFLOND>18Ø THEN DIFLOND=DIFLOND-36Ø
1630 REM *** DEGREES TO RADIANS CONVERSION
1640 MYLAT=MYLATD*DEG
1650 MYLON=MYLOND*DEG
1660 HISLAT=HISLATD*DEG
1670 HISLON=HISLOND*DEG
1680 DIFLON=DIFLOND*DEG
1690 REM *** DISTANCE CALCULATION
1700 COSB=(SIN(MYLAT)*SIN(HISLAT))+(COS(MYLAT)*COS(HISLAT)*COS(DIFLON))
1710 BETA=FNACOS(COSB)
1720 BETA2=BETA/DEG
1730 REM *** '69.Ø5' IS THE CONVERSION FACTOR FOR STATUTE MILES.
1740 REM *** FOR KILOMETERS, USE 111.2, AND FOR NAUTICAL MILES 6Ø.Ø.
1750 DIST=BETA2*69.Ø5
1760 DISTKM=BETA2*111.2
1770 REM *** BEARING CALCULATION
1780 COSA=(SIN(HISLAT)-(SIN(MYLAT)*COSB))/(COS(MYLAT)*SIN(BETA))
1790 A=FNACOS(COSA)
1800 ANGLE=A/DEG
1810 IF DIFLON<=Ø THEN 183Ø
1820 ANGLE=36Ø!-ANGLE
1830 IF ANGLE<18Ø THEN REVERSE=ANGLE+18Ø ELSE REVERSE=ANGLE-18Ø
1840 LOCATE 19, 11: PRINT "DISTANCE                BEARING"
1850 LOCATE 2Ø,6
1860 PRINT USING "#####.# ";DIST: LOCATE 2Ø, 15: PRINT "MILES"
1870 LOCATE 21,6:PRINT USING "#####.# ";DISTKM:LOCATE 21,15:PRINT "KILOMETERS"
1880 LOCATE 2Ø,4Ø:PRINT USING "###.#&";ANGLE,DEG$
1890 LOCATE 2Ø,49: PRINT "DIRECT"
1900 LOCATE 21,4Ø: PRINT USING "###.#&";REVERSE,DEG$
1910 LOCATE 21,49: PRINT "REVERSE"
1920 RETURN
1930 'CONVERT DD,MM,SS TO DECIMAL
1940 L1=L1+(L2/6Ø)+(L3/36ØØ)
1950 G1=G1+(G2/6Ø)+(G3/36ØØ)
1960 RETURN
1970 'INPUT DECIMAL LATITUDE AND LONGITUDE
1980 LOCATE 11,12: PRINT "Example: 33.4,N
1990 LOCATE 1Ø,1Ø
2000 INPUT "ENTER DECIMAL LATITUDE    ",L1,SN$
2010 IF L1>9Ø THEN BEEP: GOTO 198Ø
2020 GOSUB 238Ø ' CAPITALIZE
2030 IF SN$="N" OR SN$="S" THEN 206Ø
2040 LOCATE 2Ø,1Ø: BEEP: PRINT "LATITUDE MUST BE EITHER N OR S : TRY AGAIN"
2050 GOTO 199Ø
2060 LOCATE 14,12: PRINT "Example: 124.5,W"
2070 LOCATE 13,1Ø: INPUT "ENTER DECIMAL LONGITUDE    ",G1,WE$
2080 IF G1>18Ø THEN BEEP: GOTO 2Ø7Ø
2090 GOSUB 238Ø ' CAPITALIZE
2100 IF WE$="E" OR WE$="W" THEN 213Ø
2110 LOCATE 2Ø,1Ø: BEEP: PRINT "LONGITUDE MUST BE EITHER E OR W : TRY AGAIN"
2120 GOTO 2Ø7Ø
2130 IF SN$="S" THEN L1=-L1
2140 IF WE$="E" THEN G1=-G1
2150 RETURN 71Ø
2160 ON ERROR GOTO 246Ø
2170 LOCATE 12,12: PRINT "Example: 44,34.6,N"
2180 LOCATE 1Ø,1Ø
2190 PRINT "ENTER LATITUDE in DEGREES, DECIMAL MINUTES + N or S"
2200 LOCATE 12,5Ø: INPUT "",L1,L2,SN$
2210 GOSUB 238Ø ' CAPITALIZE
2220 IF SN$="N" OR SN$="S" THEN 225Ø
2230 LOCATE 2Ø,1Ø: BEEP: PRINT "LATITUDE MUST BE EITHER N OR S : TRY AGAIN"
```

```
2240 GOTO 2180
2250 LOCATE 16,12: PRINT "Example: 132,34.5,W"
2260 LOCATE 14,10
2270 PRINT "ENTER LONGITUDE in DEGREES, DECIMAL MINUTES + E OR W"
2280 LOCATE 16,50: INPUT "",G1,G2,WE$
2290 GOSUB 2380 ' CAPITALIZE
2300 IF WE$="E" OR WE$="W" THEN 2330
2310 LOCATE 20,10: BEEP: PRINT "LONGITUDE MUST BE EITHER E OR W : TRY AGAIN"
2320 GOTO 2260
2330 L1=L1+(L2/60)
2340 G1=G1+(G2/60)
2350 IF SN$="S" THEN L1=-L1
2360 IF WE$="E" THEN G1=-G1
2370 RETURN 710
2380 'ROUTINE TO CAPITALIZE
2390 IF WHICH$="a" THEN WHICH$="A"
2400 IF WHICH$="b" THEN WHICH$="B"
2410 IF SN$="s" THEN SN$="S"
2420 IF SN$="n" THEN SN$="N"
2430 IF WE$="e" THEN WE$="E"
2440 IF WE$="w" THEN WE$="W"
2450 RETURN
2460 'INPUT ERROR MESSAGE
2470 LOCATE 20,10: BEEP: PRINT "INPUT ERROR: TRY AGAIN"
2480 RESUME
2490 ' CHECK FOR PROPER DIRECTIONS
2500 IF SN$="N" OR SN$="n" OR SN$="S" OR SN$="s" THEN RETURN
2510 LOCATE 20,10: BEEP: PRINT "LATITUDE MUST BE EITHER N OR S : TRY AGAIN"
2520 RETURN 570
2530 IF WE$="E" OR WE$="e" OR WE$="W" OR WE$="w" THEN RETURN
2540 LOCATE 20,10: BEEP: PRINT "LONGITUDE MUST BE EITHER E OR W : TRY AGAIN"
2550 RETURN 640
```

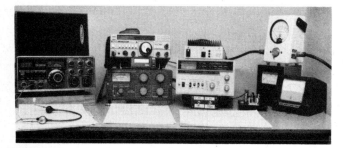

WA6YBT's elegant station layout: a 2-meter multimode transceiver, antenna tuner and HF rig, 2 meter and 70 cm transverter with a solid-state 70-cm amplifier on top of it, and antenna rotator controls, wattmeter, et al. *(WA6YBT photo)*

extent, your range on VHF is determined by your location and the quality of your station. After all, you can't expect the same performance from a 10-watt rig and a small antenna on the roof as you might from a kilowatt and stacked beams at 100 feet.

For the sake of discussion, consider a more-or-less "typical" station. On 2-meter SSB, a hypothetical typical station would probably consist of a low-powered rig, perhaps a multi-mode rig (SSB/CW/FM), followed by a 100-watt amplifier. The antenna of our typical station might be a single long Yagi at around 50 feet, fed with low-loss coax.

Using SSB or CW, how far could this station cover on an average night? Location plays a big role, but it's probably safe to estimate that you could talk to a similarly equipped station about 200 miles away almost 100% of the time. Naturally, higher-power stations with tall antennas and low-noise receive preamps will have a greater range than this, up to a practical maximum of about 350-400 miles in the Midwest (less in the hilly West and East).

On 220 MHz, a similar station might expect to cover about the same distance, and somewhat less (150 miles) on 432 MHz. This assumes normal propagation conditions and a reasonably unobstructed horizon. This range is a lot greater than you would get for noise-free communication on FM, and it represents the sort of capability the typical station should seek. Increase the height of the antenna to 80 feet and the range might increase to 250 miles, probably more, depending on your location. That's not bad for reliable communication!

Band Openings and DX

The main thrill of the VHF and UHF bands for most of us is the occasional "band opening," when signals from far away are received as if they are next door. DX of well over 1000 miles on 6 meters is commonplace during the summer, and occurs at least once or twice a year on 144, 220 and 432 MHz.

DX propagation on the VHF/UHF bands is strongly influenced by the seasons. Summer and fall are definitely the most active times in the spectrum above 50 MHz, although band openings occur at other times as well.

The following is a review of the different types of "band openings," what bands they affect and how they may be predicted. Later on, how to operate these modes will be discussed. Remember that there is a lot of variation, and that no two band openings are alike. This uncertainty is part of what makes VHF/UHF interesting! Here is a list of the main types:

1) *Tropospheric—or simply "tropo"—openings*. Tropo is the most common form of DX-producing propagation on the bands above 144 MHz. It comes in several forms, depending on local and regional weather patterns. This is because it is caused by the weather. Tropo may cover only a few hundred miles, or it may include huge areas of the country at once.

a) *"Radiation Inversion"* (mostly summer). This common type of tropo is caused by the Earth cooling off in the evening. The air just above the Earth's surface also cools, while the air a few hundred meters above remains warm. This creates an inversion that refracts VHF/UHF radio signals. If there is little or no wind, you will notice a gradual improvement in the strength of signals out to a range of 100-200 miles (less in hilly areas) as evening passes into night. This form of tropo mostly affects the bands above 144 MHz, and is seldom noticeable at 50 MHz.

b) *Broad, regional tropo openings* (late summer and early fall). These are the DXer's dream! In one of these openings, stations as far away as 1200 miles (maybe more!) are brought into range on VHF/UHF. The broad, regional type of tropo opening is caused by stagnation of a large, slow-moving high-pressure system. The stations that benefit the most are often on the south or western sides (the so-called "back side") of the system. This sort of sluggish weather system may often be forecast just by looking at a weather map.

c) *Wave-cyclone tropo* (spring). These openings don't usually last long. They are brought about by an advancing cold front that interacts with the warm sector ahead of it. The resulting contrast in air temperatures may cause thunderstorms, but sometimes don't. If conditions are just right, a band opening may result. These openings usually involve stations in the warm sector ahead of the cold front. You may feel that there's a pipeline between you and a DX station 1000 miles or more away. You may not be able to contact anybody else!

There are several other types of tropo openings, but space doesn't allow more than a brief discussion here. Coastal breezes sometimes cause long, narrow tropo openings along the East and Gulf coasts. Rarely, cold fronts may cause brief openings as they slide under warmer air. US West Coast VHFers are always on the alert for the California-to-Hawaii duct that permits 2500-mile DX. For a very complete discussion of all the major forms of weather-related VHF propagation, see "The Weather Which Brings VHF DX" by Emil Pocock, W3EP, May 1983 *QST*.

2) *The Scatter modes*. Long-distance communication on the VHF and lower UHF bands is possible using the ionosphere. Some modes are within the reach of modest stations, whereas others require very large antennas and high power. They all have one thing in common: They take advantage of the scattering of radio waves whenever there is an irregularity in the uppermost atmosphere or ionosphere. Very briefly, the main types are:

a) *Meteor scatter*. When meteors enter the Earth's atmosphere, they ionize a small trail through the E layer. This ionization doesn't last long, often only a second. However, before it dissipates, the ionization can scatter, or sometimes reflect, VHF radio waves. This sort of propagation may be identified by the fact that it's very brief, but signals may be rather strong. It is quite common on 6 meters, less so on 2 meters, and rare indeed on 220 and 432 MHz. No success-

W9IP/2 in Western New York: 8 x 19-element RIW 432-MHz EME array fed with open-wire line.

ful QSOs have been completed any higher in frequency. Operating techniques for this mode are discussed later.

b) *Tropo scatter*. This takes place in the troposphere, and is the result of wave scattering from bundles of turbulent air. Signals are always weak, with a large amount of fading. Depending on frequency, tropo scatter may be assisted by scatter in the D and E layers of the ionosphere. Contacts out to about 1000 miles are possible, but difficult; definitely the territory of well-equipped stations.

3) *Sporadic E (often called "E_s," for "E sporadic")*. This type of propagation is the most spectacular DX-producer on the 50-MHz band, where it may occur almost every day during late June and July and early August. A short E_s season also occurs during December-January. Sporadic E is most common in mid-morning and again around sunset, during the summer months, but it can occur any time, any date. E_s occurs once or twice a year on the 2-meter band in most areas.

E_s results from small patches of ionization in the E layer of the ionosphere. These patches move about, intensify or disappear altogether. The fast-moving patches of ionization appear to follow a general SE-NW track as they fly across the mid-latitude Northern Hemisphere.

Reflections from E_s patches make coast-to-coast DX possible on 6 meters. Single-hop distances of 1200-1400 miles are most common, but multi-hop openings are common on 6 meters. Multi-hop has also been observed on 2 meters. It is exceedingly rare, but it can provide the most spectacular 144-MHz DX available without moonbounce. Signals have been heard, but to date no QSOs have actually taken place on 220 MHz.

E_s signals are usually quite strong, but they may fade away altogether within a short time. On 6 meters, some days are seemingly filled with E skip. Strong signals from all parts of the country can pour in for hours and hours, or sometimes only a small area is covered. However, it is also common to go through "dry periods" when there may be no E_s for a week or more, even in the middle of the summer. That's one reason they call it "sporadic" E!

Sporadic E openings usually last only about half an hour on 2 meters, but some have lasted much longer. They are much more rare on 2 meters than on 6. Probably fewer than one in 20 strong E_s openings on 6 meters ever materialize on 2 meters.

4) *Aurora ("Au")*. The aurora borealis, or "northern lights," is a beautiful spectacle which is seen occasionally by those who live in Canada, the northern part of the USA and Europe. Similar "southern lights" are sometimes visible in the southernmost parts of South America and Africa. The aurora is caused by the Earth intercepting a massive number of charged particles thrown from the Sun during a solar "storm." These particles are funnelled into the polar regions of the Earth by its magnetic field. As the charged particles interact with the upper atmosphere, the air glows, which we see as the aurora. And, important to VHFers, these particles also provide an irregular, moving curtain of ionization which can propagate signals for many hundreds of miles.

Like sporadic E, aurora is much more common on 6 meters than on 2 meters. Au is heard often on 6 meters but not on 2 meters. Nevertheless, 2-meter Au is far more common than 2-meter sporadic E, at least above 40° N latitude. Au is also possible on 220 and 432 MHz, and many tremendous DX contacts have been made on these bands. Current record distances are over 1000 miles on 144, 220 and 432 MHz.

Aurora can be predicted by listening to radio station WWV at the National Bureau of Standards. At 18 minutes past each hour, WWV transmits a summary of the condition of the Earth's geomagnetic field. If the K index is 4 or above, you should watch for Au. However, as many VHFers have learned, a high K index is no guarantee of an aurora. Similarly, K indices of only 3 have occasionally produced spectacular radio auroras at middle latitudes. When in doubt, point the antenna north and listen!

Auroral DX signals are highly distorted. CW is the most practical mode, although SSB is sometimes used. Stations point their antennas generally northwards and listen for the telltale hissing note that is characteristic of auroral signals. More about operating Au is presented later.

5) *Transequatorial propagation (TE)*. This strange mode is responsible for almost unbelievable DX on 50 through 220 MHz. It is a strictly north-south path, crossing the equator. Stations which are equally spaced north and south of the geomagnetic equator have communicated over 4800 miles on 2 meters. Apparently some form of ionospheric ducting is responsible, although there is little agreement on the exact mechanism. TE is most common during and after periods of geomagnetic unrest. Signals have a weak, fluttery quality. This form of propagation is quite worthy of continued experimentation by amateurs.

6) *EME, or Earth-Moon-Earth, often called "moonbounce."* This is the ultimate VHF/UHF DX medium! Moonbouncers use the Moon as a passive reflector for their signals, and QSO distance is limited only by the diameter of the Earth. Any two stations who can simultaneously see the Moon may be able to work each other via EME. QSOs between the USA and Europe or Japan are commonplace on VHF and UHF by using this mode. That's DX!

Previously the territory of only the biggest and most serious VHFers, moonbounce has now become more widely popular. Thanks to the efforts of dedicated moonbouncers such as Bob Sutherland, W6PO, and Al Katz, K2UYH, hundreds of stations are active, mainly on 144 and 432 MHz. This huge increase in activity is largely the result of the "snowball" effect: More stations are EME-capable, so there's more incentive for others to "get on the Moon." Also, several individuals have assembled gigantic antenna arrays which make up for the inadequacy of smaller antennas. The result is that even modestly equipped VHF stations (150 watts and

one or two Yagis) are capable of making a few moonbounce contacts. Activity is constantly increasing. There is even an EME contest in which moonbouncers compete on an international scale.

Moonbounce requires larger antennas than most terrestrial VHF/UHF work. In addition, you must have a high-power transmitting amplifier and a low-noise receiving preamplifier if you wish to be able to work more than the biggest guns. A "typical" EME station on 144 MHz probably consists of four long-boom Yagi antennas on an azimuth-elevation mount (for pointing at the Moon), a kilowatt amplifier and a GaAs-FET preamplifier. On 432 MHz, the "average" antenna is eight long Yagis. You can make contacts with a smaller antenna, but they will be with only larger stations on the other end. Yagi antennas aren't the only type available; collinears and quagis are also widely used. Several VHFers have also built large parabolic dish antennas.

WA8NLC's 17-foot dish for 1296-MHz EME.

HOW DO I OPERATE ON VHF/UHF?
Normal Conditions

The most important rule to follow, on VHF/UHF like all other amateur bands, is to listen first. Even on the relatively uncrowded VHF bands, interference is common near the calling frequencies. The first thing to do when you switch on the radio is to tune around, listening for activity. Of course, the calling frequencies are the best place to start listening. If you listen for a few minutes, you'll probably hear someone make a call, even if the band isn't open. If you don't hear anyone, then it's time to make some noise yourself!

Stirring up activity on the lower VHF bands is usually just a matter of pointing the antenna and calling CQ. Because most VHF beams are rather narrow, you might have to call CQ in several directions before you find someone. Several short CQs are always more productive than one long-winded CQ. But don't make CQs too short; you have to give the other station time to turn the antenna toward you.

Give your rotator lots of exercise; don't point the antenna at the same place all the time. You never know if a new station or some DX might be available at some odd beam heading. VHFers in out-of-the-way locations, far from major cities, monitor the bands in the hope of hearing you.

Band Openings

How about DX? What is the best way to work DX when the band is open? There's no simple answer. Each main type of band opening or propagation mode requires its own techniques. This is natural, because the strength and duration of openings vary considerably. For example, you wouldn't expect to operate the same way during a 10-second meteor-burst QSO as during a three-day tropo opening.

The following is a review of the different main types of propagation and descriptions of the ways that most VHFers take advantage of them.

Tropo

Tropospheric propagation—usually called "tropo"—is usually most noticeable on the 144- through 1296-MHz bands. It often lasts at least several hours, and sometimes several days, so there's no big hurry to make contacts. You have time to listen carefully and determine how the opening is developing, how it is drifting and who is active. For stations on the West Coast, the openings usually extend up the coast and inland a couple of hundred miles or so. Therefore, the biggest question is how far north or south, and how far east, the opening will extend. In the middle part of the country and the East Coast, openings sometimes extend for well over a thousand miles, and in several directions.

[Recalling the ARRL VHF QSO Party in September 1979, in which the Midwest, South and East Coast experienced one of the greatest tropo openings in VHF history, my multi-operator contest log shows literally dozens of QSOs with stations more than 750 miles away, and several over 1000 miles. The VHF bands were open over a huge area; within one hour, for example, stations in southern Texas and Vermont were worked from my QTH in Illinois on 2 meters. The 432-MHz band was almost as good.]

East of the Rockies, big tropo openings occur at least once a year, although they aren't often as massive or long-lived as the September '79 opening. Unfortunately, amateurs in the mountainous West seldom experience tropo of this scale. In California, most openings are either along the coast or between the mainland and Hawaii. The latter, which accounts for the world's record tropo distances (greater than 2500 miles!) on VHF and several UHF bands, is rare.

To take best advantage of tropo openings, you need to know how they develop with time. Here are some tips: First, remember that most weather systems drift to the east with time. Large high-pressure systems tend to drift east and spread out as they mature. The long-haul DX may be available just before the opening terminates in your area. In other words, it may appear that the opening gets better and better and then all of a sudden you're out of it. If you're in the middle of the opening, sometimes stations will be "talking over your head." Be patient. Your time will come as the opening drifts east. Second, remember that most tropo-type openings improve as the evening progresses. If the opening is good and the weather is stable at sunset, then start the coffee pot, because you'll be DXing all night! In fact, the best tropo DX seems to come just around sunrise. But don't forget that if nobody is on the air, it may seem like the opening has ended. In the small hours of the morning, you might have trouble finding anyone on the air.

AAØL, Colorado, uses this 144- and 432-MHz array for both tropo and EME.

Tropo/Ionospheric Scatter

Tropo scatter usually requires prearranged schedules, a lot of patience and good equipment at each end of the path. It is often assisted by D-layer scattering, airplanes or even meteor scatter. Some form of tropo scatter is usable on all the VHF and UHF bands. The best means to take advantage of tropo scatter is to use medium-speed CW and follow a sequence in transmit-receive between the two stations; 30-second sequences seem to work best. One station transmits for the first 30 seconds of each minute, and the other station listens, then they trade off. If you're lucky, a passing airplane or some other anomaly will briefly enhance signals enough to make the QSO.

Meteor Scatter

Meteor scatter is very widely used on 50 and 144 MHz, and it has been used on 220 and 432 as well. Operation with this exciting mode of DX comes under two main headings: prearranged schedules and random contacts. Either SSB or CW may be used, although SSB is more popular in North America. European meteor-scatter enthusiasts use very high speed CW (50-100 WPM) which is recorded on tape.

Most meteor-scatter work is done during major meteor showers, and many stations "run skeds" (prearranged schedules) with others in states or grid squares that they need. In a "sked," 15-second transmit-receive sequences are the norm for North America (Europeans use longer sequences). One station, almost always the westernmost, will "take the first and third." This means that he/she transmits during the first and third 15-second interval of each minute, and the other station transmits during the other two segments. This is a very simple procedure that ensures that only one station is transmitting when a meteor falls. See accompanying sidebar.

A specific frequency, far removed from local activity centers, is chosen when the sked is set up (on the HF bands or on the telephone). It is important that both stations have

Meteor-Scatter Procedure

In a meteor-scatter QSO, neither station can hear the other except when a meteor trail exists to scatter or reflect their signals. The two stations take turns transmitting so that they can be sure of hearing the other if a meteor happens to fall. They agree beforehand on the sequence of transmission. One station agrees to transmit the 1st and 3rd 15 seconds of each minute, and the other station takes the 2nd and 4th. It is standard procedure for the western-most station to transmit during the 1st and 3rd.

It's important to have a format for transmissions so you know what the other station has heard. This format is used by most US stations:

Transmitting	Means you have copied:
Call signs	nothing, or only partial calls
Calls plus signal report (or grid square or state)	full calls—both sets
ROGER plus signal report (or grid square or state)	full calls, plus signal report (or grid square or state)
ROGER	ROGER from other station

Remember, for a valid QSO to take place, you must exchange full call signs, some piece of information, and acknowledgment. Too many meteor QSOs have not been completed for lack of ROGERs. Don't quit too soon; be sure the other station has received your acknowledgment. Often, stations will add "73" when they want to indicate that they have heard the other station's ROGER.

Until a few years ago, it was universal practice to give a signal report which indicated the length of the meteor burst. "S1" meant that you were just hearing pings, "S2" meant 1-5 second bursts, and so on. Unfortunately, virtually everyone was sending "S2," so there was no mystery at all, and no significant information being exchanged.

Grid squares have become popular as the "piece of information" in meteor-scatter QSOs. More and more stations are sending their grid-square designator instead of "S2." This is especially true on "random" meteor-scatter QSOs, where you might not know in advance where the station is located. Other stations prefer to give their state instead. For an excellent summary of meteor-scatter procedure, see "Meteor Scatter Communications" by Clarke Greene, K1JX, in January 1986 QST (pp 14-17).

accurate frequency readout and synchronized clocks, but with today's technology this is not the big problem that it once was. Schedules normally run for ½ hour or 1 hour. On 220 and 432 MHz, however, two-hour skeds are not uncommon because many fewer meteor trails are available at those frequencies. Meteor-scatter QSOs on 220 and 432 are well earned!

The best way to get the feel for the meteor-scatter QSO format is to listen to a couple of skeds between experienced operators. Then, look in *QST* or ask around for the name/call sign of veteran meteor-scatter operators in the 800-1000 mile range from you (this is the easiest distance for meteor scatter). Call them on the telephone or catch them on one of the national VHF nets and arrange a sked. After you cut your teeth on "easy" skeds, you'll be ready for more difficult DX.

A lot of stations make plenty of meteor-scatter QSOs without the help of skeds. Especially during major meteor showers, VHFers congregate near the calling frequency of each band. There they wait for meteor bursts like hunters waiting for ducks. Energetic operators call CQ while everyone else listens; when the big "Blue Whizzer" meteors blast in, the band comes alive with dozens of quick QSOs. For a brief time, normally five seconds to perhaps 30 seconds, 2

meters may sound like 20 meters! Then the band is quiet again . . . until the next meteor burst!

The quality of shower-related meteor-scatter DX depends on three factors. These three factors are well known or can be predicted. The most important is the "radiant effect." This refers to where the shower's radiant is located in the sky. The radiant is the spot in the sky from which the meteors appear to fall. If the radiant is below the horizon, or too high in the sky, you will hear very few meteors. The most productive spot for the radiant is at an elevation of about 45° and an azimuth of 90° from the path you're trying to work. The second important factor is the velocity of the meteors. Slow meteors cannot ionize sufficiently to propagate 144- or 220-MHz signals, no matter how many meteors there are. For 144 MHz, meteors slower than 50 km/s are usually inadequate (see accompanying sidebar detailing major meteor showers). Third, the shower will have a peak in the number of meteors which the Earth intercepts. However, because the peak of many meteor showers is more than a day in length, the exact time of the peak is not as important as most people think. Two interesting references are "Improving Meteor Scatter Communications" by Joe Reisert, W1JR, in June 1984 *Ham Radio,* and "VHF Meteor Scatter: an Astronomical Perspective," by M. R. Owen, W9IP, in June 1986 *QST.*

It takes a lot of persistence and a good station to be successful with "random" meteor scatter. This is mainly because you must overcome tremendous QRM in addition to the fluctuations of meteor propagation. At least 100 watts is necessary for much success in meteor scatter, and a full kilowatt will help a lot. One or two Yagis, stacked vertically, is a good antenna system.

In populated areas, it can be difficult to hear incoming meteor-scatter DX if many local stations are calling CQ. Therefore, many areas observe 15-second sequencing for random meteor-scatter QSOs, just as for skeds. Those who want to call CQ do so at the same time so everyone can listen for responses between transmissions. Sometimes a bit of peer pressure is necessary to keep everyone together, but it pays off in more QSOs for all. The same QSO format is used for scheduled and "random" meteor-scatter QSOs.

Sporadic E

Sporadic E (E_s) propagation is very common on 6 meters during the summer and very early fall, and rare at other times. It is always rare on 2 meters, but the greatest chance of it happening is during July and early August. On 6 meters, the band may open for hours on end, constantly shifting, disappearing briefly, then reappearing. Signals are often very strong, and 10-watt stations can work 1000 miles with ease. On 2 meters, the openings are equally strong but often very short; a half-hour sporadic E opening on 2 meters is rare but wonderful.

Predicting these openings in advance is very difficult, if not impossible. A very weak correlation may exist between sporadic E and large thunderstorms, but this might be because big storms and E skip are both more common during the summer. The best way to alert yourself to the possibility of sporadic E is to monitor lower frequencies, because this type of propagation usually rises in frequency. An opening may begin as "short skip" on 10 meters, then rise to include 6 meters. Then, distant FM broadcast and TV stations will become audible as the maximum frequency of the E cloud reflections goes higher. Then, if you're very lucky indeed, the E cloud will create an opening on 2 meters. If you are monitoring 6 meters, you will notice that the skip distance gets shorter and shorter as the E_s MUF (maximum usable fre-

Meteor Showers

Every day, the Earth is bombarded by billions of tiny grains of interplanetary debris, called meteors. They create short-lived trails of E-layer ionization which can be used as reflectors for VHF radio waves. On a normal morning, careful listeners can hear about 3-5 meteor pings (short bursts of meteor-reflected signal) per hour on 2 meters.

At several times during the year, the Earth passes through huge clouds of concentrated meteoric debris, and VHFers enjoy a "meteor shower." During meteor showers, 2-meter operators may hear 50 or more pings and bursts per hour. Here are some data on the major meteor showers of the year. Other showers also occur, but they are very minor.

Major Meteor Showers

Shower	Date range	Peak date	Time above quarter max	Approximate visual rate	Speed km/s	Best Path	Time (local)
Quadrantids	Jan 1-6	Jan 3/4	14 hours	40-150	41	NE-SW;SE-NW	13-15;05-07
Eta Aquarids	Apr 21-May 12	May 4/5	3 days	10-40	65	NE-SW;E-W;SE-NW	05-07;06-09;09-11
Arietids	May 29-Jun 19	Jun 7	?	60	37	N-S	06-07 & 13-14
Perseids	Jul 23-Aug 20	Aug 12	4.6 days	50-100	59	NE-SW;SE-NW	09-11;01-03
Orionids	Oct 2-Nov 7	Oct 22	2 days	10-70	66	NE-SW;N-S;NW-SE	01-03;01-02 & 07-09; 07-08
Geminids	Dec 4-16	Dec 13/14	2.6 days	50-80	34	N-S	22-24 & 05-07

Meteor-Shower Peaks

These calculated predictions of meteor-shower peak times/dates are based on the latest astronomical data. Nevertheless, meteor showers are notoriously fickle about meeting their predicted peaks. Therefore, don't put too much faith in these estimates. The time at which the shower's radiant is at a 45° elevation, perpendicular to the path you want to work, is much more important than the actual peak of the shower. Best times and paths are listed above. These values are calculated for Kansas City and are valid for the entire USA because the time indicated is local standard time. One hour before and after the given time ranges should generally be good as well.

Meteor Shower Peak Times (UTC): 1987-1991

Note: The first line indicates month-day (eg, 1-3 is January 3) and the second line (eg, 2330) indicates UTC.

Shower	1987	1988	1989	1990	1991	Shower	1987	1988	1989	1990	1991
Quadrantids	1-3 2330	1-4 0600	1-3 1200	1-3 1800	1-4 0000	Perseids	8-12 2330	8-12 0530	8-12 1200	8-12 1800	8-13 0000
Eta Aquarids	5-6 1400	5-5 2000	5-6 0230	5-6 0830	5-6 1500	Orionids	10-22 1130	10-21 1730	10-22 0000	10-22 0600	10-22 1200
Arietids	6-7 1930	6-7 0200	6-7 0800	6-7 1400	6-7 2030	Geminids	12-14 0600	12-13 1200	12-13 1800	12-14 0000	12-14 0630

quency) goes higher. If you hear loud 6-meter signals from 400 or fewer miles away, watch 2 meters for some action.

You'll need to be quick on your toes to take advantage of sporadic E. You might not even notice an E_s band opening in progress because signals are so strong ("DX is never that loud!"). Quick, brief calls and short exchanges are needed if you are to make the most of the opening. Don't spend a half hour calling a long-winded CQ. The opening will be gone by the time you're through. Once the E_s opening is in progress, most stations don't waste a lot of time hunting around. Just work everything you can hear, and quickly. Almost all stations will be DX, so you can't lose!

Don't worry if you aren't running a lot of power and a huge antenna; sporadic E is the greatest equalizer in VHF. The size of your station hardly matters at all. It's mostly location that counts! You may be having the greatest opening of all time, and your friend a few miles away might hear little or nothing. That's the way E_s is. Next week, your buddy may be the lucky one and you will be left out, so make the most of it!

Aurora

Aurora (Au) is a propagation mode which favors stations at high latitudes. It is a wonderful blessing for those who must suffer through long, cold winters because other forms of propagation are rare during the winter. Aurora can come at almost any time of the year. New England stations get Au on 2 meters about five to 10 times a year, whereas stations in Tennessee get it once a year if they're lucky. Central and Southern California almost never see aurora.

The MUF of the aurora seems to rise quickly, so don't wait for the lower-frequency VHF bands to get exhausted before moving up in frequency. Check the higher bands right away.

You'll notice Au by its characteristic hiss. Signals are distorted by reflection and scattering off the rapidly moving curtain of ionization. They sound like they are being transmitted by a leaking high-pressure steam vent rather than radio. SSB voice signals are so badly distorted that often you cannot understand them unless the speaker talks very slowly and clearly. The amount of distortion increases with frequency. Most 50-MHz Au contacts are made on SSB, where distortion is the least. On 144, 220 and 432 MHz, CW is the only really useful means of communicating via AU.

If you suspect Au, tune to the CW calling frequency (144.100) and listen with the antenna to the north. Maybe you'll hear some signals. Try swinging the antenna as much as 45° either side of due north to peak signals. In general, the longest-distance DX stations peak the farthest away from due north. Also, it is possible to work stations far south of you by using the aurora; in that case your antenna is often pointed north.

High power isn't necessary for aurora, but it helps. Ten-watt stations have made Au QSOs but it takes a lot of perseverance. If you have low power, try calling only the loudest stations you hear. The weak ones probably won't be able to hear you. Increasing your power to 100 watts will greatly improve your chances of making Au QSOs. As with most short-lived DX openings, it pays to keep transmissions brief.

Aurora openings may last only a few minutes or they may last many hours, and the opening may return the next night, too. If WWV indicates a geomagnetic storm, begin listening on 2 meters in the late afternoon. Many spectacular Au openings begin before sunset and continue all evening. If you get the feeling that the Au has faded away, don't give up too soon. Aurora has a habit of "dying" and then returning several times, often around midnight. And remember: If you experience a terrific Au opening, look for an encore performance about 27-28 days later, because of the rotation of the sun.

Propagation Indicators

Active VHFers keep a careful eye on various propagation indicators to tell if the VHF bands will be open. The kind of indicator you monitor is related to the kind of propagation you are expecting. For example, during the summer it pays to watch closely for sporadic E because openings may be very brief. If your area only gets one or two per year on 2 meters, you sure don't want to miss them!

Many forms of VHF/UHF propagation develop first at low frequency and then move upward to include the higher bands. Aurora is a good example. Usually it is heard first on 10 meters, then 6 meters, then 2 meters. Depending on your location and the intensity of the aurora, the time delay between hearing it on 6 meters and its appearance on 2 meters may be only a few minutes, to as much as an hour, later. Still, it shows up first at low frequency. The same is true of sporadic E; it will be noticed first on 10 meters, then 6, then 2.

Tropospheric propagation, particularly tropo ducting, acts in just the reverse manner. Inversions and ducts form at higher frequencies first. As the inversion layer grows in thickness, it refracts lower and lower frequencies. However, most of us don't notice this because amateur activity drops off with higher frequencies, and atmospheric losses increase. So, it may be true that the band opens high in frequency first, but it may appear to open first on 2 meters because that is where much of the country's VHF activity is. Six-meter tropo openings are very rare because few inversion layers ever develop sufficient thickness to enhance such long-wavelength signals.

How can you monitor for band openings? The best way is to take advantage of the "beacons" which are inadvertently provided by commercial TV and FM broadcast stations. Television Channels 2 through 6 (54-88 MHz) are great for catching sporadic E. As the E opening develops, you will see one (or more) stations appear on each channel. First, you may see Channel 2 get cluttered with stations 1000 miles or more away, then the higher channels may follow. If you see strong DX stations on Channel 6, better get the 2-meter rig warmed up. Channel 7 is at 175 MHz, so you rarely see any sporadic E there or on any higher channels. If you do, however, it means that a major 2-meter opening is in progress!

The gap between TV Channel 6 and 7 is occupied partly by the FM broadcast band (88-108 MHz). Monitoring that spectrum will give you a similar "feel" for propagation conditions.

Several amateurs have built converters to monitor TV video carrier frequencies. A system like this can be used to keep

WB8ART's EME station: The tower on the right is 32 feet high, with 4 Cushcraft Boomers. The other tower is 62 feet, with 4 KLM 432-LBXs and one KLM 220 LBX. *(WA6YBT photo)*

track of meteor showers. It's a way of checking to tell if meteors are plentiful or not, even if there appears to be little activity on the amateur VHF bands. It also can alert you to aurora and sporadic E. A variety of systems for monitoring propagation have been discussed by Joe Reisert, W1JR, in the June 1984 issue of *Ham Radio*.

EME: Earth-Moon-Earth

EME, or moonbounce, is available any time the Moon is in a favorable position. Fortunately, the Moon's position may be easily calculated in advance, so you always know when this form of DX will be ready. It's not actually that simple, of course, because the Earth's geomagnetic field can play havoc with EME signals as they are leaving the Earth's atmosphere and as they return. Not only can absorption (path loss) vary, but the polarization of the radio waves can rotate, causing abnormally high path losses at some times. Still, most EME activity is predictable.

You will find many more signals "off the Moon" during times when the Moon is nearest the Earth (perigee), when it is overhead at relatively high northern latitudes ("positive declination"), and when the Moon is nearly at Full phase. The one weekend per month which has the best combination of these three factors is informally designated the "activity" or "skeds" weekend, and most EMEers will be on the air then.

Moonbounce Operating Practices

After traveling 400,000 km, bouncing off a poorly reflective Moon, and returning 400,000 km, EME signals are quite weak. A large antenna, high transmitter power, low-noise preamplifier and very careful listening are all essential for EME. Nevertheless, hundreds of amateurs have made EME contacts, and their numbers are growing.

The most popular band for EME is 144 MHz, followed by 432 MHz. Other bands with regular EME activity are 220, 1296 and 2304/2320 MHz.

Moonbounce QSO procedure is different on 144 and 432 MHz. On 144 MHz, schedule transmissions are 2 minutes long, whereas on 432 they are 2½ minutes long. In addition, the meaning of signal reports is different on the two bands. This difference in procedure is somewhat confusing to newcomers. Most EMEers operate on only one band, so they grow accustomed to the procedure on that band pretty quickly.

Signal Report / Meaning

Signal Report	Meaning 144 MHz	432 MHz
T	Signal just detectable	Portions of calls copyable
M	Portions of calls copyable	Complete calls copied
O	Both calls fully copied	Good signal, easily copied
R	Both calls, and "O" signal report copied	Calls and report copied

What does this difference in reporting mean? Well, the biggest difference is that "M" reports aren't good enough for a valid QSO on 2 meters but they are good enough on 432. So long as everyone understands the system, then there is no confusion. On both bands, if signals are really good, then normal RST reports are exchanged.

The majority of EME QSOs are made without any prearranged schedules. Almost always, there is no rigid transmitting time-slot sequencing. Just as in CW QSOs on HF, you transmit when the other station turns it over to you. This is particularly true during EME contests, where time-slot transmissions slow down the exchange of information. Why take 10-15 minutes when two or three will do?

On the other hand, many QSOs on EME are made with the assistance of "skeds." This is especially the case for newcomers and rare DX stations.

On 144 MHz, each station transmits for 2 minutes, then listens as the other station transmits. Which 2-minute sequence you transmit in is agreed to in advance. The common terminology is "even" and "odd." The even station transmits the 2nd, 4th, 6th, 8th...2-minute segment of the hour, while the odd station transmits the others (1st, 3rd, 5th...). Therefore, 0030-0032 is an "even" slot, even if the sked begins on the half hour. In most cases, the westernmost station takes the odd sequence. An operating chart, similar to many which have been published, will help you keep track:

Transmitting Sequence

Odd (eastern)	Even (western)
00-02	02-04
04-06	06-08
08-10	10-12
12-14	14-16
16-18	18-20
20-22	22-24
24-26	26-28
28-30	30-32
32-34	34-36
36-38	38-40
40-42	42-44
44-46	46-48
48-50	50-52
52-54	54-56
56-58	58-00

During the schedule, at the point when you've copied portions of both calls, the last 30 seconds of your 2-minute sequence is reserved for signal reports; otherwise, call sets are transmitted for the full 2 minutes.

On 432 MHz, sequences are longer. Each transmitting slot is 2½ minutes. You either transmit "first" or "second." Naturally, "first" means that you transmit for the first 2½ minutes of each 5 minutes.

144-MHz Procedure—2-Min Sequence

Period	1½ minutes	30 seconds
1	Calls (W6XXX DE W1XXX)	
2	W1XXX DE W6XXX	T T T T
3	W6XXX DE W1XXX	O O O O
4	RO RO RO RO	DE W1XXX K
5	R R R R R	DE W6XXX K
6	QRZ? EME	DE W1XXX K

432-MHz Procedure—2½-Min Sequence

Period	2 minutes	30 seconds
1	VE7BBG DE K2UYH	
2	K2UYH DE VE7BBG	
3	VE7BBG DE K2UYH	T T T
4	K2UYH DE VE7BBG	M M M
5	RM RM RM RM	DE K2UYH K
6	R R R R R	DE VE7BBG SK

On 220 MHz, some stations use the 144-MHz procedure and some use the 432-MHz procedure. Which one is used is determined in advance. On 1296 and above, the 432 procedure is always used.

EME operation generally takes place in the lowest parts of the VHF bands: 144.000-144.070; 220.000-220.050; 432.000-432.070; 1296.000-1296.050. Terrestrial QSOs are strongly discouraged in these portions of the bands.

This is particularly true for 432 EME; 144-MHz EME is active during the week and on non-skeds weekends as well. See accompanying sidebar on EME operating practices.

Hilltopping and Portable Operation

One of the nice things about the VHF/UHF bands is that antennas are relatively small, and station equipment can be packed up and easily transported. Portable operation, commonly called "hilltopping" or "mountaintopping," is a favorite activity for many amateurs. This is especially true during VHF and UHF contests, where a station can be very popular by being located in a rare grid square. If you are on a hilltop or mountaintop as well, you will have a very competitive signal. See accompanying sidebar for further information.

Hilltopping is fun and exciting because hills elevate your antenna far above surrounding terrain and therefore your VHF/UHF range is greatly extended. If you live in a low-lying area such as a valley, a drive up to the top of a nearby hill or mountain will have the same effect as buying a new tower and antenna, a high-power amplifier and a preamplifier, all in one!

The popularity of hilltopping has grown as more stations acquire solid-state rigs and amplifiers. The "box and brick" (compact multi-mode VHF rig and solid-state 100-200 watt amplifier) combination is ideally suited for mobile and

VHF Mountaintopping for the '80s

Recently, VHF has seen a resurgence of an activity as old as VHF itself—mountaintopping. Few are blessed with a home operating location that facilitates a total command of the frequency. Consequently, ardent VHFers construct bigger and bigger arrays and amplifiers for the home station in order to produce that booming signal. Some, however, utilize the "great equalizer" to compete on an equal or superior footing with the big home stations—namely, a mountaintop location. Here, perched high above all the home stations, a small portable rig with a single Yagi antenna only a few feet off the ground suddenly sounds like a kilowatt feeding a killer antenna installed at home. Simple equipment performs amazingly well from a mountaintop QTH on VHF.

A mountaintop expedition can vary from a spur-of-the-moment Sunday afternoon picnic to a full-fledged weekend contest. Quick trips can also be conducted during band openings. Since a contest optimizes the opportunity to work a lot of stations on VHF, this sidebar is mostly a "how-to" for weekend contest operation conducted by one or two people. But this can be scaled down to a mountaintop stay of shorter duration.

The Times Are a Changin'

Old-timers will remember the drudgery of lugging "boat anchors" up rocky crevices. Dragging equipment and generators weighing hundreds of pounds up steep mountainsides was no picnic. Those who suffered sprained backs soon gave it up. The advent of solid-state equipment, however, has made mountaintopping a far less strenuous activity. With a greater selection of such compact commercially available equipment, VHF has become a hotbed of activity. Even some of the highly competitive HF types have found new worlds to conquer above 50 MHz. A key factor in this revival has been the introduction of a worldwide grid locator system—now much in vogue of VHF. The use of grid squares in the major VHF contests has tickled the innermost secret desire of every radio amateur—to be on the receiving end of a DX pileup. Now instead of going on safari to some distant DX land, you can head for the mountains—some nearby mountaintop located in a rare grid square.

Choosing a Site

Choosing a mountaintop site involves considering how far you want to travel to get there, accessibility to the top of the mountain and its all important grid-square location. Ideally, your mountain is only a short driving distance away, towers into the cirrusphere, has a six-lane interstate to the top and rises within a grid square that has never been on the air before!

Obviously, some of these considerations may have to be compromised. Your first step to finding VHF heaven involves extensive study of a road atlas. How far do you want to travel? Where are the mountains? How high are they? Can you drive to the top? Draw in the grid line boundaries so you can tell which square it is in. Ask some active VHFers which are the difficult squares to work. When you start zeroing in on a potential site, you may want to get a topographic survey map of the area to determine access roads and direction of "drop-off" from the summit.

I've never operated from a mountaintop without first scoping it out in person. Access is most important. Thus far, I've operated from sites which I have reached by car, ferry, gondola, 4-wheel drive, motor home and hiking. Unless you are going with mini-radios and gel-cells, you want to get there without backpacking it. A passable road to the top is ideal. When checking out a potential site, bring a compass and 2-meter FM hand-held. A call on 146.52-MHz simplex should tell you how good the location is. Are you blocked in any direction? Is it already "RF-city" with commercial installations—a potential source of interference? Will you be able to clear any trees with a lightweight mast? Then the prime requisite: Is there a picnic table permanently at the site? If not, plan on bringing an operating table and chair—which adds considerable bulk and weight to transport.

Once you've selected an operating site, be sure you have secured the necessary permission to use the site. This may simply require verbal permission from some authority or the owner, or, it could involve a lengthy exchange of correspondence with a state environmental agency of forests and parks and the signing of a liability release. But be sure you have permission. The last thing you want is the local sheriff shining a flashlight in your eyes at 3 AM, rousting you out of a fantastic tropo opening on 2 meters. You'll find rangers on fire watch most helpful in pointing out how to obtain necessary permission.

Power Source

Unless you are awfully lucky to find a location that will permit you to just "plug in," make plans for providing your own power. With a single-band operation from a car (with antenna mast mounted just outside the car window), the car battery will probably suffice. Run a set of heavy-duty jumper cables directly to the car battery. Even a solid-state "brick" amplifier can be run off the car battery without ill effects. Just in case, park the car facing downhill!

For a more serious effort utilizing several VHF and UHF bands, a small generator is recommended. If the word generator conjures up an image of an ugly engine block from a 1947 LaSalle, then tune in to the modern world. Small, even attractive, generators in the 500-1000 W category, that look more like American Tourister luggage, are now available. Mine is a 650-watt beauty that weighs in at 43 pounds, and runs for four hours on a half gallon of petrol. And quiet? You can hear the wings of a Monarch butterfly flutter at 50 paces. A not-filled 5-gallon gerry-can provides more than enough flammable juice for a contest weekend.

Equipment

A mountaintop location effectively places your antenna atop a natural tower of hundreds or perhaps thousands of feet. With this height advantage, compact, lightweight, low-powered radios that can be boosted up to the 50- to 100-W range with solid-state amplifiers will perform nicely. Low-powered portable transceivers are manufactured for just this purpose. The popular 10- and 25-watt multimode rigs are also quite adequate. Many discontinued models can be obtained at a substantial savings through the Ham Ads section of QST. Use of transverters operating with mobile-type HF radios should also be considered.

If you don't have any sizable trees to get over, you can use simple mast sections that fit together. I use 5-foot sections available at the popular shack of radios. They are easily transportable. Important too is the method of antenna rotation. If you can install the antenna mast right next to the operating position, do it. It will save all the hassle of installing motorized rotators. Nothing beats the "Armstrong" method for speed and simplicity. I use a cross-piece of aluminum tubing mounted with U-bolts to the mast at arm level. See Fig 12-3. This provides instantaneous antenna-peaking capability—a necessity on V/UHF. While home stations are twirling their antennas in every direction trying to peak a weak signal, I've already worked him!

Installing antennas for several bands on the same mast is recommended. They should be oriented in the same direction. Many contacts on UHF are the result of moving stations over from other bands. For example, in a contest if you move a multiplier to 432 MHz after first working on 2 meters, and both antennas are on the same mast, you will first want to peak the signal on 144 MHz. Then, when you move to 432 MHz where the antennas are probably a bit more sharp and propagation perhaps marginal, both antennas will be pointing at each other for maximum signal. This can make the difference in whether the contact is made.

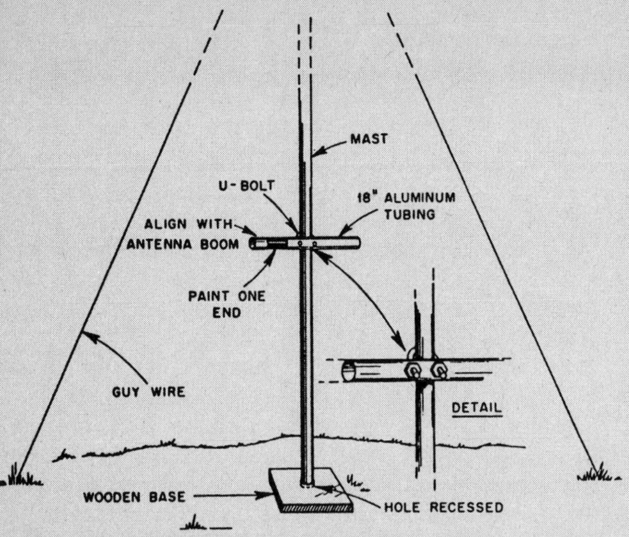

Fig 12-3—A closeup view of the "Armstrong" rotator, used for quick peaking of signals.

If your mountaintop operation involves staying overnight, additional attention must be paid to having the proper survival equipment. The most luxurious way to go is a van or RV. Otherwise, a tent will be required. For the rugged outdoors type, this can be as appealing as the radio part. I find that cooking steaks over a campfire with a canopy of stars overhead (while a programmable keyer is calling CQ) is half the fun. But keep this aspect of the operation also as simple as possible. Champagne and caviar can be held for another time. I've also found out the hard way that one can expect heavy winds on mountaintops. Large tents blow down easily in such weather.

Further on the subject of weather, just because you're topping it in July or August, don't expect it will always be T-shirt and shorts weather. No matter what the season, expect to need a heavy jacket after dark. I always bring a heavy flannel shirt and ski jacket for night, and shorts in the daytime. And bring lightweight raingear, just in case. And depending on the habitat, don't be surprised to be introduced to a critter or two, especially after dark!

Getting Started

Okay, you've read this far and are beginning to say to yourself: "Self, I think I'd like to try that." But there is a little voice of caution in you that says: "Don't go bonkers until you've sampled a little first." Good advice!

Start out by setting up on an easily accessible mountain for an afternoon during a contest period on a single band. For a first effort, I recommend 2 meters. With so many 10-W multimode rigs out there in radioland, 2 meters is your "bread and butter" band. Using a multi-element Yagi a few feet off the ground of a strategically located mountain or hill can whet your appetite. I first got hooked by operating from the side of a highway on Hogback Mountain, Vermont, with a 3-W portable 2-meter radio to a 30-W brick and 11-element Yagi. I was astounded by the results, with contacts hundreds of miles away. This launched my interest in acquiring more equipment for portable mountaintop use, each operation adding a new band or better antenna. The basic formula of keeping it lightweight and simple has prevailed, however.

Now what's holding you back from operating from Mount Everest?—*John F. Lindholm, W1XX* [originally published in Mar 1986 *QST*]

A Checklist of Typical Items Needed for a Weekend Portable Mountaintop Operation

Radio Equipment
☐ Transceivers for each band
☐ Solid-state amplifiers
☐ Antennas
☐ Coax (Belden 9913 or equiv)
☐ Keyer
☐ Paddle
☐ Antenna masts
☐ Earphones
☐ Coax connector cables
☐ Coax adapters
☐ dc cables with plugs
☐ Power supply (ies)
☐ Fuses
☐ Antenna rotator crosspiece
☐ Microphones
☐ Key line with plug
☐ Multi-ac plug outlet
☐ SWR meter
☐ VTVM
☐ Multi-dc plug box
☐ Clip leads

Tools
☐ Wrenches
☐ Pliers
☐ Screwdrivers
☐ Hammer

Power Source
☐ Generator
☐ Power cable
☐ Gasoline
☐ Oil
☐ Jumper cables (if on car battery)
☐ Gas funnel

Personal
☐ Toothbrush/toothpaste
☐ Soap
☐ Towel
☐ Suntan lotion
☐ Change of clothes
☐ Rain gear
☐ Hat
☐ Warm jacket
☐ Alarm clock
☐ Bug spray
☐ Insect repellant
☐ Electric shaver
☐ Toilet paper

Camping Equipment
☐ Matches
☐ Tent stakes (enough for tent and masts)
☐ Tarpaulin canopy
☐ Tent
☐ Ground cloth
☐ Extra rain cover
☐ Cot
☐ Lantern
☐ Flashlight
☐ Pot and pan set
☐ Pot holder
☐ Spatula
☐ Can opener
☐ Water bottle (5 gal)
☐ Rope
☐ Charcoal briquettes
☐ Table and chair (if needed)
☐ Sleeping bag
☐ Food and drink
☐ Knife and fork set
☐ Paper plates
☐ Paper towels
☐ Paper cups
☐ Cooler with ice
☐ Cook stove
☐ Stove fuel
☐ Funnel for fuel
☐ Fluorescent-type battery lantern
☐ Old newspaper
☐ Coffee cup
☐ Aluminum foil
☐ Trash bag

Miscellaneous
☐ Compass
☐ 24-hour clock
☐ Black plastic tape
☐ Masking tape
☐ Logs
☐ Grid square maps
☐ Pencils (many)
☐ Clipboard
☐ Highlighter

DJ9BV's 4 × 29 element Yagi for 432 MHz, each boom 8½ wavelengths long.

portable operation. You need no other power source than a car battery, and even with a simple antenna your signals will be outstanding.

Many VHF/UHFers drive to the top of hills or mountains and set up their station. A hilltop park, rest area or farmer's field are equally good sites, so long as they are clear of trees and obstructions. You should watch out for high-power FM or TV broadcasters who may also be taking advantage of the hill's good location; their powerful signals may cause intermodulation problems in your receiver.

Antennas, on a couple of 10-foot mast sections, may be turned by hand as the operator sits in the passenger seat of the car. A few hours of operating like this can be wonderfully enjoyable and can net you a lot of good VHF/UHF DX. In fact, some VHF enthusiasts have very modest home stations but rather elaborate hilltopping stations. When they notice that band conditions are improving, they hop in the car and "head for the hills." There, they have a really excellent site and can make many more QSOs.

In some places, there are no roads to the tops of hills or mountains where you might wish to operate. In this case, it is a simple (but sometimes strenuous) affair to hike to the top, carrying the car battery, rig and antenna. Many hilltoppers have had great fun by setting up on the top of a firetower on a hilltop, relying on a battery for power.

Some VHF/UHFers, especially contesters, like to take the entire station, high power amplifiers and all, to hilltops for extended operation. They may stay there for several days, camping out and DXing. Probably the most outstanding example of this kind of operation is put on regularly by the group at W2SZ/1. Dozens of operators and helpers assemble this multiband station, often with moonbounce capability, and

operate major VHF and UHF contests from Mount Greylock in Western Massachusetts. Their winning scores in virtually every VHF contest they have entered testifies to their skill and the effectiveness of hilltopping!

Contests

The greatest amount of activity on the VHF/UHF bands occurs during contests. VHF/UHF contests are scheduled for some of the best propagation dates during the year. Not only are propagation conditions generally good, but activity is always very high. Many stations "come out of the woodwork" just for the contest, and many individuals and groups go hilltopping to rare states or grid squares.

A VHF/UHF contest is a challenging but friendly battle between you, your station, other contesters, propagation and Murphy's Law. Your score is determined by a combination of skill and luck. It is not always the biggest or loudest station that scores well. The ability to listen, switch bands quickly and to take advantage of rapidly changing propagation conditions is more important than brute strength.

There are quite a few VHF/UHF contests. Some are for all the VHF and UHF bands, while others are for one band only. Some run for entire weekends and others for only a few hours. Despite their differences, all contests share a basic similarity. In all North American contests, your score is determined by the number of contacts (or more precisely, the number of QSO points) you make, multiplied by some "multiplier," which may be ARRL DXCC countries, grid squares, call-sign prefixes or some other factor. In most of the current major contests, you keep track of QSOs and multipliers by band. In other words, you can work the same station on each band for separate QSO and multiplier credit.

Listed below are the major contests of the VHF/UHF realm in North America. Other contests are popular in Europe and Asia, but these are not listed here because information about them is not usually available to most of us. All the listed contests are open to all licensed amateurs, regardless of their affiliation with any club or organization. Detailed rules for each contest are published in QST, CQ and the newsletters of the sponsoring organizations. For more information about contesting, see Chapter 7.

1) *ARRL VHF Sweepstakes:* This one-weekend contest occurs in January. It is favored by clubs because it permits club members to pool their individual scores for the club's total. In addition, individuals and multioperator groups compete. Scores are determined by total QSO points per band multiplied by the number of grid squares worked per band.

2) *ARRL Spring Sprints:* These contests, as their names imply, are short: Only 4 hours. There is a separate Sprint for each of the major VHF and UHF bands. Each one takes place on a different evening, a little more than a week apart. The dates vary, but are usually in the spring. The evenings for the contests are chosen to coincide with each band's activity night. So, the 2-meter Sprint takes place on a Monday, the 220-MHz Sprint on Tuesday the next week, the 432-MHz Sprint on Wednesday a week later than that, 902 MHz on the Friday after that, and the 1296-MHz Sprint on the following Thursday. The 6-meter Sprint usually is held on a Saturday at the end of the sequence, to place it closer to the sporadic E season. Your score is the number of QSO points multiplied by the number of grid squares worked in the Sprint's 4-hour time period. Fast-paced and fun!

3) *ARRL June VHF QSO Party:* This contest is the highlight of the contest season for most VHFers. Scheduled for

One of the many multiop contest efforts by N2SB, this during the VHF Sweepstakes.

the second full weekend of June, the "June Contest" sees the most activity of any North American VHF contest. Conditions on all the VHF bands are usually good, with 6 meters leading the way. This contest covers all the VHF and UHF bands, and QSOs and multipliers are accumulated for each band. Your score is determined by multiplying QSO points per band by grid squares per band. Single-band and multiband awards are given to high-scoring individuals in each ARRL section. In addition, multi-operator groups compete against each other (and the competition can be fierce!).

4) *CQ Magazine Worldwide VHF WPX Contest*: This contest is a recent addition to the contest calendar, and it is receiving a lot of attention. It is based upon call sign prefixes as multipliers and is international in scope. Your score is the number of different prefixes worked per band times the number of QSO points per band. All VHF bands and UHF bands through 1296 MHz are permitted. A multitude of awards are available, based on power level, number of operators and number of bands employed.

5) *ARRL UHF Contest*: As its name suggests, this contest is restricted to the UHF bands (plus 220 MHz). It takes place over a full weekend in August. All UHF bands are permitted, and grid squares are the multiplier. Less equipment is involved for this contest than for contests which cover all VHF/UHF bands, so many groups go hilltopping for the UHF Contest.

6) *ARRL September VHF QSO Party*: The rules for this contest are identical to those for the June VHF QSO Party. This contest is also very popular, and many multioperator groups travel to rare states and grid squares for it. By the second weekend of September, most 6-meter sporadic E has disappeared, but tropo conditions are often extremely good. Therefore, the bands above 144 MHz are the scene of tremendous activity during the September contest. Some of the best tropo openings of recent decades have taken place during this contest, and they are made even better by the high level of activity.

7) *ARRL EME Contest*: This contest is devoted to moonbounce. It takes place over two full weekends which are spaced almost a month apart. The date of the contest is different each year because of the variable phase of the Moon. The dates are usually chosen by active moonbouncers to coincide with the best combination of high lunar declination, perigee and the full phase of the Moon. This contest is international in scope. Your score is the number of QSO points made via EME per band, multiplied by the number of US call districts and ARRL countries per band. Hundreds of EMEers participate

in this challenging test of moonbounce capability.

Contests are lots of fun, whether you're actively competing or not. You don't have to be a full-time competitor to participate or to enjoy yourself! In fact, most of the participants in contests aren't really "in" the contest. Lots of operators get on the air to pass out points, to have fun for a while, and to listen for rare DX. Others try their hardest for the entire contest, keep track of their score and send their logs in for awards. Either way, the contest is a fun challenge.

If you don't plan on being a serious competitor, then you just need to know what the "exchange" is. The exchange is the minimum information that must be passed between each station to validly count the QSO in the contest. In all of the terrestrial VHF contests sponsored by the ARRL, you need to pass only your grid square designator to the other station, and receive theirs, plus acknowledgment, to count the QSO. Other contests require some different information, such as serial numbers or signal reports, so it pays to check the rules to make sure.

A serious contest effort requires dedication and effort, as well as a station that can withstand a real workout. Contests are a challenge to operators and equipment alike. A good contest score is the result of hard work, a good station and favorable propagation. A good score is something to be proud of, especially if there is lots of stiff competition. And with the popularity of VHF and UHF contests these days, competition is always stiff!

If you are a competitor in the contest, you will probably need to keep a "dupe sheet" which helps prevent duplicate QSOs. A dupe sheet is a large piece of paper on which you record each call sign that you work in a sort of a matrix so you can check it quickly. If the other station's call sign is already in the dupe sheet, working him/her again won't count for contest credit. Don't rely on your memory. After 24 hours of contesting, most of us have trouble remembering our own names! Also, if you make more than 200 QSOs in an ARRL VHF contest, you must submit your dupe sheet along with your logs. Dupe sheets, as well as log sheets, are available from ARRL HQ if you send a large self-addressed, stamped envelope (SASE) with your request.

Many operators don't feel that their stations are competitive on all the VHF and UHF bands. Not to worry; in most contests, it is possible to compete on only one band if you want to. This has the advantage of concentrating your efforts where your station is the strongest, allowing you to devote full time to just one band. You don't need to be high powered to compete as a single-band entry. Location makes a lot of difference, and hilltopping single-banders have had tremendous success, particularly if they go to a rare grid square.

Other operators like to try for all-band competition. In this case, it's a real advantage to be able to hop from one band to another. You can quickly check 6 meters for activity while also tuning 2 meters. Or, if you work a rare grid on one band, you can take advantage of the opportunity by asking the other station to switch bands right then. Some contesters work one station on all possible bands within two minutes by band hopping! This is a speedy way to increase your grids and QSOs.

Multioperator stations, in which more than one person performs operating tasks such as logging, dupe checking or making contacts, are very popular in VHF/UHF contests. In ARRL contests, all multiop stations are also multiband. For multiop stations, the ability to operate several bands simultaneously is an obvious advantage. Being able to pass messages from one operating position to another is also important. It allows one band-station to alert another to changing propagation conditons. For example, the 6-meter position may notice the beginning of an aurora several minutes before it will appear on 2 meters. If the 6-meter operator can alert the other operators, valuable time can be saved.

Similarly, it is advantageous to be able to pass "referrals" from one station to another in a multiop contest effort. A referral is an on-the-spot schedule to call another station at a particular time and frequency. It's the same as a quick band change for single-op stations. For example, if you are working the multioperator contest station W8VP on 2 meters, you might request an immediate sked on 220. If the 2-meter operator at W8VP can pass a quick message to the 220 station, they'll be looking for you, and you will probably make the QSO promptly.

During VHF/UHF contests, the above-listed times for activity nights are often observed for special activity, particularly on 220 MHz and up. In other words, look for increased activity on 220 MHz at 8 PM, 432 MHz at 9 PM, and 1296 MHz at 10 PM local during major contests. In some areas, this scheme also applies to morning (AM) as well. Bigger stations sometimes turn their antennas in accordance with the minute hand of the clock, looking north at the beginning of the hour, east at 15 minutes past, and so forth.

This is KD8SI's (Kettering, Ohio) VHF antenna system. Left tower: 4 1296-MHz Yagis, 2 432-MHz Yagis, miscellaneous FM antennas, and an HF rotatable dipole. Right tower is a 6-antenna array of KLM LBXs for 2-meter EME. *(WA6YBT photo)*

This 4 × 17 element array for 144 and 8 × 15 for 432 MHz belongs to Finland's OH5IY.

SOURCES OF INFORMATION

Many VHF/UHF operators like to keep abreast of the latest happenings on the bands such as new DX records, band openings or new designs for equipment. There are many excellent sources for current information about VHF/UHF. In addition, they provide a way to share ideas and ask questions, and for newcomers to become familiar with operation above 50 MHz.

The main sources are nets, newsletters and published columns. Each has its own use and appeal; active VHFers usually seek out at least one of them.

Nets

Several nets meet regularly on the HF and VHF bands so that VHF/UHFers can chat with each other. These are listed below. It's a good idea to listen first, before checking in the first time, so you'll know the format of the net. Some nets like to get urgent news, scheduling information and other "hot" topics out of the way early, and save questions and discussion until later. Others are more free-form. You can learn quite a bit just by tuning in to these nets for a few weeks. Regular participants in the nets are often very knowledgeable and experienced. The technical discussions which sometimes take place can be very informative.

1) *Central States VHF Net*. Meets each Sunday at 0230 UTC, 3.818 MHz. Open to all interested stations. A terrific source of information, as well as skeds for meteor scatter and EME. Operates informally during major meteor showers.

2) *144 and 432 MHz EME Nets*. These meet each Saturday and Sunday at 1700 and 1600 UTC, respectively, on 14.345 MHz. These are international nets which serve those interested in moonbounce. Each net meeting is usually occupied by a review of recent EME conditions, news from participants and setting up schedules for future EME work. The 432 EME net

is one hour long, whereas the 2-meter net often continues for several hours.

3) *Sidewinders On Two (SWOT) Net.* This net meets on 2-meter SSB, at regular intervals which vary by region. It is a good place to find active 2-meter operators. For information about the SWOT net in your area, consult the SWOT newsletter (c/o Howard Hallman, WD5DJT, 3230 Springfield, Lancaster, TX 75134).

4) *East Coast 432 Net.* This net is called each Wednesday at 9-11 PM local on 432.090 MHz. It is particularly popular in the Southeast, although they have had checkins from as many as 30 states in the past few years.

5) *Local Nets.* Lots of these nets, both formal and informal, are scattered across the country. It's impossible to tabulate their meeting times and frequencies. The best thing to do is to get on the air and ask other VHF operators about nets in your area.

Newsletters

Several VHF-related newsletters are available to interested operators. The newsletters are usually very up-to-date, with lots of tips on VHF activity, construction articles and news of VHF DX. They usually have a nominal subscription fee; check with each publisher for a sample copy and rates. The list below is culled from several published sources and may not include all the important newsletters in the USA.

VHF/UHF and Above Information Exchange: c/o Rusty Landes, KA0HPK, PO Box 270, West Terre Haute, IN 47885.

KC0W's VHF-Plus Update: 3090 Point Pleasant Dr, Hebron, KY 41048.

Midwest VHF Report: c/o Roger Cox, WB0DGF, 3451 Dudley, Lincoln, NE 68503.

2-Meter EME Bulletin: c/o Gene Shea, KB7Q, 417 Stadhauer, Bozeman, MT 59715.

220 Notes: c/o Art Reis, K9XI, 215 Villa Rd, Steamwood, IL 60103.

The Rochester VHF Group Journal: 6484 Rte 96, Victor, NY 14564.

The Pack Rats' Cheese Bits: c/o Doc Cutler, K3GAS, 7815 New Second St, Elkins Park, PA 19117.

The West Coast VHFer: 560 West Yucca St, Oxnard, CA 93033.

Northeast VHF News: c/o Lewis Collins, W1GXT, 10 Marshall Terrace, Wayland, MA 01778.

In addition, those who are truly serious about 432-and-up EME may want to subscribe to the *432 and Above EME News.* Published by Al Katz, K2UYH (326 Old Trenton Rd, RD 4, Trenton, NJ 08691), it is devoted exclusively to current news about 432, 1296 and 2304-MHz EME. Its circulation is limited, so only those who are active on 432-and-up EME should request a copy.

Columns

Several national Amateur Radio magazines include columns on VHF/UHF as regular features. *QST* has The World Above 50 MHz, a valuable record of VHF/UHF activity in North America; Bill Tynan, W3XO, receives regular reports from active VHFers and compiles a summary of band openings, new DX records and interesting observations. Also in *QST* is The New Frontier, conducted by Bob Atkins, KA1GT. This column is devoted to the higher UHF bands and microwaves. Bob includes a lot of technical information as well as summaries of American and foreign activity. In *Ham Radio*, Joe Reisert, W1JR, writes an informative series called VHF/UHF World. He covers a different general topic each month. Topics

N4GJV, North Carolina, put up this 70-cm EME array, consisting of 16 13-element Yagis each with a wood boom.

range from design of high-level mixers, to operating EME and meteor-scatter techniques. *CQ* magazine runs a monthly column entitled VHF—Principles, Practices, and Projects for the VHFer, by Steve Katz, WB2WIK. This wide-ranging column covers product reviews of some new equipment, discussion of operating techniques and technical topics. And the latest addition to this field is *73* magazine's monthly Above and Beyond, conducted by Peter Putman, KT2B.

Awards

Several awards have helped to spur activity on the VHF/UHF bands. The most popular are WAS, WAC and VUCC.

JA0JCJ is rightfully proud of this antenna system, 4 Yagis on 144 MHz and 4 Yagis on 432 MHz.

WA6YBT/8, Dayton, Ohio: A total of 14 elements on 144 MHz and 18 elements on 432 MHz. The tower is homebrew, made of 1-in angle iron on a 3-foot-square base. *(WA6YBT photo)*

It was long thought that WAS, Worked All States, was an impossible dream for VHFers. After all, it was "common knowledge" that no VHFer could work more than a thousand miles or so. To think of working coast-to-coast was silly, and to consider Alaska and Hawaii was crazy. To work all 50 states was just impossible. Fortunately, not everyone listens to "common knowledge," and WAS has been attained by several hundred amateurs on the 50-, 144-, 220- and 432-MHz bands.

WAS requires that you work each state in the United States and confirm the QSOs with QSL cards. The quest for this award has probably been responsible for much of the technical advancement of VHF/UHFers during the past three decades. Moonbounce activity has benefited most directly because WAS on any band above 144 MHz requires EME capability. (It should be noted, however, that a handful of diehards have worked 48 states on 2 meters *without* moonbounce!)

The 6-meter band was host to dozens of WAS-achievers during the peak years of the past sunspot cycle (1979-1981). Long-haul F_2 openings made transcontinental and intercontinental QSOs common for several months. Amateurs in the "lower 48" states worked Alaska, Hawaii and tons of international DX at that time. As of late 1986, there is little hope of F_2 because the sunspot cycle is at a low point. Nevertheless, multihop sporadic E openings have occurred from time to time between the continental USA and the "DX states." WAS on 6 meters is still a possibility.

Using portable moonbounce stations, several enterprising groups have mounted EME-DXpeditions to rare states. This has allowed quite a few hard-working VHF/UHFers to com-plete WAS, even when there was no resident EME activity in some states. At last count, over 80 stations had received WAS on 2 meters. At least one station, W1JR, has achieved WAS on 50, 144 and 432 MHz.

WAC, Worked All Continents, has also been achieved on 6 meters with the assistance of F_2 propagation, but now that's nearly gone (for the next 5 years or so). The chances of sporadic E of sufficient distance for catching all continents are almost nil, so we'll have to wait a while for more WAC awards on this band.

WAC is available to EMEers on 144, 432 and perhaps 1296 MHz. Stations from all continents have been active on these bands in recent years, although 1296 is not fully represented. In fact, several top-scoring stations in the annual EME contests have made QSOs on all continents during a single weekend. The rarest continent is probably South America, where only a small handful of EMEers are active. WAC is not possible on 220 MHz because this band is not authorized for amateur use outside of ITU Region 2 (North and South America).

ARRL breathed new life into the VHF and UHF bands in January 1983 when the League began sponsoring the VUCC (VHF/UHF Century Club) awards. These awards, based upon grid squares, rejuvenated the VHF and UHF bands tremendously. Somewhat later, the required exchange in all terrestrial ARRL-sponsored VHF and UHF contests was changed to include grid squares, and activity in those contests increased as well. You can do a lot of grid-hunting during a good contest!

The VUCC award is based on working and confirming a certain number of grid squares on the VHF/UHF bands. For 50 and 144 MHz, the number is 100. On 220 and 432 MHz, the number is 50, and on 902 and 1296 MHz the minimum number is 25. 2.3-GHz operators need to work 10 grids, and five grids are required on the higher microwave bands. This is quite a challenge, and qualifying for the VUCC award is a real accomplishment!

VUCC endorsement stickers are available for those who work specified numbers of additional grids above the minimum required for the basic award. Several stations have exceeded 300 grids on 6 meters and 200 on 2 meters, and a few have worked over 100 grids on 432. Impressive!

Another award should be mentioned although no one has yet received it on VHF: DXCC. The DX Century Club award, based on working 100 ARRL DXCC countries, has been a popular activity among low-band DXers for years. Many people have thought it preposterous to think that DXCC could be achieved by VHFers. So far, this is the case. However, some hard-working DXers on 6 meters were getting pretty close by the end of the last sunspot cycle. With the increasing availability of 6-meter privileges in Europe and elsewhere, 6-meter DXCC may become a reality. What's more, several 2-meter and 70-cm moonbouncers have attained very impressive totals. One British 2-meter operator is rumored to have worked over 80 countries so far, and several 432 EME stations are over 50 countries worked. Because EME is continually growing in popularity worldwide, someone may reach DXCC on either 2 meters or 432 MHz, using moonbounce, before the end of the decade. For specific details on the ARRL awards program, see Chapter 8.

Problems

With the ever-increasing sharing of the VHF and UHF spectrum by commercial, industrial and private radio services, a certain amount of interference with amateur operation is almost inevitable. Amateurs are well acquainted with interference, and so we normally solve interference problems by ourselves.

Several main types of interference are common. The first is our old friend, television interference (TVI). Fundamental overload, particularly of TV Channels 2 and 3 from 6-meter transmitters, is still common. Some Channel 12 and 13 viewing is bothered by 220-MHz transmissions. Fortunately, as modern TV manufacturers have slowly improved the quality of their sets, the amount of TVI is beginning to decline.

A more common form of TVI is called CATVI, Cable Television Interference. It results from many cable systems distributing their signals, inside shielded cables, on frequencies which are allocated to amateurs. As long as the cable remains a "closed system," in which all of their signal stays inside the cable and our signals stay out, then everything is usually okay. However, it is a sad fact that many cable systems are deteriorating because of age, and have begun to leak. When a cable system leaks, your perfectly clean and proper VHF signal can get into the cable and cause enormous amounts of mischief. By the same token, the cable company's signals can leak out and interfere with legitimate amateur reception.

RFI—radio frequency interference—is a common complaint of owners of unshielded or poorly designed electronic entertainment equipment. Amateur transmissions, especially high power, may be picked up and rectified, causing very annoying problems. RFI may include stereos, video-cassette recorders and telephones. The symptoms of RFI usually include muffled noises which coincide with keying or SSB voice peaks, or partial to total disruption of VCR pictures.

This chapter cannot cover all the complex methods which are used to track down and correct TVI/RFI problems. However, a few general principles may help in beginning your search for a solution:

1) With ordinary TVI, be sure your transmitter is clean, all coax connectors are tightened, and a good dc and RF ground is provided in your shack—before you look elsewhere for the cause of the problem. Then, find out if TVI affects all televisions or just one. If it's just one, then the problem is probably in the set and not your station.

2) With CATVI, remember that it is the cable company's responsibility to keep its system "closed" to the limits of FCC rules. Unfortunately, the FCC's limits are loose enough that in some cases there will be interference-causing leakage from the system which is still within FCC limits. In that case, there is no easy solution to the problem. However, you may be pleasantly surprised by the cooperative attitude of some cable TV operators. (Cable company technicians often have worked overtime trying to solve CATVI complaints, but not everyone is so fortunate.)

3) In dealing with RFI, the main goal is to keep your RF out of the entertainment system. This is often solved rather simply by bypassing the speaker, microphone and power leads with disc ceramic capacitors. In other cases, particularly some telephones, you must also employ RF chokes.

For further information on RFI and CATVI, consult *The Radio Frequency Interference* handbook, published by ARRL.

Several other types of interference plague VHF/UHF amateurs. These are the receive-only kinds of interference, which just affect your receiving capability. One form has already been mentioned: CATV leakage. For example, cable channel "E" is distributed on a frequency in the middle of the amateur 2-meter band. If the cable system leaks, you may experience very disruptive interference. Reducing leakage or

perhaps eliminating the use of channel "E" may be satisfactory. This problem has vexed many stalwart VHFers already, and no end is in sight.

Scanner "birdies" may be a problem in your area. All scanners are superheterodyne-type receivers which generate local oscillator signals, just as your receiving system does. Unfortunately, many scanners have inadequate shielding between the local oscillator and the scanner's antenna. The result is radiation of the oscillator's signal each time its channel is scanned—a very annoying chirp, swoosh or buzz sound every second or so. If the scanner's local oscillator frequency happens to fall near a frequency you're listening to, you'll hear the scanner instead. Amateurs can easily pick up 10-15 scanner birdies within 5 kHz of the 2-meter calling frequency (for example, a New York State Police frequency is

The WBØTEM antenna array in Akron, Iowa. The antenna system includes (at left) 24 19-element RIW Yagis for 70-cm, a 100-foot tower (in the center) with antennas for 2 and 6 meters and an HF tribander, and at right, 16 homebrew 13-element Yagis on 220 MHz and 8 16-element F9FT Yagis on 2 meters. *(WA6YBT photo)*

about 10.7 MHz above 144.200!). Very little can be done about this problem aside from installing tuned traps in everyone's scanners—an unattractive prospect to most scanner owners.

Many amateurs who operate 432 MHz have lived with radar interference for years. Radar interference is identified by a very rapid burst of noise which sounds vaguely like ignition noise, repeated on a regular basis. Although some radars are being phased out (no pun intended), many remain. Amateurs are secondary users of the 420-450 MHz band, so we must accept this interference. The only solution may be directional antennas which may null out the interference, not a very satisfying alternative in some cases. 432-MHz EMEers sometimes hear radar interference off the Moon!

N3BFL, with W3PM logging, operated 2-meter SSB during Field Day. *(WB3AMY photo)*

NEW FRONTIERS

The VHF/UHF spectrum offers amateurs a real opportunity to contribute to their own knowledge as well as the advancement of science in general. We as amateurs are capable of several aspects of personal research which are not possible for the limited resources of most scientific research organizations. Although these organizations are able to conduct costly experiments, they are limited in the time available, geographic coverage and, sometimes, by the "it's impossible" syndrome. Hams, on the air 24 hours a day, scattered across the globe, don't know that some aspects of propagation are "impossible," so they occasionally happen for us.

For example, it is commonly "known" that VHF meteor scatter is limited to frequencies below about 150 MHz, yet hams have made dozens of contacts at 220 and 432 MHz. A spirit of curiosity and a willingness to learn is all that is required to turn VHF/UHF operating into an interesting scientific investigation.

What are some of the topics amateurs can contribute to? The following list includes just a few of the possibilities:

1) *Sporadic E above 220 MHz.* Signals have been heard on 220.1 MHz, but to date no contacts have been completed.

However, sporadic-E-propagated commercial TV has been seen on TV channel 13 on several occasions. How high in frequency does sporadic E go?

2) *Aurora on 902 or 1296.* Many aurora contacts have been made on 432 MHz; is Au "impossible" on 902? 1296?

3) *Unusual forms of F-layer ionospheric propagation—* FAI, TE and what else? Little is known about their characteristics, how they relate to overall geomagnetic activity and frequency. Amateurs can discover a lot here.

4) *Multiple-hop sporadic E on 50 MHz.* Single-hop propagation on this band is a daily occurrence during the summer. In recent years, multiple-hop openings have not been uncommon. Now that many European countries are beginning to allow 6-meter amateur operation, we can find out how many hops are actually possible!

5) *Meteor scatter.* Astronomers are very interested in the orbits of comets and their swarms of debris which give rise to meteor showers. However, it is often difficult for astronomers to observe meteors (because of clouds, moonlight and so on). Amateurs can make very substantial contributions to the study of meteors by keeping accurate records of meteor-scatter contacts.

6) *Polarization of E-layer signals.* Do VHF signals rotate in polarity when reflected by sporadic-E or meteor-trail ionization? Are vertical antennas better than horizontal antennas for E-layer signals?

7) *How effective is diversity reception* in different types of VHF/UHF reception? Stations with two or more antennas could investigate.

8) *New modes of data transfer.* Packet, AMTOR and the like—what are some of the limits to their use? Who knows how many amateurs may attempt to send packet information via EME.

In addition to this list, one should consider the exciting challenges available to amateurs in ITU Region 2 on the new 902-MHz band. Its propagation characteristics should be a compromise between those of 432 MHz and 1296 MHz. High power levels are more easily obtained at 902 than 1296 because several vacuum tubes are usable at 902 but not much higher. Also, antennas are still very small, so this could be an ideal band for experimentation. The field is wide open for those who want to design equipment and antennas for this new band.

The most exciting new frontier, for Novices certainly, is Novice Enhancement. Under rules adopted by the FCC (at the urging of ARRL), Novices now have access to VHF/UHF: Novices may use up to 25 watts output in the 222.10-223.91 MHz band, with all authorized modes and up to 5 watts output in the 1270-1295 MHz band, with all authorized modes. Interesting times indeed are ahead for Novices who now can also share in the enjoyment of the VHF/UHF spectrum.

Chapter 13

Satellites

I s there such a thing as an impossible dream? Imagine relaxed ragchewing with exotic DX stations on SSB without the congestion found on 20 or 75 meters, all with only 15 watts of RF power. Not possible? Sure it is—just try your hand at amateur satellite operation.

Oh sure, you're thinking, just another lecture on amateur satellites. Hold on—don't turn the page yet! This chapter isn't strictly about satellites, but about how you can enjoy the fun of operating a super DX machine at a casual pace with no ill effects to your stamina or heart! For example, on a given morning, casual QSOs can be held with hams in Texas, Colorado, Antarctica, England, Massachusetts, Grenada and South Africa. Interested?

W1INF, the ARRL HQ laboratory station (with KA1JQW operating), has OSCAR 10 Mode B capability. The uplink is a 2-meter multimode transceiver driving a 2-meter to 70-cm transverter, while the downlink is a receiving converter and 10-meter receiver. A GaAsFET preamplifier is mounted at the antenna.

Not for Beginners Only

If you think this discussion is aimed at beginners, you are very right. Every amateur is a "beginner" at the satellite game at one time or another. Part of the beauty of satellite operation is that there is always something new to learn, even for the ham who has been operating satellites for many years. And it's very different from operating on the HF bands.

This chapter will provide virtually everything you need to know to get started. In addition, having a few of the right references handy will be helpful; it is highly recommended that every satellite user obtain and read a copy of *The Satellite Experimenter's Handbook* by Martin Davidoff, K2UBC, which is published by ARRL.[1] This fine publication is packed

with useful information that ranges far beyond the scope of this chapter. Also, another good reference source is *The AMSAT Phase III Satellite Operations Manual* published by AMSAT in conjunction with Project Oscar, Inc, and is available from AMSAT.[2]

AMSAT

This is a name you'll frequently encounter in the context of Amateur Radio satellites. You'll come to recognize it often associated with some of the leading activities and developments in Amateur Radio. Specifically, AMSAT is the official name of the Radio Amateur Satellite Corporation, a nonprofit, educational, scientific corporation founded in the greater Washington, DC, area in 1969. It has come to symbolize the highest aspirations of radio amateurs the world over through its achievements in building and operating the OSCAR series of satellites.

AMSAT's purposes include the following:

1) Providing satellites that can be used for Amateur Radio communications and experimentation by suitably equipped Amateur Radio stations throughout the world on a nondiscriminatory basis.

2) Encouraging the development of skills and the advancement of specialized knowledge in the art and practice of Amateur Radio communications and space science.

3) Fostering of international goodwill and cooperation through joint experimentation and study, and through the wide participation in these activities on a noncommercial basis by the radio amateurs of the world.

4) Facilitating communications by means of amateur satellites in times of emergency.

5) Encouraging the more effective and expanded use of the higher frequency amateur bands.

6) The dissemination of scientific, technical and operational information derived from such communications and experimentation, and the encouragement of the publication of such information in suitable public media.

AMSAT is a member-supported organization open to all radio amateurs and interested nonamateurs. AMSAT encourages the participation of all interested individuals in its activities regardless of membership and invites licensed Amateur Radio operators of all countries to engage in communications through the AMSAT-OSCAR series of satellites.

Obviously, building and launching satellites is very expensive. This is true even though much volunteer labor is used and donations of materials and services are occasionally made

[1]M. Davidoff, *The Satellite Experimenter's Handbook* (Newington: ARRL, 1984). Available from your local radio store or from ARRL HQ for $10 ($11 outside US). Add $2.50 ($3.50 UPS) per order for shipping and handling.

[2]Available for a $15 contribution from AMSAT, PO Box 27, Washington DC 20044.

by industry. Governments have helped with "free launches" in the past. AMSAT operating funds derive largely from donations, membership dues and grants from amateur organizations.

AMSAT is the body of amateurs and volunteers who provide this most important service together with other supporting functions to the amateur community. The key word in this context is "volunteer." AMSAT's paid staff has never numbered more than three. Most of the work is done by volunteers—people who want to participate in this exciting program solely for the sake of the promise of tomorrow. They work, and in doing so they are the program. In fact, AMSAT is mostly folks like us, who volunteer time and resources to support the amateur satellite program and the ideals that it represents.

So what does all this mean? In the current period of depressed sunspot activity, it means we have access to some new and valuable international DX operating frequencies on the VHF/UHF bands. We gain predictable clear communications covering nearly half the globe at any one time. Through satellite communications, friendships have been born that might never have been possible but for the amateur satellite program. International roundtable QSOs on UHF are now commonplace. These exchanges have materially helped amateur affairs at home by providing insight and ideas from amateurs across the globe.

One of the many benefits of AMSAT membership is the biweekly publication *Amateur Satellite Report*. This interesting newsletter keeps you abreast of the things you need to know to stay "tuned-in" to satellites today. Being an AMSAT member yields dividends that pay off year after year. Write or call AMSAT, PO Box 27, Washington, DC 20044, or call 301-589-6062 for free information about getting started in satellites today.

Getting Help—AMSAT Nets and Area Coordinators

The AMSAT Area Coordinator is a colleague of many talents, willing to share expertise with anyone who asks. He's the traditional "Elmer," a general-information source and spokesman all rolled into one.

For the over 100 AMSAT Area Coordinators and Assistant Area Coordinators across North America, the primary mission is to help the newcomer get started in OSCAR. The Area Coordinator is the satellite-newcomer's conduit to getting started and later becoming proficient in Amateur Radio satellites. Assigned based on state and province, these experienced satellite users are a ready source of information and guidance available to all.

Typically, the satellite newcomer will have the following questions:

- "Who else in this area is active on OSCAR?"
- "What types of equipment will I need?"
- "When can I work the satellite?"
- "Where can I get information?"
- "How do I know when OSCAR's in range?"

The Area Coordinator is the accessible person who knows many of the answers to the "Who, What, When, Where and How" questions. And on the occasion when he may be unsure of exactly what to do, he often knows how to get an authoritative answer.

Besides being a ready source of information to the newcomer, the Area Coordinator helps build the organization. He does this by presenting a positive image of achievement and pride at conventions and hamfests. In a very real

sense, the excellence in building and using OSCARs, some of the most complex systems ever built by amateurs, is reflected in the attitude and appearance of the AMSAT Area Coordinator.

In sum, the Area Coordinator is a motivator and doer. He motivates people to join in and enjoy OSCAR satellites, and he's there to help the newcomer get rolling until the newcomer can do most of it on his own. The Area Coordinator realizes that without a strong AMSAT built through dedicated members, there will be no more satellites.

Every organization has folks who are considered "spark plugs." AMSAT calls its spark plugs "Area Coordinators" and charges them to help interested amateurs begin the exciting path to one of the most advanced area of communications in Amateur Radio today—amateur satellites.

Other information paths for AMSAT are to be found on the many AMSAT-sponsored nets. A listing of these nets may be found in Table 13-1. Many of the AMSAT Area Coordinators will also be found on these nets. The name and address of your local Area Coordinator is available from AMSAT.

Table 13-1
AMSAT Information Net Schedules

Net Name	Day/Time (UTC)	Frequency (MHz)	Notes
HF Nets			
AMSAT International	Sun 1800	21.280	
AMSAT International	Sun 1900	14.282	
AMSAT European	Sat 1000	14.280	
AMSAT UK 80 m	Mon/Wed	3.780	1900 Local Time
AMSAT UK 80 m	Sun	3.780	1015 Local Time
AMSAT Asia/Pacific	Sun 1100	14.305	
AMSAT South Pacific	Sat 2200	14.282	
AMSAT South Africa	Sun 0900	14.280	
AMSAT South Africa	Sun 0900	7.080	
East Coast 75 m	Wed	3.840	2000 Local Tue
Mid-America 75 m	Wed	3.840	2100 Local Tue
West Coast 75 m	Wed	3.857	2000 Local Tue
Australian AMSAT	Sun 1000	3.685	
New Zealand	Wed 0800	3.850	
VHF Nets			
Los Angeles 2 m	Wed	144.144	2000 Local Tue
Houston Area	Tues	145.450	2200 Local Time
Chicago Area	Wed	146.880	1930 Local Time
AMSAT South Africa	Sun 0900	145.650	
AMSAT UK 2 m	Sun	144.280	1930 Local Time

THE PHASES OF THE OSCAR PROGRAM

The world of amateur satellites had its beginning in the form of what is now known as OSCAR Phase I (OSCAR stands for *O*rbiting *S*atellite *C*arrying *A*mateur *R*adio). These satellites were very low cost, low powered piggy-backed packages launched with other Government satellites. The principal aim was to prove that private parties—radio amateurs—could build a satellite. This point was well proved.

The next objective was to demonstrate that useful communications could be achieved through amateur satellites, and this effort was called Phase II. OSCARs 6, 7 and 8 were the crowning achievements of Phase II. These satellites provided lots of challenges for amateur operators, and even just finding the satellite was a major accomplishment for many. For others, the hearing of one's own signal and/or being able to hold brief QSOs with others was an achievement of note. In

Amateur Satellites: A History

OSCAR 1, the first of the "Phase I" satellites, was launched on December 12, 1961. It sent information concerning its internal temperature back to Earth on the 2-meter band. The now traditional CW identification HI was varied in code speed as the temperature changed. The 0.10-watt transmitter onboard discharged its batteries after only three weeks, but the mission was so successful that construction of a second satellite was begun.

OSCAR 2 was launched on June 2, 1962. Virtually identical to OSCAR I, Amateur Radio's second venture into space lasted 18 days.

OSCAR 3, launched March 9, 1965, was the world's first free-access communications satellite. During its two-week life, over 100 pioneering amateurs in 16 countries communicated through the 1-watt linear transponder on 2 meters.

OSCAR 4 was launched on December 21, 1965. More advanced in design than OSCAR 3, this satellite had a 2-meter-to-70-cm linear transponder with an output of 3 watts. Unfortunately, a launch-vehicle defect placed OSCAR 4 into a high elliptical orbit, which prevented widespread amateur use. A handful of hams did communicate through it, however, including the first US to USSR satellite contact of any kind.

OSCAR 5 was built by students at Melbourne University, in Australia. A new amateur organization, AMSAT, prepared the satellite for launch and coordinated ground activities. OSCAR 5 transmitted telemetry about its operating parameters on both 2 meters and 10 meters. Its batteries lasted for over a month.

OSCAR 6 was the first of the Phase II satellites. Launched on October 15, 1972, it carried a 1-watt-output 2-meter-to-10-meter Mode A linear transponder. Not only was the satellite magnetically stabilized, but the internal battery package was continuously recharged by solar cells. Tens of thousands of contacts were made during the nearly five-year lifespan of this satellite.

OSCAR 7 was built by hams from West Germany, Canada, Australia and the United States. Two elaborate linear transponders each ran 2-watts output; one was a 2-meter-to-10-meter Mode A unit like the one on OSCAR 6, the other a 70-cm-to-2-meter Mode B unit. Internal circuitry, as well as ground commands, controlled this sophisticated satellite's functions. OSCAR 7 was launched on November 15, 1974.

OSCAR 8, a cooperative effort of United States, Japanese, German and Canadian amateurs, was launched on March 5, 1978. The two transponders on board were a 2-meter-to-10-meter Mode A unit as used before, and a 2-meter-to-70-cm Mode J unit. Both had an output of 2 watts.

Radio Sputniks 1 and 2 were launched from the Soviet Union on October 26, 1978. Each satellite carried a 2-meter-to-10-meter transponder with extremely high sensitivity. Like their OSCAR counterparts, the RS satellites transmitted telemetry data relating to their well-being, and were commandable from the ground.

OSCAR Phase III-A, the first of a planned series of satellites that would finally realize the dream of reliable high-altitude long-distance satellite communications over extended periods, was launched on May 23, 1980, a date that unfortunately came to be known as "Black Friday." Phase III-A piggybacked aboard the European Space Agency Ariane LO2 rocket. A few minutes after liftoff from the ESA launch facility in Kourou, French Guiana, the Ariane rocket failed, dumping Phase III-A and the dream that rode with it into the Atlantic.

OSCAR 9, built by a group of radio amateurs and educators at the University of Surrey in England, went aloft in October 1981 as part of a secondary payload aboard a NASA Delta rocket. This is a scientific/educational low-orbit satellite containing many experiments but no amateur transponders. These include HF beacons at 7.050, 14.002, 21.002 and 29.510 MHz for propagation studies, general and engineering data beacons on 2 meters and 70 cm, and two additional beacons on 13 cm and 3 cm. An Earth-imaging camera, magnetometer, synthesized voice telemetry capability and onboard computer round out the experimental hardware. UoSAT-OSCAR 9 is fully operational as this is written.

Radio Sputniks 3-8 were launched simultaneously aboard a single vehicle in December 1981. Several of these RS satellites carried 2-meter-to-10-meter transponders that permitted communications over distances greater than 5000 miles. Also aboard two of them was a unique device nicknamed "Robot" that could automatically handle a CW QSO with terrestrial stations.

Iskra 2, was launched manually by two Soviet cosmonauts from the *Salyut 7* space station in May 1982. With the call sign RK02, Iskra 2 sported a telemetry beacon on 10-meters and a 15-meter-to-10-meter HF transponder. Iskra 2 was destroyed on reentering the atmosphere a few weeks after launch.

Iskra 3, launched in November 1982 from *Salyut 7,* was even shorter lived than its predecessor.

OSCAR 10, the second Phase III satellite, was launched on June 16, 1983, aboard an ESA Ariane rocket, and was successfully placed in its initial elliptical orbit. OSCAR 10 has proved to be a communications resource for radio amateurs throughout the world. OSCAR 10's computer (IHU) began showing the accumulated effects of near-Earth space particle radiation in mid-1986. This is a problem of intense subatomic particles trapped in the Earth's magnetic fields, irradiating the computer memory chips and either destroying the integrated-circuit chip or placing unwanted charge states on those chips, giving false memory information. Faulty operation of the IHU means that the satellite has become uncommandable and uncontrollable.

OSCAR 11, another scientific/educational low-orbit satellite like OSCAR 9, was built at the University of Surrey in England and launched on March 1, 1984. UoSAT-OSCAR 11 has also demonstrated the feasibility of store-and-forward packet digital communications and is fully operational as this is written.

OSCAR 12 or Fuji-OSCAR 12 (FO-12) is the first Japanese OSCAR and is a joint effort between the Japan Amateur Radio League (JARL) and the Japan AMSAT (JAMSAT) organization. It was launched by the Japanese National Space Development Agency (NASDA) as one of the payloads on an experimental launch vehicle H-1—truly a Japanese product in total. FO-12 is a Phase II spacecraft in a 1500-km circular orbit. It carries a Mode J transponder that serves two functions: linear SSB service (Mode JA) and an advanced packet store-and-forward global message service with a 1.5-megabyte RAM storage bank called Mode JD (for Mode J digital).

short, while some useful communications could be conducted through Phase II satellites, just the fact of being able to communicate at all was the end goal, much like working some rare DX on the HF bands.

AMSAT visionaries saw that amateur satellites could have a much more meaningful communications role if they were available for longer periods of a day. This kind of thinking led to the Phase III program. While it is still necessary to track the satellite to aim the antennas properly, this task is much easier.

The real challenge now provided by the Phase III satellites is the art of communications itself. We are so conditioned to some of the types of propagation provided by ionospheric communications that we may well not recognize this latest challenge. OSCAR 10 (see below) has shown some excellent examples of the results that clear intercontinental communications provide, forms of "cultural exchanges" not previously possible (for example, meeting foreign amateurs with interests similar to your own and having hour-long QSOs with them). This realm of discovery in turn provides bidirectional information flows useful to all. This type of global communication is very real, not just a satellite planner's dream.

Phase III satellites have their limitations, however. They are not available all of the time, and the satellite user still has to track the satellite and re-aim antennas as the satellite moves. These limitations mean that operators need graphical or computer tracking aids, antenna rotators and other peripheral equipment. Moreover, certain types of urgent communications (ie, emergency communications) cannot wait for the satellite to come into view.

The next logical AMSAT effort—Phase IV—has begun and will feature a geosynchronous stationary satellite, much like nearly all of the commercial satellites. For any given QTH, there will be only one set of antenna aiming angles, and these antennas can be fixed in position—no rotators. This means that the antenna requirements for any QTH will be well known and fixed. This then opens the opportunity for highly portable antennas and full-time access for such needs as emergency communications, since the main thrust of Phase IV will be *public service*. The Phase IV plan proposes a multi-transponder package using the 2 meter, 70 cm, 24 cm and 13-cm bands. Features would include a circuit-switched transponder capable of handling numerous voice conversations between teleports with interconnects to FM repeaters, a dedicated high-speed packet transponder for linking terrestrial regional networks into a semi-global network, a linear transponder for SSB and CW, a special facility for ATV (amateur television) using digitized video and, for UHF experimenters, a microwave-beacon experiment. Phase IV will even further enhance the art of communications beyond what we have begun to see in OSCAR 10 and Phase III.

About Orbits

Some of you may have operated through OSCARs 6, 7 and 8, or the various Russian satellites. They were in circular orbits less than 1200 miles above the Earth. Known as Phase II satellites, these spacecraft provided communications over distances up to 4500 miles. They were in range of a given point on the Earth for at the most 20 minutes or so during each orbit. Phase II satellites moved quickly, so operators had to work hard to keep their antennas pointed at the "birds," much like the fabled "one-armed-paperhanger-in-a-stiff-breeze" situation. One consequence of these 20-minute passes was that contacts tended to be short, contest-like exchanges of signal report, name and QTH. The current Fuji-OSCAR 12 spacecraft (see Fig 13-1) is in an orbit similar to the AMSAT-OSCAR Phase II series, but its mission, high-speed digital store-and-forward message handling by AX.25 packet, is suited to the short access.

Looking for something that would support communications for longer periods over greater distances, AMSAT designers developed a new generation of satellites, Phase III. As shown in Fig 13-2, these satellites have elliptical orbits rather than circular ones. At apogee, the point in its orbit where it is farthest from the Earth, OSCAR 10 is about 22,300 miles (36,000 kilometers) away. At perigee, when it is closest, OSCAR 10 is about 2230 miles (3600 km) from the Earth.

In practice, this means that maximum communications

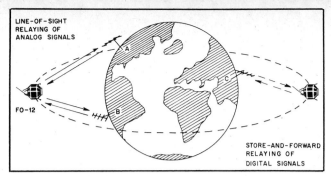

Fig 13-1—FO-12 provides for conventional amateur communications through its transponders, as well as digital store-and-forward capabilities.

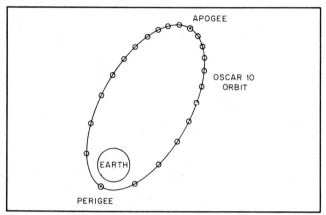

Fig 13-2—OSCAR 10 was designed for an elliptical orbit to allow it to be in view of much of the Northern Hemisphere for hours at a time. The circles of the ellipse represent the approximate position of OSCAR 10 during each half hour of its orbit. Note that the satellite "slows down" as it nears apogee. It passes through perigee (where it is out of sight of stations in the Northern Hemisphere) rather quickly.

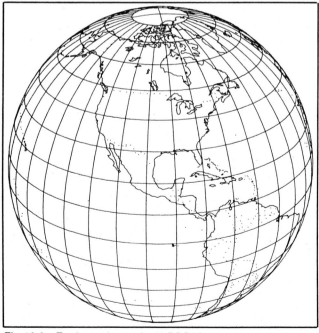

Fig 13-3—Earth as viewed from OSCAR 10 located at 35,512 km over 26:00N 95:00W, facing 0 degrees azimuth (perspective projection prepared by William D. Johnston, N5KR).

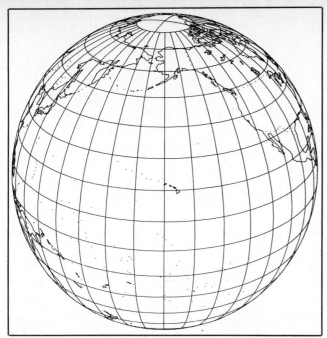

Fig 13-4—Earth as viewed from OSCAR 10 located at 35,512 km over 26:00N 160:00W, facing 0 degrees azimuth (perspective projection prepared by William D. Johnston, N5KR).

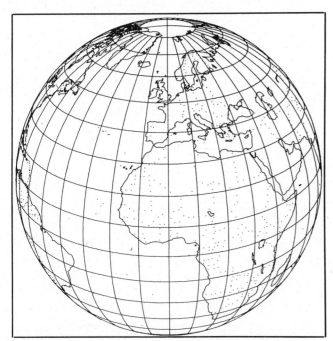

Fig 13-5—Earth as viewed from OSCAR 10 located at 35,512 km over 26:00N 0:00W, facing 0 degrees azimuth (perspective projection prepared by William D. Johnston, N5KR).

distance via OSCAR 10 is more than 10,000 miles when the satellite is at apogee. It is accessible to nearly half the Earth! Figs 13-3, 13-4 and 13-5 show views of the Earth from OSCAR 10 apogees in several positions, giving some very good ideas of the communication ranges from the continental US when the satellite is high in the orbit. Because the orbit is elliptical, OSCAR 10 is in view of stations for six to eight hours at a time. Being able to work stations half a world away for that length of time lends new meaning to amateur DX communications. These satellite views of Earth are typical

for all Phase III and Phase IV spacecraft, except the latter won't be moving relative to its position over the equator.

What is OSCAR?

OSCAR (also called AO-# which stands for AMSAT-OSCAR and the specific number designation of the satellite) is a series of amateur-communications satellites designed and built by radio amateurs for the sole purpose of supporting Amateur Radio communications and experimentation. It is very much like the satellites that allow you instant telephone access to relatives overseas and to watch televised events occurring on the other side of the world as they happen. The OSCAR satellites receive transmissions from earth stations and relay them back, allowing hams to communicate over great distances without worrying about the whims of ionospheric propagation. OSCAR 10 (AO-10) is the second Phase III satellite, but the first to be successfully placed into operation. [Unfortunately, as this material is being prepared, OSCAR 10 is suffering from technical problems. However, much of what is written here about OSCAR 10 will apply to the next Phase III satellite, which has been assembled and is scheduled for launch in 1987.]

Fig 13-6 shows OSCAR 10 pretty much as it looks today. Weighing about 200 pounds, this satellite was launched by a European Space Agency rocket in June 1983. Although it is an "amateur" satellite, there is nothing amateur about OSCAR 10's design and construction. It is built to the same

Fig 13-6—Phase III-B assembly being preflight vibration tested. (*Photograph courtesy AMSAT and W4PUJ*)

high standards as other communications satellites and was subjected to strenuous testing before launch. Historically, amateur satellites have provided reliable service well beyond their design lifespans.

Links: Up and Down

OSCAR 10 carries two linear translators or (a possible misnomer) *transponders*. A satellite transponder acts somewhat like a typical VHF FM repeater. The basic idea is the same: You transmit to it on one frequency, and it retransmits your signal back on another. There are some major differences from normal repeaters, however. Unlike FM repeaters that are designed for nonlinear modes only, OSCAR 10's transponders are linear. They faithfully reproduce all modes, including SSB. (Please do not try to use FM for OSCAR 10, however, as you will wipe out all of the other operators and unduly tax the satellite.) The other big difference is that OSCAR 10's transponders cover not just a single frequency, but a whole range of frequencies. This range of frequencies is called a *passband*; the passband may cover 20 kHz or more, even 800 kHz (as on OSCAR 10 Mode L). Amateur satellites retransmit every signal heard in the receiver passband, so many stations can use the transponder simultaneously.

The frequency on which you transmit to the satellite is called the *uplink,* while the frequency on which the satellite retransmits your signal to Earth is called the *downlink*. Uplink and downlink frequencies are on different bands. Various band combinations are called modes; see Table 13-2. For example, Phase II satellites carried a Mode A transponder that used 2 meters for the uplink and 10 meters for the downlink.

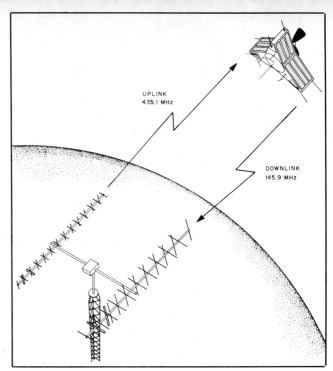

Fig 13-7—OSCAR 10 Mode B operation; stations on earth transmit to the satellite on 435 MHz and listen on 145 MHz.

Table 13-2

OSCAR Operating Modes

Mode	Uplink Band	Downlink Band
A	2 m (145 MHz)	10 m (29 MHz)
B	70 cm (435 MHz)	2 m (145 MHz)
J*	2 m (145 MHz)	70 cm (435 MHz)
L	24 cm (1269 MHz)	70 cm (435 MHz)
S	70 cm (435 MHz)	13 cm (2401 MHz)
JL	2 m/23 cm	70 cm
K	15 m	10 m

*JA = voice, JD = digital

OSCAR 10 carries a Mode B transponder and a Mode L transponder. The Mode B uplink is at 435 MHz, while the downlink is at 145 MHz. See Fig 13-7. The passband is 152 kHz wide. Mode L uses a 1269-MHz uplink and a 436-MHz downlink. The passband is 800 kHz wide—more than twice as wide as the 20-meter ham band. Most of the activity on OSCAR 10 is on Mode B, but the operating procedures described here apply to Mode L and most of all future Phase III and Phase IV satellites as well.

Phase III-C carrries a Mode B transponder, much like AO-10. It also provides a Mode JL transponder that will allow simultaneous uplinks on 145 MHz and 1268 MHz using a common downlink on 436 MHz. There will be some overlap in these passbands, allowing Mode J operators to talk with Mode L operators. If that isn't enough, Phase III-C also has a limited Mode S transponder to pave the way for future UHF satellite links and a Mode L packet digital transponder called "RUDAK." The possibilities for Phase III-C are truly exciting.

OSCAR-10 Band Plan

Since OSCAR 10 is in view of large areas of the Earth (nearly an entire hemisphere at a time) for up to 10 hours per orbit and transponder bandwidth is limited, some form of cooperation and coordination is required. A general scheme for satellite band planning, based on the downlink passband, has been followed for AO-10. This means that the downlink is the frame of reference for passband partitioning. The user is thus obliged to control the uplink frequency to assure that the resultant downlink frequency falls within the recommended allocations for that mode.

On OSCAR 10 Mode B, the downlink passband is partitioned into three major segments (see Fig 13-8) as follows:

The general communications segment is 115 kHz wide with edges at 145.845 MHz and 145.960 MHz. This segment is further divided into three sections by emission type as follows:

CW: 145.845 to 145.875 MHz
Mixed CW/SSB: 145.875 to 145.925 MHz
SSB: 145.925 to 145.960 MHz

This sectioning is essentially identical in style to Phase II satellite band plans. Specific uplink/downlink frequencies are presented in Table 13-3. These relationships need to be adjusted on the uplink for any Doppler corrections. The "special" emissions currently used—RTTY and SSTV—are allocated calling frequencies:

RTTY: 145.880 MHz
SSTV: 145.888 MHz

The second major segment within the 2-meter downlink is allocated to channelized special services. All Special Service Channels (SSC) are 4 kHz wide, and their placement is at the passband extremities just above and below the beacons and their associated guard bands. The SSC functions are explained later.

The final portion of the 2-meter downlink is assigned to the engineering beacon and the general beacon. Protecting

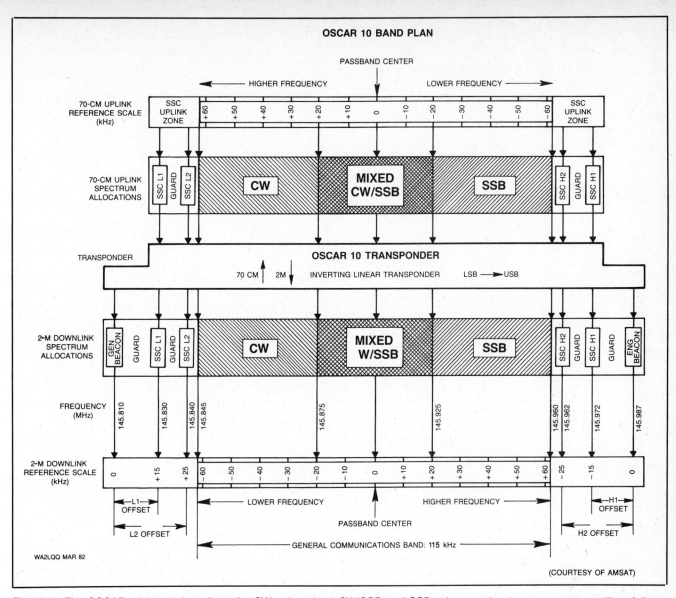

Fig 13-8—The OSCAR 10 band plan allows for CW only, mixed CW/SSB and SSB-only operation (most amateur satellites follow a similar type of plan that allows for mode separation). Courteous operators observe this voluntary band plan at all times.

these beacons from interference are 13- to 17-kHz-wide guard bands. The engineering beacon is sent in a computer compatible format, while the general beacon transmits information in RTTY, Morse code and phase-shift keying (PSK).

Special Service Channels are a new concept for amateur satellites, as they are in contrast to the usually nonchannelized communications predominating the amateur bands (with the notable exception of the VHF/UHF FM repeater subbands). The four SSCs are as follows (L = Low end of 2 meter passband, H = High end):

L1—International packet digital communications, 145.830 MHz.

L2—CW/RTTY bulletins, 145.840 MHz.

H1—Education and scientific channel, 145.972 MHz.

H2—Voice bulletin channel, 145.962 MHz.

L1 is dedicated to serving the packet digital communications needs of the worldwide amateur community. Long distance high-speed packet exchanges are an everyday operation on L1.

L2 provides amateurs all over the world with code practice and information bulletins in CW and RTTY, such as W1AW provides. Special receiving-sensitivity tests, the ZRO Memorial

Technical Achievement Tests, have also been conducted on this channel.

The education/scientific channel, H1, is available to any amateur or noncommercial group (with a ham sponsor), on a scheduled basis, to perform serious scientific experimentation.

The voice bulletin channel, H2, is being employed to disseminate information of interest to the amateur population as a whole. Information pertinent to the operation of the satellite, such as orbital data, is transmitted. Bulletins of general interest, comparable to those being sent by AMSAT, W1AW and GB2RS, are transmitted in both real time and by prerecorded messages. For those voice nets using H2 and having general amateur participation, the H2 frequency will be kept clear and checkins taken on frequencies below 145.960 MHz.

OSCAR 10 Beacon and Telemetry

Many amateurs are highly interested in the location and condition of various spacecraft orbiting the earth. Some of these operators have specialized in deciphering the orbits and

Table 13-3
AMSAT OSCAR-10 Frequency Conversion Chart

Revised Mode B Frequency Guide*
(Exclusive of Doppler Shift)

Uplink		Downlink	
		145.987	Engineering Beacon
435.0323	Scheduled Use	145.9720	SSC H1
435.0423	Scheduled Use	145.9620	SSC H2
435.0447		145.960	GCB Upper Limit
435.050		145.955	
.055		.950	
.060		.945	
.065		.940	
.070		.935	
.075		.930	
.080		.925	
.085		.920	
.090		.915	
.095		.910	
.100		.905	
435.1037	Center Band	145.901	
.105		.900	
.110		.895	
.115		.890	
.120		.885	
.125		.880	
.130		.875	
.135		.870	
.140		.865	
.145		.860	
.150		.855	
.155		.850	
.160		.845	GCB Lower Limit
435.1647	Scheduled Use	145.8400	SSC L2
435.1747	Scheduled Use	145.8300	SSC L1
		145.810	General Beacon

*Based on conversion frequency of 581.0047 MHz.
SSC—Special Service Channel
GCB—General Communications Band

Preliminary Mode L Frequency Guide*
(Exclusive of Doppler Shift)

Uplink	Downlink	
1269.050	436.950	Upper Limit
.075	.925	
.100	.900	
.125	.875	
.150	.850	
.175	.825	
.200	.800	
.225	.775	
.250	.750	
.275	.725	
.300	.700	
.325	.675	
.350	.650	
.375	.625	
.400	.600	
.425	.575	
1269.450	Center Band	436.550
.475	.525	
.500	.500	
.525	.475	
.550	.450	
.575	.425	
.600	.400	
.625	.375	
.650	.350	
.675	.325	
.700	.300	
.725	.275	
.750	.250	
.775	.225	
.800	.200	
.825	.175	
1269.850	435.150	Lower Limit
	436.040	Engineering Beacon
	436.020	General Beacon

*Based on a translation frequency of 1706.00 MHz estimated.

---

telemetry (TLM) data from numerous governmental orbiters. The OSCAR-10 beacon also provides a wealth of information. AO-10 has four beacon frequencies as follows:

- Mode B General Beacon—145.810 MHz
- Mode B Engineering Beacon—145.987 MHz
- Mode L General Beacon—436.020 MHz
- Mode L Engineering Beacon—436.040 MHz

The Mode B General Beacon and the Mode L Engineering Beacon are most routinely in use.

These beacon signals are used for general signal-strength references, but they contain a wealth of other information of interest to space-active amateurs. Most of the data is in the form of PSK-modulated signals. Available kits allow the enterprising amateur to construct a PSK decoder. All is not lost, however, for those who don't have the PSK decoders. In Mode B, CW-status information is transmitted for five minutes at 0 and 30 minutes past the hour. Fifty-baud, 170-Hz RTTY data is transmitted at 15 and 45 minutes past the hour. See Fig 13-9. In Mode L, the RTTY data is transmitted at 0, 15, 30 and 45 minutes past the hour.

According to command station VE1SAT/VE6, the so-called "Z" block of data sent as RTTY is a version of the "Y" block of telemetry sent via PSK. Thus, many of the key operating parameters are now available for general consumption to anyone who has 45-baud Baudot RTTY demodulation and display capability.

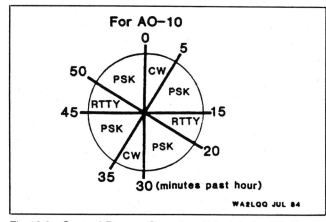

Fig 13-9—General Beacon Operating Schedule for AO-10.

The exact baud rate is 50 rather than the 45 baud (60 WPM) standard, but most electronic and mechanical machines equipped to receive 45 baud will respond well to 50 baud. The AO-10 RTTY telemetry is sent in FSK with a 170-Hz shift and is standard Baudot (5-level) coded.

The following table exemplifies the format of the RTTY as one might see it printed or displayed on a RTTY machine or computer screen. The telemetry channels will appear as six rows (lines) of 10 columns, as shown below and in Table 13-4:

	1	2	3	4	5	6	7	8	9	10
A										
B										
C										
D										
E										
F										

Each of the 60 telemetry channels are sent as a 1-, 2- or 3-digit numeric group, such as "1," "16" or "165." The entire RTTY beacon sequence begins with three lines of standard RYRY synchronizing characters, followed by the identification:

HI HI THIS IS AMSAT OSCAR 10

This is followed by the UTC time in the standard HH:MM:SS format. Next is the AMSAT day number where day 1 equals 1 January 1978. This is followed by three hexidecimal numbers denoting safety information, transponder status and command serial number. Next are seven decimal numbers which indicate the IHU multiplexer status. The block of 60 telemetry channels then follow four blank lines.

Following the data block is another AO-10 identifier and a QTC message notation, followed by a line of showing the AO-10 Mean Anomaly (in 256 parts) and the transponder AGC attenuation. The noted QTC message follows, containing information such as the important AO-10 operating schedule. A second complete presentation of the telemetry data is again transmitted, followed by a second important operating information QTC message block.

Table 13-4 shows a complete RTTY beacon data block received starting at 0245 UTC on AMSAT day 2793, or 24 August 1985. Table 13-5 provides the complete engineering information needed to identify and decode the telemetry data. Table 13-6 shows the data received on day 2793 and its reduction to engineering units.

OSCAR 10 Operating Schedule

The OSCAR-10 spacecraft, the forerunner of and role model for all future amateur satellites, has a very sophisticated onboard computer to facilitate the command and control of the entire vehicle. Like all computers, it needs to keep track of time, and, in fact, it keeps time with several different methods. One such AO-10 method is a clock based on each complete orbital revolution, which is numerically divided into 256 parts, called *Mean Anomaly* (MA), or simply "Phase." Zero MA starts at perigee, with the halfway point, MA = 128, being at apogee.

If you have used the computer approach to locate AO-10, you will have the Phase printed out for you. AO-10's onboard controller uses these MA values to turn the Mode B and L transponder on and off according to desired schedules defined by ground controllers and spacecraft needs.

The satellite does not operate Mode B and Mode L simultaneously. Rather, the transponders are turned on and off according to a fixed operating schedule defined by the control stations on Earth. This is done to make sure that the spacecraft's batteries are charged (from solar cells) and discharged (through transponder use) at a rate that will assure the longest possible battery life. The OSCAR 10 schedule

Table 13-4
OSCAR 10 RTTY Beacon Data Frame

```
RYRYRYRYRYRYRYRYRYRYRYRYRYRY
RYRYRYRYRYRYRYRYRYRYRYRYRYRY
RYRYRYRYRYRYRYRYRYRYRYRYRYRY
Z  HI. THIS IS AMSAT OSCAR 10
     02.45.34   2793

48   6    0    0   10  226   1

232  6   155  135  204  6   102  61   212  44
148  11   5    5   157 57   119   5   136  20
27   6   139  78   8    5   135  78   106   5

134  13  113  137  145 106  81   158  143  13
193 148  144  13   245 138  146  93   198 140
123  12  183  146  132 130   5   139  143  10
HI HI DE AO 10 QTC 037
AT MA 57  AGC N0
NEW SCHEDULE STARTING AUG 5
OFF MA 40 TO 189
MODE L ON AT MA 190 TO 206
MODE B ON AT MA 207 TO 39
2 METRE OMNI ANTENNA ON MA 45 TO 184
RESTRICTED SCHEDULE DUE ECLIPSES
AO 10 HI HI

Z  HI. THIS IS AMSAT OSCAR 10
     02.47.49   2793

48   6    0    0   10  226   1

233  6   155  135  204  6   102  61   212  44
148  11   5    5   157 56   122   6   136  19
17   5   139  79   5    5   135  76   106   5

134 116  112  137  145 13   81   157  142  12
201 148  143  18   245 138  146  12   198 140
123  37  191  146  132 130   5   139  143  10
HI HI DE AO 10 QTC 039
ECLIPSE DATA FOLLOWS
23 AUG 0537 TO 0723 ES 1719 TO 1903
24 AUG 0501 TO 0644 ES 1643 TO 1825
25 AUG 0425 TO 0605 ES 1607 TO 1745
26 AUG 0349 TO 0525 ES 1531 TO 1705
27 AUG 0314 TO 0445 ES 1456 TO 1625
AO 10 HI HI
```

needs to be adjusted several times each year because of the satellite's position relative to the sun and Earth. The current operating schedule will include the on and off periods for both the Mode B and Mode L transponders.

You can obtain amateur-satellite operating schedules from several sources. If you're active on HF, you can hear the schedule on the weekly 75-meter AMSAT nets listed in this chapter. Or, if you've located OSCAR 10, you can get this information on the General Beacon frequency of 145.810 MHz. Information from the operating schedule will give the on and off periods in terms of MA for both the Mode B and L translators; see Fig 13-10 for a typical schedule.

Satellite users must be aware of the operating schedule of the spacecraft so that they can set their operations accordingly. Typically, the transponders are turned off during the perigee periods, while Mode L will be turned on near apogee, or during those periods where the +Z axis of the spacecraft is most closely aimed toward the Earth. Most other periods will see the Mode B transponder turned on.

Spin Modulation

Spin modulation is a term and phenomenon that has become a significant subject for discussion with the advent

Table 13-5

OSCAR 10 Telemetry Data Format

TLM Row/Col	Meaning	Equation	Units
A1	Solar Panel output & BCR input voltage	$n \times 150$	mV
A2	70-cm transmitter average power output	$(253-n)^2/2000$	W
A3	70-cm receiver temperature	$(n-127)/1.82$	C
A4	Nutation damper temperature	$(n-127)/182$	C
A5	BCR output and main battery voltage	$(n-10) \times 75$	mV
A6	Special purpose	xxxxxxxxxxxxxx	—
A7	2-m transmitter temperature	$(n-127)/1.82$	C
A8	14 volt rail current to transponder	$(n-15) \times 20.64$	mA
A9	10 volt regulator voltage	$(n-12) \times 50$	mV
A10	He tank pressure at high pres reg	$(n-34) \times 44.46$	bar
B1	IHU temperature	$(n-127)/1.82$	C
B2	14 volt rail current to magnetorquers and antenna relay	— $(n-15) \times 4.128$	mA
B3	BCR #1 status	$0 = $ off, $n>10 = $ on	—
B4	He tank pressure at low pres reg	$(n-37) \times 0.8$	bar
B5	BCR temperature	$(n-127)/1.82$	C
B6	10 volt regulator current	$(n-15) \times 4.128$	mA
B7	BCR #2 status	$0 = $ off, $n>10 = $ on	—
B8	Not used	xxxxxxxxxxxx	—
B9	SEU temperature	$(n-127)/1.82$	C
B10	Battery charge current	$(n-15) \times 10.32$	mA
C1	Top photocell sensor	65 means sun normal to Z (spin) axis 20 – 30 nominal	— —
C2	Special purpose	xxxxxxxxxxxx	—
C3	Main battery case #1 temperature	$(n-127)/1.82$	C
C4	Active BCR output current	$(n-15) \times 20.64$	mA
C5	Bottom photocell sensor	(same as C1)	—
C6	Kick motor strut temperature	(Inoperative)	—
C7	Main battery case #2 temperature	$(n-127)/1.82$	C
C8	Active BCR input current on 28 V line	$(n-15) \times 10.32$	mA
C9	Spin rate (if $n<139$) or (if $n>=139$)	$r = (139-n) \times 0.8 + 20$ $r = 508/(n-116) - 2$	r/min r/min
C10	24-cm receiver AGC (if $n<100$) or (if $n>=100$)	$AGC = 0$ $AGC = (n-100)^2/189$	dB dB

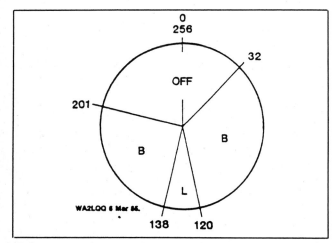

Fig 13-10—OSCAR 10 Transponder Schedule (typical); check beacon for lastest information.

of OSCAR 10. An understanding of spin modulation, and how to deal with it, is needed for effective satellite communications.

Since a satellite antenna and its gain pattern are firmly anchored to the rotating spacecraft, a ground station's position relative to the pattern will change moment by moment. Both the polarization and gain of an antenna will vary with the observer's location. A ground station will therefore see cyclical gain and polarization changes on a downlink signal resulting from satellite rotation. These changes are called spin modulation. The frequency of this spin modulation depends on the spacecraft's rotation rate which, in turn, depends on the attitude-stabilization technique employed. OSCARs 5, 6, 7 and 8 rotated at frequencies on the order of 0.01 Hz, or about one revolution every four minutes. Spin modulation at 0.01 Hz sounds much like a slow fade. Its effect on intelligibility is minor, unless the signal drops below the noise level.

The attitude scheme used on OSCAR 10 differs considerably. The spacecraft is spun at roughly 60 r/min about the Z axis. Because of the three-pointed star shape of OSCAR 10, gain and polarization variations on the links vary at three times the spin rate, or about 3 Hz. When OSCAR 10 is used in the main lobe of the antenna pattern, the effects of spin modulation are hardly noticeable. When a ground station is located on the fringes of the satellite's antenna

D1	Auxiliary battery temperature	$(n-127)/1.82$	C
D2	Solar panel #6 current	$(n-15) \times 4.128$	mA
D3	2-m transmitter average power output	$(200-n)^2/2000$	W
D4	He tank temperature	$(n-127)/1.82$	C
D5	Solar panel #1 temperature	$(n-127)/1.82$	C
D6	Solar panel #5 current	$(n-15) \times 4.128$	mA
D7	70-cm receiver AGC	$(n-83)^2/1000$	dB
D8	70-cm transmitter temperature	$(n-127)/1.82$	C
D9	Solar panel #3 temperature	$(n-127)/1.82$	C
D10	Solar panel #4 current	$(n-15) \times 4.128$	mA
E1	Special purpose	xxxxxxxxxxxx	—
E2	24-cm receiver temperature	$(n-127)/1.82$	C
E3	Solar panel #5 temperature	$(n-127)/1.82$	C
E4	Solar panel #3 current	$(n-15) \times 4.128$	mA
E5	14 volt regulator voltage	$(n-10) \times 61.5$	mV
E6	Wall temperature in arm #3	$(n-127)/1.82$	C
E7	Top surface temperature in arm #1	$(n-127)/1.82$	C
E8	Solar panel #2 current	$(n-15) \times 4.128$	mA
E9	Internal 9 volt bus from transponder	$(n-10) \times 50$	mV
E10	Wall temperature in arm #2	$(n-127)/1.82$	C
F1	Bottom surface temperature in arm #1	$(n-127)/1.82$	C
F2	Solar panel #1 current	$(n-15) \times 4.128$	mA
F3	Special purpose	xxxxxxxxxxxx	—
F4	Wall temperature in arm #1	$(n-127)/1.82$	C
F5	N_2O_4 tank temperature	$(n-127)/1.82$	C
F6	UDMH tank temperature	$(n-127)/1.82$	C
F7	Auxiliary battery voltage	$(n-10) \times 75$	mV
F8	Central cylinder temperature in arm #1	$(n-127)/1.82$	C
F9	Earth sensor temperature	$(n-127)/1.82$	C
F10	Mode L transponder 9 V regulated line	$(n-10) \times 44$	mV

Notes

B1 IHU = Integrated Housekeeping Unit, the computer
B3 BCR = Battery Charge Regulator
B4 He = Helium
B9 SEU = Sensor Electronics Unit
C8 Direct from solar panels
F5 N_2O_4 = nitrogen tetroxide, the propellant oxidizer
F6 UDMH = unsymmetrical di-methyl hydrazine, the
propellant fuel

Compiled by ZL1AOX, VE1SAT, KA9Q, DB5ER, R. Gape, WA2LQQ.

pattern, however, it may observe gain variations that exceed 10 dB. A signal affected by spin modulation at a frequency of 3 Hz has a rapid flutter that can be very annoying. At times, the flutter is so pronounced that signals are difficult, if not impossible, to copy even though they are strong. These problems are especially exacerbated by those stations using linearly polarized (LP) uplink signals, or circularly polarized (CP) signals, of the wrong sense for the particular conditions. Switchable CP antennas are a great assistance in alleviating the effects of spin moduation and are highly recommended.

Satellite Tracking

Expert HF DXers know where and when to point their beams, based on years of experience. Tracking satellites is similar in some ways, yet strikingly different in others. It does take experience to become a proficient satellite tracker and to really understand what is going on. In this way, satellite trackers and HF DXers are similar. HF DXers, however, sometimes have to shrug their shoulders at the unpredictable nature of ionospheric propagation. Predicting OSCAR access is much more precise. There is an enormous satisfaction in positioning a simple graphical tracker or dumping a bunch of numbers into a computer and being presented with the information that an object traveling at greater than 18,000 miles per hour is going to pop over your horizon in precisely 38 minutes and 22 seconds. And then, it does!

There are two fundamental reasons that you need to keep track of OSCARs. First of all, they move—some fast, others not so fast. You need to keep track of when the satellite is "in view" of your QTH. Second, since most satellite communications require some sort of directional antennas, you need to know where to point the array. Thus, the two primary functions of OSCAR-tracking efforts are position determination and scheduling. There are other functions that might be determined, but these two are the most basic.

Tracking OSCAR satellites requires information in four areas:

1) You need information about the OSCAR to be tracked—its precise location and rates of movement at a precisely defined instant.

2) You must know your own location to a reasonable degree of accuracy.

3) You must know the time of day reasonably accurately.

4) Most importantly, you need a device to coordinate the first three items. This can be a graphical tracker or a computer program. Both will be discussed here.

Glossary of Satellite Terminology

AMSAT—A registered trademark of the Radio Amateur Satellite Corporation, a nonprofit scientific/educational organization located in Washington, DC. It builds and operates Amateur Radio satellites and has sponsored the OSCAR program since the launch of OSCAR 5. (AMSAT, PO Box 27, Washington, DC 20044)

Anomalistic period—The elapsed time between two successive perigees of a satellite.

AO-#—The designator used for AMSAT OSCAR spacecraft in flight, by sequence number (AO-5 through AO-10).

AOS—Acquisition of signal—The time at which radio signals are first heard from a satellite, usually just after it rises above the horizon.

Apogee—The point in a satellite's orbit where it is farthest from Earth.

Area coordinators—An AMSAT corps of volunteers who organize and coordinate amateur-satellite-user activity in their particular state, municipality, region or country. This is the AMSAT grassroots organization set up to assist all current and prospective OSCAR users.

Argument of perigee—The polar angle that locates the perigee point of a satellite in the orbital plane; drawn between the ascending node, geocenter and perigee; and measured from the ascending node in the direction of satellite motion.

Ascending node—The point on the ground track of the satellite orbit where the subsatellite point (SSP) crosses the equator from the Southern Hemisphere into the Northern Hemisphere.

Az-el mount—An antenna mount that allows antenna positioning in both the azimuth and elevation planes.

Azimuth—Direction (side-to-side in the horizontal plane) from a given point on Earth, usually expressed in degrees. North = 0° or 360°; East = 90°; South = 180°; West = 270°.

Circular polarization (CP)—A special case of radio energy emission where the electric and magnetic field vectors rotate about the central axis of radiation. As viewed along the radiation path, the rotation directions are considered to be right-hand (RHCP) if the rotation is clockwise, and left-hand (LHCP) if the rotation is counter-clockwise.

Codestore—An onboard digital memory system that can be loaded with data and messages by ground stations for later rebroadcast in Morse, Baudot or other codes.

Descending node—The point on the ground track of the satellite orbit where the subsatellite point (SSP) crosses the equator from the Northern Hemisphere into the Southern Hemisphere.

Desense—A problem characteristic of many radio receivers in which a strong RF signal overloads the receiver, reducing sensitivity.

Doppler effect—An apparent shift in frequency caused by satellite movement toward or away from your location.

Downlink—The frequency on which radio signals originate from a satellite for reception by stations on Earth.

Eccentricity—The orbital parameter used to describe the geometric shape of an elliptical orbit; eccentricity values vary from e = 0 to e = 1, where e = 0 describes a circle and e = 1 describes a straight line.

Table 13-6

**OSCAR 10 Telemetry Data Reduction
Data from Table 13-4**

Date: 2793 = 24 August 1985
Time: 0245:34 UTC

TLM Channel	Count n	Value	TLM Channel	Count n	Value
A1	232	34800 mV	D1	134	3.8 C
A2	6	30.5 W	D2	13	0 mA
A3	155	15.4 C	D3	113	3.8 W
A4	135	4.4 C	D4	137	5.5 C
A5	204	14550 mV	D5	145	9.9 C
A6	6	—	D6	106	376 mA
A7	102	−13.7 C	D7	81	0 dB
A8	61	949 mA	D8	158	17.0 C
A9	212	10000 mV	D9	143	8.8 C
A10	44	444.6 bar	D10	13	0 mA
B1	148	11.5 C	E1	193	—
B2	11	0 mA	E2	148	36.3 C
B3	5	Off	E3	144	9.3 C
B4	5	−25.6 bar	E4	13	0 mA
B5	157	16.5 C	E5	245	14452 mV
B6	57	173 mA	E6	138	6.0 C
B7	119	On	E7	146	10.4 C
B8	5	—	E8	93	322 mA
B9	136	4.9 C	E9	198	9400 mV
B10	20	52 mA	E10	140	7.1 C
C1	27	nominal	F1	123	−2.2 C
C2	6	—	F2	12	0 mA
C3	139	6.6	F3	183	—
C4	78	1300 mA	F4	146	10.4 C
C5	8	no sun	F5	132	2.7 C
C6	5	—	F6	130	1.6 C
C7	135	4.4 C	F7	5	0 mV
C8	78	650 mA	F8	139	6.6 C
C9	106	46.4 rpm	F9	143	8.8 C
C10	5	0 dB	F10	10	0 mV

Graphical Tracking

Graphical (or manual) tracking methods generally employ a map, typically an azimuthal equidistant projection centered on the North Pole, and one or more clear overlays that allow you to use the map for changing satellite orbits and for different locations on Earth. The overlays allow you to determine which satellite orbits will bring the bird within range of your QTH. They also give beam headings for azimuth and elevation.

Most amateur operators who are new to the OSCAR-10 scene are caught up with the idea that they *must* use computer tracking methods to generate the numerical data needed to aim their antennas at the satellite. To those of us who used graphical tracking methods for years to follow the Phase II satellites such as OSCAR 8, the computer methods were such a revelation that we quickly became married to them. The notion that computers were the only hope was reinforced by the graphical tracking presentations for Phase III satellites provided at the time of the Phase III-A demise in 1980 (see accompanying sidebar on Amateur Radio satellite history). Those manual tracking methods seemed unduly complicated, so we charged ahead with our computers.

While a great many amateurs do have computers that can be used for tracking

EIRP—Effective isotropic radiated power—same as ERP except the antenna reference is an isotropic radiator.

Elliptical orbit—Those orbits in which the satellite path describes an ellipse with the Earth at one focus.

Elevation—Angle above the local horizontal plane, usually specified in degrees. (0° = plane of the Earth's surface at your location; 90° = straight up, perpendicular to the plane of the Earth)

Epoch—The reference time at which a particular set of parameters describing satellite motion (Keplerian elements) are defined.

EQX—The reference equator crossing of the ascending node of a satellite orbit, usually specified in UTC time and degrees of longitude of the crossing.

ERP—Effective radiated power—System power output after transmission-line losses and antenna gain (referenced to a dipole) are considered.

ESA—European Space Agency—A consortium of European governmental groups pooling resources for space exploration and development.

FO-12—Japanese Amateur Radio satellite, Fuji-OSCAR 12.

Geocenter—The center of the Earth.

Geostationary orbit—A satellite orbit at such an altitude (approximately 22,300 miles) over the equator that the satellite appears to be fixed above a given point.

Ground station—A radio station, on or near the surface of the Earth, designed to transmit or receive to/from a spacecraft.

Groundtrack—The imaginary line traced on the surface of the Earth by the subsatellite point (SSP).

IHU—Inflight Housekeeping Unit, the name for the central computer brain for the AO-10 spacecraft. This unit provides the complete command and communication link to all other subunits within AO-10. Uploaded ground programs and commands then provide for automated operations, a new concept with AO-10.

Inclination—The angle between the orbital plane of a satellite and the equatorial plane of the Earth.

Increment—The change in longitude of ascending node between two successive passes of a specified satellite, measured in degrees West per orbit.

Iskra—Soviet low-orbit satellites launched manually by cosmonauts aboard *Salyut* missions. Iskra means "spark" in Russian.

JAMSAT—Japan AMSAT organization.

JAS-1—The prelaunch designation for the Japanese Amateur Radio satellite Fuji-OSCAR 12 (FO-12).

Keplerian Elements—The classical set of six orbital element numbers used to define and compute satellite orbital motions. The set is comprised of inclination, Right Ascension of Ascending Node (RAAN), eccentricity, argument of perigee, mean anomaly and mean motion, all specified at a particular epoch or reference year, day and time. Additionally, a decay rate or drag factor is usually included to refine the computation.

LHCP—Left-hand circular polarization.

(and if you have one, that is the way to go), there are a large number of new satellite operators who are not so equipped. It's easy to be misled into thinking that you *must* purchase and master a computer before making even a single OSCAR contact. That's enough to make some potential satellite operators lose interest at the onset. Don't be intimidated! Today there are at least two very good, low-cost graphical tracking packages available to you. They give excellent results—finding OSCAR 10 is a snap. Best of all, the investment is downright trivial compared to the cost of even the least-expensive computer.

Try one of the graphical tracking methods, even if you already have a computer to use for satellite tracking. The introduction to, and use of, the graphical methods will expand your knowledge and understanding of the nature of the OSCAR 10 orbit and make you a wiser communicator.

The two graphical-tracking packages that should be most useful to you are the *OSCARLOCATOR* from ARRL and the *Satellipse* from ZRO Technical Products.[3,4] See Figs 13-11 through 13-16. While you are at it, obtain a copy of the ARRL's *Satellite Experimenter's Handbook* mentioned previously. This publication presents an excellent discussion of graphical tracking. As an added bonus, these publications treat Phase II spacecraft tracking as well as Phase III.

All satellite tracking methods, computer and graphical, need periodic updating of orbital parameters and other reference information. Each satellite has different characteristics, so you'll need data for each satellite of interest.

Notes

[3]The *OSCARLOCATOR* is available from your local radio store or from ARRL for $8.50. Add $2.50 ($3.50 UPS) per order for shipping and handling.

[4]The *Satellipse* is available from ZRO Technical Products, Box 11, Endicott, NY 13760, for $10.

Amateurs can obtain this information from the following sources:

• AMSAT publications, including *Amateur Satellite*

OSCAR-10 Antenna Pointing Angles

Date: _____

Apogee Time Hr	Mean Anomaly	UTC Time Hr	Azimuth deg	Range × 10³ km	Elevation deg
−5.5	7				
−5.0	18				
−4.5	29				
−4.0	40				
−3.0	62				
−2.0	84				
−1.0	106				
0.0	128				
1.0	150				
2.0	172				
3.0	194				
4.0	216				
4.5	227				
5.0	238				
5.5	249				

Fig 13-11—Tracking Data Form.

Glossary—Continued

LOS—Loss of signal—The time when a satellite passes out of range and signals from it can no longer be heard. This usually occurs just after the satellite goes below the horizon.

Mean anomaly (MA)—An angle that increases uniformly with time, starting at perigee, used to indicate where a satellite is located along its orbit. MA is usually specified at the reference epoch time where the Keplerian elements are defined. For AO-10 the orbital time is divided into 256 parts, rather than degrees of a circle, and MA (sometimes called Phase) is specified from 0 to 255. Perigee is therefore at MA = 0 with apogee at MA = 128.

Mean motion—The Keplerian element that indicates the complete number of orbits a satellite makes in a day.

NASA—National Aeronautics and Space Administration, the US space agency.

Nodal period—The amount of time between two successive ascending nodes of a satellite orbit.

Orbital elements—See Keplerian Elements.

Orbital plane—An imaginary plane, extending throughout space, that contains the satellite orbit.

OSCAR—*O*rbiting *S*atellite *C*arrying *A*mateur *R*adio.

OSCARLOCATOR—A graphical satellite tracking device used to locate a satellite in its orbit and aid in pointing antennas. Available from ARRL.

PACSAT—A proposed AMSAT packet-radio satellite with store-and-forward capability.

Pass—An orbit of a satellite.

Passband—The range of frequencies handled by a satellite translator or transponder.

Perigee—The point in a satellite's orbit where it is closest to the Earth.

Period—The time required for a satellite to make one complete revolution about the Earth. See anomalistic period and nodal period.

Phase I—The term given to the earliest, short-lived OSCAR satellites that were not equipped with solar cells. When their batteries were depleted, they ceased operating.

Phase II—Low-altitude, long-lived OSCAR satellites. Equipped with solar panels that powered the spacecraft systems and recharged their batteries, these satellites have been shown to be capable of lasting up to five years (OSCARs 6, 7 and 8, for example).

Phase III—Extended-range, high-orbit OSCAR satellites with very long-lived solar power systems (OSCAR 10 and Phase III-C).

Phase IV—Proposed OSCAR satellites in geostationary orbits.

Precession—An effect that is characteristic of AO-10 and Phase III orbits. The satellite apogee SSP will gradually change over time.

Project OSCAR—The California-based group, among the first to recognize the potential of space for Amateur Radio; responsible for OSCARs 1 through 4.

PSK—Phase-shift-keying, a digital-communications modulation technique, especially employed on the OSCAR 10 telemetry.

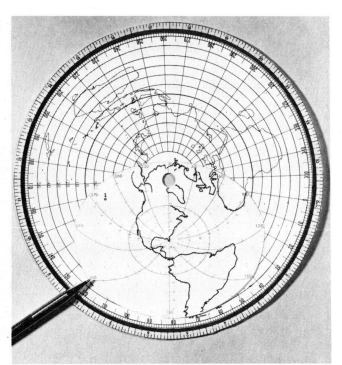

Fig 13-12—*Satellipse*: Cut out one of the circular maps according to the instructions. Mount the AZ/EL (with the range circles) overlap to the map centered over your QTH. It may be convenient to trace the continental outlines onto the overlay. *(WD4FAB photo)*

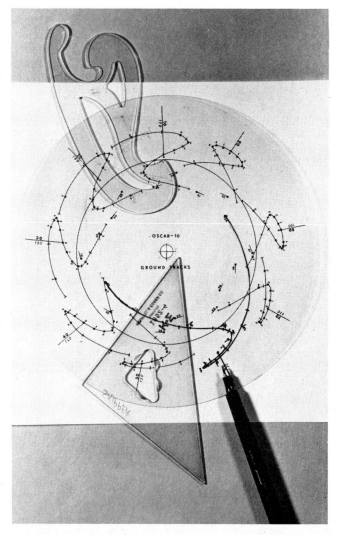

Fig 13-13—*Satellipse*, continued: Trace the AO-10 satellite ground track, corresponding to the current argument of perigee, onto the clear acetate circular overlay. Follow the *Satellipse* instructions to select the proper ground track template. Use a fine-point permanent marker for this tracing (such as the Pilot SC-UF pen). *(WD4FAB photo)*

QRP days—Special orbits set aside for very low power uplink operating through the satellites.

RAAN—Right Ascension of Ascending Node—The Keplerian element specifying the angular distance, measured eastward along the celestial equator, between the vernal equinox and the hour circle of the ascending node of a spacecraft. This can be simplified to mean roughly the longitude of the ascending node.

Radio Sputnik—Soviet Amateur Radio satellites (see RS #).

Reference orbit—The orbit of Phase II satellites beginning with the first ascending node during that UTC day.

RHCP—Right-hand circular polarization.

RS #—The designator used for most Soviet Amateur Radio satellites (RS-1 through RS-8, for example).

Satellipse—A graphical tracking device, similar to *OSCARLOCATOR,* designed to be used with satellites in elliptical orbits.

Satellite pass—Segment of orbit during which the satellite "passes" nearby and in range of a particular ground station.

Sidereal day—The amount of time required for the Earth to rotate exactly 360 degrees about its axis with respect to the "fixed" stars. The sidereal day contains 1436.07 minutes (see Solar day).

Solar day—The solar day, by definition, contains exactly 24 hours (1440 minutes). During the solar day, the Earth rotates slightly more than 360 degrees about its axis with respect to "fixed" stars (see Sidereal day).

Spin modulation—Periodic amplitude fade-and-peak resulting from the rotation of a satellite's antennas about its spin axis, rotating the antenna peaks-and-nulls.

SSC—Special service channels—Frequencies in the downlink passband of AO-10 that are set aside for authorized, scheduled use in such areas as education, data exchange, scientific experimentation, bulletins and official traffic.

SSP—Subsatellite point—Point on the surface of the Earth directly between the satellite and the geocenter.

Telemetry—Radio signals, originating at a satellite, that convey information on the performance or status of on-board subsystems. Also refers to the information itself.

Transponder—A device onboard a satellite that receives radio signals in one segment of the spectrum, amplifies them, translates (shifts) their frequency to another segment of the spectrum and retransmits them. Also called linear translator.

UoSAT-OSCAR (UO #)—An Amateur Radio satellite built under the coordination of radio amateurs and educators at the University of Surrey, England.

Uplink—The frequency at which signals are transmitted from ground stations to a satellite.

UTC—Coordinated Universal Time, the time of day corresponding to the zero meridian.

Window—Overlap region between acquisition circles of two ground stations referenced to a specific satellite. Communication between two stations is possible when the subsatellite point is within the window.

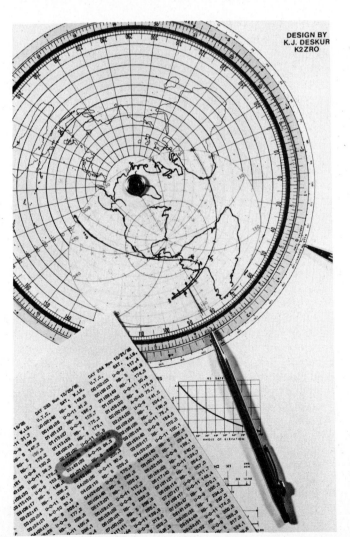

DESIGN BY
K.J. DESKUR
K2ZRO

Report, a biweekly AMSAT newsletter[5]

• *QST*

• Daily W1AW bulletins[6]

• *Project OSCAR Orbital Calendar*, a yearly publication of daily satellite reference predictions[7]

• Various AMSAT nets, especially the Tuesday evening net on 3857 kHz at 8 PM Eastern and Pacific times, and 9 PM Eastern, Central and Mountain times.

Tabulated satellite reference information, covering all current amateur satellites and good for about six weeks at a time, is available from ARRL HQ. Include a legal-size SASE with two units of First-Class postage with your request. Keep a number of SASEs on file, and you will receive the routine updates.

As mentioned before, most graphical tracking methods are based on an azimuthal equidistant projection map of the Earth, centered on the North Pole. A series of clear-plastic

[5]AMSAT, PO Box 27, Washington, DC 20044. *Amateur Satellite Report* is published bi-weekly and is available with AMSAT membership.

[6]See the W1AW Schedule in *QST* (or available from ARRL HQ for an SASE) for more information.

[7]Available annually for a $10 donation from Project OSCAR, PO Box 1136, Los Altos, CA 98510.

Fig 13-14—*Satellipse*, continued: Assemble the *Satellipse* using the snap fastener. From the ARRL table of EQXs and Apogees, locate the apogee desired for the track of interest. Place the apogee index of the ground track overlay to the desired longitude, shown here as 84.5 degrees west. Rotate the time scale to place the apogee time, 02:43:23, at the zero of the apogee time scale. Note also that the spacecraft mean anomaly figures correspond to the apogee time scale and use these values according to the AO-10 operating schedule. (WD4FAB photo)

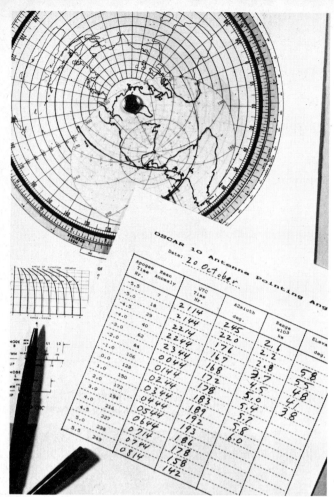

OSCAR 10 Antenna Pointing Ang

Date: 20 October

Apogee Mean Time Anomaly Hr.		UTC Time Hr.	AZimuth deg.	Range x10³ km	ELeva deg.
-5.5					
-5.0	7	2114			
-4.5	18	2144	245		
-4.0	29	2214	220		
-3.0	40	2214	220	2.6	
-2.0	62	2244	176	2.2	
-0.0	84	2344	167	2.2	
0.0	106	0044	168	2.8	
1.0	128	0144	172	3.7	58
2.0	150	0244	178	4.5	55
3.0	172	0344	183	5.0	48
4.0	194	0444	189	5.4	41
4.5	216	0544	192	5.7	38
5.0	227	0644	193	5.8	
5.5	238	0714	186	6.0	
	249	0744	178		
		0814	158		
			142		

Fig 13-15—*Satellipse*, continued: Using the antenna angle table of Fig 13-11, chart the UTC time, azimuth and range for each time point along the satellite ground track. Using the small graph just below the circular maps on the *Satellipse*, determine the antenna elevation corresponding to each apogee time and range data point. You will now have a completed set of tracking data for a desired orbit. *(WD4FAB photo)*

overlays is supplied. Different satellites usually have different orbits, which means that separate overlays are needed for each bird. Both of the graphical tracking packages mentioned here provide overlays for current satellites.

You'll need two overlays for each satellite. The ground-track overlay, which pivots around a rivet positioned at the North Pole, relates *the path of the satellite* to the map of the Earth. It shows the various locations that the satellite track can take.

The other overlay provides satellite *visibility circles* for your QTH. This overlay tells you which part of an orbit will bring the satellite in view of your QTH. It also shows the azimuth and elevation headings so you can point your antennas at the satellite.

Each of the graphical tracking packages mentioned here come with complete instructions for use. While it might seem like a lot of work to set one up, graphical trackers really don't require that much effort. They become a lot of fun and a self-satisfying achievement, once the method is learned.

A valid question is, "How good are the graphical tracking methods compared to computer-based systems?" Some test cases were run to find out. In each case, the *Satellipse* results

were within 3 degrees of the computer data in both azimuth and elevation. These differences are well inside the half-power beamwidth of highly directive crossed Yagi antennas. Even the most proficient operator would not be able to detect them.

Using a Graphical Tracker

To make the manual plotters work, you need an accurate reference point. This is simply a precise reckoning of where the satellite was at a selected instant. With the low earth-orbiting Phase II satellites, which have very nearly circular orbits, the reference used is the time and longitude of the satellite's northbound equator crossing, EQX. This is also known as the *ascending node*. More precisely, the reference orbit (the one most often used for reference purposes) is the first EQX of the UTC day.

The reference orbit consists of the time and longitude of the equator crossing. Orbit numbers are sometimes also included. To track subsequent orbits, you increment the longitude of the reference orbit by a fixed amount (provided in the bulletins). Similarly, the time increment is given. Given a reference orbit, together with the longitude and time increments, you have sufficient information for accurate tracking for several weeks, at least.

With elliptically orbiting satellites, such as OSCAR 10, things are a bit more complex. Since both the velocity and altitude of AO-10 change constantly, the shape of the ground track overlays are anything but a simple curve. The Earth is rotating at a fixed rate, but the motion of the satellite above any one point on the Earth is variable. The interaction of the two motions yields a ground track that doubles back on itself. Moreover, the shape of the ground-track overlay must be updated periodically to account for changes in orbital geometry. When used with OSCAR 10, the manual trackers most often use the time and location (both latitude and longitude) of the apogee as a reference, rather than the EQX data. Since apogee latitude and longitude are predictably related, usually only the time and longitude is given in the references. This apogee reference data, however, can also be incremented to extend the usefulness of a single data point.

Using graphical tracking methods for OSCAR 10 requires a periodic redrawing of the OSCAR 10 ground track, which changes with the argument of perigee (see glossary). You don't need the argument of perigee data for using the *OSCARLOCATOR*, as the ground track is routinely presented in the satellite column in *QST* and also available from AMSAT. Data for estimating the argument of perigee, and thus the specific ground track needed for tracing, are all contained in the package of information that comes with the *Satellipse*.

With the key pieces of information needed for daily tracking of OSCAR 10, the apogee reference time and longitude, taken from the previously noted sources, daily apogee updates can be made. The apparent daily apogee progression is very predictable, with time values of − 40.95 minutes per day and longitude progression of 9.37 degrees East per day. Keep in mind that there are two OSCAR 10 apogees per day, and when the one you are tracking passes out of view to the East, look for its companion appearing 699.52 minutes later and 175.3 degrees to the West of the last apogee. These numbers are based on the basic motion of OSCAR 10 of a Mean Motion of 2.05854 orbits per day, and an apogee longitude progression of 184.68 degrees per orbit. With this data, anyone can project, from a single reference, the OSCAR 10 apogee positions for quite a period of time.

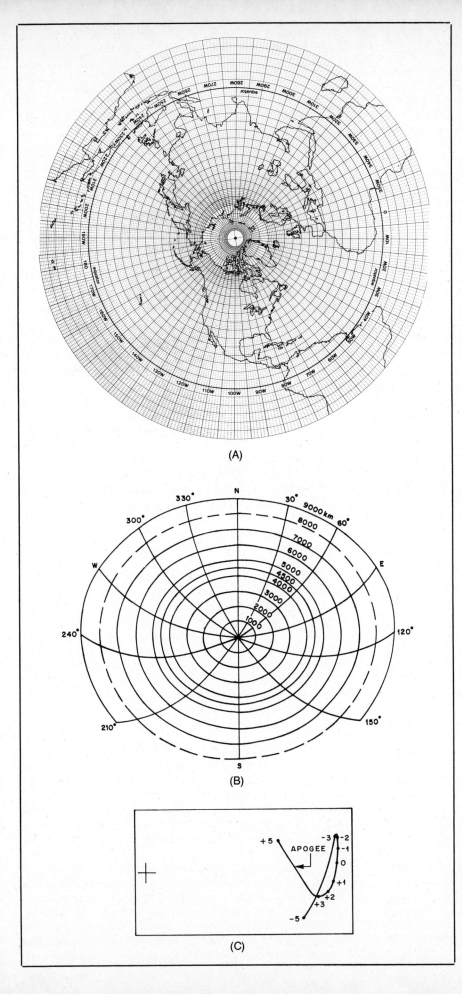

(A)

(B)

(C)

With all of the foregoing preparations, the actual tracking becomes easy, with the preferential nod going to the *Satellipse*, as it contains a "circular slide rule" calculator for setting apogee reference UTC time to the satellite "zero" time of the apogee reference and bringing out the MA or phase timing of the spacecraft. From that point, the current UTC time can be compared to the satellite times for a determination of what portion of the ground track is to be used for the current satellite location. Setting the ground-track apogee mark to the apogee reference longitude then completes the satellite positioning for the desired time. Antenna-aiming azimuth bearing can be read directly from the QTH visibility circle overlay, and aiming elevation data can be easily estimated from the information given with the *Satellipse*. Users of the *OSCARLOCATOR* are instructed to do their own estimating of the needed elevation angles through listening tests.

Computer-Based Tracking

For those of you with personal computers, there is a wealth of software available. Most operators are quite satisfied to use the computer to provide numerical data to locate all amateur satellites. They then take this information and aim their antennas accordingly. For AO-10, this is usually sufficient. Changes in antenna positions are infrequent—anywhere from every half hour up to three hours before significant repointing is needed.

Another class of software allows the computer to control antenna position automatically. This approach has some inherent technical problems that are very far afield from computer byte bashing. Unless you are extremely well versed in the software, electronics and mechanics of digitally controlled, closed-loop servo systems, you should forget automatic-antenna tracking. Once you get into tracking AO-10, you will find that you need to adjust the position of your antennas only once or twice an hour by

Fig 13-16—The *OSCARLOCATOR* from ARRL contains all the tools needed to find OSCAR 10. The polar projection map at A is used for tracking all amateur satellites. The clear acetate QTH range circle (B) is centered over your QTH. The AO-10 ground track, shown at C, shows the path the satellite will take. The ground track changes periodically, and updates are published as necessary in *QST*.

fairly small increments. It is just not worth the effort to control the antenna position automatically. In the days of OSCARs 6, 7 and 8, when the satellite was workable for 16 to 22 minutes each pass, operations in the shack were a bit busy. AO-10 is literally a world of difference.

Satellite-tracking software for a number of computers is available from AMSAT, through the AMSAT Software Exchange (ASE). Most of this software is based on the original work by Dr Tom Clark, W3IWI, that was published in *Orbit*.[8] Since the original work, software specialists have found many innovative ways to express Dr Clark's computational methods. A listing of the various versions as of early 1987 is given in Table 13-7. For a current program catalog and ordering information, write to AMSAT Software Exchange, PO Box 27, Washington, DC 20044.

Some commercial software vendors advertise satellite-tracking programs in the ham magazines. An elegant package

[8]T. Clark, "Basic Orbits," *Orbit*, Mar/Apr 1981, pp 6-11 and 19-20.

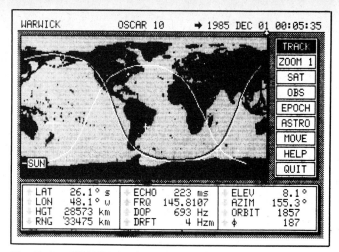

Fig 13-17—*GRAFTRAK II* from Silicon Solutions provides an elaborate map display as well as all important satellite parameters. This software runs on the IBM PC and requires an 8087 math coprocessor to help with its intensive calculations.

Table 13-7

AMSAT Software Exchange Satellite-Tracking Programs

1) Radio Shack TRS-80® Model I, Level II BASIC, 32 kbytes RAM needed. No instructions included; software manual (see item 20) recommended.

2) Radio Shack TRS-80 Model III, 32 kbytes RAM needed.

3) North Star BASIC under North Star DOS for 5¼ inch, hard sector, single- or double-density drives.

4) Microsoft BASIC, version 5.21 under CP/M®, single-density, single-sided 8-inch disk. No instructions included; software manual (see item 20) recommended.

5) Apple® II, APPLESOFT BASIC, on 13- or 16-sectored diskettes or cassette. Menu driven, output to screen or printer.

6) IBM® PC, PC-XT, or PC-AT version, by WØSL. Menu driven for tabulated output for up to eight satellites in real time. Graphics display of world map. Requires 128 kbytes RAM, DOS 2.0 or later, and BASICA.

7) IBM PCjr version by WØSL. Tracking with graphics. Requires 128 kbytes RAM, DOS 2.0 or later and BASICA as above, but modified to run on PCjr.

8) IBM PC and compatibles version by N4HY. Called QUIKTRAK, it is menu driven for tracking and scheduling and features a new "Window Track" mode for DX.

9) Texas Instruments TI 99/4A, cassette only.

10) Apple II antenna positioning and controlling software by KØRZ.

11) Radio Shack TRS-80 Model 4, for TRSDOS Version 6.0.

12) Radio Shack TRS-80 Color Computer. Requires 32 kbytes RAM and extended BASIC (cassette only).

13) Commodore 64™, AMSAT VR85. Datapoint map of 2000, 20 satellites.

14) QUIKTRAK-2064, enhanced version of AMS-2064, including machine language file, cassette or disk.

15) Atari® —disk only.

16) Timex-2068, cassette only of W3IWI program.

17) HP-41C programmable calculator, version ORBIT I of the W3IWI program (approximates real time operation).

18) HP-41C programmable calculator, version ORBIT II of the W3IWI program, converted to run with time module (real-time tracking).

19) Heathkit® H89 version of W3IWI program. CP/M version configured for H89, CP/M & MBASIC. Requires 5¼ inch H-17 single-density, single-sided, hard-sector disk.

20) *Using Microcomputer Programs for Radio Amateur Satellite Orbital Prediction* by N5AHD. This manual tells how to use the W3IWI program on Radio Shack, CP/M and S-100 bus computers.

21) UoSAT telemetry capture and decoding software for the IBM PC.

that is advertised in *QST* is *Graftrak II* from Silicon Solutions. This program, which operates on the IBM® PC, provides a sophisticated, colorful map display showing the satellite path over your QTH. See Fig 13-17. Also from Silicon Solutions is *Silicon Ephemeris*, a satellite-tracking program that has a tabular output. Both packages are available from Silicon Solutions Inc, PO Box 742546, Houston, TX 77274-2546, tel 713-661-8727.

Spectrum West offers *Autotrak: Computer Rotor Control* for several popular computers. This software and hardware package allows the computer to control azimuth and elevation rotators, so the antenna positions are updated as the computer recalculates the satellite position. For more information, contact Spectrum West, 5717 NE 56th, Seattle, WA 98105, tel 206-523-6167.

Another package, using a different computational algorithm than W3IWI employed, is available from Manfred Mueller, KG6EF and Gordon Mueller, KB6BPL. The Muellers provide a Sharp PC-1246 pocket computer and a BASIC routine that was conceived by Dr Karl Meinzer, DJ4ZC. While this computation is somewhat simpler than the W3IWI Keplerian method, it has yet to achieve a great following in the US. For more information, contact the Muellers at 4914 Commonwealth Ave, La Canada, CA 91011, tel 818-790-6695.

Station Requirements for OSCAR Operation

Most long-distance communications depend on natural propagation phenomena. Some days, a kilowatt and high Yagi are needed to establish communications; at other times, 5 W and a long wire will do the trick. Stations designed to take advantage of ionospheric propagation are largely a matter of personal preference. Bigger stations usually mean bigger signals.

Operation through a satellite is somewhat different. Fixed conditions such as transponder receiver sensitivity and power output, and variable conditions such as spacecraft attitude and path losses, are the dominating factors in establishing minimum ground-station requirements. Once the minimum requirements are met, you can have reliable communications any time the satellite is in view, regardless of changing prop-

Those Keplerian Elements

All of the current crop of computer programs that are used to compute the position of a satellite and determine the corresponding antenna pointing angles use a basic set of numerical constants as inputs. These input constants are derived from classical astronomical motions and are called the primary Keplerian elements.

Numerous articles have appeared in various publications describing the virtues of some of these computer programs. These articles usually gloss over the mechanics of orbital motion because it is not possible for most amateurs to become instant experts in celestial mechanics. Even becoming acquainted with the terms will require some reasoned study. Here is a basic description of some of the terms associated with orbital mechanics. These descriptions will provide most of the known identification references to the specific element, so that you will recognize the term no matter how it is declared. These element descriptions will also be presented in the same order as they are commonly listed and used, so that sequential associations will also be recognized. Mathematical references will also be avoided; there are enough texts around that treat these elements in those terms.

Epoch, TØ: A fancy name for a specific reference time. This is the UTC date and time for which the Keplerian elements are defined. Epoch values may be stated in conventional dating methods of day, month and year (ddmmyyyy) followed by UTC clock time. Internally, all computer programs use a numeric day-of-year date reference, and many of us are more accustomed to stating the Epoch in this manner. For instance, 11 November 1987 is day number 315 of 1987, and most programs will accept the compound number of 85315.458333 for 1100 UTC on that date. The decimal portion of the number, .458333, is that fraction of the 24-hour period corresponding to the clock time of 1100 hours of day 315.

Inclination, IØ (degrees): Satellite orbits are best described as an elliptical motion within a flat plane in inertial space. The tilt of this plane with regard to the Earth's equator is the angle called Inclination. If IØ is zero degrees, then the orbit is in the equatorial plane, typical for most geostationary communications satellites, while a value of 90 degrees describes a polar orbit. See Fig 13-24.

Right Ascension of the Ascending Node (RAAN), OØ (degrees): This angle describes the location, with respect to the Earth's longitude coordinates, of the tilted orbital plane. Convention places the OØ at that point on the Earth's longitude scale (degrees west longitude) that the satellite track crosses the equator while traveling from the Southern Hemisphere to the Northern Hemisphere. See Fig 13-24.

Eccentricity, EØ: A dimensionless number that describes the shape of an ellipse. All orbits are elliptical motions of which the circle is only a special case. Since elipses can vary in shape, we need to know the "flatness" of the orbit. If the orbit is truly circular, then EØ = 0, but if the orbit is very flat, the value goes to EØ = 1.0, which is a straight line. Beyond this point, the "elliptical" orbit is no longer elliptical, but is either parabolic or hyperbolic and is no longer a closed, or periodic, orbit. See Fig 13-25.

Argument of Perigee, WØ (degrees): This awkward title is merely a statement of the position of the perigee, or lowest altitude point of the orbit, with respect to the RAAN. Measuring orbital position, in degrees, about the center of the Earth, starting from the ascending node crossing of the equator around to the perigee position, then, is the value of WØ. You can then reason that a value of WØ = 270 degrees means that the high point of the orbit, the apogee, will be at the most northerly latitude. Conversely, a value of WØ = 90 degrees places the apogee at the most southerly latitude. See Fig 13-26.

Mean Anomaly, MA, MØ (degrees): Another "strange" term, but if you read "anomaly" to be identical to "angle," it then starts to have meaning. MA is merely a statement of the angular position of the satellite in its orbit at the very moment of the reference time is set for the Keplerian elements. MØ is expressed in degrees centered on the Earth's center, starting from the perigee. Another use of MA is in the AO-10 computer for keeping track of time, since the orbital period is a constant time unit. The computer takes the full orbital revolution, from one perigee to the next perigee, and divides that time into a computer-usable number of 256 parts, rather than the conventional angular measurement of 360 degrees. AO-10 scheduling is done in increments of this version of MA, sometimes called "Phase."

Mean Motion, NØ (orbits per day): This is a very simple notation that merely expresses the number of orbits a satellite completes around the Earth for each UTC day. Orbital period can be easily derived from NØ by dividing 24 hours by NØ.

Semi-Major Axis, SMA, AØ (kilometers): The SMA is a dimensional measurement of one half of the total "length" of the longest axis of an ellipse. In Earth orbits, if the NØ is stated, the use of the SMA is not necessary as they are numerically related in a very direct manner. Most programs will accept either NØ or SMA as input but only use one or the other. The SMA is commonly not stated in lists of Keplerian elements for OSCAR satellites. See Fig 13-25.

Decay Rate, Drag Factor, N1 (orbits/day/day): These two separate names describe the same quantity, which is the first derivative of the Mean Motion. It is, as the name states, the rate of slowing down of an orbit. This factor has been a relatively recent addition to the orbital computations and assists greatly with prediction precision. For those satellites in low orbits, N1 becomes a rather high number, and its impact rather apparent.

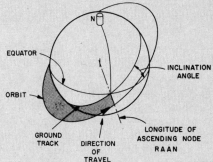

Fig 13-24—Inclination and RAAN.

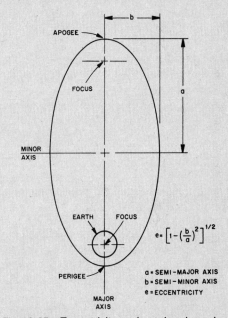

$$e = \left[1 - \left(\frac{b}{a} \right)^2 \right]^{1/2}$$

a = SEMI-MAJOR AXIS
b = SEMI-MINOR AXIS
e = ECCENTRICITY

Fig 13-25—Eccentricity and semi-major axis.

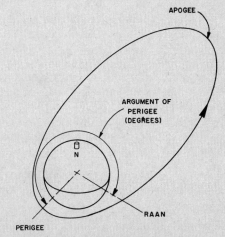

Fig 13-26—Argument of perigee.

agation phenomena. The communications requirements for OSCAR 10 have been pretty well defined since the spacecraft became operational in 1983. See Chapter 23 of *The ARRL Handbook* for detailed information on "link budgets." Unlike other areas of amateur interest, an inadequate station won't merely *limit* your ability to communicate—it may well prohibit communications *entirely*.

Satisfactory satellite communications might be possible with an existing station, but planning the ground station around certain parameters will enhance its effectiveness. Ground stations for Phase III satellites such as OSCAR 10 need to consider such subtleties as low-loss feed line, a single-transistor low-noise GaAsFET preamplifier, and high-gain VHF/UHF antennas that can be pointed accurately.

The parameters that should be considered for an OSCAR ground station are discussed in detail in *The ARRL Handbook* and are outlined below:

1) 146-MHz antenna gain of at least 13 dBi, circularly polarized. Right-hand circular polarization is a must for effective operation. More antenna gain is always desirable for better performance. Antenna gain of much more than 18 dBi, however, is not cost effective—satellite transponder noise floors will limit further benefits.

2) 435-MHz antenna gain of at least 13 dBi, circularly polarized. For this band, it is highly desirable to be able to switch between right-hand circular polarization (RHCP) and left-hand circular polarization (LHCP). Gains of 14 to 18 dBi are preferred for OSCAR 10 Mode B. For OSCAR 10 Mode L, antenna gain around 20 dBi is necessary. Future Mode L operations are expected to be satisfactory with the 14- to 18-dBi-class antenna.

3) 1269-MHz antenna gain of 20+ dBi is required for OSCAR 10 Mode L. Current AO-10 users are having success with linear antenna polarization, even though the spacecraft antenna uses RHCP. Future Phase III satellite operations will probably demand a switchable circular polarization. Future Mode L operations are also expected to be less demanding in the antenna department. Gain in the range of 16 to 20 dBi should yield satisfactory results.

4) Circularly polarized antennas must be mounted on nonmetallic booms so that the antenna performance is not degraded by a metal boom within the antenna field. Similarly, coaxial feed lines must be routed axially rearward past the antenna reflector to avoid pattern distortion.

5) Feed-line losses of less than 2 dB are needed; less than 1 dB is better. This can be achieved without expensive Hardline through the use of antenna-mounted preamplifiers and other equipment.

6) 435-MHz EIRP (effective isotropic radiated power) of no more than 500 W. Most operations can be conducted with 100-W EIRP, and at times it's possible to use only 10-W EIRP.

7) 1269-MHz EIRP of 3000 to 25,000 W for OSCAR 10 Mode L. Future satellite requirements are expected to be in the 500 to 3000 W category.

8) 146-MHz receive system noise figure no greater than 2 dB. Noise figures much less than 1 dB are not warranted because of the limitations imposed by terrestrial noise sources.

9) 435-MHz receive system noise figure of less than 1 dB. Terrestrial noise levels are not a limiting factor on this band.

10) Frequency readout resolution of 1 kHz on transmit and receive. Accurate frequency readout allows you to find your signal on the downlink without causing needless QRM.

11) An accurate means of determining antenna direction at all times during an orbit. This means knowing where the

antennas are pointed, and where to point them. Accurate graphical plotting or computer printout of pointing angles is required.

Like many other aspects of Amateur Radio, trade-offs can be made to minimize expenditure of time and money. Up to a point, antenna gain is easier to obtain than transmitter power. It is easier to put up a bigger Yagi to increase your EIRP than to build or buy an amplifier. For example, a commercially available 40-element 435-MHz crossed Yagi makes enough EIRP for AO-10, Mode B communications with only 25 W of 435-MHz RF at the antenna. Often, only 5 to 15 W are all that's necessary.

Better antennas also improve receive performance. All things considered, it is much more cost effective—and desirable from a performance standpoint—to optimize the receiving set-up before investing in higher transmit EIRP. Don't be an alligator!

AMSAT and Project OSCAR have jointly produced a very useful manual that defines the details of the satellite link characteristics. This manual is recommended reading for all interested amateurs.

Circular Polarization

In the HF bands, polarization differences between antennas are not really noticeable because of the nonlinearities of ionospheric reflections. On the VHF and UHF bands, however, there is little ionospheric reflection. Cross-polarized stations (one using a vertical antenna, the other a horizontal antenna) often find considerable difficulty, with upwards of 20-dB polarization loss. Such linearly polarized antennas are "horizontal" or "vertical" in terms of the antenna's position relative to the surface of the Earth, a reference that loses its meaning in space.

The use of circularly polarized (CP) antennas for space communications is well established. If spacecraft antennas used linear polarization, ground stations would not be able to maintain polarization alignment with the spacecraft because of changing orientation. Ground stations using CP antennas are not sensitive to the polarization motions of the spacecraft antenna, and therefore will maintain a better communications link.

All AO-10 gain antennas (for 2 meters, 70 cm and 24 cm) are configured for RHCP operation along their maximum gain direction. See Table 13-8 and Fig 13-18. Since this direction is also the main antenna lobe along the spacecraft +Z axis, the best communications with AO-10 will also be along that direction. Since the AO-10 radiations are RHCP, ground stations should also be RHCP for optimum communications.

There are times, however, when LHCP provides a better satellite link. AO-10 is designed so that the main antenna lobe is oriented toward the center of the Earth when the satellite is at apogee. It is only so oriented for up to a few hours either

---

Table 13-8

Polarization and Gain of the OSCAR 10 Antennas

Frequency (MHz)	High Gain Antennas Polarization	Gain (dBi)	Omni Antennas Polarization	Gain (dBi)
146	RHCP	9.0	Linear	0.0
436	RHCP	9.5	Linear	2.1
1269	RHCP	12.0	RHCP	0.0

(adapted from *The AMSAT-Phase III Satellite Operations Manual* prepared by AMSAT and Project OSCAR)

Fig 13-18—This diagram shows the far-field radiation pattern (undeformed) for the 2-meter high-gain antenna on OSCAR 10. (Adapted from *The AMSAT-Phase III Satellite Operations Manual* prepared by AMSAT and Project OSCAR.)

decibels. This way you can note signal strengths in decibels above the noise, which has more meaning in weak-signal work than normal S-meter readings.

Once you've found the beacon and know that you're hearing OSCAR 10, tune up through the passband (145.830 to 145.970 MHz). Note that the satellite passband is divided according to the voluntary frequency plan shown in Fig 13-8. Nearly all of the CW activity is below 145.900 MHz; nearly all of the SSB activity is above that frequency. You will be able to hear packet activity on Special Service Channel (SSC) L2, near 145.840 MHz.

It will take a while to get oriented to most OSCAR transponder operations because nearly all spacecraft translators are *inverting*, as shown in Fig 13-19. This means that a signal transmitted at the high end of the 435-MHz uplink passband will come out on the low end of the 145-MHz receiving passband. SSB signals are inverted as well: If you transmit LSB on the uplink, it will come out as USB on the downlink. Common practice is to transmit up to AO-10 on LSB, providing a USB downlink signal.

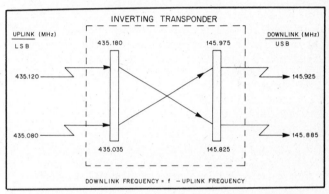

Fig 13-19—The OSCAR 10 Mode B transponder, typical for most satellite V/UHF links, is inverting. Signals transmitted to the satellite at the high end of the 435-MHz uplink come out at the low end of the 145-MHz downlink. Signals transmitted to the satellite on LSB return to Earth as USB signals. The translation frequency, f, is 581.005 MHz.

side of apogee, however. As the satellite moves away from apogee, the orientation changes and the RHCP main lobe is not centered on Earth. At other times, ground command stations have to orient the spacecraft to odd angles to maintain proper sunlight on the solar cells that are used for prime power. Again, at such times the satellite antenna's main lobe is not pointed directly at Earth. During the periods when the main lobe is directed away from Earth, communications are still possible through the use of OSCAR 10's antenna side lobes. CP antenna patterns, however, exhibit side lobes that can be polarized in the opposite sense to the main lobe. At times when only LHCP side lobes are available, ground stations must be able to switch to LHCP to maintain optimum communications.

Independently switchable RHCP/LHCP antenna circularity is necessary, especially for 70 cm. It is expected to be required for 24-cm operation on Phase III-C as well. Operators who use AO-10 Mode L have determined that the uplink can be affected by LHCP antenna lobes when the satellite is located off-axis to the ground station. Switchable RHCP/LHCP on 2 meters may be convenient, but it is rarely needed. See *The ARRL Handbook* and *The Satellite Experimenter's Handbook* for detailed information on the technical aspects of assembling an OSCAR station.

ON THE AIR THROUGH OSCAR 10

It's a good idea to tune to the beacon frequency of 145.81 MHz each time you begin satellite operation. Among other things, the beacon provides a constant signal for peaking your antenna on the satellite. Determine the beacon strength. If you have the receiving capability recommended earlier, you will have useful readings on your S-meter (normally, less than S9 values). It may be helpful to use a signal source and a switchable attenuator to calibrate your S-meter in terms of

If everything is working right, conversational QSOs can be held with signals ranging from 6 to 15 dB above the noise floor of your receiver. Typically, the beacon is 12 to 15 dB above the noise. Confirmable QSOs have been made with signals as low as 2 to 4 dB above the noise. It is amazingly different from the type of communications that you may have been doing on the HF bands. Note that there is *no* excuse for QRM. Anyone able to receive the satellite should also be able to hear *everyone else* and allow sufficient elbow room for rational QSOs without crowding.

Locating Your Signal

Find some vacant territory on the receiving passband of the satellite for testing your own signal, for example, about 145.960 MHz. There's no need to hurry—AO-10 will be around for awhile. The frequency chart shown in Table 13-3 will help you to find the right 435-MHz transmitting frequency to correspond with your chosen 145-MHz downlink frequency. If you wish, you can purchase a handy circular slide rule for the uplink-downlink frequency relationships.[9]

[9]The OSCAR 10 ''No Ditter'' is available from Dave Guimont, Jr, WB6LLO, 5030 July St, San Diego, CA 92110, for $3, postage paid.

Assuming you've tuned your receiver to 145.960 MHz, the nominal transmitting frequency is 435.045 MHz. Send a few dits and listen for them on the receiver. Headphones are very helpful here. Tune your *transmitter* frequency a bit on either side of nominal and find your own signal coming back. Note the offset from the nominal frequency; you'll need to know this number any time you want to bring your transmitter to a frequency you're listening on.

The offset is a combination of equipment calibration and Doppler shift. *Doppler shift* is caused by the relative motion between you and the satellite. As the satellite moves toward you, the frequency of the downlink signals will increase slightly. As the satellite passes overhead and moves away from you, the frequency of the downlink signals will be slightly lower than nominal. Doppler shift through a transponder becomes the sum of the Doppler shifts of both the uplink and downlink signals. Since the Mode B transponder is inverting, an increase in uplink frequency causes a decrease in downlink frequency, so the Doppler shifts tend to cancel.

If you are using the Ten-Tec 2510 Mode B Satellite Station, tuning the frequency offset is even easier. Set your HF receiver, used as a tunable IF, to 29.0 MHz, and set the 2510 to the desired receiving frequency. Tune the HF receiver a few kilohertz on either side of 29.0 MHz and you will find your signal. Once set, keep your hands off the HF receiver tuning knob! Adjust the HF rig only for very small Doppler-shift corrections, and do all of your QSYing with the 2510

tuning knob. Your transmit signal will always follow your receiver; it's that easy.

Now that you've found your signal, compare its strength with that of the 145.810-MHz General Beacon. Your signal strength on the downlink should never exceed that of the beacon. If it does, decrease your transmitter power output accordingly. The transponder must share its power among all users. If your signal is louder than the beacon, you'll activate the transponder AGC and degrade performance. It takes only one hog to make communications difficult for everyone. Don't ever feel that you have to be as strong as the beacon. Good conversations are had at 8 dB above noise when the beacon is 12-15 dB above the noise, and your lower uplink power is all the better for the spacecraft.

At Last: Operating Through OSCAR 10

Let's try a CW QSO first. There are two ways of finding someone to work on the satellite, just like on any other band or mode: You can *call* CQ and hope someone answers, or you can *answer* a CQ. If you've got a good signal through the satellite, you may want to try calling CQ at first. This way, you won't have to worry about bringing your transmitter to someone else's frequency. You'll be able to get the frequency controls set in advance, allowing you to concentrate on making the contact. After a contact or two, venture down the band and try to find another station to call.

Find a clear spot in the CW portion of the passband and

Doppler Shift

Doppler shifting of signals is caused by the relative motion between you and the satellite. As the satellite moves toward you, the frequency of the downlink signals will increase by a small amount as the velocity of the satellite adds to the velocity of the transmitted signal. As the satellite passes overhead and starts to move away from you, there will be a rapid drop in frequency of a few kilohertz, much the same way as the tone of a car horn or a train whistle drops as the vehicle moves past the observer. Martin Davidoff, K2UBC, provides a very complete and understandable discussion of Doppler shift in Chapter 10 of *The Satellite Experimenter's Handbook*. This brief outline on Doppler shifts provides highlights of the effect.

The Doppler effect is different for stations located at different distances from the satellite because the relative velocity of the satellite with respect to the observer is dependent on the observer's distance from the satellite. The result of all this is that signals passing through the satellite transponder shift slowly around the calculated downlink frequency. Locating your downlink signal is more than a simple computation, since tuning is needed to compensate for the Doppler shift.

Doppler shift through a transponder becomes the sum of the Doppler shifts of both the uplink and downlink signals. In the case of an inverting type transponder (as in OSCAR 10 Mode B), a Doppler-shifted increase in the uplink frequency causes a corresponding decrease in downlink frequency, so the resultant Doppler shift is the *difference* of the Doppler shifts, rather than the *sum*. The shifts tend to cancel.

While Doppler shifts can be observed on OSCAR 10, it's not nearly so apparent as was for OSCARs 7 and 8. The time required to observe a given change in Doppler shifts is much longer for OSCAR 10 than for the earlier low-altitude satellites, and is therefore much less of a problem. For instance, the total OSCAR 10 Doppler shift is approximately 4.5 kHz over the period of time that normal operations are conducted, approximately four hours on each side of apogee. This rate of change amounts to

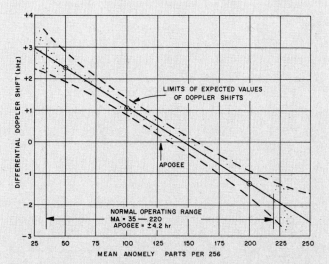

Fig 13-22—OSCAR 10 Mode B differential Doppler shift, normal for MA operating range.

approximately 9 Hz per minute. For OSCAR 8, operators experienced a total shift of up to 12 kHz over a period of 16 minutes, a rate of approximately 750 Hz per minute!

Fig 13-22 illustrates a mean-computed OSCAR 10 Mode B differential Doppler shift averaged over quite a few orbits. It must be pointed out that the complete whole-orbit AO-10 Doppler-shift curve is not a straight line. However, over the portion of the orbit that encompasses the majority of operations, a straight-line approximation will suffice quite well. These ranges cover approximately MA = 35 to MA = 220, and the total shift averages to be 4529 Hz, or 8.96 Hz/minute. Fig 13-22 can be used to estimate the expected Dopple shift for various operating conditions.

Calibrating Your S-Meter

While this book is primarily about amateur operating techniques and not about amateur equipment, there is one measurement and adjustment that needs to be noted to improve your operating abilities. Making your transceiver S-meter a truly useful measure of signal strengths, rather than just a guess-meter, is the point of reference.

Installing a low-noise preamplifier into a 2-meter receive system, preferably very near the antenna, will do wonders for satellite reception. Signals will be noticeably stronger, but so too will the noise level, sometimes to unbearable levels. Useful reception of OSCAR 10 signals does not require that the receiver noise level be brought up to the S9 level. Most low-noise receiving systems will audibly sound better when the noise level is brought back down to a much lower level, say to S1. If a convenient control to adjust the receiver to that point was available, life would be fine. That kind of control can be built and easily calibrated and provides a sizable improvement to communications through OSCAR 10.

A very useful adjunct of controlling the indicated noise level of a receiver system is that the AGC metering, the S-meter, becomes more sensitive and readable at the lower levels. In fact, knowing the value of reading a received signal, compared to your noise floor, becomes a very useful tool for the operator. Received noise floors of relatively decent OSCAR ground stations, static permitting, is really limited by the noise floor of the spacecraft transceiver itself.

Achieving all of this ability is one of the lower cost items that you can provide for your station. First of all, calibrating the receiver AGC meter provides the measurement knowledge you need. Fig 13-23 shows two ways to insert an attenuator into the receiver system for calibration, depending on your receiver system. Beg or borrow the use of a good switchable RF attenuator and set it up as shown in Fig 13-23. Locate a low-level steady carrier, birdie or beacon on 2 meters. If you cannot find such a usable signal, try your 2-meter hand-held at low power into a shielded dummy load. Often the dedicated Radio Shack Weather Alert radios will have such a beacon from its local oscillator. Some of these signals are just at 145.825 MHz, and are very convenient for routinely checking satellite receiver operation.

Start adding RF or IF attenuation to achieve a meter reading of S1 and note that value. Continue adjusting the attenuator for obtaining the value at the other significant marks of your S-meter. With this data tabulated, normalize

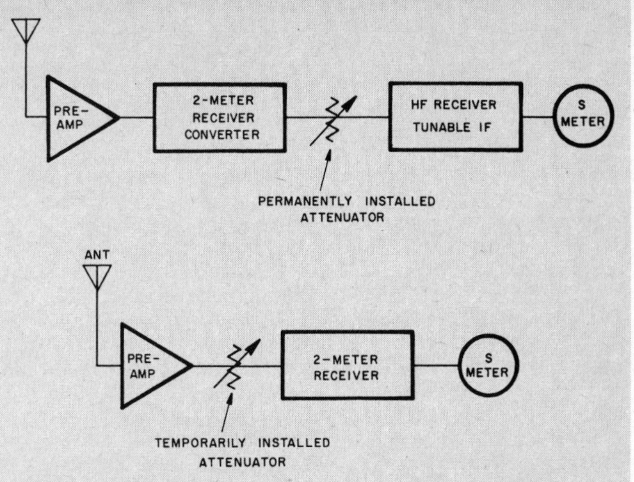

Fig 13-23—Installing the attenuator (two methods).

the data to the S1 meter reading by making that value 0 dB signal strength and note the dB values for the meter marks above S1.

With this data, the sophisticated attenuator will no longer be needed, and it can be returned to the measurement laboratory. For those using a 2-meter receive converter, it is very convenient to install a simple IF attenuator ahead of the tunable IF, the HF receiver. Using this attenuator permits setting the OSCAR 10 noise floor at your receiver S1 indication and keeps your ears from becoming desensed with the audio white noise. In addition, received signals from the spacecraft are all reportable to the other stations in terms of dB above your noise floor, a much more meaningful report rather than the wholly arbitrary S-meter readings. For those stations using 2-meter transceivers, consideration should be given to installing a switchable attenuator in the receiver IF line, as it would be a very usable improvement to the low-noise operation.

Attenuators for this operation should have independently selectable values of 5, 10 and 20 dB, which will permit selecting any value from zero to 35 dB in steps of 5 dB. Chapter 25 of *The ARRL Handbook* has the details on such switch-selectable attenuators for RF service, and they are easy to build with small toggle or slide switches.

---

bring your transmitter to the frequency. Do this the intelligent way: Look up the frequencies in Table 13-3, set your transmitter to the right frequency (remember the offset), and send a couple of dots if necessary to zero in on your signal. Resist the temptation to put a "brick" on the dot lever and crank the transmitter control until you find your signal. Such "swishing the passband" causes interference to everyone else and is considered extremely bad manners.

Now call CQ. Although you can hear your signal on the downlink, you'll probably want to use a sidetone oscillator for monitoring your sending. OSCAR 10 is quite a distance away, so it takes a little time for your signal to make the round trip. When AO-10 is in the highest part of its orbit, the delay is about ¼ second. It can be disorienting to listen to a delayed downlink signal while you're sending. It's a good idea to adjust the sidetone level so that it's loud enough to use as a monitor, yet weak enough to let you hear downlink signals between characters or words.

Since satellite operation is full duplex, don't be surprised if someone answers your CQ *before* you've finished. There's really no reason to wait—it's not like routine single-frequency

communications where the receiver is disabled until the end of a transmission. This won't be difficult to get used to, especially if you've used a full break-in CW rig before. Should Doppler change during the course of your QSO, tune your *transmitter* to keep your transmitter and receiver on the same frequency. This way, you won't "walk" down the band as things change. That's all there is to it: OSCAR 10 operation is just like what you're used to on HF, only the band is *always* open when the satellite is in view and it uses a lot less power.

Now Try SSB

SSB operation via OSCAR 10 is similar to CW. The time delay is even more difficult to get over, though. New operators usually stumble over their own return voices when they first try AO-10. You may find it more comfortable to turn down the receiver audio gain control while you're talking. The Ten-Tec 2510 has a push-button control to mute the receive audio during transmit. These approaches, while helpful, force the operator back to monolog transmission without the full-duplex features available from AO-10. With some practice, you should be able to listen to the *pitch* of your return voice

Things Even Your Best Friends May be Afraid to Tell You

The Hard Lessons

"Boy, I called that WA3URV several times. He kept calling CQ on AO-10 and he sure was loud. Doesn't that guy have a receiver or is it me?"

Sound typical? Perhaps the biggest complaint, and biggest problem for some satellite operators, is typified by the last comment. The old adage from HF hamming that "if you can't hear them, you can't work them" is very much to the point for satellite operators. If you can only manage to be a typical "alligator" (all mouth and no ears), then you should not QRM the satellite until appropriate changes have been made in your station capabilities.

Let us examine this situation more closely. To successfully operate OSCAR 10, you have to be able to hear it if you ever hope to hear your own signal. Now this sounds like a simple statement until you realize that the satellite is sometimes *36,000 km* above the Earth and moving at a relatively high rate of speed. Knowing where the satellite is in time and space is a prime prerequisite, followed by the need to be able to aim your directional antennas toward the satellite. Recovering usable RF signals from the electromagnetic energy around us then becomes the final technical challenge.

For some unexplained reason, hams invest lots of time and effort in their transmitting setups, but are willing to use almost anything for receiving. The frustrations dealt out to other hams trying to contact an "alligator" station is only a small part of the problem. In their frustration of receiving their own downlink signals poorly or not at all, a lot of hams will increase their power output, figuring that will make their downlink signal stronger.

OSCAR 10 is very different from the HF bands, and more power will *not* get you more received signal. Instead, the super strong stations, certainly those stronger than the beacon on the downlink, just compress the spacecraft AGC, reducing the spacecraft receive sensitivity and blocking out the lower power users, ie, desensing the transponder.

Is there a solution? There is a very effective remedy, which also happens to be very simple: effective receiving equipment. Most significant of all, achieving effective receiving is less expensive than the increased power.

So then how do you get an effective receiving capability? Obviously, there is no one answer. A lot depends upon the individual circumstances. There are many little things that will improve almost every station that doesn't have first-class receiving capabilities.

These improvements can be made and tested without ever transmitting any signals. The OSCAR 10 beacon is just the reference tool that is needed. The amateur station that is going to succeed in enjoying OSCAR 10 will be able to measurably receive the beacon up to 15 to 20 dB out of the noise. The following suggestions should be helpful in achieving these goals:

1) *Use switchable, circularly polarized, gain antennas.* Most of the satellite's antennas are circularly polarized, a 3-dB gain over a linear antenna. More importantly, fading will be drastically reduced, as will be spin modulation. Using linearly polarized antennas on OSCAR 10 may be a disappointing experience. Spin modulation, a form of very fast fading, will make the satellite signal sound like it's coming from the whirling blades of a helicopter. The ability to switch circular senses (from left hand to right and vice versa), from inside the shack, can also be a valuable asset.

2) *Use a 2-meter preamplifier.* One of the big factors in popularizing 2-meter sideband and CW in the past few years has been the advent and proliferation of commercially made multimode transceivers. Drawing from their experience in designing HF sideband equipment as well as VHF FM gear, manufacturers have combined the two, yielding transceivers with few compromises. One of the compromises does affect satellite users, though. The receiver front ends are designed with the best attributes of FM transceivers destined for use in a highly populated area, such as Japan. The receivers are very resistant to overload on the FM channels from both nearby stations

and not be bothered by the *content* of what is being said, thus achieving full break-in voice operation.

Although headphones can be used to enhance any weak-signal work, they are mandatory for AO-10 SSB operation. If you don't use headphones, you'll have problems with the receiver audio getting into the transmitter. The result can be anything from lousy, booming transmitted audio to a screeching, full-feedback oscillation through the satellite. Please use your headphones.

An added bonus of lightweight headphones that do not block off all external sounds is that you can hear your own acoustic voice, as well as your voice as transmitted through the satellite. You can readily tune your *transmitter* for the same voice pitch from both sources, thus ensuring that you are on "your frequency" and that any other station that hears you will be able to make contact without chasing you up and down the passband. The message here is that once you have a QSO established, *hold your receive frequency fixed* and make any Doppler adjustments with your transmitter VFO.

OSCAR 10 Mode L Operation

A great deal has been said about operating the OSCAR 10 Mode B transponder, and not very much about Mode L. The supporters of Mode L may feel they are being slighted, but that is not really the case. Nearly everything noted about the operating styles needed for Mode B are applicable for Mode L, save that there are currently a lot fewer L operators than B operators.

Mode L satellite operation is the real future for the satellite service, as there is much more frequency space to be used. OSCAR 10 Mode L is currently configured for a passband of 800 kHz, equivalent in space to the entire 15- and 20-meter bands combined. Only a minor electronic fault on OSCAR 10 has held back Mode L, as that fault requires quite a substantially greater uplink ERP than for Mode B.

As this book is going to press, the worldwide AMSAT engineering team is assembling a Phase III-C spacecraft, a structural twin to OSCAR 10, with electronic improvements. It will have features well beyond OSCAR 10. It will carry both Mode J and Mode L transponders, and will feature a new combination called Mode JL. This new system may well be the mainstay operating system.

Amateurs around the world becoming proficient in Mode B operations are really preparing themselves for using a superb Mode JL transponder in the future. One feature will be different from Mode B. The terrestrial noise that can be quite troublesome on Mode B is absent from Mode L, and it is a real pleasure to operate. If you have a good 70-cm transceiver and antenna in use for Mode B, try adding a mast-mounted GaAsFET preamp and listen to the current Mode L—you will be pleasantly surprised.

FUJI

Operations on the new Japanese OSCAR have commenced with outstanding success. Thousands of users, newcomers and veterans alike, have taken to the air to try out the latest

and from out-of-band commercial signals. Unfortunately, the noise figure of these transceivers is less than optimum. They have been measured as having noise figures between 5.5 and 10 dB. Add a couple of dB of feed-line loss to this, and your 2-meter downlink capabilities are severely limited. The solution is to use a low-noise preamplifier.

The ARRL Handbook and the various ham magazines offer simple designs, and inexpensive units are commercially available. The low-noise GaAsFET type is recommended. The caveat regarding transmitting into the preamp applies here; you will be guaranteed to be changing GaAsFETs if you transmit without proper protective devices. The preamp should be connected to the transceiver so it is only in the receive RF line. Some manufacturers have provided an alternative: preamps with built-in switching to automatically bypass themselves on transmit. Mounting the preamp at the antenna is a must (unless very short feed-lines are used) as it will circumvent the feed-line's contribution to receive noise figure; therefore, it's the recommended approach.

3) *Minimize system losses.* As mentioned, any losses between the receiver front end and the antenna degrade system performance. Transmission-line losses and connector losses all reduce effectiveness. Perhaps the soundest investment a ground-station builder can make is in quality, low-loss coaxial line and high-quality connectors. Properly installed and weatherproofed, an installation like this should provide years of trouble-free and satisfactory service. Make good, clean connections to the antennas and weatherproof them with CoaxSeal™ and vinyl electrician's tape overwrap. Anchor the feed lines to solid tie points to prevent placing undue strain on the cables. UHF connectors (PL-259/SO-239) have no place at 70 cm; use N connectors or other constant-impedance connectors. They're also worth using at 2 meters, too. Don't use cheap relays; they are false economy.

4) *Use az-el mounts for gain antennas.* Make sure that you can aim your antennas accurately to the desired points in space. This requires that you have the necessary controls, equipment and knowledge of the antenna-aiming process to be certain of the entire operation.

5) *Have accurate orbit information.* Whether you are using a plotter, a calculator, a computer printout or a computerized aiming system to point your antennas, use the most accurate information that you can get. This is even more acutely needed, since the antenna directivity increases correspondingly as the gain increases. Phase II low-orbit satellites are somewhat dependent on solar influences, so don't rely on year-old data. Phase III satellites are more tolerant of errors of a few minutes, even to an hour.

6) *Calibrate your receiver S-Meter.* You must be able to measure the strength of the OSCAR 10 beacon above your local noise level; it should be at least 10-dB S/N ratio on the meter. This will assure you that your receive sensitivity is adequate and that you know the beacon strength to properly judge your own signal strength to not exceed that of the beacon. If fact, if you are receiving the beacon strongly (12-15 dB above noise), then try even lower ERP. You do not have to be as strong as the beacon to have good communication.

7) *Use a receiver with a good noise blanker.* Two-meter terrestrial noise levels, those characteristic of civilization, can be abominable and raise havoc with reception, making OSCAR 10 operations less than enjoyable. Unfortunately, too many of the multimode VHF transceivers have inadequate noise blankers, while their HF counterparts have very good ones. One solution is to use the HF receiver and a 2-meter receive converter for OSCAR 10 reception.

8) *Use a reliable means for adjusting transmitter output.* As mentioned, nobody is pleased if you overload the satellite. On most modern SSB rigs, turning the microphone gain up or down is a suitable way to adjust output power. Most of these rigs also have controls for adjusting output power. Monitor your power levels as well as your returned signal strengths and adjust accordingly. You will be surprised as to how well low-power levels communicate through OSCAR 10. Again, you do not have to be as strong as the beacon to have excellent communications.

offering in amateur satellite technology. Known as JAS-1 prior to launch, this new satellite now follows Eastern and Western traditions by being named Fuji-OSCAR 12 (FO-12). Users of FO-12 have found it sensitive and fun. Its Mode JA transponder (Mode J-analog voice) is easily accessed using 10- to 100-watt ERP levels when transponder loading is low. The most frequent comment by users is about the high Doppler shift and rapid antenna pointing required to stay on target, reminiscent of OSCAR 8. Many veterans of OSCAR 8 Mode J, who have been only sporadically active on AO-10, are having a great deal of fun with FO-12.

FO-12 Operating Frequencies and Procedures

JARL and JAMSAT have provided the following specifications for FO-12. The uplink for FO-12 Mode JA is 145.900 through 146.000 MHz. The corresponding downlink band is 435.900 through 435.800 MHz. Like Phase III satellites, this is an inverting, linear transponder. Operating frequencies for Mode JA, used for SSB and CW, may be determined as follows: The sum of the uplink and downlink frequencies equals 581.800 MHz exclusive of Doppler shift. For example, if the uplink is 145.920 MHz, the downlink will be 435.880 MHz. Doppler shift can be as high as plus or minus 5 kHz on the downlink passband and 8 kHz on the beacon. The beacon frequency on-orbit has been measured at 435.797 MHz. The downlink power is up to 2 watts PEP EIRP. The downlink antennas are LHCP polarized while the uplink antenna is linear. About 100 watts ERP or less is the recommended uplink.

The JARL band plan is very much in accord with previous OSCARs, and recommends CW users in the lower third of the passband from 435.800 through 435.835 MHz. SSB operators should use the upper third from 435.865 to 435.900 MHz. Mixed SSB and CW are welcome in the center third. To maximize passband usefulness, JARL recommends you control your uplink frequency to maintain a constant downlink frequency as heard at your QTH. The reasons for this are fairly subtle, but seemed to work on OSCAR 8 Mode J.

Differences in Doppler shift between occupants of the passband can cause some occupants to move across the downlink passband faster than others. The result is that "collisions" between QSOs are more likely than on other current satellites. Consequently, it is recommended that the same procedure used on OSCAR 8 Mode J and OSCAR 10 should be used on FO-12 Mode J. That is, once you have found a clear spot in the downlink passband, try to remain there by adjusting your uplink frequency *only*.

Satellite DXing

DX on OSCAR 10 is a strange affair. The ham just across town is nearly as far away as the ham across the ocean. In both cases, the signals need to go up to and back from the satellite 36,000 km away. Nevertheless, there are a great many traditional DX countries represented on OSCAR 10, as shown in Table 13-9. These hams are all quite good operators, and

Table 13-9

Satellite DX Countries

(Courtesy of ASR No. 110/111)

According to Ed Steeb, WA2RDE, AMSAT DX Bureau Manager, DXCC countries heard and/or worked on OSCAR 10 is reaching impressive numbers. Here is the late 1985 tally of countries and prefixes.

A5	GW3,GW6,GW8	OA4	V3
A71	H44	OE1,OE3-8,E9	VP2D (Dominica)
A92	HA,HG2,HG5	OH2,OH5,OH7	VP2E (Anguilla)
BV	HB9	OHØ	VP2V (Br VI)
BY1	HBØ	OK1,OK3	VP5D (Turks)
C3Ø	HC1,HC2	ON1,ON4,ON6,ON7	VP8 (Falklands)
C6A	HC8	OX3	VP9
CE	HI8	OY9	VS6
CEØA	HK4	OZ1,OZ5,OZ6	VU
CO2	HL1,HL3	P29	XE1,XE2
CN8	HZ1	PA2,PA3,PA8,PE1	Y24
CT1	IØ-I8,IT9	PJ2	YB
CX	IKØ,IK1,IK6,IK8,	PJ7	YJ8
DB,DC,DD,DF,DG,	IC8,IN2-3	PY2,PY6	YO2,YO3
DJ,DK,DL	IV3,IW1-6	PZ1	YU2,YU3,YU6
DPØ	IS	SK1,SK5,SK6,SKØ	YV4,YV5
DU2,DU6	J37	SM1-3,SM5,SM7,	ZD8
EA1,EA3,EA5	JA1-4,JA6-Ø,JC,	SMØ	Z25
EA8	JE1-3,JE6-7,	SP9	ZF1
EI1,EI6,EI9	JF1-2,JF6-7,	SU	ZK1,ZK2,ZK2N,
F1,F2,F5-6,F9	JG1-2,JH1-4	SV1,SV7	ZK2I,ZK2U,ZK2E
FC,FD,FE	JH7-8,JK8	T29	ZL1-4
FT8	JD1 (Ogasawara)	TF	ZL (Kermadec)
FC	JY1	TI2	ZS2,ZS3,ZS5,ZS6
FS7	K,W,N,AA-AL	TR8	ZR1-3,ZR6
FK2,FK8	KG4	TU2	3A
FM5,FM7	KH2,KG6	TZ6	3D2
FOØ	KH6,WH6	UA1,UA3-4	4U (ITU Geneva)
FO8	KL7	UK5	4U (UN)
FR7 (Reunion)	KP2,KV4	UB5	4X4
FY7	KP4	UC2	5B4
G1,G3-6,G8,GØ	KP5 (Desecheo)	UL7	5N8
GD4	LA1,LA3,LA5-6,	UM8	6W
GI1,GI4,GI6,	LA8-9	VE1-8,VO1,VY	7P8
GI8	LU1,LU2,LU4-5	VK1-5,VK7-8	8Q7
GJ6,GJ8	LU7-9	VK9 (Willis)	9H1
GM1,GM3,GM4	LX1	VKØ (Heard)	9M2
GM6-8	LZ		9Y4
GU6			

in many cases don't operate HF (be it because of license class or preference). This, in turn, means QSOs of a very different flavor than many you might find on HF.

Indeed there are some hams who have been so conditioned to HF operations that at first they just send QRZ, signal report and 73. They quickly find out that they will exhaust the available operators to talk to. Moreover, there is no need for any pileups, since the band conditions don't change, and the satellite will be there the next day, too. The ham who just engages in signal-report exchanges on OSCAR 10 might lose interest rapidly because he is not prepared to communicate fully in the manner that is available. QSL cards from a selection of some of the stations operated on OSCAR 10 are shown in Fig 13-20.

Another form of DX operating is exchanging international grid squares with other stations. For more information on grid squares, see Chapter 12. Essentially, working grid squares is an activity somewhat like county hunting, only based on geographical rather than political boundaries. Fig 13-21 shows US and European grid-square maps.

OPERATING AWARDS

A good many of the conventional operating awards (typically with specific satellite endorsements) are available

to satellite operators. In addition, there are some very fine awards available only for satellite operations.

For openers, there is a nonendorsable ARRL DXCC satellite award that is highly prized. Originally only for low-altitude satellites, this award has been extended for contacts on *all* amateur satellites. Another traditional ARRL award is the Worked All States (WAS) via satellite. The rules for this award are the same for all other forms of WAS, and can be found in Chapter 8.

Also available through the ARRL is the IARU-sponsored Worked All Continents (WAC) award with a satellite endorsement. Amateurs who work and confirm all six continents are eligible. See Chapter 8 for further details, and Chapter 17 for sample awards application forms.

AMSAT offers some interesting and challenging awards. You can start with the Phase 3 Satellite Communicators' Club Certificate (SCC), which requires only that you report a successful two-way contact, through OSCAR 10, to AMSAT SCC Manager, PO Box 27, Washington, DC 20044. No form is needed; just send the information about the QSO along with an SASE and $1 ($2 for AMSAT nonmembers).

A unique award just for the listening is also offered by AMSAT in the form of the ZRO Memorial Station Engineering Award (in memory of K2ZRO). This is a test of

Fig 13-20—OSCAR 10 operating brings many enjoyable DX contacts. (*WD4FAB photo*)

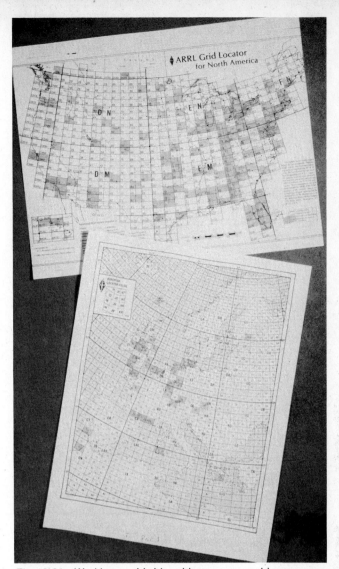

Fig 13-21—Working worldwide grid squares provides many QSOs on OSCAR 10. (*WD4FAB photo*)

your operating ability and your station receiving capability, and is conducted by a control station sending random numeric code groups at sequentially reduced levels of power, in 3 dB steps. These signals are referenced to the satellite's beacon level, and when the test signals reach -21 dBb (21 dB below the beacon), you will know that your abilities are being challenged, as the noise floor of the satellite transponder is right there for you to copy also. These tests signals have gotten to as low as -24 dBb (Z-8) which is pretty difficult copy. All of this ZRO test activity is without your transmitting any signals. Details are available from AMSAT ZRO Test, PO Box 177, Warwick, NY 10990; please enclose an SASE.

Another trio of challenging AMSAT operating awards, each with progressive difficulty, are the ASMAT OSCAR Award (AOA), the OSCAR Sexagesimal Award (OSA) and the OSCAR Century Award (OCA). They are presented for 20, 60 and 100-QSO elements respectively. The counting unit is defined as confirmed satellite contacts with different states, Canadian call areas or countries. A service charge of $3.50 per certificate ($5 for AMSAT nonmembers) should be sent with the cards and return postage to the AMSAT Awards Manager at the AMSAT Washington, DC, address. Endorsement stickers are available for 10-QSO-element intervals between awards at no extra charge, save for an SASE.

CQ Magazine offers a pair of very difficult awards via satellite. These are the *CQ* CW DX Award with OSCAR endorsement, and the *CQ* SSB DX Award with OSCAR endorsement. Each basic award is for confirmed QSOs with 100 countries via satellite, with endorsements for additional countries in increments of 50. See Chapter 8 for further details on *CQ* awards.

The Northern Alberta UHF Society offers a fine-looking certificate called the VE Satellite Award. This is available for W/VE amateur stations who submit QSL cards confirming satellite contacts with eight different Canadian call areas (VE1,2,3,4,5,6,7,8,0, VO1,2, and VY1). DX stations, including KH6, must contact only four different call areas. Application fee for this award is $1 or 4 IRCs for DX stations. Extra postage should be included for registered return mail of cards and certificate. Submit cards and certificate request to Ray J. Nadeau, VE6SF, Committee Chairman, PO Box 52, Barrhead, AB T0G 0E0, Canada.

Chapter 14

Emergency Communications

People in desperate need of help can depend on Amateur Radio to serve them. Since 1913, ham communicators have been dedicated volunteers for the public interest, convenience and necessity by handling free and reliable communications for people in disaster-stricken areas until normal communications are restored.

Emergency communication is an Amateur Radio communication directly relating to the immediate safety of human life or the immediate protection of property, and usually concerns disasters, severe weather or accidents. The ability of amateurs to respond effectively to these situations with emergency communications depends on planning and training of operators.

PLANNING

More than any other facet of Amateur Radio, emergency communications requires a plan—an orderly arrangement of time, talent and activities which insures that performance is smooth and objectives are met. Basically, a plan is a method

The Monsanto Amateur Radio Association operated from this command post to assist community-wide agencies in evacuation and disaster-relief efforts in the Miamisburg, Ohio, area, following a train derailment and toxic-chemical spill. WA8SPN (left) and N8DNG were among the 357 amateurs from a variety of neighboring clubs and ARES groups who volunteered their services during the emergency. (Curtis Cobbler photo)

of achieving a goal. A lack of an emergency-communications plan could hamper urgent operations, defer crucial decisions or delay critical supplies. So plan now to analyze what emergencies are likely to occur in your area, develop guidelines for providing communications after a disaster, know the proper contact people and inform local authorities of your group's capabilities.

Start with a small plan, such as developing a community-awareness program for severe-weather emergencies. Next, test the plan a piece at a time, but redefine the plan if it isn't satisfactory. Without a great deal of risk, testing with simulated-emergency drills teaches communicators what to do in a real emergency. Finally, prepare a few contingency plans just in case the original plan doesn't work.

Large cities usually have capable relief efforts handled by paid professionals, and there always seems to be some equipment and facilities that remain operable. Even though damage may be more concentrated outside a city, it can be remote from fire-fighting or public-works equipment and law-enforcement authorities. The rural public then, with few volunteers spread over a wide area, may be isolated, unable to call for help or incapable of reporting all of the damage. Thus, communications for city or rural emergencies each require a particular response and careful planning.

It's futile to look back regretfully at past emergencies and wish you had been better prepared. Prepare yourself *now* for emergency communications by maintaining a dependable transmitter-receiver setup, an emergency-power source, having a plan and learning proper procedures.

PROCEDURES

Besides having plans, it is also necessary to have procedures—the best methods or ways to do a job. Procedures become habits independent of a plan when everyone knows what happens next and can tell others what to do. Actually, the size of a disaster affects the size of the response, but not the procedures. Before disasters occur, there are many existing procedures; for example, how to correctly coordinate or deploy people, equipment and supplies. There are procedures to use a repeater and an autopatch, to check into a net and to format or handle traffic. Because it takes time to learn activities that are not normally used every day, excessively detailed procedures will confuse people and should be avoided.

Command Post and Emergency Operations Center

Amateur Radio emergency communications frequently use the combined concepts of a command post and an operations center. See Fig 14-1. Although a command post (CP) controls

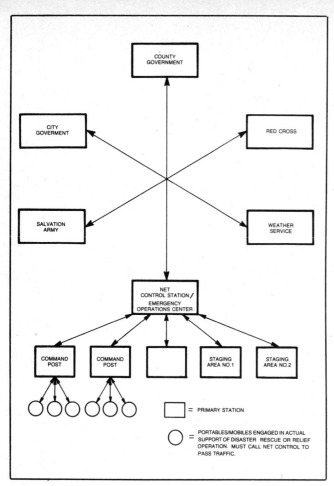

Fig 14-1—The interaction between the EOC/NCS and the command post(s) in a local emergency.

Works) staff. If the status of an accident changes—for example, a car hits a utility pole which later causes a fire—the command post keeps control until the support agencies arrive and take over their specific responsibilities: Injuries—medical; fires—fire department; disabled vehicles—tow truck; and utility poles—utility company. The EOC gets updated through communications from the CP, must anticipate needs to supply support and assistance, and may send more manpower and equipment to a staging area to be stored where it can be available almost instantly. By being outside the perimeter of dangerous activities, the EOC can use the proper type of radio communications, concentrate on gathering data from other agencies and then provide the right response.

The success of the CP and EOC procedures of directing requires communication. Whether there is a minor vehicle accident or a major disaster operation, such communications are important to understand because in many cases they are handled by Amateur Radio. The CPs, who request action and provide information, have duties like net participants checking into an amateur net with emergency or priority traffic. The EOC, who coordinates relief efforts, has functions similar to an amateur net control station (NCS). The effectiveness of the amateur effort in an emergency depends mainly on radio-communications discipline.

Tactical Traffic

Tactical traffic is first-response communications in an emergency situation involving a few people in a small area. It may be urgent instructions or inquiries such as "send an ambulance" or "who has the medical supplies?" The 146.52-MHz calling frequency or 2-meter FM repeater and net frequencies (see Fig 14-2) are typically used for tactical communications because compatible mobile, portable and fixed-station equipment is so plentiful and popular. Tactical traffic, even though unformatted and seldom written, is particularly important in localized communications when working with government and law-enforcement agencies. In some relief activities, tactical traffic takes the form of operational or administrative traffic. Use the 12-hour local-time system for time and dates when working with relief agencies, because most may not understand the 24-hour or UTC systems. Another way to make tactical traffic clear is to use tactical call signs.

initial activities in emergency and disaster situations, the CP may be unapparent because it is self-starting and automatic. Consider an automobile accident where a citizen or an amateur, first on the scene, becomes a temporary command post to call or radio for help. A law-enforcement officer is dispatched to the accident scene in a squad and, upon arriving, takes over the CP tasks. The command post's general procedure is to assess the situation, report to a dispatcher and ask for equipment and manpower. Incidentally, a command post may expand into multiple command posts or move to accommodate situations. Relief efforts, like those in this simple example of an automobile accident, begin when someone takes charge, makes a decision and directs the efforts of others.

The dispatcher at the Emergency Operations Center (EOC) responds to the CP by dispatching equipment and manpower: More police squads, an ambulance, a fire truck, tow truck, utility crew or DPW (Department of Public

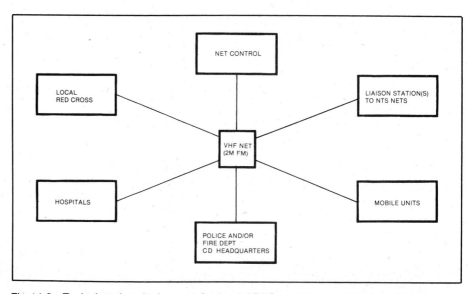

Fig 14-2—Typical station deployment for local ARES net coverage in an emergency.

Tactical call signs are words that describe a function, location or agency, and their use prevents confusing listeners or agencies who are monitoring. When operators change shifts or locations, the set of tactical calls remains the same. Amateurs may use tactical call signs such as "parade headquarters," "finish line," "Red Cross," "Net Control" or "Weather Center" to promote efficiency and coordination in public-service communication activities. However, tactical call signs are not substitutes for station call signs when fulfilling the identification requirements of §97.84 of the FCC rules; amateurs must identify their station operation with its FCC-assigned call sign at the end of a transmission or series of transmissions and at intervals not to exceed 10 minutes during a transmission.

Formal Message Traffic

Formal message traffic is long-term communications that involves many people over a large area, generally uses standard ARRL-message format and is handled on well-established National Traffic System (NTS) nets primarily on 75-meter SSB or 80-meter CW and 2-meter FM. See Fig 14-3. [In addition, there is much NTS net activity on 40 and 20 meters, including regular liaison to the International Assistance and Traffic Net, IATN, which meets on 14.303 MHz daily at 1100 UTC, to provide international traffic outlets, as suggested in Fig 14-4.] Formal messages are used for severe weather and disaster reports. These radiograms, already familiar to many agency officials and to the public, avoid message duplication while insuring accuracy from origination to delivery because they are usually written. When accuracy is more important than speed, getting the message on paper before it is transmitted is an inherent advantage of formal traffic. A message should be read to the originators before sending it, since they are responsible for its content.

The General Plan

Whether traffic is tactical or by formal message, the operations success will depend on the use of a few well thought-out plans and procedures. Before an emergency occurs, register your station with your local ARRL Emergency Coordinator (EC), to be discussed in detail later. The EC will explain to local civic and relief agencies in your community what the Amateur Service can offer in time of disaster.

Determine how to effectively activate your group. But be care-ful; a telephone alerting "tree" call-up, even if based on a current list of phone numbers, might fail if there are gaps in the calling sequence, members are not near a phone or there is no phone service. However, alerting tones and frequent announcements on a well-monitored repeater can round up many operators at once. Or you could set aside an unused 2-meter simplex frequency for alerting; instead of turning radios off, your group would monitor this frequency for alerts without the need of any equipment modifications. Since this channel is normally quiet, any activity on it would probably be an alert announcement. See accompanying sidebar on the alert-frequency concept.

During an emergency, report to the EC so that up-to-the-minute data on operators will be available. Don't rely on one leader; everyone should keep handy an emergency reference list of relief-agency officials, police, sheriff and fire departments, ambulance service and NTS nets. Be ready to help,

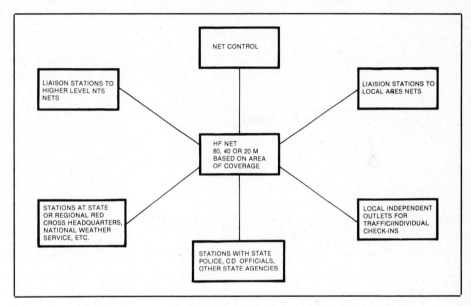

Fig 14-3—Typical structure of an HF network for emergency communications.

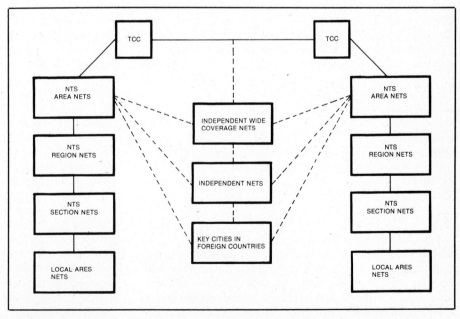

Fig 14-4—Emergency communications through liaison to NTS for handling formal message traffic.

Board Adopts Alert Frequency Concept

At their Annual Meeting in April 1983, the ARRL Board of Directors accepted an Emergency Communications Advisory Committee (ECAC) recommendation concerning the AB1Z Eastern Massachusetts emergency alerting plan, directing the General Manager to implement this recommendation. Then New England Division Vice Director K1PAD summed up the essence of the unique alerting plan:

"The method is infinitely simple and requires no modification for anyone with a 2-meter FM rig. The idea is this. Set aside a simplex frequency and, when you aren't using your rig for anything else, monitor this frequency. There will be no two-way communication on the frequency. In the event of an emergency during odd hours, people would be alerted by an alert transmit station and told which repeater to go to for more information. A group of alert transmit stations have been trained and would retransmit the alert right on the simplex frequency to spread the message over an increasingly large geographical area.

"The beauty of this system is that absolutely no hardware modifications or additions are necessary and people can participate immediately. The only requirement is the discipline to dial up the alert frequency instead of turning the rig off. Not having two-way communications on the frequency keeps it quiet and also puts into action the only proven way to cure any malicious interference that might occur—and that is 'dead silence.' This technique has been used a number of times in Eastern Massachusetts, most notably during a huge fire in Lynn, Massachusetts (see Feb 1982 *QST*, pp 83-84). I know the system works."

Readers of the Lynn fire article will remember that the Eastern Massachusetts ARRL leadership decided to use 145.695 MHz as their alerting frequency (for reasons explained below). However, the League's Emergency Communications Advisory Committee chose not to designate an official frequency for this emergency alerting plan, preferring to let the decision be made at the local level. The ECAC recommendation (as submitted by Chairman W5GHP) which the Board adopted, is as follows:

"The committee agrees with the AB1Z emergency alerting plan. This plan is very easy to implement and can be used in its present form, even though more advanced systems may be in the planning stage or even the hardware stage. No special equipment is required. Concerning the recommendation of a specific frequency, it is felt that this should be a consideration of the local ARRL Emergency Coordinator in cooperation with the ARRL Section Emergency Coordinator. The ECAC therefore recommends that as much publicity as possible be given to this plan, and that information on it be published in all manuals concerned with emergency communication."

AB1Z, the originator of this idea, has generated an extensive amount of documentation on the concept. Here's a portion of his explanation:

"The alert frequency is a new kind of OFF position for any 2-meter transceiver or scanner. The designated simplex frequency is the place to leave the knobs when you want silence. This frequency is reserved for an emergency announcement when hams are urgently needed to provide communication. The emphasis is on speed; it will bring you the word hours sooner. The alert frequency can also be used to transmit a warning when a major emergency is imminent; that could give you just the time you need to protect yourself and your family.

"Like tone signal systems, the alert frequency is a selective alert system. The receiver stays silent except when there is an alert. Unlike tone systems, the alert frequency doesn't require any hardware. No black boxes need to be built or hooked up. The frequency itself is the selective feature. This means that any ham with a synthesized rig or scanner can 'install' it immediately just by punching it in. That's what this concept is all about.

"The alert frequency idea grows out of the Eastern Massachusetts Plan's philosophy of reliability through redundancy. No single person or piece of equipment is essential for its success.

"The alert frequency isn't intended to be the only method of getting hams in an emergency. Many other methods may be appropriate. The alert frequency fills in the gaps left by the limitations of other methods. Its virtue is that it can reach many hams outside the hours when repeaters are active, without requiring any special hardware.

"It's not a distress frequency. The addressee for a distress call is not the ham community, but a public-safety

---

but stay off the air unless there is a specific job to be done which you can efficiently handle. Always listen before you transmit. Work and cooperate with the local civic and relief agencies as the EC suggests; offer these agencies your services directly in the absence of an EC. During a major emergency, copy special W1AW bulletins for the latest developments.

Afterward, let your EC or Net Manager know about your activities, so that a timely report can be submitted to League HQ. Amateur Radio has won glowing public tribute in emergencies; help maintain this record.

For a comprehensive study of emergency communications and training, see the ARRL *Emergency Coordinator's Handbook* (FSD-9). Contact the ARRL HQ Field Services Department for details on how to obtain a copy. "Experience is the worst teacher; it gives the test before presenting the lesson" (Vernon Law). Train and drill *now* so you can be prepared for an emergency.

Training

Amateurs need training in operating procedures and communications skills. In an emergency, radios don't communicate, but people do. In an emergency, amateurs with all sorts of varied interests participate, and many of those who offer to help may not even have experience in public-service activities. Rarely are there enough trained operators, especially if a crisis persists for a long time.

Proper disaster training replaces chaotic pleas with organized communications. Well-trained communicators respond during drills or actual emergencies with quick, effective and efficient communications—that is "professionally." Each understands his or her role in the emergency-communications plan and sets a good example by knowing the proper procedures to use. The ARRL recognizes the need for emergency preparedness and emergency-communications training through sponsorship of the Amateur Radio Emergency Service (ARES).

Whether training takes place with programs at club meetings, on-the-air or with a personal approach, the basic subjects should cover emergency communications, traffic handling, net or repeater operation and technical knowledge. Packet radio, one powerful technological tool, can be used for traffic handling; see Chapter 10. Prepare and edit messages off line as text files which can then be sent error-free in just seconds, an important time-saver for busy traffic channels. Public service agencies are impressed by fast and accurate printed messages. Get as many people involved as possible to learn how emergency communications should be handled. Explain what's going on, assign each participant with a useful role, and who knows? A few students might be encouraged to become active members of the local ARES/NTS groups.

Practical on-the-air activities, such as the ARRL's Field Day and Simulated Emergency Test (specific rules for which are published in *QST*) offer training opportunities on a nation-

agency. To reach them, we have autopatches...Metropolitan areas usually have many repeaters within range; most times it's possible to raise someone on at least one of them. Remember, a call on the alert frequency may wake hundreds of people, so using it should be a last-ditch response to a truly desperate situation.

"It's not an emergency operating frequency. The actual operation takes place on one of the many available repeaters, standard simplex frequencies or RACES frequencies.

"It's not an emergency check-in frequency. The alert announcement specifies the frequency to check in on. The alert frequency goes silent again as soon as the alert announcement is finished, in case a further alert might be needed later.

"Obviously, it's not a routine calling frequency. Yes, your friend might be listening. But if you use the alert frequency to call, you'll be discouraging other hams from using it as a way to shut the rig off. That's what we all want to avoid.

"To avoid confusion about any of these points, it's not called an 'emergency frequency.' Its only name is 'the alert frequency.'

"We have many operating frequencies, many methods of calling for help and many frequencies that can be used for calling; but we have only one recognized and reserved alert frequency. Therefore, its usefulness should be protected by using it only for its recognized purpose.

"Basically, just treat the designated simplex frequency as the receiver's OFF position. No changes in operating habits are needed, except to discontinue the practice of shutting off the power. Leaving the offset switch wherever it was for the last repeater used will make sure that any accidental transmission isn't on the alert frequency."

The ECAC and the Board, in adopting this concept in principle for nationwide implementation, did not name any specific 2-meter frequency, therefore leaving that decision to the local emergency communications leadership. However, it may be useful to readers to understand why Eastern Massachusetts amateurs adopted 145.695 MHz. AB1Z addressed this point:

"The basic criterion for a frequency that hundreds of hams will be listening to around the clock is that it should be completely free of anything except high-priority alerts. A channel is needed that has nothing already on it.

"Obviously, repeater channels and FM simplex frequencies can't be used. That leaves out everything above 146 MHz and the inputs and the outputs of the 145 MHz group. We don't want to come into conflict with the active CW and SSB segment at the bottom of the band or its logical expansion space. Scratch 144.9 MHz and below. Satellite frequencies are out. So everything above 145.8 MHz is eliminated. 144.9-145.1 MHz is assigned to FM weak signal work; however, it's also used for simplex operation sometimes. So is 145.52 MHz, which is a simple click down from 52/52 on a lot of rigs. Better rule out everything below 145.6 MHz.

"As for possible conflicts with the regular band plan, 145.5-145.8 MHz is listed as "miscellaneous and experimental modes" in the band plan in the latest *ARRL Repeater Directory*. There isn't a lot of that, and we're out in the middle, away from the frequencies most likely to be used. This type of activity is likely to involve weak signals, which would not ordinarily be detected by typical FM receivers with average antennas and locations. As for interfering with these activities, alert transmissions (including test transmissions) will be extremely rare. Thus, the time taken away from other possible uses is insignificant.

"As a cross-check, several stations, including one at a well-known repeater site, monitored the frequency for a couple of months without detecting any carriers. So we really do have the channel we need. The wide publicity being given to the alert function should also reduce the chance for casual use."

It should be reemphasized that the ECAC and the Board, in adopting this concept, decided to make the specific frequency choice a local option. If such an alerting plan will enhance the emergency response of your local radio amateur organization, then serious consideration should be given to implementing a simplex alert frequency in your area. Further details on this innovative approach to emergency alerting can be obtained by sending a business-sized SASE to John A. Carroll, AB1Z, 25 Evergreen Ave, Bedford, MA 01730 [originally published in July 1983 *QST*].

wide basis for individuals and groups. Participation in such events reveals weak areas where more training is needed. In addition, drills and tests can be designed specifically to check dependability of emergency equipment or to rate training in the local area.

Field Day

The League's Field Day (FD) gets more amateurs out of their cozy shacks and into tents on hilltops than any other event. You may not be operating from a tent after a disaster, but the training you will get from FD is invaluable. In the League Field Day event, a premium is placed on sharp operating skills, adapting equipment that can meet the challenges of emergency preparedness and flexible logistics. Amateurs produce portable stations capable of long-range communications at almost any place and under almost any condition. Alternatives to commercial power in the form of generators, car batteries, windmills or solar power are used to power equipment to make as many contacts as possible. FD is held on the fourth full weekend of June, but FD enthusiasts get the most out of their training by keeping preparedness programs alive during the rest of the year.

Simulated Emergency Test

The ARRL Simulated Emergency Test (SET) builds emergency-communications character. The purposes of SET are to:

1) Help amateurs gain experience in communicating using standard procedures under simulated emergency conditions and to experiment with some new concepts.

2) Determine strong points, capabilities and limitations in providing emergency communications to improve the response to a real emergency.

3) Provide a public demonstration, to served agencies and through the news media, of the value of Amateur Radio to the public, particularly in time of need.

The goals of SET are to:

1) Strengthen VHF-to-HF links at the local level, ensuring that ARES and NTS work in concert.

2) Encourage greater use of digital modes for handling high-volume traffic and point-to-point Welfare messages out of the affected simulated-disaster area.

3) Implement the Memoranda of Understanding between the League, the users and cooperative agencies.

4) Focus energies on ARES communications at the local level. Increase use and recognition of tactical communications on behalf of served agencies and less amateur-to-amateur formal radiogram traffic.

Help promote the SET on nets and repeaters, or at club meetings and publicize it with net bulletins and in club newsletters. SET is conducted on the third full weekend of October; but some groups have their SETs any time during the period of September 1 through October 31, especially if an alternate date coincides more favorably with a planned communications activity and provides greater publicity.

Drills and Tests

A drill or test that includes interest and practical value makes a group glad to participate because it seems worthy of their efforts. Formulate plans around a simulated disaster based on one described later in this chapter such as weather-caused disasters or vehicle accidents. Elaborate on the situation to develop a realistic scenario or have the drill in conjunction with a local event. Many ARRL Section Emergency Coordinators (SECs) have developed training activities that are specifically designed for your state, section or local area. During a drill:

1) Announce the emergency situation. Activate the emergency net. Dispatch mobiles to served agencies.

2) Originate messages and requests for supplies on behalf of served agencies by using tactical communications.

3) Use emergency-powered repeaters and employ digital modes.

4) As warranted by traffic loads, assign liaison stations to receive traffic on the local net and relay to your section net. Be sure there is a representative on each session of the section nets to receive traffic coming to your area.

After a drill:

1) Determine the results of the emergency communications.

2) Critique the drill.

3) Report your efforts, including any photos, clippings and other items of interest, to your SEC or ARRL HQ.

Net Operators

Training should involve as many different operators as possible in net control station (NCS) and liaison functions; don't have the same operator performing the same functions repeatedly. There should be plenty of work for everyone. Good liaison and cooperation at all levels of NTS requires versatile operators who can operate either phone or CW. Even though phone operators may not feel comfortable on CW and vice versa, encourage net operators to gain familiarity on both modes by giving them proper training. The liaison between different NTS region net cycles as well as between section nets are examples of the need for versatile operators. For more information on NTS and traffic handling, see Chapter 15.

ARRL Emergency Communications and Certification Course

The League is presently working on a certification program to provide training and formal recognition of amateur achievement in the field of emergency communications. When available, this course will probably be administered to club members by ECs or club officers (perhaps as an on-the-air activity), and each amateur who demonstrates basic knowledge and completes this training course successfully will be certified.

GOVERNMENT AND RELIEF AGENCIES

Government and relief agencies provide commendable and effective emergency management to help communities in disasters. They must rely on normal communications, but even reliable communications systems may fail, be unavailable or become overloaded in an emergency. In disaster situations, agency-to-agency radio systems may be found to be incompatible. Agency operators might be unfamiliar with the names of officials, special terms or the routine procedures of other agencies and may be reluctant to share their system or use another's system. Fortunately, Amateur Radio communicators can serve and support in these situations, not only to bridge communications gaps with survivable and compatible mobile, portable and fixed stations, but also to supply trained volunteers needed by the agencies for collection and exchange of critical emergency information.

As government and relief agencies are restrained by budget cuts, the general public becomes increasingly dependent on volunteer programs and quickly recognizes the value of efforts of radio amateurs who serve the public interest. This public recognition is important support for the continued existence and justification of Amateur Radio.

By using Amateur Radio operators in the amateur frequency bands, the ARRL has been and continues to be in the forefront of supplying emergency communications, direct to the general public and through various agencies. Increased involvement, coordination and valuable assistance with served agencies is possible, particularly with those organizations

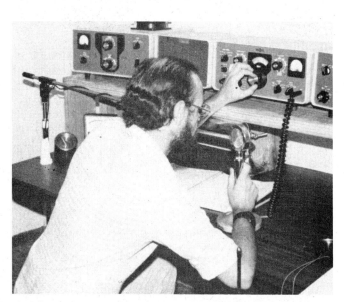

Chuck, NI5I (left), and Rick, WB5TJV, were two of the many hams who provided communications in the wake of the devastating Mexico City earthquake. (WB5TJV photos)

K7UGA Lauds Amateurs In Grenada Emergency

On October 25, 1983, US troops stunned the world when they landed on the island of Grenada. Not since Vietnam had this country sent its soldiers on a major military offensive. And not since the earliest days of wireless had much of the world been riveted on the words of an Amateur Radio operator. That ham, Mark Barettella, KA2ORK, originally of Ridgefield, New Jersey, gave via Amateur Radio an eyewitness account of those dramatic events in the Caribbean.

Mark, then a student at St George's University School of Medicine, was at times the only nonmilitary source of information from Grenada. His live ham-radio accounts also aided US troop movements. He and his radio became the primary link between worried parents and medical students, and between concerned mainlanders and American citizens then on the island. In the process, Mark and his Amateur Radio transmissions captured worldwide attention.

In the November 2, 1983, issue of the *Congressional Record,* then US Senator Barry Goldwater, K7UGA, formally praised the outstanding efforts of KA2ORK and the many other radio amateurs who provided crucial emergency communications during the Grenada operation:

"A Commendation for Amateur Radio Operators

"MR. GOLDWATER. Mr. President, I would like to speak briefly about something this Nation and its people are built on; namely traditions. In Grenada, our Rangers, Marines, and Seal teams along with those of the joint Caribbean force helped a tiny nation to rid itself of what most surely would have been a reign of tyranny and communism. The United States came to the aid of those nations to help them in a time of need and to protect our citizens in a foreign nation. Students who were evacuated from the island of Grenada spoke of our military and our country with patriotic pride—a tradition that in recent years has been missing in the United States. I hope that we can continue to strengthen these same traditions that have made this a great nation.

"Mr. President, with all due respect to our fighting forces and what they have been through in the last several weeks, there are other traditions which perhaps have been less noticed—especially in the case of Grenada—that I do not think should go unnoticed.

"As you and my other colleagues know, my hobby is Amateur Radio. It is not only a hobby, but to most hams as they are known, it is also a service. And it is this tradition of service from the Amateur Radio fraternity that I speak of today. During the first two or three days during which our forces were conducting operations in Grenada, the island was virtually cut off from the outside world communications-wise. Yes, we have spent millions of dollars on communications for our military for use in crisis and wartime situations. However, on this particular occasion, probably the most up to date accounts of what was happening in and around St. George's Medical College area, were given by ham radio operators. Mark Barettella, KA2ORK/J3, and Don Atkinson, J37AH, maintained communications throughout a very critical situation and were, at times, the only sources of information coming from Grenada. Ham radio operators here in the United States monitored frequencies used by Mark and Don and stayed in contact with them night and day. Ham radio operators provided a great service, not only to their government, but also to the people of the United States. Like hams that have gone before them, they have a tradition of service in times of local and national emergencies.

"I think it is fitting today that we should honor these amateurs, and the Amateur Radio community in general, for also being a part of the finest traditions of this country. They are a national resource that we should be proud of and should appreciate.

"Mr. President, in addition, I would like to enter into the record the call signs of at least a few amateurs that we know of at this point: N2DRA, K3RZR, W3DOS, KA3DTE, KC2PK, WD4CNR, KØIND, VE3AJN, K4MM, WA4ZHC, W4PPC, W1ISO, WA4CCK, WD4AIE, WB4CKO, WR4S, WB4FTK, N4GFQ.

"I am sure that there are many others who are known to us and who are not included in this list that should receive our thanks."

---

which the League has signed official Memoranda of Understanding:

- The American National Red Cross
- The Salvation Army
- The Federal Emergency Management Agency (FEMA)
- The National Communications System (NCS)
- The Associated Public Safety Communications Officers, Inc (APCO)

[At this writing, the League is exploring the possibility of Memoranda of Understanding with the National Weather Service (NWS) and the Civil Air Patrol (CAP).]

The American National Red Cross and the Salvation Army voluntarily extend assistance to individuals and families in need as a result of a disaster. Red Cross Chapters, for example, establish, coordinate and maintain continuity of communications during disaster-relief operations, both national and international. These agencies have long recognized that the Amateur Radio Service, because of its excellent geographical station coverage, can render valuable aid to maintain the continuity of communications during disasters and emergencies when normal communications facilities are disrupted or overloaded.

The Federal Emergency Management Agency (FEMA) is a federal agency that supports state and local civil-preparedness and emergency-management agencies. With headquarters in Washington and 10 regional offices through-

out the country, it can provide technical assistance, guidance and financial aid to state and local governments wishing to upgrade their emergency communications and warning systems. Since FEMA is in charge of the RACES program, its recognition of ARRL-sponsored emergency preparedness programs can be a powerful tool in selling ARES capability to local emergency-management officials and in the combined growth of ARES and RACES. The Radio Amateur Civil Emergency Service (RACES) was set up in 1966 as a special phase of amateur operation that provides radio communications for civil-preparedness purposes only during periods of local, regional or national civil emergencies. The RACES regulations make it simple and possible for an ARES group whose members are all enrolled in and certified by RACES to operate in an emergency with great flexibility. During a nondeclared emergency, RACES can operate under ARES. But when an emergency or disaster is officially declared by a state or federal authority, the operation can become RACES using the same operators and the same frequencies. For more information on RACES, contact your State Emergency Management or Civil-Preparedness Office or FEMA.

The National Communication System (NCS) is a confederation of government agencies and departments established by Presidential Memorandum to ensure that the critical telecommunications needs of the Federal Government can be met in a possible emergency, ranging from a routine situation to national emergencies and international crises including nuclear attack, while at the same time achieving the most effective and economical fulfillment of the day-to-day telecommunications requirements. The ARRL Field Organization continues to participate in communications tests sponsored by NCS to study telecommunications readiness in any conceivable national emergency. Through this participation, radio amateurs have received recognition at the highest levels of government.

The Associated Public Safety Communications Officers, Inc (APCO) and amateurs share common bonds of communications in the public interest. APCO is made up of law-enforcement, fire and public-safety communications personnel. These officials have primary responsibility for the management, maintenance and operation of communications facilities in the public domain, and for establishing international standards for public-safety communications, professionalism and continuity of communications through education, standardization and the exchange of information.

[Note: Although there is as yet no national agreement with the Civil Air Patrol, local ARES groups do work with CAP, chartered by Congress as an official auxiliary of the US Air Force. Emergency services, the most critical of the three CAP missions, is entirely dependent on communications for its effective operation. Amateur Radio supplements CAP emergency-services communication. Both amateurs and CAP personnel train in drills and exercises to maintain and enhance communications skills to serve the public.]

Chuck, N6CFQ, relays information on the progress of the St Patrick's Day Parade in Beverly Hills, California. *(WA6VIP photo)*

An increased awareness of radio amateur capabilities has also been fostered by ARRL's active participation as a member of NVOAD (National Volunteer Organizations Active in Disaster). This has increased amateur visibility before the volunteer sector. NVOAD coordinates the volunteer efforts of its 23 member-agencies (the Red Cross and Salvation Army are among its members). ARRL and radio amateurs have continued to accelerate their presence amongst these agencies by grassroots participation at regional conferences. It's this increased recognition of Amateur Radio that enhances the image of our hobby and provides more "customers" for amateurs and our communications skills.

PUBLIC SERVICE EVENTS

A primary responsibility of the Amateur Radio Service, as

established by Part 97 of the FCC regulations, is to render public-service communications to the general public, particularly in times of emergency when normal communications are not available. Progressive experience can be gained by helping with public-service events such as message centers, parades and sports events. Aside from being fun, the skills learned from these activities can later be applied in handling an emergency or crisis.

For Free

Amateurs, according to FCC regulations, must not become involved in moneymaking commercial activities or events whatsoever. Communications provided for the public's benefit, *yes;* for the direct benefit of the sponsor, *no.* The transmission of business communications by an Amateur Radio station is prohibited except for emergency communications. So instead, select public-service events that are held just for fun, or at most, those that raise or pledge funds for nonprofit organizations with no prizes at the finish line other than ribbons, trophies or congratulations.

Message Centers

Message centers are Amateur Radio showcase stations set up and operated in conspicuous places to send free radiogram messages for the public. Many clubs provide such radiogram-message services at shopping malls, fairs, information booths, festivals, exhibits, conferences and at special local events to demonstrate Amateur Radio while handling traffic. Message centers also afford opportunities to train operators, practice handling messages which are similar to disaster Welfare traffic, and show the public that amateurs are capable, serious and responsible communicators. A plan for a message center should include a display, equipment, public relations and operators.

Successful centers include eye-catching organized displays and attractive stations which favorably promote Amateur Radio or lure newcomers to the hobby. Cover a table with bright-colored cloth and match the color to the surroundings. Put out a few radio magazines and hang up some QSLs, maps, posters and operating aids. ARRL HQ can supply free literature, pamphlets, press handouts and various informative videotapes to help make your club display a success.

Equipment should be simple, safe and uncomplicated so as not to overwhelm people. Try only a few of the following: HF stations with CW, SSB, RTTY or ASCII capability, and VHF stations with FM, packet radio, OSCAR or ATV activities. Keep expensive portable radios secure from possible theft, out of reach from visitors or stored under a protected table. To promote a good public-relations image, consider the visitors; use an extension loudspeaker, desk microphone or computer readout of code. Clear, local contacts impress people more than long-distance, scratchy QSOs. Decide on using power from batteries, commercial power or generators, and where to connect a ground. Put the antenna as high as possible. Check prior to the event for receiver hash or public-address system interference. To instill interest, have an easy-to-operate shortwave receiver accessible for spectators to tune and listen. Be sure not to talk to the public using ham jargon.

It's a good idea to make some form of barrier to separate talking visitors and busy operators. But then be sure to have friendly and knowledgeable club members conveniently available to welcome visitors. They should clearly describe to passersby what Amateur Radio can do, answer questions, explain the message process, assist in preparing messages and maybe even invite those interested to the next club meeting. Keep explanations simple, basic and informative; because of the varying degrees of interest the public has, only a few will be interested in specific details. Discuss the variety of Amateur Radio, even if it is not demonstrated. Most visitors will want to know about the role of Amateur Radio in the community, about FCC exams and the license process. Well-planned message centers with properly scheduled operators can handle thousands of messages at one event. Some repeat spectators, address books in hand, excitedly write out bunches of messages while inquisitive crowds of people watch and wait for their turns. Handling messages for the local community is a valuable way amateurs can enhance public recognition to their traditional public-service responsibilites. Another benefit is to keep themselves in practice and their equipment ready for emergencies.

Naturally, before getting involved in message centers you should become familiar with the traffic-handling procedures as described in Chapter 15 (which also contains additional hints on exhibit-station operation). Ask your ARRL Section Traffic Manager (STM) to visit your club to conduct a training seminar and to provide any operating hints pertaining to traffic nets in your section. If no local traffic net exists, your club should consider initiating a net on an available 2-meter repeater. Coordinate these efforts with the STM and the trustee(s) of the repeater you'll use. Encourage club members to participate in traffic-handling activities, either from their home stations or as a group activity from a message center. After all, network discipline and message-handling procedures are fundamental concepts of Amateur Radio operation.

Parades

From a communications viewpoint, parades simply relocate people and equipment, with a few requests for supplies or medical aid along the way; parades simulate evacuations. The first step in handling parade communications is to select a group representative who will contact the parade organizers and officials during the planning stages. This representative can then instruct the ham group and describe what to do for the parade.

As for most public-service events, arrive early and fill vacant positions with standby operators. Wear proper identification, whether it is a special jacket, cap or badge, for official access past police lines. In most parades, the basic order for the day is to provide communications for disoriented people or those needing other assistance; leave parade decisions to officials and first-aid or medical care to Red Cross personnel.

Some operators will be assigned to find parade officials and help them exchange information using hand-held radios. Others work in pairs or teams with the parade marshall and parade specialists like the band, float, vehicle and marcher officials. These officials will be busy once the busloads of uniformed marchers with musical instruments, banners, flags or equipment arrive at the staging area and are merged to make up the parade. Set up net controls at parade headquarters where a telephone is available, at an announcement or information booth or at the staging area near the starting line. Establish a procedure for lost children, monitored by a police liaison, with which parent inquiry and lost-child descriptions can be announced on the net for matching.

Next, after initial parade staging is underway, many other operators have specific on-route assignments. Position hand-held operators at intervals along the entire parade route where assignments can be carried out immediately or questions can

get individual answers. Operators can walk, march or ride along with the parade to report any specific problems, while one amateur on top of a high building can spot general problems. Communicators near loud marching bands will need earphones to hear communications when reporting parade delays, gaps or units out of sequence. Hot weather can keep first-aid volunteers busy with victims of dehydration, sunstroke and heat exhaustion. In some cases, a request for an ambulance, paramedic or water will be needed, and parade floats may need repairs, gas or towing.

Then, at the destaging area, more officials rely on communications to prevent clogging the area and stopping the parade. Large parades can require over 75 amateurs to handle communications, and they must plan and practice many months before the event. As a result, amateurs provide reliable communications to resolve unanticipated problems, handle last-minute changes with immediate action and, despite the excitement of a parade, learn appropriate actions to be applied to next year's parade or in a real emergency.

Sporting Events

Sporting events include all types of outdoor athletic activities where there is competition or movement, or when

N3DHC keeps a close watch on runners as they pass Checkpoint 14.5 during the 1986 Pittsburgh Marathon. *(KC3ET photo)*

participants are timed. Again, it is important to assess the communication needs and to plan with event officials, law-enforcement officers, medical and first-aid personnel, and with fellow amateurs. Know the route and preview the course; everyone should understand their functions. Somewhat like parades, there are strategic locations where sport-event amateur communicators should be positioned to be most useful:

1) Stationed with or shadowing event officials and organizers or law-enforcement officers.

2) At check-in point, staging area or starting line.

3) At water stops, aid stations and checkpoints around the course or route.

4) Near road intersections or sharp turns where safety of crowds or participants is crucial.

5) In the course preview car, pace car, follow-up vehicle or roving sheriff's squad.

6) At the net-control station.

7) Alongside first-aid or Red Cross stations, medical personnel or ambulance.

8) At message traffic centers.

9) Near a telephone or at a home station with a telephone.

10) At the destaging area or finish line.

There are many types of sporting events in which amateurs serve. The list is nearly endless: an air show and balloon race, soapbox derbies, road rallies, Boy Scout hikes, bicycle tours, regattas, golf tournaments, kayak and river-raft races, football games, mini-olympics, Special Olympics, World Olympics, ski-lift operations, slalom courses and cross-country meets.

By far the most popular event is the A-Thon: a walk-athon, marathon, sport-athon, bike-athon or even a triathlon (usually running, bicycling and swimming or canoeing). Long-distance marathons require a large, disciplined force of communicators, as is the case with the New York City Marathon which is served by well over 200 ham radio operators. Good coverage allows operators, for example, to radio the identifying numbers and locations of marathon participants who drop out of the course. This information is posted at a base station to help friends and relatives locate their favorite runners.

Physical endurance is not only for the participants; hams use roller skates, bicycles, golf-carts, motorcycles or cars to make themselves mobile assistants around the routes of sport events. Some communicators cover a race at one particular point on the route until all participants have passed, then they become mobile and leapfrog down the route to repeat the procedure. Roving mobiles, sometimes called roamers, can keep track of everyone's location, determine race miles remaining, warn of potential traffic hazards, find lost people or items, report harassing spectators or post signs for riders or runners. Yet, even with this variety, keep in mind that amateurs *are not* to be used as parking-lot attendants or crowd-control police. We are *communicators*.

Portable operators, those not on wheels, can report medical emergencies, give weather reports and can handle messages for participants or spectators. Separate UHF channels for administrative purposes are useful to get supplies like drinking cups, water or bandages, and to report malfunctioning facilities or request more volunteers. Telephones may be crowded in the area, so first-aid stations or checkpoints with bulletin boards can double as NTS message centers. It then is easy for the NCS to give assignments and tell the officials that the route is ready and participants are set to go.

In major sports events, the Red Cross can dispatch their own first-aid vehicles and keep them near densely populated or hazardous areas for fast accident response. A specific medical frequency might be assigned. Be prepared to report minor medical situations from injuries, blisters and cramps that may need first-aid volunteers with bandages or ice, to serious problems like nausea, unconsciousness or cardiac difficulties, where an ambulance might be called to take the victim to a hospital for examination. If you do call an ambulance, assign someone to flag it down and also to give the driver final directions for finding the victim.

Public-service events are great teachers. Once you gain skill in working with the public in hectic and near-emergency conditions which can occur with a message center, parade or sports event, you're fairly well-prepared and experienced enough to tackle communications for the more difficult natural disaster assignments. An impending natural disaster gives various warning periods, from the slow approach of a severe-weather storm to the sudden devastation of a volcanic eruption or an earthquake, situations where advance planning and training are crucial to the success of emergency communications.

NATURAL DISASTERS

Relentlessly, Nature creates severe weather and natural calamities such as a hurricane, tornado, flood, winter storm, forest fire, earthquake and volcanic eruption that can cause human suffering and create needs which the victims cannot alleviate without assistance. Despite the spectrum of requirements desperately needed to help people in a disaster, it is generally understood and agreed that amateurs will neither seek nor accept any duties other than Amateur Radio communications. Volunteer communicators do not, for example, enforce local laws, make major decisions, work as common laborers or replace contractors, rent generators, tents or lights to the public; nor do they sandbag floods or break ice jams. Instead, amateurs simply handle radio communications.

Most needed during the first few days after a disaster occurs or until normal communications service is restored, Amateur Radio disaster communications range from spotting threatening weather and helping communities prepare for it, to handling Welfare traffic and surveying property damage. Here are several typical categories of amateur disaster communications:

1) Severe-weather spotting and reporting.
2) Evacuating people to safe areas.
3) Shelter operations.
4) Assisting government groups and agencies.
5) Victim rescue operations.
6) Medical help requests.
7) Critical supplies requests.
8) Health-and-Welfare traffic.
9) Property damage surveys and cleanup.

Severe Weather Spotting

Unfortunately, nasty weather hits somewhere every day. Long ago, amateurs exchanged simple information among themselves about the approach and progress of storms. Next, concerned hams phoned the weather offices to share a few reports they thought might be of particular interest to the public. National Weather Service (NWS) forecasters were relying on spotters: police, sheriff, highway patrol, emergency government and trained individuals who reported weather information by telephone. But when severe weather strikes, professional spotters may be burdened with law-enforcement tasks, phone lines become overloaded (slow or no dial tones), special communication circuits might go out of service or, worse yet, there can be a loss of electrical power. Because of these uncertainties, weather-center officials welcomed amateur operators and encouraged them to install their battery-powered radio equipment on-site so forecasters could monitor the weather nets, request specific area observations and maintain communications in a serious emergency.

Those first, informal ham weather nets had great potential to access perhaps hundreds of observers in a wide area for sources of timely on-the-spot reports to supplement forecasters' radar data. Many Meteorologists-in-Charge eagerly began to instruct hams in the types of information needed during severe-weather emergencies and even in radar interpreting. Eventually, Amateur Radio SKYWARN operations developed as an important part of community disaster-preparedness programs. SKYWARN is a plan sponsored by NWS to report and track destructive storms or other severe, unusual or abnormal weather conditions. Accurate observations and rapid communications during extreme weather situations now proves to be fundamental to NWS. Amateur Radio operators nationwide are a first-response group making great contributions of most critical information, reports which are invaluable to the success of an early storm-warning effort.

Weather spotting is popular because the procedures are easy to learn and reports can be given from the relative safety and convenience of a home or an auto. Reports on a severe-weather net are limited to fairly drastic weather data, unless specifically requested by the net-control operator, so most amateurs monitor net operations and transmit only when they can help.

Weather forecasters, depending on their geographical location, need reports on the following during the summer or thunderstorm season:

1) Tornadoes, funnels or wall clouds.
2) Hail.
3) Damaging winds, usually 50 miles per hour or greater.
4) Flash flooding.
5) Heavy rains, rate of 2 inches per hour or more.

During the winter or snow season report:

1) High winds.
2) Heavy, drifting snow.
3) Freezing precipitation.
4) Sleet.
5) New snow accumulations of 2 or more inches.

Here's a four-step method to describe the weather you spot:

1) What: Tornadoes, funnels, heavy rain and such.
2) Where: Direction and distance from a known location; for example, 3 miles south of Newington.
3) When: Time of observation.
4) How: Storm's direction, speed of travel, size, intensity and destructiveness. Include uncertainty as needed—"Funnel cloud, but too far away to be certain it is on the ground."

Severe Weather Alert on Net

The net control station (NCS) using a 2-meter repeater directs and maintains control over traffic being passed on the Weather Net, collates reports, relates pertinent material to the Weather Service and organizes liaison with other area repeaters. Priority Stations, those that are assigned and use tactical call signs, may call any other station without going through net control. The NCS should start the net if he/she hears a NOAA (National Oceanic and Atmospheric Administration) radio alert or upon request by NWS or the EC. The NCS should keep in mind that the general public or government officials might be listening to net operations with VHF scanners. Here are some guidelines an NCS might use to initiate and handle a severe-weather net on a repeater:

1) Activate alert tone on repeater.
2) Read weather-net activation format.
3) Appoint backup NCS to copy and log all traffic and to take over in the event the NCS goes off-the-air or needs relief.
4) Ask NWS for the current weather status, location and

expiration time, and specific information needed.
5) Check in all available operators.
6) Assign operators to priority stations and liaisons.
7) Give severe-weather report outline and updates.
8) Be apprised of situation and assignments by EC.
9) Periodically read instructions on net procedures and types of severe weather to report.
10) Acknowledge and respond to all calls immediately.
11) Require that net stations request permission to leave the net.
12) During periods of inactivity and to keep the frequency open, advise stations that may have just started listening to the repeater that a net is in progress.
13) Close the net after operations conclude.

At the Weather Service

The NCS position at the Weather Service can be manned by the EC and other ARES personnel. Operators assigned there must have with them 2-meter hand-helds with fully charged batteries capable of operating on the repeater frequency. The station located at the NWS office can be connected to similarly equipped positions located at the Emergency Government Office, Red Cross and the sheriff's department with an off-the-air intercom system which allows some traffic handling without loading up the repeater. Provide a supplementary radio channel of communications in anticipation of an overload or loss of primary communications circuits.

In the event traffic is flowing faster than can be easily written and digested, NWS personnel may request that a hand-held radio be placed at the severe-weather desk so that forecasters can monitor incoming traffic directly. Nevertheless, all traffic should be written on report forms.

If a disaster should occur during a severe-weather net, communications get equal priority. Shift to disaster-relief operations if needed during or after a severe-weather alert.

Repeater Liaisons

Assign properly equipped and located stations to act as liaisons with other repeaters. Two stations should be appointed to each liaison assignment. One monitors the weather repeater at all times and switches to the assigned repeater to pass traffic, then comes back. The second monitors the assigned repeater and switches to the weather repeater only to pass traffic. If there aren't enough qualified liaison stations, one station can be given both assignments, providing the operator is competent and properly equipped to monitor both channels simultaneously.

Shelters

A shelter or relief center is a temporary place of protection where rescuers can bring disaster victims and where supplies can be coordinated—perhaps a hospital's mass-casualty center that treats the injured, a school or church which has cots and blankets and where volunteers can serve meals, or a fire-fighter's camp where workers can seek comfort. Many displaced people can stay at the homes of friends or relatives, but those searching for family members or in need are housed in shelters. Whether a shelter is for a few stranded motorists during a snowstorm or a whole community of homeless residents after a disaster, it is an ideal location to set up an Amateur Radio communications base station, basically a combined relief-agency-support operation and a modified, but familiar, message center. In fact, the amateur station operators could share a table with shelter registration workers. Make sure you obtain permission for access to the shelter to assist and do not upset the evacuees.

Health-and-Welfare Traffic

There can be a tremendous amount of radio traffic to handle during a disaster because phone lines that remain in working order should be reserved for emergency use by those people in peril. Shortly after a major disaster, *Emergency* messages leaving the disaster area have life-and-death urgency or are for medical help and critical supplies—they take supremacy. Next, *Priority* traffic, a message with an emergency-related nature but not of the utmost urgency, is handled. Then, *Welfare* traffic originated by evacuees at shelters or by the injured at hospitals, relayed by Amateur Radio, flows one way and results in timely advisories to those waiting outside the disaster area.

It's only logical that hams inside the disaster area cannot immediately find out about Aunt Irene when their own lives are threatened, and when they are busy handling bona fide emergency messages, so for a while, leave them to the tasks of surviving. Later, amateurs patiently waiting outside the disaster area can originate their Welfare messages for concerned friends and relatives of the possible victims or ask about property damage within the disaster-affected area. Thus, the incoming traffic begins. Yet, actively soliciting traffic for an emergency crisis area should be avoided because it can severely overload an already busy system at a time when guaranteed delivery and reply is uncertain. Incoming Welfare traffic should be handled only after all emergency and priority traffic is cleared. Welfare inquiries can take time to discover their hard-to-find answers, and an advisory to the inquirer uses even more time. Meanwhile, some questions might have already been answered through restored circuits.

Shelter stations, acting as net control stations, can exchange information on the HF bands with many stations directly to destination areas as propagation permits. Or they can handle formal traffic through a few outside operators who in turn can link to NTS schedules. By having many amateurs NTS-trained, it's easy to adapt from local-tactical to long-distance-formal traffic handling as the situation requires.

Damage Surveys

Damage caused by natural disasters can be sudden and extensive. Responsible officials near the disaster area, paralyzed without communications, will need help to contact appropriate officials outside to give damage reports. Such data will be used to initiate and then coordinate disaster relief. Amateur Radio operators, good neighbors from nearby counties and states, offer to help but often are unable to cross roadblocks established to limit access by sightseers and potential looters. Call-letter license plates on the front of the car or placards inside windshields may help to gain entry into these areas. It's important for amateurs to keep complete and accurate logs or written records for use by officials to summarize the situation, survey the damage or to be used as a guide for replacement operators. Even in disasters, accuracy and thoroughness are important.

Weather Warnings

NWS policy is to issue warnings only when there is absolute certainty, for fear that warnings of nonexistent storms might cause the public to ignore later warnings for real storms.

Public confidence increases with reliable weather warnings. When NWS calls a weather alert, it will contact the local EC by phone or voice-message pager, or the EC may call NWS to check on a weather situation.

Hurricanes

A hurricane is wind of force with a velocity of 75 miles per hour or more. These strong winds may cause tidal waves along shores and flooding inland which may lead to injury or death. A hurricane can bring about devastation—flatten shopping centers, uproot trees, destroy homes, overturn autos and carry mobile homes away.

A Hurricane Watch means a hurricane may threaten coastal and inland areas, a real possibility, but not necessarily imminent. Listen for further advisories and be prepared to act promptly if a warning is issued.

A Hurricane Warning is issued when a hurricane is expected to strike an area within 24 hours. It may include an assessment of flood danger, small-craft or gale warnings for the storm's periphery, estimated storm effects and recommended emergency procedures.

Amateurs can spot and report the approach of hurricanes well ahead of any news service. In fact, indirectly their information is edited and then broadcast on the local radio or TV to keep citizens of affected areas immediately informed. The Hurricane Watch Net on 14.325 MHz, for example, serves either the Atlantic or Pacific during a watch or warning period and keeps in touch with the National Hurricane Center as amateurs report location, strength and movement of hurricanes and tropical storms. Frequent, detailed information is issued on nets when storms pose a threat to the US mainland. In addition to hurricane spotting, local communicators may announce on the nets if residents have evacuated from low-lying flood areas and coastal shores to seek safety at motels and shelters. Other amateurs across the country can help by relaying information, keeping the net frequency clear and by listening.

Tornadoes

A tornado is an intensely destructive, advancing whirlwind formed from strongly ascending air currents. With winds of up to 300 miles per hour, tornadoes appear as rotating, funnel-shaped clouds from gray to black in color, extending toward the ground from the base of a thundercloud and may sound like the roaring of an airplane or locomotive. Even though they are shortlived over a small area, tornadoes are the most violent of all atmospheric phenomena. Tornadoes that don't touch the ground are called funnels.

A Tornado Watch is issued when a tornado may occur near your area; watch the sky.

A Tornado Warning means take shelter immediately, a tornado has actually been sighted or indicated by radar and may strike in your vicinity. Protect yourself from being blown away, struck by falling objects or injured by flying debris.

"Tornado alley" runs in the 5th, 9th and 10th US call areas. Prepared amateurs in these areas receive Tornado Spotter's Training and refresher courses presented by NWS personnel, through the local government, ARES group or at radio clubs. Knowing what severe storm elements should be reported, Amateur Radio's quick response has reduced injuries and fatalities by giving early warning for residents to get into underground shelters. Operators have seen a tornado knock out telephone and electrical services just as quickly as it can flip over a truck or destroy a home in its path. Traffic lights and gas pumps won't work without electricity, creating problems for motorists. After a tornado strikes, amateurs provide communications in cooperation with local government and relief agencies who rescue victims, provide medical treatment and procure critical supplies. Welfare messages are sent from shelters, where survivors, whose homes have been damaged, receive assistance. Teams of amateurs and officials can survey and report property damage.

KB3OM was a part of the team surveying tornado damage in Lycoming County, Pennsylvania in the aftermath of a series of massive tornadoes that rumbled through eastern Ohio and western Pennsylvania. *(K3QDA photo)*

Floods, Mudslides and Tidal Waves

Floods occur when excessive rainfall causes rivers to overflow their banks, or when heavy rains and warmer than usual temperatures in combination melt snow, or if frozen ground prevents water runoff from soaking into soil, or when dams break. Floods can be minor, moderate or major. Floods or volcanic eruptions may melt mountain snow, causing mudslides.

A tidal wave or tsunami is actually a series of waves caused by a disturbance which may be associated with earthquakes, volcanoes or sometimes hurricanes.

Don't wait for the water level to rise or for officials to ask for help; sound the alarm and activate a weather net immediately. Besides handling weather data for NWS, enact the response plans necessary to relay tactical flood information to local officials, and assist their decision making by answering the following questions:

1) Which rivers and streams are affected and what are their conditions?

2) When will flooding probably begin, and where are the flood plain areas?

3) What are river-level or depth-gauge readings for comparison to flood levels?

Mobile operators may find roads flooded or closed and bridges washed out, isolating communities or themselves. Flood-rescue operations then may be handled by marine police boats with an amateur aboard. If power and telephones are out, portable radio operators can help with relief operations to evacuate families to care facilities or shelters outside the flood-sticken areas, where a fixed station should be set up. The officials will need to know the number and location of evacuees. As the river recedes, the water level drops in some

areas, but it may rise elsewhere to threaten residents. Liaison to repeaters downstream can warn others of flooding. Property-damage reports and Welfare traffic will usually be followed by a clean-up operation.

Winter Storms

A Winter Storm Watch indicates there is a threat of severe winter weather in a particular area.

A Winter Storm Warning is issued when heavy snow (6 inches or more in a 12-hour period, 8 inches or more in a 24-hour period), sleet, freezing rain or a substantial layer of ice is expected to accumulate, are forecast to occur separately or in combination.

Freezing rain or freezing drizzle is forecast when expected rain is likely to freeze as soon as it strikes the ground, coating roads and walkways with ice, or when snow falls on warm highways to form ice hazards.

Sleet is small particles of ice, usually mixed with rain. If enough sleet accumulates on the ground, it will make the roads slippery.

Travelers advisories are issued when ice and snow are expected to hinder travel, but not seriously enough to require warnings.

A blizzard is the most dangerous of all winter storms. It combines cold air, heavy snow and strong winds that blow snow about and may reduce visibility to only a few yards.

A Blizzard Warning is issued when there is considerable snow and winds of 35 miles per hour or more.

A Severe Blizzard Warning means that a heavy snowfall is expected, with winds of at least 45 miles per hour and temperatures 10 degrees F or lower.

Blizzards and snowstorms can create vehicle-traffic problems by making roads impassable, stranding motorists or drastically delaying their progress. In some areas, local snowmobile club members or 4-wheel-drive-vehicle enthusiasts cooperate with hams to coordinate and transport key medical personnel to hospitals when snowdrifts render roads almost impassable, or they may assist search teams looking for motorists. Blizzards can isolate communities, and freezing rain or ice storms can bring down wires, causing telephone and power outages. Once officials determine the location of shelters, radio operators can be assigned to set up equipment at the shelter sites. Use of repeaters and autopatches can allow Welfare phone calls for those inside the shelters to inform and reassure friends, families and relatives waiting for them to return after the storm.

Brush and Forest Fires

A dry spell of prolonged hot weather parches shrubs, brush and trees. This dry vegetation, ignited by lightning, arson or even a helicopter crash, can start a forest fire and can quickly become worse when winds spread the burning material. Amateurs, in the 6th and 7th US call areas in particular, train with local fire department or public safety officials to assist during forest fires. An amateur who spots a small fire that has the potential of being a forest fire can radio through a repeater for the Park Service to dispatch a district ranger and fire-fighting equipment to contain the fire. When fires are out of control, hams help with communications to evacuate people to shelters, report the fires' movement or radio requests for supplies and volunteers. ARES groups over a wide geographical area can set up portable repeaters and communicate various needs until Department of Forestry personnel set up their own communications.

Fire safety rules are of special importance in an emergency, but also should be observed every day to prevent disaster. Since most fires occur in the home, even an alert urban amateur can spot and report a building on fire as easily as a rural amateur can catch sight of a forest fire. Fires are extinguished by taking away the fuel or its air (smother it), or cooling it with water and fire-extinguishing chemicals. A radio call to the fire department will bring the needed control. The Red Cross generally helps find shelter for the homeless after fires make large buildings, such as apartments, uninhabitable. So even in cities, communications for routing people during and after fires may be needed.

Earthquakes and Volcanic Eruptions

Earthquakes are caused when underground volcanic forces break and shift rock beneath the surface causing a shake or tremble of the earth's crust. The actual movement of the earth is seldom a direct cause of death or injury, but can cause buildings and other structures to collapse and may annihilate communications, telephone service and electrical power. Most casualties result from falling objects and debris, splintering glass and fires.

Earthquake prediction has progressed, yet precise predictions do not exist. With no advance notice of earthquakes, spotting or evacuating is unlikely. Instead, amateurs are often the first to alert communities immediately after earthquakes and volcanic eruptions occur, and their warnings are definitely credited with lessening personal injuries in the area. Amateurs then may assist with rescue operations, getting medical help or critical supplies and appraisal of damage. The rest of the world will be trying to get information, too, and the biggest job is to handle Welfare messages. You might be overwhelmed with a sense of gratification, though, when delivering a *favorable* Welfare message to a family when they become ecstatic with the news of loved ones and cry with happiness. This is one of the rewards of helping with emergency communications.

The most difficult disaster communications to prepare for are accidents and hazardous situations because they are so unpredictable and can happen anywhere.

ACCIDENTS AND HAZARDS

An emergency autopatch system connected to a law-enforcement agency, such as a sheriff's department or state police, can dramatically enhance the public-service capabilities of Amateur Radio operators. The mobile or portable operator sends a three-digit code on a key pad attached to a 2-meter radio for quick and direct access to the sheriff to report a public-safety situation. Generally, a sheriff's emergency patch is used only to report incidents which pose a threat to life or personal safety such as vehicle accidents, disabled vehicles or debris in traffic, injured persons, criminal activities and fires. Sheriff's departments usually accept reports of such incidents anywhere in their county and will relay information to the proper agency when it pertains to adjacent areas.

Here's a typical Sheriff's patch procedure:
1) Give your call and say "emergency patch."
2) Drop your carrier momentarily.
3) Dial access code.
4) Wait for sheriff's operator to answer.
5) Give your call and your traffic.
6) After sheriff's operator acknowledges, dump the patch by dialing the dump code.
7) Give your call and say "patch clear."
8) Stay on frequency for at least three minutes after completing your call in case the sheriff's operator needs more information.

The ability to call the police or for an ambulance, without depending on another amateur to monitor the frequency, saves precious minutes. Quick reaction and minimum delay is what makes an autopatch useful in emergencies. Autopatch, when used responsibly, is a valuable asset to the community.

Vehicular Accidents

Autopatch calls concerning highway safety are permitted by the FCC in cases where there is an immediate threat to life or property. Vehicle-accident reports, by far the most common public-service activity on repeaters, can involve anything from bikes, motorcycles and automobiles, to buses, trucks, trains and airplanes.

The first activities handled by experts at a vehicle-accident scene are keyed to rescue, stabilize and transport of victims; then they ensure security, develop a perimeter, handle vehicle traffic and control fires or prevent fires from gasoline spills. Finally, routine operations restore the area with wrecking, towing and salvage operations, and clearing debris.

Distress Calling

An amateur who finds himself in a situation where immediate emergency assistance is required (at sea, in a remote location, etc) would call MAYDAY on whatever frequency seems to offer the best chance of getting a useful answer. MAYDAY is from the French *m'aider* (help me). On CW, use $\overline{SOS}$ to call for help. QRRR has been discarded by ARRL. The distress call should be repeated over and over again for as long as possible until answered. The amateur involved should be prepared to supply the following information to the stations who respond to an $\overline{SOS}$ or MAYDAY:

•The location of the emergency, with enough detail to permit rescuers to locate it without difficulty.
•The nature of the distress.
•The type of assistance required (medical aid, evacuation, food, clothing, etc).
•Any other information that might be helpful in locating the emergency area or in sending assistance.

A dedicated effort by the Santa Barbara ARC and the Santa Barbara South County ARES resulted in this mobile communications van. *(photo courtesy of WB6UNH)*

If you should have to report a vehicle accident, remain calm, and get as much information as you can. This is one time you certainly have the right to break into a conversation on a repeater. Use plain language, say exactly what you mean, and be brief and to the point. Do not guess about injuries; if you don't know, say so. Some accidents may look worse than they really are; requesting an ambulance to be sent needlessly could divert it away from a bona fide injury accident occurring at the same time elsewhere. And besides, police cruisers are generally only minutes away in an urban area. Here's what you should report for a vehicle accident:

1) Highway number (eg, I-43, I-94, US-45).
2) Direction of travel (north, south, east, west).
3) Address or street intersection, if on city streets.
4) Traffic blocked, or if accident is out of traffic.
5) Apparent injuries, number and extent.
6) Vehicles on fire, smoking or a fuel spill.

Example: "This is K1XA, reporting a two-car accident, I-94 at Edgerton, northbound, blocking lane number two, property damage only."

Freeway Warning

Many public-safety agencies recommend that you do not stop on freeways or expressways to render assistance at an accident scene unless you are involved, are a witness or have sufficient medical training to alleviate life-threatening injuries and can stabilize the injured until help arrives. Freeways are not like city streets. They are extremely dangerous because of the heavy flow of high-speed traffic. Even under ideal conditions, driver fatigue or inattentiveness, high speeds and short distances between vehicles often make it impossible to stop a vehicle from striking stationary objects in the traffic lanes. If you must stop on a freeway, pull out of traffic and onto distress lanes, and exercise extreme caution to protect yourself. Don't add to the traffic problem. Instead, radio for help.

Public-Safety Agencies

Amateur Radio affords public-safety agencies, such as local police and fire officials, with an extremely valuable resource to be used in times of emergency. Once initial acceptance by the authorities is achieved, an ongoing working relationship between amateurs and safety agencies is based on the efficiency of our performance. Officials tend to be very cautious and skeptical about those who are not members of the public-safety professions, based on experiences in which well-intended, but somewhat overzealous, volunteers have complicated or jeopardized efforts in emergencies. At times, officials may have trouble separating problem *solvers* from problem *makers*. Public-safety officials may be understaffed and have a limited budget, so they might accept communications help if it is offered in the proper spirit. Here are several image-building rules for working with safety agencies:

1) Do not take individual or independent action. Work with ARES or RACES groups or a local club through your EC as a direct chain of command with the agencies for an impression of an organized, efficient volunteer operation.

2) Be honest. If your group cannot handle a request, say so and explain why. Safety personnel often risk their lives based on a fellow disaster worker's promise to perform.

3) Have an amateur representative meet with public-safety officials in advance of major emergencies so each group will know the capabilities of the other and can gain confidence. The representative should present himself or herself professionally with a calm, businesslike manner and in conservative attire.

4) Do not have more flashing lights, signs, decals and antennas than used by the average police or fire vehicle. Safety professionals are trained against overkill and use the minimum resources necessary to get the job done.

5) In the field, wear a simple jumpsuit with an ARES patch to give a professional image and to help officials identify radio operators in the emergency.

6) Do not assume more authority or responsibility than you've been given.

7) If contacted by members of the press, restrict comments solely to the amateurs' role in the situation. Emergency status should come from the press information officer or government agency concerned.

It is important that the public-service lifeline provided by Amateur Radio be understood by public-safety agencies before the next disaster occurs. It is up to you to invite the local agencies to observe or cooperatively participate with your group. For further information, see the three-part series by Jerry Boyd, KG6LF, published in January 1982, February 1983 and June 1985 *QST*.

Police Assistance

Amateurs have been known to help citizens in distress. Act as a communicator to radio for the police when you spot criminal activities. Memorize a description of the suspects for apprehension or information about a vehicle for later recovery. Use caution in the area of criminal activities; uniformed personnel arriving on the scene do not know who you are. Be an observer; let the officers act. For instance, if you see a vehicle traveling at high speed at night without lights, report the location and direction of travel to the sheriff and let the authorities determine if the driver is under the influence of alcohol. Don't try following the vehicle. Some Amateur Radio groups provide police communications assistance involving free taxi service to citizens who do not wish to drive home on New Year's Eve, as lookouts for vandals on city/town streets and on freeway overpasses on Halloween, or as reassurance to the elderly at a senior citizens' picnic. Communicators have even reported sighting a person who is about to make a life-threatening jump into a river. Assisting the police in small ways can make a big difference to the community.

Search and Rescue

Amateurs helping search for an injured climber use repeaters to coordinate the rescue and alert authorities. A small airplane crashes, and amateurs direct their search with signals from its Emergency Locator Transmitter (ELT). It's reassuring to team up with local search-and-rescue organizations who have familiarity with the area. Once the victim is found, the communicator can radio the status, phonepatch for medical information, talk in further help and plan for a return transportation route for getting the injured to hospital treatment. If the victim is found in good condition and can talk, Amateur Radio can bolster the hopes of base-camp personnel and the family of the victim with direct communications.

Even in cities, there is sometimes a need for searches. An elderly person, out for a walk, gets lost and doesn't return home or wanders from a nursing home and is reported missing. After a reasonable time, a local search team plans and coordinates a search. Amateurs can take part by providing communications, a valuable part of any search. When the missing person is discovered, there may be a need to radio for an ambulance for transportation to a nearby hospital.

Hospitals

Hospital phones can fail, for example, when a construction crew using a backhoe accidentally cuts though the main trunk line supplying telephone service to several hundred users, including the hospital. Such major hospital telephone outages can stop emergency incoming phone calls besides preventing the hospital staff from phoning to discuss medical treatment with outside specialists. Several hand-held equipped amateurs can first handle emergency calls from nursing homes, fire departments and police stations or provide communications to temporarily replace a defective hospital paging system. Then, they can help restore critical interdepartmental hospital communications and, finally, with nearby hospitals. Preparation for hospital help begins by cooperating with administrators and public-relations personnel. Then it may require inside signal checks or mounting an outside antenna, eliminating interference to medical or security equipment and finding a good NCS location in the hospital.

An electrical problem plunges a hospital into darkness. Emergency power lights some areas to reveal that smoke inundates the nursery and intensive-care units, necessitating evacuation of patients and newborn infants to another hospital or to the Red Cross building. A real lifesaver, Amateur Radio can help relocating maneuvers with critical communications or as a backup until regular service is restored.

Working with local hospitals doesn't always require such grave extremes. It might simply be relaying information from the poison control center to a campsite victim. Or it could be a hospital drill where reports of "victims'" conditions are sent to the hospital from a disaster site. One typical drill involved a simulated cropduster plane crash on an elementary-school playground during recess. The doctors and officials depended on communications to find and help casualties who were contaminated by cropdusting chemicals. Amateurs worldwide hold emergency nets, helping to find rare treatment or medication, special consultants or care and even locating free jet transportation to a hospital. The needs differ but the response is brought by amateur communications.

Toxic-Chemical Spills and Hazardous Materials

A toxic-chemical spill suddenly appears when gasoline pours from a ruptured bulk-storage tank. A water supply is unexpectedly contaminated or a fire causes chlorine gas to escape at an apartment swimming pool. On a highway, a faulty shutoff valve lets chemicals leak from a truck or drums of chemicals on a transport fall onto the highway and rupture. Amateur communications have helped in all these situations.

In general, don't rush into a Hazardous Materials ("Haz Mat") incident area to help without knowledge or respect for what is involved. Vehicles carrying 1000 pounds or more of a Haz Mat are required by Federal regulations to display a placard bearing a four-digit identification number. Radio the placard number to the authorities, and they will decide whether to send Haz Mat experts to contain the spills. Follow directions from those in command, provide communications to help them evacuate residents in the immediate area and coordinate between the spill site and the shelter buildings. Communications also assists public-service agencies to set flares for traffic control, help reroute motorists and assess damage. Sometimes, autopatch traffic for police or fire department workers is needed until the chemical is determined to be nontoxic or the spill is contained. The National Transportation Safety Board, the Environmental Protection

Agency and many local police and fire departments continue to praise ARES volunteers in their assistance with toxic spills.

AMATEUR RADIO GROUPS

The Amateur Radio Emergency Service

In 1935, the League developed what is now called the Amateur Radio Emergency Service (ARES), an organization of radio amateurs who have voluntarily registered their capabilities and equipment for emergency communications. See Fig 14-5. They are groups of trained operators ready to serve the public when disaster strikes and regular communications fail. ARES recruits members from existing clubs rather than starting from scratch, but includes amateurs outside the club area since emergencies do not recognize boundary lines.

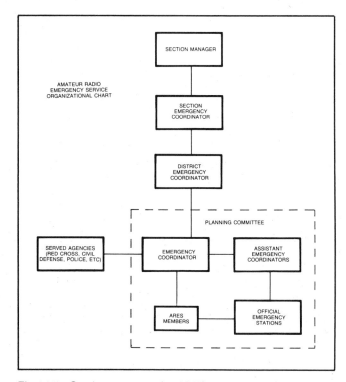

Fig 14-5—Section structure for ARES.

Interested in getting active in public-service activities or preparing for emergency communications? Then join ARES in your area and meet other amateurs to exchange ideas, ask questions and help with message centers, sports events or weather spotting. Any licensed amateur with a sincere desire to serve is eligible for ARES membership. The possession of emergency-powered equipment is recommended, but is not a requirement. Information about ARES may be obtained from your ARRL Section Manager (listed on page 8 of *QST*) or League HQ.

Official Emergency Station

Once you get some ARES training and practice, you might want to build and refine your skills in emergency communications. If you are an ARRL full member and hold at least a Technician class license, there are several opportunities available. The first is the Official Emergency Station (OES)

ARES Personal Checklist

The following represents recommendations of equipment and supplies ARES members should consider having available for use during an emergency or public-service activity.

1) ARES Identification Card
2) FCC Amateur Radio License
3) Radio Gear
   a) rig (2 meters)
   b) microphone
   c) headphones
   d) power supply/extra batteries
   e) antennas with mounts
   f) spare fuses
   g) patch cords/adapters (BNC to PL259/RCA phono to PL259)
   h) SWR meter
   i) extra coax
4) Writing Gear
   a) pen/pencil/eraser
   b) clipboard
   c) message forms
   d) logbook
   e) note paper
5) Personal Gear (short duration)
   a) snacks
   b) liquid refreshments
   c) smoking supplies/candy
   d) throat lozenges
   e) personal medicine
   f) aspirin
   g) extra pair of prescription glasses
6) Personal Gear (72-hour duration)
   a) foul-weather gear
   b) three-day supply of drinking water
   c) cooler with three-day supply of food

   d) mess kit with cleaning supplies
   e) first-aid kit
   f) personal medicine
   g) aspirin
   h) throat lozenges
   i) sleeping bag
   j) toilet articles
   k) mechanical alarm clock
   l) flashlight with batteries/lantern
   m) candles
   n) waterproof matches
   o) extra pair of prescription glasses
7) Tool Box (72-hour duration)
   a) screwdrivers
   b) pliers
   c) socket wrenches
   d) electrical tape
   e) 12/120-volt soldering iron
   f) solder
   g) volt-ohm meter
8) Other (72-hour duration)
   a) hatchet/ax
   b) saw
   c) pick
   d) shovel
   e) siphon
   f) jumper cables
   g) 3/8-inch hemp rope
   h) highway flares
   i) extra gasoline and oil

---

appointment, which requires regular participation in ARES, including drills, emergency nets and, possibly, real emergency situations. An OES aims for high standards of activity, emergency-preparedness and operating skills.

Emergency Coordinator

Next, when you feel qualified enough to become a team leader of your local ARES group, consider the Emergency Coordinator (EC) appointment, if that position is vacant in your area. An EC, usually responsible for a county, is the person who can plan, organize, maintain response-readiness and coordinate for emergency communications. See Fig 14-6. Much of the work involves promoting a working relationship with local government and served agencies. The busy EC can hold meetings, train members, keep records, encourage newcomers, determine equipment availability, lead others in drills or be first on the scene in an actual disaster. Some highly populated or emergency-prone areas may also need one or more Assistant Emergency Coordinators (AEC) to help the EC. The AEC is an unofficial appointment made by an EC and, as a specific exception,

tion, can hold any class license and need not be an ARRL member.

District Emergency Coordinator

If there are many ARES groups in an area, a District Emergency Coordinator (DEC) might be appointed. Usually responsible for several counties, the DEC coordinates emergency plans between local groups and ARES nets, directs the overall communication needs of a large area or can be

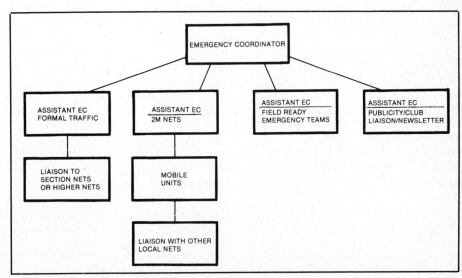

Fig 14-6—Local ARES structure for a county, city or other area of coverage.

a backup for an EC. As a model emergency communicator, the DEC trains clubs in tactical traffic, formal traffic, disaster communications and operating skills.

Section Emergency Coordinator

Finally, there is one rare individual who can qualify as top leader of each ARRL Section emergency structure, the Section Emergency Coordinator (SEC). Only the Section Manager can appoint a candidate to become the SEC. The SEC does some fairly hefty work on a section-wide level: making policies and plans and establishing goals, selecting the DECs and ECs, promoting ARES membership and keeping tabs on emergency preparedness. During an actual emergency, the SEC follows activities from behind the scenes, making sure that plans work and section communications are effective.

ARES has responded countless times to communications emergencies and must continue to introduce Amateur Radio to the ever changing stream of agency officials who have never heard of it or used it. Experience has proven that radio amateurs react and work together more capably in time of emergency when practice has been conducted in an organized group. There is no substitute for experience gained—*before* the need arises.

An ARES group needs to refresh its training with meetings, scheduled nets, drills or real emergencies. Welcome all amateurs to the ARES unit; an effective ARES group is to our benefit, as well as to the benefit of the entire community.

Section Manager

The overall leader of the ARRL Field Organization in each section is the Section Manager (SM), who is elected by the League membership in that section. The SM not only appoints the SEC to handle details of ARES and emergency-preparedness activities, but also appoints a Section Traffic Manager (STM) to handle details of the NTS and formal message traffic operation within the section. Response to emergency and public service needs combines the ARES and the NTS. [For further details on the ARRL Field Organization, see Chapter 17.]

The National Traffic System

In 1949 the League created the National Traffic System (NTS) to handle medium- and long-haul formal message traffic through networks whose operations can be expedited to meet the needs of an emergency situation.

The main function of NTS in an emergency is to link various local activities and to allow traffic destined outside of a local area to be systematically relayed to the addressee. In a few rare cases, a message can be handled by taking it directly to a net in the state where the addressee lives for rapid delivery by an amateur there within toll-free calling distance. However, NTS is set up on the basis of being able to relay large amounts of traffic systematically, efficiently and according to an established flow pattern. This proven and dependable scheme is what makes NTS so vital to emergency communications. See Chapter 15 for more information on NTS.

Additional details on ARES and NTS can be found in the *Public Service Communications Manual,* the *Emergency Communicator's Handbook* and the *ARRL Net Directory,* all published by ARRL (send an SASE for ordering information). Information on emergency communications and traffic handling also appears regularly in *QST.*

Radio Amateur Civil Emergency Service

A subpart of the US amateur regulations (Part 97, subpart F) provides for RACES, the Radio Amateur Civil Emergency Service. RACES is a special phase of amateur operation sponsored by FEMA, that provides radio communications for civil-preparedness purposes only, during periods of local, regional or national civil emergencies. These emergencies are not limited to war-related activities, but can include natural disasters such as fires, floods and earthquakes. As defined in the rules, RACES is a radiocommunication service conducted by volunteer licensed amateurs, and is designed to provide emergency communications to local or state civil-preparedness agencies. It is important to note that RACES operation is authorized by the FCC upon request of a state or federal official and is strictly limited to official civil-preparedness activity, in the event of an emergency-communications situation.

Amateurs operating in a local RACES organization must be officially enrolled in that local civil-preparedness group. RACES operation is conducted by amateurs using their own primary station licenses, and by existing RACES stations. The FCC no longer issues new RACES (WC prefix) station call signs. Operator privileges in RACES are dependent upon, and identical to, those for the class of license held in the Amateur Radio Service. All of the authorized frequencies and emissions allocated to the Amateur Radio Service are also available to RACES on a shared basis. But in the event that the President invokes his War Emergency Powers, amateurs involved with RACES would be limited to the certain frequencies (while all other amateur operation would be silenced). Please note that the RACES frequency allocation on 2 meters includes the standard FM repeater subbands. See below.

While RACES was originally based on potential use for wartime, it has evolved over the years, as has the meaning of civil defense (which is also called civil preparedness), to encompass all types of emergencies. It should be emphasized again that RACES is part of the amateur service, its regulations are part of the amateur regulations, and it operates in the amateur bands. The segments of the amateur bands it uses are shared with the rest of the amateur service in peacetime; in the event of war, its frequency segments would be exclusive.

While operating in a RACES capacity, RACES stations and amateurs registered in the local RACES organization may not communicate with amateurs not operating in a RACES capacity. (Of course, such restrictions do not apply when such stations are operating in a non-RACES—such as ARES—amateur capacity.) Only civil-preparedness communications can be transmitted, as defined in section 97.161 of the FCC regulations. Tests and drills are permitted only for a maximum of one hour per week. All test and drill messages must be clearly so identified.

Although RACES and ARES are separate entities, the League advocates dual membership and cooperative efforts between both groups whenever possible. The RACES regulations make it simple and possible for an ARES group whose members are all enrolled in and certified by RACES to operate in an emergency with great flexibility. Using the same operators and the same frequencies, an ARES group also enrolled as RACES can "switch hats" from ARES to RACES and RACES to ARES to meet the requirements of the situation as it develops. For example, during a "non-declared emergency," ARES can operate under ARES, but when an emergency or disaster is officially declared by a state or federal authority, the operation can become RACES with no change in personnel or frequencies.

This situation is still not well understood and accepted throughout the United States; both ARES and RACES still exist, separately, in many areas. League officials will have to determine the situation in their own area. Where there is currently no RACES, it would be a simple matter for an ARES group to enroll in that capacity, after a sophisticated presentation to the civil-preparedness authorities. In cases where both ARES and RACES exist, it is possible to join both or to be involved in either. As time progresses, the goal would be the merger into one strong organization, with coordination between ARES and RACES officials using the same groups of amateurs. In some sections of the US today, the ARES structure has also been accepted as the RACES structure. (For more information on RACES, contact your State Emergency Management or Civil-Preparedness Office, the Federal Emergency Management Agency, or the Federal Communications Commission.)

The FCC's decision to expand the RACES allocation on 2 meters was an important step forward for both the RACES program and ARES. This is because for years, a vexing dilemma had been facing amateur groups interested in putting a 2-meter RACES repeater on the air in their communities. That is, FCC-designated frequencies for 2-meter RACES operation were not in harmony with standard amateur repeater pairs. This had an inhibiting effect on the growth of RACES repeaters (funding for which can sometimes be obtained from local government).

While all amateur frequencies are available to RACES on a shared basis during normal times, in the event of a presidentially declared war emergency, the only permissible amateur operations would be by a RACES group, and only on those oddball frequencies. Why put a repeater on the air when it will be unusable when it's needed the most? This has not been a very workable situation, particularly since 2-meter repeaters play a crucial role in most emergencies. The realization that standard repeater fre-

quencies should be available to RACES took a substantial amount of time to filter through the maze of government agencies concerned.

Admittedly, the wheels of government move slowly. Luckily for amateurs, the climate of the FCC continues in the deregulatory mode. Through the efforts of ARRL and other agencies, the FCC announced that, effective March 26, 1984, all 2-meter repeater frequencies were available to RACES. This brings more repeaters into RACES operation, expands the potential for the use of RACES stations in the future, and eliminates the uneasiness of trying to figure out a reasonable way of complying with the nonstandard 2-meter RACES segments in the event of a national emergency.

Further, this will have a very beneficial effect on cooperation between RACES groups and the League-sponsored ARES. RACES and ARES can now work together on the same frequencies to meet the needs of any local or state emergency.

The beauty of ARES/RACES togetherness is that RACES has official government recognition (and resources), while the League's ARES has a skilled Emergency Coordinator at the helm, the backing of ARRL Field Organization personpower, and freedom from the above-indicated operating restraints. Public-service 2-meter repeaters can now be operated under joint ARES/RACES coordination, and it is completely unnecessary for anyone even to contemplate altering repeater frequencies to operate on RACES frequencies during a declared emergency. ARES can shift to RACES, or vice versa, without a shift in frequency.

As noted above, in areas where there is currently no RACES organization, it would be a relatively simple matter for an ARES group to enroll in that capacity. The first step is for the ARRL EC or DEC to make a sophisticated presentation to local civil-preparedness authorities. Perhaps the logical long-range goal is for ARES and RACES to merge into one vigorous organization. Since the all-important 2-meter frequencies have been reconciled, the concept of a unified public-service force is something worth striving for.

Chapter 15

Traffic Handling

F or just pennies a day, you can protect you and your family from all sorts of catastrophic illnesses with the Mutual of Podunk health-care policy....

Yes, at the amazing low price of $9.95, you can turn boring old potatoes, carrots and okra into culinary masterpieces with the Super Veggie-Whatchamadoodler—a modest investment for your family's mealtime happiness....

Tired of all that ugly fat on your otherwise-beautiful body? Our Exer-Torture home gym will trim those thunder thighs in 30 days or your money back....

Ah, those pitches. Everybody is trying to sell something—even the participants in the specialty modes of Amateur Radio. Our eyes light up thinking about all those wondrous gizmos that digitize, packetize and equalize. Those kinds of specialty modes are easily remembered and can quickly gain popularity. Most people, though, forget about the oldest specialty mode in Amateur Radio—traffic handling.

Admittedly, it's hard for traffic handlers to compete with other specialty modes because it just doesn't look exciting. After all, tossing messages from Great-Uncle Levi and Grandma Strauss sure doesn't put you on the leading edge of our high-tech hobby, does it? Yet, over 500 nets are members of the ARRL National Traffic System (NTS), probably one of the most highly organized special interests of Amateur Radio. Today's traffic handlers are at this very moment setting new standards for traffic handling via RTTY and packet—two of the hottest specialty modes going—as well as using traditional modes.

If you enjoy emergency-communications preparation, traffic handling is for you. Sure, it's true that over 90% of all messages handled via Amateur Radio are of the "at the state fair, wish you were here" or the "happy holidays" variety—certainly not life-and-death stuff. But consider this: Your local fire department often conducts drills without ever putting out a real fire, your civil defense goes out regularly and looks for "pink tornadoes," and many department stores hire people to come in and pretend to be shoplifters to check the alertness of their employees. Similarly, when real emergencies rear their ugly heads, traffic handlers just take it all in stride and churn messages out like they always do.

Traffic handling is also an excellent way to paint a friendly picture of Amateur Radio to the nonamateur public. What sometimes seems to be unimportant to us is usually never unimportant to that person in the address block of a message. Almost every traffic handler can relate stories of delivering a Christmas message from some long-lost friend or relative that touched the heart of the recipient such that they could hear the tears welling up at the other end of the phone. Those

happy recipients will always mentally connect Amateur Radio traffic handling with good and happy things, and often this is the most satisfying part of this hobby within a hobby.

Next time you go to a hamfest, see if you can spot the traffic handlers. They are almost always reveling in a big social cluster, sharing stories and enjoying a unique camaraderie. Traffickers always enthusiastically look forward to the next hamfest, because it means another chance to spend time with their special friends of the airwaves. These friendships often last a lifetime, transcending barriers of age, geographical distance, background, gender or physical ability.

Young or old, rural or urban, there's a place reserved for you on the traffic nets. Young people often can gain respect among a much older peer group and obtain high levels of responsibility through the traffic nets, good preparation for job opportunities and scholarships. "Nine to fivers" on a tight schedule can still manage to get a regular dose of Amateur Radio in just 15 to 30 minutes of net operation—a lot of hamming in a little time. Retired people can stay active in an important activity and provide a service to the general public.

Even if you live in the sticks, where you rarely get a delivery, you can still perform a vital function in NTS as a net control station (NCS) or as a representative to the upper echelons of NTS. You don't have to check in every night (a popular myth about traffic handling—if you can donate time just once a week you are certainly welcome on NTS). You don't need fancy antennas or huge amplifiers, and you don't even need those ARRL message pads. For the cost of a pad or paper and a pencil, you can interface with a system that covers thousands of miles and consists of tens of thousands of users. A world of fun and friendship is waiting for you in traffic handling. You only have to check into a net to become part of it.

MAKING THE BIG STEP: CHOOSING A NET AND CHECKING INTO IT

Checking into a net for the first time is a lot like making your first dive off the high board. It's not usually very pretty, but it's a start. Once you've gotten over the initial shock of hitting the water that hard, the next one comes a lot easier. But, like a beginning diver, you can do a little advance preparation that will get you emotionally prepared for your first plunge.

Before you attempt to interface with the world of NTS, you need the right "software." Get a copy of the *ARRL Net Directory*, available from ARRL ($1) and two free ARRL operating aids—FSD-3, the list of ARRL numbered radiograms, and FSD-218, the pink "Q-signals for nets and message form" card. Read all the articles in the *Net Directory*, and keep the two operating aids between the covers of your logbook for quick reference. After you have a basic understanding of the materials, you're ready to pick out a net in your ARRL section/state that suits you.

Go through the *Net Directory* and match up your time schedule with the nets in your section/state. If you're a Novice, your section/state slow-speed net is a good choice (or your neighboring section, if yours doesn't have one). But if you aren't a Novice, and still don't have the confidence to try the "big" CW section net, don't feel embarrassed about checking into a slow-speed net. Slow-speed nets are chock-full of veterans that help with NCS and NTS duties, and they're willing to help you, too. Perhaps you'd like to try the section phone net or weather net or a local 2-meter net. Some sections have also established RTTY nets. At any rate, you are the sole judge of what you want to try first.

Once you've chosen a net, it's a good idea to listen to it for a few days before you check in. Although this chapter will deal with a generalized format for net operation, each net is "fingerprinted" with its own special style of operation, and it's best to become acquainted with it before you jump in. If you have a friend who checks into the net, let your friend "take you by the hand" and tell you about the ins and outs of the net.

When the big day arrives, keep in mind that everybody on that net had to check into the net for the first time once. You aren't doing anything different than the rest of them, and this, like your first QSO as a Novice, is just another "rite of passage" in the ham world. You will discover that it doesn't hurt, and that many other folks will be pleased, even happy, that you checked in with them. (See accompanying sidebar for general recommendations on how to make your NCS love you!)

On CW

First, we'll pay a visit to a session of the Mowegia CW Net, not so long ago, on an 80-meter frequency not so far away. KØSI is calling tonight's session, using that peculiar CW net shorthand that we aren't too used to yet.

KØSI: MCWN MCWN MCWN DE KØSI KØSI KØSI GE AND PSE QNZ V V V MCWN QNI QTC?

(Translation: Calling the Mowegia CW Net, calling the Mowegia CW Net, this is KØSI. Good evening, this is a directed net, please zero beat with me. [Then, he sent some Vs to aid stations in zero beating him.] Calling the Mowegia CW Net...any checkins, any traffic?)

In the meantime, NDØN, K2ONP and WØOUD are waiting in the wings to check in. Since each is an experienced traffic handler, each listens carefully before jumping in, so as to not step on anyone.

NDØN: N

KØSI: N

(Notice that NDØN just sent the first letter of his suffix, and the NCS acknowledged it. This is common practice, but if you have the letter E or T or K as the first letter of your suffix, or if it is the same as another net operator's first letter, you might use another letter. It's not a hard-and-fast rule.)

NDØN: DE NDØN GE PETE QTC SOUTH CORNER 1 AIØO 1 K

(NDØN has one piece of traffic for South Corner, and one for AIØO.)

KØSI: NDØN DE KØSI GE JOHN QSL TU A̅S̅

(Good evening, John, I acknowledge your traffic. Thank you, please stand by.)

How to be the Kind of Net Operator the Net Control Station (NCS) Loves

As a net operator, you have a duty to be self-disciplined. A net is only as good as its worst operator. You can be an exemplary net operator by following a few easy guidelines.

1) *Zero beat the NCS.* The NCS doesn't have time to chase all over the band for you. Make sure you're on frequency, and you will never be known at the annual net picnic as "old so-and-so who's always off frequency."

2) *Don't be late.* There's no such thing as "fashionably late" on a net. Liaison stations are on a tight timetable. Don't hold them up by checking in 10 minutes late with three pieces of traffic.

3) *Speak only when spoken to by the NCS.* Unless it is a bona fide emergency situation, you don't need to "help" the NCS unless asked. If you need to contact the NCS, make it brief. Resist the urge to help clear the frequency for the NCS or to "advise" the NCS. The NCS, not you, is boss.

4) Unless otherwise instructed by the NCS, *transmit only* to the NCS. Side comments to another station in the net are out of order.

5) *Stay until you are excused.* If the NCS calls you and you don't respond because you're getting a "cold one" from the fridge, the NCS may assume you've left the net, and net business may be stymied. If you need to leave the net prematurely, contact the NCS and simply ask to be excused (QNX PSE on CW).

6) *Be brief when transmitting to the NCS.* A simple "yes" (C) or "no" (N) will usually suffice. Shaggy dog tales only waste valuable net time.

7) *Know how the net runs.* The NCS doesn't have time to explain procedure to you. After you have been on the net for a while, you should already know these things.

K2ONP: M

KØSI: M

K2ONP: DE K2ONP GE PETE TEN REP QRU K

(K2ONP is representative to the NTS Tenth Region Net tonight, and he has no traffic.)

KØSI: K2ONP DE KØSI HI GEO TU A̅S̅

WØOUD: BK

KØSI: BK

WØOUD: DE WØOUD GE PETE QRU K

KØSI: WØOUD DE KØSI GE LETHA QNU TU A̅S̅

(Good evening, Letha, the net has traffic for you. Please stand by.)

(Since Letha lives in South Corner, the NCS is going to move WØOUD and NDØN off frequency to pass the South Corner traffic.)

KØSI: ØN?

NDØN: HR

(HR [here], or C [yes] are both acceptable ways to answer the NCS, who will usually use only your suffix from here on out to address questions to you.)

KØSI: OUD?

WØOUD: C

KØSI: NDØN ES WØOUD PSE UP 4 1 SOUTH CORNER K

(Please go up 4 kHz and pass the South Corner traffic.)

NDØN: GG (going)

WØOUD: GG

(When two stations go off frequency, the receiving station always calls the transmitting station. If the NCS had said WØOUD PSE QNV NDØN UP 4 GET 1 SOUTH CORNER, WØOUD would have called NDØN first on frequency to see if she copied him. This is done often when conditions are bad. If they don't make connection, they will return to net frequency. If they

ARRL QN Signals for CW Net Use

QNA*   Answer in prearranged order.

QNB*   Act as relay Between _____ and _____.

QNC   All net stations Copy.
      I have a message for all net stations.

QND*   Net is Directed (controlled by net control station.)

QNE*   Entire net stand by.

QNF   Net is Free (not controlled.)

QNG   Take over as net control station.

QNH   Your net frequency is High.

QNI   Net stations report In.
      I am reporting into the net. (Follow with a list of traffic or QRU.)

QNJ   Can you copy me?

QNK*   Transmit messages for _____ to _____.

QNL   Your net frequency is Low.

QNM*   You are QRMing the net. Stand by.

QNN   Net control station is _____.
      What station has net control?

QNO   Station is leaving the net.

QNP   Unable to copy you.
      Unable to copy _____.

QNQ*   Move frequency to _____ and wait for _____ to finish handling traffic. Then send him traffic for _____.

QNR*   Answer _____ and Receive traffic.

QNS   Following Stations are in the net.*
      (Follow with list.)
      Request list of stations in the net.

QNT   I request permission to leave the net for _____ minutes.

QNU*   The net has traffic for you. Stand by.

QNV*   Establish contact with _____ on this frequency. If successful, move to _____ and send him traffic for _____.

QNW   How do I route messages for _____?

QNX   You are excused from the net.*
      Request to be excused from the net.

QNY*   Shift to another frequency (or to _____ kHz) to clear traffic with _____.

QNZ   Zero beat your signal with mine.

*For use only by the Net Control Station.

Notes on Use of QN Signals

The QN signals listed above are special ARRL signals for use in amateur CW nets *only*. They are not for use in casual amateur conversation. Other meanings that may be used in other services do not apply. Do not use QN signals on phone nets. *Say it with words.* QN signals need not be followed by a question mark, even though the meaning may be interrogatory.

These "Special QN Signals for New Use" originated in the late 1940s in the Michigan QMN Net, and were first known to Headquarters through the then head traffic honcho W8FX. Ev Battey, W1UE, then ARRL assistant communications manager, thought enough of them to print them in *QST* and later to make them standard for ARRL nets, with a few modifications. The original list was designed to make them easy to remember by association. For example, QNA meant "Answer in *A*lphabetical order," QNB meant "Act as relay *B*etween...," QNC meant "All *Net Copy,*" QND meant "*Net* is *D*irected," etc. Subsequent modifications have tended away from this very principle, however, in order that some of the less-used signals could be changed to another, more needed, use.

Since the QN signals started being used by amateurs, international QN signals having entirely different meanings have been adopted. Concerned that this might make our use of QN signals with our own meanings at best obsolete, at worst illegal, ARRL informally queried FCC's legal branch. The opinion then was that no difficulty was foreseen as long as we continued to use them only in amateur nets.

Occasionally, a purist will insist that our use of QN signals for net purposes is illegal. Should anyone feel so strongly about it as to make a legal test, we might find ourselves deprived of their use. Until or unless we reach such an eventuality, however, let's continue to use them, and use them right. After all, we were using them first.

---

do make connection, and pass the traffic, they will return as soon as they are done.)

AI0O: O

K0SI: O

AI0O: DE AI0O GE PETE QRU K

K0SI: AI0O DE K0SI GE ROB QNU PSE UP 4 AI0O 1 WID ND0N AFTER W0OUD THEN BOTH QNX 73 K

(Good evening, Rob, the net has traffic for you. Please go up 4 and get one for you from ND0N [WID means "with"] after he finishes with W0OUD. Then, when you are both finished, you and ND0N are both excused from the net. 73!)

AI0O: 73 GG

As you can see, it doesn't take much to say a lot on a CW net. Now that all the net business is taken care of, The NCS will start excusing other stations. Since K2ONP has a schedule to make with the Tenth Region Net, he will be excused first.

K0SI: K2ONP DE K0SI TU GEO FER QNI NW QRU QNX TU 73 K

K2ONP: TU PETE CUL 73 DE K2ONP $\overline{SK}$

W0OUD: OUD

(Letha is back from receiving her traffic.)

K0SI: OUD TU LETHA NW QRU QNX 88 K

W0OUD: GN PETE CUL 88 DE W0OUD $\overline{SK}$

Now, the NCS will close the net.

K0SI: MCWN NW QNF [the net is free] GN DE K0SI CL

When you check into a CW net for the first time, don't worry about speed. The NCS will answer you at about the speed you check into the net. You will discover that everyone on a CW net checks in with a different speed, just as everyone has a different voice on SSB. It's nothing to be self-conscious about. As the saying goes, "We're all in this together." The goal is to pass the traffic correctly, with 100 percent accuracy, not burn up the ether with our spiffy fists. Likewise, don't hesitate to slow down for someone else. Remember, when handling traffic, 100 percent is the minimum acceptable performance level!

On SSB

Now, let's journey through a world with plenty of shadow and substance, the Mowegia Single Sideband Net.

As we look in on K0PCK calling tonight's session, keep in mind these few pointers:

1) The net preamble, given at the beginning of each session, will usually give you all the information you need to survive on the net. Method of checking in varies greatly from net to net. For instance, some section nets have a prearranged net roll, some take checkins by alphabetical order, or some even take checkins by city, county or geographical area. Don't feel intimidated by a prearranged net roll. Those nets will always stand by near the end of the session to take stations not on the net roster.

2) As you listen to a net, you will find that on phone, formality also varies. Some SSB nets are strictly business, while others are "chattier." However, don't always confuse lack of formality with looseness on a net. There is still a

definite net procedure to adhere to.

3) Once, on a close play, the catcher asked umpire Bill Klem, "Well, what is it?"

"It ain't nothin' till I call it," he growled.

By the same token, you need to keep in mind that the NCS is the absolute boss when the net is in session. On CW nets, this doesn't seem to be much of a problem to the tightness of operation, but on phone nets, sometimes a group of "well-meaners" can really slow down the net. So don't "help" unless NCS tells you to.

Since net time is upon us, let's get back to the beginning of the Mowegia Single Sideband Net.

"Calling the Mowegia Single Sideband Net, calling the Mowegia Single Sideband Net. This is KØPCK, net control. The Mowegia Single Sideband Net meets on 3988 kHz nightly at 6 PM for the purpose of handling traffic in Mowegia and to provide a link for out-of-section traffic through the ARRL National Traffic System. My name is Ben, Bravo Echo November, located in Toad Lick, Mowegia. When I call for the letter corresponding to the first letter of the suffix of your call, please give your call sign only."

"Is there any emergency, priority or time value traffic for the net?"

(wait 5 seconds)

"Any relays?"

(wait 5 seconds)

"Any low-power, mobile or portable stations wishing to check in?"

(wait 5 seconds)

"Any relays?"

(wait 5 seconds)

"This is KØPCK for the Mowegia Sideband Net. Do we have any formal written traffic?"

"KØORB."

"KØORB, Good evening Bill. List your traffic, please."

"Good evening, Ben. I have two pieces to go out-of-state."

"Very good. Who is our Tenth Region Rep tonight?"

"Good evening, Ben. This is NIØR, Ten Rep."

"NIØR, this is KØPCK. Hi, Roger. Please call KØORB, move him to 3973 and get Bill's out-of-state traffic."

"KØORB, this NIØR. See you on 73."

"Going. KØORB."

Now, let's sit back and analyze this. As you can see, the format is pretty much identical except that it takes more words. Oh, yes, one other difference...you may have noticed that not one single Q-signal was used! Q-signals should not be used on phone nets; using them detracts from the "professionalism" of a net. Work hard at avoiding them, and you will reduce the "lingo barrier," making it a little less intimidating for a potential new checkin.

Two Special Cases

RTTY nets are almost identical to CW nets, except for the method of transmission, but you need to remember two extra tips:

1) It is always important to zero-beat the NCS, but on RTTY it is crucial! It doesn't take much to turn QTC Corn Shuck Holler 1 to QTHGCRRX ULUSK HXLBELZZ4.

2) Stay far away from the key that sends RYRYRYRYRYRY. If you follow rule number one, RYing isn't necessary.

The other "exception to the rule" is the local 2-meter FM net. Many 2-meter nets are designed for ragchewing or weather spotting, so often if you bring traffic to the net, be prepared to coach someone in the nuances of traffic handling.

Who Owns the Frequency?

Traffic nets sometimes have difficulties when it comes time for the call-up, and a ragchew is taking place on the published net frequency. What to do? Well, you could break in on the ragchew and ask politely if the participants would mind relinquishing the frequency. This usually works, but what if it doesn't? The net has no more right to the frequency than the stations occupying it at net time, and the ragchew stations would be perfectly within their rights to decline to relinquish it.

The best thing to do in such a case is to call the net near, but not directly on, the normal frequency—far enough (hopefully) to avoid causing QRM, but not so far that net stations can't find the net. Usually, the ragchewers will hear the net and move a bit farther away—or even if they don't, the net can usually live with the situation until the ragchew is over.

It is possible to conceive of a situation, especially on 75-meter phone, in which the net frequency is occupied and the entire segment from 3850 to 4000 kHz is wall-to-wall stations. In such a case, the net will just have to do the best it can and take its chances on QRMing and being QRMed. In any case, it is *not* productive to argue about who has the most right to the frequency. Common courtesy says that the first occupants have, but there are many extenuating circumstances. Avoid such controversies, especially on the air.

Accordingly, net frequencies should be considered "approximate," inasmuch as it may be necessary for nets to vary their frequencies according to band conditions at the time. Further, no amateur or organization has any preemptory right to any specific amateur frequency; please note FCC Rule 97.63(a), which states "An amateur station may transmit on any frequency within any authorized amateur frequency band."

A few other things to remember:

1) Unlike a "double" on CW or SSB, where the NCS might even still get both calls, a "double" on an FM repeater either captures only one station, or makes an ear-splitting squeally heterodyne. Drag your feet a little before you check in, so you are less likely to double.

2) Be especially aware to wait for the squelch tail or courtesy beep. More people seem to time out the repeater on net night than any other!

3) Remember that a lot of people have scanners, many with the local repeater programmed in on one of the channels. Design your behavior in such a way that it attracts nonhams to Amateur Radio. In other words, don't do anything you wouldn't do in front of the whole town! [For more information on RTTY operating, see Chapter 9; see Chapter 11 for additional details about FM.]

Oh, yes. One other thing will insure your success and build your reputation as a good traffic handler. It's a simple thing, really, but you'd be surprised how much mileage you can get out of saying "please" and "thank you."

MAKING IT, TAKING IT AND GIVING IT AWAY: MESSAGE HANDLING AND MESSAGE FORM

By this stage, you have probably been checking into the net for a while, and things have started to move along quite smoothly on your journey as a rookie traffic handler. But in the life of any new traffic op, the fateful day comes along when the NCS points RF at you and says those words that strike fear in almost every newcomer: "Pse up 4 get one QTC."

Now what? You could suddenly have "rig trouble" or a

"power outage" or a "telephone call," or just bump your dial and disappear. After all, it has been done before, and everyone that didn't do it sure thought of it the first time they were asked to take traffic! Of course, there is a more honorable route—go ahead and take it! Chances are you will be no worse than anyone else your first time out. The late Bob Peavler, WØBV, a well-known and respected trafficker, used to say, "Don't be afraid of making a mistake. Chances are we have already made all of them before you."

To ease the shock of your first piece of traffic, maybe it would be a good idea to go over message form "by the book"—ARRL message form, that is.

ARRL Message Form—the Right Way, and the Right Way

A common line of non-traffickers is, "Aw, why do they have to go through that ARRL message-form stuff? It just confuses people and besides, my message is just a few words or so. It's silly to go through all that rigamarole."

Well, then, let's imagine that you are going to write a letter to your best friend. What do you think would happen to your letter if you decided that the standardized method the post office used was "silly," so you signed the front, put the addressee's address where the return address is supposed to go, and stamped the inside of the letter? It would probably end up in the Dead Letter Office.

Amateur message form is standardized so it will reach its destination speedily and correctly. It is very important for every amateur to understand correct message form, because you never know when you will be called upon in a emergency. Most nonhams think all hams know how to handle messages, and it's troublesome to discover how few do. You can completely change the meaning of a piece of traffic by accident if you don't know the ARRL message form, and as you will see later, this can be a real "disaster." Learn it the right way, and this will never happen.

If you will examine the sample message in this section, you will notice that the message is essentially broken into four parts: the preamble, the address block, the text and the signature. The preamble is analogous to the return address in a letter and contains the following:

1) The number denotes the message number of the originating station. Most traffic handlers begin with number 1 on January 1, but some stations with heavy volumes of traffic begin the numbering sequence every quarter or every month.

2) The precedence indicates the relative importance of the message. Most messages are Routine (R) precedence—in fact, about 99 out of 100 are in this category. You might ask, then why use any precedence on routine messages? The reason is that operators should get used to having a precedence on messages so they will be accustomed to it and be alerted in case a message shows up with a different precedence. A Routine message is one that has no urgency aspect of any kind, such as a greeting. And that's what most amateur messages are—just greetings.

The Welfare (W) precedence refers to either an inquiry as to the health and welfare of an individual in the disaster area or an advisory from the disaster area that indicates all is well. Welfare traffic is handled only after all emergency and priority traffic is cleared. The Red Cross equivalent to an incoming Welfare message is DWI (Disaster Welfare Inquiry).

The Priority (P) precedence is getting into the category of high importance and is applicable in a number of circumstances: (1) important messages having a specific time limit, (2) official messages not covered in the emergency category, (3) press dispatches and emergency-related traffic not of the utmost urgency, and (4) notice of death or injury in a disaster area, personal or official.

The highest order of precedence is EMERGENCY (always spelled out, regardless of mode). This indicates any message having life-and-death urgency to any person or group of persons, which is transmitted by Amateur Radio in the absence of regular commercial facilities. This includes official messages of welfare agencies during emergencies requesting supplies, materials or instructions vital to relief of stricken populace in emergency areas. During normal times, it will be very rare.

3) Handling Instructions are optional cues to handle a message in a specific way. For instance, HXG tells us to cancel delivery if it requires a toll call or mail delivery, and to service it back instead. Most messages will not contain handling instructions.

4) Although the station of origin block seems self-explanatory, many new traffic handlers make the common

Handling Instructions

HXA—(Followed by number.) Collect landline delivery authorized by addressee within _____ miles. (If no number, authorization is unlimited.)

HXB—(Followed by number.) Cancel message if not delivered within _____ hours of filing time; service originating station.

HXC—Report date and time of delivery (TOD) to originating station.

HXD—Report to originating station the identity of station from which received, plus date, time and method of delivery.

HXE—Delivering station get reply from addressee, originate message back.

HXF—(Followed by number.) Hold delivery until _____ (date).

HXG—Delivery by mail or landline toll call not required. If toll or other expense involved, cancel message and service originating station.

An HX prosign (when used) will be inserted in the message preamble before the station of origin, thus: NR 207 R HXA50 W1AW 12... (etc). If more than one HX prosign is used, they can be combined if no numbers are to be inserted; otherwise the HX should be repeated, thus: NR 207 R HXAC W1AW... (etc), but: NR 207 R HXA50 HXC W1AW... (etc). On phone, use phonetics for the letter or letters following the HX, to insure accuracy.

mistake of exchanging their call sign for the station of origin after handling it. The station of origin never changes. That call serves as the return route should the message encounter trouble, and replacing it with your call will eliminate that route. A good rule of thumb is *never* to change any part of a message.

5) The check is merely the word count of the text of the message. The signature is not counted in the check. If you discover that the check is wrong, you may not change it, but you may amend it by putting a slash bar and the amended count after the original count. [See below for additional information on the message check.]

Another common mistake of new traffickers involves "ARL" checks. A check of ARL 8 merely means the text has an ARL numbered radiogram message text in it, and a word count of 8. It does not mean ARRL numbered message no. 8. This confusion has happened before, with unpleasant results. For instance, an amateur with limited traffic experience once received a message with a check of ARL 13. The message itself was an innocuous little greeting from some sort of fair, but the amateur receiving it thought the message was ARRL numbered message 13—"Medical emergency exists here." Consequently, he unknowingly put a family through a great deal of unnecessary stress. When the smoke cleared, the family was on the verge of bringing legal action against the ham, who himself developed an intense hatred for traffic of any sort and refused to ever handle another message. These kinds of episodes certainly don't help the "white hat" image of Amateur Radio!

When counting messages, don't forget that each "X-ray" (instead of periods), "Query" and initial group counts as a word. Ten-digit telephone numbers count as three words; the ARRL-recommended procedure for counting the telephone number in the text of a radiogram message is to separate the telephone number into groups, with the area code (if any) counting as one word, the three-digit exchange counting as one word, and the last four digits counting as one word. Separating the telephone number into separate groups also helps to minimize garbling. Also remember that closings such as "love" or "sincerely" (that would be in the signature of a letter) are considered part of the text in a piece of amateur traffic.

6) The place of origin can either be the location (City/State or City/Province) of the originating station or the location of the third party wishing to initiate a message through the originating station. Use standard abbreviations for state or province. ZIP or postal codes are not necessary. For messages from outside the US and Canada, city and country is usually used. If a message came from the MARS (Military Affiliate Radio System) system, place of origin may read something like "Korea via Mars."

7) The filing time is another option, usually used if speed of delivery is of significant importance. Filing times should be in UTC or "Zulu" time.

8) The final part of the preamble, the date, is the month and day the message was filed—year isn't necessary.

Next in the message is the address. Although things like ZIP code and phone number aren't entirely necessary, the more items included in the address, the better its chances of reaching its destination. To experienced traffic handlers, ZIP codes and telephone area codes can be tip-offs to what area of the state the traffic goes, and can serve as a method of verification in case of garbling. For example, all ZIP codes in Minnesota start with a 5. Therefore, if a piece of traffic

Checking Your Message

Traffic handlers don't have to dine out to fight over the check! Even good ops find much confusion when counting up the text of a message. You can eliminate some of this confusion by remembering these basic rules:

1) Punctuation ("X-rays," "Querys") count separately as a word.

2) Mixed letter-number groups (1700Z, for instance) count as one word.

3) Initial or number groups count as one word if sent together, two if sent separately.

4) The signature does not count as part of the text, but any closing lines, such as "Love" or "Best wishes" do.
Here are some examples:

- Charles J McClain—3 words
- W B Stewart—3 words
- St Louis—2 words
- 3 PM—2 words
- SASE—1 word
- ARL FORTY SIX—3 words
- 2N1601—1 word
- Seventy three—2 words
- 73—1 word

Telephone numbers count as 3 words (area code, prefix, number), and ZIP codes count as one. ZIP + 4 codes count as two words. Canadian postal codes count as two words (first three characters, last three characters.)

Although it is improper to change the text of a message, you may change the check. Always do this by following the original check with a slash bar, then the corrected check. On phone, use the words "corrected to."

Book Messages

When sending book traffic, always send the common parts first, followed by the "uncommon" parts. For example:

R NØFQW ARL 8 BETHEL MO SEP 7 $\overline{BT}$
ARL FIFTY ONE BETHEL SHEEP FESTIVAL X LOVE $\overline{BT}$
PHIL AND JANE $\overline{BT}$
NR 107 TONY AND LYN CALHOUN $\overline{AA}$
160 NORTH DOUGLAS $\overline{AA}$ SPRINGFIELD IL 62702 $\overline{BT}$
NR 108 JOE WOOD AJØX $\overline{AA}$
84 MAIN STREET $\overline{AA}$
LAUREL MS 39440 $\overline{BT}$
NR 109 JEAN WILCOX $\overline{AA}$
1243 EDGEWOOD DRIVE $\overline{AA}$
LODI CA 95240 $\overline{AR}$ N

Before sending the book traffic to another operator, announce beforehand that it is book traffic. Say "Follows book traffic." Then use the above format. On CW, a simple HR BK TFC will do.

sent as St Joseph, MO, with a ZIP of 56374 has been garbled along the way, it conceivably can be rerouted. So, when it comes to addresses, the adage "the more, the better" applies.

The text, of course, is the message itself. You can expedite the counting of the check by following this simple rule—when copying by hand, write five words to a line. When copying with a typewriter, or when sending a message via RTTY, type the message 10 words to a line. You will discover that this is a quick way to see if your message count agrees with the check. If you don't agree, nine times out of 10 you have dropped or added an "X-ray," so copy carefully. Another important thing to remember is that you never end a text with a "X-ray"—it just wastes space and makes the word count longer.

Finally, the signature. Remember, complimentary closing words like "sincerely" belong in the text, not the signature. In addition, signatures like "Dody, Vanessa, Jeremy,

MARS

Most modern-day traffic handlers don't realize that our standard message preamble is largely fashioned after the form used in the Army Amateur Radio System, which had its heyday in the '30s. The ARRL standard preamble prior to that time was quite different, but the AARS form was adopted because it had advantages and was so widely used. AARS nets were numerous in the '30s, using WL calls on two frequencies (3497.5 and 6990 kc.) outside the amateur bands and many frequencies, using amateur calls and amateur participants, inside the bands.

MARS, the post-World War II successor to AARS, encompasses all three of the US armed services. Although it operates numerous nets, all of which are outside the amateur bands on military frequencies, thousands of amateurs participate in MARS. This service performs some traffic coverage that we amateurs are not permitted to perform. MARS, which stands for "Military Affiliate Radio System," is conducted in three different organizations under the direction of the Army, Navy and Air Force. Therefore, MARS is not in the strictest sense Amateur Radio.

Nevertheless, many amateur messages find their way into MARS circuits, and MARS messages find their way into amateur nets. In fact, NTS has a semiformal liaison with MARS to handle the many messages from families in the States to their sons and daughters serving overseas.

Traffic for some points overseas at which US military personnel are stationed can be handled via MARS, provided a complete military address is given, even though some of these points cannot be covered by Amateur Radio or NTS. The traffic is originated in standard ARRL form and refiled into MARS form (now quite a bit different from ours) when it is introduced into a MARS circuit for transmission overseas. In this manner, traffic may be exchanged with military personnel in West Germany, Japan and a few other countries that do not otherwise permit the handling of international third-party traffic.

Traffic coming from MARS circuits into amateur nets for delivery by Amateur Radio are converted from MARS to amateur form and handled as any other amateur message. The *exception* occurs when traffic originates overseas. In this case, the name of the country in which it originates, followed by "via MARS," should appear as the place of origin, so it does not appear that such messages were handled illegally by Amateur Radio. There are places where US military personnel are stationed that even MARS cannot handle traffic with, presumably because of objections by the host country. This information is in the hands of MARS "gateway" stations and changes from time to time.

The amount of MARS traffic appearing on amateur nets is not great, since MARS has a pretty good system of handling it on MARS frequencies, but it is important that we maintain liaison as closely as possible since all civilian MARS members are US licensed amateurs.

Information concerning MARS may be obtained directly from the individual branches at these addresses:

Air Force MARS
Chief, Air Force MARS
HQ, AFCC/DOOCC
Scott AFB, IL 62225-6001

Navy-Marine Corps MARS
Director, Navy-Marine Corps MARS
Naval Communications Unit
Washington, DC 20390-5161

HQ Army MARS
US Army Information Systems Command
AS-OPS-OA
Ft Huachuca, AZ 85613

Ashleigh, and Uncle Porter," no matter how long, go entirely on the signature line.

At the bottom of our sample message you will see call signs next to the blanks marked "sent" and "received." These are not sent as the message, but are just bookkeeping notes for your own files. If necessary, you could help the originating station trace the path of the message.

Keeping It Legal

The following is quoted directly from the US Amateur Regulations:

§97.114 Limitations on third-party traffic

a) Subject to the limitations specified in paragraphs (b) and (c) of this section, an amateur radio station may transmit third-party traffic.

b) The transmission or delivery of the following third-party traffic is prohibited:

(1) International third-party traffic except with countries which have assented thereto;

(2) Third-party traffic involving material compensation, either tangible or intangible, direct or indirect, to a third party, a station licensee, a control operator or any other person;

(3) Except for an emergency communications as defined in this part, third-party traffic consisting of business communications on behalf of any party.

(c) The licensee of an amateur radio station may not permit any person to participate in traffic from that station as a third party if:

(1) The control operator is not present at the control point and is not continuously monitoring and supervising the third-party participation to ensure compliance with the rules;

(2) The third party is a prior amateur radio licensee whose license was revoked; suspended for less than the balance of the license term and the suspension is still in effect; suspended for the balance of the license and relicensing has not taken place; surrendered for cancellation following notice of revocation, suspension or monetary forfeiture proceedings; or who is the subject of a cease and desist order which relates to amateur operation and which is still in effect.

The FCC defines business communications as any transmission or communication the purpose of which is to facilitate the regular business or commercial affairs of any party. [§97.3(bb)]. Emergency communication is defined in §97.3(w) as "any amateur communication directly relating to the immediate safety of life of individuals or the immediate protection of property."

It's self-explanatory. Every amateur should be familiar with these rules. Also, while third-party traffic is permitted in the US and Canada, this is not so for most other nations. A special legal agreement is required in each country to make such traffic permissible, both internally and externally. A list of third-party agreements in effect when this book was published appears in Chapter 5. Check this list before agreeing to handle *any kind* of international traffic.

Such agreements specify that only unimportant, personal, nonbusiness communications be handled—things that ordinarily would not utilize commercial facilities. (In an emergency situation, amateurs generally handle traffic *first* and face the possible consequences *later*. It is not unusual for a special limited-duration third-party agreement to be instituted by the affected country during an overseas disaster.) The key point here is, particularly under routine day-to-day nonemergency conditions, if we value our privileges, we must take care not to abuse any regulations, whether it be on the national or international level.

Some Helpful Hints about Receiving Traffic

1) Once you have committed the format of ARRL message form to memory, there's no need to use the "official" message pads from ARRL except for deliveries. Traffic handlers have many varied materials on hand for message handling. Some just use scrap paper. Many buy inexpensive 200-sheet 6- × 9-inch plain tablets available at stationery stores. Those who like to copy with a typewriter often use roll paper or fan-fold computer paper to provide a continuous stream of paper, separating each message as needed. As long as you can keep track of it, anything goes in the way of writing material.

2) Don't say "QSL" or "I roger number..." unless you mean it! It's not "Roger" unless you've received the contents of the message 100 percent. It's no shame to ask for fills (ie, repeats of parts of the message). Make sure you have received the traffic correctly before going on to the next one.

3) Full- (QSK) or semi-breakin can be very useful in handling traffic on CW (or VOX on SSB). If you get behind, saying "break" or sending a string of dits will alert the other op that you need a fill.

4) You can get a fill by asking for "word before" (WB), "word after" (WA), "all before" (AB), "all after" (AA) or "between" (BN).

Sending the Traffic

Just because you've taken a few messages, don't get the notion that being good at receiving traffic makes you a good sender, too. In the "dark ages" of hamdom, when one had to journey to the FCC office to take exams, one not only had to receive code in the code test, but send it as well. It used to be amazing how many folks could copy perfect 20-WPM code only to be inadequate on the sending portion. A lot of good ops became quite amazed that they could be so ham-fisted on a straight key!

Good traffic operators know they have to learn the nuances of sending messages as well as getting them. Your ability to send can "make or break" the other operator's ability to receive traffic in poor conditions. Imagine yourself and the other operator as a pair of computers interfaced over the telephone. Your computer (your brain) must successfully transfer data through your modem (your rig) over the lines of communication (the amateur frequencies) to the other modem, and ultimately, to the other computer.

Of course, for this transfer to be successful, two major items must be just right. The modems must operate at the same baud rate, and they must operate on matching protocol. Likewise, you must be careful to send your traffic at a comfortable speed for the receiving op, and use standardized protocol (standard ARRL message form). As you will see, these are slightly different for phone and CW, with even a couple of other deviations for RTTY or FM.

Sending the Traffic by CW

Someone once remarked, "The nice thing about CW traffic handling is that you have to spell it as you go along, so

you don't usually have to spell words over." Also, the other main difference in CW traffic handling is that you tell someone when to go to the next line or section by use of the prosigns $\overline{AA}$ or $\overline{BT}$. Keeping this in mind, let's show how our sample message would be sent:

NR 133 R HXG WØMME ARL8 MOUNT PLEASANT IA 1700Z SEP 1
MR MRS JEFF HOLTZCLAW $\overline{AA}$
ROUTE 1 BOX 127 $\overline{AA}$
TONGANOXIE KS 66086 $\overline{AA}$
913 555 1212 $\overline{BT}$
ARL FIFTY ONE OLD THRESHERS REUNION X LOVE $\overline{BT}$
UNCLE CHUCKIE $\overline{AR}$ N (if you have no more messages) or $\overline{AR}$ B (if you have further messages)

Now, let's examine a few points of interest:

1) You don't need to send "preamble words" such as precedence and check. The other operator is probably as familiar with standard ARRL form as you are (maybe more!).

2) The first three letters of the month are sufficient when sending the date.

3) In the address, always spell out words like "route" and "street."

4) Do not send dashes in the body of telephone numbers; it just wastes time.

5) Always, always, always spell out each word in the text! For example, "ur" for "your" could be misconstrued as the first two letters of the next word. Abbreviations are great for ragchewing, but not for the text of a radiogram.

6) Sometimes, if you have sent a number of messages, when you get to the next-to-last message, it's a good idea to send $\overline{AR}$ 1 instead of $\overline{AR}$ B to alert the other station that you have just one more.

7) If the other operator "breaks" you with a string of dits, stop sending and wait for the last word received by the other side. Then, when you resume sending, start up with that word, and continue through the message.

Becoming a proficient CW traffic sender is tough at first, but once you've mastered the basics, it will become second nature—no kidding!

Sending the Traffic by Phone

Phone traffic handling is a lot like the infield fly rule in baseball—everyone thinks they know the rule, but in truth few really do. Correct message handling via phone can be just as efficient as via CW if and only if the two operators follow these basic rules:

1) If it's not an actual part of the message, don't say it.

2) Unless it's a very weird spelling, don't spell it.

3) Don't spell it phonetically unless it's a letter group or mixed group, or the receiving station didn't get it when you spelled it alphabetically.

Keeping these key points in mind, let's waltz through our sample message. This is how an efficient phone traffic handler would send the message:

"Number one hundred thirty three, routine, Hotel X-ray Golf, WØ Mike Mike Echo, Mount Pleasant, Iowa, seventeen hundred Zulu, September one."

"Mr and Mrs Jeff Holtzclaw, route one, box twenty seven, Tonganoxie, T-O-N, G-A-N, O-X, I-E, Kansas, six six zero eight six. Nine one three, five five five, one two, one two. Break." You would then let up on the PTT switch (or pause if using VOX) and give the operator any fills needed in the first half of the message.

"ARL FIFTY ONE Old Threshers Reunion X-ray Love

Break Uncle Chuckie. End, no more'' (if you have no more messages), ''more'' (if you have more messages).

Notice that in phone traffic handling, a pause is the counterpart for $\overline{AA}$. Also notice the lack of extraneous words. You don't need to say ''check,'' or ''signature,'' or ''Jones, common spelling.'' (If it's common spelling, why tell someone?) You only spell the uncommon. Most importantly, you speak at about half reading speed to give the other person time to write. If the receiving operator types, or if you have worked with the other op a long time and know his capabilities, you can speak faster. Always remember that any fill slows down the message more than if you had sent the message slowly to begin with!

Oh, Yes... Those Exceptions!

Once again, RTTY and VHF FM provide the exceptions to the rules. RTTY traffic is very much like CW traffic. You include the prosigns $\overline{AA}$, $\overline{BT}$, and $\overline{AR}$, as a typed AA, BT or AR. Single-space your message (you don't need to waste paper) and send ''BBBB'' or ''NNNN'' as the last line of your message depending on whether you have more traffic or not. Use three or four lines between messages. This allows you to get four or five average messages on a standard 8½ × 11-inch sheet of paper. For further details, see Chapter 9.

When sending a message over your local repeater, remember that you often will be working with someone who isn't a traffic handler. It may be necessary to break more often (between the preamble and the address, for instance). Also, always make sure they understand about ARL numbered radiogram texts, and if they don't have a list, tell them what the message means. [The complete list of ARRL numbered radiogram texts appears in Chapter 17.] FM is a quiet mode, so you can get away with less spelling than you do on SSB.

If you yourself are already into traffic, don't try to force-feed correct traffic procedure in the case of someone just starting out; ease the person into it a little at a time. You will give a nontrafficker a more positive impression of traffic handling and may even make someone more receptive to joining a net. After all, our goal as traffic handlers is to make it look fun!

Delivering a Message

Up to now, all our traffic work has been carried out in the safe, secure world of Amateur Radio. All of this changes, though, when we get a piece of traffic for delivery. Suddenly, we're thrust out among the general public in the real world with this little message and expected to give it to someone. Unfortunately, many hams don't realize the importance of this little action and miss an opportunity to engrave a favorable impression of Amateur Radio on nonhams. It's ironic that many hams can chat for hours on CW or SSB, but can't pick up the telephone and deliver a 15-word message without mumbling, stuttering or acting embarrassed. Delivering messages should be a treat, not a chore.

Let's go through a few guidelines for delivery, and if you keep these tips in mind, you and the party on the other end of the phone will enjoy the delivery.

1) Ask for the person named in the message. If he or she is not home, ask the person on the phone if they would take a message for that person.

2) Introduce yourself. Don't you hate phone calls from people you don't know and don't bother to give a name? Chances are they're trying to sell you something, and you brush them off. Most people have no idea what Amateur

Radio is about, and it's up to you to make a good first impression.

3) Tell who the message is from before you give the message. Since the signature appears at the end of the message, most hams give it last, but you will hold the deliveree's attention longer if you give it first. When you get letters in the mail, you check out the return addresses first, don't you? Then you open them in some sort of order of importance. Likewise, the party on the phone will want to know the sender of the message first.

A good way to start off a delivery is to say something like, ''Hi, Mr/Mrs/Miss so-and-so, my name is whatever, and I received a greeting message via Amateur Radio for you from wherever from such-and-such person.'' This usually gives you some credibility with your listener, because you mentioned someone they know. They will usually respond by telling such-and-such is their relative, college friend, etc. At that point, you have become less of a stranger in their eyes, and now they don't have to worry about you trying to sell them some aluminum siding or a lake lot at Casa Burrito Estates. Make sure you say it's a *greeting* message, too, to allay any fears of the addressee that some bad news is imminent.

4) When delivering the message, skip the preamble and just give the text, avoiding ARL text abbreviations. Chances are, Grandma Ollie doesn't give two hoots about the check of a message, and thinks ARL FORTY SIX is an all-purpose cleaner. Always give the ''translation'' of an ARL numbered text, even if the message is going to another ham.

5) Unless it is a ''mass origination'' message from a fair or other special event, ask the party if they would like to send a return message. Explain that it's absolutely free, and that you would be happy to send a reply if they wish. Experienced traffickers can vouch that it's easy to get a lot of return and repeat business once you've opened the door to someone. It's not uncommon for strangers to ask for your name or phone number once they discover Amateur Radio is a handy way to tell relatives ''Arriving 3 PM on Flight 202 next Wednesday.''

To Mail or Not to Mail

Suppose you get a message that doesn't have a phone number, or the message would require a toll call. Then what? If you don't know anyone on 2 meters that could deliver it, or Directory Assistance is of no help, you are faced with the decision whether to mail it or not. There is no hard and fast rule on this (unless, of course, the message has an HXG attached). Always remember that since this is a free service, you are under no obligation to shell out a stamp just because you accepted the message.

Many factors may influence your decision. If you live in a large urban area, you probably have more deliveries than most folks, and mail delivery could be a big out-of-pocket expense that you're not willing to accept. If you live out in the wide open spaces, you may be the only ham for miles around, and probably consider mail delivery more often than most. Are you a big softie on Christmas or Mother's Day? If so, you may be willing to brunt the expense of a few stamps during those times of the year when you wouldn't otherwise. At any rate, the decision is entirely up to you.

Although you may be absolved from the responsibility of mailing a message, you don't just chuck the message in the trash. You do have a duty to inform the originating station that the message could not be delivered. A simple ARL SIXTY SEVEN followed by a brief reason (no listing, no one home for three days, mail delivery returned by post office, and so forth) will suffice. This message always goes to the station of origin, not the person in the signature. The originating station will appreciate your courtesy.

Now That You're Moving Up in the World

By now, you are starting to get a grasp of the traffic world. You've been checking in to a net on a regular basis, and you're pretty good at message form. Maybe you've even delivered a few messages. Now you are ready to graduate from Basic Traffic 101 and enroll in Intermediate Traffic 102. Good for you! You have now surpassed 80 percent of your peers in a skillful specialty area of Amateur Radio. However, there's still a lot to learn, so let's move on.

Book Messages

Over the years, book messages have caused a lot of needless headaches and consternation among even the best traffic handlers. Many hams avoid booking anything just because they think it's too confusing. Truthfully, book messages are fairly simple to understand, but folks tend to make them harder than they actually are.

So, just what are book messages? Book messages are merely messages with the same text and different addresses. They come in two categories—ones with different signatures, and ones with the same signatures. Elsewhere in the chapter are examples. Often you will see book messages around holiday times and during fairs or other public events.

Counting a book message is no worse than understanding it. Since book messages are quicker to send than regular messages, they don't count as much as an ordinary message. Count every three book messages as one message. If you have one or two messages left over, count one more. For instance:

Book Messages		Regular Messages
3	counts as	1
4	counts as	2
5	counts as	2
6	counts as	2
7	counts as	3
8	counts as	3

and so on.

Oh, yes... one other thing about book messages. When you check into a net with a bunch of book messages, give the regular message count only. Don't say, "I have a book of seven for Outer Baldonia." (The NCS has enough to keep track of without having to break his train of thought to divide by three.) Say instead, "Outer Baldonia three." Then, when you and the station from Outer Baldonia go off frequency to pass the traffic, tell him that it is book traffic. When he tells you to begin sending, give the common parts first, then the "uncommon" parts (addresses and possibly signatures.) By following this procedure, you will avoid a lot of confusion.

Suppose you get a book of traffic on the NTS Region net bound for your state, but to different towns. When you take them to your section net, you will not be able to send them as a book, since they must be sent to different stations. Now what? Simply "unbook" them, and send them as individual messages. For instance, let's say you get a book of three messages (which is a count of one) for the Mowegia section from the region net. Two are for Mowegia City, and one is for Swan Valley. Simply list your traffic as Mowegia City 1 (because those two are still booked together) and Swan Valley 1, for a total count of two. Books aren't ironclad chunks of traffic, but a stepsaver that can be used to your advantage.

They can be unbooked at any time. Use them whenever you can, and don't be afraid of them.

BECOMING NCS AND LIVING TO TELL ABOUT IT

Some momentous evening in your traffic career, you may be called upon to take the net. Perhaps the NCS had a power failure, or is on vacation, or perhaps a vacancy occurred in the daily NCS rotation on your favorite net. Should this be the case, consider yourself lucky. Net Managers entrust few members with net-control duties.

Of course, you probably won't be thinking how lucky you are when the Net Manager says "QNG" and sticks your call after it. Once again, just like your first checkin or your first piece of traffic, you will just have to grit your teeth and live through it. However, you can make the jump easier by following these hints *long before* you are asked to be a net control:

1) Become familiar with the other stations on the net. Even if you never become NCS, it pays to know who you work with and where they live.

2) Pay close attention to the stations that go off frequency to pass traffic. What frequencies does the net use to move traffic? Which stations are off frequency at the moment? You will gain a feel for the net control job just by keeping track of the action.

3) Try to guess what the NCS will do next. You will discover many dilemmas when you try to second-guess the NCS. Often different amounts of traffic with equal precedence appear on the net, and a skillful NCS must rank them in order of importance. For instance, if you follow the NCS closely, you will discover that traffic for the NTS rep, such as out-of-state traffic, gets higher priority than one for the NTS rep's city. Situations like these are fun to second-guess when you are standing by on the net and will better prepare you for the day *you* might get to run the net.

Should that day arrive, just keep your cool and try to implement the techniques used by your favorite net-control stations. After a few rounds of NCS duties, you will develop your own style, and who knows? Perhaps some new hopeful for NCS will try to emulate *you* some day! See accompanying sidebars for further hints on developing a "type-NCS" personality and on proper net-control methods.

HANDLING TRAFFIC AT FAIRS OR OTHER PUBLIC EVENTS

A very special and important aspect of message handling is that of how to handle traffic at public events. If the event is of any size, like a state fair, it doesn't take long to swamp a group of operators with traffic. Only by efficient, tight organization can a handful of amateurs keep a lid on the backlog.

No matter what size your public event, the following points need to be considered for any traffic station accessible to the public:

1) Often, more nontraffic handlers than traffic handlers will be working in the booth. This means your group will have to lay out a standard operating procedure to help the non-traffickers assist the experienced ops.

Jobs such as meeting the public, filling out the message blanks, sorting the "in" and "out" piles, and keeping the booth tidied up, can all be performed by people with little or no traffic skill, and is a good way to introduce those people into the world of traffic.

Are You a Type-NCS Personality?

As net control station (NCS), it pays to remember that the net regulars are the net. Your function is to preside over the net in the most efficient, businesslike way possible so that the net participants can promptly finish their duties and go on to other ones. You must be tolerant and calm, yet confident and quick in your decisions. An ability to "take things as they come" is a must. Remember that you were appointed NCS because your Net Manager believes in you and your abilities.

1) *Be the boss, but don't be bossy.* It's your job to teach net discipline and train new net operators (and retrain some old ones!). You are the absolute boss when the net is in session, even over your Net Manager. However, you must be a "benevolent monarch" rather than a tyrant. Nets lose participation quickly one night a week when it's Captain Bligh's turn to call the net. If the net has a good turnout every night but one, that tells something about its NCS.

2) *Be punctual.* Many of the net participants have other commitments or nets to attend to; liaison stations are often on a tight schedule to make the NTS region or area net. If you, as NCS, don't care when the net starts, others will think it's okay for them to be late, too. Then traffic doesn't get passed in time, and someone may miss his NTS liaison. In short, the system is close to breaking down.

3) *Know your territory.* Your members have names—use them. They also live somewhere—by knowing their locations, you can quickly ascertain who needs to get the traffic. As NCS, it's your responsibility to know the geography of your net. You also need to understand where your net fits into the scheme of NTS.

4) *Take extra care to keep your antennas in good shape,* because an NCS can't run a net with a "wimpy" signal. Although you don't have to be the loudest one on the net, you do have to be heard. You will discover that the best way to do this is to have a good antenna system. A linear amplifier alone won't help you hear those weak checkins!

5) *The NCS establishes the net frequency.* Just because the *Net Directory* lists a certain frequency doesn't give you squatters rights to it if a QSO is already in progress there. Move to a nearby clear frequency, close enough for the net to find you. QRM is a fact of life on HF, especially on 75/80 meters, so live with it.

6) *Keep a log of every net session.* Just because the FCC dropped the logging requirements doesn't mean that you have to drop it. It's a personal decision. The Net Manager may need information about a checkin or a piece of traffic, and your log details can be helpful to him in determining what happened on a particular night.

7) *Don't hamstring the net by waiting to move the traffic.* Your duty is to get traffic moving as quickly as possible. As soon as you can get two stations moving, send them off to clear the traffic. If you have more than one station holding traffic for the same city, let the "singles" (stations with only one piece for that city) go before the ones with more than one piece for that city. The quicker the net gets the traffic moved, the sooner the net can be finished and the net operators can be free to do whatever they want.

---

2) If you plan to handle fairly large amounts of traffic, the incoming traffic needs to be sorted. A good system is to have an "in-state" pile, an "in-region" pile, and three piles for the three areas of NTS. After the traffic has been sent, it needs to be stacked in numerical order in the "out" pile. Keeping it in numerical order makes it easier to find should it need to be referred to.

Since your station will be on a number of hours, plan to check into your region and area, as well as your section, net.

Another good idea is to have "helpers" on 2 meters who can also take some of your traffic to the region and/or area net. These arrangements need to be worked out in advance.

3) Make up your radiogram blanks so that most of the preamble is already on them, and all you need to fill in is the number, check and date. In the message portion, put only about 20 word lines to discourage lengthy messages. Try to convince your "customer" to use a standard ARL text so you can book your messages.

The Net Control Sheet

Net controlling is no easy task, requiring much talent on the part of the operator. A useful prop is a set system for keeping track of net operation so that you don't get mixed up and start to lose control. This can best be effected by a sheet of paper on which you record who has what traffic, who covers what and who is on what side frequency. Trying to keep this information in your head is a losing battle unless you have a remarkable memory.

There are many methods for doing this, depending to a great extent on what net is being controlled and the exact procedure used. In general, however, the best method is to list the calls of stations reporting in vertically down the page, followed by their coverage, if known. The coverage may be unnecessary if the NCS knows his stations, but it is a good idea to leave space for it in case an unfamiliar station reports in and you have to ask his location.

Next, list horizontally across the page the traffic reported by each station coming into the net, using destination (abbreviated) followed by the number of messages. From this you can see at a glance when traffic flow can start. In most nets, it is best to start it right away, and not wait until all stations have reported in. As traffic is passed, it can be crossed off. Whenever you get a station who has no traffic and for whom there is none, that station can be excused (QNX) and crossed off the list. As stations clear traffic and there is none for them, they also can be excused.

If side frequency (QNY) procedure is used, net controlling is a bit more complicated, but the use of such frequencies vastly speeds up the process. In this case, you will need to keep track of who is on which side frequency clearing which traffic, and you will be kept busy dispatching them, sending stations up or down to meet stations already on side frequencies, checking stations back into the net as they return, etc. Both your fingers and your key or mike button will be kept going, and it can be a nightmare if not handled properly.

Probably the best method is to keep the side frequencies on a separate sheet, each side frequency utilizing a separate column, labeled up 5, up 10, down 5, down 10, or whatever spacing intervals you find practical. As two stations are dispatched to a side frequency, enter the suffixes of their calls in the appropriate column, at the same time crossing out the traffic they are sent there to handle. When they return, cross them out of the side-frequency column; this side frequency can then accommodate two other stations. Of course, if you dispatch a third station to the side frequency to wait to clear traffic to one of the stations already there, enter him in the column also, and then only one of the two originally dispatched will return, so just cross off that one. When all your listed traffic is crossed off, the net is ready to secure (QNX QNF).

There are a number of refinements to this method, but the above is basic and a good way to start. Experience will soon indicate better ways to do it. For instance, K2KIR's novel approach follows.

Net Controlling EAN

At the NTS area and region levels, the complexity of the net control task often suggests the desirability of using a matrix form of log sheet. Many of the Eastern Area Net (EAN) Net Control Stations use the form here. A similar form for region nets can be made by replacing the RN column headings with the section net designations and the CAN/PAN columns with single column titled THRU (or EAN, CAN or PAN). Although this form can be used with no other accessories such as a pencil, the use of moveable objects such as 6-32 hex nuts, buttons or push pins materially aids the visualization of which stations are on the net frequency and hence which stations are available for pairing off and passing traffic.

As stations are sent off frequency, their hex nuts are moved from the net column to the appropriate side frequency and a single diagonal line is drawn through the traffic being cleared. If, for any reason, the traffic is not cleared, a circle is drawn around the traffic total as a reminder to the NCS that it still must be cleared. (In other words, the circle overrides the slash.) Assuming the traffic is cleared on the second attempt, an opposite diagonal is drawn, thus totally crossing out the traffic quantity.

For the example shown, the following notes should explain how the various features of this NCS sheet are utilized:

Halfway through the Eastern Area Net session:
• W1EFW and VE3GOL are both clear and have been QNXed at 0055 and 0057, respectively.
• W8PMJ has cleared his CAN and PAN traffic, as well as 8RN traffic from W1TN. He has yet to clear his 3RN, plus receive 8RN from W3YQ. He is presently DOWN TEN with W2CS.
• W2CS was previously sent off to clear his 4RN with W4NTO, but they were unable to complete the pairing. The NCS will try again later, perhaps by using a relay station.
• WB4PNY has a net report for K2KIR, to be sent direct to him if he QNIs.
• N2AKZ and KW1U are UP TEN, clearing 2RN traffic.
• KQ3T, W4NTO and K4ZK are standing by on the net frequency, waiting for new assignments.

There are many techniques for determining in what order to clear the traffic listed, but a couple of fairly common ones are worth mentioning.

1) Assign highest priorities to stations having the largest totals. Thus, W1NJM (PAN RX), W8PMJ (8RN TX and RX) and W3YQ (3RN TX) should be kept waiting as little as possible. If conditions are good, QNQing other stations to these three will speed things up immensely.

2) Tackle the smaller individual destination totals first. This allows early excusing of stations having small traffic totals to clear, and avoids a last-minute panic near the end of the net.

3) Clear short-haul pairings first. In the winter near a sunspot minimum, short-haul communications are most

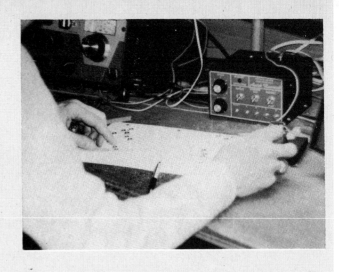

30				EASTERN AREA NET														
+30	+25	+20	+15	+10	+5	IN	STATION	OUT	1RN	2RN	3RN	4RN	8RN	ECN	CAN	PAN	OTR	−5	−10	−15	−20	−25

(FIG. 1 — matrix log sheet showing stations W1EFW, N2AKZ, KQ3T, W4NTO, W8PMJ, VE3GOL, WB4PNY, W1NJM, W3YQ, W1TN, W2CS, K4ZK, KW1U with traffic entries)

FIG. 1

likely to be successful early in the net session before the skip has gone out. At all other times, especially in the summer with nets on Daylight Savings Time, it delays long-haul pairings until band conditions have improved for those paths.

Obviously, there are times when these (or any other algorithms) will conflict with each other. In the final analysis, a good NCS bases his decisions on the specifics of a given net session: the band conditions, the operators, the traffic distribution and the time remaining.—*Bud Hippisley, K2KIR*

A real time- and headache-saver in this department is to fill out the message blank for the sender. This way, you can write in the X-rays and other jargon that the sender is unaware of.

4) Most importantly, realize that you sometimes have to work at getting "customers" as much as if you were selling something! Most people have no concept of Amateur Radio at all, and don't understand how message handling works. ("How can they get it? They don't have a radio like that" is a very common question!)

Use posters to make your booth appealing to the eye. Make sure one of the posters is of the "How your message gets to its destination" variety, such as the one shown in this section. Don't go over someone's head when answering a question—explain it simply and succinctly.

Finally, don't be afraid to "solicit" business. Get up in front of the booth and say "hi" to folks. If they say "hi" back, ask them if they would like to send a free greeting to a friend or relative anywhere in the US (grandparents and grandchildren are the easiest to convince!). Even if they decide not to send a message, your friendliness will help keep our image of "good guys in white hats" viable among the general public, which is every bit as great a service as message handling. For further hints, see accompanying sidebar.

THE ARRL NATIONAL TRAFFIC SYSTEM— MESSAGE HANDLING'S "ROAD MAP"

Although you probably never think about it, when you check into your local net or section net, you are participating in one of the most cleverly designed game plans ever written— the National Traffic System (NTS). Even though the ARRL conceived NTS way back in 1949, and it has grown from one

Handy Hints for Handling Traffic at Fairs or Other Public Events

1) Although you may only be there a day or two, don't compromise your station too much. Try to put up the most you can for an antenna system because band conditions on traffic nets in the summer can really be the pits! Usually, you will be surrounded by electrical lines at fairs, so a line filter is a must. An inboard SSB or CW filter in your rig is a definite plus, too, and may save you many headaches.

2) Don't huddle around the rigs or seat yourself in the back of the booth. Get up front and meet the people. After all, your purpose is to "show off" Amateur Radio to perk the interest of nonhams.

3) Most people will not volunteer to send a piece of traffic, nor will they believe a message is really "free." It's up to you to solicit "business." Be cheerful.

4) Always use "layman's language" when explaining about Amateur Radio to nonhams. Say "message," not "traffic." Don't ramble about the workings

2. AN AMATEUR RADIO OPERATOR TAKES YOUR MESSAGE AND RELAYS IT VIA AMATEUR RADIO TO THE CITY OF DESTINATION.

3. THE RECEIVING STATION GETS YOUR MESSAGE AND CALLS YOUR AUNT PATTY ON THE TELEPHONE.

4. AUNT PATTY GETS YOUR MESSAGE AND IS GLAD TO HEAR FROM YOU.

1. YOU SEND A MESSAGE FROM THE FAIR TO YOUR AUNT PATTY, VIA AMATEUR RADIO.

FREE RADIOGRAMS

of NTS or repeaters; your listener just wants to know how Aunt Patty will get the message. "We take the message and send it via Amateur Radio to Aunt Patty's town, and the ham there will call her on the phone and deliver it to her," will do.

5) Make sure your pencils or pens are attached to the booth with a long string, or you will be out of writing utensils in the first hour!

6) Make sure there are plenty of instructions around for hams not familiar with traffic handling to help them "get the hang of" the situation.

7) Make sure your booth is colorful and attractive. You will catch the public's eye better if you give them something to notice, such as this suggested poster idea.

NATIONAL TRAFFIC SYSTEM AREA & REGION MAP

PAN   CAN   EAN

when radio was still only a wild dream. Three of the Standard Time Zones are the basis for the three NTS areas—Eastern Area (Eastern Time Zone), Central Area (Central Time Zone) and Pacific area (Mountain and Pacific Time Zones). Within these areas are a total of 12 regions. You may wonder why NTS has 12 regions—why not just break it up into 10 regions, one for each area? Ah, but check the map. You will discover that NTS not only covers the US, but our Canadian neighbors as well. Then, of course, the region nets are linked to section/local nets.

The interconnecting lines between the boxes on the flow chart represent liaison stations to and from each level of NTS. The liaisons from area net to area net have a special name, the Transcontinental Corps (TCC). In addition to the functions shown, TCC stations also link the various cycles of NTS to each other.

The clever part about the NTS setup, though, is that in any given cycle of NTS, all nets in the same level commence at approximately the same local time. This allows time for liaisons to the next level to pick up any outgoing traffic and meet the next net. In addition, this gives the TCC stations at least an hour before their duties commence on another area net or their schedule begins with another TCC station.

The original NTS plan calls for four cycles of traffic nets, but usually two cycles are sufficient to handle a normal load of traffic on the system. However, during the holiday season, or in times of emergency, many more messages are dumped into the system, forcing NTS to expand to four cycles temporarily. The cycles of normal operation are Cycle Two, the daytime cycle, which consists primarily of phone nets, and

regular cycle to two to three, NTS hasn't outgrown itself and remains the most streamlined method of traffic handling in the world. (During this discussion, please refer to the accompanying Section/Region/Area map, the NTS Routing Guide and the NTS Flow Chart.)

Actually, the National Traffic System can trace its roots to the railroad's adoption of Standard Time back in 1883,

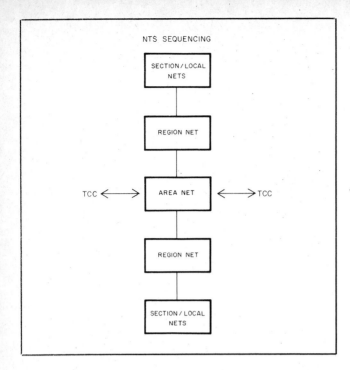

NTS SEQUENCING

SECTION/LOCAL NETS

REGION NET

TCC ← → AREA NET ← → TCC

REGION NET

SECTION/LOCAL NETS

Table 1
National Traffic System Routing Guide

State/Province	Abbrev.	Region	Area
Alaska	AK	7	PAN
Alabama	AL	5	CAN
Alberta	AB	7	PAN
Arizona	AZ	12	PAN
Arkansas	AR	5	CAN
British Columbia	BC	7	PAN
California	CA	6	PAN
Colorado	CO	12	PAN
Connecticut	CT	1	EAN
Delaware	DE	3	EAN
District of Columbia	DC	3	EAN
Florida	FL	4	EAN
Georgia	GA	4	EAN
Guam	GU	6	PAN
Hawaii	HI	6	PAN
Idaho	ID	7	PAN
Illinois	IL	9	CAN
Indiana	IN	9	CAN
Iowa	IA	10	CAN
Kansas	KS	10	CAN
Kentucky	KY	9	CAN
Labrador	LB	11	EAN
Louisiana	LA	5	CAN
Maine	ME	1	EAN
Manitoba	MB	10	CAN
Maryland	MD	3	EAN
Massachusetts	MA	1	EAN
Michigan	MI	8	EAN
Minnesota	MN	10	CAN
Mississippi	MS	5	CAN
Missouri	MO	10	CAN
Montana	MT	7	PAN
Nebraska	NE	10	CAN
Nevada	NV	6	PAN
New Brunswick	NB	11	EAN
New Hampshire	NH	1	EAN
New Jersey	NJ	2	EAN
New Mexico	NM	12	PAN
New York	NY	2	EAN
Newfoundland	NF	11	EAN
North Carolina	NC	4	EAN
North Dakota	ND	10	CAN
Nova Scotia	NS	11	EAN
Ohio	OH	8	EAN
Oklahoma	OK	5	CAN
Ontario	ON	11	EAN
Oregon	OR	7	PAN
Pennsylvania	PA	3	EAN
Prince Edward Island	PEI	11	EAN
Puerto Rico	PR	4	EAN
Quebec	PQ	11	EAN
Rhode Island	RI	1	EAN
Saskatchewan	SK	10	CAN
South Carolina	SC	4	EAN
South Dakota	SD	10	CAN
Tennessee	TN	5	CAN
Texas	TX	5	CAN
Utah	UT	12	PAN
Vermont	VT	1	EAN
Virginia	VA	4	EAN
Washington	WA	7	PAN
West Virginia	WV	8	EAN
Wisconsin	WI	9	CAN
Wyoming	WY	12	PAN
Virgin Islands	VI	4	EAN
APO New York	APO NY	2	EAN
APO San Francisco	APO SF	6	PAN

Cycle Four, the nighttime cycle, made up mostly of CW nets. In addition, Cycle One has been implemented in the Pacific and Eastern areas. Now that the rudiments of NTS have been covered, let's see where you fit in.

NTS and You

Before the adoption of NTS, upper-level traffic handlers worked a system called the "trunk line" system, where a handful of stations carried the burden of cross-country traffic, day in day out. Nowadays, no one has to be an "iron man" or "iron woman" within NTS if they choose not to. If each liaison slot and TCC slot were filled by one person, one day a week, this would allow over 1000 hams to participate in NTS! Unfortunately, many hams have to double- and triple-up duties, so there is plenty of room for any interested amateur.

An NTS liaison spot one day (or night) a week is a great way to stay active in the traffic circuit. Many hams who don't have time to make the section nets get satisfaction in the traffic world by holding a TCC slot or area liasion once a week. If you would enjoy such a post, drop a note to your STM, Net Manager or TCC Director. They will be happy to add another to their fold.

However, remember that the area and region net are very different from your section or local net in one aspect. The function of the section or local net is to "saturate" its jurisdiction, so the more checkins, the better. On the upper-level nets, though, the name of the game is to move the traffic as quickly and efficiently as possible. Therefore, additional checkins— other than specified liaisons and stations holding traffic— only slow down the net. [However, if you are a station holding traffic to be moved, you can enter NTS at any level to pass your traffic, even if you've never been on an upper-level net before. Entering the system at the section or local level is preferred.]

If you are interested in finding out more about the "brass tacks" of the workings to NTS, get a copy of the *Public Service Communications Manual* (FSD-235), available free from ARRL HQ. Every aspect of NTS is explained, as well

as information about local net operating procedure, RACES and ARES operation. The *PSCM* will also orient you with net procedure of region and area NTS nets. It takes a little more skill and savvy to become a regular part of NTS, but the rewards are worth the effort. If you have the chance, go for it!

So There You Have It

Although this chapter is by no means a complete guide to

Fame and Glory: Your Traffic Total, PSHR Report, and Traffic Awards and Appointments

Even if you handle only one message in a month's time, you should send a message to your ARRL Section Traffic Manager (STM) reporting your activity. Your report should include your total originations, messages received, messages sent and deliveries.

An *origination* is any message obtained from a third party for sending from your station. If you send a message to Uncle Filbert on his birthday, you don't get an origination. However, if your mom or your neighbor wants you to send him one with her signature, it qualifies (it counts as one originated and one sent). The origination category is essentially an "extra" credit for an off-the-air function. This is because of the critical value of contact with the general public and to motivate traffickers to be somewhat more aggressive in making their message-handling services known to the general public.

Any formal piece of traffic you get via Amateur Radio counts as a message *received*. Any message you send via Amateur Radio, even if you originated it, counts as a message *sent*. Therefore, any time you relay a message, you get two points: one received and one sent.

Any time you take a message and give it to the party it's addressed to, on a mode other than Amateur Radio, you are credited with *a delivery*. (It's okay if the addressee is a ham.) As long as you deliver it off the air (eg, telephone, mail, in person), you get a delivery point.

Your monthly report to your STM, if sent in radiogram *format,* should look something like this:

NR 111 R NIØR 15 ST JOSEPH MO NOV 2

PETE STEWART KØSI

MISSOURI STM

COLUMBIA MO 65203

OCTOBER TRAFFIC ORIG 2 RCVD 5 SENT 6 DLVD 1

TOTAL 14 X 73

ROGER NIØR

If you have a traffic total of 500 or more in any month, or have over 100 originations-plus-deliveries in a month, you are eligible for the Brass Pounders League (BPL) certificate, even if you did all that traffic on SSB! If you make BPL three times, you receive a handsome medallion for your shack.

Another mark of distinction is the Public Service Honor Roll (PSHR). You don't have to handle a single message to get PSHR, so it's a favorite among traffickers in rural areas. The categories have traditionally been listed in the Public Service column of *QST*, and they include checking in CW and phone nets, acting as net control, handling an emergency message, and so on. See the Public Service column for particulars.

Almost any station, with regular participation, can get an ARRL Net Certificate through the Net Manager. Once you have participated regularly in a net for three months, you are eligible. If you are a League member, you can also become eligible for an Official Relay Station appointment in the ARRL Field Organization (for details concerning the ARRL Field Organization, see Chapter 17). Either certificate makes a handsome addition to your shack.

traffic handling, it should serve as a good reference for verterans and newcomers alike. If you've never been involved in message handling, perhaps your interest has been piqued. Should that be the case, don't put it off. Find a net that's custom-made for you and check into it! You'll find plenty of fine folks that will soon become close friends as you begin to work with them, learn from them, and yes, even chat with them when the net is over.

The roots of traffic handling run deep into the history of Amateur Radio, yet its branches reach out toward many tomorrows. Our future lies in proving our worth to the nonham public, and what better way to ensure the continuance of our hobby than by uniting family and friends via Amateur Radio?

Sure, it takes some effort, but any trafficker will tell you he stays with it because of the satisfaction he gets from hearing those voices on the other end of the phone say, "Oh, isn't that nice!" We've plenty of room for you—come grow with us!

Tips on Handling NTS Traffic By Packet Radio

Listing Messages

* After logging on to your local NTS-supported bulletin board, type the command LT, meaning List Traffic. The BBS will sort and display an index of all NTSXX traffic awaiting delivery. The index will contain information that looks something like this:

MSG#	TR	SIZE	TO	FROM	@BBS	DATE	Title
200	TN	282	NTSCT	KY1T		870225	QTC 1 Hartford
198	TN	302	NTSRI	K1CE		870224	QTC 1 Cranston
192	TN	215	NTSIN	WF4R		870224	QTC 1 Indianapolis
190	TN	200	NTSCT	AJ6F		870224	QTC 1 Waterbury
188	TN	315	NTSCT	KH6WZ		870224	QTC 1 Newington
187	TN	320	NTSCT	K6TP		870224	QTC 1 New Haven
186	TN	300	NTSAL	WA4STO		870224	QTC 1 Birmingham
184	TN	295	NTSCA	WB4FDT		870224	QTC 1 Fresno

Receiving Messages

* To take a message off the Bulletin Board for telephone delivery to the third party, or for relay to a NTS Local or Section Net, type the R or RT command, meaning Read Traffic, and the message number. R 188 will cause the BBS to find and send the message text file containing the RADIOGRAM for the third party in Newington. The RADIOGRAM will look like any other, with preamble, address, text and signature, only some additional packet-related message header information is added. This information includes the routing path of the message for auditing purposes; e.g., to discern any excessive delays in the system.

* After the message is saved to the printer or disk, the message should be KILLED by using the KT command, meaning Kill Traffic, and the message number. In the above case, at the BBS prompt, type KT 188. This prevents the message from being delivered twice.

* At the time the message is killed, many BBS's will automatically send a message back to the station in the FROM field with information on who took the traffic, and when it was taken!

Delivering or Relaying A Message

* A downloaded RADIOGRAM should, of course, be handled expeditiously in the traditional way: telephone delivery, or relay to a phone or cw net.

Sending Messages

* To send a RADIOGRAM, use the ST command, meaning Send Traffic. The BBS will prompt you for the NTSXX address (NTSOH, for example), the message title which should contain the city in the address of the RADIOGRAM (QTC 1 Dayton), and the text of the message in RADIOGRAM format. The BBS, usually within the hour, will check its outgoing mailpouch, find the NTSOH message and automatically forward it to the next packet station in line to the NTSOH node. Note: Some states have more than one ARRL Section. If you do not know the destination ARRL Section ("Is San Angelo in the ARRL Northern, Southern or West Texas Section?"), then simply use the state designator NTSTX.

* Note: While NTS/packet radio message forwarding is evolving rapidly, there are still some gaps. When uploading an NTS message destined for a distant state, use handling instruction 'HXC' to ask the delivering station to report back to you the date and time of delivery.

Hopefully, in a few short months, national reliability will be the rule.

* Unbundle your messages please: one NTS message per BBS message. Please remember that traffic eventually will have to be broken down to the individual addressee somewhere down the line for ultimate delivery. When you place two or more NTS messages destined for different addressees within one packet message, eventually the routing will require the messages to be broken up by either the BBS SYSOP or the relay station, placing an additional, unreasonable burden on them. Therefore it is good practice for the originator to expend the extra word processing in the first place and create individual messages per city regardless if there are common parts of other messages. This means that book messages are not suitable in packet at this time unless they are going to the same city. Bottom line: Messages should be sent unbundled. (Tnx NI6A for this tip)

We Want You!

Local and Section BBS's need to be checked daily for NTS traffic. SYSOPs and STMs can't do it alone. They need your help to clear NTS RADIOGRAMS every day, seven days a week, for delivery and relay. If you are a traffic handler/packeteer, contact your Section Traffic Manager or Section Manager for information on existing NTS/packet procedures in your Section.

If you are a packeteer, and know nothing of NTS traffic handling, contact ARRL HQ, your Section Manager or Section Traffic Manager for information on how you can put your packet radio gear to use in serving the public in routine times, but especially in times of emergency!

And, if you enjoy phone/cw traffic handling, but aren't on packet yet, discover the incredible speed, and accuracy of packet radio traffic handling. You probably already have a small computer and 2-meter rig; all you need is a packet radio "black box" to connect between your 2-meter rig and computer. Cost: about $100. For more information on packet radio, contact ARRL HQ.

Chapter 16

Image Communications

A mateur Radio operators have always had a strong propensity and well-deserved reputation for resourcefulness, whether it be fashioning an envelope detector out of a safety pin, razor blade and pencil lead, or making antennas out of coat hangers and toy springs. On a larger scale, hams have successfully transformed piles of surplus parts and metal scraps into radio transmitters capable of worldwide communications. Amateur image communicators continue to uphold this venerable tradition, but now with a slightly different leaning.

This time, hams have been successfully incorporating elements of the home video and personal computer revolutions into their image systems. Components such as color video cameras and monitors, video tape recorders, computer processors, printers and large capacity solid-state memories are having a profound beneficial effect upon the operation, performance and costs related to the transmission and reception of both still and moving pictures over the airwaves. Most hams are surprised to learn that some of their home video and computer equipment can indeed serve as the foundation for building a reasonably priced, yet highly effective, image communication system.

In this chapter, the goal is to improve your image—quite literally—and you won't even have to attend charm school! The focus will be on the three main image communications systems: fast-scan amateur television (FSTV), also referred to as ATV; slow-scan television (SSTV); and facsimile (FAX). Each mode has something unique and fascinating to offer. (The technical aspects of these systems are fully covered in *The ARRL Handbook for the Radio Amateur* and on a monthly basis in *SPEC-COM Magazine,* c/o Mike Stone, WBØQCD, PO Box H, Lowden, IA 52255.) Regulatory requirements along with beneficial operating suggestions and procedures for achieving more effective communications will be discussed. Armed with this information and aided by some old-fashioned common sense and on-the-air courtesy, you'll be all set to see and be seen. After all, hams should be seen as well as heard!

FAST-SCAN AMATEUR TELEVISION

"If it looks like a duck, sounds like a duck, walks like a duck, then it must be a...," or so the saying goes. Well, fast-scan amateur television (FSTV), also referred to simply as amateur television (ATV), certainly looks and sounds like commercial-broadcast-TV quality, and the pictures move around on the screen like commercial-broadcast TV. But alas, ATV is still just "amateur." The similarity between the systems is by no means accidental since ATVers use the same basic transmission standards as does the commercial world. Amateurs are prohibited, however, from engaging in any overt

forms of broadcasting or entertainment. Also, amateurs operate at considerably lower power levels which must be compensated through the use of high-gain antennas, sensitive pre-amps and low-loss cable at the receiver. Nevertheless, there is no greater compliment that an ATVer can be paid than hearing from a receiving station that your picture is coming across "just like commercial broadcast quality." That's like getting a 5-by-9 report on the phone bands! Most folks viewing ATV for the first time are surprised by the performance levels attainable by the relatively simple and low-cost amateur TV equipment.

Because of its large bandwidth, ATV may not be operated below the 70-cm band. This, in turn, makes ATV largely a line-of-sight or local-coverage system. Exceptions do occur, though, especially when inversion layers in the atmosphere are present and ducting occurs. Under these conditions, DX of several hundred miles is possible.

In recent years, ATV has undergone some dramatic changes, most notably in performance improvements and expanded portable applications. The most evident is in the widespread use of color because of the availability of low-cost color cameras and color monitors originally aimed at the home-video entertainment market. Color transmission has also been facilitated by the expanded use of broadband solid-state transmitters and stripline amplifiers which provide the necessary frequency response and phase characteristics to faithfully transmit color. The effectiveness of the low-power solid-state equipment has been further enhanced through the expanded use of ATV repeaters throughout the country, and low-noise GaAsFET receiver preamps. Finally, the small-size ATV transmitters are making their way out of the shack and into the field for use in serious public service applications.

License Requirements and Operating Frequencies

ATV can be used by any ham holding a Technician or higher-class license. Most operations can be found in the 420- to 440-MHz segment of the 70-cm band, with operation in the 1240-1294 portion of the 23-cm band gaining in popularity. Under Novice Enhancement, Novices, too, can operate ATV in their portion of the 23-cm band. See the ARRL band plan in Chapter 2 for more information. Some amateurs are also experimenting with video transmission in the 2300-2450 band using modified equipment originally intended for commercial microwave TV distribution systems, and some testing can even be found at 10 GHz using Gunnplexers. Interim plans for the new 902-928 MHz band, as adopted by the ARRL Board of Directors, allocate 910-916 MHz and 922-928 MHz to ATV and wideband modes. It is important that you first check with your ATV club, VHF/UHF local frequency coordination organization, or the ARRL *Repeater Directory* to determine the specific ATV frequencies for your area. Generally speaking, the East Coast and Midwest use 439.25 MHz for simplex and a 439.25 MHz input/426.25 MHz or 421.25 MHz output for ATV repeaters. The West Coast predominately uses 426.25 MHz for simplex and 434 MHz in/1253 MHz out for repeater use. Another West Coast frequency for simplex or duplex work is 1289 MHz. ATVers always try to avoid interfering with the weak-signal work (DX/EME) on 432 MHz and 1296 MHz, while also steering clear of the 440-450 FM voice repeater segment. Additionally, amateurs in the cities of Seattle, Detroit, Cleveland and Buffalo are prohibited (by FCC Docket 85-113) from operating in the 420-430 MHz portion of the 70-cm band.

Antenna Polarization

Antenna polarization also varies throughout the country. Most nonrepeater areas tend to use horizontal, as do the DXers. Most repeaters use vertical, although several newer ATV repeaters are experimenting with horizontal. It's highly important that you are using the same polarization as those with whom you intend to communicate. Using a different polarization can mean the difference between receiving a good picture and none at all. Again, check with your local ATV club or frequency coordinators.

Sound Format

Another interoperability requirement which will affect your equipment selection and operation is the method of transmitting and receiving sound. As with frequencies and antenna polarization, voice format varies in different regions. There are three possibilities:

1) *4.5-MHz FM subcarrier audio;* audio can be heard on a standard TV set along with the picture. This is also the format used by commercial-broadcast TV.

2) *On-carrier audio;* requires a separate audio receiver to demodulate. This mode was an outgrowth of modifying amateur or surplus business transmitters having an existing "on-carrier" FM voice capability and modifying the unit for AM video using collector, plate or grid modulation of the final. Some ATV repeaters will convert the on-carrier audio at its input to the subcarrier format on output so that standard TV receivers can be used in the shack to receive the audio.

3) *2-meters;* although this mode requires a separate rig, it has the advantage of letting other local hams know what is happening on ATV to publicize your activities. It allows non-ATVers to participate in ATV nets and can be instrumental in picking up many new video converts.

In sum, before setting up your station and attempting to operate, be sure to check:

1) simplex and repeater frequencies
2) antenna polarization
3) sound format

Equipment

The color video camera that you purchased to videotape pictures of the kids growing up is an ideal starting point. If you don't already have one, used black-and-white cameras can be found at hamfests in the $50 range, while nice new color units with a built-in electronic viewfinder can be obtained at the neighborhood electronics TV discounter for under $400. Some of the more fancy (and more expensive) cameras even have built-in character generators to compose on-screen messages and IDs.

Even if you don't have a camera, don't despair. As a video source, you can use the composite video output from a home color-graphics-computer or even a videotape recorder output to show prerecorded tapes that a friend may have taken for you. The composite video signal is then fed into an ATV transmitter or transceiver. Transceivers are gaining in popularity because of their inherent ease of operation and small size. The antenna is connected to the rig's RF connector via low-loss cable.

On the receive side, a downconverter or transceiver is connected to the VHF antenna terminals on an unmodified home TV set to translate the received signal to an unused VHF channel (typically Channel 2 or 3). Add a mic and provide good lighting, and you've got your ATV station ready to go. A typical ATV station is shown in Fig 16-1.

Fig 16-1—Tommy Noeman, KA3MKC, operates his fast-scan ATV station using home-video equipment and a commercially available transceiver. *(WA9GVK photo)*

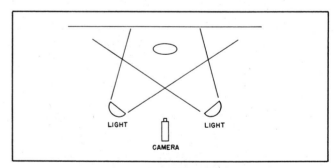

Fig 16-2—A popular lighting arrangement for ATV or SSTV pictures. Using two lights helps eliminate harsh shadows that result from a single light source. Frequently one light is set brighter or closer than the other to provide a little shading.

Lighting

Good lighting is one of the most important and underestimated aspects of ATV. If it's a choice between good picture contrast generated by good lighting or higher transmitted power, contrast wins most every time. In other words, a high-contrast picture from a low-powered transmitter will look much better through the "snow" than a low-contrast picture from a high-powered transmitter. Fig 16-2 shows one possible lighting configuration to eliminate harsh shadows. For black-and-white work, incandescent 100-watt bulbs with reflectors will do a good job. For color, 3200 degree Kelvin lights, available at your local camera store, will give you excellent contrast while maintaining the proper color hue for home cameras. Check with your local photographic camera store or *SPEC-COM Magazine,* Sept/Oct 1985, to obtain more detailed information on the art of lighting. You can experiment by monitoring your camera output on a closed-circuit local video monitor prior to transmitting. Once the lighting looks good, set the video gain control as high as possible before the picture smears or "whites out" as seen on a video monitor driven by the demodulated RF output of your transmitter or by reports from remote receivers. If you want to transmit titles or call letters, experiments have demonstrated that large black letters on a white background are most effective in cutting through noise.

Identifying

Fast-scan TV transmissions may be identified in video, or by CW or voice on the audio on-carrier or subcarrier portion. A picture of your QSL card makes an ideal video identification.

Signal Reporting

Signal reporting on ATV is markedly different than in all other aspects of Amateur Radio. Instead of the RST system used for voice and CW, image communicators use the "P" system which describes the amount of "snow" or noise observed on the screen (see Fig 16-3) and described below:

P5—Excellent; no visible noise. "Closed-circuit" or commercial-broadcast quality.

P4—Very good; some noise visible.

P3—Fairly good; noticeable or moderate noise present but picture is very usable.

P2—Passable; high noise; picture can still be seen but small details are lost.

P1—Weak picture, limited use; very high, objectionable noise.

P0—Not usable; picture is totally lost in noise; can detect sync bars or barely detect picture presence but is totally unusable.

In some cases, you can split hairs with the system. If you receive an excellent picture with very slight noise, you may want to say it is a "four point five" or P4.5.

What to Show

There are many uses for ATV around the shack to show off your equipment, prize QSLs, family, pets, friends, photographs, home videotapes and vacation movies. How about showing a videotape on how you erected your 100-foot antenna tower? (You'll be surprised how few hams really know how it's done!) How about an informative presentation showing the progress you're making on a new electronics project? You can use multiple cameras and even special-effects generators to superimpose lettering over your pictures or perform split-screen effects. If you also operate SSTV, you can retransmit the SSTV images you've received from overseas to all the local amateurs via ATV. Amateur television has been successfully transmitted from private aircraft, hot air balloons, and even radio-controlled airplanes and robots. The FCC has given permission to amateurs to retransmit the Space Shuttle video provided over NASA's public information channel on Satcom 1R. Videotape exchanges with amateurs of other countries also makes interesting viewing. You can also show off the video output of your home computer to display animated color graphics or repeat W1AW RTTY and ARRL bulletin-board messages.

Thanks to the availability of compact and portable gear, ATV has been making tremendous inroads in the public service arena nationwide. Hams have used ATV during national disaster drills to send video from hospitals back to emergency command posts; see Fig 16-4. In Oklahoma, a camera and ATV transmitter mounted atop a tower searches the skies for tornados and relays the pictures directly to the National Weather Service, providing a valuable early warning capability. Amateur TV is used annually to help monitor and coordinate the Tournament of Roses Parade, marathons and yacht races; see Fig 16-5. During the Christmas season, amateurs have taken their TV systems into hospitals to enable children to see and talk to Santa Claus over live TV. Hams have used their equipment to send Voyager 2-received pictures of Jupiter and Saturn from the Jet Propulsion Laboratory in Pasadena, California, to remote locations such as an observatory and convention center to enable more widespread viewing. Amateur TV is also always an attention-getter for Amateur Radio demonstrations held at shopping malls and Boy Scout Jamborees.

Maintaining Strong ATV Activity

Based upon experience of ATV organizations throughout the country, several recommendations have emerged as key factors for maintaining enthusiasm and steady growth of on-the-air activity and club support:

1) *Weekly on-the-air meeting.* Set aside a specific weekly time for an on-the-air get-together. This is the time when newcomers can be guaranteed that there will be a signal available for receiver testing and someone available to check the quality of their transmitted picture. It's a time to compare notes and help each other out by exchanging suggestions. (By the way, it's also the best time to show off the videotapes of your recent vacation trip to ensure it will be watched by a large audience!) The meeting is run by a control station who asks all "video rangers" to check in and then queries each one in a disciplined roundtable fashion so each person has a chance to transmit. The weekly get-together usually turns out to be one of the most busy, lively and enlightening aspects of ATV operation.

2) *Use 2 meters for audio.* Even though your ATV operations may use an on-carrier or subcarrier audio signal, 2 meters is recommended for voice to publicize your operations to the rest of the amateur world who may not be fully aware of all the ATV activity in the 70-cm and higher regions of the spectrum. ATV repeaters can be adapted to receive and transmit the audio portion on 2 meters.

3) *Promote a club project.* Public service efforts are ideal because they not only perform a valuable community service but they also generate a sense of purpose and enthusiasm among club members. It's a good feeling to work as a team and use your gear productively.

4) *Vary the picture content.* It's great to see your pretty face all the time on ATV but try to be creative in what you show. Variety is the spice of ATV life!

Self-Monitoring

Although you can usually depend on the received station to give you on-the-air signal reports, the quickest and most accurate method is to use a RF demodulator driving a video

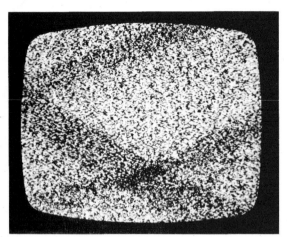

P0 Total Noise Visible. No picture at all or detectable Video Sync Bars.

P1 High noise visible. Weak picture.

P2 High noise visible. Fair picture. Fair detail.

P3 Noise Visible. Strong picture. Recognizable detail.

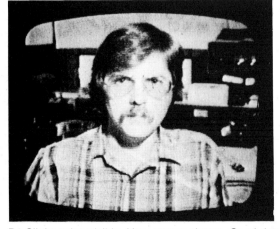

P4 Slight noise visible. Very strong picture. Good detail.

P5 No noise visible. Closed circuit picture. Excellent detail.

Fig 16-3—ATV picture quality reporting system: P5—Excellent, P4—Very Good, P3—Fairly Good, P2—Passable, P0—Weak Picture. *(Photos courtesy of* SPEC-COM *and Dave Williams, WB0ZJP)*

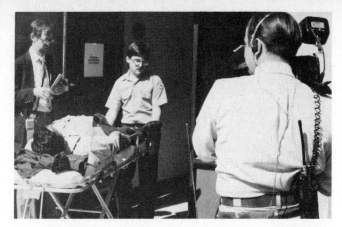

Fig 16-4—During the National Disaster Medical System drill, ATV was used to send pictures of simulated injury victims arriving at a Virginia hospital to drill coordinators at a USAF base command post in Maryland. *(WA9GVK photo)*

monitor to check the quality of your transmitted signal. You'll be able to determine contrast, focusing, framing, color level and color hue, while making sure you are not bending and rolling by excessive sync compression. Several commercial ATV transmitters have a built-in RF demodulator and monitor driver. You can also construct a device to do this function using a coupler, detector and video buffer. Making effective use of this capability will save you much on-the-air time in getting your station set up properly.

TV Audio Noise Squelch

One of the more annoying, but easily solvable, aspects of ATV operation is the audio noise generated by your TV when no signal is being received. When using your commercial TV set to receive video and audio subcarrier, you may notice that when you switch from receive to transmit, the TV set will emit an undesirable hiss because of the absence of a receive signal. That's the same sound that may have awakened you after you dozed off during the late, late show after the station went off the air! Hams have resorted to laboriously turning down the TV audio each time they transmit. However, there are two simple automated and effective alternatives for solving the problem:

1) *Noise squelch.* Communications Concepts, 2648 North Aragon Ave, Dayton, OH 45420, markets a very inexpensive noise squelch which connects to the earphone jack of your TV. An external speaker is then connected to the unit's output. Without touching any knobs or controls, the squelch will automatically turn off the TV audio to the speaker if only noise is present on the channel. When the TV receives a signal, the squelch will sense the quieting of noise and will allow the audio to pass through to the speaker.

2) *VHF TV relay.* Another technique is to use an RF relay connected to the TV's VHF antenna terminals along with an RF modulator capable of upconverting camera video to the same VHF channel as outputted by the receive converter. The relay is activated using your ATV transmitter PTT line. When you are in the transmit mode, the up-converted video from your camera is fed via a relay into the VHF terminals of the TV. On receive, the relay connects the ATV receiver downconverter output to the TV's VHF antenna input. This not only quiets the TV on transmit but allows the TV to perform a dual function of both ATV receiver display and local camera monitor. This technique is fully described in a construction article in August 1985 *QST*, pp 32-37.

(A)

(B)

Fig 16-5—Doug McKinney, KC3RL, checks out his portable 8-watt ATV station powered by a 10 ampere-hour motorcycle battery. This arrangement is ideal for public service applications, such as emergency drills, parades and marathon coordination. *(WA9GVK photo)*

SLOW-SCAN TELEVISION

Nestled among the CW, phone and RTTY operators in the Amateur Radio bands is a sizable following of hams who regularly exchange still pictures in a matter of seconds virtually anywhere on earth. They are using a system called slow-scan television (SSTV), which was originally designed by an amateur in the early 1960s. Over the years, the amateur community has been continually refining and improving the quality of SSTV. Amateur success with SSTV during the past two decades has led to its application by the military and commercial users as a reliable long-range, narrow-bandwidth transmission system. The worldwide appeal of SSTV is manifested by the many DX stations that are now equipped for this type of picture transmission. Several amateurs have even worked over 100 DXCC countries on SSTV!

Just as the name implies, SSTV is the transmission of a picture by very slowly transmitting the picture elements, while a television monitor at the receiving end reproduces it in step. An SSTV signal is a variable frequency audio tone from 1500 Hz for black to 2300 Hz for white, with 1200 Hz used for synchronization pulses. Unlike fast-scan television, which uses 30 frames per second, a single SSTV frame takes at least eight to fill the screen. Additionally, the vertical resolution of SSTV is only 120 lines (or 128 for some digital systems) compared with 525 lines for fast-scan. (Some high-resolution experimental designs are operating with 256 lines.) These disadvantages are offset by the fact that SSTV requires less than 1/2000 of a fast-scan TV's bandwidth. Thus, the FCC permits it in any amateur phone band.

The basic SSTV format represents a trade off among bandwidth, picture rate and resolution. To achieve practical HF long-distance communications, the SSTV spectrum was designed to fit into a standard 3-kHz voice bandwidth through a reduction in picture resolution and frame rate. Thus, SSTV resolution is lower than FSTV and is displayed in the form of still pictures. A sample SSTV picture is shown in Fig 16-6.

In recent years, amateurs have been actively experimenting with various forms of video-processing techniques to provide limited motion and increased resolution. The greatest advancements are currently being made in the realm of color SSTV. Unfortunately, most of this work has been done independently from each other, resulting in a multitude of different SSTV color standards. Although the 120-line/8-s format is standard for black and white, newcomers to SSTV should be cautioned that color SSTV standards are in a state of flux which may severely limit interoperability. In the coming years, it is expected that the amateur community will adopt a single color format.

License Requirements and Operating Frequencies

In the HF bands, a General- or higher-class license is required to operate SSTV. Operation is restricted to the phone portion of the bands. At UHF and above, a Technician-class amateur license or higher is needed, although the vast majority of SSTV activity occurs in the HF bands. In the US, slow-scan TV using double-sideband AM or FM on the HF bands is not permitted.

The common accepted SSTV calling frequencies are 3.845 MHz (Advanced), 7.171 MHz (Advanced), 14.230 MHz (General) and 28.680 MHz (General). Traditionally, 20 meters has been the most popular band for SSTV operations. A weekly international SSTV net is held each Saturday at 1800 UTC on 14.230 MHz. Many years ago, when SSTV was first authorized, the FCC recommended that

Fig 16-6—SSTV picture as seen on a standard TV set using a digital scan converter.

SSTVers not spread out across the band even though it was legal to do so. A "gentlemen's agreement" has remained to this day that SSTVers operate as close as possible to the above calling frequencies to maintain the problem-free operation that has existed for nearly 20 years.

Identifying

On SSTV, the legal identification must be made by voice or CW. Sending "This is W9NTP" on the screen is not sufficient. Most stations intersperse the picture with comments anyway, so voice ID is not much of a problem. Otherwise SSTV operating procedures are quite similar to those used on SSB.

Equipment

All you need to get started is an SSB station (or FM station for VHF/UHF), a monitor, scan converter and a video source. Like RTTY, SSTV is a 100%-duty-cycle transmission. Most sideband rigs will have to run considerably below their voice power ratings to avoid ruining the final amplifier or power supply. Early SSTV monitors used long-persistence CRTs much like classical radar displays. In a darkened environment, the image remained visible for a few seconds while the frame was completed. This type of reception is unusual today and has been replaced with digital scan converters which convert SSTV to FSTV to place a bright image on a conventional television monitor. Some of this older equipment is available almost for the asking and is a good way, on a temporary basis, to examine the SSTV mode without investing much money.

Motivated by the difficulty of observing SSTV pictures on the long-persistence CRTs, the scan converter is a result of recent advances in digital techniques. In receive mode, the device converts the incoming audio to a fast-scan video signal that is usable by a conventional fast-scan TV monitor. Similarly, on transmit the converter changes the fast-scan camera output to a standard slow-scan signal. A personal computer equipped with the proper software and interface makes a highly cost-effective slow-scan converter. Computers such as the TRS-80C®, TRS-80® Model III/IV, VIC 20™, Commodore 64™, Apple®, Atari® 400/800 (see August 1985 *QST*, pp 13-16), and IBM® PC (see Dec 1985 *QST,* pp 14-17) have been successfully used for SSTV. A comprehensive listing of software and interface sources is available from *SPEC-COM Magazine,* c/o Mike Stone, WBØQCD, PO Box H, Lowden, IA 52255.

The video source can be a black-and-white or color camera, video recorder or audio tape recorder, or the video output from a personal color-graphics computer. Since the signal emanating from the transmit-side of the scan converter is audio, it can be recorded on a cassette machine. The frequency response is not critical, but some consideration should be given for the wow and flutter specifications to minimize picture skew. You can have a friend pre-record some pictures from his/her camera onto your recorder and use them for on-the-air transmissions. The combinations of computer, camera and tapes (both audio and video) are endless. The superposition of graphics and images both generated by the operator and those received over the air can be recorded on either audio or video tapes for later use.

To send SSTV over the air, just couple the audio output of the scan converter directly into the microphone input of any single-sideband transmitter or FM transmitter on VHF or UHF. To receive, you tune in the signals on an SSB receiver and feed the tones into the scan converter which is connected to the monitor.

Setting Up Your Studio

Standard photographic lighting arrangements work well for TV. The set-up in Fig 16-2 is good because it reduces background shadows and harsh shadows on the subject. A pair of 100-watt bulbs with reflectors would be a minimum requirement for the typical camera. Outdoors, of course, the light level is pretty much predetermined. Care must be taken, however, to prevent direct sunlight from shining into the lens, since most TV camera pickup tubes can be damaged. Adjust the brightness and contrast controls on the camera so that the darkest screen area comes out black while the brightest areas reach maximum brightness without "blooming" on the monitor screen. This assumes that the monitor brightness and contrast controls are set correctly. (Many of the newer cameras are fully automatic and do not need any adjustment.) It is also highly useful to run a test audio tape consisting of a test pattern with various gray shades through your system to properly set up your equipment. Such a tape may be obtained by an on-the-air recording from another amateur or obtained commercially. Many computer programs and special modes of commercial scan converters also provide the test signals to adjust your equipment properly.

Camera lens adjustment is also important. Depth of field is the ability of the lens to focus sharply on both near and far objects at the same time. Since the focusing range is most limited for wide lens aperture, the most accurate focusing can be done at such a setting, the lowest f number value. While a narrow aperture (high f number) requires more light on the subject, it permits the viewing of objects at different distances without refocusing; ie, it has greater depth of field. Many cameras have fixed-aperture lenses with electronically controlled light sensitivity. Lenses may also be purchased for various focal lengths to give wide angle, telephoto or zoom capabilities.

Operating

It's customary to send two or three frames of each subject

to ensure that the other station gets at least one good-quality picture. Many operators like to send a couple of frames followed by voice comments about the picture just sent. Some of the more elaborate setups are equipped to send video and audio simultaneously with the voice on one sideband and the picture on the other. The "ISB" (independent sideband) signals can be copied with two receivers.

An SSTV signal must be tuned in properly so the picture will come out with the proper brightness, and the 1200-Hz synchronization pulses will be detected. If the signal is not "in sync," the picture will appear wildly skewed. The easiest way to tune SSTV is to wait for the transmitting operator to say something on voice and then fine tune for proper pitch while the person is talking. With experience, you may find you are able to zero in on an SSTV signal by listening to the sync pulses and by watching for proper synchronization on the screen. Many SSTV monitors are equipped with automatic or manual tuning aids.

If you want to record slow-scan pictures off the air, there are several ways of doing it. One is to tape record the audio signal for later playback. The other is to take a photographic picture of the image right from the SSTV screen. An instant print camera equipped with a closeup lens enables you to see the results shortly after the picture is taken. If you want to do this without darkening the room lights, you'll have to fabricate a light-tight hood to fit between the camera and monitor screen. It is also possible to feed the converter's fast-scan output to a videotape recorder for later viewing.

Picture Subjects

The selection of things to show is endless, but you will find that high contrast black-and-white or high-saturation color pictures, that are not too cluttered, work best. A "live" close-in shot of you in front of the operating position is probably the most desirable picture. Don't forget a couple of frames with your call, name and QTH. A closeup of your QSL card is ideal, and how about showing off your prized QSLs? Illustrating a technical conversation with a simple schematic diagram can often make clear a point that would be very difficult to get across on voice alone. There is a wealth of good SSTV material in newspapers. Home-brew cartoons are also very popular subjects.

Some SSTVers find it convenient to mount the camera in front of an easel onto which various prepared cards can be inserted. These cards can be photographs, drawings or lettered signs. A kitchen-type noteboard or menu board with press-on removable letters is handy for making headings. Today it is more convenient to use the keyboard on a computer or other specialized captioning devices to produce these letters.

It is fun to make up taped "programs" of related pictures. Some operators have dozens of programs on different cassette tapes that can be selected and put on the air in seconds. People might enjoy seeing a pictorial tour of your shack and QTH. Do you have another hobby? Maybe that person you are working would like to learn about your model train collection. A closeup of your 1862 Cannonball Express would be a lot more interesting than a few bland comments about the weather.

Spectacular pictures of Saturn and Jupiter taken by the Voyager 2 spacecraft and received at the Jet Propulsion Laboratory in California have been relayed via SSTV instantaneously to amateurs around the world. Color SSTV using the ROBOT 1200C scan-converter format came into its own when Space Shuttle *Challenger* carried aloft SSTV

Fig 16-7—SSTV picture using the ROBOT 1200C format transmitted from the Space Shuttle during the STS-51F mission.

equipment on the STS-51F mission and exchanged pictures with Amateur Radio clubs throughout the world (Fig 16-7).

Of course, there are more down-to-earth applications of SSTV. Slow-scan television has good potential for rendering service to the public by means of third-party traffic. For example, the scientists working on the Antarctic ice pack often don't see their families for months—except by amateur SSTV. This was one of the demonstrations that was used by slow-scan pioneers in 1968 to convince the FCC that slow-scan television should be permanently authorized in the ham bands. Anyone can write or call almost anywhere in the world these days, but grandma and grandpa would certainly enjoy seeing pictures of the kids from 3000 miles away by SSTV.

In whatever you show, use your imagination. But remember—never to pass up the opportunity to transmit your own image occasionally to let everyone on the frequency know with whom they are communicating!

FACSIMILE

Facsimile (FAX) is a method of transmitting highly detailed pictures onto paper via voice-bandwidth radio circuits. It has been around for a long time and has always been the prime method for newspapers to obtain photos from distant locations. Because of its narrow bandwidth, it is permitted in the HF bands, thereby making worldwide coverage possible; the majority of amateur two-way FAX operation is conducted in the HF bands. Amateur facsimile equipment may also be used to receive commercially generated weather pictures either directly from satellites on VHF in the VHF/UHF bands or by HF relay stations. These satellite weather maps are the same as shown on the TV weather report by your local weather forecaster.

Given that the transmission bandwidths of FAX and SSTV are about equal, FAX produces a much higher resolution image than SSTV but at the expense of a much longer transmission time.

With the help of the personal computer, FAX is currently undergoing a major metamorphosis within the amateur community. The older mechanical facsimile scanning equipment is being replaced by computers that process and display the pictures on a standard TV set. A graphics printer

can be used to generate the traditional hard-copy facsimile output. Also, these more up-to-date approaches are opening the door to amateur color FAX transmission. As personal computers are becoming the hardware of choice for FAX as well as for SSTV scan-conversion, the lines of difference between the two image communications systems can be expected to become blurred in the not-too-distant future. Even today, software and minor hardware interface changes permit the amateur to use both modes with a cost-effective commonality of hardware components.

Equipment

Historically, FAX transmission has been accomplished mechanically by light-scanning a picture that is placed on a rotating drum. For reception, a synchronized rotating drum with light-sensitive paper is exposed to a modulated light beam or graphite paper exposed to a wire-arc scanner to produce black or gray shades. The light-sensitive paper usually requires developing while the wire-arc technique requires none. Both techniques produce high resolution pictures. A typical mechanical unit is the surplus Western-Union Desk-FAX machine, shown in Fig 16-8, which can be purchased for about $50. Using a speed converter, the unit can also be adapted to copy all weather FAX satellites.

Fig 16-8—Surplus Western Union Telefax Transceiver. The drum accommodates a sheet of paper approximately 6 × 4 in.

In the past several years, economical memory systems and personal computer programs have been developed to allow amateurs to copy FAX pictures on a standard TV monitor. Hard copies can be made on a computer printer or video printer. The recent introduction of digitized TV with built-in memory has opened up new vistas by allowing both all-electronic FAX and SSTV copy with a minimum of additional hardware.

Computers offer some interesting advantages over the mechanical approach, particularly in the area of image processing. With the processor, selected sections of a received picture can be electronically blown-up to enable more detailed viewing. Before computers, a photographic process was required to perform the enlarging process. Additionally, contrast-enhancement and noise-suppression techniques can be implemented to futher improve the picture quality.

The reception of satellite weather pictures has been successfully accomplished on several personal computers, such as the TRS-80C®, VIC 20™ (see August 1985 *QST*, pp 25-31), and the IBM ® PC (see January 1986 *QST*, p 21).

Frequencies and Operational Procedures

Amateur FAX operators have established the following frequencies for operation and nets:

7.245 MHz
14.245 MHz
21.345 MHz (International Facsimile Net, meets every Sunday at 1600 UTC)
28.945 MHz

All are invited to check into the International Facsimile Net but be sure to use voice first. Do not transmit pictures without permission of the net control station (NCS). The NCS coordinates the net and selects individual stations based upon propagation conditions. FCC rules require a CW or voice identification prior to sending CQ by FAX. When calling CQ, use large graphics and send at either 120 or 180 lines per minute (LPM), the most commonly used amateur operating speeds. Answer all FAX CQs by voice first.

You should always establish reliable voice contact before attempting to transmit pictures. Very-high-resolution pictures should not be sent unless band conditions are very good. Amateur AM or FM FAX is now allowed in the phone portions of all amateur bands. With a general-coverage 2-30 MHz receiver, you can copy commercial press wire photos, government weather charts and Navy rebroadcasts of weather satellite pictures.

Standards

FAX machines use either AM or FM. On the HF bands, amateurs use FM, as do all commercial FAX stations operating below 30 MHz. Frequency-modulated FAX is more immune from noise, interference and fading than AM. Amplitude-modulated FAX is used by all satellite FAX stations because it is immune to Doppler frequency shift. The DESK-FAX machine in Fig 16-8 is an AM-FAX machine that can be speed-converted from 180 to 120 LPM to copy the AM-FAX weather satellites directly.

FAX standards have been established over the years to meet various requirements. FAX stations do not transmit line synchronization during picture transmission. Sync pulses are sent *before* each picture, and the receiving station "phases" these pulses to the left edge of the paper. The FAX machine must be capable of free running throughout the entire picture. This is accomplished by using a synchronous ac motor locked to the ac power lines to drive the drum. FAX speeds have been standardized in terms of lines per minute. Speeds vary from 60 LPM (one line per second) to 480 LPM (eight lines per second). In general, 60 LPM (low speed) is mainly used by the press wire-photo stations to send very high resolution pictures (approximately four times more resolution than broadcast TV), while 480 LPM (fast speed) is used to transmit lower resolution images (approximately half that of commercial TV broadcast) such as weather radar. Amateurs can use any line speed, with the final decision based upon resolution capability of the equipment, picture quality and band conditions.

Another aspect of FAX speed is lines per inch (LPI). This is usually fixed by the gear ratios in each FAX machine. The LPI varies among machines from 72 for medium resolution to about 96 LPI for high-resolution machines. Sending pictures between FAX machines with drastically different LPI capabilities will produce squashing or expanded top-to-bottom picture distortion. The LPI difference between FAX machines is called the "Index of Cooperation."

Color FAX

Color pictures may also be transmitted by FAX using a variety of methods. Press color wire-photos are sent by transmitting three different pictures: cyan, magenta and yellow. Alignment marks are provided in the corners of all three pictures. No practical method has yet been developed by amateurs to copy these pictures in color on a CRT. However, each frame can be readily copied in black and white with outstanding quality.

Another color FAX method, and the one used by amateurs with a home computer, color monitor and special software, is called "line sequential color." This system transmits the first line of the picture three times—first in red, then green and finally in blue (or RGB line sequential). The process is repeated for all remaining lines. It is not possible to receive this picture on a paper machine since each line would appear scrambled. The new generation digital television sets will be able to copy this FAX color system at twice the resolution of previous personal-computer-based systems.

Weather-Satellite FAX Reception

Amateur FAX operation can be extended for direct copy of satellite weather FAX transmissions. The United States, Soviet Union and several other countries have polar-orbiting weather satellites about 500 miles high. Each of the polar satellites transmits a continuous picture at 120 LPM. Each satellite passes in view of every part of the earth at least two times each day. The United States polar orbiting weather satellites are called the "NOAA" series. (NOAA orbital information is provided in all ARRL bulletins transmitted by W1AW.) Polar satellites transmit pictures that are approximately twice the resolution of broadcast TV. All polar

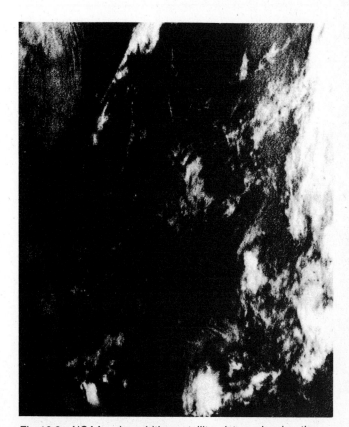

Fig 16-9—NOAA polar orbiting satellite picture showing the East Coast of the US; note Florida in the lower left.

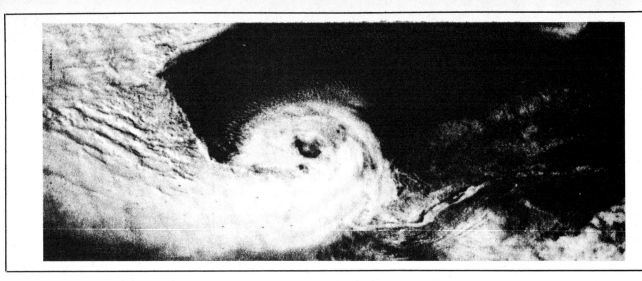

Fig 16-10—A typical GOES picture (GOES CENTRAL) showing a hurricane.

satellites operate in a VHF band from 135 to 138 MHz (typically 137.5 MHz or 137.62 MHz) using AM video modulation of a VHF-FM carrier. A wideband VHF-FM receiver such as the Vanguard Labs Model FMR-206P (15-kHz passband) with a turnstyle or directional beam can be used to copy the satellites. AMSAT provides cassettes and disk programs for various personal computers to track all of the polar satellites (as well as the OSCARs, space shuttle flights, and so on). It is always desirable to know the exact time and heading of each satellite before trying to copy it. An overhead pass will provide a copy time of 10 to 15 minutes. This will provide a high-resolution weather picture about 800 miles wide (east to west) and about 1500 miles high (north to south). Fig 16-9 shows a NOAA polar satellite picture. An AM-FAX machine or computer with an AM-FAX program is required.

It is also possible for amateurs to copy the Geostationary Weather Satellites (GOES) located above the equator in stationary orbit. They monitor North and South America with an updated weather picture about every half hour on a frequency of 1.691 GHz (S-band). The signal is also an AM-modulated FM carrier just like the polar satellites except that the speed is 240 LPM and at a lower resolution in the neighborhood of broadcast TV. The GOES transmit power is quite high so that only a four-foot dish is sufficient. A 1.691-GHz down-converter with an output of 135 MHz can be fed into the VHF polar orbiting satellite (NOAA) receiver.

There are three GOES satellites: GOES EAST (off the US East Coast), GOES WEST (off the US West Coast) and GOES CENTRAL (south of the US). GOES pictures are retransmitted every few hours in the HF bands on 8.080, 10.865 and 16.4118 MHz by US Navy station NAFAX in Virginia. See Fig 16-10 for a typical GOES picture.

Further FAX details and technical information can be found in *The ARRL Handbook for the Radio Amateur* and *SPEC-COM Magazine*.

Chapter 17

References

ARRL DXCC COUNTRIES LIST

NOTE: * INDICATES CURRENT LIST OF COUNTRIES FOR WHICH QSLs MAY BE FORWARDED BY THE ARRL MEMBERSHIP OUTGOING QSL SERVICE.

NOTE: † INDICATES COUNTRIES WITH WHICH U.S. AMATEURS MAY LEGALLY HANDLE THIRD-PARTY MESSAGE TRAFFIC.

HOW TO USE THE ARRL OUTGOING QSL SERVICE

1) Presort your DX QSLs alphabetically by call-sign prefix (AP, C6, CE, DL, F, G, JA, LU, PY, 5N, 9Y and so on).
2) Enclose the address label from your current copy of *QST*. The label shows that you are a current ARRL member.
3) Enclose payment of $1 per each pound (or less) of cards — approximately 155 cards weigh one pound. In other words, $1 is the *minimum charge* whether you send 1 card or 155 cards. Please pay by check (or money order) and write your call sign on the check. Do not send cash.
4) Include only the cards, address label and check in the package. Wrap the package securely and address it to the ARRL Outgoing QSL Service, 225 Main St., Newington, CT 06111.
5) Further details are available from the Outgoing QSL Bureau at ARRL HQ.

Prefix	Country	CONTINENT	ZONE ITU	ZONE CQ	MIXED	PHONE	CW	RTTY	SAT.		160	80	40	20	15	12	10	
A2*	Botswana	AF	57	38														
A3*	Tonga	OC	62	32														
A4*	Oman	AS	39	21														
A5	Bhutan	AS	41	22														
A6	United Arab Emirates	AS	39	21														
A7	Qatar	AS	39	21														
A9*	Bahrain	AS	39	21														
AP-AS*	Pakistan	AS	41	21														
BV	Taiwan	AS	44	24														
BY, BT*	China	AS	(A)	23,24														
C2*	Nauru	OC	65	31														
C3*	Andorra	EU	27	14														
C5†*	The Gambia	AF	46	35														
C6*	Bahamas	NA	11	08														
C8-9	Mozambique	AF	53	37														
CA-CE†*	Chile	SA	14, 16	12														
CE9/KC4▲*	Antarctica	AN	(B)	(C)														
CE0†*	Easter I.	SA	63	12														
CE0†*	San Felix	SA	14	12														
CE0†*	Juan Fernandez	SA	14	12														
CM, CO†*	Cuba	NA	11	08														
CN*	Morocco	AF	37	33														
CP†*	Bolivia	SA	12, 14	10														
CT*	Portugal	EU	37	14														
CU, CT2*	Azores	EU	36	14														
CT3*	Madeira Is.	AF	36	33														
CV-CX†*	Uruguay	SA	14	13														
CY0	Sable I.	NA	09	05														
CY0	St. Paul I.	NA	09	05														
D2-3*	Angola	AF	52	36														
D4*	Cape Verde	AF	46	35														
D6²⁶	Comoros	AF	53	39														
DA-DL²*	Fed. Rep. of Germany	EU	28	14														
DU-DZ*	Philippines	OC	50	27														
EA-EH*	Spain	EU	37	14														
EA6-EH6*	Balearic Is.	EU	37	14														
EA8-EH8*	Canary Is.	AF	36	33														
EA9-EH9*	Ceuta and Melilla	AF	37	33														
EI-EJ*	Ireland	EU	27	14														
EL†*	Liberia	AF	46	35														

Prefix	Country	CONTINENT	ZONE ITU	ZONE CQ	MIXED	PHONE	CW	RTTY	SAT.		160	80	40	20	15	12	10	
EP-EQ*	Iran	AS	40	21														
ET	Ethiopia	AF	48	37														
F*	France	EU	27	14														
FT8W*	Crozet	AF	68	39														
FT8X*	Kerguelen Is.	AF	68	39														
FT8Z*	Amsterdam & St. Paul Is.	AF	68	39														
FG*	Guadeloupe	NA	11	08														
FG, FS1*	Saint Martin	NA	11	08														
FH26*	Mayotte	AF	53	39														
FK*	New Caledonia	OC	56	32														
FM*	Martinique	NA	11	08														
FO*	Clipperton I.	NA	10	07														
FO*	Fr. Polynesia	OC	63	32														
FP*	St. Pierre & Miquelon	NA	09	05														
FR/G4*	Glorioso Is.	AF	53	39														
FR/J,E4*	Juan de Nova, Europa	AF	53	39														
FR*	Reunion	AF	53	39														
FR/T*	Tromelin	AF	53	39														
FW*	Wallis & Futuna Is.	OC	62	32														
FY*	Fr. Guiana	SA	12	09														
G*#	England	EU	27	14														
GD*	Isle of Man	EU	27	14														
GI*	Northern Ireland	EU	27	14														
GJ*	Jersey	EU	27	14														
GM*	Scotland	EU	27	14														
GU*	Guernsey & Dep.	EU	27	14														
GW*	Wales	EU	27	14														
H4*	Solomon Islands	OC	51	28														
HA, HG*	Hungary	EU	28	15														
HB*	Switzerland	EU	28	14														
HB0*	Liechtenstein	EU	28	14														
HC-HD†*	Ecuador	SA	12	10														
HC8-HD8†*	Galapagos Is.	SA	12	10														
HH†*	Haiti	NA	11	08														
HI†*	Dominican Republic	NA	11	08														
HJ-HK†*	Colombia	SA	12	09														
HK0†*	Malpelo I.	NA	12	09														
HK0†*	San Andreas & Providencia	NA	11	07														
HL*	Korea	AS	44	25														
HO-HP†*	Panama	NA	11	07														
HQ-HR†*	Honduras	NA	11	07														
HS*	Thailand	AS	49	26														
HV*	Vatican	EU	28	15														
HZ	Saudi Arabia	AS	39	21														
I*	Italy	EU	28	15														
IS0, IM0*	Sardinia	EU	28	15														
J2*	Djibouti	AF	48	37														
J3†*	Grenada	NA	11	08														
J5	Guinea-Bissau	AF	46	35														
J6†*	St. Lucia	NA	11	08														
J7†*	Dominica	NA	11	08														
J8†*	St. Vincent & Dep.	NA	11	08														

#Third-party traffic permitted with special-events stations in the United Kingdom having the prefix GB *only*, with the exception that GB3 stations are not included in this agreement.

Prefix	Country	CONTINENT	ZONE ITU	CQ	MIXED	PHONE	CW	RTTY	SAT.	160	80	40	20	15	12	10	
JA-JS*	Japan	AS	45	25													
JD1⁵*	Minami Torishima	OC	45	27													
JD1⁶*	Ogasawara	AS	45	27													
JT-JV*	Mongolia	AS	32, 33	23													
JW*	Svalbard	EU	18	40													
JX*	Jan Mayen	EU	18	40													
JY†*	Jordan	AS	39	20													
K, W, N, AA-AK	United States of America	NA	6, 7, 8	3, 4, 5													
KC6²⁸(E. Caroline Is.)	Micronesia	OC	65	27													
KC6²⁹(W. Caroline Is.)	Belau	OC	64	27													
KG4†*	Guantanamo Bay	NA	11	08													
KH1†	Baker, Howland Is.	OC	61, 62	31													
KH2†*	Guam	OC	64	27													
KH3†	Johnston I.	OC	61	31													
KH4†*	Midway Is.	OC	61	31													
KH5†	Palmyra, Jarvis Is.	OC	61	31													
KH5K†	Kingman Reef	OC	61	31													
KH6†*	Hawaiian Is.	OC	61	31													
KH7†	Kure I.	OC	61	31													
KH8†*	American Samoa	OC	62	32													
KH9†	Wake I.	OC	65	31													
KHØ†*	Mariana Is.	OC	64	27													
KL7†*	Alaska	NA	1, 2	1													
KP1†	Navassa I.	NA	11	08													
KP2†*	Virgin Is.	NA	11	08													
KP4†*	Puerto Rico	NA	11	08													
KP5²⁷†	Desecheo Is.	NA	11	08													
KX6*	Marshall Is.	OC	65	31													
LA-LN*	Norway	EU	18	14													
LO-LW†*	Argentina	SA	14, 16	13													
LX*	Luxembourg	EU	27	14													
LZ*	Bulgaria	EU	28	20													
OA-OC†*	Peru	SA	12	10													
OD*	Lebanon	AS	39	20													
OE*	Austria	EU	28	15													
OF-OI*	Finland	EU	18	15													
OHØ*	Aland Is.	EU	18	15													
OJØ*	Market Reef	EU	18	15													
OK-OM*	Czechoslovakia	EU	28	15													
ON-OT*	Belgium	EU	27	14													
OX*	Greenland	NA	5, 75	40													
OY*	Faroe Is.	EU	18	14													
OZ*	Denmark	EU	18	14													
P2⁷*	Papua New Guinea	OC	51	28													
PA-PI*	Netherlands	EU	27	14													
PJ2-4, P4*	Neth. Antilles	SA	11	09													
PJ5-8*	St.Maarten,Saba,St.Eustatius	NA	11	08													
PP-PY†*	Brazil	SA	(D)	11													
PPØ-PYØ†*	Fernando de Noronha	SA	13	11													
PPØ-PYØ†*	St. Peter & St. Paul Rocks	SA	13	11													
PPØ-PYØ†*	Trindade & Martin Vaz. Is.	SA	15	11													
PZ*	Suriname	SA	12	09													

Prefix	Country	CONTINENT	ZONE ITU	ZONE CQ	MIXED	PHONE	CW	RTTY	SAT.	160	80	40	20	15	12	10	
S2*	Bangladesh	AS	41	22													
S7*	Seychelles	AF	53	39													
S9	Sao Tome & Principe	AF	47	36													
SA-SM*	Sweden	EU	18	14													
SN-SR*	Poland	EU	28	15													
ST*	Sudan	AF	48	34													
ST0*	Southern Sudan	AF	48	34													
SU*	Egypt	AF	38	34													
SV-SZ*	Greece	EU	28	20													
SV5*	Dodecanese	EU	28	20													
SV9*	Crete	EU	28	20													
SV/A*	Mount Athos	EU	28	20													
T2[16]	Tuvalu	OC	65	31													
T30	W. Kiribati (Gilbert & Ocn Is.)	OC	65	31													
T31	C. Kiribati (Brit. Phoenix Is.)	OC	62	31													
T32	East Kiribati (Line Is.)	OC	61,63	31													
T5	Somalia	AF	48	37													
T7*	San Marino	EU	28	15													
TA-TC*	Turkey	EU/AS	39	20													
TF*	Iceland	EU	17	40													
TG, TD†*	Guatemala	NA	11	07													
TI, TE†*	Costa Rica	NA	11	07													
TI9†*	Cocos I.	NA	11	07													
TJ	Cameroon	AF	47	36													
TK*	Corsica	EU	28	15													
TL[8]	Central African Rep.	AF	47	36													
TN[9]	Congo	AF	52	36													
TR[10]	Gabon	AF	52	36													
TT[11]	Chad	AF	47	36													
TU[12]	Ivory Coast	AF	46	35													
TY[13]	Benin	AF	46	35													
TZ[14]	Mali	AF	46	35													
UA1,3,4,6*	European Russian R.S.F.S.R.	EU	(E)	16													
UA1*	Franz Josef Land	EU	75	40													
UA2*	Kaliningrad	EU	29	15													
UA8,9,0*	Asiatic R.S.F.S.R.	AS	(F)	(G)													
UB, UT, UY*	Ukraine	EU	29	16													
UC*	Byelorussia	EU	29	16													
UD*	Azerbaijan	AS	29	21													
UF*	Georgia	AS	29	21													
UG*	Armenia	AS	29	21													
UH*	Turkmenistan	AS	30	17													
UI*	Uzbekistan	AS	30	17													
UJ*	Tadzhikistan	AS	30	17													
UL*	Kazakhstan	AS	29-31	17													
UM*	Kirghizia	AS	30,31	17													
UO*	Moldavia	EU	29	16													
UP*	Lithuania	EU	29	15													
UQ*	Latvia	EU	29	15													
UR*	Estonia	EU	29	15													

Prefix	Country	Continent	ITU	CQ	MIXED	PHONE	CW	RTTY	SAT.		160	80	40	20	15	12	10	
V2†*	Antigua & Barbuda	NA	11	08														
V3†*	Belize	NA	11	07														
V4¹⁵†	St. Christopher & Nevis	NA	11	08														
V8*	Brunei	OC	54	28														
VE, VO, VY†*	Canada	NA	(H)	1-5														
VK†*	Australia	OC	(I)	29, 30														
VK†*	Lord Howe I.	OC	60	30														
VK9†*	Willis I.	OC	60	30														
VK9†*	Christmas I.	OC	54	29														
VK9†*	Cocos-Keeling Is.	OC	54	29														
VK9†*	Mellish Reef	OC	56	30														
VK9†*	Norfolk I.	OC	60	32														
VKØ†*	Heard I.	AF	68	39														
VKØ†*	Macquarie I.	OC	60	30														
VP2E¹⁵	Anguilla	NA	11	08														
VP2M¹⁵*	Montserrat	NA	11	08														
VP2V¹⁵*	Br. Virgin Is.	NA	11	08														
VP5*	Turks & Caicos Is.	NA	11	08														
VP8*	Falkland Is.	SA	16	13														
VP8, LU*	South Georgia I.	SA	73	13														
VP8, LU*	South Orkney Is.	SA	73	13														
VP8, LU*	South Sandwich Is.	SA	73	13														
VP8,CE9,HFØ,LU,4K1*	South Shetland Is.	SA	73	13														
VP9*	Bermuda	NA	11	05														
VQ9*	Chagos	AF	41	39														
VR6†	Pitcairn I.	OC	63	32														
VS6*	Hong Kong	AS	44	24														
VU*	India	AS	41	22														
VU*	Andaman & Nicobar Is.	AS	49	26														
VU*	Laccadive Is.	AS	41	22														
XA-XI†*	Mexico	NA	10	06														
XA4-XI4†*	Revilla Gigedo	NA	10	06														
XT¹⁷	Burkina Faso	AF	46	35														
XU	Kampuchea	AS	49	26														
XW	Laos	AS	49	26														
XX9	Macao	AS	44	24														
XY-XZ	Burma	AS	49	26														
Y2-9²*	German Dem. Rep.	EU	28	14														
YA	Afghanistan	AS	40	21														
YB-YH²¹*	Indonesia	OC	51,54	28														
YI*	Iraq	AS	39	21														
YJ*	Vanuatu	OC	56	32														
YK*	Syria	AS	39	20														
YN†*	Nicaragua	NA	11	07														
YO-YR*	Romania	EU	28	20														
YS†*	El Salvador	NA	11	07														
YT-YU, YZ*	Yugoslavia	EU	28	15														
YV-YY†*	Venezuela	SA	12	09														
YVØ†*	Aves I.	NA	11	08														
Z2*	Zimbabwe	AF	53	38														
ZA	Albania	EU	28	15														

Prefix	Country	CONTINENT	ZONE ITU	ZONE CQ	MIXED	PHONE	CW	RTTY	SAT.		160	80	40	20	15	12	10	
ZB2*	Gibraltar	EU	37	14														
ZC4*	UK Sov. Base Areas on Cyprus	AS	39	20														
ZD7	St. Helena	AF	66	36														
ZD8*	Ascension I.	AF	66	36														
ZD9	Tristan da Cunha & Gough I.	AF	66	38														
ZF*	Cayman Is.	NA	11	08														
ZK1*	So. Cook Is.	OC	62	32														
ZK1*	No. Cook Is.	OC	62	32														
ZK2	Niue	OC	62	32														
ZK3	Tokelau Is.	OC	62	31														
ZL-ZM*	New Zealand	OC	60	32														
ZL7*	Chatham Is.	OC	60	32														
ZL8*	Kermadec Is.	OC	60	32														
ZL9*	Auckland I. & Campbell I.	OC	60	32														
ZP†*	Paraguay	SA	14	11														
ZR-ZU*	South Africa	AF	57	38														
ZR2-ZU2*	Prince Edward & Marion Is.	AF	57	38														
ZR3-ZU3*	(Namibia) S.W. Africa	AF	57	38														
1AØ[1]	Sov. Mil. Order of Malta	EU	28	15														
1S[1]	Spratly Is.	AS	50	26														
3A*	Monaco	EU	27	14	.													
3B6, 7*	Agalega & St. Brandon	AF	53	39														
3B8*	Mauritius	AF	53	39														
3B9*	Rodriguez I.	AF	53	39														
3C	Equatorial Guinea	AF	47	36														
3CØ	Pagalu I.	AF	52	36														
3D2*	Fiji	OC	56	32														
3D6†*	Swaziland	AF	57	38														
3V	Tunisia	AF	37	33														
3W, XV	Vietnam	AS	49	26														
3X	Guinea	AF	46	35														
3Y*	Bouvet	AF	67	38														
3Y*	Peter I	AN	72	12														
4P-4S*	Sri Lanka	AS	41	22														
4U†*	ITU Geneva	EU	28	14														
4U	HQ, United Nations	NA	08	05														
4W	Yemen	AS	39	21														
4X, 4Z†*	Israel	AS	39	20														
5A	Libya	AF	38	34														
5B*	Cyprus	AS	39	20														
5H-5I	Tanzania	AF	53	37														
5N-5O*	Nigeria	AF	46	35														
5R-5S	Madagascar	AF	53	39														
5T[18]*	Mauritania	AF	46	35														
5U[19]	Niger	AF	46	35														
5V*	Togo	AF	46	35														
5W*	Western Samoa	OC	62	32														
5X	Uganda	AF	48	37														
5Y-5Z*	Kenya	AF	48	37														

Prefix	Country	CONTINENT	ZONE ITU	ZONE CQ	MIXED	PHONE		RTTY	SAT.		160	80	40	20	15	12	10	
6V-6W[20]*	Senegal	AF	46	35														
6Y[†]*	Jamaica	NA	11	08														
7O	People's Dem. Rep. of Yemen	AS	39	21														
7P*	Lesotho	AF	57	38														
7Q	Malawi	AF	53	37														
7T-7Y*	Algeria	AF	37	33														
8P*	Barbados	NA	11	08														
8Q	Maldive Is.	AS/AF	41	22														
8R[†]*	Guyana	SA	12	09														
9G[22][†]	Ghana	AF	46	35														
9H*	Malta	EU	28	15														
9I-9J*	Zambia	AF	53	36														
9K*	Kuwait	AS	39	21														
9L[†]*	Sierra Leone	AF	46	35														
9M2, 4[23]*	West Malaysia	AS	54	28														
9M6, 8[23]*	East Malaysia	OC	54	28														
9N	Nepal	AS	42	22														
9Q-9T*	Zaire	AF	52	36														
9U[24]	Burundi	AF	52	36														
9V[25]*	Singapore	AS	54	28														
9X[24]*	Rwanda	AF	52	36														
9Y-9Z[†]*	Trinidad & Tobago	SA	11	09														
J2/A*	Abu Ail, Jabal at Tair	AS	39	21														

Notes

[1]Unofficial Prefix.

[2](DA-DL) Only contacts made September 17, 1973, and after, will count for this country.

[3](DM, DT, Y2-9) Only contacts made September 17, 1973, and after, will count for this country.

[4](FR) Only contacts made June 25, 1960, and after, will count for this country.

[5](JD, KA1) Formerly Marcus Island.

[6](JD, KA1) Formerly Bonin and Volcano Islands.

[7](P2) Only contacts made September 16, 1975, and after, will count for this country.

[8](TL) Only contacts made August 13, 1960, and after, will count for this county.

[9](TN) Only contacts made August 16, 1960, and after, will count for this country.

[10](TR) Only contacts made August 17, 1960, and after, will count for this country.

[11](TT) Only contacts made August 7, 1960, and after, will count for this country.

[12](TU) Only contacts made August 7, 1960, and after, will count for this country.

[13](TY) Only contacts made August 1, 1960, and after, will count for this country.

[14](TZ) Only contacts made June 20, 1960, and after, will count for this country.

[15](VP2) For credits on QSOs made before June 1, 1958, see page 97, June 1958 QST.

[16](T2, VR8) Only contacts made January 1, 1976, and after, will count for this country.

[17](XT) Only contacts made August 6, 1960, and after, will count for this country.

[18](5T) Only contacts made June 20, 1960, and after, will count for this country.

[19](5U) Only contacts made August 3, 1960, and after, will count for this country.

[20](6W) Only contacts made June 20, 1960, and after, will count for this country.

[21](8F, YB) Only contacts made May 1, 1963, and after, will count for this country.

[22](9G) Only contacts made March 5, 1957, and after, will count for this country.

[23](9M2, 4, 6, 8) Only contacts made September 16, 1963, and after, will count for this country.

[24](9U, 9X) Only contacts made July 1, 1962, and after, will count for this country.

[25](9V) Contacts made between September 15, 1963 and August 9, 1965, count for East Malaysia.

[26](D6, FH8) Only contacts made July 5, 1975, and after, will count for this country.

[27](KP5, KP4) Only contacts made March 1, 1979, and after, will count for this country.

[28](KC6) Includes Yap Islands beginning January 1, 1981.

[29](KC6) Doesn't include Yap Is. beginning January 1, 1981.

▲Also ATØ, DPØ, FT8Y, LU, OR4, VKØ, VP8, Y8, ZL5, ZS1, ZXØ, 3Y, 4K1, 8J1, etc. QSL via country under whose auspices the particular station is operating. The availability of a third-party traffic agreement and a QSL Bureau applies to the country under whose auspices the particular station is operating.

DELETED COUNTRIES

Credit for any of these countries can be given if the date of contact with the country in question agrees with the date(s) shown in the corresponding footnote.

Prefix	Country	CONTINENT	ZONE ITU	ZONE CQ	MIXED	PHONE	CW	RTTY	SAT.	160	80	40	20	15	10		
(AC3¹)²	Sikkim	AS	41	22													
(AC4¹)³	Tibet	AS	41	23													
C9⁴	Manchuria	AS	33	24													
CN2⁵	Tangier	AF	37	33													
CR8⁶	Damao, Diu	AS	41	22													
CR8⁶	Goa	AS	41	22													
CR8³⁴	Portuguese Timor	OC	54	28													
DA,DJ,DK,DL,DM⁷	Germany	EU	28	14													
EA9⁸	Ifni	AF	37	33													
EA9³⁸	Rio de Oro	AF	37	33													
ET2⁹	Eritrea	AF	48	37													
FF¹⁰	Fr. West Africa	AF	46	35													
FH,FB8³⁵	Comoros	AF	53	39													
FI8¹¹	Fr. Indo-China	AS	49	26													
FN8¹²	French India	AS	41	22													
FQ8¹³	Fr. Equatorial Africa	AF	47, 52	36													
HKØ⁴²	Bajo Nuevo	NA	11	08													
I1¹⁴	Trieste	EU	28	15													
I5¹⁵	Italian Somaliland	AF	48	37													
JD1/7J1⁴⁰	Okino Tori-shima	AS	45	27													
JZØ¹⁶	Netherlands N. Guinea	OC	51	28													
KP3,KS4,HKØ⁴²	Serrana Bank & Roncador Cay	NA	11	07													
KR6,8,JR6,KA6¹⁷	Okinawa (Ryukyu Islands)	AS	45	25													
KS4¹⁸	Swan Islands	NA	11	07													
KZ5³⁹	Canal Zone	NA	11	07													
P2,VK9¹⁹	Papua Territory	OC	51	28													
P2,VK9¹⁹	Terr. New Guinea	OC	51	28													
PK1-3²⁰	Java	OC	54	28													
PK4²⁰	Sumatra	OC	54	28													
PK5²⁰	Netherlands Borneo	OC	54	28													
PK6²⁰	Celebe & Molucca Is.	OC	54	28													
UN1²¹	Karelo-Finish Rep.	EU	19	16													
VO²²	Newfoundland, Labrador	NA	09	02, 05													
VQ1,5H1²³	Zanzibar	AF	53	37													
VQ6²⁴	British Somaliland	AF	48	37													
VQ9³⁴	Aldabra	AF	53	39													
VQ9³⁶	Desroches	AF	53	39													
VQ9³⁶	Farquhar	AF	53	39													
VS4²⁵	Sarawak	OC	54	28													
VS9H²⁶	Kuria Muria I.	AS	39	21													
ZC5²⁵	British North Borneo	OC	54	28													
ZC6, 4X1²⁷	Palestine	AS	39	20													
ZD4²⁸	Gold Coast, Togoland	AF	46	35													
(1M¹)²⁹	Minerva Reef	OC	62	32													
7O/VS9K⁴¹	Kamaran Is.	AS	39	21													
8Z4⁴³	Saudi Arabia/Iraq Neutral Zone	AS	39	21													
9K3,8Z5³⁰	Kuwait/Saudi Arabia Neutral Zone	AS	39	21													
9M2, VS2²⁵	Malaya	AS	54	28													
9S4³¹	Saar	EU	28	14													
9U5³²	Ruanda-Urundi	AF	52	36													
³³	Blenheim Reef	AF	41	39													
³⁷	Geyser Reef	AF	53	39													

Notes

[1]Unofficial prefix.

[2](AC3) Only contacts made before May 1, 1975, will count for this country. Contacts made May 1, 1975, and after, count as India.

[3](AC4) Only contacts made before June 1, 1974, will count for this country.

[4](C9) Only contacts made before September 16, 1963, will count for this country.

[5](CN2) Only contacts made before July 1, 1960, will count for this country.

[6](CR8) Only contacts made before January 1, 1962, will count for this country.

[7](DA, DJ, DK, DL, DM) Only contacts made before September 17, 1973, will count for this country.

[8](EA9) Only contacts made May 13, 1969, and before count for this country.

[9](ET2) Only contacts made November 14, 1962, and before, will count for this country.

[10](FF) Only contacts made August 6, 1960, and before, will count for this country.

[11](FI8) Only contacts made before December 21, 1950, will count for this country.

[12](FN8) Only contacts made before November 1, 1954, will count for this country.

[13](FQ8) Only contacts made August 16, 1960, and before, will count for this country.

[14](I1) Only contacts made before April 1, 1957, will count for this country. Contacts made April 1, 1957, and after, count as Italy.

[15](I5) Only contacts made June 30, 1960, and before, will count for this country.

[16](JZØ) Only contacts made before May 1, 1963, will count for this country.

[17](KR6, JR6, KA6) Only contacts made before May 15, 1972, will count for this country. Contacts made May 15, 1972, and after, count as Japan.

[18](KS4) Only contacts made before September 1, 1972, will count for this country. Contacts made September 1, 1972, and after, count as Honduras.

[19](P2, VK9) Only contacts made before September 16, 1975, will count for this country.

[20](PK1, 2, 3, 4, 5, 6) Only contacts made before May 1, 1963, will count for this country.

[21](UN1) Only contacts made June 30, 1960, and before, will count for this country. Contacts made July 1, 1960, and after, count as European Russian S.F.S.R.

[22](VO) Only contacts made before April 1, 1949, will count for this country.

[23](VQ1, 5H1) Only contacts made before June 1, 1974, will count for this country.

[24](VQ6) Only contacts made June 30, 1960, and before, will count for this country.

[25](9M2, VS2) Only contacts made September 15, 1963, and before will count for this country.

[26](VS9H) Only contacts made November 30, 1967 and before, will count for this country. Contacts made December 1, 1967, and after, count as Oman.

[27](ZC6, 4X1) Only contacts made July 1, 1968, and before, will count for this country.

[28](ZD4) Only contacts made March 5, 1957, and before, will count for this country.

[29](1M) Only contacts made July 15, 1972, and before, will count for this country. Contacts made July 16, 1972, and after, count as Tonga.

[30](9K3, 8Z5) Only contacts made before December 15, 1969, will count for this country.

[31](9S4) Only contacts made before April 1, 1957, count for this country.

[32](9U5) Only contacts made between July 1, 1960, and July 1, 1962, will count for this country.

[33](Blenheim Reef) Only contacts made between May 4, 1967, and July 1, 1975, will count for this country. Contacts made July 1, 1975, and after, count as Chagos.

[34](CR8) Only contacts made before September 15, 1976, will count for this country.

[35](FH, FB8) Only contacts made before July 6, 1975, count for this country.

[36](VQ9) Only contacts made before June 29, 1976, count for these countries.

[37](Geyser Reef) Only contacts made between May 4, 1967 and March 1, 1978, will count for this country.

[38](EA9) Only contacts made before January 8, 1976 will count for this country.

[39](KZ5) Only contacts made before October 1, 1979 will count for this county.

[40](JD1/7J1) Only contacts made between May 30, 1976 and November 30, 1980, will count for this country. Contacts made December 1, 1980, and after, count as Ogasawara.

[41](7O/VS9K) Only contacts made before March 11, 1982, will count for this country.

[42](HKØ) (KP3, KS4, HKØ) Only contacts made before September 17, 1981, will count for these countries.

[43](8Z4) Only contacts made before December 26, 1981, will count for this country.

PREFIX CROSS REFERENCES

A8 = EL	KA1 = JD1	VP2G (before 1975) = J3	ZD3 (before 1966) = C5
AC (before 1972) = A5	KA2AA-KA8ZZ = JA	VP2K (before 1984) = V4 or	ZD4 (before 1958) = 9G
AH = KH	KB6 (before 1979) = KH1	VP2E	ZD5 (before 1969) = 3D6
AL7 = KL7	KC4 (Navassa) = KP1	VP2L (before 1980) = J6	ZD6 (before 1965) = 7Q
AM-AO = EA	KG6 (before 1979) = KH2	VP2S (before 1980) = J8	ZE (before 1981) = Z2-9
AT-AW = VU	KG6I (before 1970) = JD1	VP3 (before 1967) = 8R	ZK9 (1983) = ZK2
AX = VK	KG6R, S, T (before 1979) = KHØ	VP4 (before 1963) = 9Y	ZM6 (before 1963) = 5W
AY-AZ = LU	KJ6 (before 1979) = KH3	VP5 (Jamaica) = 6Y	ZM7 (before 1984) = ZK3
CF-CK = VE	KM6 (before 1979) = KH4	VP6 (before 1967) = 8P	ZS7 (before 1969) = 3D6
CL = CO	KP4 (Desecheo) = KP5	VP7 (before 1974) = C6	ZS8 (before 1967) = 7P
CQ-CS = CT	KP6 (before 1979) = KH5	VQ2 (before 1965) = 9J	ZS9 (before 1967) = A2
CR3 (before 1974) = J5	KS6 (before 1979) = KH8	VQ3 (before 1962) = 5H	ZV-ZZ = PY
CR4 (before 1976) = D4	KV4 (before 1979) = KP2	VQ4 (before 1964) = 5Z	3B-3C (before 1968) = VE
CR5 (before 1976) = S9	KW6 (before 1979) = KH9	VQ5 (before 1963) = 5X	3G = CE
CR6 (before 1976) = D2	L2-9 = LU	VQ8 (before 1969) = 3B	3Z = SP
CR7 (before 1976) = C9	LY = UP	VQ8 (Chagos) = VQ9	4A-4C = XE
CR9 (before 1985) = XX9	M1 (before 1984) = T7	VQ9 (Seychelles) = S7	4D-4I = DU
CY-CZ = VE	MP4B (before 1972) = A9	VR1 (before 1980) = T30/31	4J-4L = U
CY9 (before 1985) = CYØ	MP4M (before 1972) = A4	VR2 (before 1971) = 3D2	4M = YV
DM-DT (before 1980) = Y2-9	MP4Q (before 1972) = A7	VR3 (before 1980) = T32	4N-4O = YU
EAØ (before 1969) = 3C	MP4T, D (before 1972) = A6	VR4 (before 1979) = H4	4T = OA
EK, EM-EO, ER-ES, EU-EZ = U	NH = KH	VR5 (before 1971) = A3	4V = HH
FA-FF (after 1983) = F	NL7 = KL7	VR8 (before 1979) = T2	5J-5K = HK
FA (before 1963) = 7X	NP = KP	VS1 (before 1966) = 9V	5L-5M = EL
FB8 (before 1961) = 5R	OQ (before 1961) = 9Q	VS5 (before 1985) = V8	6C = YK
FB8 (before 1985) = FT	PX (before 1970) = C3	VS7 (before 1949) = 4S	6D-6J = XE
FC (before 1985) = TK	RA, RN = UA	VS9A, P, S (before 1968) = 70	6O = T5
FD8 (before 1961) = 5V	RB-RR = UB-UR	VS9M = 8Q	6T-6U = ST
FE8 (before 1961) = TJ	RS-RZ = U	VS9O (before 1961) = A4	7A-7I = YB
FL (before 1978) = J2	S4 (Ciskei) = ZS	VX-VY = CYØ/VE	7G (before 1967) = 3X
FU8 (before 1982) = YJ	S8 (Transkei) = ZS	WH = KH	7J-7N = JA, JD
GB = G	T4 = CO	WL7 = KL7	7S = SM
GC (before 1977) = GJ/GU	T4 (Venda) = ZS	WP = KP	7Z = HZ
H3 = HP	TH, TM, TO-TQ, TV-TX = F	XJ-XO = VE	8A-8I = YB
H5 (Bophutatswana) = ZS	UN, UV, UW, UZ = UA	XP = OX	8J-8N = JA
H7 = YN	V9 (Venda) = ZS	XQ-XR = CE	8O = A2
HE = HB	VA-VG = VE	XV = 3W	8S = SM
HM (before 1982) = HL	VH-VN = VK	XX7 (before 1976) = C9	9A (before 1984) = T7
HT = YN	VK9 (Nauru) = C2	YL = UQ	9B-9D = EP
HU = YS	VP1 (before 1982) = V3	ZB1 (before 1965) = 9H	9E-9F = ET
HW-HY = F	VP2A (before 1982) = V2	ZD1 (before 1962) = 9L	
J4 = SV	VP2D (before 1979) = J7	ZD2 (before 1961) = 5N	

CONTINENT	
AF = AFRICA	
AN = ANTARCTICA	
AS = ASIA	
EU = EUROPE	
NA = NORTH AMERICA	
OC = OCEANIA	
SA = SOUTH AMERICA	

ZONE NOTES

(A) 33, 42, 43, 44
(B) 67, 69-74
(C) 12, 13, 29, 30, 32, 38, 39
(D) 12, 13, 15
(E) 19, 20, 29, 30
(F) 20-26, 30-35, 75
(G) 16, 17, 18, 19, 23
(H) 2, 3, 4, 9, 75
(I) 55, 58, 59

Allocation of International Call Signs

Call Sign
Series *Allocated to*

Series	Allocated to
AAA-ALZ	United States of America
AMA-AOZ	Spain
APA-ASZ	Pakistan (Islamic Republic of)
ATA-AWZ	India (Republic of)
AXA-AXZ	Australia
AYA-AZZ	Argentine Republic
A2A-A2Z	Botswana (Republic of)
A3A-A3Z	Tonga (Kingdom of)
A4A-A4Z	Oman (Sultanate of)
A5A-A5Z	Bhutan (Kingdom of)
A6A-A6Z	United Arab Emirates
A7A-A7Z	Qatar (State of)
A8A-A8Z	Liberia (Republic of)
A9A-A9Z	Bahrain (State of)
BAA-BZZ	China (People's Republic of)
CAA-CEZ	Chile
CFA-CKZ	Canada
CLA-CMZ	Cuba
CNA-CNZ	Morocco (Kingdom of)
COA-COZ	Cuba
CPA-CPZ	Bolivia (Republic of)
CQA-CUZ	Portugal
CVA-CXZ	Uruguay (Oriental Republic of)
CYA-CZZ	Canada
C2A-C2Z	Nauru (Republic of)
C3A-C3Z	Andorra (Principality of)
C4A-C4Z	Cyprus (Republic of)
C5A-C5Z	Gambia (Republic of the)
C6A-C6Z	Bahamas (Commonwealth of the)
C7A-C7Z*	World Meteorological Organization
C8A-C9Z	Mozambique (People's Republic of)
DAA-DRZ	Germany (Federal Republic of)
DSA-DTZ	Republic of Korea
DUA-DZZ	Philippines (Republic of the)
D2A-D3Z	Angola (People's Republic of)
D4A-D4Z	Cape Verde (Republic of)
D5A-D5Z	Liberia (Republic of)
D6A-D6Z	Comoros (Federal and Islamic Republic of the)
D7A-D9Z	Republic of Korea
EAA-EHZ	Spain
EIA-EJZ	Ireland
EKA-EKZ	Union of Soviet Socialist Republics
ELA-ELZ	Liberia (Republic of)
EMA-EOZ	Union of Soviet Socialist Republics
EPA-EQZ	Iran (Islamic Republic of)
ERA-ESZ	Union of Soviet Socialist Republics
ETA-ETZ	Ethiopia
EUA-EWZ	Byelorussian Soviet Socialist Republic
EXA-EZZ	Union of Soviet Socialist Republics
FAA-FZZ	France
GAA-GZZ	United Kingdom of Great Britain and Northern Ireland
HAA-HAZ	Hungarian People's Republic
HBA-HBZ	Switzerland (Confederation of)
HCA-HDZ	Ecuador
HEA-HEZ	Switzerland (Confederation of)
HFA-HFZ	Poland (People's Republic of)
HGA-HGZ	Hungarian People's Republic
HHA-HHZ	Haiti (Republic of)
HIA-HIZ	Dominican Republic
HJA-HKZ	Colombia (Republic of)
HLA-HLZ	Republic of Korea
HMA-HMZ	Democratic People's Republic of Korea
HNA-HNZ	Iraq (Republic of)
HOA-HPZ	Panama (Republic of)
HQA-HRZ	Honduras (Republic of)
HSA-HSZ	Thailand
HTA-HTZ	Nicaragua
HUA-HUZ	El Salvador (Republic of)
HVA-HVZ	Vatican City State
HWA-HYZ	France
HZA-HZZ	Saudi Arabia (Kingdom of)
H2A-H2Z	Cyprus (Republic of)
H3A-H3Z	Panama (Republic of)
H4A-H4Z	Solomon Islands
H6A-H7Z	Nicaragua
H8A-H9Z	Panama (Republic of)
IAA-IZZ	Italy
JAA-JSZ	Japan
JTA-JVZ	Mongolian People's Republic

Series	Allocated to
JWA-JXZ	Norway
JYA-JYZ	Jordan (Hashemite Kingdom of)
JZA-JZZ	Indonesia (Republic of)
J2A-J2Z	Djibouti (Republic of)
J3A-J3Z	Grenada
J4A-J4Z	Greece
J5A-J5Z	Guinea-Bissau (Republic of)
J6A-J6Z	Saint Lucia
J7A-J7Z	Dominica
J8A-J8Z	St. Vincent and the Grenadines
KAA-KZZ	United States of America
LAA-LNZ	Norway
LOA-LWZ	Argentina (Republic of)
LXA-LXZ	Luxembourg
LYA-LYZ	Union of Soviet Socialist Republics
LZA-LZZ	Bulgaria (People's Republic of)
L2A-L9Z	Argentina (Republic of)
MAA-MZZ	United Kingdom of Great Britain and Northern Ireland
NAA-NZZ	United States of America
OAA-OCZ	Peru
ODA-ODZ	Lebanon
OEA-OEZ	Austria
OFA-OJZ	Finland
OKA-OMZ	Czechoslovak Socialist Republic
ONA-OTZ	Belgium
OUA-OZZ	Denmark
PAA-PIZ	Netherlands (Kingdom of the)
PJA-PJZ	Netherlands Antilles
PKA-POZ	Indonesia (Republic of)
PPA-PYZ	Brazil (Federative Republic of)
PZA-PZZ	Suriname (Republic of)
P2A-P2Z	Papua New Guinea
P3A-P3Z	Cyprus (Republic of)
P4A-P4Z	Netherlands Antilles
P5A-P9Z	Democratic People'e Republic of Korea
QAA-QZZ	(Service abbreviations)
RAA-RZZ	Union of Soviet Socialist Republics
SAA-SMZ	Sweden
SNA-SRZ	Poland (People's Republic of)
SSA-SSM	Egypt (Arab Republic of)
SSN-STZ	Sudan (Democratic Republic of the)
SUA-SUZ	Egypt (Arab Republic of)
SVA-SZZ	Greece
S2A-S3Z	Bangladesh (People's Republic of)
S6A-S6Z	Singapore (Republic of)
S7A-S7Z	Seychelles (Republic of)
S9A-S9Z	Sao Tome and Principe (Democratic Republic of)
TAA-TCZ	Turkey
TDA-TDZ	Guatemala (Republic of)
TEA-TEZ	Costa Rica
TFA-TFZ	Iceland
TGA-TGZ	Guatemala (Republic of)
THA-THZ	France
TIA-TIZ	Costa Rica
TJA-TJZ	Cameroon (United Republic of)
TKA-TKZ	France
TLA-TLZ	Central African Republic
TMA-TMZ	France
TNA-TNZ	Congo (People's Republic of the)
TOA-TQZ	France
TRA-TRZ	Gabon Republic
TSA-TSZ	Tunisia
TTA-TTZ	Chad (Republic of)
TUA-TUZ	Ivory Coast (Republic of the)
TVA-TXZ	France
TYA-TYZ	Benin (People's Republic of)
TZA-TZZ	Mali (Republic of)
T2A-T2Z	Tuvalu
T3A-T3Z	Kiribati Republic
T4A-T4Z	Cuba
T5A-T5Z	Somali Democratic Republic
T6A-T6Z	Afghanistan (Democratic Republic of)
T7A-T7Z	San Marino (Republic of)
UAA-UQZ	Union of Soviet Socialist Republics
URA-UTZ	Ukrainian Soviet Socialist Republic
UUA-UZZ	Union of Soviet Socialist Republics
VAA-VGZ	Canada
VHA-VNZ	Australia
VOA-VOZ	Canada

Series	Allocated to
VPA-VSZ	United Kingdom of Great Britain and Northern Ireland
VTA-VWZ	India (Republic of)
VXA-VYZ	Canada
VZA-VZZ	Australia
V2A-V2Z	Antigua and Barbuda
V3A-V3Z	Belize
V4A-V4Z	St. Christopher and Nevis
V8A-V8Z	Brunei
WAA-WZZ	United States of America
XAA-XIZ	Mexico
XJA-XOZ	Canada
XPA-XPZ	Denmark
XQA-XRZ	Chile
XSA-XSZ	China (People's Republic of)
XTA-XTZ	Burkina Faso
XUA-XUZ	Democratic Kampuchea
XVA-XVZ	Viet Nam (Socialist Republic of)
XWA-XWZ	Lao People's Democratic Republic
XXA-XXZ	Portugal
XYA-XZZ	Burma (Socialist Republic of the Union of)
YAA-YAZ	Afghanistan (Democratic Republic of)
YBA-YHZ	Indonesia (Republic of)
YIA-YIZ	Iraq (Republic of)
YJA-YJZ	New Hebrides
YKA-YKZ	Syrian Arab Republic
YLA-YLZ	Union of Soviet Socialist Republics
YMA-YMZ	Turkey
YNA-YNZ	Nicaragua
YOA-YRZ	Romania (Socialist Republic of)
YSA-YSZ	El Salvador (Republic of)
YTA-YUZ	Yugoslavia (Socialist Federal Republic of)
YVA-YYZ	Venezuela (Republic of)
YZA-YZZ	Yugoslavia (Socialist Federal Republic of)
Y2A-Y9Z	German Democratic Republic
ZAA-ZAZ	Albania (Socialist People's Republic of)
ZBA-ZJZ	United Kingdom of Great Britain and Northern Ireland
ZKA-ZMZ	New Zealand
ZNA-ZOZ	United Kingdom of Great Britain and Northern Ireland
ZPA-ZPZ	Paraguay (Republic of)
ZQA-ZQZ	United Kingdom of Great Britain and Northern Ireland
ZRA-ZUZ	South Africa (Republic of)
ZVA-ZZZ	Brazil (Federative Republic of)
Z2A-Z2Z	Zimbabwe (Republic of)
2AA-2ZZ	United Kingdom of Great Britain and Northern Ireland
3AA-3AZ	Monaco
3BA-3AZ	Mauritius
3CA-3CZ	Equatorial Guinea (Republic of)
3DA-3DM	Swaziland (Kingdom of)
3DN-3DZ	Fiji
3EA-3FZ	Panama (Republic of)
3GA-3GZ	Chile
3HA-3UZ	China (People's Republic of)
3VA-3VZ	Tunisia
3WA-3WZ	Viet Nam (Socialist Republic of)
3XA-3XZ	Guinea (People's Revolutionary Republic of)
3YA-3YZ	Norway
3ZA-3ZZ	Poland (People's Republic of)
4AA-4CA	Mexico
4DA-4IZ	Philippines (Republic of the)
4JA-4LZ	Union of Soviet Socialist Republics
4MA-4MZ	Venezuela (Republic of)
4NA-4OZ	Yugoslavia (Socialist Federal Republic of)
4PA-4SZ	Sri Lanka (Democratic Socialist Republic of)
4TA-4TZ	Peru
4UA-4UZ*	United Nations Organization
4VA-4VZ	Haiti (Republic of)
4WA-4WZ	Yemen Arab Republic
4XA-4XZ	Israel (State of)
4YA-4YZ*	International Civil Aviation Organization

4ZA-4ZZ	Israel (State of)	
5AA-5AZ	Libya (Socialist People's Libyan Arab Jamahiriya)	
5BA-5BZ	Cyprus (Republic of)	
5CA-5GZ	Morocco (Kingdom of)	
5HA-5IZ	Tanzania (United Republic of)	
5JA-5KZ	Colombia (Republic of)	
5LA-5MZ	Liberia (Republic of)	
5NA-5OZ	Nigeria (Federal Republic of)	
5PA-5QZ	Denmark	
5RA-5SZ	Madagascar (Democratic Republic of)	
5TA-5TZ	Mauritania (Islamic Republic of)	
5UA-5UZ	Niger (Republic of the)	
5VA-5VZ	Togolese Republic	
5WA-5WZ	Western Samoa	
5XA-5XZ	Uganda (Republic of)	
5YA-5ZZ	Kenya (Republic of)	
6AA-6BZ	Egypt (Arab Republic of)	
6CA-6CZ	Syrian Arab Republic	
6DA-6JZ	Mexico	
6KA-6NZ	Republic of Korea	

6OA-6OZ	Somali Democratic Republic	
6PA-6SZ	Pakistan (Islamic Republic of)	
6TA-6UZ	Sudan (Democratic Republic of the)	
6VA-6WZ	Senegal (Republic of the)	
6XA-6XZ	Madagascar (Democratic Republic of)	
6YA-6YZ	Jamaica	
6ZA-6ZZ	Liberia (Republic of)	
7AA-7IZ	Indonesia (Republic of)	
7JA-7NZ	Japan	
7OA-7OZ	Yemen (People's Democratic Republic of)	
7PA-7PZ	Lesotho (Kingdom of)	
7QA-7QZ	Malawi (Republic of)	
7RA-7RZ	Algeria (Algerian Democratic and Popular Republic)	
7SA-7SZ	Sweden	
7TA-7YZ	Algeria (Algerian Democratic and Popular Republic)	
7ZA-7ZZ	Saudi Arabia (Kingdom of)	
8AA-8IZ	Indonesia (Republic of)	
8JA-8NZ	Japan	

8OA-8OZ	Botswana (Republic of)	
8PA-8PZ	Barbados	
8QA-8QZ	Maldives (Republic of)	
8RA-8RZ	Guyana	
8SA-8SZ	Sweden	
8TA-8YZ	India (Republic of)	
8ZA-8ZZ	Saudi Arabia (Kingdom of)	
9BA-9DZ	Iran (Islamic Republic of)	
9EA-9FZ	Ethiopia	
9GA-9GZ	Ghana	
9HA-9HZ	Malta (Republic of)	
9IA-9JZ	Zambia (Republic of)	
9KA-9KZ	Kuwait (State of)	
9LA-9LZ	Sierra Leone	
9MA-9MZ	Malaysia	
9NA-9NZ	Nepal	
9OA-9TZ	Zaire (Republic of)	
9UA-9UZ	Burundi (Republic of)	
9VA-9VZ	Singapore (Republic of)	
9WA-9WZ	Malaysia	
9XA-9XZ	Rwanda (Republic of)	
9YA-9ZZ	Trinidad and Tobago	

USSR Oblasts

Prefix	No.	Name	Prefix	No.	Name	Prefix	No.	Name
UD-D	1	AZERBAIJAN	UB-G	78	HERSONSKAYA	UA3L	155	SMOLENSKAYA
UD-N	2	NAKHICHEVAN	UB-T	79	HMELNICKAYA	UA4A	156	VOLGRADSKAYA
UD-K	3	GORNO-KARABAKH	UB-C	80	CHERKASSKAYA	UA3R	157	TAMBOVSKAYA
UG-G	4	ARMENIA	UB-R	81	CHERNIGOVSKAYA	UA9H	158	TOMSKAYA
UC-L	5	BRESTSKAYA	UB-Y	82	CHERNOVICKAYA	UAOY	159	TUVA
UC-W	6	VIETEBSKAYA	UR-R	83	ESTONIA	UA3P	160	TULSKAYA
UC-O	7	GOMELSKAYA	UA9W	84	BASHKIR	UA9L	161	TIUMENSKAYA
UC-I	8	GRODNENSKAYA	UAOO	85	BURYAT	UA9J	162	KHANTY MENSYJSKY
UC-C	9	MINSKAYA	UA6W	86	DAGHESTAN	UA9K	163	YAMALO NENETSKY
UC-S	10	MOGILEVSKAYA	UA6X	87	KABARDINO-BALKARSK	UA4L	164	ULIANOVSKAYA
UC2-	11	DELETED 1960	UA1N	88	KARELIAN	UA9A	165	CHELIABINSKAYA
UF-F	12	GEORGIA	UA6I	89	KALYMK	UAOU	166	CHITINSKAYA
UF-V	13	ABKHAZIAN	UA9X	90	KOMI	UA9S	167	ORENBURGSKAYA
UF-Q	14	ADJAR	UA4S	91	MARI	UA3M	168	YAROSLAVSKAYA
UF-O	15	SOUTH OSSETIA	UA4U	92	MORDOVIAN	UA1A	169	LENINGRAD
UL-B	16	CELINOGRADSKAYA	UA6J	93	NORTH OSSETIA	UA3A	170	MOSCOW
UL-I	17	AKTUBINSKAYA	UA4P	94	TATAR	4KO-	171	DELETED 1984
UL-Q	18	ALMA-STINSKAYA	UA4W	95	UDMURT	4K1-	172	DELETED 1984
UL-J	19	EAST KAZAKHSTANSKAYA	UA6P	96	CHECHENO-INGUSH	UI-D	173	SYRDARINSKAYA
UL-O	20	GURIEVSKAYA	UA4Y	97	CHUVASH	UA8T	174	UST.ORDYNSKY BURIATSKY
UL-T	21	JAMBULSKAYA	UAOQ	98	YAKUT	UA8V	175	AGINSKY BURIATSKY
UL-M	22	URALSKAYA	UA9Y	99	ALTAI	UL-Y	176	TURGAY (1970)
UL-P	23	KARAGANDINSKAYA	UA9Z	100	GORNO-ALTAI AUT	UM-P	177	NARYNSKY (1970)
UL-K	24	KIZIL-ORDINSKAYA	UA6A	101	KRASNODAR	UL-R	178	DZHEZKAZGANSKAYA (1973)
UL-E	25	KOKCHETAVASKAYA	UA6Y	102	ADIGEI AUT	UL-A	179	MANGISHLAKSKAYA (1973)
UL-L	26	KUSTANAYSKAYA	UAOA	103	KRASNOYARSK	UH-B	180	NEBIT DAG (1983)
UL-F	27	PAVOLDARSKAYA	UAOW	104	KHAKASS AUT	UI-V	181	DZHIZAKSKAYA (1973)
UL-C	28	NORTH KAZAKHSTANSKAYA	UAOB	105	TAJMYRSKY	UJ-K	182	KULYABSKAYA (1973)
UL-D	29	SEMIPALATINSKAYA	UAOH	106	EVENKIYSKY	UJ-X	183	KURGAN-TYUBINSKAYA (1977)
UL-V	30	TALDY-KURGANSKAYA	UAOL	107	PRIMORYE	UM-T	184	TALASSKAYA (1980)
UL-N	31	CHIMKENTSKAYA	UA6H	108	STAVROPOL	UI-Q	185	NAVOYASKAYA (1982)
UM-	32	DELETED 1959	UA6E	109	KARACHAI-CHERKESS	UT-U	186	KIEV CITY (1984)
UM-Q	33	ISSYK-KUL-PRZHEVALSK	UAOC	110	KHABAROVSK	UT-J	187	SEVASTOPOL CITY (1984)
UM-N	34	OSHKAYA	UAOD	111	JEWISH	UC-A	188	MINSK CITY (1984)
UM•	35	DELETED 1959	UAOJ	112	AMURSKAYA	UI-A	189	TASHKENT CITY (1984)
UM-M	36	KIRGHIZ	UA1O	113	ARKANGELSKAYA	UL-G	190	ALMA ATA CITY (1984)
UQ-G	37	LATVIA	UA1P	114	NENETSKIY	UH-A	191	ASHKHABAD CITY (1984)
UP-B	38	LITHUANIA	UA6U	115	ASTRAKHANSKAYA			
UO-O	39	MOLDAVIA	UA4-	116	DELETED 1962			
UJ-J	40	TADZHIK	UA3Z	117	BELGORODSKAYA			
UJ-S	41	LENINABAD	UA3Y	118	BRIANSKAYA			
UJ-R	42	GORNO-BADAKHSTAN	UA3V	119	VLADIMIRSKAYA			
UH-H	43	TURKOMAN	UA1Q	120	VOLOGODSKAYA			
UH-E	44	MARYISKAYA	UA3Q	121	VORONEJSKAYA	UA1A	169	LENINGRAD
UH-W	45	TASHAUZSKAYA	UA3T	122	GORKOVSKAYA	UA1C	136	LENINGRADSKAYA
UH-Y	46	CHARDJOUSKAYA	UA3U	123	IVANOVSKAYA	UA1N	88	KARELIAN
UI-F	47	ANDIJANSKAYA	UAOS	124	IRKUTSKAYA	UA1O	113	ARKANGELSKAYA
UI-L	48	BUKHARSKAYA	UA2F	125	KALININGRADSKAYA	UA1P	114	NENETSKIY
UI-C	49	KASHKADARSKAYA	UA3I	126	KALININSKAYA	UA1Q	120	VOLOGODSKAYA
UI-O	50	NAMANGAN	UA3X	127	KALUJSKAYA	UA1T	144	NOVGORODSKAYA
UI-I	51	SAMARKANDSKAYA	UAOZ	128	KAMCHATSKAYA	UA1W	149	PSKOVSKAYA
UI-T	52	SURKHAM DARINSKAYA	UAOX	129	KORYAKSKY	UA1Z	143	MURMANSKAYA
UI-B	53	TASHKENTSKAYA	UA9U	130	KEMEROVSKAYA	UA2F	125	KALININGRADSKAYA
UI-G	54	FERGANSKAYA	UA4N	131	KIROVSKAYA	UA3A	170	MOSCOW
UI-U	55	KHOREZMSKAYA	UA3N	132	KOSTROMSKAYA	UA3D	142	MOSCOWSKAYA
UI-Z	56	KARA-KALPAK	UA4H	133	KUIBISHEVSKAYA	UA3E	147	ORLOVSKAYA
UB-N	57	VINNICKAYA	UA9Q	134	KURGANSKAYA	UA3G	137	LIPECKAYA
UB-P	58	VOLINSKAYA	UA3W	135	KURSKAYA	UA3I	126	KALININSKAYA
UB-M	59	VOROSHILOVGRADSKAYA	UA1C	136	LENINGRADSKAYA	UA3L	155	SMOLENSKAYA
UB-E	60	DNEPROPETROVSKAYA	UA3G	137	LIPECKAYA	UA3M	168	YAROSLAVSKAYA
UB5-	61	DELETED 1963	UAOI	138	MAGADANSKAYA	UA3N	132	KOSTROMSKAYA
UB-X	62	ZHITOMIRSKAYA	UAOK	139	CHUKOTSKIY	UA3P	160	TULSKAYA
UB-D	63	ZAKARPATSKAYA	UA9F	140	PERMSKAYA	UA3Q	121	VORONEJSKAYA
UB-Q	64	ZAPOROJSKAYA	UA9G	141	KOMI PERMYATSKY	UA3R	157	TAMBOVSKAYA
UB-U	65	KIEVSKAYA	UA3D	142	MOSCOWSKAYA	UA3S	151	RIASANSKAYA
UB-V	66	KIROVGRADSKAYA	UA1Z	143	MURMANSKAYA	UA3T	122	GORKOVSKAYA
UB-J	67	CRIMSKAYA	UA1T	144	NOVGORODSKAYA	UA3U	123	IVANOVSKAYA
UB-W	68	LVOVSKAYA	UA9O	145	NOVOSIBIRSKAYA	UA3V	119	VLADIMIRSKAYA
UB-Z	69	NIKOLAEVSKAYA	UA9M	146	OMSKAYA	UA3W	135	KURSKAYA
UB-F	70	ODESSKAYA	UA3E	147	ORLOVSKAYA	UA3X	127	KALUJSKAYA
UB-H	71	POLTAVSKAYA	UA4F	148	PENZENSKAYA	UA3Y	118	BRIANSKAYA
UB-K	72	ROVENSKAYA	UA1W	149	PSKOVSKAYA	UA3Z	117	BELGORODSKAYA
UB-I	73	DONECKAYA	UA6L	150	ROSTOVSKAYA	UA4A	156	VOLGRADSKAYA
UB-S	74	IVANO-FRANKOVSKAYA	UA3S	151	RIASANSKAYA	UA4C	152	SARATOVSKAYA
UB-A	75	SUMSKAYA	UA4C	152	SARATOVSKAYA	UA4F	148	PENZENSKAYA
UB-B	76	TERNOPOLSKAYA	UAOF	153	SAKHALINSKAYA	UA4H	133	KUIBISHEVSKAYA
UB-L	77	KARKOVSKAYA	UA9C	154	SVERDLOVSKAYA	UA4L	164	ULIANOVSKAYA

Prefix	No.	Oblast	Prefix	No.	Oblast	Prefix	No.	Oblast
UA4N	131	KIROVSKAYA	UAOX	129	KORYAKSKY	UI-C	49	KASHKADARSKAYA
UA4P	94	TATAR	UAOY	159	TUVA	UI-D	173	SYRDARINSKAYA
UA4S	91	MARI	UAOZ	128	KAMCHATSKAYA	UI-F	47	ANDIJANSKAYA
UA4U	92	MORDOVIAN	UB5-	61	DELETED 1963	UI-G	54	FERGANSKAYA
UA4W	95	UDMURT	UB-A	75	SUMSKAYA	UI-I	51	SAMARKANDSKAYA
UA4Y	97	CHUVASH	UB-B	76	TERNOPOLSKAYA	UI-L	48	BUKHARSKAYA
UA4-	116	DELETED 1962	UB-C	80	CHERKASSKAYA	UI-O	50	NAMANGAN
UA6A	101	KRASNODAR	UB-D	63	ZAKARPATSKAYA	UI-Q	185	NAVOYASKAYA (1982)
UA6E	109	KARACHAI-CHERKESS	UB-E	60	DNEPROPETROVSKAYA	UI-T	52	SURKHAM DARINSKAYA
UA6H	108	STAVROPOL	UB-F	70	ODESSKAYA	UI-U	55	KHOREZMSKAYA
UA6I	89	KALYMK	UB-G	78	HERSONSKAYA	UI-V	181	DZHIZAKSKAYA (1973)
UA6J	93	NORTH OSSETIA	UB-H	71	POLTAVSKAYA	UI-Z	56	KARA-KALPAK
UA6L	150	ROSTOVSKAYA	UB-I	73	DONECKAYA	UJ-J	40	TADZHIK
UA6P	96	CHECHENO-INGUSH	UB-J	67	CRIMSKAYA	UJ-K	182	KULYABSKAYA (1973)
UA6U	115	ASTRAKHANSKAYA	UB-K	72	ROVENSKAYA	UJ-R	42	GORNO-BADAKHSTAN
UA6W	86	DAGHESTAN	UB-L	77	KARKOVSKAYA	UJ-S	41	LENINABAD
UA6X	87	KABARDINO-BALKARSK	UB-M	59	VOROSHILOVGRADSKAYA	UJ-X	183	KURGAN-TYUBINSKAYA (1977)
UA6Y	102	ADIGEI AUT	UB-N	57	VINNICKAYA	UL-A	179	MANGISHLAKSKAYA (1973)
UA8T	174	UST.ORDYNSKY BURIATSKY	UB-P	58	VOLINSKAYA	UL-B	16	CELINOGRADSKAYA
UA8V	175	AGINSKY BURIATSKY	UB-Q	64	ZAPOROJSKAYA	UL-C	28	NORTH KAZAKHSTANSKAYA
UA9A	165	CHELIABINSKAYA	UB-R	81	CHERNIGOVSKAYA	UL-D	29	SEMIPALATINSKAYA
UA9C	154	SVERDLOVSKAYA	UB-S	74	IVANO-FRANKOVSKAYA	UL-E	25	KOKCHETAVASKAYA
UA9F	140	PERMSKAYA	UB-T	79	HMELNICKAYA	UL-F	27	PAVOLDARSKAYA
UA9G	141	KOMI PERMYATSKY	UB-U	65	KIEVSKAYA	UL-G	190	ALMA ATA CITY (1984)
UA9H	158	TOMSKAYA	UB-V	66	KIROVGRADSKAYA	UL-I	17	AKTUBINSKAYA
UA9J	162	KHANTY MENSYJSKY	UB-W	68	LVOVSKAYA	UL-J	19	EAST KAZAKHSTANSKAYA
UA9K	163	YAMALO NENETSKY	UB-X	62	ZHITOMIRSKAYA	UL-K	24	KIZIL-ORDINSKAYA
UA9L	161	TIUMENSKAYA	UB-Y	82	CHERNOVICKAYA	UL-L	26	KUSTANAYSKAYA
UA9M	146	OMSKAYA	UB-Z	69	NIKOLAEVSKAYA	UL-M	22	URALSKAYA
UA9O	145	NOVOSIBIRSKAYA	UC2-	11	DELETED 1960	UL-N	31	CHIMKENTSKAYA
UA9Q	134	KURGANSKAYA	UC-A	188	MINSK CITY (1984)	UL-O	20	GURIEVSKAYA
UA9S	167	ORENBURGSKAYA	UC-C	9	MINSKAYA	UL-P	23	KARAGANDINSKAYA
UA9U	130	KEMEROVSKAYA	UC-I	8	GRODNENSKAYA	UL-Q	18	ALMA-STINSKAYA
UA9W	84	BASHKIR	UC-L	5	BRESTSKAYA	UL-R	178	DZHEZKAZGANSKAYA (1973
UA9X	90	KOMI	UC-O	7	GOMELSKAYA	UL-T	21	JAMBULSKAYA
UA9Y	99	ALTAI	UC-S	10	MOGILEVSKAYA	UL-V	30	TALDY-KURGANSKAYA
UA9Z	100	GORNO-ALTAI AUT	UC-W	6	VIETEBSKAYA	UL-Y	176	TURGAY (1970)
UAOA	103	KRASNOYARSK	UD-D	1	AZERBAIJAN	UM-M	36	KIRGHIZ
UAOB	105	TAJMYRSKY	UD-K	3	GORNO-KARABAKH	UM-N	34	OSHKAYA
UAOC	110	KHABAROVSK	UD-N	2	NAKHICHEVAN	UM-P	177	NARYNSKY (1970)
UAOD	111	JEWISH	UF-F	12	GEORGIA	UM-Q	33	ISSYK-KUL-PRZHEVALSK
UAOF	153	SAKHALINSKAYA	UF-O	15	SOUTH OSSETIA	UM-T	184	TALASSKAYA (1980)
UAOH	106	EVENKIYSKY	UF-Q	14	ADJAR	UM-	32	DELETED 1959
UAOI	138	MAGADANSKAYA	UF-V	13	ABKHAZIAN	UM-	35	DELETED 1959
UAOJ	112	AMURSKAYA	UG-G	4	ARMENIA	UO-O	39	MOLDAVIA
UAOK	139	CHUKOTSKIY	UH-A	191	ASHKHABAD CITY (1984)	UP-B	38	LITHUANIA
UAOL	107	PRIMORYE	UH-B	180	NEBIT DAG (1983)	UQ-G	37	LATVIA
UAOO	85	BURYAT	UH-E	44	MARYISKAYA	UR-R	83	ESTONIA
UAOQ	98	YAKUT	UH-H	43	TURKOMAN	UT-J	187	SEVASTOPOL CITY (1984
UAOS	124	IRKUTSKAYA	UH-W	45	TASHAUZSKAYA	UT-U	186	KIEV CITY (1984)
UAOU	166	CHITINSKAYA	UH-Y	46	CHARDJOUSKAYA	4K1-	172	DELETED 1984
UAOW	104	KHAKASS AUT	UI-A	189	TASHKENT CITY (1984)	4KO-	171	DELETED 1984
			UI-B	53	TASHKENTSKAYA			

DELETIONS - 11 32 35 61 116 171 172
MISC - UA6A = UA6B   UP-B = UP-P   UR-R = UR-T
    Calls may begin with R or U - RI = UI   UZ = RZ   UB = RB, etc
    UKRAINE - UB = UT = UY
    R.S.F.S.R. - UA = UN = UV = UW = UZ
Calls issued before 1971 may not follow this pattern.
Club calls can be identified by a W,X,Y or Z in the second letter following the number.
The oblast can be determined by the letter following the number for all republics, except the RSFSR where the number and following letter are needed.

—K1KI

ITU Recommended Phonetics

A—Alfa (**AL** FAH)
B—Bravo (**BRAH** VOH)
C—Charlie (**CHAR** LEE or **SHAR** LEE)
D—Delta (**DELL** TAH)
E—Echo (**ECK** OH)
F—Foxtrot (**FOKS** TROT)
G—Golf (**GOLF**)
H—Hotel (HOH **TELL**)
I—India (**IN** DEE AH)
J—Juliett (**JEW** LEE **ETT**)
K—Kilo (**KEY** LOH)
L—Lima (**LEE** MAH)
M—Mike (**MIKE**)
N—November (NO **VEM** BER)
O—Oscar (**OSS** CAH)
P—Papa (PAH **PAH**)
Q—Quebec (KEH **BECK**)
R—Romeo (**ROW** ME OH)
S—Sierra (SEE **AIR** RAH)
T—Tango (**TANG** GO)
U—Uniform (**YOU** NEE FORM or **OO** NEE FORM)
V—Victor (**VIK** TAH)
W—Whiskey (**WISS** KEY)
X—X-Ray (**ECKS** RAY)
Y—Yankee (**YANG** KEY)
Z—Zulu (**ZOO** LOO)

Note: The **Boldfaced** syllables are emphasized.

INTERNATIONAL MORSE CODE	AMERICAN MORSE CODE
A–Z, 0–9, and punctuation in Morse code	A–Z, 0–9, and punctuation in Morse code

Punctuation:
Period (.)
Comma (,)
Interrogation (?)
Colon (:)
Semicolon (;)
Hyphen (-)
Slash (/)
Quotation marks ('')

Morse Code for Other Languages

Code	Japanese	Korean	Arabic	Hebrew	Russian	Greek
·	ヘ he	ㅏ a		vav	Е,Э E	E epsilon
—	ム mu	ㅓ ŏ	ت ta	tav	Т T	T tau
··	nigori	ㅑ ya	ي ya	yod	И I	I iota
·—	イ i	ㅗ o	ا alif	aleph	А A	A alpha
—·—	タ ta	ㅛ yo	ن noon	nun	Н N	N nu
——	ヨ yo	ㅁ m	م meem	mem	М M	M mu
···	ラ ra	ㅕ yŏ	س seen	shin	С S	Σ sigma
··—	ウ u		tet	tet	У U	ΟΥ omicron ypsilon
·—·	ナ na	ㅠ yu	ر ra	reish	Р R	Ρ rho
·——	ヤ ya	ㅂ p(b)	و waw	tzadi	В V	Ω omega
—··	ホ ho		د dal	dalet	Д D	Δ delta
—·—	ワ wa	ㅇ -ng	ك kaf	chaf	Н K	K kappa
·—·—	リ ri	ㅅ s	غ ghain	gimmel	Г G	Γ gamma
—·—·	レ re	ㅍ p'	خ kha	heh	О O	O omicron
····	ヌ nu	ㅌ r-(-l)	ح ha	chet	Х H	H eta
···—	ク ku	ㄴ n	ض dad	dad	Ж J	ΗΥ eta ypsilon
··—·	チ ti		ف fa	feh	Ф F	Φ phi
·——	ノ no				Ю yu	ΑΥ alpha ypsilon
·—··	カ ka	ㄱ k(g)	ل lam	lamed	Л L	Λ lambda
·—··	ロ ro		ع ain		Я ya	ΑΙ alpha iota
··——	ツ tu	ㅈ ch(j)		peh	П P	Π pi
·—·—	ヲ wo	ㅎ h	ج jeem	ayen	Й Y	ΤΙ ypsilon iota
—···	ハ ha	ㄷ t(d)	ب ba	bet	В B	B beta
—·——	マ ma	ㅋ k'	ص sad		ь,ъ mute	Ξ xi
—·—·	ニ ni	ㅊ ch'	ث tha	samech	Ц TS	Θ theta
—·—·	ケ ke	ㅔ e	ز za		Ы I	Τ ypsilon
—·——	フ hu	ㅌ t'	ذ dhal	zain	З Z	Z zeta
—··—·	ネ ne	ㅐ ae	ق qaf	kof	Щ SHCH	Ψ psi
—··—	ソ so		ز zay		Ч CH	ΕΤ epsilon ypsilon
————	コ ko		ش sheen		Ш SH	X khi
·—·—	ト to		ه he			
··—·	ミ mi					
·	han-nigori					
·—···	オ o					
·—·—·	ヰ (w)i					
·—·—·	ン n					
·—··—	テ te					
·—··—	ヱ (w)e					
—····—	- hyphen					
·——··	セ se					
—··—·	メ me					
—··—·	モ mo					
—··—·	ユ yu					
·—·—·	キ ki					
—·—··	サ sa					
—·—··	ル ru					
—·—··	エ e					
——·—·	ヒ hi					
——·—·	シ si					
——·—·	ア a					
·—···—	ス su					

ي lam·alif

Spanish Phonetics

America	ah-MAIR-ika
Brasil	brah-SIL
Canada	cana-DAH
Dinamarca	dina-MAR-ka
Espana	es-PAHN-yah
Francia	FRAHN-seeah
Grenada	gre-NAH-dah
Holanda	oh-LONN-dah
Italia	i-TAL-eeah
Japon	hop-OWN
Kilowatio	kilo-WAT-eeoh
Lima	LIMA
Mejico	MEH-heeco
Noruega	nor-WAY-gah
Ontario	on-TAR-eeoh
Portugal	portu-GAL
Quito	KEY-toe
Roma	ROW-mah
Santiago	santee-AH-go
Toronto	tor-ON-toe
Uniforme	oonee-FORM-eh
Victoria	vic-TOR-eeah
Washington, Wisky	washingtone, wisky
Xilofono	see-LOW-phono
Yucatan	yuca-TAN
Zelandia	see-LAND-eeah

W		DOE-bleh-vay
Ø	cero	SEH-roe
1	uno	OO-no
2	dos	DOS
3	tres	TRAYCE
4	cuatro	KWAT-roe
5	cinco	SINK-oh
6	seis	SAYCE
7	siete	see-AY-teh
8	ocho	OCH-oh
9	nueve	new-AY-veh

—John Mason Jr., EA4AXW

DX Operating Code

For W/VE Amateurs

Some amateur DXers have caused considerable confusion and interference in their efforts to work DX stations. The points below, if observed by all W/VE amateurs, will help make DX more enjoyable for all.

1) *Call* DX only after he calls CQ, QRZ? or signs SK, or voice equivalents thereof. Make your calls short.

2) Do not call a DX station:
   a) On the frequency of the station he is calling until you are sure the QSO is over (SK).
   b) Because you hear someone else calling him.
   c) When he signs KN, AR or CL.
   d) Exactly on his frequency.
   e) After he calls a directional CQ, unless of course you are in the right direction or area.

3) Keep within frequency band limits. Some DX stations can get away with working outside, but you cannot.

4) Observe calling instructions given by DX stations. Example: 15U means "call 15 kHz *up* from my frequency." 15D means *down*, etc.

5) Give honest reports. Many DX stations *depend* on W/VE reports for adjustment of station and equipment.

6) Keep your signal clean. Key clicks, ripple, feedback or splatter gives you a bad reputation and may get you a citation from FCC.

7) *Listen* and call the station you want. Calling CQ DX is not the best assurance that the rare DX will reply.

8) When there are several W or VE stations waiting, avoid asking DX to "listen for a friend." Also avoid engaging him in a ragchew against his wishes.

For Overseas Amateurs

To all overseas amateur stations:

In their eagerness to work you, many W and VE amateurs resort to practices that cause confusion and QRM. Most of this is good-intentional but ill-advised; some of it is intentional and selfish. The key to the cessation of unethical DX operating practices is in your hands. We believe that your adoption of certain operating habits will increase your enjoyment of Amateur Radio and that of amateurs on this side who are eager to work you. We recommend your adoption of the following principles:

1) Do not answer calls on your own frequency.

2) Answer calls from W/VE stations only when their signals are of good quality.

3) Refuse to answer calls from other stations when you are already in contact with someone, and do not acknowledge calls from amateurs who indicate they wish to be "next."

4) Give *everybody* a break. When many W/VE amateurs are patiently and quietly waiting to work you, avoid complying with requests to "listen for a friend."

5) Tell listeners where to call you by indicating how many kilohertz up (U) or down (D) from your frequency you are listening.

6) Use the ARRL-recommended ending signals, especially KN to indicate to impatient listeners the status of the QSO. KN means "Go ahead (specific station); all others keep out."

7) Let it be known that you avoid working amateurs who are constant violators of these principles.

Phillips Code

At some point in your amateur career, you will come across the words "Phillips Code." In the late 1800s, the telegraph was widely used by the news services. Owing to the large volume of information that had to be relayed, telegraphers made extensive use of abbreviations. In 1879, a high-speed operator named Walter P. Phillips published the first edition of the Phillips Code, a comprehensive list of abbreviations, many of which were derived from popular abbreviations in use at the time. The code became a standard for the industry, and through its use, many telegraphers of the day were able to clear 15,000 words per shift! It was to their advantage to handle as much information as possible; it was common to be paid by the word or message. Today this code is rarely used.

Some abbreviations from the Phillips Code are included in here. The complete list is many times longer. You can see by studying this list how some of our currently used abbreviations were derived.

ABB	Abbreviate	FQ	Frequent	MA	May	SG	Signify
AGR	Agree	FZ	Freeze	MD	Made	SNT	Sent
ALG	Along	GD	Good	MK	Make	TBL	Trouble
APP	Appoint	GNT	Grant	MO	Month	TH	Those
APRL	April	GVT	Government	N	Not	TKT	Ticket
BAJ	Badge	GW	Grow	ND	Need	U	You
BF	Before	HAP	Happy	NR	Near	Un	Until
BH	Both	HH	Has had	NTC	Notice	UR	Your
BLU	Blue	HRD	Hears	OB	Obtain	VA	Virginia
BWR	Beware	HRT	Hurt	OFN	Often	VU	View
C	See	IA	Iowa	OMN	Omission	VY	Very
CFM	Confirm	IFM	Inform	OT	Out	WB	Will be
CK	Check	INC	Increase	PAS	Pays	WD	Would
COL	Colonel	ITN	Intention	PB	Probable	WH	Which
CR	Care	IW	It was	PD	Paid	XAC	Exact
DA	Day	JL	Jail	POS	Possible	XM	Extreme
DED	Dead	JP	Japan	PRM	Permanent	XPT	Export
DMH	Diminish	JUN	June	QA	Qualify	XU	Exclude
DRU	Drew	KD	Kind	QNY	Quantity	Y	Year
EA	Each	KG	King	QRL	Quarrel	YD	Yield
EDU	Educate	KM	Communicate	R	Are	YOA	Years of age
EGO	Egotism	KN	Known	RF	Refer	ZA	Sea
ENR	Enter	LAK	Lake	RLY	Really	ZD	Said
EV	Ever	LGR	Longer	RPT	Repeat	ZM	Seem
FA	Fail	LIB	Liberty	SB	Subsequent		
FEV	Fever	LV	Leave				

The Origin of "73"

The traditional expression "73" goes right back to the beginning of the landline telegraph days. It is found in some of the earliest editions of the numerical codes, each with a different definition, but each with the same idea in mind—it indicated that the end, or signature, was coming up. But there are no data to prove that any of these were used.

The first authentic use of 73 is in the publication *The National Telegraphic Review and Operators' Guide*, first published in April 1857. At that time, 73 meant "My love to you"! Succeeding issues of this publication continued to use this definition of the term. Curiously enough, some of the other numerals used then had the same definition as they have now, but within a short time, the use of 73 began to change.

In the National Telegraph Convention, the numeral was changed from the Valentine-type sentiment to a vague sign of fraternalism. Here, 73 was a greeting, a friendly "word" between operators and it was so used on all wires.

In 1859, the Western Union Company set up the standard "92 Code." A list of numerals from one to 92 was compiled to indicate a series of prepared phrases for use by the operators on the wires. Here, in the 92 Code, 73 changes from a fraternal sign to a very flowery "accept my compliments," which was in keeping with the florid language of that era.

Over the years from 1859 to 1900, the many manuals of telegraphy show variations of this meaning. Dodge's *The Telegraph Instructor* shows it merely as "compliments." The *Twentieth Century Manual of Railway and Commercial Telegraphy* defines it two ways, one listing as "my compliments to you"; but in the glossary of abbreviations it is merely "compliments." Theodore A. Edison's *Telegraphy Self-Taught* shows a return to "accept my compliments." By 1908, however, a later edition of the Dodge Manual gives us today's definition of "best regards" with a backward look at the older meaning in another part of the work where it also lists it as "compliments."

"Best regards" has remained ever since as the "put-it-down-in-black-and-white" meaning of 73 but it has acquired overtones of much warmer meaning. Today, amateurs use it more in the manner that James Reid had intended that it be used—a "friendly word between operators."—*Louise Ramsey Moreau, W3WRE*

The RST System

READABILITY
1—Unreadable.
2—Barely readable, occasional words distinguishable.
3—Readable with considerable difficulty.
4—Readable with practically no difficulty.
5—Perfectly readable.
SIGNAL STRENGTH
1—Faint signals barely perceptible.
2—Very weak signals.
3—Weak signals.
4—Fair signals.
5—Fairly good signals.
6—Good signals.
7—Moderately strong signals.
8—Strong signals.
9—Extremely strong signals.
TONE
1—Sixty-cycle ac or less, very rough and broad.
2—Very rough ac, very harsh and broad.
3—Rough ac tone, rectified but not filtered.
4—Rough note, some trace of filtering.
5—Filtered rectified ac but strongly ripple-modulated.
6—Filtered tone, definite trace of ripple modulation.
7—Near pure tone, trace of ripple modulation.
8—Near perfect tone, slight trace of modulation.
9—Perfect tone, no trace of ripple or modulation of any kind.

The "tone" report refers only to the purity of the signal, and has no connection with its stability or freedom from clicks or chirps. If the signal has the characteristic steadiness of crystal control, add X to the report (e.g., RST 469X). If it has a chirp or "tail" (either on "make" or "break") add C (e.g., 469C). If it has clicks or noticeable other keying transients, add K (e.g., 469K). Of course a signal could have both chirps and clicks, in which case both C and K could be used (e.g., RST 469CK).

Q Signals

Given below are a number of Q signals whose meanings most often need to be expressed with brevity and clarity in amateur work. (Q abbreviations take the form of questions only when each is sent followed by a question mark.)

QRG Will you tell me my exact frequency (or that of ____)? Your exact frequency (or that of ____) is ____ kHz.

QRH Does my frequency vary? Your frequency varies.

QRI How is the tone of my transmission? The tone of your transmission is ____ (1. Good; 2. Variable; 3. Bad).

QRJ Are you receiving me badly? I cannot receive you. Your signals are too weak.

QRK What is the intelligibility of my signals (or those of ____)? The intelligibility of your signals (or those of ____) is ____ (1. Bad; 2. Poor; 3. Fair; 4. Good; 5. Excellent).

QRL Are you busy? I am busy (or I am busy with ____). Please do not interfere.

QRM Is my transmission being interfered with? Your transmission is being interfered with ____ (1. Nil; 2. Slightly; 3. Moderately; 4. Severely; 5. Extremely.)

QRN Are you troubled by static? I am troubled by static ____ (1-5 as under QRM).

QRO Shall I increase power? Increase power.

QRP Shall I decrease power? Decrease power.

QRQ Shall I send faster? Send faster (____ WPM).

QRS Shall I send more slowly? Send more slowly (____ WPM).

QRT Shall I stop sending? Stop sending.

QRU Have you anything for me? I have nothing for you.

QRV Are you ready? I am ready.

QRW Shall I inform ____ that you are calling on ____ kHz? Please inform ____ that I am calling on ____ kHz.

QRX When will you call me again? I will call you again at ____ hours (on ____ kHz).

QRY What is my turn? Your turn is numbered ____

QRZ Who is calling me? You are being called by ____ (on ____ kHz).

QSA What is the strength of my signals (or those of ____)? The strength of your signals (or those of ____) is ____ (1. Scarcely perceptible; 2. Weak; 3. Fairly good; 4. Good; 5. Very good).

QSB Are my signals fading? Your signals are fading.

QSD Is my keying defective? Your keying is defective.

QSG Shall I send ____ messages at a time? Send ____ messages at a time.

QSK Can you hear me between your signals and if so can I break in on your transmission? I can hear you between my signals; break in on my transmission.

QSL Can you acknowledge receipt? I am acknowledging receipt

QSM Shall I repeat the last message which I sent you, or some previous message? Repeat the last message which you sent me [or message(s) number(s) ____].

QSN Did you hear me (or ____) on ____ kHz? I did hear you (or ____) on ____ kHz.

QSO Can you communicate with ____ direct or by relay? I can communicate with ____ direct (or by relay through ____).

QSP Will you relay to ____? I will relay to ____

QST General call preceding a message addressed to all amateurs and ARRL members. This is in effect "CQ ARRL."

QSU Shall I send or reply on this frequency (or on ____ kHz)? Send a series of Vs on this frequency (or ____ kHz).

QSW Will you send on this frequency (or on ____ kHz)? I am going to send on this frequency (or on ____ kHz).

QSX Will you listen to ____ on ____ kHz? I am listening to ____ on ____ kHz.

QSY Shall I change to transmission on another frequency? Change to transmission on another frequency (or on ____ kHz).

QSZ Shall I send each word or group more than once? Send each word or group twice (or ____ times).

QTA Shall I cancel message number ____? Cancel message number ____

QTB Do you agree with my counting of words? I do not agree with your counting of words. I will repeat the first letter or digit of each word or group.

QTC How many messages have you to send? I have ____ messages for you (or for ____).

QTH What is your location? My location is ____

QTR What is the correct time? The time is ____

The following Ham Outline Maps are reprinted with the permission of FM-RTTY Publishing. For ordering information regarding these or other Outline Maps, send an SASE to Jim Labo, KØOST, FM-RTTY Publishing, PO Box 842, Denver, CO 80201-0842.

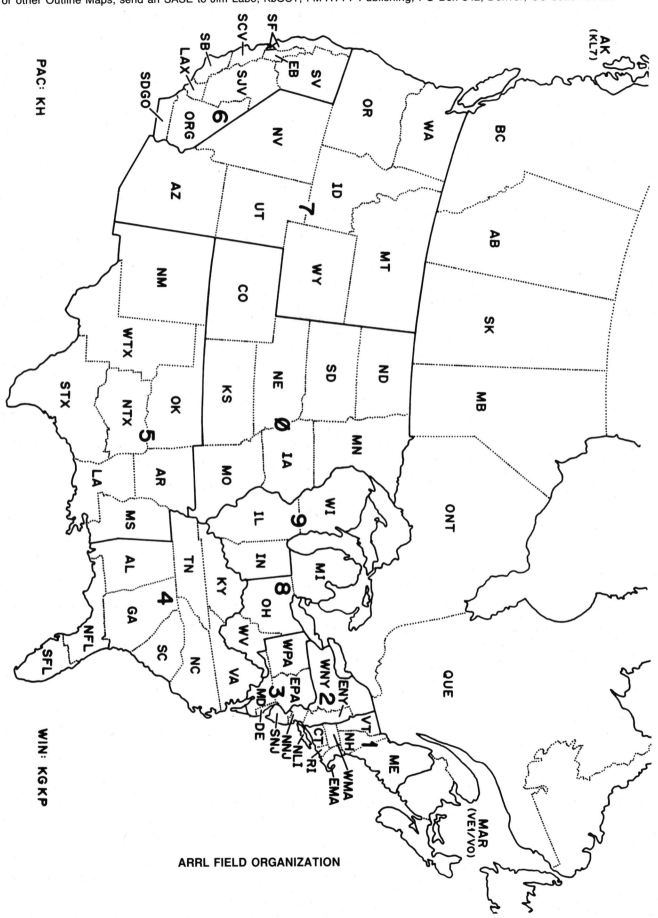

ARRL FIELD ORGANIZATION

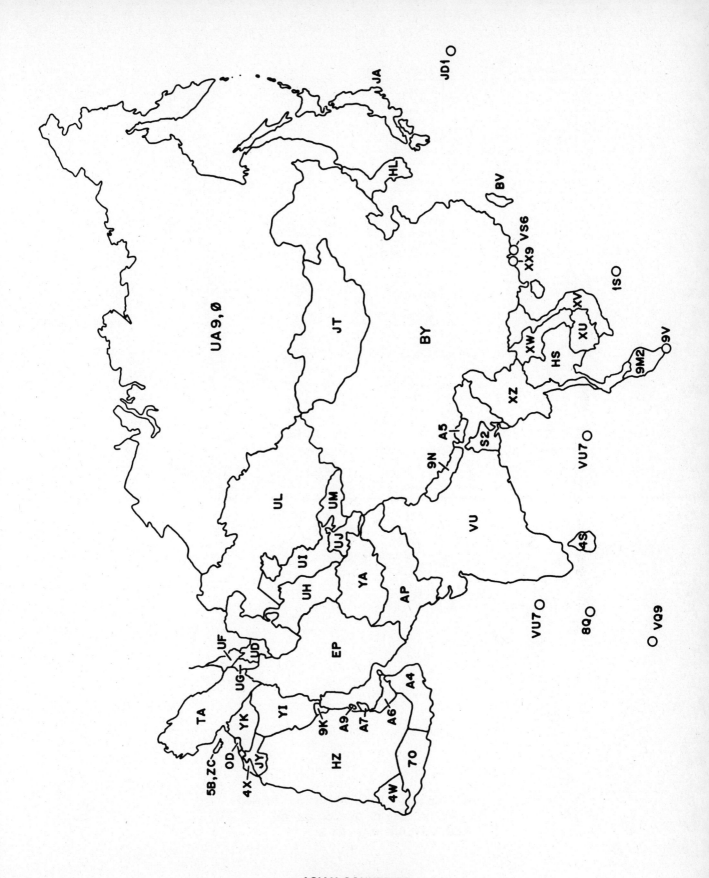

**ASIAN COUNTRIES
HAM OUTLINE MAP No 1**

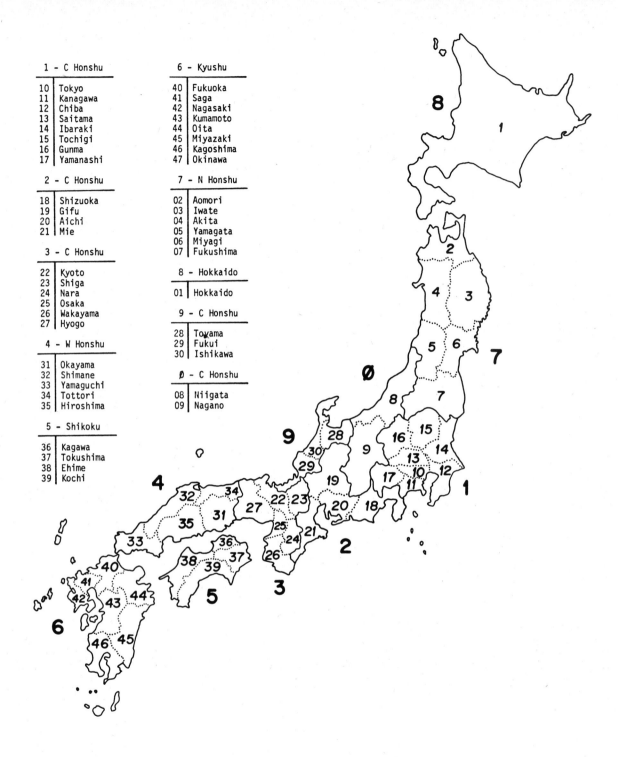

1 - C Honshu

10	Tokyo
11	Kanagawa
12	Chiba
13	Saitama
14	Ibaraki
15	Tochigi
16	Gunma
17	Yamanashi

2 - C Honshu

18	Shizuoka
19	Gifu
20	Aichi
21	Mie

3 - C Honshu

22	Kyoto
23	Shiga
24	Nara
25	Osaka
26	Wakayama
27	Hyogo

4 - W Honshu

31	Okayama
32	Shimane
33	Yamaguchi
34	Tottori
35	Hiroshima

5 - Shikoku

36	Kagawa
37	Tokushima
38	Ehime
39	Kochi

6 - Kyushu

40	Fukuoka
41	Saga
42	Nagasaki
43	Kumamoto
44	Oita
45	Miyazaki
46	Kagoshima
47	Okinawa

7 - N Honshu

02	Aomori
03	Iwate
04	Akita
05	Yamagata
06	Miyagi
07	Fukushima

8 - Hokkaido

01	Hokkaido

9 - C Honshu

28	Toyama
29	Fukui
30	Ishikawa

0 - C Honshu

08	Niigata
09	Nagano

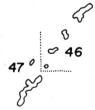

JAPANESE PREFECTURES (KENS)
 PREFIX: JA-JS 7J-7N 8J-8N
HAM OUTLINE MAP No 2

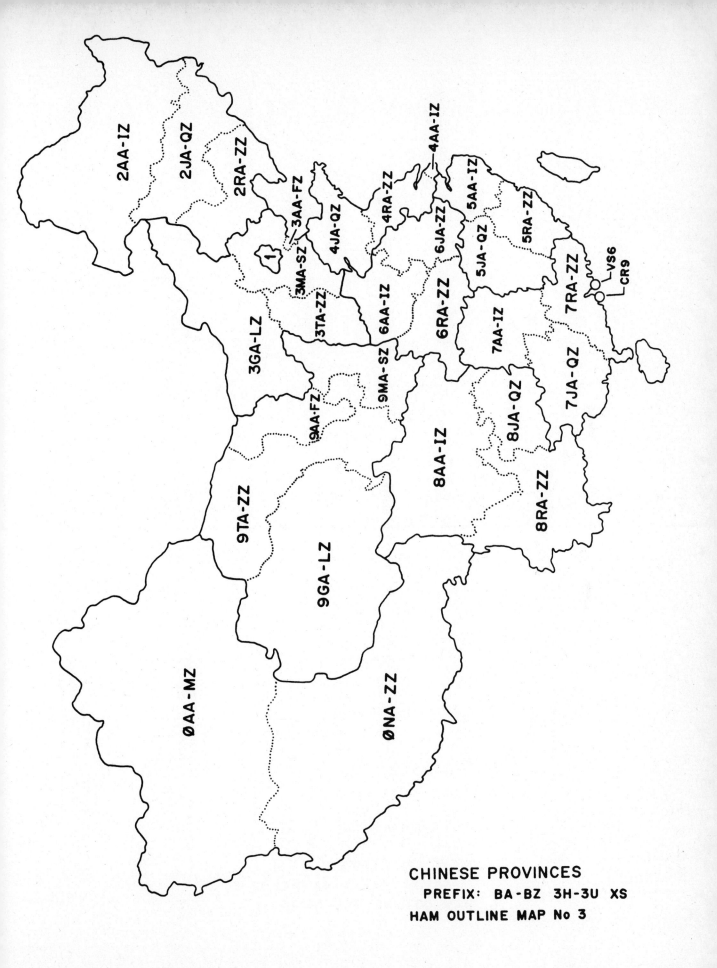

CHINESE PROVINCES
  PREFIX: BA-BZ 3H-3U XS
HAM OUTLINE MAP No 3

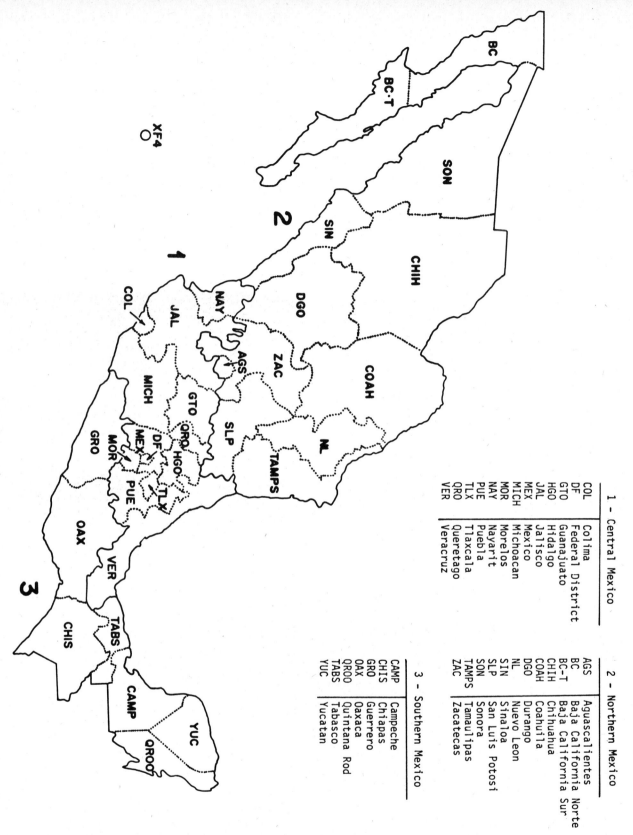

MEXICAN STATES
PREFIX: XA-XI 4A-4C 6D-6J
HAM OUTLINE MAP No 4

1 – Central Mexico

COL	Colima
DF	Federal District
GTO	Guanajuato
HGO	Hidalgo
JAL	Jalisco
MEX	Mexico
MICH	Michoacan
MOR	Morelos
NAY	Nayarit
PUE	Puebla
TLX	Tlaxcala
QRO	Queretago
VER	Veracruz

2 – Northern Mexico

AGS	Aguascalientes
BC	Baja California Norte
BC-T	Baja California Sur
CHIH	Chihuahua
COAH	Coahuila
DGO	Durango
NL	Nuevo Leon
SIN	Sinaloa
SLP	San Luis Potosi
SON	Sonora
TAMPS	Tamaulipas
ZAC	Zacatecas

3 – Southern Mexico

CAMP	Campeche
CHIS	Chiapas
GRO	Guerrero
OAX	Oaxaca
QROO	Quintana Rod
TABS	Tabasco
YUC	Yucatan

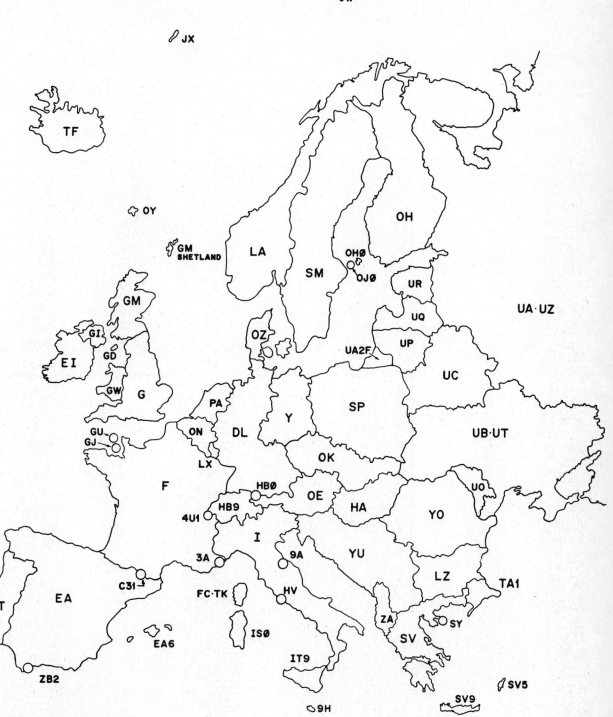

EUROPEAN COUNTRIES
HAM OUTLINE MAP No 6

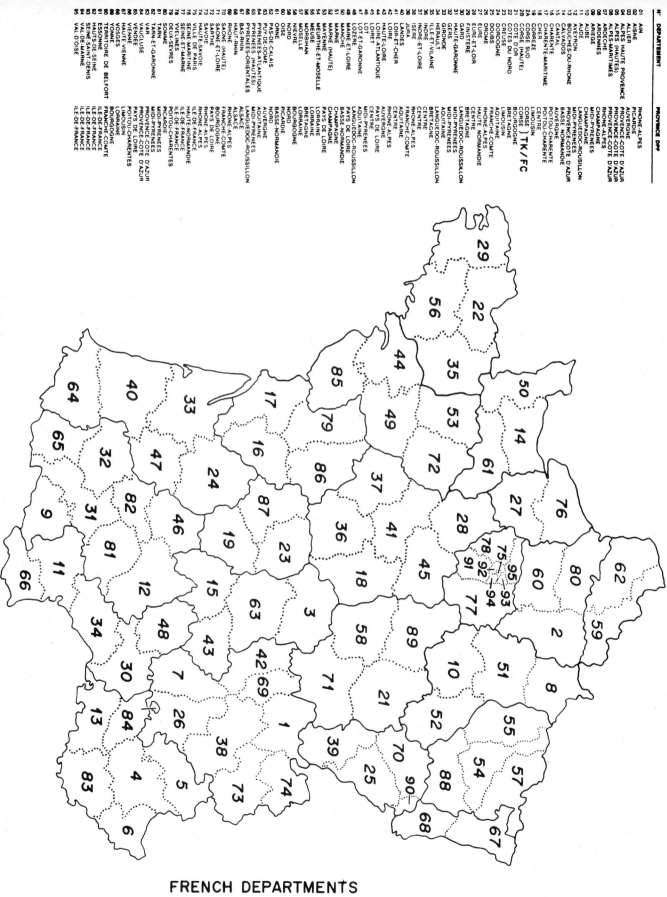

N°	DÉPARTEMENT	PROVINCE	DPF
01	AIN	RHONE-ALPES	
02	AISNE	PICARDIE	
03	ALLIER	AUVERGNE	
04	ALPES HAUTE PROVENCE	PROVENCE-COTE D'AZUR	
05	ALPES (HAUTES)	PROVENCE-COTE D'AZUR	
06	ALPES-MARITIMES	PROVENCE-COTE D'AZUR	
07	ARDECHE	RHONE-ALPES	
08	ARDENNES	CHAMPAGNE	
09	ARIEGE	MIDI-PYRENEES	
10	AUBE	CHAMPAGNE	
11	AUDE	LANGUEDOC-ROUSSILLON	
12	AVEYRON	MIDI-PYRENEES	
13	BOUCHES-DU-RHONE	PROVENCE-COTE D'AZUR	
14	CALVADOS	BASSE NORMANDIE	
15	CANTAL	AUVERGNE	
16	CHARENTE	POITOU-CHARENTE	
17	CHARENTE-MARITIME	POITOU-CHARENTE	
18	CHER	CENTRE	
19	CORREZE	LIMOUSIN	
20	CORSE (HAUTE)	CORSE	} TK/FC
21	CORSE SUD	CORSE	
22	COTE DU NORD	BRETAGNE	
23	CREUSE	LIMOUSIN	
24	DORDOGNE	AQUITAINE	
25	DOUBS	FRANCHE-COMTE	
26	DROME	RHONE-ALPES	
27	EURE	HAUTE NORMANDIE	
28	EURE-ET-LOIR	CENTRE	
29	FINISTERE	BRETAGNE	
30	GARD	LANGUEDOC-ROUSSILLON	
31	HAUTE-GARONNE	MIDI-PYRENEES	
32	GERS	MIDI-PYRENEES	
33	GIRONDE	AQUITAINE	
34	HERAULT	LANGUEDOC-ROUSSILLON	
35	ILLE-ET-VILAINE	BRETAGNE	
36	INDRE	CENTRE	
37	INDRE-ET-LOIRE	CENTRE	
38	ISERE	RHONE-ALPES	
39	JURA	FRANCHE-COMTE	
40	LANDES	AQUITAINE	
41	LOIR-ET-CHER	CENTRE	
42	LOIRE	RHONE-ALPES	
43	HAUTE-LOIRE	AUVERGNE	
44	LOIRE-ATLANTIQUE	PAYS DE LOIRE	
45	LOIRET	CENTRE	
46	LOT	MIDI-PYRENEES	
47	LOT-ET-GARONNE	AQUITAINE	
48	LOZERE	LANGUEDOC-ROUSSILLON	
49	MAINE-ET-LOIRE	PAYS DE LOIRE	
50	MANCHE	BASSE-NORMANDIE	
51	MARNE	CHAMPAGNE	
52	HAUTE-MARNE	CHAMPAGNE	
53	MAYENNE	PAYS DE LOIRE	
54	MEURTHE-ET-MOSELLE	LORRAINE	
55	MEUSE	LORRAINE	
56	MORBIHAN	BRETAGNE	
57	MOSELLE	LORRAINE	
58	NIEVRE	BOURGOGNE	
59	NORD	NORD	
60	OISE	PICARDIE	
61	ORNE	BASSE-NORMANDIE	
62	PAS-DE-CALAIS	NORD	
63	PUY-DE-DOME	AUVERGNE	
64	PYRENEES-ATLANTIQUE	AQUITAINE	
65	PYRENEES (HAUTES)	MIDI-PYRENEES	
66	PYRENEES-ORIENTALES	LANGUEDOC-ROUSSILLON	
67	BAS-RHIN	ALSACE	
68	HAUT-RHIN	ALSACE	
69	RHONE	RHONE-ALPES	
70	SAONE (HAUTE)	FRANCHE-COMTE	
71	SAONE-ET-LOIRE	BOURGOGNE	
72	SARTHE	PAYS DE LOIRE	
73	SAVOIE	RHONE-ALPES	
74	HAUTE-SAVOIE	RHONE-ALPES	
75	VILLE DE PARIS	ILE-DE-FRANCE	
76	SEINE-MARITIME	HAUTE-NORMANDIE	
77	SEINE-ET-MARNE	ILE-DE-FRANCE	
78	YVELINES	ILE-DE-FRANCE	
79	DEUX-SEVRES	POITOU-CHARENTES	
80	SOMME	PICARDIE	
81	TARN	MIDI-PYRENEES	
82	TARN-ET-GARONNE	MIDI-PYRENEES	
83	VAR	PROVENCE-COTE D'AZUR	
84	VAUCLUSE	PROVENCE-COTE D'AZUR	
85	VENDEE	PAYS DE LOIRE	
86	VIENNE	POITOU-CHARENTES	
87	HAUTE-VIENNE	LIMOUSIN	
88	VOSGES	LORRAINE	
89	YONNE	BOURGOGNE	
90	TERRITOIRE DE BELFORT	FRANCHE-COMTE	
91	ESSONNE	ILE-DE-FRANCE	
92	HAUTS-DE-SEINE	ILE-DE-FRANCE	
93	SEINE-SAINT-DENIS	ILE-DE-FRANCE	
94	VAL-DE-MARNE	ILE-DE-FRANCE	
95	VAL-D'OISE	ILE-DE-FRANCE	

FRENCH DEPARTMENTS
PREFIX: F HW-HY TH TK TM TO-TQ TV-TX
HAM OUTLINE MAP No 7

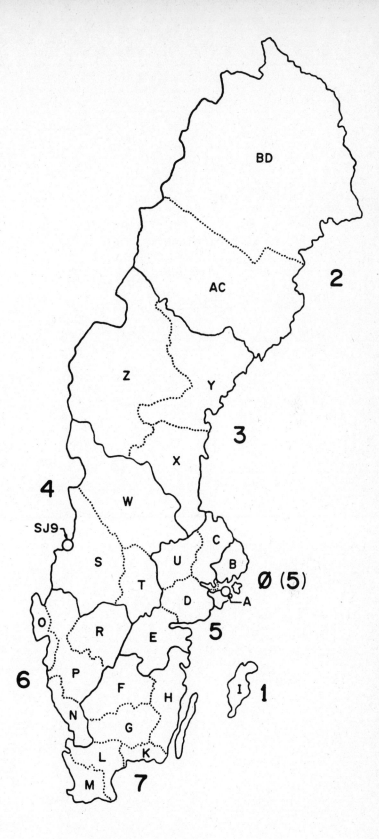

1 - Gotland

I	Gotland

2 - Northern

AC	Vasterbotten
BD	Norrbotten

3 - Northern

X	Gavleborg
Y	Vasternorrland
Z	Jamtland

4 - West Central

S	Varmland
T	Orebro
W	Kopparberg

5 - East Central

A	Stockholm City
B	Stockholm County
C	Uppsala
D	Sodermanland
E	Ostergotland
U	Vastmanland

6 - Southwestern

N	Halland
O	Goteborg och Bohus
P	Alvsborg
R	Skaraborg

7 - Southern

F	Jonkoping
G	Kronoberg
H	Kalmar
K	Blekinge
L	Kristianstad
M	Malmohus

Ø - Stockholm Area

A	Stockholm City
B	Stockholm County

SM8	Maritime Mobile
SJ9	Morokulien

SK	Club
SL	Military
SM	Individual

SWEDISH COUNTIES (LAENS)
PREFIX: SA-SM 7S 8S
HAM OUTLINE MAP No 8

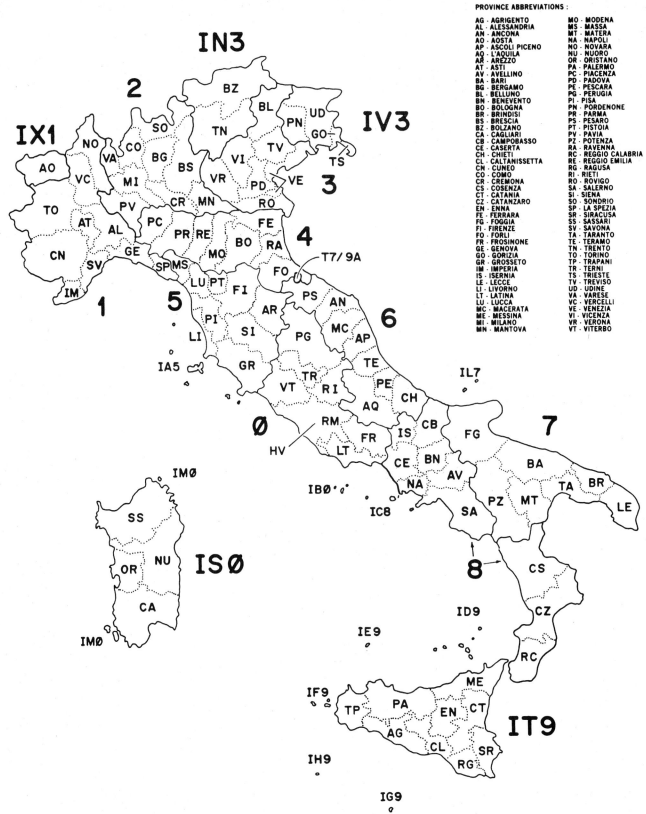

PROVINCE ABBREVIATIONS :

AG - AGRIGENTO
AL - ALESSANDRIA
AN - ANCONA
AO - AOSTA
AP - ASCOLI PICENO
AQ - L'AQUILA
AR - AREZZO
AT - ASTI
AV - AVELLINO
BA - BARI
BG - BERGAMO
BL - BELLUNO
BN - BENEVENTO
BO - BOLOGNA
BR - BRINDISI
BS - BRESCIA
BZ - BOLZANO
CA - CAGLIARI
CB - CAMPOBASSO
CE - CASERTA
CH - CHIETI
CL - CALTANISSETTA
CN - CUNEO
CO - COMO
CR - CREMONA
CS - COSENZA
CT - CATANIA
CZ - CATANZARO
EN - ENNA
FE - FERRARA
FG - FOGGIA
FI - FIRENZE
FO - FORLI
FR - FROSINONE
GE - GENOVA
GO - GORIZIA
GR - GROSSETO
IM - IMPERIA
IS - ISERNIA
LE - LECCE
LI - LIVORNO
LT - LATINA
LU - LUCCA
MC - MACERATA
ME - MESSINA
MI - MILANO
MN - MANTOVA

MO - MODENA
MS - MASSA
MT - MATERA
NA - NAPOLI
NO - NOVARA
NU - NUORO
OR - ORISTANO
PA - PALERMO
PC - PIACENZA
PD - PADOVA
PE - PESCARA
PG - PERUGIA
PI - PISA
PN - PORDENONE
PR - PARMA
PS - PESARO
PT - PISTOIA
PV - PAVIA
PZ - POTENZA
RA - RAVENNA
RC - REGGIO CALABRIA
RE - REGGIO EMILIA
RG - RAGUSA
RI - RIETI
RO - ROVIGO
SA - SALERNO
SI - SIENA
SO - SONDRIO
SP - LA SPEZIA
SR - SIRACUSA
SS - SASSARI
SV - SAVONA
TA - TARANTO
TE - TERAMO
TN - TRENTO
TO - TORINO
TP - TRAPANI
TR - TERNI
TS - TRIESTE
TV - TREVISO
UD - UDINE
VA - VARESE
VC - VERCELLI
VE - VENEZIA
VI - VICENZA
VR - VERONA
VT - VITERBO

ITALIAN PROVINCES
   PREFIX: I IA-IZ
HAM OUTLINE MAP No 9

WESTERN USSR OBLASTS
MAY 1984 SYSTEM · FORMER UA1-6 CALL AREAS
HAM OUTLINE MAP No 12 REVISED

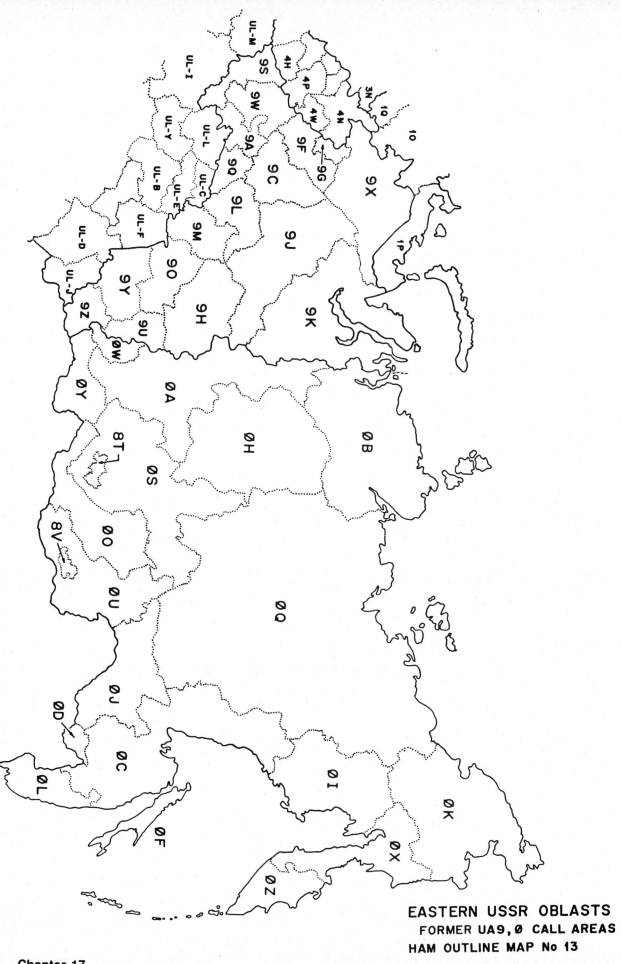

EASTERN USSR OBLASTS
FORMER UA9,Ø CALL AREAS
HAM OUTLINE MAP No 13

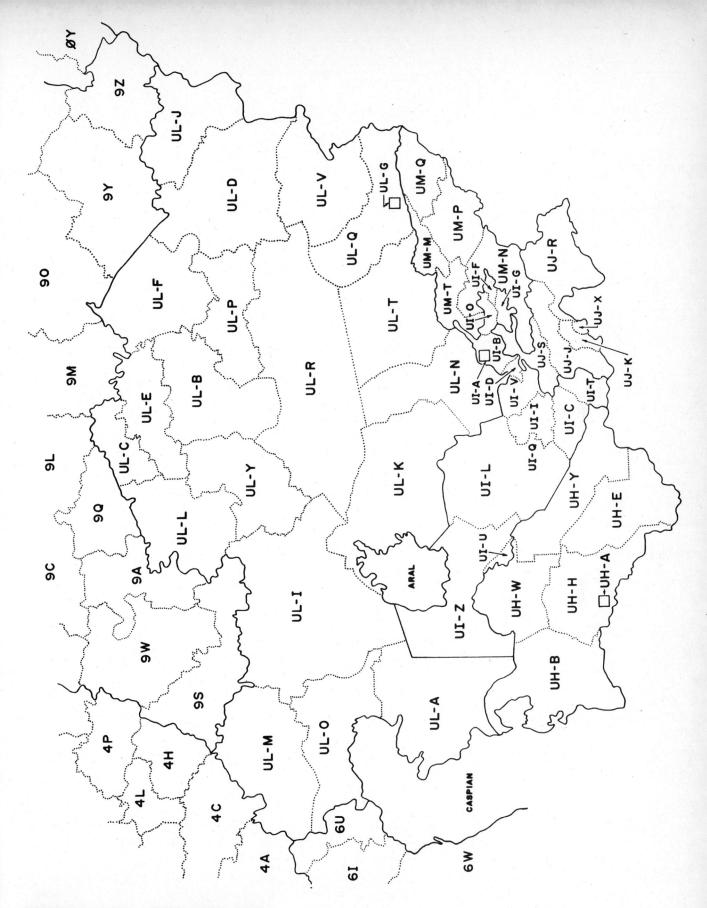

CENTRAL ASIA USSR OBLASTS
MAY 1984 SYSTEM · FORMER 7,8 CALL AREAS
HAM OUTLINE MAP No 14

SOUTH AMERICAN COUNTRIES
HAM OUTLINE MAP No 16

PP8

PV8

PU8

PT8

PY8

PW8

PR8

PS8

PT7

PS7

PY9

PY7

PR7

PP7

PP2

PY6

PT2

PP6

PT9

PY4

PY2

PY5

PP1

PP5

PY1

PY3

PP1—Espirito Santo ES
PP2—Goias GO
PP5—Santa Catarina SC
PP6—Sergipe SE
PP7—Alagoas AL
PP8—Amazonas AM
PR7—Paraiba PB
PR8—Maranhao MA
PS7—Rio Grande do Norte
PS8—Piaui PI
PT2—Distrite Federal DF (Brasilia)
PT7—Ceara CE
PT8—Acre AC
PT9—Mato Grosso do Sul MS
PU8—Amapa AP
PV8—Roraima RR
PW8—Rondonia RO
PY1—Rio de Janeiro RJ
PY2—Sao Paulo SP
PY3—Rio Gran de do Sul RS
PY4—Minas Gerais MG
PY5—Parana PR
PY6—Bahia BA
PY7—Pernambuco PE
PY8—Para PA
PY9—Mato Grosso MT
PY0T—Trinidade Island
PY0F—Fernando de Noronha Is.
PY0S—Sao Pedro/S. Paulo Rocks

BRAZILIAN STATES
PREFIX: PP-PY ZV-ZZ
HAM OUTLINE MAP No 17

References   17-33

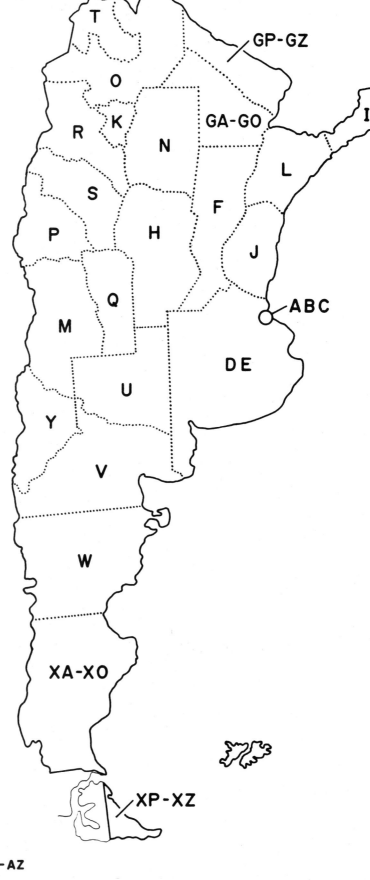

Province is indicated by letter(s)
after number in call sign

A B C   Buenos Aires City
  D E   Buenos Aires Province
    F   Santa Fe
GA-GO   Chaco
GP-GZ   Formosa
    H   Cordoba
    I   Misiones
    J   Entre Rios
    K   Tucuman
    L   Corrientes
    M   Mendoza
    N   Santiago Del Estero
    O   Salta
    P   San Juan
    Q   San Luis
    R   Catamarga
    S   La Rioja
    T   Jujuy
    U   La Pampas
    V   Rio Negro
    W   Chubut
XA-XO   Santa Cruz
XP-XZ   Tierra Del Fuego
    Y   Neuquen
    Z   Antarctica

ARGENTINE PROVINCES
  PREFIX: LO-LW L2-L9 AY-AZ
HAM OUTLINE MAP No 18

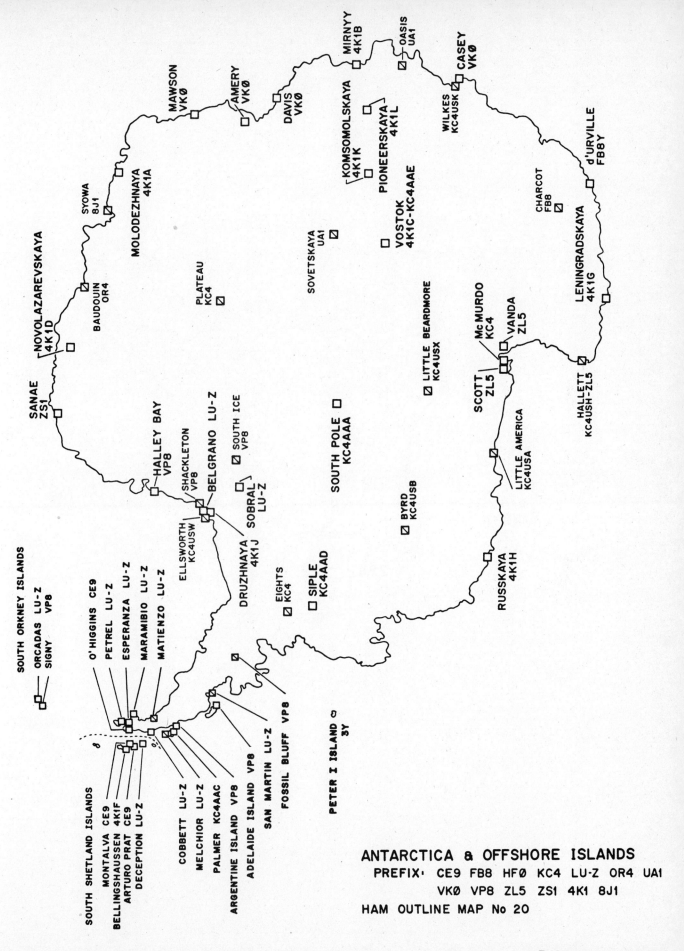

MAWSON VKØ
AMERY VKØ
DAVIS VKØ
MIRNYY 4K1B
OASIS UA1
CASEY VKØ
WILKES KC4USK
d'URVILLE FB8Y
KOMSOMOLSKAYA 4K1K
PIONEERSKAYA 4K1L
CHARCOT FB8
SYOWA 8J1
MOLODEZHNAYA 4K1A
VOSTOK 4K1C-KC4AAE
LENINGRADSKAYA 4K1G
SOVETSKAYA UA1
NOVOLAZAREVSKAYA 4K1D
BAUDOUIN OR4
PLATEAU KC4
LITTLE BEARDMORE
McMURDO KC4
VANDA ZL5
SANAE ZS1
HALLEY BAY VP8
SHACKLETON VP8
BELGRANO LU-Z
SOUTH ICE VP8
SOUTH POLE KC4AAA
SCOTT ZL5
LITTLE AMERICA KC4USA
HALLETT KC4USH-ZL5
SOUTH ORKNEY ISLANDS
ORCADAS LU-Z
SIGNY VP8
ELLSWORTH KC4USW
DRUZHNAYA 4K1J
SOBRAL LU-Z
EIGHTS KC4
SIPLE KC4AAD
BYRD KC4USB
RUSSKAYA 4K1H
O'HIGGINS CE9
PETREL LU-Z
ESPERANZA LU-Z
MARAMBIO LU-Z
MATIENZO LU-Z
SAN MARTIN LU-Z
FOSSIL BLUFF VP8
PETER I ISLAND 3Y
SOUTH SHETLAND ISLANDS
MONTALVA CE9
BELLINGSHAUSSEN 4K1F
ARTURO PRAT CE9
DECEPTION LU-Z
COBBETT LU-Z
MELCHIOR LU-Z
PALMER KC4AAC
ARGENTINE ISLAND VP8
ADELAIDE ISLAND VP8

ANTARCTICA & OFFSHORE ISLANDS
PREFIX: CE9 FB8 HFØ KC4 LU-Z OR4 UA1
VKØ VP8 ZL5 ZS1 4K1 8J1
HAM OUTLINE MAP No 20

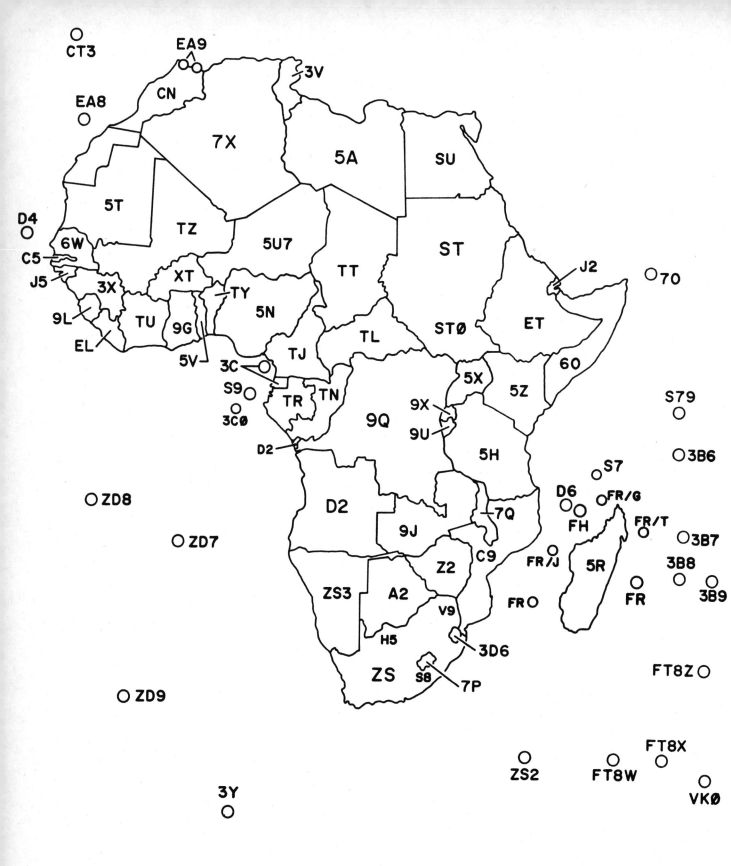

CT3
EA9
3V
EA8
CN
7X
5A
SU
5T
TZ
ST
D4
6W
5U7
J2
70
C5
J5
XT
TT
3X
TY
ET
9L
TU
9G
5N
60
EL
TL
STØ
5V
3C
TJ
5X
5Z
S9
TR
TN
S79
3CØ
9X
9Q
9U
3B6
D2
5H
S7
ZD8
D6
FR/G
FH
FR/T
ZD7
7Q
3B7
9J
3B8
Z2
C9
FR/J
5R
3B9
ZS3
A2
FR
FR
V9
3D6
H5
ZD9
ZS
S8
7P
FT8Z

ZS2
FT8W
FT8X

3Y
VKØ

AFRICAN COUNTRIES
HAM OUTLINE MAP No 22

Some Abbreviations for CW Work

Although abbreviations help to cut down unnecessary transmission, make it a rule not to abbreviate unnecessarily when working an operator of unknown experience.

AA	All after	GND	Ground	SASE	Self-addressed, stamped envelope
AB	All before	GUD	Good	SED	Said
ABT	About	HI	The telegraphic laugh; high	SIG	Signature; signal
ADR	Address	HR	Here, hear	SINE	Operator's personal initials or
AGN	Again	HV	Have		nickname
ANT	Antenna	HW	How	SKED	Schedule
BCI	Broadcast interference	LID	A poor operator	SRI	Sorry
BCL	Broadcast listener	MA, MILS	Milliamperes	SSB	Single sideband
BK	Break; break me; break in	MSG	Message; prefix to radiogram	SVC	Service; prefix to service message
BN	All between; been	N	No	T	Zero
BUG	Semi-automatic key	NCS	Net control station	TFC	Traffic
B4	Before	ND	Nothing doing	TMW	Tomorrow
C	Yes	NIL	Nothing; I have nothing for you	TNX-TKS	Thanks
CFM	Confirm; I confirm	NM	No more	TT	That
CK	Check	NR	Number	TU	Thank you
CL	I am closing my station; call	NW	Now; I resume transmission	TVI	Television interference
CLD-CLG	Called; calling	OB	Old boy	TX	Transmitter
CQ	Calling any station	OC	Old chap	TXT	Text
CUD	Could	OM	Old man	UR-URS	Your; you're; yours
CUL	See you later	OP-OPR	Operator	VFO	Variable-frequency oscillator
CW	Continuous wave	OT	Old timer; old top	VY	Very
	(i.e., radiotelegraph)	PBL	Preamble	WA	Word after
DLD-DLVD	Delivered	PSE	Please	WB	Word before
DR	Dear	PWR	Power	WD-WDS	Word; words
DX	Distance, foreign countries	PX	Press	WKD-WKG	Worked; working
ES	And, &	R	Received as transmitted; are	WL	Well; will
FB	Fine business, excellent	RCD	Received	WUD	Would
FM	Frequency modulation	RCVR (RX)	Receiver	WX	Weather
GA	Go ahead (or resume sending)	REF	Refer to; referring to; reference	XCVR	Transceiver
GB	Good-by	RFI	Radio frequency interference	XMTR (TX)	Transmitter
GBA	Give better address	RIG	Station equipment	XTAL	Crystal
GE	Good evening	RPT	Repeat; I repeat	XYL (YF)	Wife
GG	Going	RTTY	Radioteletype	YL	Young lady
GM	Good morning	RX	Receiver	73	Best regards
GN	Good night			88	Love and kisses

RETURN POSTAGE CHART FOR DXCC AND WAS QSL CARDS

(Please include sufficient postage, as follows, for the safe return of your cards)

Award	UPS	Registered	Certified	First Class
DXCC (new application)*	$ 1.75	$ 6.00 (10.50[1])	$ 3.25	$ 2.50 (7.00[1])
DXCC (endorsement)**	1.75	4.50 (6.50[1])	1.75	1.00 (3.00[1])
5BDXCC	3.00	8.50 (13.25[2])	5.75	5.00 (9.75[2])

*Initial 100 cards
**25-card endorsement
[1]Foreign small-packet Air Mail
[2]Foreign surface Mail

		UPS	Registered	Certified	First Class
WAS	(United States)	$ 1.75	$ 4.62	$ 1.82	$ 1.07
5BWAS		2.00	9.47	6.77	5.92

Foreign Small-Packet Air Mail

	Mexico, Central America, the Caribbean Islands, Colombia, Venezuela	South America (except Colombia & Venezuela), Europe & North Africa	U.S.S.R., Asia, Pacific Ocean Islands, Africa (except North Africa), The Indian Ocean & the Middle East
WAS	$ 1.82	$ 2.50	$ 3.18
5BWAS	5.54	8.98	12.42

All of the above are U.S. funds
For Registered Small-Packet Air Mail, add $3.55 to the above fees.

UPS or Registered Mail is recommended

These rates are subject to change at any time.

ARRL WAS AWARD APPLICATION

Please PRINT.

Call: _____

Name _____
(Print exactly as you want it to appear on certificate)

List any ex-calls used on
any cards submitted:

Address _____
(number and street)

_____

_____
(City, State/Country, ZIP/PC)

```
┌──────────────────────────────┐
│   FOR RETURN POSTAGE         │
│   Cash/Check/Stamps/IRCs     │
│   enclosed.                  │
│                              │
│       $ _____      │
│                              │
│   Return my cards via:       │
│   ☐ Registered               │
│   ☐ Certified • U.S. only    │
│   ☐ United Parcel Service    │
│   ☐ First-class mail         │
└──────────────────────────────┘
```

W/VEs: Membership control number
(from your **QST** wrapper)   : _____

Is this your first WAS application? ☐ YES   ☐ NO

If NO, my original WAS number held is _____

I was first licensed in (year): _____

I am applying for ONE of the following WAS awards (each is separately numbered):
Please Check: (Use MCS-225 for 5BWAS)

☐ Basic award (mixed bands and/or modes)

☐ SSTV   ☐ 144 MHz   ☐ 50 MHz
☐ OSCAR Satellite   ☐ 160 Meters   ☐ 220 MHz
☐ RTTY   ☐ 75 Meters 2-Letter Extra Class ☐ 432 MHz

I am applying for the following ENDORSEMENTS (Check **ALL** that apply):

☐ SSB   ☐ Novice

☐ CW   ☐ QRP   ☐ Single Band: _____
                                              (Meters)

The undersigned affirms that he/she has read, understood, and abided by the rules of the WAS award (MCS-264).

Signature _____ Date _____ 19 _____

- -

HF AWARDS MANAGER VERIFICATION

I hereby verify that I have personally inspected the confirmations with all 50 states and verify that this application is
correct and true. In my own handwriting, I confirm this application is for the following SPECIALITY awards or
ENDORSEMENTS: _____ (write NONE if none)

_____   _____ H- _____   _____ 19 ____
       (signature)   (call)   (Mgr. 4-digit code #)   (date)

- -

DIRECTIONS TO APPLICANT

1. Read the WAS Rules (MCS-264) carefully.

2. Fill out BOTH sides of this application.

3. Sort cards alphabetically by state.

4. Present application and cards to your ARRL Special Service Club HF Awards Manager for verification. **Applications from DX stations may be certified by the Awards Manager of your IARU member-society.**

5. Send application to ARRL HQ. Send cards also ONLY if there is no local HF Awards Manager for verification.

6. If mailing cards, enclose postage sufficient for return of cards (see attached CD-16). MAIL TO:

   ARRL WAS Award
   225 Main Street
   Newington, CT 06111 USA

STATE	CALL	DATE	BAND	MODE	
Alabama					
Alaska					
Arizona					
Arkansas					
California					
Colorado					
Connecticut					
Delaware					
Florida					
Georgia					
Hawaii					
Idaho					
Illinois					
Indiana					
Iowa					
Kansas					
Kentucky					
Louisiana					
Maine					
Maryland (D.C.)					
Massachusetts					
Michigan					
Minnesota					
Mississippi					
Missouri					
Montana					
Nebraska					
Nevada					
New Hampshire					
New Jersey					
New Mexico					
New York					
North Carolina					
North Dakota					
Ohio					
Oklahoma					
Oregon					
Pennsylvania					
Rhode Island					
South Carolina					
South Dakota					
Tennessee					
Texas					
Utah					
Vermont					
Virginia					
Washington					
West Virginia					
Wisconsin					
Wyoming					

ARRL 5BWAS AWARD APPLICATION

Please PRINT.

Call: _____

List any ex-calls used on
any cards submitted:

_____

Name _____
(Print exactly as you want it to appear on certificate)

Address _____
(number and street)

_____
(City, State/Country, ZIP/PC)

FOR RETURN POSTAGE
Cash/Check/Stamps/IRCs
enclosed.

$ _____

Return my cards via:
☐ Registered
☐ Certified • U.S. only
☐ United Parcel Service
☐ First-class mail

W/VEs: Membership control number
(from your **QST** wrapper)   : _____

Is this your first WAS application? ☐ YES   ☐ NO

If NO, my original WAS number held is _____

I was first licensed in (year): _____

☐ $20 is enclosed for a plaque

The undersigned affirms that he/she has read, understood, and abided by the
rules of the 5BWAS award printed below and the WAS Rules (MCS-264).

Signature _____ Date _____ 19 _____

- -

HF AWARDS MANAGER VERIFICATION

I hereby verify that I have personally inspected the confirmations with all 50 states on five different bands and verify that
this application is correct and true.

_____   _____ H- _____ _____ 19 _____
(signature)   (call)   (Mgr. 4-digit code #)   (date)

- -

5BWAS RULES

1. The 5BWAS certificate and plaque (see below) will be issued for having submitted confirmations with each of the 50 United States for contacts **dated January 1, 1970,** or after, on five amateur bands (10 and 24 MHz excluded). Phone and CW segments of a band do not count as separate bands.
2. WAS Rules (MCS-264), that do not conflict with these 5BWAS rules, also apply to the 5BWAS award.
3. There are no specialty 5Band awards or endorsements.
4. The 5BWAS certificate is offered free of charge. A handsome 9 × 12″ personalized walnut plaque is available for a fee of $20.00 U.S. (check or money order).

- -

DIRECTIONS TO APPLICANT

1. Read 5BWAS rules above and WAS rules (MCS-264) carefully.
2. Fill out BOTH sides of this application.
3. Sort cards alphabetically by state and band.
4. Present application and cards to your ARRL Special Service Club HF Awards Manager for verification. Applications from DX stations may be certified by the Awards Manager of your IARU member-society.

5. Send application to ARRL HQ. Send cards also ONLY if there is no local HF Awards Manager.
6. If mailing cards, enclose postage sufficient for return of cards (see attached CD-16). MAIL TO:

   ARRL 5BWAS Award
   225 Main Street
   Newington, CT 06111 USA

7. Enclose $20 (check) if you want the plaque.

MCS-225 (487)   —OVER—   Printed in U.S.A.

5B-WAS RECORD SHEET

DIRECTIONS: Enter callsigns in the appropriate boxes by state and band.

QSLs must be for contacts **on or after January 1, 1970.**

STATE	80	40	20	15	10
Alabama					
Alaska					
Arizona					
Arkansas					
California					
Colorado					
Connecticut					
Delaware					
Florida					
Georgia					
Hawaii					
Idaho					
Illinois					
Indiana					
Iowa					
Kansas					
Kentucky					
Louisiana					
Maine					
Maryland (D.C.)					
Massachusetts					
Michigan					
Minnesota					
Mississippi					
Missouri					
Montana					
Nebraska					
Nevada					
New Hampshire					
New Jersey					
New Mexico					
New York					
North Carolina					
North Dakota					
Ohio					
Oklahoma					
Oregon					
Pennsylvania					
Rhode Island					
South Carolina					
South Dakota					
Tennessee					
Texas					
Utah					
Vermont					
Virginia					
Washington					
West Virginia					
Wisconsin					
Wyoming					

MCS-225 (487)

Printed in U.S.A.

ARRL VUCC AWARD APPLICATION

1. Please PRINT.   2. Use separate application for EACH band.

Call: _____   Name _____
(Print exactly as you want it to appear on certificate)

List any ex-calls used on   Address _____
any cards submitted: (number and street)

_____   _____
                                            (City, State/Country, ZIP/PC)

FOR RETURN POSTAGE   Membership control number
Money/Stamps enclosed   (from your **QST** wrapper) : _____

$ _____   ☐ Initial application (check one) ☐ Endorsement

PLUS $1 for certificate or   BAND (check **one** only): ☐ 50 ☐ 144 ☐ 220 ☐ 432 ☐ 902 ☐ 1296
s.a.s.e. (2 units first class
postage) for endorsement.   ☐ 2.3 ☐ 3.4 ☐ 5.7 ☐ 10
Return my cards via:
                                   INITIAL applicants: _____
☐ Registered   Nr. of grid squares claimed.
☐ Certified • U.S. only
☐ United Parcel Service   ENDORSEMENTS:
☐ First-class mail
                                   _____ + _____ = _____
                                      Nr. previously   Nr. additional in   NEW TOTAL
                                       credited   this application

The undersigned affirms he/she has read, understood, and abided by the rules of the VUCC (MCS-261).

Signature _____ Date _____ 19 ____

- -

VHF AWARDS MANAGER VERIFICATION

Grid square credits claimed on this   V-
application have been verified as correct _____ _____ _____
and true. TOTAL number of grid squares   (signature)   (call)   (Mgr. 4-digit
credited =   code #)

_____ .   _____
                                          (date)

- -

DIRECTIONS TO APPLICANT

1. Fill out this application and MCS-259. Read the VUCC Rules (MCS-261) carefully.

2. Enclose postage sufficient for return of cards (if mailed), and return address label.

3. For initial applicants, enclose $1.00 U.S. for mailing certificate (W/VEs only).

4. For endorsement, enclose s.a.s.e. with 2 units of first class postage (W/VEs only).

5. Sort QSLs alphabetically by field, and numerically within the field, from 00 to 99.

6. Forward cards to nearest VHF Awards Manager (see attached information). DO NOT SEND CARDS TO ARRL HQ.

MCS-260 (186)

Field:
(first 2 letters of locator)

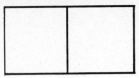

SQ.	Callsign	SQ.	Callsign
00		50	
01		51	
02		52	
03		53	
04		54	
05		55	
06		56	
07		57	
08		58	
09		59	
10		60	
11		61	
12		62	
13		63	
14		64	
15		65	
16		66	
17		67	
18		68	
19		69	
20		70	
21		71	
22		72	
23		73	
24		74	
25		75	
26		76	
27		77	
28		78	
29		79	
30		80	
31		81	
32		82	
33		83	
34		84	
35		85	
36		86	
37		87	
38		88	
39		89	
40		90	
41		91	
42		92	
43		93	
44		94	
45		95	
46		96	
47		97	
48		98	
49		99	

TOTAL number of grid squares worked in this field = [   ]

MCS-259 (186)

ARRL VUCC AWARD

CALLSIGN: _____

☐ 50 ☐ 144 ☐ 220 ☐ 432 ☐ 902 ☐ 1296 ☐ 2.3 ☐ 3.4 ☐ 5.7 ☐ 10

DIRECTIONS: This side of tally sheet good for two fields. 1. Enter your callsign and check band. 2. Enter two-letter field. 3. Enter callsigns of stations worked in appropriate grid square. 4. Total grid squares for each field.

Field:
(first 2 letters of locator)

SQ.	Callsign	SQ.	Callsign
00		50	
01		51	
02		52	
03		53	
04		54	
05		55	
06		56	
07		57	
08		58	
09		59	
10		60	
11		61	
12		62	
13		63	
14		64	
15		65	
16		66	
17		67	
18		68	
19		69	
20		70	
21		71	
22		72	
23		73	
24		74	
25		75	
26		76	
27		77	
28		78	
29		79	
30		80	
31		81	
32		82	
33		83	
34		84	
35		85	
36		86	
37		87	
38		88	
39		89	
40		90	
41		91	
42		92	
43		93	
44		94	
45		95	
46		96	
47		97	
48		98	
49		99	

TOTAL number of grid squares worked in this field = [   ]

WAC AWARD APPLICATION

From:

Name _____   Callsign _____

Mailing Address _____

City/Town _____   State _____   ZIP Code _____

This application is for:

☐ Basic certificate (mixed-mode)   ☐ QRP sticker (5 watts output
☐ CW certificate   or less)
☐ Phone certificate   ☐ 1.8-MHz sticker
☐ SSTV certificate   ☐ 3.5-MHz sticker
☐ RTTY certificate   ☐ 50-MHz sticker
☐ FAX certificate   ☐ 144-MHz sticker
☐ Satellite certificate   ☐ 430-MHz sticker
☐ 5-band certificate
☐ 6-band sticker   My 5-band certificate is dated _____

Enclosed are QSL cards from the following stations:

(Enter band(s) and callsigns)

	MHz	MHz	MHz	MHz	MHz
N. America					
S. America					
Oceania					
Asia					
Europe					
Africa					

The undersigned has abided by all rules set forth for this award.
My ARRL membership does not expire until _____

Signature _____   Callsign _____

Date _____

Only amateurs from the US and possessions, and
Puerto Rico, should use this form. ARRL membership is required. All other applicants must apply
0186   through their country's IARU member-society.

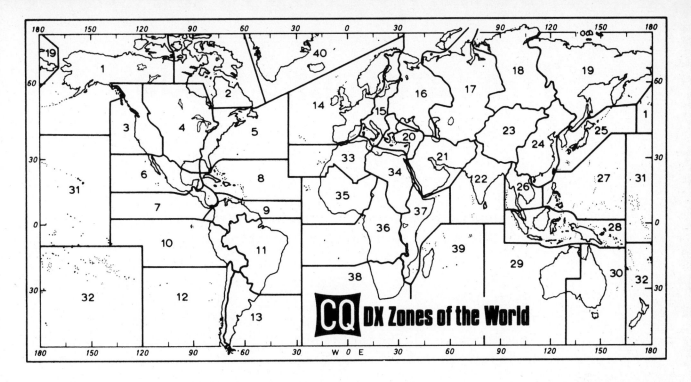

CQ DX Zones of the World

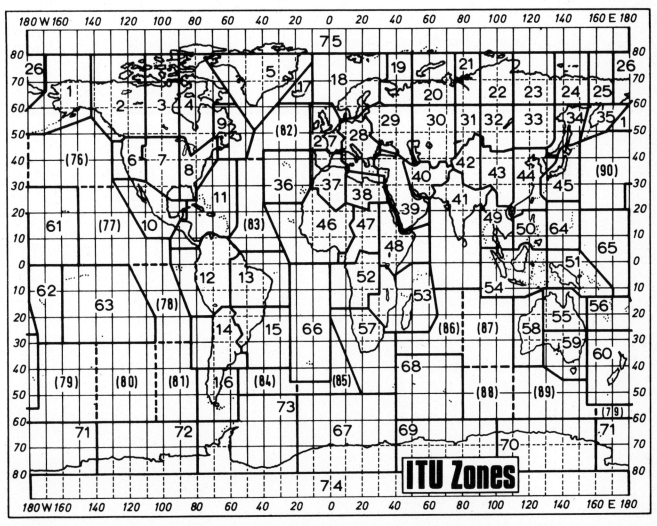

ITU Zones

The Amateur's Code

ONE

The Amateur is Considerate...He never knowingly uses the air in such a way as to lessen the pleasure of others.

TWO

The Amateur is Loyal...He offers his loyalty, encouragement and support to his fellow radio amateurs, his local club and to the American Radio Relay League, through which Amateur Radio is represented.

THREE

The Amateur is Progressive...He keeps his station abreast of science. It is well-built and efficient. His operating practice is above reproach.

FOUR

The Amateur is Friendly...Slow and patient sending when requested, friendly advice and counsel to the beginner, kindly assistance, cooperation and consideration for the interests of others; these are marks of the amateur spirit.

FIVE

The Amateur is Balanced...Radio is his hobby. He never allows it to interfere with any of the duties he owes to his home, his job, his school, or his community.

SIX

The Amateur is Patriotic...His knowledge and his station are always ready for the service of his country and his community.

— PAUL M. SEGAL

United States Counties

(information courtesy of *CQ*'s Counties Award Record Book; for information on county hunting, see p 8-9)

Alabama (67 counties)
Autauga
Baldwin
Barbour
Bibb
Blount
Bullock
Butler
Calhoun
Chambers
Cherokee
Chilton
Choctaw
Clarke
Clay
Cleburne
Coffee
Colbert
Conecuh
Coosa
Covington
Crenshaw
Cullman
Dale
Dallas
Dekalb
Elmore
Escambia
Etowah
Fayette
Franklin
Geneva
Greene
Hale
Henry
Houston
Jackson
Jefferson
Lamar
Lauderdale
Lawrence
Lee
Limestone
Lowndes
Macon
Madison
Marengo
Marion
Marshall
Mobile
Monroe
Montgomery
Morgan
Perry
Pickens
Pike
Randolph
Russel
Saint Clair
Shelby
Sumter
Talladega
Tallapoosa
Tuscaloosa
Walker
Washington
Wilcox
Winston

Alaska (4 counties)
Southeastern
Northwestern
South Central
Central

Arizona (15 counties)
Apache
Cochise
Coconino
Gila
Graham
Greenlee
Lez Paz
Maricopa
Mohave
Navajo
Pima
Pinal
Santa Cruz
Yavapai
Yuma

Arkansas (75 counties)
Arkansas
Ashley
Baxter
Benton
Boone
Bradley
Calhoun
Carroll
Chicot
Clark
Clay
Cleburne
Cleveland
Columbia
Conway
Craighead
Crawford
Crittenden
Cross
Dallas
Desha
Drew
Faulkner
Franklin
Fulton
Garland
Grant
Greene
Hempstead
Hot Spring
Howard
Independence
Izard
Jackson
Jefferson
Johnson
Lafayette
Lawrence
Lee
Lincoln
Little River
Logan
Lonoke
Madison
Marion
Miller
Mississippi
Monroe
Montgomery
Nevada
Newton
Ouachita
Perry
Phillips
Pike
Poinsett

Polk
Pope
Prairie
Pulaski
Randolph
St. Francis
Saline
Scott
Searcy
Sebastian
Sevier
Sharp
Stone
Union
Van Buren
Washington
White Woodruff
Yell

California (58 counties)
Alameda
Alpine
Amador
Butte
Calaveras
Colusa
Contra Costa
Del Norte
El Dorado
Fresno
Glenn
Humboldt
Imperial
Inyo
Kern
Kings
Lake
Lassen
Los Angeles
Madera
Marin
Mariposa
Mendocino
Merced
Modoc
Mono
Monterey
Napa
Nevada
Orange
Placer
Plumas
Riverside
Sacramento
San Benito
San Bernardino
San Diego
San Francisco
San Joaquin
San Luis Obispo
San Mateo
Santa Barbara
Santa Clara
Santa Cruz
Shasta
Sierra
Siskiyou
Solano
Sonoma
Stanislaus
Sutter
Tehama
Trinity
Tulare

Toulumne
Ventura
Yolo
Yuba

Colorado (63 counties)
Adams
Alamosa
Arapahoe
Archuleta
Baca
Bent
Boulder
Chaffee
Cheyenne
Clear Creek
Conejos
Costilla
Crowley
Custer
Delta
Denver
Dolores
Douglas
Eagle
Elbert
El Paso
Fremont
Garfield
Gilpin
Grand
Gunnison
Hinsdale
Huerfano
Jackson
Jefferson
Kiowa
Kit Carson
Lake
La Plata
Larimer
Las Animas
Lincoln
Logan
Mesa
Mineral
Moffat
Montezuma
Montrose
Morgan
Otero
Ouray
Park
Phillips
Pitkin
Prowers
Pueblo
Rio Blanco
Rio Grande
Routt
Saguache
San Juan
San Miguel
Sedgwick
Summit
Teller
Washington
Weld
Yuma

Connecticut (8 counties)
Fairfield
Hartford
Litchfield

Middlesex
New Haven
New London
Tolland
Windham

Delaware (3 counties)

Kent
New Castle
Sussex

Florida (67 counties)

Alachua
Baker
Bay
Bradford
Brevard
Broward
Calhoun
Charlotte
Citrus
Clay
Collier
Columbia
Dade
De Soto
Dixie
Duval
Escambia
Flagler
Franklin
Gadsden
Gilchrist
Glades
Gulf
Hamilton
Hardee
Hendry
Hernando
Highlands
Hillsborough
Holmes
Indian River
Jackson
Jefferson
Lafayette
Lake
Lee
Leon
Levy
Liberty
Madison
Manatee
Marion
Martin
Monroe
Nassau
Okaloosa
Okeechobee
Orange
Osceola
Palm Beach
Pasco
Pinellas
Polk
Putnam
St. Johns
St. Lucie
Santa Rosa
Sarasota
Seminole
Sumter
Suwannee
Taylor
Union
Volusia
Wakulla
Walton
Washington

Georgia (159 counties)

Appling
Atkinson
Bacon
Baker
Baldwin
Banks
Barrow
Bartow
Ben Hill
Berrien
Bibb
Bleckley
Brantley
Brooks
Bryan
Bulloch
Burke
Butts
Calhoun
Camden
Candler
Carroll
Catoosa
Charlton
Chatham
Chattahoochee
Chattooga
Cherokee
Clarke
Clay
Clayton
Clinch
Cobb
Coffee
Colquitt
Columbia
Cook
Coweta
Crawford
Crisp
Dade
Dawson
Decatur
De Kalb
Dodge
Dooly
Dougherty
Douglas
Early
Echols
Effingham
Elbert
Emanuel
Evans
Fannin
Fayette
Floyd
Forsyth
Franklin
Fulton
Gilmer
Glascock
Glynn
Gordon
Grady
Greene
Gwinnett
Habersham
Hall
Hancock
Haralson
Harris
Hart Heard
Henry
Houston
Irwin
Jackson

Jasper
Jeff Davis
Jefferson
Jenkins
Johnson
Jones
Lamar
Lanier
Laurens
Lee
Liberty
Lincoln
Long
Lowndes
Lumpkin
McDuffie
McIntosh
Macon
Madison
Marion
Meriwether
Miller
Mitchell
Monroe
Montgomery
Morgan
Murray
Muscogee
Newton
Oconee
Oglethorpe
Paulding
Peach
Pickens
Pierce
Pike
Polk
Pulaski
Putnam
Quitman
Rabun
Randolph
Richmond
Rockdale
Schley
Screven
Seminole
Spalding
Stephens
Stewart
Sumter
Talbot
Taliaferro
Tattnall
Taylor
Telfair
Terrell
Thomas
Tift
Toombs
Towns
Treutlen
Troup
Turner
Twiggs
Union
Upson
Walker
Walton
Ware
Warren
Washington
Wayne
Webster
Wheeler
White
Whitfield
Wilcox

Wilkes
Wilkinson
Worth

Hawaii (5 counties)

Hawaii
Honolulu
Kalawao
Kauai
Maui

Idaho (44 counties)

Ada
Adams
Bannock
Bear Lake
Benewah
Bingham
Blaine
Boise
Bonner
Bonneville
Boundary
Butte
Camas
Canyon
Caribou
Cassia
Clark
Clearwater
Custer
Elmore
Franklin
Fremont
Gem
Gooding
Idaho
Jefferson
Jerome
Kootenai
Latah
Lemhi
Lewis
Lincoln
Madison
Minidoka
Nez Perce
Oneida
Owyhee
Payette
Power
Shoshone
Teton
Twin Falls
Valley
Washington

Illinois (102 counties)

Adams
Alexander
Bond
Boone
Brown
Bureau
Calhoun
Carroll
Cass
Champaign
Christian
Clark
Clay
Clinton
Coles
Cook
Crawford
Cumberland
De Kalb
De Witt

Douglas
Du Page
Edgar
Edwards
Effingham
Fayette
Ford
Franklin
Fulton
Gallatin
Greene
Grundy
Hamilton
Hancock
Hardin Henderson
Henry
Iroquois
Jackson
Jasper
Jefferson
Jersey
Jo Daviess
Johnson
Kane
Kankakee
Kendall
Knox
Lake
La Salle
Lawrence
Lee
Livingston
Logan
McDonough
McHenry
McLean
Macon
Macoupin
Madison
Marion
Marshall
Mason
Massac
Menard
Mercer
Monroe
Montgomery
Morgan
Moultrie
Ogle
Peoria
Perry
Piatt
Pike
Pope
Pulaski
Putnam
Randolph
Richland
Rock Island
St. Clair
Saline
Sangamon
Schuyler
Scott
Shelby
Stark
Stephenson
Tazewell
Union
Vermilion
Wabash
Warren
Washington
Wayne
White
Whiteside
Will
Williamson

Winnebago
Woodford

Indiana (92 counties)

Adams
Allen
Bartholomew
Benton
Blackford
Boone
Brown
Carroll
Cass
Clark
Clay
Clinton
Crawford
Daviess
Dearborn
Decatur
De Kalb
Delaware
Dubois
Elkhart
Fayette
Floyd
Fountain
Franklin
Fulton
Gibson
Grant
Greene
Hamilton
Hancock
Harrison
Hendricks
Henry
Howard
Huntington
Jackson
Jasper
Jay
Jefferson
Jennings
Johnson
Knox
Kosciusko
Lagrange
Lake
La Porte
Lawrence
Madison
Marion
Marshall
Martin
Miami
Monroe
Montgomery
Morgan
Newton
Noble
Ohio
Orange
Owen
Parke
Perry
Pike
Porter
Posey
Pulaski
Putnam
Randolph
Ripley
Rush
St. Joseph
Scott
Shelby
Spencer
Starke

Steuben
Sullivan
Switzerland
Tippecanoe
Tipton
Union
Vanderburgh
Vermillion
Vigo
Wabash
Warren
Warrick
Washington
Wayne
Wells
White
Whitley

Iowa (99 counties)

Adair
Adams
Allamakee
Appanoose
Audubon
Benton
Black Hawk
Boone
Bremer
Buchanan
Buena Vista
Butler
Calhoun
Carroll
Cass
Cedar
Cerro Gordo
Cherokee
Chickasaw
Clarke
Clay
Clayton
Clinton
Crawford
Dallas
Davis
Decatur
Delaware
Des Moines
Dickinson
Dubuque
Emmet
Fayette
Floyd
Franklin
Fremont
Greene
Grundy
Guthrie
Hamilton
Hancock
Hardin
Harrison
Henry
Howard
Humboldt
Ida
Iowa
Jackson
Jasper
Jefferson
Johnson
Jones
Keokuk
Kossuth
Lee
Linn
Louisa
Lucas
Lyon

Madison
Mahaska
Marion
Marshall
Mills
Mitchell
Monona
Monroe
Montgomery
Muscatine
O'Brien
Osceola
Page
Palo Alto
Plymouth
Pocahontas
Polk
Pottawattamie
Poweshiek
Ringgold
Sac
Scott
Shelby
Sioux
Story
Tama
Taylor
Union
Van Buren
Wapello
Warren
Washington
Wayne
Webster
Winnebago
Winneshiek
Woodbury
Worth
Wright

Kansas (105 counties)

Allen
Anderson
Atchison
Barber
Barton
Bourbon
Brown
Butler
Chase
Chautauqua
Cherokee
Cheyenne
Clark
Clay
Cloud
Coffey
Comanche
Cowley
Crawford
Decatur
Dickinson
Doniphan
Douglas
Edwards
Elk
Ellis
Ellsworth
Finney
Ford
Franklin
Geary
Gove
Graham
Grant
Gray
Greeley
Greenwood
Hamilton

Harper
Harvey
Haskell
Hodgeman
Jackson
Jefferson
Jewell
Johnson
Kearny
Kingman
Kiowa
Labette
Lane
Leavenworth
Lincoln
Linn
Logan
Lyon
McPherson
Marion
Marshall
Meade
Miami
Mitchell
Montgomery
Morris
Morton
Nemaha
Neosho
Ness
Norton
Osage
Osborne
Ottawa
Pawnee
Phillips
Pottawatomie
Pratt
Rawlins
Reno
Republic
Rice
Riley
Rooks
Rush
Russell
Saline
Scott
Sedgwick
Seward
Shawnee
Sheridan
Sherman
Smith
Stafford
Stanton
Stevens
Sumner
Thomas
Trego
Wabaunsee
Wallace
Washington
Wichita
Wilson
Woodson
Wyandotte

Kentucky (120 counties)

Adair
Allen
Anderson
Ballard
Barren
Bath
Bell
Boone
Bourbon
Boyd

Boyle
Bracken
Breathitt
Breckenridge
Bullitt
Butler
Caldwell
Calloway
Campbell
Carlisle
Carroll
Carter
Casey
Christian
Clark
Clay
Clinton
Crittenden
Cumberland
Daviess
Edmonson
Elliott
Estill
Fayette
Fleming
Floyd
Franklin
Fulton
Gallatin
Garrard
Grant
Graves
Grayson
Green
Greenup
Hancock
Hardin
Harlan
Harrison
Hart
Henderson
Henry
Hickman
Hopkins
Jackson
Jefferson
Jessamine
Johnson
Kenton
Knott
Knox
Larue
Laurel
Lawrence
Lee
Leslie
Letcher
Lewis
Lincoln
Livingston
Logan
Lyon
McCracken
McCreary
McLean
Madison
Magoffin
Marion
Marshall
Martin
Mason
Meade
Menifee
Mercer
Metcalfe
Monroe
Montgomery
Morgan
Muhlenberg

Nelson
Nicholas
Ohio
Oldham
Owen
Owsley
Pendleton
Perry
Pike
Powell
Pulaski
Robertson
Rockcastle
Rowan
Russell
Scott
Shelby
Simpson
Spencer
Taylor
Todd
Trigg
Trimble
Union
Warren
Washington
Wayne
Webster
Whitley
Wolfe
Woodford

Louisiana (64 parishes)

Acadia
Allen
Ascension
Assumption
Avoyelles
Beauregard
Bienville
Bossier
Caddo
Calcasieu
Caldwell
Cameron
Catahoula
Claiborne
Concordia
De Soto
E. Baton Rouge
East Carroll
East Feliciana
Evangeline
Franklin
Grant
Iberia
Iberville
Jackson
Jefferson
Jefferson Davis
Lafayette
Lafourche
La Salle
Lincoln
Livingston
Madison
Morehouse
Natchitoches
Orleans
Ouachita
Plaquemines
Pointe Coupee
Rapides
Red River
Richland
Sabine
St. Bernard
St. Charles
St. Helena

St. James
St. John the Baptist
St. Landry
St. Martin
St. Mary
St. Tammany
Tangipahoa
Tensas
Terrebone
Union
Vermilion
Vernon
Washington
Webster
W. Baton Rouge
West Carroll
West Feliciana
Winn

Maine (16 countries)

Androscoggin
Aroostock
Cumberland
Franklin
Hancock
Kennebec
Knox
Lincoln
Oxford
Penobscot
Piscataquis
Sagadahoc
Somerset
Waldo
Washington
York

Maryland (24 counties)

Allegany
Anne Arundel
Baltimore
Baltimore City
Calvert
Caroline
Carroll
Cecil
Charles
Dorchester
Frederick
Garrett
Harford
Howard
Kent
Montgomery
Prince Georges
Queen Annes
St. Marys
Somerset
Talbot
Washington
Wicomico
Worcester

Massachusetts
  (14 counties)

Barnstable
Berkshire
Bristol
Dukes
Essex
Franklin
Hampden
Hampshire
Middlesex
Nantucket
Norfolk
Plymouth
Suffolk
Worcester

Michigan (83 counties)

Alcona
Alger
Allegan
Alpena
Antrim
Arenac
Baraga
Barry
Bay
Benzie
Berrien
Branch
Calhoun
Cass
Charlevoix
Cheboygan
Chippewa
Clare
Clinton
Crawford
Delta
Dickinson
Eaton
Emmet
Genesee
Gladwin
Gogebic
Grand Traverse
Gratiot
Hillsdale
Houghton
Huron
Ingham
Ionia
Iosco
Iron
Isabella
Jackson
Kalamazoo
Kalkaska
Kent
Keweenaw
Lake
Lapeer
Leelanau
Lenawee
Livingston
Luce
Mackinac
Macomb
Manistee
Marquette
Mason
Mecosta
Menominee
Midland
Missaukee
Monroe
Montcalm
Montmorency
Muskegon
Newaygo
Oakland
Oceana
Ogemaw
Ontonagon
Osceola
Oscoda
Otsego
Ottawa
Presque Isle
Roscommon
Saginaw
St. Clair
St. Joseph
Sanilac
Schoolcraft
Shiawassee
Tuscola
Van Buren
Washtenaw
Wayne
Wexford

Minnesota (87 counties)

Aitkin
Anoka
Becker
Beltrami
Benton
Big Stone
Blue Earth
Brown
Carlton
Carver
Cass
Chippewa
Chisago
Clay
Clearwater
Cook
Cottonwood
Crow Wing
Dakota
Dodge
Douglas
Faribault
Fillmore
Freeborn
Goodhue
Grant
Hennepin
Houston
Hubbard
Isanti
Itasa
Jackson
Kanabec
Kandiyohi
Kittson
Koochiching
Lac Qui Parle
Lake
L. of the Woods
Le Sueur
Lincoln
Lyon
McLeod
Mahnomen
Marshall
Martin
Meeker
Mille Lacs
Morrison
Mower
Murray
Nicollet
Nobles
Norman
Olmsted
Otter Trail
Pennington
Pine
Pipestone
Polk
Pope
Ramsey
Red Lake
Redwood
Renville
Rice
Rock
Roseau
St. Louis
Scott
Sherburne
Sibley
Stearns
Steele
Stevens
Swift
Todd
Traverse
Wabasha
Wadena
Waseca
Washington
Watonwan
Wilkin
Winona
Wright
Yellow Medicine

Mississippi (82 counties)

Adams
Alcorn
Amite
Attala
Benton
Bolivar
Calhoun
Carroll
Chickasaw
Choctaw
Claiborne
Clarke
Clay
Coahoma
Copiah
Covington
De Soto
Forrest
Franklin
George
Greene
Grenada
Hancock
Harrison
Hinds
Holmes
Humphreys
Issaquena
Itawamba
Jackson
Jasper
Jefferson
Jefferson Davis
Jones
Kemper
Lafayette
Lamar
Lauderdale
Lawrence
Leake
Lee
Leflore
Lincoln
Lowndes
Madison
Marion
Marshall
Monroe
Montgomery
Neshoba
Newton
Noxubee
Oktibbeha
Panola
Pearl River
Perry
Pike
Pontotoc
Prentiss
Quitman
Rankin
Scott
Sharkey
Simpson
Smith
Stone
Sunflower
Tallahatchie
Tate
Tippah
Tishomingo
Tunica
Union
Walthall
Warren
Washington
Wayne
Webster
Wilkinson
Winston
Yalobusha
Yazoo

Missouri (115 counties)

Adair
Andrew
Atchison
Audrain
Barry
Barton
Bates
Benton
Bollinger
Boone
Buchanan
Butler
Caldwell
Callaway
Camden
Cape Girardeau
Carroll
Carter
Cass
Cedar
Chariton
Christian
Clark
Clay
Clinton
Cole
Cooper
Crawford
Dade
Dallas
Daviess
De Kalb
Dent
Douglas
Dunklin
Franklin
Gasconade
Gentry
Greene
Grundy
Harrison
Henry
Hickory
Holt
Howard
Howell
Iron
Jackson
Jasper
Jefferson
Johnson
Knox
Laclede
Lafayette
Lawrence
Lewis
Lincoln

Linn
Livingston
McDonald
Macon
Madison
Maries
Marion
Mercer
Miller
Mississippi
Moniteau
Monroe
Montgomery
Morgan
New Madrid
Madrid
Newton
Nodaway
Oregon
Osage
Ozark
Pemiscot
Perry
Pettis
Phelps
Pike
Platte
Polk
Pulaski
Putnam
Ralls
Randolph
Ray
Reynolds
Ripley
St. Charles
St. Clair
St. Francois
St. Louis
St. Louis City
Ste. Genevieve
Saline
Schuyler
Scotland
Scott
Shannon
Shelby
Stoddard
Stone
Sullivan
Taney
Texas
Vernon
Warren
Washington
Wayne
Webster
Worth
Wright

Montana (56 counties)

Beaverhead
Big Horn
Blaine
Broadwater
Carbon
Carter
Cascade
Chouteau
Custer
Daniels
Dawson
Deer Lodge
Fallon
Fergus
Flathead
Gallatin
Garfield
Glacier

Golden Valley
Granite
Hill
Jefferson
Judith Basin
Lake
Lewis and Clark
Liberty
Lincoln
McCone
Madison
Meagher
Mineral
Missoula
Musselshell
Park
Petroleum
Phillips
Pondera
Powder River
Powell Prairie
Ravalli
Richland
Roosevelt
Rosebud
Sanders
Sheridan
Silver Bow
Stillwater
Sweet Grass
Teton
Toole
Treasure
Valley
Wheatland
Wibaux
Yellowstone

Nebraska (93 counties)

Adams
Antelope
Arthur
Banner
Blaine
Boone
Box Butte
Boyd
Brown
Buffalo
Burt
Butler
Cass
Cedar
Chase
Cherry
Cheyenne
Clay
Colfax
Cuming
Custer
Dakota
Dawes
Dawson
Deuel
Dixon
Dodge
Douglas
Dundy
Fillmore
Franklin
Frontier
Furnas
Gage
Garden
Garfield
Gosper
Grant
Greeley
Hall

Hamilton
Harlan
Hayes
Hitchcock
Holt
Hooker
Howard
Jefferson
Johnson
Kearney
Keith
Keya Paha
Kimball
Knox
Lancaster
Lincoln
Logan
Loup
McPherson
Madison
Merrick
Morrill
Nance
Nemaha
Nuckolls
Otoe
Pawnee
Perkins
Phelps
Pierce
Platte
Polk
Red Willow
Richardson
Rock
Saline
Sarpy
Saunders
Scotts Bluff
Seward
Sheridan
Sherman
Sioux
Stanton
Thayer
Thomas
Thurston
Valley
Washington
Wayne
Webster
Wheeler
York

Nevada (16 counties)

Churchill
Clark
Douglas
Elko
Esmeralda
Eureka
Humboldt
Lander
Lincoln
Lyon
Mineral
Nye
Pershing
Storey
Washoe
White Pine

New Hampshire (10 counties)

Belknap
Carroll
Cheshire
Coos
Grafton

Hillsboro
Merrimack
Rockingham
Strafford
Sullivan

New Jersey (21 counties)

Atlantic
Bergen
Burlington
Camden
Cape May
Cumberland
Essex
Gloucester
Hudson
Hunterdon
Mercer
Middlesex
Monmouth
Morris
Ocean
Passaic
Salem
Somerset
Sussex
Union
Warren

New Mexico (33 counties)

Bernalillo
Catron
Chaves
Cibola
Colfax
Curry
De Baca
Dona Ana
Eddy
Grant
Guadalupe
Harding
Hidalgo
Lea
Lincoln
Los Alamos
Luna
McKinley
Mora
Otero
Quay
Rio Arriba
Roosevelt
Sandoval
San Juan
San Miguel
Santa Fe
Sierra
Socorro
Taos
Torrance
Union
Valencia

New York (62 counties)

Albany
Allegany
Bronx
Broome
Cattaraugus
Cayuga
Chautauqua
Chemung
Chenango
Clinton
Columbia
Cortland
Delaware
Dutchess

Erie
Essex
Franklin
Fulton
Genesee
Greene
Hamilton
Herkimer
Jefferson
Kings
Lewis
Livingston
Madison
Monroe
Montgomery
Nassau
New York
Niagara
Oneida
Onondaga
Ontario
Orange
Orleans
Oswego
Otsego
Putnam
Queens
Rensselaer
Rockland
St. Lawrence
Saratoga
Schenectady
Schoharie
Schuyler
Seneca
Staten Island
Steuben
Suffolk
Sullivan
Tioga
Tompkins
Ulster
Warren
Washington
Wayne
Westchester
Wyoming
Yates

North Carolina (100 counties)

Alamance
Alexander
Alleghany
Anson
Ashe
Avery
Beaufort
Bertie
Bladen
Brunswick
Buncombe
Burke
Cabarrus
Caldwell
Camden
Carteret
Caswell
Catawba
Chatham
Cherokee
Chowan
Clay
Cleveland
Columbus
Craven
Cumberland
Currituck
Dare

Davidson
Davie
Duplin
Durham
Edgecombe
Forsyth
Franklin
Gaston
Gates
Graham
Granville
Greene
Guilford
Halifax
Harnett
Haywood
Henderson
Hertford
Hoke
Hyde
Iredell
Jackson
Johnston
Jones
Lee
Lenoir
Lincoln
McDowell
Macon
Madison
Martin
Mecklenburg
Mitchell
Montgomery
Moore
Nash
New Hanover
Northampton
Onslow
Orange
Pamlico
Pasquotank
Pender
Perquimans
Person
Pitt
Polk
Randolph
Richmond
Robeson
Rockingham
Rowan
Rutherford
Sampson
Scotland
Stanly
Stokes
Surry
Swain
Transylvania
Tyrrell
Union
Vance
Wake
Warren
Washington
Watauga
Wayne
Wilkes
Wilson
Yadkin
Yancey

North Dakota (53 counties)

Adams
Barnes
Benson
Billings
Bottineau

Bowman
Burke
Burleigh
Cass
Cavalier
Dickey
Divide
Dunn
Eddy
Emmons
Foster
Golden Valley
Grand Forks
Grant
Griggs
Hettinger
Kidder
La Moure
Logan
McHenry
McIntosh
McKenzie
McLean
Mercer
Morton
Mountrail
Nelson
Oliver
Pembina
Pierce
Ramsey
Ranson
Renville
Richland
Rolette
Sargent
Sheridan
Sioux
Slope
Stark
Steele
Stutsman
Towner
Traill
Walsh
Ward
Wells
Williams

Ohio (88 counties)

Adams
Allen
Ashland
Ashtabula
Athens
Auglaize
Belmont
Brown
Butler
Carroll
Champaign
Clark
Clermont
Clinton
Columbiana
Coshocton
Crawford
Cuyahoga
Darke
Defiance
Delaware
Erie
Fairfield
Fayette
Franklin
Fulton
Gallia
Geauga
Greene

Guernsey
Hamilton
Hancock
Hardin
Harrison
Henry
Highland
Hocking
Holmes
Huron
Jackson
Jefferson
Knox
Lake
Lawrence
Licking
Logan
Lorain
Lucas
Madison
Mahoning
Marion
Medina
Meigs
Mercer
Miami
Monroe
Montgomery
Morgan
Morrow
Muskingum
Noble
Ottawa
Paulding
Perry
Pickaway
Pike
Portage
Preble
Putnam
Richland
Ross
Sandusky
Scioto
Seneca
Shelby
Stark
Summit
Trumbull
Tuscarawas
Union
Van Wert
Vinton
Warren
Washington
Wayne
Williams
Wood
Wyandot

Oklahoma (77 counties)
Adair
Alfalfa
Atoka
Beaver
Beckham
Blaine
Bryan
Caddo
Canadian
Carter
Cherokee
Choctaw
Cimarron
Cleveland
Coal
Commanche
Cotton
Craig

Creek
Custer
Delaware
Dewey
Ellis
Garfield
Garvin
Grady
Grant
Greer
Harmon
Harper
Haskell
Hughes
Jackson
Jefferson
Johnston
Kay
Kingfisher
Kiowa
Latimer
Le Flore
Lincoln
Logan
Love
McClain
McCurtain
McIntosh
Major
Marshall
Mayes
Murray
Muskogee
Noble
Nowata
Okfuskee
Oklahoma
Okmulgee
Osage
Ottawa
Pawnee
Payne
Pittsburg
Pontotoc
Pottawatomie
Pushmataha
Roger Mills
Rogers
Seminole
Sequoyah
Stephens
Texas
Tillman
Tulsa
Wagoner
Washington
Washita
Woods
Woodward

Oregon (36 counties)

Baker
Benton
Clackamas
Clatsop
Columbia
Coos
Crook
Curry
Deschutes
Douglas
Gilliam
Grant
Harney
Hood River
Jackson
Jefferson
Josephine
Klamath

Lake
Lane
Lincoln
Linn
Malheur
Marion
Morrow
Multnomah
Polk
Sherman
Tillamook
Umatilla
Union
Wallowa
Wasco
Washington
Wheeler
Yamhill

Pennsylvania (67 counties)

Adams
Allegheny
Armstrong
Beaver
Bedford
Berks
Blair
Bradford
Bucks
Butler
Cambria
Cameron
Carbon
Centre
Chester
Clarion
Clearfield
Clinton
Columbia
Crawford
Cumberland
Dauphin
Delaware
Elk
Erie
Fayette
Forest
Franklin
Fulton
Greene
Huntingdon
Indiana
Jefferson
Juniata
Lackawanna
Lancaster
Lawrence
Lebanon
Lehigh
Luzerne
Lycoming
McKean
Mercer
Mifflin
Monroe
Montgomery
Montour
Northampton
Northumberland
Perry
Philadelphia
Pike Potter
Schuylkill
Snyder
Somerset
Sullivan
Susquehanna
Tioga
Union

Venango
Warren
Washington
Wayne
Westmoreland
Wyoming
York

Rhode Island (5 counties)

Bristol
Kent
Newport
Providence
Washington

South Carolina
(46 counties)

Abbeville
Aiken
Allendale
Anderson
Bamberg
Barnwell
Beaufort
Berkeley
Calhoun
Charleston
Cherokee
Chester
Chesterfield
Clarendon
Colleton
Darlington
Dillon
Dorchester
Edgefield
Fairfield
Florence
Georgetown
Greenville
Greenwood
Hampton
Horry
Jasper
Kershaw
Lancaster
Laurens
Lee
Lexington
McCormick
Marion
Marlboro
Newberry
Oconee
Orangeburg
Pickens
Richland
Saluda
Spartanburg
Sumter
Union
Williamsburg
York

South Dakota (66 counties)

Aurora
Beadle
Bennett
Bon Homme
Brookings
Brown
Brule
Buffalo
Butte
Campbell
Charles Mix
Clark
Clay
Codington

Corson
Custer
Davison
Day
Deuel
Dewey
Douglas
Edmunds
Fall River
Faulk
Grant
Gregory
Haakon
Hamlin
Hand
Hanson
Harding
Hughes
Hutchinson
Hyde
Jackson
Jerauld
Jones
Kingsbury
Lake
Lawrence
Lincoln
Lyman
McCook
McPherson
Marshall
Meade
Mellette
Miner
Minnehaha
Moody
Pennington
Perkins
Potter
Roberts
Sanborn
Shannon
Spink
Stanley
Sully
Todd
Tripp
Turner
Union
Walworth
Yankton
Ziebach

Tennessee (95 counties)

Anderson
Bedford
Benton
Bledsoe
Blount
Bradley
Campbell
Cannon
Carroll
Carter
Cheatham
Chester
Claiborne
Clay
Cocke
Coffee
Crockett
Cumberland
Davidson
Decatur
DeKalb
Dickson
Dyer
Fayette
Fentress

Franklin
Gibson
Giles
Grainger
Greene
Grundy
Hamblen
Hamilton
Hancock
Hardeman
Hardin
Hawkins
Haywood
Henderson
Henry
Hickman
Houston
Humphreys
Jackson
Jefferson
Johnson
Knox
Lake
Lauderdale
Lawrence
Lewis
Lincoln
Loudon
McMinn
McNairy
Macon
Madison
Marion
Marshall
Maury
Meigs
Monroe
Montgomery
Moore
Morgan
Obion
Overton
Perry
Pickett
Polk
Putnam
Rhea
Roane
Robertson
Rutherford
Scott
Sequatchie
Sevier
Shelby
Smith
Stewart
Sullivan
Sumner
Tipton
Trousdale
Unicoi
Union
Van Buren
Warren
Washington
Wayne
Weakley
White
Williamson
Wilson

Texas (254 counties)

Anderson
Andrews
Angelina
Aransas
Archer
Armstrong
Atascosa

Austin
Bailey
Bandera
Bastrop
Baylor
Bee
Bell
Bexar
Blanco
Borden
Bosque
Bowie
Brazoria
Brazos
Brewster
Briscoe
Brooks
Brown
Burleson
Burnet
Caldwell
Calhoun
Callahan
Cameron
Camp
Carson
Cass
Castro
Chambers
Cherokee
Childress
Clay
Cochran
Coke
Coleman
Collin
Collingsworth
Colorado
Comal
Comanche
Concho
Cooke
Coryell
Cottle
Crane
Crockett
Crosby
Culberson
Dallam
Dallas
Dawson
Deaf Smith
Delta
Denton
De Witt
Dickens
Dimmit
Donley
Duval
Eastland
Ector
Edwards
Ellis
El Paso
Erath
Falls
Fannin
Fayette
Fisher
Floyd
Foard
Fort Bend
Franklin
Freestone
Frio
Gaines
Galveston
Garza
Gillespie

Glasscock
Goliad
Gonzales
Gray
Grayson
Gregg
Grimes
Guadelupe
Hale
Hall
Hamilton
Hansford
Hardeman
Hardin
Harris
Harrison
Hartley
Haskell
Hays
Hemphill
Henderson
Hidalgo
Hill
Hockley
Hood
Hopkins
Houston
Howard
Hudspeth
Hunt
Hutchinson
Irion
Jack
Jackson
Jasper
Jeff Davis
Jefferson
Jim Hogg
Jim Wells
Johnson
Jones
Karnes
Kaufman
Kendall
Kenedy
Kent
Kerr
Kimble
King
Kinney
Kleberg
Knox
Lamar
Lamb
Lampasas
La Salle
Lavaca
Lee
Leon
Liberty
Limestone
Lipscomb
Live Oak
Llano
Loving
Lubbock
Lynn
McCulloch
McLennan
McMullen
Madison
Marion
Martin
Mason
Matagorda
Maverick
Medina
Menard
Midland

Milam
Mills
Mitchell
Montague
Montgomery
Moore
Morris
Motley
Nacogdoches
Navarro
Newton
Nolan
Nueces
Ochiltree
Oldham
Orange
Palo Pinto
Panola
Parker
Parmer
Pecos
Polk
Potter
Presidio
Rains
Randall
Reagan
Real
Red River
Reeves
Refugio
Roberts
Robertson
Rockwall
Runnels
Rusk
Sabine
San Augustine
San Jacinto
San Patricio
San Saba
Schleicher
Scurry
Shackelford
Shelby
Sherman
Smith
Somervell
Starr
Stephens
Sterling
Stonewall
Sutton
Swisher
Tarrant
Taylor
Terrell
Throckmorton
Titus
Tom Green
Travis
Trinity
Tyler
Upshur
Upton
Uvalde
Val Verde
Van Zandt
Victoria
Walker
Waller
Ward
Washington
Webb
Wharton
Wheeler
Wichita
Wilbarger
Willacy

Williamson
Wilson
Winkler
Wise
Wood
Yoakum
Young
Zapata
Zavala

Utah (29 counties)

Beaver
Boxelder
Cache
Carbon
Daggett
Davis
Duchesne
Emery
Garfield
Grand
Iron
Juab
Kane
Millard
Morgan
Piute
Rich
Salt Lake
San Juan
Sanpete
Sevier
Summit
Tooele
Uintah
Utah
Wasatch
Washington
Wayne
Weber

Vermont (14 counties)

Addison
Bennington
Caledonia
Chittendon
Essex
Franklin
Grand Isle
Lamoille
Orange
Orleans
Rutland
Washington
Windham
Windsor

Virginia (95 counties)

Accomack
Albemarle
Alleghany
Amelia
Amherst
Appomattox
Arlington
Augusta
Bath
Bedford
Bland
Botetourt
Brunswick
Buchanan
Buckingham
Campbell
Caroline
Carroll
Charles City
Charlotte
Chesterfield

Clarke
Craig
Culpeper
Cumberland
Dickenson
Dinwiddie
Essex
Fairfax
Fauquier
Floyd
Fluvanna
Franklin
Frederick
Giles
Gloucester
Goochland
Grayson
Greene
Greensville
Halifax
Hanover
Henrico
Henry
Highland
Isle of Wight
James City
King and Queen
King George
King William
Lancaster
Lee
Loudoun
Louisa
Lunenberg
Madison
Mathews
Mecklenburg
Middlesex
Montgomery
Nelson
New Kent
Northampton
Northumberland
Nottoway
Orange
Page
Patrick
Pittsylvania
Powhatan
Prince Edward
Prince George
Prince William
Pulaski
Rappahannock
Richmond
Roanoke
Rockbridge
Rockingham
Russell
Scott
Shenandoah
Smyth
Southampton
Spotsylvania
Stafford
Surry
Sussex
Tazewell
Warren
Washington
Westmoreland
Wise
Wythe
York

Washington (39 counties)

Adams
Asotin
Benton

Chelan
Clallam
Clark
Columbia
Cowlitz
Douglas
Ferry
Franklin
Garfield
Grant
Grays Harbor
Island
Jefferson
King
Kitsap
Kittitas
Klickitat
Lewis
Lincoln
Mason
Okanogan
Pacific
Pend Oreille
Pierce
San Juan
Skagit
Skamania
Snohomish
Spokane
Stevens
Thurston
Wahkiakum
Walla Walla
Whatcom
Whitman
Yakima

West Virginia (55 counties)

Barbour
Berkeley
Boone
Braxton
Brooke
Cabell
Calhoun
Clay
Doddridge
Fayette
Gilmer
Grant
Greenbrier
Hampshire
Hancock
Hardy
Harrison
Jackson
Jefferson
Kanawha
Lewis
Lincoln
Logan
McDowell
Marion
Marshall
Mason
Mercer
Mineral
Mingo
Monongalia
Monroe
Morgan
Nicholas
Ohio
Pendleton
Pleasants
Pocahontas
Preston
Putnam
Raleigh

Randolph
Ritchie
Roane
Summers
Taylor
Tucker
Tyler
Upshur
Wayne
Webster
Wetzel
Wirt
Wood
Wyoming

Wisconsin (72 counties)

Adams
Ashland
Barron
Bayfield
Brown
Buffalo
Burnett
Calumet
Chippewa
Clark
Columbia
Crawford
Dane
Dodge
Door
Douglas
Dunn
Eau Claire
Florence
Fond du Lac
Forest
Grant
Green
Green Lake
Iowa
Iron
Jackson
Jefferson
Juneau
Kenosha
Kewaunee
La Crosse
Lafayette
Langlade
Lincoln
Manitowoc
Marathon
Marinette
Marquette
Menominee
Milwaukee
Monroe
Oconto
Oneida
Outagamie
Ozaukee
Pepin
Pierce
Polk
Portage
Price
Racine
Richland
Rock
Rusk
St. Croix
Sauk
Sawyer
Shawano
Sheboygan
Taylor
Trempealeau
Vernon

Vilas	**Wyoming** (23 counties)	Hot Springs	Sublette
Walworth	Albany	Johnson	Sweetwater
Washburn	Big Horn	Laramie	Teton
Washington	Campbell	Lincoln	Uinta
Waukesha	Carbon	Natrona	Washakie
Waupaca	Converse	Niobrara	Weston
Waushara	Crook	Park	
Winnebago	Fremont	Platte	
Wood	Goshen	Sheridan	**Total US Counties = 3076**

Notes: Since USA-CA started, counties that have been absorbed include Princess Ann, Norfolk and Nansemond in Virginia; Ormsby in Nevada; Washabaugh in South Dakota. Carson City, Nevada, is now considered an independent city. Richmond County in New York is now called Staten Island County. A new county has been added to New Mexico—Cibola (used to be part of Valencia County).

The following is a list of independent cities and the county for which each can be used.

Independent City   Counts as_____ County

[VIRGINIA]

Alexandria	Arlington or Fairfax
Bedford	Bedford
Bristol	Washington
Buena Vista	Rockbridge
Charlottesville	Albermarle
Chesapeake	Isle of Wight
Clifton Forge	Alleghany
Colonial Heights	Chesterfield or Prince George
Covington	Alleghany
Danville	Pittsylvania
Emporia	Greensville
Fairfax	Fairfax
Falls Church	Fairfax
Fort Monroe	York
Franklin	Southampton
Fredericksburg	Spotsylvania
Galax	Carroll or Grayson
Hampton	York
Harrisonburg	Rockingham
Hopewell	Prince George
Lexington	Rockbridge
Lynchburg	Amherst or Bedford or Campbell
Manassas	Prince William
Manassas Park	Prince William
Martinsville	Henry
Newport News	York
Norfolk	Isle of Wight
Norton	Wise
Petersburg	Chesterfield or Dinwiddie or Prince George
Poquoson	York
Portsmouth	Isle of Wight
Radford	Montgomery
Richmond	Chesterfield or Henrico
Roanoke	Roanoke
Salem	Roanoke
South Boston	Halifax
Staunton	Augusta
Suffolk	Isle of Wight or Southampton
Virginia Beach	Isle of Wight
Waynesboro	Augusta
Williamsburg	James City
Winchester	Frederick

Carson City, Nevada	Douglas or Lyon or Story or Washoe
Washington, DC	Montgomery or Prince Georges, Maryland

Worldwide UTC Sunrise and Sunset Times

The following pages contain computer-generated tables of sunrise and sunset times at various locations throughout the world for the 7th and 21st day of each month. Times are in Coordinated Universal Time (UTC). The calculations are based on the latitude and longitude coordinates in Chapter 4 and should be accurate to a minute or two.

These calculations are based on the formulas provided in *DXing by Computer,* by Van Brollini, NS6N, and Walter Buchanan, published in *Ham Radio,* August 1984.

Sunrise and Sunset, in UTC, for 07-Jan

Pref	State/Province/Country/City	Sunrise	Sunset
VE1	New Brunswick, Saint John	12:03	20:56
	Nova Scotia, Halifax	11:51	20:49
	Prince Edward Island,		
	Charlottetown	11:54	20:41
VE2	Quebec, Montreal	12:34	21:26
	Quebec City	12:29	21:11
VE3	Ontario, London	12:56	22:05
	Ottawa	12:42	21:34
	Sudbury	13:07	21:52
	Toronto	12:51	21:55
VE4	Manitoba, Winnipeg	14:25	22:42
VE5	Saskatchewan, Regina	14:58	23:10
	Saskatoon	15:14	23:10
VE6	Alberta, Calgary	15:38	23:45
	Edmonton	15:48	23:30
VE7	British Columbia, Prince George	16:28	00:05
	Prince Rupert	17:00	00:33
	Vancouver	16:07	00:29
VE8	Northwest Territories,		
	Yellowknife	17:01	22:24
	Resolute	24 hrs night	
VY1	Yukon, Whitehorse	18:05	00:06
VO1	Newfoundland, St. John's	11:18	19:54
VO2	Labrador, Goose Bay	12:15	19:59
W1	Connecticut, Hartford	12:18	21:34
	Maine, Bangor	12:12	21:09
	Portland	12:14	21:19
	Massachusetts, Boston	12:13	21:26
	New Hampshire, Concord	12:18	21:25
	Rhode Island, Providence	12:13	21:29
	Vermont, Montpelier	12:26	21:26
W2	New Jersey, Atlantic City	12:18	21:48
	New York, Albany	12:25	21:36
	Buffalo	12:46	21:56
	New York City	12:20	21:43
	Syracuse	12:36	21:44
W3	Delaware, Wilmington	12:23	21:52
	District of Columbia,		
	Washington	12:27	22:00
	Maryland, Baltimore	12:26	21:57
	Pennsylvania, Harrisburg	12:30	21:56
	Philadelphia	12:22	21:50
	Pittsburgh	12:43	22:08
	Scranton	12:29	21:48
W4	Alabama, Montgomery	12:47	22:54
	Florida, Jacksonville	12:24	22:40
	Miami	12:08	22:44
	Pensacola	12:46	23:02
	Georgia, Atlanta	12:43	22:43
	Savannah	12:26	22:34
	Kentucky, Lexington	12:54	22:33
	Louisville	13:00	22:37
	North Carolina, Charlotte	12:32	22:25
	Raleigh	12:25	22:15
	Wilmington	12:18	22:16
	South Carolina, Columbia	12:30	22:29
	Tennessee, Knoxville	12:46	22:35
	Memphis	13:09	23:03
	Nashville	12:58	22:47
	Virginia, Norfolk	12:18	22:03
	Richmond	12:24	22:06
W5	Arkansas, Little Rock	13:17	23:12
	Louisiana, New Orleans	12:57	23:15
	Shreveport	13:17	23:23
	Mississippi, Jackson	13:02	23:10
	New Mexico, Albuquerque	14:15	00:09
	Oklahoma, Oklahoma City	13:39	23:31

Pref	State/Province/Country/City	Sunrise	Sunset
	Texas, Abilene	13:41	23:47
	Amarillo	13:56	23:49
	Dallas	13:30	23:35
	El Paso	14:06	00:16
	San Antonio	13:29	23:50
W6	California, Los Angeles	14:59	00:57
	San Francisco	15:25	01:05
W7	Arizona, Flagstaff	14:35	00:29
	Phoenix	14:33	00:35
	Idaho, Boise	15:18	00:23
	Pocatello	15:01	00:10
	Montana, Billings	14:54	23:44
	Butte	15:11	23:59
	Great Falls	15:12	23:49
	Nevada, Las Vegas	14:52	00:40
	Reno	15:19	00:50
	Oregon, Portland	15:50	00:42
	Utah, Salt Lake City	14:52	00:14
	Washington, Seattle	15:56	00:33
	Spokane	15:37	00:13
	Wyoming, Cheyenne	14:24	23:45
	Sheridan	14:45	23:42
W8	Michigan, Detroit	13:01	22:14
	Grand Rapids	13:14	22:23
	Sault Ste. Marie	13:21	22:05
	Traverse City	13:19	22:16
	Ohio, Cincinnati	12:57	22:30
	Cleveland	12:53	22:11
	Columbus	12:54	22:21
	West Virginia, Charleston	12:44	22:20
W9	Illinois, Chicago	13:18	22:34
	Indiana, Indianapolis	13:06	22:34
	Wisconsin, Green Bay	13:28	22:27
	Milwaukee	13:22	22:31
W0	Colorado, Denver	14:21	23:50
	Grand Junction	14:33	00:06
	Iowa, Des Moines	13:41	22:59
	Kansas, Pratt	13:50	23:30
	Wichita	13:44	23:25
	Minnesota, Duluth	13:53	22:35
	Minneapolis	13:51	22:46
	Missouri, Columbia	13:28	23:01
	Kansas City	13:37	23:10
	St. Louis	13:18	22:54
	Nebraska, North Platte	14:08	23:29
	Omaha	13:49	23:09
	North Dakota, Fargo	14:12	22:53
	South Dakota, Rapid City	14:27	23:29
A2	Botswana, Gaborone	03:33	17:10
A3	Tonga, Nukualofa	17:05	06:27
A4	Oman, Muscat	02:49	13:33
A5	Bhutan, Thimbu	00:53	11:22
A6	United Arab Emirates, Abu		
	Dhabi	03:08	13:49
A7	Qatar, Ad-Dawhah	03:20	13:58
A9	Bahrein, Al-Manamah	03:26	14:00
AP	Pakistan, Karachi	02:17	12:57
	Islamabad	02:12	12:13
BV	Taiwan, Taipei	22:40	09:19
BY	People's Rep. of China, Beijing	23:36	09:03
	Harbin	23:14	08:03
	Shanghai	22:53	09:06
	Fuzhou	22:51	09:25
	Xian	23:51	09:49
	Chongqing	23:50	10:09
	Chengdu	00:02	10:16

Pref	State/Province/Country/City	Sunrise	Sunset
	Lhasa	00:51	11:10
	Urumqi	01:43	10:47
	Kashi	02:16	11:46
C2	Nauru	18:53	07:02
C3	Andorra	07:23	16:35
C5	The Gambia, Banjul	07:31	18:53
C6	Bahamas, Nassau	11:56	22:34
C9	Mozambique, Maputo	03:04	16:46
	Mozambique	02:53	15:52
CE	Chile, Santiago	09:40	23:57
CE0	Easter Island	12:30	02:16
CE0	San Felix	10:34	00:17
CE0	Juan Fernandez	10:12	00:29
CM,CO	Cuba, Havana	12:12	22:58
CN	Morocco, Casablanca	07:35	17:36
CP	Bolivia, La Paz	10:07	23:11
CT	Portugal, Lisbon	07:55	17:29
CT2, CU	Azores , Ponta Delgada	08:58	18:38
CT3	Madeira Islands, Funchal	08:10	18:16
CX	Uruguay, Montevideo	08:38	23:02
D2	Angola, Luanda	04:54	17:31
D4	Cape Verde, Praia	08:01	19:18
D6	Comoros, Moroni	02:47	15:34
DA-DL	Germany, Fed Rep of (W), Bonn	07:32	15:42
DU	Philippines, Manilla	22:22	09:40
EA	Spain, Madrid	07:38	17:03
EA6	Balearic Is., Palma	07:10	16:40
EA8	Canary Is., Las Palmas	07:50	18:15
EA9	Ceuta & Melilla, Ceuta	07:32	17:21
	Melilla	07:21	17:14
EI	Ireland, Dublin	08:38	16:23
EL	Liberia, Monrovia	06:55	18:42
EP	Iran, Tehran	03:43	13:33
ET	Ethiopia, Addis Ababa	03:42	15:19
	Asmera	03:52	15:07
F	France, Paris	07:43	16:09
FG	Guadeloupe	10:36	21:49
FG, FS	St. Martin	10:45	21:50
FH	Mayotte	02:38	15:30
FK	New Caledonia, Noumia	18:16	07:42
FM	Martinique	10:30	21:48
FO	Clipperton	13:36	01:08
FO	Fr. Polynesia, Tahiti	16:09	05:17
FP	St. Pierre & Miquelon, St. Pierre	11:28	20:11
FR	Glorioso	02:33	15:19
FR	Juan de Nova	02:41	15:47
	Europa	02:41	16:07
FR	Reunion	01:42	15:04
FR	Tromelin	01:57	14:59
FT8W	Crozet	00:50	16:25
FT8X	Kerguelen	23:27	15:30
FT8Y	Antarctica, Dumont D'Urville	15:03	14:28
FT8Z	Amsterdam & St. Paul Is.,		
	Amsterdam	23:35	14:14
FW	Wallis & Futuna Is., Wallis	17:24	06:17
FY	Fr. Guiana, Cayenne	09:39	21:30
G	England, London	08:05	16:07
GD	Isle of Man	08:37	16:10
GI	Northern Ireland, Belfast	08:44	16:14
GJ	Jersey	08:03	16:25
GM	Scotland, Glasgow	08:45	16:00
	Aberdeen	08:45	15:42
GU	Guernsey	08:06	16:26
GW	Wales, Cardiff	08:17	16:19
H4	Solomon Islands, Honiara	19:06	07:45
HA	Hungary, Budapest	06:30	15:07
HB	Switzerland, Bern	07:15	15:56
HB0	Liechtenstein	07:07	15:46
HC	Ecuador, Quito	11:13	23:21
HC8	Galapagos Is.	12:03	00:12
HH	Haiti, Port-Au-Prince	11:23	22:26
HI	Dominican Republic, Santo		
	Domingo	11:13	22:17
HK	Colombia, Bogota	11:06	22:58
HK0	Malpelo Is.	11:33	23:27
HK0	San Adreas	11:49	23:15
HL, HM	Korea, Seoul	22:47	08:28
HP	Panama, Panama	11:35	23:12
HR	Honduras, Tegucigalpa	12:14	23:34
HS	Thailand, Bangkok	23:43	11:04
HV	Vatican City	06:37	15:53
HZ, 7Z	Saudi Arabia, Dharan	03:28	14:02
	Mecca	04:00	14:52
	Riyadh	03:38	14:19
I	Italy, Rome	06:37	15:53
	Trieste	06:45	15:35
	Sicily	06:19	16:00
IS	Sardinia, Cagliari	06:43	16:15
J2	Djibouti, Djibouti	03:28	14:57
J3	Grenada	10:29	21:56
J5	Guinea-Bissau, Bissau	07:24	18:51
J6	St. Lucia	10:29	21:49
J7	Dominica	10:33	21:48
J8	St. Vincent	10:29	21:52
JA-JS	Japan, Tokyo	21:51	07:41
	Nagasaki	22:23	08:28
	Sapporo	22:06	07:14
JD	Minami Torishima	20:28	07:10
JD	Ogasawara, Kazan Is.	21:27	07:56
JT	Mongolia, Ulan Bator	00:41	09:15
JW	Svalbard, Spitsbergen	24 hrs night	
JX	Jan Mayen	24 hrs night	
JY	Jordan, Amman	04:37	14:46
KC4	Antarctica, Byrd Station	24 hrs day	
	McMurdo Sound	24 hrs day	
	Palmer Station	05:55	02:48
KC6	Micronesia, Ponape	19:40	07:24
KC6	Belau, Yap	21:05	08:40
	Koror	21:16	08:59
KG4	Guantanamo Bay	11:37	22:35
KH1	Baker, Howland Is.	17:47	05:52
KH2	Guam, Agana	20:45	08:07
KH3	Johnston Is.	17:45	04:54
KH4	Midway Is.	18:42	05:08
KH5	Palmyra Is.	17:00	04:48
KH5K	Kingman Reef	17:05	04:48
KH6	Hawaii, Hilo	16:56	03:55
	Honolulu	17:10	04:04
KH7	Kure Is.	18:46	05:11
KH8	American Samoa, Pago Pago	17:00	05:56
KH9	Wake Is.	19:28	06:29
KH0	Mariana Is., Saipan	20:44	08:00
KL7	Alaska, Adak	19:52	03:51
	Anchorage	19:10	01:01
	Fairbanks	19:47	00:07
	Juneau	17:42	00:23
	Nome	20:52	01:22
KP1	Navassa Is.	11:33	22:37
KP2	Virgin Islands, Charlotte Amalie	10:53	21:57
KP4	Puerto Rico, San Juan	10:58	22:02
KP5	Desecheo Is.	11:03	22:08
KX6	Marshall Islands, Kwajalein	19:08	06:45
LA-LJ	Norway, Oslo	08:16	14:29
LU	Argentina, Buenos Aires	08:48	23:10
LX	Luxembourg	07:31	15:50
LZ	Bulgaria, Sofia	05:57	15:07
OA	Peru, Lima	10:50	23:38
OD	Lebanon, Beirut	04:43	14:43
OE	Austria, Vienna	06:44	15:16
OH	Finland, Helsinki	07:20	13:30
OH0	Aland Is.	07:40	13:50
OJ0	Market Reef	07:45	13:53
OK	Czechoslovakia, Prague	7:00	15:15
ON	Belgium, Brussels	07:44	15:52
OX, XP	Greenland, Godthaab	13:13	17:51
	Thule	24 hrs night	
OY	Faroe Islands, Torshavn	09:45	15:20
OZ	Denmark, Copenhagen	07:37	14:53
P2	Papua New Guinea, Madang	20:11	08:35
	Port Moresby	19:57	08:36
PA-PI	Netherlands, Amsterdam	07:49	15:42
PJ, P4	Netherlands Antilles, Willemstad	10:58	22:25
PJ5-8	St. Maarten and Saba,		
	St. Maarten	10:45	21:52
PY	Brazil, Brasilia	08:46	21:48
	Rio De Janeiro	08:14	21:43
	Natal	08:12	20:40
	Manaus	09:57	22:15
	Porto Alegre	08:30	22:30
PY0	Fernando De Noronha	08:05	20:25
PY0	St. Peter & St. Paul Rocks	08:01	20:05

Pref	State/Province/Country/City	Sunrise	Sunset	Pref	State/Province/Country/City	Sunrise	Sunset
PYØ	Trindade & Martin Vaz Is.,			VK	Australia, Canberra (VK1)	18:56	09:22
	Trindade	07:23	20:42		Sydney (VK2)	18:51	09:10
PZ	Suriname, Paramaribo	09:52	21:40		Melbourne (VK3)	19:05	09:45
S2	Bangladesh, Dacca	00:42	11:26		Brisbane (VK4)	18:59	08:47
S7	Seychelles, Victoria	02:12	14:35		Adelaide (VK5)	19:39	10:03
S9	Sao Tome	05:35	17:42		Perth (VK6)	21:18	11:26
SJ-SM	Sweden, Stockholm	07:40	14:06		Hobart, Tasmania (VK7)	18:40	09:52
SP	Poland, Krakow	06:37	14:53		Darwin (VK8)	20:57	09:46
	Warsaw	06:44	14:39	VK	Lord Howe Is.	18:25	08:32
ST	Sudan, Khartoum	04:18	15:32	VK9	Christmas Is.	22:41	11:24
STØ	Southern Sudan, Juba	04:04	15:54	VK9	Cocos-Keeling Is.	23:14	12:02
SU	Egypt, Cairo	04:51	15:09	VK9	Mellish Reef	19:08	08:16
SV	Greece, Athens	05:41	15:20	VK9	Norfolk Is.	07:56	07:51
SV5	Dodecanese, Rhodes	05:19	15:06	VK9	Willis Is.	19:36	08:39
SV9	Crete	05:28	15:21	VKØ	Heard Is.	22:51	15:32
SV/A	Mount Athos	05:45	15:11	VKØ	Macquarie Is.	16:59	10:01
T2	Tuvalu, Funafuti	17:53	06:29	VP2E	Anguilla	10:45	21:50
T3Ø	West Kiribati, Bonriki	18:31	06:34	VP2M	Montserrat	10:39	21:49
T31	Central Kiribati, Kanton	17:24	05:40	VP2V	British Virgin Is., Tortola	10:52	21:56
T32	East Kiribati, Christmas Is.	16:34	04:35	VP5	Turks & Caicos Islands,		
T5	Somalia, Mogadishu	03:04	15:04		Grand Turk	11:24	22:17
T7	San Marino	06:45	15:48	VP8	Falkland Islands, Stanley	07:43	00:10
TA	Turkey, Ankara	05:10	14:38	VP8	So. Georgia Is.	06:04	23:01
	Istanbul	05:29	14:50	VP8	So. Orkney Is.	05:47	00:28
TF	Iceland, Reykjavik	11:13	15:54	VP8	So. Sandwich Islands, Saunders		
TG	Guatemala, Guatemala City	12:28	23:46		Is.	04:58	22:46
TI	Costa Rica, San Jose	11:54	23:28	VP8	So. Shetland Is., King George		
TI9	Cocos Is.	11:59	23:48		Is.	06:21	01:36
TJ	Cameroon, Yaounde	05:22	17:16	VP9	Bermuda	11:20	21:28
TK	Corsica	06:52	16:07	VQ9	Chagos, Diego Garcia	01:00	13:32
TL	Central African Republic, Bangui	04:55	16:47	VR6	Pitcairn Is.	13:57	03:35
TN	Congo, Brazzaville	04:53	17:15	VS6	Hong Kong	23:03	09:53
TR	Gabon, Libreville	05:24	17:30	VU	India, Bombay	01:43	12:45
TT	Chad, N'Djamena	05:22	16:49		Calcutta	00:47	11:36
TU	Ivory Coast, Abidjan	06:26	18:16		New Delhi	01:45	12:08
TY	Benin, Porto Novo	06:02	17:48		Bangalore	01:13	12:37
TZ	Mali, Bamako	06:55	18:20	VU7	Andaman Islands, Port Blair	00:10	11:38
UA	Russia, European,			VU7	Laccadive Is.	01:26	13:00
	Leningrad (UA1)	06:57	13:12	XE	Mexico, Mexico City (XE1)	13:11	00:12
	Archangel (UA1)	07:10	11:36		Chihuahua (XE2)	13:58	00:21
	Murmansk (UA1)	24 hrs	night		Merida (XE3)	12:37	23:31
	Moscow (UA3)	05:57	13:13	XF4	Revilla Gigedo	14:00	01:02
	Kuibyshev (UA4)	04:52	12:38	XT	Burkina Faso, Onagadougou	06:29	17:54
	Rostov (UA6)	05:09	13:46	XU	Kampuchea, Phnom Penh	23:22	10:50
UA1P	Franz Josef Land	24 hrs	night	XW	Laos, Viangchan	23:42	10:48
UA2	Kaliningrad	07:01	14:26	XX9	Macao	23:06	09:56
UA9,Ø	Russia, Asiatic,			XZ	Burma, Rangoon	00:06	11:16
	Novosibirsk (UA9)	02:51	10:16	Y2-9	German Dem.Rep.(E), Berlin	07:15	15:08
	Perm (UA9)	04:57	11:43	YA	Afghanistan, Kandahar	02:35	12:49
	Omsk (UA9)	03:29	10:54		Kabul	02:30	12:27
	Norilsk (UAØ)	24 hrs	night	YB-YD	Indonesia, Jakarta	22:44	11:12
	Irkutsk (UAØ)	01:11	09:05		Medan	23:33	11:28
	Vladivostok (UAØ)	22:44	07:51		Pontianak	22:44	10:52
	Petropavlovsk (UAØ)	21:37	05:24		Jayapura	20:35	08:50
	Khabarovsk (UAØ)	22:50	07:19	YI	Iraq, Baghdad	04:05	14:09
	Krasnoyarsk (UAØ)	02:18	09:31	YJ	Vanuatu, Port Vila	18:18	07:26
	Yakutsk (UAØ)	00:39	06:14	YK	Syria, Damascus	04:39	14:41
	Wrangel Island (UAØ)	24 hrs	night	YN, HT	Nicaragua, Managua	12:06	23:33
	Kyzyl (UAØY)	01:47	09:47	YO	Romania, Bucharest	05:51	14:51
UB	Ukraine, Kiev	05:57	14:10	YS	El Salvador, San Salvador	12:22	23:43
UC	Byelorussia, Minsk	06:26	14:04	YU	Yugoslavia, Belgrade	06:15	15:11
UD	Azerbaijan, Baku	04:03	13:28	YV	Venezuela, Caracas	10:47	22:19
UF	Georgia, Tbilisi	04:28	13:45	YVØ	Aves Is.	10:43	21:57
UG	Armenia, Ierevan	04:25	13:50	Z2,ZE	Zimbabwe, Harare	03:27	16:36
UH	Turkoman, Ashkhabad	03:22	13:01	ZA	Albania, Tirane	06:06	15:26
UI	Uzbek, Bukhara	03:03	12:32	ZB	Gibraltar	07:33	17:21
	Tashkent	02:48	12:08	ZC4	British Cyprus	04:55	14:51
UJ	Tadzhik, Samarkand	02:54	12:23	ZD7	St. Helena	05:58	19:00
	Dushanbe	02:44	12:16	ZD8	Ascension Is.	06:46	19:20
UL	Kazakh, Alma-ata	02:24	11:31	ZD9	Tristan da Cunha	05:37	20:12
UM	Kirghiz, Frunze	02:32	11:42	ZF	Cayman Is.	12:00	23:00
UO	Moldavia, Kishinev	05:50	14:30	ZK1	No. Cook Is., Manihiki	16:28	05:10
UP	Lithuania, Vilna	06:38	14:09	ZK1	So. Cook Is., Rarotonga	16:04	05:25
UQ	Latvia, Riga	06:59	13:59	ZK2	Niue	16:44	05:58
UR	Estonia, Tallinn	07:14	13:38	ZK3	Tokelaus, Atafu	17:18	05:54
V2	Antigua & Barbuda, St. Johns	10:38	21:47	ZL	New Zealand, Auckland (ZL1)	17:09	07:43
V3	Belize, Belmopan	12:26	23:35		Wellington (ZL2)	16:56	07:57
V4	St. Christopher & Nevis	10:42	21:50		Christchurch (ZL3)	16:57	08:13
V8	Brunei, Bandar Seri Begawan	22:30	10:21		Dunedin (ZL4)	16:56	08:30
VE1, CY	Sable Is.	11:34	20:37	ZL5	Antarctica, Scott Base	24 hrs	day
VE1, CY	St. Paul Is.	11:46	20:25	ZL7	Chatham Is.	16:11	07:31

Pref	State/Province/Country/City	Sunrise	Sunset
ZL8	Kermadec Is.	16:59	06:55
ZL9	Auckland & Campbell Is.,		
	Auckland	16:51	09:08
	Campbell Is.	16:31	09:07
ZP	Paraguay, Asuncion	09:07	22:45
ZS	South Africa, Cape Town (ZS1)	03:42	18:01
	Port Elizabeth (ZS2)	03:13	17:32
	Bloemfontein (ZS4)	03:23	17:19
	Durban (ZS5)	03:02	17:01
	Johannesburg (ZS6)	03:22	17:04
ZS2	Prince Edward & Marion Is.,		
	Marion	01:43	17:25
ZS3	Namibia, Windhoek	04:13	17:41
1A0	S.M.O.M	06:38	15:54
1S	Spratly Is.	22:49	10:27
3A	Monaco	07:04	16:08
3B6	Agalega	01:58	14:40
3B7	St. Brandon	01:34	14:38
3B8	Mauritius	01:36	14:54
3B9	Rodriguez Is.	01:14	14:30
3C	Equatorial Guinea, Bata	05:25	17:27
	Malabo	05:33	17:27
3C0	Pagalu Is.	05:37	17:49
3D2	Fiji Is., Suva	17:37	06:47
3D6	Swaziland, Mbabane	03:10	16:52
3V	Tunisia, Tunis	06:32	16:17
3W	Vietnam, Ho Chi Minh City		
	(Saigon)	23:13	10:44
	Hanoi	23:35	10:29
3X	Republic of Guinea, Conakry	07:12	18:48
3Y	Bouvet	03:22	20:21
3Y	Peter I Is.	24 hrs day	
4S	Sri Lanka, Colombo	00:54	12:38
4U	ITU Geneva	07:17	16:04
4U	United Nations Hq.	12:20	21:43
4W	Yemen, Sanaa	03:31	14:46
4X, 4Z	Israel, Jerusalem	04:40	14:49
5A	Libya, Tripoli	06:12	16:18
	Benghazi	05:41	15:49
5B	Cyprus, Nicosia	04:55	14:48

Pref	State/Province/Country/City	Sunrise	Sunset
5H	Tanzania, Dar es Salaam	3:12	15:43
5N	Nigeria, Lagos	05:59	17:45
5R	Madagascar, Antananarivo	02:19	15:32
5T	Mauritania, Nouakchott	07:37	18:42
5U	Niger, Niamey	06:16	17:38
5V	Togo, Lome	06:07	17:54
5W	Western Samoa, Apia	17:06	05:59
5X	Uganda, Kampala	03:52	15:58
5Z	Kenya, Nairobi	04:52	17:04
6W	Senegal, Dakar	07:37	18:54
6Y	Jamaica, Kingston	11:40	22:45
7O	Yemen, People's Dem. Rep.,		
	Aden	03:23	14:47
7P	Lesotho, Maseru	03:17	17:13
7Q	Malawi, Lilongwe	03:23	16:18
	Blantyre	03:15	16:16
7X	Algeria, Algiers	07:01	16:46
8P	Barbados, Bridgetown	10:22	21:45
8Q	Maldive Is.	01:15	13:08
8R	Guyana, Georgetown	10:06	21:50
9G	Ghana, Accra	06:12	18:01
9H	Malta	06:13	16:02
9J	Zambia, Lusaka	03:42	16:42
9K	Kuwait	03:44	14:04
9L	Sierra Leone, Freetown	07:09	18:48
9M2	West Malaysia, Kuala Lumpur	23:21	11:17
9M6,8	East Malaysia, Sabah,		
	Sandakan (9M6)	22:19	10:07
	Sarawak, Kuching (9M8)	22:43	10:45
9N	Nepal, Katmandu	01:10	11:38
9Q	Zaire, Kinshasa	04:53	17:15
	Kisangani	04:22	16:27
	Lubumbashi	03:52	16:39
9U	Burundi, Bujumbura	03:59	16:17
9V	Singapore	23:09	11:12
9X	Rwanda, Kigali	03:58	16:12
9Y	Trinidad & Tobago, Port of		
	Spain	10:24	21:57
J2/A	Abu Ail	03:34	14:54

Pref	State/Province/Country/City	Sunrise	Sunset
VE1	New Brunswick, Saint John	11:57	21:13
	Nova Scotia, Halifax	11:44	21:05
	Prince Edward Island,		
	Charlottetown	11:47	20:58
VE2	Quebec, Montreal	12:27	21:43
	Quebec City	12:22	21:29
VE3	Ontario, London	12:51	22:21
	Ottawa	12:35	21:51
	Sudbury	13:00	22:09
	Toronto	12:45	22:11
VE4	Manitoba, Winnipeg	14:16	23:02
VE5	Saskatchewan, Regina	14:48	23:29
	Saskatoon	15:03	23:31
VE6	Alberta, Calgary	15:28	00:05
	Edmonton	15:37	23:52
VE7	British Columbia, Prince George	16:16	00:28
	Prince Rupert	16:47	00:56
	Vancouver	15:58	00:48
VE8	Northwest Territories,		
	Yellowknife	16:37	23:00
	Resolute	24 hrs night	
VY1	Yukon, Whitehorse	17:45	00:37
VO1	Newfoundland, St. John's	11:10	20:12
VO2	Labrador, Goose Bay	12:03	20:21
W1	Connecticut, Hartford	12:13	21:50
	Maine, Bangor	12:06	21:26
	Portland	12:09	21:35
	Massachusetts, Boston	12:08	21:42
	New Hampshire, Concord	12:12	21:41
	Rhode Island, Providence	12:08	21:45
	Vermont, Montpelier	12:20	21:42
W2	New Jersey, Atlantic City	12:13	22:03
	New York, Albany	12:20	21:52
	Buffalo	12:41	22:12
	New York City	12:15	21:58
	Syracuse	12:30	22:00
W3	Delaware, Wilmington	12:19	22:06
	District of Columbia,		
	Washington	12:23	22:14
	Maryland, Baltimore	12:22	22:12
	Pennsylvania, Harrisburg	12:26	22:11
	Philadelphia	12:18	22:05
	Pittsburgh	12:38	22:23
	Scranton	12:24	22:03
W4	Alabama, Montgomery	12:45	23:06
	Florida, Jacksonville	12:23	22:52
	Miami	12:08	22:54
	Pensacola	12:45	23:14
	Georgia, Atlanta	12:41	22:56
	Savannah	12:24	22:46
	Kentucky, Lexington	12:50	22:47
	Louisville	12:56	22:51
	North Carolina, Charlotte	12:29	22:38
	Raleigh	12:22	22:28
	Wilmington	12:15	22:29
	South Carolina, Columbia	12:27	22:42
	Tennessee, Knoxville	12:43	22:49
	Memphis	13:06	23:16
	Nashville	12:55	23:00
	Virginia, Norfolk	12:15	22:16
	Richmond	12:21	22:19
W5	Arkansas, Little Rock	13:14	23:25
	Louisiana, New Orleans	12:55	23:26
	Shreveport	13:15	23:36
	Mississippi, Jackson	13:01	23:22
	New Mexico, Albuquerque	14:13	00:22
	Oklahoma, Oklahoma City	13:37	23:44
	Texas, Abilene	13:39	00:00
	Amarillo	13:53	00:02
	Dallas	13:28	23:47
	El Paso	14:05	00:28
	San Antonio	13:28	00:01
W6	California, Los Angeles	14:56	01:10
	San Francisco	15:22	01:19
W7	Arizona, Flagstaff	14:33	00:42
	Phoenix	14:31	00:47
	Idaho, Boise	15:12	00:39
	Pocatello	14:55	00:26

Pref	State/Province/Country/City	Sunrise	Sunset
	Montana, Billings	14:48	00:01
	Butte	15:04	00:17
	Great Falls	15:04	00:07
	Nevada, Las Vegas	14:49	00:53
	Reno	15:15	01:04
	Oregon, Portland	15:44	00:59
	Utah, Salt Lake City	14:47	00:29
	Washington, Seattle	15:49	00:51
	Spokane	15:29	00:31
	Wyoming, Cheyenne	14:19	00:00
	Sheridan	14:39	23:58
W8	Michigan, Detroit	12:55	22:30
	Grand Rapids	13:08	22:38
	Sault Ste. Marie	13:14	22:23
	Traverse City	13:13	22:33
	Ohio, Cincinnati	12:53	22:44
	Cleveland	12:48	22:27
	Columbus	12:49	22:36
	West Virginia, Charleston	12:40	22:34
W9	Illinois, Chicago	13:13	22:49
	Indiana, Indianapolis	13:02	22:49
	Wisconsin, Green Bay	13:22	22:43
	Milwaukee	13:17	22:47
W0	Colorado, Denver	14:17	00:04
	Grand Junction	14:30	00:20
	Iowa, Des Moines	13:36	23:14
	Kansas, Pratt	13:47	23:44
	Wichita	13:41	23:39
	Minnesota, Duluth	13:45	22:53
	Minneapolis	13:44	23:03
	Missouri, Columbia	13:24	23:15
	Kansas City	13:34	23:24
	St. Louis	13:15	23:08
	Nebraska, North Platte	14:03	23:44
	Omaha	13:44	23:24
	North Dakota, Fargo	14:04	23:11
	South Dakota, Rapid City	14:21	23:45
A2	Botswana, Gaborone	03:44	17:10
A3	Tonga, Nukualofa	17:15	06:28
A4	Oman, Muscat	02:49	13:43
A5	Bhutan, Thimbu	00:53	11:33
A6	United Arab Emirates, Abu		
	Dhabi	03:08	13:59
A7	Qatar, Ad-Dawhah	03:20	14:09
A9	Bahrein, Al-Manamah	03:26	14:11
AP	Pakistan, Karachi	02:17	13:07
	Islamabad	02:10	12:25
BV	Taiwan, Taipei	22:40	09:29
BY	People's Rep. of China, Beijing	23:32	09:18
	Harbin	23:07	08:21
	Shanghai	22:52	09:17
	Fuzhou	22:51	09:36
	Xian	23:48	10:01
	Chongqing	23:49	10:20
	Chengdu	00:00	10:28
	Lhasa	00:50	11:22
	Urumqi	01:37	11:03
	Kashi	02:12	12:01
C2	Nauru	18:59	07:07
C3	Andorra	07:18	16:51
C5	The Gambia, Banjul	07:34	19:01
C6	Bahamas, Nassau	11:56	22:45
C9	Mozambique, Maputo	03:15	16:45
	Mozambique	03:01	15:54
CE	Chile, Santiago	09:53	23:54
CE0	Easter Island	12:41	02:15
CE0	San Felix	10:45	00:17
CE0	Juan Fernandez	10:25	00:27
CM,CO	Cuba, Havana	12:12	23:08
CN	Morocco, Casablanca	07:33	17:48
CP	Bolivia, La Paz	10:15	23:13
CT	Portugal, Lisbon	07:51	17:44
CT2, CU	Azores, Ponta Delgada	08:55	18:52
CT3	Madeira Islands, Funchal	08:08	18:28
CX	Uruguay, Montevideo	08:52	22:59
D2	Angola, Luanda	05:01	17:34
D4	Cape Verde, Praia	08:03	19:26

Pref	State/Province/Country/City	Sunrise	Sunset
D6	Comoros, Moroni	02:55	15:37
DA-DL	Germany, Fed. Rep. of (W),		
	Bonn	07:23	16:02
DU	Philippines, Manilla	22:25	09:48
EA	Spain, Madrid	07:33	17:17
EA6	Balearic Is., Palma	07:06	16:55
EA8	Canary Is., Las Palmas	07:49	18:26
EA9	Ceuta & Melilla, Ceuta	07:29	17:35
	Melilla	07:18	17:27
EI	Ireland, Dublin	08:27	16:45
EL	Liberia, Monrovia	06:59	18:48
EP	Iran, Tehran	03:40	13:46
ET	Ethiopia, Addis Ababa	03:45	15:26
	Asmera	03:54	15:16
F	France, Paris	07:34	16:28
FG	Guadeloupe	10:38	21:57
FG, FS	St. Martin	10:47	21:59
FH	Mayotte	02:46	15:32
FK	New Caledonia, Noumia	18:26	07:43
FM	Martinique	10:33	21:56
FO	Clipperton	13:39	01:16
FO	Fr. Polynesia, Tahiti	16:18	05:19
FP	St. Pierre & Miquelon, St. Pierre	11:21	20:28
FR	Glorioso	02:41	15:22
FR	Juan de Nova	02:50	15:49
	Europa	02:50	16:07
FR	Reunion	01:52	15:05
FR	Tromelin	02:05	15:01
FT8W	Crozet	01:08	16:17
FT8X	Kerguelen	23:47	15:21
FT8Y	Antarctica, Dumont D'Urville	16:41	13:00
FT8Z	Amsterdam & St. Paul Is.,		
	Amsterdam	23:50	14:11
FW	Wallis & Futuna Is., Wallis	17:32	06:19
FY	Fr. Guiana, Cayenne	09:43	21:36
G	England, London	07:54	16:27
GD	Isle of Man	08:24	16:33
GI	Northern Ireland, Belfast	08:31	16:37
GJ	Jersey	07:54	16:44
GM	Scotland, Glasgow	08:31	16:24
	Aberdeen	08:30	16:08
GU	Guernsey	07:57	16:46
GW	Wales, Cardiff	08:07	16:40
H4	Solomon Islands, Honiara	19:13	07:48
HA	Hungary, Budapest	06:23	15:25
HB	Switzerland, Bern	07:08	16:14
HB0	Liechtenstein	07:00	16:04
HC	Ecuador, Quito	11:19	23:26
HC8	Galapagos Is.	12:08	00:17
HH	Haiti, Port-Au-Prince	11:24	22:35
HI	Dominican Republic, Santo		
	Domingo	11:15	22:26
HK	Colombia, Bogota	11:10	23:04
HK0	Malpelo Is.	11:37	23:33
HK0	San Adreas	11:52	23:22
HL, HM	Korea, Seoul	22:43	08:42
HP	Panama, Panama	11:38	23:19
HR	Honduras, Tegucigalpa	12:17	23:42
HS	Thailand, Bangkok	23:46	11:12
HV	Vatican City	06:32	16:09
HZ, 7Z	Saudi Arabia, Dharan	03:28	14:13
	Mecca	04:01	15:02
	Riyadh	03:38	14:29
I	Italy, Rome	06:32	16:09
	Trieste	06:38	15:52
	Sicily	06:15	16:14
IS	Sardinia, Cagliari	06:39	16:29
J2	Djibouti, Djibouti	03:31	15:04
J3	Grenada	10:32	22:03
J5	Guinea-Bissau, Bissau	07:27	18:59
J6	St. Lucia	10:32	21:57
J7	Dominica	10:35	21:56
J8	St. Vincent	10:32	21:59
JA-JS	Japan, Tokyo	21:48	07:55
	Nagasaki	22:21	08:41
	Sapporo	22:00	07:30
JD	Minami Torishima	20:29	07:20
JD	Ogasawara, Kazan Is.	21:26	08:07
JT	Mongolia, Ulan Bator	00:33	09:33
JW	Svalbard, Spitsbergen	24 hrs night	
JX	Jan Mayen	24 hrs night	
JY	Jordan, Amman	04:36	14:58
KC4	Antarctica, Byrd Station	24 hrs day	
	McMurdo Sound	24 hrs day	
	Palmer Station	06:48	02:05
KC6	Micronesia, Ponape	19:44	07:31
KC6	Belau, Yap	21:08	08:47
	Koror	21:20	09:05
KG4	Guantanamo Bay	11:38	22:45
KH1	Baker, Howland Is.	17:52	05:57
KH2	Guam, Agana	20:48	08:15
KH3	Johnston Is.	17:47	05:02
KH4	Midway Is.	18:41	05:19
KH5	Palmyra Is.	17:04	04:54
KH5K	Kingman Reef	17:09	04:54
KH6	Hawaii, Hilo	16:57	04:05
	Honolulu	17:11	04:13
KH7	Kure Is.	18:46	05:22
KH8	American Samoa, Pago Pago	17:09	05:59
KH9	Wake Is.	19:30	06:38
KH0	Mariana Is., Saipan	20:47	08:08
KL7	Alaska, Adak	19:42	04:12
	Anchorage	18:48	01:33
	Fairbanks	19:14	00:50
	Juneau	17:26	00:51
	Nome	20:21	02:04
KP1	Navassa Is.	11:35	22:46
KP2	Virgin Islands, Charlotte Amalie	10:54	22:06
KP4	Puerto Rico, San Juan	11:00	22:11
KP5	Desecheo Is.	11:05	22:16
KX6	Marshall Islands, Kwajalein	19:11	06:51
LA-LJ	Norway, Oslo	07:56	14:59
LU	Argentina, Buenos Aires	09:01	23:07
LX	Luxembourg	07:22	16:10
LZ	Bulgaria, Sofia	05:51	15:23
OA	Peru, Lima	10:57	23:41
OD	Lebanon, Beirut	04:41	14:56
OE	Austria, Vienna	06:36	15:34
OH	Finland, Helsinki	07:01	14:00
OH0	Aland Is.	07:21	14:20
OJ0	Market Reef	07:25	14:24
OK	Czechoslovakia, Prague	06:51	15:35
ON	Belgium, Brussels	07:34	16:12
OX, XP	Greenland, Godthaab	12:43	18:32
	Thule	24 hrs night	
OY	Faroe Islands, Torshavn	09:22	15:53
OZ	Denmark, Copenhagen	07:23	15:17
P2	Papua New Guinea, Madang	20:17	08:39
	Port Moresby	20:05	08:40
PA-PI	Netherlands, Amsterdam	07:38	16:04
PJ, P4	Netherlands Antilles, Willemstad	11:01	22:32
PJ5-8	St. Maarten and Saba, St.		
	Maarten	10:46	22:00
PY	Brazil, Brasilia	08:55	21:50
	Rio De Janeiro	08:24	21:43
	Natal	08:19	20:44
	Manaus	10:03	22:19
	Porto Alegre	08:42	22:29
PY0	Fernando De Noronha	08:11	20:29
PY0	St. Peter & St. Paul Rocks	08:06	20:10
PY0	Trindade & Martin Vaz Is.,		
	Trindade	07:32	20:43
PZ	Suriname, Paramaribo	09:56	21:46
S2	Bangladesh, Dacca	00:42	11:36
S7	Seychelles, Victoria	02:18	14:39
S9	Sao Tome	05:41	17:47
SJ-SM	Sweden, Stockholm	07:22	14:34
SP	Poland, Krakow	06:28	15:13
	Warsaw	06:33	15:00
ST	Sudan, Khartoum	04:20	15:41
ST0	Southern Sudan, Juba	04:08	16:00
SU	Egypt, Cairo	04:50	15:20
SV	Greece, Athens	05:38	15:34
SV5	Dodecanese, Rhodes	05:16	15:20
SV9	Crete	05:26	15:34
SV/A	Mount Athos	05:41	15:26
T2	Tuvalu, Funafuti	18:00	06:33
T30	West Kiribati, Bonriki	18:36	06:39
T31	Central Kiribati, Kanton	17:30	05:45
T32	East Kiribati, Christmas Is.	16:39	04:41
T5	Somalia, Mogadishu	03:08	15:09
T7	San Marino	06:39	16:04

Pref	State/Province/Country/City	Sunrise	Sunset
TA	Turkey, Ankara	05:06	14:52
	Istanbul	05:25	15:05
TF	Iceland, Reykjavik	10:43	16:34
TG	Guatemala, Guatemala City	12:31	23:54
TI	Costa Rica, San Jose	11:58	23:35
TI9	Cocos Is.	12:03	23:54
TJ	Cameroon, Yaounde	05:27	17:22
TK	Corsica	06:47	16:22
TL	Central African Republic, Bangui	04:59	16:53
TN	Congo, Brazzaville	04:59	17:19
TR	Gabon, Libreville	05:30	17:35
TT	Chad, N'Djamena	05:25	16:56
TU	Ivory Coast, Abidjan	06:31	18:22
TY	Benin, Porto Novo	06:06	17:54
TZ	Mali, Bamako	06:58	18:27
UA	Russia, European,		
	Leningrad (UA1)	06:37	13:41
	Archangel (UA1)	06:38	12:19
	Murmansk (UA1)	08:21	11:36
	Moscow (UA3)	05:43	13:37
	Kuibyshev (UA4)	04:41	12:59
	Rostov (UA6)	05:01	14:04
UA1P	Franz Josef Land	24 hrs	night
UA2	Kaliningrad	06:48	14:49
UA9,0	Russia, Asiatic,		
	Novosibirsk (UA9)	02:38	10:40
	Perm (UA9)	04:41	12:10
	Omsk (UA9)	03:16	11:18
	Norilsk (UA0)	04:50	07:46
	Irkutsk (UA0)	01:00	09:27
	Vladivostok (UA0)	22:38	08:07
	Petropavlovsk (UA0)	21:26	05:46
	Khabarovsk (UA0)	22:42	07:38
	Krasnoyarsk (UA0)	02:04	09:55
	Yakutsk (UA0)	00:16	06:47
	Wrangel Island (UA0)	24 hrs	night
	Kyzyl (UA0Y)	01:37	10:08
UB	Ukraine, Kiev	05:48	14:29
UC	Byelorussia, Minsk	06:14	14:26
UD	Azerbaijan, Baku	03:59	13:43
UF	Georgia, Tbilisi	04:23	14:00
UG	Armenia, Ierevan	04:20	14:05
UH	Turkoman, Ashkhabad	03:19	13:15
UI	Uzbek, Bukhara	02:59	12:47
	Tashkent	02:43	12:23
UJ	Tadzhik, Samarkand	02:49	12:37
	Dushanbe	02:40	12:31
UL	Kazakh, Alma-ata	02:19	11:47
UM	Kirghiz, Frunze	02:27	11:58
UO	Moldavia, Kishinev	05:42	14:48
UP	Lithuania, Vilna	06:25	14:32
UQ	Latvia, Riga	06:44	14:24
UR	Estonia, Tallinn	06:56	14:07
V2	Antigua & Barbuda, St. Johns	10:40	21:56
V3	Belize, Belmopan	12:28	23:43
V4	St. Christopher & Nevis	10:43	21:58
V8	Brunei, Bandar Seri Begawan	22:35	10:27
VE1, CY	Sable Is.	11:28	20:53
VE1, CY	St. Paul Is.	11:39	20:43
VK	Australia, Canberra (VK1)	19:09	09:19
	Sydney (VK2)	19:04	09:07
	Melbourne (VK3)	19:20	09:41
	Brisbane (VK4)	19:10	08:47
	Adelaide (VK5)	19:52	10:00
	Perth (VK6)	21:30	11:24
	Hobart, Tasmania (VK7)	18:56	09:46
	Darwin (VK8)	21:05	09:49
VK	Lord Howe Is.	18:38	08:31
VK9	Christmas Is.	22:49	11:27
VK9	Cocos-Keeling Is.	23:21	12:05
VK9	Mellish Reef	19:17	08:18
VK9	Norfolk Is.	18:07	07:50
VK9	Willis Is.	19:44	08:41
VK0	Heard Is.	23:14	15:20
VK0	Macquarie Is.	17:23	09:47
VP2E	Anguilla	10:47	21:58
VP2M	Montserrat	10:41	21:58
VP2V	British Virgin Is., Tortola	10:53	22:05
VP5	Turks & Caicos Islands,		
	Grand Turk	11:25	22:26
VP8	Falkland Islands, Stanley	08:05	23:59

Pref	State/Province/Country/City	Sunrise	Sunset
VP8	So. Georgia Is.	06:28	22:48
VP8	So. Orkney Is.	06:20	00:05
VP8	So. Sandwich Islands, Saunders Is.	05:27	22:28
VP8	So. Shetland Is., King George Is.	06:59	01:09
VP9	Bermuda	11:19	21:40
VQ9	Chagos, Diego Garcia	01:07	13:35
VR6	Pitcairn Is.	14:07	03:34
VS6	Hong Kong	23:04	10:03
VU	India, Bombay	01:45	12:54
	Calcutta	00:48	11:46
	New Delhi	01:44	12:20
	Bangalore	01:16	12:44
VU7	Andaman Islands, Port Blair	00:13	11:46
VU7	Laccadive Is.	01:30	13:07
XE	Mexico, Mexico City (XE1)	13:13	00:21
	Chihuahua (XE2)	13:57	00:32
	Merida (XE3)	12:38	23:41
XF4	Revilla Gigedo	14:02	01:11
XT	Burkina Faso, Onagadougou	06:32	18:02
XU	Kampuchea, Phnom Penh	23:25	10:58
XW	Laos, Viangchan	23:44	10:56
XX9	Macao	23:07	10:06
XZ	Burma, Rangoon	00:08	11:25
Y2-9	German Dem.Rep.(E), Berlin	07:05	15:29
YA	Afghanistan, Kandahar	02:34	13:01
	Kabul	02:27	12:40
YB-YD	Indonesia, Jakarta	22:51	11:16
	Medan	23:37	11:34
	Pontianak	22:50	10:57
	Jayapura	20:40	08:55
YI	Iraq, Baghdad	04:03	14:22
YJ	Vanuatu, Port Vila	18:27	07:28
YK	Syria, Damascus	04:37	14:53
YN, HT	Nicaragua, Managua	12:09	23:40
YO	Romania, Bucharest	05:45	15:07
YS	El Salvador, San Salvador	12:24	23:50
YU	Yugoslavia, Belgrade	06:09	15:28
YV	Venezuela, Caracas	10:50	22:27
YV0	Aves Is.	10:45	22:05
Z2,ZE	Zimbabwe, Harare	03:36	16:37
ZA	Albania, Tirane	06:02	15:41
ZB	Gibraltar	07:30	17:35
ZC4	British Cyprus	04:53	15:04
ZD7	St. Helena	06:06	19:02
ZD8	Ascension Is.	06:53	19:24
ZD9	Tristan da Cunha	05:51	20:09
ZF	Cayman Is.	12:01	23:09
ZK1	No. Cook Is., Manihiki	16:35	05:14
ZK1	So. Cook Is., Rarotonga	16:13	05:26
ZK2	Niue	16:53	05:59
ZK3	Tokelaus, Atafu	17:25	05:57
ZL	New Zealand, Auckland (ZL1)	17:23	07:40
	Wellington (ZL2)	17:11	07:52
	Christchurch (ZL3)	17:13	08:07
	Dunedin (ZL4)	17:14	08:23
ZL5	Antarctica, Scott Base	24 hrs	day
ZL7	Chatham Is.	16:28	07:25
ZL8	Kermadec Is.	17:10	06:54
ZL9	Auckland & Campbell Is.,		
	Auckland	17:12	08:57
	Campbell Is.	16:53	08:55
ZP	Paraguay, Asuncion	09:17	22:45
ZS	South Africa, Cape Town (ZS1)	03:55	17:58
	Port Elizabeth (ZS2)	03:26	17:29
	Bloemfontein (ZS4)	03:35	17:18
	Durban (ZS5)	03:14	17:00
	Johannesburg (ZS6)	03:33	17:04
ZS2	Prince Edward & Marion Is.,		
	Marion	02:02	17:17
ZS3	Namibia, Windhoek	04:23	17:41
1A0	S.M.O.M	06:33	16:09
1S	Spratly Is.	22:52	10:33
3A	Monaco	06:58	16:24
3B6	Agalega	02:05	14:43
3B7	St. Brandon	01:43	14:40
3B8	Mauritius	01:46	14:56
3B9	Rodriguez Is.	01:23	14:31
3C	Equatorial Guinea, Bata	05:30	17:32
	Malabo	05:37	17:33

Pref	State/Province/Country/City	Sunrise	Sunset
3CØ	Pagalu Is.	05:42	17:54
3D2	Fiji Is., Suva	17:46	06:48
3D6	Swaziland, Mbabane	03:20	16:52
3V	Tunisia, Tunis	06:29	16:31
3W	Vietnam, Ho Chi Minh City		
	(Saigon)	23:16	10:51
	Hanoi	23:36	10:39
3X	Republic of Guinea, Conakry	07:16	18:55
3Y	Bouvet	03:46	20:08
3Y	Peter I Is.	06:48	05:38
4S	Sri Lanka, Colombo	00:58	12:44
4U	ITU Geneva	07:10	16:21
4U	United Nations Hq.	12:15	21:58
4W	Yemen, Sanaa	03:33	14:54
4X, 4Z	Israel, Jerusalem	04:38	15:01
5A	Libya, Tripoli	06:10	16:31
	Benghazi	05:39	16:02
5B	Cyprus, Nicosia	04:52	15:01
5H	Tanzania, Dar es Salaam	03:19	15:46
5N	Nigeria, Lagos	06:03	17:51
5R	Madagascar, Antananarivo	02:28	15:33
5T	Mauritania, Nouakchott	07:38	18:51
5U	Niger, Niamey	06:19	17:46
5V	Togo, Lome	06:11	18:01
5W	Western Samoa, Apia	17:14	06:02
5X	Uganda, Kampala	03:57	16:04
5Z	Kenya, Nairobi	04:58	17:09
6W	Senegal, Dakar	07:39	19:02
6Y	Jamaica, Kingston	11:41	22:54

Pref	State/Province/Country/City	Sunrise	Sunset
7O	Yemen, People's Dem. Rep.,		
	Aden	03:26	14:55
7P	Lesotho, Maseru	03:29	17:12
7Q	Malawi, Lilongwe	03:31	16:20
	Blantyre	03:23	16:18
7X	Algeria, Algiers	06:57	17:00
8P	Barbados, Bridgetown	10:25	21:53
8Q	Maldive Is.	01:20	13:14
8R	Guyana, Georgetown	10:10	21:57
9G	Ghana, Accra	06:16	18:07
9H	Malta	06:10	16:16
9J	Zambia, Lusaka	03:50	16:44
9K	Kuwait	03:43	14:16
9L	Sierra Leone, Freetown	07:12	18:54
9M2	West Malaysia, Kuala Lumpur	23:25	11:23
9M6,8	East Malaysia, Sabah,		
	Sandakan (9M6)	22:23	10:13
	Sarawak, Kuching (9M8)	22:48	10:51
9N	Nepal, Katmandu	01:10	11:49
9Q	Zaire, Kinshasa	04:59	17:19
	Kisangani	04:27	16:33
	Lubumbashi	03:59	16:42
9U	Burundi, Bujumbura	04:05	16:22
9V	Singapore	23:14	11:17
9X	Rwanda, Kigali	04:04	16:17
9Y	Trinidad & Tobago, Port of		
	Spain	10:28	22:04
J2/A	Abu Ail	03:37	15:02

Pref	State/Province/Country/City	Sunrise	Sunset
VE1	New Brunswick, Saint John	11:39	21:37
	Nova Scotia, Halifax	11:27	21:29
	Prince Edward Island,		
	Charlottetown	11:29	21:23
VE2	Quebec, Montreal	12:09	22:07
	Quebec City	12:03	21:54
VE3	Ontario, London	12:35	22:43
	Ottawa	12:17	22:16
	Sudbury	12:41	22:34
	Toronto	12:28	22:34
VE4	Manitoba, Winnipeg	13:54	23:30
VE5	Saskatchewan, Regina	14:26	23:58
	Saskatoon	14:39	00:02
VE6	Alberta, Calgary	15:05	00:35
	Edmonton	15:10	00:25
VE7	British Columbia, Prince George	15:49	01:01
	Prince Rupert	16:20	01:29
	Vancouver	15:36	01:16
VE8	Northwest Territories,		
	Yellowknife	15:53	23:50
	Resolute	17:34	19:33
VY1	Yukon, Whitehorse	17:06	01:22
VO1	Newfoundland, St. John's	10:51	20:38
VO2	Labrador, Goose Bay	11:37	20:53
W1	Connecticut, Hartford	11:58	22:11
	Maine, Bangor	11:48	21:49
	Portland	11:52	21:58
	Massachusetts, Boston	11:52	22:04
	New Hampshire, Concord	11:56	22:04
	Rhode Island, Providence	11:52	22:06
	Vermont, Montpelier	12:02	22:06
W2	New Jersey, Atlantic City	12:00	22:23
	New York, Albany	12:04	22:14
	Buffalo	12:25	22:34
	New York City	12:01	22:18
	Syracuse	12:14	22:23
W3	Delaware, Wilmington	12:05	22:26
	District of Columbia,		
	Washington	12:09	22:34
	Maryland, Baltimore	12:08	22:32
	Pennsylvania, Harrisburg	12:12	22:31
	Philadelphia	12:04	22:25
	Pittsburgh	12:24	22:43
	Scranton	12:09	22:24
W4	Alabama, Montgomery	12:36	23:22
	Florida, Jacksonville	12:14	23:07
	Miami	12:02	23:07
	Pensacola	12:36	23:29
	Georgia, Atlanta	12:30	23:12
	Savannah	12:14	23:02
	Kentucky, Lexington	12:38	23:06
	Louisville	12:43	23:10
	North Carolina, Charlotte	12:18	22:56
	Raleigh	12:10	22:46
	Wilmington	12:05	22:46
	South Carolina, Columbia	12:17	22:58
	Tennessee, Knoxville	12:32	23:07
	Memphis	12:55	23:33
	Nashville	12:44	23:18
	Virginia, Norfolk	12:03	22:35
	Richmond	12:08	22:38
W5	Arkansas, Little Rock	13:03	23:42
	Louisiana, New Orleans	12:47	23:41
	Shreveport	13:05	23:51
	Mississippi, Jackson	12:51	23:38
	New Mexico, Albuquerque	14:02	00:39
	Oklahoma, Oklahoma City	13:25	00:02
	Texas, Abilene	13:29	00:15
	Amarillo	13:42	00:20
	Dallas	13:18	00:03
	El Paso	13:56	00:44
	San Antonio	13:20	00:15
W6	California, Los Angeles	14:46	01:27
	San Francisco	15:09	01:38
W7	Arizona, Flagstaff	14:22	00:59
	Phoenix	14:21	01:04
	Idaho, Boise	14:55	01:01
	Pocatello	14:39	00:48

Pref	State/Province/Country/City	Sunrise	Sunset
	Montana, Billings	14:29	00:26
	Butte	14:46	00:41
	Great Falls	14:45	00:33
	Nevada, Las Vegas	14:37	01:11
	Reno	15:02	01:24
	Oregon, Portland	15:26	01:23
	Utah, Salt Lake City	14:32	00:50
	Washington, Seattle	15:29	01:17
	Spokane	15:10	00:57
	Wyoming, Cheyenne	14:05	00:21
	Sheridan	14:21	00:22
W8	Michigan, Detroit	12:40	22:51
	Grand Rapids	12:52	23:01
	Sault Ste. Marie	12:55	22:48
	Traverse City	12:56	22:56
	Ohio, Cincinnati	12:40	23:04
	Cleveland	12:33	22:48
	Columbus	12:35	22:56
	West Virginia, Charleston	12:27	22:53
W9	Illinois, Chicago	12:57	23:11
	Indiana, Indianapolis	12:48	23:09
	Wisconsin, Green Bay	13:05	23:07
	Milwaukee	13:01	23:10
W0	Colorado, Denver	14:03	00:24
	Grand Junction	14:16	00:40
	Iowa, Des Moines	13:21	23:35
	Kansas, Pratt	13:34	00:03
	Wichita	13:28	23:57
	Minnesota, Duluth	13:26	23:18
	Minneapolis	13:27	23:27
	Missouri, Columbia	13:11	23:35
	Kansas City	13:20	23:44
	St. Louis	13:02	23:27
	Nebraska, North Platte	13:49	00:05
	Omaha	13:29	23:45
	North Dakota, Fargo	13:45	23:36
	South Dakota, Rapid City	14:04	00:08
A2	Botswana, Gaborone	03:56	17:04
A3	Tonga, Nukualofa	17:26	06:23
A4	Oman, Muscat	02:44	13:55
A5	Bhutan, Thimbu	00:45	11:47
A6	United Arab Emirates, Abu		
	Dhabi	03:02	14:11
A7	Qatar, Ad-Dawhah	03:14	14:21
A9	Bahrein, Al-Manamah	03:19	14:23
AP	Pakistan, Karachi	02:11	13:19
	Islamabad	02:00	12:42
BV	Taiwan, Taipei	22:34	09:41
BY	People's Rep. of China, Beijing	23:18	09:38
	Harbin	22:49	08:45
	Shanghai	22:43	09:33
	Fuzhou	22:44	09:49
	Xian	23:38	10:18
	Chongqing	23:41	10:35
	Chengdu	23:51	10:43
	Lhasa	00:42	11:36
	Urumqi	01:21	11:26
	Kashi	01:58	12:21
C2	Nauru	19:02	07:10
C3	Andorra	07:02	17:13
C5	The Gambia, Banjul	07:32	19:09
C6	Bahamas, Nassau	11:50	22:57
C9	Mozambique, Maputo	03:28	16:39
	Mozambique	03:10	15:52
CE	Chile, Santiago	10:10	23:44
CE0	Easter Island	12:54	02:08
CE0	San Felix	10:58	00:10
CE0	Juan Fernandez	10:42	00:16
CM,CO	Cuba, Havana	12:07	23:20
CN	Morocco, Casablanca	07:22	18:05
CP	Bolivia, La Paz	10:25	23:10
CT	Portugal, Lisbon	07:38	18:03
CT2, CU	Azores, Ponta Delgada	08:42	19:11
CT3	Madeira Islands, Funchal	07:58	18:44
CX	Uruguay, Montevideo	09:09	22:48
D2	Angola, Luanda	05:07	17:34
D4	Cape Verde, Praia	08:01	19:34

Pref	State/Province/Country/City	Sunrise	Sunset
D6	Comoros, Moroni	03:02	15:36
DA-DL	Germany, Fed. Rep. of (W), Bonn	07:00	16:31
DU	Philippines, Manilla	22:23	09:56
EA	Spain, Madrid	07:19	17:38
EA6	Balearic Is., Palma	06:52	17:14
EA8	Canary Is., Las Palmas	07:42	18:40
EA9	Ceuta & Melilla, Ceuta	07:17	17:52
	Melilla	07:07	17:44
EI	Ireland, Dublin	08:01	17:17
EL	Liberia, Monrovia	07:00	18:53
EP	Iran, Tehran	03:29	14:04
ET	Ethiopia, Addis Ababa	03:45	15:32
	Asmera	03:52	15:24
F	France, Paris	07:14	16:55
FG	Guadeloupe	10:35	22:06
FG, FS	St. Martin	10:43	22:09
FH	Mayotte	02:54	15:31
FK	New Caledonia, Noumia	18:37	07:38
FM	Martinique	10:31	22:04
FO	Clipperton	13:39	01:22
FO	Fr. Polynesia, Tahiti	16:28	05:16
FP	St. Pierre & Miquelon, St. Pierre	11:02	20:54
FR	Glorioso	02:48	15:21
FR	Juan de Nova	02:59	15:46
	Europa	03:02	16:02
FR	Reunion	02:03	15:00
FR	Tromelin	02:14	14:58
FT8W	Crozet	01:33	15:58
FT8X	Kerguelen	00:15	14:58
FT8Y	Antarctica, Dumont D'Urville	18:01	11:47
FT8Z	Amsterdam & St. Paul Is., Amsterdam	00:09	13:58
FW	Wallis & Futuna Is., Wallis	17:40	06:18
FY	Fr. Guiana, Cayenne	09:45	21:41
G	England, London	07:31	16:57
GD	Isle of Man	07:57	17:06
GI	Northern Ireland, Belfast	08:04	17:11
GJ	Jersey	07:33	17:12
GM	Scotland, Glasgow	08:01	17:00
	Aberdeen	07:58	16:46
GU	Guernsey	07:35	17:13
GW	Wales, Cardiff	07:43	17:10
H4	Solomon Islands, Honiara	19:19	07:48
HA	Hungary, Budapest	06:03	15:51
HB	Switzerland, Bern	06:48	16:39
HBØ	Liechtenstein	06:40	16:30
HC	Ecuador, Quito	11:22	23:29
HC8	Galapagos Is.	12:12	00:20
HH	Haiti, Port-Au-Prince	11:21	22:45
HI	Dominican Republic, Santo Domingo	11:11	22:36
HK	Colombia, Bogota	11:12	23:08
HKØ	Malpelo Is.	11:39	23:37
HKØ	San Adreas	11:51	23:30
HL, HM	Korea, Seoul	22:31	09:00
HP	Panama, Panama	11:38	23:25
HR	Honduras, Tegucigalpa	12:15	23:50
HS	Thailand, Bangkok	23:44	11:19
HV	Vatican City	06:17	16:30
HZ, 7Z	Saudi Arabia, Dharan	03:22	14:26
	Mecca	03:56	15:13
	Riyadh	03:32	14:41
I	Italy, Rome	06:17	16:30
	Trieste	06:20	16:17
	Sicily	06:03	16:32
IS	Sardinia, Cagliari	06:25	16:49
J2	Djibouti, Djibouti	03:30	15:11
J3	Grenada	10:31	22:11
J5	Guinea-Bissau, Bissau	07:26	19:06
J6	St. Lucia	10:30	22:05
J7	Dominica	10:33	22:05
J8	St. Vincent	10:30	22:07
JA-JS	Japan, Tokyo	21:37	08:12
	Nagasaki	22:11	08:57
	Sapporo	21:44	07:52
JD	Minami Torishima	20:23	07:32
JD	Ogasawara, Kazan Is.	21:19	08:20
JT	Mongolia, Ulan Bator	00:13	09:59
JW	Svalbard, Spitsbergen	24 hrs night	
JX	Jan Mayen	10:07	15:26
JY	Jordan, Amman	04:26	15:14
KC4	Antarctica, Byrd Station	24 hrs	day
	McMurdo Sound	24 hrs	day
	Palmer Station	07:53	01:06
KC6	Micronesia, Ponape	19:45	07:36
KC6	Belau, Yap	21:08	08:54
	Koror	21:20	09:11
KG4	Guantanamo Bay	11:34	22:55
KH1	Baker, Howland Is.	17:55	06:01
KH2	Guam, Agana	20:46	08:23
KH3	Johnston Is.	17:44	05:12
KH4	Midway Is.	18:34	05:33
KH5	Palmyra Is.	17:05	04:59
KH5K	Kingman Reef	17:10	05:00
KH6	Hawaii, Hilo	16:53	04:15
	Honolulu	17:07	04:24
KH7	Kure Is.	18:38	05:36
KH8	American Samoa, Pago Pago	17:17	05:57
KH9	Wake Is.	19:26	06:48
KHØ	Mariana Is., Saipan	20:44	08:17
KL7	Alaska, Adak	19:17	04:43
	Anchorage	18:08	02:19
	Fairbanks	18:21	01:49
	Juneau	16:52	01:31
	Nome	19:29	03:01
KP1	Navassa Is.	11:31	22:56
KP2	Virgin Islands, Charlotte Amalie	10:51	22:16
KP4	Puerto Rico, San Juan	10:56	22:21
KP5	Desecheo Is.	11:01	22:26
KX6	Marshall Islands, Kwajalein	19:11	06:58
LA-LJ	Norway, Oslo	07:19	15:43
LU	Argentina, Buenos Aires	09:18	22:56
LX	Luxembourg	07:00	16:38
LZ	Bulgaria, Sofia	05:35	15:45
OA	Peru, Lima	11:05	23:39
OD	Lebanon, Beirut	04:31	15:12
OE	Austria, Vienna	06:16	16:01
OH	Finland, Helsinki	06:23	14:44
OHØ	Aland Is.	06:43	15:04
OJØ	Market Reef	06:47	15:08
OK	Czechoslovakia, Prague	06:29	16:03
ON	Belgium, Brussels	07:11	16:41
OX, XP	Greenland, Godthaab	11:52	19:28
	Thule	24 hrs night	
OY	Faroe Islands, Torshavn	08:39	16:42
OZ	Denmark, Copenhagen	06:54	15:53
P2	Papua New Guinea, Madang	20:22	08:41
	Port Moresby	20:11	08:39
PA-PI	Netherlands, Amsterdam	07:13	16:35
PJ, P4	Netherlands Antilles, Willemstad	11:00	22:39
PJ5-8	St. Maarten and Saba, St. Maarten	10:43	22:10
PY	Brazil, Brasilia	09:03	21:47
	Rio De Janeiro	08:35	21:38
	Natal	08:24	20:45
	Manaus	10:07	22:21
	Porto Alegre	08:57	22:20
PYØ	Fernando De Noronha	08:15	20:31
PYØ	St. Peter & St. Paul Rocks	08:09	20:14
PYØ	Trindade & Martin Vaz Is., Trindade	07:43	20:39
PZ	Suriname, Paramaribo	09:57	21:51
S2	Bangladesh, Dacca	00:36	11:48
S7	Seychelles, Victoria	02:23	14:40
S9	Sao Tome	05:44	17:50
SJ-SM	Sweden, Stockholm	06:46	15:16
SP	Poland, Krakow	06:06	15:41
	Warsaw	06:08	15:31
ST	Sudan, Khartoum	04:18	15:49
STØ	Southern Sudan, Juba	04:09	16:05
SU	Egypt, Cairo	04:41	15:35
SV	Greece, Athens	05:25	15:53
SV5	Dodecanese, Rhodes	05:04	15:38
SV9	Crete	05:14	15:51
SV/A	Mount Athos	05:27	15:46
T2	Tuvalu, Funafuti	18:06	06:33
T30	West Kiribati, Bonriki	18:39	06:43
T31	Central Kiribati, Kanton	17:34	05:47
T32	East Kiribati, Christmas Is.	16:42	04:45
T5	Somalia, Mogadishu	03:11	15:13
T7	San Marino	06:22	16:27

Pref	State/Province/Country/City	Sunrise	Sunset
TA	Turkey, Ankara	04:52	15:12
	Istanbul	05:10	15:26
TF	Iceland, Reykjavik	09:53	17:30
TG	Guatemala, Guatemala City	12:29	00:02
TI	Costa Rica, San Jose	11:57	23:42
TI9	Cocos Is.	12:04	23:59
TJ	Cameroon, Yaounde	05:29	17:27
TK	Corsica	06:31	16:44
TL	Central African Republic, Bangui	05:01	16:58
TN	Congo, Brazzaville	05:04	17:21
TR	Gabon, Libreville	05:33	17:39
TT	Chad, N'Djamena	05:24	17:03
TU	Ivory Coast, Abidjan	06:32	18:27
TY	Benin, Porto Novo	06:07	17:59
TZ	Mali, Bamako	06:57	18:35
UA	Russia, European,		
	Leningrad (UA1)	06:00	14:25
	Archangel (UA1)	05:46	13:17
	Murmansk (UA1)	06:55	13:08
	Moscow (UA3)	05:14	14:13
	Kuibyshev (UA4)	04:15	13:31
	Rostov (UA6)	04:41	14:30
UA1P	Franz Josef Land	24 hrs	night
UA2	Kaliningrad	06:19	15:24
UA9,Ø	Russia, Asiatic,		
	Novosibirsk (UA9)	02:10	11:14
	Perm (UA9)	04:08	12:49
	Omsk (UA9)	02:48	11:52
	Norilsk (UAØ)	03:18	09:24
	Irkutsk (UAØ)	00:35	09:57
	Vladivostok (UAØ)	22:22	08:30
	Petropavlovsk (UAØ)	21:00	06:18
	Khabarovsk (UAØ)	22:22	08:05
	Krasnoyarsk (UAØ)	01:34	10:31
	Yakutsk (UAØ)	23:33	07:36
	Wrangel Island (UAØ)	21:32	02:51
	Kyzyl (UAØY)	01:13	10:38
UB	Ukraine, Kiev	05:25	14:58
UC	Byelorussia, Minsk	05:47	14:59
UD	Azerbaijan, Baku	03:45	14:04
UF	Georgia, Tbilisi	04:07	14:21
UG	Armenia, Ierevan	04:06	14:25
UH	Turkoman, Ashkhabad	03:06	13:34
UI	Uzbek, Bukhara	02:45	13:07
	Tashkent	02:28	12:44
UJ	Tadzhik, Samarkand	02:36	12:57
	Dushanbe	02:26	12:50
UL	Kazakh, Alma-ata	02:02	12:10
UM	Kirghiz, Frunze	02:11	12:20
UO	Moldavia, Kishinev	05:23	15:14
UP	Lithuania, Vilna	05:58	15:06
UQ	Latvia, Riga	06:12	15:02
UR	Estonia, Tallinn	06:20	14:49
V2	Antigua & Barbuda, St. Johns	10:37	22:05
V3	Belize, Belmopan	12:25	23:52
V4	St. Christopher & Nevis	10:40	22:08
V8	Brunei, Bandar Seri Begawan	22:36	10:32
VE1, CY	Sable Is.	11:11	21:16
VE1, CY	St. Paul Is.	11:19	21:09
VK	Australia, Canberra (VK1)	19:27	09:07
	Sydney (VK2)	19:21	08:56
	Melbourne (VK3)	19:39	09:28
	Brisbane (VK4)	19:24	08:39
	Adelaide (VK5)	20:10	09:49
	Perth (VK6)	21:46	11:15
	Hobart, Tasmania (VK7)	19:19	09:30
	Darwin (VK8)	21:12	09:48
VK	Lord Howe Is.	18:53	08:21
VK9	Christmas Is.	22:55	11:26
VK9	Cocos-Keeling Is.	23:29	12:04
VK9	Mellish Reef	19:26	08:14
VK9	Norfolk Is.	18:22	07:41
VK9	Willis Is.	19:53	08:38
VKØ	Heard Is.	23:47	14:54
VKØ	Macquarie Is.	17:59	09:18
VP2E	Anguilla	10:43	22:08
VP2M	Montserrat	10:38	22:07
VP2V	British Virgin Is., Tortola	10:50	22:14
VP5	Turks & Caicos Islands, Grand Turk	11:20	22:37
VP8	Falkland Islands, Stanley	08:36	23:34

Pref	State/Province/Country/City	Sunrise	Sunset
VP8	So. Georgia Is.	07:03	22:19
VP8	So. Orkney Is.	07:08	23:23
VP8	So. Sandwich Islands, Saunders Is.	06:07	21:53
VP8	So. Shetland Is., King George Is.	07:51	00:23
VP9	Bermuda	11:09	21:56
VQ9	Chagos, Diego Garcia	01:12	13:36
VR6	Pitcairn Is.	14:20	03:28
VS6	Hong Kong	22:59	10:14
VU	India, Bombay	01:41	13:04
	Calcutta	00:43	11:57
	New Delhi	01:36	12:34
	Bangalore	01:15	12:52
VU7	Andaman Islands, Port Blair	00:12	11:53
VU7	Laccadive Is.	01:29	13:14
XE	Mexico, Mexico City (XE1)	13:09	00:31
	Chihuahua (XE2)	13:49	00:46
	Merida (XE3)	12:33	23:51
XF4	Revilla Gigedo	13:58	01:21
XT	Burkina Faso, Onagadougou	06:31	18:09
XU	Kampuchea, Phnom Penh	23:24	11:05
XW	Laos, Viangchan	23:40	11:06
XX9	Macao	23:02	10:17
XZ	Burma, Rangoon	00:05	11:34
Y2-9	German Dem.Rep.(E), Berlin	06:40	16:00
YA	Afghanistan, Kandahar	02:25	13:16
	Kabul	02:17	12:57
YB-YD	Indonesia, Jakarta	22:56	11:17
	Medan	23:39	11:38
	Pontianak	22:53	11:00
	Jayapura	20:44	08:57
YI	Iraq, Baghdad	03:53	14:38
YJ	Vanuatu, Port Vila	18:36	07:25
YK	Syria, Damascus	04:27	15:10
YN, HT	Nicaragua, Managua	12:08	23:48
YO	Romania, Bucharest	05:28	15:31
YS	El Salvador, San Salvador	12:23	23:58
YU	Yugoslavia, Belgrade	05:51	15:52
YV	Venezuela, Caracas	10:50	22:33
YVØ	Aves Is.	10:43	22:14
Z2,ZE	Zimbabwe, Harare	03:45	16:34
ZA	Albania, Tirane	05:47	16:02
ZB	Gibraltar	07:18	17:52
ZC4	British Cyprus	04:42	15:21
ZD7	St. Helena	06:15	18:59
ZD8	Ascension Is.	06:59	19:24
ZD9	Tristan da Cunha	06:09	19:56
ZF	Cayman Is.	11:58	23:19
ZK1	No. Cook Is., Manihiki	16:42	05:13
ZK1	So. Cook Is., Rarotonga	16:24	05:22
ZK2	Niue	17:03	05:55
ZK3	Tokelaus, Atafu	17:31	05:57
ZL	New Zealand, Auckland (ZL1)	17:41	07:27
	Wellington (ZL2)	17:33	07:36
	Christchurch (ZL3)	17:37	07:50
	Dunedin (ZL4)	17:39	08:04
ZL5	Antarctica, Scott Base	24 hrs day	
ZL7	Chatham Is.	16:52	07:07
ZL8	Kermadec Is.	17:25	06:45
ZL9	Auckland & Campbell Is., Auckland	17:42	08:34
	Campbell Is.	17:25	08:29
ZP	Paraguay, Asuncion	09:30	22:39
ZS	South Africa, Cape Town (ZS1)	04:12	17:48
	Port Elizabeth (ZS2)	03:43	17:19
	Bloemfontein (ZS4)	03:49	17:09
	Durban (ZS5)	03:29	16:51
	Johannesburg (ZS6)	03:46	16:57
ZS2	Prince Edward & Marion Is., Marion	02:28	16:57
ZS3	Namibia, Windhoek	04:35	17:36
1AØ	S.M.O.M	06:17	16:31
1S	Spratly Is.	22:52	10:40
3A	Monaco	06:41	16:47
3B6	Agalega	02:12	14:43
3B7	St. Brandon	01:52	14:37
3B8	Mauritius	01:56	14:51
3B9	Rodriguez Is.	01:33	14:27
3C	Equatorial Guinea, Bata	05:33	17:36
	Malabo	05:39	17:38

Pref	State/Province/Country/City	Sunrise	Sunset
3CØ	Pagalu Is.	05:46	17:56
3D2	Fiji Is., Suva	17:55	06:45
3D6	Swaziland, Mbabane	03:33	16:45
3V	Tunisia, Tunis	06:17	16:49
3W	Vietnam, Ho Chi Minh City		
	(Saigon)	23:16	10:58
	Hanoi	23:31	10:49
3X	Republic of Guinea, Conakry	07:16	19:01
3Y	Bouvet	04:21	19:39
3Y	Peter I Is.	08:57	03:35
4S	Sri Lanka, Colombo	00:58	12:50
4U	ITU Geneva	06:52	16:46
4U	United Nations Hq.	12:01	22:18
4W	Yemen, Sanaa	03:31	15:03
4X, 4Z	Israel, Jerusalem	04:29	15:17
5A	Libya, Tripoli	06:01	16:47
	Benghazi	05:30	16:17
5B	Cyprus, Nicosia	04:41	15:19
5H	Tanzania, Dar es Salaam	03:24	15:47
5N	Nigeria, Lagos	06:04	17:56
5R	Madagascar, Antananarivo	02:38	15:29
5T	Mauritania, Nouakchott	07:35	19:00
5U	Niger, Niamey	06:17	17:54
5V	Togo, Lome	06:12	18:06
5W	Western Samoa, Apia	17:22	06:00
5X	Uganda, Kampala	04:01	16:07
5Z	Kenya, Nairobi	05:02	17:11
6W	Senegal, Dakar	07:37	19:10
6Y	Jamaica, Kingston	11:38	23:04

Pref	State/Province/Country/City	Sunrise	Sunset
7O	Yemen, People's Dem. Rep.,		
	Aden	03:25	15:03
7P	Lesotho, Maseru	03:43	17:04
7Q	Malawi, Lilongwe	03:39	16:18
	Blantyre	03:32	16:15
7X	Algeria, Algiers	06:45	17:18
8P	Barbados, Bridgetown	10:23	22:01
8Q	Maldive Is.	01:21	13:19
8R	Guyana, Georgetown	10:11	22:02
9G	Ghana, Accra	06:17	18:12
9H	Malta	05:59	16:33
9J	Zambia, Lusaka	03:59	16:42
9K	Kuwait	03:35	14:30
9L	Sierra Leone, Freetown	07:13	19:00
9M2	West Malaysia, Kuala Lumpur	23:27	11:27
9M6,8	East Malaysia, Sabah,		
	Sandakan (9M6)	22:24	10:18
	Sarawak, Kuching (9M8)	22:51	10:54
9N	Nepal, Katmandu	01:02	12:03
9Q	Zaire, Kinshasa	05:04	17:21
	Kisangani	04:30	16:36
	Lubumbashi	04:07	16:40
9U	Burundi, Bujumbura	04:09	16:24
9V	Singapore	23:16	11:20
9X	Rwanda, Kigali	04:08	16:19
9Y	Trinidad & Tobago, Port of		
	Spain	10:27	22:10
J2/A	Abu Ail	03:35	15:10

Pref	State/Province/Country/City	Sunrise	Sunset
VE1	New Brunswick, Saint John	11:18	21:58
	Nova Scotia, Halifax	11:07	21:49
	Prince Edward Island, Charlottetown	11:07	21:44
VE2	Quebec, Montreal	11:48	22:27
	Quebec City	11:41	22:16
VE3	Ontario, London	12:15	23:02
	Ottawa	11:56	22:36
	Sudbury	12:19	22:55
	Toronto	12:09	22:53
VE4	Manitoba, Winnipeg	13:29	23:54
VE5	Saskatchewan, Regina	14:01	00:23
	Saskatoon	14:12	00:28
VE6	Alberta, Calgary	14:39	01:00
	Edmonton	14:42	00:53
VE7	British Columbia, Prince George	15:20	01:29
	Prince Rupert	15:51	01:58
	Vancouver	15:12	01:39
VE8	Northwest Territories, Yellowknife	15:10	00:31
	Resolute	15:16	21:51
VY1	Yukon, Whitehorse	16:27	02:00
VO1	Newfoundland, St. John's	10:28	21:00
VO2	Labrador, Goose Bay	11:09	21:21
W1	Connecticut, Hartford	11:39	22:29
	Maine, Bangor	11:28	22:09
	Portland	11:32	22:17
	Massachusetts, Boston	11:34	22:22
	New Hampshire, Concord	11:36	22:22
	Rhode Island, Providence	11:34	22:24
	Vermont, Montpelier	11:42	22:25
W2	New Jersey, Atlantic City	11:43	22:39
	New York, Albany	11:45	22:32
	Buffalo	12:06	22:52
	New York City	11:43	22:35
	Syracuse	11:55	22:41
W3	Delaware, Wilmington	11:48	22:43
	District of Columbia, Washington	11:53	22:50
	Maryland, Baltimore	11:52	22:48
	Pennsylvania, Harrisburg	11:54	22:48
	Philadelphia	11:47	22:41
	Pittsburgh	12:07	23:00
	Scranton	11:51	22:41
W4	Alabama, Montgomery	12:23	23:34
	Florida, Jacksonville	12:02	23:18
	Miami	11:52	23:16
	Pensacola	12:24	23:40
	Georgia, Atlanta	12:17	23:25
	Savannah	12:02	23:14
	Kentucky, Lexington	12:22	23:21
	Louisville	12:27	23:26
	North Carolina, Charlotte	12:04	23:09
	Raleigh	11:56	23:00
	Wilmington	11:51	22:59
	South Carolina, Columbia	12:03	23:11
	Tennessee, Knoxville	12:17	23:21
	Memphis	12:41	23:47
	Nashville	12:29	23:32
	Virginia, Norfolk	11:48	22:49
	Richmond	11:53	22:53
W5	Arkansas, Little Rock	12:49	23:56
	Louisiana, New Orleans	12:35	23:52
	Shreveport	12:52	00:04
	Mississippi, Jackson	12:38	23:50
	New Mexico, Albuquerque	13:47	00:53
	Oklahoma, Oklahoma City	13:11	00:16
	Texas, Abilene	13:16	00:28
	Amarillo	13:28	00:33
	Dallas	13:05	00:16
	El Paso	13:43	00:56
	San Antonio	13:08	00:26
W6	California, Los Angeles	14:32	01:40
	San Francisco	14:53	01:53
W7	Arizona, Flagstaff	14:07	01:13
	Phoenix	14:07	01:16
	Idaho, Boise	14:36	01:20
	Pocatello	14:20	01:07

Pref	State/Province/Country/City	Sunrise	Sunset
	Montana, Billings	14:08	00:46
	Butte	14:25	01:02
	Great Falls	14:22	00:55
	Nevada, Las Vegas	14:22	01:25
	Reno	14:45	01:40
	Oregon, Portland	15:05	01:44
	Utah, Salt Lake City	14:15	01:07
	Washington, Seattle	15:06	01:39
	Spokane	14:47	01:19
	Wyoming, Cheyenne	13:47	00:38
	Sheridan	14:01	00:42
W8	Michigan, Detroit	12:21	23:09
	Grand Rapids	12:33	23:19
	Sault Ste. Marie	12:33	23:09
	Traverse City	12:35	23:16
	Ohio, Cincinnati	12:23	23:20
	Cleveland	12:15	23:05
	Columbus	12:18	23:12
	West Virginia, Charleston	12:11	23:09
W9	Illinois, Chicago	12:39	23:28
	Indiana, Indianapolis	12:31	23:25
	Wisconsin, Green Bay	12:44	23:26
	Milwaukee	12:42	23:28
W0	Colorado, Denver	13:46	00:41
	Grand Junction	13:59	00:56
	Iowa, Des Moines	13:03	23:53
	Kansas, Pratt	13:18	00:18
	Wichita	13:13	00:12
	Minnesota, Duluth	13:04	23:39
	Minneapolis	13:06	23:47
	Missouri, Columbia	12:54	23:51
	Kansas City	13:03	00:00
	St. Louis	12:45	23:43
	Nebraska, North Platte	13:31	00:22
	Omaha	13:12	00:02
	North Dakota, Fargo	13:23	23:58
	South Dakota, Rapid City	13:45	00:28
A2	Botswana, Gaborone	04:05	16:54
A3	Tonga, Nukualofa	17:33	06:15
A4	Oman, Muscat	02:35	14:03
A5	Bhutan, Thimbu	00:35	11:57
A6	United Arab Emirates, Abu Dhabi	02:53	14:20
A7	Qatar, Ad-Dawhah	03:05	14:30
A9	Bahrein, Al-Manamah	03:09	14:33
AP	Pakistan, Karachi	02:02	13:28
	Islamabad	01:46	12:55
BV	Taiwan, Taipei	22:24	09:50
BY	People's Rep. of China, Beijing	23:01	09:55
	Harbin	22:27	09:06
	Shanghai	22:30	09:44
	Fuzhou	22:34	09:58
	Xian	23:24	10:32
	Chongqing	23:29	10:46
	Chengdu	23:39	10:54
	Lhasa	00:30	11:47
	Urumqi	01:01	11:45
	Kashi	01:42	12:37
C2	Nauru	19:02	07:10
C3	Andorra	06:44	17:31
C5	The Gambia, Banjul	07:27	19:13
C6	Bahamas, Nassau	11:40	23:06
C9	Mozambique, Maputo	03:37	16:28
	Mozambique	03:15	15:46
CE	Chile, Santiago	10:23	23:30
CE0	Easter Island	13:04	01:57
CE0	San Felix	11:08	23:59
CE0	Juan Fernandez	10:55	00:02
CM,CO	Cuba, Havana	11:58	23:28
CN	Morocco, Casablanca	07:09	18:18
CP	Bolivia, La Paz	10:30	23:04
CT	Portugal, Lisbon	07:21	18:19
CT2, CU	Azores, Ponta Delgada	08:26	19:26
CT3	Madeira Islands, Funchal	07:45	18:57
CX	Uruguay, Montevideo	09:23	22:33
D2	Angola, Luanda	05:10	17:31
D4	Cape Verde, Praia	07:56	19:39

Pref	State/Province/Country/City	Sunrise	Sunset
D6	Comoros, Moroni	03:06	15:31
DA-DL	Germany, Fed. Rep. of (W),		
	Bonn	06:35	16:56
DU	Philippines, Manilla	22:17	10:01
EA	Spain, Madrid	07:02	17:55
EA6	Balearic Is., Palma	06:35	17:31
EA8	Canary Is., Las Palmas	07:31	18:50
EA9	Ceuta & Melilla, Ceuta	07:03	18:07
	Melilla	06:53	17:58
EI	Ireland, Dublin	07:33	17:44
EL	Liberia, Monrovia	06:58	18:55
EP	Iran, Tehran	03:14	14:18
ET	Ethiopia, Addis Ababa	03:42	15:35
	Asmera	03:46	15:29
F	France, Paris	06:50	17:18
FG	Guadeloupe	10:29	22:11
FG, FS	St. Martin	10:37	22:15
FH	Mayotte	02:58	15:26
FK	New Caledonia, Noumia	18:45	07:29
FM	Martinique	10:25	22:09
FO	Clipperton	13:35	01:26
FO	Fr. Polynesia, Tahiti	16:34	05:09
FP	St. Pierre & Miquelon, St. Pierre	10:40	21:15
FR	Glorioso	02:52	15:17
FR	Juan de Nova	03:05	15:39
	Europa	03:10	15:54
FR	Reunion	02:10	14:52
FR	Tromelin	02:20	14:52
FT8W	Crozet	01:54	15:37
FT8X	Kerguelen	00:39	14:34
FT8Y	Antarctica, Dumont D'Urville	18:57	10:49
FT8Z	Amsterdam & St. Paul Is.,		
	Amsterdam	00:24	13:42
FW	Wallis & Futuna Is., Wallis	17:45	06:13
FY	Fr. Guiana, Cayenne	09:43	21:42
G	England, London	07:04	17:23
GD	Isle of Man	07:28	17:35
GI	Northern Ireland, Belfast	07:34	17:40
GJ	Jersey	07:09	17:35
GM	Scotland, Glasgow	07:30	17:31
	Aberdeen	07:25	17:18
GU	Guernsey	07:11	17:37
GW	Wales, Cardiff	07:17	17:35
H4	Solomon Islands, Honiara	19:23	07:44
HA	Hungary, Budapest	05:41	16:13
HB	Switzerland, Bern	06:26	17:00
HB0	Liechtenstein	06:18	16:52
HC	Ecuador, Quito	11:22	23:29
HC8	Galapagos Is.	12:12	00:19
HH	Haiti, Port-Au-Prince	11:14	22:51
HI	Dominican Republic, Santo		
	Domingo	11:05	22:42
HK	Colombia, Bogota	11:10	23:10
HK0	Malpelo Is.	11:37	23:38
HK0	San Adreas	11:46	23:34
HL, HM	Korea, Seoul	22:15	09:15
HP	Panama, Panama	11:35	23:28
HR	Honduras, Tegucigalpa	12:10	23:54
HS	Thailand, Bangkok	23:39	11:24
HV	Vatican City	05:59	16:48
HZ, 7Z	Saudi Arabia, Dharan	03:11	14:35
	Mecca	03:48	15:20
	Riyadh	03:23	14:50
I	Italy, Rome	05:59	16:48
	Trieste	05:59	16:37
	Sicily	05:47	16:47
IS	Sardinia, Cagliari	06:09	17:05
J2	Djibouti, Djibouti	03:26	15:15
J3	Grenada	10:26	22:15
J5	Guinea-Bissau, Bissau	07:22	19:10
J6	St. Lucia	10:25	22:10
J7	Dominica	10:27	22:10
J8	St. Vincent	10:26	22:12
JA-JS	Japan, Tokyo	21:22	08:26
	Nagasaki	21:58	09:09
	Sapporo	21:25	08:11
JD	Minami Torishima	20:14	07:41
JD	Ogasawara, Kazan Is.	21:09	08:30
JT	Mongolia, Ulan Bator	23:50	10:21
JW	Svalbard, Spitsbergen	09:27	12:52
JX	Jan Mayen	08:51	16:42

Pref	State/Province/Country/City	Sunrise	Sunset
JY	Jordan, Amman	04:13	15:26
KC4	Antarctica, Byrd Station	24 hrs day	
	McMurdo Sound	14:14	11:59
	Palmer Station	08:43	00:15
KC6	Micronesia, Ponape	19:42	07:38
KC6	Belau, Yap	21:05	08:57
	Koror	21:18	09:13
KG4	Guantanamo Bay	11:27	23:02
KH1	Baker, Howland Is.	17:54	06:00
KH2	Guam, Agana	20:41	08:27
KH3	Johnston Is.	17:37	05:17
KH4	Midway Is.	18:23	05:43
KH5	Palmyra Is.	17:03	05:01
KH5K	Kingman Reef	17:07	05:02
KH6	Hawaii, Hilo	16:46	04:21
	Honolulu	16:59	04:31
KH7	Kure Is.	18:27	05:47
KH8	American Samoa, Pago Pago	17:22	05:51
KH9	Wake Is.	19:19	06:55
KH0	Mariana Is., Saipan	20:39	08:22
KL7	Alaska, Adak	18:51	05:08
	Anchorage	17:28	02:58
	Fairbanks	17:33	02:36
	Juneau	16:17	02:05
	Nome	18:42	03:48
KP1	Navassa Is.	11:25	23:02
KP2	Virgin Islands, Charlotte Amalie	10:44	22:22
KP4	Puerto Rico, San Juan	10:49	22:27
KP5	Desecheo Is.	10:54	22:32
KX6	Marshall Islands, Kwajalein	19:08	07:00
LA-LJ	Norway, Oslo	06:42	16:19
LU	Argentina, Buenos Aires	09:32	22:42
LX	Luxembourg	06:36	17:01
LZ	Bulgaria, Sofia	05:17	16:04
OA	Peru, Lima	11:09	23:35
OD	Lebanon, Beirut	04:17	15:26
OE	Austria, Vienna	05:53	16:23
OH	Finland, Helsinki	05:45	15:22
OH0	Aland Is.	06:05	15:42
OJ0	Market Reef	06:09	15:45
OK	Czechoslovakia, Prague	06:04	16:28
ON	Belgium, Brussels	06:45	17:06
OX, XP	Greenland, Godthaab	11:06	20:14
	Thule	14:02	19:35
OY	Faroe Islands, Torshavn	07:58	17:23
OZ	Denmark, Copenhagen	06:23	16:23
P2	Papua New Guinea, Madang	20:23	08:38
	Port Moresby	20:14	08:36
PA-PI	Netherlands, Amsterdam	06:46	17:01
PJ, P4	Netherlands Antilles, Willemstad	10:55	22:43
PJ5-8	St. Maarten and Saba, St.		
	Maarten	10:37	22:16
PY	Brazil, Brasilia	09:09	21:41
	Rio De Janeiro	08:44	21:29
	Natal	08:26	20:42
	Manaus	10:08	22:20
	Porto Alegre	09:08	22:08
PY0	Fernando De Noronha	08:16	20:29
PY0	St. Peter & St.Paul Rocks	08:08	20:14
PY0	Trindade & Martin Vaz Is.,		
	Trindade	07:50	20:31
PZ	Suriname, Paramaribo	09:55	21:53
S2	Bangladesh, Dacca	00:27	11:56
S7	Seychelles, Victoria	02:24	14:38
S9	Sao Tome	05:43	17:50
SJ-SM	Sweden, Stockholm	06:10	15:52
SP	Poland, Krakow	05:41	16:05
	Warsaw	05:41	15:57
ST	Sudan, Khartoum	04:12	15:54
ST0	Southern Sudan, Juba	04:07	16:06
SU	Egypt, Cairo	04:29	15:46
SV	Greece, Athens	05:09	16:08
SV5	Dodecanese, Rhodes	04:49	15:52
SV9	Crete	05:00	16:05
SV/A	Mount Athos	05:09	16:03
T2	Tuvalu, Funafuti	18:09	06:29
T30	West Kiribati, Bonriki	18:38	06:43
T31	Central Kiribati, Kanton	17:35	05:46
T32	East Kiribati, Christmas Is.	16:41	04:45
T5	Somalia, Mogadishu	03:10	15:13
T7	San Marino	06:02	16:46

Pref	State/Province/Country/City	Sunrise	Sunset
TA	Turkey, Ankara	04:34	15:29
	Istanbul	04:52	15:43
TF	Iceland, Reykjavik	09:07	18:16
TG	Guatemala, Guatemala City	12:23	00:07
TI	Costa Rica, San Jose	11:54	23:45
TI9	Cocos Is.	12:02	00:00
TJ	Cameroon, Yaounde	05:27	17:28
TK	Corsica	06:13	17:02
TL	Central African Republic, Bangui	04:59	16:59
TN	Congo, Brazzaville	05:05	17:19
TR	Gabon, Libreville	05:32	17:38
TT	Chad, N'Djamena	05:19	17:07
TU	Ivory Coast, Abidjan	06:30	18:29
TY	Benin, Porto Novo	06:05	18:01
TZ	Mali, Bamako	06:52	18:39
UA	Russia, European,		
	Leningrad (UA1)	05:23	15:01
	Archangel (UA1)	04:59	14:04
	Murmansk (UA1)	05:51	14:11
	Moscow (UA3)	04:43	14:43
	Kuibyshev (UA4)	03:47	13:59
	Rostov (UA6)	04:19	14:52
UA1P	Franz Josef Land	24 hrs night	
UA2	Kaliningrad	05:49	15:53
UA9,Ø	Russia, Asiatic,		
	Novosibirsk (UA9)	01:40	11:44
	Perm (UA9)	03:34	13:23
	Omsk (UA9)	02:18	12:22
	Norilsk (UAØ)	02:13	10:29
	Irkutsk (UAØ)	00:08	10:24
	Vladivostok (UAØ)	22:03	08:49
	Petropavlovsk (UAØ)	20:32	06:45
	Khabarovsk (UAØ)	21:58	08:28
	Krasnoyarsk (UAØ)	01:03	11:02
	Yakutsk (UAØ)	22:52	08:17
	Wrangel Island (UAØ)	20:16	04:07
	Kyzyl (UAØY)	00:46	11:04
UB	Ukraine, Kiev	05:00	15:23
UC	Byelorussia, Minsk	05:19	15:27
UD	Azerbaijan, Baku	03:27	14:20
UF	Georgia, Tbilisi	03:49	14:39
UG	Armenia, Ierevan	03:49	14:42
UH	Turkoman, Ashkhabad	02:50	13:49
UI	Uzbek, Bukhara	02:28	13:23
	Tashkent	02:11	13:02
UJ	Tadzhik, Samarkand	02:19	13:14
	Dushanbe	02:10	13:06
UL	Kazakh, Alma-ata	01:43	12:28
UM	Kirghiz, Frunze	01:52	12:38
UO	Moldavia, Kishinev	05:01	15:35
UP	Lithuania, Vilna	05:28	15:34
UQ	Latvia, Riga	05:40	15:34
UR	Estonia, Tallinn	05:44	15:25
V2	Antigua & Barbuda, St. Johns	10:31	22:10
V3	Belize, Belmopan	12:19	23:58
V4	St. Christopher & Nevis	10:34	22:13
V8	Brunei, Bandar Seri Begawan	22:34	10:33
VE1, CY	Sable Is.	10:51	21:35
VE1, CY	St. Paul Is.	10:57	21:31
VK	Australia, Canberra (VK1)	19:41	08:53
	Sydney (VK2)	19:34	08:43
	Melbourne (VK3)	19:54	09:12
	Brisbane (VK4)	19:34	08:28
	Adelaide (VK5)	20:24	09:34
	Perth (VK6)	21:58	11:02
	Hobart, Tasmania (VK7)	19:38	09:10
	Darwin (VK8)	21:16	09:43
VK	Lord Howe Is.	19:06	08:08
VK9	Christmas Is.	22:59	11:22
VK9	Cocos-Keeling Is.	23:33	11:59
VK9	Mellish Reef	19:32	08:08
VK9	Norfolk Is.	18:33	07:30
VK9	Willis Is.	19:59	08:32
VKØ	Heard Is.	00:14	14:25
VKØ	Macquarie Is.	18:29	08:48
VP2E	Anguilla	10:36	22:14
VP2M	Montserrat	10:32	22:12
VP2V	British Virgin Is., Tortola	10:43	22:21
VP5	Turks & Caicos Islands,		
	Grand Turk	11:12	22:44
VP8	Falkland Islands, Stanley	09:03	23:07

Pref	State/Province/Country/City	Sunrise	Sunset
VP8	So. Georgia Is.	07:32	21:49
VP8	So. Orkney Is.	07:48	22:43
VP8	So. Sandwich Islands, Saunders Is.	06:42	21:18
VP8	So. Shetland Is., King George Is.	08:33	23:40
VP9	Bermuda	10:56	22:08
VQ9	Chagos, Diego Garcia	01:15	13:33
VR6	Pitcairn Is.	14:29	03:18
VS6	Hong Kong	22:51	10:22
VU	India, Bombay	01:34	13:10
	Calcutta	00:35	12:05
	New Delhi	01:25	12:44
	Bangalore	01:10	12:56
VU7	Andaman Islands, Port Blair	00:08	11:56
VU7	Laccadive Is.	01:26	13:17
XE	Mexico, Mexico City (XE1)	13:02	00:38
	Chihuahua (XE2)	13:38	00:57
	Merida (XE3)	12:26	23:59
XF4	Revilla Gigedo	13:51	01:28
XT	Burkina Faso, Onagadougou	06:26	18:13
XU	Kampuchea, Phnom Penh	23:20	11:08
XW	Laos, Viangchan	23:34	11:12
XX9	Macao	22:53	10:24
XZ	Burma, Rangoon	23:59	11:39
Y2-9	German Dem.Rep.(E), Berlin	06:12	16:27
YA	Afghanistan, Kandahar	02:13	13:27
	Kabul	02:03	13:10
YB-YD	Indonesia, Jakarta	22:58	11:14
	Medan	23:38	11:39
	Pontianak	22:53	11:00
	Jayapura	20:45	08:56
YI	Iraq, Baghdad	03:40	14:51
YJ	Vanuatu, Port Vila	18:42	07:18
YK	Syria, Damascus	04:13	15:23
YN, HT	Nicaragua, Managua	12:03	23:51
YO	Romania, Bucharest	05:08	15:50
YS	El Salvador, San Salvador	12:17	00:03
YU	Yugoslavia, Belgrade	05:31	16:12
YV	Venezuela, Caracas	10:46	22:37
YVØ	Aves Is.	10:37	22:19
Z2,ZE	Zimbabwe, Harare	03:51	16:27
ZA	Albania, Tirane	05:29	16:20
ZB	Gibraltar	07:03	18:07
ZC4	British Cyprus	04:28	15:35
ZD7	St. Helena	06:21	18:53
ZD8	Ascension Is.	07:01	19:21
ZD9	Tristan da Cunha	06:25	19:40
ZF	Cayman Is.	11:50	23:26
ZK1	No. Cook Is., Manihiki	16:46	05:09
ZK1	So. Cook Is., Rarotonga	16:32	05:13
ZK2	Niue	17:10	05:48
ZK3	Tokelaus, Atafu	17:34	05:54
ZL	New Zealand, Auckland (ZL1)	17:56	07:12
	Wellington (ZL2)	17:50	07:18
	Christchurch (ZL3)	17:56	07:30
	Dunedin (ZL4)	18:00	07:42
ZL5	Antarctica, Scott Base	14:01	12:15
ZL7	Chatham Is.	17:11	06:47
ZL8	Kermadec Is.	17:36	06:34
ZL9	Auckland & Campbell Is.,		
	Auckland	18:07	08:08
	Campbell Is.	17:53	08:01
ZP	Paraguay, Asuncion	09:39	22:29
ZS	South Africa, Cape Town (ZS1)	04:26	17:34
	Port Elizabeth (ZS2)	03:56	17:05
	Bloemfontein (ZS4)	04:00	16:58
	Durban (ZS5)	03:40	16:39
	Johannesburg (ZS6)	03:55	16:47
ZS2	Prince Edward & Marion Is.,		
	Marion	02:49	16:35
ZS3	Namibia, Windhoek	04:43	17:27
1AØ	S.M.O.M	05:59	16:48
1S	Spratly Is.	22:49	10:42
3A	Monaco	06:22	17:06
3B6	Agalega	02:15	14:39
3B7	St. Brandon	01:58	14:31
3B8	Mauritius	02:03	14:43
3B9	Rodriguez Is.	01:40	14:19
3C	Equatorial Guinea, Bata	05:32	17:36
	Malabo	05:38	17:39

Pref	State/Province/Country/City	Sunrise	Sunset
3CØ	Pagalu Is.	05:46	17:55
3D2	Fiji Is., Suva	18:02	06:38
3D6	Swaziland, Mbabane	03:43	16:35
3V	Tunisia, Tunis	06:02	17:03
3W	Vietnam, Ho Chi Minh City		
	(Saigon)	23:12	11:02
	Hanoi	23:24	10:57
3X	Republic of Guinea, Conakry	07:12	19:04
3Y	Bouvet	04:51	19:09
3Y	Peter I Is.	10:05	02:26
4S	Sri Lanka, Colombo	00:56	12:52
4U	ITU Geneva	06:30	17:07
4U	United Nations Hq.	11:43	22:35
4W	Yemen, Sanaa	03:25	15:08
4X, 4Z	Israel, Jerusalem	04:16	15:29
5A	Libya, Tripoli	05:48	16:59
	Benghazi	05:17	16:29
5B	Cyprus, Nicosia	04:27	15:33
5H	Tanzania, Dar es Salaam	03:26	15:44
5N	Nigeria, Lagos	06:01	17:58
5R	Madagascar, Antananarivo	02:45	15:22
5T	Mauritania, Nouakchott	07:28	19:06
5U	Niger, Niamey	06:12	17:58
5V	Togo, Lome	06:10	18:07
5W	Western Samoa, Apia	17:26	05:55
5X	Uganda, Kampala	04:00	16:06
5Z	Kenya, Nairobi	05:02	17:11
6W	Senegal, Dakar	07:31	19:15
6Y	Jamaica, Kingston	11:31	23:10

Pref	State/Province/Country/City	Sunrise	Sunset
7O	Yemen, People's Dem. Rep.,		
	Aden	03:20	15:07
7P	Lesotho, Maseru	03:54	16:52
7Q	Malawi, Lilongwe	03:44	16:13
	Blantyre	03:37	16:09
7X	Algeria, Algiers	06:30	17:32
8P	Barbados, Bridgetown	10:19	22:05
8Q	Maldive Is.	01:20	13:20
8R	Guyana, Georgetown	10:08	22:04
9G	Ghana, Accra	06:15	18:13
9H	Malta	05:44	16:48
9J	Zambia, Lusaka	04:04	16:36
9K	Kuwait	03:23	14:41
9L	Sierra Leone, Freetown	07:09	19:03
9M2	West Malaysia, Kuala Lumpur	23:26	11:28
9M6,8	East Malaysia, Sabah,		
	Sandakan (9M6)	22:22	10:20
	Sarawak, Kuching (9M8)	22:50	10:54
9N	Nepal, Katmandu	00:52	12:13
9Q	Zaire, Kinshasa	05:05	17:19
	Kisangani	04:30	16:36
	Lubumbashi	04:11	16:36
9U	Burundi, Bujumbura	04:10	16:22
9V	Singapore	23:16	11:21
9X	Rwanda, Kigali	04:08	16:18
9Y	Trinidad & Tobago, Port of		
	Spain	10:23	22:14
J2/A	Abu Ail	03:30	15:14

Pref	State/Province/Country/City	Sunrise	Sunset
VE1	New Brunswick, Saint John	10:52	22:18
	Nova Scotia, Halifax	10:41	22:09
	Prince Edward Island,		
	Charlottetown	10:40	22:06
VE2	Quebec, Montreal	11:22	22:48
	Quebec City	11:13	22:38
VE3	Ontario, London	11:51	23:21
	Ottawa	11:30	22:57
	Sudbury	11:52	23:17
	Toronto	11:44	23:12
VE4	Manitoba, Winnipeg	12:59	00:19
VE5	Saskatchewan, Regina	13:30	00:48
	Saskatoon	13:39	00:55
VE6	Alberta, Calgary	14:08	01:26
	Edmonton	14:08	01:21
VE7	British Columbia, Prince George	14:45	01:58
	Prince Rupert	15:16	02:28
	Vancouver	14:43	02:03
VE8	Northwest Territories,		
	Yellowknife	14:22	01:14
	Resolute	13:37	23:25
VY1	Yukon, Whitehorse	15:42	02:40
VO1	Newfoundland, St. John's	10:00	21:23
VO2	Labrador, Goose Bay	10:35	21:49
W1	Connecticut, Hartford	11:16	22:47
	Maine, Bangor	11:02	22:29
	Portland	11:08	22:36
	Massachusetts, Boston	11:10	22:40
	New Hampshire, Concord	11:12	22:41
	Rhode Island, Providence	11:11	22:42
	Vermont, Montpelier	11:17	22:45
W2	New Jersey, Atlantic City	11:21	22:55
	New York, Albany	11:21	22:51
	Buffalo	11:41	23:11
	New York City	11:21	22:53
	Syracuse	11:31	23:00
W3	Delaware, Wilmington	11:26	22:59
	District of Columbia,		
	Washington	11:32	23:06
	Maryland, Baltimore	11:30	23:04
	Pennsylvania, Harrisburg	11:32	23:05
	Philadelphia	11:25	22:58
	Pittsburgh	11:44	23:17
	Scranton	11:28	22:59
W4	Alabama, Montgomery	12:05	23:46
	Florida, Jacksonville	11:46	23:29
	Miami	11:38	23:25
	Pensacola	12:08	23:51
	Georgia, Atlanta	11:58	23:38
	Savannah	11:44	23:26
	Kentucky, Lexington	12:01	23:36
	Louisville	12:06	23:41
	North Carolina, Charlotte	11:45	23:23
	Raleigh	11:36	23:14
	Wilmington	11:33	23:12
	South Carolina, Columbia	11:45	23:24
	Tennessee, Knoxville	11:58	23:35
	Memphis	12:22	00:00
	Nashville	12:09	23:46
	Virginia, Norfolk	11:28	23:04
	Richmond	11:32	23:08
W5	Arkansas, Little Rock	12:31	00:09
	Louisiana, New Orleans	12:19	00:03
	Shreveport	12:35	00:16
	Mississippi, Jackson	12:21	00:02
	New Mexico, Albuquerque	13:28	01:07
	Oklahoma, Oklahoma City	12:52	00:30
	Texas, Abilene	12:59	00:40
	Amarillo	13:09	00:47
	Dallas	12:48	00:28
	El Paso	13:26	01:07
	San Antonio	12:53	00:36
W6	California, Los Angeles	14:14	01:53
	San Francisco	14:33	02:08
W7	Arizona, Flagstaff	13:48	01:27
	Phoenix	13:49	01:29
	Idaho, Boise	14:11	01:40
	Pocatello	13:56	01:25

Pref	State/Province/Country/City	Sunrise	Sunset
	Montana, Billings	13:42	01:08
	Butte	13:58	01:23
	Great Falls	13:54	01:18
	Nevada, Las Vegas	14:03	01:40
	Reno	14:23	01:57
	Oregon, Portland	14:38	02:05
	Utah, Salt Lake City	13:52	01:24
	Washington, Seattle	14:38	02:01
	Spokane	14:19	01:42
	Wyoming, Cheyenne	13:24	00:56
	Sheridan	13:35	01:02
W8	Michigan, Detroit	11:58	23:28
	Grand Rapids	12:09	23:38
	Sault Ste. Marie	12:06	23:31
	Traverse City	12:10	23:37
	Ohio, Cincinnati	12:02	23:36
	Cleveland	11:52	23:23
	Columbus	11:56	23:29
	West Virginia, Charleston	11:50	23:24
W9	Illinois, Chicago	12:16	23:46
	Indiana, Indianapolis	12:09	23:42
	Wisconsin, Green Bay	12:19	23:46
	Milwaukee	12:18	23:47
W0	Colorado, Denver	13:24	00:57
	Grand Junction	13:38	01:12
	Iowa, Des Moines	12:40	00:11
	Kansas, Pratt	12:58	00:33
	Wichita	12:52	00:28
	Minnesota, Duluth	12:37	00:01
	Minneapolis	12:40	00:07
	Missouri, Columbia	12:33	00:07
	Kansas City	12:42	00:16
	St. Louis	12:24	23:59
	Nebraska, North Platte	13:08	00:40
	Omaha	12:49	00:20
	North Dakota, Fargo	12:56	00:20
	South Dakota, Rapid City	13:19	00:47
A2	Botswana, Gaborone	04:14	16:41
A3	Tonga, Nukualofa	17:40	06:03
A4	Oman, Muscat	02:22	14:11
A5	Bhutan, Thimbu	00:20	12:06
A6	United Arab Emirates, Abu		
	Dhabi	02:40	14:28
A7	Qatar, Ad-Dawhah	02:51	14:38
A9	Bahrein, Al-Manamah	02:55	14:41
AP	Pakistan, Karachi	01:48	13:36
	Islamabad	01:28	13:08
BV	Taiwan, Taipei	22:11	09:58
BY	People's Rep. of China, Beijing	22:39	10:12
	Harbin	22:01	09:27
	Shanghai	22:14	09:56
	Fuzhou	22:20	10:07
	Xian	23:06	10:45
	Chongqing	23:13	10:56
	Chengdu	23:23	11:05
	Lhasa	00:14	11:58
	Urumqi	00:36	12:04
	Kashi	01:20	12:53
C2	Nauru	19:00	07:07
C3	Andorra	06:20	17:50
C5	The Gambia, Banjul	07:19	19:16
C6	Bahamas, Nassau	11:27	23:14
C9	Mozambique, Maputo	03:46	16:14
	Mozambique	03:19	15:37
CE	Chile, Santiago	10:36	23:12
CE0	Easter Island	13:14	01:43
CE0	San Felix	11:17	23:45
CE0	Juan Fernandez	11:08	23:44
CM,CO	Cuba, Havana	11:46	23:35
CN	Morocco, Casablanca	06:51	18:31
CP	Bolivia, La Paz	10:34	22:54
CT	Portugal, Lisbon	07:00	18:35
CT2, CU	Azores, Ponta Delgada	08:06	19:41
CT3	Madeira Islands, Funchal	07:28	19:09
CX	Uruguay, Montevideo	09:36	22:14
D2	Angola, Luanda	05:11	17:25
D4	Cape Verde, Praia	07:47	19:42

Pref	State/Province/Country/City	Sunrise	Sunset
D6	Comoros, Moroni	03:08	15:24
DA-DL	Germany, Fed. Rep. of (W), Bonn	06:03	17:22
DU	Philippines, Manilla	22:09	10:05
EA	Spain, Madrid	06:39	18:12
EA6	Balearic Is., Palma	06:13	17:47
EA8	Canary Is., Las Palmas	07:16	19:00
EA9	Ceuta & Melilla, Ceuta	06:43	18:21
	Melilla	06:34	18:12
EI	Ireland, Dublin	06:59	18:13
EL	Liberia, Monrovia	06:53	18:55
EP	Iran, Tehran	02:55	14:32
ET	Ethiopia, Addis Ababa	03:36	15:36
	Asmera	03:37	15:33
F	France, Paris	06:21	17:42
FG	Guadeloupe	10:20	22:15
FG, FS	St. Martin	10:26	22:20
FH	Mayotte	03:01	15:18
FK	New Caledonia, Noumia	18:52	07:17
FM	Martinique	10:17	22:13
FO	Clipperton	13:28	01:27
FO	Fr. Polynesia, Tahiti	16:38	04:59
FP	St. Pierre & Miquelon, St. Pierre	10:12	21:37
FR	Glorioso	02:54	15:09
FR	Juan de Nova	03:09	15:29
	Europa	03:17	15:41
FR	Reunion	02:16	14:40
FR	Tromelin	02:23	14:43
FT8W	Crozet	02:16	15:10
FT8X	Kerguelen	01:04	14:04
FT8Y	Antarctica, Dumont D'Urville	19:52	09:49
FT8Z	Amsterdam & St. Paul Is., Amsterdam	00:40	13:21
FW	Wallis & Futuna Is., Wallis	17:47	06:04
FY	Fr. Guiana, Cayenne	09:38	21:41
G	England, London	06:32	17:50
GD	Isle of Man	06:53	18:05
GI	Northern Ireland, Belfast	06:59	18:10
GJ	Jersey	06:39	18:00
GM	Scotland, Glasgow	06:53	18:02
	Aberdeen	06:46	17:52
GU	Guernsey	06:41	18:02
GW	Wales, Cardiff	06:45	18:02
H4	Solomon Islands, Honiara	19:24	07:38
HA	Hungary, Budapest	05:13	16:36
HB	Switzerland, Bern	05:59	17:23
HBØ	Liechtenstein	05:50	17:14
HC	Ecuador, Quito	11:19	23:26
HC8	Galapagos Is.	12:09	00:16
HH	Haiti, Port-Au-Prince	11:03	22:56
HI	Dominican Republic, Santo Domingo	10:54	22:47
HK	Colombia, Bogota	11:05	23:09
HKØ	Malpelo Is.	11:33	23:37
HKØ	San Adreas	11:39	23:36
HL, HM	Korea, Seoul	21:55	09:31
HP	Panama, Panama	11:29	23:29
HR	Honduras, Tegucigalpa	12:01	23:58
HS	Thailand, Bangkok	23:30	11:27
HV	Vatican City	05:35	17:06
HZ, 7Z	Saudi Arabia, Dharan	02:57	14:44
	Mecca	03:36	15:27
	Riyadh	03:10	14:58
I	Italy, Rome	05:35	17:06
	Trieste	05:33	16:58
	Sicily	05:27	17:03
IS	Sardinia, Cagliari	05:47	17:21
J2	Djibouti, Djibouti	03:19	15:17
J3	Grenada	10:19	22:17
J5	Guinea-Bissau, Bissau	07:14	19:12
J6	St. Lucia	10:16	22:13
J7	Dominica	10:18	22:13
J8	St. Vincent	10:17	22:14
JA-JS	Japan, Tokyo	21:03	08:40
	Nagasaki	21:41	09:21
	Sapporo	21:00	08:30
JD	Minami Torishima	20:01	07:49
JD	Ogasawara, Kazan Is.	20:54	08:39
JT	Mongolia, Ulan Bator	23:22	10:44
JW	Svalbard, Spitsbergen	06:40	15:34
JX	Jan Mayen	07:36	17:52

Pref	State/Province/Country/City	Sunrise	Sunset
JY	Jordan, Amman	03:56	15:38
KC4	Antarctica, Byrd Station	11:40	04:41
	McMurdo Sound	17:05	09:03
	Palmer Station	09:33	23:20
KC6	Micronesia, Ponape	19:37	07:38
KC6	Belau, Yap	20:58	08:58
	Koror	21:12	09:13
KG4	Guantanamo Bay	11:16	23:07
KH1	Baker, Howland Is.	17:51	05:58
KH2	Guam, Agana	20:33	08:30
KH3	Johnston Is.	17:28	05:22
KH4	Midway Is.	18:08	05:53
KH5	Palmyra Is.	16:58	05:00
KH5K	Kingman Reef	17:01	05:02
KH6	Hawaii, Hilo	16:35	04:27
	Honolulu	16:47	04:38
KH7	Kure Is.	18:12	05:57
KH8	American Samoa, Pago Pago	17:25	05:43
KH9	Wake Is.	19:08	07:00
KHØ	Mariana Is., Saipan	20:30	08:25
KL7	Alaska, Adak	18:19	05:35
	Anchorage	16:43	03:39
	Fairbanks	16:40	03:25
	Juneau	15:36	02:40
	Nome	17:49	04:35
KP1	Navassa Is.	11:14	23:07
KP2	Virgin Islands, Charlotte Amalie	10:34	22:27
KP4	Puerto Rico, San Juan	10:39	22:32
KP5	Desecheo Is.	10:44	22:37
KX6	Marshall Islands, Kwajalein	19:01	07:01
LA-LJ	Norway, Oslo	05:58	16:58
LU	Argentina, Buenos Aires	09:45	22:23
LX	Luxembourg	06:06	17:26
LZ	Bulgaria, Sofia	04:53	16:22
OA	Peru, Lima	11:11	23:27
OD	Lebanon, Beirut	03:59	15:38
OE	Austria, Vienna	05:24	16:47
OH	Finland, Helsinki	05:01	16:00
OHØ	Aland Is.	05:21	16:20
OJØ	Market Reef	05:25	16:24
OK	Czechoslovakia, Prague	05:33	16:53
ON	Belgium, Brussels	06:14	17:32
OX, XP	Greenland, Godthaab	10:14	21:01
	Thule	12:02	21:29
OY	Faroe Islands, Torshavn	07:11	18:05
OZ	Denmark, Copenhagen	05:46	16:55
P2	Papua New Guinea, Madang	20:23	08:34
	Port Moresby	20:15	08:29
PA-PI	Netherlands, Amsterdam	06:13	17:29
PJ, P4	Netherlands Antilles, Willemstad	10:48	22:46
PJ5-8	St. Maarten and Saba, St. Maarten	10:27	22:20
PY	Brazil, Brasilia	09:13	21:32
	Rio De Janeiro	08:51	21:16
	Natal	08:26	20:37
	Manaus	10:07	22:16
	Porto Alegre	09:19	21:52
PYØ	Fernando De Noronha	08:15	20:25
PYØ	St. Peter & St. Paul Rocks	08:05	20:11
PYØ	Trindade & Martin Vaz Is., Trindade	07:56	20:19
PZ	Suriname, Paramaribo	09:50	21:53
S2	Bangladesh, Dacca	00:15	12:03
S7	Seychelles, Victoria	02:24	14:34
S9	Sao Tome	05:41	17:47
SJ-SM	Sweden, Stockholm	05:28	16:29
SP	Poland, Krakow	05:11	16:30
	Warsaw	05:09	16:25
ST	Sudan, Khartoum	04:03	15:58
STØ	Southern Sudan, Juba	04:03	16:06
SU	Egypt, Cairo	04:13	15:57
SV	Greece, Athens	04:48	16:23
SV5	Dodecanese, Rhodes	04:29	16:06
SV9	Crete	04:41	16:19
SV/A	Mount Athos	04:47	16:20
T2	Tuvalu, Funafuti	18:10	06:23
T3Ø	West Kiribati, Bonriki	18:35	06:41
T31	Central Kiribati, Kanton	17:33	05:42
T32	East Kiribati, Christmas Is.	16:38	04:43
T5	Somalia, Mogadishu	03:06	15:12
T7	San Marino	05:37	17:06

Pref	State/Province/Country/City	Sunrise	Sunset
TA	Turkey, Ankara	04:13	15:46
	Istanbul	04:29	16:01
TF	Iceland, Reykjavik	08:15	19:02
TG	Guatemala, Guatemala City	12:15	00:11
TI	Costa Rica, San Jose	11:47	23:46
TI9	Cocos Is.	11:57	00:00
TJ	Cameroon, Yaounde	05:23	17:27
TK	Corsica	05:49	17:20
TL	Central African Republic, Bangui	04:55	16:58
TN	Congo, Brazzaville	05:04	17:14
TR	Gabon, Libreville	05:29	17:36
TT	Chad, N'Djamena	05:12	17:10
TU	Ivory Coast, Abidjan	06:25	18:28
TY	Benin, Porto Novo	05:59	18:01
TZ	Mali, Bamako	06:44	18:41
UA	Russia, European,		
	Leningrad (UA1)	04:40	15:39
	Archangel (UA1)	04:06	14:51
	Murmansk (UA1)	04:45	15:12
	Moscow (UA3)	04:06	15:15
	Kuibyshev (UA4)	03:13	14:27
	Rostov (UA6)	03:51	15:14
UA1P	Franz Josef Land	04:24	12:53
UA2	Kaliningrad	05:13	16:24
UA9,Ø	Russia, Asiatic,		
	Novosibirsk (UA9)	01:04	12:14
	Perm (UA9)	02:53	13:58
	Omsk (UA9)	01:42	12:52
	Norilsk (UAØ)	01:06	11:31
	Irkutsk (UAØ)	23:36	10:51
	Vladivostok (UAØ)	21:38	09:08
	Petropavlovsk (UAØ)	19:59	07:13
	Khabarovsk (UAØ)	21:29	08:51
	Krasnoyarsk (UAØ)	00:25	11:34
	Yakutsk (UAØ)	22:05	08:59
	Wrangel Island (UAØ)	19:01	05:17
	Kyzyl (UAØY)	00:14	11:31
UB	Ukraine, Kiev	04:29	15:48
UC	Byelorussia, Minsk	04:44	15:57
UD	Azerbaijan, Baku	03:05	14:37
UF	Georgia, Tbilisi	03:26	14:57
UG	Armenia, Ierevan	03:26	14:59
UH	Turkoman, Ashkhabad	02:29	14:05
UI	Uzbek, Bukhara	02:06	13:40
	Tashkent	01:48	13:19
UJ	Tadzhik, Samarkand	01:57	13:30
	Dushanbe	01:48	13:22
UL	Kazakh, Alma-ata	01:19	12:48
UM	Kirghiz, Frunze	01:27	12:57
UO	Moldavia, Kishinev	04:33	15:57
UP	Lithuania, Vilna	04:53	16:04
UQ	Latvia, Riga	05:01	16:08
UR	Estonia, Tallinn	05:01	16:02
V2	Antigua & Barbuda, St. Johns	10:21	22:15
V3	Belize, Belmopan	12:09	00:03
V4	St. Christopher & Nevis	10:24	22:18
V8	Brunei, Bandar Seri Begawan	22:30	10:33
VE1, CY	Sable Is.	10:26	21:55
VE1, CY	St. Paul Is.	10:29	21:53
VK	Australia, Canberra (VK1)	19:55	08:33
	Sydney (VK2)	19:47	08:24
	Melbourne (VK3)	20:10	08:51
	Brisbane (VK4)	19:44	08:14
	Adelaide (VK5)	20:37	09:15
	Perth (VK6)	22:10	10:45
	Hobart, Tasmania (VK7)	19:57	08:46
	Darwin (VK8)	21:19	09:35
VK	Lord Howe Is.	19:17	07:51
VK9	Christmas Is.	23:00	11:15
VK9	Cocos-Keeling Is.	23:35	11:51
VK9	Mellish Reef	19:37	07:58
VK9	Norfolk Is.	18:43	07:14
VK9	Willis Is.	20:03	08:22
VKØ	Heard Is.	00:43	13:51
VKØ	Macquarie Is.	18:59	08:11
VP2E	Anguilla	10:26	22:19
VP2M	Montserrat	10:22	22:17
VP2V	British Virgin Is., Tortola	10:33	22:25
VP5	Turks & Caicos Islands,		
	Grand Turk	11:00	22:51
VP8	Falkland Islands, Stanley	09:30	22:35

Pref	State/Province/Country/City	Sunrise	Sunset
VP8	So. Georgia Is.	08:02	21:13
VP8	So. Orkney Is.	08:28	21:58
VP8	So. Sandwich Islands, Saunders Is.	07:17	20:38
VP8	So. Shetland Is., King George Is.	09:16	22:51
VP9	Bermuda	10:39	22:20
VQ9	Chagos, Diego Garcia	01:15	13:27
VR6	Pitcairn Is.	14:37	03:05
VS6	Hong Kong	22:39	10:28
VU	India, Bombay	01:23	13:16
	Calcutta	00:22	12:12
	New Delhi	01:10	12:54
	Bangalore	01:02	12:59
VU7	Andaman Islands, Port Blair	00:00	11:58
VU7	Laccadive Is.	01:19	13:18
XE	Mexico, Mexico City (XE1)	12:51	00:43
	Chihuahua (XE2)	13:23	01:07
	Merida (XE3)	12:14	00:05
XF4	Revilla Gigedo	13:40	01:33
XT	Burkina Faso, Onagadougou	06:18	18:16
XU	Kampuchea, Phnom Penh	23:12	11:10
XW	Laos, Viangchan	23:24	11:17
XX9	Macao	22:41	10:31
XZ	Burma, Rangoon	23:50	11:44
Y2-9	German Dem.Rep.(E), Berlin	05:39	16:55
YA	Afghanistan, Kandahar	01:56	13:39
	Kabul	01:44	13:23
YB-YD	Indonesia, Jakarta	22:58	11:09
	Medan	23:34	11:38
	Pontianak	22:50	10:57
	Jayapura	20:44	08:52
YI	Iraq, Baghdad	03:22	15:03
YJ	Vanuatu, Port Vila	18:47	07:08
YK	Syria, Damascus	03:56	15:35
YN, HT	Nicaragua, Managua	11:56	23:54
YO	Romania, Bucharest	04:42	16:10
YS	El Salvador, San Salvador	12:09	00:06
YU	Yugoslavia, Belgrade	05:05	16:32
YV	Venezuela, Caracas	10:39	22:38
YVØ	Aves Is.	10:28	22:23
Z2,ZE	Zimbabwe, Harare	03:56	16:17
ZA	Albania, Tirane	05:06	16:37
ZB	Gibraltar	06:44	18:21
ZC4	British Cyprus	04:09	15:48
ZD7	St. Helena	06:25	18:44
ZD8	Ascension Is.	07:02	19:15
ZD9	Tristan da Cunha	06:40	19:20
ZF	Cayman Is.	11:39	23:31
ZK1	No.Cook Is., Manihiki	16:47	05:02
ZK1	So. Cook Is., Rarotonga	16:38	05:02
ZK2	Niue	17:15	05:37
ZK3	Tokelaus, Atafu	17:35	05:48
ZL	New Zealand, Auckland (ZL1)	18:11	06:52
	Wellington (ZL2)	18:08	06:55
	Christchurch (ZL3)	18:15	07:05
	Dunedin (ZL4)	18:22	07:16
ZL5	Antarctica, Scott Base	17:04	09:06
ZL7	Chatham Is.	17:31	06:22
ZL8	Kermadec Is.	17:46	06:18
ZL9	Auckland & Campbell Is.,		
	Auckland	18:33	07:36
	Campbell Is.	18:21	07:28
ZP	Paraguay, Asuncion	09:48	22:15
ZS	South Africa, Cape Town (ZS1)	04:39	17:15
	Port Elizabeth (ZS2)	04:09	16:46
	Bloemfontein (ZS4)	04:10	16:42
	Durban (ZS5)	03:51	16:23
	Johannesburg (ZS6)	04:04	16:32
ZS2	Prince Edward & Marion Is., Marion	03:12	16:07
ZS3	Namibia, Windhoek	04:50	17:15
1AØ	S.M.O.M	05:36	17:06
1S	Spratly Is.	22:43	10:43
3A	Monaco	05:57	17:25
3B6	Agalega	02:17	14:32
3B7	St. Brandon	02:02	14:21
3B8	Mauritius	02:09	14:32
3B9	Rodriguez Is.	01:46	14:08
3C	Equatorial Guinea, Bata	05:29	17:34
	Malabo	05:34	17:37

Pref	State/Province/Country/City	Sunrise	Sunset	Pref	State/Province/Country/City	Sunrise	Sunset
3CØ	Pagalu Is.	05:44	17:52	7O	Yemen, People's Dem. Rep.,		
3D2	Fiji Is., Suva	18:07	06:28		Aden	03:12	15:09
3D6	Swaziland, Mbabane	03:52	16:21	7P	Lesotho, Maseru	04:05	16:36
3V	Tunisia, Tunis	05:42	17:18	7Q	Malawi, Lilongwe	03:47	16:04
3W	Vietnam, Ho Chi Minh City				Blantyre	03:41	16:00
	(Saigon)	23:05	11:03	7X	Algeria, Algiers	06:10	17:47
	Hanoi	23:12	11:03	8P	Barbados, Bridgetown	10:11	22:08
3X	Republic of Guinea, Conakry	07:06	19:05	8Q	Maldive Is.	01:15	13:19
3Y	Bouvet	05:21	18:33	8R	Guyana, Georgetown	10:03	22:04
3Y	Peter I Is.	11:08	01:18	9G	Ghana, Accra	06:10	18:13
4S	Sri Lanka, Colombo	00:50	12:52	9H	Malta	05:24	17:02
4U	ITU Geneva	06:03	17:28	9J	Zambia, Lusaka	04:08	16:27
4U	United Nations Hq.	11:21	22:53	9K	Kuwait	03:08	14:51
4W	Yemen, Sanaa	03:16	15:11	9L	Sierra Leone, Freetown	07:03	19:04
4X, 4Z	Israel, Jerusalem	03:59	15:41	9M2	West Malaysia, Kuala Lumpur	23:22	11:26
5A	Libya, Tripoli	05:30	17:11	9M6,8	East Malaysia, Sabah,		
	Benghazi	05:00	16:41		Sandakan (9M6)	22:17	10:19
5B	Cyprus, Nicosia	04:08	15:46		Sarawak, Kuching (9M8)	22:47	10:52
5H	Tanzania, Dar es Salaam	03:27	15:39	9N	Nepal, Katmandu	00:37	12:22
5N	Nigeria, Lagos	05:56	17:58	9Q	Zaire, Kinshasa	05:04	17:14
5R	Madagascar, Antananarivo	02:50	15:11		Kisangani	04:27	16:33
5T	Mauritania, Nouakchott	07:18	19:11		Lubumbashi	04:13	16:28
5U	Niger, Niamey	06:04	18:01	9U	Burundi, Bujumbura	04:09	16:18
5V	Togo, Lome	06:05	18:07	9V	Singapore	23:13	11:18
5W	Western Samoa, Apia	17:29	05:46	9X	Rwanda, Kigali	04:06	16:14
5X	Uganda, Kampala	03:57	16:04	9Y	Trinidad & Tobago, Port of		
5Z	Kenya, Nairobi	05:00	17:07		Spain	10:16	22:15
6W	Senegal, Dakar	07:23	19:19	J2/A	Abu Ail	03:21	15:18
6Y	Jamaica, Kingston	11:21	23:14				

Sunrise and Sunset, in UTC, for 21-Mar

Pref	State/Province/Country/City	Sunrise	Sunset
VE1	New Brunswick, Saint John	10:26	22:37
	Nova Scotia, Halifax	10:16	22:27
	Prince Edward Island, Charlottetown	10:14	22:25
VE2	Quebec, Montreal	10:56	23:07
	Quebec City	10:46	22:57
VE3	Ontario, London	11:27	23:37
	Ottawa	11:04	23:15
	Sudbury	11:25	23:36
	Toronto	11:19	23:30
VE4	Manitoba, Winnipeg	12:29	00:41
VE5	Saskatchewan, Regina	12:59	01:11
	Saskatoon	13:07	01:20
VE6	Alberta, Calgary	13:37	01:49
	Edmonton	13:34	01:47
VE7	British Columbia, Prince George	14:11	02:25
	Prince Rupert	14:41	02:55
	Vancouver	14:13	02:25
VE8	Northwest Territories, Yellowknife	13:36	01:53
	Resolute	12:11	00:42
VY1	Yukon, Whitehorse	14:59	03:15
VO1	Newfoundland, St. John's	09:32	21:43
VO2	Labrador, Goose Bay	10:02	22:15
W1	Connecticut, Hartford	10:52	23:03
	Maine, Bangor	10:37	22:47
	Portland	10:43	22:53
	Massachusetts, Boston	10:46	22:56
	New Hampshire, Concord	10:47	22:58
	Rhode Island, Providence	10:47	22:58
	Vermont, Montpelier	10:52	23:03
W2	New Jersey, Atlantic City	10:59	23:09
	New York, Albany	10:57	23:07
	Buffalo	11:17	23:28
	New York City	10:58	23:08
	Syracuse	11:06	23:17
W3	Delaware, Wilmington	11:04	23:14
	District of Columbia, Washington	11:10	23:20
	Maryland, Baltimore	11:08	23:18
	Pennsylvania, Harrisburg	11:09	23:19
	Philadelphia	11:03	23:13
	Pittsburgh	11:22	23:32
	Scranton	11:04	23:15
W4	Alabama, Montgomery	11:48	23:56
	Florida, Jacksonville	11:29	23:38
	Miami	11:24	23:32
	Pensacola	11:51	00:00
	Georgia, Atlanta	11:40	23:49
	Savannah	11:27	23:36
	Kentucky, Lexington	11:40	23:50
	Louisville	11:45	23:55
	North Carolina, Charlotte	11:25	23:35
	Raleigh	11:17	23:26
	Wilmington	11:14	23:23
	South Carolina, Columbia	11:26	23:35
	Tennessee, Knoxville	11:38	23:47
	Memphis	12:03	00:12
	Nashville	11:49	23:59
	Virginia, Norfolk	11:07	23:17
	Richmond	11:12	23:21
W5	Arkansas, Little Rock	12:11	00:21
	Louisiana, New Orleans	12:03	00:12
	Shreveport	12:17	00:26
	Mississippi, Jackson	12:03	00:12
	New Mexico, Albuquerque	13:09	01:18
	Oklahoma, Oklahoma City	12:32	00:41
	Texas, Abilene	12:41	00:50
	Amarillo	12:49	00:59
	Dallas	12:30	00:38
	El Paso	13:08	01:17
	San Antonio	12:37	00:45
W6	California, Los Angeles	13:55	02:04
	San Francisco	14:12	02:21
W7	Arizona, Flagstaff	13:29	01:38
	Phoenix	13:31	01:40
	Idaho, Boise	13:46	01:57
	Pocatello	13:32	01:42

Pref	State/Province/Country/City	Sunrise	Sunset
	Montana, Billings	13:15	01:26
	Butte	13:31	01:42
	Great Falls	13:26	01:38
	Nevada, Las Vegas	13:43	01:52
	Reno	14:01	02:11
	Oregon, Portland	14:12	02:23
	Utah, Salt Lake City	13:29	01:40
	Washington, Seattle	14:10	02:22
	Spokane	13:51	02:02
	Wyoming, Cheyenne	13:01	01:11
	Sheridan	13:09	01:20
W8	Michigan, Detroit	11:34	23:44
	Grand Rapids	11:44	23:55
	Sault Ste. Marie	11:39	23:50
	Traverse City	11:44	23:55
	Ohio, Cincinnati	11:40	23:50
	Cleveland	11:28	23:39
	Columbus	11:34	23:44
	West Virginia, Charleston	11:28	23:38
W9	Illinois, Chicago	11:52	00:02
	Indiana, Indianapolis	11:47	23:57
	Wisconsin, Green Bay	11:53	00:04
	Milwaukee	11:53	00:04
W0	Colorado, Denver	13:02	01:12
	Grand Junction	13:16	01:26
	Iowa, Des Moines	12:16	00:26
	Kansas, Pratt	12:37	00:46
	Wichita	12:31	00:41
	Minnesota, Duluth	12:09	00:21
	Minneapolis	12:15	00:26
	Missouri, Columbia	12:11	00:21
	Kansas City	12:20	00:30
	St. Louis	12:03	00:13
	Nebraska, North Platte	12:45	00:55
	Omaha	12:25	00:36
	North Dakota, Fargo	12:28	00:40
	South Dakota, Rapid City	12:54	01:05
A2	Botswana, Gaborone	04:20	16:27
A3	Tonga, Nukualofa	17:44	05:51
A4	Oman, Muscat	02:08	14:16
A5	Bhutan, Thimbu	00:05	12:13
A6	United Arab Emirates, Abu Dhabi	02:26	14:34
A7	Qatar, Ad-Dawhah	02:37	14:45
A9	Bahrein, Al-Manamah	02:40	14:49
AP	Pakistan, Karachi	01:34	13:42
	Islamabad	01:09	13:19
BV	Taiwan, Taipei	21:57	10:05
BY	People's Rep. of China, Beijing	22:16	10:26
	Harbin	21:34	09:46
	Shanghai	21:56	10:05
	Fuzhou	22:06	10:14
	Xian	22:47	10:56
	Chongqing	22:57	11:05
	Chengdu	23:06	11:15
	Lhasa	23:58	12:06
	Urumqi	00:11	12:22
	Kashi	00:58	13:08
C2	Nauru	18:56	07:03
C3	Andorra	05:56	18:06
C5	The Gambia, Banjul	07:10	19:17
C6	Bahamas, Nassau	11:12	23:20
C9	Mozambique, Maputo	03:53	16:00
	Mozambique	03:21	15:27
CE	Chile, Santiago	10:47	22:53
CE0	Easter Island	13:21	01:28
CE0	San Felix	11:24	23:31
CE0	Juan Fernandez	11:19	23:25
CM,CO	Cuba, Havana	11:32	23:40
CN	Morocco, Casablanca	06:32	18:41
CP	Bolivia, La Paz	10:37	22:44
CT	Portugal, Lisbon	06:39	18:49
CT2, CU	Azores, Ponta Delgada	07:45	19:54
CT3	Madeira Islands, Funchal	07:10	19:19
CX	Uruguay, Montevideo	09:48	21:55
D2	Angola, Luanda	05:11	17:17
D4	Cape Verde, Praia	07:37	19:44

Pref	State/Province/Country/City	Sunrise	Sunset
D6	Comoros, Moroni	03:09	15:15
DA-DL	Germany, Fed. Rep. of (W),		
	Bonn	05:33	17:45
DU	Philippines, Manilla	21:59	10:06
EA	Spain, Madrid	06:17	18:27
EA6	Balearic Is., Palma	05:51	18:01
EA8	Canary Is., Las Palmas	07:00	19:08
EA9	Ceuta & Melilla, Ceuta	06:23	18:33
	Melilla	06:14	18:23
EI	Ireland, Dublin	06:25	18:39
EL	Liberia, Monrovia	06:47	18:53
EP	Iran, Tehran	02:35	14:44
ET	Ethiopia, Addis Ababa	03:29	15:36
	Asmera	03:28	15:35
F	France, Paris	05:52	18:04
FG	Guadeloupe	10:10	22:17
FG, FS	St. Martin	10:15	22:23
FH	Mayotte	03:02	15:09
FK	New Caledonia, Noumia	18:58	07:04
FM	Martinique	10:07	22:14
FO	Clipperton	13:20	01:27
FO	Fr. Polynesia, Tahiti	16:42	04:48
FP	St. Pierre & Miquelon, St. Pierre	09:45	21:57
FR	Glorioso	02:54	15:01
FR	Juan de Nova	03:12	15:19
	Europa	03:22	15:28
FR	Reunion	02:21	14:28
FR	Tromelin	02:26	14:32
FT8W	Crozet	02:35	14:43
FT8X	Kerguelen	01:26	13:34
FT8Y	Antarctica, Dumont D'Urville	20:40	08:53
FT8Z	Amsterdam & St. Paul Is.,		
	Amsterdam	00:53	13:00
FW	Wallis & Futuna Is., Wallis	17:49	05:55
FY	Fr. Guiana, Cayenne	09:33	21:39
G	England, London	06:01	18:14
GD	Isle of Man	06:18	18:32
GI	Northern Ireland, Belfast	06:24	18:37
GJ	Jersey	06:10	18:22
GM	Scotland, Glasgow	06:17	18:31
	Aberdeen	06:08	18:23
GU	Guernsey	06:12	18:24
GW	Wales, Cardiff	06:13	18:26
H4	Solomon Islands, Honiara	19:24	07:30
HA	Hungary, Budapest	04:45	16:56
HB	Switzerland, Bern	05:31	17:43
HB0	Liechtenstein	05:23	17:34
HC	Ecuador, Quito	11:15	23:22
HC8	Galapagos Is.	12:05	00:12
HH	Haiti, Port-Au-Prince	10:52	23:00
HI	Dominican Republic, Santo		
	Domingo	10:43	22:51
HK	Colombia, Bogota	11:00	23:07
HK0	Malpelo Is.	11:28	23:35
HK0	San Adreas	11:30	23:37
HL, HM	Korea, Seoul	21:34	09:44
HP	Panama, Panama	11:21	23:28
HR	Honduras, Tegucigalpa	11:52	23:59
HS	Thailand, Bangkok	23:21	11:28
HV	Vatican City	05:12	17:22
HZ, 7Z	Saudi Arabia, Dharan	02:43	14:51
	Mecca	03:24	15:32
	Riyadh	02:56	15:04
I	Italy, Rome	05:12	17:22
	Trieste	05:06	17:17
	Sicily	05:06	17:16
IS	Sardinia, Cagliari	05:25	17:35
J2	Djibouti, Djibouti	03:10	15:18
J3	Grenada	10:10	22:18
J5	Guinea-Bissau, Bissau	07:06	19:13
J6	St. Lucia	10:07	22:14
J7	Dominica	10:08	22:16
J8	St. Vincent	10:08	22:16
JA-JS	Japan, Tokyo	20:43	08:52
	Nagasaki	21:23	09:32
	Sapporo	20:36	08:47
JD	Minami Torishima	19:47	07:55
JD	Ogasawara, Kazan Is.	20:39	08:47
JT	Mongolia, Ulan Bator	22:53	11:05
JW	Svalbard, Spitsbergen	04:42	17:24
JX	Jan Mayen	06:27	18:53

Pref	State/Province/Country/City	Sunrise	Sunset
JY	Jordan, Amman	03:39	15:48
KC4	Antarctica, Byrd Station	13:52	02:22
	McMurdo Sound	18:48	07:12
	Palmer Station	10:17	22:29
KC6	Micronesia, Ponape	19:30	07:37
KC6	Belau, Yap	20:51	08:58
	Koror	21:05	09:12
KG4	Guantanamo Bay	11:04	23:11
KH1	Baker, Howland Is.	17:47	05:54
KH2	Guam, Agana	20:24	08:31
KH3	Johnston Is.	17:17	05:25
KH4	Midway Is.	17:52	06:01
KH5	Palmyra Is.	16:52	04:59
KH5K	Kingman Reef	16:55	05:01
KH6	Hawaii, Hilo	16:23	04:31
	Honolulu	16:35	04:42
KH7	Kure Is.	17:56	06:05
KH8	American Samoa, Pago Pago	17:27	05:33
KH9	Wake Is.	18:57	07:04
KH0	Mariana Is., Saipan	20:20	08:27
KL7	Alaska, Adak	17:47	06:00
	Anchorage	15:58	04:15
	Fairbanks	15:49	04:08
	Juneau	14:57	03:12
	Nome	16:59	05:18
KP1	Navassa Is.	11:03	23:11
KP2	Virgin Islands, Charlotte Amalie	10:23	22:30
KP4	Puerto Rico, San Juan	10:28	22:35
KP5	Desecheo Is.	10:33	22:41
KX6	Marshall Islands, Kwajalein	18:54	07:01
LA-LJ	Norway, Oslo	05:16	17:32
LU	Argentina, Buenos Aires	09:57	22:04
LX	Luxembourg	05:36	17:48
LZ	Bulgaria, Sofia	04:28	16:39
OA	Peru, Lima	11:12	23:18
OD	Lebanon, Beirut	03:40	15:49
OE	Austria, Vienna	04:56	17:08
OH	Finland, Helsinki	04:19	16:35
OH0	Aland Is.	04:39	16:55
OJ0	Market Reef	04:43	16:59
OK	Czechoslovakia, Prague	05:03	17:15
ON	Belgium, Brussels	05:43	17:55
OX, XP	Greenland, Godthaab	09:24	21:43
	Thule	10:24	23:00
OY	Faroe Islands, Torshavn	06:25	18:43
OZ	Denmark, Copenhagen	05:09	17:23
P2	Papua New Guinea, Madang	20:21	08:28
	Port Moresby	20:15	08:22
PA-PI	Netherlands, Amsterdam	05:41	17:54
PJ, P4	Netherlands Antilles, Willemstad	10:39	22:46
PJ5-8	St. Maarten and Saba,		
	St. Maarten	10:16	22:23
PY	Brazil, Brasilia	09:15	21:22
	Rio De Janeiro	08:56	21:03
	Natal	08:24	20:31
	Manaus	10:04	22:11
	Porto Alegre	09:28	21:35
PY0	Fernando De Noronha	08:13	20:20
PY0	St. Peter & St. Paul Rocks	08:01	20:08
PY0	Trindade & Martin Vaz Is.,		
	Trindade	08:01	20:07
PZ	Suriname, Paramaribo	09:44	21:51
S2	Bangladesh, Dacca	00:01	12:09
S7	Seychelles, Victoria	02:22	14:28
S9	Sao Tome	05:37	17:43
SJ-SM	Sweden, Stockholm	04:47	17:02
SP	Poland, Krakow	04:41	16:53
	Warsaw	04:36	16:49
ST	Sudan, Khartoum	03:53	16:00
ST0	Southern Sudan, Juba	03:57	16:04
SU	Egypt, Cairo	03:57	16:06
SV	Greece, Athens	04:27	16:37
SV5	Dodecanese, Rhodes	04:09	16:19
SV9	Crete	04:21	16:31
SV/A	Mount Athos	04:25	16:35
T2	Tuvalu, Funafuti	18:09	06:16
T30	West Kiribati, Bonriki	18:31	06:37
T31	Central Kiribati, Kanton	17:30	05:37
T32	East Kiribati, Christmas Is.	16:33	04:40
T5	Somalia, Mogadishu	03:02	15:09
T7	San Marino	05:12	17:23

Pref	State/Province/Country/City	Sunrise	Sunset
TA	Turkey, Ankara	03:50	16:00
	Istanbul	04:06	16:16
TF	Iceland, Reykjavik	07:26	19:44
TG	Guatemala, Guatemala City	12:05	00:12
TI	Costa Rica, San Jose	11:39	23:46
TI9	Cocos Is.	11:51	23:58
TJ	Cameroon, Yaounde	05:17	17:24
TK	Corsica	05:26	17:36
TL	Central African Republic, Bangui	04:49	16:56
TN	Congo, Brazzaville	05:02	17:09
TR	Gabon, Libreville	05:25	17:32
TT	Chad, N'Djamena	05:03	17:10
TU	Ivory Coast, Abidjan	06:19	18:26
TY	Benin, Porto Novo	05:53	18:00
TZ	Mali, Bamako	06:35	18:42
UA	Russia, European,		
	Leningrad (UA1)	03:58	16:14
	Archangel (UA1)	03:15	15:34
	Murmansk (UA1)	03:43	16:06
	Moscow (UA3)	03:29	15:44
	Kuibyshev (UA4)	02:40	14:53
	Rostov (UA6)	03:23	15:35
UA1P	Franz Josef Land	02:11	14:59
UA2	Kaliningrad	04:38	16:52
UA9,Ø	Russia, Asiatic,		
	Novosibirsk (UA9)	00:28	12:42
	Perm (UA9)	02:14	14:29
	Omsk (UA9)	01:06	13:20
	Norilsk (UAØ)	00:03	12:26
	Irkutsk (UAØ)	23:03	11:16
	Vladivostok (UAØ)	21:14	09:25
	Petropavlovsk (UAØ)	19:25	07:39
	Khabarovsk (UAØ)	21:00	09:12
	Krasnoyarsk (UAØ)	23:48	12:03
	Yakutsk (UAØ)	21:19	09:37
	Wrangel Island (UAØ)	17:52	06:17
	Kyzyl (UAØY)	23:42	11:55
UB	Ukraine, Kiev	03:59	16:11
UC	Byelorussia, Minsk	04:10	16:23
UD	Azerbaijan, Baku	02:42	14:52
UF	Georgia, Tbilisi	03:02	15:13
UG	Armenia, Ierevan	03:04	15:14
UH	Turkoman, Ashkhabad	02:08	14:18
UI	Uzbek, Bukhara	01:44	13:54
	Tashkent	01:24	13:35
UJ	Tadzhik, Samarkand	01:35	13:45
	Dushanbe	01:27	13:37
UL	Kazakh, Alma-ata	00:54	13:05
UM	Kirghiz, Frunze	01:03	13:14
UO	Moldavia, Kishinev	04:06	16:17
UP	Lithuania, Vilna	04:18	16:32
UQ	Latvia, Riga	04:23	16:38
UR	Estonia, Tallinn	04:20	16:36
V2	Antigua & Barbuda, St. Johns	10:10	22:18
V3	Belize, Belmopan	11:58	00:06
V4	St. Christopher & Nevis	10:13	22:21
V8	Brunei, Bandar Seri Begawan	22:24	10:31
VE1, CY	Sable Is.	10:01	22:12
VE1, CY	St. Paul Is.	10:01	22:13
VK	Australia, Canberra (VK1)	20:07	08:14
	Sydney (VK2)	19:59	08:05
	Melbourne (VK3)	20:23	08:30
	Brisbane (VK4)	19:51	07:58
	Adelaide (VK5)	20:49	08:56
	Perth (VK6)	22:20	10:27
	Hobart, Tasmania (VK7)	20:14	08:21
	Darwin (VK8)	21:20	09:26
VK	Lord Howe Is.	19:27	07:34
VK9	Christmas Is.	23:01	11:07
VK9	Cocos-Keeling	23:36	11:43
VK9	Mellish Reef	19:40	07:47
VK9	Norfolk Is.	18:51	06:58
VK9	Willis Is.	20:06	08:12
VKØ	Heard Is.	01:09	13:18
VKØ	Macquarie Is.	19:27	07:36
VP2E	Anguilla	10:15	22:23
VP2M	Montserrat	10:12	22:19
VP2V	British Virgin Is., Tortola	10:21	22:29
VP5	Turks & Caicos Islands, Grand Turk	10:48	22:56
VP8	Falkland Islands, Stanley	09:54	22:03
VP8	So. Georgia Is.	08:29	20:39
VP8	So. Orkney Is.	09:03	21:14
VP8	So. Sandwich Islands, Saunders Is.	07:49	19:59
VP8	So. Shetland Is., King George Is.	09:54	22:06
VP9	Bermuda	10:21	22:30
VQ9	Chagos, Diego Garcia	01:14	13:20
VR6	Pitcairn Is.	14:44	02:51
VS6	Hong Kong	22:26	10:34
VU	India, Bombay	01:12	13:19
	Calcutta	00:09	12:17
	New Delhi	00:54	13:02
	Bangalore	00:53	13:00
VU7	Andaman Islands, Port Blair	23:52	11:59
VU7	Laccadive Is.	01:11	13:18
XE	Mexico, Mexico City (XE1)	12:39	00:47
	Chihuahua (XE2)	13:07	01:15
	Merida (XE3)	12:02	00:09
XF4	Revilla Gigedo	13:29	01:37
XT	Burkina Faso, Onagadougou	06:10	18:17
XU	Kampuchea, Phnom Penh	23:04	11:11
XW	Laos, Viangchan	23:13	11:20
XX9	Macao	22:28	10:36
XZ	Burma, Rangoon	23:39	11:47
Y2-9	German Dem.Rep.(E), Berlin	05:07	17:20
YA	Afghanistan, Kandahar	01:39	13:48
	Kabul	01:25	13:35
YB-YD	Indonesia, Jakarta	22:56	11:03
	Medan	23:29	11:35
	Pontianak	22:46	10:53
	Jayapura	20:41	08:47
YI	Iraq, Baghdad	03:04	15:13
YJ	Vanuatu, Port Vila	18:50	06:57
YK	Syria, Damascus	03:37	15:46
YN, HT	Nicaragua, Managua	11:47	23:54
YO	Romania, Bucharest	04:17	16:28
YS	El Salvador, San Salvador	12:00	00:07
YU	Yugoslavia, Belgrade	04:39	16:50
YV	Venezuela, Caracas	10:31	22:38
YVØ	Aves Is.	10:18	22:25
Z2,ZE	Zimbabwe, Harare	04:00	16:06
ZA	Albania, Tirane	04:42	16:53
ZB	Gibraltar	06:24	18:33
ZC4	British Cyprus	03:50	15:59
ZD7	St. Helena	06:27	18:34
ZD8	Ascension Is.	07:01	19:08
ZD9	Tristan da Cunha	06:52	19:00
ZF	Cayman Is.	11:28	23:35
ZK1	No. Cook Is., Manihiki	16:48	04:54
ZK1	So. Cook Is., Rarotonga	16:43	04:49
ZK2	Niue	17:19	05:26
ZK3	Tokelaus, Atafu	17:34	05:41
ZL	New Zealand, Auckland (ZL1)	18:24	06:31
	Wellington (ZL2)	18:24	06:31
	Christchurch (ZL3)	18:33	06:40
	Dunedin (ZL4)	18:41	06:49
ZL5	Antarctica, Scott Base	18:49	07:13
ZL7	Chatham Is.	17:49	05:57
ZL8	Kermadec Is.	17:55	06:02
ZL9	Auckland & Campbell Is., Auckland	18:57	07:05
	Campbell Is.	18:46	06:55
ZP	Paraguay, Asuncion	09:54	22:01
ZS	South Africa, Cape Town (ZS1)	04:50	16:57
	Port Elizabeth (ZS2)	04:21	16:27
	Bloemfontein (ZS4)	04:19	16:26
	Durban (ZS5)	04:00	16:07
	Johannesburg (ZS6)	04:11	16:18
ZS2	Prince Edward & Marion Is., Marion	03:32	15:40
ZS3	Namibia, Windhoek	04:55	17:02
1AØ	S.M.O.M	05:12	17:22
1S	Spratly Is.	22:36	10:43
3A	Monaco	05:32	17:43
3B6	Agalega	02:17	14:24
3B7	St. Brandon	02:04	14:11
3B8	Mauritius	02:14	14:20
3B9	Rodriguez Is.	01:50	13:56
3C	Equatorial Guinea, Bata	05:24	17:31
	Malabo	05:28	17:35

Pref	State/Province/Country/City	Sunrise	Sunset
3CØ	Pagalu Is.	05:41	17:48
3D2	Fiji Is., Suva	18:10	06:16
3D6	Swaziland, Mbabane	03:59	16:06
3V	Tunisia, Tunis	05:21	17:31
3W	Vietnam, Ho Chi Minh City		
	(Saigon)	22:56	11:04
	Hanoi	23:00	11:07
3X	Republic of Guinea, Conakry	06:58	19:05
3Y	Bouvet	05:49	17:58
3Y	Peter I Is.	12:02	00:16
4S	Sri Lanka, Colombo	00:44	12:51
4U	ITU Geneva	05:36	17:48
4U	United Nations Hq.	10:58	23:08
4W	Yemen, Sanaa	03:06	15:14
4X, 4Z	Israel, Jerusalem	03:42	15:50
5A	Libya, Tripoli	05:12	17:21
	Benghazi	04:42	16:51
5B	Cyprus, Nicosia	03:49	15:58
5H	Tanzania, Dar es Salaam	03:26	15:32
5N	Nigeria, Lagos	05:50	17:57
5R	Madagascar, Antananarivo	02:54	15:00
5T	Mauritania, Nouakchott	07:07	19:15
5U	Niger, Niamey	05:55	18:02
5V	Togo, Lome	05:59	18:05
5W	Western Samoa, Apia	17:31	05:37
5X	Uganda, Kampala	03:53	16:00
5Z	Kenya, Nairobi	04:56	17:03
6W	Senegal, Dakar	07:13	19:20
6Y	Jamaica, Kingston	11:10	23:18

Pref	State/Province/Country/City	Sunrise	Sunset
7O	Yemen, People's Dem. Rep.,		
	Aden	03:03	15:10
7P	Lesotho, Maseru	04:13	16:20
7Q	Malawi, Lilongwe	03:48	15:55
	Blantyre	03:44	15:50
7X	Algeria, Algiers	05:50	18:00
8P	Barbados, Bridgetown	10:02	22:09
8Q	Maldive Is.	01:10	13:17
8R	Guyana, Georgetown	09:56	22:03
9G	Ghana, Accra	06:04	18:11
9H	Malta	05:05	17:14
9J	Zambia, Lusaka	04:10	16:17
9K	Kuwait	02:51	15:00
9L	Sierra Leone, Freetown	06:56	19:03
9M2	West Malaysia, Kuala Lumpur	23:17	11:24
9M6,8	East Malaysia, Sabah,		
	Sandakan(9M6)	22:11	10:18
	Sarawak, Kuching (9M8)	22:42	10:49
9N	Nepal, Katmandu	00:21	12:30
9Q	Zaire, Kinshasa	05:02	17:09
	Kisangani	04:23	16:29
	Lubumbashi	04:14	16:20
9U	Burundi, Bujumbura	04:06	16:13
9V	Singapore	23:08	11:15
9X	Rwanda, Kigali	04:03	16:10
9Y	Trinidad & Tobago, Port of		
	Spain	10:08	22:16
J2/A	Abu Ail	03:12	15:19

Pref	State/Province/Country/City	Sunrise	Sunset	Pref	State/Province/Country/City	Sunrise	Sunset
VE1	New Brunswick, Saint John	09:54	22:59		Montana, Billings	12:43	01:49
	Nova Scotia, Halifax	09:44	22:48		Butte	12:59	02:05
	Prince Edward Island,				Great Falls	12:52	02:02
	Charlottetown	09:41	22:48		Nevada, Las Vegas	13:18	02:06
VE2	Quebec, Montreal	10:23	23:29		Reno	13:34	02:28
	Quebec City	10:12	23:21		Oregon, Portland	13:40	02:45
VE3	Ontario, London	10:57	23:57		Utah, Salt Lake City	13:01	01:57
	Ottawa	10:32	23:37		Washington, Seattle	13:36	02:46
	Sudbury	10:52	24:00		Spokane	13:16	02:26
	Toronto	10:49	23:50		Wyoming, Cheyenne	12:33	01:29
VE4	Manitoba, Winnipeg	11:52	01:08		Sheridan	12:38	01:42
VE5	Saskatchewan, Regina	12:22	01:39	W8	Michigan, Detroit	11:04	00:03
	Saskatoon	12:28	01:49		Grand Rapids	11:14	00:15
VE6	Alberta, Calgary	12:59	02:17		Sault Ste. Marie	11:06	00:13
	Edmonton	12:53	02:19		Traverse City	11:12	00:16
VE7	British Columbia, Prince George	13:30	02:56		Ohio, Cincinnati	11:13	00:06
	Prince Rupert	13:59	03:27		Cleveland	11:00	23:57
	Vancouver	13:37	02:51		Columbus	11:06	00:01
VE8	Northwest Territories,				West Virginia, Charleston	11:02	23:54
	Yellowknife	12:39	02:40	W9	Illinois, Chicago	11:23	00:21
	Resolute	10:24	02:19		Indiana, Indianapolis	11:19	00:14
VY1	Yukon, Whitehorse	14:06	03:58		Wisconsin, Green Bay	11:22	00:25
VO1	Newfoundland, St. John's	08:58	22:08		Milwaukee	11:23	00:23
VO2	Labrador, Goose Bay	09:21	22:46	W0	Colorado, Denver	12:35	01:29
W1	Connecticut, Hartford	10:24	23:22		Grand Junction	12:50	01:43
	Maine, Bangor	10:05	23:09		Iowa, Des Moines	11:47	00:45
	Portland	10:12	23:14		Kansas, Pratt	12:11	01:02
	Massachusetts, Boston	10:17	23:16		Wichita	12:06	00:56
	New Hampshire, Concord	10:17	23:18		Minnesota, Duluth	11:36	00:44
	Rhode Island, Providence	10:18	23:16		Minneapolis	11:43	00:47
	Vermont, Montpelier	10:21	23:24		Missouri, Columbia	11:44	00:37
W2	New Jersey, Atlantic City	10:33	23:26		Kansas City	11:54	00:47
	New York, Albany	10:27	23:27		St. Louis	11:36	00:29
	Buffalo	10:47	23:47		Nebraska, North Platte	12:17	01:13
	New York City	10:30	23:26		Omaha	11:57	00:54
	Syracuse	10:36	23:37		North Dakota, Fargo	11:55	01:03
W3	Delaware, Wilmington	10:37	23:31		South Dakota, Rapid City	12:23	01:26
	District of Columbia,						
	Washington	10:43	23:36	A2	Botswana, Gaborone	04:27	16:09
	Maryland, Baltimore	10:41	23:35	A3	Tonga, Nukualofa	17:49	05:36
	Pennsylvania, Harrisburg	10:42	23:37	A4	Oman, Muscat	01:52	14:23
	Philadelphia	10:35	23:30	A5	Bhutan, Thimbu	23:46	12:22
	Pittsburgh	10:54	23:49	A6	United Arab Emirates, Abu		
	Scranton	10:36	23:33		Dhabi	02:09	14:41
W4	Alabama, Montgomery	11:26	00:08	A7	Qatar, Ad-Dawhah	02:19	14:52
	Florida, Jacksonville	11:09	23:48	A9	Bahrein, Al-Manamah	02:22	14:57
	Miami	11:06	23:39	AP	Pakistan, Karachi	01:17	13:50
	Pensacola	11:31	00:10		Islamabad	00:47	13:31
	Georgia, Atlanta	11:17	00:02	BV	Taiwan, Taipei	21:39	10:12
	Savannah	11:05	23:47	BY	People's Rep. of China, Beijing	21:49	10:43
	Kentucky, Lexington	11:14	00:05		Harbin	21:02	10:08
	Louisville	11:19	00:11		Shanghai	21:35	10:16
	North Carolina, Charlotte	11:02	23:48		Fuzhou	21:47	10:22
	Raleigh	10:52	23:40		Xian	22:23	11:09
	Wilmington	10:51	23:36		Chongqing	22:36	11:15
	South Carolina, Columbia	11:03	23:48		Chengdu	22:45	11:25
	Tennessee, Knoxville	11:13	00:01		Lhasa	23:37	12:16
	Memphis	11:39	00:25		Urumqi	23:40	12:42
	Nashville	11:25	00:13		Kashi	00:31	13:25
	Virginia, Norfolk	10:42	23:32	C2	Nauru	18:51	06:57
	Richmond	10:46	23:37	C3	Andorra	05:26	18:25
W5	Arkansas, Little Rock	11:48	00:34	C5	The Gambia, Banjul	06:59	19:19
	Louisiana, New Orleans	11:43	00:22	C6	Bahamas, Nassau	10:55	23:28
	Shreveport	11:55	00:38	C9	Mozambique, Maputo	04:01	15:42
	Mississippi, Jackson	11:41	00:24		Mozambique	03:23	15:15
	New Mexico, Albuquerque	12:45	01:32	CE	Chile, Santiago	10:59	22:31
	Oklahoma, Oklahoma City	12:08	00:55	CE0	Easter Island	13:30	01:09
	Texas, Abilene	12:19	01:02	CE0	San Felix	11:32	23:12
	Amarillo	12:26	01:12	CE0	Juan Fernandez	11:31	23:03
	Dallas	12:07	00:51	CM,CO	Cuba, Havana	11:16	23:47
	El Paso	12:47	01:29	CN	Morocco, Casablanca	06:10	18:54
	San Antonio	12:16	00:55	CP	Bolivia, La Paz	10:40	22:31
W6	California, Los Angeles	13:32	02:17	CT	Portugal, Lisbon	06:12	19:05
	San Francisco	13:46	02:37	CT2, CU	Azores, Ponta Delgada	07:19	20:10
W7	Arizona, Flagstaff	13:05	01:52	CT3	Madeira Islands, Funchal	06:48	19:31
	Phoenix	13:08	01:52	CX	Uruguay, Montevideo	10:02	21:32
	Idaho, Boise	13:16	02:17	D2	Angola, Luanda	05:10	17:08
	Pocatello	13:02	02:02	D4	Cape Verde, Praia	07:25	19:46

Pref	State/Province/Country/City	Sunrise	Sunset
D6	Comoros, Moroni	03:09	15:05
DA-DL	Germany, Fed. Rep. of (W), Bonn	04:55	18:13
DU	Philippines, Manilla	21:47	10:08
EA	Spain, Madrid	05:49	18:44
EA6	Balearic Is., Palma	05:24	18:18
EA8	Canary Is., Las Palmas	06:40	19:18
EA9	Ceuta & Melilla, Ceuta	05:59	18:47
	Melilla	05:50	18:37
EI	Ireland, Dublin	05:44	19:09
EL	Liberia, Monrovia	06:39	18:51
EP	Iran, Tehran	02:11	14:58
ET	Ethiopia, Addis Ababa	03:19	15:35
	Asmera	03:15	15:37
F	France, Paris	05:16	18:29
FG	Guadeloupe	09:57	22:20
FG, FS	St. Martin	10:02	22:27
FH	Mayotte	03:03	14:58
FK	New Caledonia, Noumia	19:03	06:48
FM	Martinique	09:55	22:16
FO	Clipperton	13:10	01:27
FO	Fr. Polynesia, Tahiti	16:45	04:35
FP	St. Pierre & Miquelon, St. Pierre	09:12	22:20
FR	Glorioso	02:55	14:50
FR	Juan de Nova	03:15	15:06
	Europa	03:28	15:13
FR	Reunion	02:26	14:12
FR	Tromelin	02:28	14:20
FT8W	Crozet	02:57	14:10
FT8X	Kerguelen	01:52	12:58
FT8Y	Antarctica, Dumont D'Urville	21:37	07:47
FT8Z	Amsterdam & St. Paul Is., Amsterdam	01:08	12:34
FW	Wallis & Futuna Is., Wallis	17:50	05:44
FY	Fr. Guiana, Cayenne	09:25	21:37
G	England, London	05:22	18:42
GD	Isle of Man	05:36	19:04
GI	Northern Ireland, Belfast	05:41	19:10
GJ	Jersey	05:34	18:48
GM	Scotland, Glasgow	05:33	19:05
	Aberdeen	05:21	18:59
GU	Guernsey	05:35	18:50
GW	Wales, Cardiff	05:35	18:54
H4	Solomon Islands, Honiara	19:23	07:21
HA	Hungary, Budapest	04:10	17:20
HB	Switzerland, Bern	04:57	18:06
HBØ	Liechtenstein	04:49	17:58
HC	Ecuador, Quito	11:10	23:17
HC8	Galapagos Is.	12:01	00:07
HH	Haiti, Port-Au-Prince	10:38	23:04
HI	Dominican Republic, Santo Domingo	10:29	22:54
HK	Colombia, Bogota	10:53	23:04
HKØ	Malpelo Is.	11:21	23:31
HKØ	San Adreas	11:19	23:38
HL, HM	Korea, Seoul	21:09	09:59
HP	Panama, Panama	11:12	23:27
HR	Honduras, Tegucigalpa	11:40	00:01
HS	Thailand, Bangkok	23:10	11:30
HV	Vatican City	04:43	17:41
HZ, 7Z	Saudi Arabia, Dharan	02:25	14:59
	Mecca	03:08	15:37
	Riyadh	02:39	15:11
I	Italy, Rome	04:43	17:41
	Trieste	04:34	17:39
	Sicily	04:41	17:31
IS	Sardinia, Cagliari	04:59	17:52
J2	Djibouti, Djibouti	03:00	15:18
J3	Grenada	10:00	22:18
J5	Guinea-Bissau, Bissau	06:55	19:13
J6	St. Lucia	09:56	22:16
J7	Dominica	09:56	22:18
J8	St. Vincent	09:57	22:17
JA-JS	Japan, Tokyo	20:19	09:06
	Nagasaki	21:01	09:44
	Sapporo	20:06	09:06
JD	Minami Torishima	19:30	08:02
JD	Ogasawara, Kazan Is.	20:20	08:56
JT	Mongolia, Ulan Bator	22:19	11:29
JW	Svalbard, Spitsbergen	02:07	19:48
JX	Jan Mayen	05:03	20:07

Pref	State/Province/Country/City	Sunrise	Sunset
JY	Jordan, Amman	03:17	15:59
KC4	Antarctica, Byrd Station	16:27	23:37
	McMurdo Sound	20:49	05:01
	Palmer Station	11:08	21:27
KC6	Micronesia, Ponape	19:22	07:35
KC6	Belau, Yap	20:41	08:57
	Koror	20:57	09:11
KG4	Guantanamo Bay	10:49	23:16
KH1	Baker, Howland Is.	17:42	05:49
KH2	Guam, Agana	20:13	08:33
KH3	Johnston Is.	17:04	05:28
KH4	Midway Is.	17:33	06:10
KH5	Palmyra Is.	16:44	04:56
KH5K	Kingman Reef	16:46	05:00
KH6	Hawaii, Hilo	16:09	04:35
	Honolulu	16:19	04:48
KH7	Kure Is.	17:37	06:14
KH8	American Samoa, Pago Pago	17:28	05:21
KH9	Wake Is.	18:42	07:08
KHØ	Mariana Is., Saipan	20:08	08:29
KL7	Alaska, Adak	17:08	06:28
	Anchorage	15:05	04:59
	Fairbanks	14:47	05:00
	Juneau	14:08	03:50
	Nome	15:57	06:09
KP1	Navassa Is.	10:49	23:14
KP2	Virgin Islands, Charlotte Amalie	10:09	22:34
KP4	Puerto Rico, San Juan	10:14	22:39
KP5	Desecheo Is.	10:19	22:44
KX6	Marshall Islands, Kwajalein	18:45	07:00
LA-LJ	Norway, Oslo	04:24	18:13
LU	Argentina, Buenos Aires	10:10	21:41
LX	Luxembourg	05:00	18:14
LZ	Bulgaria, Sofia	03:59	16:58
OA	Peru, Lima	11:13	23:08
OD	Lebanon, Beirut	03:17	16:02
OE	Austria, Vienna	04:21	17:32
OH	Finland, Helsinki	03:27	17:17
OHØ	Aland Is.	03:47	17:37
OJØ	Market Reef	03:51	17:41
OK	Czechoslovakia, Prague	04:26	17:42
ON	Belgium, Brussels	05:05	18:23
OX, XP	Greenland, Godthaab	08:24	22:34
	Thule	08:20	00:54
OY	Faroe Islands, Torshavn	05:30	19:28
OZ	Denmark, Copenhagen	04:25	17:58
P2	Papua New Guinea, Madang	20:18	08:20
	Port Moresby	20:14	08:12
PA-PI	Netherlands, Amsterdam	05:01	18:23
PJ, P4	Netherlands Antilles, Willemstad	10:28	22:47
PJ5-8	St. Maarten and Saba, St. Maarten	10:02	22:27
PY	Brazil, Brasilia	09:18	21:09
	Rio De Janeiro	09:02	20:47
	Natal	08:22	20:23
	Manaus	10:01	22:04
	Porto Alegre	09:38	21:15
PYØ	Fernando De Noronha	08:10	20:13
PYØ	St. Peter & St. Paul Rocks	07:56	20:03
PYØ	Trindade & Martin Vaz Is., Trindade	08:06	19:52
PZ	Suriname, Paramaribo	09:36	21:49
S2	Bangladesh, Dacca	23:44	12:16
S7	Seychelles, Victoria	02:19	14:21
S9	Sao Tome	05:31	17:38
SJ-SM	Sweden, Stockholm	03:56	17:42
SP	Poland, Krakow	04:04	17:20
	Warsaw	03:57	17:19
ST	Sudan, Khartoum	03:41	16:03
STØ	Southern Sudan, Juba	03:50	16:01
SU	Egypt, Cairo	03:36	16:16
SV	Greece, Athens	04:01	16:53
SV5	Dodecanese, Rhodes	03:45	16:33
SV9	Crete	03:57	16:45
SV/A	Mount Athos	03:57	16:52
T2	Tuvalu, Funafuti	18:08	06:07
T30	West Kiribati, Bonriki	18:25	06:33
T31	Central Kiribati, Kanton	17:27	05:31
T32	East Kiribati, Christmas Is.	16:27	04:36
T5	Somalia, Mogadishu	02:56	15:05
T7	San Marino	04:42	17:44

Pref	State/Province/Country/City	Sunrise	Sunset
TA	Turkey, Ankara	03:23	16:17
	Istanbul	03:37	16:34
TF	Iceland, Reykjavik	06:25	20:34
TG	Guatemala, Guatemala City	11:53	00:14
TI	Costa Rica, San Jose	11:30	23:46
TI9	Cocos Is.	11:44	23:56
TJ	Cameroon, Yaounde	05:11	17:21
TK	Corsica	04:57	17:55
TL	Central African Republic, Bangui	04:42	16:53
TN	Congo, Brazzaville	04:59	17:02
TR	Gabon, Libreville	05:20	17:27
TT	Chad, N'Djamena	04:52	17:11
TU	Ivory Coast, Abidjan	06:12	18:24
TY	Benin, Porto Novo	05:45	17:58
TZ	Mali, Bamako	06:24	18:43
UA	Russia, European,		
	Leningrad (UA1)	03:06	16:55
	Archangel (UA1)	02:14	16:26
	Murmansk (UA1)	02:27	17:11
	Moscow (UA3)	02:45	16:18
	Kuibyshev (UA4)	01:59	15:24
	Rostov (UA6)	02:49	15:59
UA1P	Franz Josef Land	23:12	17:47
UA2	Kaliningrad	03:55	17:25
UA9,Ø	Russia, Asiatic,		
	Novosibirsk (UA9)	23:45	13:15
	Perm (UA9)	01:26	15:07
	Omsk (UA9)	00:23	13:53
	Norilsk (UAØ)	22:46	13:33
	Irkutsk (UAØ)	22:24	11:46
	Vladivostok (UAØ)	20:44	09:44
	Petropavlovsk (UAØ)	18:45	08:09
	Khabarovsk (UAØ)	20:25	09:37
	Krasnoyarsk (UAØ)	23:04	12:37
	Yakutsk (UAØ)	20:24	10:22
	Wrangel Island (UAØ)	16:28	07:32
	Kyzyl (UAØY)	23:04	12:24
UB	Ukraine, Kiev	03:21	16:38
UC	Byelorussia, Minsk	03:28	16:55
UD	Azerbaijan, Baku	02:14	15:10
UF	Georgia, Tbilisi	02:34	15:31
UG	Armenia, Ierevan	02:36	15:31
UH	Turkoman, Ashkhabad	01:43	14:34
UI	Uzbek, Bukhara	01:17	14:11
	Tashkent	00:56	13:53
UJ	Tadzhik, Samarkand	01:07	14:02
	Dushanbe	01:00	13:53
UL	Kazakh, Alma-ata	00:24	13:25
UM	Kirghiz, Frunze	00:33	13:33
UO	Moldavia, Kishinev	03:32	16:41
UP	Lithuania, Vilna	03:35	17:04
UQ	Latvia, Riga	03:37	17:14
UR	Estonia, Tallinn	03:29	17:16
V2	Antigua & Barbuda, St. Johns	09:57	22:21
V3	Belize, Belmopan	11:45	00:09
V4	St. Christopher & Nevis	10:00	22:24
V8	Brunei, Bandar Seri Begawan	22:16	10:28
VE1, CY	Sable Is.	09:31	22:33
VE1, CY	St. Paul Is.	09:28	22:37
VK	Australia, Canberra (VK1)	20:21	07:50
	Sydney (VK2)	20:11	07:43
	Melbourne (VK3)	20:39	08:05
	Brisbane (VK4)	20:00	07:39
	Adelaide (VK5)	21:02	08:32
	Perth (VK6)	22:32	10:06
	Hobart, Tasmania (VK7)	20:33	07:52
	Darwin (VK8)	21:21	09:16
VK	Lord Howe Is.	19:38	07:12
VK9	Christmas Is.	23:01	10:57
VK9	Cocos-Keeling Is.	23:37	11:32
VK9	Mellish Reef	19:44	07:33
VK9	Norfolk Is.	19:01	06:38
VK9	Willis Is.	20:08	07:59
VKØ	Heard Is.	01:39	12:37
VKØ	Macquarie Is.	19:59	06:54
VP2E	Anguilla	10:01	22:26
VP2M	Montserrat	09:59	22:22
VP2V	British Virgin Is., Tortola	10:08	22:33
VP5	Turks & Caicos Islands,		
	Grand Turk	10:32	23:01
VP8	Falkland Islands, Stanley	10:23	21:24

Pref	State/Province/Country/City	Sunrise	Sunset
VP8	So. Georgia Is.	09:01	19:57
VP8	So. Orkney Is.	09:46	20:22
VP8	So. Sandwich Islands, Saunders Is.	08:26	19:11
VP8	So. Shetland Is., King George Is.	10:39	21:10
VP9	Bermuda	09:59	22:42
VQ9	Chagos, Diego Garcia	01:12	13:12
VR6	Pitcairn Is.	14:51	02:33
VS6	Hong Kong	22:10	10:39
VU	India, Bombay	00:58	13:24
	Calcutta	23:53	12:23
	New Delhi	00:34	13:12
	Bangalore	00:42	13:01
VU7	Andaman Islands, Port Blair	23:41	12:00
VU7	Laccadive Is.	01:02	13:18
XE	Mexico, Mexico City (XE1)	12:25	00:51
	Chihuahua (XE2)	12:47	01:25
	Merida (XE3)	11:46	00:15
XF4	Revilla Gigedo	13:15	01:41
XT	Burkina Faso, Onagadougou	05:59	18:18
XU	Kampuchea, Phnom Penh	22:53	11:12
XW	Laos, Viangchan	22:59	11:24
XX9	Macao	22:13	10:42
XZ	Burma, Rangoon	23:26	11:49
Y2-9	German Dem.Rep.(E), Berlin	04:27	17:49
YA	Afghanistan, Kandahar	01:18	13:59
	Kabul	01:02	13:48
YB-YD	Indonesia, Jakarta	22:54	10:55
	Medan	23:22	11:32
	Pontianak	22:41	10:48
	Jayapura	20:37	08:41
YI	Iraq, Baghdad	02:42	15:26
YJ	Vanuatu, Port Vila	18:54	06:43
YK	Syria, Damascus	03:14	15:59
YN, HT	Nicaragua, Managua	11:37	23:55
YO	Romania, Bucharest	03:46	16:49
YS	El Salvador, San Salvador	11:48	00:09
YU	Yugoslavia, Belgrade	04:08	17:12
YV	Venezuela, Caracas	10:21	22:38
YVØ	Aves Is.	10:05	22:28
Z2,ZE	Zimbabwe, Harare	04:03	15:53
ZA	Albania, Tirane	04:14	17:11
ZB	Gibraltar	05:59	18:47
ZC4	British Cyprus	03:27	16:13
ZD7	St. Helena	06:30	18:21
ZD8	Ascension Is.	07:00	18:59
ZD9	Tristan da Cunha	07:07	18:35
ZF	Cayman Is.	11:13	23:40
ZK1	No. Cook Is., Manihiki	16:47	04:44
ZK1	So. Cook Is., Rarotonga	16:48	04:34
ZK2	Niue	17:23	05:12
ZK3	Tokelaus, Atafu	17:33	05:32
ZL	New Zealand, Auckland (ZL1)	18:39	06:06
	Wellington (ZL2)	18:42	06:03
	Christchurch (ZL3)	18:53	06:10
	Dunedin (ZL4)	19:03	06:16
ZL5	Antarctica, Scott Base	20:52	05:00
ZL7	Chatham Is.	18:09	05:26
ZL8	Kermadec Is.	18:05	05:42
ZL9	Auckland & Campbell Is.,		
	Auckland	19:24	06:28
	Campbell Is.	19:15	06:15
ZP	Paraguay, Asuncion	10:02	21:43
ZS	South Africa, Cape Town (ZS1)	05:02	16:34
	Port Elizabeth (ZS2)	04:33	16:05
	Bloemfontein (ZS4)	04:29	16:06
	Durban (ZS5)	04:10	15:46
	Johannesburg (ZS6)	04:10	15:46
ZS2	Prince Edward & Marion Is.,		
	Marion	03:55	15:06
ZS3	Namibia, Windhoek	05:01	16:46
1AØ	S.M.O.M	04:43	17:41
1S	Spratly Is.	22:27	10:42
3A	Monaco	05:01	18:03
3B6	Agalega	02:17	14:14
3B7	St. Brandon	02:07	13:58
3B8	Mauritius	02:18	14:05
3B9	Rodriguez Is.	01:54	13:42
3C	Equatorial Guinea, Bata	05:18	17:27
	Malabo	05:21	17:32

Pref	State/Province/Country/City	Sunrise	Sunset
3CØ	Pagalu Is.	05:37	17:42
3D2	Fiji Is., Suva	18:14	06:03
3D6	Swaziland, Mbabane	04:07	15:48
3V	Tunisia, Tunis	04:56	17:46
3W	Vietnam, Ho Chi Minh City		
	(Saigon)	22:46	11:04
	Hanoi	22:44	11:13
3X	Republic of Guinea, Conakry	06:49	19:05
3Y	Bouvet	06:21	17:16
3Y	Peter I Is.	13:06	23:02
4S	Sri Lanka, Colombo	00:35	12:49
4U	ITU Geneva	05:04	18:10
4U	United Nations Hq.	10:30	23:26
4W	Yemen, Sanaa	02:54	15:16
4X, 4Z	Israel, Jerusalem	03:20	16:02
5A	Libya, Tripoli	04:50	17:33
	Benghazi	04:21	17:03
5B	Cyprus, Nicosia	03:25	16:12
5H	Tanzania, Dar es Salaam	03:24	15:24
5N	Nigeria, Lagos	05:42	17:55
5R	Madagascar, Antananarivo	02:58	14:46
5T	Mauritania, Nouakchott	06:53	19:18
5U	Niger, Niamey	05:44	18:04
5V	Togo, Lome	05:51	18:03
5W	Western Samoa, Apia	17:32	05:26
5X	Uganda, Kampala	03:48	15:55
5Z	Kenya, Nairobi	04:52	16:57
6W	Senegal, Dakar	07:01	19:22
6Y	Jamaica, Kingston	10:57	23:21

Pref	State/Province/Country/City	Sunrise	Sunset
7O	Yemen, People's Dem. Rep.,		
	Aden	02:52	15:11
7P	Lesotho, Maseru	04:23	16:00
7Q	Malawi, Lilongwe	03:50	15:43
	Blantyre	03:46	15:38
7X	Algeria, Algiers	05:25	18:14
8P	Barbados, Bridgetown	09:50	22:10
8Q	Maldive Is.	01:03	13:14
8R	Guyana, Georgetown	09:48	22:01
9G	Ghana, Accra	05:57	18:09
9H	Malta	04:40	17:28
9J	Zambia, Lusaka	04:13	16:05
9K	Kuwait	02:31	15:10
9L	Sierra Leone, Freetown	06:47	19:02
9M2	West Malaysia, Kuala Lumpur	23:10	11:20
9M6,8	East Malaysia, Sabah,		
	Sandakan (9M6)	22:03	10:15
	Sarawak, Kuching (9M8)	22:36	10:45
9N	Nepal, Katmandu	00:02	12:39
9Q	Zaire, Kinshasa	04:59	17:02
	Kisangani	04:17	16:25
	Lubumbashi	04:14	16:10
9U	Burundi, Bujumbura	04:03	16:06
9V	Singapore	23:03	11:11
9X	Rwanda, Kigali	03:59	16:04
9Y	Trinidad & Tobago, Port of		
	Spain	09:58	22:15
J2/A	Abu Ail	03:00	15:21

Pref	State/Province/Country/City	Sunrise	Sunset
VE1	New Brunswick, Saint John	09:29	23:16
	Nova Scotia, Halifax	09:20	23:05
	Prince Edward Island, Charlottetown	09:15	23:06
VE2	Quebec, Montreal	09:58	23:47
	Quebec City	09:46	23:40
VE3	Ontario, London	10:34	00:13
	Ottawa	10:07	23:55
	Sudbury	10:26	00:18
	Toronto	10:25	00:07
VE4	Manitoba, Winnipeg	11:24	01:30
VE5	Saskatchewan, Regina	11:52	02:01
	Saskatoon	11:57	02:13
VE6	Alberta, Calgary	12:29	02:40
	Edmonton	12:21	02:44
VE7	British Columbia, Prince George	12:57	03:22
	Prince Rupert	13:26	03:53
	Vancouver	13:09	03:12
VE8	Northwest Territories, Yellowknife	11:53	03:19
	Resolute	08:37	03:59
VY1	Yukon, Whitehorse	13:24	04:34
VO1	Newfoundland, St. John's	08:31	22:27
VO2	Labrador, Goose Bay	08:49	23:11
W1	Connecticut, Hartford	10:01	23:37
	Maine, Bangor	09:41	23:26
	Portland	09:48	23:30
	Massachusetts, Boston	09:54	23:31
	New Hampshire, Concord	09:54	23:34
	Rhode Island, Providence	09:56	23:32
	Vermont, Montpelier	09:57	23:41
W2	New Jersey, Atlantic City	10:12	23:40
	New York, Albany	10:04	23:43
	Buffalo	10:24	00:03
	New York City	10:08	23:40
	Syracuse	10:13	23:53
W3	Delaware, Wilmington	10:16	23:45
	District of Columbia, Washington	10:23	23:50
	Maryland, Baltimore	10:21	23:49
	Pennsylvania, Harrisburg	10:20	23:51
	Philadelphia	10:14	23:44
	Pittsburgh	10:33	00:04
	Scranton	10:14	23:48
W4	Alabama, Montgomery	11:09	00:18
	Florida, Jacksonville	10:53	23:57
	Miami	10:52	23:46
	Pensacola	11:15	00:19
	Georgia, Atlanta	11:00	00:12
	Savannah	10:48	23:57
	Kentucky, Lexington	10:54	00:18
	Louisville	10:59	00:24
	North Carolina, Charlotte	10:43	00:00
	Raleigh	10:34	23:52
	Wilmington	10:33	23:47
	South Carolina, Columbia	10:46	23:59
	Tennessee, Knoxville	10:55	00:13
	Memphis	11:21	00:37
	Nashville	11:06	00:25
	Virginia, Norfolk	10:23	23:44
	Richmond	10:27	23:49
W5	Arkansas, Little Rock	11:30	00:45
	Louisiana, New Orleans	11:27	00:30
	Shreveport	11:38	00:48
	Mississippi, Jackson	11:25	00:34
	New Mexico, Albuquerque	12:27	01:43
	Oklahoma, Oklahoma City	11:50	01:07
	Texas, Abilene	12:02	01:12
	Amarillo	12:07	01:24
	Dallas	11:50	01:01
	El Paso	12:30	01:38
	San Antonio	12:01	01:03
W6	California, Los Angeles	13:14	02:28
	San Francisco	13:26	02:50
W7	Arizona, Flagstaff	12:47	02:03
	Phoenix	12:51	02:03
	Idaho, Boise	12:52	02:34
	Pocatello	12:39	02:18

Pref	State/Province/Country/City	Sunrise	Sunset
	Montana, Billings	12:18	02:07
	Butte	12:33	02:23
	Great Falls	12:25	02:21
	Nevada, Las Vegas	12:59	02:18
	Reno	13:13	02:42
	Oregon, Portland	13:15	03:03
	Utah, Salt Lake City	12:40	02:12
	Washington, Seattle	13:09	03:06
	Spokane	12:49	02:46
	Wyoming, Cheyenne	12:11	01:44
	Sheridan	12:13	01:59
W8	Michigan, Detroit	10:42	00:19
	Grand Rapids	10:51	00:31
	Sault Ste. Marie	10:40	00:32
	Traverse City	10:48	00:34
	Ohio, Cincinnati	10:53	00:20
	Cleveland	10:38	00:12
	Columbus	10:45	00:15
	West Virginia, Charleston	10:42	00:07
W9	Illinois, Chicago	11:01	00:37
	Indiana, Indianapolis	10:58	00:28
	Wisconsin, Green Bay	10:58	00:43
	Milwaukee	11:00	00:40
WØ	Colorado, Denver	12:14	01:43
	Grand Junction	12:29	01:56
	Iowa, Des Moines	11:25	01:00
	Kansas, Pratt	11:51	01:15
	Wichita	11:46	01:09
	Minnesota, Duluth	11:10	01:03
	Minneapolis	11:18	01:05
	Missouri, Columbia	11:24	00:51
	Kansas City	11:33	01:00
	St. Louis	11:16	00:42
	Nebraska, North Platte	11:55	01:28
	Omaha	11:35	01:09
	North Dakota, Fargo	11:29	01:22
	South Dakota, Rapid City	11:59	01:43

Pref	State/Province/Country/City	Sunrise	Sunset
A2	Botswana, Gaborone	04:33	15:56
A3	Tonga, Nukualofa	17:54	05:24
A4	Oman, Muscat	01:39	14:29
A5	Bhutan, Thimbu	23:32	12:29
A6	United Arab Emirates, Abu Dhabi	01:56	14:47
A7	Qatar, Ad-Dawhah	02:06	14:59
A9	Bahrein, Al-Manamah	02:08	15:03
AP	Pakistan, Karachi	01:04	13:56
	Islamabad	00:29	13:42
BV	Taiwan, Taipei	21:26	10:19
BY	People's Rep. of China, Beijing	21:28	10:58
	Harbin	20:37	10:26
	Shanghai	21:19	10:25
	Fuzhou	21:34	10:28
	Xian	22:06	11:20
	Chongqing	22:21	11:24
	Chengdu	22:29	11:34
	Lhasa	23:22	12:25
	Urumqi	23:17	12:59
	Kashi	00:10	13:39
C2	Nauru	18:48	06:54
C3	Andorra	05:03	18:41
C5	The Gambia, Banjul	06:50	19:20
C6	Bahamas, Nassau	10:42	23:34
C9	Mozambique, Maputo	04:08	15:28
	Mozambique	03:25	15:06
CE	Chile, Santiago	11:09	22:14
CEØ	Easter Island	13:37	00:55
CEØ	San Felix	11:39	22:59
CEØ	Juan Fernandez	11:41	22:46
CM,CO	Cuba, Havana	11:04	23:52
CN	Morocco, Casablanca	05:52	19:04
CP	Bolivia, La Paz	10:43	22:21
CT	Portugal, Lisbon	05:52	19:18
CT2, CU	Azores, Ponta Delgada	06:59	20:23
CT3	Madeira Islands, Funchal	06:31	19:41
CX	Uruguay, Montevideo	10:12	21:14
D2	Angola, Luanda	05:09	17:01
D4	Cape Verde, Praia	07:16	19:49

Pref	State/Province/Country/City	Sunrise	Sunset
D6	Comoros, Moroni	03:10	14:57
DA-DL	Germany, Fed. Rep. of (W), Bonn	04:25	18:35
DU	Philippines, Manilla	21:38	10:10
EA	Spain, Madrid	05:27	18:59
EA6	Balearic Is., Palma	05:04	18:32
EA8	Canary Is., Las Palmas	06:26	19:25
EA9	Ceuta & Melilla, Ceuta	05:40	18:59
	Melilla	05:32	18:49
EI	Ireland, Dublin	05:12	19:35
EL	Liberia, Monrovia	06:33	18:50
EP	Iran, Tehran	01:52	15:10
ET	Ethiopia, Addis Ababa	03:12	15:35
	Asmera	03:06	15:39
F	France, Paris	04:48	18:50
FG	Guadeloupe	09:48	22:22
FG, FS	St. Martin	09:51	22:30
FH	Mayotte	03:05	14:49
FK	New Caledonia, Noumia	19:08	06:36
FM	Martinique	09:46	22:18
FO	Clipperton	13:03	01:27
FO	Fr. Polynesia, Tahiti	16:48	04:25
FP	St. Pierre & Miquelon, St. Pierre	08:46	22:39
FR	Glorioso	02:55	14:43
FR	Juan de Nova	03:18	14:56
	Europa	03:33	15:01
FR	Reunion	02:31	14:01
FR	Tromelin	02:31	14:11
FT8W	Crozet	03:15	13:45
FT8X	Kerguelen	02:12	12:31
FT8Y	Antarctica, Dumont D'Urville	22:24	06:52
FT8Z	Amsterdam & St. Paul Is., Amsterdam	01:21	12:15
FW	Wallis & Futuna Is., Wallis	17:51	05:36
FY	Fr. Guiana, Cayenne	09:20	21:35
G	England, London	04:52	19:05
GD	Isle of Man	05:02	19:30
GI	Northern Ireland, Belfast	05:07	19:37
GJ	Jersey	05:05	19:09
GM	Scotland, Glasgow	04:57	19:34
	Aberdeen	04:44	19:29
GU	Guernsey	05:07	19:11
GW	Wales, Cardiff	05:04	19:18
H4	Solomon Islands, Honiara	19:23	07:14
HA	Hungary, Budapest	03:44	17:40
HB	Switzerland, Bern	04:31	18:25
HBØ	Liechtenstein	04:22	18:17
HC	Ecuador, Quito	11:07	23:14
HC8	Galapagos Is.	11:57	00:03
HH	Haiti, Port-Au-Prince	10:28	23:07
HI	Dominican Republic, Santo Domingo	10:19	22:58
HK	Colombia, Bogota	10:47	23:02
HKØ	Malpelo Is.	11:16	23:29
HKØ	San Adreas	11:11	23:39
HL, HM	Korea, Seoul	20:49	10:12
HP	Panama, Panama	11:05	23:27
HR	Honduras, Tegucigalpa	11:31	00:03
HS	Thailand, Bangkok	23:01	11:32
HV	Vatican City	04:20	17:56
HZ, 7Z	Saudi Arabia, Dharan	02:11	15:06
	Mecca	02:56	15:42
	Riyadh	02:26	15:17
I	Italy, Rome	04:20	17:56
	Trieste	04:09	17:58
	Sicily	04:21	17:44
IS	Sardinia, Cagliari	04:38	18:06
J2	Djibouti, Djibouti	02:52	15:19
J3	Grenada	09:52	22:19
J5	Guinea-Bissau, Bissau	06:47	19:14
J6	St. Lucia	09:47	22:18
J7	Dominica	09:47	22:20
J8	St. Vincent	09:49	22:18
JA-JS	Japan, Tokyo	20:00	09:18
	Nagasaki	20:44	09:54
	Sapporo	19:43	09:23
JD	Minami Torishima	19:17	08:08
JD	Ogasawara, Kazan Is.	20:05	09:03
JT	Mongolia, Ulan Bator	21:52	11:49
JW	Svalbard, Spitsbergen	24 hrs day	
JX	Jan Mayen	03:48	21:15

Pref	State/Province/Country/City	Sunrise	Sunset
JY	Jordan, Amman	03:01	16:09
KC4	Antarctica, Byrd Station	24 hrs night	
	McMurdo Sound	23:03	02:40
	Palmer Station	11:51	20:37
KC6	Micronesia, Ponape	19:16	07:34
KC6	Belau, Yap	20:34	08:57
	Koror	20:51	09:10
KG4	Guantanamo Bay	10:38	23:20
KH1	Baker, Howland Is.	17:38	05:46
KH2	Guam, Agana	20:04	08:34
KH3	Johnston Is.	16:54	05:31
KH4	Midway Is.	17:18	06:18
KH5	Palmyra Is.	16:38	04:55
KH5K	Kingman Reef	16:40	04:59
KH6	Hawaii, Hilo	15:58	04:39
	Honolulu	16:07	04:52
KH7	Kure Is.	17:22	06:22
KH8	American Samoa, Pago Pago	17:30	05:13
KH9	Wake Is.	18:31	07:12
KHØ	Mariana Is., Saipan	19:58	08:32
KL7	Alaska, Adak	16:37	06:52
	Anchorage	14:21	05:35
	Fairbanks	13:55	05:45
	Juneau	13:30	04:22
	Nome	15:07	06:53
KP1	Navassa Is.	10:39	23:18
KP2	Virgin Islands, Charlotte Amalie	09:58	22:37
KP4	Puerto Rico, San Juan	10:03	22:43
KP5	Desecheo Is.	10:09	22:48
KX6	Marshall Islands, Kwajalein	18:38	07:00
LA-LJ	Norway, Oslo	03:43	18:48
LU	Argentina, Buenos Aires	10:21	21:23
LX	Luxembourg	04:31	18:36
LZ	Bulgaria, Sofia	03:36	17:14
OA	Peru, Lima	11:13	23:00
OD	Lebanon, Beirut	03:00	16:13
OE	Austria, Vienna	03:54	17:52
OH	Finland, Helsinki	02:45	17:51
OHØ	Aland Is.	03:05	18:11
OJØ	Market Reef	03:09	18:16
OK	Czechoslovakia, Prague	03:57	18:04
ON	Belgium, Brussels	04:35	18:46
OX, XP	Greenland, Godthaab	07:34	23:16
	Thule	05:51	03:16
OY	Faroe Islands, Torshavn	04:45	20:06
OZ	Denmark, Copenhagen	03:50	18:26
P2	Papua New Guinea, Madang	20:17	08:15
	Port Moresby	20:14	08:05
PA-PI	Netherlands, Amsterdam	04:30	18:48
PJ, P4	Netherlands Antilles, Willemstad	10:20	22:48
PJ5-8	St. Maarten and Saba, St. Maarten	09:52	22:30
PY	Brazil, Brasilia	09:20	21:00
	Rio De Janeiro	09:08	20:34
	Natal	08:21	20:17
	Manaus	09:58	22:00
	Porto Alegre	09:47	20:59
PYØ	Fernando De Noronha	08:08	20:08
PYØ	St. Peter & St. Paul Rocks	07:52	20:00
PYØ	Trindade & Martin Vaz Is., Trindade	08:10	19:41
PZ	Suriname, Paramaribo	09:31	21:47
S2	Bangladesh, Dacca	23:32	12:21
S7	Seychelles, Victoria	02:17	14:16
S9	Sao Tome	05:28	17:35
SJ-SM	Sweden, Stockholm	03:16	18:16
SP	Poland, Krakow	03:35	17:42
	Warsaw	03:26	17:43
ST	Sudan, Khartoum	03:31	16:05
STØ	Southern Sudan, Juba	03:44	15:59
SU	Egypt, Cairo	03:21	16:24
SV	Greece, Athens	03:41	17:06
SV5	Dodecanese, Rhodes	03:26	16:45
SV9	Crete	03:39	16:56
SV/A	Mount Athos	03:36	17:06
T2	Tuvalu, Funafuti	18:08	06:00
T30	West Kiribati, Bonriki	18:21	06:30
T31	Central Kiribati, Kanton	17:24	05:26
T32	East Kiribati, Christmas Is.	16:23	04:33
T5	Somalia, Mogadishu	02:52	15:02
T7	San Marino	04:18	18:00

Pref	State/Province/Country/City	Sunrise	Sunset
TA	Turkey, Ankara	03:02	16:32
	Istanbul	03:16	16:49
TF	Iceland, Reykjavik	05:35	21:17
TG	Guatemala, Guatemala City	11:44	00:16
TI	Costa Rica, San Jose	11:22	23:46
TI9	Cocos Is.	11:38	23:54
TJ	Cameroon, Yaounde	05:06	17:19
TK	Corsica	04:34	18:10
TL	Central African Republic, Bangui	04:37	16:51
TN	Congo, Brazzaville	04:57	16:57
TR	Gabon, Libreville	05:17	17:24
TT	Chad, N'Djamena	04:44	17:12
TU	Ivory Coast, Abidjan	06:06	18:22
TY	Benin, Porto Novo	05:39	17:57
TZ	Mali, Bamako	06:16	18:45
UA	Russia, European,		
	Leningrad (UA1)	02:25	17:29
	Archangel (UA1)	01:23	17:10
	Murmansk (UA1)	01:22	18:09
	Moscow (UA3)	02:10	16:46
	Kuibyshev (UA4)	01:27	15:49
	Rostov (UA6)	02:22	16:18
UA1P	Franz Josef Land	24 hrs day	
UA2	Kaliningrad	03:20	17:52
UA9,Ø	Russia, Asiatic,		
	Novosibirsk (UA9)	23:11	13:42
	Perm (UA9)	00:48	15:38
	Omsk (UA9)	23:49	14:20
	Norilsk (UAØ)	21:40	14:32
	Irkutsk (UAØ)	21:52	12:10
	Vladivostok (UAØ)	20:21	10:01
	Petropavlovsk (UAØ)	18:13	08:34
	Khabarovsk (UAØ)	19:58	09:58
	Krasnoyarsk (UAØ)	22:28	13:06
	Yakutsk (UAØ)	19:39	11:00
	Wrangel Island (UAØ)	15:13	08:40
	Kyzyl (UAØY)	22:33	12:47
UB	Ukraine, Kiev	02:52	17:00
UC	Byelorussia, Minsk	02:55	17:21
UD	Azerbaijan, Baku	01:53	15:24
UF	Georgia, Tbilisi	02:11	15:47
UG	Armenia, Ierevan	02:15	15:46
UH	Turkoman, Ashkhabad	01:23	14:47
UI	Uzbek, Bukhara	00:56	14:25
	Tashkent	00:34	14:08
UJ	Tadzhik, Samarkand	00:47	14:16
	Dushanbe	00:39	14:07
UL	Kazakh, Alma-ata	00:00	13:41
UM	Kirghiz, Frunze	00:10	13:49
UO	Moldavia, Kishinev	03:06	17:00
UP	Lithuania, Vilna	03:02	17:31
UQ	Latvia, Riga	03:00	17:44
UR	Estonia, Tallinn	02:49	17:49
V2	Antigua & Barbuda, St. Johns	09:47	22:24
V3	Belize, Belmopan	11:35	00:12
V4	St. Christopher & Nevis	09:50	22:27
V8	Brunei, Bandar Seri Begawan	22:11	10:26
VE1, CY	Sable Is.	09:07	22:49
VE1, CY	St. Paul Is.	09:01	22:56
VK	Australia, Canberra (VK1)	20:32	07:32
	Sydney (VK2)	20:22	07:25
	Melbourne (VK3)	20:51	07:45
	Brisbane (VK4)	20:07	07:25
	Adelaide (VK5)	21:13	08:14
	Perth (VK6)	22:41	09:49
	Hobart, Tasmania (VK7)	20:49	07:29
	Darwin (VK8)	21:22	09:08
VK	Lord Howe Is.	19:47	06:56
VK9	Christmas Is.	23:01	10:50
VK9	Cocos-Keeling Is.	23:38	11:24
VK9	Mellish Reef	09:47	07:23
VK9	Norfolk Is.	09:09	06:24
VK9	Willis Is.	20:11	07:50
VKØ	Heard Is.	02:03	2:06
VKØ	Macquarie Is.	20:26	06:20
VP2E	Anguilla	09:51	22:30
VP2M	Montserrat	09:49	22:25
VP2V	British Virgin Is., Tortola	09:57	22:36
VP5	Turks & Caicos Islands, Grand Turk	00:21	23:06
VP8	Falkland Islands, Stanley	00:46	20:54

Pref	State/Province/Country/City	Sunrise	Sunset
VP8	So. Georgia Is.	09:27	09:24
VP8	So. Orkney Is.	00:20	09:40
VP8	So. Sandwich Islands, Saunders Is.	08:561	08:34
VP8	So. Shetland Is., King George Is.	01:16	20:27
VP9	Bermuda	09:43	22:52
VQ9	Chagos, Diego Garcia	01:11	03:06
VR6	Pitcairn Is.	04:57	02:20
VS6	Hong Kong	21:58	00:44
VU	India, Bombay	00:47	03:27
	Calcutta	23:41	02:28
	New Delhi	00:19	03:20
	Bangalore	00:33	03:02
VU7	Andaman Islands, Port Blair	23:34	02:00
VU7	Laccadive Is.	00:54	03:18
XE	Mexico, Mexico City (XE1)	02:14	00:55
	Chihuahua (XE2)	02:32	01:33
	Merida (XE3)	01:35	00:19
XF4	Revilla Gigedo	03:04	01:44
XT	Burkina Faso, Onagadougou	05:51	08:19
XU	Kampuchea, Phnom Penh	22:46	01:12
XW	Laos, Viangchan	22:49	01:27
XX9	Macao	22:01	00:47
XZ	Burma, Rangoon	23:16	01:52
Y2-9	German Dem.Rep.(E), Berlin	03:56	08:14
YA	Afghanistan, Kandahar	01:02	04:08
	Kabul	00:44	03:59
YB-YD	Indonesia, Jakarta	22:53	00:49
	Medan	23:17	01:30
	Pontianak	22:38	00:44
	Jayapura	20:34	08:37
YI	Iraq, Baghdad	02:25	05:36
YJ	Vanuatu, Port Vila	08:57	06:33
YK	Syria, Damascus	02:57	06:09
YN, HT	Nicaragua, Managua	01:29	23:56
YO	Romania, Bucharest	03:22	07:06
YS	El Salvador, San Salvador	01:40	00:10
YU	Yugoslavia, Belgrade	03:43	07:29
YV	Venezuela, Caracas	00:14	22:39
YVØ	Aves Is.	09:56	22:30
Z2,ZE	Zimbabwe, Harare	04:06	05:42
ZA	Albania, Tirane	03:52	07:26
ZB	Gibraltar	05:40	08:59
ZC4	British Cyprus	03:09	06:24
ZD7	St. Helena	06:32	08:12
ZD8	Ascension Is.	06:59	08:53
ZD9	Tristan da Cunha	07:20	08:15
ZF	Cayman Is.	01:02	23:44
ZK1	No. Cook Is., Manihiki	06:48	04:37
ZK1	So. Cook Is., Rarotonga	06:52	04:23
ZK2	Niue	07:27	05:01
ZK3	Tokelaus, Atafu	07:33	05:25
ZL	New Zealand, Auckland (ZL1)	08:51	05:47
	Wellington (ZL2)	08:57	05:41
	Christchurch (ZL3)	09:09	05:47
	Dunedin (ZL4)	09:21	05:51
ZL5	Antarctica, Scott Base	23:12	02:33
ZL7	Chatham Is.	08:26	05:02
ZL8	Kermadec Is.	08:13	05:27
ZL9	Auckland & Campbell Is., Auckland	09:46	05:59
	Campbell Is.	09:39	05:44
ZP	Paraguay, Asuncion	00:08	21:30
ZS	South Africa, Cape Town (ZS1)	05:13	06:16
	Port Elizabeth (ZS2)	04:44	05:47
	Bloemfontein (ZS4)	04:37	05:51
	Durban (ZS5)	04:18	05:31
	Johannesburg (ZS6)	04:26	15:46
ZS2	Prince Edward & Marion Is., Marion	04:14	14:41
ZS3	Namibia, Windhoek	05:06	16:34
1AØ	S.M.O.M	04:21	17:57
1S	Spratly Is.	22:20	10:42
3A	Monaco	04:38	18:20
3B6	Agalega	02:17	14:07
3B7	St. Brandon	02:10	13:49
3B8	Mauritius	02:22	13:54
3B9	Rodriguez Is.	01:58	13:31
3C	Equatorial Guinea, Bata	05:14	17:24
	Malabo	05:16	17:30

Pref	State/Province/Country/City	Sunrise	Sunset
3CØ	Pagalu Is.	05:34	17:38
3D2	Fiji Is., Suva	18:17	05:53
3D6	Swaziland, Mbabane	04:14	15:34
3V	Tunisia, Tunis	04:37	17:58
3W	Vietnam, Ho Chi Minh City		
	(Saigon)	22:39	11:04
	Hanoi	22:33	11:17
3X	Republic of Guinea, Conakry	06:42	19:05
3Y	Bouvet	06:47	16:43
3Y	Peter I Is.	14:00	22:01
4S	Sri Lanka, Colombo	00:29	12:48
4U	ITU Geneva	04:38	18:29
4U	United Nations Hq.	10:08	23:40
4W	Yemen, Sanaa	02:45	15:18
4X, 4Z	Israel, Jerusalem	03:04	16:11
5A	Libya, Tripoli	04:34	17:43
	Benghazi	04:04	17:13
5B	Cyprus, Nicosia	03:06	16:23
5H	Tanzania, Dar es Salaam	03:23	15:18
5N	Nigeria, Lagos	05:36	17:54
5R	Madagascar, Antananarivo	03:01	14:35
5T	Mauritania, Nouakchott	06:43	19:22
5U	Niger, Niamey	05:35	18:05
5V	Togo, Lome	05:45	18:02
5W	Western Samoa, Apia	17:33	05:17
5X	Uganda, Kampala	03:45	15:52
5Z	Kenya, Nairobi	04:49	16:53
6W	Senegal, Dakar	06:52	19:24
6Y	Jamaica, Kingston	10:46	23:25

Pref	State/Province/Country/City	Sunrise	Sunset
7O	Yemen, People's Dem. Rep.,		
	Aden	02:44	15:13
7P	Lesotho, Maseru	04:31	15:45
7Q	Malawi, Lilongwe	03:52	15:35
	Blantyre	03:48	15:28
7X	Algeria, Algiers	05:06	18:26
8P	Barbados, Bridgetown	09:42	22:11
8Q	Maldive Is.	00:58	13:12
8R	Guyana, Georgetown	09:42	22:00
9G	Ghana, Accra	05:51	18:07
9H	Malta	04:21	17:40
9J	Zambia, Lusaka	04:15	15:55
9K	Kuwait	02:16	15:18
9L	Sierra Leone, Freetown	06:40	19:02
9M2	West Malaysia, Kuala Lumpur	23:06	11:18
9M6,8	East Malaysia, Sabah,		
	Sandakan (9M6)	21:58	10:14
	Sarawak, Kuching (9M8)	22:32	10:42
9N	Nepal, Katmandu	23:48	12:46
9Q	Zaire, Kinshasa	04:57	16:57
	Kisangani	04:14	16:21
	Lubumbashi	04:15	16:02
9U	Burundi, Bujumbura	04:00	16:02
9V	Singapore	22:59	11:08
9X	Rwanda, Kigali	03:56	16:00
9Y	Trinidad & Tobago, Port of		
	Spain	09:51	22:16
J2/A	Abu Ail	02:51	15:23

Pref	State/Province/Country/City	Sunrise	Sunset
VE1	New Brunswick, Saint John	09:04	23:37
	Nova Scotia, Halifax	08:56	23:25
	Prince Edward Island,		
	Charlottetown	08:50	23:27
VE2	Quebec, Montreal	09:34	00:07
	Quebec City	09:20	00:01
VE3	Ontario, London	10:11	00:32
	Ottawa	09:42	00:15
	Sudbury	10:01	00:40
	Toronto	10:02	00:26
VE4	Manitoba, Winnipeg	10:55	01:54
VE5	Saskatchewan, Regina	11:23	02:26
	Saskatoon	11:25	02:40
VE6	Alberta, Calgary	11:59	03:06
	Edmonton	11:47	03:13
VE7	British Columbia, Prince George	12:23	03:52
	Prince Rupert	12:51	04:23
	Vancouver	12:41	03:36
VE8	Northwest Territories,		
	Yellowknife	11:03	04:05
	Resolute	24 hrs day	
VY1	Yukon, Whitehorse	12:38	05:15
VO1	Newfoundland, St. John's	08:04	22:49
VO2	Labrador, Goose Bay	08:16	23:40
W1	Connecticut, Hartford	09:39	23:54
	Maine, Bangor	09:16	23:46
	Portland	09:25	23:49
	Massachusetts, Boston	09:32	23:49
	New Hampshire, Concord	09:31	23:53
	Rhode Island, Providence	09:34	23:49
	Vermont, Montpelier	09:33	00:00
W2	New Jersey, Atlantic City	09:52	23:56
	New York, Albany	09:42	00:01
	Buffalo	10:02	00:22
	New York City	09:47	23:57
	Syracuse	09:50	00:11
W3	Delaware, Wilmington	09:55	00:01
	District of Columbia,		
	Washington	10:03	00:05
	Maryland, Baltimore	10:01	00:04
	Pennsylvania, Harrisburg	10:00	00:08
	Philadelphia	09:54	00:00
	Pittsburgh	10:12	00:20
	Scranton	09:52	00:05
W4	Alabama, Montgomery	10:53	00:30
	Florida, Jacksonville	10:38	00:08
	Miami	10:39	23:54
	Pensacola	11:00	00:30
	Georgia, Atlanta	10:43	00:25
	Savannah	10:33	00:08
	Kentucky, Lexington	10:35	00:33
	Louisville	10:40	00:36
	North Carolina, Charlotte	10:26	00:13
	Raleigh	10:16	00:05
	Wilmington	10:16	23:59
	South Carolina, Columbia	10:29	00:11
	Tennessee, Knoxville	10:37	00:27
	Memphis	11:03	00:50
	Nashville	10:48	00:39
	Virginia, Norfolk	10:04	23:58
	Richmond	10:08	00:04
W5	Arkansas, Little Rock	11:13	00:58
	Louisiana, New Orleans	11:12	00:40
	Shreveport	11:22	00:59
	Mississippi, Jackson	11:09	00:45
	New Mexico, Albuquerque	12:10	01:56
	Oklahoma, Oklahoma City	11:32	01:20
	Texas, Abilene	11:46	01:23
	Amarillo	11:50	01:37
	Dallas	11:34	01:12
	El Paso	12:15	01:49
	San Antonio	11:47	01:13
W6	California, Los Angeles	12:57	02:40
	San Francisco	13:07	03:04
W7	Arizona, Flagstaff	12:29	02:16
	Phoenix	12:34	02:15
	Idaho, Boise	12:29	02:53
	Pocatello	12:16	02:36

Pref	State/Province/Country/City	Sunrise	Sunset
	Montana, Billings	11:52	02:28
	Butte	12:08	02:44
	Great Falls	11:59	02:44
	Nevada, Las Vegas	12:41	02:32
	Reno	12:53	02:57
	Oregon, Portland	12:50	03:24
	Utah, Salt Lake City	12:18	02:29
	Washington, Seattle	12:43	03:28
	Spokane	12:23	03:09
	Wyoming, Cheyenne	11:49	02:01
	Sheridan	11:49	02:19
W8	Michigan, Detroit	10:19	00:37
	Grand Rapids	10:29	00:49
	Sault Ste. Marie	10:14	00:53
	Traverse City	10:24	00:53
	Ohio, Cincinnati	10:33	00:35
	Cleveland	10:16	00:30
	Columbus	10:25	00:31
	West Virginia, Charleston	10:23	00:22
W9	Illinois, Chicago	10:39	00:54
	Indiana, Indianapolis	10:38	00:44
	Wisconsin, Green Bay	10:34	01:02
	Milwaukee	10:37	00:58
W0	Colorado, Denver	11:53	01:59
	Grand Junction	12:09	02:12
	Iowa, Des Moines	11:03	01:17
	Kansas, Pratt	11:32	01:29
	Wichita	11:27	01:24
	Minnesota, Duluth	10:44	01:25
	Minneapolis	10:54	01:25
	Missouri, Columbia	11:04	01:06
	Kansas City	11:13	01:16
	St. Louis	10:57	00:57
	Nebraska, North Platte	11:33	01:45
	Omaha	11:13	01:26
	North Dakota, Fargo	11:03	01:44
	South Dakota, Rapid City	11:36	02:02
A2	Botswana, Gaborone	04:41	15:44
A3	Tonga, Nukualofa	18:00	05:14
A4	Oman, Muscat	01:28	14:36
A5	Bhutan, Thimbu	23:19	12:38
A6	United Arab Emirates, Abu		
	Dhabi	01:44	14:55
A7	Qatar, Ad-Dawhah	01:53	15:07
A9	Bahrein, Al-Manamah	01:56	15:12
AP	Pakistan, Karachi	00:52	14:04
	Islamabad	00:13	13:54
BV	Taiwan, Taipei	21:14	10:26
BY	People's Rep. of China, Beijing	21:07	11:14
	Harbin	20:12	10:47
	Shanghai	21:04	10:36
	Fuzhou	21:21	10:37
	Xian	21:49	11:32
	Chongqing	22:06	11:34
	Chengdu	22:14	11:45
	Lhasa	23:08	12:35
	Urumqi	22:53	13:18
	Kashi	23:50	13:54
C2	Nauru	18:46	06:51
C3	Andorra	04:41	18:59
C5	The Gambia, Banjul	06:43	19:23
C6	Bahamas, Nassau	10:29	23:42
C9	Mozambique, Maputo	04:16	15:16
	Mozambique	03:28	14:58
CE	Chile, Santiago	11:21	21:57
CE0	Easter Island	13:45	00:42
CE0	San Felix	11:47	22:46
CE0	Juan Fernandez	11:53	22:29
CM,CO	Cuba, Havana	10:52	23:59
CN	Morocco, Casablanca	05:36	19:17
CP	Bolivia, La Paz	10:46	22:13
CT	Portugal, Lisbon	05:32	19:33
CT2, CU	Azores , Ponta Delgada	06:40	20:37
CT3	Madeira Islands, Funchal	06:15	19:52
CX	Uruguay, Montevideo	10:25	20:57
D2	Angola, Luanda	05:10	16:56
D4	Cape Verde, Praia	07:08	19:52

Pref	State/Province/Country/City	Sunrise	Sunset
D6	Comoros, Moroni	03:12	14:50
DA-DL	Germany, Fed. Rep. of (W),		
	Bonn	03:56	19:00
DU	Philippines, Manila	21:30	10:14
EA	Spain, Madrid	05:07	19:15
EA6	Balearic Is., Palma	04:43	18:48
EA8	Canary Is., Las Palmas	06:12	19:35
EA9	Ceuta & Melilla, Ceuta	05:22	19:12
	Melilla	05:14	19:02
EI	Ireland, Dublin	04:39	20:03
EL	Liberia, Monrovia	06:28	18:50
EP	Iran, Tehran	01:34	15:24
ET	Ethiopia, Addis Ababa	03:07	15:36
	Asmera	02:58	15:43
F	France, Paris	04:21	19:13
FG	Guadeloupe	09:39	22:26
FG, FS	St. Martin	09:42	22:35
FH	Mayotte	03:07	14:42
FK	New Caledonia, Noumia	19:15	06:25
FM	Martinique	09:38	22:22
FO	Clipperton	12:57	01:29
FO	Fr. Polynesia, Tahiti	16:52	04:16
FP	St. Pierre & Miquelon, St. Pierre	08:20	23:00
FR	Glorioso	02:57	14:36
FR	Juan de Nova	03:22	14:47
	Europa	03:39	14:50
FR	Reunion	02:37	13:51
FR	Tromelin	02:35	14:02
FT8W	Crozet	03:36	13:20
FT8X	Kerguelen	02:36	12:03
FT8Y	Antarctica, Dumont D'Urville	23:21	05:51
FT8Z	Amsterdam & St. Paul Is.,		
	Amsterdam	01:35	11:56
FW	Wallis & Futuna Is., Wallis	17:54	05:28
FY	Fr. Guiana, Cayenne	09:16	21:35
G	England, London	04:21	19:32
GD	Isle of Man	04:28	20:00
GI	Northern Ireland, Belfast	04:32	20:07
GJ	Jersey	04:37	19:33
GM	Scotland, Glasgow	04:21	20:06
	Aberdeen	04:05	20:03
GU	Guernsey	04:38	19:35
GW	Wales, Cardiff	04:34	19:44
H4	Solomon Islands, Honiara	19:24	07:08
HA	Hungary, Budapest	03:17	18:02
HB	Switzerland, Bern	04:05	18:47
HBØ	Liechtenstein	03:56	18:39
HC	Ecuador, Quito	11:05	23:11
HC8	Galapagos Is.	11:55	00:01
HH	Haiti, Port-Au-Prince	10:18	23:12
HI	Dominican Republic, Santo		
	Domingo	10:09	23:03
HK	Colombia, Bogota	10:43	23:01
HKØ	Malpelo Is.	11:12	23:29
HKØ	San Adreas	11:04	23:42
HL, HM	Korea, Seoul	20:30	10:26
HP	Panama, Panama	11:00	23:29
HR	Honduras, Tegucigalpa	11:24	00:06
HS	Thailand, Bangkok	22:53	11:35
HV	Vatican City	03:58	18:14
HZ, 7Z	Saudi Arabia, Dharan	01:58	15:14
	Mecca	02:46	15:48
	Riyadh	02:14	15:25
I	Italy, Rome	03:58	18:14
	Trieste	03:44	18:18
	Sicily	04:02	17:58
IS	Sardinia, Cagliari	04:18	18:21
J2	Djibouti, Djibouti	02:45	15:21
J3	Grenada	09:45	22:22
J5	Guinea-Bissau, Bissau	06:40	19:17
J6	St. Lucia	09:39	22:21
J7	Dominica	09:39	22:24
J8	St. Vincent	09:41	22:21
JA-JS	Japan, Tokyo	19:42	09:31
	Nagasaki	20:27	10:06
	Sapporo	19:20	09:41
JD	Minami Torishima	19:05	08:15
JD	Ogasawara, Kazan Is.	19:52	09:12
JT	Mongolia, Ulan Bator	21:25	12:12
JW	Svalbard, Spitsbergen	24 hrs day	
JX	Jan Mayen	02:00	22:59

Pref	State/Province/Country/City	Sunrise	Sunset
JY	Jordan, Amman	02:45	16:20
KC4	Antarctica, Byrd Station	24 hrs night	
	McMurdo Sound	24 hrs night	
	Palmer Station	12:41	19:43
KC6	Micronesia, Ponape	19:11	07:35
KC6	Belau, Yap	20:28	08:58
	Koror	20:46	09:10
KG4	Guantanamo Bay	10:28	23:26
KH1	Baker, Howland Is.	17:36	05:44
KH2	Guam, Agana	19:57	08:37
KH3	Johnston Is.	16:45	05:35
KH4	Midway Is.	17:04	06:27
KH5	Palmyra Is.	16:34	04:55
KH5K	Kingman Reef	16:35	05:00
KH6	Hawaii, Hilo	15:48	04:45
	Honolulu	15:57	04:58
KH7	Kure Is.	17:08	06:31
KH8	American Samoa, Pago Pago	17:33	05:05
KH9	Wake Is.	18:22	07:18
KHØ	Mariana Is., Saipan	19:50	08:35
KL7	Alaska, Adak	16:06	07:19
	Anchorage	13:34	06:18
	Fairbanks	12:57	06:38
	Juneau	12:49	04:58
	Nome	14:10	07:45
KP1	Navassa Is.	10:29	23:23
KP2	Virgin Islands, Charlotte Amalie	09:49	22:42
KP4	Puerto Rico, San Juan	09:54	22:48
KP5	Desecheo Is.	09:59	22:53
KX6	Marshall Islands, Kwajalein	18:32	07:01
LA-LJ	Norway, Oslo	02:59	19:27
LU	Argentina, Buenos Aires	10:33	21:06
LX	Luxembourg	04:02	19:00
LZ	Bulgaria, Sofia	03:13	17:32
OA	Peru, Lima	11:16	22:53
OD	Lebanon, Beirut	02:43	16:25
OE	Austria, Vienna	03:26	18:15
OH	Finland, Helsinki	02:01	18:32
OHØ	Aland Is.	02:21	18:52
OJØ	Market Reef	02:24	18:56
OK	Czechoslovakia, Prague	03:28	18:29
ON	Belgium, Brussels	04:05	19:12
OX, XP	Greenland, Godthaab	06:38	00:08
	Thule	24 hrs day	
OY	Faroe Islands, Torshavn	03:56	20:51
OZ	Denmark, Copenhagen	03:14	18:58
P2	Papua New Guinea, Madang	20:16	08:11
	Port Moresby	20:16	08:00
PA-PI	Netherlands, Amsterdam	03:58	19:15
PJ, P4	Netherlands Antilles, Willemstad	10:14	22:50
PJ5-8	St. Maarten and Saba,		
	St. Maarten	09:43	22:35
PY	Brazil, Brasilia	09:24	20:52
	Rio De Janeiro	09:14	20:23
	Natal	08:21	20:13
	Manaus	09:57	21:57
	Porto Alegre	09:57	20:45
PYØ	Fernando De Noronha	08:07	20:04
PYØ	St. Peter & St. Paul Rocks	07:49	19:58
PYØ	Trindade & Martin Vaz Is.,		
	Trindade	08:15	19:31
PZ	Suriname, Paramaribo	09:26	21:47
S2	Bangladesh, Dacca	23:20	12:29
S7	Seychelles, Victoria	02:16	14:12
S9	Sao Tome	05:25	17:33
SJ-SM	Sweden, Stockholm	02:34	18:54
SP	Poland, Krakow	03:06	18:06
	Warsaw	02:54	18:10
ST	Sudan, Khartoum	03:23	16:09
STØ	Southern Sudan, Juba	03:40	15:59
SU	Egypt, Cairo	03:06	16:35
SV	Greece, Athens	03:22	17:20
SV5	Dodecanese, Rhodes	03:07	16:59
SV9	Crete	03:21	17:09
SV/A	Mount Athos	03:15	17:23
T2	Tuvalu, Funafuti	18:09	05:55
T30	West Kiribati, Bonriki	18:18	06:28
T31	Central Kiribati, Kanton	17:23	05:23
T32	East Kiribati, Christmas Is.	16:20	04:31
T5	Somalia, Mogadishu	02:48	15:00
T7	San Marino	03:54	18:19

Pref	State/Province/Country/City	Sunrise	Sunset
TA	Turkey, Ankara	02:41	16:48
	Istanbul	02:54	17:06
TF	Iceland, Reykjavik	04:40	22:08
TG	Guatemala, Guatemala City	11:36	00:20
TI	Costa Rica, San Jose	11:16	23:48
TI9	Cocos Is.	11:34	23:54
TJ	Cameroon, Yaounde	05:02	17:18
TK	Corsica	04:12	18:28
TL	Central African Republic, Bangui	04:33	16:50
TN	Congo, Brazzaville	04:57	16:53
TR	Gabon, Libreville	05:14	17:22
TT	Chad, N'Djamena	04:38	17:14
TU	Ivory Coast, Abidjan	06:02	18:22
TY	Benin, Porto Novo	05:34	17:57
TZ	Mali, Bamako	06:09	18:47
UA	Russia, European,		
	Leningrad (UA1)	01:41	18:08
	Archangel (UA1)	00:26	18:03
	Murmansk (UA1)	00:01	19:27
	Moscow (UA3)	01:33	17:18
	Kuibyshev (UA4)	00:54	16:17
	Rostov (UA6)	01:56	16:40
UA1P	Franz Josef Land	24 hrs day	
UA2	Kaliningrad	02:45	18:23
UA9,Ø	Russia, Asiatic,		
	Novosibirsk (UA9)	22:36	14:13
	Perm (UA9)	00:08	16:14
	Omsk (UA9)	23:14	14:51
	Norilsk (UAØ)	20:15	15:53
	Irkutsk (UAØ)	21:21	12:37
	Vladivostok (UAØ)	19:58	10:19
	Petropavlovsk (UAØ)	17:40	09:02
	Khabarovsk (UAØ)	19:30	10:21
	Krasnoyarsk (UAØ)	21:52	13:38
	Yakutsk (UA1Ø)	18:50	11:45
	Wrangel Island (UAØ)	13:25	10:23
	Kyzyl (UAØY)	22:02	13:14
UB	Ukraine, Kiev	02:23	17:25
UC	Byelorussia, Minsk	02:21	17:50
UD	Azerbaijan, Baku	01:32	15:41
UF	Georgia, Tbilisi	01:50	16:04
UG	Armenia, Ierevan	01:54	16:02
UH	Turkoman, Ashkhabad	01:03	15:02
UI	Uzbek, Bukhara	00:36	14:41
	Tashkent	00:13	14:25
UJ	Tadzhik, Samarkand	00:26	14:32
	Dushanbe	00:19	14:22
UL	Kazakh, Alma-ata	23:37	13:59
UM	Kirghiz, Frunze	23:48	14:08
UO	Moldavia, Kishinev	02:40	17:22
UP	Lithuania, Vilna	02:27	18:01
UQ	Latvia, Riga	02:22	18:18
UR	Estonia, Tallinn	02:06	18:28
V2	Antigua & Barbuda, St. Johns	09:38	22:28
V3	Belize, Belmopan	11:26	00:16
V4	St. Christopher & Nevis	09:41	22:32
V8	Brunei, Bandar Seri Begawan	22:07	10:26
VE1, CY	Sable Is.	08:44	23:08
VE1, CY	St. Paul Is.	08:35	23:18
VK	Australia, Canberra (VK1)	20:45	07:15
	Sydney (VK2)	20:34	07:09
	Melbourne (VK3)	21:06	07:26
	Brisbane (VK4)	20:16	07:12
	Adelaide (VK5)	21:26	07:57
	Perth (VK6)	22:52	09:34
	Hobart, Tasmania (VK7)	21:07	07:07
	Darwin (VK8)	21:24	09:01
VK	Lord Howe Is.	19:58	06:41
VK9	Christmas Is.	23:03	10:44
VK9	Cocos-Keeling Is.	23:40	11:17
VK9	Mellish Reef	19:51	07:15
VK9	Norfolk Is.	19:19	06:10
VK9	Willis Is.	20:15	07:41
VKØ	Heard Is.	02:31	11:34
VKØ	Macquarie Is.	20:55	05:47
VP2E	Anguilla	09:41	22:35
VP2M	Montserrat	09:40	22:29
VP2V	British Virgin Is., Tortola	09:48	22:41
VP5	Turks & Caicos Islands,		
	Grand Turk	10:10	23:12
VP8	Falkland Islands, Stanley	11:11	20:24

Pref	State/Province/Country/City	Sunrise	Sunset
VP8	So. Georgia Is.	09:56	18:50
VP8	So. Orkney Is.	10:59	18:57
VP8	So. Sandwich Islands, Saunders Is.	09:30	17:56
VP8	So. Shetland Is., King George Is.	11:59	19:40
VP9	Bermuda	09:27	23:03
VQ9	Chagos, Diego Garcia	01:12	13:01
VR6	Pitcairn Is.	15:05	02:08
VS6	Hong Kong	21:47	10:51
VU	India, Bombay	00:37	13:32
	Calcutta	23:30	12:35
	New Delhi	00:05	13:29
	Bangalore	00:26	13:05
VU7	Andaman Islands, Port Blair	23:27	12:03
VU7	Laccadive Is.	00:48	13:20
XE	Mexico, Mexico City (XE1)	12:04	01:01
	Chihuahua (XE2)	12:18	01:42
	Merida (XE3)	11:24	00:25
XF4	Revilla Gigedo	12:54	01:50
XT	Burkina Faso, Onagadougou	05:44	18:21
XU	Kampuchea, Phnom Penh	22:39	11:15
XW	Laos, Viangchan	22:39	11:32
XX9	Macao	21:50	10:54
XZ	Burma, Rangoon	23:07	11:57
Y2-9	German Dem.Rep.(E), Berlin	03:24	18:41
YA	Afghanistan, Kandahar	00:47	14:19
	Kabul	00:27	14:11
YB-YD	Indonesia, Jakarta	22:53	10:45
	Medan	23:13	11:29
	Pontianak	22:35	10:42
	Jayapura	20:33	08:34
YI	Iraq, Baghdad	02:09	15:47
YJ	Vanuatu, Port Vila	19:01	06:24
YK	Syria, Damascus	02:41	16:21
YN, HT	Nicaragua, Managua	11:22	23:58
YO	Romania, Bucharest	02:58	17:25
YS	El Salvador, San Salvador	11:32	00:13
YU	Yugoslavia, Belgrade	03:19	17:49
YV	Venezuela, Caracas	10:08	22:40
YVØ	Aves Is.	09:48	22:34
Z2,ZE	Zimbabwe, Harare	04:11	15:33
ZA	Albania, Tirane	03:31	17:43
ZB	Gibraltar	05:22	19:13
ZC4	British Cyprus	02:52	16:36
ZD7	St. Helena	06:36	18:03
ZD8	Ascension Is.	07:00	18:47
ZD9	Tristan da Cunha	07:34	17:57
ZF	Cayman Is.	10:53	23:49
ZK1	No. Cook Is., Manihiki	16:49	04:31
ZK1	So. Cook Is., Rarotonga	16:58	04:12
ZK2	Niue	17:32	04:51
ZK3	Tokelaus, Atafu	17:34	05:20
ZL	New Zealand, Auckland (ZL1)	19:05	05:29
	Wellington (ZL2)	19:14	05:20
	Christchurch (ZL3)	19:27	05:24
	Dunedin (ZL4)	19:41	05:27
ZL5	Antarctica, Scott Base	24 hrs night	
ZL7	Chatham Is.	18:45	04:39
ZL8	Kermadec Is.	18:23	05:13
ZL9	Auckland & Campbell Is.,		
	Auckland	20:11	05:29
	Campbell Is.	20:06	05:13
ZP	Paraguay, Asuncion	10:16	21:18
ZS	South Africa, Cape Town (ZS1)	05:25	16:00
	Port Elizabeth (ZS2)	04:56	15:31
	Bloemfontein (ZS4)	04:46	15:37
	Durban (ZS5)	04:28	15:17
	Johannesburg (ZS6)	04:34	15:33
ZS2	Prince Edward & Marion Is., Marion	04:34	14:15
ZS3	Namibia, Windhoek	05:13	16:23
1AØ	S.M.O.M	03:59	18:14
1S	Spratly Is.	22:14	10:43
3A	Monaco	04:14	18:38
3B6	Agalega	02:19	14:00
3B7	St. Brandon	02:13	13:40
3B8	Mauritius	02:28	13:44
3B9	Rodriguez Is.	02:04	13:21
3C	Equatorial Guinea, Bata	05:11	17:22
	Malabo	05:13	17:29

Pref	State/Province/Country/City	Sunrise	Sunset
3CØ	Pagalu Is.	05:32	17:35
3D2	Fiji Is., Suva	18:21	05:43
3D6	Swaziland, Mbabane	04:22	15:21
3V	Tunisia, Tunis	04:19	18:12
3W	Vietnam, Ho Chi Minh City		
	(Saigon)	22:32	11:06
	Hanoi	22:22	11:23
3X	Republic of Guinea, Conakry	06:36	19:06
3Y	Bouvet	07:16	16:09
3Y	Peter I Is.	15:08	20:49
4S	Sri Lanka, Colombo	00:24	12:48
4U	ITU Geneva	04:13	18:50
4U	United Nations Hq.	09:47	23:57
4W	Yemen, Sanaa	02:37	15:22
4X, 4Z	Israel, Jerusalem	02:48	16:23
5A	Libya, Tripoli	04:18	17:55
	Benghazi	03:48	17:24
5B	Cyprus, Nicosia	02:49	16:36
5H	Tanzania, Dar es Salaam	03:23	15:13
5N	Nigeria, Lagos	05:31	17:54
5R	Madagascar, Antananarivo	03:06	14:26
5T	Mauritania, Nouakchott	06:34	19:26
5U	Niger, Niamey	05:28	18:08
5V	Togo, Lome	05:41	18:02
5W	Western Samoa, Apia	17:36	05:10
5X	Uganda, Kampala	03:42	15:50
5Z	Kenya, Nairobi	04:47	16:51
6W	Senegal, Dakar	06:44	19:28
6Y	Jamaica, Kingston	10:37	23:29

Pref	State/Province/Country/City	Sunrise	Sunset
7O	Yemen, People's Dem. Rep.,		
	Aden	02:37	15:15
7P	Lesotho, Maseru	04:41	15:31
7Q	Malawi, Lilongwe	03:54	15:27
	Blantyre	03:52	15:20
7X	Algeria, Algiers	04:48	18:40
8P	Barbados, Bridgetown	09:35	22:14
8Q	Maldive Is.	00:54	13:11
8R	Guyana, Georgetown	09:37	22:01
9G	Ghana, Accra	05:47	18:07
9H	Malta	04:03	17:53
9J	Zambia, Lusaka	04:18	15:47
9K	Kuwait	02:01	15:28
9L	Sierra Leone, Freetown	06:35	19:03
9M2	West Malaysia, Kuala Lumpur	23:02	11:17
9M6,8	East Malaysia, Sabah,		
	Sandakan (9M6)	21:53	10:14
	Sarawak, Kuching (9M8)	22:29	10:40
9N	Nepal, Katmandu	23:34	12:55
9Q	Zaire, Kinshasa	04:57	16:53
	Kisangani	04:11	16:19
	Lubumbashi	04:17	15:55
9U	Burundi, Bujumbura	03:59	15:58
9V	Singapore	22:56	11:06
9X	Rwanda, Kigali	03:55	15:57
9Y	Trinidad & Tobago, Port of		
	Spain	09:45	22:18
J2/A	Abu Ail	02:44	15:26

Pref	State/Province/Country/City	Sunrise	Sunset
VE1	New Brunswick, Saint John	08:48	23:53
	Nova Scotia, Halifax	08:40	23:41
	Prince Edward Island, Charlottetown	08:33	23:44
VE2	Quebec, Montreal	09:17	00:24
	Quebec City	09:03	00:19
VE3	Ontario, London	09:56	00:47
	Ottawa	09:26	00:32
	Sudbury	09:43	00:57
	Toronto	09:46	00:41
VE4	Manitoba, Winnipeg	10:35	02:14
VE5	Saskatchewan, Regina	11:02	02:47
	Saskatoon	11:04	03:02
VE6	Alberta, Calgary	11:38	03:27
	Edmonton	11:24	03:36
VE7	British Columbia, Prince George	11:59	04:15
	Prince Rupert	12:27	04:47
	Vancouver	12:21	03:56
VE8	Northwest Territories, Yellowknife	10:23	04:44
	Resolute	24 hrs day	
VY1	Yukon, Whitehorse	12:03	05:50
VO1	Newfoundland, St. John's	07:46	23:07
VO2	Labrador, Goose Bay	07:53	00:03
W1	Connecticut, Hartford	09:25	00:09
	Maine, Bangor	09:00	00:02
	Portland	09:10	00:05
	Massachusetts, Boston	09:17	00:04
	New Hampshire, Concord	09:16	00:08
	Rhode Island, Providence	09:20	00:03
	Vermont, Montpelier	09:17	00:16
W2	New Jersey, Atlantic City	09:39	00:09
	New York, Albany	09:27	00:16
	Buffalo	09:47	00:37
	New York City	09:33	00:11
	Syracuse	09:35	00:26
W3	Delaware, Wilmington	09:42	00:14
	District of Columbia, Washington	09:50	00:18
	Maryland, Baltimore	09:48	00:17
	Pennsylvania, Harrisburg	09:46	00:21
	Philadelphia	09:40	00:13
	Pittsburgh	09:58	00:34
	Scranton	09:38	00:19
W4	Alabama, Montgomery	10:43	00:39
	Florida, Jacksonville	10:29	00:17
	Miami	10:32	00:02
	Pensacola	10:51	00:39
	Georgia, Atlanta	10:32	00:35
	Savannah	10:23	00:18
	Kentucky, Lexington	10:23	00:45
	Louisville	10:27	00:51
	North Carolina, Charlotte	10:15	00:24
	Raleigh	10:04	00:16
	Wilmington	10:05	00:10
	South Carolina, Columbia	10:18	00:22
	Tennessee, Knoxville	10:25	00:38
	Memphis	10:52	01:01
	Nashville	10:36	00:50
	Virginia, Norfolk	09:53	00:10
	Richmond	09:55	00:16
W5	Arkansas, Little Rock	11:02	01:09
	Louisiana, New Orleans	11:03	00:49
	Shreveport	11:12	01:09
	Mississippi, Jackson	10:59	00:55
	New Mexico, Albuquerque	11:58	02:07
	Oklahoma, Oklahoma City	11:21	01:31
	Texas, Abilene	11:36	01:33
	Amarillo	11:39	01:48
	Dallas	11:24	01:22
	El Paso	12:05	01:59
	San Antonio	11:38	01:22
W6	California, Los Angeles	12:47	02:51
	San Francisco	12:55	03:17
W7	Arizona, Flagstaff	12:18	02:27
	Phoenix	12:24	02:25
	Idaho, Boise	12:14	03:08
	Pocatello	12:01	02:51

Pref	State/Province/Country/City	Sunrise	Sunset
	Montana, Billings	11:36	02:44
	Butte	11:51	03:01
	Great Falls	11:41	03:01
	Nevada, Las Vegas	12:29	02:43
	Reno	12:40	03:11
	Oregon, Portland	12:34	03:40
	Utah, Salt Lake City	12:05	02:43
	Washington, Seattle	12:25	03:46
	Spokane	12:05	03:27
	Wyoming, Cheyenne	11:35	02:15
	Sheridan	11:33	02:35
W8	Michigan, Detroit	10:05	00:51
	Grand Rapids	10:14	01:04
	Sault Ste. Marie	09:57	01:10
	Traverse City	10:07	01:09
	Ohio, Cincinnati	10:20	00:48
	Cleveland	10:02	00:44
	Columbus	10:11	00:45
	West Virginia, Charleston	10:10	00:35
W9	Illinois, Chicago	10:24	01:08
	Indiana, Indianapolis	10:25	00:57
	Wisconsin, Green Bay	10:18	01:18
	Milwaukee	10:22	01:13
WØ	Colorado, Denver	11:40	02:12
	Grand Junction	11:56	02:25
	Iowa, Des Moines	10:49	01:32
	Kansas, Pratt	11:20	01:41
	Wichita	11:15	01:36
	Minnesota, Duluth	10:27	01:42
	Minneapolis	10:38	01:41
	Missouri, Columbia	10:51	01:19
	Kansas City	11:00	01:29
	St. Louis	10:44	01:10
	Nebraska, North Platte	11:19	01:59
	Omaha	10:59	01:40
	North Dakota, Fargo	10:45	02:01
	South Dakota, Rapid City	11:20	02:18
A2	Botswana, Gaborone	04:48	15:37
A3	Tonga, Nukualofa	18:06	05:08
A4	Oman, Muscat	01:21	14:42
A5	Bhutan, Thimbu	23:11	12:46
A6	United Arab Emirates, Abu Dhabi	01:37	15:02
A7	Qatar, Ad-Dawhah	01:46	15:14
A9	Bahrein, Al-Manamah	01:48	15:19
AP	Pakistan, Karachi	00:44	14:11
	Islamabad	00:02	14:04
BV	Taiwan, Taipei	21:06	10:34
BY	People's Rep. of China, Beijing	20:54	11:27
	Harbin	19:55	11:04
	Shanghai	20:54	10:46
	Fuzhou	21:13	10:44
	Xian	21:38	11:43
	Chongqing	21:57	11:43
	Chengdu	22:05	11:54
	Lhasa	22:59	12:44
	Urumqi	22:38	13:33
	Kashi	23:37	14:07
C2	Nauru	18:46	06:51
C3	Andorra	04:26	19:14
C5	The Gambia, Banjul	06:39	19:27
C6	Bahamas, Nassau	10:22	23:49
C9	Mozambique, Maputo	04:23	15:08
	Mozambique	03:32	14:54
CE	Chile, Santiago	11:31	21:47
CEØ	Easter Island	13:53	00:34
CEØ	San Felix	11:54	22:39
CEØ	Juan Fernandez	12:03	22:19
CM,CO	Cuba, Havana	10:46	00:06
CN	Morocco, Casablanca	05:25	19:27
CP	Bolivia, La Paz	10:51	22:08
CT	Portugal, Lisbon	05:20	19:46
CT2, CU	Azores, Ponta Delgada	06:28	20:49
CT3	Madeira Islands, Funchal	06:05	20:02
CX	Uruguay, Montevideo	10:36	20:46
D2	Angola, Luanda	05:13	16:54
D4	Cape Verde, Praia	07:04	19:56

Pref	State/Province/Country/City	Sunrise	Sunset
D6	Comoros, Moroni	03:15	14:47
DA-DL	Germany, Fed. Rep. of (W),		
	Bonn	03:35	19:21
DU	Philippines, Manilla	21:26	10:18
EA	Spain, Madrid	04:53	19:29
EA6	Balearic Is., Palma	04:30	19:01
EA8	Canary Is., Las Palmas	06:03	19:43
EA9	Ceuta & Melilla, Ceuta	05:11	19:23
	Melilla	05:03	19:13
EI	Ireland, Dublin	04:16	20:26
EL	Liberia, Monrovia	06:26	18:52
EP	Iran, Tehran	01:23	15:35
ET	Ethiopia, Addis Ababa	03:04	15:38
	Asmera	02:54	15:47
F	France, Paris	04:02	19:32
FG	Guadeloupe	09:35	22:31
FG, FS	St. Martin	09:37	22:40
FH	Mayotte	03:11	14:39
FK	New Caledonia, Noumia	19:21	06:19
FM	Martinique	09:34	22:26
FO	Clipperton	12:54	01:32
FO	Fr. Polynesia, Tahiti	16:57	04:11
FP	St. Pierre & Miquelon, St. Pierre	08:03	23:17
FR	Glorioso	03:00	14:33
FR	Juan de Nova	03:27	14:43
	Europa	03:45	14:44
FR	Reunion	02:42	13:45
FR	Tromelin	02:39	13:58
FT8W	Crozet	03:52	13:04
FT8X	Kerguelen	02:54	11:44
FT8Y	Antarctica, Dumont D'Urville	00:12	05:00
FT8Z	Amsterdam & St. Paul Is.,		
	Amsterdam	01:47	11:44
FW	Wallis & Futuna Is., Wallis	17:57	05:25
FY	Fr. Guiana, Cayenne	09:14	21:36
G	England, London	04:00	19:53
GD	Isle of Man	04:04	20:24
GI	Northern Ireland, Belfast	04:08	20:31
GJ	Jersey	04:18	19:52
GM	Scotland, Glasgow	03:55	20:32
	Aberdeen	03:37	20:31
GU	Guernsey	04:19	19:55
GW	Wales, Cardiff	04:12	20:05
H4	Solomon Islands, Honiara	19:26	07:06
HA	Hungary, Budapest	02:59	18:20
HB	Switzerland, Bern	03:48	19:04
HB0	Liechtenstein	03:38	18:57
HC	Ecuador, Quito	11:05	23:11
HC8	Galapagos Is.	11:55	00:01
HH	Haiti, Port-Au-Prince	10:13	23:17
HI	Dominican Republic, Santo		
	Domingo	10:04	23:08
HK	Colombia, Bogota	10:42	23:03
HK0	Malpelo Is.	11:11	23:30
HK0	San Adreas	11:00	23:45
HL, HM	Korea, Seoul	20:18	10:38
HP	Panama, Panama	10:57	23:31
HR	Honduras, Tegucigalpa	11:20	00:10
HS	Thailand, Bangkok	22:50	11:38
HV	Vatican City	03:44	18:28
HZ, 7Z	Saudi Arabia, Dharan	01:50	15:22
	Mecca	02:40	15:54
	Riyadh	02:07	15:32
I	Italy, Rome	03:44	18:28
	Trieste	03:27	18:35
	Sicily	03:50	18:10
IS	Sardinia, Cagliari	04:05	18:34
J2	Djibouti, Djibouti	02:42	15:24
J3	Grenada	09:42	22:25
J5	Guinea-Bissau, Bissau	06:37	19:20
J6	St. Lucia	09:35	22:25
J7	Dominica	09:34	22:28
J8	St. Vincent	09:38	22:25
JA-JS	Japan, Tokyo	19:31	09:43
	Nagasaki	20:17	10:16
	Sapporo	19:05	09:56
JD	Minami Torishima	18:58	08:22
JD	Ogasawara, Kazan Is.	19:44	09:20
JT	Mongolia, Ulan Bator	21:07	12:30
JW	Svalbard, Spitsbergen	24 hrs day	
JX	Jan Mayen	24 hrs day	

Pref	State/Province/Country/City	Sunrise	Sunset
JY	Jordan, Amman	02:35	16:30
KC4	Antarctica, Byrd Station	24 hrs night	
	McMurdo Sound	24 hrs night	
	Palmer Station	13:24	19:00
KC6	Micronesia, Ponape	19:09	07:37
KC6	Belau, Yap	20:26	09:01
	Koror	20:44	09:12
KG4	Guantanamo Bay	10:22	23:31
KH1	Baker, Howland Is.	17:36	05:44
KH2	Guam, Agana	19:53	08:41
KH3	Johnston Is.	16:40	05:40
KH4	Midway Is.	16:56	06:35
KH5	Palmyra Is.	16:32	04:57
KH5K	Kingman Reef	16:33	05:02
KH6	Hawaii, Hilo	15:42	04:50
	Honolulu	15:51	05:04
KH7	Kure Is.	17:00	06:40
KH8	American Samoa, Pago Pago	17:37	05:01
KH9	Wake Is.	18:16	07:23
KH0	Mariana Is., Saipan	19:46	08:39
KL7	Alaska, Adak	15:45	07:40
	Anchorage	12:58	06:54
	Fairbanks	12:09	07:27
	Juneau	12:19	05:28
	Nome	13:23	08:32
KP1	Navassa Is.	10:24	23:28
KP2	Virgin Islands, Charlotte Amalie	09:44	22:47
KP4	Puerto Rico, San Juan	09:49	22:53
KP5	Desecheo Is.	09:54	22:58
KX6	Marshall Islands, Kwajalein	18:30	07:04
LA-LJ	Norway, Oslo	02:26	20:01
LU	Argentina, Buenos Aires	10:44	20:55
LX	Luxembourg	03:43	19:20
LZ	Bulgaria, Sofia	02:58	17:47
OA	Peru, Lima	11:19	22:50
OD	Lebanon, Beirut	02:32	16:36
OE	Austria, Vienna	03:08	18:34
OH	Finland, Helsinki	01:27	19:05
OH0	Aland Is.	01:47	19:25
OJ0	Market Reef	01:50	19:30
OK	Czechoslovakia, Prague	03:08	18:49
ON	Belgium, Brussels	03:45	19:32
OX, XP	Greenland, Godthaab	05:52	00:54
	Thule	24 hrs day	
OY	Faroe Islands, Torshavn	03:18	21:29
OZ	Denmark, Copenhagen	02:48	19:23
P2	Papua New Guinea, Madang	20:18	08:10
	Port Moresby	20:18	07:57
PA-PI	Netherlands, Amsterdam	03:36	19:37
PJ, P4	Netherlands Antilles, Willemstad	10:10	22:54
PJ5-8	St. Maarten and Saba,		
	St. Maarten	09:38	22:39
PY	Brazil, Brasilia	09:28	20:47
	Rio De Janeiro	09:21	20:17
	Natal	08:22	20:12
	Manaus	09:58	21:56
	Porto Alegre	10:06	20:36
PY0	Fernando De Noronha	08:08	20:03
PY0	St. Peter & St. Paul Rocks	07:49	19:59
PY0	Trindade & Martin Vaz Is.,		
	Trindade	08:21	19:25
PZ	Suriname, Paramaribo	09:25	21:49
S2	Bangladesh, Dacca	23:13	12:35
S7	Seychelles, Victoria	02:17	14:11
S9	Sao Tome	05:25	17:33
SJ-SM	Sweden, Stockholm	02:02	19:25
SP	Poland, Krakow	02:46	18:26
	Warsaw	02:32	18:32
ST	Sudan, Khartoum	03:19	16:13
ST0	Southern Sudan, Juba	03:39	16:01
SU	Egypt, Cairo	02:57	16:44
SV	Greece, Athens	03:10	17:33
SV5	Dodecanese, Rhodes	02:56	17:11
SV9	Crete	03:10	17:20
SV/A	Mount Athos	03:02	17:36
T2	Tuvalu, Funafuti	18:11	05:52
T30	West Kiribati, Bonriki	18:18	06:29
T31	Central Kiribati, Kanton	17:23	05:22
T32	East Kiribati, Christmas Is.	16:19	04:32
T5	Somalia, Mogadishu	02:48	15:01
T7	San Marino	03:39	18:35

Pref	State/Province/Country/City	Sunrise	Sunset
TA	Turkey, Ankara	02:28	17:01
	Istanbul	02:40	17:20
TF	Iceland, Reykjavik	03:55	22:53
TG	Guatemala, Guatemala City	11:32	00:24
TI	Costa Rica, San Jose	11:14	23:50
TI9	Cocos Is.	11:32	23:56
TJ	Cameroon, Yaounde	05:01	17:19
TK	Corsica	03:58	18:42
TL	Central African Republic, Bangui	04:32	16:52
TN	Congo, Brazzaville	04:58	16:52
TR	Gabon, Libreville	05:14	17:22
TT	Chad, N'Djamena	04:34	17:18
TU	Ivory Coast, Abidjan	06:01	18:23
TY	Benin, Porto Novo	05:32	17:59
TZ	Mali, Bamako	06:05	18:51
UA	Russia, European,		
	Leningrad (UA1)	01:08	18:41
	Archangel (UA1)	23:38	18:50
	Murmansk (UA1)	21:57	21:31
	Moscow (UA3)	01:07	17:44
	Kuibyshev (UA4)	00:31	16:40
	Rostov (UA6)	01:38	16:58
UA1P	Franz Josef Land	24 hrs day	
UA2	Kaliningrad	02:20	18:48
UA9,Ø	Russia, Asiatic,		
	Novosibirsk (UA9)	22:11	14:38
	Perm (UA9)	23:38	16:43
	Omsk (UA9)	22:49	15:16
	Norilsk (UAØ)	24 hrs day	
	Irkutsk (UAØ)	20:59	12:59
	Vladivostok (UAØ)	19:43	10:34
	Petropavlovsk (UAØ)	17:18	09:25
	Khabarovsk (UAØ)	19:12	10:40
	Krasnoyarsk (UAØ)	21:25	14:04
	Yakutsk (UAØ)	18:12	12:23
	Wrangel Island (UAØ)	24 hrs day	
	Kyzyl (UAØY)	21:41	13:35
UB	Ukraine, Kiev	02:02	17:46
UC	Byelorussia, Minsk	01:58	18:14
UD	Azerbaijan, Baku	01:19	15:54
UF	Georgia, Tbilisi	01:35	16:18
UG	Armenia, Ierevan	01:41	16:16
UH	Turkoman, Ashkhabad	00:51	15:14
UI	Uzbek, Bukhara	00:22	14:55
	Tashkent	23:59	14:39
UJ	Tadzhik, Samarkand	00:13	14:45
	Dushanbe	00:07	14:35
UL	Kazakh, Alma-ata	23:22	14:15
UM	Kirghiz, Frunze	23:33	14:23
UO	Moldavia, Kishinev	02:22	17:39
UP	Lithuania, Vilna	02:03	18:25
UQ	Latvia, Riga	01:54	18:45
UR	Estonia, Tallinn	01:34	18:59
V2	Antigua & Barbuda, St. Johns	09:34	22:33
V3	Belize, Belmopan	11:21	00:21
V4	St. Christopher & Nevis	09:36	22:36
V8	Brunei, Bandar Seri Begawan	22:06	10:27
VE1, CY	Sable Is.	08:28	23:24
VE1, CY	St. Paul Is.	08:17	23:36
VK	Australia, Canberra (VK1)	20:55	07:04
	Sydney (VK2)	20:44	06:58
	Melbourne (VK3)	21:18	07:14
	Brisbane (VK4)	20:24	07:04
	Adelaide (VK5)	21:37	07:47
	Perth (VK6)	23:01	09:24
	Hobart, Tasmania (VK7)	21:22	06:52
	Darwin (VK8)	21:27	08:57
VK	Lord Howe Is.	20:08	06:32
VK9	Christmas Is.	23:05	10:41
VK9	Cocos-Keeling Is.	23:43	11:14
VK9	Mellish Reef	19:56	07:10
VK9	Norfolk Is.	19:27	06:01
VK9	Willis Is.	20:19	07:37
VKØ	Heard Is.	02:52	11:12
VKØ	Macquarie Is.	21:19	05:23
VP2E	Anguilla	09:36	22:40
VP2M	Montserrat	09:36	22:34
VP2V	British Virgin Is., Tortola	09:43	22:46
VP5	Turks & Caicos Islands,		
	Grand Turk	10:04	23:18
VP8	Falkland Islands, Stanley	11:32	20:03

Pref	State/Province/Country/City	Sunrise	Sunset
VP8	So. Georgia Is.	10:19	18:27
VP8	So. Orkney Is.	11:31	18:25
VP8	So. Sandwich Islands, Saunders Is.	09:57	17:28
VP8	So. Shetland Is., King George Is.	12:34	19:05
VP9	Bermuda	09:17	23:13
VQ9	Chagos, Diego Garcia	01:14	12:59
VR6	Pitcairn Is.	15:12	02:01
VS6	Hong Kong	21:40	10:57
VU	India, Bombay	00:32	13:38
	Calcutta	23:23	12:41
	New Delhi	23:57	13:38
	Bangalore	00:23	13:09
VU7	Andaman Islands, Port Blair	23:24	12:06
VU7	Laccadive Is.	00:46	13:22
XE	Mexico, Mexico City (XE1)	11:59	01:06
	Chihuahua (XE2)	12:09	01:51
	Merida (XE3)	11:19	00:31
XF4	Revilla Gigedo	12:49	01:55
XT	Burkina Faso, Onagadougou	05:40	18:25
XU	Kampuchea, Phnom Penh	22:36	11:18
XW	Laos, Viangchan	22:34	11:37
XX9	Macao	21:43	11:00
XZ	Burma, Rangoon	23:03	12:01
Y2-9	German Dem.Rep.(E), Berlin	03:01	19:03
YA	Afghanistan, Kandahar	00:38	14:28
	Kabul	00:17	14:22
YB-YD	Indonesia, Jakarta	22:54	10:43
	Medan	23:12	11:30
	Pontianak	22:35	10:42
	Jayapura	20:33	08:33
YI	Iraq, Baghdad	01:58	15:58
YJ	Vanuatu, Port Vila	19:06	06:20
YK	Syria, Damascus	02:30	16:31
YN, HT	Nicaragua, Managua	11:18	00:02
YO	Romania, Bucharest	02:42	17:41
YS	El Salvador, San Salvador	11:29	00:17
YU	Yugoslavia, Belgrade	03:03	18:05
YV	Venezuela, Caracas	10:05	22:43
YVØ	Aves Is.	09:43	22:38
Z2,ZE	Zimbabwe, Harare	04:15	15:29
ZA	Albania, Tirane	03:17	17:57
ZB	Gibraltar	05:11	19:24
ZC4	British Cyprus	02:41	16:47
ZD7	St. Helena	06:40	17:59
ZD8	Ascension Is.	07:02	18:45
ZD9	Tristan da Cunha	07:45	17:45
ZF	Cayman Is.	10:47	23:55
ZK1	No. Cook Is., Manihiki	16:52	04:28
ZK1	So. Cook Is., Rarotonga	17:04	04:06
ZK2	Niue	17:37	04:46
ZK3	Tokelaus, Atafu	17:36	05:18
ZL	New Zealand, Auckland (ZL1)	19:16	05:17
	Wellington (ZL2)	19:27	05:07
	Christchurch (ZL3)	19:42	05:09
	Dunedin (ZL4)	19:58	05:10
ZL5	Antarctica, Scott Base	24 hrs night	
ZL7	Chatham Is.	19:00	04:24
ZL8	Kermadec Is.	18:31	05:04
ZL9	Auckland & Campbell Is.,		
	Auckland	20:30	05:10
	Campbell Is.	20:27	04:52
ZP	Paraguay, Asuncion	10:23	21:11
ZS	South Africa, Cape Town (ZS1)	05:35	15:50
	Port Elizabeth (ZS2)	05:06	15:20
	Bloemfontein (ZS4)	04:55	15:28
	Durban (ZS5)	04:37	15:08
	Johannesburg (ZS6)	04:41	15:26
ZS2	Prince Edward & Marion Is.,		
	Marion	04:51	13:58
ZS3	Namibia, Windhoek	05:19	16:16
1AØ	S.M.O.M	03:44	18:28
1S	Spratly Is.	22:12	10:45
3A	Monaco	03:59	18:54
3B6	Agalega	02:21	13:58
3B7	St. Brandon	02:18	13:36
3B8	Mauritius	02:33	13:39
3B9	Rodriguez Is.	02:09	13:16
3C	Equatorial Guinea, Bata	05:11	17:23
	Malabo	05:12	17:30

Pref	State/Province/Country/City	Sunrise	Sunset
3CØ	Pagalu Is.	05:32	17:35
3D2	Fiji Is., Suva	18:26	05:39
3D6	Swaziland, Mbabane	04:29	15:14
3V	Tunisia, Tunis	04:07	18:24
3W	Vietnam, Ho Chi Minh City		
	(Saigon)	22:30	11:09
	Hanoi	22:17	11:29
3X	Republic of Guinea, Conakry	06:33	19:09
3Y	Bouvet	07:39	15:46
3Y	Peter I Is.	16:17	19:40
4S	Sri Lanka, Colombo	00:22	12:50
4U	ITU Geneva	03:56	19:07
4U	United Nations Hq.	09:33	00:11
4W	Yemen, Sanaa	02:32	15:26
4X, 4Z	Israel, Jerusalem	02:38	16:32
5A	Libya, Tripoli	04:08	18:04
	Benghazi	03:38	17:34
5B	Cyprus, Nicosia	02:38	16:47
5H	Tanzania, Dar es Salaam	03:25	15:11
5N	Nigeria, Lagos	05:29	17:56
5R	Madagascar, Antananarivo	03:11	14:21
5T	Mauritania, Nouakchott	06:29	19:31
5U	Niger, Niamey	05:24	18:12
5V	Togo, Lome	05:39	18:03
5W	Western Samoa, Apia	17:40	05:07
5X	Uganda, Kampala	03:42	15:50
5Z	Kenya, Nairobi	04:47	16:50
6W	Senegal, Dakar	06:40	19:32
6Y	Jamaica, Kingston	10:32	23:34

Pref	State/Province/Country/City	Sunrise	Sunset
7O	Yemen, People's Dem. Rep.,		
	Aden	02:33	15:19
7P	Lesotho, Maseru	04:49	15:23
7Q	Malawi, Lilongwe	03:58	15:23
	Blantyre	03:56	15:16
7X	Algeria, Algiers	04:36	18:52
8P	Barbados, Bridgetown	09:31	22:18
8Q	Maldive Is.	00:52	13:12
8R	Guyana, Georgetown	09:35	22:02
9G	Ghana, Accra	05:45	18:08
9H	Malta	03:52	18:05
9J	Zambia, Lusaka	04:22	15:43
9K	Kuwait	01:53	15:37
9L	Sierra Leone, Freetown	06:33	19:05
9M2	West Malaysia, Kuala Lumpur	23:01	11:18
9M6,8	East Malaysia, Sabah,		
	Sandakan (9M6)	21:51	10:16
	Sarawak, Kuching (9M8)	22:29	10:41
9N	Nepal, Katmandu	23:26	13:03
9Q	Zaire, Kinshasa	04:58	16:52
	Kisangani	04:11	16:20
	Lubumbashi	04:20	15:52
9U	Burundi, Bujumbura	04:00	15:58
9V	Singapore	22:55	11:06
9X	Rwanda, Kigali	03:55	15:56
9Y	Trinidad & Tobago, Port of		
	Spain	09:42	22:20
J2/A	Abu Ail	02:40	15:30

Pref	State/Province/Country/City	Sunrise	Sunset
VE1	New Brunswick, Saint John	08:37	00:09
	Nova Scotia, Halifax	08:29	23:56
	Prince Edward Island,		
	Charlottetown	08:21	00:00
VE2	Quebec, Montreal	09:06	00:39
	Quebec City	08:51	00:35
VE3	Ontario, London	09:46	01:01
	Ottawa	09:15	00:47
	Sudbury	09:31	01:13
	Toronto	09:36	00:56
VE4	Manitoba, Winnipeg	10:21	02:33
VE5	Saskatchewan, Regina	10:48	03:06
	Saskatoon	10:48	03:22
VE6	Alberta, Calgary	11:23	03:46
	Edmonton	11:07	03:58
VE7	British Columbia, Prince George	11:41	04:37
	Prince Rupert	12:09	05:10
	Vancouver	12:08	04:14
VE8	Northwest Territories,		
	Yellowknife	09:47	05:24
	Resolute	24 hrs day	
VY1	Yukon, Whitehorse	11:34	06:23
VO1	Newfoundland, St. John's	07:34	23:24
VO2	Labrador, Goose Bay	07:36	00:24
W1	Connecticut, Hartford	09:16	00:22
	Maine, Bangor	08:49	00:17
	Portland	08:59	00:19
	Massachusetts, Boston	09:07	00:18
	New Hampshire, Concord	09:06	00:22
	Rhode Island, Providence	09:11	00:17
	Vermont, Montpelier	09:06	00:31
W2	New Jersey, Atlantic City	09:30	00:21
	New York, Albany	09:17	00:30
	Buffalo	09:37	00:51
	New York City	09:24	00:24
	Syracuse	09:25	00:41
W3	Delaware, Wilmington	09:34	00:27
	District of Columbia,		
	Washington	09:42	00:30
	Maryland, Baltimore	09:39	00:30
	Pennsylvania, Harrisburg	09:37	00:34
	Philadelphia	09:32	00:26
	Pittsburgh	09:50	00:47
	Scranton	09:29	00:33
W4	Alabama, Montgomery	10:37	00:50
	Florida, Jacksonville	10:24	00:26
	Miami	10:28	00:10
	Pensacola	10:46	00:48
	Georgia, Atlanta	10:26	00:46
	Savannah	10:17	00:28
	Kentucky, Lexington	10:15	00:58
	Louisville	10:19	01:03
	North Carolina, Charlotte	10:08	00:35
	Raleigh	09:58	00:28
	Wilmington	09:59	00:21
	South Carolina, Columbia	10:12	00:33
	Tennessee, Knoxville	10:18	00:49
	Memphis	10:45	01:12
	Nashville	10:29	01:02
	Virginia, Norfolk	09:45	00:22
	Richmond	09:48	00:28
W5	Arkansas, Little Rock	10:55	01:20
	Louisiana, New Orleans	10:59	00:59
	Shreveport	11:07	01:19
	Mississippi, Jackson	10:53	01:05
	New Mexico, Albuquerque	11:52	02:18
	Oklahoma, Oklahoma City	11:14	01:43
	Texas, Abilene	11:31	01:43
	Amarillo	11:32	01:59
	Dallas	11:18	01:33
	El Paso	12:00	02:09
	San Antonio	11:33	01:31
W6	California, Los Angeles	12:41	03:02
	San Francisco	12:47	03:29
W7	Arizona, Flagstaff	12:12	02:38
	Phoenix	12:18	02:36
	Idaho, Boise	12:03	03:23
	Pocatello	11:51	03:05

Pref	State/Province/Country/City	Sunrise	Sunset
	Montana, Billings	11:24	03:00
	Butte	11:40	03:17
	Great Falls	11:28	03:18
	Nevada, Las Vegas	12:22	02:55
	Reno	12:32	03:23
	Oregon, Portland	12:22	03:56
	Utah, Salt Lake City	11:56	02:56
	Washington, Seattle	12:12	04:03
	Spokane	11:52	03:44
	Wyoming, Cheyenne	11:26	02:28
	Sheridan	11:22	02:50
W8	Michigan, Detroit	09:55	01:05
	Grand Rapids	10:04	01:18
	Sault Ste. Marie	09:45	01:27
	Traverse City	09:57	01:25
	Ohio, Cincinnati	10:12	01:01
	Cleveland	09:53	00:57
	Columbus	10:03	00:58
	West Virginia, Charleston	10:02	00:47
W9	Illinois, Chicago	10:15	01:22
	Indiana, Indianapolis	10:16	01:10
	Wisconsin, Green Bay	10:07	01:33
	Milwaukee	10:12	01:27
W0	Colorado, Denver	11:32	02:25
	Grand Junction	11:48	02:37
	Iowa, Des Moines	10:40	01:45
	Kansas, Pratt	11:13	01:54
	Wichita	11:07	01:48
	Minnesota, Duluth	10:15	01:59
	Minneapolis	10:27	01:56
	Missouri, Columbia	10:43	01:32
	Kansas City	10:52	01:41
	St. Louis	10:36	01:22
	Nebraska, North Platte	11:10	02:12
	Omaha	10:50	01:53
	North Dakota, Fargo	10:33	02:18
	South Dakota, Rapid City	11:10	02:32
A2	Botswana, Gaborone	04:55	15:34
A3	Tonga, Nukualofa	18:12	05:06
A4	Oman, Muscat	01:18	14:50
A5	Bhutan, Thimbu	23:06	12:55
A6	United Arab Emirates, Abu		
	Dhabi	01:33	15:09
A7	Qatar, Ad-Dawhah	01:42	15:22
A9	Bahrein, Al-Manamah	01:44	15:28
AP	Pakistan, Karachi	00:41	14:19
	Islamabad	23:56	14:15
BV	Taiwan, Taipei	21:03	10:42
BY	People's Rep. of China, Beijing	20:45	11:40
	Harbin	19:44	11:19
	Shanghai	20:49	10:55
	Fuzhou	21:10	10:53
	Xian	21:32	11:54
	Chongqing	21:52	11:52
	Chengdu	22:00	12:04
	Lhasa	22:54	12:53
	Urumqi	22:28	13:48
	Kashi	23:28	14:20
C2	Nauru	18:48	06:53
C3	Andorra	04:17	19:28
C5	The Gambia, Banjul	06:38	19:32
C6	Bahamas, Nassau	10:18	23:57
C9	Mozambique, Maputo	04:31	15:05
	Mozambique	03:38	14:53
CE	Chile, Santiago	11:41	21:41
CE0	Easter Island	14:01	00:30
CE0	San Felix	12:02	22:35
CE0	Juan Fernandez	12:14	22:13
CM,CO	Cuba, Havana	10:43	00:13
CN	Morocco, Casablanca	05:19	19:37
CP	Bolivia, La Paz	10:57	22:07
CT	Portugal, Lisbon	05:12	19:58
CT2, CU	Azores, Ponta Delgada	06:21	21:02
CT3	Madeira Islands, Funchal	05:59	20:13
CX	Uruguay, Montevideo	10:46	20:40
D2	Angola, Luanda	05:17	16:54
D4	Cape Verde, Praia	07:03	20:02

Pref	State/Province/Country/City	Sunrise	Sunset
D6	Comoros, Moroni	03:20	14:47
DA-DL	Germany, Fed. Rep. of (W),		
	Bonn	03:20	19:40
DU	Philippines, Manilla	21:25	10:23
EA	Spain, Madrid	04:44	19:42
EA6	Balearic Is., Palma	04:22	19:14
EA8	Canary Is., Las Palmas	05:59	19:52
EA9	Ceuta & Melilla, Ceuta	05:04	19:35
	Melilla	04:57	19:24
EI	Ireland, Dublin	03:59	20:48
EL	Liberia, Monrovia	06:27	18:56
EP	Iran, Tehran	01:16	15:46
ET	Ethiopia, Addis Ababa	03:05	15:42
	Asmera	02:53	15:53
F	France, Paris	03:48	19:50
FG	Guadeloupe	09:34	22:36
FG, FS	St. Martin	09:35	22:46
FH	Mayotte	03:16	14:39
FK	New Caledonia, Noumia	19:28	06:17
FM	Martinique	09:33	22:31
FO	Clipperton	12:54	01:36
FO	Fr. Polynesia, Tahiti	17:03	04:09
FP	St. Pierre & Miquelon, St. Pierre	07:51	23:34
FR	Glorioso	03:05	14:33
FR	Juan de Nova	03:33	14:41
	Europa	03:52	14:41
FR	Reunion	02:49	13:43
FR	Tromelin	02:44	13:57
FT8W	Crozet	04:07	12:53
FT8X	Kerguelen	03:12	11:31
FT8Y	Antarctica, Dumont D'Urville	01:11	04:05
FT8Z	Amsterdam & St. Paul Is.,		
	Amsterdam	01:59	11:37
FW	Wallis & Futuna Is., Wallis	18:02	05:24
FY	Fr. Guiana, Cayenne	09:16	21:39
G	England, London	03:45	20:13
GD	Isle of Man	03:46	20:47
GI	Northern Ireland, Belfast	03:49	20:54
GJ	Jersey	04:04	20:10
GM	Scotland, Glasgow	03:35	20:56
	Aberdeen	03:15	20:58
GU	Guernsey	04:05	20:13
GW	Wales, Cardiff	03:57	20:25
H4	Solomon Islands, Honiara	19:30	07:06
HA	Hungary, Budapest	02:47	18:37
HB	Switzerland, Bern	03:35	19:21
HBØ	Liechtenstein	03:26	19:14
HC	Ecuador, Quito	11:07	23:14
HC8	Galapagos Is.	11:57	00:03
HH	Haiti, Port-Au-Prince	10:11	23:24
HI	Dominican Republic, Santo		
	Domingo	10:02	23:14
HK	Colombia, Bogota	10:43	23:06
HKØ	Malpelo Is.	11:12	23:33
HKØ	San Adreas	11:00	23:50
HL, HM	Korea, Seoul	20:10	10:50
HP	Panama, Panama	10:57	23:35
HR	Honduras, Tegucigalpa	11:19	00:15
HS	Thailand, Bangkok	22:49	11:44
HV	Vatican City	03:35	18:42
HZ, 7Z	Saudi Arabia, Dharan	01:46	15:30
	Mecca	02:37	16:01
	Riyadh	02:03	15:40
I	Italy, Rome	03:35	18:42
	Trieste	03:16	18:51
	Sicily	03:42	18:22
IS	Sardinia, Cagliari	03:57	18:47
J2	Djibouti, Djibouti	02:42	15:29
J3	Grenada	09:41	22:30
J5	Guinea-Bissau, Bissau	06:37	19:25
J6	St. Lucia	09:35	22:30
J7	Dominica	09:33	22:34
J8	St. Vincent	09:37	22:30
JA-JS	Japan, Tokyo	19:24	09:54
	Nagasaki	20:11	10:26
	Sapporo	18:55	10:10
JD	Minami Torishima	18:55	08:30
JD	Ogasawara, Kazan Is.	19:40	09:29
JT	Mongolia, Ulan Bator	20:54	12:47
JW	Svalbard, Spitsbergen	24 hrs day	
JX	Jan Mayen	24 hrs day	

Pref	State/Province/Country/City	Sunrise	Sunset
JY	Jordan, Amman	02:29	16:40
KC4	Antarctica, Byrd Station	24 hrs night	
	McMurdo Sound	24 hrs night	
	Palmer Station	14:09	18:20
KC6	Micronesia, Ponape	19:10	07:40
KC6	Belau, Yap	20:26	09:05
	Koror	20:44	09:16
KG4	Guantanamo Bay	10:20	23:38
KH1	Baker, Howland Is.	17:38	05:47
KH2	Guam, Agana	19:52	08:46
KH3	Johnston Is.	16:39	05:46
KH4	Midway Is.	16:52	06:44
KH5	Palmyra Is.	16:33	05:00
KH5K	Kingman Reef	16:33	05:06
KH6	Hawaii, Hilo	15:40	04:57
	Honolulu	15:48	05:11
KH7	Kure Is.	16:55	06:48
KH8	American Samoa, Pago Pago	17:42	05:01
KH9	Wake Is.	18:14	07:29
KHØ	Mariana Is., Saipan	19:45	08:45
KL7	Alaska, Adak	15:29	08:00
	Anchorage	12:28	07:29
	Fairbanks	11:17	08:22
	Juneau	11:55	05:56
	Nome	12:35	09:25
KP1	Navassa Is.	10:22	23:34
KP2	Virgin Islands, Charlotte Amalie	09:42	22:54
KP4	Puerto Rico, San Juan	09:47	22:59
KP5	Desecheo Is.	09:53	23:04
KX6	Marshall Islands, Kwajalein	18:30	07:08
LA-LJ	Norway, Oslo	01:58	20:32
LU	Argentina, Buenos Aires	10:55	20:49
LX	Luxembourg	03:29	19:38
LZ	Bulgaria, Sofia	02:49	18:01
OA	Peru, Lima	11:24	22:50
OD	Lebanon, Beirut	02:26	16:46
OE	Austria, Vienna	02:55	18:51
OH	Finland, Helsinki	00:59	19:37
OHØ	Aland Is.	01:19	19:57
OJØ	Market Reef	01:22	20:03
OK	Czechoslovakia, Prague	02:54	19:08
ON	Belgium, Brussels	03:30	19:52
OX, XP	Greenland, Godthaab	05:07	01:43
	Thule	24 hrs day	
OY	Faroe Islands, Torshavn	02:44	22:07
OZ	Denmark, Copenhagen	02:28	19:48
P2	Papua New Guinea, Madang	20:21	08:11
	Port Moresby	20:22	07:58
PA-PI	Netherlands, Amsterdam	03:20	19:58
PJ, P4	Netherlands Antilles, Willemstad	10:10	22:59
PJ5-8	St. Maarten and Saba,		
	St. Maarten	09:36	22:46
PY	Brazil, Brasilia	09:33	20:46
	Rio De Janeiro	09:28	20:14
	Natal	08:26	20:13
	Manaus	10:01	21:57
	Porto Alegre	10:15	20:31
PYØ	Fernando De Noronha	08:11	20:05
PYØ	St. Peter & St. Paul Rocks	07:51	20:01
PYØ	Trindade & Martin Vaz Is.,		
	Trindade	08:28	19:23
PZ	Suriname, Paramaribo	09:26	21:52
S2	Bangladesh, Dacca	23:10	12:43
S7	Seychelles, Victoria	02:20	14:12
S9	Sao Tome	05:27	17:36
SJ-SM	Sweden, Stockholm	01:36	19:56
SP	Poland, Krakow	02:32	18:45
	Warsaw	02:16	18:52
ST	Sudan, Khartoum	03:18	16:19
STØ	Southern Sudan, Juba	03:40	16:04
SU	Egypt, Cairo	02:52	16:53
SV	Greece, Athens	03:02	17:45
SV5	Didecanese, Rhodes	02:49	17:22
SV9	Crete	03:03	17:31
SV/A	Mount Athos	02:53	17:49
T2	Tuvalu, Funafuti	18:15	05:53
T3Ø	West Kiribati, Bonriki	18:19	06:31
T31	Central Kiribati, Kanton	17:26	05:24
T32	East Kiribati, Christmas Is.	16:21	04:35
T5	Somalia, Mogadishu	02:49	15:04
T7	San Marino	03:28	18:50

Pref	State/Province/Country/City	Sunrise	Sunset
TA	Turkey, Ankara	02:20	17:14
	Istanbul	02:31	17:34
TF	Iceland, Reykjavik	03:10	23:43
TG	Guatemala, Guatemala City	11:31	00:29
TI	Costa Rica, San Jose	11:14	23:55
TI9	Cocos Is.	11:33	23:59
TJ	Cameroon, Yaounde	05:02	17:22
TK	Corsica	03:48	18:56
TL	Central African Republic, Bangui	04:33	16:55
TN	Congo, Brazzaville	05:01	16:53
TR	Gabon, Libreville	05:16	17:25
TT	Chad, N'Djamena	04:34	17:23
TU	Ivory Coast, Abidjan	06:02	18:27
TY	Benin, Porto Novo	05:33	18:02
TZ	Mali, Bamako	06:05	18:56
UA	Russia, European,		
	Leningrad (UA1)	00:41	19:13
	Archangel (UA1)	22:49	19:44
	Murmansk (UA1)	24 hrs day	
	Moscow (UA3)	00:47	18:08
	Kuibyshev (UA4)	00:14	17:02
	Rostov (UA6)	01:25	17:15
UA1P	Franz Josef Land	24 hrs day	
UA2	Kaliningrad	02:01	19:11
UA9,Ø	Russia, Asiatic,		
	Novosibirsk (UA9)	21:52	15:02
	Perm (UA9)	23:15	17:11
	Omsk (UA9)	22:30	15:40
	Norilsk (UA9)	24 hrs day	
	Irkutsk (UAØ)	20:43	13:20
	Vladivostok (UAØ)	19:32	10:49
	Petropavlovsk (UAØ)	17:01	09:46
	Khabarovsk (UAØ)	18:58	10:57
	Krasnoyarsk (UAØ)	21:05	14:29
	Yakutsk (UAØ)	17:38	13:01
	Wrangel Island (UAØ)	24 hrs day	
	Kyzyl (UAØY)	21:25	13:55
UB	Ukraine, Kiev	01:48	18:05
UC	Byelorussia, Minsk	01:40	18:36
UD	Azerbaijan, Baku	01:10	16:07
UF	Georgia, Tbilisi	01:26	16:32
UG	Armenia, Ierevan	01:32	16:29
UH	Turkoman, Ashkhabad	00:43	15:26
UI	Uzbek, Bukhara	00:14	15:07
	Tashkent	23:50	14:52
UJ	Tadzhik, Samarkand	00:05	14:58
	Dushanbe	23:58	14:48
UL	Kazakh, Alma-ata	23:12	14:29
UM	Kirghiz, Frunze	23:23	14:37
UO	Moldavia, Kishinev	02:10	17:56
UP	Lithuania, Vilna	01:44	18:48
UQ	Latvia, Riga	01:32	19:12
UR	Estonia, Tallinn	01:08	19:30
V2	Antigua & Barbuda, St. Johns	09:32	22:39
V3	Belize, Belmopan	11:20	00:27
V4	St. Christopher & Nevis	09:35	22:42
V8	Brunei, Bandar Seri Begawan	22:07	10:31
VE1, CY	Sable Is.	08:18	23:39
VE1, CY	St. Paul Is.	08:05	23:52
VK	Australia, Canberra (VK1)	21:06	06:57
	Sydney (VK2)	20:54	06:53
	Melbourne (VK3)	21:29	07:07
	Brisbane (VK4)	20:32	07:00
	Adelaide (VK5)	21:47	07:40
	Perth (VK6)	23:11	09:19
	Hobart, Tasmania (VK7)	21:35	06:43
	Darwin (VK8)	21:32	08:57
VK	Lord Howe Is.	20:17	06:26
VK9	Christmas Is.	23:10	10:41
VK9	Cocos-Keeling Is.	23:48	11:14
VK9	Mellish Reef	20:02	07:08
VK9	Norfolk Is.	19:36	05:57
VK9	Willis Is.	20:25	07:36
VKØ	Heard Is.	03:13	10:57
VKØ	Macquarie Is.	21:41	05:05
VP2E	Anguilla	09:35	22:46
VP2M	Montserrat	09:34	22:40
VP2V	British Virgin Is., Tortola	09:41	22:53
VP5	Turks & Caicos Islands,		
	Grand Turk	10:01	23:25
VP8	Falkland Islands, Stanley	11:51	19:48

Pref	State/Province/Country/City	Sunrise	Sunset
VP8	So. Georgia Is.	10:41	18:10
VP8	So. Orkney Is.	12:02	17:59
VP8	So. Sandwich Islands, Saunders Is.	10:23	17:07
VP8	So. Shetland Is., King George Is.	13:07	18:35
VP9	Bermuda	09:11	23:23
VQ9	Chagos, Diego Garcia	01:17	13:00
VR6	Pitcairn Is.	15:20	01:57
VS6	Hong Kong	21:37	11:05
VU	India, Bombay	00:30	13:44
	Calcutta	23:20	12:49
	New Delhi	23:52	13:47
	Bangalore	00:22	13:14
VU7	Andaman Islands, Port Blair	23:23	12:11
VU7	Laccadive Is.	00:46	13:27
XE	Mexico, Mexico City (XE1)	11:57	01:12
	Chihuahua (XE2)	12:05	02:00
	Merida (XE3)	11:16	00:38
XF4	Revilla Gigedo	12:47	02:01
XT	Burkina Faso, Onagadougou	05:40	18:30
XU	Kampuchea, Phnom Penh	22:35	11:23
XW	Laos, Viangchan	22:33	11:43
XX9	Macao	21:40	11:07
XZ	Burma, Rangoon	23:01	12:07
Y2-9	German Dem.Rep.(E), Berlin	02:45	19:24
YA	Afghanistan, Kandahar	00:32	14:38
	Kabul	00:10	14:33
YB-YD	Indonesia, Jakarta	22:58	10:44
	Medan	23:14	11:33
	Pontianak	22:37	10:45
	Jayapura	20:36	08:35
YI	Iraq, Baghdad	01:53	16:08
YJ	Vanuatu, Port Vila	19:12	06:18
YK	Syria, Damascus	02:24	16:42
YN, HT	Nicaragua, Managua	11:18	00:06
YO	Romania, Bucharest	02:31	17:56
YS	El Salvador, San Salvador	11:28	00:22
YU	Yugoslavia, Belgrade	02:52	18:21
YV	Venezuela, Caracas	10:05	22:48
YVØ	Aves Is.	09:42	22:44
Z2,ZE	Zimbabwe, Harare	04:21	15:27
ZA	Albania, Tirane	03:07	18:11
ZB	Gibraltar	05:04	19:36
ZC4	British Cyprus	02:34	16:58
ZD7	St. Helena	06:46	17:58
ZD8	Ascension Is.	07:06	18:46
ZD9	Tristan da Cunha	07:57	17:38
ZF	Cayman Is.	10:45	00:01
ZK1	No. Cook Is., Manihiki	16:56	04:28
ZK1	So. Cook Is., Rarotonga	17:11	04:04
ZK2	Niue	17:43	04:44
ZK3	Tokelaus, Atafu	17:40	05:19
ZL	New Zealand, Auckland (ZL1)	19:28	05:10
	Wellington (ZL2)	19:40	04:58
	Christchurch (ZL3)	19:56	04:59
	Dunedin (ZL4)	20:13	04:59
ZL5	Antarctica, Scott Base	24 hrs night	
ZL7	Chatham Is.	19:14	04:14
ZL8	Kermadec Is.	18:40	05:00
ZL9	Auckland & Campbell Is.,		
	Auckland	20:49	04:56
	Campbell Is.	20:47	04:36
ZP	Paraguay, Asuncion	10:31	21:07
ZS	South Africa, Cape Town (ZS1)	05:46	15:44
	Port Elizabeth (ZS2)	05:17	15:14
	Bloemfontein (ZS4)	05:04	15:24
	Durban (ZS5)	04:46	15:03
	Johannesburg (ZS6)	04:49	15:22
ZS2	Prince Edward & Marion Is.,		
	Marion	05:07	13:47
ZS3	Namibia, Windhoek	05:26	16:14
1AØ	S.M.O.M	03:35	18:42
1S	Spratly Is.	22:12	10:49
3A	Monaco	03:49	19:09
3B6	Agalega	02:26	13:58
3B7	St. Brandon	02:23	13:35
3B8	Mauritius	02:40	13:36
3B9	Rodriguez Is.	02:15	13:14
3C	Equatorial Guinea, Bata	05:12	17:26
	Malabo	05:13	17:33

Pref	State/Province/Country/City	Sunrise	Sunset
3CØ	Pagalu Is.	05:35	17:37
3D2	Fiji Is., Suva	18:32	05:37
3D6	Swaziland, Mbabane	04:38	15:10
3V	Tunisia, Tunis	04:00	18:35
3W	Vietnam, Ho Chi Minh City		
	(Saigon)	22:29	11:13
	Hanoi	22:14	11:36
3X	Republic of Guinea, Conakry	06:33	19:13
3Y	Bouvet	08:01	15:28
3Y	Peter I Is.	24 hrs night	
4S	Sri Lanka, Colombo	00:23	12:54
4U	ITU Geneva	03:44	19:23
4U	United Nations Hq.	09:24	00:24
4W	Yemen, Sanaa	02:31	15:32
4X, 4Z	Israel, Jerusalem	02:33	16:42
5A	Libya, Tripoli	04:02	18:15
	Benghazi	03:33	17:44
5B	Cyprus, Nicosia	02:31	16:58
5H	Tanzania, Dar es Salaam	03:28	15:12
5N	Nigeria, Lagos	05:30	17:59
5R	Madagascar, Antananarivo	03:17	14:19
5T	Mauritania, Nouakchott	06:27	19:38
5U	Niger, Niamey	05:23	18:17
5V	Togo, Lome	05:40	18:07
5W	Western Samoa, Apia	17:45	05:06
5X	Uganda, Kampala	03:44	15:52
5Z	Kenya, Nairobi	04:50	16:52
6W	Senegal, Dakar	06:39	19:37
6Y	Jamaica, Kingston	10:30	23:41

Pref	State/Province/Country/City	Sunrise	Sunset
7O	Yemen, People's Dem. Rep.,		
	Aden	02:33	15:24
7P	Lesotho, Maseru	04:58	15:18
7Q	Malawi, Lilongwe	04:03	15:23
	Blantyre	04:02	15:15
7X	Algeria, Algiers	04:29	19:04
8P	Barbados, Bridgetown	09:31	22:23
8Q	Maldive Is.	00:54	13:16
8R	Guyana, Georgetown	09:36	22:06
9G	Ghana, Accra	05:46	18:12
9H	Malta	03:45	18:16
9J	Zambia, Lusaka	04:28	15:42
9K	Kuwait	01:48	15:46
9L	Sierra Leone, Freetown	06:33	19:09
9M2	West Malaysia, Kuala Lumpur	23:03	11:21
9M6,8	East Malaysia, Sabah,		
	Sandakan (9M6)	21:52	10:19
	Sarawak, Kuching (9M8)	22:31	10:43
9N	Nepal, Katmandu	23:22	13:12
9Q	Zaire, Kinshasa	05:01	16:53
	Kisangani	04:13	16:22
	Lubumbashi	04:24	15:52
9U	Burundi, Bujumbura	04:03	15:59
9V	Singapore	22:57	11:09
9X	Rwanda, Kigali	03:58	15:58
9Y	Trinidad & Tobago, Port of		
	Spain	09:42	22:25
J2/A	Abu Ail	02:39	15:35

Pref	State/Province/Country/City	Sunrise	Sunset
VE1	New Brunswick, Saint John	08:36	00:15
	Nova Scotia, Halifax	08:29	00:02
	Prince Edward Island, Charlottetown	08:20	00:07
VE2	Quebec, Montreal	09:05	00:46
	Quebec City	08:50	00:42
VE3	Ontario, London	09:45	01:07
	Ottawa	09:14	00:54
	Sudbury	09:30	01:20
	Toronto	09:35	01:02
VE4	Manitoba, Winnipeg	10:19	02:40
VE5	Saskatchewan, Regina	10:46	03:13
	Saskatoon	10:45	03:31
VE6	Alberta, Calgary	11:21	03:54
	Edmonton	11:04	04:06
VE7	British Columbia, Prince George	11:39	04:46
	Prince Rupert	12:06	05:19
	Vancouver	12:06	04:21
VE8	Northwest Territories, Yellowknife	09:38	05:39
	Resolute	24 hrs day	
VY1	Yukon, Whitehorse	11:27	06:36
VO1	Newfoundland, St. John's	07:32	23:32
VO2	Labrador, Goose Bay	07:33	00:33
W1	Connecticut, Hartford	09:15	00:28
	Maine, Bangor	08:49	00:24
	Portland	08:59	00:26
	Massachusetts, Boston	09:07	00:24
	New Hampshire, Concord	09:06	00:29
	Rhode Island, Providence	09:10	00:23
	Vermont, Montpelier	09:06	00:37
W2	New Jersey, Atlantic City	09:30	00:27
	New York, Albany	09:17	00:36
	Buffalo	09:36	00:57
	New York City	09:24	00:30
	Syracuse	09:25	00:47
W3	Delaware, Wilmington	09:34	00:33
	District of Columbia, Washington	09:42	00:36
	Maryland, Baltimore	09:39	00:36
	Pennsylvania, Harrisburg	09:37	00:40
	Philadelphia	09:32	00:32
	Pittsburgh	09:49	00:53
	Scranton	09:29	00:39
W4	Alabama, Montgomery	10:38	00:55
	Florida, Jacksonville	10:25	00:31
	Miami	10:29	00:14
	Pensacola	10:47	00:53
	Georgia, Atlanta	10:27	00:51
	Savannah	10:18	00:33
	Kentucky, Lexington	10:15	01:03
	Louisville	10:20	01:09
	North Carolina, Charlotte	10:08	00:40
	Raleigh	09:58	00:33
	Wilmington	10:00	00:26
	South Carolina, Columbia	10:13	00:38
	Tennessee, Knoxville	10:19	00:55
	Memphis	10:46	01:17
	Nashville	10:30	01:07
	Virginia, Norfolk	09:46	00:27
	Richmond	09:48	00:33
W5	Arkansas, Little Rock	10:56	01:25
	Louisiana, New Orleans	10:59	01:04
	Shreveport	11:07	01:25
	Mississippi, Jackson	10:54	01:10
	New Mexico, Albuquerque	11:52	02:24
	Oklahoma, Oklahoma City	11:14	01:48
	Texas, Abilene	11:31	01:49
	Amarillo	11:32	02:04
	Dallas	11:19	01:38
	El Paso	12:00	02:14
	San Antonio	11:34	01:36
W6	California, Los Angeles	12:41	03:07
	San Francisco	12:47	03:34
W7	Arizona, Flagstaff	12:12	02:44
	Phoenix	12:18	02:41
	Idaho, Boise	12:03	03:29
	Pocatello	11:51	03:12

Pref	State/Province/Country/City	Sunrise	Sunset
	Montana, Billings	11:23	03:07
	Butte	11:39	03:24
	Great Falls	11:27	03:25
	Nevada, Las Vegas	12:23	03:00
	Reno	12:32	03:29
	Oregon, Portland	12:21	04:02
	Utah, Salt Lake City	11:56	03:02
	Washington, Seattle	12:11	04:10
	Spokane	11:51	03:51
	Wyoming, Cheyenne	11:26	02:34
	Sheridan	11:21	02:57
W8	Michigan, Detroit	09:55	01:11
	Grand Rapids	10:03	01:25
	Sault Ste. Marie	09:44	01:33
	Traverse City	09:56	01:31
	Ohio, Cincinnati	10:12	01:07
	Cleveland	09:52	01:03
	Columbus	10:03	01:04
	West Virginia, Charleston	10:02	00:53
W9	Illinois, Chicago	10:15	01:28
	Indiana, Indianapolis	10:16	01:16
	Wisconsin, Green Bay	10:07	01:40
	Milwaukee	10:12	01:34
W0	Colorado, Denver	11:32	02:31
	Grand Junction	11:48	02:43
	Iowa, Des Moines	10:40	01:51
	Kansas, Pratt	11:13	01:59
	Wichita	11:07	01:54
	Minnesota, Duluth	10:14	02:06
	Minneapolis	10:26	02:03
	Missouri, Columbia	10:43	01:38
	Kansas City	10:52	01:47
	St. Louis	10:36	01:28
	Nebraska, North Platte	11:10	02:18
	Omaha	10:50	02:00
	North Dakota, Fargo	10:32	02:25
	South Dakota, Rapid City	11:09	02:39
A2	Botswana, Gaborone	05:00	15:35
A3	Tonga, Nukualofa	18:17	05:07
A4	Oman, Muscat	01:19	14:54
A5	Bhutan, Thimbu	23:08	12:59
A6	United Arab Emirates, Abu Dhabi	01:35	15:14
A7	Qatar, Ad-Dawhah	01:44	15:27
A9	Bahrein, Al-Manamah	01:45	15:32
AP	Pakistan, Karachi	00:42	14:23
	Islamabad	23:57	14:20
BV	Taiwan, Taipei	21:04	10:46
BY	People's Rep. of China, Beijing	20:45	11:46
	Harbin	19:43	11:26
	Shanghai	20:50	11:00
	Fuzhou	21:11	10:57
	Xian	21:32	11:59
	Chongqing	21:53	11:57
	Chengdu	22:01	12:09
	Lhasa	22:55	12:58
	Urumqi	22:27	13:55
	Kashi	23:28	14:26
C2	Nauru	18:51	06:56
C3	Andorra	04:16	19:34
C5	The Gambia, Banjul	06:40	19:36
C6	Bahamas, Nassau	10:20	00:02
C9	Mozambique, Maputo	04:35	15:06
	Mozambique	03:41	14:55
CE	Chile, Santiago	11:47	21:42
CE0	Easter Island	14:06	00:32
CE0	San Felix	12:07	22:36
CE0	Juan Fernandez	12:19	22:14
CM,CO	Cuba, Havana	10:44	00:17
CN	Morocco, Casablanca	05:20	19:43
CP	Bolivia, La Paz	11:00	22:09
CT	Portugal, Lisbon	05:12	20:04
CT2, CU	Azores, Ponta Delgada	06:21	21:07
CT3	Madeira Islands, Funchal	06:00	20:18
CX	Uruguay, Montevideo	10:52	20:40
D2	Angola, Luanda	05:20	16:57
D4	Cape Verde, Praia	07:05	20:05

Pref	State/Province/Country/City	Sunrise	Sunset
D6	Comoros, Moroni	03:23	14:49
DA-DL	Germany, Fed. Rep. of (W), Bonn	03:18	19:48
DU	Philippines, Manila	21:27	10:27
EA	Spain, Madrid	04:44	19:48
EA6	Balearic Is., Palma	04:22	19:20
EA8	Canary Is., Las Palmas	06:00	19:57
EA9	Ceuta & Melilla, Ceuta	05:04	19:40
	Melilla	04:57	19:29
EI	Ireland, Dublin	03:56	20:56
EL	Liberia, Monrovia	06:30	18:59
EP	Iran, Tehran	01:16	15:52
ET	Ethiopia, Addis Ababa	03:07	15:46
	Asmera	02:54	15:57
F	France, Paris	03:47	19:57
FG	Guadeloupe	09:36	22:40
FG, FS	St. Martin	09:37	22:50
FH	Mayotte	03:19	14:41
FK	New Caledonia, Noumia	19:32	06:18
FM	Martinique	09:35	22:35
FO	Clipperton	12:56	01:40
FO	Fr. Polynesia, Tahiti	17:07	04:11
FP	St. Pierre & Miquelon, St. Pierre	07:50	23:41
FR	Glorioso	03:08	14:35
FR	Juan de Nova	03:37	14:43
	Europa	03:57	14:43
FR	Reunion	02:53	13:44
FR	Tromelin	02:48	13:59
FT8W	Crozet	04:14	12:52
FT8X	Kerguelen	03:19	11:30
FT8Y	Antarctica, Dumont D'Urville	01:38	03:44
FT8Z	Amsterdam & St. Paul Is., Amsterdam	02:04	11:37
FW	Wallis & Futuna Is., Wallis	18:06	05:27
FY	Fr. Guiana, Cayenne	09:18	21:43
G	England, London	03:42	20:21
GD	Isle of Man	03:43	20:56
GI	Northern Ireland, Belfast	03:46	21:03
GJ	Jersey	04:02	20:18
GM	Scotland, Glasgow	03:31	21:05
	Aberdeen	03:11	21:08
GU	Guernsey	04:03	20:21
GW	Wales, Cardiff	03:55	20:33
H4	Solomon Islands, Honiara	19:34	07:08
HA	Hungary, Budapest	02:46	18:44
HB	Switzerland, Bern	03:34	19:28
HB0	Liechtenstein	03:25	19:21
HC	Ecuador, Quito	11:10	23:16
HC8	Galapagos Is.	12:00	00:06
HH	Haiti, Port-Au-Prince	10:13	23:28
HI	Dominican Republic, Santo Domingo	10:04	23:18
HK	Colombia, Bogota	10:46	23:09
HK0	Malpelo Is.	11:15	23:36
HK0	San Adreas	11:02	23:54
HL, HM	Korea, Seoul	20:11	10:56
HP	Panama, Panama	11:00	23:39
HR	Honduras, Tegucigalpa	11:21	00:19
HS	Thailand, Bangkok	22:51	11:47
HV	Vatican City	03:34	18:48
HZ, 7Z	Saudi Arabia, Dharan	01:47	15:35
	Mecca	02:39	16:05
	Riyadh	02:04	15:44
I	Italy, Rome	03:34	18:48
	Trieste	03:15	18:57
	Sicily	03:43	18:28
IS	Sardinia, Cagliari	03:57	18:53
J2	Djibouti, Djibouti	02:44	15:32
J3	Grenada	09:43	22:33
J5	Guinea-Bissau, Bissau	06:39	19:28
J6	St. Lucia	09:37	22:34
J7	Dominica	09:35	22:38
J8	St. Vincent	09:39	22:34
JA-JS	Japan, Tokyo	19:25	09:59
	Nagasaki	20:12	10:31
	Sapporo	18:54	10:17
JD	Minami Torishima	18:56	08:34
JD	Ogasawara, Kazan Is.	19:41	09:34
JT	Mongolia, Ulan Bator	20:53	12:54
JW	Svalbard, Spitsbergen	24 hrs day	
JX	Jan Mayen	24 hrs day	

Pref	State/Province/Country/City	Sunrise	Sunset
JY	Jordan, Amman	02:30	16:45
KC4	Antarctica, Byrd Station	24 hrs night	
	McMurdo Sound	24 hrs night	
	Palmer Station	14:26	18:09
KC6	Micronesia, Ponape	19:12	07:44
KC6	Belau, Yap	20:28	09:09
	Koror	20:47	09:20
KG4	Guantanamo Bay	10:22	23:42
KH1	Baker, Howland Is.	17:41	05:50
KH2	Guam, Agana	19:54	08:50
KH3	Johnston Is.	16:41	05:49
KH4	Midway Is.	16:53	06:49
KH5	Palmyra Is.	16:36	05:03
KH5K	Kingman Reef	16:36	05:09
KH6	Hawaii, Hilo	15:42	05:01
	Honolulu	15:50	05:16
KH7	Kure Is.	16:56	06:53
KH8	American Samoa, Pago Pago	17:46	05:03
KH9	Wake Is.	18:16	07:34
KH0	Mariana Is., Saipan	19:47	08:49
KL7	Alaska, Adak	15:27	08:08
	Anchorage	12:20	07:42
	Fairbanks	10:59	08:47
	Juneau	11:50	06:07
	Nome	12:19	09:47
KP1	Navassa Is.	10:24	23:38
KP2	Virgin Islands, Charlotte Amalie	09:44	22:58
KP4	Puerto Rico, San Juan	09:49	23:03
KP5	Desecheo Is.	09:54	23:08
KX6	Marshall Islands, Kwajalein	18:32	07:12
LA-LJ	Norway, Oslo	01:52	20:44
LU	Argentina, Buenos Aires	11:00	20:50
LX	Luxembourg	03:27	19:45
LZ	Bulgaria, Sofia	02:48	18:08
OA	Peru, Lima	11:27	22:52
OD	Lebanon, Beirut	02:27	16:51
OE	Austria, Vienna	02:54	18:58
OH	Finland, Helsinki	00:53	19:50
OH0	Aland Is.	01:13	20:10
OJ0	Market Reef	01:16	20:15
OK	Czechoslovakia, Prague	02:52	19:15
ON	Belgium, Brussels	03:28	19:59
OX, XP	Greenland, Godthaab	04:52	02:04
	Thule	24 hrs day	
OY	Faroe Islands, Torshavn	02:36	22:21
OZ	Denmark, Copenhagen	02:24	19:57
P2	Papua New Guinea, Madang	20:24	08:13
	Port Moresby	20:26	08:00
PA-PI	Netherlands, Amsterdam	03:17	20:06
PJ, P4	Netherlands Antilles, Willemstad	10:12	23:02
PJ5-8	St. Maarten and Saba, St. Maarten	09:38	22:50
PY	Brazil, Brasilia	09:37	20:48
	Rio De Janeiro	09:32	20:16
	Natal	08:29	20:15
	Manaus	10:04	22:00
	Porto Alegre	10:20	20:32
PY0	Fernando De Noronha	08:14	20:08
PY0	St. Peter & St. Paul Rocks	07:53	20:04
PY0	Trindade & Martin Vaz Is., Trindade	08:32	19:25
PZ	Suriname, Paramaribo	09:28	21:56
S2	Bangladesh, Dacca	23:12	12:47
S7	Seychelles, Victoria	02:23	14:15
S9	Sao Tome	05:30	17:38
SJ-SM	Sweden, Stockholm	01:30	20:07
SP	Poland, Krakow	02:30	18:52
	Warsaw	02:14	19:00
ST	Sudan, Khartoum	03:20	16:23
ST0	Southern Sudan, Juba	03:42	16:07
SU	Egypt, Cairo	02:53	16:58
SV	Greece, Athens	03:02	17:50
SV5	Dodecanese, Rhodes	02:49	17:28
SV9	Crete	03:04	17:37
SV/A	Mount Athos	02:53	17:55
T2	Tuvalu, Funafuti	18:18	05:55
T30	West Kiribati, Bonriki	18:22	06:34
T31	Central Kiribati, Kanton	17:29	05:27
T32	East Kiribati, Christmas Is.	16:24	04:38
T5	Somalia, Mogadishu	02:52	15:07
T7	San Marino	03:28	18:56

Pref	State/Province/Country/City	Sunrise	Sunset
TA	Turkey, Ankara	02:19	17:20
	Istanbul	02:31	17:40
TF	Iceland, Reykjavik	02:56	00:02
TG	Guatemala, Guatemala City	11:33	00:33
TI	Costa Rica, San Jose	11:16	23:58
TI9	Cocos Is.	11:36	00:03
TJ	Cameroon, Yaounde	05:05	17:26
TK	Corsica	03:48	19:02
TL	Central African Republic, Bangui	04:35	16:58
TN	Congo, Brazzaville	05:04	16:56
TR	Gabon, Libreville	05:19	17:27
TT	Chad, N'Djamena	04:36	17:26
TU	Ivory Coast, Abidjan	06:04	18:30
TY	Benin, Porto Novo	05:36	18:06
TZ	Mali, Bamako	06:07	18:59
UA	Russia, European,		
	Leningrad (UA1)	00:35	19:25
	Archangel (UA1)	22:32	20:06
	Murmansk (UA1)	24 hrs day	
	Moscow (UA3)	00:44	18:18
	Kuibyshev (UA4)	00:12	17:10
	Rostov (UA6)	01:24	17:22
UA1P	Franz Josef Land	24 hrs day	
UA2	Kaliningrad	01:58	19:20
UA9,Ø	Russia, Asiatic,		
	Novosibirsk (UA9)	21:48	15:11
	Perm (UA9)	23:10	17:21
	Omsk (UA9)	22:26	15:49
	Norilsk (UAØ)	24 hrs day	
	Irkutsk (UAØ)	20:40	13:28
	Vladivostok (UAØ)	19:32	10:55
	Petropavlovsk (UAØ)	16:58	09:54
	Khabarovsk (UAØ)	18:57	11:04
	Krasnoyarsk (UAØ)	21:01	14:39
	Yakutsk (UAØ)	17:30	13:15
	Wrangel Island (UAØ)	24 hrs day	
	Kyzyl (UAØY)	21:23	14:03
UB	Ukraine, Kiev	01:46	18:12
UC	Byelorussia, Minsk	01:37	18:44
UD	Azerbaijan, Baku	01:10	16:13
UF	Georgia, Tbilisi	01:26	16:38
UG	Armenia, Ierevan	01:32	16:35
UH	Turkoman, Ashkhabad	00:43	15:32
UI	Uzbek, Bukhara	00:14	15:13
	Tashkent	23:49	14:58
UJ	Tadzhik, Samarkand	00:04	15:03
	Dushanbe	23:58	14:53
UL	Kazakh, Alma-ata	23:12	14:36
UM	Kirghiz, Frunze	23:22	14:43
UO	Moldavia, Kishinev	02:09	18:03
UP	Lithuania, Vilna	01:41	18:57
UQ	Latvia, Riga	01:28	19:21
UR	Estonia, Tallinn	01:03	19:41
V2	Antigua & Barbuda, St. Johns	09:34	22:43
V3	Belize, Belmopan	11:21	00:31
V4	St. Christopher & Nevis	09:37	22:46
V8	Brunei, Bandar Seri Begawan	22:09	10:34
VE1, CY	Sable Is.	08:17	23:45
VE1, CY	St. Paul Is.	08:04	23:59
VK	Australia, Canberra (VK1)	21:12	06:58
	Sydney (VK2)	21:00	06:53
	Melbourne (VK3)	21:35	07:07
	Brisbane (VK4)	20:37	07:01
	Adelaide (VK5)	21:53	07:41
	Perth (VK6)	23:16	09:20
	Hobart, Tasmania (VK7)	21:42	06:42
	Darwin (VK8)	21:36	08:59
VK	Lord Howe Is.	20:22	06:27
VK9	Christmas Is.	23:13	10:44
VK9	Cocos-Keeling Is.	23:52	11:16
VK9	Mellish Reef	20:06	07:10
VK9	Norfolk Is.	19:41	05:58
VK9	Willis Is.	20:28	07:38
VKØ	Heard Is.	03:21	10:54
VKØ	Macquarie Is.	21:49	05:03
VP2E	Anguilla	09:36	22:50
VP2M	Montserrat	09:36	22:44
VP2V	British Virgin Is., Tortola	09:42	22:57
VP5	Turks & Caicos Islands,		
	Grand Turk	10:03	23:29
VP8	Falkland Islands, Stanley	11:59	19:46

Pref	State/Province/Country/City	Sunrise	Sunset
VP8	So. Georgia Is.	10:49	18:08
VP8	So. Orkney Is.	12:13	17:53
VP8	So. Sandwich Islands, Saunders Is.	10:33	17:03
VP8	So. Shetland Is., King George Is.	13:20	18:29
VP9	Bermuda	09:12	23:28
VQ9	Chagos, Diego Garcia	01:21	13:02
VR6	Pitcairn Is.	15:24	01:59
VS6	Hong Kong	21:39	11:09
VU	India, Bombay	00:32	13:48
	Calcutta	23:22	12:53
	New Delhi	23:53	13:51
	Bangalore	00:24	13:17
VU7	Andaman Islands, Port Blair	23:26	12:14
VU7	Laccadive Is.	00:48	13:30
XE	Mexico, Mexico City (XE1)	11:59	01:17
	Chihuahua (XE2)	12:06	02:04
	Merida (XE3)	11:18	00:42
XF4	Revilla Gigedo	12:49	02:05
XT	Burkina Faso, Onagadougou	05:42	18:33
XU	Kampuchea, Phnom Penh	22:38	11:26
XW	Laos, Viangchan	22:34	11:47
XX9	Macao	21:42	11:12
XZ	Burma, Rangoon	23:03	12:11
Y2-9	German Dem.Rep.(E), Berlin	02:43	19:32
YA	Afghanistan, Kandahar	00:33	14:43
	Kabul	00:11	14:38
YB-YD	Indonesia, Jakarta	23:01	10:47
	Medan	23:16	11:36
	Pontianak	22:40	10:48
	Jayapura	20:39	08:37
YI	Iraq, Baghdad	01:53	16:13
YJ	Vanuatu, Port Vila	19:16	06:20
YK	Syria, Damascus	02:25	16:47
YN, HT	Nicaragua, Managua	11:20	00:10
YO	Romania, Bucharest	02:31	18:03
YS	El Salvador, San Salvador	11:30	00:26
YU	Yugoslavia, Belgrade	02:51	18:27
YV	Venezuela, Caracas	10:07	22:51
YVØ	Aves Is.	09:44	22:48
Z2,ZE	Zimbabwe, Harare	04:25	15:29
ZA	Albania, Tirane	03:07	18:17
ZB	Gibraltar	05:04	19:41
ZC4	British Cyprus	02:35	17:03
ZD7	St. Helena	06:50	18:00
ZD8	Ascension Is.	07:09	18:48
ZD9	Tristan da Cunha	08:02	17:39
ZF	Cayman Is.	10:47	00:05
ZK1	No. Cook Is., Manihiki	17:00	04:31
ZK1	So. Cook Is., Rarotonga	17:15	04:06
ZK2	Niue	17:47	04:46
ZK3	Tokelaus, Atafu	17:43	05:21
ZL	New Zealand, Auckland (ZL1)	19:33	05:11
	Wellington (ZL2)	19:46	04:58
	Christchurch (ZL3)	20:02	04:59
	Dunedin (ZL4)	20:20	04:59
ZL5	Antarctica, Scott Base	24 hrs night	
ZL7	Chatham Is.	19:21	04:14
ZL8	Kermadec Is.	18:45	05:01
ZL9	Auckland & Campbell Is.,		
	Auckland	20:56	04:54
	Campbell Is.	20:55	04:34
ZP	Paraguay, Asuncion	10:35	21:09
ZS	South Africa, Cape Town (ZS1)	05:51	15:44
	Port Elizabeth (ZS2)	05:22	15:15
	Bloemfontein (ZS4)	05:09	15:25
	Durban (ZS5)	04:51	15:04
	Johannesburg (ZS6)	04:54	15:24
ZS2	Prince Edward & Marion Is.,		
	Marion	05:14	13:46
ZS3	Namibia, Windhoek	05:30	16:15
1AØ	S.M.O.M	03:35	18:48
1S	Spratly Is.	22:14	10:53
3A	Monaco	03:48	19:15
3B6	Agalega	02:29	14:00
3B7	St. Brandon	02:27	13:37
3B8	Mauritius	02:44	13:38
3B9	Rodriguez Is.	02:19	13:16
3C	Equatorial Guinea, Bata	05:15	17:29
	Malabo	05:16	17:36

Pref	State/Province/Country/City	Sunrise	Sunset
3CØ	Pagalu Is.	05:38	17:40
3D2	Fiji Is., Suva	18:36	05:39
3D6	Swaziland, Mbabane	04:42	15:11
3V	Tunisia, Tunis	04:00	18:41
3W	Vietnam, Ho Chi Minh City		
	(Saigon)	22:32	11:17
	Hanoi	22:16	11:40
3X	Republic of Guinea, Conakry	06:36	19:16
3Y	Bouvet	08:10	15:26
3Y	Peter I Is.	24 hrs night	
4S	Sri Lanka, Colombo	00:26	12:57
4U	ITU Geneva	03:43	19:30
4U	United Nations Hq.	09:24	00:30
4W	Yemen, Sanaa	02:33	15:36
4X, 4Z	Israel, Jerusalem	02:34	16:47
5A	Libya, Tripoli	04:03	18:20
	Benghazi	03:34	17:49
5B	Cyprus, Nicosia	02:32	17:03
5H	Tanzania, Dar es Salaam	03:32	15:15
5N	Nigeria, Lagos	05:33	18:03
5R	Madagascar, Antananarivo	03:21	14:21
5T	Mauritania, Nouakchott	06:29	19:42
5U	Niger, Niamey	05:25	18:21
5V	Togo, Lome	05:43	18:10
5W	Western Samoa, Apia	17:48	05:08
5X	Uganda, Kampala	03:47	15:55
5Z	Kenya, Nairobi	04:53	16:55
6W	Senegal, Dakar	06:41	19:41
6Y	Jamaica, Kingston	10:32	23:45

Pref	State/Province/Country/City	Sunrise	Sunset
7O	Yemen, People's Dem. Rep.,		
	Aden	02:35	15:28
7P	Lesotho, Maseru	05:03	15:19
7Q	Malawi, Lilongwe	04:07	15:25
	Blantyre	04:06	15:17
7X	Algeria, Algiers	04:29	19:09
8P	Barbados, Bridgetown	09:33	22:26
8Q	Maldive Is.	00:56	13:19
8R	Guyana, Georgetown	09:38	22:09
9G	Ghana, Accra	05:49	18:15
9H	Malta	03:45	18:22
9J	Zambia, Lusaka	04:32	15:44
9K	Kuwait	01:49	15:51
9L	Sierra Leone, Freetown	06:35	19:12
9M2	West Malaysia, Kuala Lumpur	23:06	11:24
9M6,8	East Malaysia, Sabah,		
	Sandakan (9M6)	21:55	10:22
	Sarawak, Kuching (9M8)	22:34	10:46
9N	Nepal, Katmandu	23:23	13:17
9Q	Zaire, Kinshasa	05:04	16:56
	Kisangani	04:16	16:25
	Lubumbashi	04:28	15:54
9U	Burundi, Bujumbura	04:06	16:02
9V	Singapore	23:00	11:12
9X	Rwanda, Kigali	04:01	16:01
9Y	Trinidad & Tobago, Port of		
	Spain	09:44	22:28
J2/A	Abu Ail	02:41	15:39

Pref	State/Province/Country/City	Sunrise	Sunset
VE1	New Brunswick, Saint John	08:44	00:14
	Nova Scotia, Halifax	08:36	00:01
	Prince Edward Island,		
	Charlottetown	08:28	00:05
VE2	Quebec, Montreal	09:13	00:45
	Quebec City	08:58	00:40
VE3	Ontario, London	09:53	01:06
	Ottawa	09:22	00:53
	Sudbury	09:38	01:18
	Toronto	09:43	01:01
VE4	Manitoba, Winnipeg	10:28	02:38
VE5	Saskatchewan, Regina	10:55	03:11
	Saskatoon	10:55	03:27
VE6	Alberta, Calgary	11:30	03:51
	Edmonton	11:14	04:03
VE7	British Columbia, Prince George	11:49	04:42
	Prince Rupert	12:16	05:15
	Vancouver	12:15	04:19
VE8	Northwest Territories,		
	Yellowknife	09:56	05:28
	Resolute	24 hrs day	
VY1	Yukon, Whitehorse	11:42	06:28
VO1	Newfoundland, St. John's	07:41	23:30
VO2	Labrador, Goose Bay	07:43	00:29
W1	Connecticut, Hartford	09:22	00:28
	Maine, Bangor	08:56	00:23
	Portland	09:06	00:25
	Massachusetts, Boston	09:14	00:23
	New Hampshire, Concord	09:13	00:28
	Rhode Island, Providence	09:17	00:23
	Vermont, Montpelier	09:13	00:36
W2	New Jersey, Atlantic City	09:37	00:27
	New York, Albany	09:24	00:35
	Buffalo	09:44	00:56
	New York City	09:31	00:30
	Syracuse	09:32	00:46
W3	Delaware, Wilmington	09:40	00:32
	District of Columbia,		
	Washington	09:49	00:36
	Maryland, Baltimore	09:46	00:35
	Pennsylvania, Harrisburg	09:44	00:40
	Philadelphia	09:39	00:32
	Pittsburgh	09:56	00:52
	Scranton	09:36	00:38
W4	Alabama, Montgomery	10:44	00:55
	Florida, Jacksonville	10:30	00:32
	Miami	10:35	00:16
	Pensacola	10:52	00:54
	Georgia, Atlanta	10:33	00:51
	Savannah	10:24	00:34
	Kentucky, Lexington	10:22	01:03
	Louisville	10:26	01:09
	North Carolina, Charlotte	10:15	00:40
	Raleigh	10:04	00:33
	Wilmington	10:06	00:26
	South Carolina, Columbia	10:19	00:38
	Tennessee, Knoxville	10:25	00:55
	Memphis	10:52	01:17
	Nashville	10:36	01:07
	Virginia, Norfolk	09:52	00:27
	Richmond	09:55	00:33
W5	Arkansas, Little Rock	11:02	01:25
	Louisiana, New Orleans	11:05	01:04
	Shreveport	11:13	01:25
	Mississippi, Jackson	11:00	01:11
	New Mexico, Albuquerque	11:58	02:24
	Oklahoma, Oklahoma City	11:21	01:48
	Texas, Abilene	11:37	01:49
	Amarillo	11:39	02:04
	Dallas	11:25	01:38
	El Paso	12:06	02:15
	San Antonio	11:40	01:37
W6	California, Los Angeles	12:47	03:07
	San Francisco	12:54	03:34
W7	Arizona, Flagstaff	12:18	02:44
	Phoenix	12:24	02:41
	Idaho, Boise	12:10	03:28
	Pocatello	11:58	03:11

Pref	State/Province/Country/City	Sunrise	Sunset
	Montana, Billings	11:31	03:05
	Butte	11:46	03:22
	Great Falls	11:35	03:24
	Nevada, Las Vegas	12:29	03:00
	Reno	12:38	03:29
	Oregon, Portland	12:29	04:01
	Utah, Salt Lake City	12:03	03:01
	Washington, Seattle	12:19	04:08
	Spokane	11:59	03:49
	Wyoming, Cheyenne	11:33	02:34
	Sheridan	11:29	02:56
W8	Michigan, Detroit	10:02	01:11
	Grand Rapids	10:10	01:24
	Sault Ste. Marie	09:52	01:32
	Traverse City	10:03	01:30
	Ohio, Cincinnati	10:18	01:06
	Cleveland	09:59	01:03
	Columbus	10:09	01:03
	West Virginia, Charleston	10:09	00:53
W9	Illinois, Chicago	10:22	01:28
	Indiana, Indianapolis	10:23	01:15
	Wisconsin, Green Bay	10:14	01:38
	Milwaukee	10:19	01:33
W0	Colorado, Denver	11:38	02:30
	Grand Junction	11:55	02:43
	Iowa, Des Moines	10:47	01:51
	Kansas, Pratt	11:19	01:59
	Wichita	11:14	01:53
	Minnesota, Duluth	10:22	02:04
	Minneapolis	10:33	02:02
	Missouri, Columbia	10:50	01:37
	Kansas City	10:59	01:47
	St. Louis	10:43	01:28
	Nebraska, North Platte	11:17	02:18
	Omaha	10:57	01:59
	North Dakota, Fargo	10:40	02:23
	South Dakota, Rapid City	11:16	02:38

Pref	State/Province/Country/City	Sunrise	Sunset
A2	Botswana, Gaborone	05:01	15:40
A3	Tonga, Nukualofa	18:18	05:12
A4	Oman, Muscat	01:24	14:56
A5	Bhutan, Thimbu	23:13	13:00
A6	United Arab Emirates, Abu		
	Dhabi	01:40	15:15
A7	Qatar, Ad-Dawhah	01:49	15:28
A9	Bahrein, Al-Manamah	01:51	15:33
AP	Pakistan, Karachi	00:47	14:25
	Islamabad	00:03	14:20
BV	Taiwan, Taipei	21:09	10:47
BY	People's Rep. of China, Beijing	20:52	11:46
	Harbin	19:50	11:25
	Shanghai	20:56	11:01
	Fuzhou	21:16	10:58
	Xian	21:38	11:59
	Chongqing	21:59	11:58
	Chengdu	22:06	12:09
	Lhasa	23:00	12:59
	Urumqi	22:34	13:53
	Kashi	23:35	14:26
C2	Nauru	18:54	06:59
C3	Andorra	04:23	19:33
C5	The Gambia, Banjul	06:44	19:38
C6	Bahamas, Nassau	10:25	00:03
C9	Mozambique, Maputo	04:37	15:11
	Mozambique	03:44	14:59
CE	Chile, Santiago	11:47	21:48
CE0	Easter Island	14:07	00:37
CE0	San Felix	12:08	22:41
CE0	Juan Fernandez	12:19	22:20
CM,CO	Cuba, Havana	10:49	00:19
CN	Morocco, Casablanca	05:26	19:43
CP	Bolivia, La Paz	11:02	22:13
CT	Portugal, Lisbon	05:18	20:04
CT2, CU	Azores, Ponta Delgada	06:27	21:07
CT3	Madeira Islands, Funchal	06:06	20:18
CX	Uruguay, Montevideo	10:52	20:46
D2	Angola, Luanda	05:23	17:00
D4	Cape Verde, Praia	07:09	20:07

Pref	State/Province/Country/City	Sunrise	Sunset
D6	Comoros, Moroni	03:26	14:53
DA-DL	Germany, Fed. Rep. of (W), Bonn	03:27	19:45
DU	Philippines, Manilla	21:32	10:29
EA	Spain, Madrid	04:51	19:47
EA6	Balearic Is., Palma	04:29	19:19
EA8	Canary Is., Las Palmas	06:05	19:58
EA9	Ceuta & Melilla, Ceuta	05:11	19:40
	Melilla	05:03	19:30
EI	Ireland, Dublin	04:06	20:53
EL	Liberia, Monrovia	06:33	19:02
EP	Iran, Tehran	01:23	15:52
ET	Ethiopia, Addis Ababa	03:11	15:48
	Asmera	02:59	15:59
F	France, Paris	03:55	19:55
FG	Guadeloupe	09:40	22:42
FG, FS	St. Martin	09:42	22:52
FH	Mayotte	03:21	14:45
FK	New Caledonia, Noumia	19:34	06:23
FM	Martinique	09:40	22:37
FO	Clipperton	13:00	01:42
FO	Fr. Polynesia, Tahiti	17:09	04:16
FP	St. Pierre & Miquelon, St. Pierre	07:58	23:39
FR	Glorioso	03:11	14:39
FR	Juan de Nova	03:39	14:48
	Europa	03:58	14:47
FR	Reunion	02:55	13:49
FR	Tromelin	02:50	14:03
FT8W	Crozet	04:13	13:00
FT8X	Kerguelen	03:17	11:38
FT8Y	Antarctica, Dumont D'Urville	01:13	04:15
FT8Z	Amsterdam & St. Paul Is., Amsterdam	02:04	11:44
FW	Wallis & Futuna Is., Wallis	18:08	05:31
FY	Fr. Guiana, Cayenne	09:22	21:45
G	England, London	03:52	20:18
GD	Isle of Man	03:53	20:52
GI	Northern Ireland, Belfast	03:57	20:59
GJ	Jersey	04:11	20:15
GM	Scotland, Glasgow	03:42	21:01
	Aberdeen	03:23	21:02
GU	Guernsey	04:12	20:18
GW	Wales, Cardiff	04:04	20:30
H4	Solomon Islands, Honiara	19:36	07:12
HA	Hungary, Budapest	02:54	18:42
HB	Switzerland, Bern	03:42	19:26
HBØ	Liechtenstein	03:33	19:19
HC	Ecuador, Quito	11:13	23:20
HC8	Galapagos Is.	12:04	00:09
HH	Haiti, Port-Au-Prince	10:18	23:29
HI	Dominican Republic, Santo Domingo	10:08	23:20
HK	Colombia, Bogota	10:49	23:12
HKØ	Malpelo Is.	11:18	23:39
HKØ	San Adreas	11:06	23:56
HL, HM	Korea, Seoul	20:17	10:56
HP	Panama, Panama	11:04	23:41
HR	Honduras, Tegucigalpa	11:25	00:21
HS	Thailand, Bangkok	22:55	11:50
HV	Vatican City	03:41	18:47
HZ, 7Z	Saudi Arabia, Dharan	01:53	15:36
	Mecca	02:43	16:07
	Riyadh	02:10	15:45
I	Italy, Rome	03:41	18:47
	Trieste	03:22	18:56
	Sicily	03:49	18:28
IS	Sardinia, Cagliari	04:04	18:52
J2	Djibouti, Djibouti	02:48	15:35
J3	Grenada	09:47	22:36
J5	Guinea-Bissau, Bissau	06:43	19:31
J6	St. Lucia	09:41	22:36
J7	Dominica	09:39	22:40
J8	St. Vincent	09:43	22:36
JA-JS	Japan, Tokyo	19:31	09:59
	Nagasaki	20:18	10:31
	Sapporo	19:02	10:16
JD	Minami Torishima	19:01	08:36
JD	Ogasawara, Kazan Is.	19:46	09:35
JT	Mongolia, Ulan Bator	21:01	12:53
JW	Svalbard, Spitsbergen	24 hrs day	
JX	Jan Mayen	24 hrs day	

Pref	State/Province/Country/City	Sunrise	Sunset
JY	Jordan, Amman	02:36	16:45
KC4	Antarctica, Byrd Station	24 hrs night	
	McMurdo Sound	24 hrs night	
	Palmer Station	14:12	18:29
KC6	Micronesia, Ponape	19:16	07:46
KC6	Belau, Yap	20:32	09:11
	Koror	20:50	09:22
KG4	Guantanamo Bay	10:27	23:44
KH1	Baker, Howland Is.	17:44	05:53
KH2	Guam, Agana	19:58	08:52
KH3	Johnston Is.	16:45	05:51
KH4	Midway Is.	16:58	06:50
KH5	Palmyra Is.	16:39	05:06
KH5K	Kingman Reef	16:39	05:12
KH6	Hawaii, Hilo	15:47	05:03
	Honolulu	15:55	05:17
KH7	Kure Is.	17:02	06:54
KH8	American Samoa, Pago Pago	17:48	05:07
KH9	Wake Is.	18:21	07:35
KHØ	Mariana Is., Saipan	19:51	08:51
KL7	Alaska, Adak	15:36	08:05
	Anchorage	12:36	07:33
	Fairbanks	11:27	08:24
	Juneau	12:03	06:01
	Nome	12:44	09:27
KP1	Navassa Is.	10:29	23:40
KP2	Virgin Islands, Charlotte Amalie	09:48	22:59
KP4	Puerto Rico, San Juan	09:53	23:05
KP5	Desecheo Is.	09:59	23:10
KX6	Marshall Islands, Kwajalein	18:36	07:14
LA-LJ	Norway, Oslo	02:06	20:37
LU	Argentina, Buenos Aires	11:00	20:56
LX	Luxembourg	03:36	19:43
LZ	Bulgaria, Sofia	02:55	18:07
OA	Peru, Lima	11:29	22:56
OD	Lebanon, Beirut	02:33	16:52
OE	Austria, Vienna	03:02	18:56
OH	Finland, Helsinki	01:07	19:42
OHØ	Aland Is.	01:27	20:02
OJØ	Market Reef	01:30	20:07
OK	Czechoslovakia, Prague	03:01	19:13
ON	Belgium, Brussels	03:37	19:57
OX, XP	Greenland, Godthaab	05:16	01:46
	Thule	24 hrs day	
OY	Faroe Islands, Torshavn	02:53	22:10
OZ	Denmark, Copenhagen	02:35	19:52
P2	Papua New Guinea, Madang	20:27	08:17
	Port Moresby	20:28	08:04
PA-PI	Netherlands, Amsterdam	03:27	20:03
PJ, P4	Netherlands Antilles, Willemstad	10:16	23:04
PJ5-8	St. Maarten and Saba, St. Maarten	09:43	22:51
PY	Brazil, Brasilia	09:39	20:53
	Rio De Janeiro	09:34	20:20
	Natal	08:32	20:19
	Manaus	10:07	22:04
	Porto Alegre	10:21	20:38
PYØ	Fernando De Noronha	08:17	20:11
PYØ	St. Peter & St. Paul Rocks	07:57	20:07
PYØ	Trindade & Martin Vaz Is., Trindade	08:33	19:30
PZ	Suriname, Paramaribo	09:32	21:58
S2	Bangladesh, Dacca	23:17	12:49
S7	Seychelles, Victoria	02:26	14:18
S9	Sao Tome	05:33	17:42
SJ-SM	Sweden, Stockholm	01:44	20:00
SP	Poland, Krakow	02:39	18:50
	Warsaw	02:23	18:57
ST	Sudan, Khartoum	03:24	16:25
STØ	Southern Sudan, Juba	03:46	16:10
SU	Egypt, Cairo	02:59	16:59
SV	Greece, Athens	03:09	17:50
SV5	Dodecanese, Rhodes	02:55	17:28
SV9	Crete	03:10	17:37
SV/A	Mount Athos	03:00	17:55
T2	Tuvalu, Funafuti	18:21	05:59
T3Ø	West Kiribati, Bonriki	18:26	06:37
T31	Central Kiribati, Kanton	17:32	05:30
T32	East Kiribati, Christmas Is.	16:27	04:41
T5	Somalia, Mogadishu	02:56	15:10
T7	San Marino	03:35	18:55

Pref	State/Province/Country/City	Sunrise	Sunset
TA	Turkey, Ankara	02:26	17:19
	Istanbul	02:38	17:39
TF	Iceland, Reykjavik	03:19	23:45
TG	Guatemala, Guatemala City	11:38	00:35
TI	Costa Rica, San Jose	11:20	00:01
TI9	Cocos Is.	11:39	00:05
TJ	Cameroon, Yaounde	05:08	17:28
TK	Corsica	03:55	19:02
TL	Central African Republic, Bangui	04:39	17:01
TN	Congo, Brazzaville	05:07	17:00
TR	Gabon, Libreville	05:22	17:31
TT	Chad, N'Djamena	04:40	17:28
TU	Ivory Coast, Abidjan	06:08	18:33
TY	Benin, Porto Novo	05:39	18:08
TZ	Mali, Bamako	06:11	19:02
UA	Russia, European,		
	Leningrad (UA1)	00:49	19:17
	Archangel (UA1)	22:58	19:46
	Murmansk (UA1)	24 hrs day	
	Moscow (UA3)	00:55	18:13
	Kuibyshev (UA4)	00:21	17:06
	Rostov (UA6)	01:32	17:20
UA1P	Franz Josef Land	24 hrs day	
UA2	Kaliningrad	02:09	19:16
UA9,Ø	Russia, Asiatic,		
	Novosibirsk (UA9)	21:59	15:07
	Perm (UA9)	23:23	17:15
	Omsk (UA9)	22:37	15:45
	Norilsk (UAØ)	24 hrs day	
	Irkutsk (UAØ)	20:50	13:25
	Vladivostok (UAØ)	19:39	10:54
	Petropavlovsk (UAØ)	17:08	09:51
	Khabarovsk (UAØ)	19:05	11:02
	Krasnoyarsk (UAØ)	21:12	14:34
	Yakutsk (UAØ)	17:47	13:04
	Wrangel Island (UAØ)	24 hrs day	
	Kyzyl (UAØY)	21:32	14:00
UB	Ukraine, Kiev	01:55	18:10
UC	Byelorussia, Minsk	01:47	18:41
UD	Azerbaijan, Baku	01:17	16:13
UF	Georgia, Tbilisi	01:33	16:37
UG	Armenia, Ierevan	01:39	16:34
UH	Turkoman, Ashkhabad	00:50	15:31
UI	Uzbek, Bukhara	00:21	15:13
	Tashkent	23:56	14:58
UJ	Tadzhik, Samarkand	00:11	15:03
	Dushanbe	00:05	14:53
UL	Kazakh, Alma-ata	23:19	14:34
UM	Kirghiz, Frunze	23:30	14:42
UO	Moldavia, Kishinev	02:17	18:01
UP	Lithuania, Vilna	01:52	18:53
UQ	Latvia, Riga	01:40	19:16
UR	Estonia, Tallinn	01:16	19:34
V2	Antigua & Barbuda, St. Johns	09:38	22:45
V3	Belize, Belmopan	11:26	00:33
V4	St. Christopher & Nevis	09:41	22:48
V8	Brunei, Bandar Seri Begawan	22:13	10:37
VE1, CY	Sable Is.	08:25	23:44
VE1, CY	St. Paul Is.	08:12	23:58
VK	Australia, Canberra (VK1)	21:12	07:04
	Sydney (VK2)	21:00	06:59
	Melbourne (VK3)	21:35	07:14
	Brisbane (VK4)	20:38	07:06
	Adelaide (VK5)	21:53	07:47
	Perth (VK6)	23:17	09:25
	Hobart, Tasmania (VK7)	21:41	06:49
	Darwin (VK8)	21:38	09:03
VK	Lord Howe Is.	20:23	06:33
VK9	Christmas Is.	23:16	10:47
VK9	Cocos-Keeling Is.	23:54	11:20
VK9	Mellish Reef	20:08	07:15
VK9	Norfolk Is.	19:41	06:03
VK9	Willis Is.	20:30	07:42
VKØ	Heard Is.	03:18	11:04
VKØ	Macquarie Is.	21:45	05:13
VP2E	Anguilla	09:41	22:52
VP2M	Montserrat	09:40	22:46
VP2V	British Virgin Is., Tortola	09:47	22:58
VP5	Turks & Caicos Islands,		
	Grand Turk	10:08	23:31
VP8	Falkland Islands, Stanley	11:56	19:55
VP8	So. Georgia Is.	10:46	18:17
VP8	So. Orkney Is.	12:06	18:06
VP8	So. Sandwich Islands, Saunders Is.	10:28	17:15
VP8	So. Shetland Is., King George Is.	13:12	18:43
VP9	Bermuda	09:18	23:29
VQ9	Chagos, Diego Garcia	01:23	13:06
VR6	Pitcairn Is.	15:26	02:04
VS6	Hong Kong	21:44	11:10
VU	India, Bombay	00:36	13:50
	Calcutta	23:27	12:55
	New Delhi	23:59	13:52
	Bangalore	00:28	13:20
VU7	Andaman Islands, Port Blair	23:30	12:17
VU7	Laccadive Is.	00:52	13:33
XE	Mexico, Mexico City (XE1)	12:03	01:18
	Chihuahua (XE2)	12:11	02:05
	Merida (XE3)	11:22	00:44
XF4	Revilla Gigedo	12:53	02:07
XT	Burkina Faso, Onagadougou	05:46	18:35
XU	Kampuchea, Phnom Penh	22:42	11:29
XW	Laos, Viangchan	22:39	11:49
XX9	Macao	21:47	11:13
XZ	Burma, Rangoon	23:08	12:13
Y2-9	German Dem.Rep.(E), Berlin	02:52	19:29
YA	Afghanistan, Kandahar	00:39	14:43
	Kabul	00:17	14:38
YB-YD	Indonesia, Jakarta	23:04	10:50
	Medan	23:20	11:39
	Pontianak	22:43	10:51
	Jayapura	20:42	08:41
YI	Iraq, Baghdad	01:59	16:14
YJ	Vanuatu, Port Vila	19:18	06:24
YK	Syria, Damascus	02:31	16:48
YN, HT	Nicaragua, Managua	11:24	00:12
YO	Romania, Bucharest	02:38	18:02
YS	El Salvador, San Salvador	11:34	00:28
YU	Yugoslavia, Belgrade	02:59	18:26
YV	Venezuela, Caracas	10:11	22:54
YVØ	Aves Is.	09:48	22:50
Z2,ZE	Zimbabwe, Harare	04:27	15:33
ZA	Albania, Tirane	03:14	18:16
ZB	Gibraltar	05:11	19:41
ZC4	British Cyprus	02:41	17:04
ZD7	St. Helena	06:52	18:04
ZD8	Ascension Is.	07:12	18:52
ZD9	Tristan da Cunha	08:02	17:45
ZF	Cayman Is.	10:51	00:07
ZK1	No. Cook Is., Manihiki	17:02	04:34
ZK1	So. Cook Is., Rarotonga	17:17	04:10
ZK2	Niue	17:49	04:51
ZK3	Tokelaus, Atafu	17:46	05:25
ZL	New Zealand, Auckland (ZL1)	19:33	05:17
	Wellington (ZL2)	19:46	05:04
	Christchurch (ZL3)	20:02	05:06
	Dunedin (ZL4)	20:18	05:06
ZL5	Antarctica, Scott Base	24 hrs night	
ZL7	Chatham Is.	19:20	04:21
ZL8	Kermadec Is.	18:46	05:06
ZL9	Auckland & Campbell Is.,		
	Auckland	20:54	05:03
	Campbell Is.	20:52	04:44
ZP	Paraguay, Asuncion	10:36	21:14
ZS	South Africa, Cape Town (ZS1)	05:51	15:50
	Port Elizabeth (ZS2)	05:22	15:21
	Bloemfontein (ZS4)	05:09	15:30
	Durban (ZS5)	04:52	15:10
	Johannesburg (ZS6)	04:55	15:29
ZS2	Prince Edward & Marion Is.,		
	Marion	05:12	13:54
ZS3	Namibia, Windhoek	05:32	16:20
1AØ	S.M.O.M.	03:42	18:48
1S	Spratly Is.	22:18	10:55
3A	Monaco	03:55	19:14
3B6	Agalega	02:32	14:04
3B7	St. Brandon	02:29	13:41
3B8	Mauritius	02:46	13:43
3B9	Rodriguez Is.	02:21	13:20
3C	Equatorial Guinea, Bata	05:18	17:32
	Malabo	05:19	17:39

Pref	State/Province/Country/City	Sunrise	Sunset
3CØ	Pagalu Is.	05:41	17:43
3D2	Fiji Is., Suva	18:38	05:43
3D6	Swaziland, Mbabane	04:43	15:17
3V	Tunisia, Tunis	04:06	18:41
3W	Vietnam, Ho Chi Minh City		
	(Saigon)	22:36	11:19
	Hanoi	22:20	11:42
3X	Republic of Guinea, Conakry	06:39	19:19
3Y	Bouvet	08:06	15:35
3Y	Peter I Is.	24 hrs night	
4S	Sri Lanka, Colombo	00:29	13:00
4U	ITU Geneva	03:51	19:28
4U	United Nations Hq.	09:31	00:30
4W	Yemen, Sanaa	02:37	15:38
4X, 4Z	Israel, Jerusalem	02:39	16:48
5A	Libya, Tripoli	04:08	18:20
	Benghazi	03:39	17:49
5B	Cyprus, Nicosia	02:38	17:04
5H	Tanzania, Dar es Salaam	03:34	15:18
5N	Nigeria, Lagos	05:36	18:05
5R	Madagascar, Antananarivo	03:23	14:25
5T	Mauritania, Nouakchott	06:33	19:43
5U	Niger, Niamey	05:30	18:23
5V	Togo, Lome	05:46	18:13
5W	Western Samoa, Apia	17:51	05:12
5X	Uganda, Kampala	03:50	15:58
5Z	Kenya, Nairobi	04:56	16:59
6W	Senegal, Dakar	06:45	19:43
6Y	Jamaica, Kingston	10:37	23:46

Pref	State/Province/Country/City	Sunrise	Sunset
7O	Yemen, People's Dem. Rep.,		
	Aden	02:39	15:30
7P	Lesotho, Maseru	05:04	15:25
7Q	Malawi, Lilongwe	04:09	15:29
	Blantyre	04:08	15:21
7X	Algeria, Algiers	04:35	19:09
8P	Barbados, Bridgetown	09:37	22:29
8Q	Maldive Is.	01:00	13:22
8R	Guyana, Georgetown	09:42	22:12
9G	Ghana, Accra	05:52	18:18
9H	Malta	03:52	18:22
9J	Zambia, Lusaka	04:34	15:49
9K	Kuwait	01:54	15:52
9L	Sierra Leone, Freetown	06:39	19:15
9M2	West Malaysia, Kuala Lumpur	23:09	11:27
9M6,8	East Malaysia, Sabah,		
	Sandakan (9M6)	21:59	10:25
	Sarawak, Kuching (9M8)	22:37	10:49
9N	Nepal, Katmandu	23:28	13:18
9Q	Zaire, Kinshasa	05:07	17:00
	Kisangani	04:19	16:28
	Lubumbashi	04:30	15:58
9U	Burundi, Bujumbura	04:09	16:05
9V	Singapore	23:03	11:15
9X	Rwanda, Kigali	04:04	16:04
9Y	Trinidad & Tobago, Port of		
	Spain	09:48	22:31
J2/A	Abu Ail	02:45	15:41

Pref	State/Province/Country/City	Sunrise	Sunset
VE1	New Brunswick, Saint John	08:56	00:05
	Nova Scotia, Halifax	08:48	23:52
	Prince Edward Island,		
	Charlottetown	08:41	23:56
VE2	Quebec, Montreal	09:25	00:35
	Quebec City	09:11	00:30
VE3	Ontario, London	10:04	00:58
	Ottawa	09:34	00:43
	Sudbury	09:51	01:08
	Toronto	09:54	00:52
VE4	Manitoba, Winnipeg	10:43	02:26
VE5	Saskatchewan, Regina	11:10	02:58
	Saskatoon	11:11	03:14
VE6	Alberta, Calgary	11:46	03:39
	Edmonton	11:31	03:48
VE7	British Columbia, Prince George	12:07	04:27
	Prince Rupert	12:34	05:00
	Vancouver	12:29	04:07
VE8	Northwest Territories,		
	Yellowknife	10:29	04:58
	Resolute	24 hrs day	
VY1	Yukon, Whitehorse	12:09	06:03
VO1	Newfoundland, St. John's	07:54	23:19
VO2	Labrador, Goose Bay	08:00	00:15
W1	Connecticut, Hartford	09:33	00:20
	Maine, Bangor	09:08	00:14
	Portland	09:18	00:16
	Massachusetts, Boston	09:25	00:15
	New Hampshire, Concord	09:24	00:19
	Rhode Island, Providence	09:28	00:15
	Vermont, Montpelier	09:25	00:27
W2	New Jersey, Atlantic City	09:47	00:20
	New York, Albany	09:35	00:27
	Buffalo	09:55	00:48
	New York City	09:42	00:22
	Syracuse	09:44	00:38
W3	Delaware, Wilmington	09:51	00:25
	District of Columbia,		
	Washington	09:59	00:29
	Maryland, Baltimore	09:56	00:28
	Pennsylvania, Harrisburg	09:55	00:32
	Philadelphia	09:49	00:24
	Pittsburgh	10:07	00:45
	Scranton	09:47	00:31
W4	Alabama, Montgomery	10:52	00:50
	Florida, Jacksonville	10:38	00:27
	Miami	10:41	00:12
	Pensacola	11:00	00:50
	Georgia, Atlanta	10:41	00:46
	Savannah	10:32	00:29
	Kentucky, Lexington	10:31	00:57
	Louisville	10:36	01:02
	North Carolina, Charlotte	10:23	00:35
	Raleigh	10:13	00:27
	Wilmington	10:14	00:21
	South Carolina, Columbia	10:27	00:33
	Tennessee, Knoxville	10:34	00:49
	Memphis	11:01	01:12
	Nashville	10:45	01:01
	Virginia, Norfolk	10:01	00:21
	Richmond	10:04	00:27
W5	Arkansas, Little Rock	11:11	01:20
	Louisiana, New Orleans	11:12	01:00
	Shreveport	11:21	01:20
	Mississippi, Jackson	11:08	01:06
	New Mexico, Albuquerque	12:07	02:18
	Oklahoma, Oklahoma City	11:29	01:42
	Texas, Abilene	11:45	01:44
	Amarillo	11:47	01:59
	Dallas	11:33	01:33
	El Paso	12:14	02:10
	San Antonio	11:47	01:33
W6	California, Los Angeles	12:56	03:02
	San Francisco	13:03	03:28
W7	Arizona, Flagstaff	12:27	02:38
	Phoenix	12:33	02:36
	Idaho, Boise	12:22	03:19
	Pocatello	12:09	03:02

Pref	State/Province/Country/City	Sunrise	Sunset
	Montana, Billings	11:44	02:56
	Butte	11:59	03:13
	Great Falls	11:49	03:13
	Nevada, Las Vegas	12:38	02:54
	Reno	12:48	03:22
	Oregon, Portland	12:42	03:52
	Utah, Salt Lake City	12:13	02:54
	Washington, Seattle	12:33	03:58
	Spokane	12:13	03:38
	Wyoming, Cheyenne	11:44	02:26
	Sheridan	11:41	02:46
W8	Michigan, Detroit	10:13	01:03
	Grand Rapids	10:22	01:15
	Sault Ste. Marie	10:05	01:22
	Traverse City	10:16	01:21
	Ohio, Cincinnati	10:28	00:59
	Cleveland	10:10	00:55
	Columbus	10:20	00:56
	West Virginia, Charleston	10:19	00:46
W9	Illinois, Chicago	10:33	01:20
	Indiana, Indianapolis	10:33	01:08
	Wisconsin, Green Bay	10:26	01:29
	Milwaukee	10:31	01:24
WØ	Colorado, Denver	11:49	02:23
	Grand Junction	12:05	02:36
	Iowa, Des Moines	10:58	01:43
	Kansas, Pratt	11:29	01:53
	Wichita	11:23	01:47
	Minnesota, Duluth	10:35	01:54
	Minneapolis	10:46	01:52
	Missouri, Columbia	11:00	01:30
	Kansas City	11:09	01:40
	St. Louis	10:52	01:21
	Nebraska, North Platte	11:28	02:10
	Omaha	11:08	01:51
	North Dakota, Fargo	10:53	02:13
	South Dakota, Rapid City	11:28	02:29
A2	Botswana, Gaborone	04:58	15:46
A3	Tonga, Nukualofa	18:16	05:17
A4	Oman, Muscat	01:30	14:53
A5	Bhutan, Thimbu	23:20	12:57
A6	United Arab Emirates, Abu		
	Dhabi	01:46	15:12
A7	Qatar, Ad-Dawhah	01:55	15:25
A9	Bahrein, Al-Manamah	01:57	15:30
AP	Pakistan, Karachi	00:53	14:21
	Islamabad	00:11	14:15
BV	Taiwan, Taipei	21:16	10:44
BY	People's Rep. of China, Beijing	21:02	11:38
	Harbin	20:03	11:15
	Shanghai	21:03	10:56
	Fuzhou	21:22	10:55
	Xian	21:47	11:54
	Chongqing	22:06	11:53
	Chengdu	22:14	12:05
	Lhasa	23:08	12:54
	Urumqi	22:46	13:45
	Kashi	23:45	14:18
C2	Nauru	18:55	07:01
C3	Andorra	04:35	19:25
C5	The Gambia, Banjul	06:48	19:37
C6	Bahamas, Nassau	10:31	00:00
C9	Mozambique, Maputo	04:33	15:17
	Mozambique	03:43	15:04
CE	Chile, Santiago	11:42	21:56
CEØ	Easter Island	14:03	00:43
CEØ	San Felix	12:05	22:48
CEØ	Juan Fernandez	12:14	22:28
CM,CO	Cuba, Havana	10:55	00:16
CN	Morocco, Casablanca	05:34	19:38
CP	Bolivia, La Paz	11:01	22:18
CT	Portugal, Lisbon	05:28	19:57
CT2, CU	Azores, Ponta Delgada	06:37	21:01
CT3	Madeira Islands, Funchal	06:14	20:13
CX	Uruguay, Montevideo	10:47	20:55
D2	Angola, Luanda	05:23	17:03
D4	Cape Verde, Praia	07:13	20:06

Pref	State/Province/Country/City	Sunrise	Sunset
D6	Comoros, Moroni	03:25	14:57
DA-DL	Germany, Fed. Rep. of (W),		
	Bonn	03:43	19:33
DU	Philippines, Manilla	21:36	10:28
EA	Spain, Madrid	05:02	19:40
EA6	Balearic Is., Palma	04:39	19:12
EA8	Canary Is., Las Palmas	06:12	19:54
EA9	Ceuta & Melilla, Ceuta	05:20	19:34
	Melilla	05:12	19:24
EI	Ireland, Dublin	04:24	20:38
EL	Liberia, Monrovia	06:36	19:02
EP	Iran, Tehran	01:32	15:46
ET	Ethiopia, Addis Ababa	03:14	15:48
	Asmera	03:03	15:57
F	France, Paris	04:10	19:44
FG	Guadeloupe	09:44	22:41
FG, FS	St. Martin	09:46	22:50
FH	Mayotte	03:21	14:49
FK	New Caledonia, Noumia	19:31	06:28
FM	Martinique	09:44	22:36
FO	Clipperton	13:03	01:42
FO	Fr. Polynesia, Tahiti	17:07	04:20
FP	St. Pierre & Miquelon, St. Pierre	08:11	23:29
FR	Glorioso	03:10	14:43
FR	Juan de Nova	03:37	14:52
	Europa	03:56	14:53
FR	Reunion	02:53	13:54
FR	Tromelin	02:49	14:08
FT8W	Crozet	04:03	13:12
FT8X	Kerguelen	03:06	11:52
FT8Y	Antarctica, Dumont D'Urville	00:28	05:04
FT8Z	Amsterdam & St. Paul Is.,		
	Amsterdam	01:58	11:53
FW	Wallis & Futuna Is., Wallis	18:08	05:34
FY	Fr. Guiana, Cayenne	09:24	21:46
G	England, London	04:08	20:05
GD	Isle of Man	04:11	20:36
GI	Northern Ireland, Belfast	04:15	20:44
GJ	Jersey	04:26	20:04
GM	Scotland, Glasgow	04:02	20:44
	Aberdeen	03:44	20:44
GU	Guernsey	04:27	20:07
GW	Wales, Cardiff	04:20	20:17
H4	Solomon Islands, Honiara	19:36	07:15
HA	Hungary, Budapest	03:07	18:32
HB	Switzerland, Bern	03:56	19:16
HB0	Liechtenstein	03:46	19:08
HC	Ecuador, Quito	11:15	23:21
HC8	Galapagos Is.	12:05	00:11
HH	Haiti, Port-Au-Prince	10:23	23:28
HI	Dominican Republic, Santo		
	Domingo	10:13	23:18
HK	Colombia, Bogota	10:52	23:13
HK0	Malpelo Is.	11:21	23:40
HK0	San Adreas	11:10	23:55
HL, HM	Korea, Seoul	20:26	10:49
HP	Panama, Panama	11:07	23:41
HR	Honduras, Tegucigalpa	11:29	00:20
HS	Thailand, Bangkok	22:59	11:49
HV	Vatican City	03:52	18:39
HZ, 7Z	Saudi Arabia, Dharan	01:59	15:32
	Mecca	02:49	16:04
	Riyadh	02:16	15:42
I	Italy, Rome	03:52	18:39
	Trieste	03:35	18:46
	Sicily	03:58	18:21
IS	Sardinia, Cagliari	04:14	18:45
J2	Djibouti, Djibouti	02:52	15:34
J3	Grenada	09:51	22:35
J5	Guinea-Bissau, Bissau	06:47	19:30
J6	St. Lucia	09:45	22:35
J7	Dominica	09:44	22:38
J8	St. Vincent	09:47	22:35
JA-JS	Japan, Tokyo	19:40	09:54
	Nagasaki	20:26	10:26
	Sapporo	19:13	10:07
JD	Minami Torishima	19:07	08:33
JD	Ogasawara, Kazan Is.	19:53	09:31
JT	Mongolia, Ulan Bator	21:15	12:42
JW	Svalbard, Spitsbergen	24 hrs day	
JX	Jan Mayen	24 hrs day	

Pref	State/Province/Country/City	Sunrise	Sunset
JY	Jordan, Amman	02:44	16:41
KC4	Antarctica, Byrd Station	24 hrs night	
	McMurdo Sound	24 hrs night	
	Palmer Station	13:39	19:05
KC6	Micronesia, Ponape	19:19	07:47
KC6	Belau, Yap	20:35	09:11
	Koror	20:53	09:22
KG4	Guantanamo Bay	10:32	23:42
KH1	Baker, Howland Is.	17:46	05:54
KH2	Guam, Agana	20:02	08:51
KH3	Johnston Is.	16:50	05:50
KH4	Midway Is.	17:05	06:46
KH5	Palmyra Is.	16:42	05:07
KH5K	Kingman Reef	16:42	05:12
KH6	Hawaii, Hilo	15:52	05:01
	Honolulu	16:00	05:15
KH7	Kure Is.	17:09	06:50
KH8	American Samoa, Pago Pago	17:47	05:11
KH9	Wake Is.	18:26	07:33
KH0	Mariana Is., Saipan	19:56	08:50
KL7	Iaska, Adak	15:52	07:52
	Anchorage	13:04	07:07
	Fairbanks	12:13	07:42
	Juneau	12:26	05:41
	Nome	13:27	08:48
KP1	Navassa Is.	10:33	23:38
KP2	Virgin Islands, Charlotte Amalie	09:53	22:58
KP4	Puerto Rico, San Juan	09:58	23:03
KP5	Desecheo Is.	10:04	23:08
KX6	Marshall Islands, Kwajalein	18:39	07:14
LA-LJ	Norway, Oslo	02:32	20:14
LU	Argentina, Buenos Aires	10:55	21:04
LX	Luxembourg	03:51	19:31
LZ	Bulgaria, Sofia	03:07	17:59
OA	Peru, Lima	11:29	23:00
OD	Lebanon, Beirut	02:41	16:46
OE	Austria, Vienna	03:16	18:45
OH	Finland, Helsinki	01:33	19:18
OH0	Aland Is.	01:53	19:38
OJ0	Market Reef	01:56	19:43
OK	Czechoslovakia, Prague	03:16	19:01
ON	Belgium, Brussels	03:52	19:44
OX, XP	Greenland, Godthaab	05:57	01:09
	Thule	24 hrs day	
OY	Faroe Islands, Torshavn	03:23	21:43
OZ	Denmark, Copenhagen	02:55	19:36
P2	Papua New Guinea, Madang	20:28	08:19
	Port Moresby	20:28	08:07
PA-PI	Netherlands, Amsterdam	03:43	19:49
PJ, P4	Netherlands Antilles, Willemstad	10:20	23:04
PJ5-8	St. Maarten and Saba,		
	St. Maarten	09:47	22:50
PY	Brazil, Brasilia	09:38	20:57
	Rio De Janeiro	09:31	20:26
	Natal	08:32	20:21
	Manaus	10:08	22:06
	Porto Alegre	10:16	20:45
PY0	Fernando De Noronha	08:18	20:13
PY0	St. Peter & St. Paul Rocks	07:58	20:09
PY0	Trindade & Martin Vaz Is.,		
	Trindade	08:31	19:35
PZ	Suriname, Paramaribo	09:34	21:59
S2	Bangladesh, Dacca	23:23	12:46
S7	Seychelles, Victoria	02:27	14:21
S9	Sao Tome	05:35	17:43
SJ-SM	Sweden, Stockholm	02:08	19:39
SP	Poland, Krakow	02:54	18:38
	Warsaw	02:40	18:44
ST	Sudan, Khartoum	03:28	16:24
ST0	Southern Sudan, Juba	03:48	16:11
SU	Egypt, Cairo	03:06	16:54
SV	Greece, Athens	03:18	17:44
SV5	Dodecanese, Rhodes	03:04	17:22
SV9	Crete	03:19	17:31
SV/A	Mount Athos	03:10	17:47
T2	Tuvalu, Funafuti	18:21	06:02
T30	West Kiribati, Bonriki	18:27	06:39
T31	Central Kiribati, Kanton	17:33	05:32
T32	East Kiribati, Christmas Is.	16:29	04:42
T5	Somalia, Mogadishu	02:58	15:11
T7	San Marino	03:47	18:46

Pref	State/Province/Country/City	Sunrise	Sunset
TA	Turkey, Ankara	02:37	17:12
	Istanbul	02:49	17:31
TF	Iceland, Reykjavik	03:59	23:08
TG	Guatemala, Guatemala City	11:42	00:34
TI	Costa Rica, San Jose	11:23	00:00
TI9	Cocos Is.	11:42	00:06
TJ	Cameroon, Yaounde	05:10	17:29
TK	Corsica	04:06	18:54
TL	Central African Republic, Bangui	04:41	17:02
TN	Congo, Brazzaville	05:08	17:02
TR	Gabon, Libreville	05:24	17:32
TT	Chad, N'Djamena	04:44	17:28
TU	Ivory Coast, Abidjan	06:10	18:33
TY	Benin, Porto Novo	05:42	18:09
TZ	Mali, Bamako	06:15	19:01
UA	Russia, European,		
	Leningrad (UA1)	01:15	18:55
	Archangel (UA1)	23:42	19:06
	Murmansk (UA1)	24 hrs day	
	Moscow (UA3)	01:14	17:57
	Kuibyshev (UA4)	00:39	16:52
	Rostov (UA6)	01:46	17:10
UA1P	Franz Josef Land	24 hrs day	
UA2	Kaliningrad	02:27	19:00
UA9,Ø	Russia, Asiatic,		
	Novosibirsk (UA9)	22:18	14:51
	Perm (UA9)	23:45	16:56
	Omsk (UA9)	22:56	15:29
	Norilsk (UAØ)	24 hrs day	
	Irkutsk (UAØ)	21:06	13:11
	Vladivostok (UAØ)	19:51	10:46
	Petropavlovsk (UAØ)	17:25	09:37
	Khabarovsk (UAØ)	19:20	10:51
	Krasnoyarsk (UAØ)	21:32	14:17
	Yakutsk (UAØ)	18:17	12:37
	Wrangel Island (UAØ)	24 hrs day	
	Kyzyl (UAØY)	21:48	13:47
UB	Ukraine, Kiev	02:10	17:58
UC	Byelorussia, Minsk	02:05	18:26
UD	Azerbaijan, Baku	01:27	16:05
UF	Georgia, Tbilisi	01:44	16:30
UG	Armenia, Ierevan	01:49	16:27
UH	Turkoman, Ashkhabad	01:00	15:25
UI	Uzbek, Bukhara	00:31	15:06
	Tashkent	00:07	14:50
UJ	Tadzhik, Samarkand	00:21	14:56
	Dushanbe	00:15	14:46
UL	Kazakh, Alma-ata	23:31	14:26
UM	Kirghiz, Frunze	23:41	14:34
UO	Moldavia, Kishinev	02:30	17:51
UP	Lithuania, Vilna	02:10	18:38
UQ	Latvia, Riga	02:01	18:58
UR	Estonia, Tallinn	01:41	19:13
V2	Antigua & Barbuda, St. Johns	09:43	22:43
V3	Belize, Belmopan	11:31	00:32
V4	St. Christopher & Nevis	09:46	22:47
V8	Brunei, Bandar Seri Begawan	22:15	10:37
VE1, CY	Sable Is.	08:37	23:35
VE1, CY	St. Paul Is.	08:25	23:47
VK	Australia, Canberra (VK1)	21:06	07:13
	Sydney (VK2)	20:55	07:07
	Melbourne (VK3)	21:29	07:23
	Brisbane (VK4)	20:35	07:13
	Adelaide (VK5)	21:47	07:55
	Perth (VK6)	23:12	09:33
	Hobart, Tasmania (VK7)	21:33	07:01
	Darwin (VK8)	21:38	09:07
VK	Lord Howe Is.	20:18	06:41
VK9	Christmas Is.	23:15	10:51
VK9	Cocos-Keeling Is.	23:54	11:24
VK9	Mellish Reef	20:06	07:19
VK9	Norfolk Is.	19:38	06:10
VK9	Willis Is.	20:29	07:46
VKØ	Heard Is.	03:04	11:20
VKØ	Macquarie Is.	21:31	05:30
VP2E	Anguilla	09:46	22:50
VP2M	Montserrat	09:45	22:44
VP2V	British Virgin Is., Tortola	09:52	22:57
VP5	Turks & Caicos Islands,		
	Grand Turk	10:13	23:28
VP8	Falkland Islands, Stanley	11:44	20:11

Pref	State/Province/Country/City	Sunrise	Sunset
VP8	So. Georgia Is.	10:31	18:35
VP8	So. Orkney Is.	11:45	18:31
VP8	So. Sandwich Islands, Saunders Is.	10:10	17:35
VP8	So. Shetland Is., King George Is.	12:47	19:11
VP9	Bermuda	09:26	23:24
VQ9	Chagos, Diego Garcia	01:24	13:09
VR6	Pitcairn Is.	15:23	02:10
VS6	Hong Kong	21:49	11:08
VU	India, Bombay	00:41	13:48
	Calcutta	23:33	12:52
	New Delhi	00:06	13:48
	Bangalore	00:32	13:19
VU7	Andaman Islands, Port Blair	23:33	12:16
VU7	Laccadive Is.	00:55	13:33
XE	Mexico, Mexico City (XE1)	12:08	01:16
	Chihuahua (XE2)	12:18	02:01
	Merida (XE3)	11:28	00:42
XF4	Revilla Gigedo	12:58	02:05
XT	Burkina Faso, Onagadougou	05:50	18:35
XU	Kampuchea, Phnom Penh	22:45	11:28
XW	Laos, Viangchan	22:44	11:47
XX9	Macao	21:52	11:10
XZ	Burma, Rangoon	23:12	12:12
Y2-9	German Dem.Rep.(E), Berlin	03:09	19:16
YA	Afghanistan, Kandahar	00:46	14:39
	Kabul	00:25	14:33
YB-YD	Indonesia, Jakarta	23:04	10:53
	Medan	23:22	11:40
	Pontianak	22:45	10:52
	Jayapura	20:43	08:43
YI	Iraq, Baghdad	02:07	16:08
YJ	Vanuatu, Port Vila	19:16	06:29
YK	Syria, Damascus	02:39	16:42
YN, HT	Nicaragua, Managua	11:28	00:12
YO	Romania, Bucharest	02:50	17:53
YS	El Salvador, San Salvador	11:38	00:27
YU	Yugoslavia, Belgrade	03:11	18:17
YV	Venezuela, Caracas	10:14	22:53
YVØ	Aves Is.	09:53	22:48
Z2,ZE	Zimbabwe, Harare	04:26	15:38
ZA	Albania, Tirane	03:25	18:08
ZB	Gibraltar	05:20	19:35
ZC4	British Cyprus	02:50	16:58
ZD7	St. Helena	06:50	18:09
ZD8	Ascension Is.	07:12	18:55
ZD9	Tristan da Cunha	07:56	17:54
ZF	Cayman Is.	10:56	00:05
ZK1	No. Cook Is., Manihiki	17:02	04:38
ZK1	So. Cook Is., Rarotonga	17:15	04:16
ZK2	Niue	17:47	04:56
ZK3	Tokelaus, Atafu	17:46	05:28
ZL	New Zealand, Auckland (ZL1)	19:27	05:26
	Wellington (ZL2)	19:38	05:15
	Christchurch (ZL3)	19:53	05:18
	Dunedin (ZL4)	20:09	05:19
ZL5	Antarctica, Scott Base	24 hrs night	
ZL7	Chatham Is.	19:11	04:32
ZL8	Kermadec Is.	18:42	05:13
ZL9	Auckland & Campbell Is.,		
	Auckland	20:42	05:17
	Campbell Is.	20:39	04:59
ZP	Paraguay, Asuncion	10:33	21:20
ZS	South Africa, Cape Town (ZS1)	05:46	15:58
	Port Elizabeth (ZS2)	05:17	15:29
	Bloemfontein (ZS4)	05:06	15:37
	Durban (ZS5)	04:48	15:17
	Johannesburg (ZS6)	04:52	15:35
ZS2	Prince Edward & Marion Is.,		
	Marion	05:03	14:06
ZS3	Namibia, Windhoek	05:29	16:26
1AØ	S.M.O.M	03:53	18:40
1S	Spratly Is.	22:21	10:55
3A	Monaco	04:07	19:05
3B6	Agalega	02:32	14:07
3B7	St. Brandon	02:28	13:45
3B8	Mauritius	02:44	13:48
3B9	Rodriguez Is.	02:19	13:25
3C	Equatorial Guinea, Bata	05:20	17:33
	Malabo	05:21	17:40

Pref	State/Province/Country/City	Sunrise	Sunset
3CØ	Pagalu Is.	05:42	17:45
3D2	Fiji Is., Suva	18:37	05:48
3D6	Swaziland, Mbabane	04:40	15:23
3V	Tunisia, Tunis	04:15	18:35
3W	Vietnam, Ho Chi Minh City		
	(Saigon)	22:39	11:19
	Hanoi	22:26	11:40
3X	Republic of Guinea, Conakry	06:43	19:19
3Y	Bouvet	07:52	15:53
3Y	Peter I Is.	16:36	19:40
4S	Sri Lanka, Colombo	00:32	13:00
4U	ITU Geneva	04:04	19:18
4U	United Nations Hq.	09:42	00:22
4W	Yemen, Sanaa	02:42	15:36
4X, 4Z	Israel, Jerusalem	02:47	16:43
5A	Libya, Tripoli	04:16	18:15
	Benghazi	03:47	17:44
5B	Cyprus, Nicosia	02:47	16:58
5H	Tanzania, Dar es Salaam	03:35	15:21
5N	Nigeria, Lagos	05:39	18:06
5R	Madagascar, Antananarivo	03:21	14:30
5T	Mauritania, Nouakchott	06:38	19:42
5U	Niger, Niamey	05:34	18:22
5V	Togo, Lome	05:49	18:13
5W	Western Samoa, Apia	17:50	05:16
5X	Uganda, Kampala	03:52	16:00
5Z	Kenya, Nairobi	04:57	17:00
6W	Senegal, Dakar	06:50	19:42
6Y	Jamaica, Kingston	10:41	23:45

Pref	State/Province/Country/City	Sunrise	Sunset
7O	Yemen, People's Dem. Rep.,		
	Aden	02:43	15:29
7P	Lesotho, Maseru	05:00	15:32
7Q	Malawi, Lilongwe	04:08	15:33
	Blantyre	04:06	15:25
7X	Algeria, Algiers	04:45	19:03
8P	Barbados, Bridgetown	09:41	22:28
8Q	Maldive Is.	01:02	13:22
8R	Guyana, Georgetown	09:45	22:12
9G	Ghana, Accra	05:55	18:18
9H	Malta	04:01	18:16
9J	Zambia, Lusaka	04:33	15:53
9K	Kuwait	02:02	15:48
9L	Sierra Leone, Freetown	06:42	19:15
9M2	West Malaysia, Kuala Lumpur	23:11	11:28
9M6,8	East Malaysia, Sabah,		
	Sandakan (9M6)	22:01	10:26
	Sarawak, Kuching (9M8)	22:39	10:51
9N	Nepal, Katmandu	23:35	13:14
9Q	Zaire, Kinshasa	05:08	17:02
	Kisangani	04:21	16:29
	Lubumbashi	04:30	16:02
9U	Burundi, Bujumbura	04:10	16:07
9V	Singapore	23:05	11:16
9X	Rwanda, Kigali	04:05	16:06
9Y	Trinidad & Tobago, Port of		
	Spain	09:52	22:31
J2/A	Abu Ail	02:49	15:40

Pref	State/Province/Country/City	Sunrise	Sunset
VE1	New Brunswick, Saint John	09:15	23:44
	Nova Scotia, Halifax	09:07	23:33
	Prince Edward Island, Charlottetown	09:01	23:35
VE2	Quebec, Montreal	09:44	00:15
	Quebec City	09:31	00:09
VE3	Ontario, London	10:22	00:39
	Ottawa	09:53	00:23
	Sudbury	10:11	00:47
	Toronto	10:12	00:33
VE4	Manitoba, Winnipeg	11:06	02:02
VE5	Saskatchewan, Regina	11:34	02:34
	Saskatoon	11:37	02:47
VE6	Alberta, Calgary	12:10	03:13
	Edmonton	11:59	03:20
VE7	British Columbia, Prince George	12:34	03:59
	Prince Rupert	13:03	04:30
	Vancouver	12:52	03:44
VE8	Northwest Territories, Yellowknife	11:15	04:10
	Resolute	24 hrs day	
VY1	Yukon, Whitehorse	12:50	05:21
VO1	Newfoundland, St. John's	08:15	22:57
VO2	Labrador, Goose Bay	08:27	23:47
W1	Connecticut, Hartford	09:50	00:02
	Maine, Bangor	09:27	23:54
	Portland	09:36	23:57
	Massachusetts, Boston	09:42	23:57
	New Hampshire, Concord	09:42	00:01
	Rhode Island, Providence	09:45	23:57
	Vermont, Montpelier	09:44	00:08
W2	New Jersey, Atlantic City	10:02	00:04
	New York, Albany	09:52	00:09
	Buffalo	10:12	00:30
	New York City	09:57	00:05
	Syracuse	10:01	00:19
W3	Delaware, Wilmington	10:06	00:09
	District of Columbia, Washington	10:14	00:13
	Maryland, Baltimore	10:11	00:12
	Pennsylvania, Harrisburg	10:10	00:16
	Philadelphia	10:04	00:08
	Pittsburgh	10:22	00:28
	Scranton	10:03	00:13
W4	Alabama, Montgomery	11:03	00:38
	Florida, Jacksonville	10:48	00:16
	Miami	10:49	00:03
	Pensacola	11:10	00:38
	Georgia, Atlanta	10:53	00:33
	Savannah	10:43	00:17
	Kentucky, Lexington	10:45	00:41
	Louisville	10:50	00:47
	North Carolina, Charlotte	10:36	00:21
	Raleigh	10:26	00:13
	Wilmington	10:26	00:07
	South Carolina, Columbia	10:39	00:20
	Tennessee, Knoxville	10:47	00:35
	Memphis	11:13	00:58
	Nashville	10:58	00:47
	Virginia, Norfolk	10:15	00:06
	Richmond	10:18	00:12
W5	Arkansas, Little Rock	11:23	01:06
	Louisiana, New Orleans	11:22	00:49
	Shreveport	11:32	01:08
	Mississippi, Jackson	11:19	00:53
	New Mexico, Albuquerque	12:20	02:04
	Oklahoma, Oklahoma City	11:42	01:28
	Texas, Abilene	11:56	01:32
	Amarillo	12:00	01:45
	Dallas	11:44	01:21
	El Paso	12:25	01:58
	San Antonio	11:57	01:22
W6	California, Los Angeles	13:08	02:49
	San Francisco	13:17	03:12
W7	Arizona, Flagstaff	12:40	02:25
	Phoenix	12:44	02:23
	Idaho, Boise	12:40	03:00
	Pocatello	12:27	02:44

Pref	State/Province/Country/City	Sunrise	Sunset
	Montana, Billings	12:03	02:35
	Butte	12:19	02:52
	Great Falls	12:10	02:51
	Nevada, Las Vegas	12:51	02:40
	Reno	13:03	03:06
	Oregon, Portland	13:01	03:31
	Utah, Salt Lake City	12:29	02:37
	Washington, Seattle	12:54	03:35
	Spokane	12:34	03:16
	Wyoming, Cheyenne	12:00	02:09
	Sheridan	12:00	02:27
W8	Michigan, Detroit	10:30	00:45
	Grand Rapids	10:39	00:57
	Sault Ste. Marie	10:25	01:01
	Traverse City	10:34	01:01
	Ohio, Cincinnati	10:43	00:43
	Cleveland	10:27	00:37
	Columbus	10:35	00:39
	West Virginia, Charleston	10:33	00:30
W9	Illinois, Chicago	10:49	01:02
	Indiana, Indianapolis	10:48	00:52
	Wisconsin, Green Bay	10:45	01:10
	Milwaukee	10:48	01:06
W0	Colorado, Denver	12:04	02:07
	Grand Junction	12:19	02:20
	Iowa, Des Moines	11:14	01:25
	Kansas, Pratt	11:43	01:37
	Wichita	11:37	01:32
	Minnesota, Duluth	10:55	01:32
	Minneapolis	11:05	01:32
	Missouri, Columbia	11:14	01:14
	Kansas City	11:23	01:24
	St. Louis	11:07	01:05
	Nebraska, North Platte	11:44	01:53
	Omaha	11:24	01:34
	North Dakota, Fargo	11:13	01:51
	South Dakota, Rapid City	11:46	02:10
A2	Botswana, Gaborone	04:49	15:54
A3	Tonga, Nukualofa	18:09	05:24
A4	Oman, Muscat	01:37	14:44
A5	Bhutan, Thimbu	23:29	12:47
A6	United Arab Emirates, Abu Dhabi	01:54	15:03
A7	Qatar, Ad-Dawhah	02:03	15:15
A9	Bahrein, Al-Manamah	02:06	15:20
AP	Pakistan, Karachi	01:01	14:12
	Islamabad	00:23	14:02
BV	Taiwan, Taipei	21:24	10:35
BY	People's Rep. of China, Beijing	21:18	11:22
	Harbin	20:22	10:54
	Shanghai	21:14	10:45
	Fuzhou	21:31	10:45
	Xian	21:59	11:40
	Chongqing	22:16	11:42
	Chengdu	22:24	11:53
	Lhasa	23:18	12:43
	Urumqi	23:04	13:26
	Kashi	00:00	14:02
C2	Nauru	18:55	07:01
C3	Andorra	04:52	19:07
C5	The Gambia, Banjul	06:52	19:32
C6	Bahamas, Nassau	10:39	23:51
C9	Mozambique, Maputo	04:24	15:26
	Mozambique	03:37	15:08
CE	Chile, Santiago	11:29	22:08
CEØ	Easter Island	13:54	00:52
CEØ	San Felix	11:55	22:56
CEØ	Juan Fernandez	12:02	22:39
CM,CO	Cuba, Havana	11:02	00:08
CN	Morocco, Casablanca	05:46	19:25
CP	Bolivia, La Paz	10:55	22:22
CT	Portugal, Lisbon	05:43	19:41
CT2, CU	Azores, Ponta Delgada	06:51	20:45
CT3	Madeira Islands, Funchal	06:25	20:01
CX	Uruguay, Montevideo	10:33	21:07
D2	Angola, Luanda	05:19	17:06
D4	Cape Verde, Praia	07:18	20:01

Pref	State/Province/Country/City	Sunrise	Sunset
D6	Comoros, Moroni	03:21	15:00
DA-DL	Germany, Fed. Rep. of (W),		
	Bonn	04:07	19:08
DU	Philippines, Manilla	21:40	10:23
EA	Spain, Madrid	05:17	19:23
EA6	Balearic Is., Palma	04:54	18:56
EA8	Canary Is., Las Palmas	06:22	19:43
EA9	Ceuta & Melilla, Ceuta	05:33	19:20
	Melilla	05:25	19:10
EI	Ireland, Dublin	04:51	20:10
EL	Liberia, Monrovia	06:37	18:59
EP	Iran, Tehran	01:44	15:32
ET	Ethiopia, Addis Ababa	03:16	15:45
	Asmera	03:08	15:52
F	France, Paris	04:32	19:21
FG	Guadeloupe	09:49	22:35
FG, FS	St. Martin	09:52	22:44
FH	Mayotte	03:16	14:52
FK	New Caledonia, Noumia	19:23	06:35
FM	Martinique	09:48	22:31
FO	Clipperton	13:06	01:38
FO	Fr. Polynesia, Tahiti	17:01	04:25
FP	St. Pierre & Miquelon, St. Pierre	08:31	23:08
FR	Glorioso	03:06	14:46
FR	Juan de Nova	03:31	14:57
	Europa	03:48	15:00
FR	Reunion	02:45	14:00
FR	Tromelin	02:43	14:12
FT8W	Crozet	03:43	13:31
FT8X	Kerguelen	02:43	12:14
FT8Y	Antarctica, Dumont D'Urville	23:26	06:05
FT8Z	Amsterdam & St. Paul Is.,		
	Amsterdam	01:43	12:07
FW	Wallis & Futuna Is., Wallis	18:03	05:38
FY	Fr. Guiana, Cayenne	09:25	21:44
G	England, London	04:32	19:39
GD	Isle of Man	04:40	20:07
GI	Northern Ireland, Belfast	04:44	20:14
GJ	Jersey	04:48	19:40
GM	Scotland, Glasgow	04:33	20:12
	Aberdeen	04:17	20:10
GU	Guernsey	04:49	19:43
GW	Wales, Cardiff	04:45	19:51
H4	Solomon Islands, Honiara	19:33	07:18
HA	Hungary, Budapest	03:28	18:10
HB	Switzerland, Bern	04:16	18:55
HBØ	Liechtenstein	04:07	18:47
HC	Ecuador, Quito	11:14	23:21
HC8	Galapagos Is.	12:04	00:10
HH	Haiti, Port-Au-Prince	10:28	23:21
HI	Dominican Republic, Santo		
	Domingo	10:19	23:12
HK	Colombia, Bogota	10:53	23:11
HKØ	Malpelo Is.	11:21	23:38
HKØ	San Adreas	11:13	23:51
HL, HM	Korea, Seoul	20:40	10:34
HP	Panama, Panama	11:09	23:38
HR	Honduras, Tegucigalpa	11:33	00:15
HS	Thailand, Bangkok	23:03	11:44
HV	Vatican City	04:09	18:22
HZ, 7Z	Saudi Arabia, Dharan	02:08	15:23
	Mecca	02:56	15:57
	Riyadh	02:24	15:33
I	Italy, Rome	04:09	18:22
	Trieste	03:54	18:26
	Sicily	04:12	18:06
IS	Sardinia, Cagliari	04:28	18:29
J2	Djibouti, Djibouti	02:55	15:30
J3	Grenada	09:55	22:30
J5	Guinea-Bissau, Bissau	06:50	19:26
J6	St. Lucia	09:49	22:30
J7	Dominica	09:48	22:33
J8	St. Vincent	09:51	22:30
JA-JS	Japan, Tokyo	19:53	09:39
	Nagasaki	20:38	10:14
	Sapporo	19:31	09:49
JD	Minami Torishima	19:15	08:24
JD	Ogasawara, Kazan Is.	20:02	09:21
JT	Mongolia, Ulan Bator	21:36	12:20
JW	Svalbard, Spitsbergen	24 hrs day	
JX	Jan Mayen	02:19	22:57

Pref	State/Province/Country/City	Sunrise	Sunset
JY	Jordan, Amman	02:55	16:28
KC4	Antarctica, Byrd Station	24 hrs night	
	McMurdo Sound	24 hrs night	
	Palmer Station	12:47	19:56
KC6	Micronesia, Ponape	19:20	07:44
KC6	Belau, Yap	20:38	09:07
	Koror	20:55	09:19
KG4	Guantanamo Bay	10:38	23:34
KH1	Baker, Howland Is.	17:45	05:53
KH2	Guam, Agana	20:06	08:46
KH3	Johnston Is.	16:55	05:44
KH4	Midway Is.	17:14	06:35
KH5	Palmyra Is.	16:43	05:04
KH5K	Kingman Reef	16:44	05:09
KH6	Hawaii, Hilo	15:58	04:54
	Honolulu	16:07	05:07
KH7	Kure Is.	17:18	06:40
KH8	American Samoa, Pago Pago	17:42	05:15
KH9	Wake Is.	18:31	07:26
KHØ	Mariana Is., Saipan	20:00	08:44
KL7	Alaska, Adak	16:17	07:26
	Anchorage	13:47	06:24
	Fairbanks	13:11	06:43
	Juneau	13:01	05:05
	Nome	14:23	07:50
KP1	Navassa Is.	10:39	23:32
KP2	Virgin Islands, Charlotte Amalie	09:59	22:51
KP4	Puerto Rico, San Juan	10:04	22:56
KP5	Desecheo Is.	10:09	23:01
KX6	Marshall Islands, Kwajalein	18:42	07:10
LA-LJ	Norway, Oslo	03:11	19:34
LU	Argentina, Buenos Aires	10:42	21:16
LX	Luxembourg	04:13	19:07
LZ	Bulgaria, Sofia	03:24	17:40
OA	Peru, Lima	11:25	23:03
OD	Lebanon, Beirut	02:53	16:33
OE	Austria, Vienna	03:37	18:23
OH	Finland, Helsinki	02:13	18:38
OHØ	Aland Is.	02:33	18:58
OJ0	Market Reef	02:36	19:02
OK	Czechoslovakia, Prague	03:39	18:36
ON	Belgium, Brussels	04:16	19:19
OX, XP	Greenland, Godthaab	06:51	00:13
	Thule	24 hrs day	
OY	Faroe Islands, Torshavn	04:08	20:56
OZ	Denmark, Copenhagen	03:25	19:04
P2	Papua New Guinea, Madang	20:26	08:20
	Port Moresby	20:25	08:09
PA-PI	Netherlands, Amsterdam	04:09	19:22
PJ, P4	Netherlands Antilles, Willemstad	10:23	22:59
PJ5-8	St. Maarten and Saba,		
	St. Maarten	09:53	22:43
PY	Brazil, Brasilia	09:32	21:01
	Rio De Janeiro	09:23	20:33
	Natal	08:30	20:22
	Manaus	10:06	22:06
	Porto Alegre	10:05	20:55
PYØ	Fernando De Noronha	08:16	20:14
PYØ	St. Peter & St. Paul Rocks	07:58	20:08
PYØ	Trindade & Martin Vaz Is.,		
	Trindade	08:24	19:41
PZ	Suriname, Paramaribo	09:36	21:56
S2	Bangladesh, Dacca	23:30	12:37
S7	Seychelles, Victoria	02:25	14:21
S9	Sao Tome	05:35	17:42
SJ-SM	Sweden, Stockholm	02:46	19:00
SP	Poland, Krakow	03:17	18:14
	Warsaw	03:06	18:17
ST	Sudan, Khartoum	03:33	16:18
STØ	Southern Sudan, Juba	03:49	16:08
SU	Egypt, Cairo	03:16	16:43
SV	Greece, Athens	03:33	17:28
SV5	Dodecanese, Rhodes	03:18	17:07
SV9	Crete	03:32	17:17
SV/A	Mount Athos	03:26	17:31
T2	Tuvalu, Funafuti	18:18	06:04
T30	West Kiribati, Bonriki	18:27	06:38
T31	Central Kiribati, Kanton	17:32	05:32
T32	East Kiribati, Christmas Is.	16:29	04:41
T5	Somalia, Mogadishu	02:58	15:10
T7	San Marino	04:05	18:27

Pref	State/Province/Country/City	Sunrise	Sunset
TA	Turkey, Ankara	02:52	16:56
	Istanbul	03:05	17:14
TF	Iceland, Reykjavik	04:53	22:13
TG	Guatemala, Guatemala City	11:46	00:29
TI	Costa Rica, San Jose	11:26	23:57
TI9	Cocos Is.	11:43	00:03
TJ	Cameroon, Yaounde	05:11	17:27
TK	Corsica	04:23	18:36
TL	Central African Republic, Bangui	04:42	17:00
TN	Congo, Brazzaville	05:06	17:02
TR	Gabon, Libreville	05:23	17:31
TT	Chad, N'Djamena	04:47	17:23
TU	Ivory Coast, Abidjan	06:12	18:31
TY	Benin, Porto Novo	05:44	18:06
TZ	Mali, Bamako	06:18	18:56
UA	Russia, European,		
	Leningrad (UA1)	01:53	18:15
	Archangel (UA1)	00:39	18:08
	Murmansk (UA1)	00:16	19:29
	Moscow (UA3)	01:45	17:25
	Kuibyshev (UA4)	01:05	16:24
	Rostov (UA6)	02:07	16:48
UA1P	Franz Josef Land	24 hrs day	
UA2	Kaliningrad	02:57	18:30
UA9,Ø	Russia, Asiatic,		
	Novosibirsk (UA9)	22:47	14:20
	Perm (UA9)	00:20	16:20
	Omsk (UA9)	23:25	14:58
	Norilsk (UAØ)	20:31	15:55
	Irkutsk (UAØ)	21:32	12:44
	Vladivostok (UAØ)	20:08	10:27
	Petropavlovsk (UAØ)	17:52	09:09
	Khabarovsk (UAØ)	19:41	10:28
	Krasnoyarsk (UAØ)	22:03	13:45
	Yakutsk (UAØ)	19:02	11:50
	Wrangel Island (UAØ)	13:44	10:22
	Kyzyl (UAØY)	22:13	13:21
UB	Ukraine, Kiev	02:34	17:33
UC	Byelorussia, Minsk	02:33	17:57
UD	Azerbaijan, Baku	01:43	15:49
UF	Georgia, Tbilisi	02:00	16:12
UG	Armenia, Ierevan	02:05	16:10
UH	Turkoman, Ashkhabad	01:14	15:10
UI	Uzbek, Bukhara	00:46	14:49
	Tashkent	00:23	14:33
UJ	Tadzhik, Samarkand	00:37	14:40
	Dushanbe	00:30	14:30
UL	Kazakh, Alma-ata	23:48	14:07
UM	Kirghiz, Frunze	23:58	14:16
UO	Moldavia, Kishinev	02:51	17:29
UP	Lithuania, Vilna	02:39	18:08
UQ	Latvia, Riga	02:33	18:24
UR	Estonia, Tallinn	02:18	18:34
V2	Antigua & Barbuda, St. Johns	09:48	22:37
V3	Belize, Belmopan	11:36	00:25
V4	St. Christopher & Nevis	09:51	22:40
V8	Brunei, Bandar Seri Begawan	22:16	10:35
VE1, CY	Sable Is.	08:54	23:16
VE1, CY	St. Paul Is.	08:46	23:26
VK	Australia, Canberra (VK1)	20:53	07:25
	Sydney (VK2)	20:42	07:19
	Melbourne (VK3)	21:14	07:37
	Brisbane (VK4)	20:25	07:22
	Adelaide (VK5)	21:34	08:08
	Perth (VK6)	23:00	09:44
	Hobart, Tasmania (VK7)	21:15	07:17
	Darwin (VK8)	21:33	09:10
VK	Lord Howe Is.	20:07	06:51
VK9	Christmas Is.	23:11	10:53
VK9	Cocos-Keeling Is.	23:49	11:27
VK9	Mellish Reef	20:00	07:24
VK9	Norfolk Is.	19:27	06:20
VK9	Willis Is.	20:23	07:51
VKØ	Heard Is.	02:38	11:46
VK0	Macquarie Is.	21:02	05:58
VP2E	Anguilla	09:51	22:43
VP2M	Montserrat	09:50	22:38
VP2V	British Virgin Is., Tortola	09:57	22:50
VP5	Turks & Caicos Islands,		
	Grand Turk	10:20	23:20
VP8	Falkland Islands, Stanley	11:19	20:35

Pref	State/Province/Country/City	Sunrise	Sunset
VP8	So. Georgia Is.	10:03	19:02
VP8	So. Orkney Is.	11:06	19:09
VP8	So. Sandwich Islands, Saunders Is.	09:37	18:07
VP8	So. Shetland Is., King George Is.	12:05	19:52
VP9	Bermuda	09:37	23:11
VQ9	Chagos, Diego Garcia	01:21	13:11
VR6	Pitcairn Is.	15:14	02:18
VS6	Hong Kong	21:56	11:00
VU	India, Bombay	00:47	13:41
	Calcutta	23:40	12:44
	New Delhi	00:15	13:38
	Bangalore	00:36	13:14
VU7	Andaman Islands, Port Blair	23:36	12:12
VU7	Laccadive Is.	00:58	13:29
XE	Mexico, Mexico City (XE1)	12:14	01:09
	Chihuahua (XE2)	12:28	01:51
	Merida (XE3)	11:34	00:34
XF4	Revilla Gigedo	13:04	01:58
XT	Burkina Faso, Onagadougou	05:53	18:30
XU	Kampuchea, Phnom Penh	22:48	11:24
XW	Laos, Viangchan	22:49	11:41
XX9	Macao	21:59	11:02
XZ	Burma, Rangoon	23:17	12:05
Y2-9	German Dem.Rep.(E), Berlin	03:35	18:48
YA	Afghanistan, Kandahar	00:57	14:27
	Kabul	00:37	14:19
YB-YD	Indonesia, Jakarta	23:02	10:54
	Medan	23:23	11:38
	Pontianak	22:45	10:52
	Jayapura	20:42	08:43
YI	Iraq, Baghdad	02:19	15:56
YJ	Vanuatu, Port Vila	19:10	06:34
YK	Syria, Damascus	02:51	16:29
YN, HT	Nicaragua, Managua	11:31	00:07
YO	Romania, Bucharest	03:09	17:33
YS	El Salvador, San Salvador	11:42	00:22
YU	Yugoslavia, Belgrade	03:30	17:57
YV	Venezuela, Caracas	10:17	22:49
YVØ	Aves Is.	09:57	22:43
Z2,ZE	Zimbabwe, Harare	04:19	15:43
ZA	Albania, Tirane	03:41	17:51
ZB	Gibraltar	05:33	19:21
ZC4	British Cyprus	03:02	16:45
ZD7	St. Helena	06:45	18:13
ZD8	Ascension Is.	07:09	18:57
ZD9	Tristan da Cunha	07:42	18:07
ZF	Cayman Is.	11:02	23:58
ZK1	No. Cook Is., Manihiki	16:58	04:40
ZK1	So. Cook Is., Rarotonga	17:07	04:22
ZK2	Niue	17:41	05:01
ZK3	Tokelaus, Atafu	17:43	05:30
ZL	New Zealand, Auckland (ZL1)	19:13	05:39
	Wellington (ZL2)	19:21	05:31
	Christchurch (ZL3)	19:35	05:35
	Dunedin (ZL4)	19:49	05:38
ZL5	Antarctica, Scott Base	24 hrs night	
ZL7	Chatham Is.	18:53	04:50
ZL8	Kermadec Is.	18:31	05:23
ZL9	Auckland & Campbell Is.,		
	Auckland	20:18	05:41
	Campbell Is.	20:13	05:24
ZP	Paraguay, Asuncion	10:24	21:28
ZS	South Africa, Cape Town (ZS1)	05:33	16:10
	Port Elizabeth (ZS2)	05:04	15:41
	Bloemfontein (ZS4)	04:55	15:47
	Durban (ZS5)	04:37	15:27
	Johannesburg (ZS6)	04:42	15:43
ZS2	Prince Edward & Marion Is.,		
	Marion	04:42	14:26
ZS3	Namibia, Windhoek	05:21	16:32
1AØ	S.M.O.M	04:09	18:22
1S	Spratly Is.	22:24	10:52
3A	Monaco	04:25	18:46
3B6	Agalega	02:28	14:10
3B7	St. Brandon	02:22	13:50
3B8	Mauritius	02:37	13:54
3B9	Rodriguez Is.	02:12	13:31
3C	Equatorial Guinea, Bata	05:20	17:32
	Malabo	05:22	17:38

Pref	State/Province/Country/City	Sunrise	Sunset
3CØ	Pagalu Is.	05:41	17:45
3D2	Fiji Is., Suva	18:30	05:53
3D6	Swaziland, Mbabane	04:31	15:31
3V	Tunisia, Tunis	04:29	18:20
3W	Vietnam, Ho Chi Minh City		
	(Saigon)	22:42	11:15
	Hanoi	22:32	11:32
3X	Republic of Guinea, Conakry	06:45	19:15
3Y	Bouvet	07:23	16:20
3Y	Peter I Is.	15:12	21:04
4S	Sri Lanka, Colombo	00:34	12:57
4U	ITU Geneva	04:23	18:58
4U	United Nations Hq.	09:57	00:05
4W	Yemen, Sanaa	02:46	15:31
4X, 4Z	Israel, Jerusalem	02:58	16:31
5A	Libya, Tripoli	04:28	18:03
	Benghazi	03:58	17:32
5B	Cyprus, Nicosia	02:59	16:44
5H	Tanzania, Dar es Salaam	03:32	15:22
5N	Nigeria, Lagos	05:40	18:03
5R	Madagascar, Antananarivo	03:15	14:36
5T	Mauritania, Nouakchott	06:43	19:35
5U	Niger, Niamey	05:37	18:17
5V	Togo, Lome	05:50	18:11
5W	Western Samoa, Apia	17:45	05:20
5X	Uganda, Kampala	03:51	15:59
5Z	Kenya, Nairobi	04:56	17:00
6W	Senegal, Dakar	06:54	19:37
6Y	Jamaica, Kingston	10:47	23:38

Pref	State/Province/Country/City	Sunrise	Sunset
7O	Yemen, People's Dem. Rep.,		
	Aden	02:46	15:24
7P	Lesotho, Maseru	04:49	15:41
7Q	Malawi, Lilongwe	04:03	15:37
	Blantyre	04:01	15:30
7X	Algeria, Algiers	04:58	18:49
8P	Barbados, Bridgetown	09:44	22:23
8Q	Maldive Is.	01:03	13:20
8R	Guyana, Georgetown	09:46	22:10
9G	Ghana, Accra	05:56	18:16
9H	Malta	04:14	18:02
9J	Zambia, Lusaka	04:27	15:57
9K	Kuwait	02:12	15:37
9L	Sierra Leone, Freetown	06:44	19:12
9M2	West Malaysia, Kuala Lumpur	23:12	11:26
9M6,8	East Malaysia, Sabah,		
	Sandakan (9M6)	22:03	10:23
	Sarawak, Kuching (9M8)	22:39	10:49
9N	Nepal, Katmandu	23:44	13:04
9Q	Zaire, Kinshasa	05:06	17:02
	Kisangani	04:20	16:29
	Lubumbashi	04:26	16:05
9U	Burundi, Bujumbura	04:08	16:08
9V	Singapore	23:05	11:15
9X	Rwanda, Kigali	04:04	16:06
9Y	Trinidad & Tobago, Port of		
	Spain	09:54	22:27
J2/A	Abu Ail	02:53	15:35

Pref	State/Province/Country/City	Sunrise	Sunset
VE1	New Brunswick, Saint John	09:32	23:22
	Nova Scotia, Halifax	09:23	23:11
	Prince Edward Island,		
	Charlottetown	09:18	23:12
VE2	Quebec, Montreal	10:01	23:53
	Quebec City	09:49	23:46
VE3	Ontario, London	10:37	00:19
	Ottawa	10:10	00:01
	Sudbury	10:29	00:24
	Toronto	10:28	00:13
VE4	Manitoba, Winnipeg	11:26	01:36
VE5	Saskatchewan, Regina	11:55	02:07
	Saskatoon	11:59	02:20
VE6	Alberta, Calgary	12:32	02:46
	Edmonton	12:23	02:50
VE7	British Columbia, Prince George	12:59	03:29
	Prince Rupert	13:28	04:00
	Vancouver	13:12	03:19
VE8	Northwest Territories,		
	Yellowknife	11:54	03:26
	Resolute	08:33	04:13
VY1	Yukon, Whitehorse	13:25	04:41
VO1	Newfoundland, St. John's	08:34	22:33
VO2	Labrador, Goose Bay	08:51	23:17
W1	Connecticut, Hartford	10:04	23:43
	Maine, Bangor	09:44	23:32
	Portland	09:52	23:36
	Massachusetts, Boston	09:57	23:37
	New Hampshire, Concord	09:57	23:40
	Rhode Island, Providence	09:59	23:37
	Vermont, Montpelier	10:00	23:47
W2	New Jersey, Atlantic City	10:15	23:46
	New York, Albany	10:07	23:48
	Buffalo	10:27	00:09
	New York City	10:11	23:46
	Syracuse	10:16	23:59
W3	Delaware, Wilmington	10:19	23:50
	District of Columbia,		
	Washington	10:26	23:55
	Maryland, Baltimore	10:24	23:54
	Pennsylvania, Harrisburg	10:24	23:57
	Philadelphia	10:17	23:50
	Pittsburgh	10:36	00:09
	Scranton	10:17	23:54
W4	Alabama, Montgomery	11:12	00:23
	Florida, Jacksonville	10:57	00:02
	Miami	10:56	23:51
	Pensacola	11:18	00:25
	Georgia, Atlanta	11:03	00:18
	Savannah	10:52	00:02
	Kentucky, Lexington	10:58	00:24
	Louisville	11:02	00:29
	North Carolina, Charlotte	10:47	00:05
	Raleigh	10:37	23:57
	Wilmington	10:36	23:52
	South Carolina, Columbia	10:49	00:04
	Tennessee, Knoxville	10:58	00:19
	Memphis	11:24	00:42
	Nashville	11:09	00:30
	Virginia, Norfolk	10:26	23:49
	Richmond	10:30	23:55
W5	Arkansas, Little Rock	11:33	00:50
	Louisiana, New Orleans	11:31	00:36
	Shreveport	11:42	00:53
	Mississippi, Jackson	11:28	00:39
	New Mexico, Albuquerque	12:30	01:49
	Oklahoma, Oklahoma City	11:53	01:12
	Texas, Abilene	12:06	01:17
	Amarillo	12:11	01:29
	Dallas	11:54	01:06
	El Paso	12:34	01:43
	San Antonio	12:05	01:09
W6	California, Los Angeles	13:18	02:33
	San Francisco	13:29	02:55
W7	Arizona, Flagstaff	12:50	02:09
	Phoenix	12:54	02:08
	Idaho, Boise	12:55	02:40
	Pocatello	12:42	02:24

Pref	State/Province/Country/City	Sunrise	Sunset
	Montana, Billings	12:20	02:13
	Butte	12:36	02:29
	Great Falls	12:28	02:27
	Nevada, Las Vegas	13:03	02:24
	Reno	13:16	02:47
	Oregon, Portland	13:18	03:09
	Utah, Salt Lake City	12:43	02:18
	Washington, Seattle	13:12	03:12
	Spokane	12:52	02:52
	Wyoming, Cheyenne	12:14	01:50
	Sheridan	12:16	02:05
W8	Michigan, Detroit	10:45	00:25
	Grand Rapids	10:54	00:37
	Sault Ste. Marie	10:43	00:38
	Traverse City	10:51	00:39
	Ohio, Cincinnati	10:56	00:25
	Cleveland	10:41	00:18
	Columbus	10:49	00:21
	West Virginia, Charleston	10:45	00:13
W9	Illinois, Chicago	11:04	00:42
	Indiana, Indianapolis	11:02	00:33
	Wisconsin, Green Bay	11:01	00:48
	Milwaukee	11:03	00:45
WØ	Colorado, Denver	12:17	01:48
	Grand Junction	12:32	02:02
	Iowa, Des Moines	11:28	01:06
	Kansas, Pratt	11:55	01:20
	Wichita	11:49	01:15
	Minnesota, Duluth	11:13	01:09
	Minneapolis	11:21	01:11
	Missouri, Columbia	11:27	00:57
	Kansas City	11:36	01:06
	St. Louis	11:19	00:48
	Nebraska, North Platte	11:58	01:34
	Omaha	11:38	01:15
	North Dakota, Fargo	11:32	01:28
	South Dakota, Rapid City	12:02	01:49
A2	Botswana, Gaborone	04:38	16:00
A3	Tonga, Nukualofa	17:59	05:28
A4	Oman, Muscat	01:43	14:34
A5	Bhutan, Thimbu	23:36	12:34
A6	United Arab Emirates, Abu		
	Dhabi	02:00	14:52
A7	Qatar, Ad-Dawhah	02:10	15:04
A9	Bahrein, Al-Manamah	02:12	15:08
AP	Pakistan, Karachi	01:08	14:01
	Islamabad	00:33	13:47
BV	Taiwan, Taipei	21:30	10:24
BY	People's Rep. of China, Beijing	21:31	11:03
	Harbin	20:40	10:32
	Shanghai	21:23	10:31
	Fuzhou	21:37	10:34
	Xian	22:09	11:25
	Chongqing	22:24	11:29
	Chengdu	22:33	11:40
	Lhasa	23:26	12:30
	Urumqi	23:20	13:05
	Kashi	00:13	13:44
C2	Nauru	18:52	06:58
C3	Andorra	05:06	18:47
C5	The Gambia, Banjul	06:54	19:25
C6	Bahamas, Nassau	10:45	23:39
C9	Mozambique, Maputo	04:13	15:32
	Mozambique	03:30	15:10
CE	Chile, Santiago	11:15	22:17
CEØ	Easter Island	13:42	00:59
CEØ	San Felix	11:44	23:02
CEØ	Juan Fernandez	11:47	22:49
CM,CO	Cuba, Havana	11:07	23:57
CN	Morocco, Casablanca	05:56	19:10
CP	Bolivia, La Paz	10:47	22:25
CT	Portugal, Lisbon	05:55	19:24
CT2, CU	Azores, Ponta Delgada	07:03	20:28
CT3	Madeira Islands, Funchal	06:35	19:46
CX	Uruguay, Montevideo	10:18	21:17
D2	Angola, Luanda	05:14	17:06
D4	Cape Verde, Praia	07:20	19:53

Pref	State/Province/Country/City	Sunrise	Sunset
D6	Comoros, Moroni	03:15	15:01
DA-DL	Germany, Fed. Rep. of (W),		
	Bonn	04:28	18:41
DU	Philippines, Manilla	21:42	10:15
EA	Spain, Madrid	05:31	19:04
EA6	Balearic Is., Palma	05:07	18:38
EA8	Canary Is., Las Palmas	06:29	19:31
EA9	Ceuta & Melilla, Ceuta	05:44	19:04
	Melilla	05:35	18:54
EI	Ireland, Dublin	05:15	19:41
EL	Liberia, Monrovia	06:37	18:55
EP	Iran, Tehran	01:55	15:15
ET	Ethiopia, Addis Ababa	03:17	15:39
	Asmera	03:10	15:44
F	France, Paris	04:51	18:56
FG	Guadeloupe	09:52	22:27
FG, FS	St. Martin	09:55	22:35
FH	Mayotte	03:09	14:54
FK	New Caledonia, Noumia	19:13	06:40
FM	Martinique	09:50	22:23
FO	Clipperton	13:07	01:32
FO	Fr. Polynesia, Tahiti	16:53	04:29
FP	St. Pierre & Miquelon, St. Pierre	08:49	22:45
FR	Glorioso	03:00	14:47
FR	Juan de Nova	03:23	15:00
	Europa	03:38	15:05
FR	Reunion	02:36	14:05
FR	Tromelin	02:36	14:15
FT8W	Crozet	03:21	13:48
FT8X	Kerguelen	02:19	12:33
FT8Y	Antarctica, Dumont D'Urville	22:32	06:53
FT8Z	Amsterdam & St. Paul Is.,		
	Amsterdam	01:26	12:18
FW	Wallis & Futuna Is., Wallis	17:56	05:40
FY	Fr. Guiana, Cayenne	09:24	21:40
G	England, London	04:55	19:12
GD	Isle of Man	05:05	19:37
GI	Northern Ireland, Belfast	05:10	19:43
GJ	Jersey	05:08	19:15
GM	Scotland, Glasgow	05:00	19:40
	Aberdeen	04:46	19:36
GU	Guernsey	05:10	19:17
GW	Wales, Cardiff	05:07	19:24
H4	Solomon Islands, Honiara	19:27	07:18
HA	Hungary, Budapest	03:47	17:46
HB	Switzerland, Bern	04:34	18:31
HB0	Liechtenstein	04:25	18:23
HC	Ecuador, Quito	11:11	23:18
HC8	Galapagos Is.	12:02	00:08
HH	Haiti, Port-Au-Prince	10:32	23:12
HI	Dominican Republic, Santo		
	Domingo	10:23	23:03
HK	Colombia, Bogota	10:52	23:06
HK0	Malpelo Is.	11:20	23:34
HK0	San Adreas	11:15	23:44
HL, HM	Korea, Seoul	20:52	10:17
HP	Panama, Panama	11:09	23:32
HR	Honduras, Tegucigalpa	11:36	00:07
HS	Thailand, Bangkok	23:05	11:36
HV	Vatican City	04:23	18:02
HZ, 7Z	Saudi Arabia, Dharan	02:14	15:11
	Mecca	03:00	15:47
	Riyadh	02:29	15:22
I	Italy, Rome	04:23	18:02
	Trieste	04:11	18:04
	Sicily	04:24	17:49
IS	Sardinia, Cagliari	04:41	18:11
J2	Djibouti, Djibouti	02:56	15:24
J3	Grenada	09:56	22:24
J5	Guinea-Bissau, Bissau	06:51	19:19
J6	St. Lucia	09:51	22:22
J7	Dominica	09:51	22:25
J8	St. Vincent	09:53	22:23
JA-JS	Japan, Tokyo	20:04	09:23
	Nagasaki	20:47	09:59
	Sapporo	19:46	09:28
JD	Minami Torishima	19:21	08:13
JD	Ogasawara, Kazan Is.	20:09	09:08
JT	Mongolia, Ulan Bator	21:55	11:55
JW	Svalbard, Spitsbergen	24 hrs day	
JX	Jan Mayen	03:47	21:25

Pref	State/Province/Country/City	Sunrise	Sunset
JY	Jordan, Amman	03:04	16:14
KC4	Antarctica, Byrd Station	24 hrs night	
	McMurdo Sound	23:22	02:29
	Palmer Station	11:59	20:39
KC6	Micronesia, Ponape	19:20	07:39
KC6	Belau, Yap	20:38	09:02
	Koror	20:55	09:14
KG4	Guantanamo Bay	10:42	23:25
KH1	Baker, Howland Is.	17:43	05:51
KH2	Guam, Agana	20:08	08:39
KH3	Johnston Is.	16:58	05:35
KH4	Midway Is.	17:22	06:23
KH5	Palmyra Is.	16:43	05:00
KH5K	Kingman Reef	16:44	05:04
KH6	Hawaii, Hilo	16:02	04:44
	Honolulu	16:11	04:57
KH7	Kure Is.	17:26	06:27
KH8	American Samoa, Pago Pago	17:35	05:17
KH9	Wake Is.	18:35	07:17
KH0	Mariana Is., Saipan	20:03	08:36
KL7	Alaska, Adak	16:40	06:58
	Anchorage	14:23	05:43
	Fairbanks	13:56	05:52
	Juneau	13:32	04:29
	Nome	15:08	07:01
KP1	Navassa Is.	10:43	23:23
KP2	Virgin Islands, Charlotte Amalie	10:02	22:42
KP4	Puerto Rico, San Juan	10:07	22:48
KP5	Desecheo Is.	10:13	22:53
KX6	Marshall Islands, Kwajalein	18:42	07:05
LA-LJ	Norway, Oslo	03:45	18:55
LU	Argentina, Buenos Aires	10:26	21:26
LX	Luxembourg	04:34	18:42
LZ	Bulgaria, Sofia	03:39	17:20
OA	Peru, Lima	11:18	23:04
OD	Lebanon, Beirut	03:03	16:18
OE	Austria, Vienna	03:56	17:59
OH	Finland, Helsinki	02:47	17:59
OH0	Aland Is.	03:07	18:19
OJ0	Market Reef	03:10	18:23
OK	Czechoslovakia, Prague	04:00	18:10
ON	Belgium, Brussels	04:38	18:52
OX, XP	Greenland, Godthaab	07:35	23:24
	Thule	05:36	03:40
OY	Faroe Islands, Torshavn	04:46	20:13
OZ	Denmark, Copenhagen	03:52	18:32
P2	Papua New Guinea, Madang	20:21	08:19
	Port Moresby	20:19	08:10
PA-PI	Netherlands, Amsterdam	04:32	18:54
PJ, P4	Netherlands Antilles, Willemstad	10:25	22:53
PJ5-8	St. Maarten and Saba,		
	St. Maarten	09:56	22:35
PY	Brazil, Brasilia	09:25	21:04
	Rio De Janeiro	09:13	20:38
	Natal	08:25	20:22
	Manaus	10:03	22:04
	Porto Alegre	09:52	21:03
PY0	Fernando De Noronha	08:12	20:12
PY0	St. Peter & St. Paul Rocks	07:56	20:05
PY0	Trindade & Martin Vaz Is.,		
	Trindade	08:15	19:45
PZ	Suriname, Paramaribo	09:35	21:52
S2	Bangladesh, Dacca	23:36	12:27
S7	Seychelles, Victoria	02:21	14:20
S9	Sao Tome	05:32	17:40
SJ-SM	Sweden, Stockholm	03:18	18:23
SP	Poland, Krakow	03:38	17:48
	Warsaw	03:28	17:49
ST	Sudan, Khartoum	03:35	16:10
ST0	Southern Sudan, Juba	03:49	16:04
SU	Egypt, Cairo	03:25	16:30
SV	Greece, Athens	03:45	17:11
SV5	Dodecanese, Rhodes	03:29	16:51
SV9	Crete	03:42	17:01
SV/A	Mount Athos	03:39	17:12
T2	Tuvalu, Funafuti	18:12	06:04
T30	West Kiribati, Bonriki	18:25	06:35
T31	Central Kiribati, Kanton	17:28	05:30
T32	East Kiribati, Christmas Is.	16:27	04:37
T5	Somalia, Mogadishu	02:56	15:06
T7	San Marino	04:21	18:06

Pref	State/Province/Country/City	Sunrise	Sunset
TA	Turkey, Ankara	03:05	16:37
	Istanbul	03:19	16:55
TF	Iceland, Reykjavik	05:37	21:25
TG	Guatemala, Guatemala City	11:48	00:21
TI	Costa Rica, San Jose	11:27	23:51
TI9	Cocos Is.	11:42	23:59
TJ	Cameroon, Yaounde	05:10	17:23
TK	Corsica	04:37	18:16
TL	Central African Republic, Bangui	04:41	16:56
TN	Congo, Brazzaville	05:02	17:01
TR	Gabon, Libreville	05:21	17:28
TT	Chad, N'Djamena	04:49	17:17
TU	Ivory Coast, Abidjan	06:11	18:27
TY	Benin, Porto Novo	05:43	18:01
TZ	Mali, Bamako	06:20	18:49
UA	Russia, European,		
	Leningrad (UA1)	02:27	17:36
	Archangel (UA1)	01:24	17:18
	Murmansk (UA1)	01:22	18:18
	Moscow (UA3)	02:12	16:53
	Kuibyshev (UA4)	01:29	15:55
	Rostov (UA6)	02:25	16:24
UA1P	Franz Josef Land	24 hrs day	
UA2	Kaliningrad	03:23	17:59
UA9,Ø	Russia, Asiatic,		
	Novosibirsk (UA9)	23:13	13:49
	Perm (UA9)	00:50	15:45
	Omsk (UA9)	23:51	14:27
	Norilsk (UAØ)	21:39	14:41
	Irkutsk (UAØ)	21:55	12:16
	Vladivostok (UAØ)	20:24	10:07
	Petropavlovsk (UAØ)	18:16	08:40
	Khabarovsk (UAØ)	20:01	10:04
	Krasnoyarsk (UAØ)	22:31	13:12
	Yakutsk (UAØ)	19:40	11:07
	Wrangel Island (UAØ)	15:12	08:49
	Kyzyl (UAØY)	22:36	12:54
UB	Ukraine, Kiev	02:55	17:07
UC	Byelorussia, Minsk	02:57	17:27
UD	Azerbaijan, Baku	01:56	15:30
UF	Georgia, Tbilisi	02:15	15:52
UG	Armenia, Ierevan	02:18	15:51
UH	Turkoman, Ashkhabad	01:26	14:52
UI	Uzbek, Bukhara	00:59	14:31
	Tashkent	00:37	14:14
UJ	Tadzhik, Samarkand	00:50	14:21
	Dushanbe	00:43	14:12
UL	Kazakh, Alma-ata	00:03	13:47
UM	Kirghiz, Frunze	00:13	13:55
UO	Moldavia, Kishinev	03:09	17:06
UP	Lithuania, Vilna	03:04	17:37
UQ	Latvia, Riga	03:02	17:50
UR	Estonia, Tallinn	02:51	17:56
V2	Antigua & Barbuda, St. Johns	09:51	22:29
V3	Belize, Belmopan	11:39	00:17
V4	St. Christopher & Nevis	09:54	22:32
V8	Brunei, Bandar Seri Begawan	22:15	10:31
VE1, CY	Sable Is.	09:10	22:55
VE1, CY	St. Paul Is.	09:04	23:02
VK	Australia, Canberra (VK1)	20:37	07:35
	Sydney (VK2)	20:27	07:29
	Melbourne (VK3)	20:57	07:49
	Brisbane (VK4)	20:13	07:29
	Adelaide (VK5)	21:19	08:18
	Perth (VK6)	22:46	09:53
	Hobart, Tasmania (VK7)	20:55	07:32
	Darwin (VK8)	21:27	09:12
VK	Lord Howe Is.	19:53	07:00
VK9	Christmas Is.	23:06	10:54
VK9	Cocos-Keeling Is.	23:43	11:28
VK9	Mellish Reef	19:52	07:27
VK9	Norfolk Is.	19:14	06:27
VK9	Willis Is.	20:16	07:54
VKØ	Heard Is.	02:10	12:09
VKØ	Macquarie Is.	20:32	06:23
VP2E	Anguilla	09:55	22:35
VP2M	Montserrat	09:53	22:30
VP2V	British Virgin Is., Tortola	10:01	22:41
VP5	Turks & Caicos Islands, Grand Turk	10:24	23:11
VP8	Falkland Islands, Stanley	10:52	20:57

Pref	State/Province/Country/City	Sunrise	Sunset
VP8	So. Georgia Is.	09:34	19:26
VP8	So. Orkney Is.	10:27	19:42
VP8	So. Sandwich Islands, Saunders Is.	09:03	18:36
VP8	So. Shetland Is., King George Is.	11:24	20:28
VP9	Bermuda	09:46	22:57
VQ9	Chagos, Diego Garcia	01:16	13:10
VR6	Pitcairn Is.	15:02	02:24
VS6	Hong Kong	22:02	10:49
VU	India, Bombay	00:51	13:32
	Calcutta	23:45	12:33
	New Delhi	00:23	13:25
	Bangalore	00:37	13:07
VU7	Andaman Islands, Port Blair	23:38	12:05
VU7	Laccadive Is.	00:59	13:23
XE	Mexico, Mexico City (XE1)	12:18	01:00
	Chihuahua (XE2)	12:36	01:38
	Merida (XE3)	11:39	00:24
XF4	Revilla Gigedo	13:08	01:49
XT	Burkina Faso, Onagadougou	05:55	18:23
XU	Kampuchea, Phnom Penh	22:50	11:17
XW	Laos, Viangchan	22:53	11:32
XX9	Macao	22:04	10:52
XZ	Burma, Rangoon	23:20	11:57
Y2-9	German Dem.Rep.(E), Berlin	03:58	18:20
YA	Afghanistan, Kandahar	01:06	14:13
	Kabul	00:48	14:04
YB-YD	Indonesia, Jakarta	22:57	10:54
	Medan	23:21	11:34
	Pontianak	22:42	10:49
	Jayapura	20:39	08:41
YI	Iraq, Baghdad	02:28	15:41
YJ	Vanuatu, Port Vila	19:02	06:37
YK	Syria, Damascus	03:01	16:14
YN, HT	Nicaragua, Managua	11:33	00:01
YO	Romania, Bucharest	03:25	17:12
YS	El Salvador, San Salvador	11:44	00:15
YU	Yugoslavia, Belgrade	03:46	17:35
YV	Venezuela, Caracas	10:18	22:43
YVØ	Aves Is.	10:00	22:35
Z2,ZE	Zimbabwe, Harare	04:11	15:46
ZA	Albania, Tirane	03:55	17:32
ZB	Gibraltar	05:44	19:05
ZC4	British Cyprus	03:12	16:29
ZD7	St. Helena	06:37	18:16
ZD8	Ascension Is.	07:04	18:57
ZD9	Tristan da Cunha	07:25	18:19
ZF	Cayman Is.	11:06	23:49
ZK1	No. Cook Is., Manihiki	16:52	04:41
ZK1	So. Cook Is., Rarotonga	16:57	04:26
ZK2	Niue	17:32	05:05
ZK3	Tokelaus, Atafu	17:37	05:30
ZL	New Zealand, Auckland (ZL1)	18:56	05:51
	Wellington (ZL2)	19:03	05:44
	Christchurch (ZL3)	19:15	05:50
	Dunedin (ZL4)	19:27	05:54
ZL5	Antarctica, Scott Base	23:33	02:21
ZL7	Chatham Is.	18:32	05:05
ZL8	Kermadec Is.	18:18	05:30
ZL9	Auckland & Campbell Is., Auckland	19:52	06:01
	Campbell Is.	19:46	05:47
ZP	Paraguay, Asuncion	10:13	21:34
ZS	South Africa, Cape Town (ZS1)	05:18	16:20
	Port Elizabeth (ZS2)	04:49	15:51
	Bloemfontein (ZS4)	04:42	15:55
	Durban (ZS5)	04:24	15:35
	Johannesburg (ZS6)	04:31	15:50
ZS2	Prince Edward & Marion Is., Marion	04:19	14:44
ZS3	Namibia, Windhoek	05:11	16:37
1AØ	S.M.O.M	04:24	18:02
1S	Spratly Is.	22:24	10:46
3A	Monaco	04:41	18:25
3B6	Agalega	02:22	14:11
3B7	St. Brandon	02:14	13:53
3B8	Mauritius	02:27	13:58
3B9	Rodriguez Is.	02:03	13:35
3C	Equatorial Guinea, Bata	05:19	17:28
	Malabo	05:21	17:34

Pref	State/Province/Country/City	Sunrise	Sunset
3CØ	Pagalu Is.	05:38	17:42
3D2	Fiji Is., Suva	18:22	05:57
3D6	Swaziland, Mbabane	04:19	15:38
3V	Tunisia, Tunis	04:40	18:03
3W	Vietnam, Ho Chi Minh City		
	(Saigon)	22:43	11:09
	Hanoi	22:37	11:22
3X	Republic of Guinea, Conakry	06:46	19:09
3Y	Bouvet	06:53	16:45
3Y	Peter I Is.	14:09	22:01
4S	Sri Lanka, Colombo	00:34	12:53
4U	ITU Geneva	04:41	18:35
4U	United Nations Hq.	10:11	23:46
4W	Yemen, Sanaa	02:49	15:23
4X, 4Z	Israel, Jerusalem	03:07	16:17
5A	Libya, Tripoli	04:37	17:48
	Benghazi	04:08	17:18
5B	Cyprus, Nicosia	03:10	16:28
5H	Tanzania, Dar es Salaam	03:27	15:22
5N	Nigeria, Lagos	05:40	17:58
5R	Madagascar, Antananarivo	03:06	14:39
5T	Mauritania, Nouakchott	06:47	19:26
5U	Niger, Niamey	05:39	18:10
5V	Togo, Lome	05:49	18:06
5W	Western Samoa, Apia	17:38	05:22
5X	Uganda, Kampala	03:49	15:56
5Z	Kenya, Nairobi	04:53	16:58
6W	Senegal, Dakar	06:56	19:29
6Y	Jamaica, Kingston	10:50	23:30

Pref	State/Province/Country/City	Sunrise	Sunset
7O	Yemen, People's Dem. Rep.,		
	Aden	02:48	15:17
7P	Lesotho, Maseru	04:37	15:49
7Q	Malawi, Lilongwe	03:56	15:39
	Blantyre	03:53	15:32
7X	Algeria, Algiers	05:09	18:32
8P	Barbados, Bridgetown	09:46	22:16
8Q	Maldive Is.	01:02	13:16
8R	Guyana, Georgetown	09:46	22:05
9G	Ghana, Accra	05:55	18:12
9H	Malta	04:25	17:45
9J	Zambia, Lusaka	04:20	15:59
9K	Kuwait	02:20	15:23
9L	Sierra Leone, Freetown	06:45	19:06
9M2	West Malaysia, Kuala Lumpur	23:10	11:22
9M6,8	East Malaysia, Sabah,		
	Sandakan (9M6)	22:02	10:19
	Sarawak, Kuching (9M8)	22:37	10:46
9N	Nepal, Katmandu	23:52	12:51
9Q	Zaire, Kinshasa	05:02	17:01
	Kisangani	04:18	16:26
	Lubumbashi	04:19	16:06
9U	Burundi, Bujumbura	04:05	16:06
9V	Singapore	23:03	11:12
9X	Rwanda, Kigali	04:01	16:04
9Y	Trinidad & Tobago, Port of		
	Spain	09:55	22:21
J2/A	Abu Ail	02:56	15:27

Pref	State/Province/Country/City	Sunrise	Sunset
VE1	New Brunswick, Saint John	09:52	22:52
	Nova Scotia, Halifax	09:43	22:41
	Prince Edward Island, Charlottetown	09:39	22:41
VE2	Quebec, Montreal	10:22	23:22
	Quebec City	10:11	23:14
VE3	Ontario, London	10:55	23:50
	Ottawa	10:31	23:30
	Sudbury	10:51	23:52
	Toronto	10:47	23:43
VE4	Manitoba, Winnipeg	11:52	01:00
VE5	Saskatchewan, Regina	12:21	01:31
	Saskatoon	12:27	01:42
VE6	Alberta, Calgary	12:58	02:10
	Edmonton	12:53	02:11
VE7	British Columbia, Prince George	13:29	02:48
	Prince Rupert	13:59	03:19
	Vancouver	13:36	02:44
VE8	Northwest Territories, Yellowknife	12:40	02:30
	Resolute	10:32	02:04
VY1	Yukon, Whitehorse	14:07	03:49
VO1	Newfoundland, St. John's	08:56	22:00
VO2	Labrador, Goose Bay	09:21	22:38
W1	Connecticut, Hartford	10:22	23:15
	Maine, Bangor	10:04	23:02
	Portland	10:11	23:07
	Massachusetts, Boston	10:15	23:09
	New Hampshire, Concord	10:16	23:11
	Rhode Island, Providence	10:17	23:10
	Vermont, Montpelier	10:19	23:17
W2	New Jersey, Atlantic City	10:31	23:20
	New York, Albany	10:26	23:20
	Buffalo	10:46	23:41
	New York City	10:28	23:19
	Syracuse	10:35	23:30
W3	Delaware, Wilmington	10:35	23:24
	District of Columbia, Washington	10:41	23:30
	Maryland, Baltimore	10:40	23:28
	Pennsylvania, Harrisburg	10:40	23:30
	Philadelphia	10:33	23:23
	Pittsburgh	10:52	23:43
	Scranton	10:34	23:27
W4	Alabama, Montgomery	11:23	00:02
	Florida, Jacksonville	11:06	23:43
	Miami	11:03	23:34
	Pensacola	11:28	00:05
	Georgia, Atlanta	11:15	23:56
	Savannah	11:03	23:41
	Kentucky, Lexington	11:12	23:59
	Louisville	11:17	00:04
	North Carolina, Charlotte	10:59	23:42
	Raleigh	10:50	23:34
	Wilmington	10:48	23:30
	South Carolina, Columbia	11:01	23:42
	Tennessee, Knoxville	11:11	23:55
	Memphis	11:37	00:19
	Nashville	11:23	00:07
	Virginia, Norfolk	10:40	23:25
	Richmond	10:44	23:30
W5	Arkansas, Little Rock	11:46	00:28
	Louisiana, New Orleans	11:40	00:16
	Shreveport	11:53	00:32
	Mississippi, Jackson	11:39	00:18
	New Mexico, Albuquerque	12:43	01:26
	Oklahoma, Oklahoma City	12:06	00:49
	Texas, Abilene	12:17	00:56
	Amarillo	12:23	01:06
	Dallas	12:05	00:45
	El Paso	12:44	01:23
	San Antonio	12:14	00:49
W6	California, Los Angeles	13:30	02:11
	San Francisco	13:44	02:31
W7	Arizona, Flagstaff	13:03	01:46
	Phoenix	13:06	01:46
	Idaho, Boise	13:14	02:10
	Pocatello	13:00	01:55
	Montana, Billings	12:41	01:42
	Butte	12:57	01:58
	Great Falls	12:51	01:55
	Nevada, Las Vegas	13:16	02:00
	Reno	13:32	02:21
	Oregon, Portland	13:39	02:38
	Utah, Salt Lake City	13:00	01:51
	Washington, Seattle	13:35	02:39
	Spokane	13:15	02:19
	Wyoming, Cheyenne	12:31	01:23
	Sheridan	12:36	01:35
W8	Michigan, Detroit	11:03	23:57
	Grand Rapids	11:13	00:08
	Sault Ste. Marie	11:04	00:06
	Traverse City	11:11	00:09
	Ohio, Cincinnati	11:11	00:00
	Cleveland	10:58	23:51
	Columbus	11:05	23:55
	West Virginia, Charleston	11:00	23:48
W9	Illinois, Chicago	11:21	00:15
	Indiana, Indianapolis	11:18	00:07
	Wisconsin, Green Bay	11:21	00:18
	Milwaukee	11:22	00:17
W0	Colorado, Denver	12:33	01:22
	Grand Junction	12:48	01:36
	Iowa, Des Moines	11:46	00:38
	Kansas, Pratt	12:09	00:56
	Wichita	12:04	00:50
	Minnesota, Duluth	11:35	00:37
	Minneapolis	11:41	00:40
	Missouri, Columbia	11:43	00:31
	Kansas City	11:52	00:40
	St. Louis	11:34	00:22
	Nebraska, North Platte	12:15	01:07
	Omaha	11:55	00:47
	North Dakota, Fargo	11:54	00:56
	South Dakota, Rapid City	12:22	01:19
A2	Botswana, Gaborone	04:22	16:06
A3	Tonga, Nukualofa	17:44	05:33
A4	Oman, Muscat	01:49	14:18
A5	Bhutan, Thimbu	23:44	12:16
A6	United Arab Emirates, Abu Dhabi	02:06	14:36
A7	Qatar, Ad-Dawhah	02:16	14:47
A9	Bahrein, Al-Manamah	02:19	14:51
AP	Pakistan, Karachi	01:14	13:44
	Islamabad	00:44	13:25
BV	Taiwan, Taipei	21:36	10:07
BY	People's Rep. of China, Beijing	21:47	10:37
	Harbin	21:01	10:01
	Shanghai	21:33	10:10
	Fuzhou	21:45	10:16
	Xian	22:21	11:03
	Chongqing	22:34	11:10
	Chengdu	22:43	11:20
	Lhasa	23:35	12:11
	Urumqi	23:39	12:35
	Kashi	00:29	13:18
C2	Nauru	18:47	06:53
C3	Andorra	05:25	18:19
C5	The Gambia, Banjul	06:55	19:14
C6	Bahamas, Nassau	10:52	23:22
C9	Mozambique, Maputo	03:55	15:39
	Mozambique	03:18	15:12
CE	Chile, Santiago	10:53	22:29
CE0	Easter Island	13:24	01:06
CE0	San Felix	11:26	23:10
CE0	Juan Fernandez	11:25	23:01
CM,CO	Cuba, Havana	11:13	23:41
CN	Morocco, Casablanca	06:07	18:48
CP	Bolivia, La Paz	10:35	22:27
CT	Portugal, Lisbon	06:10	18:58
CT2, CU	Azores, Ponta Delgada	07:17	20:04
CT3	Madeira Islands, Funchal	06:46	19:25
CX	Uruguay, Montevideo	09:55	21:29
D2	Angola, Luanda	05:05	17:04
D4	Cape Verde, Praia	07:22	19:42

Pref	State/Province/Country/City	Sunrise	Sunset
D6	Comoros, Moroni	03:04	15:01
DA-DL	Germany, Fed. Rep. of (W), Bonn	04:54	18:05
DU	Philippines, Manilla	21:44	10:03
EA	Spain, Madrid	05:47	18:38
EA6	Balearic Is., Palma	05:23	18:12
EA8	Canary Is., Las Palmas	06:38	19:12
EA9	Ceuta & Melilla, Ceuta	05:57	18:41
	Melilla	05:48	18:31
EI	Ireland, Dublin	05:44	19:01
EL	Liberia, Monrovia	06:35	18:47
EP	Iran, Tehran	02:09	14:52
ET	Ethiopia, Addis Ababa	03:16	15:30
	Asmera	03:12	15:32
F	France, Paris	05:15	18:22
FG	Guadeloupe	09:54	22:15
FG, FS	St. Martin	09:58	22:22
FH	Mayotte	02:59	14:54
FK	New Caledonia, Noumia	18:58	06:45
FM	Martinique	09:52	22:11
FO	Clipperton	13:07	01:22
FO	Fr. Polynesia, Tahiti	16:40	04:31
FP	St. Pierre & Miquelon, St. Pierre	09:11	22:13
FR	Glorioso	02:50	14:47
FR	Juan de Nova	03:10	15:02
	Europa	03:22	15:10
FR	Reunion	02:21	14:09
FR	Tromelin	02:24	14:16
FT8W	Crozet	02:50	14:09
FT8X	Kerguelen	01:44	12:57
FT8Y	Antarctica, Dumont D'Urville	21:26	07:49
FT8Z	Amsterdam & St. Paul Is., Amsterdam	01:02	12:32
FW	Wallis & Futuna Is., Wallis	17:45	05:40
FY	Fr. Guiana, Cayenne	09:21	21:32
G	England, London	05:22	18:34
GD	Isle of Man	05:36	18:56
GI	Northern Ireland, Belfast	05:41	19:02
GJ	Jersey	05:33	18:40
GM	Scotland, Glasgow	05:33	18:57
	Aberdeen	05:21	18:51
GU	Guernsey	05:34	18:42
GW	Wales, Cardiff	05:34	18:47
H4	Solomon Islands, Honiara	19:18	07:17
HA	Hungary, Budapest	04:09	17:13
HB	Switzerland, Bern	04:56	17:59
HBØ	Liechtenstein	04:48	17:51
HC	Ecuador, Quito	11:06	23:13
HC8	Galapagos Is.	11:56	00:03
HH	Haiti, Port-Au-Prince	10:35	22:59
HI	Dominican Republic, Santo Domingo	10:26	22:49
HK	Colombia, Bogota	10:49	22:59
HKØ	Malpelo Is.	11:17	23:27
HKØ	San Adreas	11:16	23:33
HL, HM	Korea, Seoul	21:07	09:53
HP	Panama, Panama	11:08	23:23
HR	Honduras, Tegucigalpa	11:37	23:56
HS	Thailand, Bangkok	23:06	11:25
HV	Vatican City	04:41	17:34
HZ, 7Z	Saudi Arabia, Dharan	02:22	14:53
	Mecca	03:05	15:32
	Riyadh	02:36	15:06
I	Italy, Rome	04:41	17:34
	Trieste	04:32	17:32
	Sicily	04:39	17:25
IS	Sardinia, Cagliari	04:57	17:46
J2	Djibouti, Djibouti	02:56	15:13
J3	Grenada	09:56	22:13
J5	Guinea-Bissau, Bissau	06:51	19:09
J6	St. Lucia	09:52	22:11
J7	Dominica	09:53	22:13
J8	St. Vincent	09:54	22:12
JA-JS	Japan, Tokyo	20:17	09:00
	Nagasaki	20:58	09:38
	Sapporo	20:04	09:00
JD	Minami Torishima	19:27	07:56
JD	Ogasawara, Kazan Is.	20:17	08:50
JT	Mongolia, Ulan Bator	22:18	11:22
JW	Svalbard, Spitsbergen	02:22	19:26
JX	Jan Mayen	05:07	19:54

Pref	State/Province/Country/City	Sunrise	Sunset
JY	Jordan, Amman	03:15	15:53
KC4	Antarctica, Byrd Station	16:04	23:52
	McMurdo Sound	20:31	05:11
	Palmer Station	10:58	21:29
KC6	Micronesia, Ponape	19:18	07:31
KC6	Belau, Yap	20:37	08:52
	Koror	20:53	09:06
KG4	Guantanamo Bay	10:46	23:11
KH1	Baker, Howland Is.	17:38	05:45
KH2	Guam, Agana	20:09	08:28
KH3	Johnston Is.	17:01	05:23
KH4	Midway Is.	17:30	06:04
KH5	Palmyra Is.	16:40	04:52
KH5K	Kingman Reef	16:42	04:55
KH6	Hawaii, Hilo	16:06	04:30
	Honolulu	16:16	04:42
KH7	Kure Is.	17:34	06:08
KH8	American Samoa, Pago Pago	17:24	05:18
KH9	Wake Is.	18:39	07:03
KHØ	Mariana Is., Saipan	20:04	08:25
KL7	Alaska, Adak	17:07	06:21
	Anchorage	15:06	04:50
	Fairbanks	14:49	04:50
	Juneau	14:09	03:42
	Nome	15:59	05:59
KP1	Navassa Is.	10:46	23:09
KP2	Virgin Islands, Charlotte Amalie	10:06	22:29
KP4	Puerto Rico, San Juan	10:11	22:34
KP5	Desecheo Is.	10:16	22:39
KX6	Marshall Islands, Kwajalein	18:41	06:56
LA-LJ	Norway, Oslo	04:25	18:04
LU	Argentina, Buenos Aires	10:04	21:38
LX	Luxembourg	04:59	18:07
LZ	Bulgaria, Sofia	03:57	16:52
OA	Peru, Lima	11:08	23:04
OD	Lebanon, Beirut	03:15	15:56
OE	Austria, Vienna	04:20	17:25
OH	Finland, Helsinki	03:28	17:08
OHØ	Aland Is.	03:48	17:28
OJØ	Market Reef	03:51	17:32
OK	Czechoslovakia, Prague	04:25	17:35
ON	Belgium, Brussels	05:04	18:16
OX, XP	Greenland, Godthaab	08:25	22:23
	Thule	08:30	00:36
OY	Faroe Islands, Torshavn	05:31	19:19
OZ	Denmark, Copenhagen	04:25	17:49
P2	Papua New Guinea, Madang	20:14	08:16
	Port Moresby	20:10	08:09
PA-PI	Netherlands, Amsterdam	05:00	18:16
PJ, P4	Netherlands Antilles, Willemstad	10:25	22:42
PJ5-8	St. Maarten and Saba, St. Maarten	09:59	22:22
PY	Brazil, Brasilia	09:13	21:06
	Rio De Janeiro	08:57	20:44
	Natal	08:18	20:19
	Manaus	09:56	22:00
	Porto Alegre	09:33	21:12
PYØ	Fernando De Noronha	08:06	20:09
PYØ	St. Peter & St. Paul Rocks	07:51	19:59
PYØ	Trindade & Martin Vaz Is., Trindade	08:00	19:49
PZ	Suriname, Paramaribo	09:33	21:44
S2	Bangladesh, Dacca	23:42	12:10
S7	Seychelles, Victoria	02:14	14:17
S9	Sao Tome	05:27	17:34
SJ-SM	Sweden, Stockholm	03:57	17:33
SP	Poland, Krakow	04:03	17:12
	Warsaw	03:56	17:11
ST	Sudan, Khartoum	03:37	15:58
STØ	Southern Sudan, Juba	03:46	15:57
SU	Egypt, Cairo	03:34	16:10
SV	Greece, Athens	03:59	16:46
SV5	Dodecanese, Rhodes	03:43	16:27
SV9	Crete	03:55	16:38
SV/A	Mount Athos	03:55	16:46
T2	Tuvalu, Funafuti	18:04	06:03
T30	West Kiribati, Bonriki	18:21	06:29
T31	Central Kiribati, Kanton	17:22	05:27
T32	East Kiribati, Christmas Is.	16:23	04:31
T5	Somalia, Mogadishu	02:52	15:00
T7	San Marino	04:40	17:37

Pref	State/Province/Country/City	Sunrise	Sunset
TA	Turkey, Ankara	03:21	16:11
	Istanbul	03:36	16:28
TF	Iceland, Reykjavik	06:27	20:24
TG	Guatemala, Guatemala City	11:50	00:09
TI	Costa Rica, San Jose	11:26	23:41
TI9	Cocos Is.	11:40	23:51
TJ	Cameroon, Yaounde	05:07	17:17
TK	Corsica	04:55	17:48
TL	Central African Republic, Bangui	04:38	16:48
TN	Congo, Brazzaville	04:55	16:58
TR	Gabon, Libreville	05:16	17:23
TT	Chad, N'Djamena	04:49	17:06
TU	Ivory Coast, Abidjan	06:08	18:19
TY	Benin, Porto Novo	05:41	17:53
TZ	Mali, Bamako	06:21	18:39
UA	Russia, European,		
	Leningrad (UA1)	03:07	16:46
	Archangel (UA1)	02:15	16:16
	Murmansk (UA1)	02:31	17:00
	Moscow (UA3)	02:45	16:09
	Kuibyshev (UA4)	01:59	15:16
	Rostov (UA6)	02:48	15:51
UA1P	Franz Josef Land	23:31	17:20
UA2	Kaliningrad	03:55	17:17
UA9,Ø	Russia, Asiatic,		
	Novosibirsk (UA9)	23:45	13:07
	Perm (UA9)	01:27	14:58
	Omsk (UA9)	00:23	13:45
	Norilsk (UAØ)	22:49	13:21
	Irkutsk (UAØ)	22:23	11:38
	Vladivostok (UAØ)	20:42	09:38
	Petropavlovsk (UAØ)	18:45	08:01
	Khabarovsk (UAØ)	20:24	09:30
	Krasnoyarsk (UAØ)	23:04	12:29
	Yakutsk (UAØ)	20:25	10:13
	Wrangel Island (UAØ)	16:32	07:19
	Kyzyl (UAØY)	23:03	12:16
UB	Ukraine, Kiev	03:21	16:31
UC	Byelorussia, Minsk	03:28	16:47
UD	Azerbaijan, Baku	02:13	15:03
UF	Georgia, Tbilisi	02:32	15:25
UG	Armenia, Ierevan	02:34	15:25
UH	Turkoman, Ashkhabad	01:41	14:27
UI	Uzbek, Bukhara	01:15	14:05
	Tashkent	00:54	13:46
UJ	Tadzhik, Samarkand	01:06	13:55
	Dushanbe	00:58	13:47
UL	Kazakh, Alma-ata	00:22	13:18
UM	Kirghiz, Frunze	00:32	13:27
UO	Moldavia, Kishinev	03:31	16:34
UP	Lithuania, Vilna	03:35	16:56
UQ	Latvia, Riga	03:37	17:05
UR	Estonia, Tallinn	03:30	17:07
V2	Antigua & Barbuda, St. Johns	09:54	22:16
V3	Belize, Belmopan	11:42	00:04
V4	St. Christopher & Nevis	09:57	22:19
V8	Brunei, Bandar Seri Begawan	22:13	10:23
VE1, CY	Sable Is.	09:29	22:26
VE1, CY	St. Paul Is.	09:26	22:30
VK	Australia, Canberra (VK1)	20:14	07:48
	Sydney (VK2)	20:05	07:40
	Melbourne (VK3)	20:32	08:03
	Brisbane (VK4)	19:55	07:37
	Adelaide (VK5)	20:56	08:30
	Perth (VK6)	22:26	10:03
	Hobart, Tasmania (VK7)	20:27	07:50
	Darwin (VK8)	21:16	09:12
VK	Lord Howe Is.	19:32	07:10
VK9	Christmas Is.	22:56	10:54
VK9	Cocos-Keeling Is.	23:32	11:29
VK9	Mellish Reef	19:39	07:30
VK9	Norfolk Is.	18:55	06:36
VK9	Willis Is.	20:03	07:56
VKØ	Heard Is.	01:31	12:37
VKØ	Macquarie Is.	19:51	06:53
VP2E	Anguilla	09:58	22:21
VP2M	Montserrat	09:55	22:17
VP2V	British Virgin Is., Tortola	10:04	22:28
VP5	Turks & Caicos Islands,		
	Grand Turk	10:29	22:56
VP8	Falkland Islands, Stanley	10:15	21:23

Pref	State/Province/Country/City	Sunrise	Sunset
VP8	So. Georgia Is.	08:53	19:56
VP8	So. Orkney Is.	09:36	20:23
VP8	So. Sandwich Islands, Saunders Is.	08:17	19:12
VP8	So. Shetland Is., King George Is.	10:30	21:12
VP9	Bermuda	09:57	22:36
VQ9	Chagos, Diego Garcia	01:08	13:08
VR6	Pitcairn Is.	14:46	02:30
VS6	Hong Kong	22:07	10:34
VU	India, Bombay	00:54	13:18
	Calcutta	23:50	12:18
	New Delhi	00:32	13:06
	Bangalore	00:38	12:56
VU7	Andaman Islands, Port Blair	23:38	11:55
VU7	Laccadive Is.	00:58	13:13
XE	Mexico, Mexico City (XE1)	12:22	00:46
	Chihuahua (XE2)	12:44	01:19
	Merida (XE3)	11:43	00:09
XF4	Revilla Gigedo	13:12	01:36
XT	Burkina Faso, Onagadougou	05:55	18:13
XU	Kampuchea, Phnom Penh	22:50	11:07
XW	Laos, Viangchan	22:56	11:19
XX9	Macao	22:10	10:37
XZ	Burma, Rangoon	23:23	11:45
Y2-9	German Dem.Rep.(E), Berlin	04:26	17:42
YA	Afghanistan, Kandahar	01:16	13:53
	Kabul	01:00	13:42
YB-YD	Indonesia, Jakarta	22:50	10:51
	Medan	23:18	11:28
	Pontianak	22:37	10:44
	Jayapura	20:33	08:37
YI	Iraq, Baghdad	02:40	15:20
YJ	Vanuatu, Port Vila	18:49	06:40
YK	Syria, Damascus	03:12	15:53
YN, HT	Nicaragua, Managua	11:33	23:50
YO	Romania, Bucharest	03:44	16:42
YS	El Salvador, San Salvador	11:45	00:04
YU	Yugoslavia, Belgrade	04:06	17:05
YV	Venezuela, Caracas	10:18	22:34
YVØ	Aves Is.	10:02	22:23
Z2,ZE	Zimbabwe, Harare	03:58	15:49
ZA	Albania, Tirane	04:12	17:04
ZB	Gibraltar	05:57	18:41
ZC4	British Cyprus	03:25	16:07
ZD7	St. Helena	06:25	18:18
ZD8	Ascension Is.	06:55	18:55
ZD9	Tristan da Cunha	07:01	18:32
ZF	Cayman Is.	11:10	23:35
ZK1	No. Cook Is., Manihiki	16:43	04:41
ZK1	So. Cook Is., Rarotonga	16:43	04:31
ZK2	Niue	17:18	05:08
ZK3	Tokelaus, Atafu	17:29	05:28
ZL	New Zealand, Auckland (ZL1)	18:33	06:04
	Wellington (ZL2)	18:35	06:01
	Christchurch (ZL3)	18:46	06:09
	Dunedin (ZL4)	18:56	06:15
ZL5	Antarctica, Scott Base	20:34	05:10
ZL7	Chatham Is.	18:03	05:25
ZL8	Kermadec Is.	17:59	05:39
ZL9	Auckland & Campbell Is.,		
	Auckland	19:16	06:27
	Campbell Is.	19:08	06:15
ZP	Paraguay, Asuncion	09:56	21:41
ZS	South Africa, Cape Town (ZS1)	04:56	16:32
	Port Elizabeth (ZS2)	04:27	16:02
	Bloemfontein (ZS4)	04:23	16:03
	Durban (ZS5)	04:04	15:44
	Johannesburg (ZS6)	04:14	15:57
ZS2	Prince Edward & Marion Is., Marion	03:48	15:05
ZS3	Namibia, Windhoek	04:56	16:43
1AØ	S.M.O.M	04:41	17:35
1S	Spratly Is.	22:23	10:37
3A	Monaco	05:00	17:56
3B6	Agalega	02:12	14:10
3B7	St. Brandon	02:02	13:55
3B8	Mauritius	02:13	14:02
3B9	Rodriguez Is.	01:49	13:39
3C	Equatorial Guinea, Bata	05:14	17:23
	Malabo	05:17	17:27

Pref	State/Province/Country/City	Sunrise	Sunset
3CØ	Pagalu Is.	05:32	17:38
3D2	Fiji Is., Suva	18:08	06:00
3D6	Swaziland, Mbabane	04:02	15:45
3V	Tunisia, Tunis	04:54	17:39
3W	Vietnam, Ho Chi Minh City		
	(Saigon)	22:43	10:59
	Hanoi	22:41	11:07
3X	Republic of Guinea, Conakry	06:45	19:00
3Y	Bouvet	06:13	17:15
3Y	Peter I Is.	12:54	23:06
4S	Sri Lanka, Colombo	00:32	12:44
4U	ITU Geneva	05:02	18:03
4U	United Nations Hq.	10:28	23:19
4W	Yemen, Sanaa	02:51	15:11
4X, 4Z	Israel, Jerusalem	03:18	15:56
5A	Libya, Tripoli	04:48	17:27
	Benghazi	04:18	16:57
5B	Cyprus, Nicosia	03:23	16:05
5H	Tanzania, Dar es Salaam	03:19	15:20
5N	Nigeria, Lagos	05:38	17:50
5R	Madagascar, Antananarivo	02:52	14:43
5T	Mauritania, Nouakchott	06:50	19:13
5U	Niger, Niamey	05:40	17:59
5V	Togo, Lome	05:47	17:59
5W	Western Samoa, Apia	17:27	05:22
5X	Uganda, Kampala	03:44	15:51
5Z	Kenya, Nairobi	04:48	16:53
6W	Senegal, Dakar	06:58	19:18
6Y	Jamaica, Kingston	10:53	23:16

Pref	State/Province/Country/City	Sunrise	Sunset
7O	Yemen, People's Dem. Rep.,		
	Aden	02:49	15:07
7P	Lesotho, Maseru	04:18	15:58
7Q	Malawi, Lilongwe	03:45	15:40
	Blantyre	03:41	15:34
7X	Algeria, Algiers	05:23	18:08
8P	Barbados, Bridgetown	09:47	22:05
8Q	Maldive Is.	00:59	13:09
8R	Guyana, Georgetown	09:44	21:57
9G	Ghana, Accra	05:53	18:04
9H	Malta	04:38	17:22
9J	Zambia, Lusaka	04:08	16:01
9K	Kuwait	02:29	15:04
9L	Sierra Leone, Freetown	06:43	18:57
9M2	West Malaysia, Kuala Lumpur	23:06	11:16
9M6,8	East Malaysia, Sabah,		
	Sandakan (9M6)	21:59	10:11
	Sarawak, Kuching (9M8)	22:32	10:40
9N	Nepal, Katmandu	00:00	12:33
9Q	Zaire, Kinshasa	04:55	16:58
	Kisangani	04:13	16:20
	Lubumbashi	04:09	16:06
9U	Burundi, Bujumbura	03:58	16:02
9V	Singapore	22:58	11:06
9X	Rwanda, Kigali	03:55	16:00
9Y	Trinidad & Tobago, Port of		
	Spain	09:55	22:11
J2/A	Abu Ail	02:57	15:16

Pref	State/Province/Country/City	Sunrise	Sunset
VE1	New Brunswick, Saint John	10:09	22:25
	Nova Scotia, Halifax	09:59	22:15
	Prince Edward Island, Charlottetown	09:57	22:13
VE2	Quebec, Montreal	10:39	22:55
	Quebec City	10:29	22:46
VE3	Ontario, London	11:11	23:25
	Ottawa	10:48	23:03
	Sudbury	11:09	23:25
	Toronto	11:03	23:18
VE4	Manitoba, Winnipeg	12:12	00:30
VE5	Saskatchewan, Regina	12:42	01:00
	Saskatoon	12:50	01:09
VE6	Alberta, Calgary	13:20	01:38
	Edmonton	13:17	01:36
VE7	British Columbia, Prince George	13:54	02:14
	Prince Rupert	14:24	02:44
	Vancouver	13:56	02:14
VE8	Northwest Territories, Yellowknife	13:17	01:43
	Resolute	11:49	00:36
VY1	Yukon, Whitehorse	14:41	03:05
VO1	Newfoundland, St. John's	09:15	21:32
VO2	Labrador, Goose Bay	09:45	22:04
W1	Connecticut, Hartford	10:36	22:51
	Maine, Bangor	10:20	22:36
	Portland	10:26	22:41
	Massachusetts, Boston	10:30	22:44
	New Hampshire, Concord	10:31	22:46
	Rhode Island, Providence	10:31	22:45
	Vermont, Montpelier	10:36	22:51
W2	New Jersey, Atlantic City	10:44	22:57
	New York, Albany	10:41	22:55
	Buffalo	11:01	23:16
	New York City	10:42	22:56
	Syracuse	10:50	23:05
W3	Delaware, Wilmington	10:48	23:01
	District of Columbia, Washington	10:54	23:07
	Maryland, Baltimore	10:52	23:06
	Pennsylvania, Harrisburg	10:53	23:07
	Philadelphia	10:47	23:00
	Pittsburgh	11:06	23:20
	Scranton	10:48	23:03
W4	Alabama, Montgomery	11:32	23:44
	Florida, Jacksonville	11:14	23:25
	Miami	11:08	23:19
	Pensacola	11:36	23:47
	Georgia, Atlanta	11:24	23:36
	Savannah	11:11	23:23
	Kentucky, Lexington	11:24	23:37
	Louisville	11:29	23:42
	North Carolina, Charlotte	11:10	23:22
	Raleigh	11:01	23:13
	Wilmington	10:58	23:10
	South Carolina, Columbia	11:11	23:23
	Tennessee, Knoxville	11:22	23:35
	Memphis	11:47	23:59
	Nashville	11:34	23:46
	Virginia, Norfolk	10:51	23:04
	Richmond	10:56	23:09
W5	Arkansas, Little Rock	11:56	00:08
	Louisiana, New Orleans	11:48	23:59
	Shreveport	12:02	00:13
	Mississippi, Jackson	11:48	23:59
	New Mexico, Albuquerque	12:53	01:06
	Oklahoma, Oklahoma City	12:16	00:29
	Texas, Abilene	12:26	00:37
	Amarillo	12:34	00:46
	Dallas	12:14	00:26
	El Paso	12:53	01:04
	San Antonio	12:21	00:32
W6	California, Los Angeles	13:39	01:52
	San Francisco	13:56	02:09
W7	Arizona, Flagstaff	13:13	01:26
	Phoenix	13:15	01:27
	Idaho, Boise	13:30	01:45
	Pocatello	13:15	01:30

Pref	State/Province/Country/City	Sunrise	Sunset
	Montana, Billings	12:59	01:15
	Butte	13:15	01:31
	Great Falls	13:10	01:26
	Nevada, Las Vegas	13:27	01:39
	Reno	13:45	01:59
	Oregon, Portland	13:56	02:11
	Utah, Salt Lake City	13:13	01:27
	Washington, Seattle	13:54	02:10
	Spokane	13:34	01:51
	Wyoming, Cheyenne	12:45	00:59
	Sheridan	12:53	01:08
W8	Michigan, Detroit	11:17	23:32
	Grand Rapids	11:28	23:43
	Sault Ste. Marie	11:22	23:38
	Traverse City	11:27	23:43
	Ohio, Cincinnati	11:24	23:37
	Cleveland	11:12	23:27
	Columbus	11:18	23:32
	West Virginia, Charleston	11:12	23:26
W9	Illinois, Chicago	11:36	23:50
	Indiana, Indianapolis	11:31	23:44
	Wisconsin, Green Bay	11:37	23:52
	Milwaukee	11:37	23:52
W0	Colorado, Denver	12:46	00:59
	Grand Junction	13:00	01:14
	Iowa, Des Moines	12:00	00:14
	Kansas, Pratt	12:21	00:34
	Wichita	12:15	00:28
	Minnesota, Duluth	11:53	00:09
	Minneapolis	11:58	00:14
	Missouri, Columbia	11:55	00:09
	Kansas City	12:04	00:18
	St. Louis	11:47	00:00
	Nebraska, North Platte	12:29	00:43
	Omaha	12:09	00:23
	North Dakota, Fargo	12:12	00:28
	South Dakota, Rapid City	12:38	00:53

Pref	State/Province/Country/City	Sunrise	Sunset
A2	Botswana, Gaborone	04:07	16:11
A3	Tonga, Nukualofa	17:31	05:36
A4	Oman, Muscat	01:53	14:03
A5	Bhutan, Thimbu	23:50	12:00
A6	United Arab Emirates, Abu Dhabi	02:11	14:21
A7	Qatar, Ad-Dawhah	02:22	14:32
A9	Bahrein, Al-Manamah	02:25	14:35
AP	Pakistan, Karachi	01:19	13:29
	Islamabad	00:54	13:06
BV	Taiwan, Taipei	21:42	09:52
BY	People's Rep. of China, Beijing	22:00	10:14
	Harbin	21:18	09:34
	Shanghai	21:41	09:52
	Fuzhou	21:50	10:01
	Xian	22:31	10:43
	Chongqing	22:41	10:52
	Chengdu	22:51	11:02
	Lhasa	23:42	11:53
	Urumqi	23:55	12:10
	Kashi	00:42	12:55
C2	Nauru	18:42	06:48
C3	Andorra	05:39	17:54
C5	The Gambia, Banjul	06:55	19:04
C6	Bahamas, Nassau	10:57	23:07
C9	Mozambique, Maputo	03:40	15:45
	Mozambique	03:07	15:13
CE	Chile, Santiago	10:34	22:38
CE0	Easter Island	13:08	01:12
CE0	San Felix	11:11	23:15
CE0	Juan Fernandez	11:06	23:10
CM,CO	Cuba, Havana	11:17	23:27
CN	Morocco, Casablanca	06:17	18:29
CP	Bolivia, La Paz	10:24	22:29
CT	Portugal, Lisbon	06:23	18:36
CT2, CU	Azores, Ponta Delgada	07:29	19:42
CT3	Madeira Islands, Funchal	06:54	19:06
CX	Uruguay, Montevideo	09:35	21:39
D2	Angola, Luanda	04:57	17:03
D4	Cape Verde, Praia	07:22	19:31

Pref	State/Province/Country/City	Sunrise	Sunset
D6	Comoros, Moroni	02:55	15:01
DA-DL	Germany, Fed. Rep. of (W), Bonn	05:16	17:34
DU	Philippines, Manilla	21:44	09:53
EA	Spain, Madrid	06:01	18:14
EA6	Balearic Is., Palma	05:35	17:49
EA8	Canary Is., Las Palmas	06:44	18:55
EA9	Ceuta & Melilla, Ceuta	06:08	18:20
	Melilla	05:58	18:11
EI	Ireland, Dublin	06:08	18:27
EL	Liberia, Monrovia	06:32	18:40
EP	Iran, Tehran	02:19	14:32
ET	Ethiopia, Addis Ababa	03:14	15:22
	Asmera	03:13	15:21
F	France, Paris	05:35	17:52
FG	Guadeloupe	09:55	22:04
FG, FS	St. Martin	10:01	22:10
FH	Mayotte	02:49	14:54
FK	New Caledonia, Noumia	18:44	06:49
FM	Martinique	09:52	22:01
FO	Clipperton	13:06	01:13
FO	Fr. Polynesia, Tahiti	16:28	04:33
FP	St. Pierre & Miquelon, St. Pierre	09:29	21:45
FR	Glorioso	02:41	14:46
FR	Juan de Nova	02:59	15:04
	Europa	03:09	15:13
FR	Reunion	02:08	14:13
FR	Tromelin	02:12	14:18
FT8W	Crozet	02:23	14:26
FT8X	Kerguelen	01:14	13:18
FT8Y	Antarctica, Dumont D'Urville	20:31	08:34
FT8Z	Amsterdam & St. Paul Is., Amsterdam	00:40	12:44
FW	Wallis & Futuna Is., Wallis	17:35	05:41
FY	Fr. Guiana, Cayenne	09:18	21:25
G	England, London	05:44	18:02
GD	Isle of Man	06:01	18:21
GI	Northern Ireland, Belfast	06:06	18:26
GJ	Jersey	05:53	18:10
GM	Scotland, Glasgow	06:00	18:20
	Aberdeen	05:50	18:12
GU	Guernsey	05:55	18:12
GW	Wales, Cardiff	05:56	18:15
H4	Solomon Islands, Honiara	19:10	07:16
HA	Hungary, Budapest	04:28	16:44
HB	Switzerland, Bern	05:15	17:31
HBØ	Liechtenstein	05:06	17:22
HC	Ecuador, Quito	11:01	23:08
HC8	Galapagos Is.	11:51	23:58
HH	Haiti, Port-Au-Prince	10:37	22:46
HI	Dominican Republic, Santo Domingo	10:28	22:37
HK	Colombia, Bogota	10:45	22:53
HKØ	Malpelo Is.	11:14	23:21
HKØ	San Adreas	11:15	23:24
HL, HM	Korea, Seoul	21:18	09:31
HP	Panama, Panama	11:07	23:15
HR	Honduras, Tegucigalpa	11:37	23:46
HS	Thailand, Bangkok	23:07	11:15
HV	Vatican City	04:56	17:10
HZ, 7Z	Saudi Arabia, Dharan	02:27	14:38
	Mecca	03:09	15:18
	Riyadh	02:41	14:51
I	Italy, Rome	04:56	17:10
	Trieste	04:50	17:05
	Sicily	04:50	17:03
IS	Sardinia, Cagliari	05:10	17:23
J2	Djibouti, Djibouti	02:56	15:04
J3	Grenada	09:56	22:04
J5	Guinea-Bissau, Bissau	06:51	18:59
J6	St. Lucia	09:53	22:01
J7	Dominica	09:54	22:02
J8	St. Vincent	09:54	22:02
JA-JS	Japan, Tokyo	20:27	08:40
	Nagasaki	21:07	09:19
	Sapporo	20:20	08:34
JD	Minami Torishima	19:32	07:42
JD	Ogasawara, Kazan Is.	20:23	08:34
JT	Mongolia, Ulan Bator	22:37	10:53
JW	Svalbard, Spitsbergen	04:16	17:21
JX	Jan Mayen	06:07	18:45
JY	Jordan, Amman	03:23	15:35
KC4	Antarctica, Byrd Station	13:50	01:55
	McMurdo Sound	18:44	06:48
	Palmer Station	10:07	22:10
KC6	Micronesia, Ponape	19:16	07:23
KC6	Belau, Yap	20:36	08:44
	Koror	20:51	08:58
KG4	Guantanamo Bay	10:49	22:58
KH1	Baker, Howland Is.	17:33	05:40
KH2	Guam, Agana	20:09	08:18
KH3	Johnston Is.	17:02	05:11
KH4	Midway Is.	17:37	05:48
KH5	Palmyra Is.	16:37	04:45
KH5K	Kingman Reef	16:40	04:48
KH6	Hawaii, Hilo	16:08	04:18
	Honolulu	16:20	04:29
KH7	Kure Is.	17:41	05:52
KH8	American Samoa, Pago Pago	17:13	05:19
KH9	Wake Is.	18:42	06:51
KHØ	Mariana Is., Saipan	20:05	08:14
KL7	Alaska, Adak	17:30	05:48
	Anchorage	15:40	04:05
	Fairbanks	15:30	03:58
	Juneau	14:39	03:01
	Nome	16:40	05:08
KP1	Navassa Is.	10:48	22:57
KP2	Virgin Islands, Charlotte Amalie	10:08	22:17
KP4	Puerto Rico, San Juan	10:13	22:22
KP5	Desecheo Is.	10:18	22:27
KX6	Marshall Islands, Kwajalein	18:40	06:47
LA-LJ	Norway, Oslo	04:58	17:22
LU	Argentina, Buenos Aires	09:44	21:48
LX	Luxembourg	05:19	17:37
LZ	Bulgaria, Sofia	04:12	16:27
OA	Peru, Lima	10:58	23:04
OD	Lebanon, Beirut	03:25	15:37
OE	Austria, Vienna	04:39	16:56
OH	Finland, Helsinki	04:01	16:25
OHØ	Aland Is.	04:21	16:45
OJØ	Market Reef	04:25	16:49
OK	Czechoslovakia, Prague	04:46	17:04
ON	Belgium, Brussels	05:26	17:44
OX, XP	Greenland, Godthaab	09:06	21:33
	Thule	10:01	22:55
OY	Faroe Islands, Torshavn	06:07	18:33
OZ	Denmark, Copenhagen	04:52	17:13
P2	Papua New Guinea, Madang	20:07	08:13
	Port Moresby	20:01	08:07
PA-PI	Netherlands, Amsterdam	05:24	17:42
PJ, P4	Netherlands Antilles, Willemstad	10:25	22:33
PJ5-8	St. Maarten and Saba, St. Maarten	10:01	22:10
PY	Brazil, Brasilia	09:02	21:07
	Rio De Janeiro	08:43	20:48
	Natal	08:10	20:17
	Manaus	09:50	21:57
	Porto Alegre	09:15	21:20
PYØ	Fernando De Noronha	07:59	20:05
PYØ	St. Peter & St. Paul Rocks	07:47	19:54
PYØ	Trindade & Martin Vaz Is., Trindade	07:47	19:52
PZ	Suriname, Paramaribo	09:30	21:37
S2	Bangladesh, Dacca	23:46	11:56
S7	Seychelles, Victoria	02:08	14:14
S9	Sao Tome	05:23	17:29
SJ-SM	Sweden, Stockholm	04:29	16:52
SP	Poland, Krakow	04:24	16:41
	Warsaw	04:19	16:38
ST	Sudan, Khartoum	03:38	15:47
STØ	Southern Sudan, Juba	03:43	15:50
SU	Egypt, Cairo	03:42	15:53
SV	Greece, Athens	04:11	16:24
SV5	Dodecanese, Rhodes	03:54	16:06
SV9	Crete	04:06	16:18
SV/A	Mount Athos	04:09	16:22
T2	Tuvalu, Funafuti	17:55	06:01
T30	West Kiribati, Bonriki	18:16	06:23
T31	Central Kiribati, Kanton	17:16	05:23
T32	East Kiribati, Christmas Is.	16:19	04:26
T5	Somalia, Mogadishu	02:48	14:55
T7	San Marino	04:56	17:11

Pref	State/Province/Country/City	Sunrise	Sunset
TA	Turkey, Ankara	03:34	15:48
	Istanbul	03:50	16:04
TF	Iceland, Reykjavik	07:07	19:34
TG	Guatemala, Guatemala City	11:50	23:59
TI	Costa Rica, San Jose	11:25	23:33
TI9	Cocos Is.	11:37	23:44
TJ	Cameroon, Yaounde	05:03	17:10
TK	Corsica	05:10	17:24
TL	Central African Republic, Bangui	04:35	16:42
TN	Congo, Brazzaville	04:48	16:55
TR	Gabon, Libreville	05:11	17:18
TT	Chad, N'Djamena	04:49	16:57
TU	Ivory Coast, Abidjan	06:05	18:12
TY	Benin, Porto Novo	05:39	17:46
TZ	Mali, Bamako	06:21	18:29
UA	Russia, European,		
	Leningrad (UA1)	03:40	16:03
	Archangel (UA1)	02:57	15:25
	Murmansk (UA1)	03:23	15:57
	Moscow (UA3)	03:12	15:33
	Kuibyshev (UA4)	02:23	14:42
	Rostov (UA6)	03:06	15:23
UA1P	Franz Josef Land	01:44	14:57
UA2	Kaliningrad	04:21	16:41
UA9,Ø	Russia, Asiatic,		
	Novosibirsk (UA9)	00:11	12:31
	Perm (UA9)	01:56	14:19
	Omsk (UA9)	00:49	13:09
	Norilsk (UAØ)	23:43	12:18
	Irkutsk (UAØ)	22:46	11:05
	Vladivostok (UAØ)	20:58	09:12
	Petropavlovsk (UAØ)	19:08	07:27
	Khabarovsk (UAØ)	20:44	09:01
	Krasnoyarsk (UAØ)	23:31	11:52
	Yakutsk (UAØ)	21:01	09:27
	Wrangel Island (UAØ)	17:32	06:10
	Kyzyl (UAØY)	23:26	11:44
UB	Ukraine, Kiev	03:42	16:00
UC	Byelorussia, Minsk	03:53	16:12
UD	Azerbaijan, Baku	02:26	14:40
UF	Georgia, Tbilisi	02:46	15:01
UG	Armenia, Ierevan	02:48	15:02
UH	Turkoman, Ashkhabad	01:53	14:06
UI	Uzbek, Bukhara	01:28	13:42
	Tashkent	01:08	13:23
UJ	Tadzhik, Samarkand	01:19	13:32
	Dushanbe	01:11	13:24
UL	Kazakh, Alma-ata	00:38	12:52
UM	Kirghiz, Frunze	00:47	13:02
UO	Moldavia, Kishinev	03:49	16:06
UP	Lithuania, Vilna	04:01	16:21
UQ	Latvia, Riga	04:06	16:27
UR	Estonia, Tallinn	04:02	16:25
V2	Antigua & Barbuda, St. Johns	09:55	22:04
V3	Belize, Belmopan	11:43	23:52
V4	St. Christopher & Nevis	09:59	22:07
V8	Brunei, Bandar Seri Begawan	22:09	10:17
VE1, CY	Sable Is.	09:45	22:00
VE1, CY	St. Paul Is.	09:45	22:01
VK	Australia, Canberra (VK1)	19:54	07:58
	Sydney (VK2)	19:46	07:50
	Melbourne (VK3)	20:11	08:15
	Brisbane (VK4)	19:38	07:43
	Adelaide (VK5)	20:36	08:40
	Perth (VK6)	22:07	10:12
	Hobart, Tasmania (VK7)	20:02	08:05
	Darwin (VK8)	21:06	09:12
VK	Lord Howe Is.	19:14	07:18
VK9	Christmas Is.	22:47	10:53
VK9	Cocos-Keeling is. 23:23	11:28	
VK9	Mellish Reef	19:27	07:32
VK9	Norfolk Is.	18:39	06:43
VK9	Willis Is.	19:52	07:57
VKØ	Heard Is.	00:57	13:01
VKØ	Macquarie Is.	19:16	07:19
VP2E	Anguilla	10:00	22:09
VP2M	Montserrat	09:57	22:06
VP2V	British Virgin Is., Tortola	10:07	22:16
VP5	Turks & Caicos Islands,		
	Grand Turk	10:33	22:42
VP8	Falkland Islands, Stanley	09:43	21:46

Pref	State/Province/Country/City	Sunrise	Sunset
VP8	So. Georgia Is.	08:18	20:21
VP8	So. Orkney Is.	08:53	20:56
VP8	So. Sandwich Islands, Saunders Is.	07:38	19:41
VP8	So. Shetland Is., King George Is.	09:44	21:47
VP9	Bermuda	10:06	22:17
VQ9	Chagos, Diego Garcia	01:00	13:06
VR6	Pitcairn Is.	14:31	02:35
VS6	Hong Kong	22:11	10:20
VU	India, Bombay	00:57	13:06
	Calcutta	23:54	12:04
	New Delhi	00:38	12:49
	Bangalore	00:38	12:46
VU7	Andaman Islands, Port Blair	23:37	11:45
VU7	Laccadive Is.	00:57	13:05
XE	Mexico, Mexico City (XE1)	12:24	00:34
	Chihuahua (XE2)	12:51	01:02
	Merida (XE3)	11:47	23:56
XF4	Revilla Gigedo	13:14	01:23
XT	Burkina Faso, Onagadougou	05:55	18:03
XU	Kampuchea, Phnom Penh	22:49	10:57
XW	Laos, Viangchan	22:58	11:07
XX9	Macao	22:13	10:23
XZ	Burma, Rangoon	23:24	11:33
Y2-9	German Dem.Rep.(E), Berlin	04:50	17:08
YA	Afghanistan, Kandahar	01:24	13:35
	Kabul	01:10	13:22
YB-YD	Indonesia, Jakarta	22:42	10:49
	Medan	23:14	11:21
	Pontianak	22:32	10:39
	Jayapura	20:27	08:33
YI	Iraq, Baghdad	02:49	15:01
YJ	Vanuatu, Port Vila	18:37	06:42
YK	Syria, Damascus	03:22	15:33
YN, HT	Nicaragua, Managua	11:33	23:41
YO	Romania, Bucharest	04:01	16:16
YS	El Salvador, San Salvador	11:45	23:54
YU	Yugoslavia, Belgrade	04:23	16:38
YV	Venezuela, Caracas	10:17	22:25
YVØ	Aves Is.	10:03	22:12
Z2,ZE	Zimbabwe, Harare	03:46	15:51
ZA	Albania, Tirane	04:26	16:41
ZB	Gibraltar	06:08	18:21
ZC4	British Cyprus	03:35	15:47
ZD7	St. Helena	06:14	18:19
ZD8	Ascension Is.	06:47	18:53
ZD9	Tristan da Cunha	06:40	18:44
ZF	Cayman Is.	11:13	23:22
ZK1	No. Cook Is., Manihiki	16:34	04:40
ZK1	So. Cook Is., Rarotonga	16:29	04:34
ZK2	Niue	17:06	05:11
ZK3	Tokelaus, Atafu	17:21	05:26
ZL	New Zealand, Auckland (ZL1)	18:12	06:15
	Wellington (ZL2)	18:12	06:15
	Christchurch (ZL3)	18:21	06:24
	Dunedin (ZL4)	18:29	06:32
ZL5	Antarctica, Scott Base	18:45	06:49
ZL7	Chatham Is.	17:37	05:40
ZL8	Kermadec Is.	17:42	05:46
ZL9	Auckland & Campbell Is.,		
	Auckland	18:45	06:48
	Campbell Is.	18:35	06:38
ZP	Paraguay, Asuncion	09:41	21:46
ZS	South Africa, Cape Town (ZS1)	04:37	16:41
	Port Elizabeth (ZS2)	04:08	16:12
	Bloemfontein (ZS4)	04:06	16:10
	Durban (ZS5)	03:47	15:51
	Johannesburg (ZS6)	03:58	16:03
ZS2	Prince Edward & Marion Is.,		
	Marion	03:20	15:23
ZS3	Namibia, Windhoek	04:42	16:47
1AØ	S.M.O.M	04:56	17:10
1S	Spratly Is.	22:21	10:29
3A	Monaco	05:16	17:31
3B6	Agalega	02:03	14:09
3B7	St. Brandon	01:51	13:56
3B8	Mauritius	02:00	14:05
3B9	Rodriguez Is.	01:37	13:42
3C	Equatorial Guinea, Bata	05:10	17:17
	Malabo	05:14	17:21

Pref	State/Province/Country/City	Sunrise	Sunset
3CØ	Pagalu Is.	05:27	17:34
3D2	Fiji Is., Suva	17:57	06:02
3D6	Swaziland, Mbabane	03:46	15:51
3V	Tunisia, Tunis	05:05	17:18
3W	Vietnam, Ho Chi Minh City		
	(Saigon)	22:42	10:50
	Hanoi	22:45	10:54
3X	Republic of Guinea, Conakry	06:44	18:51
3Y	Bouvet	05:38	17:41
3Y	Peter I Is.	11:54	23:57
4S	Sri Lanka, Colombo	00:29	12:37
4U	ITU Geneva	05:20	17:36
4U	United Nations Hq.	10:42	22:56
4W	Yemen, Sanaa	02:52	15:00
4X, 4Z	Israel, Jerusalem	03:26	15:38
5A	Libya, Tripoli	04:57	17:09
	Benghazi	04:27	16:38
5B	Cyprus, Nicosia	03:33	15:45
5H	Tanzania, Dar es Salaam	03:12	15:18
5N	Nigeria, Lagos	05:35	17:43
5R	Madagascar, Antananarivo	02:40	14:45
5T	Mauritania, Nouakchott	06:52	19:01
5U	Niger, Niamey	05:41	17:49
5V	Togo, Lome	05:44	17:52
5W	Western Samoa, Apia	17:17	05:23
5X	Uganda, Kampala	03:39	15:46
5Z	Kenya, Nairobi	04:42	16:49
6W	Senegal, Dakar	06:58	19:07
6Y	Jamaica, Kingston	10:55	23:04

Pref	State/Province/Country/City	Sunrise	Sunset
7O	Yemen, People's Dem. Rep.,		
	Aden	02:49	14:57
7P	Lesotho, Maseru	04:01	16:05
7Q	Malawi, Lilongwe	03:35	15:40
	Blantyre	03:30	15:35
7X	Algeria, Algiers	05:34	17:47
8P	Barbados, Bridgetown	09:47	21:55
8Q	Maldive Is.	00:56	13:03
8R	Guyana, Georgetown	09:42	21:49
9G	Ghana, Accra	05:50	17:57
9H	Malta	04:49	17:01
9J	Zambia, Lusaka	03:57	16:02
9K	Kuwait	02:36	14:47
9L	Sierra Leone, Freetown	06:42	18:49
9M2	West Malaysia, Kuala Lumpur	23:03	11:10
9M6,8	East Malaysia, Sabah,		
	Sandakan (9M6)	21:57	10:04
	Sarawak, Kuching (9M8)	22:28	10:35
9N	Nepal, Katmandu	00:06	12:17
9Q	Zaire, Kinshasa	04:48	16:55
	Kisangani	04:09	16:15
	Lubumbashi	04:00	16:05
9U	Burundi, Bujumbura	03:52	15:59
9V	Singapore	22:54	11:01
9X	Rwanda, Kigali	03:49	15:56
9Y	Trinidad & Tobago, Port of		
	Spain	09:54	22:02
J2/A	Abu Ail	02:57	15:06

Pref	State/Province/Country/City	Sunrise	Sunset
VE1	New Brunswick, Saint John	10:29	21:54
	Nova Scotia, Halifax	10:19	21:45
	Prince Edward Island, Charlottetown	10:18	21:42
VE2	Quebec, Montreal	10:59	22:24
	Quebec City	10:51	22:14
VE3	Ontario, London	11:28	22:57
	Ottawa	11:08	22:33
	Sudbury	11:30	22:53
	Toronto	11:21	22:49
VE4	Manitoba, Winnipeg	12:37	23:55
VE5	Saskatchewan, Regina	13:07	00:24
	Saskatoon	13:17	00:32
VE6	Alberta, Calgary	13:46	01:02
	Edmonton	13:45	00:57
VE7	British Columbia, Prince George	14:23	01:34
	Prince Rupert	14:53	02:04
	Vancouver	14:20	01:39
VE8	Northwest Territories, Yellowknife	14:00	00:50
	Resolute	13:16	22:59
VY1	Yukon, Whitehorse	15:20	02:16
VO1	Newfoundland, St. John's	09:37	20:59
VO2	Labrador, Goose Bay	10:13	21:25
W1	Connecticut, Hartford	10:53	22:23
	Maine, Bangor	10:40	22:06
	Portland	10:45	22:12
	Massachusetts, Boston	10:47	22:16
	New Hampshire, Concord	10:49	22:17
	Rhode Island, Providence	10:48	22:18
	Vermont, Montpelier	10:55	22:21
W2	New Jersey, Atlantic City	10:59	22:31
	New York, Albany	10:58	22:27
	Buffalo	11:19	22:47
	New York City	10:58	22:29
	Syracuse	11:08	22:36
W3	Delaware, Wilmington	11:03	22:36
	District of Columbia, Washington	11:09	22:42
	Maryland, Baltimore	11:07	22:40
	Pennsylvania, Harrisburg	11:09	22:41
	Philadelphia	11:02	22:34
	Pittsburgh	11:22	22:53
	Scranton	11:05	22:35
W4	Alabama, Montgomery	11:43	23:23
	Florida, Jacksonville	11:23	23:05
	Miami	11:15	23:01
	Pensacola	11:45	23:27
	Georgia, Atlanta	11:36	23:14
	Savannah	11:22	23:02
	Kentucky, Lexington	11:38	23:13
	Louisville	11:44	23:18
	North Carolina, Charlotte	11:22	22:59
	Raleigh	11:14	22:50
	Wilmington	11:10	22:48
	South Carolina, Columbia	11:22	23:01
	Tennessee, Knoxville	11:35	23:11
	Memphis	11:59	23:37
	Nashville	11:47	23:23
	Virginia, Norfolk	11:05	22:40
	Richmond	11:10	22:44
W5	Arkansas, Little Rock	12:08	23:46
	Louisiana, New Orleans	11:57	23:39
	Shreveport	12:12	23:52
	Mississippi, Jackson	11:58	23:38
	New Mexico, Albuquerque	13:06	00:43
	Oklahoma, Oklahoma City	12:29	00:06
	Texas, Abilene	12:36	00:16
	Amarillo	12:46	00:23
	Dallas	12:25	00:04
	El Paso	13:03	00:44
	San Antonio	12:30	00:13
W6	California, Los Angeles	13:51	01:29
	San Francisco	14:10	01:44
W7	Arizona, Flagstaff	13:26	01:03
	Phoenix	13:26	01:05
	Idaho, Boise	13:48	01:16
	Pocatello	13:33	01:02

Pref	State/Province/Country/City	Sunrise	Sunset
	Montana, Billings	13:19	00:44
	Butte	13:35	01:00
	Great Falls	13:32	00:54
	Nevada, Las Vegas	13:40	01:16
	Reno	14:00	01:33
	Oregon, Portland	14:16	01:41
	Utah, Salt Lake City	13:30	01:01
	Washington, Seattle	14:16	01:38
	Spokane	13:56	01:18
	Wyoming, Cheyenne	13:01	00:32
	Sheridan	13:12	00:38
W8	Michigan, Detroit	11:35	23:04
	Grand Rapids	11:46	23:14
	Sault Ste. Marie	11:43	23:07
	Traverse City	11:47	23:13
	Ohio, Cincinnati	11:39	23:12
	Cleveland	11:29	22:59
	Columbus	11:33	23:05
	West Virginia, Charleston	11:27	23:01
W9	Illinois, Chicago	11:53	23:23
	Indiana, Indianapolis	11:46	23:18
	Wisconsin, Green Bay	11:56	23:23
	Milwaukee	11:55	23:23
W0	Colorado, Denver	13:01	00:34
	Grand Junction	13:15	00:48
	Iowa, Des Moines	12:17	23:47
	Kansas, Pratt	12:35	00:10
	Wichita	12:29	00:04
	Minnesota, Duluth	12:14	23:37
	Minneapolis	12:18	23:43
	Missouri, Columbia	12:10	23:43
	Kansas City	12:19	23:52
	St. Louis	12:01	23:35
	Nebraska, North Platte	12:45	00:16
	Omaha	12:26	23:56
	North Dakota, Fargo	12:33	23:56
	South Dakota, Rapid City	12:57	00:24
A2	Botswana, Gaborone	03:50	16:18
A3	Tonga, Nukualofa	17:16	05:40
A4	Oman, Muscat	01:59	13:47
A5	Bhutan, Thimbu	23:57	11:42
A6	United Arab Emirates, Abu Dhabi	02:17	14:04
A7	Qatar, Ad-Dawhah	02:28	14:15
A9	Bahrein, Al-Manamah	02:32	14:18
AP	Pakistan, Karachi	01:26	13:13
	Islamabad	01:05	12:44
BV	Taiwan, Taipei	21:48	09:35
BY	People's Rep. of China, Beijing	22:16	09:48
	Harbin	21:38	09:03
	Shanghai	21:51	09:32
	Fuzhou	21:57	09:43
	Xian	22:43	10:21
	Chongqing	22:50	10:33
	Chengdu	23:00	10:42
	Lhasa	23:51	11:34
	Urumqi	00:13	11:41
	Kashi	00:57	12:30
C2	Nauru	18:36	06:43
C3	Andorra	05:57	17:26
C5	The Gambia, Banjul	06:56	18:52
C6	Bahamas, Nassau	11:04	22:50
C9	Mozambique, Maputo	03:23	15:51
	Mozambique	02:55	15:14
CE	Chile, Santiago	10:12	22:49
CE0	Easter Island	12:50	01:20
CE0	San Felix	10:53	23:22
CE0	Juan Fernandez	10:44	23:21
CM,CO	Cuba, Havana	11:23	23:11
CN	Morocco, Casablanca	06:28	18:07
CP	Bolivia, La Paz	10:11	22:31
CT	Portugal, Lisbon	06:38	18:11
CT2, CU	Azores, Ponta Delgada	07:43	19:18
CT3	Madeira Islands, Funchal	07:05	18:45
CX	Uruguay, Montevideo	09:13	21:52
D2	Angola, Luanda	04:48	17:01
D4	Cape Verde, Praia	07:24	19:19

Pref	State/Province/Country/City	Sunrise	Sunset
D6	Comoros, Moroni	02:45	15:01
DA-DL	Germany, Fed. Rep. of (W), Bonn	05:41	16:58
DU	Philippines, Manilla	21:46	09:41
EA	Spain, Madrid	06:16	17:48
EA6	Balearic Is., Palma	05:51	17:23
EA8	Canary Is., Las Palmas	06:53	18:37
EA9	Ceuta & Melilla, Ceuta	06:20	17:57
	Melilla	06:11	17:48
EI	Ireland, Dublin	06:36	17:49
EL	Liberia, Monrovia	06:30	18:32
EP	Iran, Tehran	02:32	14:09
ET	Ethiopia, Addis Ababa	03:13	15:13
	Asmera	03:14	15:09
F	France, Paris	05:58	17:18
FG	Guadeloupe	09:57	21:51
FG, FS	St. Martin	10:03	21:56
FH	Mayotte	02:38	14:55
FK	New Caledonia, Noumia	18:29	06:54
FM	Martinique	09:54	21:49
FO	Clipperton	13:05	01:04
FO	Fr. Polynesia, Tahiti	16:15	04:36
FP	St. Pierre & Miquelon, St. Pierre	09:50	21:13
FR	Glorioso	02:30	14:46
FR	Juan de Nova	02:46	15:06
	Europa	02:53	15:18
FR	Reunion	01:53	14:17
FR	Tromelin	02:00	14:20
FT8W	Crozet	01:52	14:47
FT8X	Kerguelen	00:40	13:41
FT8Y	Antarctica, Dumont D'Urville	19:28	09:27
FT8Z	Amsterdam & St. Paul Is., Amsterdam	00:16	12:58
FW	Wallis & Futuna Is., Wallis	17:24	05:41
FY	Fr. Guiana, Cayenne	09:15	21:18
G	England, London	06:10	17:26
GD	Isle of Man	06:30	17:41
GI	Northern Ireland, Belfast	06:36	17:46
GJ	Jersey	06:17	17:36
GM	Scotland, Glasgow	06:31	17:38
	Aberdeen	06:24	17:28
GU	Guernsey	06:19	17:38
GW	Wales, Cardiff	06:22	17:38
H4	Solomon Islands, Honiara	19:00	07:14
HA	Hungary, Budapest	04:50	16:12
HB	Switzerland, Bern	05:36	16:59
HB0	Liechtenstein	05:28	16:50
HC	Ecuador, Quito	10:56	23:03
HC8	Galapagos Is.	11:46	23:53
HH	Haiti, Port-Au-Prince	10:40	22:33
HI	Dominican Republic, Santo Domingo	10:31	22:24
HK	Colombia, Bogota	10:42	22:45
HK0	Malpelo Is.	11:10	23:14
HK0	San Adreas	11:16	23:13
HL, HM	Korea, Seoul	21:32	09:07
HP	Panama, Panama	11:06	23:05
HR	Honduras, Tegucigalpa	11:38	23:34
HS	Thailand, Bangkok	23:07	11:04
HV	Vatican City	05:13	16:42
HZ, 7Z	Saudi Arabia, Dharan	02:35	14:20
	Mecca	03:13	15:03
	Riyadh	02:47	14:34
I	Italy, Rome	05:13	16:42
	Trieste	05:10	16:35
	Sicily	05:04	16:39
IS	Sardinia, Cagliari	05:25	16:57
J2	Djibouti, Djibouti	02:56	14:54
J3	Grenada	09:56	21:53
J5	Guinea-Bissau, Bissau	06:51	18:49
J6	St. Lucia	09:53	21:49
J7	Dominica	09:55	21:50
J8	St. Vincent	09:54	21:51
JA-JS	Japan, Tokyo	20:40	08:17
	Nagasaki	21:18	08:58
	Sapporo	20:38	08:06
JD	Minami Torishima	19:38	07:25
JD	Ogasawara, Kazan Is.	20:31	08:16
JT	Mongolia, Ulan Bator	22:59	10:21
JW	Svalbard, Spitsbergen	06:20	15:07
JX	Jan Mayen	07:14	17:27
JY	Jordan, Amman	03:34	15:14
KC4	Antarctica, Byrd Station	11:12	04:22
	McMurdo Sound	16:39	08:42
	Palmer Station	09:09	22:58
KC6	Micronesia, Ponape	19:14	07:15
KC6	Belau, Yap	20:35	08:34
	Koror	20:49	08:50
KG4	Guantanamo Bay	10:53	22:44
KH1	Baker, Howland Is.	17:28	05:35
KH2	Guam, Agana	20:10	08:06
KH3	Johnston Is.	17:05	04:58
KH4	Midway Is.	17:45	05:29
KH5	Palmyra Is.	16:35	04:37
KH5K	Kingman Reef	16:38	04:39
KH6	Hawaii, Hilo	16:12	04:04
	Honolulu	16:24	04:14
KH7	Kure Is.	17:49	05:33
KH8	American Samoa, Pago Pago	17:02	05:20
KH9	Wake Is.	18:45	06:37
KH0	Mariana Is., Saipan	20:07	08:02
KL7	Alaska, Adak	17:56	05:11
	Anchorage	16:20	03:14
	Fairbanks	16:18	03:00
	Juneau	15:14	02:16
	Nome	17:27	04:11
KP1	Navassa Is.	10:51	22:44
KP2	Virgin Islands, Charlotte Amalie	10:11	22:03
KP4	Puerto Rico, San Juan	10:16	22:08
KP5	Desecheo Is.	10:21	22:14
KX6	Marshall Islands, Kwajalein	18:38	06:38
LA-LJ	Norway, Oslo	05:36	16:33
LU	Argentina, Buenos Aires	09:22	22:00
LX	Luxembourg	05:43	17:02
LZ	Bulgaria, Sofia	04:30	15:59
OA	Peru, Lima	10:48	23:04
OD	Lebanon, Beirut	03:36	15:15
OE	Austria, Vienna	05:02	16:23
OH	Finland, Helsinki	04:39	15:36
OH0	Aland Is.	04:59	15:56
OJ0	Market Reef	05:03	16:00
OK	Czechoslovakia, Prague	05:11	16:29
ON	Belgium, Brussels	05:52	17:08
OX, XP	Greenland, Godthaab	09:52	20:36
	Thule	11:42	21:04
OY	Faroe Islands, Torshavn	06:49	17:40
OZ	Denmark, Copenhagen	05:23	16:31
P2	Papua New Guinea, Madang	20:00	08:10
	Port Moresby	19:52	08:06
PA-PI	Netherlands, Amsterdam	05:51	17:05
PJ, P4	Netherlands Antilles, Willemstad	10:25	22:22
PJ5-8	St. Maarten and Saba, St. Maarten	10:04	21:57
PY	Brazil, Brasilia	08:49	21:09
	Rio De Janeiro	08:27	20:53
	Natal	08:03	20:14
	Manaus	09:44	21:53
	Porto Alegre	08:56	21:29
PY0	Fernando De Noronha	07:52	20:02
PY0	St. Peter & St. Paul Rocks	07:42	19:48
PY0	Trindade & Martin Vaz Is., Trindade	07:33	19:56
PZ	Suriname, Paramaribo	09:27	21:29
S2	Bangladesh, Dacca	23:52	11:40
S7	Seychelles, Victoria	02:00	14:11
S9	Sao Tome	05:17	17:24
SJ-SM	Sweden, Stockholm	05:05	16:05
SP	Poland, Krakow	04:48	16:07
	Warsaw	04:46	16:01
ST	Sudan, Khartoum	03:40	15:35
ST0	Southern Sudan, Juba	03:40	15:42
SU	Egypt, Cairo	03:51	15:33
SV	Greece, Athens	04:26	16:00
SV5	Dodecanese, Rhodes	04:07	15:43
SV9	Crete	04:18	15:55
SV/A	Mount Athos	04:24	15:56
T2	Tuvalu, Funafuti	17:46	06:00
T30	West Kiribati, Bonriki	18:12	06:17
T31	Central Kiribati, Kanton	17:10	05:19
T32	East Kiribati, Christmas Is.	16:14	04:20
T5	Somalia, Mogadishu	02:43	14:48
T7	San Marino	05:15	16:42

Pref	State/Province/Country/City	Sunrise	Sunset
TA	Turkey, Ankara	03:50	15:22
	Istanbul	04:06	15:37
TF	Iceland, Reykjavik	07:53	18:38
TG	Guatemala, Guatemala City	11:52	23:47
TI	Costa Rica, San Jose	11:24	23:23
TI9	Cocos Is.	11:34	23:37
TJ	Cameroon, Yaounde	05:00	17:03
TK	Corsica	05:27	16:56
TL	Central African Republic, Bangui	04:31	16:35
TN	Congo, Brazzaville	04:41	16:51
TR	Gabon, Libreville	05:06	17:13
TT	Chad, N'Djamena	04:49	16:46
TU	Ivory Coast, Abidjan	06:02	18:05
TY	Benin, Porto Novo	05:36	17:38
TZ	Mali, Bamako	06:21	18:18
UA	Russia, European,		
	Leningrad (UA1)	04:17	15:15
	Archangel (UA1)	03:44	14:27
	Murmansk (UA1)	04:23	14:47
	Moscow (UA3)	03:43	14:51
	Kuibyshev (UA4)	02:51	14:03
	Rostov (UA6)	03:28	14:50
UA1P	Franz Josef Land	04:04	12:26
UA2	Kaliningrad	04:51	16:00
UA9,Ø	Russia, Asiatic,		
	Novosibirsk (UA9)	00:41	11:50
	Perm (UA9)	02:31	13:33
	Omsk (UA9)	01:19	12:28
	Norilsk (UAØ)	00:44	11:06
	Irkutsk (UAØ)	23:13	10:27
	Vladivostok (UAØ)	21:16	08:44
	Petropavlovsk (UAØ)	19:36	06:49
	Khabarovsk (UAØ)	21:07	08:27
	Krasnoyarsk (UAØ)	00:03	11:10
	Yakutsk (UAØ)	21:43	08:34
	Wrangel Island (UAØ)	18:39	04:52
	Kyzyl (UAØY)	23:52	11:07
UB	Ukraine, Kiev	04:07	15:24
UC	Byelorussia, Minsk	04:21	15:33
UD	Azerbaijan, Baku	02:42	14:14
UF	Georgia, Tbilisi	03:03	14:33
UG	Armenia, Ierevan	03:04	14:35
UH	Turkoman, Ashkhabad	02:07	13:41
UI	Uzbek, Bukhara	01:44	13:16
	Tashkent	01:25	12:56
UJ	Tadzhik, Samarkand	01:34	13:06
	Dushanbe	01:26	12:59
UL	Kazakh, Alma-ata	00:56	12:24
UM	Kirghiz, Frunze	01:05	12:33
UO	Moldavia, Kishinev	04:11	15:34
UP	Lithuania, Vilna	04:30	15:40
UQ	Latvia, Riga	04:39	15:43
UR	Estonia, Tallinn	04:39	15:38
V2	Antigua & Barbuda, St. Johns	09:58	21:51
V3	Belize, Belmopan	11:46	23:39
V4	St. Christopher & Nevis	10:01	21:55
V8	Brunei, Bandar Seri Begawan	22:06	10:09
VE1, CY	Sable Is.	10:04	21:31
VE1, CY	St. Paul Is.	10:07	21:29
VK	Australia, Canberra (VK1)	19:31	08:11
	Sydney (VK2)	19:24	08:01
	Melbourne (VK3)	19:46	08:29
	Brisbane (VK4)	19:20	07:51
	Adelaide (VK5)	20:14	08:52
	Perth (VK6)	21:47	10:22
	Hobart, Tasmania (VK7)	19:33	08:23
	Darwin (VK8)	20:56	09:12
VK	Lord Howe Is.	18:54	07:28
VK9	Christmas Is.	22:37	10:52
VK9	Cocos-Keeling Is.	23:12	11:28
VK9	Mellish Reef	19:14	07:35
VK9	Norfolk Is.	18:19	06:51
VK9	Willis Is.	19:40	07:59
VKØ	Heard Is.	00:19	13:29
VKØ	Macquarie Is.	18:35	07:49
VP2E	Anguilla	10:03	21:56
VP2M	Montserrat	09:59	21:53
VP2V	British Virgin Is., Tortola	10:10	22:02
VP5	Turks & Caicos Islands,		
	Grand Turk	10:37	22:27
VP8	Falkland Islands, Stanley	09:06	22:12
VP8	So. Georgia Is.	07:38	20:51
VP8	So. Orkney Is.	08:03	21:35
VP8	So. Sandwich Islands, Saunders Is.	06:53	20:16
VP8	So. Shetland Is., King George Is.	08:52	22:29
VP9	Bermuda	10:16	21:56
VQ9	Chagos, Diego Garcia	00:52	13:04
VR6	Pitcairn Is.	14:14	02:42
VS6	Hong Kong	22:16	10:05
VU	India, Bombay	01:00	12:52
	Calcutta	23:59	11:48
	New Delhi	00:47	12:31
	Bangalore	00:39	12:35
VU7	Andaman Islands, Port Blair	23:37	11:35
VU7	Laccadive Is.	00:56	12:55
XE	Mexico, Mexico City (XE1)	12:28	00:20
	Chihuahua (XE2)	13:00	00:43
	Merida (XE3)	11:51	23:41
XF4	Revilla Gigedo	13:17	01:09
XT	Burkina Faso, Onagadougou	05:55	17:52
XU	Kampuchea, Phnom Penh	22:49	10:47
XW	Laos, Viangchan	23:01	10:53
XX9	Macao	22:18	10:08
XZ	Burma, Rangoon	23:27	11:20
Y2-9	German Dem.Rep.(E), Berlin	05:17	16:31
YA	Afghanistan, Kandahar	01:33	13:15
	Kabul	01:22	13:00
YB-YD	Indonesia, Jakarta	22:34	10:46
	Medan	23:11	11:15
	Pontianak	22:27	10:34
	Jayapura	20:20	08:29
YI	Iraq, Baghdad	03:00	14:39
YJ	Vanuatu, Port Vila	18:24	06:45
YK	Syria, Damascus	03:33	15:12
YN, HT	Nicaragua, Managua	11:33	23:30
YO	Romania, Bucharest	04:20	15:46
YS	El Salvador, San Salvador	11:46	23:42
YU	Yugoslavia, Belgrade	04:43	16:08
YV	Venezuela, Caracas	10:16	22:15
YVØ	Aves Is.	10:05	22:00
Z2,ZE	Zimbabwe, Harare	03:33	15:54
ZA	Albania, Tirane	04:43	16:13
ZB	Gibraltar	06:21	17:57
ZC4	British Cyprus	03:46	15:24
ZD7	St. Helena	06:01	18:21
ZD8	Ascension Is.	06:39	18:51
ZD9	Tristan da Cunha	06:16	18:57
ZF	Cayman Is.	11:16	23:08
ZK1	No. Cook Is., Manihiki	16:24	04:39
ZK1	So. Cook Is., Rarotonga	16:15	04:39
ZK2	Niue	16:52	05:14
ZK3	Tokelaus, Atafu	17:12	05:25
ZL	New Zealand, Auckland (ZL1)	17:48	06:29
	Wellington (ZL2)	17:44	06:32
	Christchurch (ZL3)	17:52	06:42
	Dunedin (ZL4)	17:58	06:53
ZL5	Antarctica, Scott Base	16:38	08:46
ZL7	Chatham Is.	17:08	05:59
ZL8	Kermadec Is.	17:23	05:55
ZL9	Auckland & Campbell Is.,		
	Auckland	18:09	07:14
	Campbell Is.	17:57	07:05
ZP	Paraguay, Asuncion	09:24	21:52
ZS	South Africa, Cape Town (ZS1)	04:15	16:53
	Port Elizabeth (ZS2)	03:46	16:23
	Bloemfontein (ZS4)	03:47	16:19
	Durban (ZS5)	03:27	16:00
	Johannesburg (ZS6)	03:41	16:10
ZS2	Prince Edward & Marion Is.,		
	Marion	02:48	15:45
ZS3	Namibia, Windhoek	04:26	16:52
1AØ	S.M.O.M	05:13	16:43
1S	Spratly Is.	22:20	10:20
3A	Monaco	05:34	17:02
3B6	Agalega	01:54	14:08
3B7	St. Brandon	01:38	13:58
3B8	Mauritius	01:46	14:09
3B9	Rodriguez Is.	01:22	13:45
3C	Equatorial Guinea, Bata	05:06	17:11
	Malabo	05:10	17:14

Pref	State/Province/Country/City	Sunrise	Sunset
3CØ	Pagalu Is.	05:21	17:29
3D2	Fiji Is., Suva	17:43	06:05
3D6	Swaziland, Mbabane	03:28	15:58
3V	Tunisia, Tunis	05:19	16:54
3W	Vietnam, Ho Chi Minh City		
	(Saigon)	22:41	10:40
	Hanoi	22:49	10:39
3X	Republic of Guinea, Conakry	06:43	18:42
3Y	Bouvet	04:57	18:10
3Y	Peter I Is.	10:43	00:56
4S	Sri Lanka, Colombo	00:27	12:29
4U	ITU Geneva	05:41	17:05
4U	United Nations Hq.	10:58	22:29
4W	Yemen, Sanaa	02:53	14:48
4X, 4Z	Israel, Jerusalem	03:36	15:17
5A	Libya, Tripoli	05:07	16:47
	Benghazi	04:37	16:18
5B	Cyprus, Nicosia	03:45	15:22
5H	Tanzania, Dar es Salaam	03:03	15:15
5N	Nigeria, Lagos	05:33	17:35
5R	Madagascar, Antananarivo	02:26	14:48
5T	Mauritania, Nouakchott	06:55	18:48
5U	Niger, Niamey	05:41	17:38
5V	Togo, Lome	05:41	17:44
5W	Western Samoa, Apia	17:06	05:23
5X	Uganda, Kampala	03:34	15:41
5Z	Kenya, Nairobi	04:36	16:44
6W	Senegal, Dakar	07:00	18:55
6Y	Jamaica, Kingston	10:58	22:51

Pref	State/Province/Country/City	Sunrise	Sunset
7O	Yemen, People's Dem. Rep.,		
	Aden	02:49	14:46
7P	Lesotho, Maseru	03:41	16:14
7Q	Malawi, Lilongwe	03:23	15:41
	Blantyre	03:18	15:37
7X	Algeria, Algiers	05:48	17:23
8P	Barbados, Bridgetown	09:47	21:44
8Q	Maldive Is.	00:52	12:56
8R	Guyana, Georgetown	09:39	21:41
9G	Ghana, Accra	05:47	17:49
9H	Malta	05:02	16:38
9J	Zambia, Lusaka	03:45	16:04
9K	Kuwait	02:45	14:28
9L	Sierra Leone, Freetown	06:40	18:40
9M2	West Malaysia, Kuala Lumpur	22:59	11:03
9M6,8	East Malaysia, Sabah,		
	Sandakan (9M6)	21:54	09:56
	Sarawak, Kuching (9M8)	22:23	10:29
9N	Nepal, Katmandu	00:14	11:59
9Q	Zaire, Kinshasa	04:41	16:51
	Kisangani	04:03	16:10
	Lubumbashi	03:49	16:05
9U	Burundi, Bujumbura	03:46	15:55
9V	Singapore	22:49	10:55
9X	Rwanda, Kigali	03:43	15:51
9Y	Trinidad & Tobago, Port of		
	Spain	09:53	21:52
J2/A	Abu Ail	02:58	14:54

Pref	State/Province/Country/City	Sunrise	Sunset
VE1	New Brunswick, Saint John	10:48	21:30
	Nova Scotia, Halifax	10:37	21:21
	Prince Edward Island,		
	Charlottetown	10:37	21:16
VE2	Quebec, Montreal	11:18	21:59
	Quebec City	11:10	21:48
VE3	Ontario, London	11:45	22:34
	Ottawa	11:26	22:08
	Sudbury	11:49	22:27
	Toronto	11:38	22:25
VE4	Manitoba, Winnipeg	12:59	23:26
VE5	Saskatchewan, Regina	13:30	23:55
	Saskatoon	13:41	00:01
VE6	Alberta, Calgary	14:09	00:32
	Edmonton	14:11	00:25
VE7	British Columbia, Prince George	14:49	01:02
	Prince Rupert	15:20	01:31
	Vancouver	14:42	01:11
VE8	Northwest Territories,		
	Yellowknife	14:39	00:05
	Resolute	14:41	21:27
VY1	Yukon, Whitehorse	15:56	01:34
VO1	Newfoundland, St. John's	09:57	20:33
VO2	Labrador, Goose Bay	10:38	20:53
W1	Connecticut, Hartford	11:09	22:01
	Maine, Bangor	10:58	21:41
	Portland	11:02	21:49
	Massachusetts, Boston	11:04	21:54
	New Hampshire, Concord	11:06	21:54
	Rhode Island, Providence	11:04	21:56
	Vermont, Montpelier	11:12	21:57
W2	New Jersey, Atlantic City	11:13	22:11
	New York, Albany	11:15	22:04
	Buffalo	11:35	22:24
	New York City	11:13	22:07
	Syracuse	11:25	22:13
W3	Delaware, Wilmington	11:18	22:15
	District of Columbia,		
	Washington	11:23	22:22
	Maryland, Baltimore	11:22	22:20
	Pennsylvania, Harrisburg	11:24	22:20
	Philadelphia	11:17	22:13
	Pittsburgh	11:37	22:32
	Scranton	11:21	22:13
W4	Alabama, Montgomery	11:53	23:06
	Florida, Jacksonville	11:32	22:50
	Miami	11:22	22:48
	Pensacola	11:54	23:12
	Georgia, Atlanta	11:47	22:57
	Savannah	11:32	22:46
	Kentucky, Lexington	11:52	22:53
	Louisville	11:57	22:58
	North Carolina, Charlotte	11:34	22:41
	Raleigh	11:26	22:32
	Wilmington	11:21	22:31
	South Carolina, Columbia	11:33	22:43
	Tennessee, Knoxville	11:47	22:53
	Memphis	12:11	23:18
	Nashville	11:59	23:04
	Virginia, Norfolk	11:18	22:21
	Richmond	11:23	22:25
W5	Arkansas, Little Rock	12:19	23:28
	Louisiana, New Orleans	12:06	23:24
	Shreveport	12:22	23:36
	Mississippi, Jackson	12:08	23:22
	New Mexico, Albuquerque	13:17	00:25
	Oklahoma, Oklahoma City	12:41	23:48
	Texas, Abilene	12:46	00:00
	Amarillo	12:58	00:05
	Dallas	12:35	23:48
	El Paso	13:13	00:27
	San Antonio	12:39	23:58
W6	California, Los Angeles	14:02	01:12
	San Francisco	14:23	01:25
W7	Arizona, Flagstaff	13:37	00:45
	Phoenix	13:37	00:48
	Idaho, Boise	14:06	00:52
	Pocatello	13:50	00:39

Pref	State/Province/Country/City	Sunrise	Sunset
	Montana, Billings	13:38	00:19
	Butte	13:54	00:34
	Great Falls	13:52	00:27
	Nevada, Las Vegas	13:52	00:57
	Reno	14:15	01:12
	Oregon, Portland	14:34	01:16
	Utah, Salt Lake City	13:45	00:39
	Washington, Seattle	14:36	01:11
	Spokane	14:16	00:51
	Wyoming, Cheyenne	13:17	00:10
	Sheridan	13:30	00:14
W8	Michigan, Detroit	11:51	22:41
	Grand Rapids	12:03	22:51
	Sault Ste. Marie	12:03	22:41
	Traverse City	12:05	22:48
	Ohio, Cincinnati	11:53	22:51
	Cleveland	11:45	22:37
	Columbus	11:48	22:44
	West Virginia, Charleston	11:41	22:41
W9	Illinois, Chicago	12:09	23:00
	Indiana, Indianapolis	12:01	22:57
	Wisconsin, Green Bay	12:14	22:58
	Milwaukee	12:12	23:00
W0	Colorado, Denver	13:16	00:13
	Grand Junction	13:29	00:28
	Iowa, Des Moines	12:32	23:25
	Kansas, Pratt	12:48	23:50
	Wichita	12:43	23:44
	Minnesota, Duluth	12:34	23:11
	Minneapolis	12:36	23:19
	Missouri, Columbia	12:24	23:23
	Kansas City	12:33	23:32
	St. Louis	12:15	23:15
	Nebraska, North Platte	13:01	23:54
	Omaha	12:41	23:34
	North Dakota, Fargo	12:53	23:30
	South Dakota, Rapid City	13:14	00:00

Pref	State/Province/Country/City	Sunrise	Sunset
A2	Botswana, Gaborone	03:37	16:24
A3	Tonga, Nukualofa	17:05	05:45
A4	Oman, Muscat	02:05	13:35
A5	Bhutan, Thimbu	00:05	11:28
A6	United Arab Emirates, Abu		
	Dhabi	02:23	13:51
A7	Qatar, Ad-Dawhah	02:35	14:02
A9	Bahrein, Al-Manamah	02:39	14:04
AP	Pakistan, Karachi	01:32	13:00
	Islamabad	01:16	12:27
BV	Taiwan, Taipei	21:55	09:22
BY	People's Rep. of China, Beijing	22:30	09:27
	Harbin	21:57	08:38
	Shanghai	22:00	09:16
	Fuzhou	22:04	09:30
	Xian	22:54	10:03
	Chongqing	22:59	10:17
	Chengdu	23:09	10:26
	Lhasa	00:00	11:19
	Urumqi	00:31	11:17
	Kashi	01:11	12:09
C2	Nauru	18:33	06:40
C3	Andorra	06:13	17:03
C5	The Gambia, Banjul	06:58	18:44
C6	Bahamas, Nassau	11:10	22:37
C9	Mozambique, Maputo	03:09	15:59
	Mozambique	02:46	15:17
CE	Chile, Santiago	09:55	23:00
CE0	Easter Island	12:36	01:28
CE0	San Felix	10:39	23:30
CE0	Juan Fernandez	10:27	23:32
CM,CO	Cuba, Havana	11:29	22:59
CN	Morocco, Casablanca	06:39	17:50
CP	Bolivia, La Paz	10:01	22:34
CT	Portugal, Lisbon	06:51	17:51
CT2, CU	Azores, Ponta Delgada	07:56	18:58
CT3	Madeira Islands, Funchal	07:15	18:28
CX	Uruguay, Montevideo	08:55	22:03
D2	Angola, Luanda	04:41	17:02
D4	Cape Verde, Praia	07:26	19:10

Pref	State/Province/Country/City	Sunrise	Sunset
D6	Comoros, Moroni	02:37	15:02
DA-DL	Germany, Fed. Rep. of (W),		
	Bonn	06:04	16:29
DU	Philippines, Manilla	21:48	09:32
EA	Spain, Madrid	06:31	17:27
EA6	Balearic Is., Palma	06:05	17:03
EA8	Canary Is., Las Palmas	07:01	18:22
EA9	Ceuta & Melilla, Ceuta	06:32	17:38
	Melilla	06:23	17:30
EI	Ireland, Dublin	07:02	17:17
EL	Liberia, Monrovia	06:29	18:26
EP	Iran, Tehran	02:44	13:50
ET	Ethiopia, Addis Ababa	03:13	15:06
	Asmera	03:17	15:00
F	France, Paris	06:19	16:51
FG	Guadeloupe	10:00	21:42
FG, FS	St. Martin	10:07	21:46
FH	Mayotte	02:30	14:56
FK	New Caledonia, Noumia	18:17	07:00
FM	Martinique	09:56	21:40
FO	Clipperton	13:05	00:57
FO	Fr. Polynesia, Tahiti	16:05	04:39
FP	St. Pierre & Miquelon, St. Pierre	10:09	20:47
FR	Glorioso	02:23	14:47
FR	Juan de Nova	02:36	15:10
	Europa	02:41	15:24
FR	Reunion	01:42	14:22
FR	Tromelin	01:51	14:22
FT8W	Crozet	01:26	15:06
FT8X	Kerguelen	00:12	14:03
FT8Y	Antarctica, Dumont D'Urville	18:32	10:17
FT8Z	Amsterdam & St. Paul Is.,		
	Amsterdam	23:56	13:12
FW	Wallis & Futuna Is., Wallis	17:16	05:43
FY	Fr. Guiana, Cayenne	09:14	21:13
G	England, London	06:34	16:55
GD	Isle of Man	06:57	17:07
GI	Northern Ireland, Belfast	07:03	17:12
GJ	Jersey	06:38	17:08
GM	Scotland, Glasgow	06:59	17:03
	Aberdeen	06:54	16:51
GU	Guernsey	06:41	17:09
GW	Wales, Cardiff	06:46	17:08
H4	Solomon Islands, Honiara	18:54	07:15
HA	Hungary, Budapest	05:10	15:46
HB	Switzerland, Bern	05:56	16:33
HBØ	Liechtenstein	05:48	16:24
HC	Ecuador, Quito	10:53	23:00
HC8	Galapagos Is.	11:42	23:50
HH	Haiti, Port-Au-Prince	10:44	22:23
HI	Dominican Republic, Santo		
	Domingo	10:35	22:13
HK	Colombia, Bogota	10:41	22:41
HKØ	Malpelo Is.	11:08	23:09
HKØ	San Adreas	11:17	23:05
HL, HM	Korea, Seoul	21:45	08:47
HP	Panama, Panama	11:06	22:59
HR	Honduras, Tegucigalpa	11:40	23:26
HS	Thailand, Bangkok	23:09	10:55
HV	Vatican City	05:28	16:20
HZ, 7Z	Saudi Arabia, Dharan	02:42	14:07
	Mecca	03:18	14:52
	Riyadh	02:53	14:21
I	Italy, Rome	05:28	16:20
	Trieste	05:29	16:09
	Sicily	05:17	16:19
IS	Sardinia, Cagliari	05:39	16:37
J2	Djibouti, Djibouti	02:57	14:46
J3	Grenada	09:57	21:46
J5	Guinea-Bissau, Bissau	06:52	18:41
J6	St. Lucia	09:55	21:41
J7	Dominica	09:58	21:41
J8	St. Vincent	09:56	21:43
JA-JS	Japan, Tokyo	20:52	07:58
	Nagasaki	21:28	08:41
	Sapporo	20:54	07:43
JD	Minami Torishima	19:44	07:12
JD	Ogasawara, Kazan Is.	20:39	08:02
JT	Mongolia, Ulan Bator	23:20	09:54
JW	Svalbard, Spitsbergen	08:44	12:36
JX	Jan Mayen	08:18	16:17

Pref	State/Province/Country/City	Sunrise	Sunset
JY	Jordan, Amman	03:44	14:58
KC4	Antarctica, Byrd Station	24 hrs day	
	McMurdo Sound	14:02	11:13
	Palmer Station	08:17	23:43
KC6	Micronesia, Ponape	19:13	07:09
KC6	Belau, Yap	20:35	08:28
	Koror	20:48	08:44
KG4	Guantanamo Bay	10:57	22:33
KH1	Baker, Howland Is.	17:25	05:31
KH2	Guam, Agana	20:12	07:58
KH3	Johnston Is.	17:08	04:49
KH4	Midway Is.	17:53	05:15
KH5	Palmyra Is.	16:34	04:32
KH5K	Kingman Reef	16:38	04:33
KH6	Hawaii, Hilo	16:16	03:53
	Honolulu	16:29	04:03
KH7	Kure Is.	17:57	05:18
KH8	American Samoa, Pago Pago	16:53	05:22
KH9	Wake Is.	18:49	06:26
KHØ	Mariana Is., Saipan	20:09	07:53
KL7	Alaska, Adak	18:20	04:41
	Anchorage	16:57	02:32
	Fairbanks	17:01	02:10
	Juneau	15:46	01:38
	Nome	18:10	03:21
KP1	Navassa Is.	10:55	22:33
KP2	Virgin Islands, Charlotte Amalie	10:14	21:53
KP4	Puerto Rico, San Juan	10:20	21:58
KP5	Desecheo Is.	10:25	22:04
KX6	Marshall Islands, Kwajalein	18:38	06:32
LA-LJ	Norway, Oslo	06:10	15:52
LU	Argentina, Buenos Aires	09:04	22:12
LX	Luxembourg	06:05	16:34
LZ	Bulgaria, Sofia	04:46	15:36
OA	Peru, Lima	10:40	23:05
OD	Lebanon, Beirut	03:47	14:57
OE	Austria, Vienna	05:22	15:56
OH	Finland, Helsinki	05:14	14:55
OHØ	Aland Is.	05:34	15:15
OJØ	Market Reef	05:38	15:18
OK	Czechoslovakia, Prague	05:33	16:00
ON	Belgium, Brussels	06:15	16:39
OX, XP	Greenland, Godthaab	10:34	19:48
	Thule	13:26	19:13
OY	Faroe Islands, Torshavn	07:27	16:56
OZ	Denmark, Copenhagen	05:52	15:56
P2	Papua New Guinea, Madang	19:54	08:09
	Port Moresby	19:45	08:06
PA-PI	Netherlands, Amsterdam	06:15	16:34
PJ, P4	Netherlands Antilles, Willemstad	10:26	22:14
PJ5-8	St. Maarten and Saba,		
	St. Maarten	10:07	21:47
PY	Brazil, Brasilia	08:40	21:12
	Rio De Janeiro	08:15	20:59
	Natal	07:57	20:13
	Manaus	09:39	21:51
	Porto Alegre	08:40	21:38
PYØ	Fernando De Noronha	07:47	20:00
PYØ	St. Peter & St. Paul Rocks	07:39	19:44
PYØ	Trindade & Martin Vaz Is.,		
	Trindade	07:22	20:01
PZ	Suriname, Paramaribo	09:26	21:24
S2	Bangladesh, Dacca	23:58	11:27
S7	Seychelles, Victoria	01:55	14:09
S9	Sao Tome	05:14	17:21
SJ-SM	Sweden, Stockholm	05:39	15:25
SP	Poland, Krakow	05:11	15:38
	Warsaw	05:11	15:30
ST	Sudan, Khartoum	03:43	15:26
STØ	Southern Sudan, Juba	03:38	15:37
SU	Egypt, Cairo	04:00	15:18
SV	Greece, Athens	04:39	15:40
SV5	Dodecanese, Rhodes	04:19	15:24
SV9	Crete	04:30	15:37
SV/A	Mount Athos	04:39	15:35
T2	Tuvalu, Funafuti	17:40	06:00
T3Ø	West Kiribati, Bonriki	18:09	06:14
T31	Central Kiribati, Kanton	17:06	05:17
T32	East Kiribati, Christmas Is.	16:12	04:16
T5	Somalia, Mogadishu	02:41	14:44
T7	San Marino	05:32	16:18

Pref	State/Province/Country/City	Sunrise	Sunset
TA	Turkey, Ankara	04:04	15:01
	Istanbul	04:22	15:15
TF	Iceland, Reykjavik	08:35	17:49
TG	Guatemala, Guatemala City	11:54	23:38
TI	Costa Rica, San Jose	11:24	23:16
TI9	Cocos Is.	11:33	23:31
TJ	Cameroon, Yaounde	04:58	16:59
TK	Corsica	05:43	16:34
TL	Central African Republic, Bangui	04:30	16:30
TN	Congo, Brazzaville	04:36	16:50
TR	Gabon, Libreville	05:03	17:09
TT	Chad, N'Djamena	04:50	16:38
TU	Ivory Coast, Abidjan	06:01	18:00
TY	Benin, Porto Novo	05:35	17:32
TZ	Mali, Bamako	06:22	18:10
UA	Russia, European,		
	Leningrad (UA1)	04:52	14:34
	Archangel (UA1)	04:27	13:37
	Murmansk (UA1)	05:18	13:45
	Moscow (UA3)	04:12	14:16
	Kuibyshev (UA4)	03:16	13:31
	Rostov (UA6)	03:48	14:24
UA1P	Franz Josef Land	07:29	08:56
UA2	Kaliningrad	05:18	15:26
UA9,Ø	Russia, Asiatic,		
	Novosibirsk (UA9)	01:09	11:16
	Perm (UA9)	03:02	12:56
	Omsk (UA9)	01:47	11:54
	Norilsk (UAØ)	01:40	10:03
	Irkutsk (UAØ)	23:38	09:56
	Vladivostok (UAØ)	21:33	08:21
	Petropavlovsk (UAØ)	20:01	06:17
	Khabarovsk (UAØ)	21:28	08:00
	Krasnoyarsk (UAØ)	00:32	10:34
	Yakutsk (UAØ)	22:21	07:50
	Wrangel Island (UAØ)	19:42	03:42
	Kyzyl (UAØY)	00:16	10:37
UB	Ukraine, Kiev	04:29	14:55
UC	Byelorussia, Minsk	04:48	15:00
UD	Azerbaijan, Baku	02:57	13:52
UF	Georgia, Tbilisi	03:19	14:11
UG	Armenia, Ierevan	03:18	14:14
UH	Turkoman, Ashkhabad	02:20	13:21
UI	Uzbek, Bukhara	01:58	12:55
	Tashkent	01:40	12:34
UJ	Tadzhik, Samarkand	01:49	12:46
	Dushanbe	01:40	12:38
UL	Kazakh, Alma-ata	01:13	12:00
UM	Kirghiz, Frunze	01:21	12:10
UO	Moldavia, Kishinev	04:31	15:08
UP	Lithuania, Vilna	04:57	15:07
UQ	Latvia, Riga	05:09	15:07
UR	Estonia, Tallinn	05:12	14:58
V2	Antigua & Barbuda, St. Johns	10:01	21:42
V3	Belize, Belmopan	11:49	23:30
V4	St. Christopher & Nevis	10:04	21:45
V8	Brunei, Bandar Seri Begawan	22:05	10:04
VE1, CY	Sable Is.	10:21	21:07
VE1, CY	St. Paul Is.	10:26	21:03
VK	Australia, Canberra (VK1)	19:13	08:23
	Sydney (VK2)	19:06	08:13
	Melbourne (VK3)	19:26	08:42
	Brisbane (VK4)	19:06	07:59
	Adelaide (VK5)	19:55	09:04
	Perth (VK6)	21:30	10:32
	Hobart, Tasmania (VK7)	19:10	08:40
	Darwin (VK8)	20:48	09:14
VK	Lord Howe Is.	18:37	07:38
VK9	Christmas Is.	22:30	10:53
VK9	Cocos-Keeling Is.	23:04	11:30
VK9	Mellish Reef	19:04	07:38
VK9	Norfolk Is.	18:04	07:00
VK9	Willis Is.	19:30	08:02
VKØ	Heard Is.	23:47	13:54
VKØ	Macquarie Is.	18:01	08:17
VP2E	Anguilla	10:07	21:46
VP2M	Montserrat	10:02	21:44
VP2V	British Virgin Is., Tortola	10:13	21:52
VP5	Turks & Caicos Islands,		
	Grand Turk	10:42	22:16
VP8	Falkland Islands, Stanley	08:35	22:37

Pref	State/Province/Country/City	Sunrise	Sunset
VP8	So. Georgia Is.	07:05	21:18
VP8	So. Orkney Is.	07:21	22:12
VP8	So. Sandwich Islands, Saunders Is.	06:15	20:47
VP8	So. Shetland Is., King George Is.	08:07	23:08
VP9	Bermuda	10:26	21:40
VQ9	Chagos, Diego Garcia	00:46	13:04
VR6	Pitcairn Is.	14:01	02:49
VS6	Hong Kong	22:21	09:53
VU	India, Bombay	01:04	12:42
	Calcutta	00:05	11:36
	New Delhi	00:55	12:16
	Bangalore	00:40	12:27
VU7	Andaman Islands, Port Blair	23:38	11:28
VU7	Laccadive Is.	00:56	12:48
XE	Mexico, Mexico City (XE1)	12:32	00:09
	Chihuahua (XE2)	13:08	00:28
	Merida (XE3)	11:56	23:30
XF4	Revilla Gigedo	13:21	00:59
XT	Burkina Faso, Onagadougou	05:57	17:45
XU	Kampuchea, Phnom Penh	22:50	10:40
XW	Laos, Viangchan	23:04	10:43
XX9	Macao	22:24	09:56
XZ	Burma, Rangoon	23:30	11:11
Y2-9	German Dem.Rep.(E), Berlin	05:42	16:00
YA	Afghanistan, Kandahar	01:43	12:59
	Kabul	01:33	12:42
YB-YD	Indonesia, Jakarta	22:29	10:45
	Medan	23:09	11:10
	Pontianak	22:24	10:30
	Jayapura	20:16	08:27
YI	Iraq, Baghdad	03:10	14:22
YJ	Vanuatu, Port Vila	18:14	06:48
YK	Syria, Damascus	03:44	14:54
YN, HT	Nicaragua, Managua	11:34	23:23
YO	Romania, Bucharest	04:37	15:22
YS	El Salvador, San Salvador	11:48	23:34
YU	Yugoslavia, Belgrade	05:01	15:44
YV	Venezuela, Caracas	10:17	22:08
YVØ	Aves Is.	10:08	21:50
Z2,ZE	Zimbabwe, Harare	03:23	15:58
ZA	Albania, Tirane	04:58	15:52
ZB	Gibraltar	06:33	17:38
ZC4	British Cyprus	03:58	15:07
ZD7	St. Helena	05:52	18:24
ZD8	Ascension Is.	06:32	18:51
ZD9	Tristan da Cunha	05:56	19:10
ZF	Cayman Is.	11:21	22:57
ZK1	No. Cook Is., Manihiki	16:17	04:40
ZK1	So. Cook Is., Rarotonga	16:03	04:44
ZK2	Niue	16:41	05:18
ZK3	Tokelaus, Atafu	17:05	05:25
ZL	New Zealand, Auckland (ZL1)	17:28	06:42
	Wellington (ZL2)	17:22	06:48
	Christchurch (ZL3)	17:28	07:00
	Dunedin (ZL4)	17:32	07:12
ZL5	Antarctica, Scott Base	13:53	11:25
ZL7	Chatham Is.	16:44	06:17
ZL8	Kermadec Is.	17:08	06:04
ZL9	Auckland & Campbell Is.,		
	Auckland	17:40	07:37
	Campbell Is.	17:25	07:30
ZP	Paraguay, Asuncion	09:11	21:59
ZS	South Africa, Cape Town (ZS1)	03:57	17:04
	Port Elizabeth (ZS2)	03:28	16:35
	Bloemfontein (ZS4)	03:32	16:28
	Durban (ZS5)	03:12	16:09
	Johannesburg (ZS6)	03:27	16:17
ZS2	Prince Edward & Marion Is., Marion	02:22	16:04
ZS3	Namibia, Windhoek	04:14	16:57
1AØ	S.M.O.M	05:29	16:20
1S	Spratly Is.	22:20	10:13
3A	Monaco	05:51	16:38
3B6	Agalega	01:46	14:09
3B7	St. Brandon	01:29	14:01
3B8	Mauritius	01:35	14:14
3B9	Rodriguez Is.	01:12	13:50
3C	Equatorial Guinea, Bata	05:03	17:07
	Malabo	05:08	17:10

Pref	State/Province/Country/City	Sunrise	Sunset
3CØ	Pagalu Is.	05:17	17:26
3D2	Fiji Is., Suva	17:33	06:08
3D6	Swaziland, Mbabane	03:15	16:05
3V	Tunisia, Tunis	05:32	16:35
3W	Vietnam, Ho Chi Minh City		
	(Saigon)	22:42	10:33
	Hanoi	22:54	10:28
3X	Republic of Guinea, Conakry	06:43	18:35
3Y	Bouvet	04:23	18:38
3Y	Peter I Is.	09:40	01:54
4S	Sri Lanka, Colombo	00:26	12:23
4U	ITU Geneva	06:00	16:39
4U	United Nations Hq.	11:13	22:07
4W	Yemen, Sanaa	02:56	14:39
4X, 4Z	Israel, Jerusalem	03:46	15:01
5A	Libya, Tripoli	05:18	16:31
	Benghazi	04:47	16:01
5B	Cyprus, Nicosia	03:57	15:04
5H	Tanzania, Dar es Salaam	02:57	15:15
5N	Nigeria, Lagos	05:32	17:29
5R	Madagascar, Antananarivo	02:16	14:53
5T	Mauritania, Nouakchott	06:59	18:38
5U	Niger, Niamey	05:43	17:29
5V	Togo, Lome	05:40	17:38
5W	Western Samoa, Apia	16:58	05:25
5X	Uganda, Kampala	03:31	15:37
5Z	Kenya, Nairobi	04:33	16:41
6W	Senegal, Dakar	07:02	18:46
6Y	Jamaica, Kingston	11:02	22:41

Pref	State/Province/Country/City	Sunrise	Sunset
7O	Yemen, People's Dem. Rep.,		
	Aden	02:51	14:38
7P	Lesotho, Maseru	03:26	16:22
7Q	Malawi, Lilongwe	03:15	15:43
	Blantyre	03:08	15:40
7X	Algeria, Algiers	06:00	17:04
8P	Barbados, Bridgetown	09:49	21:36
8Q	Maldive Is.	00:51	12:51
8R	Guyana, Georgetown	09:39	21:35
9G	Ghana, Accra	05:46	17:44
9H	Malta	05:14	16:19
9J	Zambia, Lusaka	03:36	16:06
9K	Kuwait	02:54	14:12
9L	Sierra Leone, Freetown	06:40	18:34
9M2	West Malaysia, Kuala Lumpur	22:57	10:59
9M6,8	East Malaysia, Sabah,		
	Sandakan (9M6)	21:53	09:51
	Sarawak, Kuching (9M8)	22:21	10:25
9N	Nepal, Katmandu	00:22	11:44
9Q	Zaire, Kinshasa	04:36	16:50
	Kisangani	04:00	16:06
	Lubumbashi	03:42	16:07
9U	Burundi, Bujumbura	03:41	15:53
9V	Singapore	22:47	10:51
9X	Rwanda, Kigali	03:39	15:49
9Y	Trinidad & Tobago, Port of		
	Spain	09:54	21:45
J2/A	Abu Ail	03:00	14:46

Pref	State/Province/Country/City	Sunrise	Sunset
VE1	New Brunswick, Saint John	11:11	21:04
	Nova Scotia, Halifax	10:59	20:56
	Prince Edward Island, Charlottetown	11:01	20:50
VE2	Quebec, Montreal	11:41	21:34
	Quebec City	11:35	21:21
VE3	Ontario, London	12:07	22:10
	Ottawa	11:50	21:42
	Sudbury	12:13	22:01
	Toronto	12:00	22:01
VE4	Manitoba, Winnipeg	13:27	22:56
VE5	Saskatchewan, Regina	13:59	23:25
	Saskatoon	14:12	23:28
VE6	Alberta, Calgary	14:38	00:01
	Edmonton	14:44	23:51
VE7	British Columbia, Prince George	15:22	00:27
	Prince Rupert	15:54	00:55
	Vancouver	15:09	00:42
VE8	Northwest Territories, Yellowknife	15:28	23:14
	Resolute	24 hrs night	
VY1	Yukon, Whitehorse	16:41	00:47
VO1	Newfoundland, St. John's	10:23	20:05
VO2	Labrador, Goose Bay	11:11	20:19
W1	Connecticut, Hartford	11:30	21:38
	Maine, Bangor	11:21	21:16
	Portland	11:24	21:25
	Massachusetts, Boston	11:24	21:31
	New Hampshire, Concord	11:28	21:31
	Rhode Island, Providence	11:24	21:33
	Vermont, Montpelier	11:35	21:33
W2	New Jersey, Atlantic City	11:32	21:50
	New York, Albany	11:36	21:41
	Buffalo	11:57	22:01
	New York City	11:33	21:46
	Syracuse	11:46	21:50
W3	Delaware, Wilmington	11:37	21:54
	District of Columbia, Washington	11:41	22:01
	Maryland, Baltimore	11:40	21:59
	Pennsylvania, Harrisburg	11:43	21:58
	Philadelphia	11:36	21:52
	Pittsburgh	11:56	22:10
	Scranton	11:41	21:51
W4	Alabama, Montgomery	12:07	22:50
	Florida, Jacksonville	11:45	22:35
	Miami	11:33	22:35
	Pensacola	12:07	22:57
	Georgia, Atlanta	12:02	22:40
	Savannah	11:46	22:30
	Kentucky, Lexington	12:09	22:33
	Louisville	12:15	22:38
	North Carolina, Charlotte	11:50	22:23
	Raleigh	11:42	22:13
	Wilmington	11:36	22:13
	South Carolina, Columbia	11:48	22:26
	Tennessee, Knoxville	12:03	22:34
	Memphis	12:27	23:01
	Nashville	12:15	22:46
	Virginia, Norfolk	11:35	22:02
	Richmond	11:40	22:06
W5	Arkansas, Little Rock	12:35	23:10
	Louisiana, New Orleans	12:18	23:09
	Shreveport	12:37	23:19
	Mississippi, Jackson	12:22	23:06
	New Mexico, Albuquerque	13:33	00:07
	Oklahoma, Oklahoma City	12:57	23:30
	Texas, Abilene	13:01	23:43
	Amarillo	13:14	23:47
	Dallas	12:49	23:31
	El Paso	13:27	00:12
	San Antonio	12:51	23:43
W6	California, Los Angeles	14:17	00:55
	San Francisco	14:41	01:05
W7	Arizona, Flagstaff	13:53	00:27
	Phoenix	13:52	00:31
	Idaho, Boise	14:27	00:29
	Pocatello	14:11	00:15

Pref	State/Province/Country/City	Sunrise	Sunset
	Montana, Billings	14:02	23:53
	Butte	14:18	00:08
	Great Falls	14:17	00:00
	Nevada, Las Vegas	14:08	00:39
	Reno	14:33	00:51
	Oregon, Portland	14:58	00:50
	Utah, Salt Lake City	14:04	00:17
	Washington, Seattle	15:01	00:43
	Spokane	14:42	00:23
	Wyoming, Cheyenne	13:37	23:48
	Sheridan	13:53	23:49
W8	Michigan, Detroit	12:12	22:19
	Grand Rapids	12:24	22:28
	Sault Ste. Marie	12:27	22:15
	Traverse City	12:28	22:23
	Ohio, Cincinnati	12:11	22:31
	Cleveland	12:05	22:15
	Columbus	12:07	22:23
	West Virginia, Charleston	11:58	22:21
W9	Illinois, Chicago	12:29	22:38
	Indiana, Indianapolis	12:20	22:36
	Wisconsin, Green Bay	12:37	22:34
	Milwaukee	12:33	22:37
W0	Colorado, Denver	13:35	23:52
	Grand Junction	13:48	00:07
	Iowa, Des Moines	12:53	23:02
	Kansas, Pratt	13:06	23:30
	Wichita	13:00	23:25
	Minnesota, Duluth	12:59	22:45
	Minneapolis	12:59	22:54
	Missouri, Columbia	12:42	23:02
	Kansas City	12:52	23:11
	St. Louis	12:33	22:55
	Nebraska, North Platte	13:21	23:32
	Omaha	13:01	23:12
	North Dakota, Fargo	13:18	23:03
	South Dakota, Rapid City	13:37	23:35

Pref	State/Province/Country/City	Sunrise	Sunset
A2	Botswana, Gaborone	03:25	16:34
A3	Tonga, Nukualofa	16:54	05:54
A4	Oman, Muscat	02:14	13:23
A5	Bhutan, Thimbu	00:16	11:15
A6	United Arab Emirates, Abu Dhabi	02:33	13:40
A7	Qatar, Ad-Dawhah	02:45	13:49
A9	Bahrein, Al-Manamah	02:50	13:52
AP	Pakistan, Karachi	01:42	12:48
	Islamabad	01:31	12:10
BV	Taiwan, Taipei	22:05	09:10
BY	People's Rep. of China, Beijing	22:50	09:06
	Harbin	22:21	08:12
	Shanghai	22:14	09:01
	Fuzhou	22:15	09:17
	Xian	23:09	09:46
	Chongqing	23:12	10:03
	Chengdu	23:23	10:11
	Lhasa	00:13	11:04
	Urumqi	00:53	10:53
	Kashi	01:30	11:48
C2	Nauru	18:32	06:40
C3	Andorra	06:34	16:40
C5	The Gambia, Banjul	07:02	18:38
C6	Bahamas, Nassau	11:20	22:25
C9	Mozambique, Maputo	02:56	16:09
	Mozambique	02:39	15:22
CE	Chile, Santiago	09:38	23:15
CE0	Easter Island	12:23	01:39
CE0	San Felix	10:27	23:41
CE0	Juan Fernandez	10:09	23:47
CM,CO	Cuba, Havana	11:38	22:48
CN	Morocco, Casablanca	06:54	17:33
CP	Bolivia, La Paz	09:53	22:40
CT	Portugal, Lisbon	07:09	17:31
CT2, CU	Azores, Ponta Delgada	08:14	18:38
CT3	Madeira Islands, Funchal	07:30	18:12
CX	Uruguay, Montevideo	08:37	22:19
D2	Angola, Luanda	04:37	17:04
D4	Cape Verde, Praia	07:31	19:03

Pref	State/Province/Country/City	Sunrise	Sunset
D6	Comoros, Moroni	02:31	15:06
DA-DL	Germany, Fed. Rep. of (W), Bonn	06:33	15:58
DU	Philippines, Manilla	21:53	09:25
EA	Spain, Madrid	06:51	17:05
EA6	Balearic Is., Palma	06:24	16:42
EA8	Canary Is., Las Palmas	07:13	18:08
EA9	Ceuta & Melilla, Ceuta	06:49	17:20
	Melilla	06:38	17:12
EI	Ireland, Dublin	07:34	16:43
EL	Liberia, Monrovia	06:30	18:23
EP	Iran, Tehran	03:00	13:32
ET	Ethiopia, Addis Ababa	03:15	15:01
	Asmera	03:22	14:53
F	France, Paris	06:46	16:22
FG	Guadeloupe	10:06	21:34
FG, FS	St. Martin	10:14	21:37
FH	Mayotte	02:23	15:01
FK	New Caledonia, Noumia	18:06	07:08
FM	Martinique	10:01	21:33
FO	Clipperton	13:09	00:51
FO	Fr. Polynesia, Tahiti	15:56	04:46
FP	St. Pierre & Miquelon, St. Pierre	10:34	20:20
FR	Glorioso	02:17	14:51
FR	Juan de Nova	02:28	15:16
	Europa	02:30	15:33
FR	Reunion	01:31	14:30
FR	Tromelin	01:43	14:28
FT8W	Crozet	01:00	15:31
FT8X	Kerguelen	23:41	14:31
FT8Y	Antarctica, Dumont D'Urville	17:22	11:25
FT8Z	Amsterdam & St. Paul Is., Amsterdam	23:36	13:29
FW	Wallis & Futuna Is., Wallis	17:09	05:48
FY	Fr. Guiana, Cayenne	09:15	21:10
G	England, London	07:04	16:24
GD	Isle of Man	07:30	16:32
GI	Northern Ireland, Belfast	07:37	16:36
GJ	Jersey	07:06	16:38
GM	Scotland, Glasgow	07:35	16:26
	Aberdeen	07:32	16:11
GU	Guernsey	07:08	16:40
GW	Wales, Cardiff	07:16	16:36
H4	Solomon Islands, Honiara	18:49	07:18
HA	Hungary, Budapest	05:36	15:18
HB	Switzerland, Bern	06:21	16:06
HB0	Liechtenstein	06:13	15:57
HC	Ecuador, Quito	10:51	22:59
HC8	Galapagos Is.	11:41	23:49
HH	Haiti, Port-Au-Prince	10:51	22:14
HI	Dominican Republic, Santo Domingo	10:42	22:04
HK	Colombia, Bogota	10:41	22:38
HK0	Malpelo Is.	11:09	23:06
HK0	San Adreas	11:21	22:59
HL, HM	Korea, Seoul	22:02	08:28
HP	Panama, Panama	11:08	22:54
HR	Honduras, Tegucigalpa	11:45	23:19
HS	Thailand, Bangkok	23:14	10:48
HV	Vatican City	05:49	15:57
HZ, 7Z	Saudi Arabia, Dharan	02:52	13:54
	Mecca	03:27	14:41
	Riyadh	03:03	14:10
I	Italy, Rome	05:49	15:57
	Trieste	05:52	15:44
	Sicily	05:34	16:00
IS	Sardinia, Cagliari	05:57	16:16
J2	Djibouti, Djibouti	03:01	14:40
J3	Grenada	10:01	21:40
J5	Guinea-Bissau, Bissau	06:56	18:35
J6	St. Lucia	10:00	21:34
J7	Dominica	10:03	21:34
J8	St. Vincent	10:01	21:36
JA-JS	Japan, Tokyo	21:08	07:40
	Nagasaki	21:43	08:24
	Sapporo	21:16	07:19
JD	Minami Torishima	19:54	07:01
JD	Ogasawara, Kazan Is.	20:50	07:48
JT	Mongolia, Ulan Bator	23:45	09:26
JW	Svalbard, Spitsbergen	24 hrs night	
JX	Jan Mayen	09:49	14:44

Pref	State/Province/Country/City	Sunrise	Sunset
JY	Jordan, Amman	03:57	14:42
KC4	Antarctica, Byrd Station	24 hrs day	
	McMurdo Sound	24 hrs day	
	Palmer Station	07:16	00:43
KC6	Micronesia, Ponape	19:15	07:05
KC6	Belau, Yap	20:38	08:23
	Koror	20:50	08:40
KG4	Guantanamo Bay	11:05	22:23
KH1	Baker, Howland Is.	17:24	05:30
KH2	Guam, Agana	20:16	07:52
KH3	Johnston Is.	17:14	04:40
KH4	Midway Is.	18:05	05:01
KH5	Palmyra Is.	16:35	04:28
KH5K	Kingman Reef	16:40	04:29
KH6	Hawaii, Hilo	16:24	03:43
	Honolulu	16:37	03:52
KH7	Kure Is.	18:09	05:05
KH8	American Samoa, Pago Pago	16:46	05:27
KH9	Wake Is.	18:57	06:17
KH0	Mariana Is., Saipan	20:15	07:45
KL7	Alaska, Adak	18:51	04:09
	Anchorage	17:43	01:44
	Fairbanks	17:58	01:12
	Juneau	16:26	00:56
	Nome	19:05	02:24
KP1	Navassa Is.	11:02	22:25
KP2	Virgin Islands, Charlotte Amalie	10:21	21:44
KP4	Puerto Rico, San Juan	10:27	21:49
KP5	Desecheo Is.	10:32	21:55
KX6	Marshall Islands, Kwajalein	18:41	06:27
LA-LJ	Norway, Oslo	06:54	15:07
LU	Argentina, Buenos Aires	08:46	22:27
LX	Luxembourg	06:33	16:04
LZ	Bulgaria, Sofia	05:08	15:12
OA	Peru, Lima	10:34	23:09
OD	Lebanon, Beirut	04:02	14:40
OE	Austria, Vienna	05:49	15:27
OH	Finland, Helsinki	05:58	14:09
OH0	Aland Is.	06:18	14:29
OJ0	Market Reef	06:22	14:32
OK	Czechoslovakia, Prague	06:01	15:30
ON	Belgium, Brussels	06:44	16:07
OX, XP	Greenland, Godthaab	11:28	18:52
	Thule	24 hrs night	
OY	Faroe Islands, Torshavn	08:15	16:06
OZ	Denmark, Copenhagen	06:27	15:18
P2	Papua New Guinea, Madang	19:51	08:10
	Port Moresby	19:40	08:09
PA-PI	Netherlands, Amsterdam	06:46	16:01
PJ, P4	Netherlands Antilles, Willemstad	10:30	22:08
PJ5-8	St. Maarten and Saba, St. Maarten	10:14	21:38
PY	Brazil, Brasilia	08:32	21:17
	Rio De Janeiro	08:04	21:08
	Natal	07:53	20:15
	Manaus	09:37	21:51
	Porto Alegre	08:25	21:51
PY0	Fernando De Noronha	07:45	20:01
PY0	St. Peter & St. Paul Rocks	07:38	19:43
PY0	Trindade & Martin Vaz Is., Trindade	07:12	20:09
PZ	Suriname, Paramaribo	09:27	21:21
S2	Bangladesh, Dacca	00:07	11:16
S7	Seychelles, Victoria	01:52	14:10
S9	Sao Tome	05:13	17:20
SJ-SM	Sweden, Stockholm	06:21	14:41
SP	Poland, Krakow	05:39	15:08
	Warsaw	05:4*	14:57
ST	Sudan, Khartoum	03:48	15:18
ST0	Southern Sudan, Juba	03:39	15:34
SU	Egypt, Cairo	04:12	15:03
SV	Greece, Athens	04:57	15:20
SV5	Dodecanese, Rhodes	04:36	15:05
SV9	Crete	04:46	15:19
SV/A	Mount Athos	04:58	15:14
T2	Tuvalu, Funafuti	17:35	06:03
T30	West Kiribati, Bonriki	18:09	06:12
T31	Central Kiribati, Kanton	17:03	05:17
T32	East Kiribati, Christmas Is.	16:11	04:14
T5	Somalia, Mogadishu	02:41	14:43
T7	San Marino	05:54	15:54

Pref	State/Province/Country/City	Sunrise	Sunset
TA	Turkey, Ankara	04:23	14:40
	Istanbul	04:42	14:53
TF	Iceland, Reykjavik	09:29	16:53
TG	Guatemala, Guatemala City	11:59	23:31
TI	Costa Rica, San Jose	11:27	23:11
TI9	Cocos Is.	11:34	23:28
TJ	Cameroon, Yaounde	04:58	16:56
TK	Corsica	06:03	16:11
TL	Central African Republic, Bangui	04:30	16:27
TN	Congo, Brazzaville	04:33	16:50
TR	Gabon, Libreville	05:02	17:08
TT	Chad, N'Djamena	04:54	16:32
TU	Ivory Coast, Abidjan	06:02	17:56
TY	Benin, Porto Novo	05:37	17:29
TZ	Mali, Bamako	06:27	18:04
UA	Russia, European,		
	Leningrad (UA1)	05:35	13:49
	Archangel (UA1)	05:23	12:40
	Murmansk (UA1)	06:34	12:28
	Moscow (UA3)	04:48	13:38
	Kuibyshev (UA4)	03:48	12:57
	Rostov (UA6)	04:14	13:56
UA1P	Franz Josef Land	24 hrs night	
UA2	Kaliningrad	05:53	14:49
UA9,Ø	Russia, Asiatic,		
	Novosibirsk (UA9)	01:43	10:40
	Perm (UA9)	03:42	12:14
	Omsk (UA9)	02:21	11:18
	Norilsk (UAØ)	02:58	08:44
	Irkutsk (UAØ)	00:09	09:23
	Vladivostok (UAØ)	21:54	07:57
	Petropavlovsk (UAØ)	20:33	05:44
	Khabarovsk (UAØ)	21:54	07:31
	Krasnoyarsk (UAØ)	01:08	09:56
	Yakutsk (UAØ)	23:09	07:00
	Wrangel Island (UAØ)	21:14	02:09
	Kyzyl (UAØY)	00:46	10:05
UB	Ukraine, Kiev	04:58	14:24
UC	Byelorussia, Minsk	05:21	14:25
UD	Azerbaijan, Baku	03:16	13:31
UF	Georgia, Tbilisi	03:39	13:49
UG	Armenia, Ierevan	03:38	13:53
UH	Turkoman, Ashkhabad	02:38	13:01
UI	Uzbek, Bukhara	02:17	12:34
	Tashkent	02:00	12:12
UJ	Tadzhik, Samarkand	02:07	12:25
	Dushanbe	01:58	12:18
UL	Kazakh, Alma-ata	01:34	11:37
UM	Kirghiz, Frunze	01:43	11:47
UO	Moldavia, Kishinev	04:56	14:40
UP	Lithuania, Vilna	05:31	14:31
UQ	Latvia, Riga	05:46	14:27
UR	Estonia, Tallinn	05:54	14:14
V2	Antigua & Barbuda, St. Johns	10:07	21:33
V3	Belize, Belmopan	11:56	23:21
V4	St. Christopher & Nevis	10:11	21:36
V8	Brunei, Bandar Seri Begawan	22:06	10:01
VE1, CY	Sable Is.	10:43	20:43
VE1, CY	St. Paul Is.	10:52	20:36
VK	Australia, Canberra (VK1)	18:55	08:39
	Sydney (VK2)	18:49	08:28
	Melbourne (VK3)	19:06	09:00
	Brisbane (VK4)	18:52	08:10
	Adelaide (VK5)	19:38	09:20
	Perth (VK6)	21:14	10:46
	Hobart, Tasmania (VK7)	18:46	09:02
	Darwin (VK8)	20:41	09:18
VK	Lord Howe Is.	18:21	07:52
VK9	Christmas Is.	22:24	10:56
VK9	Cocos-Keeling Is.	22:58	11:34
VK9	Mellish Reef	18:55	07:45
VK9	Norfolk Is.	17:50	07:12
VK9	Willis Is.	19:22	08:08
VKØ	Heard Is.	23:12	14:27
VKØ	Macquarie Is.	17:24	08:52
VP2E	Anguilla	10:14	21:37
VP2M	Montserrat	10:09	21:35
VP2V	British Virgin Is., Tortola	10:20	21:43
VP5	Turks & Caicos Islands,		
	Grand Turk	10:50	22:06
VP8	Falkland Islands, Stanley	08:02	23:07

Pref	State/Province/Country/City	Sunrise	Sunset
VP8	So. Georgia Is.	06:28	21:53
VP8	So. Orkney Is.	06:32	22:58
VP8	So. Sandwich Islands, Saunders		
	Is.	05:32	21:28
VP8	So. Shetland Is., King George		
	Is.	07:14	23:58
VP9	Bermuda	10:40	21:24
VQ9	Chagos, Diego Garcia	00:42	13:06
VR6	Pitcairn Is.	13:48	02:59
VS6	Hong Kong	22:30	09:42
VU	India, Bombay	01:11	12:33
	Calcutta	00:14	11:26
	New Delhi	01:07	12:02
	Bangalore	00:45	12:21
VU7	Andaman Islands, Port Blair	23:42	11:22
VU7	Laccadive Is.	00:59	12:43
XE	Mexico, Mexico City (XE1)	12:39	00:00
	Chihuahua (XE2)	13:20	00:14
	Merida (XE3)	12:04	23:20
XF4	Revilla Gigedo	13:29	00:50
XT	Burkina Faso, Onagadougou	06:01	17:38
XU	Kampuchea, Phnom Penh	22:54	10:34
XW	Laos, Viangchan	23:11	10:35
XX9	Macao	22:32	09:45
XZ	Burma, Rangoon	23:36	11:03
Y2-9	German Dem.Rep.(E), Berlin	06:13	15:26
YA	Afghanistan, Kandahar	01:56	12:44
	Kabul	01:48	12:25
YB-YD	Indonesia, Jakarta	22:25	10:47
	Medan	23:09	11:08
	Pontianak	22:23	10:29
	Jayapura	20:14	08:27
YI	Iraq, Baghdad	03:25	14:06
YJ	Vanuatu, Port Vila	18:05	06:55
YK	Syria, Damascus	03:58	14:38
YN, HT	Nicaragua, Managua	11:38	23:17
YO	Romania, Bucharest	05:00	14:58
YS	El Salvador, San Salvador	11:53	23:27
YU	Yugoslavia, Belgrade	05:24	15:19
YV	Venezuela, Caracas	10:20	22:02
YVØ	Aves Is.	10:13	21:43
Z2,ZE	Zimbabwe, Harare	03:14	16:04
ZA	Albania, Tirane	05:19	15:29
ZB	Gibraltar	06:49	17:20
ZC4	British Cyprus	04:13	14:49
ZD7	St. Helena	05:44	18:30
ZD8	Ascension Is.	06:28	18:54
ZD9	Tristan da Cunha	05:37	19:28
ZF	Cayman Is.	11:28	22:48
ZK1	No. Cook Is., Manihiki	16:11	04:43
ZK1	So. Cook Is., Rarotonga	15:53	04:52
ZK2	Niue	16:32	05:26
ZK3	Tokelaus, Atafu	17:01	05:27
ZL	New Zealand, Auckland (ZL1)	17:09	06:59
	Wellington (ZL2)	17:00	07:08
	Christchurch (ZL3)	17:04	07:22
	Dunedin (ZL4)	17:06	07:36
ZL5	Antarctica, Scott Base	24 hrs day	
ZL7	Chatham Is.	16:19	06:40
ZL8	Kermadec Is.	16:53	06:16
ZL9	Auckland & Campbell Is.,		
	Auckland	17:08	08:07
	Campbell Is.	16:51	08:02
ZP	Paraguay, Asuncion	08:58	22:10
ZS	South Africa, Cape Town (ZS1)	03:40	17:19
	Port Elizabeth (ZS2)	03:11	16:50
	Bloemfontein (ZS4)	03:17	16:40
	Durban (ZS5)	02:57	16:22
	Johannesburg (ZS6)	03:14	16:28
ZS2	Prince Edward & Marion Is.,		
	Marion	01:54	16:30
ZS3	Namibia, Windhoek	04:03	17:06
1AØ	S.M.O.M	05:49	15:58
1S	Spratly Is.	22:22	10:09
3A	Monaco	06:13	16:14
3B6	Agalega	01:41	14:13
3B7	St. Brandon	01:21	14:07
3B8	Mauritius	01:25	14:22
3B9	Rodriguez Is.	01:02	13:57
3C	Equatorial Guinea, Bata	05:03	17:05
	Malabo	05:09	17:07

Pref	State/Province/Country/City	Sunrise	Sunset
3CØ	Pagalu Is.	05:16	17:26
3D2	Fiji Is., Suva	17:24	06:15
3D6	Swaziland, Mbabane	03:02	16:16
3V	Tunisia, Tunis	05:48	16:16
3W	Vietnam, Ho Chi Minh City		
	(Saigon)	22:46	10:27
	Hanoi	23:02	10:18
3X	Republic of Guinea, Conakry	06:46	18:30
3Y	Bouvet	03:47	19:13
3Y	Peter I Is.	08:16	03:16
4S	Sri Lanka, Colombo	00:28	12:19
4U	ITU Geneva	06:24	16:13
4U	United Nations Hq.	11:33	21:46
4W	Yemen, Sanaa	03:01	14:32
4X, 4Z	Israel, Jerusalem	04:00	14:45
5A	Libya, Tripoli	05:32	16:15
	Benghazi	05:01	15:45
5B	Cyprus, Nicosia	04:13	14:46
5H	Tanzania, Dar es Salaam	02:53	15:17
5N	Nigeria, Lagos	05:34	17:25
5R	Madagascar, Antananarivo	02:07	15:00
5T	Mauritania, Nouakchott	07:05	18:29
5U	Niger, Niamey	05:48	17:23
5V	Togo, Lome	05:42	17:35
5W	Western Samoa, Apia	16:51	05:30
5X	Uganda, Kampala	03:30	15:36
5Z	Kenya, Nairobi	04:31	16:41
6W	Senegal, Dakar	07:07	18:39
6Y	Jamaica, Kingston	11:08	22:32

Pref	State/Province/Country/City	Sunrise	Sunset
7O	Yemen, People's Dem. Rep.,		
	Aden	02:55	14:32
7P	Lesotho, Maseru	03:12	16:35
7Q	Malawi, Lilongwe	03:08	15:48
	Blantyre	03:01	15:46
7X	Algeria, Algiers	06:17	16:45
8P	Barbados, Bridgetown	09:54	21:30
8Q	Maldive Is.	00:51	12:48
8R	Guyana, Georgetown	09:40	21:32
9G	Ghana, Accra	05:47	17:41
9H	Malta	05:30	16:01
9J	Zambia, Lusaka	03:28	16:12
9K	Kuwait	03:06	13:58
9L	Sierra Leone, Freetown	06:42	18:30
9M2	West Malaysia, Kuala Lumpur	22:57	10:57
9M6,8	East Malaysia, Sabah,		
	Sandakan (9M6)	21:54	09:47
	Sarawak, Kuching (9M8)	22:20	10:24
9N	Nepal, Katmandu	00:33	11:31
9Q	Zaire, Kinshasa	04:33	16:50
	Kisangani	04:00	16:05
	Lubumbashi	03:36	16:11
9U	Burundi, Bujumbura	03:39	15:53
9V	Singapore	22:46	10:50
9X	Rwanda, Kigali	03:37	15:49
9Y	Trinidad & Tobago, Port of		
	Spain	09:57	21:40
J2/A	Abu Ail	03:05	14:39

Pref	State/Province/Country/City	Sunrise	Sunset
VE1	New Brunswick, Saint John	11:30	20:49
	Nova Scotia, Halifax	11:18	20:41
	Prince Edward Island, Charlottetown	11:21	20:35
VE2	Quebec, Montreal	12:01	21:19
	Quebec City	11:55	21:05
VE3	Ontario, London	12:24	21:57
	Ottawa	12:09	21:27
	Sudbury	12:33	21:45
	Toronto	12:19	21:47
VE4	Manitoba, Winnipeg	13:49	22:38
VE5	Saskatchewan, Regina	14:22	23:06
	Saskatoon	14:36	23:08
VE6	Alberta, Calgary	15:02	23:42
	Edmonton	15:10	23:29
VE7	British Columbia, Prince George	15:49	00:04
	Prince Rupert	16:21	00:33
	Vancouver	15:31	00:24
VE8	Northwest Territories, Yellowknife	16:09	22:37
	Resolute	24 hrs night	
VY1	Yukon, Whitehorse	17:17	00:14
VO1	Newfoundland, St. John's	10:44	19:48
VO2	Labrador, Goose Bay	11:36	19:57
W1	Connecticut, Hartford	11:47	21:26
	Maine, Bangor	11:39	21:02
	Portland	11:42	21:11
	Massachusetts, Boston	11:42	21:18
	New Hampshire, Concord	11:46	21:17
	Rhode Island, Providence	11:41	21:20
	Vermont, Montpelier	11:53	21:18
W2	New Jersey, Atlantic City	11:47	21:39
	New York, Albany	11:53	21:28
	Buffalo	12:14	21:47
	New York City	11:49	21:33
	Syracuse	12:04	21:36
W3	Delaware, Wilmington	11:52	21:42
	District of Columbia, Washington	11:56	21:50
	Maryland, Baltimore	11:56	21:48
	Pennsylvania, Harrisburg	12:00	21:46
	Philadelphia	11:52	21:41
	Pittsburgh	12:12	21:58
	Scranton	11:58	21:39
W4	Alabama, Montgomery	12:19	22:42
	Florida, Jacksonville	11:57	22:27
	Miami	11:42	22:30
	Pensacola	12:19	22:49
	Georgia, Atlanta	12:15	22:31
	Savannah	11:58	22:22
	Kentucky, Lexington	12:24	22:22
	Louisville	12:30	22:27
	North Carolina, Charlotte	12:03	22:14
	Raleigh	11:56	22:04
	Wilmington	11:49	22:04
	South Carolina, Columbia	12:01	22:17
	Tennessee, Knoxville	12:17	22:24
	Memphis	12:40	22:51
	Nashville	12:29	22:36
	Virginia, Norfolk	11:49	21:52
	Richmond	11:55	21:55
W5	Arkansas, Little Rock	12:48	23:01
	Louisiana, New Orleans	12:30	23:02
	Shreveport	12:49	23:11
	Mississippi, Jackson	12:35	22:58
	New Mexico, Albuquerque	13:47	23:58
	Oklahoma, Oklahoma City	13:11	23:20
	Texas, Abilene	13:13	23:35
	Amarillo	13:27	23:38
	Dallas	13:02	23:23
	El Paso	13:39	00:04
	San Antonio	13:02	23:36
W6	California, Los Angeles	14:30	00:46
	San Francisco	14:55	00:54
W7	Arizona, Flagstaff	14:07	00:17
	Phoenix	14:05	00:23
	Idaho, Boise	14:46	00:15
	Pocatello	14:29	00:02

Pref	State/Province/Country/City	Sunrise	Sunset
	Montana, Billings	14:21	23:37
	Butte	14:38	23:53
	Great Falls	14:38	23:43
	Nevada, Las Vegas	14:23	00:29
	Reno	14:49	00:40
	Oregon, Portland	15:17	00:35
	Utah, Salt Lake City	14:21	00:05
	Washington, Seattle	15:22	00:27
	Spokane	15:03	00:07
	Wyoming, Cheyenne	13:53	23:36
	Sheridan	14:12	23:34
W8	Michigan, Detroit	12:29	22:05
	Grand Rapids	12:42	22:14
	Sault Ste. Marie	12:47	21:59
	Traverse City	12:47	22:09
	Ohio, Cincinnati	12:27	22:20
	Cleveland	12:22	22:02
	Columbus	12:23	22:11
	West Virginia, Charleston	12:14	22:10
W9	Illinois, Chicago	12:47	22:25
	Indiana, Indianapolis	12:36	22:25
	Wisconsin, Green Bay	12:55	22:19
	Milwaukee	12:51	22:23
W0	Colorado, Denver	13:50	23:40
	Grand Junction	14:03	23:56
	Iowa, Des Moines	13:10	22:50
	Kansas, Pratt	13:20	23:20
	Wichita	13:15	23:14
	Minnesota, Duluth	13:19	22:29
	Minneapolis	13:18	22:39
	Missouri, Columbia	12:58	22:51
	Kansas City	13:07	23:00
	St. Louis	12:49	22:44
	Nebraska, North Platte	13:37	23:20
	Omaha	13:18	23:00
	North Dakota, Fargo	13:38	22:47
	South Dakota, Rapid City	13:55	23:21
A2	Botswana, Gaborone	03:19	16:44
A3	Tonga, Nukualofa	16:50	06:02
A4	Oman, Muscat	02:23	13:18
A5	Bhutan, Thimbu	00:27	11:09
A6	United Arab Emirates, Abu Dhabi	02:42	13:34
A7	Qatar, Ad-Dawhah	02:55	13:44
A9	Bahrein, Al-Manamah	03:00	13:46
AP	Pakistan, Karachi	01:52	12:42
	Islamabad	01:44	12:01
BV	Taiwan, Taipei	22:14	09:04
BY	People's Rep. of China, Beijing	23:06	08:54
	Harbin	22:40	07:57
	Shanghai	22:26	08:53
	Fuzhou	22:25	09:11
	Xian	23:22	09:37
	Chongqing	23:23	09:56
	Chengdu	23:34	10:04
	Lhasa	00:24	10:57
	Urumqi	01:11	10:39
	Kashi	01:46	11:37
C2	Nauru	18:33	06:42
C3	Andorra	06:52	16:27
C5	The Gambia, Banjul	07:08	18:36
C6	Bahamas, Nassau	11:30	22:20
C9	Mozambique, Maputo	02:50	16:20
	Mozambique	02:36	15:29
CE	Chile, Santiago	09:29	23:28
CE0	Easter Island	12:16	01:50
CE0	San Felix	10:20	23:51
CE0	Juan Fernandez	10:00	00:01
CM,CO	Cuba, Havana	11:47	22:43
CN	Morocco, Casablanca	07:07	17:24
CP	Bolivia, La Paz	09:51	22:47
CT	Portugal, Lisbon	07:25	17:19
CT2, CU	Azores, Ponta Delgada	08:28	18:28
CT3	Madeira Islands, Funchal	07:42	18:04
CX	Uruguay, Montevideo	08:27	22:33
D2	Angola, Luanda	04:36	17:09
D4	Cape Verde, Praia	07:38	19:01

Pref	State/Province/Country/City	Sunrise	Sunset
D6	Comoros, Moroni	02:30	15:12
DA-DL	Germany, Fed. Rep. of (W),		
	Bonn	06:56	15:39
DU	Philippines, Manilla	21:59	09:23
EA	Spain, Madrid	07:07	16:53
EA6	Balearic Is., Palma	06:40	16:30
EA8	Canary Is., Las Palmas	07:24	18:01
EA9	Ceuta & Melilla, Ceuta	07:03	17:10
	Melilla	06:52	17:02
EI	Ireland, Dublin	08:00	16:21
EL	Liberia, Monrovia	06:34	18:23
EP	Iran, Tehran	03:14	13:22
ET	Ethiopia, Addis Ababa	03:20	15:01
	Asmera	03:29	14:51
F	France, Paris	07:08	16:04
FG	Guadeloupe	10:12	21:32
FG, FS	St. Martin	10:21	21:34
FH	Mayotte	02:21	15:07
FK	New Caledonia, Noumia	18:01	07:17
FM	Martinique	10:07	21:31
FO	Clipperton	13:14	00:51
FO	Fr. Polynesia, Tahiti	15:53	04:53
FP	St. Pierre & Miquelon, St. Pierre	10:54	20:05
FR	Glorioso	02:16	14:57
FR	Juan de Nova	02:25	15:23
	Europa	02:26	15:42
FR	Reunion	01:27	14:39
FR	Tromelin	01:40	14:35
FT8W	Crozet	00:44	15:51
FT8X	Kerguelen	23:23	14:54
FT8Y	Antarctica, Dumont D'Urville	16:21	12:30
FT8Z	Amsterdam & St. Paul Is.,		
	Amsterdam	23:25	13:45
FW	Wallis & Futuna Is., Wallis	17:07	05:54
FY	Fr. Guiana, Cayenne	09:18	21:11
G	England, London	07:28	16:04
GD	Isle of Man	07:57	16:09
GI	Northern Ireland, Belfast	08:04	16:14
GJ	Jersey	07:28	16:21
GM	Scotland, Glasgow	08:04	16:01
	Aberdeen	08:03	15:45
GU	Guernsey	07:30	16:22
GW	Wales, Cardiff	07:40	16:16
H4	Solomon Islands, Honiara	18:48	07:23
HA	Hungary, Budapest	05:56	15:02
HB	Switzerland, Bern	06:41	15:50
HB0	Liechtenstein	06:33	15:41
HC	Ecuador, Quito	10:53	23:01
HC8	Galapagos Is.	11:43	23:52
HH	Haiti, Port-Au-Prince	10:59	22:10
HI	Dominican Republic, Santo		
	Domingo	10:49	22:01
HK	Colombia, Bogota	10:45	22:39
HK0	Malpelo Is.	11:12	23:07
HK0	San Adreas	11:27	22:57
HL, HM	Korea, Seoul	22:17	08:17
HP	Panama, Panama	11:13	22:54
HR	Honduras, Tegucigalpa	11:51	23:17
HS	Thailand, Bangkok	23:20	10:47
HV	Vatican City	06:06	15:45
HZ, 7Z	Saudi Arabia, Dharan	03:03	13:48
	Mecca	03:35	14:37
	Riyadh	03:13	14:04
I	Italy, Rome	06:06	15:45
	Trieste	06:12	15:29
	Sicily	05:49	15:49
IS	Sardinia, Cagliari	06:13	16:05
J2	Djibouti, Djibouti	03:06	14:39
J3	Grenada	10:07	21:39
J5	Guinea-Bissau, Bissau	07:02	18:34
J6	St. Lucia	10:06	21:32
J7	Dominica	10:10	21:31
J8	St. Vincent	10:07	21:35
JA-JS	Japan, Tokyo	21:22	07:30
	Nagasaki	21:55	08:16
	Sapporo	21:34	07:06
JD	Minami Torishima	20:03	06:56
JD	Ogasawara, Kazan Is.	21:01	07:42
JT	Mongolia, Ulan Bator	00:06	09:09
JW	Svalbard, Spitsbergen	24 hrs night	
JX	Jan Mayen	24 hrs night	

Pref	State/Province/Country/City	Sunrise	Sunset
JY	Jordan, Amman	04:10	14:34
KC4	Antarctica, Byrd Station	24 hrs day	
	McMurdo Sound	24 hrs day	
	Palmer Station	06:27	01:36
KC6	Micronesia, Ponape	19:19	07:06
KC6	Belau, Yap	20:43	08:22
	Koror	20:54	08:40
KG4	Guantanamo Bay	11:12	22:20
KH1	Baker, Howland Is.	17:27	05:32
KH2	Guam, Agana	20:22	07:50
KH3	Johnston Is.	17:21	04:38
KH4	Midway Is.	18:16	04:54
KH5	Palmyra Is.	16:39	04:29
KH5K	Kingman Reef	16:44	04:29
KH6	Hawaii, Hilo	16:32	03:40
	Honolulu	16:46	03:48
KH7	Kure Is.	18:20	04:58
KH8	American Samoa, Pago Pago	16:44	05:33
KH9	Wake Is.	19:04	06:14
KH0	Mariana Is., Saipan	20:21	07:43
KL7	Alaska, Adak	19:15	03:49
	Anchorage	18:20	01:10
	Fairbanks	18:45	00:28
	Juneau	16:58	00:28
	Nome	19:52	01:42
KP1	Navassa Is.	11:09	22:21
KP2	Virgin Islands, Charlotte Amalie	10:29	21:41
KP4	Puerto Rico, San Juan	10:34	21:46
KP5	Desecheo Is.	10:39	21:52
KX6	Marshall Islands, Kwajalein	18:46	06:26
LA-LJ	Norway, Oslo	07:29	14:36
LU	Argentina, Buenos Aires	08:37	22:41
LX	Luxembourg	06:55	15:46
LZ	Bulgaria, Sofia	05:25	14:59
OA	Peru, Lima	10:32	23:15
OD	Lebanon, Beirut	04:15	14:31
OE	Austria, Vienna	06:10	15:10
OH	Finland, Helsinki	06:33	13:38
OH0	Aland Is.	06:53	13:58
OJ0	Market Reef	06:58	14:01
OK	Czechoslovakia, Prague	06:24	15:11
ON	Belgium, Brussels	07:07	15:48
OX, XP	Greenland, Godthaab	12:14	18:10
	Thule	24 hrs night	
OY	Faroe Islands, Torshavn	08:54	15:31
OZ	Denmark, Copenhagen	06:56	14:54
P2	Papua New Guinea, Madang	19:52	08:14
	Port Moresby	19:40	08:14
PA-PI	Netherlands, Amsterdam	07:11	15:40
PJ, P4	Netherlands Antilles, Willemstad	10:35	22:07
PJ5-8	St. Maarten and Saba,		
	St. Maarten	10:21	21:35
PY	Brazil, Brasilia	08:30	21:24
	Rio De Janeiro	07:59	21:17
	Natal	07:54	20:18
	Manaus	09:38	21:54
	Porto Alegre	08:17	22:03
PY0	Fernando De Noronha	07:46	20:04
PY0	St. Peter & St. Paul Rocks	07:41	19:45
PY0	Trindade & Martin Vaz Is.,		
	Trindade	07:08	20:17
PZ	Suriname, Paramaribo	09:31	21:21
S2	Bangladesh, Dacca	00:16	11:11
S7	Seychelles, Victoria	01:53	14:14
S9	Sao Tome	05:15	17:22
SJ-SM	Sweden, Stockholm	06:55	14:11
SP	Poland, Krakow	06:01	14:49
	Warsaw	06:06	14:37
ST	Sudan, Khartoum	03:55	15:16
ST0	Southern Sudan, Juba	03:43	15:35
SU	Egypt, Cairo	04:24	14:56
SV	Greece, Athens	05:12	15:10
SV5	Dodecanese, Rhodes	04:50	14:55
SV9	Crete	05:00	15:09
SV/A	Mount Athos	05:15	15:02
T2	Tuvalu, Funafuti	17:35	06:07
T30	West Kiribati, Bonriki	18:11	06:14
T31	Central Kiribati, Kanton	17:05	05:20
T32	East Kiribati, Christmas Is.	16:14	04:16
T5	Somalia, Mogadishu	02:43	14:44
T7	San Marino	06:12	15:40

Pref	State/Province/Country/City	Sunrise	Sunset
TA	Turkey, Ankara	04:39	14:28
	Istanbul	04:58	14:41
TF	Iceland, Reykjavik	10:14	16:12
TG	Guatemala, Guatemala City	12:05	23:29
TI	Costa Rica, San Jose	11:32	23:10
TI9	Cocos Is.	11:38	23:29
TJ	Cameroon, Yaounde	05:01	16:57
TK	Corsica	06:20	15:58
TL	Central African Republic, Bangui	04:34	16:28
TN	Congo, Brazzaville	04:34	16:54
TR	Gabon, Libreville	05:04	17:10
TT	Chad, N'Djamena	04:59	16:31
TU	Ivory Coast, Abidjan	06:05	17:57
TY	Benin, Porto Novo	05:41	17:29
TZ	Mali, Bamako	06:32	18:02
UA	Russia, European,		
	Leningrad (UA1)	06:10	13:19
	Archangel (UA1)	06:10	11:57
	Murmansk (UA1)	07:49	11:17
	Moscow (UA3)	05:16	13:14
	Kuibyshev (UA4)	04:14	12:36
	Rostov (UA6)	04:35	13:40
UA1P	Franz Josef Land	24 hrs night	
UA2	Kaliningrad	06:21	14:26
UA9,Ø	Russia, Asiatic,		
	Novosibirsk (UA9)	02:11	10:16
	Perm (UA9)	04:14	11:47
	Omsk (UA9)	02:49	10:54
	Norilsk (UAØ)	04:17	07:29
	Irkutsk (UAØ)	00:33	09:03
	Vladivostok (UAØ)	22:12	07:43
	Petropavlovsk (UAØ)	20:59	05:22
	Khabarovsk (UAØ)	22:16	07:14
	Krasnoyarsk (UAØ)	01:36	09:32
	Yakutsk (UAØ)	23:48	06:25
	Wrangel Island (UAØ)	24 hrs night	
	Kyzyl (UAØY)	01:10	09:45
UB	Ukraine, Kiev	05:21	14:06
UC	Byelorussia, Minsk	05:47	14:03
UD	Azerbaijan, Baku	03:33	13:19
UF	Georgia, Tbilisi	03:56	13:36
UG	Armenia, Ierevan	03:54	13:41
UH	Turkoman, Ashkhabad	02:53	12:51
UI	Uzbek, Bukhara	02:33	12:22
	Tashkent	02:17	11:59
UJ	Tadzhik, Samarkand	02:23	12:13
	Dushanbe	02:14	12:06
UL	Kazakh, Alma-ata	01:52	11:23
UM	Kirghiz, Frunze	02:00	11:33
UO	Moldavia, Kishinev	05:16	14:24
UP	Lithuania, Vilna	05:58	14:08
UQ	Latvia, Riga	06:17	14:01
UR	Estonia, Tallinn	06:28	13:44
V2	Antigua & Barbuda, St. Johns	10:14	21:31
V3	Belize, Belmopan	12:03	23:18
V4	St. Christopher & Nevis	10:18	21:34
V8	Brunei, Bandar Seri Begawan	22:09	10:02
VE1, CY	Sable Is.	11:01	20:29
VE1, CY	St. Paul Is.	11:12	20:19
VK	Australia, Canberra (VK1)	18:45	08:53
	Sydney (VK2)	18:40	08:41
	Melbourne (VK3)	18:55	09:15
	Brisbane (VK4)	18:46	08:21
	Adelaide (VK5)	19:28	09:34
	Perth (VK6)	21:06	10:58
	Hobart, Tasmania (VK7)	18:32	09:20
	Darwin (VK8)	20:40	09:24
VK	Lord Howe Is.	18:13	08:05
VK9	Christmas Is.	22:24	11:02
VK9	Cocos-Keeling Is.	22:57	11:40
VK9	Mellish Reef	18:52	07:52
VK9	Norfolk Is.	17:43	07:24
VK9	Willis Is.	19:19	08:15
VKØ	Heard Is.	22:50	14:53
VKØ	Macquarie Is.	17:00	09:20
VP2E	Anguilla	10:21	21:34
VP2M	Montserrat	10:16	21:33
VP2V	British Virgin Is., Tortola	10:28	21:40
VP5	Turks & Caicos Islands,		
	Grand Turk	10:59	22:01
VP8	Falkland Islands, Stanley	07:41	23:33
VP8	So. Georgia Is.	06:04	22:21
VP8	So. Orkney Is.	05:58	23:37
VP8	So. Sandwich Islands, Saunders Is.	05:04	22:01
VP8	So. Shetland Is., King George Is.	06:36	00:41
VP9	Bermuda	10:53	21:16
VQ9	Chagos, Diego Garcia	00:42	13:10
VR6	Pitcairn Is.	13:43	03:09
VS6	Hong Kong	22:38	09:38
VU	India, Bombay	01:19	12:29
	Calcutta	00:23	11:21
	New Delhi	01:18	11:55
	Bangalore	00:50	12:19
VU7	Andaman Islands, Port Blair	23:48	11:21
VU7	Laccadive Is.	01:04	12:42
XE	Mexico, Mexico City (XE1)	12:47	23:56
	Chihuahua (XE2)	13:31	00:08
	Merida (XE3)	12:12	23:16
XF4	Revilla Gigedo	13:36	00:46
XT	Burkina Faso, Onagadougou	06:06	17:37
XU	Kampuchea, Phnom Penh	23:00	10:33
XW	Laos, Viangchan	23:18	10:32
XX9	Macao	22:41	09:41
XZ	Burma, Rangoon	23:43	11:00
Y2-9	German Dem.Rep.(E), Berlin	06:38	15:06
YA	Afghanistan, Kandahar	02:08	12:36
	Kabul	02:01	12:16
YB-YD	Indonesia, Jakarta	22:26	10:51
	Medan	23:12	11:09
	Pontianak	22:25	10:32
	Jayapura	20:15	08:30
YI	Iraq, Baghdad	03:37	13:57
YJ	Vanuatu, Port Vila	18:02	07:02
YK	Syria, Damascus	04:11	14:29
YN, HT	Nicaragua, Managua	11:43	23:15
YO	Romania, Bucharest	05:19	14:43
YS	El Salvador, San Salvador	11:59	23:26
YU	Yugoslavia, Belgrade	05:43	15:04
YV	Venezuela, Caracas	10:25	22:02
YVØ	Aves Is.	10:20	21:40
Z2,ZE	Zimbabwe, Harare	03:11	16:12
ZA	Albania, Tirane	05:35	15:17
ZB	Gibraltar	07:04	17:10
ZC4	British Cyprus	04:27	14:40
ZD7	St. Helena	05:41	18:36
ZD8	Ascension Is.	06:28	18:58
ZD9	Tristan da Cunha	05:26	19:43
ZF	Cayman Is.	11:36	22:44
ZK1	No. Cook Is., Manihiki	16:10	04:48
ZK1	So. Cook Is., Rarotonga	15:48	05:01
ZK2	Niue	16:29	05:33
ZK3	Tokelaus, Atafu	17:00	05:32
ZL	New Zealand, Auckland (ZL1)	16:58	07:14
	Wellington (ZL2)	16:47	07:25
	Christchurch (ZL3)	16:49	07:40
	Dunedin (ZL4)	16:50	07:57
ZL5	Antarctica, Scott Base	24 hrs day	
ZL7	Chatham Is.	16:04	06:58
ZL8	Kermadec Is.	16:46	06:28
ZL9	Auckland & Campbell Is.,		
	Auckland	16:48	08:31
	Campbell Is.	16:30	08:28
ZP	Paraguay, Asuncion	08:53	22:20
ZS	South Africa, Cape Town (ZS1)	03:31	17:32
	Port Elizabeth (ZS2)	03:02	17:03
	Bloemfontein (ZS4)	03:10	16:52
	Durban (ZS5)	02:50	16:34
	Johannesburg (ZS6)	03:08	16:38
ZS2	Prince Edward & Marion Is., Marion	01:38	16:50
ZS3	Namibia, Windhoek	03:58	17:15
1AØ	S.M.O.M	06:07	15:45
1S	Spratly Is.	22:27	10:08
3A	Monaco	06:31	16:00
3B6	Agalega	01:40	14:18
3B7	St. Brandon	01:18	14:14
3B8	Mauritius	01:21	14:30
3B9	Rodriguez Is.	00:58	14:05
3C	Equatorial Guinea, Bata	05:05	17:07
	Malabo	05:12	17:08

Pref	State/Province/Country/City	Sunrise	Sunset	Pref	State/Province/Country/City	Sunrise	Sunset
3CØ	Pagalu Is.	05:17	17:29	7O	Yemen, People's Dem. Rep.,		
3D2	Fiji Is., Suva	17:21	06:23		Aden	03:01	14:30
3D6	Swaziland, Mbabane	02:56	16:26	7P	Lesotho, Maseru	03:04	16:46
3V	Tunisia, Tunis	06:03	16:06	7Q	Malawi, Lilongwe	03:06	15:55
3W	Vietnam, Ho Chi Minh City				Blantyre	02:58	15:53
	(Saigon)	22:51	10:26	7X	Algeria, Algiers	06:31	16:35
	Hanoi	23:10	10:14	8P	Barbados, Bridgetown	09:59	21:28
3X	Republic of Guinea, Conakry	06:50	18:30	8Q	Maldive Is.	00:55	12:49
3Y	Bouvet	03:23	19:41	8R	Guyana, Georgetown	09:44	21:32
3Y	Peter I Is.	06:38	04:57	9G	Ghana, Accra	05:51	17:42
4S	Sri Lanka, Colombo	00:32	12:19	9H	Malta	05:44	15:51
4U	ITU Geneva	06:44	15:57	9J	Zambia, Lusaka	03:26	16:19
4U	United Nations Hq.	11:49	21:33	9K	Kuwait	03:17	13:51
4W	Yemen, Sanaa	03:08	14:29	9L	Sierra Leone, Freetown	06:47	18:29
4X, 4Z	Israel, Jerusalem	04:12	14:37	9M2	West Malaysia, Kuala Lumpur	23:00	10:58
5A	Libya, Tripoli	05:44	16:06	9M6,8	East Malaysia, Sabah,		
	Benghazi	05:14	15:37		Sandakan (9M6)	21:58	09:48
5B	Cyprus, Nicosia	04:26	14:37		Sarawak, Kuching (9M8)	22:23	10:25
5H	Tanzania, Dar es Salaam	02:54	15:21	9N	Nepal, Katmandu	00:44	11:24
5N	Nigeria, Lagos	05:38	17:26	9Q	Zaire, Kinshasa	04:34	16:54
5R	Madagascar, Antananarivo	02:03	15:08		Kisangani	04:02	16:07
5T	Mauritania, Nouakchott	07:13	18:26		Lubumbashi	03:35	16:16
5U	Niger, Niamey	05:54	17:21	9U	Burundi, Bujumbura	03:40	15:56
5V	Togo, Lome	05:45	17:36	9V	Singapore	22:48	10:52
5W	Western Samoa, Apia	16:49	05:36	9X	Rwanda, Kigali	03:38	15:51
5X	Uganda, Kampala	03:32	15:38	9Y	Trinidad & Tobago, Port of		
5Z	Kenya, Nairobi	04:33	16:44		Spain	10:02	21:39
6W	Senegal, Dakar	07:13	18:37	J2/A	Abu Ail	03:11	14:37
6Y	Jamaica, Kingston	11:16	22:29				

Pref	State/Province/Country/City	Sunrise	Sunset
VE1	New Brunswick, Saint John	11:49	20:41
	Nova Scotia, Halifax	11:37	20:34
	Prince Edward Island,		
	Charlottetown	11:40	20:26
VE2	Quebec, Montreal	12:20	21:11
	Quebec City	12:15	20:56
VE3	Ontario, London	12:42	21:50
	Ottawa	12:28	21:19
	Sudbury	12:53	21:37
	Toronto	12:37	21:40
VE4	Manitoba, Winnipeg	14:11	22:27
VE5	Saskatchewan, Regina	14:44	22:54
	Saskatoon	15:00	22:55
VE6	Alberta, Calgary	15:24	23:30
	Edmonton	15:35	23:15
VE7	British Columbia, Prince George	16:14	23:50
	Prince Rupert	16:46	00:18
	Vancouver	15:53	00:14
VE8	Northwest Territories,		
	Yellowknife	16:48	22:09
	Resolute	24 hrs night	
VY1	Yukon, Whitehorse	17:52	23:50
VO1	Newfoundland, St. John's	11:04	19:39
VO2	Labrador, Goose Bay	12:01	19:44
W1	Connecticut, Hartford	12:04	21:19
	Maine, Bangor	11:58	20:54
	Portland	12:00	21:04
	Massachusetts, Boston	11:59	21:11
	New Hampshire, Concord	12:04	21:10
	Rhode Island, Providence	11:59	21:14
	Vermont, Montpelier	12:12	21:11
W2	New Jersey, Atlantic City	12:04	21:33
	New York, Albany	12:11	21:21
	Buffalo	12:32	21:41
	New York City	12:06	21:28
	Syracuse	12:22	21:29
W3	Delaware, Wilmington	12:09	21:37
	District of Columbia,		
	Washington	12:12	21:45
	Maryland, Baltimore	12:12	21:43
	Pennsylvania, Harrisburg	12:16	21:41
	Philadelphia	12:08	21:35
	Pittsburgh	12:29	21:53
	Scranton	12:15	21:33
W4	Alabama, Montgomery	12:33	22:39
	Florida, Jacksonville	12:10	22:26
	Miami	11:54	22:29
	Pensacola	12:32	22:47
	Georgia, Atlanta	12:29	22:28
	Savannah	12:11	22:19
	Kentucky, Lexington	12:40	22:18
	Louisville	12:46	22:22
	North Carolina, Charlotte	12:18	22:10
	Raleigh	12:11	22:00
	Wilmington	12:04	22:01
	South Carolina, Columbia	12:16	22:14
	Tennessee, Knoxville	12:32	22:21
	Memphis	12:55	22:48
	Nashville	12:44	22:32
	Virginia, Norfolk	12:04	21:48
	Richmond	12:10	21:51
W5	Arkansas, Little Rock	13:03	22:58
	Louisiana, New Orleans	12:42	23:00
	Shreveport	13:03	23:09
	Mississippi, Jackson	12:48	22:55
	New Mexico, Albuquerque	14:01	23:54
	Oklahoma, Oklahoma City	13:25	23:16
	Texas, Abilene	13:27	23:33
	Amarillo	13:42	23:34
	Dallas	13:16	23:20
	El Paso	13:52	00:01
	San Antonio	13:15	23:35
W6	California, Los Angeles	14:45	00:43
	San Francisco	15:11	00:50
W7	Arizona, Flagstaff	14:21	00:14
	Phoenix	14:19	00:20
	Idaho, Boise	15:04	00:08
	Pocatello	14:47	23:55

Pref	State/Province/Country/City	Sunrise	Sunset
	Montana, Billings	14:41	23:29
	Butte	14:57	23:44
	Great Falls	14:58	23:34
	Nevada, Las Vegas	14:38	00:25
	Reno	15:05	00:35
	Oregon, Portland	15:36	00:27
	Utah, Salt Lake City	14:38	23:59
	Washington, Seattle	15:43	00:18
	Spokane	15:23	23:58
	Wyoming, Cheyenne	14:10	23:30
	Sheridan	14:31	23:27
W8	Michigan, Detroit	12:47	21:59
	Grand Rapids	13:00	22:08
	Sault Ste. Marie	13:07	21:50
	Traverse City	13:05	22:01
	Ohio, Cincinnati	12:43	22:15
	Cleveland	12:39	21:56
	Columbus	12:40	22:06
	West Virginia, Charleston	12:29	22:05
W9	Illinois, Chicago	13:04	22:19
	Indiana, Indianapolis	12:52	22:20
	Wisconsin, Green Bay	13:14	22:12
	Milwaukee	13:09	22:16
W0	Colorado, Denver	14:07	23:35
	Grand Junction	14:19	23:51
	Iowa, Des Moines	13:27	22:44
	Kansas, Pratt	13:36	23:15
	Wichita	13:30	23:10
	Minnesota, Duluth	13:39	22:20
	Minneapolis	13:37	22:31
	Missouri, Columbia	13:14	22:46
	Kansas City	13:23	22:55
	St. Louis	13:04	22:39
	Nebraska, North Platte	13:54	23:14
	Omaha	13:35	22:54
	North Dakota, Fargo	13:58	22:38
	South Dakota, Rapid City	14:13	23:14
A2	Botswana, Gaborone	03:19	16:56
A3	Tonga, Nukualofa	16:51	06:13
A4	Oman, Muscat	02:34	13:19
A5	Bhutan, Thimbu	00:39	11:08
A6	United Arab Emirates, Abu		
	Dhabi	02:54	13:34
A7	Qatar, Ad-Dawhah	03:06	13:44
A9	Bahrein, Al-Manamah	03:12	13:45
AP	Pakistan, Karachi	02:03	12:42
	Islamabad	01:58	11:58
BV	Taiwan, Taipei	22:26	09:04
BY	People's Rep. of China, Beijing	23:22	08:49
	Harbin	23:00	07:48
	Shanghai	22:39	08:51
	Fuzhou	22:37	09:11
	Xian	23:37	09:34
	Chongqing	23:36	09:54
	Chengdu	23:47	10:02
	Lhasa	00:37	10:55
	Urumqi	01:29	10:32
	Kashi	02:02	11:32
C2	Nauru	18:39	06:48
C3	Andorra	07:09	16:20
C5	The Gambia, Banjul	07:17	18:38
C6	Bahamas, Nassau	11:41	22:20
C9	Mozambique, Maputo	02:50	16:31
	Mozambique	02:39	15:38
CE	Chile, Santiago	09:26	23:43
CE0	Easter Island	12:15	02:02
CE0	San Felix	10:20	00:03
CE0	Juan Fernandez	09:57	00:15
CM,CO	Cuba, Havana	11:57	22:44
CN	Morocco, Casablanca	07:21	17:21
CP	Bolivia, La Paz	09:52	22:57
CT	Portugal, Lisbon	07:41	17:15
CT2, CU	Azores, Ponta Delgada	08:44	18:23
CT3	Madeira Islands, Funchal	07:56	18:01
CX	Uruguay, Montevideo	08:24	22:48
D2	Angola, Luanda	04:40	17:17
D4	Cape Verde, Praia	07:47	19:03

Pref	State/Province/Country/City	Sunrise	Sunset
D6	Comoros, Moroni	02:32	15:20
DA-DL	Germany, Fed. Rep. of (W), Bonn	07:19	15:27
DU	Philippines, Manilla	22:08	09:26
EA	Spain, Madrid	07:24	16:48
EA6	Balearic Is., Palma	06:56	16:25
EA8	Canary Is., Las Palmas	07:36	18:00
EA9	Ceuta & Melilla, Ceuta	07:18	17:07
	Melilla	07:07	16:59
EI	Ireland, Dublin	08:25	16:08
EL	Liberia, Monrovia	06:41	18:27
EP	Iran, Tehran	03:29	13:18
ET	Ethiopia, Addis Ababa	03:28	15:05
	Asmera	03:38	14:53
F	France, Paris	07:29	15:54
FG	Guadeloupe	10:21	21:34
FG, FS	St. Martin	10:31	21:36
FH	Mayotte	02:24	15:15
FK	New Caledonia, Noumia	18:02	07:28
FM	Martinique	10:16	21:34
FO	Clipperton	13:21	00:54
FO	Fr. Polynesia, Tahiti	15:55	05:03
FP	St. Pierre & Miquelon, St. Pierre	11:14	19:56
FR	Glorioso	02:19	15:05
FR	Juan de Nova	02:27	15:33
	Europa	02:26	15:53
FR	Reunion	01:28	14:49
FR	Tromelin	01:42	14:44
FT8W	Crozet	00:35	16:11
FT8X	Kerguelen	23:12	15:16
FT8Y	Antarctica, Dumont D'Urville	24 hrs day	
FT8Z	Amsterdam & St. Paul Is., Amsterdam	23:21	14:00
FW	Wallis & Futuna Is., Wallis	17:10	06:02
FY	Fr. Guiana, Cayenne	09:25	21:16
G	England, London	07:51	15:52
GD	Isle of Man	08:23	15:55
GI	Northern Ireland, Belfast	08:30	15:59
GJ	Jersey	07:49	16:10
GM	Scotland, Glasgow	08:31	15:45
	Aberdeen	08:32	15:27
GU	Guernsey	07:52	16:11
GW	Wales, Cardiff	08:03	16:04
H4	Solomon Islands, Honiara	18:51	07:30
HA	Hungary, Budapest	06:17	14:52
HB	Switzerland, Bern	07:01	15:41
HB0	Liechtenstein	06:53	15:32
HC	Ecuador, Quito	10:59	23:07
HC8	Galapagos Is.	11:48	23:57
HH	Haiti, Port-Au-Prince	11:08	22:12
HI	Dominican Republic, Santo Domingo	10:59	22:03
HK	Colombia, Bogota	10:51	22:43
HK0	Malpelo Is.	11:18	23:12
HK0	San Adreas	11:35	23:00
HL, HM	Korea, Seoul	22:33	08:13
HP	Panama, Panama	11:20	22:57
HR	Honduras, Tegucigalpa	12:00	23:19
HS	Thailand, Bangkok	23:29	10:49
HV	Vatican City	06:23	15:38
HZ, 7Z	Saudi Arabia, Dharan	03:14	13:48
	Mecca	03:46	14:38
	Riyadh	03:24	14:04
I	Italy, Rome	06:23	15:38
	Trieste	06:31	15:20
	Sicily	06:05	15:45
IS	Sardinia, Cagliari	06:29	16:00
J2	Djibouti, Djibouti	03:14	14:42
J3	Grenada	10:15	21:41
J5	Guinea-Bissau, Bissau	07:10	18:37
J6	St. Lucia	10:15	21:35
J7	Dominica	10:19	21:34
J8	St. Vincent	10:15	21:37
JA-JS	Japan, Tokyo	21:37	07:27
	Nagasaki	22:09	08:13
	Sapporo	21:52	06:59
JD	Minami Torishima	20:14	06:56
JD	Ogasawara, Kazan Is.	21:13	07:41
JT	Mongolia, Ulan Bator	00:27	09:00
JW	Svalbard, Spitsbergen	24 hrs night	
JX	Jan Mayen	24 hrs night	

Pref	State/Province/Country/City	Sunrise	Sunset
JY	Jordan, Amman	04:23	14:31
KC4	Antarctica, Byrd Station	24 hrs day	
	McMurdo Sound	24 hrs day	
	Palmer Station	05:38	02:35
KC6	Micronesia, Ponape	19:26	07:10
KC6	Belau, Yap	20:50	08:26
	Koror	21:01	08:44
KG4	Guantanamo Bay	11:22	22:21
KH1	Baker, Howland Is.	17:32	05:38
KH2	Guam, Agana	20:31	07:52
KH3	Johnston Is.	17:30	04:39
KH4	Midway Is.	18:28	04:53
KH5	Palmyra Is.	16:46	04:33
KH5K	Kingman Reef	16:51	04:33
KH6	Hawaii, Hilo	16:42	03:41
	Honolulu	16:56	03:49
KH7	Kure Is.	18:32	04:57
KH8	American Samoa, Pago Pago	16:46	05:42
KH9	Wake Is.	19:14	06:15
KH0	Mariana Is., Saipan	20:30	07:45
KL7	Alaska, Adak	19:38	03:36
	Anchorage	18:57	00:45
	Fairbanks	19:34	23:51
	Juneau	17:29	00:08
	Nome	20:39	01:06
KP1	Navassa Is.	11:19	22:23
KP2	Virgin Islands, Charlotte Amalie	10:38	21:43
KP4	Puerto Rico, San Juan	10:44	21:48
KP5	Desecheo Is.	10:49	21:53
KX6	Marshall Islands, Kwajalein	18:53	06:30
LA-LJ	Norway, Oslo	08:02	14:14
LU	Argentina, Buenos Aires	08:33	22:56
LX	Luxembourg	07:17	15:35
LZ	Bulgaria, Sofia	05:43	14:53
OA	Peru, Lima	10:35	23:23
OD	Lebanon, Beirut	04:29	14:28
OE	Austria, Vienna	06:31	15:01
OH	Finland, Helsinki	07:07	13:15
OH0	Aland Is.	07:27	13:35
OJ0	Market Reef	07:32	13:38
OK	Czechoslovakia, Prague	06:46	15:00
ON	Belgium, Brussels	07:30	15:37
OX, XP	Greenland, Godthaab	13:00	17:35
	Thule	24 hrs night	
OY	Faroe Islands, Torshavn	09:32	15:04
OZ	Denmark, Copenhagen	07:23	14:38
P2	Papua New Guinea, Madang	19:56	08:21
	Port Moresby	19:43	08:22
PA-PI	Netherlands, Amsterdam	07:35	15:27
PJ, P4	Netherlands Antilles, Willemstad	10:44	22:10
PJ5-8	St. Maarten and Saba, St. Maarten	10:30	21:37
PY	Brazil, Brasilia	08:32	21:33
	Rio De Janeiro	07:59	21:28
	Natal	07:58	20:25
	Manaus	09:43	22:00
	Porto Alegre	08:16	22:16
PY0	Fernando De Noronha	07:50	20:11
PY0	St. Peter & St. Paul Rocks	07:47	19:50
PY0	Trindade & Martin Vaz Is., Trindade	07:08	20:28
PZ	Suriname, Paramaribo	09:38	21:26
S2	Bangladesh, Dacca	00:27	11:11
S7	Seychelles, Victoria	01:58	14:20
S9	Sao Tome	05:21	17:27
SJ-SM	Sweden, Stockholm	07:27	13:50
SP	Poland, Krakow	06:23	14:38
	Warsaw	06:30	14:24
ST	Sudan, Khartoum	04:04	15:18
ST0	Southern Sudan, Juba	03:49	15:40
SU	Egypt, Cairo	04:37	14:54
SV	Greece, Athens	05:27	15:05
SV5	Dodecanese, Rhodes	05:05	14:51
SV9	Crete	05:14	15:06
SV/A	Mount Athos	05:31	14:56
T2	Tuvalu, Funafuti	17:38	06:15
T30	West Kiribati, Bonriki	18:17	06:19
T31	Central Kiribati, Kanton	17:09	05:26
T32	East Kiribati, Christmas Is.	16:20	04:21
T5	Somalia, Mogadishu	02:49	14:49
T7	San Marino	06:31	15:33

Pref	State/Province/Country/City	Sunrise	Sunset
TA	Turkey, Ankara	04:56	14:23
	Istanbul	05:15	14:35
TF	Iceland, Reykjavik	11:00	15:38
TG	Guatemala, Guatemala City	12:14	23:32
TI	Costa Rica, San Jose	11:40	23:14
TI9	Cocos Is.	11:45	23:33
TJ	Cameroon, Yaounde	05:08	17:02
TK	Corsica	06:38	15:52
TL	Central African Republic, Bangui	04:40	16:33
TN	Congo, Brazzaville	04:39	17:00
TR	Gabon, Libreville	05:10	17:16
TT	Chad, N'Djamena	05:08	16:34
TU	Ivory Coast, Abidjan	06:12	18:02
TY	Benin, Porto Novo	05:48	17:33
TZ	Mali, Bamako	06:41	18:05
UA	Russia, European,		
	Leningrad (UA1)	06:43	12:56
	Archangel (UA1)	06:57	11:20
	Murmansk (UA1)	24 hrs night	
	Moscow (UA3)	05:44	12:57
	Kuibyshev (UA4)	04:39	12:22
	Rostov (UA6)	04:55	13:31
UA1P	Franz Josef Land	24 hrs night	
UA2	Kaliningrad	06:47	14:11
UA9,Ø	Russia, Asiatic,		
	Novosibirsk (UA9)	02:37	10:01
	Perm (UA9)	04:44	11:28
	Omsk (UA9)	03:15	10:39
	Norilsk (UAØ)	24 hrs night	
	Irkutsk (UAØ)	00:57	08:50
	Vladivostok (UAØ)	22:30	07:37
	Petropavlovsk (UAØ)	21:23	05:09
	Khabarovsk (UAØ)	22:37	07:04
	Krasnoyarsk (UAØ)	02:04	09:15
	Yakutsk (UAØ)	00:26	05:58
	Wrangel Island (UAØ)	24 hrs night	
	Kyzyl (UAØY)	01:33	09:32
UB	Ukraine, Kiev	05:43	13:55
UC	Byelorussia, Minsk	06:12	13:49
UD	Azerbaijan, Baku	03:49	13:13
UF	Georgia, Tbilisi	04:14	13:30
UG	Armenia, Ierevan	04:11	13:35
UH	Turkoman, Ashkhabad	03:08	12:46
UI	Uzbek, Bukhara	02:49	12:17
	Tashkent	02:34	11:53
UJ	Tadzhik, Samarkand	02:40	12:08
	Dushanbe	02:30	12:02
UL	Kazakh, Alma-ata	02:10	11:16
UM	Kirghiz, Frunze	02:18	11:27
UO	Moldavia, Kishinev	05:36	14:16
UP	Lithuania, Vilna	06:24	13:54
UQ	Latvia, Riga	06:45	13:44
UR	Estonia, Tallinn	07:01	13:23
V2	Antigua & Barbuda, St. Johns	10:24	21:32
V3	Belize, Belmopan	12:12	23:20
V4	St. Christopher & Nevis	10:27	21:35
V8	Brunei, Bandar Seri Begawan	22:16	10:07
VE1, CY	Sable Is.	11:20	20:22
VE1, CY	St. Paul Is.	11:32	20:10
VK	Australia, Canberra (VK1)	18:41	09:08
	Sydney (VK2)	18:37	08:56
	Melbourne (VK3)	18:51	09:31
	Brisbane (VK4)	18:45	08:33
	Adelaide (VK5)	19:24	09:49
	Perth (VK6)	21:03	11:12
	Hobart, Tasmania (VK7)	18:25	09:38
	Darwin (VK8)	20:42	09:32
VK	Lord Howe Is.	18:11	08:18
VK9	Christmas Is.	22:27	11:09
VK9	Cocos-Keeling Is.	22:59	11:48
VK9	Mellish Reef	18:54	08:02
VK9	Norfolk Is.	17:41	07:36
VK9	Willis Is.	19:21	08:25
VKØ	Heard Is.	22:36	15:19
VKØ	Macquarie Is.	16:44	09:48
VP2E	Anguilla	10:31	21:35
VP2M	Montserrat	10:25	21:35
VP2V	British Virgin Is., Tortola	10:37	21:41
VP5	Turks & Caicos Islands, Grand Turk	11:09	22:02
VP8	Falkland Islands, Stanley	07:28	23:57

Pref	State/Province/Country/City	Sunrise	Sunset
VP8	So. Georgia Is.	05:49	22:47
VP8	So. Orkney Is.	05:31	00:15
VP8	So. Sandwich Islands, Saunders Is.	04:43	22:32
VP8	So. Shetland Is., King George Is.	06:05	01:23
VP9	Bermuda	11:06	21:13
VQ9	Chagos, Diego Garcia	00:45	13:17
VR6	Pitcairn Is.	13:42	03:20
VS6	Hong Kong	22:49	09:38
VU	India, Bombay	01:29	12:31
	Calcutta	00:33	11:21
	New Delhi	01:30	11:54
	Bangalore	00:59	12:22
VU7	Andaman Islands, Port Blair	23:56	11:24
VU7	Laccadive Is.	01:12	12:46
XE	Mexico, Mexico City (XE1)	12:57	23:57
	Chihuahua (XE2)	13:43	00:06
	Merida (XE3)	12:23	23:17
XF4	Revilla Gigedo	13:46	00:48
XT	Burkina Faso, Onagadougou	06:15	17:40
XU	Kampuchea, Phnom Penh	23:08	10:36
XW	Laos, Viangchan	23:28	10:33
XX9	Macao	22:52	09:41
XZ	Burma, Rangoon	23:52	11:02
Y2-9	German Dem.Rep.(E), Berlin	07:02	14:53
YA	Afghanistan, Kandahar	02:21	12:34
	Kabul	02:16	12:12
YB-YD	Indonesia, Jakarta	22:30	10:58
	Medan	23:18	11:14
	Pontianak	22:30	10:37
	Jayapura	20:20	08:36
YI	Iraq, Baghdad	03:51	13:55
YJ	Vanuatu, Port Vila	18:03	07:12
YK	Syria, Damascus	04:25	14:26
YN, HT	Nicaragua, Managua	11:51	23:18
YO	Romania, Bucharest	05:37	14:36
YS	El Salvador, San Salvador	12:07	23:28
YU	Yugoslavia, Belgrade	06:01	14:56
YV	Venezuela, Caracas	10:33	22:05
YVØ	Aves Is.	10:29	21:43
Z2,ZE	Zimbabwe, Harare	03:12	16:21
ZA	Albania, Tirane	05:52	15:11
ZB	Gibraltar	07:18	17:06
ZC4	British Cyprus	04:41	14:37
ZD7	St. Helena	05:43	18:46
ZD8	Ascension Is.	06:31	19:06
ZD9	Tristan da Cunha	05:22	19:58
ZF	Cayman Is.	11:46	22:46
ZK1	No. Cook Is., Manihiki	16:14	04:56
ZK1	So. Cook Is., Rarotonga	15:49	05:11
ZK2	Niue	16:30	05:43
ZK3	Tokelaus, Atafu	17:04	05:39
ZL	New Zealand, Auckland (ZL1)	16:54	07:29
	Wellington (ZL2)	16:41	07:43
	Christchurch (ZL3)	16:42	07:59
	Dunedin (ZL4)	16:41	08:16
ZL5	Antarctica, Scott Base	24 hrs day	
ZL7	Chatham Is.	15:57	07:17
ZL8	Kermadec Is.	16:44	06:41
ZL9	Auckland & Campbell Is., Auckland	16:36	08:54
	Campbell Is.	16:16	08:53
ZP	Paraguay, Asuncion	08:52	22:31
ZS	South Africa, Cape Town (ZS1)	03:28	17:47
	Port Elizabeth (ZS2)	02:58	17:18
	Bloemfontein (ZS4)	03:08	17:05
	Durban (ZS5)	02:48	16:47
	Johannesburg (ZS6)	03:07	16:50
ZS2	Prince Edward & Marion Is., Marion	01:29	17:11
ZS3	Namibia, Windhoek	03:59	17:26
1AØ	S.M.O.M	06:24	15:39
1S	Spratly Is.	22:34	10:12
3A	Monaco	06:50	15:53
3B6	Agalega	01:43	14:26
3B7	St. Brandon	01:20	14:23
3B8	Mauritius	01:22	14:40
3B9	Rodriguez Is.	03:05	14:15
3C	Equatorial Guinea, Bata	05:11	17:12
	Malabo	05:18	17:13

Pref	State/Province/Country/City	Sunrise	Sunset
3CØ	Pagalu Is.	05:22	17:35
3D2	Fiji Is., Suva	17:22	06:32
3D6	Swaziland, Mbabane	02:55	16:38
3V	Tunisia, Tunis	06:18	16:02
3W	Vietnam, Ho Chi Minh City		
	(Saigon)	22:59	10:30
	Hanoi	23:21	10:15
3X	Republic of Guinea, Conakry	06:58	18:33
3Y	Bouvet	03:07	20:08
3Y	Peter I Is.	24 hrs day	
4S	Sri Lanka, Colombo	00:39	12:23
4U	ITU Geneva	07:03	15:49
4U	United Nations Hq.	12:06	21:28
4W	Yemen, Sanaa	03:17	14:32
4X, 4Z	Israel, Jerusalem	04:26	14:35
5A	Libya, Tripoli	05:58	16:04
	Benghazi	05:27	15:35
5B	Cyprus, Nicosia	04:41	14:34
5H	Tanzania, Dar es Salaam	02:58	15:28
5N	Nigeria, Lagos	05:45	17:30
5R	Madagascar, Antananarivo	02:04	15:18
5T	Mauritania, Nouakchott	07:22	18:27
5U	Niger, Niamey	06:02	17:24
5V	Togo, Lome	05:52	17:40
5W	Western Samoa, Apia	16:51	05:45
5X	Uganda, Kampala	03:38	15:44
5Z	Kenya, Nairobi	04:38	16:49
6W	Senegal, Dakar	07:22	18:40
6Y	Jamaica, Kingston	11:25	22:31

Pref	State/Province/Country/City	Sunrise	Sunset
7O	Yemen, People's Dem. Rep.,		
	Aden	03:09	14:33
7P	Lesotho, Maseru	03:03	16:59
7Q	Malawi, Lilongwe	03:08	16:03
	Blantyre	03:00	16:02
7X	Algeria, Algiers	06:46	16:31
8P	Barbados, Bridgetown	10:08	21:31
8Q	Maldive Is.	01:01	12:54
8R	Guyana, Georgetown	09:51	21:36
9G	Ghana, Accra	05:57	17:46
9H	Malta	05:59	15:48
9J	Zambia, Lusaka	03:28	16:28
9K	Kuwait	03:30	13:49
9L	Sierra Leone, Freetown	06:54	18:33
9M2	West Malaysia, Kuala Lumpur	23:06	11:03
9M6,8	East Malaysia, Sabah,		
	Sandakan (9M6)	22:05	09:52
	Sarawak, Kuching (9M8)	22:29	10:31
9N	Nepal, Katmandu	00:56	11:23
9Q	Zaire, Kinshasa	04:39	17:00
	Kisangani	04:07	16:13
	Lubumbashi	03:37	16:24
9U	Burundi, Bujumbura	03:45	16:03
9V	Singapore	22:54	10:57
9X	Rwanda, Kigali	03:44	15:57
9Y	Trinidad & Tobago, Port of		
	Spain	10:10	21:42
J2/A	Abu Ail	03:20	14:39

Pref	State/Province/Country/City	Sunrise	Sunset
VE1	New Brunswick, Saint John	12:00	20:44
	Nova Scotia, Halifax	11:48	20:36
	Prince Edward Island, Charlottetown	11:52	20:28
VE2	Quebec, Montreal	12:31	21:13
	Quebec City	12:26	20:58
VE3	Ontario, London	12:53	21:53
	Ottawa	12:39	21:22
	Sudbury	13:04	21:39
	Toronto	12:47	21:43
VE4	Manitoba, Winnipeg	14:23	22:28
VE5	Saskatchewan, Regina	14:56	22:56
	Saskatoon	15:13	22:56
VE6	Alberta, Calgary	15:37	23:31
	Edmonton	15:48	23:15
VE7	British Columbia, Prince George	16:27	23:50
	Prince Rupert	16:59	00:18
	Vancouver	16:05	00:15
VE8	Northwest Territories, Yellowknife	17:07	22:03
	Resolute	24 hrs night	
VY1	Yukon, Whitehorse	18:09	23:47
VO1	Newfoundland, St. John's	11:16	19:41
VO2	Labrador, Goose Bay	12:14	19:44
W1	Connecticut, Hartford	12:14	21:22
	Maine, Bangor	12:09	20:56
	Portland	12:11	21:06
	Massachusetts, Boston	12:10	21:14
	New Hampshire, Concord	12:14	21:13
	Rhode Island, Providence	12:09	21:17
	Vermont, Montpelier	12:22	21:13
W2	New Jersey, Atlantic City	12:14	21:37
	New York, Albany	12:22	21:24
	Buffalo	12:43	21:43
	New York City	12:16	21:31
	Syracuse	12:33	21:32
W3	Delaware, Wilmington	12:19	21:40
	District of Columbia, Washington	12:22	21:49
	Maryland, Baltimore	12:22	21:46
	Pennsylvania, Harrisburg	12:26	21:44
	Philadelphia	12:18	21:38
	Pittsburgh	12:39	21:56
	Scranton	12:25	21:36
W4	Alabama, Montgomery	12:42	22:43
	Florida, Jacksonville	12:19	22:30
	Miami	12:03	22:34
	Pensacola	12:41	22:52
	Georgia, Atlanta	12:38	22:32
	Savannah	12:21	22:23
	Kentucky, Lexington	12:50	22:21
	Louisville	12:56	22:26
	North Carolina, Charlotte	12:27	22:14
	Raleigh	12:20	22:04
	Wilmington	12:13	22:05
	South Carolina, Columbia	12:25	22:18
	Tennessee, Knoxville	12:42	22:24
	Memphis	13:04	22:52
	Nashville	12:54	22:35
	Virginia, Norfolk	12:14	21:51
	Richmond	12:20	21:54
W5	Arkansas, Little Rock	13:12	23:01
	Louisiana, New Orleans	12:51	23:04
	Shreveport	13:12	23:13
	Mississippi, Jackson	12:57	22:59
	New Mexico, Albuquerque	14:11	23:58
	Oklahoma, Oklahoma City	13:35	23:20
	Texas, Abilene	13:36	23:37
	Amarillo	13:51	23:38
	Dallas	13:25	23:24
	El Paso	14:01	00:06
	San Antonio	13:24	23:39
W6	California, Los Angeles	14:54	00:47
	San Francisco	15:21	00:53
W7	Arizona, Flagstaff	14:31	00:18
	Phoenix	14:28	00:24
	Idaho, Boise	15:14	00:10
	Pocatello	14:57	23:58
	Montana, Billings	14:52	23:31
	Butte	15:08	23:47
	Great Falls	15:10	23:36
	Nevada, Las Vegas	14:47	00:29
	Reno	15:15	00:38
	Oregon, Portland	15:47	00:29
	Utah, Salt Lake City	14:48	00:02
	Washington, Seattle	15:54	00:19
	Spokane	15:35	23:59
	Wyoming, Cheyenne	14:20	23:33
	Sheridan	14:42	23:29
W8	Michigan, Detroit	12:57	22:02
	Grand Rapids	13:10	22:10
	Sault Ste. Marie	13:18	21:52
	Traverse City	13:16	22:04
	Ohio, Cincinnati	12:53	22:18
	Cleveland	12:49	21:59
	Columbus	12:50	22:09
	West Virginia, Charleston	12:39	22:08
W9	Illinois, Chicago	13:14	22:22
	Indiana, Indianapolis	13:02	22:23
	Wisconsin, Green Bay	13:25	22:14
	Milwaukee	13:19	22:19
W0	Colorado, Denver	14:17	23:38
	Grand Junction	14:29	23:54
	Iowa, Des Moines	13:37	22:47
	Kansas, Pratt	13:46	23:19
	Wichita	13:40	23:13
	Minnesota, Duluth	13:50	22:22
	Minneapolis	13:48	22:34
	Missouri, Columbia	13:24	22:49
	Kansas City	13:33	22:58
	St. Louis	13:14	22:42
	Nebraska, North Platte	14:04	23:17
	Omaha	13:45	22:57
	North Dakota, Fargo	14:09	22:40
	South Dakota, Rapid City	14:24	23:16
A2	Botswana, Gaborone	03:24	17:04
A3	Tonga, Nukualofa	16:56	06:21
A4	Oman, Muscat	02:43	13:23
A5	Bhutan, Thimbu	00:47	11:12
A6	United Arab Emirates, Abu Dhabi	03:02	13:39
A7	Qatar, Ad-Dawhah	03:15	13:48
A9	Bahrein, Al-Manamah	03:20	13:50
AP	Pakistan, Karachi	02:11	12:47
	Islamabad	02:07	12:02
BV	Taiwan, Taipei	22:34	09:09
BY	People's Rep. of China, Beijing	23:32	08:52
	Harbin	23:11	07:51
	Shanghai	22:48	08:55
	Fuzhou	22:45	09:15
	Xian	23:46	09:38
	Chongqing	23:45	09:58
	Chengdu	23:56	10:06
	Lhasa	00:46	11:00
	Urumqi	01:40	10:34
	Kashi	02:12	11:35
C2	Nauru	18:45	06:54
C3	Andorra	07:20	16:23
C5	The Gambia, Banjul	07:24	18:44
C6	Bahamas, Nassau	11:50	22:24
C9	Mozambique, Maputo	02:54	16:40
	Mozambique	02:44	15:45
CE	Chile, Santiago	09:29	23:52
CE0	Easter Island	12:20	02:11
CE0	San Felix	10:24	00:12
CE0	Juan Fernandez	10:01	00:24
CM,CO	Cuba, Havana	12:06	22:48
CN	Morocco, Casablanca	07:30	17:25
CP	Bolivia, La Paz	09:58	23:04
CT	Portugal, Lisbon	07:51	17:18
CT2, CU	Azores, Ponta Delgada	08:54	18:27
CT3	Madeira Islands, Funchal	08:05	18:05
CX	Uruguay, Montevideo	08:27	22:57
D2	Angola, Luanda	04:46	17:24
D4	Cape Verde, Praia	07:54	19:09

Pref	State/Province/Country/City	Sunrise	Sunset
D6	Comoros, Moroni	02:38	15:27
DA-DL	Germany, Fed. Rep. of (W),		
	Bonn	07:31	15:28
DU	Philippines, Manilla	22:16	09:31
EA	Spain, Madrid	07:34	16:51
EA6	Balearic Is., Palma	07:06	16:28
EA8	Canary Is., Las Palmas	07:45	18:05
EA9	Ceuta & Melilla, Ceuta	07:27	17:10
	Melilla	07:16	17:03
EI	Ireland, Dublin	08:38	16:08
EL	Liberia, Monrovia	06:48	18:33
EP	Iran, Tehran	03:38	13:22
ET	Ethiopia, Addis Ababa	03:35	15:11
	Asmera	03:45	14:58
F	France, Paris	07:41	15:56
FG	Guadeloupe	10:29	21:40
FG, FS	St. Martin	10:39	21:41
FH	Mayotte	02:30	15:23
FK	New Caledonia, Noumia	18:07	07:36
FM	Martinique	10:24	21:39
FO	Clipperton	13:29	01:00
FO	Fr. Polynesia, Tahiti	16:00	05:11
FP	St. Pierre & Miquelon, St. Pierre	11:25	19:58
FR	Glorioso	02:24	15:12
FR	Juan de Nova	02:32	15:41
	Europa	02:31	16:01
FR	Reunion	01:33	14:58
FR	Tromelin	01:48	14:52
FT8W	Crozet	00:37	16:22
FT8X	Kerguelen	23:13	15:28
FT8Y	Antarctica, Dumont D'Urville	24 hrs day	
FT8Z	Amsterdam & St. Paul Is.,		
	Amsterdam	23:24	14:10
FW	Wallis & Futuna Is., Wallis	17:15	06:10
FY	Fr. Guiana, Cayenne	09:32	21:22
G	England, London	08:03	15:53
GD	Isle of Man	08:36	15:55
GI	Northern Ireland, Belfast	08:44	15:59
GJ	Jersey	08:01	16:12
GM	Scotland, Glasgow	08:45	15:44
	Aberdeen	08:46	15:26
GU	Guernsey	08:04	16:13
GW	Wales, Cardiff	08:16	16:05
H4	Solomon Islands, Honiara	18:57	07:38
HA	Hungary, Budapest	06:28	14:54
HB	Switzerland, Bern	07:12	15:43
HBØ	Liechtenstein	07:05	15:33
HC	Ecuador, Quito	11:06	23:13
HC8	Galapagos Is.	11:55	00:04
HH	Haiti, Port-Au-Prince	11:16	22:17
HI	Dominican Republic, Santo		
	Domingo	11:07	22:08
HK	Colombia, Bogota	10:58	22:50
HKØ	Malpelo Is.	11:25	23:19
HKØ	San Adreas	11:43	23:06
HL, HM	Korea, Seoul	22:42	08:17
HP	Panama, Panama	11:28	23:03
HR	Honduras, Tegucigalpa	12:08	23:25
HS	Thailand, Bangkok	23:36	10:55
HV	Vatican City	06:34	15:41
HZ, 7Z	Saudi Arabia, Dharan	03:23	13:52
	Mecca	03:54	14:43
	Riyadh	03:32	14:09
I	Italy, Rome	06:34	15:41
	Trieste	06:42	15:23
	Sicily	06:14	15:49
IS	Sardinia, Cagliari	06:39	16:03
J2	Djibouti, Djibouti	03:21	14:48
J3	Grenada	10:22	21:47
J5	Guinea-Bissau, Bissau	07:17	18:43
J6	St. Lucia	10:22	21:41
J7	Dominica	10:26	21:39
J8	St. Vincent	10:22	21:43
JA-JS	Japan, Tokyo	21:46	07:30
	Nagasaki	22:18	08:17
	Sapporo	22:02	07:02
JD	Minami Torishima	20:23	07:00
JD	Ogasawara, Kazan Is.	21:21	07:46
JT	Mongolia, Ulan Bator	00:38	09:01
JW	Svalbard, Spitsbergen	24 hrs night	
JX	Jan Mayen	24 hrs night	

Pref	State/Province/Country/City	Sunrise	Sunset
JY	Jordan, Amman	04:32	14:35
KC4	Antarctica, Byrd Station	24 hrs day	
	McMurdo Sound	24 hrs day	
	Palmer Station	05:20	03:07
KC6	Micronesia, Ponape	19:33	07:16
KC6	Belau, Yap	20:58	08:32
	Koror	21:09	08:50
KG4	Guantanamo Bay	11:30	22:26
KH1	Baker, Howland Is.	17:39	05:44
KH2	Guam, Agana	20:38	07:58
KH3	Johnston Is.	17:38	04:45
KH4	Midway Is.	18:37	04:58
KH5	Palmyra Is.	16:53	04:39
KH5K	Kingman Reef	16:58	04:39
KH6	Hawaii, Hilo	16:50	03:46
	Honolulu	17:04	03:54
KH7	Kure Is.	18:41	05:01
KH8	American Samoa, Pago Pago	16:52	05:50
KH9	Wake Is.	19:22	06:20
KHØ	Mariana Is., Saipan	20:38	07:51
KL7	Alaska, Adak	19:51	03:37
	Anchorage	19:14	00:41
	Fairbanks	19:58	23:41
	Juneau	17:44	00:06
	Nome	21:02	00:56
KP1	Navassa Is.	11:27	22:28
KP2	Virgin Islands, Charlotte Amalie	10:46	21:48
KP4	Puerto Rico, San Juan	10:52	21:53
KP5	Desecheo Is.	10:57	21:58
KX6	Marshall Islands, Kwajalein	19:01	06:36
LA-LJ	Norway, Oslo	08:19	14:11
LU	Argentina, Buenos Aires	08:37	23:05
LX	Luxembourg	07:29	15:37
LZ	Bulgaria, Sofia	05:53	14:55
OA	Peru, Lima	10:41	23:31
OD	Lebanon, Beirut	04:39	14:32
OE	Austria, Vienna	06:42	15:02
OH	Finland, Helsinki	07:23	13:12
OHØ	Aland Is.	07:43	13:32
OJØ	Market Reef	07:48	13:35
OK	Czechoslovakia, Prague	06:58	15:02
ON	Belgium, Brussels	07:42	15:38
OX, XP	Greenland, Godthaab	13:22	17:27
	Thule	24 hrs night	
OY	Faroe Islands, Torshavn	09:50	14:59
OZ	Denmark, Copenhagen	07:37	14:37
P2	Papua New Guinea, Madang	20:02	08:28
	Port Moresby	19:49	08:29
PA-PI	Netherlands, Amsterdam	07:48	15:28
PJ, P4	Netherlands Antilles, Willemstad	10:51	22:16
PJ5-8	St. Maarten and Saba,		
	St. Maarten	10:38	21:42
PY	Brazil, Brasilia	08:37	21:41
	Rio De Janeiro	08:04	21:37
	Natal	08:04	20:32
	Manaus	09:49	22:07
	Porto Alegre	08:20	22:25
PYØ	Fernando De Noronha	07:57	20:18
PYØ	St. Peter & St. Paul Rocks	07:53	19:57
PYØ	Trindade & Martin Vaz Is.,		
	Trindade	07:13	20:36
PZ	Suriname, Paramaribo	09:45	21:32
S2	Bangladesh, Dacca	00:36	11:16
S7	Seychelles, Victoria	02:04	14:27
S9	Sao Tome	05:28	17:34
SJ-SM	Sweden, Stockholm	07:43	13:48
SP	Poland, Krakow	06:35	14:40
	Warsaw	06:42	14:25
ST	Sudan, Khartoum	04:12	15:23
STØ	Southern Sudan, Juba	03:56	15:46
SU	Egypt, Cairo	04:46	14:58
SV	Greece, Athens	05:37	15:08
SV5	Dodecanese, Rhodes	05:15	14:55
SV9	Crete	05:24	15:10
SV/A	Mount Athos	05:41	14:59
T2	Tuvalu, Funafuti	17:44	06:22
T30	West Kiribati, Bonriki	18:23	06:26
T31	Central Kiribati, Kanton	17:16	05:33
T32	East Kiribati, Christmas Is.	16:27	04:27
T5	Somalia, Mogadishu	02:56	14:56
T7	San Marino	06:41	15:35

Pref	State/Province/Country/City	Sunrise	Sunset
TA	Turkey, Ankara	05:06	14:26
	Istanbul	05:26	14:38
TF	Iceland, Reykjavik	11:21	15:30
TG	Guatemala, Guatemala City	12:22	23:37
TI	Costa Rica, San Jose	11:47	23:20
TI9	Cocos Is.	11:52	23:39
TJ	Cameroon, Yaounde	05:15	17:08
TK	Corsica	06:48	15:55
TL	Central African Republic, Bangui	04:47	16:39
TN	Congo, Brazzaville	04:45	17:07
TR	Gabon, Libreville	05:17	17:22
TT	Chad, N'Djamena	05:15	16:40
TU	Ivory Coast, Abidjan	06:19	18:08
TY	Benin, Porto Novo	05:55	17:39
TZ	Mali, Bamako	06:48	18:11
UA	Russia, European,		
	Leningrad (UA1)	06:59	12:53
	Archangel (UA1)	07:20	11:11
	Murmansk (UA1)	24 hrs night	
	Moscow (UA3)	05:57	12:57
	Kuibyshev (UA4)	04:51	12:23
	Rostov (UA6)	05:06	13:33
UA1P	Franz Josef Land	24 hrs night	
UA2	Kaliningrad	07:01	14:10
UA9,Ø	Russia, Asiatic,		
	Novosibirsk (UA9)	02:51	10:01
	Perm (UA9)	04:59	11:26
	Omsk (UA9)	03:29	10:39
	Norilsk (UAØ)	24 hrs night	
	Irkutsk (UAØ)	01:10	08:51
	Vladivostok (UAØ)	22:41	07:39
	Petropavlovsk (UAØ)	21:36	05:09
	Khabarovsk (UAØ)	22:48	07:06
	Krasnoyarsk (UAØ)	02:18	09:15
	Yakutsk (UAØ)	00:44	05:53
	Wrangel Island (UAØ)	24 hrs night	
	Kyzyl (UAØY)	01:46	09:33
UB	Ukraine, Kiev	05:55	13:56
UC	Byelorussia, Minsk	06:25	13:49
UD	Azerbaijan, Baku	03:59	13:16
UF	Georgia, Tbilisi	04:24	13:33
UG	Armenia, Ierevan	04:21	13:38
UH	Turkoman, Ashkhabad	03:18	12:50
UI	Uzbek, Bukhara	03:00	12:20
	Tashkent	02:44	11:56
UJ	Tadzhik, Samarkand	02:50	12:11
	Dushanbe	02:40	12:05
UL	Kazakh, Alma-ata	02:21	11:19
UM	Kirghiz, Frunze	02:29	11:29
UO	Moldavia, Kishinev	05:47	14:17
UP	Lithuania, Vilna	06:37	13:54
UQ	Latvia, Riga	07:00	13:42
UR	Estonia, Tallinn	07:17	13:20
V2	Antigua & Barbuda, St. Johns	10:32	21:38
V3	Belize, Belmopan	12:20	23:25
V4	St. Christopher & Nevis	10:35	21:41
V8	Brunei, Bandar Seri Begawan	22:23	10:13
VE1, CY	Sable Is.	11:30	20:25
VE1, CY	St. Paul Is.	11:44	20:12
VK	Australia, Canberra (VK1)	18:45	09:17
	Sydney (VK2)	18:40	09:05
	Melbourne (VK3)	18:54	09:41
	Brisbane (VK4)	18:49	08:42
	Adelaide (VK5)	19:28	09:58
	Perth (VK6)	21:07	11:21
	Hobart, Tasmania (VK7)	18:28	09:49
	Darwin (VK8)	20:48	09:40
VK	Lord Howe Is.	18:15	08:27
VK9	Christmas Is.	22:33	11:17
VK9	Cocos-Keeling Is.	23:05	11:56
VK9	Mellish Reef	18:59	08:10
VK9	Norfolk Is.	17:46	07:45
VK9	Willis Is.	19:27	08:32
VKØ	Heard Is.	22:36	15:32
VKØ	Macquarie Is.	16:43	10:01
VP2E	Anguilla	10:39	21:40
VP2M	Montserrat	10:33	21:40
VP2V	British Virgin Is., Tortola	10:45	21:47
VP5	Turks & Caicos Islands,		
	Grand Turk	11:17	22:07
VP8	Falkland Islands, Stanley	07:29	00:09

Pref	State/Province/Country/City	Sunrise	Sunset
VP8	So. Georgia Is.	05:48	23:01
VP8	So. Orkney Is.	05:26	00:33
VP8	So. Sandwich Islands, Saunders		
	Is.	04:41	22:48
VP8	So. Shetland Is., King George		
	Is.	05:58	01:43
VP9	Bermuda	11:15	21:17
VQ9	Chagos, Diego Garcia	00:51	13:24
VR6	Pitcairn Is.	13:47	03:29
VS6	Hong Kong	22:57	09:43
VU	India, Bombay	01:37	12:36
	Calcutta	00:41	11:26
	New Delhi	01:39	11:58
	Bangalore	01:06	12:28
VU7	Andaman Islands, Port Blair	00:03	11:29
VU7	Laccadive Is.	01:19	12:52
XE	Mexico, Mexico City (XE1)	13:05	00:03
	Chihuahua (XE2)	13:52	00:11
	Merida (XE3)	12:31	23:22
XF4	Revilla Gigedo	13:54	00:53
XT	Burkina Faso, Onagadougou	06:22	17:46
XU	Kampuchea, Phnom Penh	23:15	10:41
XW	Laos, Viangchan	23:36	10:39
XX9	Macao	23:00	09:46
XZ	Burma, Rangoon	00:00	11:07
Y2-9	German Dem.Rep.(E), Berlin	07:14	14:53
YA	Afghanistan, Kandahar	02:30	12:38
	Kabul	02:25	12:16
YB-YD	Indonesia, Jakarta	22:36	11:05
	Medan	23:25	11:20
	Pontianak	22:37	10:44
	Jayapura	20:27	08:43
YI	Iraq, Baghdad	04:00	13:59
YJ	Vanuatu, Port Vila	18:09	07:20
YK	Syria, Damascus	04:34	14:30
YN, HT	Nicaragua, Managua	11:59	23:24
YO	Romania, Bucharest	05:48	14:38
YS	El Salvador, San Salvador	12:15	23:34
YU	Yugoslavia, Belgrade	06:12	14:59
YV	Venezuela, Caracas	10:40	22:11
YVØ	Aves Is.	10:36	21:48
Z2,ZE	Zimbabwe, Harare	03:18	16:29
ZA	Albania, Tirane	06:03	15:14
ZB	Gibraltar	07:28	17:10
ZC4	British Cyprus	04:50	14:40
ZD7	St. Helena	05:49	18:53
ZD8	Ascension Is.	06:37	19:13
ZD9	Tristan da Cunha	05:25	20:08
ZF	Cayman Is.	11:54	22:51
ZK1	No. Cook Is., Manihiki	16:20	05:03
ZK1	So. Cook Is., Rarotonga	15:54	05:19
ZK2	Niue	16:35	05:51
ZK3	Tokelaus, Atafu	17:10	05:47
ZL	New Zealand, Auckland (ZL1)	16:57	07:39
	Wellington (ZL2)	16:43	07:53
	Christchurch (ZL3)	16:44	08:10
	Dunedin (ZL4)	16:43	08:28
ZL5	Antarctica, Scott Base	24 hrs day	
ZL7	Chatham Is.	15:59	07:28
ZL8	Kermadec Is.	16:48	06:50
ZL9	Auckland & Campbell Is.,		
	Auckland	16:37	09:06
	Campbell Is.	16:16	09:06
ZP	Paraguay, Asuncion	08:57	22:40
ZS	South Africa, Cape Town (ZS1)	03:32	17:56
	Port Elizabeth (ZS2)	03:02	17:27
	Bloemfontein (ZS4)	03:13	17:13
	Durban (ZS5)	02:52	16:56
	Johannesburg (ZS6)	03:12	16:59
ZS2	Prince Edward & Marion Is.,		
	Marion	01:30	17:22
ZS3	Namibia, Windhoek	04:03	17:35
1AØ	S.M.O.M	06:34	15:42
1S	Spratly Is.	22:42	10:18
3A	Monaco	07:00	15:55
3B6	Agalega	01:49	14:33
3B7	St. Brandon	01:25	14:31
3B8	Mauritius	01:27	14:48
3B9	Rodriguez Is.	01:04	14:24
3C	Equatorial Guinea, Bata	05:18	17:19
	Malabo	05:25	17:19

Pref	State/Province/Country/City	Sunrise	Sunset	Pref	State/Province/Country/City	Sunrise	Sunset
3CØ	Pagalu Is.	05:29	17:41	7O	Yemen, People's Dem. Rep.,		
3D2	Fiji Is., Suva	17:27	06:40		Aden	03:16	14:39
3D6	Swaziland, Mbabane	02:59	16:47	7P	Lesotho, Maseru	03:07	17:08
3V	Tunisia, Tunis	06:28	16:06	7Q	Malawi, Lilongwe	03:14	16:11
3W	Vietnam, Ho Chi Minh City				Blantyre	03:05	16:09
	(Saigon)	23:06	10:35	7X	Algeria, Algiers	06:56	16:35
	Hanoi	23:29	10:20	8P	Barbados, Bridgetown	10:15	21:36
3X	Republic of Guinea, Conakry	07:05	18:39	8Q	Maldive Is.	01:08	13:00
3Y	Bouvet	03:06	20:22	8R	Guyana, Georgetown	09:58	21:42
3Y	Peter I Is.	24 hrs day		9G	Ghana, Accra	06:04	17:52
4S	Sri Lanka, Colombo	00:46	12:29	9H	Malta	06:09	15:51
4U	ITU Geneva	07:14	15:51	9J	Zambia, Lusaka	03:33	16:36
4U	United Nations Hq.	12:16	21:31	9K	Kuwait	03:39	13:54
4W	Yemen, Sanaa	03:24	14:37	9L	Sierra Leone, Freetown	07:01	18:39
4X, 4Z	Israel, Jerusalem	04:35	14:39	9M2	West Malaysia, Kuala Lumpur	23:13	11:09
5A	Libya, Tripoli	06:07	16:08	9M6,8	East Malaysia, Sabah,		
	Benghazi	05:36	15:39		Sandakan (9M6)	22:12	09:59
5B	Cyprus, Nicosia	04:50	14:37		Sarawak, Kuching (9M8)	22:35	10:37
5H	Tanzania, Dar es Salaam	03:04	15:35	9N	Nepal, Katmandu	01:05	11:28
5N	Nigeria, Lagos	05:52	17:36	9Q	Zaire, Kinshasa	04:45	17:07
5R	Madagascar, Antananarivo	02:09	15:26		Kisangani	04:14	16:19
5T	Mauritania, Nouakchott	07:30	18:33		Lubumbashi	03:43	16:32
5U	Niger, Niamey	06:10	17:29	9U	Burundi, Bujumbura	03:51	16:10
5V	Togo, Lome	05:59	17:46	9V	Singapore	23:01	11:04
5W	Western Samoa, Apia	16:57	05:52	9X	Rwanda, Kigali	03:50	16:04
5X	Uganda, Kampala	03:44	15:51	9Y	Trinidad & Tobago, Port of		
5Z	Kenya, Nairobi	04:44	16:56		Spain	10:17	21:48
6W	Senegal, Dakar	07:30	18:45	J2/A	Abu Ail	03:28	14:45
6Y	Jamaica, Kingston	11:33	22:36				

W1AW

ARRL operates a Headquarters station using W1AW, the original call of its founding father, Hiram Percy Maxim; thus, it is often known as the Maxim Memorial Station. Many services are performed for the operating amateur over W1AW every day, including bulletins of information and latest news, code practice at speeds from 5 through 35 WPM, and certificate-qualifying runs, etc.

IN A COMMUNICATIONS EMERGENCY monitor W1AW for bulletins as follows: *Phone*, on the hour. *RTTY*, 15 minutes past the hour. *CW*, on the half hour.

Abbreviated W1AW Schedule

(Times are in Eastern Time.)

Slow Code Practice 5-15 WPM	Mon, Wed, Fri at 9 AM and 7 PM. Tues, Thurs, Sat, Sun at 4 PM and 10 PM.
Fast Code Practice 35 WPM and above	Mon, Wed, Fri at 4 PM and 10 PM. Tues, Thurs, Sat, Sun at 7 PM; Tues, Thurs at 9 AM.
CW Bulletins 18 WPM	Daily at 5, 8 and 11 PM; Mon thru Fri at 10 AM.
Teleprinter Bulletins Baudot/ASCII	Daily at 6, 9 and 12 PM; Mon thru Fri at 11 AM.
AMTOR as time permits	
Voice Bulletins	Daily at 9:30 PM and 12:30 AM.

Code Practice and CW bulletin frequencies:
1.818, 3.580, 7.08, 14.07, 21.08, 28.08 and 147.555 MHz.

Teleprinter bulletin frequencies:
3.625, 7.095, 14.095, 21.095, 28.095 and 147.555 MHz.

Voice bulletin frequencies:
1.89, 3.99, 7.29, 14.29, 21.39, 28.59 and 147.555 MHz.

The complete W1AW operating schedule appears in *QST*. The schedule is also available from ARRL HQ for an SASE.

ARRL Field Organization

The United States and Canada are divided into 16 ARRL Divisions. Every two years the ARRL full members in each of these divisions elect a director and a vice director to represent them on the League's Board of Directors. The Board determines the policies of the League, which are carried out by the Headquarters staff. A director's function is principally policymaking at the highest level, but the Board of Directors is all-powerful in the conduct of League affairs.

The 16 divisions are further broken down into 74 sections, and the ARRL full members in each section elect a Section Manager (SM). The SM is the senior elected ARRL official in the section, and in cooperation with the director, fosters and encourages all ARRL activities within the section. A breakdown of sections within each division (and counties within each split-state section) follows:

ATLANTIC DIVISION: *Delaware; Eastern Pennsylvania* (Adams, Berks, Bradford, Bucks, Carbon, Chester, Columbia, Cumberland, Dauphin, Delaware, Juniata, Lackawanna, Lancaster, Lebanon, Lehigh, Luzerne, Lycoming, Monroe, Montgomery, Montour, Northampton, Northumberland, Perry, Philadelphia, Pike, Schuylkill, Snyder, Sullivan, Susquehanna, Tioga, Union, Wayne, Wyoming, York); *Maryland-D.C.; Southern New Jersey* (Atlantic, Burlington, Camden, Cape May, Cumberland, Gloucester, Mercer, Ocean, Salem); *Western New York* (Allegany, Broome, Cattaraugus, Cayuga, Chautauqua, Chemung, Chenango, Clinton, Cortland, Delaware, Erie, Essex, Franklin, Fulton, Genesee, Hamilton, Herkimer, Jefferson, Lewis, Livingston, Madison, Monroe, Montgomery, Niagara, Oneida, Onondaga, Ontario, Orleans, Oswego, Otsego, St. Lawrence, Schoharie, Schuyler, Seneca, Steuben, Tioga, Tompkins, Wayne, Wyoming, Yates); *Western Pennsylvania* (those counties not listed under Eastern Pennsylvania).

CENTRAL DIVISION: *Illinois; Indiana; Wisconsin.*

DAKOTA DIVISION: *Minnesota; North Dakota; South Dakota.*

DELTA DIVISION: *Arkansas; Louisiana; Mississippi; Tennessee.*

GREAT LAKES DIVISION: *Kentucky; Michigan; Ohio.*

HUDSON DIVISION: *Eastern New York* (Albany, Columbia, Dutchess, Greene, Orange, Putnam, Rensselaer, Rockland, Saratoga, Schenectady, Sullivan, Ulster, Warren, Washington, Westchester); *N.Y.C.-L.I.* (Bronx, Kings, Nassau, New York, Queens, Richmond, Suffolk); *Northern New Jersey* (Bergen, Essex, Hudson, Hunterdon, Middlesex, Monmouth, Morris, Passaic, Somerset, Sussex, Union, Warren).

MIDWEST DIVISION: *Iowa; Kansas; Missouri; Nebraska.*

NEW ENGLAND DIVISION: *Connecticut; Maine, Eastern Massachusetts* (Barnstable, Bristol, Dukes, Essex, Middlesex, Nantucket, Norfolk, Plymouth, Suffolk); *New Hampshire; Rhode Island; Vermont; Western Massachusetts* (those counties not listed under Eastern Massachusetts).

NORTHWESTERN DIVISION: *Alaska; Idaho; Montana; Oregon; Washington.*

PACIFIC DIVISION: *East Bay* (Alameda, Contra Costa, Napa, Solano); *Nevada; Pacific* (Hawaii and U.S. possessions in the Pacific); *Sacramento Valley* (Alpine, Amador, Butte, Colusa, El Dorado, Glenn, Lassen, Modoc, Nevada, Placer, Plumas, Sacramento, Shasta, Sierra, Siskiyou, Sutter, Tehama, Trinity, Yolo, Yuba); *San Francisco* (Del Norte, Humboldt, Lake, Marin, Mendocino, San Francisco, Sonoma); *San Joaquin Valley* (Calaveras, Fresno, Kern, Kings, Madera, Mariposa, Merced, Mono, San Joaquin, Stanislaus, Tulare, Tuolumne); *Santa Clara Valley* (Monterey, San Benito, San Mateo, Santa Clara, Santa Cruz).

ROANOKE DIVISION: *North Carolina; South Carolina; Virginia; West Virginia.*

ROCKY MOUNTAIN DIVISION: *Colorado; Utah; New Mexico; Wyoming.*

SOUTHEASTERN DIVISION: *Alabama; Georgia; Northern Florida* (Alachua, Baker, Bay, Bradford, Calhoun, Citrus, Clay, Columbia, Dixie, Duval, Escambia, Flagler, Franklin, Gadsden, Gilchrist, Gulf, Hamilton, Hernando, Holmes, Jackson, Jefferson, Lafayette, Lake, Leon, Levy, Liberty, Madison, Marion, Nassau, Okaloosa, Orange, Pasco, Putnam, Santa Rosa, Seminole, St. Johns, Sumter, Suwanee, Taylor, Union, Volusia, Wakulla, Walton, Washington); *Southern Florida* (those counties not listed under Northern Florida); *West Indies* (Puerto Rico and U.S. possessions in the Caribbean).

SOUTHWESTERN DIVISION: *Arizona; Los Angeles; Orange* (Inyo, Orange, Riverside, San Bernardino); *San Diego* (Imperial, San Diego); *Santa Barbara* (San Luis Obispo, Santa Barbara, Ventura).

WEST GULF DIVISION: *North Texas* (Anderson, Archer, Bell, Bosque, Bowie, Brown, Callahan, Camp, Cass, Cherokee, Clay, Coleman, Collin, Comanche, Cooke, Coryell, Dallas, Delta, Denton, Eastland, Ellis, Erath, Falls, Fannin, Franklin, Freestone, Grayson, Gregg, Hamilton, Harrison, Henderson, Hill, Hood, Hopkins, Hunt, Jack, Johnson, Kaufman, Lamar, Lampasas, Limestone, McLennan, Marion, Mills, Montague, Morris, Nacogdoches, Navarro, Palo Pinto, Panola, Parker, Rains, Red River, Rockwall, Rusk, Shackelford, Shelby, Smith, Somervell, Stephens, Tarrant, Throckmorton, Titus, Upshur, Van Zandt, Wichita, Wise, Wood, Young); *Oklahoma; South Texas* (Angelina, Aransas, Atacosa, Austin, Bandera, Bastrop, Bee, Bexar, Blanco, Brazoria, Brazos, Brooks, Burleson, Burnet, Caldwell, Calhoun, Cameron, Chambers, Colorado, Comal, Concho, DeWitt, Dimmitt, Duval, Edwards, Fayette, Fort Bend, Frio, Galveston, Gillespie, Goliad, Gonzales, Grimes, Guadalupe, Hardin, Harris, Hays, Hidalgo, Houston, Jackson, Jasper, Jefferson, Jim Hogg, Jim Wells, Karnes, Kendall, Kenedy, Kerr, Kimble, Kinney, Kleberg, LaSalle, Lavaca, Lee, Leon, Liberty, Live Oak, Llano, Madison, Mason, Matagorda, Maverick, McCulloch, McMullen, Medina, Menard, Milam, Montgomery, Newton, Nueces, Orange, Polk, Real, Refugio, Robertson, Sabine, San Augustine, San Jacinto, San Patricio, San Saba, Starr, Travis, Trinity, Tyler, Uvalde, Val Verde, Victoria, Walker, Waller, Washington, Webb,

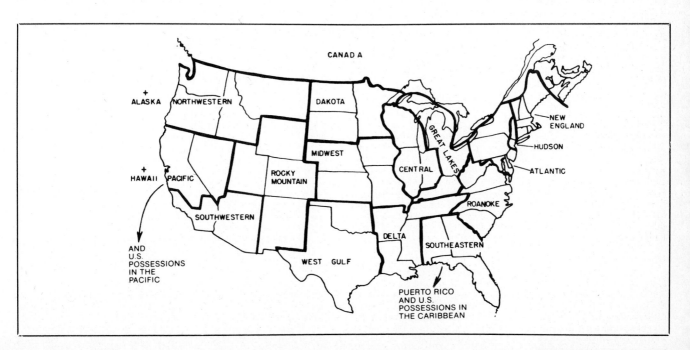

Wharton, Willacy, Williamson, Wilson, Zapata, Zavala); *West Texas* (Andrews, Armstrong, Bailey, Baylor, Bordon, Brewster, Briscoe, Carson, Castro, Childress, Cochran, Coke, Collingsworth, Cottle, Crane, Crockett, Crosby, Culberson, Dallam, Dawson, Deaf Smith, Dickens, Donley, Ector, El Paso, Fischer, Floyd, Foard, Gaines, Garza, Glasscock, Gray, Hale, Hall, Hansford, Hardeman, Hartley, Haskell, Hemphill, Hockley, Howard, Hudspeth, Hutchinson, Irion, Jeff Davis, Jones, Kent, King, Knox, Lamb, Lipscomb, Loving, Lubbock, Lynn, Martin, Midland, Mitchell, Moore, Motley, Nolan, Ochiltree, Oldham, Parmer, Pecos, Potter, Presidio, Randall, Reagan, Reeves, Roberts, Runnels, Schleicher, Scurry, Sherman, Sterling, Stonewall, Sutton, Swisher, Taylor, Terrell, Terry, Tom Green, Upton, Ward, Wheeler, Wilbarger, Winkler, Yoakum).

CANADA: *Alberta; British Columbia; Manitoba; Maritime* (Nova Scotia, New Brunswick, Prince Edward Island, Labrador, Newfoundland); *Ontario; Quebec; Saskatchewan.*

SECTION MANAGER

In each ARRL section there is an elected Section Manager (SM) who will have authority over the Field Organization in his or her section, and in cooperation with his Director, foster and encourage ARRL activities and programs within the section. Details regarding the election procedures for SMs are contained in "Rules and Regulations of the ARRL Field Organization," available on request to any ARRL member. Election notices are posted regularly in the Happenings and Canadian NewsFronts section of *QST*.

Any candidate for the office of Section Manager must be a resident of the section, a licensed amateur of Technician class or higher, and a full member of the League for a continuous term of at least two years immediately preceding receipt of a petition for nomination at Hq. If elected, he or she must maintain membership throughout the term of office.

The following is a detailed resume of the duties of the Section Manager. In discharging his responsibilities, he:

a) Recruits and appoints nine section-level assistants to serve under his general supervision and to administer the following ARRL programs in the section: emergency communications, message traffic, official observers, affiliated clubs, public information, state government liaison, technical activities and on-the-air bulletins.

b) Supervises the activities of these assistants to ensure continuing progress in accordance with overall ARRL policies and objectives.

c) Appoints qualified ARRL members in the section to volunteer positions of responsibility in support of section programs, or authorizes the respective section-level assistants to make such appointments.

d) Maintains liaison with the Division Director and makes periodic reports to him regarding the status of section activities; receives from him information and guidance pertaining to matters of mutual concern and interest; serves on the Division Cabinet and renders advice as requested by the Division Director; keeps informed on matters of policy that affect section-level programs.

e) Conducts correspondence or other communications, including personal visits to clubs, hamfests and conventions, with ARRL members and affiliated clubs in the section; either responds to their questions or concerns or refers them to the appropriate person or office in the League organization; maintains liaison with, and provides support to, representative repeater frequency-coordinating bodies having jurisdiction in the section.

f) Writes, or supervises preparation of, the monthly Section News column in *QST* to encourage member participation in the ARRL program in the section.

[Note: Move of permanent residence outside the section from which elected will be grounds for declaring the office vacant.]

ARRL LEADERSHIP APPOINTMENTS

Field Organization leadership appointments from the Section Manager are available to qualified ARRL full members in each section. These appointments are as follows: Assistant Section Manager, Section Emergency Coordinator, Section Traffic Manager, Official Observer Coordinator, State Government Liaison, Technical Coordinator, Affiliated Club Coordinator, Public Information Officer, Bulletin Manager, District Emergency Coordinator, Emergency Coordinator and Net Manager. Holders of such appointments may wear the League emblem pin with the distinctive deep-green background. Functions of these leadership officials are described below.

Assistant Section Manager

The ASM is an ARRL section-level official appointed by the Section Manager, in addition to the Section Manager's eight section-level assistants. An ASM may be appointed if the Section Manager believes such an appointment is desirable to meet the goals of the ARRL Field Organization in that section. Thus, the ASM is appointed at the complete discretion of the Section Manager, and serves at the pleasure of the Section Manager.

1) The ASM may serve as a general or as a specialized assistant to the Section Manager. That is, the ASM may assist the Section Manager with general leadership matters as the Section Manager's general understudy, or the ASM may be assigned to handle a specific important function not within the scope of the duties of the Section Manager's eight assistants.

2) At the Section Manager's discretion, the ASM may be designated as the recommended successor to the incumbent Section Manager, in case the Section Manager resigns or is otherwise unable to finish the term of office.

3) The ASM should be familiar with FSD-128, "Guidelines for the ARRL Section Manager," which contains the fundamentals of general section management.

4) The ASM must be an ARRL full member, holding at least a Technician class license.

Section Emergency Coordinator

The SEC must hold a Technician class license or higher and is appointed by the Section Manager to take care of all matters pertaining to emergency communications and the Amateur Radio Emergency Service (ARES) on a sectionwide basis. The duties of the SEC include the following:

1) The encouragement of all groups of community amateurs to establish a local emergency organization.

2) Recommendations to the SM on all section emergency policy and planning, including the development of a section emergency communications plan.

3) Cooperation and coordination with the Section Traffic Manager so that emergency nets and traffic nets in the section present a unified public service front. Cooperation and coordination should also be maintained with other section leadership officials as appropriate, particularly the State Government Liaison and the Public Information Officer.

4) Recommendations of candidates for Emergency Coordinator and District Emergency Coordinator appointments (and cancellations) to the Section Manager and determinations of areas of jurisdiction of each amateur so appointed. At the SM's discretion, the SEC may be directly in charge of making (and cancelling) such appointments. In the same way, the SEC can handle the Official Emergency Station program.

5) Promotion of ARES membership drives, meetings, activities, tests, procedures, etc., at the section level.

6) Collection and consolidation of Emergency Coordinator (or District Emergency Coordinator) monthly reports and submission of monthly progress summaries to ARRL Hq.

7) Maintenance of contact with other communication services and liaison at the section level with all agencies served in the public interest, particularly in connection with state and local government, civil preparedness, Red Cross, Salvation Army and the National Weather Service. Such contact is maintained in cooperation with the State Government Liaison.

Section Traffic Manager

The STM is appointed by the Section Manager to supervise traffic handling organization at the section level—that is, of coordinating the activities of all traffic nets, both National Traffic System-affiliated and independents, so that routings within the section and connections with other nets to effect orderly and efficient traffic flow are maintained. The STM should be a person at home and familiar with traffic handling on all modes, must have at least a Technician class license, and should possess the willingness and ability to devote equal consideration and time to all section traffic matters. The duties of the STM include the following:

1) Establishment, administration and promotion of a traffic-handling program at the section level, mainly but not restricted to NTS nets.

2) Cooperation and coordination with the Section Emergency Coordinator so that traffic nets and emergency nets in the section present a united public service front.

3) Recommendations of candidates for Net Manager and Official Relay Section appointment to the SM. At the SM's discretion, the STM may directly make (or cancel) such appointments.

4) Ability and familiarity with proper traffic-handling procedures in two or more different modes.

5) Collection of monthly net reports and submitting them to ARRL Hq.

Affiliated Club Coordinator

The ACC is the primary contact and resource person for each Amateur Radio club in the section, specializing in providing assistance to clubs. The ACC is appointed by, and reports to, the Section Manager. Duties and qualifications of the ACC include:

1) Encourage affiliated clubs in the section to become more active and, if they are interested, to apply as a Special Service Club (SSC).

2) Supply interested clubs with SSC application forms.

3) Assist clubs in completing SSC application forms, if requested.

4) Help clubs establish workable programs to use as SSCs.

5) Approve SSC application forms and pass them to the SM.

6) Work with other section leadership officials (Section Emergency Coordinator, Public Information Officer, Technical Coordinator, State Government Liaison, etc.) to insure that clubs are involved in the mainstream of ARRL Field Organization activities.

7) Encourage new clubs to become ARRL affiliated.

8) Insure that annual progress reports are forthcoming from all affiliated clubs.

9) Desire to volunteer time for Amateur Radio, specifically for strengthening Amateur Radio Clubs.

10) Technician class license required.

Bulletin Manager

Rapid dissemination of information is the lifeblood of an active, progressive organization. The ARRL Official Bulletin Station network provides a vital communications link for informing the amateur community of the latest developments in Amateur Radio and the ARRL. The ARRL Bulletin Manager is responsible for recruiting and supervising a team of Official Bulletin Stations to disseminate such news and information of interest to amateurs in the section and to provide a means of getting the news and information to all OBS appointees. The bulletins should include the content of ARRL bulletins (transmitted by W1AW), but should also include items of local, section and regional interest from other sources, such as ARRL section leadership officials, as well as information provided by the Division Director.

A special effort should be made to recruit an OBS for each major repeater in the section. This is where the greatest "audience" is to be found, many of whom are not sufficiently informed about the latest news of Amateur Radio and the League. Such bulletins should be transmitted regularly, perhaps in conjunction with a repeater net or on a repeater "bulletin board" (tone-accessed recorded announcements for repeater club members).

While the primary focus of the bulletin program in most sections will be 2-meter repeaters and 75/80 meter nets, other outlets may be appropriate. The Bulletin Manager should be sensitive to the opportunities presented by new technology; in some cases it may even be desirable to supplement on-the-air outlets with outlets provided by computer "mailboxes" and similar techniques.

Although the primary mission of OBS appointees is to copy ARRL bulletins directly from W1AW, in some sections the Bulletin Manager may take on the responsibility of retransmitting ARRL bulletins (as well as other information) for the benefit of OBS appointees, on a regularly scheduled day, time and frequency. An agreed-upon schedule should be worked out in advance. Time is of the essence when conveying news; therefore a successful Bulletin Manager will develop ways of communicating with the OBS appointees quickly and efficiently.

Bulletin Managers should be familiar with the position description of the Official Bulletin Station, which appears below. The duties of the Bulletin Manager include the following:

1) The Bulletin Manager must have a Technician class license or higher.

2) The Bulletin Manager is appointed by the SM and is required to report regularly to the SM concerning the section's bulletin program.

3) The Bulletin Manager is responsible for recruiting (and, at the discretion of the SM, appointing) and supervising a team of Official Bulletin Stations in the section. A special effort should be made to recruit OBSs for each major repeater in the section.

4) The Bulletin Manager must be capable of copying ARRL bulletins directly from W1AW on the mode(s) necessary. The Bulletin Manager may, in some cases, be required to retransmit ARRL bulletins for OBS appointees who might be unable to copy them directly from W1AW.

5) The Bulletin Manager is also responsible for funneling news and information of a local, section and regional nature to OBS appointees. In so doing, the Bulletin Manager must maintain close contact with other section-level officials, and the Division Director, to maintain an organized and unified information flow within the section.

Official Observer Coordinator

The Official Observer Coordinator is an ARRL section-level leadership official appointed by the Section Manager to supervise the Official Observer program in the section. The OO Coordinator must hold a Technician class (or higher) amateur license.

The Official Observer program has operated for more than half a century, and in that time, OO appointees have assisted thousands of amateurs whose signals, or operating procedures, were not in compliance with the regulations. The function of the OO is to *listen* for amateurs who might otherwise come to the attention of the FCC and to advise them by mail of the irregularity observed. The OO program is, in essence, for the benefit of amateurs who *want* to be helped. Official Observers must meet high standards of expertise and

experience. It is the job of the OO Coordinator to recruit, supervise and direct the efforts of OOs in the section, and to report their activity monthly to the Section Manager and to ARRL Hq.

The OO Coordinator is a key figure in the Amateur Auxiliary to the FCC's Field Operations Bureau, the foundation of which is an enhanced OO program. Jointly created by the FCC and ARRL in response to federal enabling legislation (Public Law 97-259), the Auxiliary permits a close relationship between FCC and ARRL Field Organization volunteers in monitoring the amateur airwaves for potential rules discrepancies/violations. (Contact your Section Manager for further details on the Amateur Auxiliary program.)

Public Information Officer

The ARRL Public Information Officer is a section-level official appointed by the Section Manager to be the section's expert on public information and public relations matters. The Public Information Officer is also responsible for organizing, guiding and coordinating the activities of the Public Information Assistants within the section. Usually, the PIO will be located in, or have access to, significant mass media centers within the section. The PIO must regularly report to the Section Manager. Duties of the Public Information Officer include the following:

1) Act as a public relations advisor to the Section Manager, keeping the SM informed on specific needs and opportunities, developing and recommending appropriate programs for the section.

2) Develop and coordinate the public information function through recruitment of a team of Public Information Assistants and through liaison to ARRL-affiliated clubs.

3) Assist the Section Manager and the Division Director in the preparation and dissemination of communications within the section.

4) Seek out and develop significant stories in the section and promote their dissemination.

5) Establish and maintain personal contact with those in the media (in conjunction with efforts of Public Information Assistants).

6) Promote the distribution and airing of ARRL public service announcements and other audio/visual material.

7) Encourage, organize and conduct public information/public relations sessions at conventions and hamfests.

8) Maintain contact with other section-level League officials, particularly the Section Emergency Coordinator, State Government Liaison, Affiliated Club Coordinator and Bulletin Manager, on matters appropriate to their attention, to insure a unified ARRL Field Organization.

9) Maintain the section's speaker's bureau for club and convention programs.

Local-level public relations is accomplished primarily through local affiliated clubs, which are encouraged to establish permanent public relations programs as part of club activities. Upon recommendation of a club, the Public Information Officer may—with the approval of the Section Manager—appoint the individual who handles that function for the club a Public Information Assistant (PIA). Where the responsibility cannot or will not be assumed by a club, the section PIO is authorized to seek a qualified League member who is willing to accept the responsibility of the PIA appointment.

In either case, each PIA in the section must report regularly to the PIO to maintain the appointment. Information flow is, of course, a two-way street, and the PIO is responsible for passing along news from the section or national level to the PIAs in the section. Time is of the essence when conveying news; therefore, a successful section PIO will develop ways to communicate with the PIAs instantaneously. A tie-in with the Official Bulletin Station program in the section can be very worthwhile; the Public Information Officer may wish to work closely with the Bulletin Manager in this regard.

State Government Liaison

The State Government Liaison (SGL) shall be an amateur who is aware (at a minimum) of state legislative proposals in the normal course of events and who can watch for those proposals having the potential to affect Amateur Radio without creating a conflict of interest.

The SGL shall collect and promulgate information on state ordinances affecting Amateur Radio and work (with the assistance of other ARRL members) toward assuring that they work to the mutual benefit of society and the Amateur Radio Service.

The SGL shall guide, encourage and support ARRL members in representing the interests of the Amateur Radio Service at all levels. Accordingly, the SGL shall cooperate closely with other section-level League officials, particularly the Section Emergency Coordinator and the Public Information Officer.

The SGL is appointed by the Section Manager and shall report to the SM (and keep ARRL Hq. advised) concerning all appropriate activities and developments involving the interface of Amateur Radio and government regulatory matters, particularly those which have policy implications.

In those states where there is more than one section, the Section

Managers whose territory does not encompass the state capital may simply defer to the SGL appointed by their counterpart in the section where the state capital is located. In this case, the SGL is expected to communicate equally with all Section Managers (and Section Emergency Coordinators and other section-level League officials). In sections where there is more than one government entity, i.e., Maryland-DC, West Indies, Pacific and Maritime-Nfld (where the title, of course, would be Provincial Government Liaison), there may be a Liaison appointed for each entity.

Technical Coordinator

The ARRL Technical Coordinator (TC) is a section-level official appointed by the Section Manager to coordinate all technical activities within the section. The Technical Coordinator must hold a Technician class or higher amateur license. The Technical Coordinator reports to the Section Manager and is expected to maintain contact with other section-level appointees as appropriate to insure a unified ARRL Field Organization within the section. The duties of the Technical Coordinator are as follows:

1) Encourage amateurs in the section to share their technical achievements with others through the pages of *QST*, and at club meetings, hamfests and conventions.

2) Promote technical advances and experimentation at VHF/UHF and with specialized modes, and work closely with enthusiasts in these fields within the section.

3) Serve as an advisor to radio clubs that sponsor training programs for obtaining amateur licenses or upgraded licenses, in cooperation with the ARRL Affiliated Club Coordinator.

4) In times of emergency or disaster, function as the coordinator for establishing an array of equipment for communications use and be available to supply technical expertise to government and relief agencies to set up emergency communications networks, in cooperation with the Section Emergency Coordinator.

5) Provide assistance to amateurs in the section who need technical advice.

6) Serve on RFI and TVI committees in the section for the purpose of rendering technical assistance as needed.

7) Be available to assist local technical program committees in arranging suitable programs for ARRL hamfests and conventions.

8) Convey the views of section amateurs to ARRL Hq., respective to their views about the technical contents of *QST* and ARRL books. Suggestions for improvements should also be called to the attention of the ARRL Hq. technical staff.

9) Work with appointed ARRL TAs (technical advisors) when called upon.

10) Coordinate technical information service with appointed Assistant Technical Coordinators.

11) Be available to give technical talks at club meetings, hamfests and conventions in the section.

District Emergency Coordinator

The DEC is an ARRL full member of at least Technician class experienced in emergency communications who can assist the SEC by taking charge in the area of jurisdiction especially during an emergency.
The DEC shall:

1) Coordinate the training, organization and emergency participation of Emergency Coordinators in the area of jurisdiction.

2) Make local decisions in the absence of the SEC or through coordination with the SEC concerning the allotment of available amateurs and equipment during an emergency.

3) Coordinate the interrelationship between local emergency plans and between communications networks within the area of jurisdiction.

4) Act as backup for local areas without an Emergency Coordinator and assist in maintaining contact with governmental and other agencies in the area of jurisdiction.

5) Provide direction in the routing and handling of emergency communications of either a formal or tactical nature.

6) Recommend EC appointments to the SEC and advise on OES appointments.

7) Coordinate the reporting and documentation of ARES activities in the area of jurisdiction.

8) Act as a model emergency communicator as evidenced by dedication to purpose, reliability and understanding of emergency communications.

Emergency Coordinator

Appointment of an Emergency Coordinator can be made by the SM, on recommendation of the SEC and DEC (or directly by the SEC), in every community where a qualified amateur of Technician class or higher holding ARRL full membership can be found, since it is on his work and that of the local amateurs working with him that the entire ARES organization is based. While his duties are varied and manifold,

they can be generally described as promotion and enhancement of activities of the Amateur Radio Emergency Service for public service at the local level.
The EC shall:

1) Coordinate the training, organization and emergency participation of interested amateurs in the area of jurisdiction.

2) Establish an emergency communications plan for the area that will effectively utilize the group to cover the needs for both tactical and formal message traffic requirements.

3) Establish a viable working relationship with governmental and other agencies who might need the service of ARES (e.g., Red Cross, Salvation Army, Rescue Squads, Weather Bureau, Hospitals, Police Department, Fire Department, etc.).

4) Establish local communications networks run on a regular basis and periodically test these networks by drills.

5) Establish an emergency traffic plan utilizing NTS to the extent possible and coordinate liaison with NTS nets.

6) In times of disaster, evaluate the communications needs of the jurisdiction and respond quickly to those needs.

7) Do all that is possible to further the image of Amateur Radio by dedication to purpose, reliability and a thorough understanding of the mission of Amateur Radio.

Every EC, on appointment, receives a copy of the *Emergency Coordinator's Handbook*, containing full details on how to organize and the basic principles of ARES organization, plus many samples and illustrations. Briefly, a newly appointed EC should undertake the following steps (or review each phase if he is succeeding someone else):

1) A general meeting of local amateurs.

2) Recommend candidates for Official Emergency Station appointment.

3) Designate Assistant ECs to serve specialized areas on a Planning Committee.

4) Have a meeting with the committee to discuss initial plans.

5) Contact local "to-be-served" agencies to determine requirements.

6) Draft detailed plans to serve community requirements.

7) Establish training procedures to insure that the plans can be carried out in any emergency.

The EC also has certain routine administrative duties as follows:

a) Maintain full and current information on OES appointees and ARES registrants. Keep the DEC and SEC informed on the former.

b) Issue certificates to assistant ECs (not an SM appointment) and assign each a specific subleadership role in the local ARES organization.

c) Report the status of the ARES unit to the DEC each month on the form provided (FSD-212).

d) Coordinate and cooperate with the DEC.

The area of jurisdiction of each EC is determined by the SM and SEC/DEC depending on the ARES organizational plan of the section.

Net Manager

For coordinating and supervising traffic-handling activities in the section, the SM may appoint one or more Net Managers, usually on recommendation of the Section Traffic Manager. The number of NMs appointed may depend on a section's geographical size, the number of nets operating in the section, or other factors having to do with the way the section is organized. In some cases, there may be only one Net Manager in charge of the one section net, or one NM for the phone net, one for the CW net. In larger or more traffic-active sections there may be several, including NMs for the VHF net or nets, for the RTTY net, or NTS local nets not controlled by ECs. All ARRL NMs should work under the STM in a coordinated section traffic plan.

Some nets cover more than one section but operate in NTS at the section level. In this case, the Net Manager is selected by agreement among the STMs concerned and the NM appointment conferred on him by his resident SM.

NMs may conduct any testing of candidates for ORS appointment (see below) that they consider necessary before making appointment recommendations to the STM. Net Managers also have the function of requiring that all traffic handling in ARRL recognized nets is conducted in proper ARRL form.

Remember: All appointees or appointee candidates must be ARRL full members.

ARRL STATION APPOINTMENTS

Field Organization station and individual appointments from the Section Manager are available to qualified ARRL full members in each section. These appointments are as follows: Official Relay Station, Official Emergency Station, Official Bulletin Station, Public Information Assistant, Official Observer and Assistant Technical Coordinator. All appointees receive handsome certificates from the SM and are entitled to wear ARRL membership pins with the distinctive blue background.

All appointees are required to submit regular reports to maintain appointments and to remain active in their area of specialty.

The report is the criterion of activity. An appointee who misses three consecutive monthly reports is subject to cancellation by the SM or the appropriate section leadership official, who cannot know what or how much you are doing unless you report. An appointee whose appointment is cancelled for this or other reasons must earn reinstatement by demonstrating activity and adherence to the requirements. Reinstatement of cancelled appointments, and indeed judgment of whether or not a candidate meets the requirements, is at the discretion of the SM and the section leadership.

The detailed qualifications of the six individual "station" appointments are given below. If you are interested, your SM will be glad to receive your application. Use application form FSD-187, reproduced nearby.

Official Relay Station

This is a traffic-handling appointment that is open to all licenses. This appointment applies equally to all modes and all parts of the spectrum. It is for traffic handlers, regardless of how or in what part of the spectrum they do it.

The potential value of the operator who has traffic know-how to his country and community is enhanced by his ability and the readiness of his station to function in the community interest in case of emergency. Traffic awareness and experience are often the signs by which mature amateurs may be distinguished.

Traditionally, there have been considerable differences between procedures for traffic handling by CW, phone, RTTY, ASCII and other modes. Appointment requirements for ORS do not deal with these, but with factors equally applicable to all modes. The appointed ORS may confine activities to one mode or one part of the spectrum if he wishes although versatility does indeed make it possible to perform a more complete public service. The expectation is that the ORS will set the example in traffic handling, however it is done. To the degree that he is deficient in performing traffic functions by any mode, to that extent he does not meet the qualifications for the appointment. Here are the basic requirements:

1) Full ARRL membership and Novice class license or higher.

2) Code and/or voice transmission.

3) Transmission quality, by whatever mode, must be of the highest quality, both technically and operationally. For example, CW signals must be pure, chirpless and clickless, and code sending must be well spaced and properly formed. Voice transmission must be of proper modulation percentage or deviation, precisely enunciated with minimum distortion. RTTY must be clickless, proper shift, etc.

4) All ORSs are expected to follow standard ARRL operating practices (message form, ending signals, abbreviations or prowords, courtesy, etc.).

5) Regular participation in traffic activities, either freelance or ARRL-sponsored. The latter is encouraged, but not required.

6) Handle all record communications speedily and reliably and set the example in efficient operating procedures. All traffic is relayed or delivered promptly after receipt.

7) Report monthly to the STM, including a breakdown of traffic handled during the past calendar month.

Official Emergency Station

Amateur operators of Technician class and above may be appointed OES by their SEC or SM at the recommendation of their ECs or DECs (if no EC) if they are ARRL members and interested in setting high standards of emergency preparedness and operating. In addition to candidates for this appointment operating within their own local EC jurisdictional areas, we want to recruit OESs who can set similar examples in "offshore" emergencies, such as those that frequently occur in foreign countries. Here are the standard qualifications and functions of this appointment.

1) Possession of full ARRL membership and a Technician class license or higher.

2) Regular participation in the local ARES, if any, including all drills and tests, emergency nets and, of course, real emergency situations.

3) Ability to operate independent of commercial mains including at least one-band mobile capability.

4) Must be fully acquainted with standard ARRL message form and capable of using it in handling any third-party messages.

5) Report monthly to the EC/DEC or SEC.

Official Bulletin Station

Rapid dissemination of information is the lifeblood of an active, progressive organization. The ARRL Official Bulletin Station network provides a vital communications link for informing the amateur community of the latest developments in Amateur Radio and the League. ARRL bulletins, containing up-to-the-minute news and

information of Amateur Radio, are issued by League Hq. as soon as such news breaks. These bulletins are transmitted on a regular schedule by ARRL Hq. station W1AW.

The primary mission of OBS appointees is to copy these bulletins directly off the air from W1AW—on voice, CW or RTTY/ASCII—and retransmit them locally for the benefit of amateurs in the particular coverage area, many of whom may not be equipped to receive bulletins directly from W1AW.

ARRL bulletins of major importance or of wide-ranging scope are mailed from Hq. to each Bulletin Manager and OBS appointee. However, some bulletins, such as the ARRL DX Bulletin (transmitted on Fridays UTC), are disseminated only by W1AW because of time value. Thus, it is advantageous for each OBS to copy W1AW directly. In some sections, the Bulletin Manager may assume the responsibility of copying the bulletins from W1AW; therefore, individual OBSs should be sure to meet the Bulletin Manager on a regular, agreed-upon schedule to receive the latest bulletins.

The Canadian Radio Relay League, CRRL, has a separate and well-organized OBS program, but CRRL bulletins of greatest importance are also incorporated into the regular W1AW bulletin transmissions and should be copied by Canadian OBS appointees.

Inasmuch as W1AW operates on all bands (160-2 meters), the need for OBSs on HF has lessened somewhat in recent times. However, OBS appointments for HF operation can be conferred by the Section Manager (or the Bulletin Manager, depending on how the SM organizes the section) if the need is apparent. More important, to serve the greatest possible "audience," OBS appointees who can send ARRL bulletins over VHF repeaters are of maximum usefulness and are much in demand. If possible, an OBS who can copy bulletins directly from W1AW (or the Bulletin Manager) should be assigned to each major repeater in the section. Bulletins should be transmitted regularly, perhaps in conjunction with a VHF repeater net, on a repeater bulletin board (tone-accessed recorded announcements for repeater club members), or via a local RTTY (computer) mailbox. Duties and requirements of the OBS include the following:

1) OBS candidates must have a Technician class license or higher.

2) Retransmission of ARRL bulletins must be made at least once per week to maintain appointment.

3) OBS candidates are appointed by the Section Manager (or by the Bulletin Manager, if the SM so desires) and must adhere to a schedule that is mutually agreeable, as indicated on appointment application form FSD-187.

4) OBS appointees should send a monthly activity report (such as FSD-210, under "Schedules and Net Affiliations") to the Bulletin Manager, indicating bulletin transmissions made and generally updating the Bulletin Manager to any OBS-related activities. This reporting arrangement may be modified by the Bulletin Manager as he/she sees fit.

5) As directed by the Bulletin Manager, OBSs will include in their bulletin transmissions news of local, section and regional interest.

TO SM IN THE...SECTION OF ARRL. (Address   on page 8, QST)

I feel that I can qualify and wish to apply for the following ARRL station appointment(s):

○ ORS   ○ OES   ○ PIA   ○ OBS   ○ OO

Full name_____Call_____

The following address should be the same as your ARRL/QST mailing address.

Number and Street_____

City, State/Province, ZIP/Postal Code_____

ARRL membership expiration date_____ Current age _____ Home Telephone_____

Class of present FCC/DOC license_____Date of FIRST license_____

Please provide a statement below describing your qualifications for the appointment, along with any background information that would be helpful to the Section Manager and/or the section leadership.

I understand that my appointment requires continuing activity, regular reports and maintenance of my ARRL membership as prerequisite to endorsement, and that it may be suspended or cancelled at the discretion of the SM for violation of any of the bases for appointment.

DATE_____SIGNATURE_____CALL_____

Use this form to apply for an ARRL station appointment.

Public Information Assistant

Public Information Assistants are recruited by the section Public Information Officer to provide local, "grassroots" public information and public relations. Public Information Assistants are usually prominent members of local Amateur Radio clubs, and (as with all appointments) must be ARRL full members. Good grassroots public relations involves the regular and frequent dissemination of information about amateur activities through local media as well as community activities, school programs, presentations to service clubs and community organizations, exhibits and demonstrations, and other efforts that provide visibility to the public. Functions of the PIA include the following:

1) Seek out and develop significant stories in the local area and promote their dissemination.

2) Establish and maintain personal contact with local media.

3) Develop and promote good ideas for community projects and special events to display Amateur Radio to the public.

4) Maintain regular reporting to the section PIO.

5) Distribute locally press releases and announcements concerning Amateur Radio.

6) Promote the distribution and airing of ARRL public service announcements and other audio/visual material.

7) Promote public awareness of club recruiting and training activities.

Official Observer

The Official Observer (OO) program has been sponsored by the League for over 50 years to help amateurs help each other. Official Observer appointees have aided thousands of amateurs to maintain their transmitting equipment and operating procedures in compliance with the regulations. The object of the OO program is to notify amateurs by mail of operating/technical irregularities before they come to the attention of the FCC.

The ARRL commitment to volunteer monitoring has been greatly enhanced by the creation of the Amateur Auxiliary to the FCC Field Operations Bureau, designed to enable amateurs to play a more active and direct role in upholding the traditional high standard of conduct on the amateur bands. The OO is the foundation of the Amateur Auxiliary, carrying out the all-important day-to-day maintenance monitoring of the amateur airwaves. Following recommendation by the Section Manager, potential members of the Amateur Auxiliary are provided with training materials, and all applicants must successfully complete a written examination to be enrolled as Official Observers. For further information, please contact your SM.

The OO performs his function by *listening* rather than transmitting, keeping a watchful ear out for such things as frequency instability, harmonics, hum, key clicks, broad signals, distorted audio, over-deviation, out-of-band operation, etc. The OO completes his task once the notification card is sent. Reimbursement for postage expenses are provided for through the SM. The OO:

1) Must be an ARRL full member and have been a licensee of Technician class or higher for at least four years.

2) Must undergo and complete successfully the Amateur Auxiliary training and certification procedure.

3) Must report to the OO Coordinator regularly on FSD-23.

4) Maintain regular activity in sending out notices as observed.

The OO program is one of the most important functions of the League. A sincere dedication to helping our brother and sister amateurs is required for appointment. Only the "very best" are sought.

Assistant Technical Coordinator

Appointed by the SM, or TC under delegated authority from the SM, the ATC supports the TC in two main areas of responsibility: Radio Frequency Interference, and Technical Information. ATCs can specialize in certain specific technical areas, or can be generalists. Here is a list of specific job duties:

1) Serve as a technical oracle to local hams and clubs. Correspond by telephone and letter on tech topics. Refer correspondents to other sources if specific topic is outside ATC's knowledge.

2) Serve as advisor in radio frequency interference issues. RFI can drive a wedge in neighbor and city relations. It will be the ATC with a cool head who will resolve problems. Local hams will come to you for guidance in dealing with interference problems.

3) Speak at local clubs on popular tech topics. Let local clubs know you're available and willing.

4) Represent ARRL at technical symposiums in industry; serve on CATV advisory committees; advise municipal governments on technical matters.

5) Work with other ARRL officials and appointees when called upon for technical advice especially in emergency communications situations where technical prowess can mean the difference in getting a communications system up and running, the difference between life and death.

6) Handle other miscellaneous technically-related tasks assigned by the Technical Coordinator.

ARRL Numbered Radiograms

The letters ARL are inserted in the preamble in the check and in the text before spelled out numbers, which represent texts from this list. Note that some ARL texts include insertion of numerals. *Example:* NR 1 R W1AW ARL 5 NEWINGTON CONN $\overline{DEC}$ 25 DONALD R SMITH $\overline{AA}$ 164 EAST SIXTH AVE $\overline{AA}$ NORTH RIVER CITY MO $\overline{AA}$ PHONE 733 3968 $\overline{BT}$ ARL FIFTY ARL SIXTY ONE $\overline{BT}$ DIANA $\overline{AR}$.

Group One—For possible "Relief Emergency" Use

ONE	Everyone safe here. Please don't worry.
TWO	Coming home as soon as possible.
THREE	Am in ____ hospital. Receiving excellent care and recovering fine.
FOUR	Only slight property damage here. Do not be concerned about disaster reports.
FIVE	Am moving to new location. Send no further mail or communication. Will inform you of new address when relocated.
SIX	Will contact you as soon as possible.
SEVEN	Please reply by Amateur Radio through the amateur delivering this message. This is a free public service.
EIGHT	Need additional ____ mobile or portable equipment for immediate emergency use.
NINE	Additional ____ radio operators needed to assist with emergency at this location.
TEN	Please contact ____. Advise to standby and provide further emergency information, instructions or assistance.
ELEVEN	Establish Amateur Radio emergency communications with ____ on ____ MHz.
TWELVE	Anxious to hear from you. No word in some time. Please contact me as soon as possible.
THIRTEEN	Medical emergency situation exists here.
FOURTEEN	Situation here becoming critical. Losses and damage from ____ increasing.
FIFTEEN	Please advise your condition and what help is needed.
SIXTEEN	Property damage very severe in this area.
SEVENTEEN	REACT communications services also available. Establish REACT communications with ____ on channel ____.
EIGHTEEN	Please contact me as soon as possible at ____.
NINETEEN	Request health and welfare report on ____. (State name, address and telephone number.)
TWENTY	Temporarily stranded. Will need some assistance. Please contact me at ____.
TWENTY ONE	Search and Rescue assistance is needed by local authorities here. Advise availability.
TWENTY TWO	Need accurate information on the extent and type of conditions now existing at your location. Please furnish this information and reply without delay.
TWENTY THREE	Report at once the accessibility and best way to reach your location.
TWENTY FOUR	Evacuation of residents from this area urgently needed. Advise plans for help.
TWENTY FIVE	Furnish as soon as possible the weather conditions at your location.
TWENTY SIX	Help and care for evacuation of sick and injured from this location needed at once.

Emergency/priority messages originating from official sources must carry the signature of the originating official.

Group Two—Routine messages

FORTY SIX	Greetings on your birthday and best wishes for many more to come.
FIFTY	Greetings by Amateur Radio.
FIFTY ONE	Greetings by Amateur Radio. This message is sent as a free public service by ham radio operators here at ____. Am having a wonderful time.
FIFTY TWO	Really enjoyed being with you. Looking forward to getting together again.
FIFTY THREE	Received your ____. It's appreciated; many thanks.
FIFTY FOUR	Many thanks for your good wishes.
FIFTY FIVE	Good news is always welcome. Very delighted to hear about yours.
FIFTY SIX	Congratulations on your ____, a most worthy and deserved achievement.
FIFTY SEVEN	Wish we could be together.
FIFTY EIGHT	Have a wonderful time. Let us know when you return.
FIFTY NINE	Congratulations on the new arrival. Hope mother and child are well.
*SIXTY	Wishing you the best of everything on ____.
SIXTY ONE	Wishing you a very merry Christmas and a happy New Year.
*SIXTY TWO	Greetings and best wishes to you for a pleasant ____ holiday season.
SIXTY THREE	Victory or defeat, our best wishes are with you. Hope you win.
SIXTY FOUR	Arrived safely at ____.
SIXTY FIVE	Arriving ____ on ____. Please arrange to meet me there.
SIXTY SIX	DX QSLs are on hand for you at the ____ QSL Bureau. Send ____ self-addressed envelopes.
SIXTY SEVEN	Your message number ____ undeliverable because of ____. Please advise.
SIXTY EIGHT	Sorry to hear you are ill. Best wishes for a speedy recovery.

*Can be used for all holidays.

ARL NUMBERS SHOULD BE SPELLED OUT AT ALL TIMES.

ARRL Recommended Message Precedences

Please observe the following ARRL provisions for PRECEDENCES in connection with written message traffic. These provisions are designed to increase the efficiency of our service both in normal times and in emergency.

Precedences

EMERGENCY Any message having life and death urgency to any person or group of persons, which is transmitted by Amateur Radio in the absence of regular commercial facilities. This includes official messages of welfare agencies during emergencies requesting supplies, materials or instructions vital to relief of stricken populace in emergency areas. During normal times, it will be *very rare*. On CW/RTTY, this designation will *always* be spelled out. When in doubt, do not use it.

PRIORITY Use abbreviation P on CW/RTTY. This classification is for (a) important messages having a specific time limit, (b) official messages not covered in the emergency category, (c) press dispatches and emergency-related traffic not of the *utmost* urgency, and (d) notice of death or injury in a disaster area, personal or official.

WELFARE This classification, abbreviated W on CW/RTTY, refers to either an inquiry as to the health and welfare of an individual in the disaster area or an advisory from the disaster area that indicates all is well. Welfare traffic is handled only after all emergency and priority traffic is cleared. The Red Cross equivalent to an incoming Welfare message is DWI (Disaster Welfare Inquiry).

ROUTINE Most traffic in normal times will bear this designation. In disaster situations, traffic labeled Routine (R on CW/RTTY) should be handled last, or not at all when circuits are busy with higher precedence traffic.

Note—the precedence always follows the message number. For example, a message number may be 207 R on CW and "Two Zero Seven Routine" on phone.

N5KR AZIMUTHAL EQUIDISTANT MAPS

The following computer-generated azimuthal maps centered on 21 different world locations were produced especially for *The ARRL Operating Manual* by William D. Johnston, N5KR, 1808 Pomona Dr., Las Cruces, NM 88001. Similar maps centered on your QTH or other locations can be ordered from N5KR (please include an SASE with your request for information). Note that these 21 locations are the subject of the 204 propagation tables that follow immediately after the maps.

N5KR AZIMUTHAL EQUIDISTANT MAP CENTERED ON
W1AW

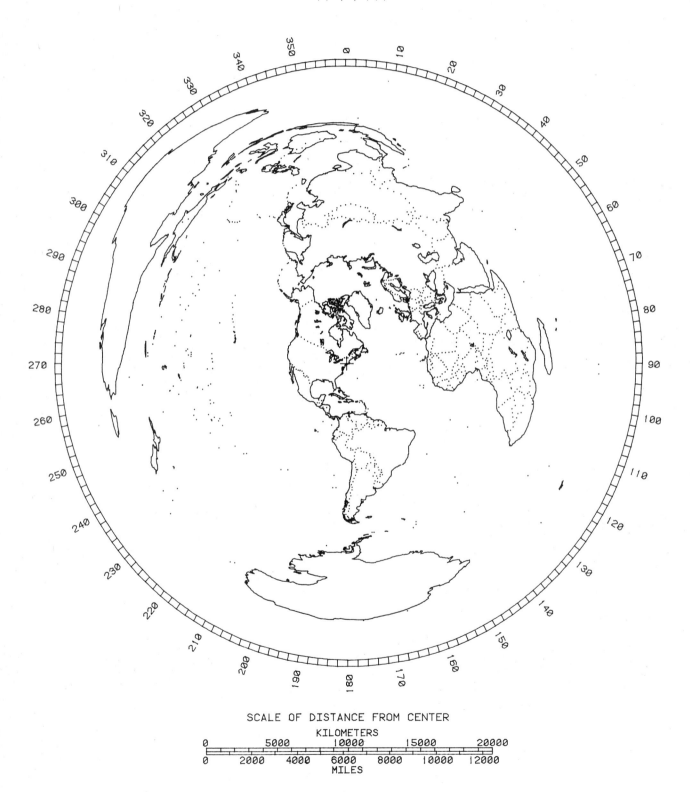

SCALE OF DISTANCE FROM CENTER

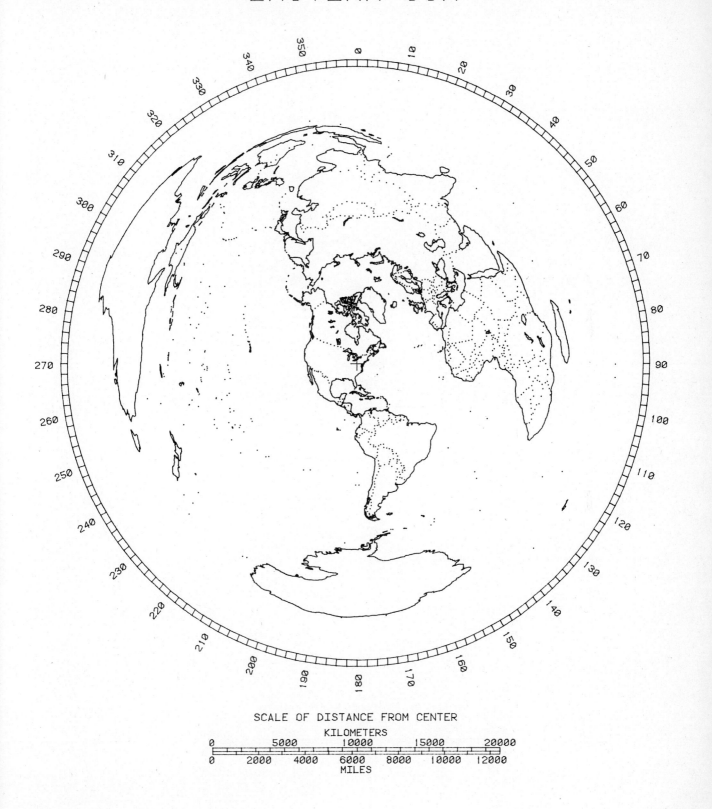

SCALE OF DISTANCE FROM CENTER
KILOMETERS

0	5000	10000	15000	20000
0	2000	4000	6000	8000	10000	12000

MILES

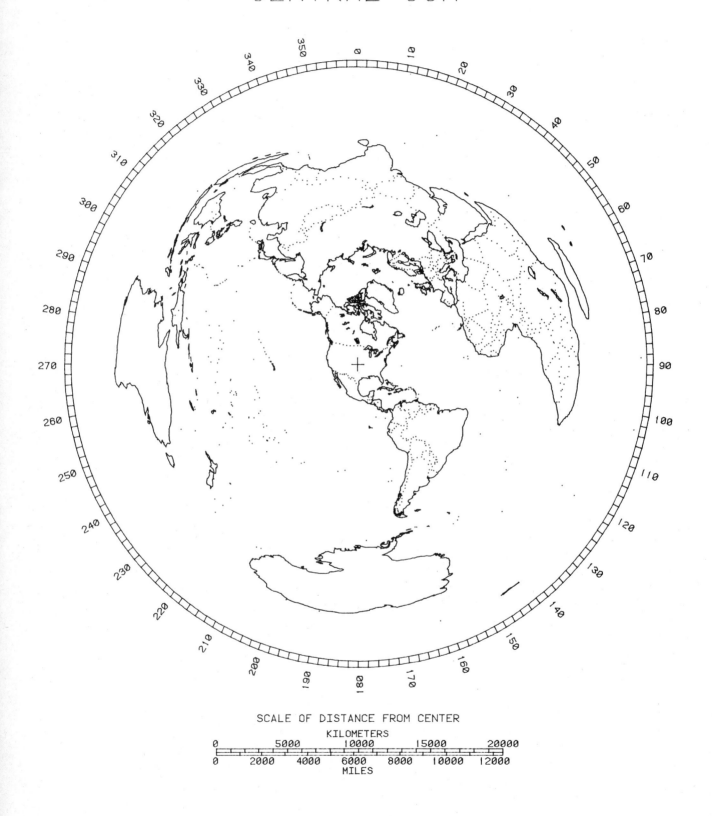

N5KR AZIMUTHAL EQUIDISTANT MAP CENTERED ON
CENTRAL USA

SCALE OF DISTANCE FROM CENTER
KILOMETERS

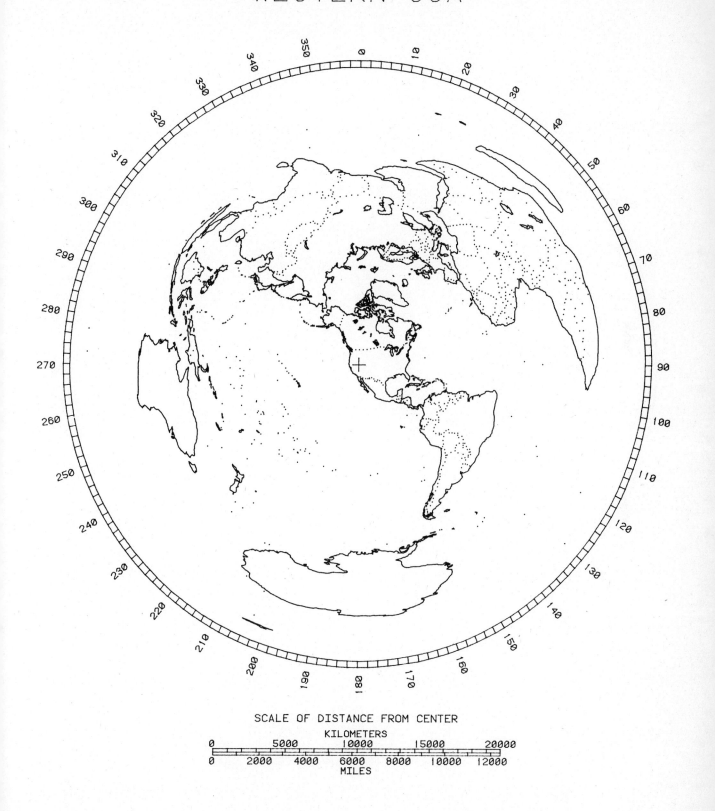

SCALE OF DISTANCE FROM CENTER
KILOMETERS

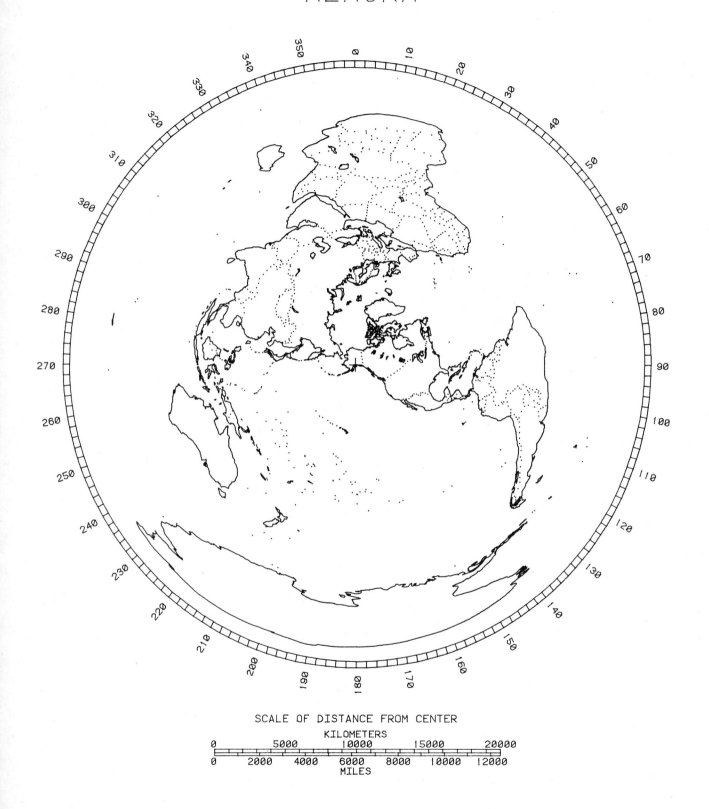

N5KR AZIMUTHAL EQUIDISTANT MAP CENTERED ON
ALASKA

SCALE OF DISTANCE FROM CENTER
KILOMETERS

0   5000   10000   15000   20000

0   2000   4000   6000   8000   10000   12000
MILES

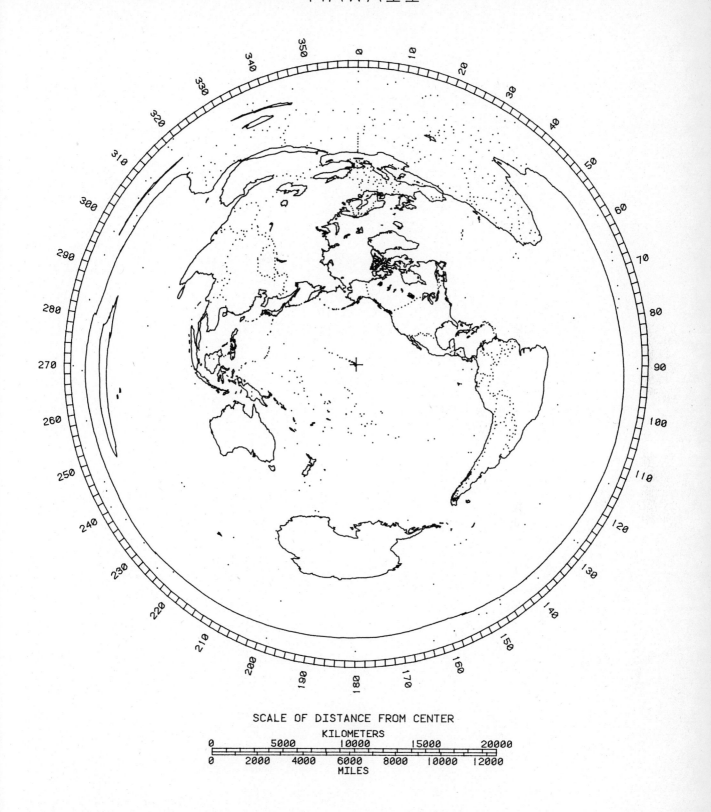

SCALE OF DISTANCE FROM CENTER

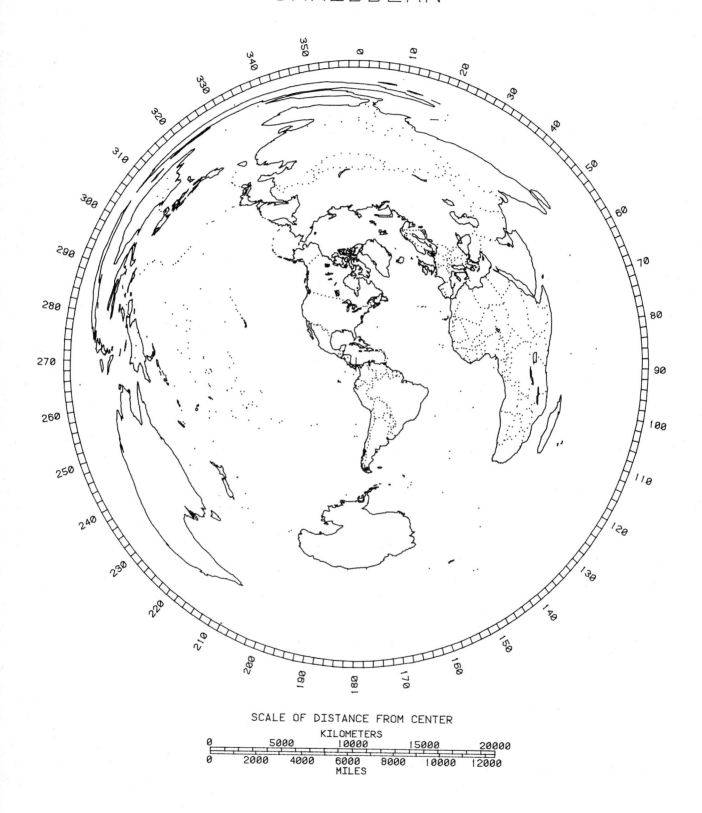

N5KR AZIMUTHAL EQUIDISTANT MAP CENTERED ON
CARIBBEAN

SCALE OF DISTANCE FROM CENTER
KILOMETERS

| 0 | 5000 | 10000 | 15000 | 20000 |

| 0 | 2000 | 4000 | 6000 | 8000 | 10000 | 12000 |
MILES

N5KR AZIMUTHAL EQUIDISTANT MAP CENTERED ON
E. SOUTH AM.

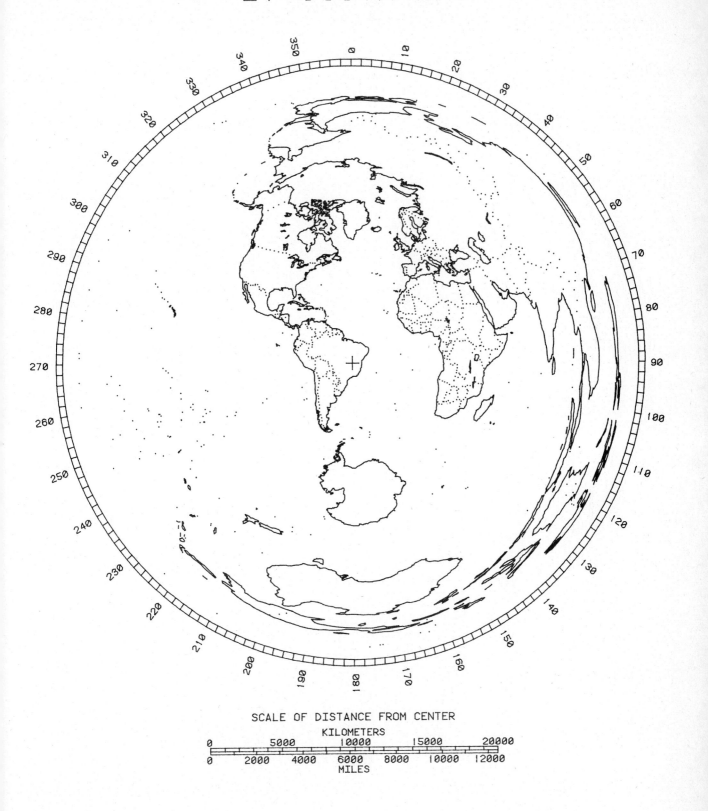

SCALE OF DISTANCE FROM CENTER

KILOMETERS

0   5000   10000   15000   20000

0   2000   4000   6000   8000   10000   12000

MILES

N5KR AZIMUTHAL EQUIDISTANT MAP CENTERED ON
S. SOUTH AM.

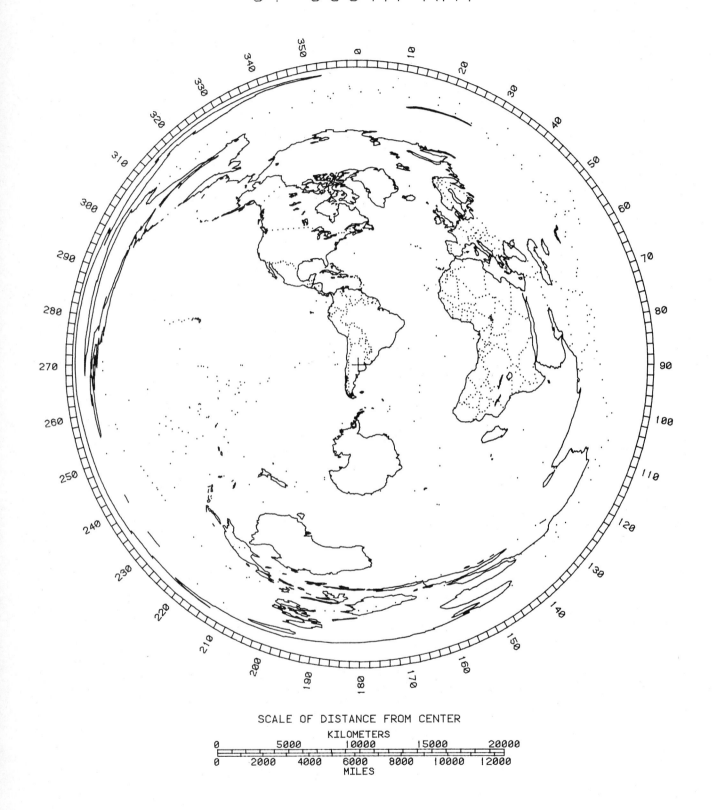

SCALE OF DISTANCE FROM CENTER

KILOMETERS

| 0 | 5000 | 10000 | 15000 | 20000 |

| 0 | 2000 | 4000 | 6000 | 8000 | 10000 | 12000 |

MILES

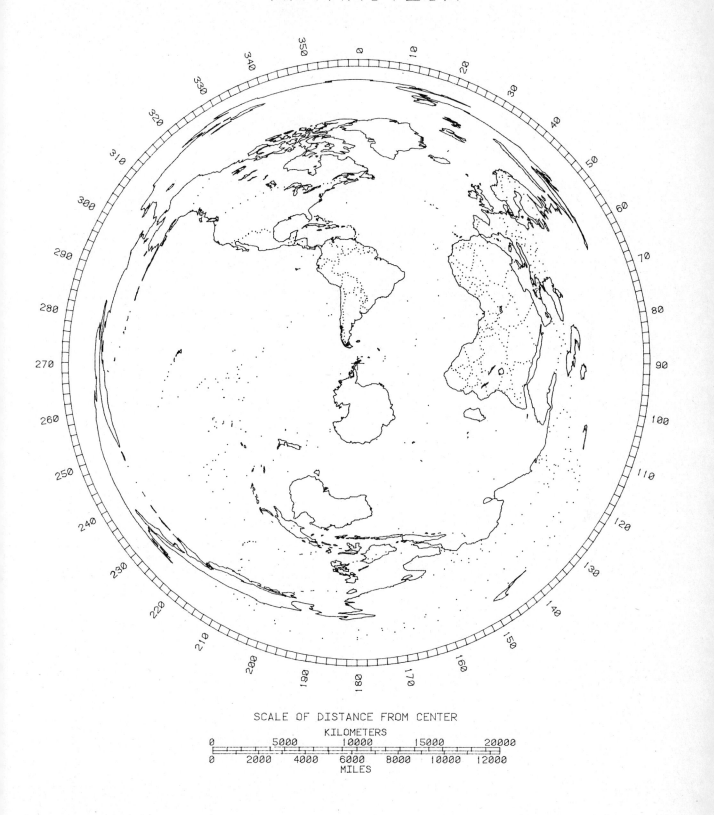

SCALE OF DISTANCE FROM CENTER
KILOMETERS

0   5000   10000   15000   20000

0   2000   4000   6000   8000   10000   12000
MILES

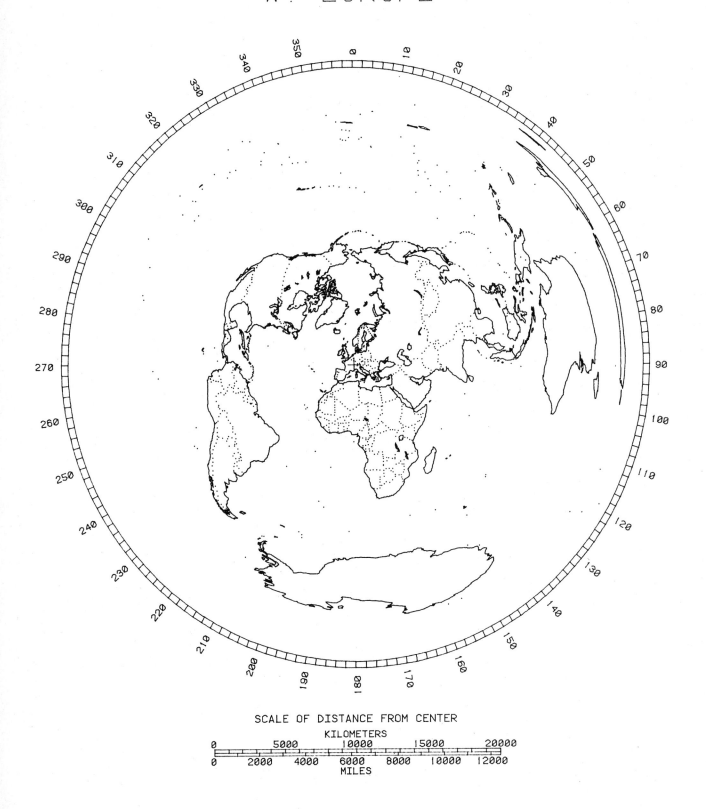

SCALE OF DISTANCE FROM CENTER
KILOMETERS
0   5000   10000   15000   20000
0   2000   4000   6000   8000   10000   12000
MILES

E. EUROPE

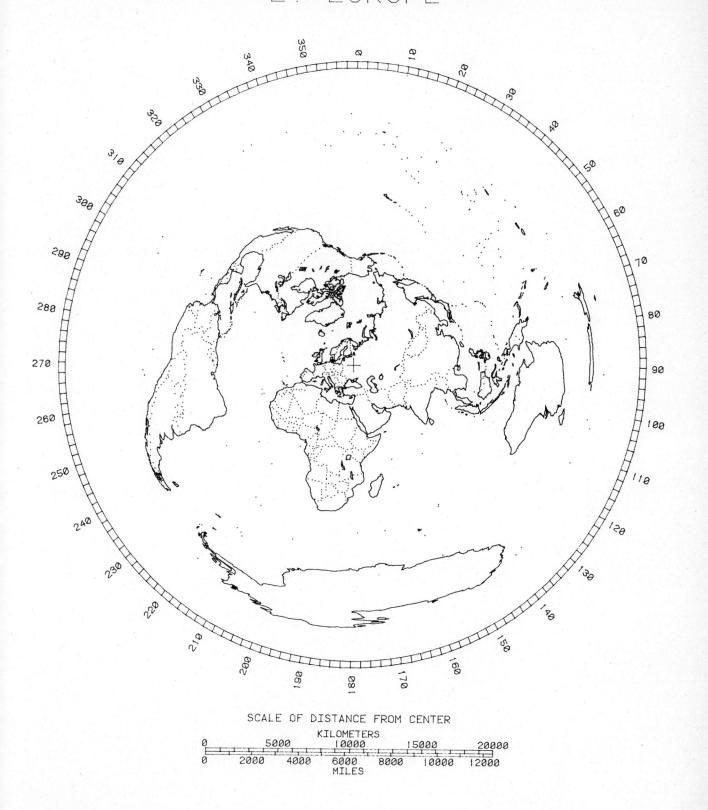

SCALE OF DISTANCE FROM CENTER
KILOMETERS

0	5000	10000	15000	20000

0	2000	4000	6000	8000	10000	12000

MILES

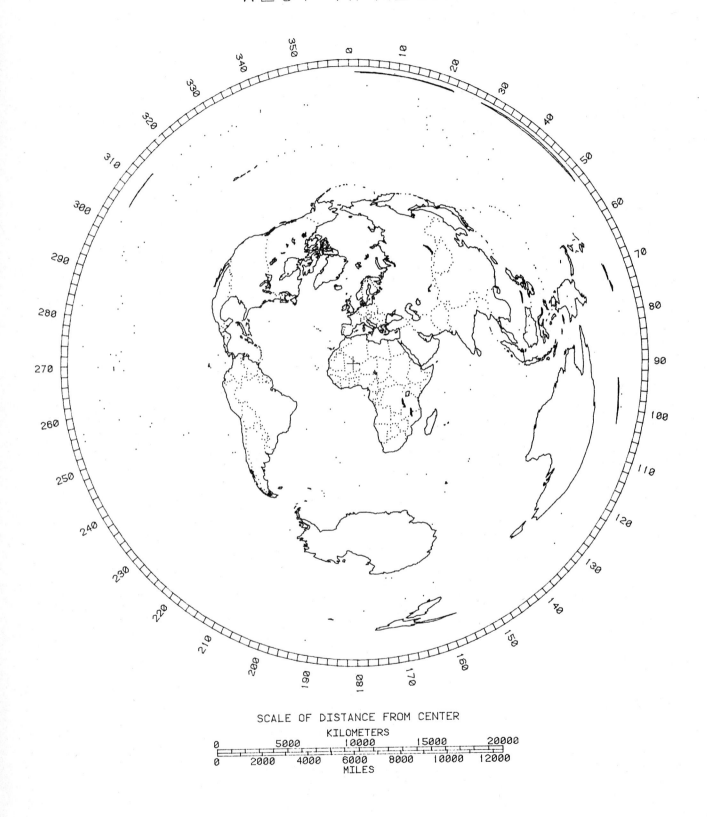

N5KR AZIMUTHAL EQUIDISTANT MAP CENTERED ON
WEST AFRICA

SCALE OF DISTANCE FROM CENTER
KILOMETERS

EAST AFRICA

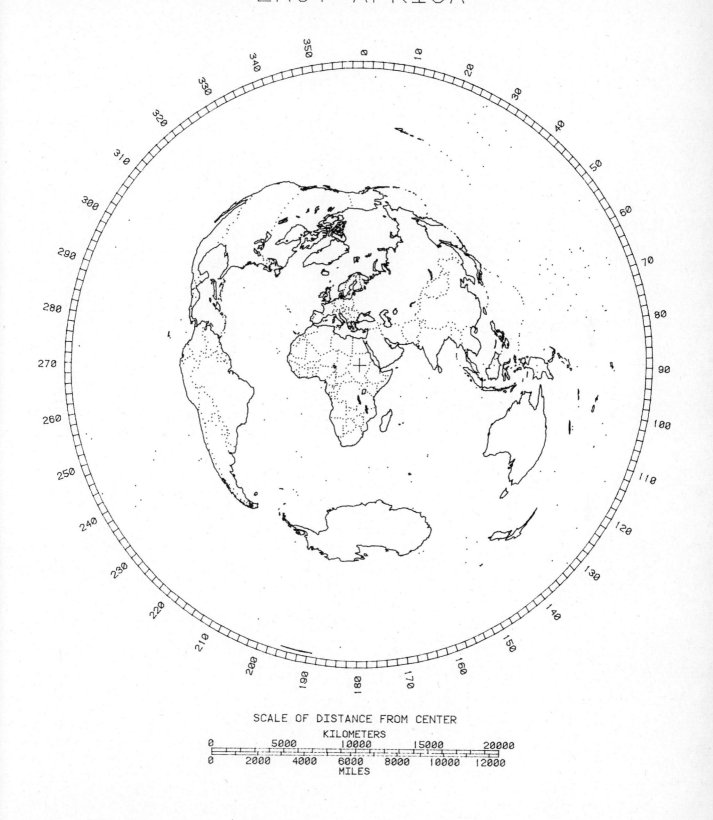

SCALE OF DISTANCE FROM CENTER

KILOMETERS

0	5000	10000	15000	20000

0	2000	4000	6000	8000	10000	12000

MILES

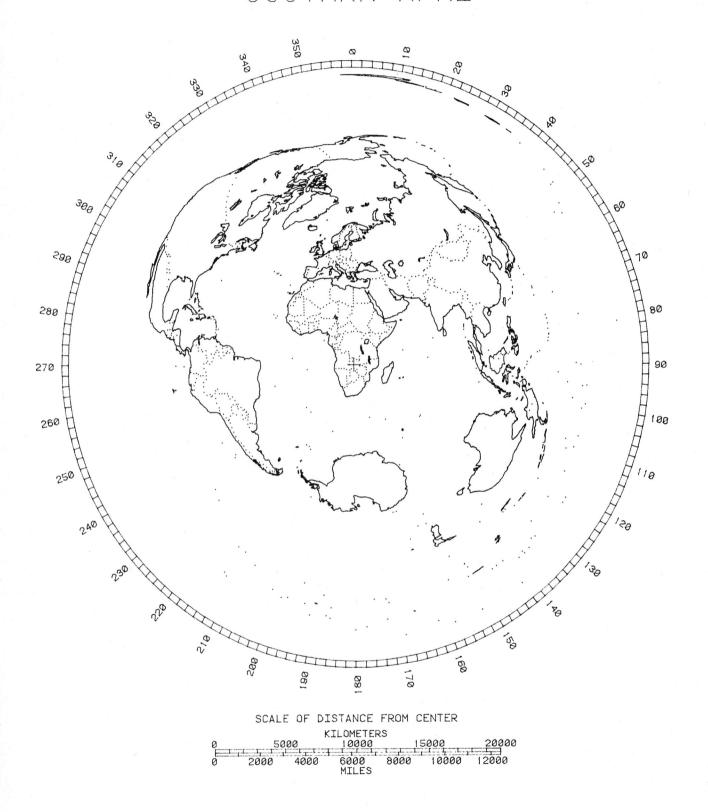

SCALE OF DISTANCE FROM CENTER

NEAR EAST

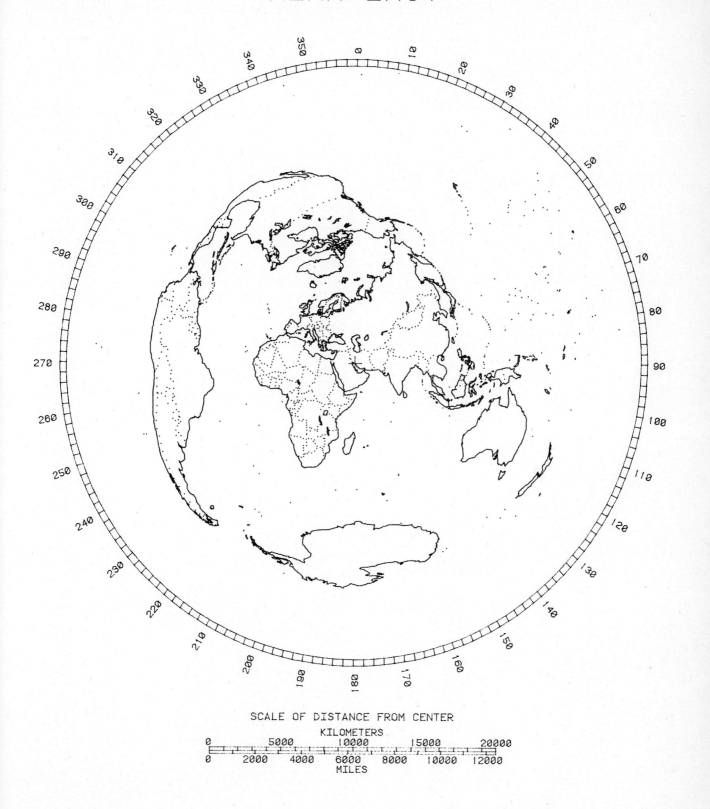

SCALE OF DISTANCE FROM CENTER

KILOMETERS

0   5000   10000   15000   20000

0   2000   4000   6000   8000   10000   12000

MILES

SOUTH ASIA

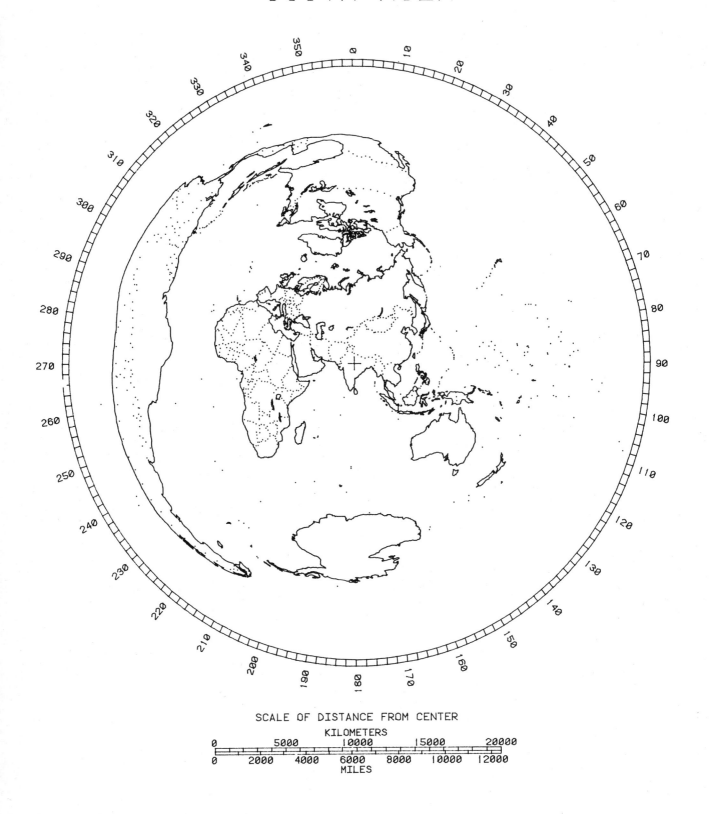

SCALE OF DISTANCE FROM CENTER
KILOMETERS

0   5000   10000   15000   20000

0   2000   4000   6000   8000   10000   12000
MILES

N5KR AZIMUTHAL EQUIDISTANT MAP CENTERED ON
S.E. ASIA

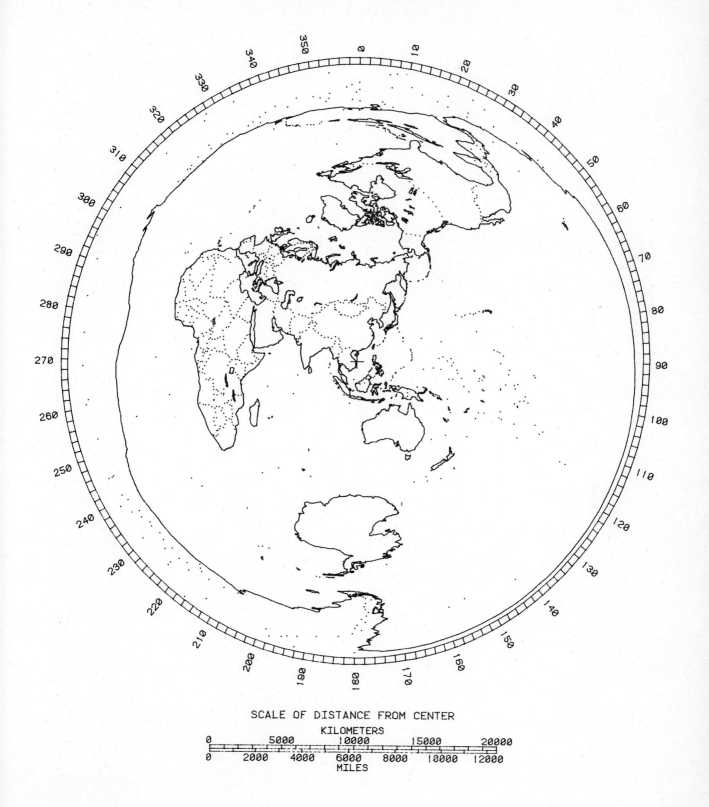

SCALE OF DISTANCE FROM CENTER

KILOMETERS

0   5000   10000   15000   20000

0   2000   4000   6000   8000   10000   12000

MILES

N5KR AZIMUTHAL EQUIDISTANT MAP CENTERED ON
FAR EAST

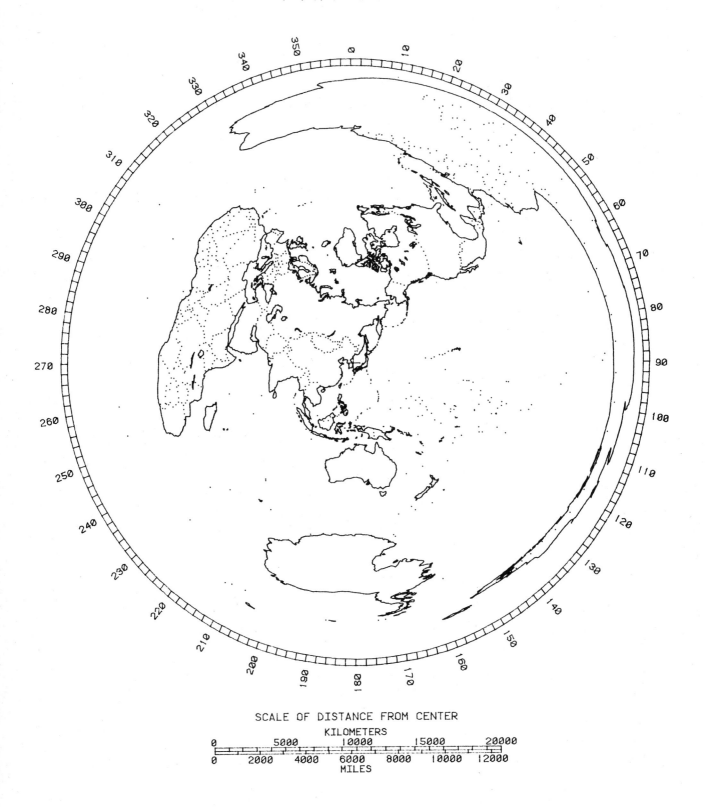

SCALE OF DISTANCE FROM CENTER

AUSTRALIA

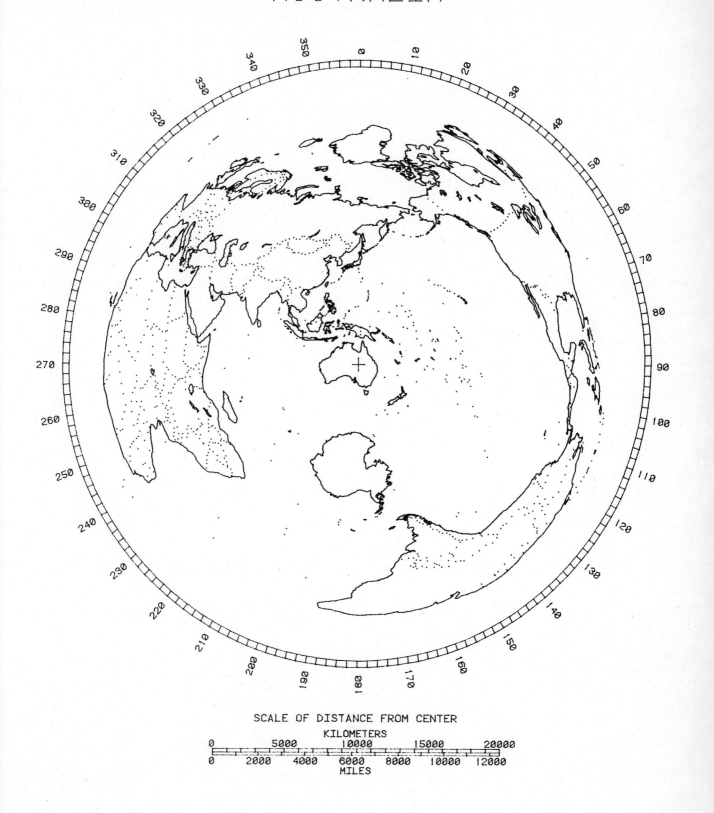

SCALE OF DISTANCE FROM CENTER

KILOMETERS

| 0 | 5000 | 10000 | 15000 | 20000 |

| 0 | 2000 | 4000 | 6000 | 8000 | 10000 | 12000 |

MILES

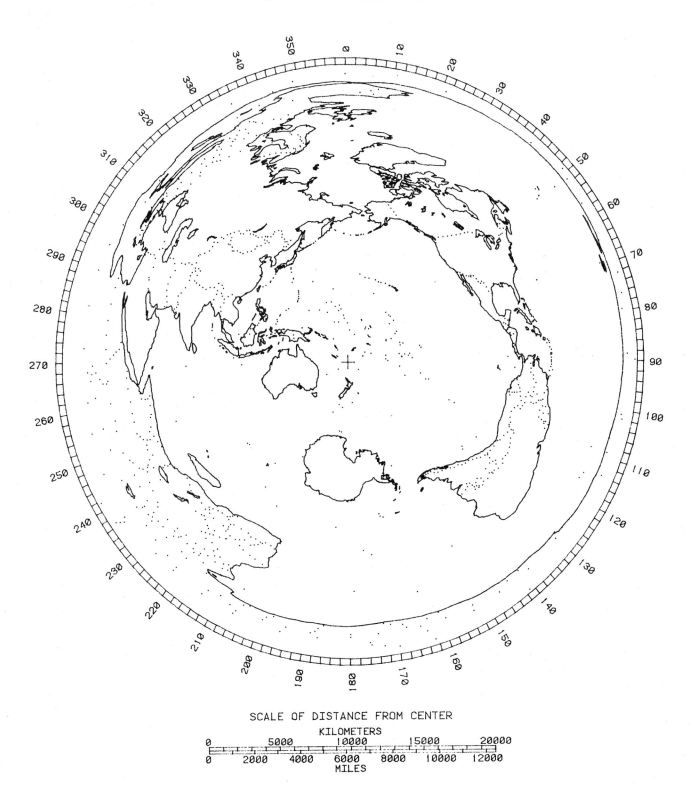

SCALE OF DISTANCE FROM CENTER

Institute for Telecommunications Sciences Propagation-Prediction Information

The following pages contain propagation-prediction information for a number of paths from the continental United States to popular DX locations around the world. The charts and tables presented here were prepared by the Institute for Telecommunications Sciences in Boulder, Colorado, using the computer program IONCAP. The calculations used to produce these charts and tables assumed average Amateur Radio stations (with regard to power level, receiver sensitivity, antenna configuration and local noise) at each end of the path.

The information presented here includes the maximum usable frequency (MUF), lowest usable frequency (LUF) and optimum traffic frequency (FOT) for each path at different times of year and with different smoothed sunspot numbers (SSNs). The curves are for the most reliable propagation mode (for example one-hop F), but other modes may be available. The MUF and LUF values shown are median numbers. Half of the time the MUF and LUF are higher (and half the time lower) for a given set of conditions.

This data is not intended as an exact prediction of when various bands will be open over various paths. Rather, it is intended to give you an idea of what bands *could* be open and to increase your odds of working the DX station you're looking for. These curves are timeless: The MUF and LUF values are tied to month and SSN, not calendar year. For example, the curves given for the path between the eastern USA and Alaska for January will be as valid in January 2001 as they are in January of this year.

Required Information

To use the charts presented here, you need to know some basic information:

1) Your general location (eastern, central or western United States)

2) The general location of the DX station you want to work. The propagation predictions presented here are for the continental United States to the following locations:
- Alaska
- Hawaii
- Caribbean
- Eastern South America
- Southern South America
- Antarctica
- Western Europe
- Eastern Europe
- Western Africa
- Eastern Africa
- Southern Africa
- Near East
- Southern Asia
- Southeast Asia
- Far East
- Australia
- South Pacific

3) The month of the year and time of day that you want to make the contact.

4) The current smoothed sunspot number. This number varies radically throughout the sunspot cycle, as shown in Fig 17-1. You can determine the current SSN from the *10.7-cm solar flux*. WWV broadcasts the latest solar flux number at 18 minutes past each hour. Use Fig 17-2 or Table 17-1 to convert the solar flux number into an SSN.

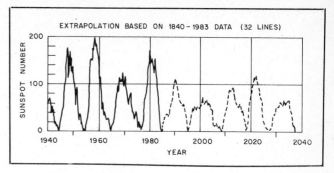

Fig 17-1—Solar sunspot number prediction for the next five sunspot cycles, from the Naval Ocean Systems Center.

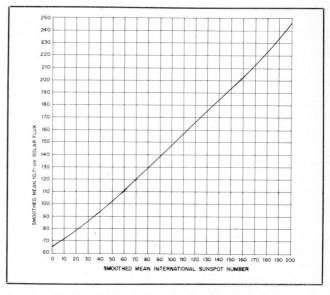

Fig 17-2—Relationship between the smoothed sunspot number (SSN) and the 10.7-cm solar flux. In the low years of a solar cycle, flux values run between about 66 and 86. Intermediate years may see values of 85 to 150. The peak years of Cycle 21 brought readings between 140 and 380.

Using the Charts

The charts on the following pages are broken down according to path, month of year, smoothed sunspot number and time of day. All of the eastern-United-States-to-DX paths are given first, then central United States, then western United States. There is one path per page. For example, the first page shows Eastern USA to Alaska. The heading for each page gives the following information:

1) Path name

2) The longitude and latitude of the point at each end of the path (used for the calculations)

3) Azimuth (beam heading) information to and from the DX location

4) The distance in miles (and kilometers) between the path end points

Each page is divided into four parts, by month, with propagation information calculated for January, April, July and October. Each monthly grouping is further divided into three

Table 17-1

Smoothed Sunspot Numbers (SSNs) Versus Solar Flux Values

SSN	Solar Flux	SSN	Solar Flux
0	64	105	150
5	67	110	155
10	71	115	159
15	75	120	164
20	79	125	169
25	83	130	173
30	86	135	178
35	90	140	183
40	94	145	188
45	98	150	193
50	102	155	198
55	107	160	203
60	111	165	208
65	115	170	213
70	119	175	218
75	123	180	224
80	128	185	229
85	132	190	234
90	137	195	240
95	141	200	245
100	146		

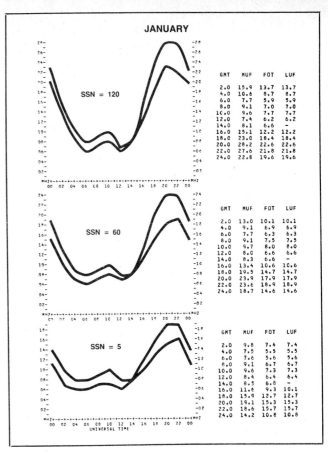

Fig 17-3—Each path is broken down into four parts (by month), and each month three SSN values are given for each month. The MUF and LUF values are presented graphically in the charts, as well as in tabular form.

parts for different SSNs: SSN = 120, SSN = 60 and SSN = 5. Each chart shows the MUF and LUF through a 24-hour period. Along with each chart is a table showing time of day, MUF and LUF in tabular form. The tables also give the FOT for the path. See Fig 17-3.

The information presented here covers a broad range of times and sunspot numbers. If you want to work a station in Alaska in January when the SSN is 5, you can predict the right band to use directly from the chart. Chances are, though, that the circumstances on a given day won't correspond exactly to the information given here. For example, what if you want to work Alaska in February, and the SSN is 20? Fortunately, you can look at the information for SSNs of 5 and 60 in January and April and use linear interpolation to make a good estimate. Remember that propagation prediction is not an exact science and that the information here is

intended to help you increase your odds of being on the right band at the right time for a given path.

For more information on propagation and propagation prediction, see *The ARRL Handbook* and *The ARRL Antenna Book*.

Eastern USA to Alaska
40°N 80°W to 61°N 150°W

JANUARY

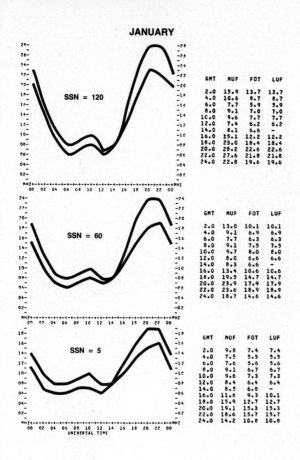

GMT	MUF	FOT	LUF
SSN = 120			
2.0	15.9	13.7	13.7
4.0	10.6	8.7	8.7
6.0	7.7	5.9	5.9
8.0	9.1	7.0	7.0
10.0	9.6	7.7	7.7
12.0	7.4	6.2	6.2
14.0	8.1	6.6	-
16.0	15.1	12.2	12.2
18.0	23.0	18.4	18.4
20.0	28.2	22.6	22.6
22.0	27.6	21.8	21.8
24.0	22.8	19.6	19.6

GMT	MUF	FOT	LUF
SSN = 60			
2.0	13.0	10.1	10.1
4.0	9.1	6.9	6.9
6.0	7.7	6.3	6.3
8.0	9.1	7.5	7.5
10.0	9.7	8.0	8.0
12.0	8.0	6.6	6.6
14.0	8.3	6.6	-
16.0	13.4	10.6	10.6
18.0	19.5	14.7	14.7
20.0	23.9	17.9	17.9
22.0	23.6	18.9	18.9
24.0	18.7	14.6	14.6

GMT	MUF	FOT	LUF
SSN = 5			
2.0	9.8	7.4	7.4
4.0	7.5	5.5	5.5
6.0	7.6	5.6	5.6
8.0	9.1	6.7	6.7
10.0	9.6	7.3	7.3
12.0	8.4	6.4	6.4
14.0	8.5	6.8	-
16.0	11.6	9.3	10.1
18.0	15.9	12.7	12.7
20.0	19.1	15.3	15.3
22.0	18.6	15.7	15.7
24.0	14.2	10.8	10.8

APRIL

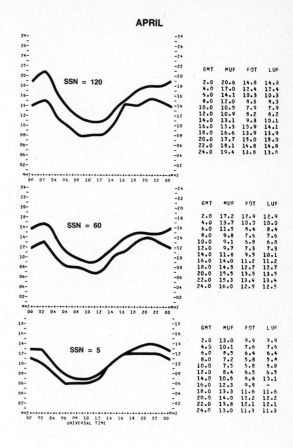

GMT	MUF	FOT	LUF
SSN = 120			
2.0	20.6	14.8	14.8
4.0	17.0	12.4	12.4
6.0	14.1	10.3	10.3
8.0	12.0	8.3	8.3
10.0	10.5	7.9	7.9
12.0	10.9	8.2	8.2
14.0	13.1	9.3	10.1
16.0	15.3	10.9	14.1
18.0	16.6	13.9	13.9
20.0	17.7	15.0	15.0
22.0	18.1	14.8	14.8
24.0	19.4	13.8	13.8

GMT	MUF	FOT	LUF
SSN = 60			
2.0	17.2	12.9	12.9
4.0	14.7	10.0	10.0
6.0	11.5	8.4	8.4
8.0	9.8	7.6	7.6
10.0	9.1	5.8	6.8
12.0	9.7	7.3	7.3
14.0	11.0	9.5	10.1
16.0	14.0	11.2	11.2
18.0	14.5	12.7	12.7
20.0	15.5	13.5	13.5
22.0	15.3	13.4	13.4
24.0	16.0	12.5	12.5

GMT	MUF	FOT	LUF
SSN = 5			
2.0	13.0	9.9	9.9
4.0	10.1	7.6	7.5
6.0	8.5	6.4	6.4
8.0	7.2	5.8	5.8
10.0	7.5	5.8	5.8
12.0	8.4	6.5	6.5
14.0	10.0	8.4	13.1
16.0	12.3	9.8	-
18.0	13.3	11.6	11.6
20.0	14.0	12.2	12.2
22.0	13.8	12.1	12.1
24.0	13.0	11.3	11.3

JULY

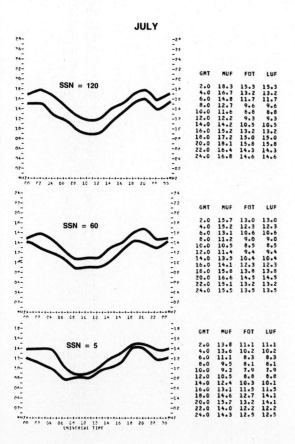

GMT	MUF	FOT	LUF
SSN = 120			
2.0	18.3	15.3	15.3
4.0	16.7	13.2	13.2
6.0	14.8	11.7	11.7
8.0	12.7	9.6	9.6
10.0	11.6	8.8	8.8
12.0	12.2	9.3	9.3
14.0	14.2	10.5	10.5
16.0	15.2	13.2	13.2
18.0	17.2	15.0	15.0
20.0	18.1	15.8	15.8
22.0	16.4	14.3	14.3
24.0	16.8	14.6	14.6

GMT	MUF	FOT	LUF
SSN = 60			
2.0	15.7	13.0	13.0
4.0	15.2	12.3	12.3
6.0	13.1	10.6	10.6
8.0	11.2	9.0	9.0
10.0	10.5	8.5	8.5
12.0	11.4	9.4	9.4
14.0	13.5	10.4	10.4
16.0	14.1	12.3	12.3
18.0	15.8	13.8	13.8
20.0	16.6	14.5	14.5
22.0	15.1	13.2	13.2
24.0	15.5	13.5	13.5

GMT	MUF	FOT	LUF
SSN = 5			
2.0	13.8	11.1	11.1
4.0	13.6	10.2	10.2
6.0	11.1	8.3	8.3
8.0	9.5	8.1	8.1
10.0	9.3	7.9	7.9
12.0	10.5	8.8	8.8
14.0	12.4	10.3	10.1
16.0	13.1	11.5	11.5
18.0	14.6	12.7	14.1
20.0	15.2	13.2	14.1
22.0	14.0	12.2	12.2
24.0	14.3	12.5	12.5

OCTOBER

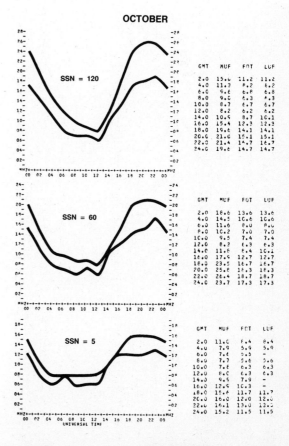

GMT	MUF	FOT	LUF
SSN = 120			
2.0	15.0	11.2	11.2
4.0	11.3	8.2	8.2
6.0	9.6	6.8	6.8
8.0	9.0	6.3	6.3
10.0	8.7	6.7	6.7
12.0	8.2	6.2	6.2
14.0	10.5	8.7	10.1
16.0	15.4	12.3	12.3
18.0	19.6	14.1	14.1
20.0	21.0	15.1	15.1
22.0	21.4	14.7	16.7
24.0	19.6	14.7	14.7

GMT	MUF	FOT	LUF
SSN = 60			
2.0	18.6	13.6	13.6
4.0	14.5	10.6	10.6
6.0	11.6	8.0	8.0
8.0	10.2	7.0	7.0
10.0	9.5	7.4	7.4
12.0	8.3	6.3	6.3
14.0	11.8	8.4	10.1
16.0	17.9	12.7	12.7
18.0	23.5	16.7	16.7
20.0	25.8	18.3	18.3
22.0	26.4	18.7	18.7
24.0	23.7	17.3	17.3

GMT	MUF	FOT	LUF
SSN = 5			
2.0	11.0	8.4	8.4
4.0	7.9	5.9	5.9
6.0	7.0	5.6	-
8.0	7.7	5.6	5.6
10.0	7.6	6.3	6.3
12.0	8.0	6.3	6.3
14.0	9.9	7.9	-
16.0	12.9	10.3	-
18.0	15.4	11.7	11.7
20.0	16.0	12.0	12.0
22.0	16.1	13.0	13.0
24.0	15.2	11.5	11.5

Eastern USA to Hawaii
40°N 80°W to 20°N 155°W

JANUARY

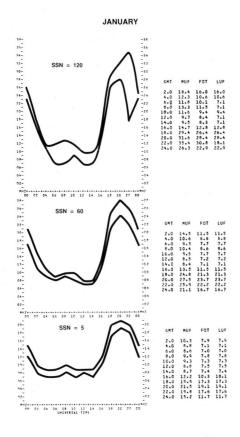

APRIL

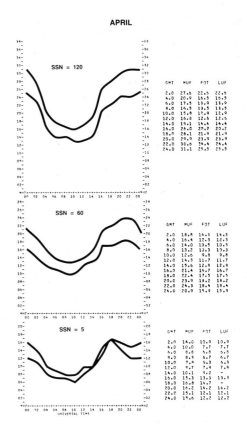

JULY

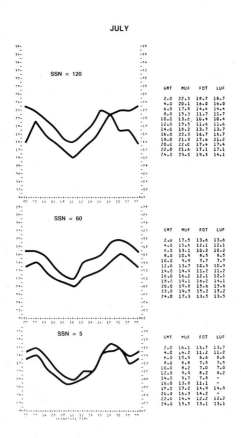

OCTOBER

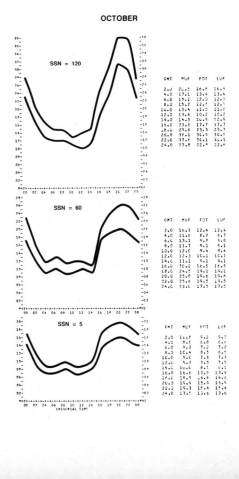

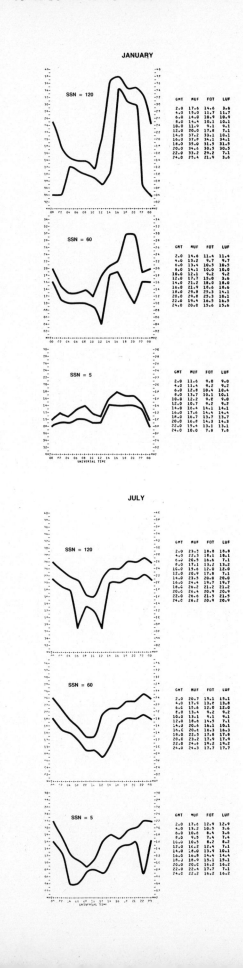

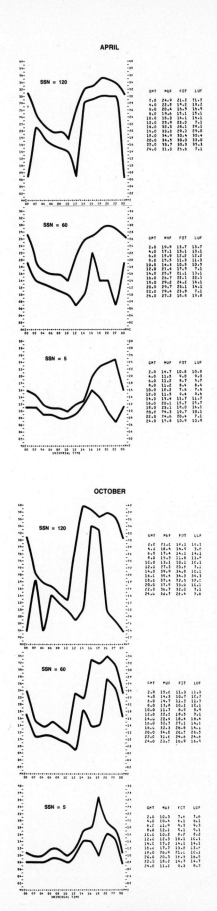

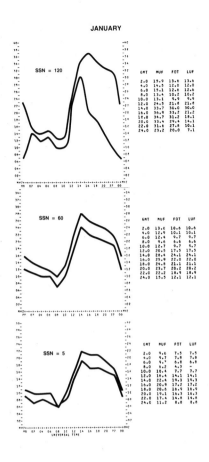

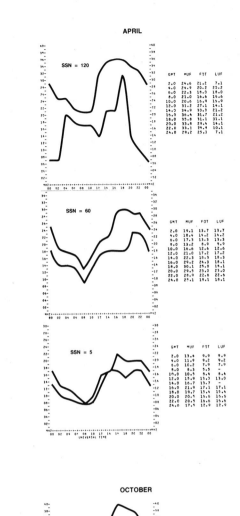

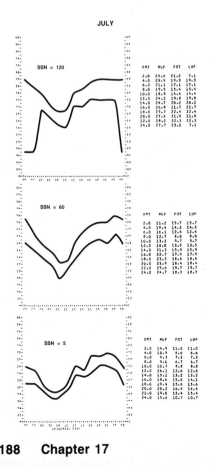

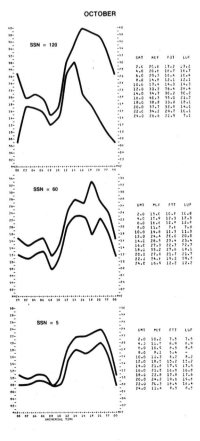

JANUARY

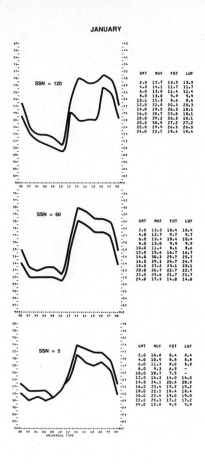

APRIL

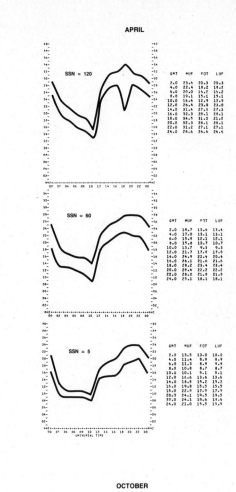

JULY

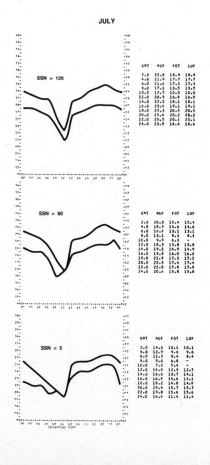

OCTOBER

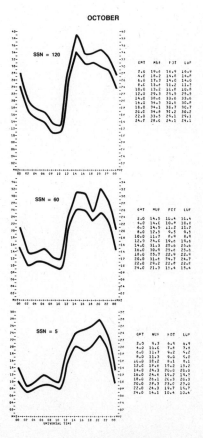

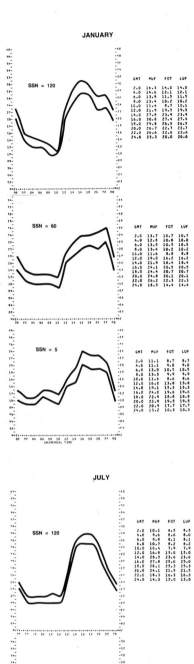

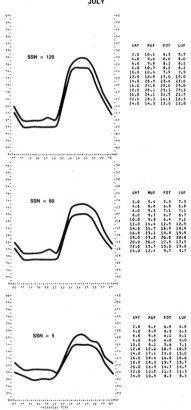

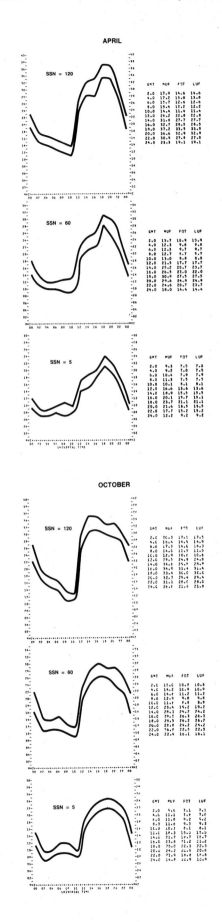

JANUARY

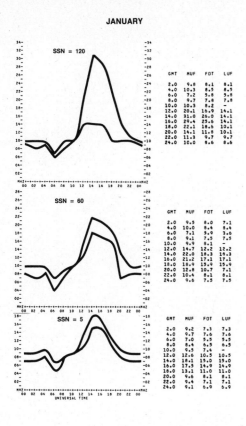

SSN = 120

GMT	MUF	FOT	LUF
2.0	9.8	8.1	8.1
4.0	10.3	8.5	8.5
6.0	7.2	5.8	5.8
8.0	9.7	7.8	7.8
10.0	10.3	8.2	–
12.0	20.1	16.9	14.1
14.0	31.0	26.0	14.1
16.0	29.4	25.6	14.1
18.0	22.1	18.5	10.1
20.0	14.1	11.8	10.1
22.0	11.3	9.7	9.7
24.0	10.0	8.6	8.6

SSN = 60

GMT	MUF	FOT	LUF
2.0	9.5	8.0	7.1
4.0	10.0	8.4	8.4
6.0	7.1	5.9	3.6
8.0	9.1	7.5	7.5
10.0	9.9	8.1	–
12.0	14.7	12.2	12.2
14.0	22.0	18.3	18.3
16.0	21.2	17.1	17.1
18.0	18.9	15.9	15.9
20.0	12.8	10.7	7.1
22.0	10.4	8.1	8.1
24.0	9.6	7.5	7.5

SSN = 5

GMT	MUF	FOT	LUF
2.0	9.2	7.3	7.3
4.0	9.7	7.6	7.6
6.0	7.0	5.5	5.5
8.0	8.4	6.5	6.5
10.0	9.5	7.4	–
12.0	12.6	10.5	10.5
14.0	18.1	15.0	15.0
16.0	17.5	14.9	14.9
18.0	13.1	11.0	11.0
20.0	9.6	8.1	8.1
22.0	9.4	7.1	7.1
24.0	9.1	6.9	6.9

UNIVERSAL TIME

APRIL

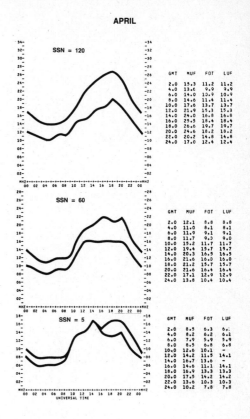

SSN = 120

GMT	MUF	FOT	LUF
2.0	15.3	11.2	11.2
4.0	13.6	9.9	9.9
6.0	14.0	13.9	10.9
8.0	14.6	11.4	11.4
10.0	17.6	13.7	13.7
12.0	21.9	15.3	15.3
14.0	24.0	16.8	16.8
16.0	25.5	18.4	18.4
18.0	26.6	19.7	19.7
20.0	24.6	18.2	18.2
22.0	20.2	14.8	14.8
24.0	17.0	12.4	12.4

SSN = 60

GMT	MUF	FOT	LUF
2.0	12.1	8.8	8.8
4.0	11.0	8.1	8.1
6.0	11.9	9.1	9.1
8.0	11.7	9.3	9.0
10.0	15.2	11.7	11.7
12.0	19.4	15.7	15.7
14.0	20.3	16.5	16.5
16.0	21.6	16.0	16.0
18.0	21.2	15.7	15.7
20.0	21.6	16.4	16.4
22.0	17.1	12.9	12.9
24.0	13.8	10.4	10.4

SSN = 5

GMT	MUF	FOT	LUF
2.0	8.5	6.3	6.
4.0	8.2	6.2	6.2
6.0	7.9	5.9	5.9
8.0	8.5	6.8	6.8
10.0	12.6	10.1	–
12.0	14.2	11.5	14.1
14.0	16.7	13.6	–
16.0	14.6	11.1	14.1
18.0	16.9	13.3	13.3
20.0	17.5	14.2	14.2
22.0	13.6	10.3	10.3
24.0	10.2	7.8	7.8

UNIVERSAL TIME

JULY

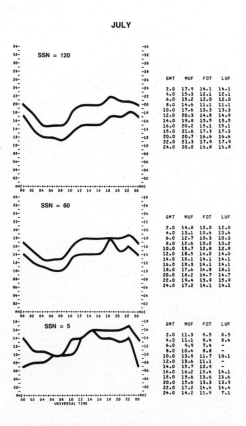

SSN = 120

GMT	MUF	FOT	LUF
2.0	17.9	14.1	14.1
4.0	15.3	12.1	12.1
6.0	15.2	12.0	12.0
8.0	14.6	11.1	11.1
10.0	17.6	13.3	13.3
12.0	20.3	14.8	14.8
14.0	19.8	15.5	15.5
16.0	20.2	15.1	15.1
18.0	21.6	17.3	17.3
20.0	20.7	16.6	16.6
22.0	21.3	17.9	17.9
24.0	20.0	16.8	15.8

SSN = 60

GMT	MUF	FOT	LUF
2.0	14.8	12.0	12.0
4.0	13.1	10.6	10.6
6.0	12.7	10.3	10.3
8.0	12.6	10.2	10.2
10.0	15.7	12.8	12.8
12.0	18.5	14.0	14.0
14.0	18.1	14.1	14.1
16.0	18.3	14.1	14.1
18.0	17.6	14.8	18.1
20.0	18.2	14.7	14.7
22.0	19.4	15.9	15.9
24.0	17.2	14.1	14.1

SSN = 5

GMT	MUF	FOT	LUF
2.0	11.3	8.5	8.5
4.0	11.1	8.4	8.4
6.0	9.9	7.4	–
8.0	10.4	8.8	–
10.0	13.8	11.7	10.1
12.0	13.6	11.1	–
14.0	15.7	12.9	–
16.0	16.2	13.4	14.1
18.0	15.6	13.6	13.6
20.0	15.6	13.3	13.3
22.0	17.2	14.4	14.4
24.0	14.2	11.9	7.1

UNIVERSAL TIME

OCTOBER

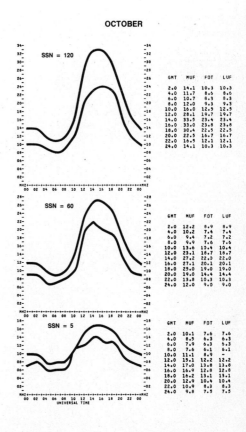

SSN = 120

GMT	MUF	FOT	LUF
2.0	14.1	10.3	10.3
4.0	11.7	8.6	8.6
6.0	10.7	8.3	8.3
8.0	12.0	9.3	9.3
10.0	16.0	12.5	12.5
12.0	28.1	19.7	19.7
14.0	33.5	23.4	23.4
16.0	33.0	23.8	23.8
18.0	30.4	22.5	22.5
20.0	22.5	16.7	16.7
22.0	16.5	12.1	12.1
24.0	14.1	10.3	10.3

SSN = 60

GMT	MUF	FOT	LUF
2.0	12.2	8.9	8.9
4.0	10.2	7.4	7.4
6.0	9.4	7.2	7.2
8.0	9.9	7.6	7.6
10.0	13.6	13.4	10.4
12.0	23.1	18.7	18.7
14.0	27.2	22.0	22.0
16.0	27.1	20.1	20.1
18.0	25.0	19.0	19.0
20.0	19.0	14.4	14.4
22.0	13.8	10.3	10.3
24.0	12.0	9.0	9.0

SSN = 5

GMT	MUF	FOT	LUF
2.0	10.1	7.6	7.6
4.0	8.5	6.3	6.3
6.0	7.9	6.3	5.3
8.0	7.6	6.1	6.1
10.0	11.1	8.9	–
12.0	15.1	12.2	12.2
14.0	17.0	13.8	13.8
16.0	16.9	12.8	12.8
18.0	16.2	13.1	13.1
20.0	12.9	10.4	10.4
22.0	10.9	8.3	8.3
24.0	9.8	7.5	7.5

UNIVERSAL TIME

JANUARY

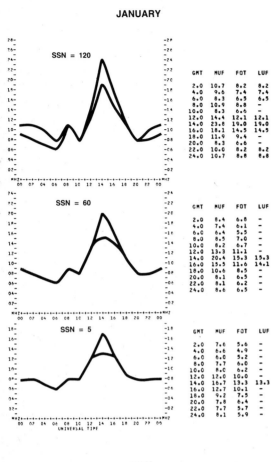

APRIL

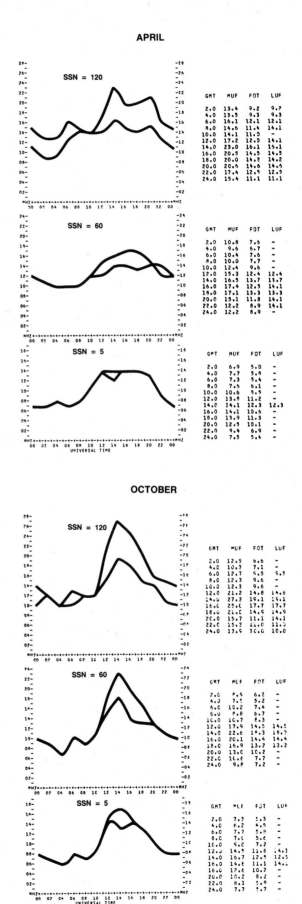

JULY

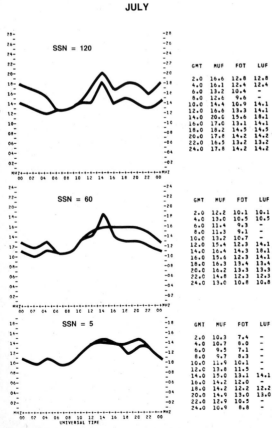

OCTOBER

JANUARY

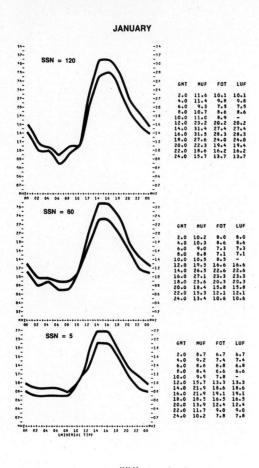

SSN = 120

GMT	MUF	FOT	LUF
2.0	11.6	10.1	10.1
4.0	11.4	9.8	9.8
6.0	9.3	7.5	7.5
8.0	10.7	8.6	8.6
10.0	11.0	8.9	-
12.0	23.2	20.2	20.2
14.0	31.4	27.4	27.4
16.0	31.5	28.3	28.3
18.0	27.6	24.0	24.0
20.0	22.3	19.4	19.4
22.0	18.6	16.2	16.2
24.0	15.7	13.7	13.7

SSN = 60

GMT	MUF	FOT	LUF
2.0	10.2	8.0	8.0
4.0	10.3	8.6	8.6
6.0	9.0	7.3	7.3
8.0	8.8	7.1	7.1
10.0	10.5	8.5	-
12.0	19.5	16.6	16.6
14.0	26.5	22.6	22.6
16.0	27.1	23.3	23.3
18.0	23.6	20.3	20.3
20.0	18.4	15.8	15.8
22.0	15.3	12.1	12.1
24.0	13.4	10.6	10.6

SSN = 5

GMT	MUF	FOT	LUF
2.0	8.7	6.7	6.7
4.0	9.2	7.4	7.4
6.0	8.6	6.8	6.8
8.0	8.4	6.6	6.6
10.0	9.9	7.8	-
12.0	15.7	13.3	13.3
14.0	21.9	18.6	18.6
16.0	21.9	19.1	19.1
18.0	18.5	16.5	16.5
20.0	13.9	12.4	12.4
22.0	11.7	9.0	9.0
24.0	10.2	7.8	7.8

APRIL

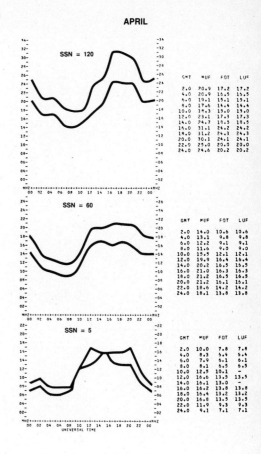

SSN = 120

GMT	MUF	FOT	LUF
2.0	20.9	17.2	17.2
4.0	20.9	16.5	16.5
6.0	19.1	15.1	15.1
8.0	17.6	14.4	14.4
10.0	19.3	15.0	15.0
12.0	23.1	17.3	17.3
14.0	24.7	19.5	18.5
16.0	31.1	24.2	24.2
18.0	31.2	24.3	24.3
20.0	30.1	24.1	24.1
22.0	25.0	20.0	20.0
24.0	24.6	20.2	20.2

SSN = 60

GMT	MUF	FOT	LUF
2.0	14.0	10.6	10.6
4.0	13.1	9.8	9.8
6.0	12.2	9.1	9.1
8.0	11.6	9.0	9.0
10.0	15.5	12.1	12.1
12.0	19.9	16.4	16.4
14.0	20.2	16.5	16.5
16.0	21.0	16.3	16.3
18.0	21.2	16.5	16.5
20.0	21.2	16.1	16.1
22.0	18.6	14.2	14.2
24.0	18.1	13.8	13.8

SSN = 5

GMT	MUF	FOT	LUF
2.0	10.0	7.8	7.8
4.0	8.3	6.4	6.4
6.0	7.9	6.1	6.1
8.0	8.1	6.5	6.5
10.0	12.5	10.1	-
12.0	16.6	13.5	13.5
14.0	16.1	13.0	-
16.0	16.2	13.8	13.8
18.0	16.4	13.2	13.2
20.0	16.8	13.5	13.5
22.0	11.9	9.3	9.3
24.0	9.1	7.1	7.1

JULY

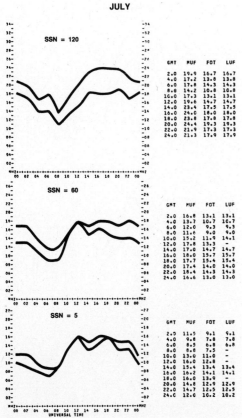

SSN = 120

GMT	MUF	FOT	LUF
2.0	19.9	16.7	16.7
4.0	17.2	13.8	13.8
6.0	17.8	14.3	14.3
8.8	14.2	10.8	10.8
10.0	17.3	13.1	13.1
12.0	19.6	14.7	14.7
14.0	23.4	17.5	17.5
16.0	24.0	18.0	18.0
18.0	23.8	17.8	17.8
20.0	24.4	19.3	19.3
22.0	21.9	17.3	17.3
24.0	21.3	17.9	17.9

SSN = 60

GMT	MUF	FOT	LUF
2.0	16.8	13.1	13.1
4.0	13.7	10.7	10.7
6.0	12.0	9.3	9.3
8.0	11.6	9.0	9.0
10.0	15.2	11.9	14.1
12.0	17.8	13.3	-
14.0	17.0	14.7	14.7
16.0	18.0	15.7	15.7
18.0	17.7	15.4	15.4
20.0	17.4	14.0	14.0
22.0	18.4	14.3	14.3
24.0	16.6	13.0	13.0

SSN = 5

GMT	MUF	FOT	LUF
2.0	11.5	9.1	9.1
4.0	9.8	7.8	7.8
6.0	8.5	6.8	6.8
8.0	8.8	7.5	-
10.0	13.0	11.0	-
12.0	16.0	12.8	-
14.0	15.4	13.4	13.4
16.0	16.2	14.1	14.1
18.0	16.0	13.9	-
20.0	14.8	12.9	12.9
22.0	14.7	12.5	12.5
24.0	12.0	10.2	10.2

OCTOBER

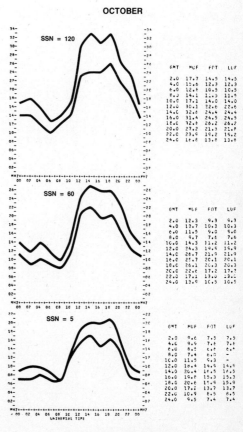

SSN = 120

GMT	MUF	FOT	LUF
2.0	17.7	14.5	14.5
4.0	15.6	12.3	12.3
6.0	12.8	10.5	10.5
8.0	14.1	11.5	11.5
10.0	17.1	14.0	14.0
12.0	30.1	22.6	22.6
14.0	32.6	24.4	24.4
16.0	31.4	24.5	24.5
18.0	32.8	26.2	26.2
20.0	27.2	21.8	21.8
22.0	23.9	19.2	19.2
24.0	16.8	13.8	13.8

SSN = 60

GMT	MUF	FOT	LUF
2.0	12.3	9.3	9.3
4.0	12.7	10.3	10.3
6.0	11.5	9.0	9.0
8.0	9.7	7.8	7.8
10.0	14.3	11.2	11.2
12.0	24.3	19.9	19.9
14.0	26.7	21.9	21.9
16.0	25.7	20.1	20.1
18.0	26.1	20.3	20.3
20.0	22.6	17.2	17.2
22.0	17.1	13.0	13.0
24.0	13.9	10.5	10.5

SSN = 5

GMT	MUF	FOT	LUF
2.0	9.6	7.5	7.5
4.0	9.9	7.6	7.6
6.0	8.5	6.4	-
8.0	7.4	6.0	-
10.0	11.5	9.3	-
12.0	18.4	14.5	14.5
14.0	20.4	16.5	16.5
16.0	19.8	15.3	15.3
18.0	20.6	15.9	15.9
20.0	17.1	13.7	13.7
22.0	10.9	8.5	8.5
24.0	9.5	7.4	7.4

JANUARY

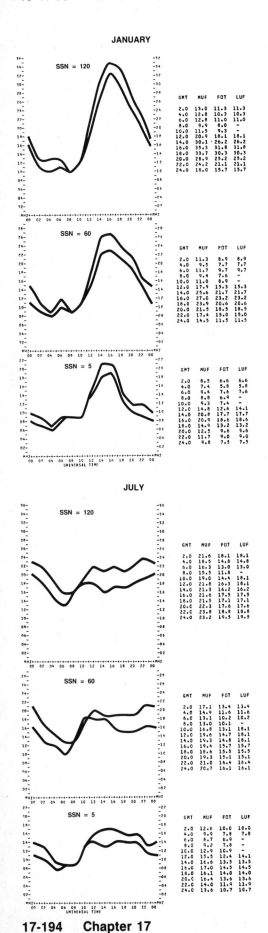

SSN = 120

GMT	MUF	FOT	LUF
2.0	13.0	11.3	11.3
4.0	12.8	10.3	10.3
6.0	12.8	11.0	11.0
8.0	9.9	8.0	-
10.0	11.5	9.3	-
12.0	20.9	18.1	18.1
14.0	30.1	26.2	26.2
16.0	35.3	31.8	31.8
18.0	33.7	30.3	30.3
20.0	28.9	25.2	25.2
22.0	24.2	21.1	21.1
24.0	18.0	15.7	15.7

SSN = 60

GMT	MUF	FOT	LUF
2.0	11.3	8.9	8.9
4.0	9.5	7.7	7.7
6.0	11.7	9.7	9.7
8.0	9.4	7.6	-
10.0	11.0	8.9	-
12.0	17.9	15.3	15.3
14.0	25.6	21.7	21.7
16.0	27.0	23.2	23.2
18.0	23.9	20.6	20.6
20.0	21.5	18.5	18.5
22.0	17.4	15.0	15.0
24.0	14.5	11.5	11.5

SSN = 5

GMT	MUF	FOT	LUF
2.0	8.5	6.6	6.6
4.0	7.4	5.8	5.8
6.0	9.4	7.6	7.6
8.0	8.8	6.9	-
10.0	9.3	7.4	-
12.0	14.8	12.6	14.1
14.0	20.8	17.7	17.7
16.0	20.9	18.6	18.6
18.0	14.9	13.2	13.2
20.0	12.5	9.6	9.6
22.0	11.7	9.0	9.0
24.0	9.8	7.5	7.5

UNIVERSAL TIME

APRIL

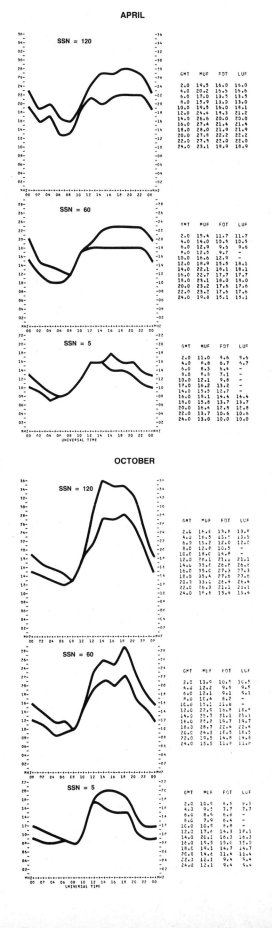

SSN = 120

GMT	MUF	FOT	LUF
2.0	19.5	16.0	16.0
4.0	20.2	16.5	16.5
6.0	17.0	13.5	13.5
8.0	15.9	13.0	13.0
10.0	19.5	16.0	19.1
12.0	24.4	19.3	21.2
14.0	26.6	20.0	20.0
16.0	27.4	21.4	21.4
18.0	28.0	21.9	21.9
20.0	27.8	22.2	22.2
22.0	27.5	22.2	22.2
24.0	23.1	18.9	18.9

SSN = 60

GMT	MUF	FOT	LUF
2.0	15.4	11.7	11.7
4.0	14.0	10.5	10.5
6.0	12.9	9.5	9.6
8.0	12.5	9.7	-
10.0	16.6	12.9	-
12.0	18.9	15.5	18.1
14.0	22.1	19.1	18.1
16.0	22.7	17.7	17.7
18.0	23.1	18.0	19.0
20.0	23.2	17.6	17.6
22.0	23.2	17.6	17.6
24.0	19.8	15.1	15.1

SSN = 5

GMT	MUF	FOT	LUF
2.0	11.0	9.6	9.5
4.0	8.8	6.7	4.7
6.0	8.3	6.4	-
8.0	8.3	7.1	-
10.0	12.1	9.8	-
12.0	16.2	13.2	-
14.0	15.5	12.7	-
16.0	18.1	14.4	14.4
18.0	15.8	13.7	13.7
20.0	16.4	12.9	12.8
22.0	13.7	10.6	10.6
24.0	13.0	10.0	10.0

UNIVERSAL TIME

JULY

SSN = 120

GMT	MUF	FOT	LUF
2.0	21.6	18.1	18.1
4.0	18.5	14.8	14.8
6.0	16.3	13.0	13.0
8.0	15.5	11.8	-
10.0	19.0	14.4	18.1
12.0	21.8	16.3	18.1
14.0	21.3	16.2	16.2
16.0	21.6	17.5	17.5
18.0	21.5	17.1	17.1
20.0	22.3	17.6	17.6
22.0	23.8	18.8	18.8
24.0	23.2	19.5	19.5

SSN = 60

GMT	MUF	FOT	LUF
2.0	17.1	13.4	13.4
4.0	14.9	11.6	11.6
6.0	13.1	10.2	10.2
8.0	13.0	10.1	-
10.0	16.8	13.1	18.1
12.0	19.6	14.7	18.1
14.0	19.3	14.8	18.1
16.0	19.4	15.7	15.7
18.0	18.6	15.5	15.5
20.0	19.3	15.1	15.1
22.0	21.0	16.4	16.4
24.0	20.7	16.1	16.1

SSN = 5

GMT	MUF	FOT	LUF
2.0	12.6	10.0	10.0
4.0	9.9	7.8	7.8
6.0	8.7	6.9	-
8.0	9.2	7.8	-
10.0	12.9	10.9	-
12.0	15.5	12.4	14.1
14.0	16.6	13.5	13.5
16.0	17.0	14.3	14.5
18.0	16.1	14.0	14.0
20.0	16.4	13.6	13.6
22.0	14.0	11.9	11.9
24.0	13.6	10.7	10.7

UNIVERSAL TIME

OCTOBER

SSN = 120

GMT	MUF	FOT	LUF
2.0	16.6	13.8	13.8
4.0	16.5	13.5	13.5
6.0	15.2	12.0	12.0
8.0	12.8	10.5	-
10.0	18.0	14.8	-
12.0	28.1	24.3	21.1
14.0	35.6	26.8	26.0
16.0	35.0	27.3	27.3
18.0	35.4	27.6	27.6
20.0	33.0	26.4	26.4
22.0	26.3	21.1	21.1
24.0	18.8	15.4	15.4

SSN = 60

GMT	MUF	FOT	LUF
2.0	13.9	10.5	10.5
4.0	12.2	9.5	9.5
6.0	12.1	9.1	9.1
8.0	10.4	8.2	-
10.0	15.1	11.8	-
12.0	22.6	18.8	18.0
14.0	25.7	21.1	21.1
16.0	25.2	19.7	19.7
18.0	28.7	22.4	22.4
20.0	24.3	18.5	18.5
22.0	19.5	14.8	14.8
24.0	15.5	11.8	11.8

SSN = 5

GMT	MUF	FOT	LUF
2.0	10.9	8.5	8.5
4.0	9.1	7.7	7.7
6.0	8.4	6.6	-
8.0	7.9	6.4	-
10.0	10.5	6.8	-
12.0	17.6	14.3	18.1
14.0	20.1	16.3	16.3
16.0	19.5	15.0	15.0
18.0	19.1	14.7	14.7
20.0	14.6	11.4	11.4
22.0	12.1	9.4	9.4
24.0	12.1	9.4	9.4

UNIVERSAL TIME

JANUARY

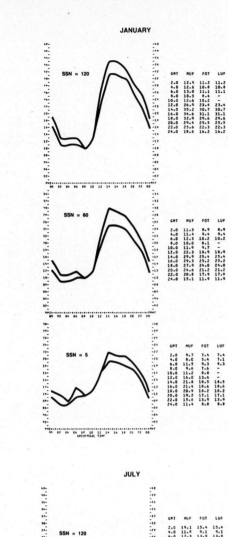

APRIL

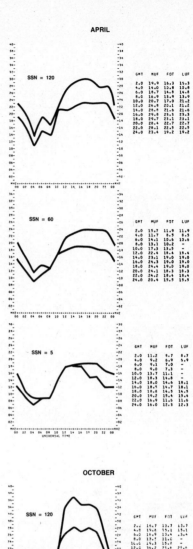

JULY

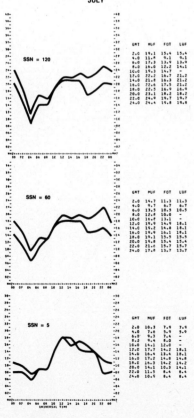

OCTOBER

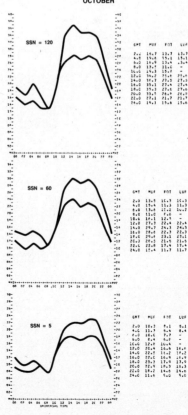

JANUARY

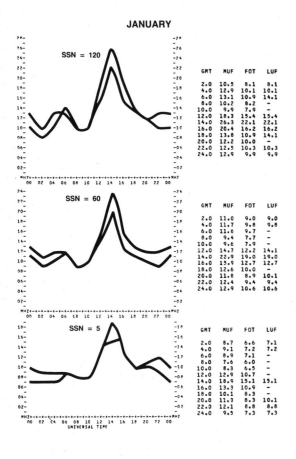

GMT	MUF	FOT	LUF
2.0	10.5	8.1	8.1
4.0	12.9	10.1	10.1
6.0	13.1	10.9	14.1
8.0	10.2	8.2	–
10.0	9.9	7.9	–
12.0	18.3	15.4	15.4
14.0	26.3	22.1	22.1
16.0	20.4	16.2	16.2
18.0	13.8	10.9	14.1
20.0	12.2	10.0	–
22.0	12.5	10.3	10.3
24.0	12.9	9.9	9.9

GMT	MUF	FOT	LUF
2.0	11.0	9.0	9.0
4.0	11.7	9.8	9.8
6.0	11.6	9.7	–
8.0	9.4	7.7	–
10.0	9.6	7.9	–
12.0	14.7	12.2	14.1
14.0	22.9	19.0	19.0
16.0	15.9	12.7	12.7
18.0	12.6	10.0	–
20.0	11.8	8.9	10.1
22.0	12.4	9.4	9.4
24.0	12.9	10.6	10.6

GMT	MUF	FOT	LUF
2.0	8.7	6.6	7.1
4.0	9.1	7.2	7.2
6.0	8.9	7.1	–
8.0	7.6	6.0	–
10.0	8.3	6.5	–
12.0	12.9	10.7	–
14.0	18.9	15.1	15.1
16.0	13.3	10.9	–
18.0	10.1	8.3	–
20.0	11.3	8.3	10.1
22.0	12.1	8.8	8.8
24.0	9.5	7.3	7.3

APRIL

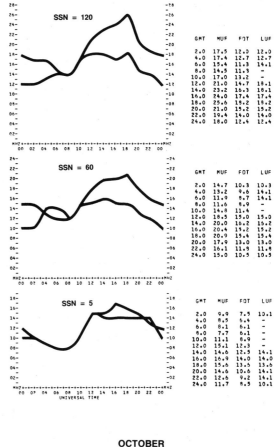

GMT	MUF	FOT	LUF
2.0	17.5	12.0	12.0
4.0	17.4	12.7	12.7
6.0	15.4	11.3	14.1
8.0	14.5	11.3	–
10.0	17.0	13.2	–
12.0	21.0	14.7	18.1
14.0	23.2	16.3	18.1
16.0	24.0	17.4	17.4
18.0	25.6	18.2	18.2
20.0	21.0	15.2	15.2
22.0	19.4	14.0	14.0
24.0	18.0	12.4	12.4

GMT	MUF	FOT	LUF
2.0	14.7	10.3	10.3
4.0	13.2	9.6	14.1
6.0	11.9	8.7	14.1
8.0	11.6	8.9	–
10.0	14.8	11.4	–
12.0	18.5	15.0	15.0
14.0	20.0	16.2	15.2
16.0	20.4	15.2	15.2
18.0	20.9	15.4	15.4
20.0	17.9	13.0	13.0
22.0	16.1	11.9	11.9
24.0	15.0	10.5	10.5

GMT	MUF	FOT	LUF
2.0	9.9	7.5	10.1
4.0	8.5	6.4	–
6.0	8.1	6.1	–
8.0	7.7	6.1	–
10.0	11.1	8.9	–
12.0	15.1	12.3	–
14.0	16.6	12.5	14.1
16.0	16.9	14.0	14.0
18.0	15.6	13.5	13.6
20.0	14.6	10.6	14.1
22.0	12.6	9.2	14.1
24.0	11.7	8.5	10.1

JULY

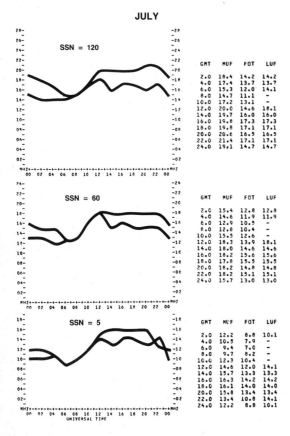

GMT	MUF	FOT	LUF
2.0	18.4	14.2	14.2
4.0	17.4	13.7	13.7
6.0	15.3	12.0	14.1
8.0	14.7	11.1	–
10.0	17.2	13.1	–
12.0	20.0	14.6	18.1
14.0	19.7	16.0	16.0
16.0	19.8	17.3	17.3
18.0	19.8	17.1	17.1
20.0	20.6	16.5	16.5
22.0	21.4	17.1	17.1
24.0	19.1	14.7	14.7

GMT	MUF	FOT	LUF
2.0	15.4	12.8	12.8
4.0	14.6	11.9	11.9
6.0	12.9	10.5	–
8.0	12.8	10.4	–
10.0	15.5	12.6	–
12.0	18.3	13.9	18.1
14.0	18.0	14.6	14.6
16.0	19.2	15.6	15.6
18.0	17.8	15.5	15.3
20.0	18.2	14.8	14.8
22.0	18.2	15.1	15.1
24.0	15.7	13.0	13.0

GMT	MUF	FOT	LUF
2.0	12.2	8.8	10.1
4.0	10.5	7.9	–
6.0	9.4	7.0	–
8.0	9.7	8.2	–
10.0	12.3	10.4	–
12.0	14.6	12.0	14.1
14.0	15.7	13.3	13.3
16.0	16.3	14.2	14.2
18.0	16.1	14.0	14.0
20.0	15.8	13.4	14.2
22.0	13.4	10.8	14.1
24.0	12.2	8.8	10.1

OCTOBER

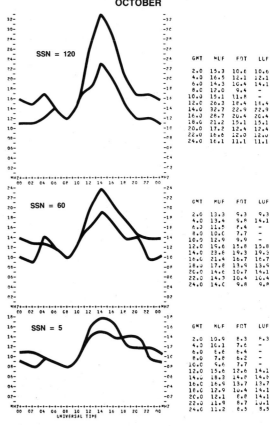

GMT	MUF	FOT	LUF
2.0	15.3	10.6	10.6
4.0	16.5	12.1	12.1
6.0	14.3	10.4	14.1
8.0	12.0	9.4	–
10.0	15.1	11.8	–
12.0	26.3	18.4	18.4
14.0	32.7	22.9	22.9
16.0	28.7	20.4	20.4
18.0	21.2	15.1	15.1
20.0	17.2	12.4	12.4
22.0	16.6	12.0	12.0
24.0	16.1	11.1	11.1

GMT	MUF	FOT	LUF
2.0	13.3	9.3	9.3
4.0	13.4	9.8	14.1
6.0	11.5	8.4	–
8.0	10.0	7.7	–
10.0	12.9	9.9	–
12.0	19.6	15.8	15.8
14.0	23.8	19.3	19.3
16.0	21.4	16.7	16.7
18.0	17.8	13.9	14.8
20.0	14.6	10.7	14.1
22.0	14.3	10.4	10.4
24.0	14.0	9.8	9.8

GMT	MUF	FOT	LUF
2.0	10.9	8.3	8.3
4.0	10.1	7.6	–
6.0	8.6	6.4	–
8.0	7.8	6.2	–
10.0	9.6	7.7	–
12.0	15.6	12.6	14.1
14.0	18.3	14.8	14.8
16.0	16.9	13.7	13.7
18.0	12.9	10.4	14.1
20.0	12.1	8.8	14.1
22.0	11.9	8.7	10.1
24.0	11.2	8.5	8.5

JANUARY

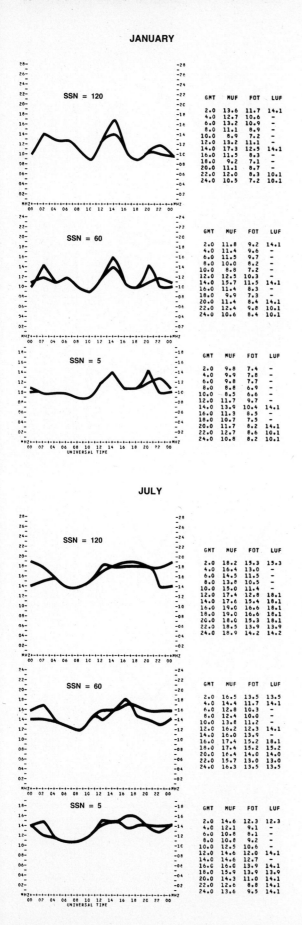

SSN = 120

GMT	MUF	FOT	LUF
2.0	13.6	11.7	14.1
4.0	12.7	10.6	-
6.0	13.2	10.9	-
8.0	11.1	8.9	-
10.0	8.9	7.2	-
12.0	13.2	11.1	-
14.0	17.3	12.5	14.1
16.0	11.5	8.3	-
18.0	9.2	7.1	-
20.0	11.1	8.7	-
22.0	12.0	8.3	10.1
24.0	10.5	7.2	10.1

SSN = 60

GMT	MUF	FOT	LUF
2.0	11.8	9.2	14.1
4.0	11.4	9.6	-
6.0	11.5	9.7	-
8.0	10.0	8.2	-
10.0	8.8	7.2	-
12.0	12.5	10.3	-
14.0	15.7	11.5	14.1
16.0	11.4	8.3	-
18.0	9.9	7.3	-
20.0	11.4	8.4	14.1
22.0	12.4	9.8	10.1
24.0	10.6	8.4	10.1

SSN = 5

GMT	MUF	FOT	LUF
2.0	9.8	7.4	-
4.0	9.9	7.8	-
6.0	9.8	7.7	-
8.0	8.5	6.6	-
10.0	8.5	6.6	-
12.0	11.7	9.7	-
14.0	13.9	10.4	14.1
16.0	11.3	8.5	-
18.0	10.7	7.5	-
20.0	11.7	8.2	14.1
22.0	12.7	8.6	10.1
24.0	10.8	8.2	10.1

UNIVERSAL TIME

APRIL

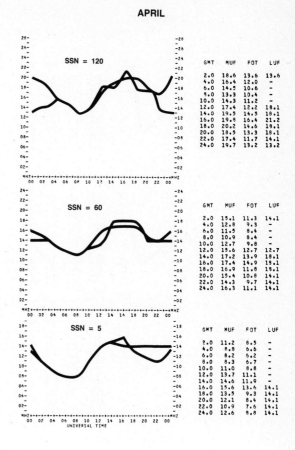

SSN = 120

GMT	MUF	FOT	LUF
2.0	18.6	13.6	13.6
4.0	16.4	12.0	-
6.0	14.5	10.6	-
8.0	13.3	10.4	-
10.0	14.3	11.2	-
12.0	17.4	12.2	18.1
14.0	19.3	14.5	18.1
16.0	19.8	16.4	21.2
18.0	20.2	14.6	19.1
20.0	18.5	13.3	18.1
22.0	17.4	11.7	14.1
24.0	19.7	13.2	13.2

SSN = 60

GMT	MUF	FOT	LUF
2.0	15.1	11.3	14.1
4.0	12.8	9.3	-
6.0	11.5	8.4	-
8.0	10.9	8.4	-
10.0	12.7	9.8	-
12.0	15.6	12.7	12.7
14.0	17.2	13.9	18.1
16.0	17.4	14.9	19.1
18.0	16.9	11.8	19.1
20.0	15.4	10.8	14.1
22.0	14.3	9.7	14.1
24.0	16.3	11.1	14.1

SSN = 5

GMT	MUF	FOT	LUF
2.0	11.2	8.5	-
4.0	8.8	6.6	-
6.0	8.2	5.2	-
8.0	8.3	6.7	-
10.0	11.0	8.8	-
12.0	13.7	11.1	-
14.0	14.6	11.9	-
16.0	15.6	13.6	14.1
18.0	13.5	9.3	14.1
20.0	12.1	8.4	14.1
22.0	10.9	7.6	14.1
24.0	12.6	8.8	14.1

UNIVERSAL TIME

JULY

SSN = 120

GMT	MUF	FOT	LUF
2.0	18.2	15.3	15.3
4.0	16.4	13.0	-
6.0	14.5	11.5	-
8.0	13.8	10.5	-
10.0	15.0	11.4	-
12.0	17.4	12.8	18.1
14.0	17.6	15.4	18.1
16.0	19.0	16.6	18.1
18.0	19.0	16.6	18.1
20.0	18.0	15.3	18.1
22.0	18.5	13.9	13.9
24.0	18.9	14.2	14.2

SSN = 60

GMT	MUF	FOT	LUF
2.0	16.5	13.5	13.5
4.0	14.4	11.7	14.1
6.0	12.8	10.3	-
8.0	12.4	10.0	-
10.0	13.8	11.2	-
12.0	16.2	12.3	14.1
14.0	16.0	13.9	-
16.0	17.4	15.2	18.1
18.0	17.4	15.2	15.2
20.0	16.4	14.0	14.0
22.0	15.7	13.0	13.0
24.0	16.3	13.5	13.5

SSN = 5

GMT	MUF	FOT	LUF
2.0	14.6	12.3	12.3
4.0	12.1	9.4	-
6.0	10.8	8.1	-
8.0	10.8	9.2	-
10.0	12.5	10.6	-
12.0	14.6	12.0	14.1
14.0	14.6	12.7	-
16.0	16.0	13.9	14.1
18.0	15.9	13.9	13.9
20.0	14.3	11.0	14.1
22.0	13.8	8.8	14.1
24.0	13.6	9.5	14.1

UNIVERSAL TIME

OCTOBER

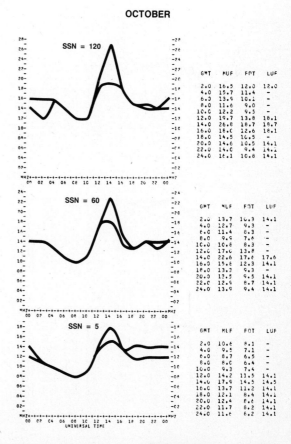

SSN = 120

GMT	MUF	FOT	LUF
2.0	16.5	12.0	12.0
4.0	15.7	11.4	-
6.0	13.9	10.1	-
8.0	11.6	9.0	-
10.0	12.2	9.5	-
12.0	19.7	13.8	18.1
14.0	26.8	18.7	18.7
16.0	14.6	12.6	18.1
18.0	14.5	10.5	-
20.0	14.6	10.5	14.1
22.0	14.0	9.4	14.1
24.0	16.1	10.8	14.1

SSN = 60

GMT	MUF	FOT	LUF
2.0	13.7	10.3	14.1
4.0	12.7	9.3	-
6.0	11.4	8.3	-
8.0	9.9	7.6	-
10.0	10.8	8.3	-
12.0	17.0	13.8	-
14.0	22.6	17.6	17.6
16.0	15.8	12.3	14.1
18.0	13.2	9.3	-
20.0	13.5	9.5	14.1
22.0	12.5	8.7	14.1
24.0	13.9	9.4	14.1

SSN = 5

GMT	MUF	FOT	LUF
2.0	10.6	8.1	-
4.0	9.5	7.1	-
6.0	8.7	6.5	-
8.0	8.0	6.4	-
10.0	9.3	7.4	-
12.0	14.2	11.5	14.1
14.0	17.9	14.5	14.5
16.0	13.7	11.2	14.1
18.0	12.1	8.4	14.1
20.0	12.4	8.8	14.1
22.0	11.7	8.2	14.1
24.0	11.6	8.2	14.1

UNIVERSAL TIME

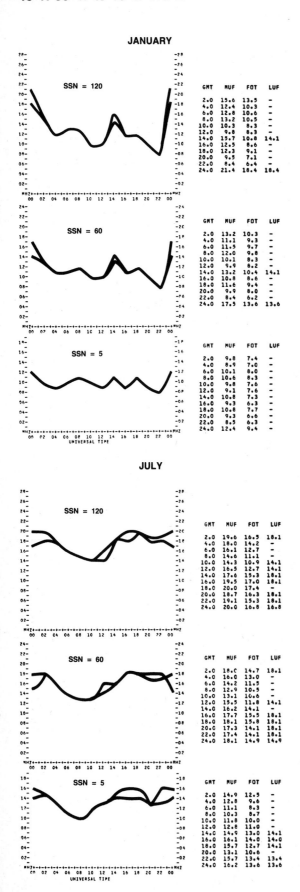

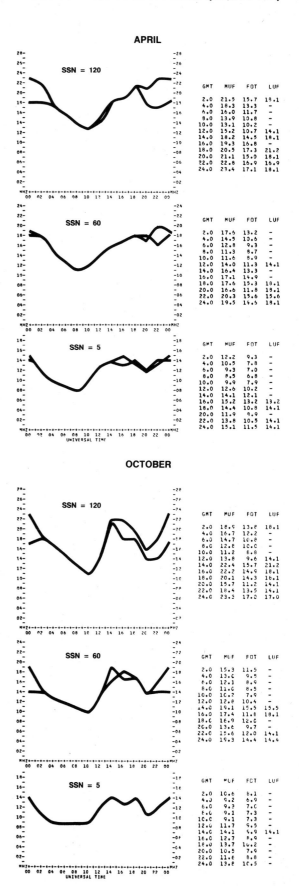

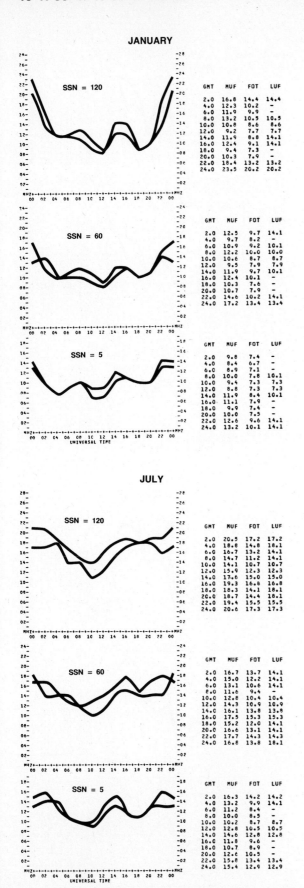

JANUARY

SSN = 120

GMT	MUF	FOT	LUF
2.0	16.8	14.4	14.4
4.0	12.3	10.2	–
6.0	11.9	9.9	–
8.0	13.2	10.5	10.5
10.0	10.8	8.6	8.6
12.0	9.2	7.7	7.7
14.0	11.9	8.8	14.1
16.0	12.4	9.1	14.1
18.0	9.4	7.3	–
20.0	10.3	7.9	–
22.0	18.4	13.2	13.2
24.0	23.5	20.2	20.2

SSN = 60

GMT	MUF	FOT	LUF
2.0	12.5	9.7	14.1
4.0	9.7	8.2	–
6.0	10.9	9.2	10.1
8.0	12.2	10.0	10.0
10.0	10.6	8.7	8.7
12.0	9.5	7.9	7.9
14.0	11.9	9.7	10.1
16.0	12.4	10.1	–
18.0	10.3	7.6	–
20.0	10.7	7.9	–
22.0	14.6	10.2	14.1
24.0	17.2	13.4	13.4

SSN = 5

GMT	MUF	FOT	LUF
2.0	9.8	7.4	–
4.0	8.4	6.7	–
6.0	8.9	7.1	–
8.0	10.0	7.8	10.1
10.0	9.4	7.3	7.3
12.0	8.8	7.3	7.3
14.0	11.9	8.4	10.1
16.0	11.1	7.9	–
18.0	9.9	7.4	–
20.0	10.0	7.5	–
22.0	12.6	9.6	14.1
24.0	13.2	10.1	14.1

UNIVERSAL TIME

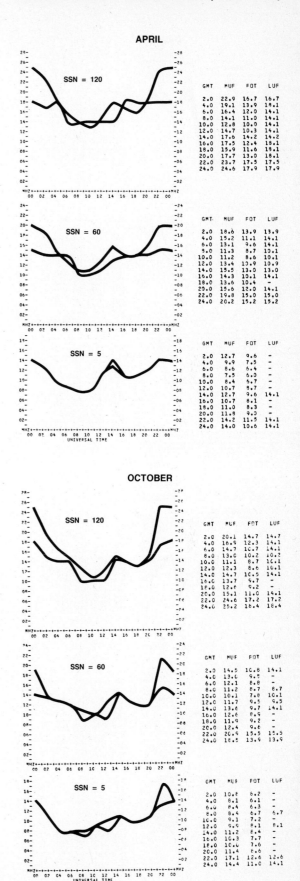

APRIL

SSN = 120

GMT	MUF	FOT	LUF
2.0	22.9	16.7	16.7
4.0	19.1	13.9	18.1
6.0	16.4	12.0	14.1
8.0	14.1	11.0	14.1
10.0	12.8	10.0	14.1
12.0	14.7	10.3	14.1
14.0	17.6	14.2	14.2
16.0	17.5	12.4	18.1
18.0	15.9	11.6	18.1
20.0	17.7	13.0	18.1
22.0	23.7	17.5	17.5
24.0	24.6	17.9	17.9

SSN = 60

GMT	MUF	FOT	LUF
2.0	18.6	13.9	13.9
4.0	15.2	11.1	14.1
6.0	13.1	9.6	14.1
8.0	11.3	8.7	10.1
10.0	11.2	8.6	10.1
12.0	13.4	13.9	10.9
14.0	15.5	13.0	13.0
16.0	14.3	10.1	14.1
18.0	13.6	10.4	–
20.0	15.6	12.0	14.1
22.0	19.8	15.0	15.0
24.0	20.2	15.2	15.2

SSN = 5

GMT	MUF	FOT	LUF
2.0	12.7	9.6	–
4.0	9.9	7.5	–
6.0	8.6	6.4	–
8.0	7.5	6.0	–
10.0	8.4	5.7	–
12.0	10.7	8.7	–
14.0	12.7	9.6	14.1
16.0	10.7	8.1	–
18.0	11.0	8.3	–
20.0	11.8	9.0	–
22.0	14.2	11.5	14.1
24.0	14.0	10.6	14.1

UNIVERSAL TIME

JULY

SSN = 120

GMT	MUF	FOT	LUF
2.0	20.5	17.2	17.2
4.0	18.8	14.8	18.1
6.0	16.7	13.2	14.1
8.0	14.7	11.2	14.1
10.0	14.1	10.7	10.7
12.0	15.9	12.3	12.3
14.0	17.6	15.0	15.0
16.0	19.3	16.8	16.8
18.0	18.3	14.1	18.1
20.0	18.7	14.4	18.1
22.0	19.4	15.5	15.5
24.0	20.6	17.3	17.3

SSN = 60

GMT	MUF	FOT	LUF
2.0	16.7	13.7	14.1
4.0	15.0	12.2	14.1
6.0	13.1	10.6	14.1
8.0	11.6	9.4	–
10.0	12.8	10.4	10.4
12.0	14.3	10.9	10.9
14.0	16.1	13.8	13.8
16.0	17.5	15.3	15.3
18.0	15.2	12.0	14.1
20.0	16.6	13.1	14.1
22.0	17.7	14.3	14.3
24.0	16.8	13.8	18.1

SSN = 5

GMT	MUF	FOT	LUF
2.0	16.3	14.2	14.2
4.0	13.2	9.9	14.1
6.0	11.2	8.4	–
8.0	10.0	8.5	–
10.0	10.2	8.7	8.7
12.0	12.8	10.5	10.5
14.0	14.6	12.8	12.8
16.0	11.8	9.6	–
18.0	10.7	8.9	–
20.0	12.6	10.5	–
22.0	15.8	13.4	13.4
24.0	15.4	12.9	12.9

UNIVERSAL TIME

OCTOBER

SSN = 120

GMT	MUF	FOT	LUF
2.0	20.1	14.7	14.7
4.0	16.9	12.3	14.1
6.0	14.7	10.7	14.1
8.0	13.0	10.2	10.2
10.0	11.1	8.7	14.1
12.0	12.3	8.6	10.1
14.0	14.7	10.5	14.1
16.0	13.7	9.7	–
18.0	12.6	9.6	–
20.0	15.1	11.0	14.1
22.0	24.6	17.2	17.2
24.0	25.2	18.4	18.4

SSN = 60

GMT	MUF	FOT	LUF
2.0	14.5	10.8	14.1
4.0	13.0	9.5	–
6.0	12.1	8.8	–
8.0	11.2	8.7	8.7
10.0	10.1	7.8	10.1
12.0	11.7	9.5	9.5
14.0	13.6	9.7	14.1
16.0	12.6	8.9	–
18.0	11.9	9.2	–
20.0	12.4	9.6	–
22.0	20.9	15.5	15.5
24.0	18.5	13.9	13.9

SSN = 5

GMT	MUF	FOT	LUF
2.0	10.8	8.2	–
4.0	8.1	6.1	–
6.0	8.4	6.3	–
8.0	8.4	6.7	6.7
10.0	9.1	7.2	–
12.0	9.6	8.1	8.1
14.0	11.2	8.4	–
16.0	10.3	7.7	–
18.0	10.0	7.6	–
20.0	11.4	8.6	–
22.0	17.1	12.6	12.6
24.0	14.4	11.0	14.1

UNIVERSAL TIME

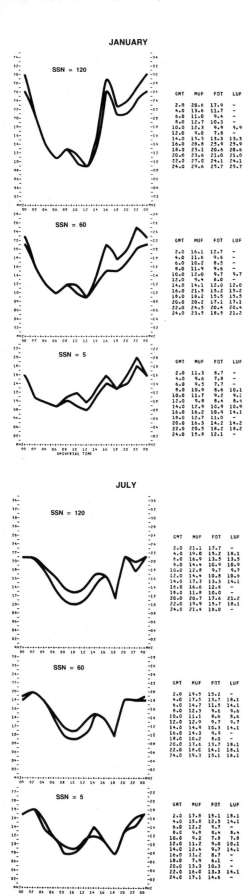

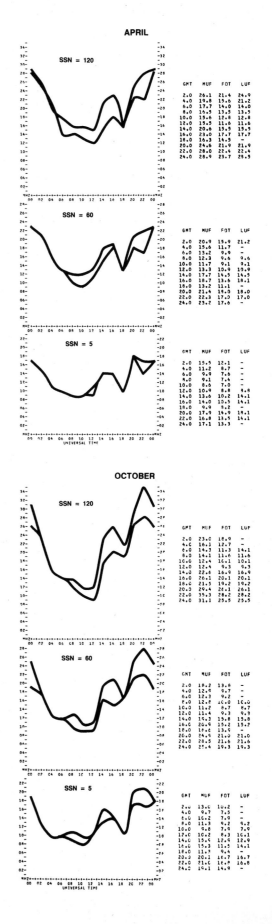

JANUARY

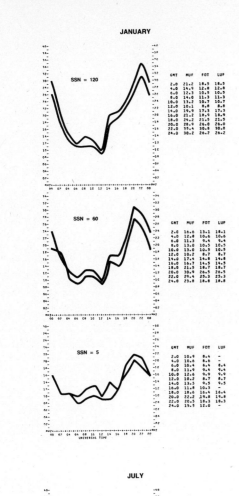

APRIL

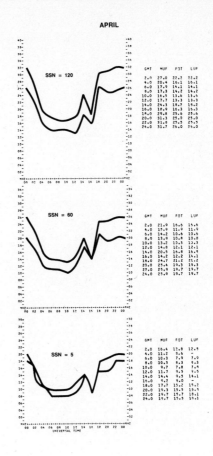

JULY

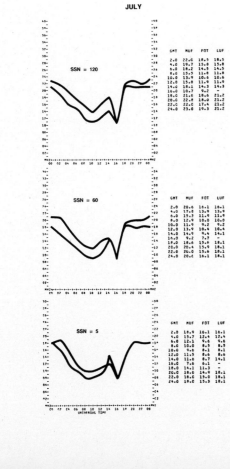

OCTOBER

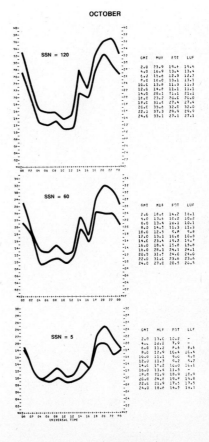

JANUARY

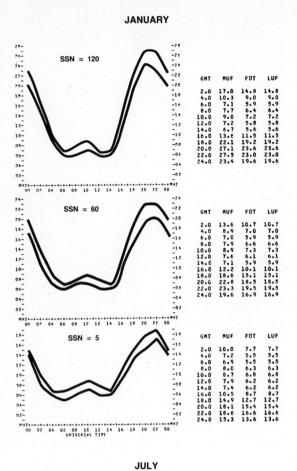

SSN = 120

GMT	MUF	FOT	LUF
2.0	17.0	14.8	14.8
4.0	10.3	9.0	9.0
6.0	7.1	5.9	5.9
8.0	7.7	6.4	6.4
10.0	9.0	7.2	7.2
12.0	7.2	5.8	5.8
14.0	6.7	5.6	5.6
16.0	13.6	11.5	11.5
18.0	22.1	19.2	19.2
20.0	27.1	23.6	23.6
22.0	27.3	23.0	23.0
24.0	23.4	19.6	19.6

SSN = 60

GMT	MUF	FOT	LUF
2.0	13.6	10.7	10.7
4.0	8.9	7.0	7.0
6.0	7.0	5.9	5.9
8.0	7.9	6.6	6.6
10.0	8.9	7.3	7.3
12.0	7.6	6.1	6.1
14.0	7.1	5.9	5.9
16.0	12.2	10.1	10.1
18.0	18.6	15.1	15.1
20.0	22.8	18.5	18.5
22.0	23.3	19.5	19.5
24.0	19.6	16.9	16.9

SSN = 5

GMT	MUF	FOT	LUF
2.0	10.0	7.7	7.7
4.0	7.2	5.5	5.5
6.0	6.9	5.5	5.5
8.0	8.0	6.3	6.3
10.0	8.7	6.8	6.8
12.0	7.9	6.2	6.2
14.0	7.4	6.2	6.2
16.0	10.5	8.7	8.7
18.0	14.9	12.7	12.7
20.0	18.1	15.4	15.4
22.0	18.6	16.6	16.6
24.0	15.3	13.6	13.6

APRIL

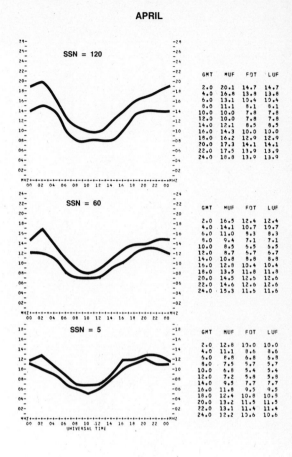

SSN = 120

GMT	MUF	FOT	LUF
2.0	20.1	14.7	14.7
4.0	16.8	13.8	13.8
6.0	13.1	10.4	10.4
8.0	11.1	8.1	8.1
10.0	10.0	7.8	7.8
12.0	10.0	7.8	7.8
14.0	12.1	8.5	8.5
16.0	14.3	10.0	10.0
18.0	16.2	12.9	12.9
20.0	17.3	14.1	14.1
22.0	17.5	13.9	13.9
24.0	18.8	13.9	13.9

SSN = 60

GMT	MUF	FOT	LUF
2.0	16.5	12.4	12.4
4.0	14.1	10.7	10.7
6.0	11.0	8.3	8.3
8.0	9.4	7.1	7.1
10.0	8.5	6.5	6.5
12.0	8.7	6.7	6.7
14.0	10.8	8.8	8.8
16.0	12.8	10.4	10.4
18.0	13.5	11.8	11.8
20.0	14.5	12.5	12.5
22.0	14.6	12.6	12.6
24.0	15.3	11.5	11.6

SSN = 5

GMT	MUF	FOT	LUF
2.0	12.8	10.0	10.0
4.0	11.1	8.6	8.6
6.0	8.8	6.8	6.8
8.0	7.5	5.7	5.7
10.0	6.8	5.4	5.4
12.0	7.2	5.8	5.8
14.0	9.5	7.7	7.7
16.0	11.8	9.5	9.5
18.0	12.4	10.8	10.8
20.0	13.2	11.5	11.5
22.0	13.1	11.4	11.4
24.0	12.2	13.6	10.6

JULY

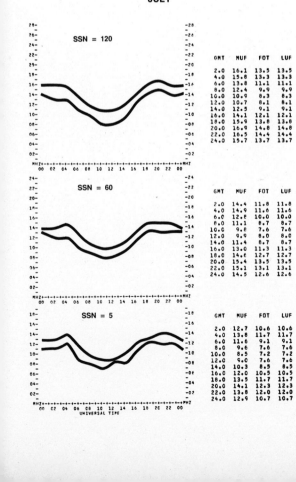

SSN = 120

GMT	MUF	FOT	LUF
2.0	16.1	13.5	13.5
4.0	15.8	13.3	13.3
6.0	13.8	11.1	11.1
8.0	12.4	9.9	9.9
10.0	10.9	8.3	8.3
12.0	10.7	8.1	8.1
14.0	12.5	9.1	9.1
16.0	14.1	12.1	12.1
18.0	15.9	13.8	13.8
20.0	16.9	14.8	14.8
22.0	16.5	14.4	14.4
24.0	15.7	13.7	13.7

SSN = 60

GMT	MUF	FOT	LUF
2.0	14.4	11.8	11.8
4.0	14.9	11.6	11.6
6.0	12.8	10.0	10.0
8.0	11.1	8.7	8.7
10.0	9.8	7.6	7.6
12.0	9.9	8.0	8.0
14.0	11.4	8.7	8.7
16.0	13.0	11.3	11.3
18.0	14.7	12.7	12.7
20.0	15.4	13.5	13.5
22.0	15.1	13.1	13.1
24.0	14.5	12.6	12.6

SSN = 5

GMT	MUF	FOT	LUF
2.0	12.7	10.6	10.6
4.0	13.8	11.7	11.7
6.0	11.6	9.1	9.1
8.0	9.6	7.6	7.6
10.0	8.5	7.2	7.2
12.0	9.0	7.6	7.6
14.0	10.3	8.5	8.5
16.0	12.0	10.5	10.5
18.0	13.5	11.7	11.7
20.0	14.1	12.3	12.3
22.0	13.8	12.0	12.0
24.0	12.9	10.7	10.7

OCTOBER

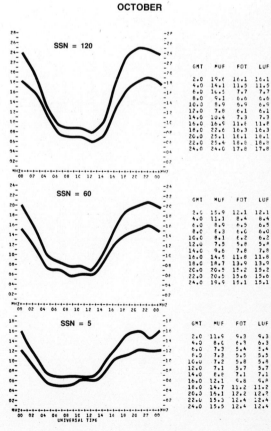

SSN = 120

GMT	MUF	FOT	LUF
2.0	19.6	16.1	16.1
4.0	14.1	11.5	11.5
6.0	10.5	7.7	7.7
8.0	9.1	6.6	6.6
10.0	8.9	6.9	6.9
12.0	7.8	6.1	6.1
14.0	10.4	7.3	7.3
16.0	16.9	11.8	11.8
18.0	22.6	16.3	16.3
20.0	25.1	18.1	18.1
22.0	25.4	18.5	18.5
24.0	24.0	17.8	17.8

SSN = 60

GMT	MUF	FOT	LUF
2.0	15.9	12.1	12.1
4.0	11.1	8.4	8.4
6.0	8.9	6.5	6.5
8.0	8.3	6.6	6.0
10.0	8.1	6.2	6.2
12.0	7.5	5.9	5.9
14.0	9.6	7.8	7.8
16.0	14.5	11.8	11.8
18.0	18.7	13.9	13.9
20.0	20.5	15.2	15.2
22.0	20.5	15.6	15.6
24.0	19.9	15.1	15.1

SSN = 5

GMT	MUF	FOT	LUF
2.0	11.9	9.3	9.3
4.0	8.0	6.3	6.3
6.0	7.3	5.4	5.4
8.0	7.3	5.5	5.5
10.0	7.2	5.8	5.8
12.0	7.1	5.7	5.7
14.0	8.8	7.1	7.1
16.0	12.1	5.8	9.8
18.0	14.7	11.2	11.2
20.0	15.2	12.2	12.2
22.0	15.3	12.4	12.4
24.0	15.5	12.4	12.4

JANUARY

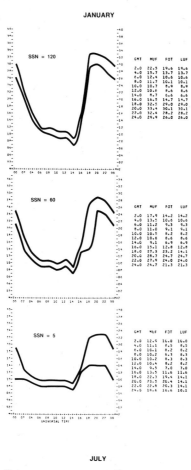

APRIL

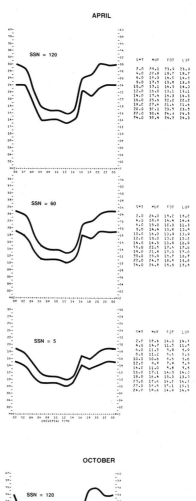

JULY

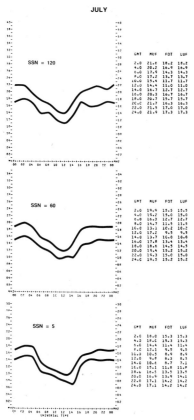

OCTOBER

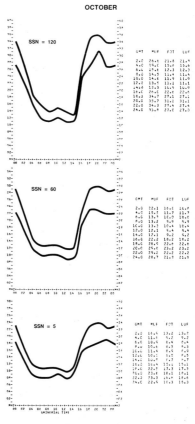

Central USA to Caribbean
38°N 98°W to 10°N 80°W

JANUARY

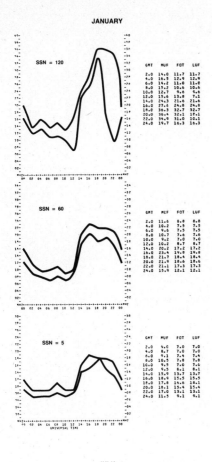

APRIL

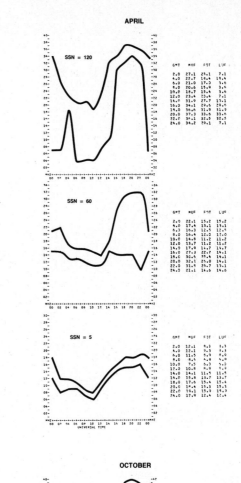

JULY

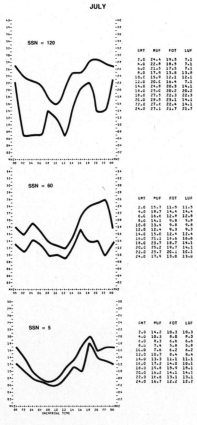

OCTOBER

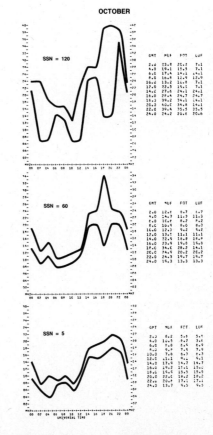

JANUARY

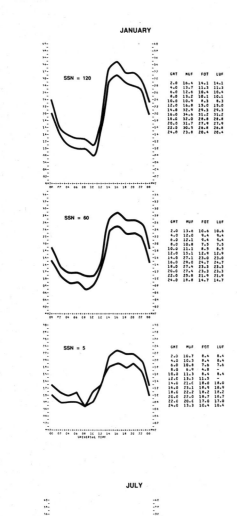

APRIL

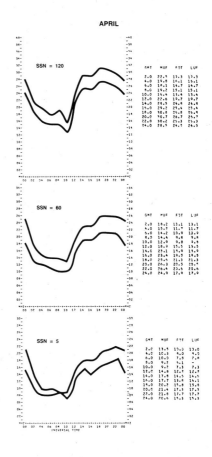

JULY

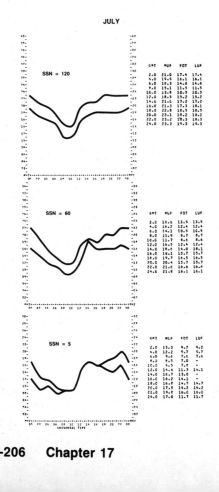

OCTOBER

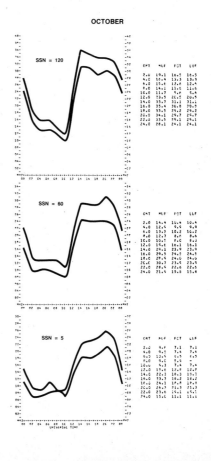

JANUARY

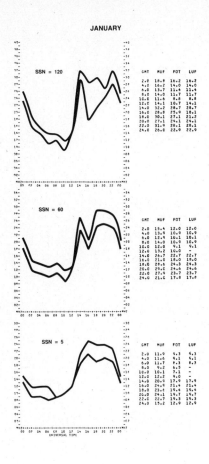

APRIL

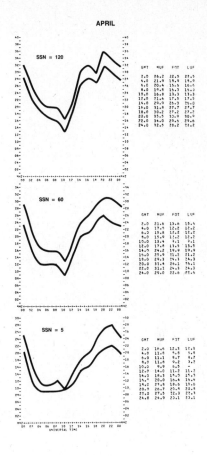

JULY

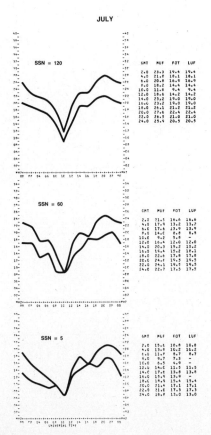

OCTOBER

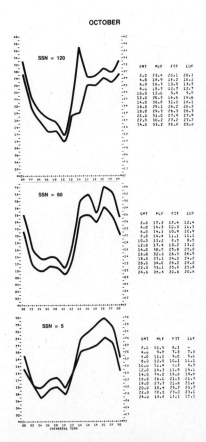

JANUARY

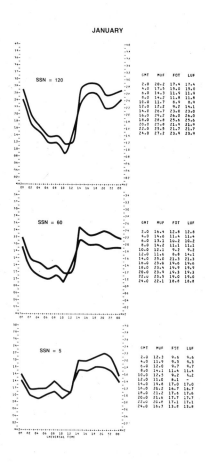

APRIL

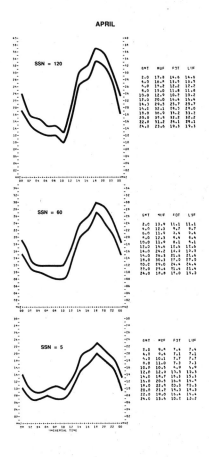

JULY

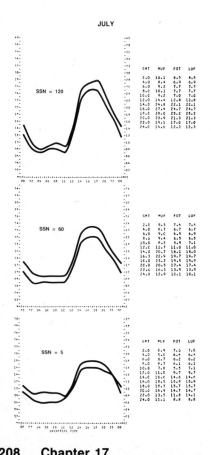

OCTOBER

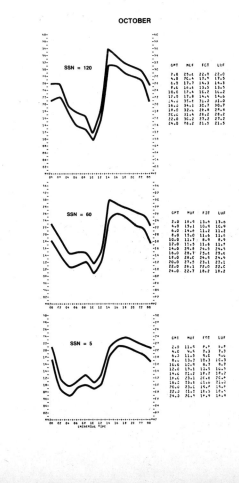

JANUARY

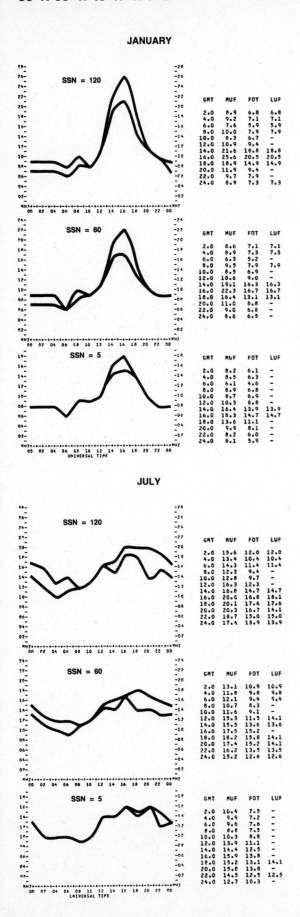

SSN = 120

GMT	MUF	FOT	LUF
2.0	8.9	6.8	6.8
4.0	9.2	7.1	7.1
6.0	7.6	5.9	5.9
8.0	10.0	7.9	7.9
10.0	8.3	6.7	–
12.0	10.9	9.4	–
14.0	21.6	18.8	18.8
16.0	25.6	20.5	20.5
18.0	18.9	14.9	14.9
20.0	11.9	9.4	–
22.0	9.7	7.9	–
24.0	8.9	7.3	7.3

SSN = 60

GMT	MUF	FOT	LUF
2.0	8.6	7.1	7.1
4.0	8.9	7.3	7.3
6.0	6.3	5.2	–
8.0	9.5	7.9	7.9
10.0	8.5	6.9	–
12.0	10.6	9.0	–
14.0	19.1	16.3	16.3
16.0	22.3	16.7	16.7
18.0	16.4	13.1	13.1
20.0	11.0	8.8	–
22.0	9.0	6.8	–
24.0	8.6	6.5	–

SSN = 5

GMT	MUF	FOT	LUF
2.0	8.2	6.1	–
4.0	8.5	6.3	–
6.0	6.1	4.6	–
8.0	8.9	6.8	–
10.0	8.7	6.9	–
12.0	10.3	8.8	–
14.0	16.4	13.9	13.9
16.0	18.3	14.7	14.7
18.0	13.6	11.1	–
20.0	9.9	8.1	–
22.0	8.2	6.0	–
24.0	8.1	5.9	–

UNIVERSAL TIME

APRIL

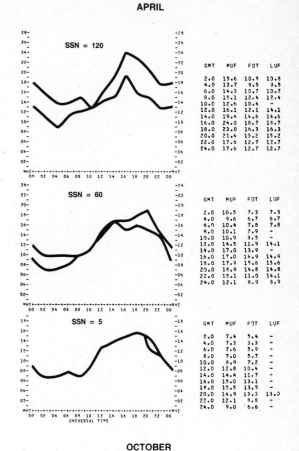

SSN = 120

GMT	MUF	FOT	LUF
2.0	15.6	10.8	10.8
4.0	13.7	9.5	9.5
6.0	14.3	10.7	10.7
8.0	15.1	12.4	12.4
10.0	12.6	10.4	–
12.0	16.1	12.1	14.1
14.0	19.4	14.6	14.6
16.0	24.0	18.7	18.7
18.0	23.0	16.3	16.3
20.0	21.4	15.2	15.2
22.0	17.5	12.7	12.7
24.0	17.6	12.7	12.7

SSN = 60

GMT	MUF	FOT	LUF
2.0	10.5	7.3	7.3
4.0	9.6	6.7	5.7
6.0	10.4	7.8	7.8
8.0	10.1	7.9	–
10.0	10.9	8.5	–
12.0	14.5	11.9	14.1
14.0	17.0	13.9	–
16.0	17.0	14.9	14.9
18.0	17.9	15.6	15.6
20.0	18.9	14.8	14.8
22.0	15.1	11.0	14.1
24.0	12.1	8.9	8.9

SSN = 5

GMT	MUF	FOT	LUF
2.0	7.4	5.4	–
4.0	7.3	5.3	–
6.0	7.6	5.9	–
8.0	7.0	5.7	–
10.0	8.9	7.2	–
12.0	12.8	10.4	–
14.0	14.4	11.7	–
16.0	15.0	13.1	–
18.0	15.5	13.5	–
20.0	14.9	13.0	13.0
22.0	12.1	9.8	–
24.0	9.0	6.6	–

UNIVERSAL TIME

JULY

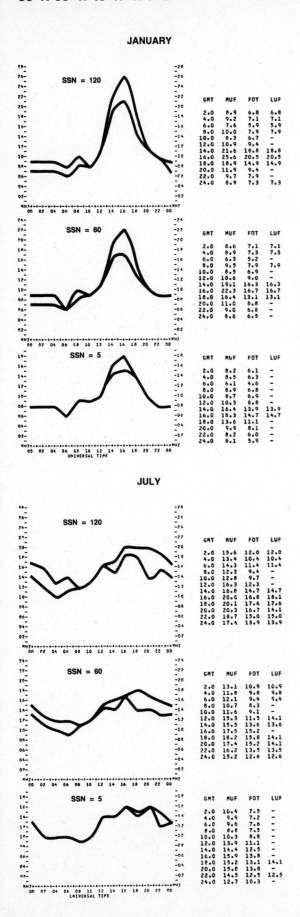

SSN = 120

GMT	MUF	FOT	LUF
2.0	15.6	12.0	12.0
4.0	13.4	10.4	10.4
6.0	14.3	11.4	11.4
8.0	12.3	9.4	–
10.0	12.8	9.7	–
12.0	16.3	12.3	–
14.0	16.8	14.7	14.7
16.0	20.0	16.8	18.1
18.0	20.1	17.6	17.6
20.0	20.3	16.7	14.1
22.0	18.7	15.0	15.0
24.0	17.4	13.9	13.9

SSN = 60

GMT	MUF	FOT	LUF
2.0	13.1	10.9	10.9
4.0	11.8	9.8	9.8
6.0	12.1	9.4	9.4
8.0	10.7	8.3	–
10.0	11.6	9.1	–
12.0	15.3	11.5	14.1
14.0	15.5	13.6	13.6
16.0	17.5	15.2	–
18.0	18.2	15.8	14.1
20.0	17.4	15.2	14.1
22.0	16.2	13.5	13.5
24.0	15.2	12.6	12.6

SSN = 5

GMT	MUF	FOT	LUF
2.0	10.4	7.5	–
4.0	9.9	7.2	–
6.0	9.6	7.6	–
8.0	8.8	7.5	–
10.0	10.3	8.8	–
12.0	13.9	11.1	–
14.0	14.4	12.5	–
16.0	15.9	13.8	–
18.0	15.2	13.1	14.1
20.0	15.8	13.8	–
22.0	14.5	12.5	12.5
24.0	12.7	10.3	–

UNIVERSAL TIME

OCTOBER

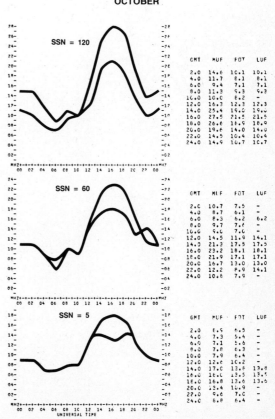

SSN = 120

GMT	MUF	FOT	LUF
2.0	14.6	10.1	10.1
4.0	11.7	8.1	8.1
6.0	9.4	7.1	7.1
8.0	11.3	9.3	9.3
10.0	10.0	8.2	–
12.0	16.3	12.3	12.3
14.0	25.4	19.0	19.0
16.0	27.5	21.5	21.5
18.0	26.6	16.9	18.9
20.0	19.6	14.0	14.0
22.0	14.5	10.4	10.4
24.0	14.9	10.7	10.7

SSN = 60

GMT	MUF	FOT	LUF
2.0	10.7	7.5	–
4.0	8.7	6.1	–
6.0	8.3	6.2	6.2
8.0	9.7	7.6	–
10.0	9.7	7.0	–
12.0	14.5	11.9	14.1
14.0	21.3	17.5	17.5
16.0	23.2	18.1	18.1
18.0	21.9	17.1	17.1
20.0	16.7	13.0	13.0
22.0	12.2	8.9	14.1
24.0	10.6	7.9	–

SSN = 5

GMT	MUF	FOT	LUF
2.0	8.9	6.5	–
4.0	7.3	5.4	–
6.0	7.1	5.6	–
8.0	7.8	6.3	–
10.0	7.9	6.4	–
12.0	12.6	10.2	–
14.0	17.0	13.8	13.8
16.0	18.0	13.5	13.5
18.0	16.8	13.6	13.5
20.0	12.4	10.9	–
22.0	9.6	7.0	–
24.0	6.8	6.4	–

UNIVERSAL TIME

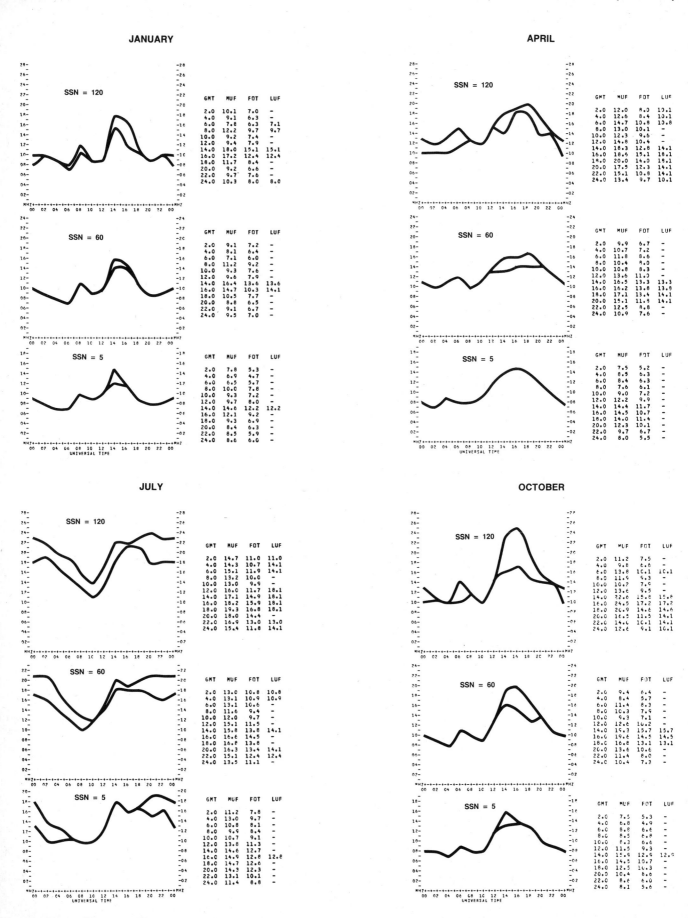

JANUARY

APRIL

JULY

OCTOBER

JANUARY

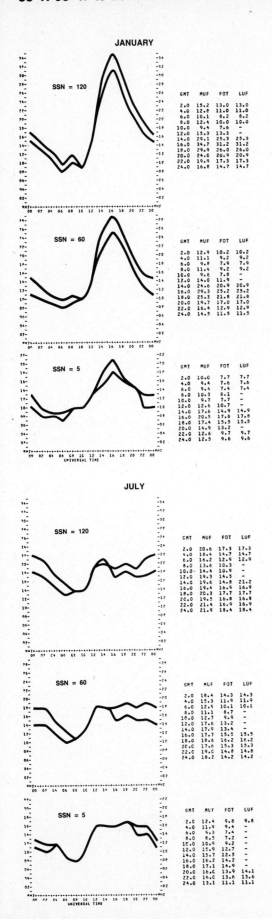

SSN = 120

GMT	MUF	FOT	LUF
2.0	15.2	13.0	13.0
4.0	12.8	11.0	11.0
6.0	10.1	8.2	8.2
8.0	12.4	10.0	10.0
10.0	9.4	7.6	-
12.0	15.3	13.3	-
14.0	29.1	25.3	25.3
16.0	34.7	31.2	31.2
18.0	29.9	26.0	26.0
20.0	24.0	20.9	20.9
22.0	19.9	17.3	17.3
24.0	16.8	14.7	14.7

SSN = 60

GMT	MUF	FOT	LUF
2.0	12.9	10.2	10.2
4.0	11.1	9.2	9.2
6.0	9.8	7.9	7.9
8.0	11.4	9.2	9.2
10.0	9.6	7.8	-
12.0	14.0	11.9	-
14.0	24.6	20.9	20.9
16.0	29.3	25.2	25.2
18.0	25.3	21.8	21.8
20.0	19.7	17.0	17.0
22.0	16.4	12.9	12.9
24.0	14.5	11.5	11.5

SSN = 5

GMT	MUF	FOT	LUF
2.0	10.0	7.7	7.7
4.0	9.4	7.6	7.6
6.0	9.4	7.4	7.4
8.0	10.3	8.1	-
10.0	9.7	7.7	-
12.0	12.6	10.7	-
14.0	17.6	14.9	14.9
16.0	20.5	17.8	17.8
18.0	17.4	15.5	15.5
20.0	14.9	13.2	-
22.0	12.6	9.7	9.7
24.0	12.5	9.6	9.6

JULY

SSN = 120

GMT	MUF	FOT	LUF
2.0	20.6	17.3	17.3
4.0	18.4	14.7	14.7
6.0	16.2	12.9	12.9
8.0	13.6	10.3	-
10.0	14.4	10.9	-
12.0	19.3	14.5	-
14.0	19.6	14.8	21.2
16.0	19.6	16.9	16.9
18.0	20.3	17.7	17.7
20.0	19.5	16.8	16.8
22.0	21.4	16.9	16.9
24.0	21.9	18.4	18.4

SSN = 60

GMT	MUF	FOT	LUF
2.0	18.4	14.3	14.3
4.0	15.3	11.9	11.9
6.0	12.6	10.1	10.1
8.0	11.1	8.7	-
10.0	12.7	9.9	-
12.0	17.6	13.2	-
14.0	17.9	13.4	-
16.0	17.7	15.5	15.5
18.0	18.6	16.2	16.2
20.0	17.6	15.3	15.3
22.0	19.0	14.8	14.8
24.0	18.2	14.2	14.2

SSN = 5

GMT	MUF	FOT	LUF
2.0	12.4	9.8	9.8
4.0	11.9	9.4	-
6.0	9.3	7.4	-
8.0	8.5	7.2	-
10.0	10.9	9.2	-
12.0	15.9	12.7	-
14.0	15.7	12.5	-
16.0	16.2	14.2	-
18.0	17.1	14.9	-
20.0	16.6	13.9	14.1
22.0	16.0	13.6	13.6
24.0	13.1	11.1	11.1

APRIL

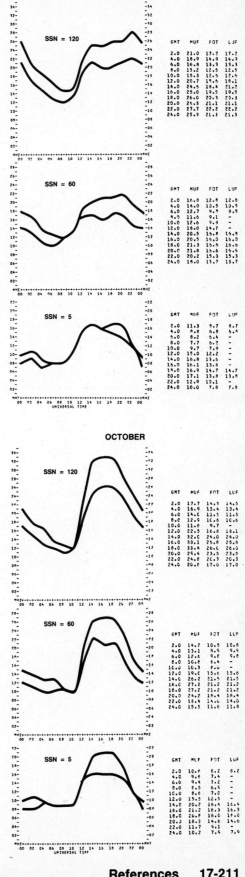

SSN = 120

GMT	MUF	FOT	LUF
2.0	21.0	17.2	17.2
4.0	18.4	14.9	14.9
6.0	16.8	13.3	13.3
8.0	15.2	12.5	12.5
10.0	15.3	12.5	12.6
12.0	20.7	15.6	18.1
14.0	24.5	18.4	21.2
16.0	25.0	19.5	19.5
18.0	26.0	20.3	20.3
20.0	24.3	21.1	21.1
22.0	27.7	22.2	22.2
24.0	25.9	21.3	21.3

SSN = 60

GMT	MUF	FOT	LUF
2.0	16.8	12.8	12.8
4.0	14.0	13.5	13.5
6.0	12.7	9.5	9.5
8.0	11.6	9.1	-
10.0	12.6	9.9	-
12.0	18.0	14.7	-
14.0	20.5	15.8	15.8
16.0	20.5	15.3	15.0
18.0	21.3	15.6	15.6
20.0	21.8	15.6	15.6
22.0	20.2	15.3	15.3
24.0	18.0	13.7	13.7

SSN = 5

GMT	MUF	FOT	LUF
2.0	11.3	9.7	8.7
4.0	8.8	6.8	6.8
6.0	8.2	6.4	-
8.0	7.7	6.2	-
10.0	9.7	7.9	-
12.0	15.0	12.2	-
14.0	16.8	13.6	-
16.0	16.1	13.8	-
18.0	16.9	14.7	14.7
20.0	17.1	13.8	13.8
22.0	12.9	10.1	-
24.0	10.0	7.8	7.8

OCTOBER

SSN = 120

GMT	MUF	FOT	LUF
2.0	17.7	14.5	14.5
4.0	16.9	13.4	13.4
6.0	14.0	11.5	11.5
8.0	12.9	10.6	10.6
10.0	11.8	9.7	-
12.0	22.5	16.8	18.1
14.0	32.6	24.0	24.0
16.0	33.1	25.8	25.8
18.0	33.4	26.0	26.0
20.0	26.4	23.5	23.5
22.0	24.8	20.3	20.3
24.0	20.8	17.0	17.0

SSN = 60

GMT	MUF	FOT	LUF
2.0	14.2	10.9	10.8
4.0	13.1	9.5	9.5
6.0	12.6	9.8	9.8
8.0	10.8	8.4	-
10.0	10.3	8.0	-
12.0	19.0	15.6	15.6
14.0	26.2	21.5	21.5
16.0	27.2	21.2	21.2
18.0	27.2	21.2	21.2
20.0	24.2	18.4	18.4
22.0	18.4	14.0	14.0
24.0	15.3	11.8	11.8

SSN = 5

GMT	MUF	FOT	LUF
2.0	10.6	8.2	8.2
4.0	9.6	7.4	-
6.0	9.4	7.2	-
8.0	8.5	6.5	-
10.0	8.6	7.0	-
12.0	15.5	12.5	-
14.0	20.2	16.4	16.4
16.0	21.2	16.3	16.3
18.0	20.8	16.0	16.0
20.0	18.3	14.6	14.6
22.0	11.7	9.1	-
24.0	10.2	7.9	7.9

JANUARY

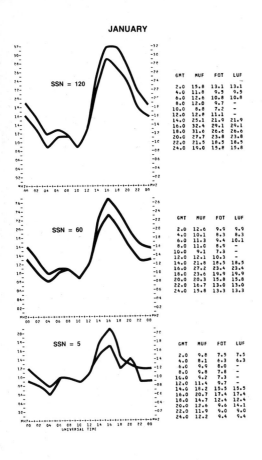

SSN = 120

GMT	MUF	FOT	LUF
2.0	15.8	13.1	13.1
4.0	11.8	9.5	9.5
6.0	12.6	10.8	10.8
8.0	12.0	9.7	–
10.0	8.8	7.2	–
12.0	12.8	11.1	–
14.0	25.1	21.9	21.9
16.0	32.4	29.1	29.1
18.0	31.6	26.6	26.6
20.0	27.7	23.8	23.8
22.0	21.5	18.5	18.5
24.0	19.0	15.8	15.8

SSN = 60

GMT	MUF	FOT	LUF
2.0	12.6	9.9	9.9
4.0	10.1	8.3	8.3
6.0	11.3	9.4	10.1
8.0	11.0	8.9	–
10.0	9.1	7.3	–
12.0	12.1	10.3	–
14.0	21.8	18.5	18.5
16.0	27.2	23.4	23.4
18.0	23.6	19.9	19.9
20.0	20.3	15.8	15.8
22.0	16.7	13.0	13.0
24.0	15.8	13.3	13.3

SSN = 5

GMT	MUF	FOT	LUF
2.0	9.8	7.5	7.5
4.0	8.1	6.3	6.3
6.0	9.9	8.0	–
8.0	9.8	7.8	–
10.0	9.2	7.3	–
12.0	11.4	9.7	–
14.0	18.2	15.5	15.5
16.0	20.7	17.4	17.4
18.0	14.7	12.4	12.4
20.0	12.6	9.6	14.1
22.0	11.9	9.0	9.0
24.0	12.2	9.4	9.4

APRIL

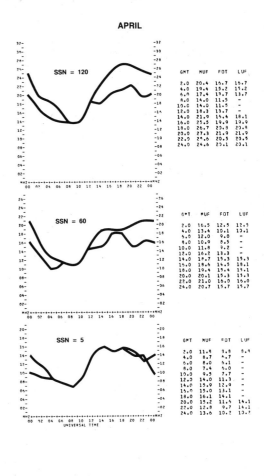

SSN = 120

GMT	MUF	FOT	LUF
2.0	20.4	16.7	16.7
4.0	19.4	15.2	15.2
6.0	17.4	13.7	13.7
8.0	14.0	11.5	–
10.0	14.0	11.5	–
12.0	18.3	13.7	–
14.0	21.9	15.4	18.1
16.0	25.5	19.9	19.9
18.0	26.7	21.9	20.9
20.0	27.3	21.9	21.9
22.0	25.6	20.5	20.5
24.0	24.6	20.1	20.1

SSN = 60

GMT	MUF	FOT	LUF
2.0	16.5	12.5	12.5
4.0	13.4	10.1	13.1
6.0	12.0	9.0	–
8.0	10.9	8.5	–
10.0	11.8	9.2	–
12.0	16.2	13.3	–
14.0	18.7	15.3	15.3
16.0	18.6	14.5	18.1
18.0	19.4	19.1	19.1
20.0	20.1	15.3	15.3
22.0	21.0	16.0	16.0
24.0	20.7	15.7	15.7

SSN = 5

GMT	MUF	FOT	LUF
2.0	11.9	9.9	8.9
4.0	8.7	6.7	–
6.0	8.0	6.1	–
8.0	7.4	5.0	–
10.0	9.5	7.7	–
12.0	14.0	11.3	–
14.0	15.9	12.9	–
16.0	15.0	13.1	–
18.0	16.1	14.1	–
20.0	15.2	11.4	14.1
22.0	12.8	9.7	14.1
24.0	13.6	10.2	13.2

JULY

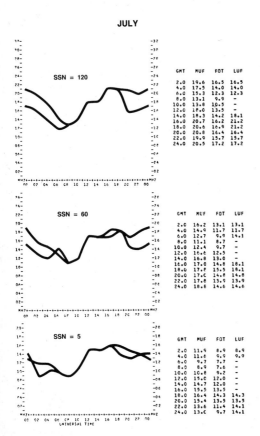

SSN = 120

GMT	MUF	FOT	LUF
2.0	19.6	16.5	16.5
4.0	17.5	14.0	14.0
6.0	15.3	12.3	12.3
8.0	13.1	9.9	–
10.0	13.8	10.5	–
12.0	18.0	13.5	–
14.0	18.3	14.2	18.1
16.0	20.7	16.2	21.2
18.0	20.6	16.9	21.2
20.0	20.8	16.4	16.4
22.0	19.9	15.7	15.7
24.0	20.5	17.2	17.2

SSN = 60

GMT	MUF	FOT	LUF
2.0	16.2	13.1	13.1
4.0	14.9	11.7	11.7
6.0	12.7	9.9	14.1
8.0	11.1	8.7	–
10.0	12.4	9.7	–
12.0	16.6	12.5	–
14.0	16.8	13.0	–
16.0	17.0	14.8	18.1
18.0	17.8	15.5	18.1
20.0	17.0	14.8	14.8
22.0	17.8	13.9	13.9
24.0	18.8	14.6	14.6

SSN = 5

GMT	MUF	FOT	LUF
2.0	11.9	8.9	8.9
4.0	11.6	9.9	9.9
6.0	9.7	7.7	–
8.0	11.1	8.7	–
10.0	10.8	9.2	–
12.0	15.0	12.0	–
14.0	14.7	12.0	–
16.0	13.5	13.0	–
18.0	16.4	14.3	14.3
20.0	15.4	13.5	14.8
22.0	13.6	11.4	14.1
24.0	13.0	9.7	14.1

OCTOBER

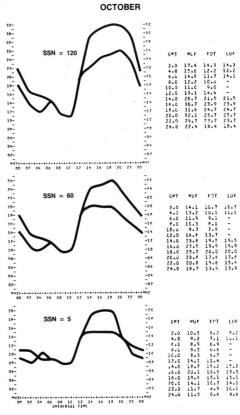

SSN = 120

GMT	MUF	FOT	LUF
2.0	17.4	14.3	14.3
4.0	15.6	12.2	12.2
6.0	14.8	11.7	14.1
8.0	12.2	10.0	–
10.0	11.0	9.0	–
12.0	19.1	14.4	–
14.0	28.7	21.5	21.5
16.0	30.7	23.9	23.9
18.0	31.6	24.7	24.7
20.0	32.1	25.7	25.7
22.0	29.7	23.7	23.7
24.0	22.4	18.4	18.4

SSN = 60

GMT	MUF	FOT	LUF
2.0	14.1	10.7	10.7
4.0	13.2	10.1	10.1
6.0	11.4	9.5	–
8.0	10.3	8.0	–
10.0	11.6	8.6	–
12.0	16.6	13.7	–
14.0	23.6	19.2	19.5
16.0	25.5	19.9	19.9
18.0	25.7	20.0	20.0
20.0	23.8	17.6	17.6
22.0	20.8	15.6	15.6
24.0	18.3	13.9	13.9

SSN = 5

GMT	MUF	FOT	LUF
2.0	10.5	8.2	8.2
4.0	9.2	7.1	10.1
6.0	8.9	6.9	–
8.0	8.2	6.6	–
10.0	8.3	6.7	–
12.0	14.1	11.4	–
14.0	18.7	15.2	15.2
16.0	20.1	15.5	15.5
18.0	19.6	15.1	15.1
20.0	14.1	10.7	14.1
22.0	11.7	8.9	10.1
24.0	11.5	8.6	8.6

Central USA to Southern Africa
38°N 98°W to 15°S 25°E

Azimuths: 82° 306°
Distance: 7506 nautical miles (13,890 km)

JANUARY

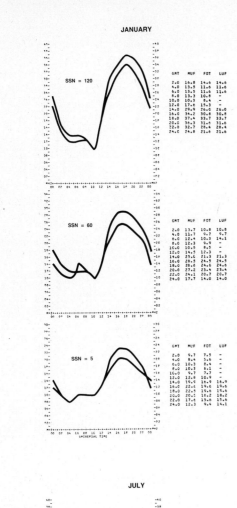

SSN = 120

GMT	MUF	FOT	LUF
2.0	16.8	14.6	14.6
4.0	13.5	11.6	11.6
6.0	13.5	11.6	11.6
8.0	13.3	10.8	-
10.0	10.3	8.4	-
12.0	17.6	15.3	-
14.0	29.4	26.0	26.0
16.0	34.2	30.8	30.8
18.0	37.4	33.7	33.7
20.0	36.3	31.6	31.6
22.0	32.7	28.4	28.4
24.0	24.8	21.6	21.6

SSN = 60

GMT	MUF	FOT	LUF
2.0	13.7	10.8	10.8
4.0	11.7	9.7	9.7
6.0	12.4	10.3	14.1
8.0	12.3	9.9	-
10.0	10.5	8.5	-
12.0	14.5	12.3	-
14.0	23.6	21.3	21.3
16.0	28.5	24.5	24.5
18.0	28.6	24.6	24.6
20.0	27.2	23.4	23.4
22.0	24.1	20.7	20.7
24.0	17.7	14.0	14.0

SSN = 5

GMT	MUF	FOT	LUF
2.0	9.7	7.5	-
4.0	8.4	5.6	-
6.0	10.3	8.4	-
8.0	10.3	8.1	-
16.0	9.7	7.7	-
12.0	12.8	10.9	-
14.0	19.6	16.9	16.9
16.0	22.6	19.6	19.6
20.0	20.5	18.2	18.2
22.0	17.6	15.6	15.6
24.0	12.3	9.4	14.1

APRIL

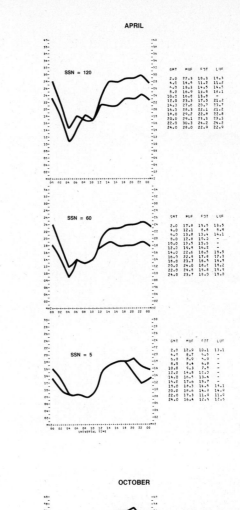

SSN = 120

GMT	MUF	FOT	LUF
2.0	22.3	18.3	19.3
4.0	14.5	11.2	11.2
6.0	18.3	14.5	14.5
8.0	16.9	13.8	18.1
10.0	16.8	13.8	-
12.0	23.3	17.5	21.7
14.0	27.6	23.7	23.7
16.0	28.3	22.1	22.1
18.0	29.7	22.8	22.8
20.0	29.1	23.3	23.3
22.0	30.3	24.2	24.2
24.0	28.0	22.9	22.9

SSN = 60

GMT	MUF	FOT	LUF
2.0	17.8	13.5	13.5
4.0	12.1	8.8	9.4
6.0	13.8	10.3	14.1
8.0	13.5	10.5	-
10.0	13.5	10.5	-
12.0	19.4	14.2	-
14.0	22.6	18.5	19.5
16.0	23.7	18.4	17.5
18.0	23.9	17.8	17.5
20.0	24.0	18.2	18.2
22.0	24.8	18.8	18.9
24.0	23.7	19.5	19.0

SSN = 5

GMT	MUF	FOT	LUF
2.0	12.9	10.1	13.1
4.0	8.7	5.5	-
6.0	8.9	4.9	-
8.0	8.4	5.8	-
10.0	9.3	7.5	-
12.0	14.8	12.3	-
14.0	16.5	13.4	-
16.0	17.6	13.7	-
18.0	18.3	14.5	14.1
20.0	18.6	14.3	14.0
22.0	17.3	11.4	11.4
24.0	16.4	12.5	12.5

JULY

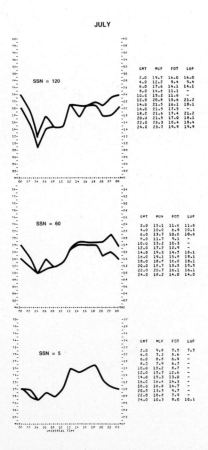

SSN = 120

GMT	MUF	FOT	LUF
2.0	19.7	16.0	16.0
4.0	12.2	9.4	9.4
6.0	17.6	14.1	14.1
8.0	14.8	11.1	-
10.0	15.2	11.8	-
12.0	20.8	13.6	21.2
14.0	21.5	16.1	18.1
16.0	21.5	17.5	-
18.0	21.6	17.4	21.2
20.0	21.5	17.0	18.1
22.0	23.3	18.4	18.4
24.0	23.7	19.9	19.9

SSN = 60

GMT	MUF	FOT	LUF
2.0	15.1	11.6	11.6
4.0	10.0	6.9	10.1
6.0	13.7	10.6	10.6
8.0	11.7	9.1	-
10.0	13.2	10.3	-
12.0	17.2	12.9	-
14.0	19.3	14.5	18.1
16.0	19.1	15.8	18.1
18.0	18.9	16.0	18.1
20.0	18.7	15.5	15.5
22.0	20.7	16.1	18.1
24.0	18.2	14.0	14.0

SSN = 5

GMT	MUF	FOT	LUF
2.0	9.8	7.5	7.5
4.0	7.2	3.6	-
6.0	8.0	6.9	-
8.0	7.9	6.7	-
10.0	10.2	8.7	-
12.0	15.7	12.6	-
14.0	15.3	13.0	-
16.0	16.4	14.3	-
18.0	16.8	14.7	-
20.0	13.3	9.7	-
22.0	10.8	7.9	-
24.0	10.3	8.0	10.1

OCTOBER

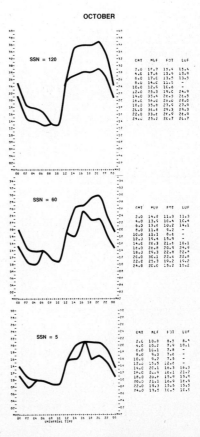

SSN = 120

GMT	MUF	FOT	LUF
2.0	17.7	15.4	15.4
4.0	17.6	13.4	13.9
6.0	17.0	13.5	13.5
8.0	14.6	11.5	-
10.0	12.9	10.6	-
12.0	25.3	19.6	24.9
14.0	35.4	26.5	26.5
16.0	38.0	28.0	28.0
18.0	35.8	29.3	29.3
20.0	34.0	29.3	26.3
22.0	33.6	28.9	26.9
24.0	25.2	20.7	20.7

SSN = 60

GMT	MUF	FOT	LUF
2.0	14.6	11.3	11.3
4.0	13.6	10.4	10.4
6.0	13.6	10.2	14.1
8.0	11.8	9.2	-
10.0	11.1	8.6	-
12.0	16.4	15.4	-
14.0	28.3	21.4	18.1
16.0	26.8	20.4	24.9
18.0	26.3	22.8	22.5
20.0	30.1	22.4	22.8
22.0	25.3	19.2	19.2
24.0	20.0	15.2	15.2

SSN = 5

GMT	MUF	FOT	LUF
2.0	10.8	8.5	8.5
4.0	10.2	7.0	10.1
6.0	10.1	7.8	-
8.0	9.3	7.6	-
10.0	9.7	7.5	-
14.0	20.1	16.3	16.3
16.0	26.9	18.1	21.2
18.0	20.7	15.9	15.9
20.0	21.1	16.9	16.9
22.0	19.3	15.5	15.5
24.0	19.5	10.5	10.5

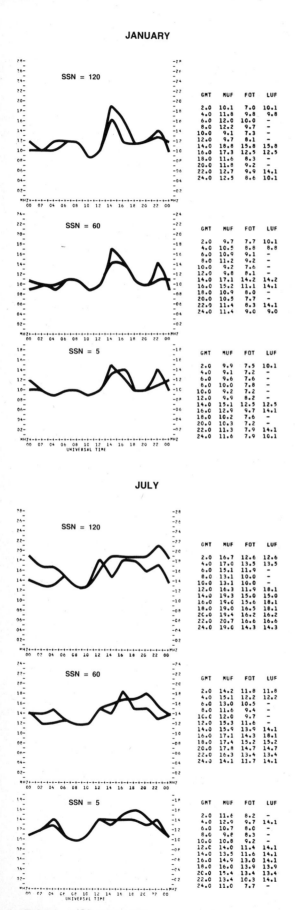

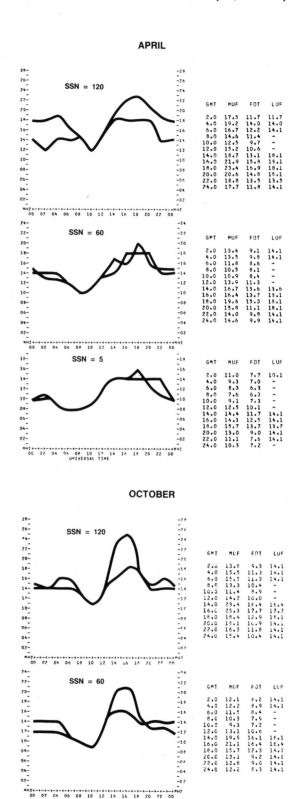

JANUARY

SSN = 120

GMT	MUF	FOT	LUF
2.0	10.1	7.0	10.1
4.0	11.8	9.8	9.8
6.0	12.0	10.0	-
8.0	12.2	9.7	-
10.0	9.1	7.3	-
12.0	9.7	8.1	-
14.0	18.8	15.8	15.8
16.0	17.3	12.5	12.5
18.0	11.6	8.3	-
20.0	11.8	9.2	-
22.0	12.7	9.9	14.1
24.0	12.5	8.6	10.1

SSN = 60

GMT	MUF	FOT	LUF
2.0	9.7	7.7	10.1
4.0	10.5	8.8	8.8
6.0	10.9	9.1	-
8.0	11.2	9.2	-
10.0	9.2	7.6	-
12.0	9.8	8.1	-
14.0	17.1	14.2	14.2
16.0	15.2	11.1	14.1
18.0	10.9	8.0	-
20.0	10.5	7.7	-
22.0	11.4	8.3	14.1
24.0	11.4	9.0	9.0

SSN = 5

GMT	MUF	FOT	LUF
2.0	9.9	7.5	10.1
4.0	9.1	7.2	-
6.0	9.6	7.6	-
8.0	10.0	7.8	-
10.0	9.2	7.2	-
12.0	9.9	8.2	-
14.0	15.1	12.5	12.5
16.0	12.9	9.7	14.1
18.0	10.2	7.6	-
20.0	10.3	7.2	-
22.0	11.3	7.9	14.1
24.0	11.6	7.9	10.1

APRIL

SSN = 120

GMT	MUF	FOT	LUF
2.0	17.5	11.7	11.7
4.0	19.2	14.0	14.0
6.0	16.7	12.2	14.1
8.0	14.6	11.4	-
10.0	12.5	9.7	-
12.0	15.2	10.6	-
14.0	18.7	13.1	18.1
16.0	21.9	15.8	19.1
18.0	23.4	16.9	18.1
20.0	20.6	14.8	19.1
22.0	18.8	13.5	13.5
24.0	17.7	11.8	14.1

SSN = 60

GMT	MUF	FOT	LUF
2.0	13.4	9.1	14.1
4.0	13.5	9.8	14.1
6.0	11.8	8.6	-
8.0	10.5	8.1	-
10.0	10.9	8.4	-
12.0	13.9	11.3	-
14.0	16.7	13.6	13.6
16.0	16.4	13.7	19.1
18.0	19.6	15.0	18.1
20.0	15.8	11.1	18.1
22.0	14.0	9.8	14.1
24.0	14.6	9.9	14.1

SSN = 5

GMT	MUF	FOT	LUF
2.0	11.0	7.7	10.1
4.0	9.3	7.0	-
6.0	8.3	6.3	-
8.0	7.6	6.0	-
10.0	9.1	7.3	-
12.0	12.5	10.1	-
14.0	14.4	11.7	14.1
16.0	14.3	12.5	14.1
18.0	15.7	13.7	13.7
20.0	13.0	9.0	14.1
22.0	11.1	7.6	14.1
24.0	10.3	7.2	-

JULY

SSN = 120

GMT	MUF	FOT	LUF
2.0	16.7	12.6	12.6
4.0	17.0	13.5	13.5
6.0	15.1	11.9	-
8.0	13.1	10.0	-
10.0	13.1	10.0	-
12.0	16.3	11.9	18.1
14.0	19.3	15.0	15.0
16.0	19.0	15.6	18.1
18.0	19.0	16.5	18.1
20.0	19.4	16.2	16.2
22.0	20.7	16.6	16.6
24.0	19.0	14.3	14.3

SSN = 60

GMT	MUF	FOT	LUF
2.0	14.2	11.8	11.8
4.0	15.1	12.2	12.2
6.0	13.0	10.5	-
8.0	11.6	9.4	-
10.0	12.0	9.7	-
12.0	15.3	11.6	-
14.0	15.9	13.9	14.1
16.0	17.1	14.3	18.1
18.0	17.4	15.2	15.2
20.0	17.8	14.7	14.7
22.0	16.3	13.4	13.4
24.0	14.1	11.7	14.1

SSN = 5

GMT	MUF	FOT	LUF
2.0	11.6	8.2	-
4.0	12.9	9.7	14.1
6.0	12.0	8.0	-
8.0	9.8	8.3	-
10.0	10.8	9.2	-
12.0	14.0	11.4	14.1
14.0	13.5	11.6	14.1
16.0	14.9	13.0	14.1
18.0	16.0	13.9	13.9
20.0	15.4	13.4	14.1
22.0	13.4	10.3	14.1
24.0	11.0	7.7	-

OCTOBER

SSN = 120

GMT	MUF	FOT	LUF
2.0	13.8	9.3	14.1
4.0	15.5	11.3	14.1
6.0	15.5	11.3	14.1
8.0	13.3	10.4	-
10.0	11.4	8.9	-
12.0	14.2	10.0	-
14.0	23.4	16.4	16.4
16.0	25.3	17.7	17.7
18.0	18.4	12.9	18.1
20.0	15.1	10.9	14.1
22.0	12.8	11.8	14.1
24.0	15.4	10.4	14.1

SSN = 60

GMT	MUF	FOT	LUF
2.0	12.1	8.2	14.1
4.0	12.2	8.9	14.1
6.0	11.5	8.4	-
8.0	10.3	7.9	-
10.0	9.3	7.2	-
12.0	13.1	10.6	-
14.0	19.5	16.1	16.1
16.0	21.1	16.4	16.4
18.0	15.7	12.3	14.1
20.0	13.1	9.2	14.1
22.0	12.8	9.0	14.1
24.0	12.2	8.3	14.1

SSN = 5

GMT	MUF	FOT	LUF
2.0	10.7	8.2	10.1
4.0	8.7	6.5	-
6.0	8.8	6.6	-
8.0	8.4	6.7	-
10.0	8.2	6.6	-
12.0	11.8	9.6	-
14.0	16.3	13.2	13.2
16.0	16.7	13.7	13.7
18.0	13.0	10.6	14.1
20.0	11.1	7.7	-
22.0	10.9	7.5	14.1
24.0	10.5	7.4	10.1

JANUARY

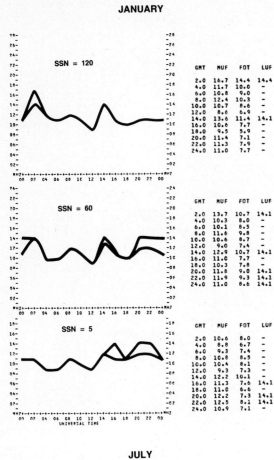

SSN = 120

GMT	MUF	FOT	LUF
2.0	16.7	14.4	14.4
4.0	11.7	10.0	-
6.0	10.8	9.0	-
8.0	12.4	10.3	-
10.0	10.7	8.6	-
12.0	8.6	6.9	-
14.0	13.6	11.4	14.1
16.0	10.6	7.7	-
18.0	9.5	5.9	-
20.0	11.4	7.1	-
22.0	11.3	7.9	-
24.0	11.0	7.7	-

SSN = 60

GMT	MUF	FOT	LUF
2.0	13.7	10.7	14.1
4.0	10.3	8.0	-
6.0	10.1	8.5	-
8.0	11.6	9.8	-
10.0	10.6	8.7	-
12.0	9.0	7.4	-
14.0	12.9	10.7	14.1
16.0	11.0	7.7	-
18.0	10.3	7.8	-
20.0	11.8	9.0	14.1
22.0	11.9	9.3	14.1
24.0	11.0	8.6	14.1

SSN = 5

GMT	MUF	FOT	LUF
2.0	10.6	8.0	-
4.0	8.8	6.7	-
6.0	9.3	7.4	-
8.0	10.8	8.5	-
10.0	10.4	8.1	-
12.0	9.3	7.3	-
14.0	12.2	10.1	-
16.0	11.3	7.6	14.1
18.0	11.0	6.6	-
20.0	12.2	7.3	14.1
22.0	12.5	8.1	14.1
24.0	10.9	7.1	-

APRIL

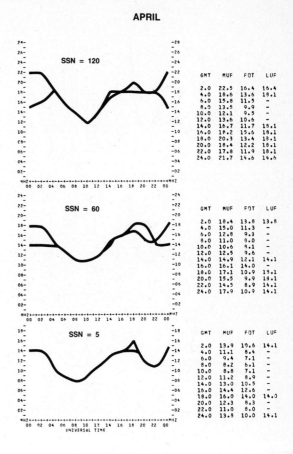

SSN = 120

GMT	MUF	FOT	LUF
2.0	22.5	16.4	16.4
4.0	18.6	13.6	18.1
6.0	15.8	11.5	-
8.0	13.5	9.9	-
10.0	12.1	9.5	-
12.0	13.6	10.6	-
14.0	16.7	11.7	18.1
16.0	18.2	15.6	18.1
18.0	20.3	13.4	18.1
20.0	18.4	12.2	18.1
22.0	17.8	11.9	18.1
24.0	21.7	14.6	14.6

SSN = 60

GMT	MUF	FOT	LUF
2.0	18.4	13.8	13.8
4.0	15.0	11.3	-
6.0	12.8	9.3	-
8.0	11.0	8.0	-
10.0	10.6	8.1	-
12.0	12.5	9.6	-
14.0	14.9	12.1	14.1
16.0	16.1	14.0	-
18.0	17.1	10.9	18.1
20.0	15.5	9.9	18.1
22.0	14.5	8.9	14.1
24.0	17.9	10.9	14.1

SSN = 5

GMT	MUF	FOT	LUF
2.0	13.9	10.6	14.1
4.0	11.1	8.4	-
6.0	9.4	7.1	-
8.0	8.2	6.1	-
10.0	8.8	7.1	-
12.0	11.2	8.9	-
14.0	13.0	10.5	-
16.0	14.4	12.6	-
18.0	16.0	14.0	14.0
20.0	12.3	8.3	-
22.0	11.0	8.0	-
24.0	13.9	10.0	14.1

JULY

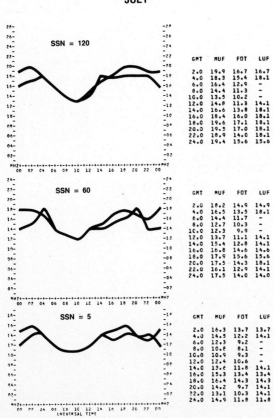

SSN = 120

GMT	MUF	FOT	LUF
2.0	19.9	16.7	16.7
4.0	18.3	15.4	18.1
6.0	16.4	12.9	-
8.0	14.4	11.3	-
10.0	13.5	10.2	-
12.0	14.8	11.3	14.1
14.0	16.6	13.8	18.1
16.0	18.4	16.0	18.1
18.0	19.6	17.1	18.1
20.0	19.5	17.0	18.1
22.0	18.9	14.0	18.1
24.0	19.4	15.6	15.6

SSN = 60

GMT	MUF	FOT	LUF
2.0	18.2	14.9	14.9
4.0	16.5	13.5	18.1
6.0	14.4	11.7	-
8.0	12.7	10.3	-
10.0	12.3	9.9	-
12.0	13.7	11.1	14.1
14.0	15.4	12.8	14.1
16.0	16.8	14.6	14.6
18.0	17.9	15.6	15.6
20.0	17.5	14.3	18.1
22.0	16.1	12.9	14.1
24.0	17.5	14.0	14.0

SSN = 5

GMT	MUF	FOT	LUF
2.0	16.3	13.7	13.7
4.0	14.5	12.2	14.1
6.0	12.3	9.2	-
8.0	10.8	8.1	-
10.0	10.9	9.3	-
12.0	12.4	10.6	-
14.0	13.6	11.8	14.1
16.0	15.3	13.4	13.4
18.0	16.4	14.3	14.3
20.0	14.2	9.7	14.1
22.0	13.1	10.3	14.1
24.0	14.9	11.8	11.8

OCTOBER

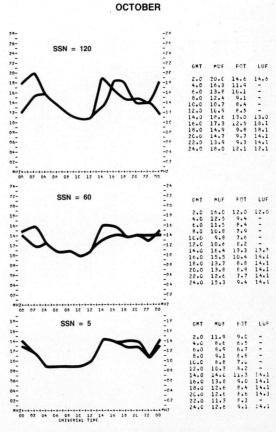

SSN = 120

GMT	MUF	FOT	LUF
2.0	20.0	14.6	14.6
4.0	16.3	11.9	-
6.0	13.8	10.1	-
8.0	12.4	9.1	-
10.0	12.9	8.5	-
12.0	14.9	9.4	-
14.0	18.6	13.0	13.0
16.0	17.3	12.5	18.1
18.0	14.9	9.8	18.1
20.0	14.7	9.7	14.1
22.0	13.9	9.3	14.1
24.0	18.0	12.1	12.1

SSN = 60

GMT	MUF	FOT	LUF
2.0	16.0	12.0	12.0
4.0	12.5	9.4	-
6.0	11.5	8.4	-
8.0	10.8	7.9	-
10.0	9.8	7.6	-
12.0	10.6	8.2	-
14.0	16.4	13.3	13.3
16.0	15.5	10.4	14.1
18.0	13.7	8.8	14.1
20.0	13.8	8.8	14.1
22.0	12.6	7.7	14.1
24.0	15.3	9.4	14.1

SSN = 5

GMT	MUF	FOT	LUF
2.0	11.9	9.0	-
4.0	8.6	6.5	-
6.0	8.9	6.7	-
8.0	9.1	6.8	-
10.0	8.8	7.0	-
12.0	10.3	8.2	-
14.0	14.0	11.3	14.1
16.0	13.8	9.0	14.1
18.0	12.6	8.8	14.1
20.0	12.6	8.6	14.1
22.0	11.3	8.1	-
24.0	12.6	9.1	14.1

JANUARY

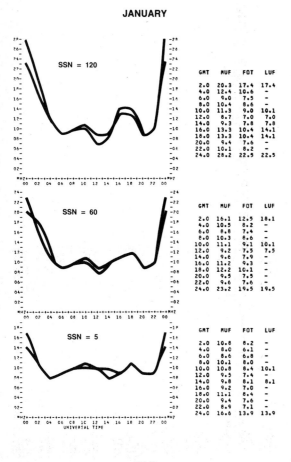

SSN = 120

GMT	MUF	FOT	LUF
2.0	20.3	17.4	17.4
4.0	12.4	10.6	-
6.0	9.0	7.5	-
8.0	10.4	8.6	-
10.0	11.3	9.0	10.1
12.0	8.7	7.0	7.0
14.0	9.3	7.8	7.8
16.0	13.3	10.4	14.1
18.0	13.3	10.4	14.1
20.0	9.4	7.6	-
22.0	10.1	8.2	-
24.0	28.2	22.5	22.5

SSN = 60

GMT	MUF	FOT	LUF
2.0	16.1	12.5	18.1
4.0	10.5	8.2	-
6.0	8.8	7.4	-
8.0	10.3	8.6	-
10.0	11.1	9.1	10.1
12.0	9.2	7.5	7.5
14.0	9.6	7.9	-
16.0	11.2	9.3	-
18.0	12.2	10.1	-
20.0	9.5	7.5	-
22.0	9.6	7.6	-
24.0	23.2	19.5	19.5

SSN = 5

GMT	MUF	FOT	LUF
2.0	10.8	8.2	-
4.0	8.0	6.1	-
6.0	8.6	6.8	-
8.0	10.1	8.0	-
10.0	10.8	8.4	10.1
12.0	9.5	7.4	-
14.0	9.8	8.1	8.1
16.0	9.2	7.0	-
18.0	11.1	8.4	-
20.0	9.4	7.6	-
22.0	8.9	7.1	-
24.0	16.6	13.9	13.9

UNIVERSAL TIME

APRIL

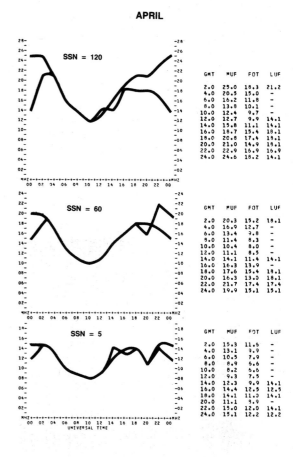

SSN = 120

GMT	MUF	FOT	LUF
2.0	25.0	18.3	21.2
4.0	20.5	15.0	-
6.0	16.2	11.8	-
8.0	13.8	10.1	-
10.0	12.4	9.7	-
12.0	12.7	9.9	14.1
14.0	15.8	11.1	14.1
16.0	18.7	15.4	18.1
18.0	20.8	17.4	18.1
20.0	21.0	14.9	18.1
22.0	22.9	16.9	16.9
24.0	24.6	18.2	14.1

SSN = 60

GMT	MUF	FOT	LUF
2.0	20.3	15.2	18.1
4.0	16.9	12.7	-
6.0	13.4	9.8	-
8.0	11.4	8.3	-
10.0	11.4	8.0	-
12.0	11.1	8.5	-
14.0	14.1	11.4	14.1
16.0	16.3	13.9	-
18.0	17.6	15.4	18.1
20.0	16.3	13.0	18.1
22.0	21.7	17.4	17.4
24.0	19.9	15.1	14.1

SSN = 5

GMT	MUF	FOT	LUF
2.0	15.3	11.6	-
4.0	13.1	9.9	-
6.0	10.5	7.9	-
8.0	8.9	6.6	-
10.0	8.2	6.6	-
12.0	9.3	7.5	-
14.0	12.3	9.9	14.1
16.0	14.4	12.5	12.5
18.0	14.1	11.0	14.1
20.0	11.1	9.9	-
22.0	15.0	12.0	14.1
24.0	15.1	12.2	12.2

UNIVERSAL TIME

JULY

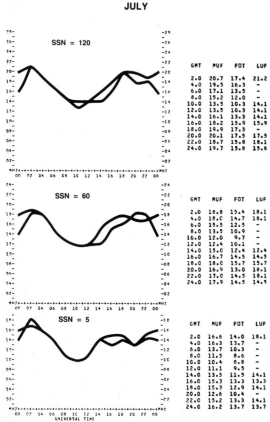

SSN = 120

GMT	MUF	FOT	LUF
2.0	20.7	17.4	21.2
4.0	19.5	16.3	-
6.0	17.1	13.5	-
8.0	15.2	12.0	-
10.0	13.5	10.3	14.1
12.0	13.5	10.3	14.1
14.0	16.1	13.3	14.1
16.0	18.2	15.9	15.9
18.0	19.9	17.3	-
20.0	20.1	17.5	17.5
22.0	18.7	15.8	18.1
24.0	19.7	15.8	15.8

SSN = 60

GMT	MUF	FOT	LUF
2.0	18.8	15.4	18.1
4.0	18.0	14.7	18.1
6.0	15.5	12.5	-
8.0	13.5	10.9	-
10.0	12.0	9.7	-
12.0	12.4	10.1	-
14.0	15.0	12.4	12.4
16.0	16.7	14.5	14.5
18.0	18.0	15.7	15.7
20.0	16.9	13.0	18.1
22.0	17.0	14.5	18.1
24.0	17.9	14.5	14.5

SSN = 5

GMT	MUF	FOT	LUF
2.0	16.6	14.0	18.1
4.0	16.3	13.7	-
6.0	13.7	10.3	-
8.0	11.5	8.6	-
10.0	10.4	8.8	-
12.0	11.1	9.5	-
14.0	13.5	11.5	14.1
16.0	15.3	13.3	13.3
18.0	15.3	12.9	14.1
20.0	12.6	10.4	-
22.0	15.2	13.3	14.1
24.0	16.2	13.7	13.7

UNIVERSAL TIME

OCTOBER

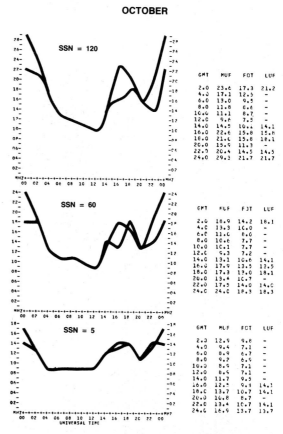

SSN = 120

GMT	MUF	FOT	LUF
2.0	23.6	17.3	21.2
4.0	17.1	12.5	-
6.0	13.0	9.5	-
8.0	11.8	8.6	-
10.0	11.1	8.7	-
12.0	9.6	7.5	-
14.0	14.5	10.4	14.1
16.0	22.6	15.8	15.8
18.0	21.6	15.8	18.1
20.0	15.9	11.3	-
22.0	20.4	14.5	14.5
24.0	29.3	21.7	21.7

SSN = 60

GMT	MUF	FOT	LUF
2.0	18.9	14.2	18.1
4.0	13.3	10.0	-
6.0	11.0	8.0	-
8.0	10.6	7.7	-
10.0	10.1	7.7	-
12.0	9.3	7.2	-
14.0	13.1	10.6	14.1
16.0	17.9	13.5	12.5
18.0	17.3	13.0	18.1
20.0	13.4	10.7	-
22.0	17.5	14.0	14.0
24.0	24.0	18.3	18.3

SSN = 5

GMT	MUF	FOT	LUF
2.0	12.9	9.8	-
4.0	9.4	7.1	-
6.0	8.9	6.7	-
8.0	9.2	6.9	-
10.0	8.9	7.1	-
12.0	8.9	7.1	-
14.0	11.7	9.5	-
16.0	12.5	9.3	14.1
18.0	13.7	10.7	14.1
20.0	10.8	8.7	-
22.0	13.4	10.7	14.1
24.0	16.9	13.7	13.7

UNIVERSAL TIME

JANUARY

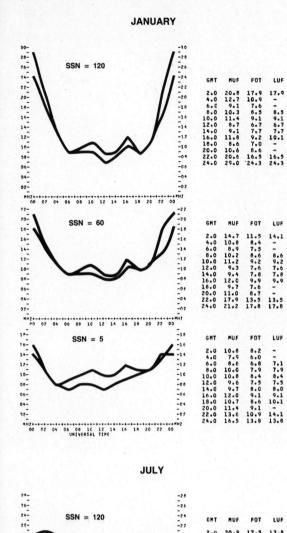

SSN = 120

GMT	MUF	FOT	LUF
2.0	20.8	17.9	17.9
4.0	12.7	10.9	-
6.0	9.1	7.6	-
8.0	10.3	8.5	8.5
10.0	11.4	9.1	9.1
12.0	8.7	6.7	6.7
14.0	9.1	7.7	7.7
16.0	11.8	9.2	10.1
18.0	8.6	7.0	-
20.0	10.6	8.6	-
22.0	20.6	16.5	16.5
24.0	29.0	24.3	24.3

SSN = 60

GMT	MUF	FOT	LUF
2.0	14.7	11.5	14.1
4.0	10.8	8.4	-
6.0	8.9	7.5	-
8.0	10.2	8.6	8.6
10.0	11.2	9.2	9.2
12.0	9.3	7.6	7.6
14.0	9.4	7.8	7.8
16.0	12.0	9.9	9.9
18.0	9.7	7.6	-
20.0	11.0	8.7	-
22.0	17.9	13.5	13.5
24.0	21.2	17.8	17.8

SSN = 5

GMT	MUF	FOT	LUF
2.0	10.8	8.2	-
4.0	7.9	6.0	-
6.0	8.6	6.8	7.1
8.0	10.0	7.9	7.9
10.0	10.8	8.4	8.4
12.0	9.6	7.5	7.5
14.0	9.7	8.0	8.0
16.0	12.0	9.1	9.1
18.0	10.7	8.6	10.1
20.0	11.4	9.1	-
22.0	13.6	10.9	14.1
24.0	16.5	13.8	13.8

APRIL

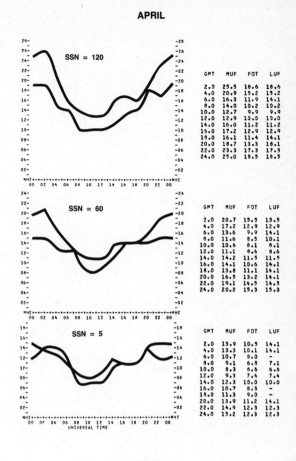

SSN = 120

GMT	MUF	FOT	LUF
2.0	25.5	18.6	18.6
4.0	20.9	15.2	15.2
6.0	16.3	11.9	14.1
8.0	14.0	10.2	10.2
10.0	12.7	9.9	9.9
12.0	12.9	10.0	10.0
14.0	16.0	11.2	11.2
16.0	17.2	12.9	12.9
18.0	16.1	11.4	14.1
20.0	18.7	13.3	18.1
22.0	23.3	17.3	17.3
24.0	25.0	18.5	18.5

SSN = 60

GMT	MUF	FOT	LUF
2.0	20.7	15.5	15.5
4.0	17.2	12.9	12.9
6.0	13.6	9.9	14.1
8.0	11.6	8.5	10.1
10.0	10.6	8.1	8.1
12.0	11.1	8.6	8.6
14.0	14.2	11.5	11.5
16.0	14.1	10.6	14.1
18.0	13.8	11.1	14.1
20.0	16.5	13.2	14.1
22.0	19.1	14.5	14.5
24.0	20.2	15.3	15.3

SSN = 5

GMT	MUF	FOT	LUF
2.0	13.9	10.5	14.1
4.0	13.3	10.1	14.1
6.0	10.7	8.0	-
8.0	9.1	6.8	7.1
10.0	8.3	6.6	6.6
12.0	9.3	7.4	7.4
14.0	12.3	10.0	10.0
16.0	10.0	8.3	-
18.0	11.3	9.0	-
20.0	13.9	11.2	14.1
22.0	14.9	12.3	12.3
24.0	15.2	12.3	12.3

JULY

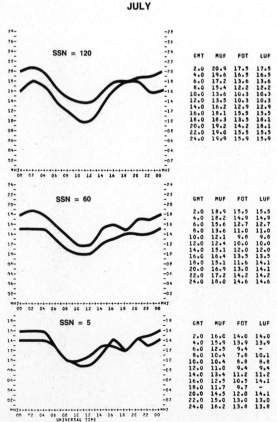

SSN = 120

GMT	MUF	FOT	LUF
2.0	20.9	17.5	17.5
4.0	19.6	16.5	16.5
6.0	17.2	13.6	13.6
8.0	15.4	12.2	12.2
10.0	13.6	10.3	10.3
12.0	13.5	10.3	10.3
14.0	16.2	12.9	12.9
16.0	18.1	15.5	15.5
18.0	18.3	13.5	18.1
20.0	19.2	14.2	18.1
22.0	19.0	15.5	15.5
24.0	19.9	15.9	15.9

SSN = 60

GMT	MUF	FOT	LUF
2.0	18.9	15.5	15.5
4.0	18.2	14.9	14.9
6.0	15.6	12.7	12.7
8.0	13.6	11.0	11.0
10.0	12.1	9.8	9.8
12.0	12.4	10.0	10.0
14.0	15.1	12.0	12.0
16.0	16.4	13.5	13.5
18.0	15.1	11.6	14.1
20.0	16.9	13.0	14.1
22.0	17.0	14.3	14.2
24.0	18.0	14.6	14.6

SSN = 5

GMT	MUF	FOT	LUF
2.0	16.0	14.0	14.0
4.0	15.9	13.9	13.9
6.0	12.5	9.4	-
8.0	10.4	7.8	10.1
10.0	10.4	8.8	8.8
12.0	11.0	9.4	9.4
14.0	13.4	11.2	11.2
16.0	12.5	10.5	14.1
18.0	11.7	9.7	-
20.0	14.5	12.0	14.1
22.0	15.0	13.0	14.1
24.0	16.2	13.8	13.8

OCTOBER

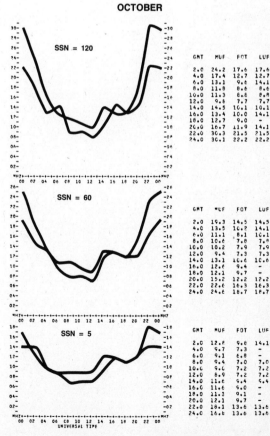

SSN = 120

GMT	MUF	FOT	LUF
2.0	24.2	17.6	17.6
4.0	17.4	12.7	12.7
6.0	13.1	9.6	14.1
8.0	11.8	8.6	8.6
10.0	11.3	8.6	8.8
12.0	9.6	7.7	7.7
14.0	14.5	10.1	10.1
16.0	13.4	10.0	14.1
18.0	12.7	9.0	-
20.0	16.7	11.9	14.1
22.0	30.3	21.5	21.5
24.0	30.1	22.2	22.2

SSN = 60

GMT	MUF	FOT	LUF
2.0	19.3	14.5	14.5
4.0	13.5	10.2	14.1
6.0	11.1	8.1	10.1
8.0	10.6	7.8	7.8
10.0	10.2	7.9	7.9
12.0	9.4	7.3	7.3
14.0	13.1	10.0	10.0
16.0	12.6	9.4	-
18.0	12.1	9.7	-
20.0	15.2	12.2	12.2
22.0	22.6	16.3	16.3
24.0	24.6	18.7	18.7

SSN = 5

GMT	MUF	FOT	LUF
2.0	12.6	9.8	14.1
4.0	9.7	7.3	-
6.0	9.1	6.8	-
8.0	9.4	7.0	7.0
10.0	9.6	7.2	7.2
12.0	8.9	7.2	7.2
14.0	11.6	9.4	9.4
16.0	11.6	9.0	-
18.0	12.1	9.7	-
20.0	13.3	10.7	14.1
22.0	18.1	13.6	13.6
24.0	16.6	13.6	13.6

JANUARY

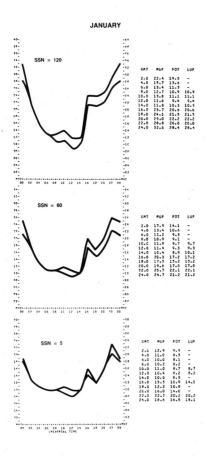

APRIL

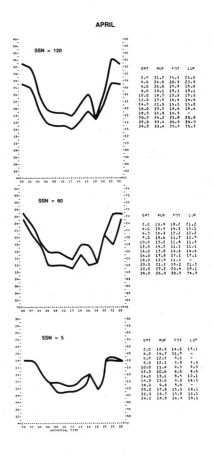

JULY

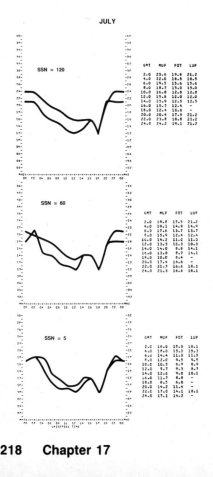

OCTOBER

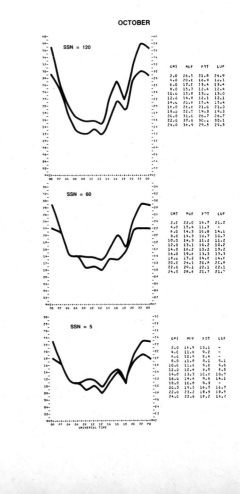

JANUARY

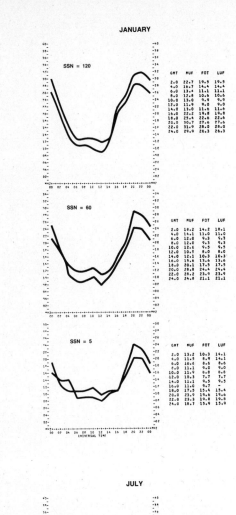

APRIL

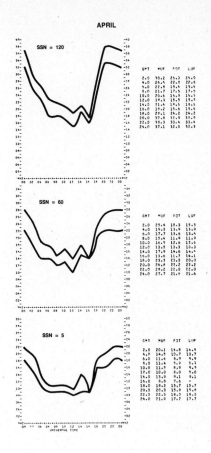

JULY

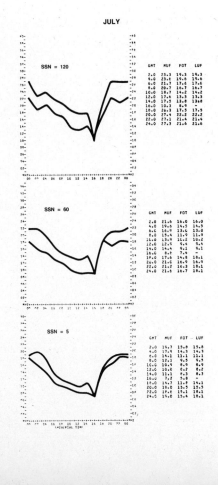

OCTOBER

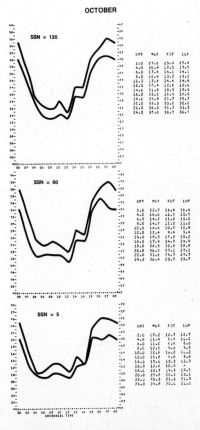

Western USA to Alaska
40°N 116°W to 61°N 150°W

JANUARY

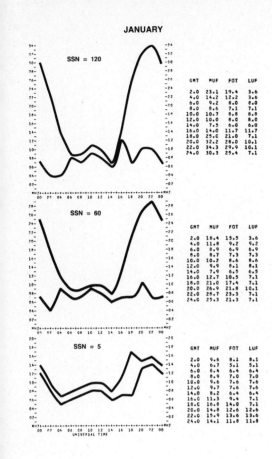

SSN = 120

GMT	MUF	FOT	LUF
2.0	23.1	19.4	3.6
4.0	14.2	12.2	3.6
6.0	9.2	8.0	8.0
8.0	8.6	7.1	7.1
10.0	10.7	8.8	8.8
12.0	10.0	8.0	8.0
14.0	7.5	6.0	6.0
16.0	14.0	11.7	11.7
18.0	25.0	21.0	7.1
20.0	32.2	28.0	10.1
22.0	34.3	29.9	10.1
24.0	30.3	25.4	7.1

SSN = 60

GMT	MUF	FOT	LUF
2.0	18.4	15.5	3.6
4.0	11.8	9.2	9.2
6.0	8.9	6.9	6.9
8.0	8.7	7.3	7.3
10.0	10.2	8.6	8.6
12.0	9.9	8.1	8.1
14.0	7.9	6.5	6.5
16.0	12.7	10.5	7.1
18.0	21.0	17.4	7.1
20.0	26.9	21.8	10.1
22.0	28.7	23.3	7.1
24.0	25.3	21.3	7.1

SSN = 5

GMT	MUF	FOT	LUF
2.0	9.6	8.1	8.1
4.0	6.7	5.1	5.1
6.0	8.4	6.4	6.4
8.0	8.9	7.0	7.0
10.0	9.6	7.6	7.6
12.0	9.7	7.6	7.6
14.0	8.2	6.4	6.4
16.0	11.3	9.4	7.1
18.0	16.8	14.0	7.1
20.0	14.8	12.6	12.6
22.0	15.9	13.6	13.6
24.0	14.1	11.8	11.8

UNIVERSAL TIME

APRIL

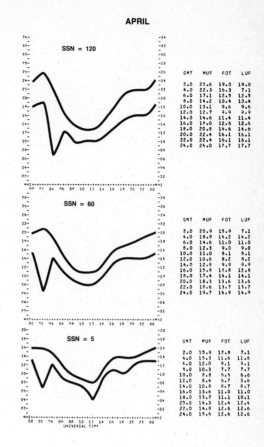

SSN = 120

GMT	MUF	FOT	LUF
2.0	23.1	19.0	19.0
4.0	22.3	16.3	7.1
6.0	17.1	12.5	12.5
8.0	14.2	10.4	10.4
10.0	13.1	9.6	9.6
12.0	12.7	9.0	9.0
14.0	14.6	11.4	11.4
16.0	16.0	12.6	12.6
18.0	20.8	14.6	14.6
20.0	22.4	16.1	16.1
22.0	22.4	16.1	16.1
24.0	24.0	17.7	17.7

SSN = 60

GMT	MUF	FOT	LUF
2.0	20.9	15.9	7.1
4.0	18.9	14.2	14.2
6.0	14.6	11.0	11.0
8.0	12.3	9.0	9.0
10.0	11.0	8.1	8.1
12.0	10.6	8.2	8.2
14.0	12.8	9.9	9.9
16.0	15.8	12.8	12.8
18.0	17.4	14.1	14.1
20.0	18.3	13.6	13.6
22.0	18.6	13.7	13.7
24.0	19.7	14.9	14.9

SSN = 5

GMT	MUF	FOT	LUF
2.0	15.9	12.8	7.1
4.0	15.3	11.6	11.6
6.0	12.0	9.1	9.1
8.0	10.3	7.7	7.7
10.0	8.8	5.5	6.6
12.0	8.4	5.7	3.6
14.0	10.8	8.7	8.7
16.0	13.6	11.0	11.0
18.0	13.7	11.1	10.1
20.0	14.8	12.4	12.4
22.0	14.8	12.6	12.6
24.0	15.6	12.6	12.6

UNIVERSAL TIME

JULY

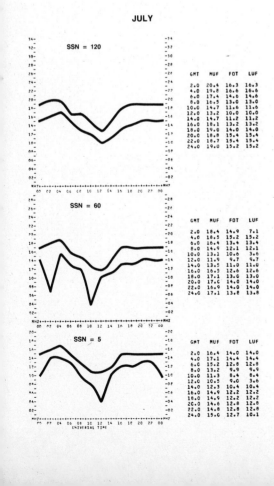

SSN = 120

GMT	MUF	FOT	LUF
2.0	20.4	16.3	16.3
4.0	19.8	16.6	16.6
6.0	17.4	14.6	14.6
8.0	16.5	13.0	13.0
10.0	14.7	11.6	11.6
12.0	13.2	10.0	10.0
14.0	14.7	11.2	11.2
16.0	18.1	13.2	13.2
18.0	19.0	14.0	14.0
20.0	18.8	15.4	15.4
22.0	18.7	15.4	15.4
24.0	19.0	15.2	15.2

SSN = 60

GMT	MUF	FOT	LUF
2.0	18.4	14.9	7.1
4.0	18.5	15.2	15.2
6.0	16.4	13.4	13.4
8.0	14.9	12.1	12.1
10.0	13.1	10.6	3.6
12.0	11.9	9.7	9.7
14.0	13.5	11.0	11.0
16.0	16.5	12.6	12.6
18.0	17.1	13.0	13.0
20.0	17.0	14.0	14.0
22.0	16.9	14.0	14.0
24.0	17.1	13.8	13.8

SSN = 5

GMT	MUF	FOT	LUF
2.0	16.4	14.0	14.0
4.0	17.1	14.4	14.4
6.0	15.2	12.8	12.8
8.0	13.2	9.9	9.9
10.0	11.3	8.4	8.4
12.0	10.5	9.0	3.6
14.0	12.3	10.4	10.4
16.0	14.9	12.2	12.2
18.0	14.9	12.2	12.2
20.0	14.6	12.8	12.8
22.0	14.8	12.8	12.8
24.0	15.0	12.7	10.1

UNIVERSAL TIME

OCTOBER

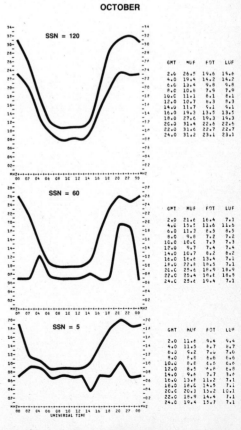

SSN = 120

GMT	MUF	FOT	LUF
2.0	26.5	19.6	19.6
4.0	19.4	14.2	14.2
6.0	13.4	9.8	9.8
8.0	10.6	7.9	7.9
10.0	10.6	8.1	8.1
12.0	10.7	8.3	8.3
14.0	11.7	9.1	9.1
16.0	19.3	13.5	13.5
18.0	27.6	19.3	19.3
20.0	31.4	22.6	22.6
22.0	31.6	22.7	22.7
24.0	31.2	23.1	23.1

SSN = 60

GMT	MUF	FOT	LUF
2.0	21.6	16.4	7.1
4.0	15.5	11.6	11.6
6.0	11.3	8.5	8.5
8.0	9.8	7.2	7.2
10.0	10.6	7.3	7.3
12.0	9.7	7.4	7.4
14.0	10.7	8.2	8.2
16.0	16.6	13.4	7.1
18.0	22.8	18.5	7.1
20.0	25.6	18.4	18.9
22.0	25.4	18.6	18.3
24.0	25.6	19.4	7.1

SSN = 5

GMT	MUF	FOT	LUF
2.0	11.6	9.4	9.4
4.0	11.5	8.7	8.7
6.0	9.2	7.0	7.0
8.0	8.6	6.6	6.6
10.0	8.8	6.6	6.6
12.0	8.5	7.8	8.3
14.0	9.6	7.7	3.6
16.0	13.6	11.2	7.1
18.0	18.6	14.5	7.1
20.0	20.1	15.2	10.1
22.0	18.9	14.4	7.1
24.0	19.4	15.7	7.1

UNIVERSAL TIME

JANUARY

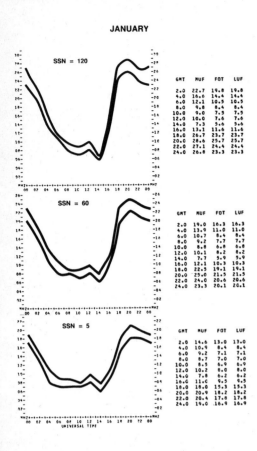

GMT	MUF	FOT	LUF
2.0	22.7	19.8	19.8
4.0	16.6	14.4	14.4
6.0	12.1	10.5	10.5
8.0	9.8	8.4	8.4
10.0	9.0	7.5	7.5
12.0	10.0	7.6	7.6
14.0	7.3	5.6	5.6
16.0	13.1	11.6	11.6
18.0	26.7	23.7	23.7
20.0	28.6	25.7	25.7
22.0	27.1	24.4	24.4
24.0	26.8	23.3	23.3

SSN = 120

SSN = 60

GMT	MUF	FOT	LUF
2.0	19.0	16.3	16.3
4.0	13.9	11.0	11.0
6.0	10.7	8.4	8.4
8.0	9.2	7.7	7.7
10.0	8.8	6.8	6.8
12.0	10.1	8.2	8.2
14.0	7.7	5.9	5.9
16.0	12.1	10.3	10.3
18.0	22.5	19.1	19.1
20.0	25.0	21.5	21.5
22.0	24.0	20.6	20.6
24.0	23.3	20.1	20.1

SSN = 5

GMT	MUF	FOT	LUF
2.0	14.6	13.0	13.0
4.0	10.9	8.4	8.4
6.0	9.2	7.1	7.1
8.0	8.7	7.0	7.0
10.0	8.5	6.9	6.9
12.0	10.2	8.0	8.0
14.0	7.8	6.2	6.2
16.0	11.0	9.5	9.5
18.0	18.0	15.3	15.3
20.0	20.9	18.2	18.2
22.0	20.4	17.8	17.8
24.0	19.0	16.9	16.9

APRIL

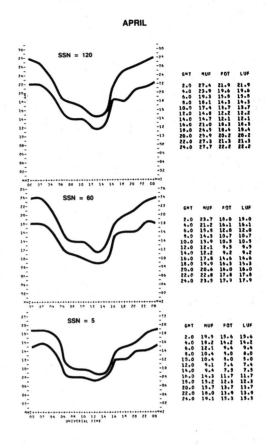

GMT	MUF	FOT	LUF
2.0	27.4	21.9	21.9
4.0	23.9	19.6	19.6
6.0	19.3	15.8	15.8
8.0	18.1	14.3	14.3
10.0	17.4	13.7	13.7
12.0	14.8	12.2	12.2
14.0	14.7	12.1	12.1
16.0	21.0	18.3	18.3
18.0	24.5	18.4	18.4
20.0	25.9	20.2	20.2
22.0	27.3	21.3	21.3
24.0	27.7	22.2	22.2

SSN = 120

SSN = 60

GMT	MUF	FOT	LUF
2.0	23.7	18.0	18.0
4.0	21.2	16.1	16.1
6.0	15.8	12.0	12.0
8.0	14.3	10.7	10.7
10.0	13.9	10.5	10.5
12.0	12.1	9.5	9.5
14.0	12.2	9.2	9.2
16.0	17.8	14.6	14.6
18.0	19.9	16.3	16.3
20.0	20.6	16.0	16.0
22.0	22.8	17.8	17.8
24.0	23.5	17.9	17.9

SSN = 5

GMT	MUF	FOT	LUF
2.0	19.5	15.6	15.6
4.0	18.2	14.2	14.2
6.0	12.1	9.4	9.4
8.0	10.4	8.0	8.0
10.0	10.4	8.0	8.0
12.0	9.1	7.4	7.4
14.0	9.4	7.5	7.5
16.0	14.3	11.7	11.7
18.0	15.2	12.3	12.3
20.0	15.7	13.7	13.7
22.0	18.0	13.9	13.9
24.0	19.1	15.3	15.3

JULY

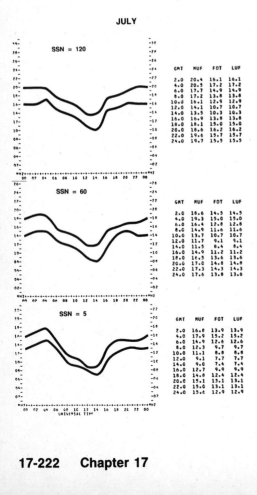

GMT	MUF	FOT	LUF
2.0	20.4	16.1	16.1
4.0	20.5	17.2	17.2
6.0	17.7	14.9	14.9
8.0	17.2	13.8	13.8
10.0	16.1	12.9	12.9
12.0	14.1	10.7	10.7
14.0	13.5	10.3	10.3
16.0	16.9	13.8	13.8
18.0	18.1	15.0	15.0
20.0	18.6	16.2	16.2
22.0	19.6	15.7	15.7
24.0	19.7	15.5	15.5

SSN = 120

SSN = 60

GMT	MUF	FOT	LUF
2.0	18.6	14.5	14.5
4.0	19.3	15.0	15.0
6.0	16.4	12.8	12.8
8.0	14.9	11.6	11.6
10.0	13.7	10.7	10.7
12.0	11.7	9.1	9.1
14.0	11.5	8.4	8.4
16.0	14.9	11.2	11.2
18.0	16.5	13.6	13.6
20.0	17.0	14.8	14.8
22.0	17.3	14.3	14.3
24.0	17.6	13.8	13.8

SSN = 5

GMT	MUF	FOT	LUF
2.0	16.8	13.9	13.9
4.0	17.9	15.2	15.2
6.0	14.9	12.6	12.6
8.0	12.3	9.7	9.7
10.0	11.1	8.8	8.8
12.0	9.1	7.7	7.7
14.0	9.0	7.4	7.4
16.0	12.7	9.9	9.9
18.0	14.8	12.4	12.4
20.0	15.1	13.1	13.1
22.0	15.0	13.1	13.1
24.0	15.6	12.9	12.9

OCTOBER

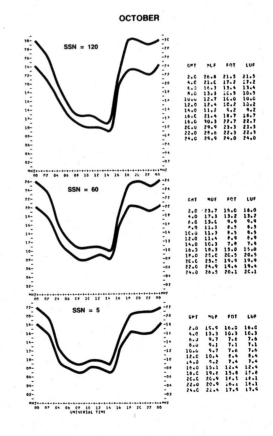

GMT	MUF	FOT	LUF
2.0	26.8	21.5	21.5
4.0	21.0	17.2	17.2
6.0	16.3	13.4	13.4
8.0	13.3	10.5	10.5
10.0	12.7	10.0	10.0
12.0	12.4	10.2	10.2
14.0	11.2	9.2	9.2
16.0	21.4	18.7	18.7
18.0	30.3	22.7	22.7
20.0	29.9	23.3	23.3
22.0	28.6	22.3	22.3
24.0	26.9	24.0	24.0

SSN = 120

SSN = 60

GMT	MUF	FOT	LUF
2.0	23.7	18.0	18.0
4.0	17.3	13.2	13.2
6.0	13.6	9.9	9.9
8.0	11.3	8.5	8.5
10.0	11.3	8.5	8.5
12.0	11.4	9.4	9.4
14.0	10.3	7.8	7.8
16.0	18.3	15.0	15.0
18.0	25.0	20.5	20.5
20.0	25.2	19.9	19.9
22.0	24.9	19.4	19.4
24.0	26.5	20.1	20.1

SSN = 5

GMT	MUF	FOT	LUF
2.0	16.9	16.0	16.0
4.0	13.3	10.3	10.3
6.0	9.7	7.6	7.6
8.0	9.1	7.1	7.1
10.0	9.7	7.6	7.6
12.0	10.4	8.4	8.4
14.0	9.2	7.4	7.4
16.0	15.1	12.4	12.4
18.0	19.6	15.8	15.8
20.0	20.9	16.1	16.1
22.0	20.9	16.1	16.1
24.0	22.4	17.4	17.4

JANUARY

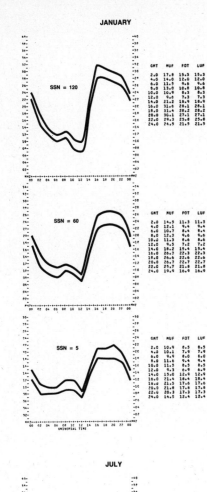

APRIL

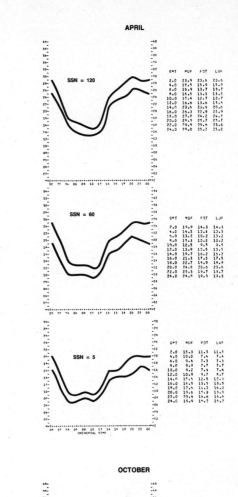

JULY

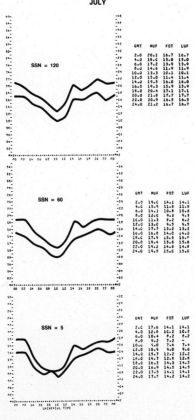

OCTOBER

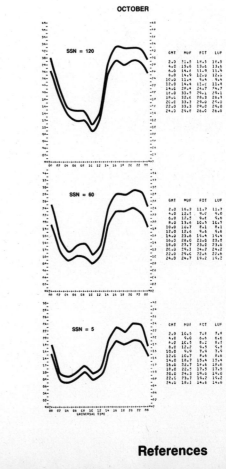

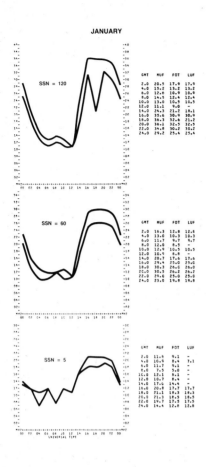

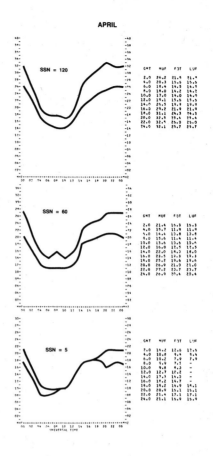

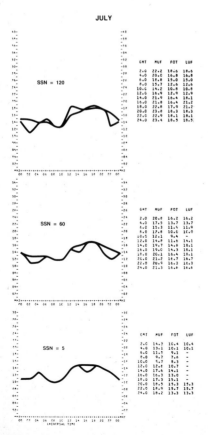

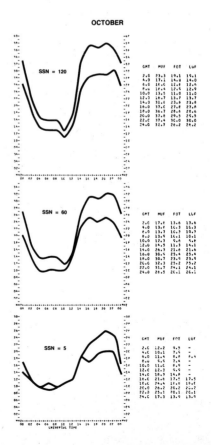

JANUARY

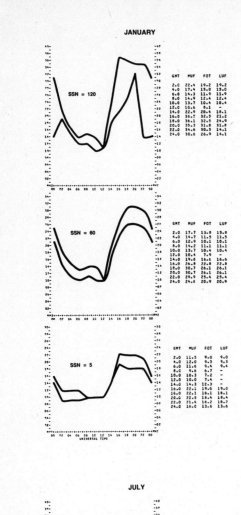

APRIL

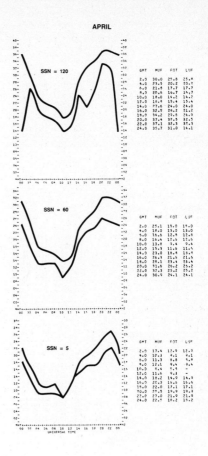

JULY

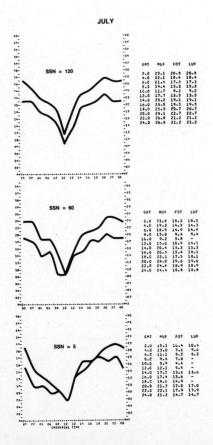

OCTOBER

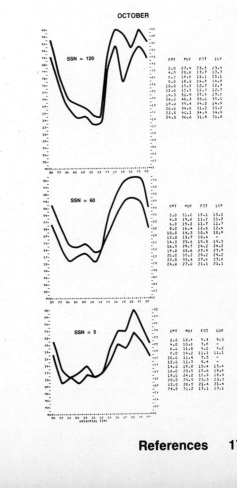

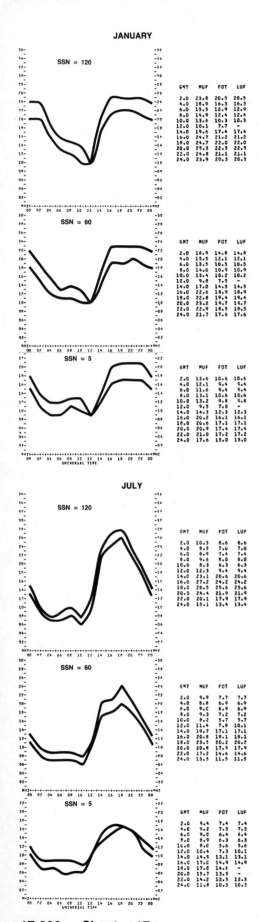

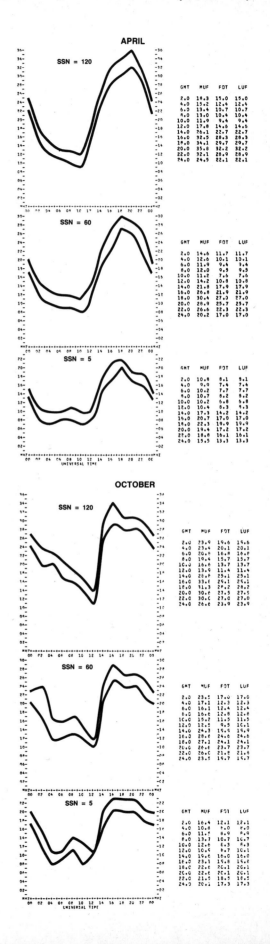

Western USA to Western Europe
40°N 116°W to 47°N 7°E

Azimuths: 36° 319°
Distance: 4761 nautical miles (8817 km)

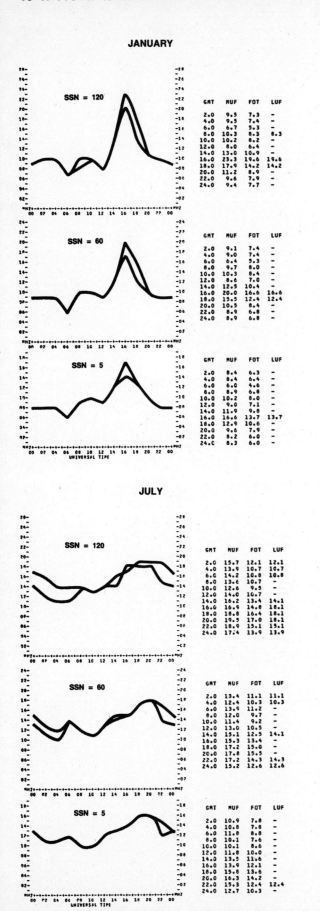

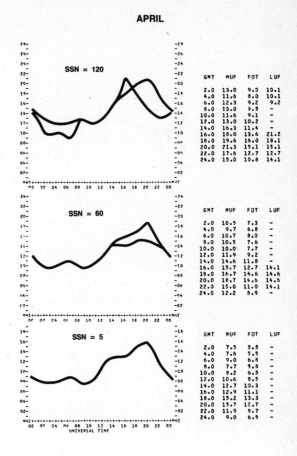

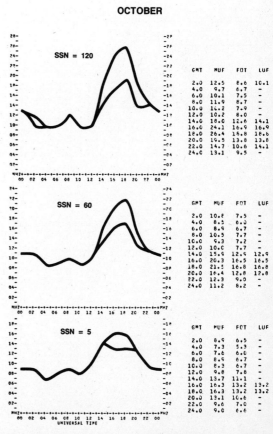

References   17-227

JANUARY

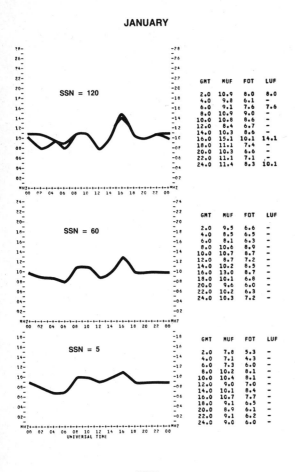

SSN = 120

GMT	MUF	FOT	LUF
2.0	10.9	8.0	8.0
4.0	9.8	6.1	-
6.0	9.1	7.6	7.6
8.0	10.9	9.0	-
10.0	10.8	8.6	-
12.0	8.4	6.7	-
14.0	10.3	8.6	-
16.0	15.1	10.1	14.1
18.0	11.1	7.4	-
20.0	10.3	6.6	-
22.0	11.1	7.1	-
24.0	11.4	8.3	10.1

SSN = 60

GMT	MUF	FOT	LUF
2.0	9.5	6.6	-
4.0	8.5	6.3	-
6.0	8.1	6.3	-
8.0	10.6	8.9	-
10.0	10.7	8.7	-
12.0	8.7	7.2	-
14.0	10.2	8.5	-
16.0	13.0	8.7	-
18.0	10.1	6.8	-
20.0	9.6	6.0	-
22.0	10.2	6.3	-
24.0	10.3	7.2	-

SSN = 5

GMT	MUF	FOT	LUF
2.0	7.8	5.3	-
4.0	7.1	4.3	-
6.0	7.3	6.0	-
8.0	10.2	8.1	-
10.0	10.4	8.1	-
12.0	9.0	7.0	-
14.0	10.1	8.4	-
16.0	10.7	7.7	-
18.0	9.1	6.5	-
20.0	8.9	6.1	-
22.0	9.1	6.2	-
24.0	9.0	6.0	-

UNIVERSAL TIME

APRIL

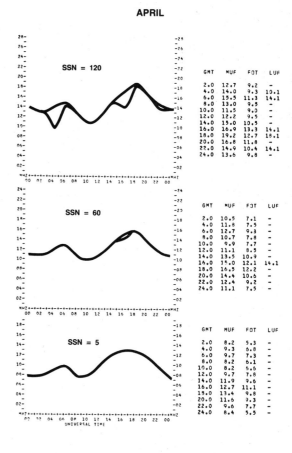

SSN = 120

GMT	MUF	FOT	LUF
2.0	12.7	9.2	-
4.0	14.0	9.3	10.1
6.0	15.5	11.3	14.1
8.0	13.0	9.5	-
10.0	11.5	9.0	-
12.0	12.2	9.5	-
14.0	15.0	10.5	-
16.0	16.9	13.3	14.1
18.0	19.2	12.7	18.1
20.0	16.8	11.8	-
22.0	14.9	10.4	14.1
24.0	13.6	9.8	-

SSN = 60

GMT	MUF	FOT	LUF
2.0	10.5	7.1	-
4.0	11.8	7.5	-
6.0	12.7	9.3	-
8.0	10.7	7.8	-
10.0	9.9	7.7	-
12.0	11.1	8.5	-
14.0	13.5	10.9	-
16.0	15.0	12.1	14.1
18.0	16.5	12.2	-
20.0	14.4	10.6	-
22.0	12.4	9.2	-
24.0	11.1	7.5	-

SSN = 5

GMT	MUF	FOT	LUF
2.0	8.2	5.3	-
4.0	9.3	5.8	-
6.0	9.7	7.3	-
8.0	8.2	6.1	-
10.0	8.2	5.6	-
12.0	9.7	7.8	-
14.0	11.9	9.6	-
16.0	12.7	11.1	-
18.0	13.4	9.8	-
20.0	11.6	9.3	-
22.0	9.6	7.7	-
24.0	8.4	5.5	-

UNIVERSAL TIME

JULY

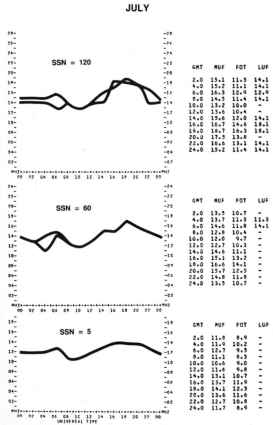

SSN = 120

GMT	MUF	FOT	LUF
2.0	15.1	11.3	14.1
4.0	15.2	11.1	14.1
6.0	16.3	12.9	12.9
8.0	14.5	11.4	14.1
10.0	13.2	10.0	-
12.0	13.6	10.4	-
14.0	15.6	12.0	14.1
16.0	16.7	14.6	18.1
18.0	18.7	16.3	18.1
20.0	17.5	13.8	-
22.0	16.6	13.1	14.1
24.0	15.2	11.4	14.1

SSN = 60

GMT	MUF	FOT	LUF
2.0	13.5	10.7	-
4.0	13.7	11.3	11.3
6.0	14.6	11.8	14.1
8.0	12.8	10.4	-
10.0	12.0	9.7	-
12.0	12.7	10.3	-
14.0	14.6	11.1	-
16.0	15.1	13.2	-
18.0	16.6	14.1	-
20.0	15.7	12.5	-
22.0	14.8	11.8	-
24.0	13.5	10.7	-

SSN = 5

GMT	MUF	FOT	LUF
2.0	11.8	8.9	-
4.0	11.9	10.2	-
6.0	12.7	9.5	-
8.0	11.1	8.3	-
10.0	10.6	9.0	-
12.0	11.6	9.8	-
14.0	13.1	10.7	-
16.0	13.7	11.9	-
18.0	14.1	11.6	-
20.0	13.6	11.6	-
22.0	12.7	10.8	-
24.0	11.7	8.9	-

UNIVERSAL TIME

OCTOBER

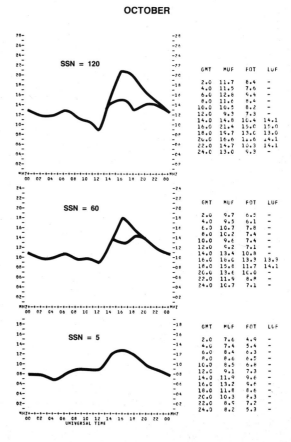

SSN = 120

GMT	MUF	FOT	LUF
2.0	11.7	8.4	-
4.0	11.5	7.6	-
6.0	12.6	9.4	-
8.0	11.6	8.4	-
10.0	10.5	8.2	-
12.0	9.3	7.3	-
14.0	14.6	10.4	14.1
16.0	21.4	15.0	15.0
18.0	19.7	13.0	13.0
20.0	16.6	14.6	14.1
22.0	14.7	10.3	14.1
24.0	13.0	9.3	-

SSN = 60

GMT	MUF	FOT	LUF
2.0	9.7	6.5	-
4.0	9.5	6.1	-
6.0	10.7	7.8	-
8.0	10.2	7.4	-
10.0	9.6	7.4	-
12.0	9.2	7.1	-
14.0	13.6	10.8	-
16.0	18.0	13.3	13.3
18.0	15.8	11.7	14.1
20.0	13.6	10.0	-
22.0	11.9	8.8	-
24.0	10.7	7.1	-

SSN = 5

GMT	MUF	FOT	LUF
2.0	7.6	4.9	-
4.0	7.4	5.4	-
6.0	8.4	6.3	-
8.0	8.6	6.5	-
10.0	8.5	6.8	-
12.0	9.1	7.3	-
14.0	11.9	9.6	-
16.0	13.2	9.6	-
18.0	11.8	8.6	-
20.0	10.3	8.3	-
22.0	8.9	7.2	-
24.0	8.2	5.3	-

UNIVERSAL TIME

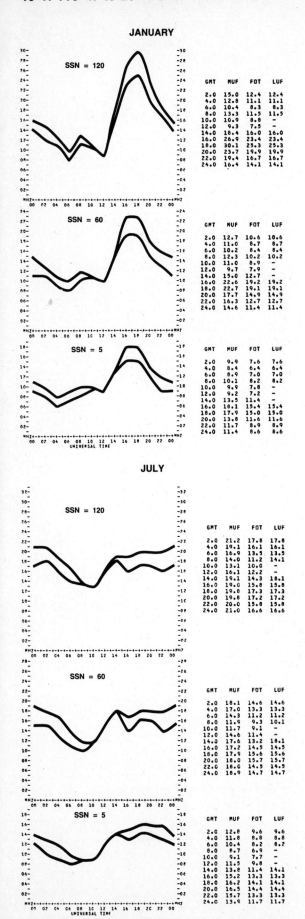

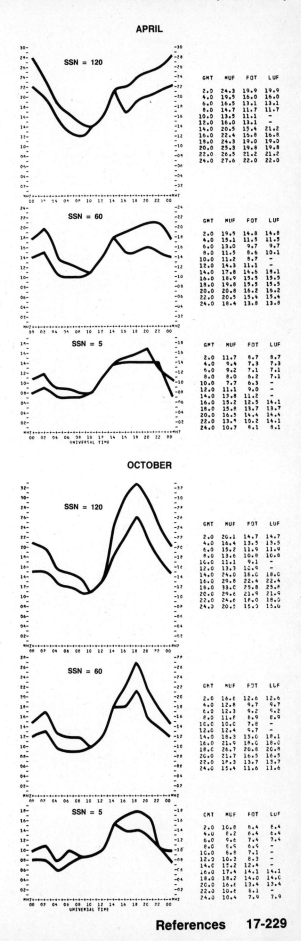

JANUARY

SSN = 120

GMT	MUF	FOT	LUF
2.0	15.0	12.4	12.4
4.0	12.8	11.1	11.1
6.0	10.4	8.3	8.3
8.0	13.3	11.5	11.5
10.0	10.9	8.8	-
12.0	9.3	7.5	-
14.0	18.4	16.0	16.0
16.0	26.9	23.4	23.4
18.0	30.1	25.3	25.3
20.0	23.7	19.9	19.9
22.0	19.4	16.7	16.7
24.0	16.4	14.1	14.1

SSN = 60

GMT	MUF	FOT	LUF
2.0	12.7	10.6	10.6
4.0	11.0	8.7	8.7
6.0	10.2	8.4	8.4
8.0	12.3	10.2	10.2
10.0	11.0	8.9	-
12.0	9.7	7.9	-
14.0	15.0	12.7	-
16.0	22.6	19.2	19.2
18.0	22.7	19.1	19.1
20.0	17.7	14.9	14.9
22.0	16.3	12.7	12.7
24.0	14.6	11.4	11.4

SSN = 5

GMT	MUF	FOT	LUF
2.0	9.9	7.6	7.6
4.0	8.4	6.4	6.4
6.0	8.9	7.0	7.0
8.0	10.1	8.2	8.2
10.0	9.9	7.8	-
12.0	9.2	7.2	-
14.0	13.5	11.4	-
16.0	18.1	15.4	15.4
18.0	17.9	15.0	15.0
20.0	13.8	11.6	11.6
22.0	11.7	8.9	8.9
24.0	11.4	8.6	8.6

UNIVERSAL TIME

APRIL

SSN = 120

GMT	MUF	FOT	LUF
2.0	24.3	19.9	19.9
4.0	19.5	16.0	16.0
6.0	16.5	13.1	13.1
8.0	14.7	11.7	11.7
10.0	13.5	11.1	-
12.0	16.0	13.1	-
14.0	20.5	15.4	21.2
16.0	22.4	15.8	16.8
18.0	24.3	19.0	19.0
20.0	25.3	19.8	19.8
22.0	26.5	21.2	21.2
24.0	27.6	22.0	22.0

SSN = 60

GMT	MUF	FOT	LUF
2.0	19.5	14.8	14.8
4.0	15.1	11.5	11.5
6.0	13.0	9.7	9.7
8.0	11.5	8.6	10.1
10.0	11.2	8.7	-
12.0	14.3	11.1	-
14.0	17.8	14.6	18.1
16.0	18.9	15.5	15.5
18.0	19.8	15.5	15.5
20.0	20.8	16.2	16.2
22.0	20.5	15.4	15.4
24.0	18.4	13.8	13.8

SSN = 5

GMT	MUF	FOT	LUF
2.0	11.7	8.7	8.7
4.0	9.4	7.3	7.3
6.0	9.2	7.1	7.1
8.0	8.0	6.2	7.1
10.0	7.7	6.3	-
12.0	11.1	9.0	-
14.0	13.8	11.2	-
16.0	15.2	12.5	14.1
18.0	15.8	13.7	13.7
20.0	16.5	14.4	14.4
22.0	13.4	10.2	14.1
24.0	10.7	8.1	8.1

UNIVERSAL TIME

JULY

SSN = 120

GMT	MUF	FOT	LUF
2.0	21.2	17.8	17.8
4.0	19.1	16.1	16.1
6.0	16.9	13.5	13.5
8.0	14.0	11.2	14.1
10.0	13.1	10.0	-
12.0	16.1	12.2	-
14.0	19.1	14.3	18.1
16.0	19.0	15.8	15.8
18.0	19.8	17.3	17.3
20.0	19.8	17.2	17.2
22.0	20.0	15.8	15.8
24.0	21.0	16.6	16.6

SSN = 60

GMT	MUF	FOT	LUF
2.0	18.1	14.6	14.6
4.0	17.0	13.3	13.3
6.0	14.3	11.2	11.2
8.0	11.9	9.3	10.1
10.0	11.7	9.1	-
12.0	14.6	11.4	-
14.0	17.6	13.2	18.1
16.0	17.2	14.5	14.5
18.0	17.9	15.6	15.6
20.0	18.0	15.7	15.7
22.0	18.0	14.5	14.5
24.0	18.9	14.7	14.7

SSN = 5

GMT	MUF	FOT	LUF
2.0	12.8	9.6	9.6
4.0	11.8	8.8	8.8
6.0	10.4	8.2	8.2
8.0	8.7	6.9	-
10.0	9.1	7.7	-
12.0	11.3	9.8	-
14.0	13.8	11.4	14.1
16.0	15.2	13.3	13.3
18.0	16.2	14.1	14.1
20.0	16.5	14.4	14.4
22.0	15.7	13.3	13.3
24.0	13.9	11.7	11.7

UNIVERSAL TIME

OCTOBER

SSN = 120

GMT	MUF	FOT	LUF
2.0	20.1	14.7	14.7
4.0	16.4	13.5	13.5
6.0	15.2	11.9	11.9
8.0	13.6	10.8	10.8
10.0	11.1	9.1	-
12.0	13.3	12.9	-
14.0	24.0	16.0	18.0
16.0	29.8	22.4	22.4
18.0	33.0	25.8	25.8
20.0	29.6	21.9	21.9
22.0	24.6	18.0	18.0
24.0	20.5	15.0	15.0

SSN = 60

GMT	MUF	FOT	LUF
2.0	16.6	12.6	12.6
4.0	12.8	9.7	9.7
6.0	12.3	9.2	9.2
8.0	11.6	8.9	8.9
10.0	10.0	7.8	-
12.0	12.6	9.7	-
14.0	18.3	15.0	18.1
16.0	21.6	18.0	18.0
18.0	26.7	20.8	20.8
20.0	21.7	16.5	16.5
22.0	18.3	13.7	13.7
24.0	15.4	11.6	11.6

SSN = 5

GMT	MUF	FOT	LUF
2.0	10.8	8.4	8.4
4.0	8.2	6.4	6.4
6.0	9.6	7.4	7.4
8.0	8.5	6.9	-
10.0	8.8	7.1	-
12.0	10.3	8.3	-
14.0	15.2	12.4	-
16.0	17.4	14.1	14.1
18.0	18.2	14.0	14.0
20.0	16.4	13.4	13.4
22.0	10.6	8.1	-
24.0	10.4	7.9	7.9

UNIVERSAL TIME

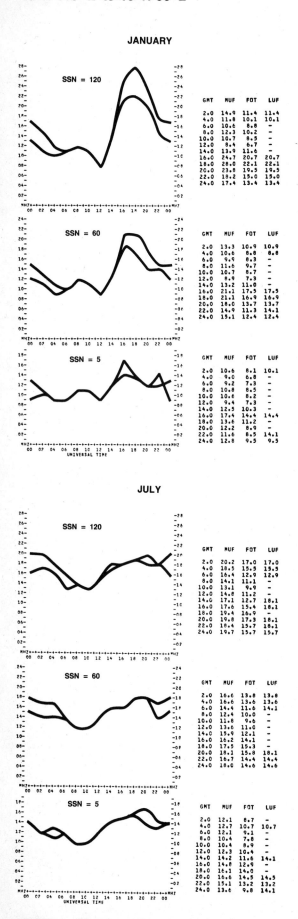

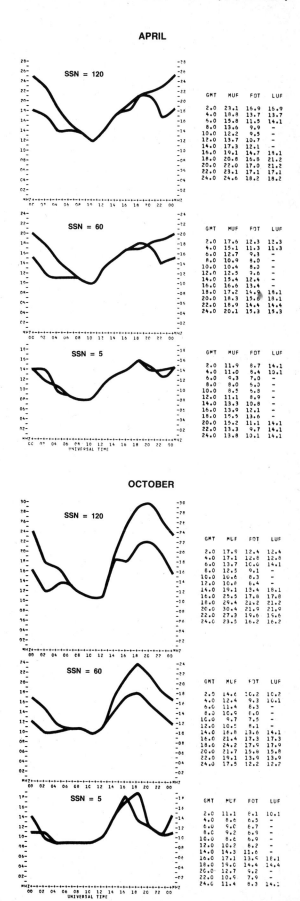

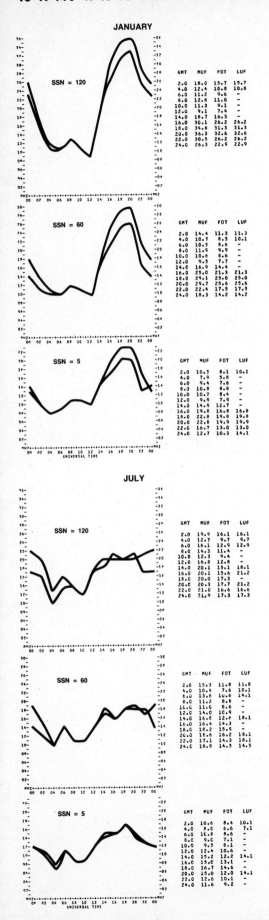

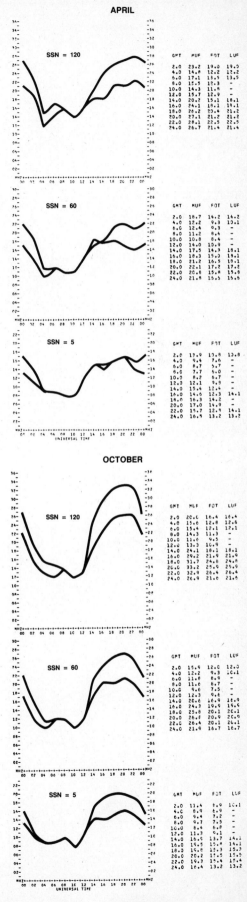

JANUARY

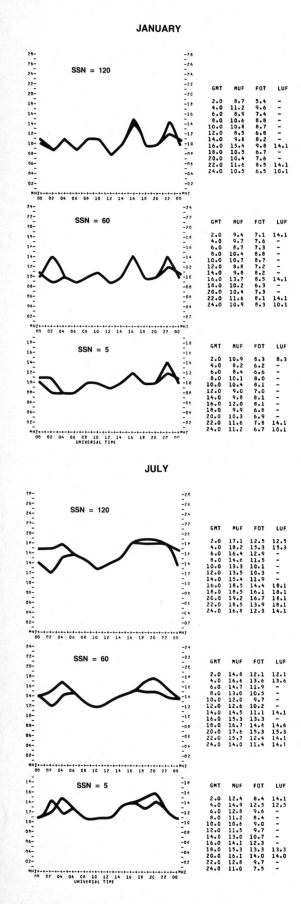

SSN = 120

GMT	MUF	FOT	LUF
2.0	8.7	5.4	-
4.0	11.2	9.6	-
6.0	8.9	7.4	-
8.0	10.6	6.8	-
10.0	10.8	8.7	-
12.0	8.5	6.8	-
14.0	9.8	8.2	-
16.0	15.4	9.8	14.1
18.0	10.5	6.7	-
20.0	10.4	7.6	-
22.0	11.6	8.5	14.1
24.0	10.5	6.5	10.1

SSN = 60

GMT	MUF	FOT	LUF
2.0	9.4	7.1	14.1
4.0	9.7	7.6	-
6.0	8.7	7.3	-
8.0	10.4	8.8	-
10.0	10.7	8.7	-
12.0	8.8	7.2	-
14.0	9.8	8.2	-
16.0	13.7	8.5	14.1
18.0	10.2	6.3	-
20.0	10.4	7.3	-
22.0	11.6	8.1	14.1
24.0	10.9	8.3	10.1

SSN = 5

GMT	MUF	FOT	LUF
2.0	10.9	8.3	8.3
4.0	8.2	6.2	-
6.0	8.4	6.6	-
8.0	10.1	8.0	-
10.0	10.4	8.1	-
12.0	9.0	7.0	-
14.0	9.8	8.1	-
16.0	12.0	8.1	-
18.0	9.9	6.8	-
20.0	10.3	6.9	-
22.0	11.6	7.8	14.1
24.0	11.2	6.7	10.1

UNIVERSAL TIME

APRIL

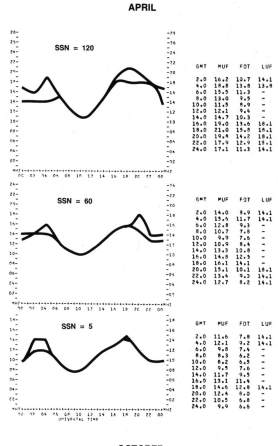

SSN = 120

GMT	MUF	FOT	LUF
2.0	16.2	10.7	14.1
4.0	18.8	13.8	13.8
6.0	15.5	11.3	-
8.0	13.0	9.5	-
10.0	11.5	8.9	-
12.0	12.1	9.4	-
14.0	14.7	10.3	-
16.0	19.0	13.6	18.1
18.0	21.0	15.8	18.1
20.0	19.8	14.2	18.1
22.0	17.9	12.9	18.1
24.0	17.1	11.3	14.1

SSN = 60

GMT	MUF	FOT	LUF
2.0	14.0	8.9	14.1
4.0	15.5	11.7	14.1
6.0	12.8	9.3	-
8.0	10.7	7.8	-
10.0	9.9	7.5	-
12.0	10.9	8.4	-
14.0	13.3	10.8	-
16.0	14.8	12.5	-
18.0	16.1	14.1	-
20.0	15.1	10.1	18.1
22.0	13.4	9.3	14.1
24.0	12.7	8.2	14.1

SSN = 5

GMT	MUF	FOT	LUF
2.0	11.6	7.8	14.1
4.0	12.1	9.2	14.1
6.0	9.8	7.4	-
8.0	8.3	6.2	-
10.0	8.2	6.5	-
12.0	9.5	7.6	-
14.0	11.7	9.5	-
16.0	13.1	11.4	-
18.0	14.6	12.8	14.1
20.0	12.4	8.0	-
22.0	10.5	6.8	-
24.0	9.9	6.6	-

UNIVERSAL TIME

JULY

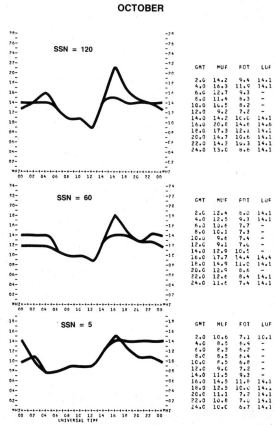

SSN = 120

GMT	MUF	FOT	LUF
2.0	17.1	12.5	12.5
4.0	18.2	15.3	15.3
6.0	16.4	12.9	-
8.0	14.6	11.5	-
10.0	13.3	10.1	-
12.0	13.5	10.3	-
14.0	15.4	11.9	-
16.0	18.5	14.4	18.1
18.0	18.5	16.1	18.1
20.0	19.2	16.7	18.1
22.0	18.5	13.9	18.1
24.0	16.8	12.3	14.1

SSN = 60

GMT	MUF	FOT	LUF
2.0	14.8	12.1	12.1
4.0	16.6	13.6	13.6
6.0	14.7	11.9	-
8.0	13.0	10.5	-
10.0	12.0	9.7	-
12.0	12.6	10.2	-
14.0	14.5	11.1	14.1
16.0	15.3	13.3	-
18.0	16.7	14.6	14.6
20.0	17.6	15.3	15.3
22.0	15.7	12.4	14.1
24.0	14.0	11.4	14.1

SSN = 5

GMT	MUF	FOT	LUF
2.0	12.4	8.4	14.1
4.0	14.9	12.5	12.5
6.0	12.8	9.6	-
8.0	11.2	8.4	-
10.0	10.6	9.0	-
12.0	11.5	9.7	-
14.0	13.0	10.7	-
16.0	14.1	12.3	-
18.0	15.3	13.3	13.3
20.0	16.1	14.0	14.0
22.0	12.8	9.7	-
24.0	11.0	7.5	-

UNIVERSAL TIME

OCTOBER

SSN = 120

GMT	MUF	FOT	LUF
2.0	14.2	9.4	14.1
4.0	16.3	11.9	14.1
6.0	12.7	9.3	-
8.0	11.4	8.3	-
10.0	14.5	8.2	-
12.0	9.2	7.2	-
14.0	14.2	10.0	14.1
16.0	20.8	14.6	14.6
18.0	17.3	12.1	14.1
20.0	14.7	10.6	14.1
22.0	14.3	10.3	14.1
24.0	13.0	8.6	14.1

SSN = 60

GMT	MUF	FOT	LUF
2.0	12.4	8.0	14.1
4.0	12.5	9.3	14.1
6.0	10.6	7.7	-
8.0	10.1	7.3	-
10.0	9.6	7.4	-
12.0	9.1	7.6	-
14.0	12.9	10.5	-
16.0	17.7	14.4	14.4
18.0	14.9	11.0	14.1
20.0	12.9	8.6	-
22.0	12.6	8.4	14.1
24.0	11.6	7.4	14.1

SSN = 5

GMT	MUF	FOT	LUF
2.0	10.6	7.1	10.1
4.0	10.5	6.4	-
6.0	8.3	6.2	-
8.0	8.5	6.4	-
10.0	8.5	6.8	-
12.0	9.0	7.2	-
14.0	11.5	9.3	-
16.0	14.5	11.8	14.1
18.0	12.5	10.0	14.1
20.0	11.1	7.2	14.1
22.0	10.8	7.0	14.1
24.0	10.0	6.7	14.1

UNIVERSAL TIME

JANUARY

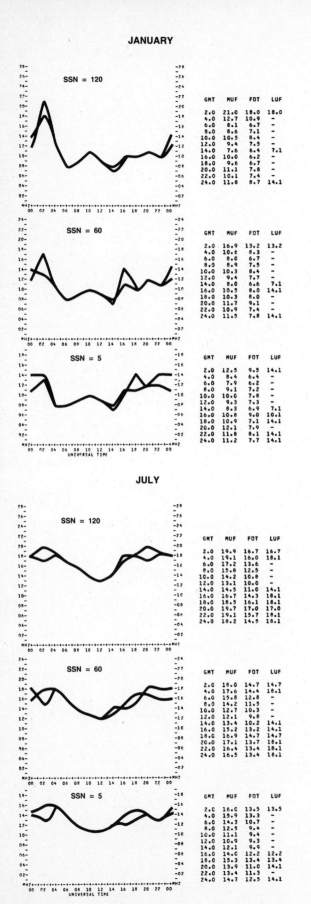

SSN = 120

GMT	MUF	FOT	LUF
2.0	21.0	18.0	18.0
4.0	12.7	10.9	-
6.0	8.1	6.7	-
8.0	8.6	7.1	-
10.0	10.5	8.4	-
12.0	9.4	7.5	-
14.0	7.6	6.4	7.1
16.0	10.0	6.2	-
18.0	9.6	6.7	-
20.0	11.1	7.8	-
22.0	10.1	7.4	-
24.0	11.8	8.7	14.1

SSN = 60

GMT	MUF	FOT	LUF
2.0	16.9	13.2	13.2
4.0	10.6	8.3	-
6.0	8.0	6.7	-
8.0	8.9	7.5	-
10.0	10.3	8.4	-
12.0	9.4	7.7	-
14.0	8.0	6.6	7.1
16.0	10.5	8.0	14.1
18.0	10.3	8.0	-
20.0	11.7	9.1	-
22.0	10.9	7.4	-
24.0	11.5	7.8	14.1

SSN = 5

GMT	MUF	FOT	LUF
2.0	12.5	9.5	14.1
4.0	8.4	6.4	-
6.0	7.9	6.2	-
8.0	9.1	7.2	-
10.0	10.0	7.8	-
12.0	9.3	7.3	-
14.0	8.3	6.9	7.1
16.0	10.8	9.0	10.1
18.0	10.9	7.1	14.1
20.0	12.1	7.9	-
22.0	11.8	8.1	14.1
24.0	11.2	7.7	14.1

APRIL

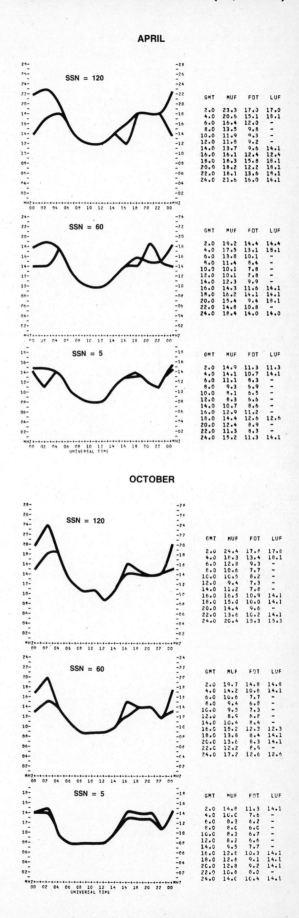

SSN = 120

GMT	MUF	FOT	LUF
2.0	23.3	17.0	17.0
4.0	20.6	15.1	18.1
6.0	16.4	12.0	-
8.0	13.5	9.8	-
10.0	11.9	9.3	-
12.0	11.8	9.2	-
14.0	13.7	9.6	14.1
16.0	16.1	12.4	12.4
18.0	18.3	15.8	18.1
20.0	18.2	12.2	18.1
22.0	18.1	13.6	18.1
24.0	21.6	16.0	14.1

SSN = 60

GMT	MUF	FOT	LUF
2.0	19.2	14.4	14.4
4.0	17.5	13.1	18.1
6.0	13.8	10.1	-
8.0	11.4	8.4	-
10.0	10.1	7.8	-
12.0	10.1	7.8	-
14.0	12.3	9.9	-
16.0	14.3	11.6	14.1
18.0	16.2	14.1	14.1
20.0	15.4	9.4	18.1
22.0	14.8	10.8	-
24.0	18.4	14.0	14.0

SSN = 5

GMT	MUF	FOT	LUF
2.0	14.9	11.3	11.3
4.0	14.1	10.7	14.1
6.0	11.1	8.3	-
8.0	9.3	6.9	-
10.0	8.1	6.5	-
12.0	8.3	5.6	-
14.0	10.7	8.6	-
16.0	12.9	11.2	-
18.0	14.4	12.6	12.6
20.0	12.4	8.9	-
22.0	11.3	8.3	-
24.0	15.2	11.3	14.1

JULY

SSN = 120

GMT	MUF	FOT	LUF
2.0	19.9	16.7	16.7
4.0	19.1	16.0	18.1
6.0	17.2	13.6	-
8.0	15.8	12.5	-
10.0	14.2	10.8	-
12.0	13.1	10.0	-
14.0	14.5	11.0	14.1
16.0	16.7	14.3	18.1
18.0	18.5	16.1	18.1
20.0	19.7	17.0	17.0
22.0	19.1	15.7	18.1
24.0	18.2	14.5	18.1

SSN = 60

GMT	MUF	FOT	LUF
2.0	18.0	14.7	14.7
4.0	17.6	14.4	18.1
6.0	15.8	12.8	-
8.0	14.2	11.5	-
10.0	12.7	10.3	-
12.0	12.1	9.8	-
14.0	13.4	10.2	14.1
16.0	15.2	13.2	14.1
18.0	16.9	14.7	14.7
20.0	17.1	13.7	18.1
22.0	16.4	13.4	18.1
24.0	16.5	13.4	18.1

SSN = 5

GMT	MUF	FOT	LUF
2.0	16.0	13.5	13.5
4.0	15.9	13.3	-
6.0	14.3	10.7	-
8.0	12.5	9.4	-
10.0	11.1	9.4	-
12.0	10.9	9.3	-
14.0	12.1	9.9	-
16.0	14.0	12.2	12.2
18.0	15.3	13.4	13.4
20.0	13.9	11.0	14.1
22.0	13.4	11.3	-
24.0	14.7	12.5	14.1

OCTOBER

SSN = 120

GMT	MUF	FOT	LUF
2.0	24.4	17.8	17.8
4.0	18.3	13.4	18.1
6.0	12.8	9.3	-
8.0	10.6	7.7	-
10.0	10.5	8.2	-
12.0	9.4	7.3	-
14.0	11.2	7.8	-
16.0	16.5	10.9	14.1
18.0	15.0	10.0	14.1
20.0	14.4	9.6	-
22.0	13.6	10.2	14.1
24.0	20.4	15.3	15.3

SSN = 60

GMT	MUF	FOT	LUF
2.0	19.7	14.8	14.8
4.0	14.2	10.6	14.1
6.0	10.6	7.7	-
8.0	9.4	6.8	-
10.0	9.5	7.3	-
12.0	8.9	6.8	-
14.0	10.4	8.4	-
16.0	15.2	12.3	12.3
18.0	13.8	8.4	14.1
20.0	13.6	8.3	14.1
22.0	12.2	8.5	-
24.0	17.2	12.6	12.6

SSN = 5

GMT	MUF	FOT	LUF
2.0	14.8	11.3	14.1
4.0	10.0	7.6	-
6.0	8.3	6.2	-
8.0	8.0	6.0	-
10.0	8.3	6.7	-
12.0	8.0	6.6	-
14.0	9.5	7.7	-
16.0	12.8	10.3	14.1
18.0	12.6	9.1	14.1
20.0	12.8	9.2	14.1
22.0	10.8	8.0	-
24.0	14.0	10.4	14.1

JANUARY

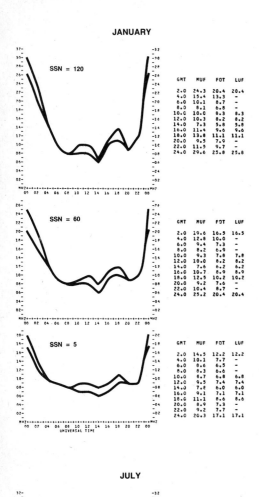

GMT	MUF	FOT	LUF
2.0	24.3	20.4	20.4
4.0	15.4	13.3	-
6.0	10.1	8.7	-
8.0	8.1	6.8	-
10.0	10.0	8.3	8.3
12.0	10.3	8.2	8.2
14.0	7.3	5.8	5.8
16.0	11.4	9.6	9.6
18.0	13.8	11.1	11.1
20.0	9.5	7.9	-
22.0	11.5	9.7	-
24.0	29.6	25.8	25.8

SSN = 120

SSN = 60

GMT	MUF	FOT	LUF
2.0	19.6	16.5	16.5
4.0	12.8	10.0	-
6.0	9.4	7.3	-
8.0	8.2	6.9	-
10.0	9.3	7.8	7.8
12.0	10.0	8.2	8.2
14.0	7.6	6.2	6.2
16.0	10.7	8.9	8.9
18.0	12.5	10.2	10.2
20.0	9.2	7.6	-
22.0	10.4	8.7	-
24.0	25.2	20.4	20.4

SSN = 5

GMT	MUF	FOT	LUF
2.0	14.5	12.2	12.2
4.0	10.1	7.7	-
6.0	8.6	6.5	-
8.0	8.3	6.6	-
10.0	8.7	6.8	6.8
12.0	9.5	7.4	7.4
14.0	7.6	6.0	6.0
16.0	9.1	7.1	7.1
18.0	11.1	8.6	8.6
20.0	8.9	7.3	-
22.0	9.2	7.7	-
24.0	20.3	17.1	17.1

UNIVERSAL TIME

APRIL

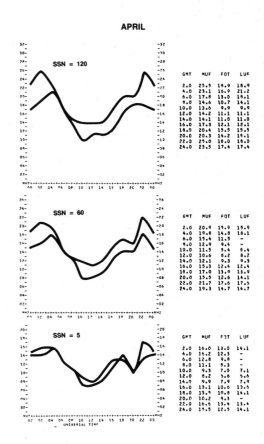

SSN = 120

GMT	MUF	FOT	LUF
2.0	25.5	18.9	18.9
4.0	23.1	16.9	21.2
6.0	17.8	13.0	19.1
8.0	14.6	10.7	14.1
10.0	13.6	9.9	9.9
12.0	14.2	11.1	11.1
14.0	14.1	11.0	11.0
16.0	17.3	12.1	12.1
18.0	20.4	15.5	15.5
20.0	20.3	14.2	19.1
22.0	25.0	18.0	18.0
24.0	23.5	17.4	17.4

SSN = 60

GMT	MUF	FOT	LUF
2.0	20.9	15.9	15.9
4.0	19.8	14.8	18.1
6.0	15.4	11.5	-
8.0	12.9	9.6	-
10.0	11.5	9.4	9.4
12.0	10.6	8.2	8.2
14.0	12.1	9.3	9.3
16.0	15.3	12.4	12.4
18.0	17.0	13.9	13.9
20.0	15.5	12.6	14.1
22.0	21.7	17.6	17.5
24.0	19.3	14.7	14.7

SSN = 5

GMT	MUF	FOT	LUF
2.0	16.0	13.0	14.1
4.0	16.2	12.3	-
6.0	12.8	9.8	-
8.0	11.1	8.3	-
10.0	9.3	7.0	7.1
12.0	8.2	5.6	5.6
14.0	9.9	7.0	7.0
16.0	13.1	10.6	13.5
18.0	13.5	10.8	14.1
20.0	10.2	9.3	-
22.0	16.6	13.4	13.4
24.0	15.5	12.5	14.1

UNIVERSAL TIME

JULY

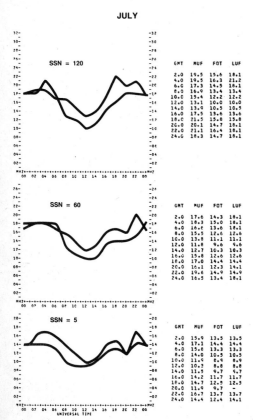

SSN = 120

GMT	MUF	FOT	LUF
2.0	19.5	15.6	18.1
4.0	19.5	16.3	21.2
6.0	17.3	14.5	18.1
8.0	16.9	13.4	13.4
10.0	15.4	12.2	12.2
12.0	13.1	10.0	10.0
14.0	13.0	10.5	10.5
16.0	17.5	13.6	13.6
18.0	21.5	15.8	15.8
20.0	20.1	14.7	18.1
22.0	21.1	16.4	18.1
24.0	18.3	14.7	18.1

SSN = 60

GMT	MUF	FOT	LUF
2.0	17.6	14.3	18.1
4.0	18.3	15.0	18.1
6.0	15.5	12.6	18.1
8.0	15.5	12.6	12.6
10.0	13.8	11.1	11.1
12.0	11.8	9.6	9.6
14.0	12.7	10.3	10.3
16.0	15.8	12.6	12.6
18.0	17.0	14.4	14.4
20.0	16.1	12.3	14.1
22.0	19.6	14.9	14.9
24.0	16.5	13.4	18.1

SSN = 5

GMT	MUF	FOT	LUF
2.0	15.9	13.5	13.5
4.0	17.1	14.4	14.4
6.0	15.8	13.3	13.3
8.0	14.0	10.5	10.5
10.0	11.9	8.9	8.9
12.0	10.3	8.8	8.8
14.0	11.5	9.7	9.7
16.0	14.2	11.7	11.7
18.0	14.7	12.5	12.5
20.0	11.9	9.7	-
22.0	16.7	13.7	13.7
24.0	14.4	12.4	14.1

UNIVERSAL TIME

OCTOBER

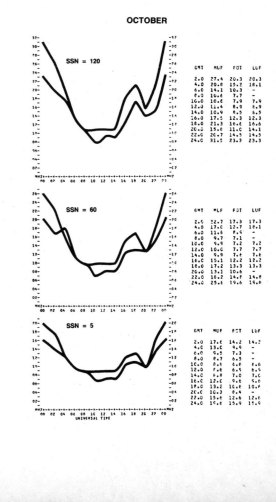

SSN = 120

GMT	MUF	FOT	LUF
2.0	27.4	20.3	20.3
4.0	20.0	15.2	18.1
6.0	14.1	10.3	-
8.0	10.6	7.7	-
10.0	10.6	7.9	7.9
12.0	11.4	8.9	8.9
14.0	10.9	8.5	8.5
16.0	17.5	12.3	12.3
18.0	21.3	16.6	16.6
20.0	15.6	11.0	14.1
22.0	20.7	14.5	14.5
24.0	31.5	23.3	23.3

SSN = 60

GMT	MUF	FOT	LUF
2.0	22.7	17.3	17.3
4.0	17.0	12.7	18.1
6.0	11.6	8.4	-
8.0	9.7	7.1	-
10.0	9.9	7.2	7.2
12.0	10.0	7.7	7.7
14.0	9.9	7.6	7.6
16.0	15.1	12.2	12.2
18.0	17.2	13.3	13.3
20.0	13.1	10.6	-
22.0	16.2	14.8	14.8
24.0	25.6	19.6	19.6

SSN = 5

GMT	MUF	FOT	LUF
2.0	17.6	14.2	14.2
4.0	13.0	9.9	-
6.0	9.5	7.3	-
8.0	8.7	6.5	-
10.0	8.6	6.6	6.6
12.0	7.6	6.5	6.5
14.0	8.8	7.0	7.0
16.0	12.6	9.6	9.6
18.0	13.6	10.6	10.6
20.0	10.3	8.4	-
22.0	15.6	12.6	12.6
24.0	19.6	15.9	15.9

UNIVERSAL TIME

Western USA to Far East
40°N 116°W to 38°N 130°E

JANUARY

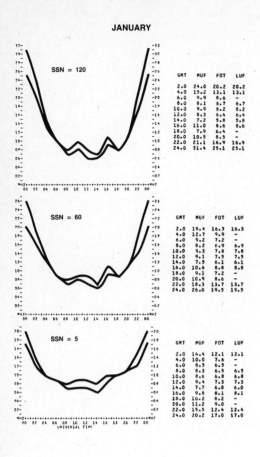

SSN = 120

GMT	MUF	FOT	LUF
2.0	24.0	20.2	20.2
4.0	15.2	13.1	13.1
6.0	9.9	8.6	-
8.0	8.1	6.7	6.7
10.0	9.9	8.2	8.2
12.0	8.3	6.4	6.4
14.0	7.2	5.8	5.8
16.0	11.0	8.6	8.6
18.0	7.9	6.4	-
20.0	10.5	8.5	-
22.0	21.1	16.9	16.9
24.0	31.4	25.1	25.1

SSN = 60

GMT	MUF	FOT	LUF
2.0	19.4	16.3	16.3
4.0	12.7	9.9	-
6.0	9.2	7.2	-
8.0	8.2	6.9	6.9
10.0	9.3	7.8	7.8
12.0	9.1	7.5	7.5
14.0	7.5	6.1	6.1
16.0	10.6	8.8	8.8
18.0	9.1	7.2	-
20.0	10.9	8.6	-
22.0	18.3	13.7	13.7
24.0	26.0	19.5	19.5

SSN = 5

GMT	MUF	FOT	LUF
2.0	14.4	12.1	12.1
4.0	10.0	7.6	-
6.0	8.5	6.5	-
8.0	8.3	6.5	6.5
10.0	8.6	6.8	6.8
12.0	9.4	7.3	7.3
14.0	7.7	6.0	6.0
16.0	9.8	8.1	8.1
18.0	10.2	8.2	-
20.0	11.2	9.0	-
22.0	15.5	12.4	12.4
24.0	20.2	17.0	17.0

APRIL

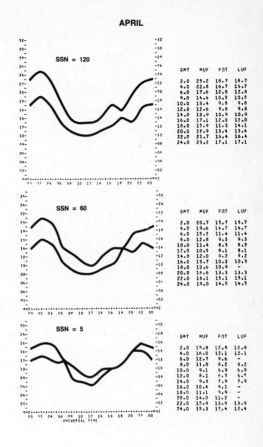

SSN = 120

GMT	MUF	FOT	LUF
2.0	25.2	18.7	18.7
4.0	22.8	16.7	15.7
6.0	17.6	12.8	12.8
8.0	14.4	10.5	10.5
10.0	13.4	9.8	9.8
12.0	12.6	9.8	9.8
14.0	13.0	10.9	10.9
16.0	17.1	12.0	12.0
18.0	15.9	11.3	14.1
20.0	16.9	13.4	13.4
22.0	21.7	15.4	16.4
24.0	23.2	17.1	17.1

SSN = 60

GMT	MUF	FOT	LUF
2.0	20.7	15.7	15.7
4.0	19.6	14.7	14.7
6.0	15.2	11.4	11.4
8.0	12.8	9.3	9.3
10.0	11.4	8.3	8.3
12.0	10.9	8.1	8.1
14.0	12.0	9.2	7.2
16.0	13.7	10.3	10.3
18.0	13.6	10.9	-
20.0	16.6	13.3	13.3
22.0	18.1	15.1	15.1
24.0	19.0	14.5	14.5

SSN = 5

GMT	MUF	FOT	LUF
2.0	15.8	12.8	12.8
4.0	16.0	12.1	12.1
6.0	12.7	9.6	-
8.0	11.0	8.2	8.2
10.0	9.1	5.9	5.9
12.0	8.1	6.5	6.5
14.0	9.8	7.9	7.9
16.0	10.4	8.1	-
18.0	11.1	9.9	-
20.0	14.0	11.2	-
22.0	15.9	13.9	13.9
24.0	15.3	12.4	12.4

JULY

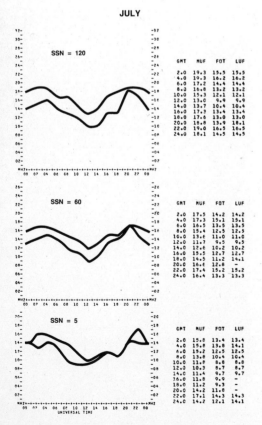

SSN = 120

GMT	MUF	FOT	LUF
2.0	19.3	15.5	15.5
4.0	19.3	16.2	16.2
6.0	17.2	14.4	14.4
8.0	16.8	13.2	13.2
10.0	15.3	12.1	12.1
12.0	13.0	9.9	9.9
14.0	13.7	10.4	10.4
16.0	17.3	13.4	13.4
18.0	17.6	13.0	13.0
20.0	18.8	13.9	18.1
22.0	19.0	16.5	16.5
24.0	18.1	14.5	14.5

SSN = 60

GMT	MUF	FOT	LUF
2.0	17.5	14.2	14.2
4.0	17.3	15.1	15.1
6.0	16.5	13.5	13.5
8.0	15.4	12.5	12.5
10.0	13.6	11.0	11.0
12.0	11.7	9.5	9.5
14.0	12.6	10.2	10.2
16.0	15.5	12.7	12.7
18.0	14.5	11.2	14.1
20.0	16.6	12.8	-
22.0	17.4	15.2	15.2
24.0	16.4	13.3	13.3

SSN = 5

GMT	MUF	FOT	LUF
2.0	15.8	13.4	13.4
4.0	15.8	13.8	14.1
6.0	15.2	12.5	12.5
8.0	13.8	10.4	10.4
10.0	11.8	8.8	8.8
12.0	10.3	8.7	8.7
14.0	11.4	9.7	9.7
16.0	11.8	9.9	-
18.0	11.2	9.9	-
20.0	14.2	11.8	-
22.0	17.1	14.3	14.3
24.0	14.2	12.1	14.1

OCTOBER

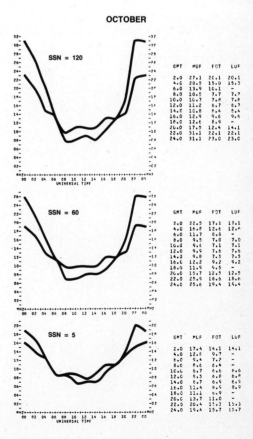

SSN = 120

GMT	MUF	FOT	LUF
2.0	27.1	20.1	20.1
4.0	20.5	15.0	15.0
6.0	13.9	10.1	-
8.0	10.5	7.7	7.7
10.0	10.7	7.8	7.8
12.0	11.2	8.7	8.7
14.0	10.8	8.4	8.4
16.0	12.9	9.6	9.6
18.0	12.6	8.9	-
20.0	17.5	12.4	14.1
22.0	31.1	22.1	22.1
24.0	31.1	23.0	23.0

SSN = 60

GMT	MUF	FOT	LUF
2.0	22.5	17.1	17.1
4.0	16.8	12.6	12.6
6.0	11.7	8.8	-
8.0	9.5	7.0	7.0
10.0	9.6	7.1	7.1
12.0	9.9	7.6	7.6
14.0	9.8	7.5	7.5
16.0	12.2	9.2	9.2
18.0	11.9	9.5	-
20.0	15.7	12.5	12.5
22.0	25.9	18.6	18.0
24.0	25.6	19.4	19.4

SSN = 5

GMT	MUF	FOT	LUF
2.0	17.4	14.1	14.1
4.0	12.8	9.7	-
6.0	9.4	7.2	-
8.0	8.6	6.4	-
10.0	8.7	6.6	6.6
12.0	9.5	6.8	6.8
14.0	8.7	6.9	6.9
16.0	11.4	8.9	8.9
18.0	11.1	8.9	-
20.0	13.7	11.0	-
22.0	20.4	15.3	15.3
24.0	19.4	15.7	15.7

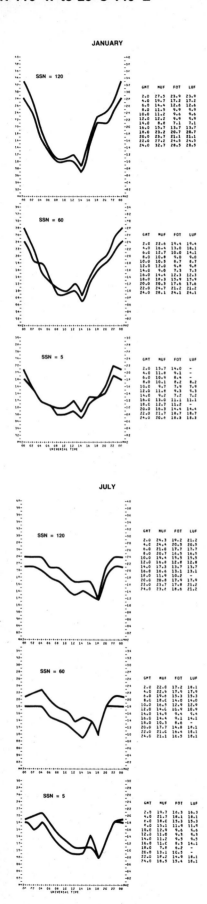

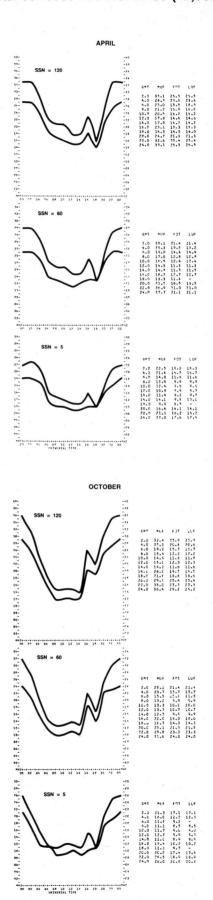

JANUARY

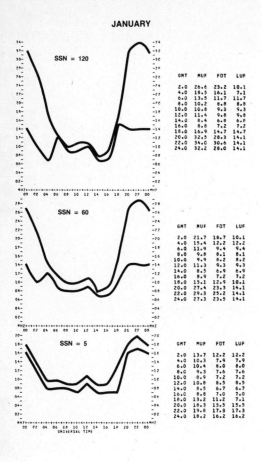

SSN = 120

GMT	MUF	FOT	LUF
2.0	26.6	23.2	10.1
4.0	18.5	16.1	7.1
6.0	13.5	11.7	11.7
8.0	10.2	8.8	8.8
10.0	10.8	9.3	9.3
12.0	11.4	9.8	9.8
14.0	8.4	6.8	6.8
16.0	8.8	7.2	7.2
18.0	16.9	14.7	14.7
20.0	32.5	28.3	14.1
22.0	34.0	30.6	14.1
24.0	32.2	28.0	14.1

SSN = 60

GMT	MUF	FOT	LUF
2.0	21.7	18.7	10.1
4.0	15.4	12.2	12.2
6.0	11.9	9.4	9.4
8.0	9.8	8.1	8.1
10.0	9.9	8.2	8.2
12.0	11.3	9.3	9.3
14.0	8.5	6.9	6.9
16.0	8.9	7.2	7.2
18.0	15.1	12.9	10.1
20.0	27.4	23.3	14.1
22.0	29.3	25.2	14.1
24.0	27.3	23.5	14.1

SSN = 5

GMT	MUF	FOT	LUF
2.0	13.7	12.2	12.2
4.0	10.3	7.9	7.9
6.0	10.4	8.0	8.0
8.0	9.3	7.6	7.6
10.0	8.9	7.2	7.2
12.0	10.8	8.5	8.5
14.0	8.5	6.7	6.7
16.0	8.8	7.0	7.0
18.0	13.2	11.2	7.1
20.0	18.3	15.5	15.5
22.0	19.8	17.3	17.3
24.0	18.2	16.2	16.2

APRIL

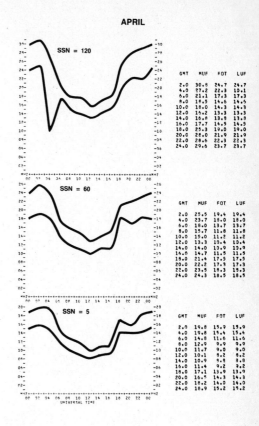

SSN = 120

GMT	MUF	FOT	LUF
2.0	30.8	24.7	24.7
4.0	27.2	22.3	10.1
6.0	21.1	17.3	17.3
8.0	18.5	14.6	14.6
10.0	18.0	14.3	14.3
12.0	15.2	13.3	13.3
14.0	16.8	13.9	13.9
16.0	17.7	14.5	14.5
18.0	25.3	19.0	19.0
20.0	28.0	21.9	21.9
22.0	28.6	22.3	22.3
24.0	29.6	23.7	23.7

SSN = 60

GMT	MUF	FOT	LUF
2.0	25.5	19.4	19.4
4.0	23.7	18.0	18.0
6.0	16.0	13.7	13.7
8.0	15.7	11.8	11.8
10.0	15.0	11.2	11.2
12.0	13.3	10.4	13.4
14.0	14.0	10.9	10.9
16.0	14.7	11.5	11.5
18.0	21.4	17.5	17.5
20.0	22.2	17.9	17.9
22.0	23.5	18.3	13.3
24.0	24.3	18.5	18.5

SSN = 5

GMT	MUF	FOT	LUF
2.0	19.8	15.9	15.9
4.0	19.8	15.4	15.4
6.0	14.8	11.6	11.6
8.0	12.9	9.9	9.9
10.0	11.7	9.0	9.0
12.0	10.1	8.2	8.2
14.0	10.9	9.8	8.8
16.0	11.4	9.2	9.2
18.0	17.1	13.9	13.9
20.0	16.5	14.3	14.3
22.0	18.2	14.0	14.0
24.0	18.9	15.2	15.2

JULY

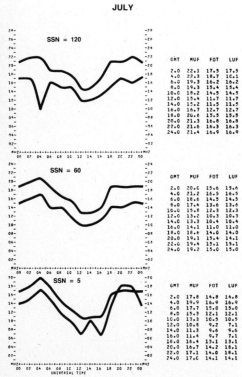

SSN = 120

GMT	MUF	FOT	LUF
2.0	22.1	17.5	17.5
4.0	22.3	18.7	10.1
6.0	19.3	16.2	16.2
8.0	19.3	15.4	15.4
10.0	18.2	14.5	14.5
12.0	15.4	11.7	11.7
14.0	15.2	11.5	11.5
16.0	16.7	12.7	12.7
18.0	20.0	15.5	15.5
20.0	21.3	16.8	16.8
22.0	21.6	16.3	16.3
24.0	21.4	16.9	16.9

SSN = 60

GMT	MUF	FOT	LUF
2.0	20.0	15.6	15.6
4.0	21.2	16.5	16.5
6.0	18.6	14.5	14.5
8.0	17.4	13.6	13.6
10.0	15.8	12.3	12.3
12.0	13.2	10.3	10.3
14.0	13.3	10.4	10.4
16.0	14.1	11.0	11.0
18.0	18.6	14.0	14.0
20.0	19.1	15.4	14.1
22.0	19.4	15.1	15.1
24.0	19.2	15.0	15.0

SSN = 5

GMT	MUF	FOT	LUF
2.0	17.8	14.8	14.8
4.0	19.9	16.9	16.9
6.0	17.7	15.0	15.0
8.0	15.3	12.1	12.1
10.0	13.3	10.5	10.5
12.0	10.8	9.2	7.1
14.0	11.3	9.6	9.6
16.0	11.4	9.7	7.1
18.0	16.4	13.1	13.1
20.0	16.7	14.2	14.1
22.0	17.1	14.0	14.1
24.0	17.0	14.1	14.1

OCTOBER

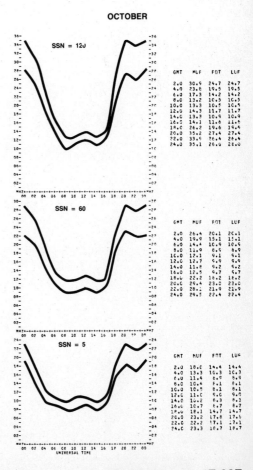

SSN = 120

GMT	MUF	FOT	LUF
2.0	30.9	24.7	24.7
4.0	23.8	19.5	19.5
6.0	17.3	14.2	14.2
8.0	13.2	10.5	10.5
10.0	13.3	10.5	10.5
12.0	14.3	11.7	11.7
14.0	13.3	10.9	10.9
16.0	14.1	11.6	11.6
18.0	26.2	19.6	19.6
20.0	35.2	27.4	27.4
22.0	33.6	26.4	26.4
24.0	35.1	28.0	28.0

SSN = 60

GMT	MUF	FOT	LUF
2.0	26.6	20.1	20.1
4.0	19.9	15.1	12.1
6.0	14.4	10.9	10.9
8.0	11.9	8.9	8.9
10.0	12.1	9.1	9.1
12.0	12.7	9.9	9.9
14.0	11.8	9.2	9.2
16.0	12.5	9.7	9.7
18.0	22.2	16.2	16.2
20.0	29.4	23.0	23.0
22.0	28.1	21.9	21.9
24.0	29.5	22.4	22.4

SSN = 5

GMT	MUF	FOT	LUF
2.0	18.0	14.4	14.4
4.0	13.3	10.3	10.3
6.0	11.4	8.9	8.9
8.0	10.4	8.1	8.1
10.0	10.5	8.1	8.1
12.0	11.0	9.0	9.0
14.0	10.2	8.3	8.3
16.0	10.7	8.7	8.7
18.0	18.1	14.7	14.7
20.0	23.2	17.8	17.5
22.0	22.2	17.1	17.1
24.0	23.3	18.7	18.7

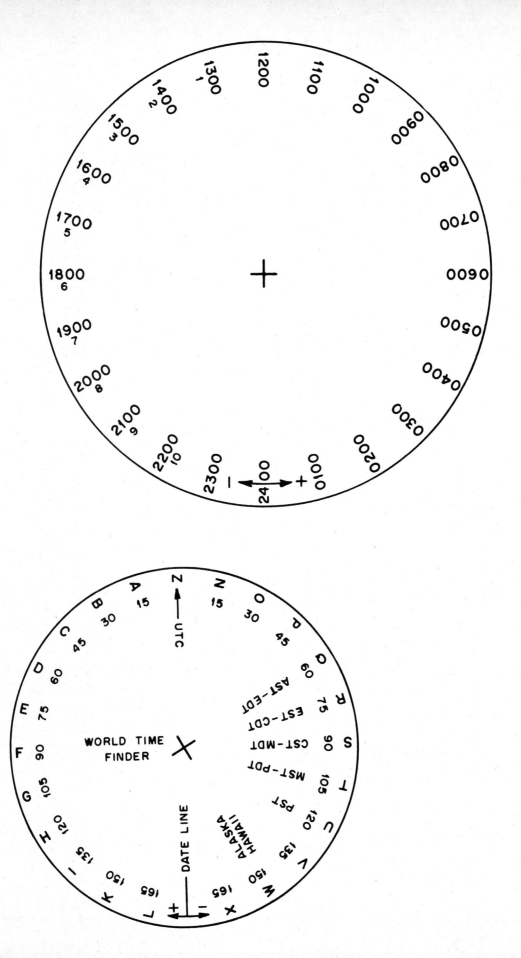

Index

OPERATING
MANUAL
PROOF OF
PURCHASE

ARRL MEMBERS

This proof of purchase may be used as a $1.50 credit on your next ARRL purchase or renewal 1 credit per member. Validate by entering your membership number — the first 7 digits on your *QST* label — below:

Please use this form to give us your comments on this book and what you'd like to see in future editions.

Name _____ Call sign _____

Address _____ Daytime Phone ( ) _____

City _____ State/Province _____ ZIP/Postal Code _____

Please affix
postage. Post
office will not
deliver without
sufficient postage

From _____

_____

_____

Editor, The Operating Manual
American Radio Relay League
225 Main Street
Newington, CT USA 06111

··· please fold and tape ···